# Probability and Random Processes

Fortuna, goddess of chance and luck

Engraving by Hans Sebald Beham (1541)

# Probability and Random Processes

FOURTH EDITION

**GEOFFREY R. GRIMMETT**
*Statistical Laboratory, Cambridge University*

and

**DAVID R. STIRZAKER**
*Mathematical Institute, Oxford University*

OXFORD
UNIVERSITY PRESS

# OXFORD
## UNIVERSITY PRESS

Great Clarendon Street, Oxford, OX2 6DP,
United Kingdom

Oxford University Press is a department of the University of Oxford.
It furthers the University's objective of excellence in research, scholarship,
and education by publishing worldwide. Oxford is a registered trade mark of
Oxford University Press in the UK and in certain other countries

Fourth Edition published in 2020

Impression: 1

Published in the United States of America by Oxford University Press
198 Madison Avenue, New York, NY 10016, United States of America

British Library Cataloguing in Publication Data
Data available

Library of Congress Control Number: 2020939178

ISBN 978–0–19–884760–1 (hbk)
ISBN 978–0–19–884759–5 (pbk)
ISBN 978–0–19–884762–5 (set with One Thousand Exercises in Probability, 3e)

Printed in Great Britain by
Bell & Bain Ltd., Glasgow

Lastly, numbers are applicable even to such things as seem to be governed by no rule, I mean such as depend on chance: the quantity of probability and proportion of it in any two proposed cases being subject to calculation as much as anything else. Upon this depend the principles of game. We find sharpers know enough of this to cheat some men that would take it very ill to be thought bubbles; and one gamester exceeds another, as he has a greater sagacity and readiness in calculating his probability to win or lose in any particular case. To understand the theory of chance thoroughly, requires a great knowledge of numbers, and a pretty competent one of Algebra.

John Arbuthnot
*An essay on the usefulness of mathematical learning*
25 November 1700

To this may be added, that some of the problems about chance having a great appearance of simplicity, the mind is easily drawn into a belief, that their solution may be attained by the mere strength of natural good sense; which generally proving otherwise, and the mistakes occasioned thereby being not infrequent, it is presumed that a book of this kind, which teaches to distinguish truth from what seems so nearly to resemble it, will be looked on as a help to good reasoning.

Abraham de Moivre
*The Doctrine of Chances*
1717

# Preface to the Fourth Edition

This book provides an extensive introduction to probability and random processes. It is intended for those working in the many and varied applications of the subject as well as for those studying more theoretical aspects. We hope it will be found suitable for mathematics undergraduates at all levels, as well as for graduate students and others with interests in these fields.

In particular, we aim:

- to give a rigorous introduction to probability theory while limiting the amount of measure theory in the early chapters;

- to discuss the most important random processes in some depth, with many examples;

- to include various topics which are suitable for undergraduate courses, but are not routinely taught;

- to impart to the beginner the flavour of more advanced work, thereby whetting the appetite for more.

The ordering and numbering of material in this fourth edition has for the most part been preserved from the third. However, a good many minor alterations and additions have been made in the pursuit of clearer exposition. Furthermore, we have revised extensively the sections on Markov chains in continuous time, and added new sections on coupling from the past, Lévy processes, self-similarity and stability, and time changes.

In an immoderate manifestation of millennial mania, the number of exercises and problems has been increased beyond the statutory 1000 to a total of 1322. Moreover, many of the existing exercises have been refreshed by additional parts, making a total of more than 3000 challenges for the diligent reader. These are frequently far from being merely drill exercises, but they complement and illustrate the text, or are entertaining, or (usually, we hope) both. In a companion volume, *One Thousand Exercises in Probability* (Oxford University Press, third edition, 2020), we give worked solutions to almost all exercises and problems.

The basic layout of the book remains unchanged. Chapters 1–5 begin with the foundations of probability theory, move through the elementary properties of random variables, and finish with the weak law of large numbers and the central limit theorem; on route, the reader meets random walks, branching processes, and characteristic functions. This material is suitable for about two lecture courses at a moderately elementary level.

The second part of the book is devoted to the theory of random processes. Chapter 6 deals with Markov chains in discrete and continuous time. The treatment of discrete-time chains is quite detailed and includes an easy proof of the limit theorem for chains with countably infinite state spaces. The sections on continuous-time chains provide a significant amplification of those of the third edition, and constitute an approachable but rigorous account of the principal theory and applications. Chapter 7 contains a general discussion of convergence, together with simple, rigorous accounts of the strong law of large numbers, and of martingale convergence. Each of these two chapters could be used as a basis for a lecture course.

Chapter 8 is an introduction to stochastic processes in their various types; most of these are studied in detail in later chapters, but we have also aspired there to engage the reader with wider aspects of probability by including new essays on a number of topics such as Lévy processes, self-similarity, and stability. Chapters 8–13 provide suitable material for about five shorter lecture courses on: stationary processes and ergodic theory; renewal processes; queues; martingales; diffusions and stochastic integration with applications to finance.

We thank those who have read and commented upon sections of this and earlier editions, and also those readers who have taken the trouble to write to us with notice of imperfections. Further help in thinning any remaining errors will be greatly appreciated.

*Cambridge and Oxford*                                                                  G.R.G.
April 2020                                                                              D.R.S.

*Note on the Frontispiece*

The iconography associated with Fortuna, the goddess of chance and luck, (Tyche or Agatho-daemon to the Greeks), has accumulated over more than two millennia, and is correspondingly complex; we give no more than a brief and much simplified account here of the various allegories involved. The goddess Fortuna was originally associated with fertility, hence the sheaf of corn shown in her right hand. Later associations with the uncertainty of sea voyages are indicated by the ship in full sail seen in the background. The sphere by her feet may initially have represented the instability of life, and this interpretation is sometimes strengthened by depicting the sphere as a bubble. (By contrast, the goddess Virtue is frequently depicted on or by a cube, representing stability.) Subsequently the sphere comes to represent the entire world, over which chance reigns supreme. The wheel carried by Fortuna in her left hand represents the fickleness and uncertainty entailed by the passage of time; that is, the inevitable ups and downs of life. 'The wheel of fortune' is a metaphor for chance and uncertainty still today, as exemplified by the title of a recent television game show.

A further discussion of the iconography is provided in Roberts 1998.

# Contents

*Contents*

# 4  Continuous random variables

# 5  Generating functions and their applications

# 6  Markov chains

# 1

# Events and their probabilities

*Summary.* Any experiment involving randomness can be modelled as a probability space. Such a space comprises a set $\Omega$ of possible outcomes of the experiment, a set $\mathcal{F}$ of events, and a probability measure $\mathbb{P}$. The definition and basic properties of a probability space are explored, and the concepts of conditional probability and independence are introduced. Many examples involving modelling and calculation are included.

## 1.1 Introduction

Much of our life is based on the belief that the future is largely unpredictable. For example, games of chance such as dice or roulette would have few adherents if their outcomes were known in advance. We express this belief in chance behaviour by the use of words such as 'random' or 'probability', and we seek, by way of gaming and other experience, to assign quantitative as well as qualitative meanings to such usages. Our main acquaintance with statements about probability relies on a wealth of concepts, some more reasonable than others. A mathematical theory of probability will incorporate those concepts of chance which are expressed and implicit in common rational understanding. Such a theory will formalize these concepts as a collection of axioms, which should lead directly to conclusions in agreement with practical experimentation. This chapter contains the essential ingredients of this construction.

## 1.2 Events as sets

Many everyday statements take the form 'the chance (or probability) of $A$ is $p$', where $A$ is some event (such as 'the sun shining tomorrow', 'Cambridge winning the Boat Race', ... ) and $p$ is a number or adjective describing quantity (such as 'one-eighth', 'low', ... ). The occurrence or non-occurrence of $A$ depends upon the chain of circumstances involved. This chain is called an *experiment* or *trial*; the result of an experiment is called its *outcome*. In general, we cannot predict with certainty the outcome of an experiment in advance of its completion; we can only list the collection of possible outcomes.

**(1) Definition.** The set of all possible outcomes of an experiment is called the **sample space** and is denoted by $\Omega$.

**(2) Example.**  A coin is tossed. There are two possible outcomes, heads (denoted by H) and tails (denoted by T), so that $\Omega = \{$H, T$\}$. We may be interested in the possible occurrences of the following events:

   (a)  the outcome is a head;

   (b)  the outcome is either a head or a tail;

   (c)  the outcome is both a head and a tail (this seems very unlikely to occur);

   (d)  the outcome is not a head.                                                                   ●

**(3) Example.**  A die is thrown once. There are six possible outcomes depending on which of the numbers 1, 2, 3, 4, 5, or 6 is uppermost. Thus $\Omega = \{1, 2, 3, 4, 5, 6\}$. We may be interested in the following events:

   (a)  the outcome is the number 1;

   (b)  the outcome is an even number;

   (c)  the outcome is even but does not exceed 3;

   (d)  the outcome is not even.                                                                    ●

We see immediately that each of the events of these examples can be specified as a subset $A$ of the appropriate sample space $\Omega$. In the first example they can be rewritten as

  (a)  $A = \{$H$\}$,                         (b)  $A = \{$H$\} \cup \{$T$\}$,

  (c)  $A = \{$H$\} \cap \{$T$\}$,              (d)  $A = \{$H$\}^c$,

whilst those of the second example become

  (a)  $A = \{1\}$,                          (b)  $A = \{2, 4, 6\}$,

  (c)  $A = \{2, 4, 6\} \cap \{1, 2, 3\}$,        (d)  $A = \{2, 4, 6\}^c$.

The *complement* of a subset $A$ of $\Omega$ is denoted here and subsequently by $A^c$; from now on, subsets of $\Omega$ containing a single member, such as $\{$H$\}$, will usually be written without the containing braces.

Henceforth we think of *events* as subsets of the sample space $\Omega$. Whenever $A$ and $B$ are events in which we are interested, then we can reasonably concern ourselves also with the events $A \cup B, A \cap B$, and $A^c$, representing '$A$ or $B$', '$A$ and $B$', and 'not $A$' respectively. Events $A$ and $B$ are called *disjoint* if their intersection is the empty set $\varnothing$; $\varnothing$ is called the *impossible event*. The set $\Omega$ is called the *certain event*, since some member of $\Omega$ will certainly occur.

Thus events are subsets of $\Omega$, but need all the subsets of $\Omega$ be events? The answer is *no*, but some of the reasons for this are too difficult to be discussed here. It suffices for us to think of the collection of events as a subcollection $\mathcal{F}$ of the set of all subsets of $\Omega$. This subcollection should have certain properties in accordance with the earlier discussion:

   (a)  if $A, B \in \mathcal{F}$ then $A \cup B \in \mathcal{F}$ and $A \cap B \in \mathcal{F}$;

   (b)  if $A \in \mathcal{F}$ then $A^c \in \mathcal{F}$;

   (c)  the empty set $\varnothing$ belongs to $\mathcal{F}$.

Any collection $\mathcal{F}$ of subsets of $\Omega$ which satisfies these three conditions is called a *field*. It follows from the properties of a field $\mathcal{F}$ that

$$\text{if}\quad A_1, A_2, \ldots, A_n \in \mathcal{F} \quad \text{then} \quad \bigcup_{i=1}^{n} A_i \in \mathcal{F};$$

| Typical notation | Set jargon | Probability jargon |
|---|---|---|
| $\Omega$ | Collection of objects | Sample space |
| $\omega$ | Member of $\Omega$ | Elementary event, outcome |
| $A$ | Subset of $\Omega$ | Event that some outcome in $A$ occurs |
| $A^c$ | Complement of $A$ | Event that no outcome in $A$ occurs |
| $A \cap B$ | Intersection | Both $A$ and $B$ |
| $A \cup B$ | Union | Either $A$ or $B$ or both |
| $A \setminus B$ | Difference | $A$, but not $B$ |
| $A \triangle B$ | Symmetric difference | Either $A$ or $B$, but not both |
| $A \subseteq B$ | Inclusion | If $A$, then $B$ |
| $\varnothing$ | Empty set | Impossible event |
| $\Omega$ | Whole space | Certain event |

Table 1.1. The jargon of set theory and probability theory.

that is to say, $\mathcal{F}$ is closed under finite unions and hence under finite intersections also (see Problem (1.8.3)). This is fine when $\Omega$ is a finite set, but we require slightly more to deal with the common situation when $\Omega$ is infinite, as the following example indicates.

**(4) Example.** A coin is tossed repeatedly until the first head turns up; we are concerned with the number of tosses before this happens. The set of all possible outcomes is the set $\Omega = \{\omega_1, \omega_2, \omega_3, \ldots\}$, where $\omega_i$ denotes the outcome when the first $i - 1$ tosses are tails and the $i$th toss is a head. We may seek to assign a probability to the event $A$, that the first head occurs after an even number of tosses, that is, $A = \{\omega_2, \omega_4, \omega_6, \ldots\}$. This is an infinite countable union of members of $\Omega$ and we require that such a set belong to $\mathcal{F}$ in order that we can discuss its probability.  ●

Thus we also require that the collection of events be closed under the operation of taking countable unions. Any collection of subsets of $\Omega$ with these properties is called a $\sigma$-*field* (or sometimes a $\sigma$-*algebra*).

**(5) Definition.** A collection $\mathcal{F}$ of subsets of $\Omega$ is called a $\sigma$**-field** if it satisfies the following conditions:

(a) $\varnothing \in \mathcal{F}$;

(b) if $A_1, A_2, \ldots \in \mathcal{F}$ then $\bigcup_{i=1}^{\infty} A_i \in \mathcal{F}$;

(c) if $A \in \mathcal{F}$ then $A^c \in \mathcal{F}$.

It follows from Problem (1.8.3) that $\sigma$-fields are closed under the operation of taking countable intersections. Here are some examples of $\sigma$-fields.

**(6) Example.** The smallest $\sigma$-field associated with $\Omega$ is the collection $\mathcal{F} = \{\varnothing, \Omega\}$.  ●

**(7) Example.** If $A$ is any subset of $\Omega$ then $\mathcal{F} = \{\varnothing, A, A^c, \Omega\}$ is a $\sigma$-field.  ●

**(8) Example.** The *power set* of $\Omega$, which is written $\{0, 1\}^{\Omega}$ and contains all subsets of $\Omega$, is obviously a $\sigma$-field. For reasons beyond the scope of this book, when $\Omega$ is infinite, its power set is too large a collection for probabilities to be assigned reasonably to all its members.  ●

To recapitulate, with any experiment we may associate a pair $(\Omega, \mathcal{F})$, where $\Omega$ is the set of all possible outcomes or *elementary events* and $\mathcal{F}$ is a $\sigma$-field of subsets of $\Omega$ which contains all the events in whose occurrences we may be interested; henceforth, to call a set $A$ an *event* is equivalent to asserting that $A$ belongs to the $\sigma$-field in question. We usually translate statements about combinations of events into set-theoretic jargon; for example, the event that both $A$ and $B$ occur is written as $A \cap B$. Table 1.1 is a translation chart.

## Exercises for Section 1.2

**1.**   Let $\{A_i : i \in I\}$ be a collection of sets. Prove 'De Morgan's Laws'†:

$$\left(\bigcup_i A_i\right)^c = \bigcap_i A_i^c, \qquad \left(\bigcap_i A_i\right)^c = \bigcup_i A_i^c.$$

**2.**   Let $A$ and $B$ belong to some $\sigma$-field $\mathcal{F}$. Show that $\mathcal{F}$ contains the sets $A \cap B$, $A \setminus B$, and $A \triangle B$.

**3.**   A conventional knock-out tournament (such as that at Wimbledon) begins with $2^n$ competitors and has $n$ rounds. There are no play-offs for the positions $2, 3, \ldots, 2^n - 1$, and the initial table of draws is specified. Give a concise description of the sample space of all possible outcomes.

**4.**   Let $\mathcal{F}$ be a $\sigma$-field of subsets of $\Omega$ and suppose that $B \in \mathcal{F}$. Show that $\mathcal{G} = \{A \cap B : A \in \mathcal{F}\}$ is a $\sigma$-field of subsets of $B$.

**5.**   Which of the following are identically true? For those that are not, say when they are true.
(a) $A \cup (B \cap C) = (A \cup B) \cap (A \cup C)$;
(b) $A \cap (B \cap C) = (A \cap B) \cap C$;
(c) $(A \cup B) \cap C = A \cup (B \cap C)$;
(d) $A \setminus (B \cap C) = (A \setminus B) \cup (A \setminus C)$.

## 1.3 Probability

We wish to be able to discuss the likelihoods of the occurrences of events. Suppose that we repeat an experiment a large number $N$ of times, keeping the initial conditions as equal as possible, and in such a way that no outcome is influenced by the other outcomes. Let $A$ be some event which may or may not occur on each repetition. Our experience of most scientific experimentation is that the proportion of times that $A$ occurs settles down to some value as $N$ becomes larger and larger; that is to say, writing $N(A)$ for the number of occurrences of $A$ in the $N$ trials, the ratio $N(A)/N$ appears to converge to a constant limit as $N$ increases‡. We can think of the ultimate value of this ratio as being the probability $\mathbb{P}(A)$ that $A$ occurs on any particular trial*; it may happen that the empirical ratio does not behave in a coherent manner and our intuition fails us at this level, but we shall not discuss this here. In practice, $N$ may be

---

†Augustus De Morgan is well known for having given the first clear statement of the principle of mathematical induction. He applauded probability theory with the words: "The tendency of our study is to substitute the satisfaction of mental exercise for the pernicious enjoyment of an immoral stimulus".

‡This property of practical experiments is called the *stability of statistical ratios* by physicists and statisticians, or the *empirical law of large numbers* by probabilists.

*This superficial discussion of probabilities is inadequate in many ways; questioning readers may care to discuss the philosophical and empirical aspects of the subject amongst themselves (see Appendix III). We do not define probability by such limits in this text.

taken to be large but finite, and the ratio $N(A)/N$ may be taken as an approximation to $\mathbb{P}(A)$. Clearly, the ratio is a number between zero and one; if $A = \varnothing$ then $N(\varnothing) = 0$ and the ratio is 0, whilst if $A = \Omega$ then $N(\Omega) = N$ and the ratio is 1. Furthermore, suppose that $A$ and $B$ are two disjoint events, each of which may or may not occur at each trial. Then

$$N(A \cup B) = N(A) + N(B)$$

and so the ratio $N(A \cup B)/N$ is the sum of the two ratios $N(A)/N$ and $N(B)/N$. We now think of these ratios as representing the probabilities of the appropriate events. The above relations become

$$\mathbb{P}(A \cup B) = \mathbb{P}(A) + \mathbb{P}(B), \qquad \mathbb{P}(\varnothing) = 0, \qquad \mathbb{P}(\Omega) = 1.$$

This discussion suggests that the probability function $\mathbb{P}$ should be *finitely additive*, which is to say that

$$\text{if } A_1, A_2, \ldots, A_n \text{ are disjoint events, then } \mathbb{P}\left(\bigcup_{i=1}^{n} A_i\right) = \sum_{i=1}^{n} \mathbb{P}(A_i).$$

A glance at Example (1.2.4) suggests the more extensive property that $\mathbb{P}$ be *countably additive*, in that the corresponding property should hold for countable collections $A_1, A_2, \ldots$ of disjoint events.

These relations are sufficient to specify the desirable properties of a probability function $\mathbb{P}$ applied to the set of events. Any such assignment of likelihoods to the members of $\mathcal{F}$ is called a *probability measure*. Some individuals refer informally to $\mathbb{P}$ as a 'probability distribution', especially when the sample space is finite or countably infinite; this practice is best avoided since the term 'probability distribution' is reserved for another purpose to be encountered in Chapter 2.

**(1) Definition.** A **probability measure** $\mathbb{P}$ on $(\Omega, \mathcal{F})$ is a function $\mathbb{P} : \mathcal{F} \rightarrow [0, 1]$ satisfying
(a) $\mathbb{P}(\varnothing) = 0, \quad \mathbb{P}(\Omega) = 1;$
(b) if $A_1, A_2, \ldots$ is a collection of disjoint members of $\mathcal{F}$, in that $A_i \cap A_j = \varnothing$ for all pairs $i, j$ satisfying $i \neq j$, then

$$\mathbb{P}\left(\bigcup_{i=1}^{\infty} A_i\right) = \sum_{i=1}^{\infty} \mathbb{P}(A_i).$$

The triple $(\Omega, \mathcal{F}, \mathbb{P})$, comprising a set $\Omega$, a $\sigma$-field $\mathcal{F}$ of subsets of $\Omega$, and a probability measure $\mathbb{P}$ on $(\Omega, \mathcal{F})$, is called a **probability space**.

A probability measure is a special example of what is called a *measure* on the pair $(\Omega, \mathcal{F})$. A measure is a function $\mu : \mathcal{F} \rightarrow [0, \infty)$ satisfying $\mu(\varnothing) = 0$ together with (b) above. A measure $\mu$ is a probability measure if $\mu(\Omega) = 1$.

We can associate a probability space $(\Omega, \mathcal{F}, \mathbb{P})$ with any experiment, and all questions associated with the experiment can be reformulated in terms of this space. It may seem natural to ask for the numerical value of the probability $\mathbb{P}(A)$ of some event $A$. The answer to such a question must be contained in the description of the experiment in question. For example, the assertion that a *fair* coin is tossed once is equivalent to saying that heads and tails have an equal probability of occurring; actually, this is the definition of fairness.

**(2) Example.** A coin, possibly biased, is tossed once. We can take $\Omega = \{H, T\}$ and $\mathcal{F} = \{\varnothing, H, T, \Omega\}$, and a possible probability measure $\mathbb{P} : \mathcal{F} \to [0, 1]$ is given by

$$\mathbb{P}(\varnothing) = 0, \quad \mathbb{P}(H) = p, \quad \mathbb{P}(T) = 1 - p, \quad \mathbb{P}(\Omega) = 1,$$

where $p$ is a fixed real number in the interval $[0, 1]$. If $p = \frac{1}{2}$, then we say that the coin is *fair*, or *unbiased*.
● 

**(3) Example.** A die is thrown once. We can take $\Omega = \{1, 2, 3, 4, 5, 6\}$, $\mathcal{F} = \{0, 1\}^{\Omega}$, and the probability measure $\mathbb{P}$ given by

$$\mathbb{P}(A) = \sum_{i \in A} p_i \quad \text{for any } A \subseteq \Omega,$$

where $p_1, p_2, \ldots, p_6$ are specified numbers from the interval $[0, 1]$ having unit sum. The probability that $i$ turns up is $p_i$. The die is fair if $p_i = \frac{1}{6}$ for each $i$, in which case

$$\mathbb{P}(A) = \tfrac{1}{6}|A| \quad \text{for any } A \subseteq \Omega,$$

where $|A|$ denotes the cardinality of $A$.
● 

The triple $(\Omega, \mathcal{F}, \mathbb{P})$ denotes a typical probability space. We now give some of its simple but important properties.

**(4) Lemma.**
(a) $\mathbb{P}(A^c) = 1 - \mathbb{P}(A)$,
(b) *if $B \supseteq A$ then* $\mathbb{P}(B) = \mathbb{P}(A) + \mathbb{P}(B \setminus A) \geq \mathbb{P}(A)$,
(c) $\mathbb{P}(A \cup B) = \mathbb{P}(A) + \mathbb{P}(B) - \mathbb{P}(A \cap B)$,
(d) **(inclusion–exclusion principle)** *more generally, if $A_1, A_2, \ldots, A_n$ are events, then*

$$\mathbb{P}\left(\bigcup_{i=1}^{n} A_i\right) = \sum_i \mathbb{P}(A_i) - \sum_{i<j} \mathbb{P}(A_i \cap A_j) + \sum_{i<j<k} \mathbb{P}(A_i \cap A_j \cap A_k) - \cdots$$
$$+ (-1)^{n+1} \mathbb{P}(A_1 \cap A_2 \cap \cdots \cap A_n)$$

*where, for example, $\sum_{i<j}$ sums over all unordered pairs $(i, j)$ with $i \neq j$.*

**Proof.**
(a) $A \cup A^c = \Omega$ and $A \cap A^c = \varnothing$, so $\mathbb{P}(A \cup A^c) = \mathbb{P}(A) + \mathbb{P}(A^c) = 1$.
(b) $B = A \cup (B \setminus A)$. This is the union of disjoint sets and therefore

$$\mathbb{P}(B) = \mathbb{P}(A) + \mathbb{P}(B \setminus A).$$

(c) $A \cup B = A \cup (B \setminus A)$, which is a disjoint union. Therefore, by (b),

$$\mathbb{P}(A \cup B) = \mathbb{P}(A) + \mathbb{P}(B \setminus A) = \mathbb{P}(A) + \mathbb{P}(B \setminus (A \cap B))$$
$$= \mathbb{P}(A) + \mathbb{P}(B) - \mathbb{P}(A \cap B).$$

(d) The proof is by induction on $n$, and is left as an *exercise* (see Exercise (1.3.4)).   ∎

In Lemma (4b), $B \setminus A$ denotes the set of members of $B$ which are not in $A$. In order to write down the quantity $\mathbb{P}(B \setminus A)$, we require that $B \setminus A$ belongs to $\mathcal{F}$, the domain of $\mathbb{P}$; this is always true when $A$ and $B$ belong to $\mathcal{F}$, and to prove this was part of Exercise (1.2.2). Notice that each proof proceeded by expressing an event in terms of disjoint unions and then applying $\mathbb{P}$. It is sometimes easier to calculate the probabilities of intersections of events rather than their unions; part (d) of the lemma is useful then, as we shall discover soon. The next property of $\mathbb{P}$ is more technical, and says that $\mathbb{P}$ is a *continuous* set function; this property is essentially equivalent to the condition that $\mathbb{P}$ is countably additive rather than just finitely additive (see Problem (1.8.16) also).

**(5) Theorem. Continuity of probability measures.** *Let $A_1, A_2, \ldots$ be an increasing sequence of events, so that $A_1 \subseteq A_2 \subseteq A_3 \subseteq \cdots$, and write $A$ for their limit:*

$$A = \bigcup_{i=1}^{\infty} A_i = \lim_{i \to \infty} A_i.$$

*Then $\mathbb{P}(A) = \lim_{i \to \infty} \mathbb{P}(A_i)$.*

*Similarly, if $B_1, B_2, \ldots$ is a decreasing sequence of events, so that $B_1 \supseteq B_2 \supseteq B_3 \supseteq \cdots$, then*

$$B = \bigcap_{i=1}^{\infty} B_i = \lim_{i \to \infty} B_i$$

*satisfies $\mathbb{P}(B) = \lim_{i \to \infty} \mathbb{P}(B_i)$.*

**Proof.** $A = A_1 \cup (A_2 \setminus A_1) \cup (A_3 \setminus A_2) \cup \cdots$ is the union of a disjoint family of events. Thus, by Definition (1),

$$\mathbb{P}(A) = \mathbb{P}(A_1) + \sum_{i=1}^{\infty} \mathbb{P}(A_{i+1} \setminus A_i)$$

$$= \mathbb{P}(A_1) + \lim_{n \to \infty} \sum_{i=1}^{n-1} \left[ \mathbb{P}(A_{i+1}) - \mathbb{P}(A_i) \right]$$

$$= \lim_{n \to \infty} \mathbb{P}(A_n).$$

To show the result for decreasing families of events, take complements and use the first part (*exercise*).   ∎

To recapitulate, statements concerning chance are implicitly related to experiments or trials, the outcomes of which are not entirely predictable. With any such experiment we can associate a probability space $(\Omega, \mathcal{F}, \mathbb{P})$ the properties of which are consistent with our shared and reasonable conceptions of the notion of chance.

**(6) Example.** In many cases, and especially in games of chance (as in (3) above), it is natural to assume that all outcomes are equally likely. In such a case, the probability of an event $A$ is

$$\mathbb{P}(A) = \frac{|A|}{|\Omega|}.$$   ●

Here is some final jargon. An event $A$ is called *null* if $\mathbb{P}(A) = 0$. If $\mathbb{P}(A) = 1$, we say that $A$ occurs *almost surely* (frequently abbreviated to a.s.). Null events should not be confused with the impossible event $\varnothing$. Null events are happening all around us, even though they have zero probability; after all, what is the chance that a dart strikes any given point of the target at which it is thrown? That is, the impossible event is null, but null events need not be impossible.

## Exercises for Section 1.3

**1.** Let $A$ and $B$ be events with probabilities $\mathbb{P}(A) = \frac{3}{4}$ and $\mathbb{P}(B) = \frac{1}{3}$. Show that $\frac{1}{12} \le \mathbb{P}(A \cap B) \le \frac{1}{3}$, and give examples to show that both extremes are possible. Find corresponding bounds for $\mathbb{P}(A \cup B)$.

**2.** A fair coin is tossed repeatedly. Show that, with probability one, a head turns up sooner or later. Show similarly that any given finite sequence of heads and tails occurs eventually with probability one. Explain the connection with Murphy's Law.

**3.** Six cups and saucers come in pairs: there are two cups and saucers which are red, two white, and two with stars on. If the cups are placed randomly onto the saucers (one each), find the probability that no cup is upon a saucer of the same pattern.

**4.** Let $A_1, A_2, \ldots, A_n$ be events where $n \ge 2$, and prove that

$$\mathbb{P}\left( \bigcup_{i=1}^{n} A_i \right) = \sum_i \mathbb{P}(A_i) - \sum_{i<j} \mathbb{P}(A_i \cap A_j) + \sum_{i<j<k} \mathbb{P}(A_i \cap A_j \cap A_k)$$

$$- \cdots + (-1)^{n+1} \mathbb{P}(A_1 \cap A_2 \cap \cdots \cap A_n).$$

In each packet of Corn Flakes may be found a plastic bust of one of the last five Vice-Chancellors of Cambridge University, the probability that any given packet contains any specific Vice-Chancellor being $\frac{1}{5}$, independently of all other packets. Show that the probability that each of the last three Vice-Chancellors is obtained in a bulk purchase of six packets is $1 - 3(\frac{4}{5})^6 + 3(\frac{3}{5})^6 - (\frac{2}{5})^6$.

**5.** Let $A_r, r \ge 1$, be events such that $\mathbb{P}(A_r) = 1$ for all $r$. Show that $\mathbb{P}(\bigcap_{r=1}^{\infty} A_r) = 1$.

**6.** You are given that at least one of the events $A_r, 1 \le r \le n$, is certain to occur, but certainly no more than two occur. If $\mathbb{P}(A_r) = p$, and $\mathbb{P}(A_r \cap A_s) = q, r \ne s$, show that $p \ge 1/n$ and $q \le 2/n$.

**7.** You are given that at least one, but no more than three, of the events $A_r, 1 \le r \le n$, occur, where $n \ge 3$. The probability of at least two occurring is $\frac{1}{2}$. If $\mathbb{P}(A_r) = p$, $\mathbb{P}(A_r \cap A_s) = q, r \ne s$, and $\mathbb{P}(A_r \cap A_s \cap A_t) = x, r < s < t$, show that $p \ge 3/(2n)$, and $q \le 4/n$.

## 1.4 Conditional probability

Many statements about chance take the form 'if $B$ occurs, then the probability of $A$ is $p$', where $B$ and $A$ are events (such as 'it rains tomorrow' and 'the bus being on time' respectively) and $p$ is a likelihood as before. To include this in our theory, we return briefly to the discussion about proportions at the beginning of the previous section. An experiment is repeated $N$ times, and on each occasion we observe the occurrences or non-occurrences of two events $A$ and $B$. Now, suppose we only take an interest in those outcomes for which $B$ occurs; all other experiments are disregarded. In this smaller collection of trials the proportion of times that $A$ occurs is $N(A \cap B)/N(B)$, since $B$ occurs at each of them. However,

$$\frac{N(A \cap B)}{N(B)} = \frac{N(A \cap B)/N}{N(B)/N}.$$

If we now think of these ratios as probabilities, we see that the probability that $A$ occurs, given that $B$ occurs, should be reasonably defined as $\mathbb{P}(A \cap B)/\mathbb{P}(B)$.

Probabilistic intuition leads to the same conclusion. Given that an event $B$ occurs, it is the case that $A$ occurs if and only if $A \cap B$ occurs. Thus the conditional probability of $A$ given $B$ should be proportional to $\mathbb{P}(A \cap B)$, which is to say that it equals $\alpha \mathbb{P}(A \cap B)$ for some constant $\alpha = \alpha(B)$. The conditional probability of $\Omega$ given $B$ must equal 1, and thus $\alpha \mathbb{P}(\Omega \cap B) = 1$, yielding $\alpha = 1/\mathbb{P}(B)$.

We formalize these notions as follows.

**(1) Definition.** If $\mathbb{P}(B) > 0$ then the **conditional probability** that $A$ occurs given that $B$ occurs is defined to be

$$\mathbb{P}(A \mid B) = \frac{\mathbb{P}(A \cap B)}{\mathbb{P}(B)}.$$

We denote this conditional probability by $\mathbb{P}(A \mid B)$, pronounced 'the probability of $A$ given $B$', or sometimes 'the probability of $A$ conditioned (or conditional) on $B$'.

**(2) Example.** Two fair dice are thrown. Given that the first shows 3, what is the probability that the total exceeds 6? The answer is obviously $\frac{1}{2}$, since the second must show 4, 5, or 6. However, let us labour the point. Clearly $\Omega = \{1, 2, 3, 4, 5, 6\}^2$, the set† of all ordered pairs $(i, j)$ for $i, j \in \{1, 2, \ldots, 6\}$, and we can take $\mathcal{F}$ to be the set of all subsets of $\Omega$, with $\mathbb{P}(A) = |A|/36$ for any $A \subseteq \Omega$. Let $B$ be the event that the first die shows 3, and $A$ be the event that the total exceeds 6. Then

$$B = \{(3, b) : 1 \leq b \leq 6\}, \quad A = \{(a, b) : a + b > 6\}, \quad A \cap B = \{(3, 4), (3, 5), (3, 6)\},$$

and

$$\mathbb{P}(A \mid B) = \frac{\mathbb{P}(A \cap B)}{\mathbb{P}(B)} = \frac{|A \cap B|}{|B|} = \frac{3}{6}. \qquad \bullet$$

**(3) Example. Boys and girls.** A family has two children. What is the probability that both are boys, given that at least one is a boy? The older and younger child may each be male or female, so there are four possible combinations of sexes, which we assume to be equally likely. Hence we can represent the sample space in the obvious way as

$$\Omega = \{GG, GB, BG, BB\}$$

where $\mathbb{P}(GG) = \mathbb{P}(BB) = \mathbb{P}(GB) = \mathbb{P}(BG) = \frac{1}{4}$. From the definition of conditional probability,

$$\mathbb{P}(BB \mid \text{one boy at least}) = \mathbb{P}(BB \mid GB \cup BG \cup BB)$$

$$= \frac{\mathbb{P}(BB \cap (GB \cup BG \cup BB))}{\mathbb{P}(GB \cup BG \cup BB)}$$

$$= \frac{\mathbb{P}(BB)}{\mathbb{P}(GB \cup BG \cup BB)} = \frac{1}{3}.$$

---

†Remember that $A \times B = \{(a, b) : a \in A, \ b \in B\}$ and that $A \times A = A^2$.

A popular but incorrect answer to the question is $\frac{1}{2}$. This is the correct answer to another question: for a family with two children, what is the probability that both are boys given that the younger is a boy? In this case,

$$\mathbb{P}(\text{BB} \mid \text{younger is a boy}) = \mathbb{P}(\text{BB} \mid \text{GB} \cup \text{BB})$$
$$= \frac{\mathbb{P}(\text{BB} \cap (\text{GB} \cup \text{BB}))}{\mathbb{P}(\text{GB} \cup \text{BB})} = \frac{\mathbb{P}(\text{BB})}{\mathbb{P}(\text{GB} \cup \text{BB})} = \frac{1}{2}.$$

The usual dangerous argument contains the assertion

$$\mathbb{P}(\text{BB} \mid \text{one child is a boy}) = \mathbb{P}(\text{other child is a boy}).$$

Why is this meaningless? [Hint: Consider the sample space.]                    ●

The next lemma is crucially important in probability theory. A family $B_1, B_2, \ldots, B_n$ of events is called a *partition* of the set $\Omega$ if

$$B_i \cap B_j = \varnothing \quad \text{when} \quad i \neq j, \quad \text{and} \quad \bigcup_{i=1}^{n} B_i = \Omega.$$

Each elementary event $\omega \in \Omega$ belongs to exactly one set in a partition of $\Omega$.

**(4) Lemma.** *For any events $A$ and $B$ such that $0 < \mathbb{P}(B) < 1$,*

$$\mathbb{P}(A) = \mathbb{P}(A \mid B)\mathbb{P}(B) + \mathbb{P}(A \mid B^c)\mathbb{P}(B^c).$$

*More generally, let $B_1, B_2, \ldots, B_n$ be a partition of $\Omega$ such that $\mathbb{P}(B_i) > 0$ for all $i$. Then*

$$\mathbb{P}(A) = \sum_{i=1}^{n} \mathbb{P}(A \mid B_i)\mathbb{P}(B_i).$$

Lemma (4) is often called the *law of total probability* or the *partition theorem for probabilities*.

**Proof.** $A = (A \cap B) \cup (A \cap B^c)$. This is a disjoint union and so

$$\mathbb{P}(A) = \mathbb{P}(A \cap B) + \mathbb{P}(A \cap B^c)$$
$$= \mathbb{P}(A \mid B)\mathbb{P}(B) + \mathbb{P}(A \mid B^c)\mathbb{P}(B^c).$$

The second part is similar (see Problem (1.8.10)).                    ■

**(5) Example.** We are given two urns, each containing a collection of coloured balls. Urn I contains two white and three blue balls, whilst urn II contains three white and four blue balls. A ball is drawn at random from urn I and put into urn II, and then a ball is picked at random from urn II and examined. What is the probability that it is blue? We assume unless otherwise specified that a ball picked randomly from any urn is equally likely to be any of those present. The reader will be relieved to know that we no longer need to describe $(\Omega, \mathcal{F}, \mathbb{P})$ in detail; we are confident that we could do so if necessary. Clearly, the colour of the final ball depends on the colour of the ball picked from urn I. So let us 'condition' on this. Let $A$ be the event that the final ball is blue, and let $B$ be the event that the first one picked was blue. Then, by Lemma (4),

$$\mathbb{P}(A) = \mathbb{P}(A \mid B)\mathbb{P}(B) + \mathbb{P}(A \mid B^c)\mathbb{P}(B^c).$$

We can easily find all these probabilities:

$$\mathbb{P}(A \mid B) = \mathbb{P}(A \mid \text{urn II contains three white and five blue balls}) = \tfrac{5}{8},$$

$$\mathbb{P}(A \mid B^c) = \mathbb{P}(A \mid \text{urn II contains four white and four blue balls}) = \tfrac{1}{2},$$

$$\mathbb{P}(B) = \tfrac{3}{5}, \qquad \mathbb{P}(B^c) = \tfrac{2}{5}.$$

Hence

$$\mathbb{P}(A) = \tfrac{5}{8} \cdot \tfrac{3}{5} + \tfrac{1}{2} \cdot \tfrac{2}{5} = \tfrac{23}{40}. \qquad \bullet$$

Unprepared readers may have been surprised by the sudden appearance of urns in this book. In the seventeenth and eighteenth centuries, lotteries often involved the drawing of slips from urns, and voting was often a matter of putting slips or balls into urns. In France today, *aller aux urnes* is synonymous with voting. It was therefore not unnatural for the numerous Bernoullis and others to model births, marriages, deaths, fluids, gases, and so on, using urns containing balls of varied hue.†

**(6) Example. Zoggles.** Only two factories manufacture zoggles. 20 per cent of the zoggles from factory I and 5 per cent from factory II are defective. Factory I produces twice as many zoggles as factory II each week. What is the probability that a zoggle, randomly chosen from a week's production, is satisfactory? Clearly this satisfaction depends on the factory of origin. Let $A$ be the event that the chosen zoggle is satisfactory, and let $B$ be the event that it was made in factory I. Arguing as before,

$$\mathbb{P}(A) = \mathbb{P}(A \mid B)\mathbb{P}(B) + \mathbb{P}(A \mid B^c)\mathbb{P}(B^c)$$
$$= \tfrac{4}{5} \cdot \tfrac{2}{3} + \tfrac{19}{20} \cdot \tfrac{1}{3} = \tfrac{51}{60}.$$

If the chosen zoggle is defective, what is the probability that it came from factory I? In our notation this is just $\mathbb{P}(B \mid A^c)$. However,

$$\mathbb{P}(B \mid A^c) = \frac{\mathbb{P}(B \cap A^c)}{\mathbb{P}(A^c)} = \frac{\mathbb{P}(A^c \mid B)\mathbb{P}(B)}{\mathbb{P}(A^c)} = \frac{\tfrac{1}{5} \cdot \tfrac{2}{3}}{1 - \tfrac{51}{60}} = \frac{8}{9}. \qquad \bullet$$

This section is terminated with a cautionary example. It is not untraditional to perpetuate errors of logic in calculating conditional probabilities. Lack of unambiguous definitions and notation has led astray many probabilists, including even Boole, who was credited by Russell with the discovery of pure mathematics and by others for some of the logical foundations of computing. The well-known 'prisoners' paradox' also illustrates some of the dangers here.

**(7) Example. Prisoners' paradox.** In a dark country, three prisoners have been incarcerated without trial. Their warder tells them that the country's dictator has decided arbitrarily to free

---

†Readers favouring applications over thought experiments may wish to consider fish in lakes rather than balls in urns.

one of them and to shoot the other two, but he is not permitted to reveal to any prisoner the fate of that prisoner. Prisoner A knows therefore that his chance of survival is $\frac{1}{3}$. In order to gain information, he asks the warder to tell him in secret the name of some prisoner (but not himself) who will be killed, and the warder names prisoner B. What now is prisoner A's assessment of the chance that he will survive? Could it be $\frac{1}{2}$: after all, he knows now that the survivor will be either A or C, and he has no information about which? Could it be $\frac{1}{3}$: after all, according to the rules, at least one of B and C has to be killed, and thus the extra information cannot reasonably affect A's earlier calculation of the odds? What does the reader think about this? The resolution of the paradox lies in the situation when either response (B or C) is possible.

An alternative formulation of this paradox has become known as the Monty Hall problem, the controversy associated with which has been provoked by Marilyn vos Savant (and many others) in *Parade* magazine in 1990; see Exercise (1.4.5).                     ●

---

## Exercises for Section 1.4

**1.** Prove that $\mathbb{P}(A \mid B) = \mathbb{P}(B \mid A)\mathbb{P}(A)/\mathbb{P}(B)$ whenever $\mathbb{P}(A)\mathbb{P}(B) \neq 0$. Show that, if $\mathbb{P}(A \mid B) > \mathbb{P}(A)$, then $\mathbb{P}(B \mid A) > \mathbb{P}(B)$.

**2.** For events $A_1, A_2, \ldots, A_n$ satisfying $\mathbb{P}(A_1 \cap A_2 \cap \cdots \cap A_{n-1}) > 0$, prove that

$$\mathbb{P}(A_1 \cap A_2 \cap \cdots \cap A_n) = \mathbb{P}(A_1)\mathbb{P}(A_2 \mid A_1)\mathbb{P}(A_3 \mid A_1 \cap A_2) \cdots \mathbb{P}(A_n \mid A_1 \cap A_2 \cap \cdots \cap A_{n-1}).$$

**3.** A man possesses five coins, two of which are double-headed, one is double-tailed, and two are normal. He shuts his eyes, picks a coin at random, and tosses it. What is the probability that the lower face of the coin is a head?

He opens his eyes and sees that the coin is showing heads; what is the probability that the lower face is a head? He shuts his eyes again, and tosses the coin again. What is the probability that the lower face is a head? He opens his eyes and sees that the coin is showing heads; what is the probability that the lower face is a head?

He discards this coin, picks another at random, and tosses it. What is the probability that it shows heads?

**4.** What do you think of the following 'proof' by Lewis Carroll that an urn cannot contain two balls of the same colour? Suppose that the urn contains two balls, each of which is either black or white; thus, in the obvious notation, $\mathbb{P}(BB) = \mathbb{P}(BW) = \mathbb{P}(WB) = \mathbb{P}(WW) = \frac{1}{4}$. We add a black ball, so that $\mathbb{P}(BBB) = \mathbb{P}(BBW) = \mathbb{P}(BWB) = \mathbb{P}(BWW) = \frac{1}{4}$. Next we pick a ball at random; the chance that the ball is black is (using conditional probabilities) $1 \cdot \frac{1}{4} + \frac{2}{3} \cdot \frac{1}{4} + \frac{2}{3} \cdot \frac{1}{4} + \frac{1}{3} \cdot \frac{1}{4} = \frac{2}{3}$. However, if there is probability $\frac{2}{3}$ that a ball, chosen randomly from three, is black, then there must be two black and one white, which is to say that originally there was one black and one white ball in the urn.

**5. The Monty Hall problem: goats and cars.** (a) In a game show; you have to choose one of three doors. One conceals a new car, two conceal old goats. You choose, but your chosen door is not opened immediately. Instead the presenter opens another door, which reveals a goat. He offers you the opportunity to change your choice to the third door (unopened and so far unchosen). Let $p$ be the (conditional) probability that the third door conceals the car. The presenter's protocol is:

(i) he is determined to show you a goat; with a choice of two, he picks one at random. Show $p = \frac{2}{3}$.

(ii) he is determined to show you a goat; with a choice of two goats (Bill and Nan, say) he shows you Bill with probability $b$. Show that, given you see Bill, the probability is $1/(1+b)$.

(iii) he opens a door chosen at random irrespective of what lies behind. Show $p = \frac{1}{2}$.

(b) Show that, for $\alpha \in [\frac{1}{2}, \frac{2}{3}]$, there exists a protocol such that $p = \alpha$. Are you well advised to change your choice to the third door?

(c) In a variant of this question, the presenter is permitted to open the first door chosen, and to reward you with whatever lies behind. If he chooses to open another door, then this door invariably conceals a goat. Let $p$ be the probability that the unopened door conceals the car, conditional on the presenter having chosen to open a second door. Devise protocols to yield the values $p = 0$, $p = 1$, and deduce that, for any $\alpha \in [0, 1]$, there exists a protocol with $p = \alpha$.

**6. The prosecutor's fallacy†.** Let $G$ be the event that an accused is guilty, and $T$ the event that some testimony is true. Some lawyers have argued on the assumption that $\mathbb{P}(G \mid T) = \mathbb{P}(T \mid G)$. Show that this holds if and only if $\mathbb{P}(G) = \mathbb{P}(T)$.

**7. Urns.** There are $n$ urns of which the $r$th contains $r - 1$ red balls and $n - r$ magenta balls. You pick an urn at random and remove two balls at random without replacement. Find the probability that:
(a) the second ball is magenta;
(b) the second ball is magenta, given that the first is magenta.

**8. Boys and girls, Example (1.4.3) revisited.** Consider a family of two children in a population in which each child is equally likely to be male as female; each child has red hair with probability $r$; these characteristics are independent of each other and occur independently between children.

What is the probability that both children are boys given that the family contains at least one red-haired boy? Show that the probability that both are boys, given that one is a boy born on a Monday, is $13/27$.

## 1.5 Independence

In general, the occurrence of some event $B$ changes the probability that another event $A$ occurs, the original probability $\mathbb{P}(A)$ being replaced by $\mathbb{P}(A \mid B)$. If the probability remains unchanged, that is to say $\mathbb{P}(A \mid B) = \mathbb{P}(A)$, then we call $A$ and $B$ 'independent'. This is well defined only if $\mathbb{P}(B) > 0$. Definition (1.4.1) of conditional probability leads us to the following.

**(1) Definition.** Events $A$ and $B$ are called **independent** if

$$\mathbb{P}(A \cap B) = \mathbb{P}(A)\mathbb{P}(B).$$

More generally, a family $\{A_i : i \in I\}$ is called **independent** if

$$\mathbb{P}\left(\bigcap_{i \in J} A_i\right) = \prod_{i \in J} \mathbb{P}(A_i)$$

for all finite subsets $J$ of $I$.

**Remark.** A common student error is to make the fallacious statement that $A$ and $B$ are independent if $A \cap B = \varnothing$.

It is straightforward to prove that two events $A$, $B$ are independent if and only if $A$, $B^c$ are independent. Similarly, for $J \subseteq I$, a family $\{A_i : i \in I\}$ is independent if and only if the family $\{A_i : i \in J\} \cup \{A_k : k \in I \setminus J\}$ is independent. See Exercise (1.5.10).

---

†The prosecution made this error in the famous Dreyfus case of 1894.

If the family $\{A_i : i \in I\}$ has the property that

$$\mathbb{P}(A_i \cap A_j) = \mathbb{P}(A_i)\mathbb{P}(A_j) \qquad \text{for all } i \neq j$$

then it is called *pairwise independent*. Pairwise-independent families are not necessarily independent, as the following example shows.

**(2) Example.** Suppose $\Omega = \{abc, acb, cab, cba, bca, bac, aaa, bbb, ccc\}$, and each of the nine elementary events in $\Omega$ occurs with equal probability $\frac{1}{9}$. Let $A_k$ be the event that the $k$th letter is $a$. It is left as an *exercise* to show that the family $\{A_1, A_2, A_3\}$ is pairwise independent but not independent. ●

**(3) Example (1.4.6) revisited.** The events $A$ and $B$ of this example are clearly dependent because $\mathbb{P}(A \mid B) = \frac{4}{5}$ and $\mathbb{P}(A) = \frac{51}{60}$. ●

**(4) Example.** Choose a card at random from a pack of 52 playing cards, each being picked with equal probability $\frac{1}{52}$. We claim that the suit of the chosen card is independent of its rank. For example,

$$\mathbb{P}(\text{king}) = \tfrac{4}{52}, \quad \mathbb{P}(\text{king} \mid \text{spade}) = \tfrac{1}{13}.$$

Alternatively,

$$\mathbb{P}(\text{spade king}) = \tfrac{1}{52} = \tfrac{1}{4} \cdot \tfrac{1}{13} = \mathbb{P}(\text{spade})\mathbb{P}(\text{king}).$$ ●

Let $C$ be an event with $\mathbb{P}(C) > 0$. To the conditional probability measure $\mathbb{P}(\cdot \mid C)$ corresponds the idea of *conditional independence*. Two events $A$ and $B$ are called *conditionally independent given $C$* if

**(5)** $$\mathbb{P}(A \cap B \mid C) = \mathbb{P}(A \mid C)\mathbb{P}(B \mid C);$$

there is a natural extension to families of events. [However, note Exercise (1.5.5).]

---

### Exercises for Section 1.5

**1.** Let $A$ and $B$ be independent events; show that $A^c$, $B$ are independent, and deduce that $A^c$, $B^c$ are independent.

**2.** We roll a die $n$ times. Let $A_{ij}$ be the event that the $i$th and $j$th rolls produce the same number. Show that the events $\{A_{ij} : 1 \leq i < j \leq n\}$ are pairwise independent but not independent.

**3.** A fair coin is tossed repeatedly. Show that the following two statements are equivalent:
(a) the outcomes of different tosses are independent,
(b) for any given finite sequence of heads and tails, the chance of this sequence occurring in the first $m$ tosses is $2^{-m}$, where $m$ is the length of the sequence.

**4.** Let $\Omega = \{1, 2, \ldots, p\}$ where $p$ is prime, $\mathcal{F}$ be the set of all subsets of $\Omega$, and $\mathbb{P}(A) = |A|/p$ for all $A \in \mathcal{F}$. Show that, if $A$ and $B$ are independent events, then at least one of $A$ and $B$ is either $\varnothing$ or $\Omega$.

**5.** Show that the conditional independence of $A$ and $B$ given $C$ neither implies, nor is implied by, the independence of $A$ and $B$. For which events $C$ is it the case that, for all $A$ and $B$, the events $A$ and $B$ are independent if and only if they are conditionally independent given $C$?

**6. Safe or sorry?** Some form of prophylaxis is said to be 90 per cent effective at prevention during one year's treatment. If the degrees of effectiveness in different years are independent, show that the treatment is more likely than not to fail within 7 years.

**7.    Families.** Jane has three children, each of which is equally likely to be a boy or a girl independently of the others. Define the events:

$$A = \{\text{all the children are of the same sex}\},$$
$$B = \{\text{there is at most one boy}\},$$
$$C = \{\text{the family includes a boy and a girl}\}.$$

(a) Show that $A$ is independent of $B$, and that $B$ is independent of $C$.
(b) Is $A$ independent of $C$?
(c) Do these results hold if boys and girls are not equally likely?
(d) Do these results hold if Jane has four children?

**8.    Galton's paradox.** You flip three fair coins. At least two are alike, and it is an evens chance that the third is a head or a tail. Therefore $\mathbb{P}(\text{all alike}) = \frac{1}{2}$. Do you agree?

**9.    Two fair dice are rolled. Show that the event that their sum is 7 is independent of the score shown by the first die.

**10.** Let $X$ and $Y$ be the scores on two fair dice taking values in the set $\{1, 2, \ldots, 6\}$. Let $A_1 = \{X + Y = 9\}$, $A_2 = \{X \in \{1, 2, 3\}\}$, and $A_3 = \{X \in \{3, 4, 5\}\}$. Show that

$$\mathbb{P}(A_1 \cap A_2 \cap A_3) = \mathbb{P}(A_1)\mathbb{P}(A_2)\mathbb{P}(A_3).$$

Are these three events independent?

---

# 1.6  Completeness and product spaces

This section should be omitted at the first reading, but we shall require its contents later. It contains only a sketch of complete probability spaces and product spaces; the reader should look elsewhere for a more detailed treatment (see Billingsley 1995). We require the following result.

**(1) Lemma.** *If $\mathcal{F}$ and $\mathcal{G}$ are two $\sigma$-fields of subsets of $\Omega$ then their intersection $\mathcal{F} \cap \mathcal{G}$ is a $\sigma$-field also. More generally, if $\{\mathcal{F}_i : i \in I\}$ is a family of $\sigma$-fields of subsets of $\Omega$ then $\mathcal{G} = \bigcap_{i \in I} \mathcal{F}_i$ is a $\sigma$-field also.*

The proof is not difficult and is left as an *exercise*. Note that the union $\mathcal{F} \cup \mathcal{G}$ may not be a $\sigma$-field, although it may be extended to a unique smallest $\sigma$-field written $\sigma(\mathcal{F} \cup \mathcal{G})$, as follows. Let $\{\mathcal{G}_i : i \in I\}$ be the collection of all $\sigma$-fields which contain both $\mathcal{F}$ and $\mathcal{G}$ as subsets; this collection is non-empty since it contains the set of all subsets of $\Omega$. Then $\mathcal{G} = \bigcap_{i \in I} \mathcal{G}_i$ is the unique smallest $\sigma$-field which contains $\mathcal{F} \cup \mathcal{G}$.

**(A) Completeness.** Let $(\Omega, \mathcal{F}, \mathbb{P})$ be a probability space. Any event $A$ which has zero probability, that is $\mathbb{P}(A) = 0$, is called *null*. It may seem reasonable to suppose that any subset $B$ of a null set $A$ will itself be null, but this may be without meaning since $B$ may not be an event, and thus $\mathbb{P}(B)$ may not be defined.

**(2) Definition.** A probability space $(\Omega, \mathcal{F}, \mathbb{P})$ is called **complete** if all subsets of null sets are events.

Any incomplete space can be completed thus. Let $\mathcal{N}$ be the collection of all subsets of null sets in $\mathcal{F}$ and let $\mathcal{G} = \sigma(\mathcal{F} \cup \mathcal{N})$ be the smallest $\sigma$-field which contains all sets in $\mathcal{F}$

and $\mathcal{N}$. It can be shown that the domain of $\mathbb{P}$ may be extended in an obvious way from $\mathcal{F}$ to $\mathcal{G}$; $(\Omega, \mathcal{G}, \mathbb{P})$ is called the *completion* of $(\Omega, \mathcal{F}, \mathbb{P})$.

**(B) Product spaces.** The probability spaces discussed in this chapter have usually been constructed around the outcomes of one experiment, but instances occur naturally when we need to combine the outcomes of several independent experiments into one space (see Examples (1.2.4) and (1.4.2)). How should we proceed in general?

Suppose two experiments have associated probability spaces $(\Omega_1, \mathcal{F}_1, \mathbb{P}_1)$ and $(\Omega_2, \mathcal{F}_2, \mathbb{P}_2)$ respectively. The sample space of the pair of experiments, considered jointly, is the collection $\Omega_1 \times \Omega_2 = \{(\omega_1, \omega_2) : \omega_1 \in \Omega_1, \omega_2 \in \Omega_2\}$ of ordered pairs. The appropriate $\sigma$-field of events is more complicated to construct. Certainly it should contain all subsets of $\Omega_1 \times \Omega_2$ of the form $A_1 \times A_2 = \{(a_1, a_2) : a_1 \in A_1, a_2 \in A_2\}$ where $A_1$ and $A_2$ are typical members of $\mathcal{F}_1$ and $\mathcal{F}_2$ respectively. However, the family of all such sets, $\mathcal{F}_1 \times \mathcal{F}_2 = \{A_1 \times A_2 : A_1 \in \mathcal{F}_1, A_2 \in \mathcal{F}_2\}$, is not in general a $\sigma$-field. By the discussion after (1), there exists a unique smallest $\sigma$-field $\mathcal{G} = \sigma(\mathcal{F}_1 \times \mathcal{F}_2)$ of subsets of $\Omega_1 \times \Omega_2$ which contains $\mathcal{F}_1 \times \mathcal{F}_2$. All we require now is a suitable probability function on $(\Omega_1 \times \Omega_2, \mathcal{G})$. Let $\mathbb{P}_{12} : \mathcal{F}_1 \times \mathcal{F}_2 \to [0, 1]$ be given by:

(3)                    $$\mathbb{P}_{12}(A_1 \times A_2) = \mathbb{P}_1(A_1)\mathbb{P}_2(A_2) \quad \text{for } A_1 \in \mathcal{F}_1, \ A_2 \in \mathcal{F}_2.$$

It can be shown that the domain of $\mathbb{P}_{12}$ can be extended from $\mathcal{F}_1 \times \mathcal{F}_2$ to the whole of $\mathcal{G} = \sigma(\mathcal{F}_1 \times \mathcal{F}_2)$. The ensuing probability space $(\Omega_1 \times \Omega_2, \mathcal{G}, \mathbb{P}_{12})$ is called the *product space* of $(\Omega_1, \mathcal{F}_1, \mathbb{P}_1)$ and $(\Omega_2, \mathcal{F}_2, \mathbb{P}_2)$. Products of larger numbers of spaces are constructed similarly. The measure $\mathbb{P}_{12}$ is sometimes called the 'product measure' since its defining equation (3) assumed that two experiments are independent. There are of course many other measures that can be applied to $(\Omega_1 \times \Omega_2, \mathcal{G})$.

In many simple cases this technical discussion is unnecessary. Suppose that $\Omega_1$ and $\Omega_2$ are finite, and that their $\sigma$-fields contain all their subsets; this is the case in Examples (1.2.4) and (1.4.2). Then $\mathcal{G}$ contains all subsets of $\Omega_1 \times \Omega_2$.

---

## 1.7 Worked examples

Here are some more examples to illustrate the ideas of this chapter. The reader is now equipped to try his or her hand at a substantial number of those problems which exercised the pioneers in probability. These frequently involved experiments having equally likely outcomes, such as dealing whist hands, putting balls of various colours into urns and taking them out again, throwing dice, and so on. In many such instances, the reader will be pleasantly surprised to find that it is not necessary to write down $(\Omega, \mathcal{F}, \mathbb{P})$ explicitly, but only to think of $\Omega$ as being a collection $\{\omega_1, \omega_2, \ldots, \omega_N\}$ of possibilities, each of which may occur with probability $1/N$. (Recall Example (1.3.6).) Thus, $\mathbb{P}(A) = |A|/N$ for any $A \subseteq \Omega$. The basic tools used in such problems are as follows.

  (a) Combinatorics: remember that the number of permutations of $n$ objects is $n!$ and that the number of ways of choosing $r$ objects from $n$ is $\binom{n}{r}$.
  (b) Set theory: to obtain $\mathbb{P}(A)$ we can compute $\mathbb{P}(A^c) = 1 - \mathbb{P}(A)$ or we can partition $A$ by conditioning on events $B_i$, and then use Lemma (1.4.4).
  (c) Use of independence.

**(1) Example.** Consider a series of hands dealt at bridge. Let $A$ be the event that in a given deal each player has one ace. Show that the probability that $A$ occurs at least once in seven deals is approximately $\frac{1}{2}$.

**Solution.** The number of ways of dealing 52 cards into four equal hands is $52!/(13!)^4$. There are 4! ways of distributing the aces so that each hand holds one, and there are $48!/(12!)^4$ ways of dealing the remaining cards. Thus

$$\mathbb{P}(A) = \frac{4!\,48!/(12!)^4}{52!/(13!)^4} \simeq \frac{1}{10}.$$

Now let $B_i$ be the event that $A$ occurs for the first time on the $i$th deal. Clearly $B_i \cap B_j = \varnothing$, $i \neq j$. Thus

$$\mathbb{P}(A \text{ occurs in seven deals}) = \mathbb{P}(B_1 \cup \cdots \cup B_7) = \sum_1^7 \mathbb{P}(B_i) \quad \text{using Definition (1.3.1).}$$

Since successive deals are independent, we have

$$\mathbb{P}(B_i) = \mathbb{P}\big(A^c \text{ occurs on deal 1,} \ A^c \text{ occurs on deal 2,}$$
$$\ldots, \ A^c \text{ occurs on deal } i-1, \ A \text{ occurs on deal } i\big)$$
$$= \mathbb{P}(A^c)^{i-1}\mathbb{P}(A) \quad \text{using Definition (1.5.1)}$$
$$\simeq \left(1 - \tfrac{1}{10}\right)^{i-1} \tfrac{1}{10}.$$

Thus

$$\mathbb{P}(A \text{ occurs in seven deals}) = \sum_1^7 \mathbb{P}(B_i) \simeq \sum_1^7 \left(\tfrac{9}{10}\right)^{i-1} \tfrac{1}{10} \simeq \tfrac{1}{2}.$$

Can you see an easier way of obtaining this answer?                                 ●

**(2) Example.** There are two roads from A to B and two roads from B to C. Each of the four roads has probability $p$ of being blocked by snow, independently of all the others. What is the probability that there is an open road from A to C?

**Solution.**

$$\mathbb{P}(\text{open road}) = \mathbb{P}\big((\text{open road from A to B}) \cap (\text{open road from B to C})\big)$$
$$= \mathbb{P}(\text{open road from A to B})\mathbb{P}(\text{open road from B to C})$$

using the independence. However, $p$ is the same for all roads; thus, using Lemma (1.3.4),

$$\mathbb{P}(\text{open road}) = \big(1 - \mathbb{P}(\text{no road from A to B})\big)^2$$
$$= \big\{1 - \mathbb{P}\big((\text{first road blocked}) \cap (\text{second road blocked})\big)\big\}^2$$
$$= \big\{1 - \mathbb{P}(\text{first road blocked})\mathbb{P}(\text{second road blocked})\big\}^2$$

using the independence. Thus

**(3)**                                $$\mathbb{P}(\text{open road}) = (1 - p^2)^2.$$

Further suppose that there is also a direct road from A to C, which is independently blocked with probability $p$. Then, by Lemma (1.4.4) and equation (3),

$$\mathbb{P}(\text{open road}) = \mathbb{P}(\text{open road} \mid \text{direct road blocked}) \cdot p$$
$$+ \mathbb{P}(\text{open road} \mid \text{direct road open}) \cdot (1 - p)$$
$$= (1 - p^2)^2 \cdot p + 1 \cdot (1 - p). \qquad \bullet$$

**(4) Example. Symmetric random walk (or 'Gambler's ruin').** A man is saving up to buy a new Jaguar at a cost of $N$ units of money. He starts with $k$ units where $0 < k < N$, and tries to win the remainder by the following gamble with his bank manager. He tosses a fair coin repeatedly; if it comes up heads then the manager pays him one unit, but if it comes up tails then he pays the manager one unit. He plays this game repeatedly until one of two events occurs: either he runs out of money and is bankrupted or he wins enough to buy the Jaguar. What is the probability that he is ultimately bankrupted?

**Solution.** This is one of many problems the solution to which proceeds by the construction of a linear difference equation subject to certain boundary conditions. Let $A$ denote the event that he is eventually bankrupted, and let $B$ be the event that the first toss of the coin shows heads. By Lemma (1.4.4),

$$(5) \qquad \mathbb{P}_k(A) = \mathbb{P}_k(A \mid B)\mathbb{P}(B) + \mathbb{P}_k(A \mid B^c)\mathbb{P}(B^c),$$

where $\mathbb{P}_k$ denotes probabilities calculated relative to the starting point $k$. We want to find $\mathbb{P}_k(A)$. Consider $\mathbb{P}_k(A \mid B)$. If the first toss is a head then his capital increases to $k + 1$ units and the game starts afresh from a different starting point. Thus $\mathbb{P}_k(A \mid B) = \mathbb{P}_{k+1}(A)$ and similarly $\mathbb{P}_k(A \mid B^c) = \mathbb{P}_{k-1}(A)$. So, writing $p_k = \mathbb{P}_k(A)$, (5) becomes

$$(6) \qquad p_k = \tfrac{1}{2}(p_{k+1} + p_{k-1}) \quad \text{if} \quad 0 < k < N,$$

which is a linear difference equation subject to the boundary conditions $p_0 = 1$, $p_N = 0$. The analytical solution to such equations is routine, and we shall return later to the general method of solution. In this case we can proceed directly. We put $b_k = p_k - p_{k-1}$ to obtain $b_k = b_{k-1}$ and hence $b_k = b_1$ for all $k$. Thus

$$p_k = b_1 + p_{k-1} = 2b_1 + p_{k-2} = \cdots = kb_1 + p_0$$

is the general solution to (6). The boundary conditions imply that $p_0 = 1$, $b_1 = -1/N$, giving

$$(7) \qquad \mathbb{P}_k(A) = 1 - \frac{k}{N}.$$

As the price of the Jaguar rises, that is as $N \to \infty$, ultimate bankruptcy becomes very likely. This is the problem of the 'symmetric random walk with two absorbing barriers' to which we shall return in more generality later. $\qquad \bullet$

**Remark.** Our experience of student calculations leads us to stress that probabilities lie between zero and one; any calculated probability which violates this must be incorrect.

**(8) Example. Testimony.**  A court is investigating the possible occurrence of an unlikely event $T$. The reliability of two independent witnesses called Alf and Bob is known to the court: Alf tells the truth with probability $\alpha$ and Bob with probability $\beta$, and there is no collusion between the two of them. Let $A$ and $B$ be the events that Alf and Bob assert (respectively) that $T$ occurred, and let $\tau = \mathbb{P}(T)$. What is the probability that $T$ occurred given that both Alf and Bob declare that $T$ occurred?

**Solution.**  We are asked to calculate $\mathbb{P}(T \mid A \cap B)$, which is equal to $\mathbb{P}(T \cap A \cap B)/\mathbb{P}(A \cap B)$. Now $\mathbb{P}(T \cap A \cap B) = \mathbb{P}(A \cap B \mid T)\mathbb{P}(T)$ and

$$\mathbb{P}(A \cap B) = \mathbb{P}(A \cap B \mid T)\mathbb{P}(T) + \mathbb{P}(A \cap B \mid T^c)\mathbb{P}(T^c).$$

We have from the independence of the witnesses that $A$ and $B$ are conditionally independent given either $T$ or $T^c$. Therefore

$$\mathbb{P}(A \cap B \mid T) = \mathbb{P}(A \mid T)\mathbb{P}(B \mid T) = \alpha\beta,$$
$$\mathbb{P}(A \cap B \mid T^c) = \mathbb{P}(A \mid T^c)\mathbb{P}(B \mid T^c) = (1 - \alpha)(1 - \beta),$$

so that

$$\mathbb{P}(T \mid A \cap B) = \frac{\alpha\beta\tau}{\alpha\beta\tau + (1 - \alpha)(1 - \beta)(1 - \tau)}.$$

As an example, suppose that $\alpha = \beta = \frac{9}{10}$ and $\tau = 1/1000$. Then $\mathbb{P}(T \mid A \cap B) = 81/1080$, which is somewhat small as a basis for a judicial conclusion.

This calculation may be informative. However, it is generally accepted that such an application of the axioms of probability is inappropriate to questions of truth and belief.    ●

**(9) Example. Zoggles revisited.**  A new process for the production of zoggles is invented, and both factories of Example (1.4.6) install extra production lines using it. The new process is cheaper but produces fewer reliable zoggles, only 75 per cent of items produced in this new way being reliable.

Factory I fails to implement its new production line efficiently, and only 10 per cent of its output is made in this manner. Factory II does better: it produces 20 per cent of its output by the new technology, and now produces twice as many zoggles in all as Factory I.

Is the new process beneficial to the consumer?

**Solution.**  Both factories now produce a higher proportion of unreliable zoggles than before, and so it might seem at first sight that there is an increased proportion of unreliable zoggles on the market.

Let $A$ be the event that a randomly chosen zoggle is satisfactory, $B$ the event that it came from factory I, and $C$ the event that it was made by the new method. Then

$$\begin{aligned}
\mathbb{P}(A) &= \tfrac{1}{3}\mathbb{P}(A \mid B) + \tfrac{2}{3}\mathbb{P}(A \mid B^c) \\
&= \tfrac{1}{3}\left( \tfrac{1}{10}\mathbb{P}(A \mid B \cap C) + \tfrac{9}{10}\mathbb{P}(A \mid B \cap C^c) \right) \\
&\quad + \tfrac{2}{3}\left( \tfrac{1}{5}\mathbb{P}(A \mid B^c \cap C) + \tfrac{4}{5}\mathbb{P}(A \mid B^c \cap C^c) \right) \\
&= \tfrac{1}{3}\left( \tfrac{1}{10}\cdot\tfrac{3}{4} + \tfrac{9}{10}\cdot\tfrac{4}{5} \right) + \tfrac{2}{3}\left( \tfrac{1}{5}\cdot\tfrac{3}{4} + \tfrac{4}{5}\cdot\tfrac{19}{20} \right) = \tfrac{523}{600} > \tfrac{51}{60},
\end{aligned}$$

so that the proportion of satisfactory zoggles has been increased.    ●

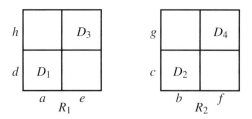

Figure 1.1. Two unions of rectangles illustrating Simpson's paradox.

**(10) Example. Simpson's paradox†.**    A doctor has performed clinical trials to determine the relative efficacies of two drugs, with the following results.

|         | Women |         | Men    |         |
|---------|-------|---------|--------|---------|
|         | Drug I | Drug II | Drug I | Drug II |
| Success | 200   | 10      | 19     | 1000    |
| Failure | 1800  | 190     | 1      | 1000    |

Which drug is the better? Here are two conflicting responses.

1. Drug I was given to 2020 people, of whom 219 were cured.  The success rate was 219/2020, which is much smaller than the corresponding figure, 1010/2200, for drug II. Therefore drug II is better than drug I.
2. Amongst women the success rates of the drugs are 1/10 and 1/20, and amongst men 19/20 and 1/2. Drug I wins in both cases.

This well-known statistical paradox may be reformulated in the following more general way.  Given three events $A, B, C$, it is possible to allocate probabilities such that

**(11)**       $\mathbb{P}(A \mid B \cap C) > \mathbb{P}(A \mid B^c \cap C)$    and    $\mathbb{P}(A \mid B \cap C^c) > \mathbb{P}(A \mid B^c \cap C^c)$

but

**(12)**                                $\mathbb{P}(A \mid B) < \mathbb{P}(A \mid B^c).$

We may think of $A$ as the event that treatment is successful, $B$ as the event that drug I is given to a randomly chosen individual, and $C$ as the event that this individual is female. The above inequalities imply that $B$ is preferred to $B^c$ when $C$ occurs and when $C^c$ occurs, but $B^c$ is preferred to $B$ overall.

---

†This paradox, named after Simpson (1951), was remarked by Yule in 1903, having been observed previously by Pearson in 1899. The nomenclature is an instance of Stigler's law of eponymy: "No law, theorem, or discovery is named after its originator". This law applies to many eponymous statements in this book, including the law itself. As remarked by A. N. Whitehead, "Everything of importance has been said before, by somebody who did not discover it".

Setting

$$a = \mathbb{P}(A \cap B \cap C), \qquad b = \mathbb{P}(A^c \cap B \cap C),$$
$$c = \mathbb{P}(A \cap B^c \cap C), \qquad d = \mathbb{P}(A^c \cap B^c \cap C),$$
$$e = \mathbb{P}(A \cap B \cap C^c), \qquad f = \mathbb{P}(A^c \cap B \cap C^c),$$
$$g = \mathbb{P}(A \cap B^c \cap C^c), \qquad h = \mathbb{P}(A^c \cap B^c \cap C^c),$$

and expanding (11)–(12), we arrive at the (equivalent) inequalities

**(13)** $\qquad ad > bc, \quad eh > fg, \quad (a+e)(d+h) < (b+f)(c+g),$

subject to the conditions $a, b, c, \ldots, h \geq 0$ and $a + b + c + \cdots + h = 1$. Inequalities (13) are equivalent to the existence of two rectangles $R_1$ and $R_2$, as in Figure 1.1, satisfying

$$\text{area}(D_1) > \text{area}(D_2), \quad \text{area}(D_3) > \text{area}(D_4), \quad \text{area}(R_1) < \text{area}(R_2).$$

Many such rectangles may be found, by inspection, as for example those with $a = \frac{3}{30}, b = \frac{1}{30}, c = \frac{8}{30}, d = \frac{3}{30}, e = \frac{3}{30}, f = \frac{8}{30}, g = \frac{1}{30}, h = \frac{3}{30}$. Similar conclusions are valid for finer partitions $\{C_i : i \in I\}$ of the sample space, though the corresponding pictures are harder to draw.

Simpson's paradox has arisen many times in practical situations. There are many well-known cases, including the admission of graduate students to the University of California at Berkeley and a clinical trial comparing treatments for kidney stones.   ●

**(14) Example. False positives.** A rare disease affects one person in $10^5$. A test for the disease shows positive with probability $\frac{99}{100}$ when applied to an ill person, and with probability $\frac{1}{100}$ when applied to a healthy person. What is the probability that you have the disease given that the test shows positive?

**Solution.** In the obvious notation,

$$\mathbb{P}(\text{ill} \mid +) = \frac{\mathbb{P}(+ \mid \text{ill})\mathbb{P}(\text{ill})}{\mathbb{P}(+ \mid \text{ill})\mathbb{P}(\text{ill}) + \mathbb{P}(+ \mid \text{healthy})\mathbb{P}(\text{healthy})}$$
$$= \frac{\frac{99}{100} \cdot 10^{-5}}{\frac{99}{100} \cdot 10^{-5} + \frac{1}{100}(1 - 10^{-5})} = \frac{99}{99 + 10^5 - 1} \simeq \frac{1}{1011}.$$

The chance of being ill is rather small. Indeed it is more likely that the test was incorrect.   ●

---

## Exercises for Section 1.7

**1.** There are two roads from A to B and two roads from B to C. Each of the four roads is blocked by snow with probability $p$, independently of the others. Find the probability that there is an open road from A to B given that there is no open route from A to C.

If, in addition, there is a direct road from A to C, this road being blocked with probability $p$ independently of the others, find the required conditional probability.

**2.** Calculate the probability that a hand of 13 cards dealt from a normal shuffled pack of 52 contains exactly two kings and one ace. What is the probability that it contains exactly one ace given that it contains exactly two kings?

**3.** A symmetric random walk takes place on the integers $0, 1, 2, \ldots, N$ with absorbing barriers at $0$ and $N$, starting at $k$. Show that the probability that the walk is never absorbed is zero.

**4.** The so-called 'sure thing principle' asserts that if you prefer $x$ to $y$ given $C$, and also prefer $x$ to $y$ given $C^c$, then you surely prefer $x$ to $y$. Agreed?

**5.** A pack contains $m$ cards, labelled $1, 2, \ldots, m$. The cards are dealt out in a random order, one by one. Given that the label of the $k$th card dealt is the largest of the first $k$ cards dealt, what is the probability that it is also the largest in the pack?

**6.** A group of $2b$ friends meet for a bridge soirée. There are $m$ men and $2b - m$ women where $2 \le m \le b$. The group divides into $b$ teams of pairs, formed uniformly at random. What is the probability that no pair comprises 2 men?

---

# 1.8 Problems

**1.** A traditional fair die is thrown twice. What is the probability that:
(a) a six turns up exactly once?
(b) both numbers are odd?
(c) the sum of the scores is 4?
(d) the sum of the scores is divisible by 3?

**2.** A fair coin is thrown repeatedly. What is the probability that on the $n$th throw:
(a) a head appears for the first time?
(b) the numbers of heads and tails to date are equal?
(c) exactly two heads have appeared altogether to date?
(d) at least two heads have appeared to date?

**3.** Let $\mathcal{F}$ and $\mathcal{G}$ be $\sigma$-fields of subsets of $\Omega$.
(a) Use elementary set operations to show that $\mathcal{F}$ is closed under countable intersections; that is, if $A_1, A_2, \ldots$ are in $\mathcal{F}$, then so is $\bigcap_i A_i$.
(b) Let $\mathcal{H} = \mathcal{F} \cap \mathcal{G}$ be the collection of subsets of $\Omega$ lying in both $\mathcal{F}$ and $\mathcal{G}$. Show that $\mathcal{H}$ is a $\sigma$-field.
(c) Show that $\mathcal{F} \cup \mathcal{G}$, the collection of subsets of $\Omega$ lying in either $\mathcal{F}$ or $\mathcal{G}$, is not necessarily a $\sigma$-field.

**4.** Describe the underlying probability spaces for the following experiments:
(a) a biased coin is tossed three times;
(b) two balls are drawn without replacement from an urn which originally contained two ultramarine and two vermilion balls;
(c) a biased coin is tossed repeatedly until a head turns up.

**5.** Show that the probability that *exactly* one of the events $A$ and $B$ occurs is

$$\mathbb{P}(A) + \mathbb{P}(B) - 2\mathbb{P}(A \cap B).$$

**6.** Prove that $\mathbb{P}(A \cup B \cup C) = 1 - \mathbb{P}(A^c \mid B^c \cap C^c)\mathbb{P}(B^c \mid C^c)\mathbb{P}(C^c)$.

**7.** (a) If $A$ is independent of itself, show that $\mathbb{P}(A)$ is 0 or 1.
(b) If $\mathbb{P}(A)$ is 0 or 1, show that $A$ is independent of all events $B$.

**8.** Let $\mathcal{F}$ be a $\sigma$-field of subsets of $\Omega$, and suppose $\mathbb{P} : \mathcal{F} \to [0, 1]$ satisfies: (i) $\mathbb{P}(\Omega) = 1$, and (ii) $\mathbb{P}$ is additive, in that $\mathbb{P}(A \cup B) = \mathbb{P}(A) + \mathbb{P}(B)$ whenever $A \cap B = \varnothing$. Show that $\mathbb{P}(\varnothing) = 0$.

**9.** Suppose $(\Omega, \mathcal{F}, \mathbb{P})$ is a probability space and $B \in \mathcal{F}$ satisfies $\mathbb{P}(B) > 0$. Let $\mathbb{Q} : \mathcal{F} \to [0, 1]$ be defined by $\mathbb{Q}(A) = \mathbb{P}(A \mid B)$. Show that $(\Omega, \mathcal{F}, \mathbb{Q})$ is a probability space. If $C \in \mathcal{F}$ and $\mathbb{Q}(C) > 0$, show that $\mathbb{Q}(A \mid C) = \mathbb{P}(A \mid B \cap C)$; discuss.

**10.** Let $B_1, B_2, \ldots$ be a partition of the sample space $\Omega$, each $B_i$ having positive probability, and show that

$$\mathbb{P}(A) = \sum_{j=1}^{\infty} \mathbb{P}(A \mid B_j)\mathbb{P}(B_j).$$

**11.** Prove **Boole's inequalities**:

$$\mathbb{P}\left(\bigcup_{i=1}^{n} A_i\right) \le \sum_{i=1}^{n} \mathbb{P}(A_i), \qquad \mathbb{P}\left(\bigcap_{i=1}^{n} A_i\right) \ge 1 - \sum_{i=1}^{n} \mathbb{P}(A_i^c).$$

**12.** Prove that

$$\mathbb{P}\left(\bigcap_{1}^{n} A_i\right) = \sum_{i} \mathbb{P}(A_i) - \sum_{i<j} \mathbb{P}(A_i \cup A_j) + \sum_{i<j<k} \mathbb{P}(A_i \cup A_j \cup A_k)$$

$$- \cdots - (-1)^{n} \mathbb{P}(A_1 \cup A_2 \cup \cdots \cup A_n).$$

**13.** Let $A_1, A_2, \ldots, A_n$ be events, and let $N_k$ be the event that exactly $k$ of the $A_i$ occur. Prove the result sometimes referred to as **Waring's theorem**:

$$\mathbb{P}(N_k) = \sum_{i=0}^{n-k} (-1)^i \binom{k+i}{k} S_{k+i}, \text{ where } S_j = \sum_{i_1 < i_2 < \cdots < i_j} \mathbb{P}(A_{i_1} \cap A_{i_2} \cap \cdots \cap A_{i_j}).$$

Use this result to find an expression for the probability that a purchase of six packets of Corn Flakes yields exactly three distinct busts (see Exercise (1.3.4)).

**14.** Prove **Bayes's formula**: if $A_1, A_2, \ldots, A_n$ is a partition of $\Omega$, each $A_i$ having positive probability, then

$$\mathbb{P}(A_j \mid B) = \frac{\mathbb{P}(B \mid A_j)\mathbb{P}(A_j)}{\sum_{1}^{n} \mathbb{P}(B \mid A_i)\mathbb{P}(A_i)}.$$

**15.** A random number $N$ of dice is thrown. Let $A_i$ be the event that $N = i$, and assume that $\mathbb{P}(A_i) = 2^{-i}$, $i \ge 1$. The sum of the scores is $S$. Find the probability that:
(a) $N = 2$ given $S = 4$;
(b) $S = 4$ given $N$ is even;
(c) $N = 2$, given that $S = 4$ and the first die showed 1;
(d) the largest number shown by any die is $r$, where $S$ is unknown.

**16.** Let $A_1, A_2, \ldots$ be a sequence of events. Define

$$B_n = \bigcup_{m=n}^{\infty} A_m, \qquad C_n = \bigcap_{m=n}^{\infty} A_m.$$

Clearly $C_n \subseteq A_n \subseteq B_n$. The sequences $\{B_n\}$ and $\{C_n\}$ are decreasing and increasing respectively with limits

$$\lim B_n = B = \bigcap_n B_n = \bigcap_n \bigcup_{m \ge n} A_m, \qquad \lim C_n = C = \bigcup_n C_n = \bigcup_n \bigcap_{m \ge n} A_m.$$

The events $B$ and $C$ are denoted $\limsup_{n \to \infty} A_n$ and $\liminf_{n \to \infty} A_n$ respectively. Show that
(a) $B = \{\omega \in \Omega : \omega \in A_n \text{ for infinitely many values of } n\}$,
(b) $C = \{\omega \in \Omega : \omega \in A_n \text{ for all but finitely many values of } n\}$.

We say that the sequence $\{A_n\}$ converges to a limit $A = \lim A_n$ if $B$ and $C$ are the same set $A$. Suppose that $A_n \to A$ and show that

(c) $A$ is an event, in that $A \in \mathcal{F}$,

(d) $\mathbb{P}(A_n) \to \mathbb{P}(A)$.

**17.** In Problem (1.8.16) above, show that $B$ and $C$ are independent whenever $B_n$ and $C_n$ are independent for all $n$. Deduce that if this holds and furthermore $A_n \to A$, then $\mathbb{P}(A)$ equals either zero or one.

**18.** Show that the assumption that $\mathbb{P}$ is *countably* additive is equivalent to the assumption that $\mathbb{P}$ is continuous. That is to say, show that if a function $\mathbb{P} : \mathcal{F} \to [0, 1]$ satisfies $\mathbb{P}(\varnothing) = 0$, $\mathbb{P}(\Omega) = 1$, and $\mathbb{P}(A \cup B) = \mathbb{P}(A) + \mathbb{P}(B)$ whenever $A, B \in \mathcal{F}$ and $A \cap B = \varnothing$, then $\mathbb{P}$ is countably additive (in the sense of satisfying Definition (1.3.1b)) if and only if $\mathbb{P}$ is continuous (in the sense of Lemma (1.3.5)).

**19.** Anne, Betty, Chloë, and Daisy were all friends at school. Subsequently each of the $\binom{4}{2} = 6$ subpairs meet up; at each of the six meetings the pair involved quarrel with some fixed probability $p$, or become firm friends with probability $1 - p$. Quarrels take place independently of each other. In future, if any of the four hears a rumour, then she tells it to her firm friends only. If Anne hears a rumour, what is the probability that:

(a) Daisy hears it?

(b) Daisy hears it if Anne and Betty have quarrelled?

(c) Daisy hears it if Betty and Chloë have quarrelled?

(d) Daisy hears it if she has quarrelled with Anne?

**20.** A biased coin is tossed repeatedly. Each time there is a probability $p$ of a head turning up. Let $p_n$ be the probability that an even number of heads has occurred after $n$ tosses (zero is an even number). Show that $p_0 = 1$ and that $p_n = p(1 - p_{n-1}) + (1 - p)p_{n-1}$ if $n \geq 1$. Solve this difference equation.

**21.** A biased coin is tossed repeatedly. Find the probability that there is a run of $r$ heads in a row before there is a run of $s$ tails, where $r$ and $s$ are positive integers.

**22.** (a) A bowl contains twenty cherries, exactly fifteen of which have had their stones removed. A greedy pig eats five whole cherries, picked at random, without remarking on the presence or absence of stones. Subsequently, a cherry is picked randomly from the remaining fifteen.

    (i) What is the probability that this cherry contains a stone?

    (ii) Given that this cherry contains a stone, what is the probability that the pig consumed at least one stone?

(b) 100 contestants buy numbered lottery tickets for a reverse raffle, in which the last ticket drawn from an urn is the winner. Halfway through the draw, the Mistress of Ceremonies discovers that 10 tickets have inadvertently not been added to the urn, so she adds them, and continues the draw. Is the lottery fair?

**23.** The '**ménages**' problem poses the following question. Some consider it to be desirable that men and women alternate when seated at a circular table. If $n$ heterosexual couples are seated randomly according to this rule, show that the probability that nobody sits next to his or her partner is

$$\frac{1}{n!} \sum_{k=0}^{n} (-1)^k \frac{2n}{2n - k} \binom{2n - k}{k} (n - k)!$$

You may find it useful to show first that the number of ways of selecting $k$ non-overlapping pairs of adjacent seats is $\binom{2n-k}{k} 2n (2n - k)^{-1}$.

**24.** An urn contains $b$ blue balls and $r$ red balls. They are removed at random and not replaced. Show that the probability that the first red ball drawn is the $(k + 1)$th ball drawn equals $\binom{r+b-k-1}{r-1} \big/ \binom{r+b}{b}$. Find the probability that the last ball drawn is red.

**25.** An urn contains $a$ azure balls and $c$ carmine balls, where $ac \neq 0$. Balls are removed at random and discarded until the first time that a ball ($B$, say) is removed having a different colour from its predecessor. The ball $B$ is now replaced and the procedure restarted. This process continues until the last ball is drawn from the urn. Show that this last ball is equally likely to be azure or carmine.

**26. Protocols.** A pack of four cards contains one spade, one club, and the two red aces. You deal two cards faces downwards at random in front of a truthful friend. She inspects them and tells you that one of them is the ace of hearts. What is the chance that the other card is the ace of diamonds? Perhaps $\frac{1}{3}$?

Suppose that your friend's protocol was:

(a) with no red ace, say "no red ace",

(b) with the ace of hearts, say "ace of hearts",

(c) with the ace of diamonds but not the ace of hearts, say "ace of diamonds".

Show that the probability in question is $\frac{1}{3}$.

Devise a possible protocol for your friend such that the probability in question is zero.

**27. Eddington's controversy.** Four witnesses, A, B, C, and D, at a trial each speak the truth with probability $\frac{1}{3}$ independently of each other. In their testimonies, A claimed that B denied that C declared that D lied. What is the (conditional) probability that D told the truth? [This problem seems to have appeared first as a parody in a university magazine of the 'typical' Cambridge Philosophy Tripos question.]

**28. The probabilistic method.** 10 per cent of the surface of a sphere is coloured blue, the rest is red. Show that, irrespective of the manner in which the colours are distributed, it is possible to inscribe a cube in $S$ with all its vertices red.

**29. Repulsion.** The event $A$ is said to be repelled by the event $B$ if $\mathbb{P}(A \mid B) < \mathbb{P}(A)$, and to be attracted by $B$ if $\mathbb{P}(A \mid B) > \mathbb{P}(A)$. Show that if $B$ attracts $A$, then $A$ attracts $B$, and $B^c$ repels $A$.

If $A$ attracts $B$, and $B$ attracts $C$, does $A$ attract $C$?

**30. Birthdays.** At a lecture, there a $m$ students born on independent days in 2007.

(a) With $2 \le m \le 365$, show that the probability that at least two of them share a birthday is $p = 1 - (365)!/\{(365 - m)!\,365^m\}$. Show that $p > \frac{1}{2}$ when $m = 23$.

(b) With $2 \le m \le 366$, find the probability $p_1$ that exactly one pair of individuals share a birthday, with no others sharing.

(c) Suppose $m$ students are born on independent random days on the planet Magrathea, whose year has $M \gg m$ days. Show that the probability $p_0$ that no two students share a birthday is approximately $\exp\left(-\frac{1}{2}m(m-1)/M\right)$ for large $M$.

**31. Lottery.** You choose $r$ of the first $n$ positive integers, and a lottery chooses a random subset $L$ of the same size. What is the probability that:

(a) $L$ includes no consecutive integers?

(b) $L$ includes exactly one pair of consecutive integers?

(c) the numbers in $L$ are drawn in increasing order?

(d) your choice of numbers is the same as $L$?

(e) there are exactly $k$ of your numbers matching members of $L$?

**32. Bridge.** During a game of bridge, you are dealt at random a hand of thirteen cards. With an obvious notation, show that $\mathbb{P}(4S, 3H, 3D, 3C) \simeq 0.026$ and $\mathbb{P}(4S, 4H, 3D, 2C) \simeq 0.018$. However if suits are not specified, so numbers denote the shape of your hand, show that $\mathbb{P}(4, 3, 3, 3) \simeq 0.11$ and $\mathbb{P}(4, 4, 3, 2) \simeq 0.22$.

**33. Poker.** During a game of poker, you are dealt a five-card hand at random. With the convention that aces may count high or low, show that:

$$\mathbb{P}(1 \text{ pair}) \simeq 0.423, \qquad \mathbb{P}(2 \text{ pairs}) \simeq 0.0475, \qquad \mathbb{P}(3 \text{ of a kind}) \simeq 0.021,$$
$$\mathbb{P}(\text{straight}) \simeq 0.0039, \qquad \mathbb{P}(\text{flush}) \simeq 0.0020, \qquad \mathbb{P}(\text{full house}) \simeq 0.0014,$$
$$\mathbb{P}(4 \text{ of a kind}) \simeq 0.00024, \qquad \mathbb{P}(\text{straight flush}) \simeq 0.000015.$$

**34. Poker dice.** There are five dice each displaying 9, 10, J, Q, K, A. Show that, when rolled:

$$\mathbb{P}(1 \text{ pair}) \simeq 0.46, \qquad \mathbb{P}(2 \text{ pairs}) \simeq 0.23, \qquad \mathbb{P}(3 \text{ of a kind}) \simeq 0.15,$$

$$\mathbb{P}(\text{no 2 alike}) \simeq 0.093, \qquad \mathbb{P}(\text{full house}) \simeq 0.039, \qquad \mathbb{P}(4 \text{ of a kind}) \simeq 0.019,$$

$$\mathbb{P}(5 \text{ of a kind}) \simeq 0.0008.$$

**35.** You are lost in the National Park of **Bandrika**†. Tourists comprise two-thirds of the visitors to the park, and give a correct answer to requests for directions with probability $\frac{3}{4}$. (Answers to repeated questions are independent, even if the question and the person are the same.) If you ask a Bandrikan for directions, the answer is always false.

(a) You ask a passer-by whether the exit from the Park is East or West. The answer is East. What is the probability this is correct?

(b) You ask the same person again, and receive the same reply. Show the probability that it is correct is $\frac{1}{2}$.

(c) You ask the same person again, and receive the same reply. What is the probability that it is correct?

(d) You ask for the fourth time, and receive the answer East. Show that the probability it is correct is $\frac{27}{70}$.

(e) Show that, had the fourth answer been West instead, the probability that East is nevertheless correct is $\frac{9}{10}$.

**36. Mr Bayes goes to Bandrika.** Tom is in the same position as you were in the previous problem, but he has reason to believe that, with probability $\epsilon$, East is the correct answer. Show that:

(a) whatever answer first received, Tom continues to believe that East is correct with probability $\epsilon$,

(b) if the first two replies are the same (that is, either WW or EE), Tom continues to believe that East is correct with probability $\epsilon$,

(c) after three like answers, Tom will calculate as follows, in the obvious notation:

$$\mathbb{P}(\text{East correct} \mid \text{EEE}) = \frac{9\epsilon}{11 - 2\epsilon}, \qquad \mathbb{P}(\text{East correct} \mid \text{WWW}) = \frac{11\epsilon}{9 + 2\epsilon}.$$

Evaluate these when $\epsilon = \frac{9}{20}$.

**37. Bonferroni's inequality.** Show that

$$\mathbb{P}\left(\bigcup_{r=1}^{n} A_r\right) \geq \sum_{r=1}^{n} \mathbb{P}(A_r) - \sum_{r < k} \mathbb{P}(A_r \cap A_k).$$

**38. Kounias's inequality.** Show that

$$\mathbb{P}\left(\bigcup_{r=1}^{n} A_r\right) \leq \min_{k} \left\{ \sum_{r=1}^{n} \mathbb{P}(A_r) - \sum_{r:r \neq k} \mathbb{P}(A_r \cap A_k) \right\}.$$

**39. The lost boarding pass‡.** The $n$ passengers for a Bell-Air flight in an airplane with $n$ seats have been told their seat numbers. They get on the plane one by one. The first person sits in the wrong seat. Subsequent passengers sit in their assigned seats whenever they find them available, or otherwise in a

---

†A fictional country made famous in the Hitchcock film 'The Lady Vanishes'.

‡The authors learned of this problem from David Bell in 2000 or earlier.

randomly chosen empty seat. What is the probability that the last passenger finds his or her assigned seat to be free?

What is the answer if the first person sits in a seat chosen uniformly at random from the $n$ available?

**40. Flash's problem.** A number $n$ of spaceships land independently and uniformly at random on the surface of planet Mongo. Each ship controls the hemisphere of which it is the centre. Show that the probability that every point on Mongo is controlled by at least one ship is $1 - 2^{-n}(n^2 - n + 2)$. [Hint: $n$ great circles almost surely partition the surface of the sphere into $n^2 - n + 2$ disjoint regions.]

**41.** Let $X$ be uniformly distributed on $\{1, 2, \ldots, n-1\}$, where $n \geq 2$. Given $X$, a team of size $X$ is selected at random from a pool of $n$ players (including you), each such subset of size $X$ being equally likely. Call the selected team A, and the remainder team B.

(a) What is the probability that your team has size $k$?

(b) Each team picks a captain uniformly at random from its members. What is the probability your team has size $k$ given that you are chosen as captain?

**42.** Alice and Bob flip a fair coin in turn. A wins if she gets a head, provided her preceding flip was a tail; B wins if he gets a tail, provided his preceding flip was a head. Let $n \geq 3$. Show that the probability the game ends on the $n$th flip is $(n+1)(n-1)/2^{n+2}$ if $n$ is odd, and $(n+2)(n-2)/2^{n+2}$ if even.

What is the probability that A wins the game?

**43.** A coin comes up heads with probability $p \in (0, 1)$. Let $k \geq 1$, and let $\rho_m$ be the probability that, in $m$ $(\geq 1)$ coin flips, the longest run of consecutive heads has length strictly less than $k$. Show that

$$\rho_m - \rho_{m-1} + (1 - p)p^k \rho_{m-k-1} = 0, \qquad m \geq k + 1,$$

and find $\rho_m$ when $k = 2$.

# 2

---

# Random variables and their distributions

*Summary.* Quantities governed by randomness correspond to functions on the probability space called random variables. The value taken by a random variable is subject to chance, and the associated likelihoods are described by a function called the distribution function. Two important classes of random variables are discussed, namely discrete variables and continuous variables. The law of averages, known also as the law of large numbers, states that the proportion of successes in a long run of independent trials converges to the probability of success in any one trial. This result provides a mathematical basis for a philosophical view of probability based on repeated experimentation. Worked examples involving random variables and their distributions are included, and the chapter terminates with sections on random vectors and on Monte Carlo simulation.

## 2.1 Random variables

We shall not always be interested in an experiment itself, but rather in some consequence of its random outcome. For example, many gamblers are more concerned with their losses than with the games which give rise to them. Such consequences, when real valued, may be thought of as functions which map $\Omega$ into the real line $\mathbb{R}$, and these functions are called 'random† variables'.

**(1) Example.** A fair coin is tossed twice: $\Omega = \{HH, HT, TH, TT\}$. For $\omega \in \Omega$, let $X(\omega)$ be the number of heads, so that

$$X(HH) = 2, \quad X(HT) = X(TH) = 1, \quad X(TT) = 0.$$

Now suppose that a gambler wagers his fortune of £1 on the result of this experiment. He gambles cumulatively so that his fortune is doubled each time a head appears, and is annihilated on the appearance of a tail. His subsequent fortune $W$ is a random variable given by

$$W(HH) = 4, \quad W(HT) = W(TH) = W(TT) = 0. \qquad \bullet$$

---

†Derived from the Old French word *randon* meaning 'haste'.

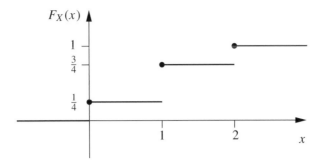

Figure 2.1. The distribution function $F_X$ of the random variable $X$ of Examples (1) and (5).

After the experiment is done and the outcome $\omega \in \Omega$ is known, a random variable $X : \Omega \to \mathbb{R}$ takes some value. In general this numerical value is more likely to lie in certain subsets of $\mathbb{R}$ than in certain others, depending on the probability space $(\Omega, \mathcal{F}, \mathbb{P})$ and the function $X$ itself. We wish to be able to describe the distribution of the likelihoods of possible values of $X$. Example (1) above suggests that we might do this through the function $f : \mathbb{R} \to [0, 1]$ defined by

$$f(x) = \text{probability that } X \text{ is equal to } x,$$

but this turns out to be inappropriate in general. Rather, we use the *distribution function* $F : \mathbb{R} \to \mathbb{R}$ defined by

$$F(x) = \text{probability that } X \text{ does not exceed } x.$$

More rigorously, this is

(2) $$F(x) = \mathbb{P}(A(x))$$

where $A(x) \subseteq \Omega$ is given by $A(x) = \{\omega \in \Omega : X(\omega) \leq x\}$. However, $\mathbb{P}$ is a function on the collection $\mathcal{F}$ of events; we cannot discuss $\mathbb{P}(A(x))$ unless $A(x)$ belongs to $\mathcal{F}$, and so we are led to the following definition.

**(3) Definition. A random variable** is a function $X : \Omega \to \mathbb{R}$ with the property that $\{\omega \in \Omega : X(\omega) \leq x\} \in \mathcal{F}$ for each $x \in \mathbb{R}$. Such a function is said to be $\mathcal{F}$-**measurable**.

If you so desire, you may pay no attention to the technical condition in the definition and think of random variables simply as functions mapping $\Omega$ into $\mathbb{R}$. We shall always use upper-case letters, such as $X$, $Y$, and $Z$, to represent generic random variables, whilst lower-case letters, such as $x$, $y$, and $z$, will be used to represent possible numerical values of these variables. Do not confuse this notation in your written work.

Every random variable has a distribution function, given by (2); distribution functions are *very* important and useful.

**(4) Definition.** The **distribution function** of a random variable $X$ is the function $F : \mathbb{R} \to [0, 1]$ given by $F(x) = \mathbb{P}(X \leq x)$.

This is the obvious abbreviation of equation (2). Events written as $\{\omega \in \Omega : X(\omega) \leq x\}$ are commonly abbreviated to $\{\omega : X(\omega) \leq x\}$ or $\{X \leq x\}$. We write $F_X$ where it is necessary to emphasize the role of $X$.

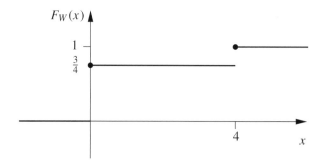

Figure 2.2. The distribution function $F_W$ of the random variable $W$ of Examples (1) and (5).

**(5) Example (1) revisited.** The distribution function $F_X$ of $X$ is given by

$$F_X(x) = \begin{cases} 0 & \text{if } x < 0, \\ \frac{1}{4} & \text{if } 0 \le x < 1, \\ \frac{3}{4} & \text{if } 1 \le x < 2, \\ 1 & \text{if } x \ge 2, \end{cases}$$

and is sketched in Figure 2.1. The distribution function $F_W$ of $W$ is given by

$$F_W(x) = \begin{cases} 0 & \text{if } x < 0, \\ \frac{3}{4} & \text{if } 0 \le x < 4, \\ 1 & \text{if } x \ge 4, \end{cases}$$

and is sketched in Figure 2.2. This illustrates the important point that the distribution function of a random variable $X$ tells us about the values taken by $X$ and their relative likelihoods, rather than about the sample space and the collection of events.        ●

**(6) Lemma.**  *A distribution function $F$ has the following properties*:
(a)  $\lim\limits_{x \to -\infty} F(x) = 0, \quad \lim\limits_{x \to \infty} F(x) = 1,$
(b)  *if $x < y$ then $F(x) \le F(y)$,*
(c)  *$F$ is right-continuous, that is, $F(x + h) \to F(x)$ as $h \downarrow 0$.*

**Proof.**
(a) Let $B_n = \{\omega \in \Omega : X(\omega) \le -n\} = \{X \le -n\}$. The sequence $B_1, B_2, \dots$ is decreasing with the empty set as limit. Thus, by Theorem (1.3.5), $\mathbb{P}(B_n) \to \mathbb{P}(\varnothing) = 0$. The other part is similar.
(b) Let $A(x) = \{X \le x\}$, $A(x, y) = \{x < X \le y\}$. Then $A(y) = A(x) \cup A(x, y)$ is a disjoint union, and so by Definition (1.3.1),

$$\mathbb{P}(A(y)) = \mathbb{P}(A(x)) + \mathbb{P}(A(x, y))$$

giving

$$F(y) = F(x) + \mathbb{P}(x < X \le y) \ge F(x).$$

(c) This is an *exercise*. Use Theorem (1.3.5).        ■

Actually, this lemma characterizes distribution functions. That is to say, $F$ is the distribution function of some random variable if and only if it satisfies (6a), (6b), and (6c).

For the time being we can forget all about probability spaces and concentrate on random variables and their distribution functions. The distribution function $F$ of $X$ contains a great deal of information about $X$.

**(7) Example. Constant variables.** The simplest random variable takes a constant value on the whole domain $\Omega$. Let $c \in \mathbb{R}$ and define $X : \Omega \to \mathbb{R}$ by

$$X(\omega) = c \quad \text{for all} \quad \omega \in \Omega.$$

The distribution function $F(x) = \mathbb{P}(X \leq x)$ is the step function

$$F(x) = \begin{cases} 0 & x < c, \\ 1 & x \geq c. \end{cases}$$

Slightly more generally, we call $X$ *constant* (*almost surely*) if there exists $c \in \mathbb{R}$ such that $\mathbb{P}(X = c) = 1$. A constant random variable is sometimes said to be *degenerate* or *trivial*. ●

**(8) Example. Bernoulli variables.** Consider Example (1.3.2). Let $X : \Omega \to \mathbb{R}$ be given by

$$X(H) = 1, \quad X(T) = 0.$$

Then $X$ is the simplest non-trivial random variable, having two possible values, 0 and 1. Its distribution function $F(x) = \mathbb{P}(X \leq x)$ is

$$F(x) = \begin{cases} 0 & x < 0, \\ 1 - p & 0 \leq x < 1, \\ 1 & x \geq 1. \end{cases}$$

$X$ is said to have the *Bernoulli distribution* sometimes denoted Bern($p$). ●

**(9) Example. Indicator functions.** A particular class of Bernoulli variables is very useful in probability theory. Let $A$ be an event and let $I_A : \Omega \to \mathbb{R}$ be the *indicator function* of $A$; that is,

$$I_A(\omega) = \begin{cases} 1 & \text{if } \omega \in A, \\ 0 & \text{if } \omega \in A^c. \end{cases}$$

Then $I_A$ is a Bernoulli random variable taking the values 1 and 0 with probabilities $\mathbb{P}(A)$ and $\mathbb{P}(A^c)$ respectively. Suppose $\{B_i : i \in I\}$ is a family of disjoint events with $A \subseteq \bigcup_{i \in I} B_i$. Then

**(10)** $$I_A = \sum_i I_{A \cap B_i},$$

an identity which is often useful. We sometimes write $I(A)$ for $I_A$. ●

**(11) Lemma.** *Let $F$ be the distribution function of $X$. Then*

(a) $\mathbb{P}(X > x) = 1 - F(x)$,

(b) $\mathbb{P}(x < X \le y) = F(y) - F(x)$,

(c) $\mathbb{P}(X = x) = F(x) - \lim_{y \uparrow x} F(y)$.

**Proof.** (a) and (b) are *exercises*.

(c) Let $B_n = \{x - 1/n < X \le x\}$ and use the method of proof of Lemma (6). ∎

Note one final piece of jargon for future use. A random variable $X$ with distribution function $F$ is said to have two 'tails' given by

$$T_1(x) = \mathbb{P}(X > x) = 1 - F(x), \quad T_2(x) = \mathbb{P}(X \le -x) = F(-x),$$

where $x$ is large and positive. We shall see later that the rates at which the $T_i$ decay to zero as $x \to \infty$ have a substantial effect on the existence or non-existence of certain associated quantities called the 'moments' of the distribution.

## Exercises for Section 2.1

**1.** Let $X$ be a random variable on a given probability space, and let $a \in \mathbb{R}$. Show that
(a) $aX$ is a random variable,
(b) $X - X = 0$, the random variable taking the value 0 always, and $X + X = 2X$.

**2.** A random variable $X$ has distribution function $F$. What is the distribution function of $Y = aX + b$, where $a$ and $b$ are real constants?

**3.** A fair coin is tossed $n$ times. Show that, under reasonable assumptions, the probability of exactly $k$ heads is $\binom{n}{k}(\frac{1}{2})^n$. What is the corresponding quantity when heads appears with probability $p$ on each toss?

**4.** Show that if $F$ and $G$ are distribution functions and $0 \le \lambda \le 1$ then $\lambda F + (1-\lambda)G$ is a distribution function. Is the product $FG$ a distribution function?

**5.** Let $F$ be a distribution function and $r$ a positive integer. Show that the following are distribution functions:   (a) $F(x)^r$,   (b) $1 - \{1 - F(x)\}^r$,   (c) $F(x) + \{1 - F(x)\} \log\{1 - F(x)\}$,
(d) $\{F(x) - 1\}e + \exp\{1 - F(x)\}$.

**6. Uniform distribution.** A random variable that is equally likely to take any value in a finite set $S$ is said to have the *uniform distribution* on $S$. If $U$ is such a random variable and $\varnothing \ne R \subseteq S$, show that the distribution of $U$ conditional on $\{U \in R\}$ is uniform on $R$.

## 2.2 The law of averages

We may recall the discussion in Section 1.3 of repeated experimentation. In each of $N$ repetitions of an experiment, we observe whether or not a given event $A$ occurs, and we write $N(A)$ for the total number of occurrences of $A$. One possible philosophical underpinning of probability theory requires that the proportion $N(A)/N$ settles down as $N \to \infty$ to some limit interpretable as the 'probability of $A$'. Is our theory to date consistent with such a requirement?

With this question in mind, let us suppose that $A_1, A_2, \ldots$ is a sequence of independent events having equal probability $\mathbb{P}(A_i) = p$, where $0 < p < 1$; such an assumption requires of course the existence of a corresponding probability space $(\Omega, \mathcal{F}, \mathbb{P})$, but we do not plan to get bogged down in such matters here. We think of $A_i$ as being the event 'that $A$ occurs on the $i$th experiment'. We write $S_n = \sum_{i=1}^{n} I_{A_i}$, the sum of the indicator functions of $A_1, A_2, \ldots, A_n$; $S_n$ is a random variable which counts the number of occurrences of $A_i$ for $1 \leq i \leq n$ (certainly $S_n$ is a function of $\Omega$, since it is the sum of such functions, and it is left as an *exercise* to show that $S_n$ is $\mathcal{F}$-measurable). The following result concerning the ratio $n^{-1}S_n$ was proved by James Bernoulli before 1692.

**(1) Theorem.** *It is the case that $n^{-1}S_n$ converges to $p$ as $n \to \infty$ in the sense that, for all* $\epsilon > 0$,

$$\mathbb{P}\left(p - \epsilon \leq \frac{1}{n}S_n \leq p + \epsilon\right) \to 1 \quad as \quad n \to \infty.$$

There are certain technicalities involved in the study of the convergence of random variables (see Chapter 7), and this is the reason for the careful statement of the theorem. For the time being, we encourage the reader to interpret the theorem as asserting simply that the proportion $n^{-1}S_n$ of times that the events $A_1, A_2, \ldots, A_n$ occur converges as $n \to \infty$ to their common probability $p$. We shall see later how important it is to be careful when making such statements.

Interpreted in terms of tosses of a fair coin, the theorem implies that the proportion of heads is (with large probability) near to $\frac{1}{2}$. As a caveat regarding the difficulties inherent in studying the convergence of random variables, we remark that it is *not* true that, in a 'typical' sequence of tosses of a fair coin, heads outnumber tails about one-half of the time.

**Proof.** Suppose that we toss a coin repeatedly, and heads occurs on each toss with probability $p$. The random variable $S_n$ has the same probability distribution as the number $H_n$ of heads which occur during the first $n$ tosses, which is to say that $\mathbb{P}(S_n = k) = \mathbb{P}(H_n = k)$ for all $k$. It follows that, for small positive values of $\epsilon$,

$$\mathbb{P}\left(\frac{1}{n}S_n \geq p + \epsilon\right) = \sum_{k \geq n(p+\epsilon)} \mathbb{P}(H_n = k).$$

We have from Exercise (2.1.3) that

$$\mathbb{P}(H_n = k) = \binom{n}{k}p^k(1-p)^{n-k} \quad \text{for} \quad 0 \leq k \leq n,$$

and hence

(2) $$\mathbb{P}\left(\frac{1}{n}S_n \geq p + \epsilon\right) = \sum_{k=m}^{n}\binom{n}{k}p^k(1-p)^{n-k}$$

where $m = \lceil n(p + \epsilon) \rceil$, the least integer not less than $n(p + \epsilon)$. The following argument is standard in probability theory. Let $\lambda > 0$ and note that $e^{\lambda k} \geq e^{\lambda n(p+\epsilon)}$ if $k \geq m$. Writing $q = 1 - p$, we have that

$$\mathbb{P}\left(\frac{1}{n}S_n \geq p + \epsilon\right) \leq \sum_{k=m}^{n} e^{\lambda[k-n(p+\epsilon)]}\binom{n}{k}p^k q^{n-k}$$

$$\leq e^{-\lambda n\epsilon}\sum_{k=0}^{n}\binom{n}{k}(pe^{\lambda q})^k(qe^{-\lambda p})^{n-k}$$

$$= e^{-\lambda n\epsilon}(pe^{\lambda q} + qe^{-\lambda p})^n,$$

by the binomial theorem. It is a simple *exercise* to show that $e^x \leq x + e^{x^2}$ for $x \in \mathbb{R}$. With the aid of this inequality, we obtain

$$\text{(3)} \qquad \mathbb{P}\left(\frac{1}{n}S_n \geq p + \epsilon\right) \leq e^{-\lambda n \epsilon}\left[pe^{\lambda^2 q^2} + qe^{\lambda^2 p^2}\right]^n$$

$$\leq e^{\lambda^2 n - \lambda n \epsilon}.$$

We can pick $\lambda$ to minimize the right-hand side, namely $\lambda = \frac{1}{2}\epsilon$, giving

$$\text{(4)} \qquad \mathbb{P}\left(\frac{1}{n}S_n \geq p + \epsilon\right) \leq e^{-\frac{1}{4}n\epsilon^2} \qquad \text{for} \quad \epsilon > 0,$$

an inequality that is commonly known as 'Bernstein's inequality'. It follows immediately that $\mathbb{P}(n^{-1}S_n \geq p + \epsilon) \to 0$ as $n \to \infty$. An exactly analogous argument shows that $\mathbb{P}(n^{-1}S_n \leq p - \epsilon) \to 0$ as $n \to \infty$, and thus the theorem is proved. ∎

Bernstein's inequality (4) is rather powerful, asserting that the chance that $S_n$ deviates from $np$ by a quantity of order $n$ tends to zero *exponentially fast* as $n \to \infty$; such an inequality is known as a 'large-deviation estimate'†. We may use the inequality to prove rather more than the conclusion of the theorem. Instead of estimating the chance that, for a specific value of $n$, $S_n$ lies between $n(p - \epsilon)$ and $n(p + \epsilon)$, let us estimate the chance that this occurs *for all large n*. Writing $A_n = \{p - \epsilon \leq n^{-1}S_n \leq p + \epsilon\}$, we wish to estimate $\mathbb{P}(\bigcap_{n=m}^{\infty} A_n)$. Now the complement of this intersection is the event $\bigcup_{n=m}^{\infty} A_n^c$, and the probability of this union satisfies, by the inequalities of Boole and Bernstein,

$$\text{(5)} \qquad \mathbb{P}\left(\bigcup_{n=m}^{\infty} A_n^c\right) \leq \sum_{n=m}^{\infty} \mathbb{P}(A_n^c) \leq \sum_{n=m}^{\infty} 2e^{-\frac{1}{4}n\epsilon^2} \to 0 \quad \text{as} \quad m \to \infty,$$

giving that, as required,

$$\text{(6)} \qquad \mathbb{P}\left(p - \epsilon \leq \frac{1}{n}S_n \leq p + \epsilon \text{ for all } n \geq m\right) \to 1 \quad \text{as} \quad m \to \infty.$$

## Exercises for Section 2.2

**1.** You wish to ask each of a large number of people a question to which the answer "yes" is embarrassing. The following procedure is proposed in order to determine the embarrassed fraction of the population. As the question is asked, a coin is tossed out of sight of the questioner. If the answer would have been "no" and the coin shows heads, then the answer "yes" is given. Otherwise people respond truthfully. What do you think of this procedure?

**2.** A coin is tossed repeatedly and heads turns up on each toss with probability $p$. Let $H_n$ and $T_n$ be the numbers of heads and tails in $n$ tosses. Show that, for $\epsilon > 0$,

$$\mathbb{P}\left(2p - 1 - \epsilon \leq \frac{1}{n}(H_n - T_n) \leq 2p - 1 + \epsilon\right) \to 1 \quad \text{as } n \to \infty.$$

**3.** Let $\{X_r : r \geq 1\}$ be observations which are independent and identically distributed with unknown distribution function $F$. Describe and justify a method for estimating $F(x)$.

†The quantity $np$ is of course the mean of $S_n$. See Example (3.3.7).

## 2.3 Discrete and continuous variables

Much of the study of random variables is devoted to distribution functions, characterized by Lemma (2.1.6). The general theory of distribution functions and their applications is quite difficult and abstract and is best omitted at this stage. It relies on a rigorous treatment of the construction of the Lebesgue–Stieltjes integral; this is sketched in Section 5.6. However, things become much easier if we are prepared to restrict our attention to certain subclasses of random variables specified by properties which make them tractable. We shall consider in depth the collection of 'discrete' random variables and the collection of 'continuous' random variables.

**(1) Definition.** The random variable $X$ is called **discrete** if it takes values in some countable subset $\{x_1, x_2, \dots\}$, only, of $\mathbb{R}$. The discrete random variable $X$ has **(probability) mass function** $f : \mathbb{R} \to [0, 1]$ given by $f(x) = \mathbb{P}(X = x)$.

We shall see that the distribution function of a discrete variable has jump discontinuities at the values $x_1, x_2, \dots$ and is constant in between; such a distribution is called *atomic*. this contrasts sharply with the other important class of distribution functions considered here.

**(2) Definition.** The random variable $X$ is called **continuous** if its distribution function can be expressed as

$$F(x) = \int_{-\infty}^{x} f(u)\, du \qquad x \in \mathbb{R},$$

for some integrable function $f : \mathbb{R} \to [0, \infty)$ called the **(probability) density function** of $X$.

The distribution function of a continuous random variable is certainly continuous (actually it is 'absolutely continuous'). For the moment we are concerned only with discrete variables and continuous variables. There is another sort of random variable, called 'singular', for a discussion of which the reader should look elsewhere. A common example of this phenomenon is based upon the Cantor ternary set† (see Grimmett and Welsh 2014, or Billingsley 1995). Other variables are 'mixtures' of discrete, continuous, and singular variables. Note that the word 'continuous' is a misnomer when used in this regard: in describing $X$ as continuous, we are referring to a property of its distribution function rather than of the random variable (function) $X$ itself.

**(3) Example. Discrete variables.** The variables $X$ and $W$ of Example (2.1.1) take values in the sets $\{0, 1, 2\}$ and $\{0, 4\}$ respectively; they are both discrete.                    ●

**(4) Example. Continuous variables.** A straight rod is flung down at random onto a horizontal plane and the angle $\omega$ between the rod and true north is measured. The result is a number in $\Omega = [0, 2\pi)$. Never mind about $\mathcal{F}$ for the moment; we can suppose that $\mathcal{F}$ contains all nice subsets of $\Omega$, including the collection of open subintervals such as $(a, b)$, where $0 \le a < b < 2\pi$. The implicit symmetry suggests the probability measure $\mathbb{P}$ which satisfies $\mathbb{P}((a, b)) = (b - a)/(2\pi)$; that is to say, the probability that the angle lies in some interval is

---

†The first such example of a nowhere dense subset of $[0, 1]$ with strictly positive Lebesgue measure is in fact due to H. S. S. Smith, whose discovery predated that of Cantor by eight years. This is an imperfect example of Stigler's law of eponymy since Smith's set was to base 4 and Cantor's to base 3.

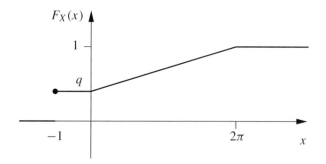

Figure 2.3. The distribution function $F_X$ of the random variable $X$ in Example (5).

directly proportional to the length of the interval. Here are two random variables $X$ and $Y$:

$$X(\omega) = \omega, \qquad Y(\omega) = \omega^2.$$

Notice that $Y$ is a function of $X$ in that $Y = X^2$. The distribution functions of $X$ and $Y$ are

$$F_X(x) = \begin{cases} 0 & x \le 0, \\ x/(2\pi) & 0 \le x < 2\pi, \\ 1 & x \ge 2\pi, \end{cases} \qquad F_Y(y) = \begin{cases} 0 & y \le 0, \\ \sqrt{y}/(2\pi) & 0 \le y < 4\pi^2, \\ 1 & y \ge 4\pi^2. \end{cases}$$

To see this, let $0 \le x < 2\pi$ and $0 \le y < 4\pi^2$. Then

$$F_X(x) = \mathbb{P}\big(\{\omega \in \Omega : 0 \le X(\omega) \le x\}\big)$$
$$= \mathbb{P}\big(\{\omega \in \Omega : 0 \le \omega \le x\}\big) = x/(2\pi),$$
$$F_Y(y) = \mathbb{P}\big(\{\omega : Y(\omega) \le y\}\big)$$
$$= \mathbb{P}\big(\{\omega : \omega^2 \le y\}\big) = \mathbb{P}\big(\{\omega : 0 \le \omega \le \sqrt{y}\}\big) = \mathbb{P}(X \le \sqrt{y})$$
$$= \sqrt{y}/(2\pi).$$

The random variables $X$ and $Y$ are continuous because

$$F_X(x) = \int_{-\infty}^{x} f_X(u)\, du, \qquad F_Y(y) = \int_{-\infty}^{y} f_Y(u)\, du,$$

where

$$f_X(u) = \begin{cases} 1/(2\pi) & \text{if } 0 \le u \le 2\pi, \\ 0 & \text{otherwise,} \end{cases}$$
$$f_Y(u) = \begin{cases} u^{-\frac{1}{2}}/(4\pi) & \text{if } 0 \le u \le 4\pi^2, \\ 0 & \text{otherwise.} \end{cases}$$
●

**(5) Example. A random variable which is neither continuous nor discrete.** A coin is tossed, and a head turns up with probability $p\,(= 1-q)$. If a head turns up then a rod is flung on the ground and the angle measured as in Example (4). Then $\Omega = \{T\}\cup\{(H, x) : 0 \le x < 2\pi\}$, in the obvious notation. Let $X : \Omega \to \mathbb{R}$ be given by

$$X(T) = -1, \qquad X((H, x)) = x.$$

The random variable $X$ takes values in $\{-1\} \cup [0, 2\pi)$ (see Figure 2.3 for a sketch of its distribution function). We say that $X$ is continuous with the exception of a 'point mass (or *atom*) at $-1$'.
●

## Exercises for Section 2.3

**1.** Let $X$ be a random variable with distribution function $F$, and let $a = (a_m : -\infty < m < \infty)$ be a strictly increasing sequence of real numbers satisfying $a_{-m} \to -\infty$ and $a_m \to \infty$ as $m \to \infty$. Define $G(x) = \mathbb{P}(X \le a_m)$ when $a_{m-1} \le x < a_m$, so that $G$ is the distribution function of a discrete random variable. How does the function $G$ behave as the sequence $a$ is chosen in such a way that $\sup_m |a_m - a_{m-1}|$ becomes smaller and smaller?

**2.** Let $X$ be a random variable and let $g : \mathbb{R} \to \mathbb{R}$ be continuous and strictly increasing. Show that $Y = g(X)$ is a random variable.

**3.** Let $X$ be a random variable with distribution function

$$\mathbb{P}(X \le x) = \begin{cases} 0 & \text{if } x \le 0, \\ x & \text{if } 0 < x \le 1, \\ 1 & \text{if } x > 1. \end{cases}$$

Let $F$ be a distribution function which is continuous and strictly increasing. Show that $Y = F^{-1}(X)$ is a random variable having distribution function $F$. Is it necessary that $F$ be continuous and/or strictly increasing?

**4.** Show that, if $f$ and $g$ are density functions, and $0 \le \lambda \le 1$, then $\lambda f + (1-\lambda)g$ is a density. Is the product $fg$ a density function?

**5.** Which of the following are density functions? Find $c$ and the corresponding distribution function $F$ for those that are.
(a) $f(x) = \begin{cases} cx^{-d} & x > 1, \\ 0 & \text{otherwise.} \end{cases}$
(b) $f(x) = ce^x(1+e^x)^{-2}, x \in \mathbb{R}$.

## 2.4 Worked examples

**(1) Example. Darts.** A dart is flung at a circular target of radius 3. We can think of the hitting point as the outcome of a random experiment; we shall suppose for simplicity that the player is guaranteed to hit the target somewhere. Setting the centre of the target at the origin of $\mathbb{R}^2$, we see that the sample space of this experiment is

$$\Omega = \{(x, y) : x^2 + y^2 < 9\}.$$

Never mind about the collection $\mathcal{F}$ of events. Let us suppose that, roughly speaking, the probability that the dart lands in some region $A$ is proportional to its area $|A|$. Thus

**(2)** $$\mathbb{P}(A) = |A|/(9\pi).$$

The scoring system is as follows. The target is partitioned by three concentric circles $C_1$, $C_2$, and $C_3$, centred at the origin with radii 1, 2, and 3. These circles divide the target into three annuli $A_1$, $A_2$, and $A_3$, where

$$A_k = \{(x, y) : k - 1 \le \sqrt{x^2 + y^2} < k\}.$$

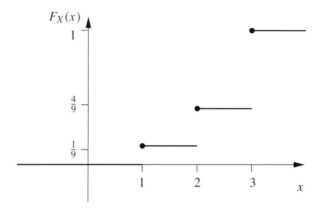

Figure 2.4. The distribution function $F_X$ of $X$ in Example (1).

We suppose that the player scores an amount $k$ if and only if the dart hits $A_k$. The resulting score $X$ is the random variable given by

$$X(\omega) = k \quad \text{whenever} \quad \omega \in A_k.$$

What is its distribution function?
**Solution.** Clearly

$$\mathbb{P}(X = k) = \mathbb{P}(A_k) = |A_k|/(9\pi) = \tfrac{1}{9}(2k - 1), \quad \text{for} \quad k = 1, 2, 3,$$

and so the distribution function of $X$ is given by

$$F_X(r) = \mathbb{P}(X \le r) = \begin{cases} 0 & \text{if } r < 1, \\ \tfrac{1}{9}\lfloor r \rfloor^2 & \text{if } 1 \le r < 3, \\ 1 & \text{if } r \ge 3, \end{cases}$$

where $\lfloor r \rfloor$ denotes the largest integer not larger than $r$ (see Figure 2.4).            ●

**(3) Example. Continuation of (1).** Let us consider a revised method of scoring in which the player scores an amount equal to the distance between the hitting point $\omega$ and the centre of the target. This time the score $Y$ is a random variable given by

$$Y(\omega) = \sqrt{x^2 + y^2}, \quad \text{if} \quad \omega = (x, y).$$

What is the distribution function of $Y$?
**Solution.** For any real $r$ let $C_r$ denote the disc with centre $(0, 0)$ and radius $r$, that is

$$C_r = \{(x, y) : x^2 + y^2 \le r\}.$$

Then

$$F_Y(r) = \mathbb{P}(Y \le r) = \mathbb{P}(C_r) = \tfrac{1}{9}r^2 \quad \text{if} \quad 0 \le r \le 3.$$

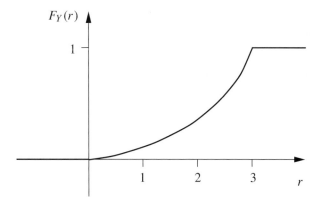

Figure 2.5. The distribution function $F_Y$ of $Y$ in Example (3).

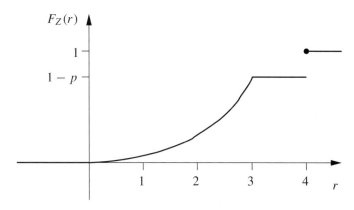

Figure 2.6. The distribution function $F_Z$ of $Z$ in Example (4).

This distribution function is sketched in Figure 2.5.                    ●

**(4) Example. Continuation of (1).**  Now suppose that the player fails to hit the target with
fixed probability $p$; if he is successful then we suppose that the distribution of the hitting point
is described by equation (2). His score is specified as follows. If he hits the target then he
scores an amount equal to the distance between the hitting point and the centre; if he misses
then he scores 4. What is the distribution function of his score $Z$?

**Solution.**  Clearly $Z$ takes values in the interval $[0, 4]$. Use Lemma (1.4.4) to see that

$$F_Z(r) = \mathbb{P}(Z \leq r)$$
$$= \mathbb{P}(Z \leq r \mid \text{hits target})\mathbb{P}(\text{hits target}) + \mathbb{P}(Z \leq r \mid \text{misses target})\mathbb{P}(\text{misses target})$$
$$= \begin{cases} 0 & \text{if } r < 0, \\ (1-p)F_Y(r) & \text{if } 0 \leq r < 4, \\ 1 & \text{if } r \geq 4, \end{cases}$$

where $F_Y$ is given in Example (3) (see Figure 2.6 for a sketch of $F_Z$).            ●

## Exercises for Section 2.4

**1.** Let $X$ be a random variable with a continuous distribution function $F$. Find expressions for the distribution functions of the following random variables:

(a) $X^2$,  
(b) $\sqrt{X}$,  
(c) $\sin X$,  
(d) $G^{-1}(X)$,  
(e) $F(X)$,  
(f) $G^{-1}(F(X))$,  

where $G$ is a continuous and strictly increasing function.

**2. Truncation**. Let $X$ be a random variable with distribution function $F$, and let $a < b$. Sketch the distribution functions of the 'truncated' random variables $Y$ and $Z$ given by

$$Y = \begin{cases} a & \text{if } X < a, \\ X & \text{if } a \le X \le b, \\ b & \text{if } X > b, \end{cases} \qquad Z = \begin{cases} X & \text{if } |X| \le b, \\ 0 & \text{if } |X| > b. \end{cases}$$

Indicate how these distribution functions behave as $a \to -\infty$, $b \to \infty$.

## 2.5  Random vectors

Suppose that $X$ and $Y$ are random variables on the probability space $(\Omega, \mathcal{F}, \mathbb{P})$. Their distribution functions, $F_X$ and $F_Y$, contain information about their associated probabilities. But how may we encapsulate information about their properties *relative to each other*? The key is to think of $X$ and $Y$ as being the components of a 'random vector' $(X, Y)$ taking values in $\mathbb{R}^2$, rather than being unrelated random variables each taking values in $\mathbb{R}$.

**(1) Example. Tontine** is a scheme wherein subscribers to a common fund each receive an annuity from the fund during his or her lifetime, this annuity increasing as the other subscribers die. When all the subscribers are dead, the fund passes to the French government (this was the case in the first such scheme designed by Lorenzo Tonti around 1653). The performance of the fund depends on the lifetimes $L_1, L_2, \ldots, L_n$ of the subscribers (as well as on their wealths), and we may record these as a vector $(L_1, L_2, \ldots, L_n)$ of random variables.   ●

**(2) Example. Darts.** A dart is flung at a conventional dartboard. The point of striking determines a distance $R$ from the centre, an angle $\Theta$ with the upward vertical (measured clockwise, say), and a score $S$. With this experiment we may associate the random vector $(R, \Theta, S)$, and we note that $S$ is a function of the pair $(R, \Theta)$.   ●

**(3) Example. Coin tossing.** Suppose that we toss a coin $n$ times, and set $X_i$ equal to 0 or 1 depending on whether the $i$th toss results in a tail or a head. We think of the vector $\mathbf{X} = (X_1, X_2, \ldots, X_n)$ as describing the result of this composite experiment. The total number of heads is the sum of the entries in $\mathbf{X}$.   ●

An individual random variable $X$ has a distribution function $F_X$ defined by $F_X(x) = \mathbb{P}(X \le x)$ for $x \in \mathbb{R}$. The corresponding 'joint' distribution function of a random vector $(X_1, X_2, \ldots, X_n)$ is the quantity $\mathbb{P}(X_1 \le x_1, X_2 \le x_2, \ldots, X_n \le x_n)$, a function of $n$ real variables $x_1, x_2, \ldots, x_n$. In order to aid the notation, we introduce an ordering of vectors of real numbers: for vectors $\mathbf{x} = (x_1, x_2, \ldots, x_n)$ and $\mathbf{y} = (y_1, y_2, \ldots, y_n)$ we write $\mathbf{x} \leq \mathbf{y}$ if $x_i \leq y_i$ for each $i = 1, 2, \ldots, n$.

**(4) Definition.**   The **joint distribution function** of a random vector $\mathbf{X} = (X_1, X_2, \ldots, X_n)$ on the probability space $(\Omega, \mathcal{F}, \mathbb{P})$ is the function $F_{\mathbf{X}} : \mathbb{R}^n \to [0, 1]$ given by $F_{\mathbf{X}}(\mathbf{x}) = \mathbb{P}(\mathbf{X} \leq \mathbf{x})$ for $\mathbf{x} \in \mathbb{R}^n$.

As before, the expression $\{\mathbf{X} \leq \mathbf{x}\}$ is an abbreviation for the event $\{\omega \in \Omega : \mathbf{X}(\omega) \leq \mathbf{x}\}$. Joint distribution functions have properties similar to those of ordinary distribution functions. For example, Lemma (2.1.6) becomes the following.

**(5) Lemma.**   *The joint distribution function $F_{X,Y}$ of the random vector $(X, Y)$ has the following properties*:

(a)  $\lim_{x,y \to -\infty} F_{X,Y}(x, y) = 0,$ $\lim_{x,y \to \infty} F_{X,Y}(x, y) = 1,$

(b)  *if $(x_1, y_1) \leq (x_2, y_2)$ then $F_{X,Y}(x_1, y_1) \leq F_{X,Y}(x_2, y_2),$*

(c)  *$F_{X,Y}$ is continuous from above, in that*

$$F_{X,Y}(x + u, y + v) \to F_{X,Y}(x, y) \quad as \quad u, v \downarrow 0.$$

We state this lemma for a random vector with only two components $X$ and $Y$, but the corresponding result for $n$ components is valid also. The proof of the lemma is left as an *exercise*. Rather more is true. It may be seen without great difficulty that

**(6)**                           $$\lim_{y \to \infty} F_{X,Y}(x, y) = F_X(x) \ (= \mathbb{P}(X \leq x))$$

and similarly

**(7)**                           $$\lim_{x \to \infty} F_{X,Y}(x, y) = F_Y(y) \ (= \mathbb{P}(Y \leq y)).$$

This more refined version of part (a) of the lemma tells us that we may recapture the individual distribution functions of $X$ and $Y$ from a knowledge of their joint distribution function. The converse is false: it is not generally possible to calculate $F_{X,Y}$ from a knowledge of $F_X$ and $F_Y$ alone. The functions $F_X$ and $F_Y$ are called the 'marginal' distribution functions of $F_{X,Y}$.

**(8) Example.**   A schoolteacher asks each member of his or her class to flip a fair coin twice and to record the outcomes. The diligent pupil $D$ does this and records a pair $(X_D, Y_D)$ of outcomes. The lazy pupil $L$ flips the coin only once and writes down the result twice, recording thus a pair $(X_L, Y_L)$ where $X_L = Y_L$. Clearly $X_D, Y_D, X_L$, and $Y_L$ are random variables with the same distribution functions. However, the pairs $(X_D, Y_D)$ and $(X_L, Y_L)$ have different *joint* distribution functions. In particular, $\mathbb{P}(X_D = Y_D = \text{heads}) = \frac{1}{4}$ since only one of the four possible pairs of outcomes contains heads only, whereas $\mathbb{P}(X_L = Y_L = \text{heads}) = \frac{1}{2}.$ ●

Once again there are two classes of random vectors which are particularly interesting: the 'discrete' and the 'continuous'.

**(9) Definition.**   The random variables $X$ and $Y$ on the probability space $(\Omega, \mathcal{F}, \mathbb{P})$ are called **(jointly) discrete** if the vector $(X, Y)$ takes values in some countable subset of $\mathbb{R}^2$ only. The jointly discrete random variables $X, Y$ have **joint (probability) mass function** $f : \mathbb{R}^2 \to [0, 1]$ given by $f(x, y) = \mathbb{P}(X = x, Y = y)$.

**(10) Definition.** The random variables $X$ and $Y$ on the probability space $(\Omega, \mathcal{F}, \mathbb{P})$ are called (**jointly**) **continuous** if their joint distribution function can be expressed as

$$F_{X,Y}(x, y) = \int_{u=-\infty}^{x} \int_{v=-\infty}^{y} f(u, v)\, du\, dv, \qquad x, y \in \mathbb{R},$$

for some integrable function $f : \mathbb{R}^2 \to [0, \infty)$ called the **joint (probability) density function** of the pair $(X, Y)$.

We shall return to such questions in later chapters. Meanwhile here are two concrete examples.

**(11) Example. Three-sided coin.** We are provided with a special three-sided coin, each toss of which results in one of the possibilities H (heads), T (tails), E (edge), each having probability $\frac{1}{3}$. Let $H_n$, $T_n$, and $E_n$ be the numbers of such outcomes in $n$ tosses of the coin. The vector $(H_n, T_n, E_n)$ is a vector of random variables satisfying $H_n + T_n + E_n = n$. If the outcomes of different tosses have no influence on each other, it is not difficult to see that

$$\mathbb{P}\big((H_n, T_n, E_n) = (h, t, e)\big) = \frac{n!}{h!\, t!\, e!} \left(\frac{1}{3}\right)^n$$

for any triple $(h, t, e)$ of non-negative integers with sum $n$. The random variables $H_n$, $T_n$, $E_n$ are (jointly) discrete and are said to have (jointly) the *trinomial* distribution.    ●

**(12) Example. Darts.** Returning to the flung dart of Example (2), let us assume that no region of the dartboard is preferred unduly over any other region of equal area. It may then be shown (see Example (2.4.3)) that

$$\mathbb{P}(R \leq r) = \frac{r^2}{\rho^2}, \quad \mathbb{P}(\Theta \leq \theta) = \frac{\theta}{2\pi}, \quad \text{for} \quad 0 \leq r \leq \rho,\ 0 \leq \theta \leq 2\pi,$$

where $\rho$ is the radius of the board, and furthermore

$$\mathbb{P}(R \leq r,\ \Theta \leq \theta) = \mathbb{P}(R \leq r)\mathbb{P}(\Theta \leq \theta).$$

It follows that

$$F_{R,\Theta}(r, \theta) = \int_{u=0}^{r} \int_{v=0}^{\theta} f(u, v)\, du\, dv$$

where

$$f(u, v) = \frac{u}{\pi \rho^2}, \quad 0 \leq u \leq \rho,\ 0 \leq v \leq 2\pi.$$

The pair $(R, \Theta)$ is (jointly) continuous.    ●

## Exercises for Section 2.5

**1.**   A fair coin is tossed twice. Let $X$ be the number of heads, and let $W$ be the indicator function of the event $\{X = 2\}$. Find $\mathbb{P}(X = x, W = w)$ for all appropriate values of $x$ and $w$.

**2.**   Let $X$ be a Bernoulli random variable, so that $\mathbb{P}(X = 0) = 1 - p$, $\mathbb{P}(X = 1) = p$. Let $Y = 1 - X$ and $Z = XY$. Find $\mathbb{P}(X = x, Y = y)$ and $\mathbb{P}(X = x, Z = z)$ for $x, y, z \in \{0, 1\}$.

**3.**   The random variables $X$ and $Y$ have joint distribution function

$$F_{X,Y}(x, y) = \begin{cases} 0 & \text{if } x < 0, \\ (1 - e^{-x}) \left( \dfrac{1}{2} + \dfrac{1}{\pi} \tan^{-1} y \right) & \text{if } x \ge 0. \end{cases}$$

Show that $X$ and $Y$ are (jointly) continuously distributed.

**4.**   Let $X$ and $Y$ have joint distribution function $F$. Show that

$$\mathbb{P}(a < X \le b, \, c < Y \le d) = F(b, d) - F(a, d) - F(b, c) + F(a, c)$$

whenever $a < b$ and $c < d$.

**5.**   Let $X, Y$ be discrete random variables taking values in the integers, with joint mass function $f$. Show that, for integers $x, y$,

$$f(x, y) = \mathbb{P}(X \ge x, \, Y \le y) - \mathbb{P}(X \ge x + 1, \, Y \le y)$$
$$- \mathbb{P}(X \ge x, \, Y \le y - 1) + \mathbb{P}(X \ge x + 1, \, Y \le y - 1).$$

Hence find the joint mass function of the smallest and largest numbers shown in $r$ rolls of a fair die.

**6.**   Is the function $F(x, y) = 1 - e^{-xy}$, $0 \le x, y < \infty$, the joint distribution function of some pair of random variables?

## 2.6 Monte Carlo simulation

It is presumably the case that the physical shape of a coin is one of the major factors relevant to whether or not it will fall with heads uppermost. In principle, the shape of the coin may be determined by direct examination, and hence we may arrive at an estimate for the chance of heads. Unfortunately, such a calculation would be rather complicated, and it is easier to estimate this chance by simulation, which is to say that we may toss the coin many times and record the proportion of successes. Similarly, roulette players are well advised to observe the behaviour of the wheel with care in advance of placing large bets, in order to discern its peculiarities (unfortunately, casinos are now wary of such observation, and change their wheels at regular intervals).

Here is a related question. Suppose that we know that our coin is fair (so that the chance of heads is $\frac{1}{2}$ on each toss), and we wish to know the chance that a sequence of 50 tosses contains a run of outcomes of the form HTHHT. In principle, this probability may be calculated explicitly and exactly. If we require only an estimate of its value, then another possibility is to simulate the experiment: toss the coin $50N$ times for some $N$, divide the result into $N$ runs of 50, and find the proportion of such runs which contain HTHHT.

It is not unusual in real life for a specific calculation to be possible in principle but extremely difficult in practice, often owing to limitations on the operating speed or the size of the memory of a computer. Simulation can provide a way around such a problem. Here are some examples.

**(1) Example. Gambler's ruin revisited.** The gambler of Example (1.7.4) eventually won his Jaguar after a long period devoted to tossing coins, and he has now decided to save up for a yacht. His bank manager has suggested that, in order to speed things up, the stake on each gamble should not remain constant but should vary as a certain prescribed function of the gambler's current fortune. The gambler would like to calculate the chance of winning the yacht in advance of embarking on the project, but he finds himself incapable of doing so.

Fortunately, he has kept a record of the extremely long sequence of heads and tails encountered in his successful play for the Jaguar. He calculates his sequence of hypothetical fortunes based on this information, until the point when this fortune reaches either zero or the price of the yacht. He then starts again, and continues to repeat the procedure until he has completed it a total of $N$ times, say. He estimates the probability that he will actually win the yacht by the proportion of the $N$ calculations which result in success.

Why might this method make him overconfident? Should he retoss the coins?          ●

**(2) Example. A dam.** It is proposed to build a dam in order to regulate the water supply, and in particular to prevent seasonal flooding downstream. How high should the dam be? Dams are expensive to construct, and some compromise between cost and risk is necessary. It is decided to build a dam which is just high enough to ensure that the chance of a flood of some given extent within ten years is less than $10^{-2}$, say. No one knows exactly how high such a dam need be, and a young probabilist proposes the following scheme. Through examination of existing records of rainfall and water demand we may arrive at an acceptable model for the pattern of supply and demand. This model includes, for example, estimates for the distributions of rainfall on successive days over long periods. With the aid of a computer, the 'real world' situation is simulated many times in order to study the likely consequences of building dams of various heights. In this way we may arrive at an accurate estimate of the height required.          ●

**(3) Example. Integration.** Let $g : [0, 1] \rightarrow [0, 1]$ be a continuous but nowhere differentiable function. How may we calculate its integral $I = \int_0^1 g(x)\, dx$? The following experimental technique is known as the 'hit or miss Monte Carlo technique'.

Let $(X, Y)$ be a random vector having the uniform distribution on the unit square. That is, we assume that $\mathbb{P}\big((X, Y) \in A\big) = |A|$, the area of $A$, for any nice subset $A$ of the unit square $[0, 1]^2$; we leave the assumption of niceness somewhat up in the air for the moment, and shall return to such matters in Chapter 4. We declare $(X, Y)$ to be 'successful' if $Y \leq g(X)$. The chance that $(X, Y)$ is successful equals $I$, the area under the curve $y = g(x)$. We now repeat this experiment a large number $N$ of times, and calculate the proportion of times that the experiment is successful. Following the law of averages, Theorem (2.2.1), we may use this value as an estimate of $I$.

Clearly it is desirable to know the accuracy of this estimate. This is a harder problem to which we shall return later.          ●

Simulation is a dangerous game, and great caution is required in interpreting the results. There are two major reasons for this. First, a computer simulation is limited by the degree to which its so-called 'pseudo-random number generator' may be trusted. It has been said for example that the summon-according-to-birthday principle of conscription to the United States armed forces may have been marred by a pseudo-random number generator with a bias for some numbers over others. Secondly, in estimating a given quantity, one may in some circumstances have little or no idea how many repetitions are necessary in order to achieve an estimate within a specified accuracy.

We have made no remark about the methods by which computers calculate 'pseudo-random numbers'. Needless to say they do not flip coins, but rely instead on operations of sufficient numerical complexity that the outcome, although deterministic, is apparently unpredictable except by an exact repetition of the calculation.

These techniques were named in honour of Monte Carlo by Metropolis, von Neumann, and Ulam, while they were involved in the process of building bombs at Los Alamos in the 1940s.

---

## 2.7  Problems

**1.**   Each toss of a coin results in a head with probability $p$. The coin is tossed until the first head appears. Let $X$ be the total number of tosses. What is $\mathbb{P}(X > m)$? Find the distribution function of the random variable $X$.

**2.**   (a) Show that any discrete random variable may be written as a linear combination of indicator variables.

(b)  Show that any random variable may be expressed as the limit of an increasing sequence of discrete random variables.

(c)  Show that the limit of any increasing convergent sequence of random variables is a random variable.

**3.**   (a) Show that, if $X$ and $Y$ are random variables on a probability space $(\Omega, \mathcal{F}, \mathbb{P})$, then so are $X + Y$, $XY$, and $\min\{X, Y\}$.

(b)  Show that the set of all random variables on a given probability space $(\Omega, \mathcal{F}, \mathbb{P})$ constitutes a vector space over the reals. If $\Omega$ is finite, write down a basis for this space.

**4.**   Let $X$ have distribution function

$$
F(x) = \begin{cases} 0 & \text{if } x < 0, \\ \frac{1}{2}x & \text{if } 0 \le x \le 2, \\ 1 & \text{if } x > 2, \end{cases}
$$

and let $Y = X^2$. Find
(a) $\mathbb{P}\left(\frac{1}{2} \le X \le \frac{3}{2}\right)$,   (b) $\mathbb{P}(1 \le X < 2)$,
(c) $\mathbb{P}(Y \le X)$,   (d) $\mathbb{P}(X \le 2Y)$,
(e) $\mathbb{P}\left(X + Y \le \frac{3}{4}\right)$,   (f) the distribution function of $Z = \sqrt{X}$.

**5.**   Let $X$ have distribution function

$$
F(x) = \begin{cases} 0 & \text{if } x < -1, \\ 1 - p & \text{if } -1 \le x < 0, \\ 1 - p + \frac{1}{2}xp & \text{if } 0 \le x \le 2, \\ 1 & \text{if } x > 2. \end{cases}
$$

Sketch this function, and find: (a) $\mathbb{P}(X = -1)$,   (b) $\mathbb{P}(X = 0)$,   (c) $\mathbb{P}(X \ge 1)$.

**6.** Buses arrive at ten minute intervals starting at noon. A man arrives at the bus stop a random number $X$ minutes after noon, where $X$ has distribution function

$$\mathbb{P}(X \le x) = \begin{cases} 0 & \text{if } x < 0, \\ x/60 & \text{if } 0 \le x \le 60, \\ 1 & \text{if } x > 60. \end{cases}$$

What is the probability that he waits less than five minutes for a bus?

**7.** Airlines find that each passenger who reserves a seat fails to turn up with probability $\frac{1}{10}$ independently of the other passengers. EasyPeasy Airlines always sell 10 tickets for their 9 seat aeroplane while RyeLoaf Airways always sell 20 tickets for their 18 seat aeroplane. Which is more often over-booked?

**8.** A fairground performer claims the power of telekinesis. The crowd throws coins and he wills them to fall heads up. He succeeds five times out of six. What chance would he have of doing at least as well if he had no supernatural powers?

**9.** Express the distribution functions of

$$X^+ = \max\{0, X\}, \quad X^- = -\min\{0, X\}, \quad |X| = X^+ + X^-, \quad -X,$$

in terms of the distribution function $F$ of the random variable $X$.

**10.** Show that $F_X(x)$ is continuous at $x = x_0$ if and only if $\mathbb{P}(X = x_0) = 0$.

**11.** The real number $m$ is called a *median* of the distribution function $F$ whenever $\lim_{y \uparrow m} F(y) \le \frac{1}{2} \le F(m)$.
(a) Show that every distribution function $F$ has at least one median, and that the set of medians of $F$ is a closed interval of $\mathbb{R}$.
(b) Show, if $F$ is continuous, that $F(m) = \frac{1}{2}$ for any median $m$.

**12. Loaded dice.**
(a) Show that it is not possible to weight two dice in such a way that the sum of the two numbers shown by these loaded dice is equally likely to take any value between 2 and 12 (inclusive).
(b) Given a fair die and a loaded die, show that the sum of their scores, modulo 6, has the same distribution as a fair die, irrespective of the loading.

**13.** A function $d : S \times S \to \mathbb{R}$ is called a *metric* on $S$ if:
(i) $d(s, t) = d(t, s) \ge 0$ for all $s, t \in S$,
(ii) $d(s, t) = 0$ if and only if $s = t$, and
(iii) $d(s, t) \le d(s, u) + d(u, t)$ for all $s, t, u \in S$.
(a) **Lévy metric.** Let $F$ and $G$ be distribution functions and define the *Lévy metric*

$$d_L(F, G) = \inf\left\{\epsilon > 0 : G(x - \epsilon) - \epsilon \le F(x) \le G(x + \epsilon) + \epsilon \text{ for all } x\right\}.$$

Show that $d_L$ is indeed a metric on the space of distribution functions.
(b) **Total variation distance.** Let $X$ and $Y$ be integer-valued random variables, and let

$$d_{TV}(X, Y) = \sum_k \left|\mathbb{P}(X = k) - \mathbb{P}(Y = k)\right|.$$

Show that $d_{TV}$ satisfies (i) and (iii) with $S$ the space of integer-valued random variables, and that $d_{TV}(X, Y) = 0$ if and only if $X$ and $Y$ have the same distribution. Thus $d_{TV}$ is a metric on the space of equivalence classes of $S$ with equivalence relation given by $X \sim Y$ if $X$ and $Y$ have the same distribution. We call $d_{TV}$ the *total variation distance*.

Show that

$$d_{\text{TV}}(X, Y) = 2 \sup_{A \subseteq \mathbb{Z}} \left| \mathbb{P}(X \in A) - \mathbb{P}(Y \in A) \right|.$$

**14.** Ascertain in the following cases whether or not $F$ is the joint distribution function of some pair $(X, Y)$ of random variables. If your conclusion is affirmative, find the distribution functions of $X$ and $Y$ separately.

(a)
$$F(x, y) = \begin{cases} 1 - e^{-x-y} & \text{if } x, y \geq 0, \\ 0 & \text{otherwise.} \end{cases}$$

(b)
$$F(x, y) = \begin{cases} 1 - e^{-x} - xe^{-y} & \text{if } 0 \leq x \leq y, \\ 1 - e^{-y} - ye^{-y} & \text{if } 0 \leq y \leq x, \\ 0 & \text{otherwise.} \end{cases}$$

**15.** It is required to place in order $n$ books $B_1, B_2, \ldots, B_n$ on a library shelf in such a way that readers searching from left to right waste as little time as possible on average. Assuming that each reader requires book $B_i$ with probability $p_i$, find the ordering of the books which minimizes $\mathbb{P}(T \geq k)$ for all $k$, where $T$ is the (random) number of titles examined by a reader before discovery of the required book.

**16. Transitive coins.** Three coins each show heads with probability $\frac{3}{5}$ and tails otherwise. The first counts 10 points for a head and 2 for a tail, the second counts 4 points for both head and tail, and the third counts 3 points for a head and 20 for a tail.

You and your opponent each choose a coin; you cannot choose the same coin. Each of you tosses your coin and the person with the larger score wins £$10^{10}$. Would you prefer to be the first to pick a coin or the second?

**17.** Before the development of radar, inertial navigation, and GPS, flying to isolated islands (for example, from Los Angeles to Hawaii) was somewhat 'hit or miss'. In heavy cloud or at night it was necessary to fly by dead reckoning, and then to search the surface. With the aid of a radio, the pilot had a good idea of the correct great circle along which to search, but could not be sure which of the two directions along this great circle was correct (since a strong tailwind could have carried the plane over its target). When you are the pilot, you calculate that you can make $n$ searches before your plane will run out of fuel. On each search you will discover the island with probability $p$ (if it is indeed in the direction of the search) independently of the results of other searches; you estimate initially that there is probability $\alpha$ that the island is ahead of you. What policy should you adopt in deciding the directions of your various searches in order to maximize the probability of locating the island?

**18.** Eight pawns are placed randomly on a chessboard, no more than one to a square. What is the probability that:
(a) they are in a straight line (do not forget the diagonals)?
(b) no two are in the same row or column?

**19.** Which of the following are distribution functions? For those that are, give the corresponding density function $f$.

(a) $F(x) = \begin{cases} 1 - e^{-x^2} & x \geq 0, \\ 0 & \text{otherwise.} \end{cases}$

(b) $F(x) = \begin{cases} e^{-1/x} & x > 0, \\ 0 & \text{otherwise.} \end{cases}$

(c) $F(x) = e^x / (e^x + e^{-x}), \ x \in \mathbb{R}$.

(d) $F(x) = e^{-x^2} + e^x / (e^x + e^{-x}), \ x \in \mathbb{R}$.

**20.** (a) If $U$ and $V$ are jointly continuous, show that $\mathbb{P}(U = V) = 0$.
(b) Let $X$ be uniformly distributed on $(0, 1)$, and let $Y = X$. Then $X$ and $Y$ are continuous, and $\mathbb{P}(X = Y) = 1$. Is there a contradiction here?

**21. Continued fractions.** Let $X$ be uniformly distributed on the interval $[0, 1]$, and express $X$ as a continued fraction thus:

$$X = \cfrac{1}{Y_1 + \cfrac{1}{Y_2 + \cfrac{1}{Y_3 + \cdots}}}.$$

Show that the joint mass function of $Y_1$ and $Y_2$ is

$$f(u, v) = \frac{1}{(uv + 1)(uv + u + 1)}, \qquad u, v = 1, 2, \ldots.$$

**22.** Let $V$ be a vector space of dimension $n$ over a finite field $\mathbb{F}$ with $q$ elements. Let $X_1, X_2, \ldots, X_m$ be independent random variables, each uniformly distributed on $V$.

(a)  Let $a_i \in \mathbb{F}$, $i = 1, 2, \ldots, m$, be not all zero. Show that the linear combination $Z = \sum_i a_i X_i$ is uniformly distributed on $V$.

(b)  Let $p_m$ be the probability that $X_1, X_2, \ldots, X_m$ are linearly dependent. Show that, if $m \leq n + 1$,

$$q^{-(n-m-1)} \leq p_m \leq q^{-(n-m)}, \qquad m = 1, 2, \ldots, n + 1.$$

**23. Modes.** A random variable $X$ with distribution function $F$ is said to be *unimodal*† about a mode $M$ if $F$ is convex on $(-\infty, M)$ and concave on $(M, \infty)$. Show that, if $F$ is unimodal about $M$, then the following hold.

(a)  $F$ is absolutely continuous, except possibly for an atom at $M$.

(b)  If $F$ is differentiable, then it has a density that is non-decreasing on $(-\infty, M)$ and non-increasing on $(M, \infty)$, and furthermore, the set of modes of $F$ is a closed bounded interval. [Cf. Problem (2.7.11).]

(c)  If the distribution functions $F$ and $G$ are unimodal about the same mode $M$, then $aF + (1 - a)G$ is unimodal about $M$ for any $0 < a < 1$.

---

†It is a source of potential confusion that the word 'mode' is used in several contexts. A function is sometimes said to be unimodal if it has a unique maximum. The word mode is also used for the value(s) of $x$ at which a mass function (or density function) $f(x)$ is maximized, and even on occasion the locations of its local maxima.

# 3

# Discrete random variables

*Summary.* The distribution of a discrete random variable may be specified via its probability mass function. The key notion of independence for discrete random variables is introduced. The concept of expectation, or mean value, is defined for discrete variables, leading to a definition of the variance and the moments of a discrete random variable. Joint distributions, conditional distributions, and conditional expectation are introduced, together with the ideas of covariance and correlation. The Cauchy–Schwarz inequality is presented. The analysis of sums of random variables leads to the convolution formula for mass functions. Random walks are studied in some depth, including the reflection principle, the ballot theorem, the hitting time theorem, and the arc sine laws for visits to the origin and for sojourn times.

## 3.1 Probability mass functions

Recall that a random variable $X$ is *discrete* if it takes values only in some countable set $\{x_1, x_2, \dots\}$. Its distribution function $F(x) = \mathbb{P}(X \leq x)$ is a jump function; just as important as its distribution function is its mass function.

**(1) Definition.** The **(probability) mass function**†of a discrete random variable $X$ is the function $f : \mathbb{R} \to [0, 1]$ given by $f(x) = \mathbb{P}(X = x)$.

The distribution and mass functions are related by

$$F(x) = \sum_{i : x_i \leq x} f(x_i), \qquad f(x) = F(x) - \lim_{y \uparrow x} F(y).$$

**(2) Lemma.** *The probability mass function* $f : \mathbb{R} \to [0, 1]$ *satisfies:*
(a) *the set of $x$ such that $f(x) \neq 0$ is countable,*
(b) $\sum_i f(x_i) = 1$, *where $x_1, x_2, \dots$ are the values of $x$ such that $f(x) \neq 0$.*

**Proof.** The proof is obvious. ∎

This lemma characterizes probability mass functions.

---

†Some refer loosely to the mass function of $X$ as its distribution.

**(3) Example. Binomial distribution.**  A coin is tossed $n$ times, and a head turns up each time with probability $p\ (= 1 - q)$. Then $\Omega = \{H, T\}^n$. The total number $X$ of heads takes values in the set $\{0, 1, 2, \ldots, n\}$ and is a discrete random variable. Its probability mass function $f(x) = \mathbb{P}(X = x)$ satisfies

$$f(x) = 0 \quad \text{if} \quad x \notin \{0, 1, 2, \ldots, n\}.$$

Let $0 \le k \le n$, and consider $f(k)$. Exactly $\binom{n}{k}$ points in $\Omega$ give a total of $k$ heads; each of these points occurs with probability $p^k q^{n-k}$, and so

$$f(k) = \binom{n}{k} p^k q^{n-k} \quad \text{if} \quad 0 \le k \le n.$$

The random variable $X$ is said to have the *binomial distribution* with parameters $n$ and $p$, written $\mathrm{bin}(n, p)$. It is the sum $X = Y_1 + Y_2 + \cdots + Y_n$ of $n$ Bernoulli variables (see Example (2.1.8)).                                                                                   ●

**(4) Example. Poisson distribution.**  If a random variable $X$ takes values in the set $\{0, 1, 2, \ldots\}$ with mass function

$$f(k) = \frac{\lambda^k}{k!} e^{-\lambda}, \qquad k = 0, 1, 2, \ldots,$$

where $\lambda > 0$, then $X$ is said to have the *Poisson distribution* with parameter $\lambda$.                     ●

## Exercises for Section 3.1

**1.**   For what values of the constant $C$ do the following define mass functions on the positive integers $1, 2, \ldots$?
(a) Geometric: $f(x) = C 2^{-x}$.
(b) Logarithmic: $f(x) = C 2^{-x}/x$.
(c) Inverse square: $f(x) = C x^{-2}$.
(d) 'Modified' Poisson: $f(x) = C 2^x/x!$.

**2.**   For a random variable $X$ having (in turn) each of the four mass functions of Exercise (3.1.1), find:
(i) $\mathbb{P}(X > 1)$,
(ii) the most probable value of $X$,
(iii) the probability that $X$ is even.

**3.**   We toss $n$ coins, and each one shows heads with probability $p$, independently of each of the others. Each coin which shows heads is tossed again. What is the mass function of the number of heads resulting from the second round of tosses?

**4.**   Let $S_k$ be the set of positive integers whose base-10 expansion contains exactly $k$ elements (so that, for example, $1024 \in S_4$). A fair coin is tossed until the first head appears, and we write $T$ for the number of tosses required. We pick a random element, $N$ say, from $S_T$, each such element having equal probability. What is the mass function of $N$?

**5.**   **Log-convexity.** (a) Show that, if $X$ is a binomial or Poisson random variable, then the mass function $f(k) = \mathbb{P}(X = k)$ has the property that $f(k - 1)f(k + 1) \le f(k)^2$.
(b) Show that, if $f(k) = 90/(\pi k)^4$, $k \ge 1$, then $f(k - 1)f(k + 1) \ge f(k)^2$.
(c) Find a mass function $f$ such that $f(k)^2 = f(k - 1)f(k + 1)$, $k \ge 1$.

## 3.2 Independence

Remember that events $A$ and $B$ are called 'independent' if the occurrence of $A$ does not change the subsequent probability of $B$ occurring. More rigorously, $A$ and $B$ are independent if and only if $\mathbb{P}(A \cap B) = \mathbb{P}(A)\mathbb{P}(B)$. Similarly, we say that discrete variables $X$ and $Y$ are 'independent' if the numerical value of $X$ does not affect the distribution of $Y$. With this in mind we make the following definition.

**(1) Definition.** Discrete variables $X$ and $Y$ are **independent** if the events $\{X = x\}$ and $\{Y = y\}$ are independent for all $x$ and $y$.

Suppose $X$ takes values in the set $\{x_1, x_2, \dots\}$ and $Y$ takes values in the set $\{y_1, y_2, \dots\}$. Let

$$A_i = \{X = x_i\}, \qquad B_j = \{Y = y_j\}.$$

Notice (see Problem (2.7.2)) that $X$ and $Y$ are linear combinations of the indicator variables $I_{A_i}$, $I_{B_j}$, in that

$$X = \sum_i x_i I_{A_i} \qquad \text{and} \qquad Y = \sum_j y_j I_{B_j}.$$

The random variables $X$ and $Y$ are independent if and only if $A_i$ and $B_j$ are independent for all pairs $i$, $j$. A similar definition holds for collections $\{X_1, X_2, \dots, X_n\}$ of discrete variables. (Recall Definition (1.5.1).)

**(2) Example. Poisson flips.** A coin is tossed once and heads turns up with probability $p = 1 - q$. Let $X$ and $Y$ be the numbers of heads and tails respectively. It is no surprise that $X$ and $Y$ are not independent. After all,

$$\mathbb{P}(X = Y = 1) = 0, \qquad \mathbb{P}(X = 1)\mathbb{P}(Y = 1) = p(1 - p).$$

Suppose now that the coin is tossed a random number $N$ of times, where $N$ has the Poisson distribution with parameter $\lambda$. It is a remarkable fact that the resulting numbers $X$ and $Y$ of heads and tails *are* independent, since

$$\mathbb{P}(X = x, \ Y = y) = \mathbb{P}\big(X = x, Y = y \mid N = x + y\big)\mathbb{P}(N = x + y)$$

$$= \binom{x + y}{x} p^x q^y \frac{\lambda^{x+y}}{(x + y)!} e^{-\lambda} = \frac{(\lambda p)^x (\lambda q)^y}{x!\, y!} e^{-\lambda}.$$

However, by Lemma (1.4.4),

$$\mathbb{P}(X = x) = \sum_{n \geq x} \mathbb{P}(X = x \mid N = n)\mathbb{P}(N = n)$$

$$= \sum_{n \geq x} \binom{n}{x} p^x q^{n-x} \frac{\lambda^n}{n!} e^{-\lambda} = \frac{(\lambda p)^x}{x!} e^{-\lambda p};$$

a similar result holds for $Y$, and so

$$\mathbb{P}(X = x, \ Y = y) = \mathbb{P}(X = x)\mathbb{P}(Y = y). \qquad \bullet$$

If $X$ is a random variable and $g : \mathbb{R} \to \mathbb{R}$, then $Z = g(X)$, defined by $Z(\omega) = g(X(\omega))$, is a random variable also. We shall need the following.

**(3) Theorem.** *If $X$ and $Y$ are independent and $g, h : \mathbb{R} \to \mathbb{R}$, then $g(X)$ and $h(Y)$ are independent also.*

**Proof.** *Exercise.* See Problem (3.11.1).                                                    ∎

More generally, we say that a family $\{X_i : i \in I\}$ of (discrete) random variables is *independent* if the events $\{X_i = x_i\}$, $i \in I$, are independent for all possible choices of the set $\{x_i : i \in I\}$ of the values of the $X_i$. That is to say, $\{X_i : i \in I\}$ is an independent family if and only if

$$\mathbb{P}(X_i = x_i \text{ for all } i \in J) = \prod_{i \in J} \mathbb{P}(X_i = x_i)$$

for all sets $\{x_i : i \in I\}$ and for all finite subsets $J$ of $I$. The conditional independence of a family of random variables, given an event $C$, is defined similarly to the conditional independence of events; see equation (1.5.5).

Independent families of random variables are very much easier to study than dependent families, as we shall see soon. Note that pairwise-independent families are not necessarily independent.

---

## Exercises for Section 3.2

**1.** Let $X$ and $Y$ be independent random variables, each taking the values $-1$ or $1$ with probability $\frac{1}{2}$, and let $Z = XY$. Show that $X$, $Y$, and $Z$ are pairwise independent. Are they independent?

**2.** Let $X$ and $Y$ be independent random variables taking values in the positive integers and having the same mass function $f(x) = 2^{-x}$ for $x = 1, 2, \ldots$. Find:
   (a) $\mathbb{P}(\min\{X, Y\} \le x)$,  (b) $\mathbb{P}(Y > X)$,
   (c) $\mathbb{P}(X = Y)$,              (d) $\mathbb{P}(X \ge kY)$, for a given positive integer $k$,
   (e) $\mathbb{P}(X \text{ divides } Y)$,  (f) $\mathbb{P}(X = rY)$, for a given positive rational $r$.

**3.** Let $X_1, X_2, X_3$ be independent random variables taking values in the positive integers and having mass functions given by $\mathbb{P}(X_i = x) = (1 - p_i)p_i^{x-1}$ for $x = 1, 2, \ldots$, and $i = 1, 2, 3$.
   (a) Show that

$$\mathbb{P}(X_1 < X_2 < X_3) = \frac{(1 - p_1)(1 - p_2)p_2 p_3^2}{(1 - p_2 p_3)(1 - p_1 p_2 p_3)}.$$

   (b) Find $\mathbb{P}(X_1 \le X_2 \le X_3)$.

**4.** Three players, A, B, and C, take turns to roll a die; they do this in the order ABCABCA....
   (a) Show that the probability that, of the three players, A is the first to throw a 6, B the second, and C the third, is $216/1001$.
   (b) Show that the probability that the first 6 to appear is thrown by A, the second 6 to appear is thrown by B, and the third 6 to appear is thrown by C, is $46656/753571$.

**5.** Let $X_r$, $1 \le r \le n$, be independent random variables which are symmetric about 0; that is, $X_r$ and $-X_r$ have the same distributions. Show that, for all $x$, $\mathbb{P}(S_n \ge x) = \mathbb{P}(S_n \le -x)$ where $S_n = \sum_{r=1}^{n} X_r$.
   Is the conclusion necessarily true without the assumption of independence?

## 3.3 Expectation

Let $x_1, x_2, \ldots, x_N$ be the numerical outcomes of $N$ repetitions of some experiment. The average of these outcomes is

$$m = \frac{1}{N} \sum_i x_i.$$

In advance of performing these experiments we can represent their outcomes by a sequence $X_1, X_2, \ldots, X_N$ of random variables, and we shall suppose that these variables are discrete with a common mass function $f$. Then, roughly speaking (see the beginning of Section 1.3), for each possible value $x$, about $Nf(x)$ of the $X_i$ will take that value $x$. Therefore, for large $N$, the average $m$ is approximately

$$m \simeq \frac{1}{N} \sum_x x Nf(x) = \sum_x xf(x)$$

where the summation here is over all possible values of the $X_i$. This average is called the 'expectation' or 'mean value' of the underlying distribution with mass function $f$.

**(1) Definition.** The **mean value**, or **expectation**, or **expected value** of the random variable $X$ with mass function $f$ is defined to be

$$\mathbb{E}(X) = \sum_{x:f(x)>0} xf(x)$$

whenever this sum is absolutely convergent.

We require *absolute* convergence in order that $\mathbb{E}(X)$ be unchanged by reordering the $x_i$. We can, for notational convenience, write $\mathbb{E}(X) = \sum_x xf(x)$. This appears to be an uncountable sum; however, all but countably many of its contributions are zero. If the numbers $f(x)$ are regarded as masses $f(x)$ at points $x$ then $\mathbb{E}(X)$ is just the position of the centre of gravity; we can speak of $X$ as having an 'atom' or 'point mass' of size $f(x)$ at $x$. We sometimes omit the parentheses and simply write $\mathbb{E}X$.

**(2) Example (2.1.5) revisited.** The random variables $X$ and $W$ of this example have mean values

$$\mathbb{E}(X) = \sum_x x\mathbb{P}(X = x) = 0 \cdot \tfrac{1}{4} + 1 \cdot \tfrac{1}{2} + 2 \cdot \tfrac{1}{4} = 1,$$

$$\mathbb{E}(W) = \sum_x x\mathbb{P}(W = x) = 0 \cdot \tfrac{3}{4} + 4 \cdot \tfrac{1}{4} = 1. \qquad \bullet$$

If $X$ is a random variable and $g : \mathbb{R} \to \mathbb{R}$, then $Y = g(X)$, given formally by $Y(\omega) = g(X(\omega))$, is a random variable also. To calculate its expectation we need first to find its probability mass function $f_Y$. This process can be complicated, and it is avoided by the following lemma (called by some the 'law of the unconscious statistician'!).

**(3) Lemma. Change of variable formula.** *If $X$ has mass function $f$ and $g : \mathbb{R} \to \mathbb{R}$, then*

$$\mathbb{E}(g(X)) = \sum_x g(x) f(x)$$

*whenever this sum is absolutely convergent.*

**Proof.** This is Problem (3.11.3).                                                ∎

**(4) Example.** Suppose that $X$ takes values $-2, -1, 1, 3$ with probabilities $\frac{1}{4}, \frac{1}{8}, \frac{1}{4}, \frac{3}{8}$ respectively. The random variable $Y = X^2$ takes values $1, 4, 9$ with probabilities $\frac{3}{8}, \frac{1}{4}, \frac{3}{8}$ respectively, and so

$$\mathbb{E}(Y) = \sum_x x \mathbb{P}(Y = x) = 1 \cdot \tfrac{3}{8} + 4 \cdot \tfrac{1}{4} + 9 \cdot \tfrac{3}{8} = \tfrac{19}{4}.$$

Alternatively, use the law of the unconscious statistician to find that

$$\mathbb{E}(Y) = \mathbb{E}(X^2) = \sum_x x^2 \mathbb{P}(X = x) = 4 \cdot \tfrac{1}{4} + 1 \cdot \tfrac{1}{8} + 1 \cdot \tfrac{1}{4} + 9 \cdot \tfrac{3}{8} = \tfrac{19}{4}.$$   ●

Lemma (3) provides a method for calculating the 'moments' of a distribution; these are defined as follows.

**(5) Definition.** If $k$ is a positive integer, the $k$th **moment** $m_k$ of $X$ is defined to be $m_k = \mathbb{E}(X^k)$. The $k$th **central moment** $\sigma_k$ is $\sigma_k = \mathbb{E}((X - m_1)^k)$.

The two moments of most use are $m_1 = \mathbb{E}(X)$ and $\sigma_2 = \mathbb{E}((X - \mathbb{E}X)^2)$, called the *mean* (or *expectation*) and *variance* of $X$. These two quantities are measures of the mean and dispersion of $X$; that is, $m_1$ is the average value of $X$, and $\sigma_2$ measures the amount by which $X$ tends to deviate from this average. The mean $m_1$ is often denoted $\mu$, and the variance of $X$ is often denoted var$(X)$. The positive square root $\sigma = \sqrt{\text{var}(X)}$ is called the *standard deviation*, and in this notation $\sigma_2 = \sigma^2$. The central moments $\{\sigma_i\}$ can be expressed in terms of the ordinary moments $\{m_i\}$. For example, $\sigma_1 = 0$ and

$$\sigma_2 = \sum_x (x - m_1)^2 f(x)$$

$$= \sum_x x^2 f(x) - 2m_1 \sum_x x f(x) + m_1^2 \sum_x f(x)$$

$$= m_2 - m_1^2,$$

which may be written as

$$\text{var}(X) = \mathbb{E}\big((X - \mathbb{E}X)^2\big) = \mathbb{E}(X^2) - (\mathbb{E}X)^2.$$

**Remark.** Experience with student calculations of variances causes us to stress the following elementary fact: *variances cannot be negative*. We sometimes omit the parentheses and write simply var $X$. The expression $\mathbb{E}(X)^2$ means $(\mathbb{E}(X))^2$ and must not be confused with $\mathbb{E}(X^2)$.

**(6) Example. Bernoulli variables.** Let $X$ be a Bernoulli variable, taking the value 1 with probability $p$ $(= 1 - q)$. Then

$$\mathbb{E}(X) = \sum_x xf(x) = 0 \cdot q + 1 \cdot p = p,$$

$$\mathbb{E}(X^2) = \sum_x x^2 f(x) = 0 \cdot q + 1 \cdot p = p,$$

$$\mathrm{var}(X) = \mathbb{E}(X^2) - \mathbb{E}(X)^2 = pq.$$

Thus the indicator variable $I_A$ has expectation $\mathbb{P}(A)$ and variance $\mathbb{P}(A)\mathbb{P}(A^c)$.    ●

**(7) Example. Binomial variables.** Let $X$ be $\mathrm{bin}(n, p)$. Then

$$\mathbb{E}(X) = \sum_{k=0}^{n} kf(k) = \sum_{k=0}^{n} k\binom{n}{k} p^k q^{n-k}.$$

To calculate this, differentiate the identity

$$\sum_{k=0}^{n} \binom{n}{k} x^k = (1+x)^n,$$

multiply by $x$ to obtain

$$\sum_{k=0}^{n} k\binom{n}{k} x^k = nx(1+x)^{n-1},$$

and substitute $x = p/q$ to obtain $\mathbb{E}(X) = np$. A similar argument shows that the variance of $X$ is given by $\mathrm{var}(X) = npq$, although it is faster to use the forthcoming Theorem (11).    ●

We can think of the process of calculating expectations as a linear operator on the space of random variables.

**(8) Theorem.** *The expectation operator $\mathbb{E}$ has the following properties:*
 (a) *if $X \geq 0$ then $\mathbb{E}(X) \geq 0$,*
 (b) *if $a, b \in \mathbb{R}$ then $\mathbb{E}(aX + bY) = a\mathbb{E}(X) + b\mathbb{E}(Y)$,*
 (c) *the random variable 1, taking the value 1 always, has expectation $\mathbb{E}(1) = 1$.*

**Proof.** (a) and (c) are obvious.
(b) Let $A_x = \{X = x\}$, $B_y = \{Y = y\}$. Then

$$aX + bY = \sum_{x,y} (ax + by) I_{A_x \cap B_y}$$

and the solution of the first part of Problem (3.11.3) shows that

$$\mathbb{E}(aX + bY) = \sum_{x,y} (ax + by)\mathbb{P}(A_x \cap B_y).$$

However,

$$\sum_y \mathbb{P}(A_x \cap B_y) = \mathbb{P}\left(A_x \cap \left(\bigcup_y B_y\right)\right) = \mathbb{P}(A_x \cap \Omega) = \mathbb{P}(A_x)$$

and similarly $\sum_x \mathbb{P}(A_x \cap B_y) = \mathbb{P}(B_y)$, which gives

$$\mathbb{E}(aX + bY) = \sum_x ax \sum_y \mathbb{P}(A_x \cap B_y) + \sum_y by \sum_x \mathbb{P}(A_x \cap B_y)$$

$$= a \sum_x x\mathbb{P}(A_x) + b \sum_y y\mathbb{P}(B_y)$$

$$= a\mathbb{E}(X) + b\mathbb{E}(Y). \qquad\blacksquare$$

**Remark.** It is not in general true that $\mathbb{E}(XY)$ is the same as $\mathbb{E}(X)\mathbb{E}(Y)$.

**(9) Lemma.** *If $X$ and $Y$ are independent then $\mathbb{E}(XY) = \mathbb{E}(X)\mathbb{E}(Y)$.*

**Proof.** Let $A_x$ and $B_y$ be as in the proof of (8). Then

$$XY = \sum_{x,y} xy I_{A_x \cap B_y}$$

and so

$$\mathbb{E}(XY) = \sum_{x,y} xy\mathbb{P}(A_x)\mathbb{P}(B_y) \qquad \text{by independence}$$

$$= \sum_x x\mathbb{P}(A_x) \sum_y y\mathbb{P}(B_y) = \mathbb{E}(X)\mathbb{E}(Y). \qquad\blacksquare$$

**(10) Definition.** $X$ and $Y$ are called **uncorrelated** if $\mathbb{E}(XY) = \mathbb{E}(X)\mathbb{E}(Y)$.

Lemma (9) asserts that independent variables are uncorrelated. The converse is not true, as Problem (3.11.16) indicates.

**(11) Theorem.** *For random variables $X$ and $Y$,*
(a) $\operatorname{var}(aX) = a^2 \operatorname{var}(X)$ *for $a \in \mathbb{R}$,*
(b) $\operatorname{var}(X + Y) = \operatorname{var}(X) + \operatorname{var}(Y)$ *if $X$ and $Y$ are uncorrelated.*

**Proof.** (a) Using the linearity of $\mathbb{E}$,

$$\operatorname{var}(aX) = \mathbb{E}\{(aX - \mathbb{E}(aX))^2\} = \mathbb{E}\{a^2(X - \mathbb{E}X)^2\}$$

$$= a^2\mathbb{E}\{(X - \mathbb{E}X)^2\} = a^2 \operatorname{var}(X).$$

(b) We have when $X$ and $Y$ are uncorrelated that

$$\operatorname{var}(X + Y) = \mathbb{E}\{(X + Y - \mathbb{E}(X + Y))^2\}$$

$$= \mathbb{E}\left[(X - \mathbb{E}X)^2 + 2(XY - \mathbb{E}(X)\mathbb{E}(Y)) + (Y - \mathbb{E}Y)^2\right]$$

$$= \operatorname{var}(X) + 2[\mathbb{E}(XY) - \mathbb{E}(X)\mathbb{E}(Y)] + \operatorname{var}(Y)$$

$$= \operatorname{var}(X) + \operatorname{var}(Y). \qquad\blacksquare$$

Theorem (11a) shows that the variance operator 'var' is *not* a linear operator, even when it is applied only to uncorrelated variables.

Sometimes the sum $S = \sum x f(x)$ does not converge absolutely, and the mean of the distribution does not exist. If $S = -\infty$ or $S = +\infty$, then we can sometimes speak of the mean as taking these values also. Of course, there exist distributions which do not have a mean value.

**(12) Example. A distribution without a mean.** Let $X$ have mass function

$$f(k) = Ak^{-2} \quad \text{for} \quad k = \pm 1, \pm 2, \ldots$$

where $A$ is chosen so that $\sum f(k) = 1$. The sum $\sum_k k f(k) = A \sum_{k \neq 0} k^{-1}$ does not converge absolutely, because both the positive and the negative parts diverge.                                    ●

This is a suitable opportunity to point out that we can base probability theory upon the expectation operator $\mathbb{E}$ rather than upon the probability measure $\mathbb{P}$. After all, our intuitions about the notion of 'average' are probably just as well developed as those about quantitative chance. Roughly speaking, the way we proceed is to postulate axioms, such as (a), (b), and (c) of Theorem (8), for a so-called 'expectation operator' $\mathbb{E}$ acting on a space of 'random variables'. The probability of an event can then be recaptured by defining $\mathbb{P}(A) = \mathbb{E}(I_A)$. Whittle (2000) is an able advocate of this approach.

This method can be easily and naturally adapted to deal with probabilistic questions in quantum theory. In this major branch of theoretical physics, questions arise which cannot be formulated entirely within the usual framework of probability theory. However, there still exists an expectation operator $\mathbb{E}$, which is applied to linear operators known as observables (such as square matrices) rather than to random variables. There does not exist a sample space $\Omega$, and nor therefore are there any indicator functions, but nevertheless there exist analogues of other concepts in probability theory. For example, the *variance* of an operator $X$ is defined by $\text{var}(X) = \mathbb{E}(X^2) - \mathbb{E}(X)^2$. Furthermore, it can be shown that $\mathbb{E}(X) = \text{tr}(UX)$ where tr denotes *trace* and $U$ is a non-negative definite operator with unit trace.

**(13) Example. Wagers.** Historically, there has been confusion amongst probabilists between the price that an individual may be willing to pay in order to play a game, and her expected return from this game. For example, I conceal $£2$ in one hand and nothing in the other, and then invite a friend to pay a fee which entitles her to choose a hand at random and keep the contents. Other things being equal (my friend is neither a compulsive gambler, nor particularly busy), it would seem that $£1$ would be a 'fair' fee to ask, since $£1$ is the expected return to the player. That is to say, faced with a modest (but random) gain, then a fair 'entrance fee' would seem to be the expected value of the gain. However, suppose that I conceal $£2^{10}$ in one hand and nothing in the other; what now is a 'fair' fee? Few persons of modest means can be expected to offer $£2^9$ for the privilege of playing. There is confusion here between fairness and reasonableness: we do not generally treat large payoffs or penalties in the same way as small ones, even though the relative odds may be unquestionable. The customary resolution of this paradox is to introduce the notion of 'utility'. Writing $u(x)$ for the 'utility' to an individual of $£x$, it would be fairer to charge a fee of $\frac{1}{2}(u(0) + u(2^{10}))$ for the above prospect. Of course, different individuals have different utility functions, although such functions have presumably various features in common: $u(0) = 0$, $u$ is non-decreasing, $u(x)$ is near to $x$ for small positive $x$, and $u$ is concave, so that in particular $u(x) \leq xu(1)$ when $x \geq 1$.

The use of expectation to assess a 'fair fee' may be convenient but is sometimes inappropriate. For example, a more suitable criterion in the finance market would be absence of arbitrage; see Exercise (3.3.7) and Section 13.10. And, in a rather general model of financial markets, there is a criterion commonly expressed as 'no free lunch with vanishing risk'.    ●

---

## Exercises for Section 3.3

**1.**   Is it generally true that $\mathbb{E}(1/X) = 1/\mathbb{E}(X)$? Is it ever true that $\mathbb{E}(1/X) = 1/\mathbb{E}(X)$?

**2.   Coupons.** Every package of some intrinsically dull commodity includes a small and exciting plastic object. There are $c$ different types of object, and each package is equally likely to contain any given type. You buy one package each day.

(a) Find the mean number of days which elapse between the acquisitions of the $j$th new type of object and the $(j + 1)$th new type.

(b) Find the mean number of days which elapse before you have a full set of objects.

**3.**   Each member of a group of $n$ players rolls a die.

(a) For any pair of players who throw the same number, the group scores 1 point. Find the mean and variance of the total score of the group.

(b) Find the mean and variance of the total score if any pair of players who throw the same number scores that number.

**4.   St Petersburg paradox†.** A fair coin is tossed repeatedly. Let $T$ be the number of tosses until the first head. You are offered the following prospect, which you may accept on payment of a fee. If $T = k$, say, then you will receive £$2^k$. What would be a 'fair' fee to ask of you?

**5.**   Let $X$ have mass function

$$f(x) = \begin{cases} \{x(x+1)\}^{-1} & \text{if } x = 1, 2, \ldots, \\ 0 & \text{otherwise,} \end{cases}$$

and let $\alpha \in \mathbb{R}$. For what values of $\alpha$ is it the case‡ that $\mathbb{E}(X^\alpha) < \infty$?

**6.**   Show that $\text{var}(a + X) = \text{var}(X)$ for any random variable $X$ and constant $a$.

**7.   Arbitrage.** Suppose you find a warm-hearted bookmaker offering payoff odds of $\pi(k)$ against the $k$th horse in an $n$-horse race where $\sum_{k=1}^{n} \{\pi(k) + 1\}^{-1} < 1$. Show that you can distribute your bets in such a way as to ensure you win.

**8.**   You roll a conventional fair die repeatedly. If it shows 1, you must stop, but you may choose to stop at any prior time. Your score is the number shown by the die on the final roll. What stopping strategy yields the greatest expected score? What strategy would you use if your score were the square of the final roll?

**9.**   Continuing with Exercise (3.3.8), suppose now that you lose $c$ points from your score each time you roll the die. What strategy maximizes the expected final score if $c = \frac{1}{3}$? What is the best strategy if $c = 1$?

**10.   Random social networks.** Let $G = (V, E)$ be a random graph with $m = |V|$ vertices and edge-set $E$. Write $d_v$ for the degree of vertex $v$, that is, the number of edges meeting at $v$. Let $Y$ be a uniformly chosen vertex, and $Z$ a uniformly chosen neighbour of $Y$.

---

†This problem was mentioned by Nicholas Bernoulli in 1713, and Daniel Bernoulli wrote about the question for the Academy of St Petersburg.

‡If $\alpha$ is not integral, than $\mathbb{E}(X^\alpha)$ is called the *fractional moment of order $\alpha$* of $X$. A point concerning notation: for real $\alpha$ and complex $x = re^{i\theta}$, $x^\alpha$ should be interpreted as $r^\alpha e^{i\theta\alpha}$, so that $|x^\alpha| = r^\alpha$. In particular, $\mathbb{E}(|X^\alpha|) = \mathbb{E}(|X|^\alpha)$.

(a) Show that $\mathbb{E}d_Z \geq \mathbb{E}d_Y$.

(b) Interpret this inequality when the vertices represent people, and the edges represent friendship.

**11.** The gambler Lester Savage makes up to three successive bets that a fair coin flip will show heads. He places a stake on each bet, which, if a head is shown, pays him back twice the stake. If a tail is shown, he loses his stake.

His stakes are determined as follows. Let $x > y > z > 0$. He bets $x$ on the first flip; if it shows heads he quits, otherwise he continues. If he continues, he bets $y$ on the second flip; if it shows heads he quits, otherwise he continues. If he continues, he bets $z$ on the third flip.

Let $G$ be his accumulated gain (positive or negative). List the possible values of $G$ and their probabilities. Show that $\mathbb{E}(G) = 0$, and find var$(G)$ and $\mathbb{P}(G < 0)$.

Lester decides to stick with the three numbers $x$, $y$, $z$ but to vary their order. How should he place his bets in order to simultaneously minimize both $\mathbb{P}(G < 0)$ and var$(G)$? Explain.

**12. Quicksort†.** A set of $n$ different words is equally likely to be in any of the $n!$ possible orders. It is decided to place them in lexicographic order using the following algorithm.

    (i) Compare the first word $w$ with the others, and find the set of earlier words and the set of later words.

    (ii) Iterate the procedure on each of the two sets thus obtained.

    (iii) Continue until the final ordering is achieved.

(a) Give an expression for the mean number $c_n$ of comparisons required.

(b) Show that $c_n = 2n(\log n + \gamma - 2) + \mathrm{O}(1)$ as $n \to \infty$, where $\gamma$ is Euler's constant.

(c) Let $n$ be replaced by a random variable $N$ with mass function

$$\mathbb{P}(N = n) = \frac{A}{(n-1)n(n+1)}, \qquad n \geq 2,$$

for suitable $A$. Show that the mean number of comparisons is 4.

---

## 3.4 Indicators and matching

This section contains light entertainment, in the guise of some illustrations of the uses of indicator functions. These were defined in Example (2.1.9) and have appeared occasionally since. Recall that

$$I_A(\omega) = \begin{cases} 1 & \text{if } \omega \in A, \\ 0 & \text{if } \omega \in A^c, \end{cases}$$

and $\mathbb{E}I_A = \mathbb{P}(A)$.

**(1) Example. Proofs of Lemma (1.3.4c, d).** Note that

$$I_A + I_{A^c} = I_{A \cup A^c} = I_\Omega = 1$$

and that $I_{A \cap B} = I_A I_B$. Thus

$$\begin{aligned} I_{A \cup B} &= 1 - I_{(A \cup B)^c} = 1 - I_{A^c \cap B^c} \\ &= 1 - I_{A^c} I_{B^c} = 1 - (1 - I_A)(1 - I_B) \\ &= I_A + I_B - I_A I_B. \end{aligned}$$

---

†Invented by C. A. R. Hoare in 1959.

Take expectations to obtain

$$\mathbb{P}(A \cup B) = \mathbb{P}(A) + \mathbb{P}(B) - \mathbb{P}(A \cap B).$$

More generally, if $B = \bigcup_{i=1}^{n} A_i$ then

$$I_B = 1 - \prod_{i=1}^{n}(1 - I_{A_i});$$

multiply this out and take expectations to obtain

$$(2) \qquad \mathbb{P}\left(\bigcup_{i=1}^{n} A_i\right) = \sum_{i} \mathbb{P}(A_i) - \sum_{i<j} \mathbb{P}(A_i \cap A_j) + \cdots + (-1)^{n+1} \mathbb{P}(A_1 \cap \cdots \cap A_n).$$

This useful identity is known as the *inclusion–exclusion formula*. Recall Lemma (1.3.4d). ●

**(3) Example. Montmort's matching problem.** A number of melodramatic applications of (2) are available, of which the following is typical. A secretary types $n$ different letters together with matching envelopes, drops the pile down the stairs, and then places the letters randomly in the envelopes. Each arrangement is equally likely, and we ask for the probability that exactly $r$ are in their correct envelopes. Rather than using (2), we shall proceed directly by way of indicator functions. (Another approach is presented in Exercise (3.4.9).)
**Solution.** Let $L_1, L_2, \ldots, L_n$ denote the letters. Call a letter *good* if it is correctly addressed and *bad* otherwise; write $X$ for the number of good letters. Let $A_i$ be the event that $L_i$ is good, and let $I_i$ be the indicator function of $A_i$. Let $j_1, \ldots, j_r, k_{r+1}, \ldots, k_n$ be a permutation of the numbers $1, 2, \ldots, n$, and define

$$(4) \qquad S = \sum_{\pi} I_{j_1} \cdots I_{j_r} (1 - I_{k_{r+1}}) \cdots (1 - I_{k_n})$$

where the sum is taken over all such permutations $\pi$. Then

$$S = \begin{cases} 0 & \text{if } X \neq r, \\ r!\,(n-r)! & \text{if } X = r. \end{cases}$$

To see this, let $L_{i_1}, \ldots, L_{i_m}$ be the good letters. If $m \neq r$ then each summand in (4) equals 0. If $m = r$ then the summand in (4) equals 1 if and only if $j_1, \ldots, j_r$ is a permutation of $i_1, \ldots, i_r$ and $k_{r+1}, \ldots, k_n$ is a permutation of the remaining numbers; there are $r!\,(n-r)!$ such pairs of permutations. It follows that $I$, given by

$$(5) \qquad I = \frac{1}{r!\,(n-r)!} S,$$

is the indicator function of the event $\{X = r\}$ that exactly $r$ letters are good. We take expectations of (4) and multiply out to obtain

$$\mathbb{E}(S) = \sum_{\pi} \sum_{s=0}^{n-r} (-1)^s \binom{n-r}{s} \mathbb{E}(I_{j_1} \cdots I_{j_r} I_{k_{r+1}} \cdots I_{k_{r+s}})$$

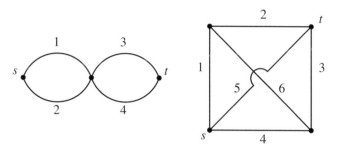

Figure 3.1. Two networks with source $s$ and sink $t$.

by a symmetry argument. However,

$$
\textbf{(6)} \qquad\qquad \mathbb{E}(I_{j_1} \cdots I_{j_r} I_{k_{r+1}} \cdots I_{k_{r+s}}) = \frac{(n-r-s)!}{n!}
$$

since there are $n!$ possible permutations, only $(n-r-s)!$ of which allocate $L_{i_1}, \ldots, L_{j_r}$, $L_{k_{r+1}}, \ldots, L_{k_{r+s}}$ to their correct envelopes. We combine (4), (5), and (6) to obtain

$$
\mathbb{P}(X = r) = \mathbb{E}(I) = \frac{1}{r!\,(n-r)!}\,\mathbb{E}(S)
$$

$$
= \frac{1}{r!\,(n-r)!} \sum_{s=0}^{n-r} (-1)^s \binom{n-r}{s} n! \frac{(n-r-s)!}{n!}
$$

$$
= \frac{1}{r!} \sum_{s=0}^{n-r} (-1)^s \frac{1}{s!}
$$

$$
= \frac{1}{r!} \left( \frac{1}{2!} - \frac{1}{3!} + \cdots + \frac{(-1)^{n-r}}{(n-r)!} \right) \qquad \text{for } r \le n-2 \text{ and } n \ge 2.
$$

In particular, as the number $n$ of letters tends to infinity, we obtain the possibly surprising result that the probability that no letter is put into its correct envelope approaches $e^{-1}$. It is left as an *exercise* to prove this without using indicators.                                  ●

**(7) Example. Reliability.** When you telephone your friend in Cambridge, your call is routed through the telephone network in a way which depends on the current state of the traffic. For example, if all lines into the Ascot switchboard are in use, then your call may go through the switchboard at Newmarket. Sometimes you may fail to get through at all, owing to a combination of faulty and occupied equipment in the system. We may think of the network as comprising nodes joined by edges, drawn as 'graphs' in the manner of the examples of Figure 3.1. In each of these examples there is a designated 'source' $s$ and 'sink' $t$, and we wish to find a path through the network from $s$ to $t$ which uses available channels. As a simple model for such a system in the presence of uncertainty, we suppose that each edge $e$ is 'working' with probability $p_e$, independently of all other edges. We write $\mathbf{p}$ for the vector of edge probabilities $p_e$, and define the *reliability* $R(\mathbf{p})$ of the network to be the probability that there is a path from $s$ to $t$ using only edges which are working. Denoting the network by $G$, we write $R_G(\mathbf{p})$ for $R(\mathbf{p})$ when we wish to emphasize the role of $G$.

We have encountered questions of reliability already. In Example (1.7.2) we were asked for the reliability of the first network in Figure 3.1 and in Problem (1.8.19) of the second, assuming on each occasion that the value of $p_e$ does not depend on the choice of $e$.

Let us write

$$X_e = \begin{cases} 1 & \text{if edge } e \text{ is working,} \\ 0 & \text{otherwise,} \end{cases}$$

the indicator function of the event that $e$ is working, so that $X_e$ takes the values 0 and 1 with probabilities $1 - p_e$ and $p_e$ respectively. Each realization $X$ of the $X_e$ either includes a working connection from $s$ to $t$ or does not. Thus, there exists a *structure function* $\zeta$ taking values 0 and 1 such that

(8)
$$\zeta(X) = \begin{cases} 1 & \text{if such a working connection exists,} \\ 0 & \text{otherwise;} \end{cases}$$

thus $\zeta(X)$ is the indicator function of the event that a working connection exists. It is immediately seen that $R(\mathbf{p}) = \mathbb{E}(\zeta(X))$. The function $\zeta$ may be expressed as

(9)
$$\zeta(X) = 1 - \prod_{\pi} I_{\{\pi \text{ not working}\}} = 1 - \prod_{\pi} \left(1 - \prod_{e \in \pi} X_e\right)$$

where $\pi$ is a typical path in $G$ from $s$ to $t$, and we say that $\pi$ is working if and only if every edge in $\pi$ is working.

For instance, in the case of the first example of Figure 3.1, there are four different paths from $s$ to $t$. Numbering the edges as indicated, we have that the structure function is given by

(10)
$$\zeta(X) = 1 - (1 - X_1 X_3)(1 - X_1 X_4)(1 - X_2 X_3)(1 - X_2 X_4).$$

As an *exercise*, expand this and take expectations to calculate the reliability of the network when $p_e = p$ for all edges $e$. $\bullet$

**(11) Example. The probabilistic method†.** Probability may be used to derive non-trivial results not involving probability. Here is an example. There are 17 fenceposts around the perimeter of a field, exactly 5 of which are rotten. Show that, irrespective of which these 5 are, there necessarily exists a run of 7 consecutive posts at least 3 of which are rotten.

Our solution involves probability. We label the posts $1, 2, \ldots, 17$, and let $I_k$ be the indicator function that post $k$ is rotten. Let $R_k$ be the number of rotten posts amongst those labelled $k + 1, k + 2, \ldots, k + 7$, all taken modulo 17. We now pick a random post labelled $K$, each being equally likely. We have that

$$\mathbb{E}(R_K) = \sum_{k=1}^{17} \frac{1}{17}(I_{k+1} + I_{k+2} + \cdots + I_{k+7}) = \sum_{j=1}^{17} \frac{7}{17} I_j = \frac{7}{17} \cdot 5.$$

Now $\frac{35}{17} > 2$, implying that $\mathbb{P}(R_K > 2) > 0$. Since $R_K$ is integer valued, it must be the case that $\mathbb{P}(R_K \geq 3) > 0$, implying that $R_k \geq 3$ for some $k$. $\bullet$

---

†Generally credited to Erdős.

## Exercises for Section 3.4

**1.**   (a) A biased coin is tossed $n$ times, and heads shows with probability $p$ on each toss. A *run* is a sequence of throws which result in the same outcome, so that, for example, the sequence HHTHTTTH contains five runs. Show that the expected number of runs is $1 + 2(n-1)p(1-p)$. Find the variance of the number of runs.

(b) Let $h$ heads and $t$ tails be arranged randomly in a line. Find the mean and variance of the number of runs of heads

**2.**   An urn contains $n$ balls numbered $1, 2, \ldots, n$. We remove $k$ balls at random (without replacement) and add up their numbers. Find the mean and variance of the total.

**3.**   Of the $2n$ people in a given collection of $n$ couples, exactly $m$ die. Assuming that the $m$ have been picked at random, find the mean number of surviving couples. This problem was formulated by Daniel Bernoulli in 1768.

**4.**   Urn R contains $n$ red balls and urn B contains $n$ blue balls. At each stage, a ball is selected at random from each urn, and they are swapped. Show that the mean number of red balls in urn R after stage $k$ is $\frac{1}{2}n\{1 + (1 - 2/n)^k\}$. This 'diffusion model' was described by Daniel Bernoulli in 1769.

**5.**   Consider a square with diagonals, with distinct source and sink. Each edge represents a component which is working correctly with probability $p$, independently of all other components. Write down an expression for the Boolean function which equals 1 if and only if there is a working path from source to sink, in terms of the indicator functions $X_i$ of the events {edge $i$ is working} as $i$ runs over the set of edges. Hence calculate the reliability of the network.

**6.**   A system is called a '$k$ out of $n$' system if it contains $n$ components and it works whenever $k$ or more of these components are working. Suppose that each component is working with probability $p$, independently of the other components, and let $X_c$ be the indicator function of the event that component $c$ is working. Find, in terms of the $X_c$, the indicator function of the event that the system works, and deduce the reliability of the system.

**7.**   **The probabilistic method.** Let $G = (V, E)$ be a finite graph. For any set $W$ of vertices and any edge $e \in E$, define the indicator function

$$I_W(e) = \begin{cases} 1 & \text{if } e \text{ connects } W \text{ and } W^c, \\ 0 & \text{otherwise.} \end{cases}$$

Set $N_W = \sum_{e \in E} I_W(e)$. Show that there exists $W \subseteq V$ such that $N_W \geq \frac{1}{2}|E|$.

**8.**   A total of $n$ bar magnets are placed end to end in a line with random independent orientations. Adjacent like poles repel, ends with opposite polarities join to form blocks. Let $X$ be the number of blocks of joined magnets. Find $\mathbb{E}(X)$ and var$(X)$.

**9.**   **Matching.** (a) Use the inclusion–exclusion formula (3.4.2) to derive the result of Example (3.4.3), namely: in a random permutation of the first $n$ integers, the probability that exactly $r$ retain their original positions is

$$\frac{1}{r!}\left(\frac{1}{2!} - \frac{1}{3!} + \cdots + \frac{(-1)^{n-r}}{(n-r)!}\right).$$

(b) Let $d_n$ be the number of derangements of the first $n$ integers (that is, rearrangements with no integers in their original positions). Show that $d_{n+1} = nd_n + nd_{n-1}$ for $n \geq 2$. Deduce the result of part (a).

(c) Given that exactly $m$ integers retain their original positions, find the probability that the integer 1 remains in first place.

**10. Birthdays.** In a lecture audience, there are $n$ students born in 2011, and they were born on independent, uniformly distributed days. Calculate the mean of the number $B$ of pairs of students sharing a birthday, and show that $\mathbb{E}(B) > 1$ if and only if $n \geq 28$. Compare this with the result of Problem (1.8.30). Find the variance of $B$.

**11. Inaba's theorem.** Show that any set of 10 points in the plane $\mathbb{R}^2$ can be covered by a suitable placement of disjoint open unit disks. [Hint: Consider an infinite array of unit disks whose centres form a triangular lattice.]

**12.** Days are either wet or dry, and, given today's weather, tomorrow's is the same as today's with probability $p$, and different otherwise. Let $w_n$ be the probability that the weather $n$ days into the future from today will be wet. Show that $w_{n+1} = 1 - p + (2p - 1)w_{n-1}$, and find $w_n$. What is the mean number of wet days in the next week?

**13.** An urn contains $b$ balls of which $g$ are green. Balls are sampled from the urn at random, one by one. After a ball is sampled, its colour is noted, and it is discarded. Find the mean and variance of the number of green balls in a sample of size $n$ ($\leq b$).

**14. Ménages (1.8.23) revisited.** Let $n$ ($\geq 2$) heterosexual couples be seated randomly at a circular table, subject only to the rule that the sexes alternate. There is no requirement that couples sit together. Let $X$ be the number of couples seated adjacently. Show that $\mathbb{E}(X) = 2$ and $\text{var}(X) = 2 - 2/(n - 1)$.

## 3.5 Examples of discrete variables

**(1) Bernoulli trials.** A random variable $X$ takes values 1 and 0 with probabilities $p$ and $q$ ($= 1 - p$), respectively. Sometimes we think of these values as representing the 'success' or the 'failure' of a trial. The mass function is

$$f(0) = 1 - p, \qquad f(1) = p,$$

and it follows that $\mathbb{E}X = p$ and $\text{var}(X) = p(1 - p)$.                                   ●

**(2) Binomial distribution.** We perform $n$ independent Bernoulli trials $X_1, X_2, \ldots, X_n$ and count the total number of successes $Y = X_1 + X_2 + \cdots + X_n$. As in Example (3.1.3), the mass function of $Y$ is

$$f(k) = \binom{n}{k} p^k (1 - p)^{n-k}, \qquad k = 0, 1, \ldots, n.$$

Application of Theorems (3.3.8) and (3.3.11) yields immediately

$$\mathbb{E}Y = np, \qquad \text{var}(Y) = np(1 - p);$$

the method of Example (3.3.7) provides a more lengthy derivation of this.                          ●

**(3) Trinomial distribution.** More generally, suppose we conduct $n$ trials, each of which results in one of three outcomes (red, white, or blue, say), where red occurs with probability $p$, white with probability $q$, and blue with probability $1 - p - q$. The probability of $r$ reds, $w$ whites, and $n - r - w$ blues is

$$\frac{n!}{r!\, w!\, (n - r - w)!} p^r q^w (1 - p - q)^{n-r-w}.$$

This is the *trinomial distribution*, with parameters $n$, $p$, and $q$. The 'multinomial distribution' is the obvious generalization of this distribution to the case of some number, say $t$, of possible outcomes.   ●

**(4) Poisson distribution.** A *Poisson* variable is a random variable with the Poisson mass function

$$f(k) = \frac{\lambda^k}{k!} e^{-\lambda}, \qquad k = 0, 1, 2, \ldots$$

for some $\lambda > 0$. It can be obtained in practice in the following way. Let $Y$ be a bin$(n, p)$ variable, and suppose that $n$ is very large and $p$ is very small (an example might be the number $Y$ of misprints on the front page of the *Grauniad*, where $n$ is the total number of characters and $p$ is the probability for each character that the typesetter has made an error). Now, let $n \to \infty$ and $p \to 0$ in such a way that $\mathbb{E}(Y) = np$ approaches a non-zero constant $\lambda$. Then, for $k = 0, 1, 2, \ldots$,

$$\mathbb{P}(Y = k) = \binom{n}{k} p^k (1-p)^{n-k} \sim \frac{1}{k!} \left(\frac{np}{1-p}\right)^k (1-p)^n \to \frac{\lambda^k}{k!} e^{-\lambda}.$$

Check that both the mean and the variance of this distribution are equal to $\lambda$. Now do Problem (2.7.7) again (*exercise*).   ●

**(5) Geometric distribution.** A *geometric* variable is a random variable with the geometric mass function

$$f(k) = p(1-p)^{k-1}, \qquad k = 1, 2, \ldots$$

for some number $p$ in $(0, 1)$. This distribution arises in the following way. Suppose that independent Bernoulli trials (parameter $p$) are performed at times $1, 2, \ldots$. Let $W$ be the time which elapses before the first success; $W$ is called a *waiting time*. Then $\mathbb{P}(W > k) = (1-p)^k$ and thus

$$\mathbb{P}(W = k) = \mathbb{P}(W > k-1) - \mathbb{P}(W > k) = p(1-p)^{k-1}.$$

The reader should check, preferably at this point, that the mean and variance are $p^{-1}$ and $(1-p)p^{-2}$ respectively.   ●

**(6) Negative binomial distribution.** More generally, in the previous example, let $W_r$ be the waiting time for the $r$th success. Check that $W_r$ has mass function

$$\mathbb{P}(W_r = k) = \binom{k-1}{r-1} p^r (1-p)^{k-r}, \qquad k = r, r+1, \ldots;$$

it is said to have the *negative binomial distribution* with parameters $r$ and $p$. The random variable $W_r$ is the sum of $r$ independent geometric variables. To see this, let $X_1$ be the waiting time for the first success, $X_2$ the *further* waiting time for the second success, $X_3$ the *further* waiting time for the third success, and so on. Then $X_1, X_2, \ldots$ are independent and geometric, and

$$W_r = X_1 + X_2 + \cdots + X_r.$$

Apply Theorems (3.3.8) and (3.3.11) to find the mean and the variance of $W_r$.   ●

## Exercises for Section 3.5

**1.    De Moivre trials.** Each trial may result in any of $t$ given outcomes, the $i$th outcome having probability $p_i$. Let $N_i$ be the number of occurrences of the $i$th outcome in $n$ independent trials. Show that

$$\mathbb{P}(N_i = n_i \text{ for } 1 \leq i \leq t) = \frac{n!}{n_1! n_2! \cdots n_t!} p_1^{n_1} p_2^{n_2} \cdots p_t^{n_t}$$

for any collection $n_1, n_2, \ldots, n_t$ of non-negative integers with sum $n$. The vector $N$ is said to have the *multinomial distribution*.

**2.    In** your pocket is a random number $N$ of coins, where $N$ has the Poisson distribution with parameter $\lambda$. You toss each coin once, with heads showing with probability $p$ each time. Show that the total number of heads has the Poisson distribution with parameter $\lambda p$.

**3.    Let** $X$ be Poisson distributed where $\mathbb{P}(X = n) = p_n(\lambda) = \lambda^n e^{-\lambda}/n!$ for $n \geq 0$. Show that $\mathbb{P}(X \leq n) = 1 - \int_0^\lambda p_n(x) \, dx$.

**4.    Capture–recapture.** A population of $b$ animals has had a number $a$ of its members captured, marked, and released. Let $X$ be the number of animals it is necessary to recapture (without re-release) in order to obtain $m$ marked animals. Show that

$$\mathbb{P}(X = n) = \frac{a}{b} \binom{a-1}{m-1} \binom{b-a}{n-m} \bigg/ \binom{b-1}{n-1},$$

and find $\mathbb{E}X$. This distribution has been called *negative hypergeometric*.

**5.    Compound Poisson distribution.** Let $\Lambda$ be a positive random variable with density function $f$ and distribution function $F$, and let $Y$ have the Poisson distribution with parameter $\Lambda$. Show for $n = 0, 1, 2, \ldots$ that

$$\mathbb{P}(Y \leq n) = \int_0^\infty p_n(\lambda) F(\lambda) \, d\lambda, \qquad \mathbb{P}(Y > n) = \int_0^\infty p_n(\lambda)[1 - F(\lambda)] \, d\lambda,$$

where $p_n(\lambda) = e^{-\lambda} \lambda^n / n!$.

# 3.6 Dependence

Probability theory is largely concerned with families of random variables; these families will not in general consist entirely of independent variables.

**(1) Example.** Suppose that we back three horses to win as an accumulator. If our stake is £1 and the starting prices are $\alpha$, $\beta$, and $\gamma$, then our total profit is

$$W = (\alpha + 1)(\beta + 1)(\gamma + 1) I_1 I_2 I_3 - 1$$

where $I_i$ denotes the indicator of a win in the $i$th race by our horse. (In checking this expression remember that a bet of £$B$ on a horse with starting price $\alpha$ brings a return of £$B(\alpha + 1)$, should this horse win.) We lose £1 if some backed horse fails to win. It seems clear that the random variables $W$ and $I_1$ are *not* independent. If the races are run independently, then

$$\mathbb{P}(W = -1) = \mathbb{P}(I_1 I_2 I_3 = 0),$$

but

$$\mathbb{P}(W = -1 \mid I_1 = 1) = \mathbb{P}(I_2 I_3 = 0)$$

which are different from each other unless the first backed horse is guaranteed victory.   ●

We require a tool for studying collections of dependent variables. Knowledge of their individual mass functions is little help by itself. Just as the main tools for studying a random variable is its distribution function, so the study of, say, a pair of random variables is based on its 'joint' distribution function and mass function.

**(2) Definition.** The **joint distribution function** $F : \mathbb{R}^2 \to [0, 1]$ of $X$ and $Y$, where $X$ and $Y$ are discrete variables, is given by

$$F(x, y) = \mathbb{P}(X \leq x \text{ and } Y \leq y).$$

Their **joint mass function** $f : \mathbb{R}^2 \to [0, 1]$ is given by

$$f(x, y) = \mathbb{P}(X = x \text{ and } Y = y).$$

Joint distribution functions and joint mass functions of larger collections of variables are defined similarly. The functions $F$ and $f$ can be characterized in much the same way (Lemmas (2.1.6) and (3.1.2)) as the corresponding functions of a single variable. We omit the details. We write $F_{X,Y}$ and $f_{X,Y}$ when we need to stress the role of $X$ and $Y$. You may think of the joint mass function in the following way. If $A_x = \{X = x\}$ and $B_y = \{Y = y\}$, then

$$f(x, y) = \mathbb{P}(A_x \cap B_y).$$

The definition of independence can now be reformulated in a lemma.

**(3) Lemma.** *The discrete random variables $X$ and $Y$ are independent if and only if*

(4)                 $f_{X,Y}(x, y) = f_X(x) f_Y(y)$   *for all $x, y \in \mathbb{R}$.*

*More generally, $X$ and $Y$ are independent if and only if $f_{X,Y}(x, y)$ can be factorized as the product $g(x)h(y)$ of a function of $x$ alone and a function of $y$ alone.*

**Proof.** This is Problem (3.11.1).                                              ■

Suppose that $X$ and $Y$ have joint mass function $f_{X,Y}$ and we wish to check whether or not (4) holds. First we need to calculate the *marginal mass functions* $f_X$ and $f_Y$ from our knowledge of $f_{X,Y}$. These are found in the following way:

$$f_X(x) = \mathbb{P}(X = x) = \mathbb{P}\left(\bigcup_y (\{X = x\} \cap \{Y = y\})\right)$$

$$= \sum_y \mathbb{P}(X = x, Y = y) = \sum_y f_{X,Y}(x, y),$$

and similarly $f_Y(y) = \sum_x f_{X,Y}(x, y)$. Having found the marginals, it is a trivial matter to see whether (4) holds or not.

**Remark.** We stress that the factorization (4) must hold for *all* $x$ and $y$ in order that $X$ and $Y$ be independent.

**(5) Example. Calculation of marginals.** In Example (3.2.2) we encountered a pair $X, Y$ of variables with a joint mass function

$$f(x, y) = \frac{\alpha^x \beta^y}{x! \, y!} e^{-\alpha - \beta} \quad \text{for} \quad x, y = 0, 1, 2, \dots$$

where $\alpha, \beta > 0$. The marginal mass function of $X$ is

$$f_X(x) = \sum_y f(x, y) = \frac{\alpha^x}{x!} e^{-\alpha} \sum_{y=0}^{\infty} \frac{\beta^y}{y!} e^{-\beta} = \frac{\alpha^x}{x!} e^{-\alpha}$$

and so $X$ has the Poisson distribution with parameter $\alpha$. Similarly $Y$ has the Poisson distribution with parameter $\beta$. It is easy to check that (4) holds, whence $X$ and $Y$ are independent.    ●

For any discrete pair $X, Y$, a real function $g(X, Y)$ is a random variable. We shall often need to find its expectation. To avoid explicit calculation of its mass function, we shall use the following more general form of the law of the unconscious statistician, Lemma (3.3.3).

**(6) Lemma.** $\mathbb{E}(g(X, Y)) = \sum_{x,y} g(x, y) f_{X,Y}(x, y)$.

**Proof.** As for Lemma (3.3.3).    ∎

For example, $\mathbb{E}(XY) = \sum_{x,y} xy f_{X,Y}(x, y)$. This formula is particularly useful to statisticians who may need to find simple ways of explaining dependence to laymen. For instance, suppose that the government wishes to announce that the dependence between defence spending and the cost of living is very small. It should *not* publish an estimate of the joint mass function unless its object is obfuscation alone. Most members of the public would prefer to find that this dependence can be represented in terms of a single number on a prescribed scale. Towards this end we make the following definition†.

**(7) Definition.** The **covariance** of $X$ and $Y$ is

$$\text{cov}(X, Y) = \mathbb{E}\big[(X - \mathbb{E}X)(Y - \mathbb{E}Y)\big].$$

The **correlation (coefficient)** of $X$ and $Y$ is

$$\rho(X, Y) = \frac{\text{cov}(X, Y)}{\sqrt{\text{var}(X) \cdot \text{var}(Y)}}$$

as long as the variances are non-zero.

Note that the concept of covariance generalizes that of variance in that $\text{cov}(X, X) = \text{var}(X)$. Expanding the covariance gives

$$\text{cov}(X, Y) = \mathbb{E}(XY) - \mathbb{E}(X)\mathbb{E}(Y).$$

---

†The concepts and terminology in this definition were formulated by Francis Galton in the late 1880s.

Remember, Definition (3.3.10), that $X$ and $Y$ are called *uncorrelated* if $\text{cov}(X, Y) = 0$. Also, independent variables are always uncorrelated, although the converse is not true. Covariance itself is not a satisfactory measure of dependence because the scale of values which $\text{cov}(X, Y)$ may take contains no points which are clearly interpretable in terms of the relationship between $X$ and $Y$. The following lemma shows that this is not the case for correlations.

**(8) Lemma.** *The correlation coefficient $\rho$ satisfies $|\rho(X, Y)| \leq 1$ with equality if and only if* $\mathbb{P}(aX + bY = c) = 1$ *for some $a, b, c \in \mathbb{R}$.*

The proof is an application of the following important inequality.

**(9) Theorem. Cauchy–Schwarz inequality.** *For random variables $X$ and $Y$,*

$$\{\mathbb{E}(XY)\}^2 \leq \mathbb{E}(X^2)\mathbb{E}(Y^2)$$

*with equality if and only if $\mathbb{P}(aX = bY) = 1$ for some real $a$ and $b$, at least one of which is non-zero.*

**Proof.** We can assume that $\mathbb{E}(X^2)$ and $\mathbb{E}(Y^2)$ are strictly positive, since otherwise the result follows immediately from Problem (3.11.2). For $a, b \in \mathbb{R}$, let $Z = aX - bY$. Then

$$0 \leq \mathbb{E}(Z^2) = a^2\mathbb{E}(X^2) - 2ab\mathbb{E}(XY) + b^2\mathbb{E}(Y^2).$$

Thus the right-hand side is a quadratic in the variable $a$ with at most one real root. Its discriminant must be non-positive. That is to say, if $b \neq 0$,

$$\mathbb{E}(XY)^2 - \mathbb{E}(X^2)\mathbb{E}(Y^2) \leq 0.$$

The discriminant is zero if and only if the quadratic has a real root. This occurs if and only if

$$\mathbb{E}\big((aX - bY)^2\big) = 0 \quad \text{for some } a \text{ and } b,$$

which, by Problem (3.11.2), completes the proof.                                        ∎

**Proof of (8).** Apply (9) to the variables $X - \mathbb{E}X$ and $Y - \mathbb{E}Y$.                   ∎

A more careful treatment than this proof shows that $\rho = +1$ if and only if $Y$ *increases* linearly with $X$ and $\rho = -1$ if and only if $Y$ *decreases* linearly as $X$ increases.

**(10) Example.** Here is a tedious numerical example of the use of joint mass functions. Let $X$ and $Y$ take values in $\{1, 2, 3\}$ and $\{-1, 0, 2\}$ respectively, with joint mass function $f$ where $f(x, y)$ is the appropriate entry in Table 3.1.

|         | $y = -1$ | $y = 0$ | $y = 2$ | $f_X$ |
|---------|----------|---------|---------|-------|
| $x = 1$ | $\frac{1}{18}$ | $\frac{3}{18}$ | $\frac{2}{18}$ | $\frac{6}{18}$ |
| $x = 2$ | $\frac{2}{18}$ | $0$ | $\frac{3}{18}$ | $\frac{5}{18}$ |
| $x = 3$ | $0$ | $\frac{4}{18}$ | $\frac{3}{18}$ | $\frac{7}{18}$ |
| $f_Y$   | $\frac{3}{18}$ | $\frac{7}{18}$ | $\frac{8}{18}$ | |

Table 3.1. The joint mass function of the random variables $X$ and $Y$. The indicated row and column sums are the marginal mass functions $f_X$ and $f_Y$.

A quick calculation gives

$$\mathbb{E}(XY) = \sum_{x,y} xy f(x, y) = 29/18,$$

$$\mathbb{E}(X) = \sum_{x} x f_X(x) = 37/18, \quad \mathbb{E}(Y) = 13/18,$$

$$\mathrm{var}(X) = \mathbb{E}(X^2) - \mathbb{E}(X)^2 = 233/324, \quad \mathrm{var}(Y) = 461/324,$$

$$\mathrm{cov}(X, Y) = 41/324, \quad \rho(X, Y) = 41/\sqrt{107413}. \qquad\qquad \bullet$$

---

## Exercises for Section 3.6

**1.** Show that the collection of random variables on a given probability space and having finite variance forms a vector space over the reals.

**2.** Find the marginal mass functions of the multinomial distribution of Exercise (3.5.1).

**3.** Let $X$ and $Y$ be discrete random variables with joint mass function

$$f(x, y) = \frac{C}{(x + y - 1)(x + y)(x + y + 1)}, \qquad x, y = 1, 2, 3, \ldots.$$

Find the marginal mass functions of $X$ and $Y$, calculate $C$, and also the covariance of $X$ and $Y$.

**4.** Let $X$ and $Y$ be discrete random variables with mean 0, variance 1, and covariance $\rho$. Show that $\mathbb{E}\left(\max\{X^2, Y^2\}\right) \le 1 + \sqrt{1 - \rho^2}$.

**5.** **Mutual information.** Let $X$ and $Y$ be discrete random variables with joint mass function $f$.
(a) Show that $\mathbb{E}(\log f_X(X)) \ge \mathbb{E}(\log f_Y(X))$.
(b) Show that the *mutual information*

$$I = \mathbb{E}\left(\log\left\{\frac{f(X, Y)}{f_X(X) f_Y(Y)}\right\}\right)$$

satisfies $I \ge 0$, with equality if and only if $X$ and $Y$ are independent.

**6.** **Voter paradox.** Let $X$, $Y$, $Z$ be discrete random variables with the property that their values are distinct with probability 1. Let $a = \mathbb{P}(X > Y)$, $b = \mathbb{P}(Y > Z)$, $c = \mathbb{P}(Z > X)$.
(a) Show that $\min\{a, b, c\} \le \frac{2}{3}$, and give an example where this bound is attained.
(b) Show that, if $X$, $Y$, $Z$ are independent and identically distributed, then $a = b = c = \frac{1}{2}$.
(c) Find $\min\{a, b, c\}$ and $\sup_p \min\{a, b, c\}$ when $\mathbb{P}(X = 0) = 1$, and $Y$, $Z$ are independent with $\mathbb{P}(Z = 1) = \mathbb{P}(Y = -1) = p$, $\mathbb{P}(Z = -2) = \mathbb{P}(Y = 2) = 1 - p$. Here, $\sup_p$ denotes the supremum as $p$ varies over $[0, 1]$.
[Part (a) is related to de Condorcet's observation that, in an election, it is possible for more than half of the voters to prefer candidate A to candidate B, more than half B to C, and more than half C to A.]

**7.** **Benford's distribution, or the law of anomalous numbers.** If one picks a numerical entry at random from an almanac, or the annual accounts of a corporation, the first two significant digits, $X$, $Y$, are found to have approximately the joint mass function

$$f(x, y) = \log_{10}\left(1 + \frac{1}{10x + y}\right), \qquad 1 \le x \le 9, \; 0 \le y \le 9.$$

Find the mass function of $X$ and an approximation to its mean. [A heuristic explanation for this phenomenon may be found in the second of Feller's volumes published in 1971. See also Berger and Hill 2015.]

**8.**   Let $X$ and $Y$ have joint mass function

$$f(j,k) = \frac{c(j+k)a^{j+k}}{j!\,k!}, \qquad j,k \geq 0,$$

where $a$ is a constant. Find $c$, $\mathbb{P}(X = j)$, $\mathbb{P}(X + Y = r)$, and $\mathbb{E}(X)$.

**9.   Correlation.** Let $X, Y, Z$ be non-degenerate and independent random variables. By considering $U = X + Y$, $V = Y + Z$, $W = Z - X$, or otherwise, show that having positive correlation is not a transitive relation.

**10.   Cauchy–Schwarz inequality.** Use the identity $a^2d^2 + b^2c^2 - 2abcd = (ad - bc)^2$ to prove the Cauchy–Schwarz inequality.

**11.   Cantelli, or one-sided Chebyshov inequality.** Show that

$$\mathbb{P}\big(X - \mathbb{E}(X) > t\big) \leq \frac{\text{var}(X)}{t^2 + \text{var}(X)}, \qquad t > 0.$$

---

## 3.7   Conditional distributions and conditional expectation

In Section 1.4 we discussed the conditional probability $\mathbb{P}(B \mid A)$. This may be set in the more general context of the conditional distribution of one variable $Y$ given the value of another variable $X$; this reduces to the definition of the conditional probabilities of events $A$ and $B$ if $X = I_A$ and $Y = I_B$.

Let $X$ and $Y$ be two discrete variables on $(\Omega, \mathcal{F}, \mathbb{P})$.

**(1) Definition.**   The **conditional distribution function** of $Y$ given $X = x$, written $F_{Y|X}(\cdot \mid x)$, is defined by

$$F_{Y|X}(y \mid x) = \mathbb{P}(Y \leq y \mid X = x)$$

for any $x$ such that $\mathbb{P}(X = x) > 0$. The **conditional (probability) mass function** of $Y$ given $X = x$, written $f_{Y|X}(\cdot \mid x)$, is defined by

**(2)**                                  $$f_{Y|X}(y \mid x) = \mathbb{P}(Y = y \mid X = x)$$

for any $x$ such that $\mathbb{P}(X = x) > 0$.

Formula (2) is easy to remember as $f_{Y|X} = f_{X,Y}/f_X$. Conditional distributions and mass functions are undefined at values of $x$ for which $\mathbb{P}(X = x) = 0$. Clearly $X$ and $Y$ are independent if and only if $f_{Y|X} = f_Y$.

Suppose we are told that $X = x$. Conditional upon this, the new distribution of $Y$ has mass function $f_{Y|X}(y \mid x)$, which we think of as a function of $y$. The expected value of this distribution, $\sum_y y f_{Y|X}(y \mid x)$, is called the *conditional expectation* of $Y$ given $X = x$ and is written $\psi(x) = \mathbb{E}(Y \mid X = x)$. Now, we observe that the conditional expectation depends on the value $x$ taken by $X$, and can be thought of as a function $\psi(X)$ of $X$ itself.

**(3) Definition.** Let $\psi(x) = \mathbb{E}(Y \mid X = x)$. Then $\psi(X)$ is called the **conditional expectation** of $Y$ given $X$, written as $\mathbb{E}(Y \mid X)$.

Although 'conditional expectation' sounds like a number, it is actually a random variable. It has the following important property.

**(4) Theorem.** *The conditional expectation* $\psi(X) = \mathbb{E}(Y \mid X)$ *satisfies*

$$\mathbb{E}(\psi(X)) = \mathbb{E}(Y).$$

**Proof.** By Lemma (3.3.3),

$$\mathbb{E}(\psi(X)) = \sum_x \psi(x) f_X(x) = \sum_{x,y} y f_{Y|X}(y \mid x) f_X(x)$$

$$= \sum_{x,y} y f_{X,Y}(x, y) = \sum_y y f_Y(y) = \mathbb{E}(Y). \qquad \blacksquare$$

This is an extremely useful theorem, to which we shall make repeated reference. It often provides a useful method for calculating $\mathbb{E}(Y)$, since it asserts that

$$\mathbb{E}(Y) = \sum_x \mathbb{E}(Y \mid X = x)\mathbb{P}(X = x).$$

For example, let $\{A_i : i \in I\}$ be a countable partition of the sample space $\Omega$, and let $X : \Omega \to \mathbb{R}$ be given by $X(\omega) = i$ if $\omega \in A_i$. The above equation becomes

$$\mathbb{E}(Y) = \sum_{i \in I} \mathbb{E}(Y \mid A_i)\mathbb{P}(A_i).$$

These equations are sometimes called the *partition theorem*, or the *law of total probability*.

**(5) Example.** A hen lays $N$ eggs, where $N$ has the Poisson distribution with parameter $\lambda$. Each egg hatches with probability $p$ ($= 1 - q$) independently of the other eggs. Let $K$ be the number of chicks. Find $\mathbb{E}(K \mid N)$, $\mathbb{E}(K)$, and $\mathbb{E}(N \mid K)$.
**Solution.** We are given that

$$f_N(n) = \frac{\lambda^n}{n!} e^{-\lambda}, \qquad f_{K|N}(k \mid n) = \binom{n}{k} p^k (1 - p)^{n-k}.$$

Therefore

$$\psi(n) = \mathbb{E}(K \mid N = n) = \sum_k k f_{K|N}(k \mid n) = pn.$$

Thus $\mathbb{E}(K \mid N) = \psi(N) = pN$ and

$$\mathbb{E}(K) = \mathbb{E}(\psi(N)) = p\mathbb{E}(N) = p\lambda.$$

To find $\mathbb{E}(N \mid K)$ we need to know the conditional mass function $f_{N\mid K}$ of $N$ given $K$. However,

$$
\begin{aligned}
f_{N\mid K}(n \mid k) &= \mathbb{P}(N = n \mid K = k) \\
&= \frac{\mathbb{P}(K = k \mid N = n)\mathbb{P}(N = n)}{\mathbb{P}(K = k)} \\
&= \frac{\binom{n}{k}p^k(1-p)^{n-k}(\lambda^n/n!)e^{-\lambda}}{\sum_{m\geq k}\binom{m}{k}p^k(1-p)^{m-k}(\lambda^m/m!)e^{-\lambda}} \quad \text{if } n \geq k \\
&= \frac{(q\lambda)^{n-k}}{(n-k)!}e^{-q\lambda}.
\end{aligned}
$$

Hence

$$
\mathbb{E}(N \mid K = k) = \sum_{n\geq k} n\frac{(q\lambda)^{n-k}}{(n-k)!}e^{-q\lambda} = k + q\lambda,
$$

giving $\mathbb{E}(N \mid K) = K + q\lambda$. ●

There is a more general version of Theorem (4), and this will be of interest later.

**(6) Theorem.** *The conditional expectation $\psi(X) = \mathbb{E}(Y \mid X)$ satisfies*

(7) $$\mathbb{E}\big(\psi(X)g(X)\big) = \mathbb{E}\big(Yg(X)\big)$$

*for any function g for which both expectations exist.*

Setting $g(x) = 1$ for all $x$, we obtain the result of (4). Whilst Theorem (6) is useful in its own right, we shall see later that its principal interest lies elsewhere. The conclusion of the theorem may be taken as a *definition* of conditional expectation—as a function $\psi(X)$ of $X$ such that (7) holds for all suitable functions $g$. Such a definition is convenient when working with a notion of conditional expectation more general than that dealt with here.

**Proof.** As in the proof of (4),

$$
\begin{aligned}
\mathbb{E}\big(\psi(X)g(X)\big) &= \sum_x \psi(x)g(x)f_X(x) = \sum_{x,y} yg(x)f_{Y\mid X}(y \mid x)f_X(x) \\
&= \sum_{x,y} yg(x)f_{X,Y}(x, y) = \mathbb{E}\big(Yg(X)\big).
\end{aligned}
$$
∎

---

## Exercises for Section 3.7

**1.** Show the following:
(a) $\mathbb{E}(aY + bZ \mid X) = a\mathbb{E}(Y \mid X) + b\mathbb{E}(Z \mid X)$ for $a, b \in \mathbb{R}$,
(b) $\mathbb{E}(Y \mid X) \geq 0$ if $Y \geq 0$,
(c) $\mathbb{E}(1 \mid X) = 1$,
(d) if $X$ and $Y$ are independent then $\mathbb{E}(Y \mid X) = \mathbb{E}(Y)$,
(e) ('pull-through property') $\mathbb{E}(Yg(X) \mid X) = g(X)\mathbb{E}(Y \mid X)$ for any suitable function $g$,
(f) ('tower property') $\mathbb{E}\{\mathbb{E}(Y \mid X, Z) \mid X\} = \mathbb{E}(Y \mid X) = \mathbb{E}\{\mathbb{E}(Y \mid X) \mid X, Z\}$.

**2.    Uniqueness of conditional expectation.** Suppose that $X$ and $Y$ are discrete random variables, and that $\phi(X)$ and $\psi(X)$ are two functions of $X$ satisfying

$$\mathbb{E}\big(\phi(X)g(X)\big) = \mathbb{E}\big(\psi(X)g(X)\big) = \mathbb{E}\big(Yg(X)\big)$$

for any function $g$ for which all the expectations exist. Show that $\phi(X)$ and $\psi(X)$ are almost surely equal, in that $\mathbb{P}(\phi(X) = \psi(X)) = 1$.

**3.** Suppose that the conditional expectation of $Y$ given $X$ is defined as the (almost surely) unique function $\psi(X)$ such that $\mathbb{E}(\psi(X)g(X)) = \mathbb{E}(Yg(X))$ for all functions $g$ for which the expectations exist. Show (a)–(f) of Exercise (3.7.1) above (with the occasional addition of the expression 'with probability 1').

**4.    Conditional variance formula.** How should we define var$(Y \mid X)$, the conditional variance of $Y$ given $X$? Show that var$(Y) = \mathbb{E}(\text{var}(Y \mid X)) + \text{var}(\mathbb{E}(Y \mid X))$.

**5.** The lifetime of a machine (in days) is a random variable $T$ with mass function $f$. Given that the machine is working after $t$ days, what is the mean subsequent lifetime of the machine when:
(a) $f(x) = (N + 1)^{-1}$ for $x \in \{0, 1, \ldots, N\}$,
(b) $f(x) = 2^{-x}$ for $x = 1, 2, \ldots$.
(The first part of Problem (3.11.13) may be useful.)

**6.** Let $X_1, X_2, \ldots$ be identically distributed random variables with mean $\mu$ and variance $\sigma^2$, and let $N$ be a random variable taking values in the non-negative integers and independent of the $X_i$. Let $S = X_1 + X_2 + \cdots + X_N$. Show that $\mathbb{E}(S \mid N) = \mu N$, and deduce that $\mathbb{E}(S) = \mu\mathbb{E}(N)$. Find var $S$ in terms of the first two moments of $N$, using the conditional variance formula of Exercise (3.7.4).

**7.** A factory has produced $n$ robots, each of which is faulty with probability $\phi$. To each robot a test is applied which detects the fault (if present) with probability $\delta$. Let $X$ be the number of faulty robots, and $Y$ the number detected as faulty. Assuming the usual independence, show that

$$\mathbb{E}(X \mid Y) = \{n\phi(1 - \delta) + (1 - \phi)Y\}/(1 - \phi\delta).$$

**8.    Families.** Each child is equally likely to be male or female, independently of all other children.
(a) Show that, in a family of predetermined size, the expected number of boys equals the expected number of girls. Was the assumption of independence necessary?
(b) A randomly selected child is male; does the expected number of his brothers equal the expected number of his sisters? What happens if you do not require independence?

**9.** Let $X$ and $Y$ be independent with mean $\mu$. Explain the error in the following equation:

$$`\mathbb{E}(X \mid X + Y = z) = \mathbb{E}(X \mid X = z - Y) = \mathbb{E}(z - Y) = z - \mu`.$$

**10.** A coin shows heads with probability $p$. Let $X_n$ be the number of flips required to obtain a run of $n$ consecutive heads. Show that $\mathbb{E}(X_n) = \sum_{k=1}^{n} p^{-k}$.

**11.    Conditional covariance.** Give a definition of the *conditional covariance* cov$(X, Y \mid Z)$. Show that

$$\text{cov}(X, Y) = \mathbb{E}\big(\text{cov}(X, Y \mid Z)\big) + \text{cov}\big(\mathbb{E}(X \mid Z), \mathbb{E}(Y \mid Z)\big).$$

**12.** An urn contains initially $b$ blue balls and $r$ red balls, where $b, r \geq 2$. Balls are drawn one by one without replacement. Show that the mean number of draws until the first colour drawn is first repeated equals 3.

**13.** (a) Let $X$ be uniformly distributed on $\{0, 1, \ldots, n\}$. Show that var$(X) = \frac{1}{12}n(n + 2)$.
(b) A student sits two examinations, gaining $X$ and $Y$ marks, respectively. In the interests of economy and fairness, the examiner determines that $X$ shall be uniformly distributed on $\{0, 1, \ldots, n\}$, and that, conditional on $X = k$, $Y$ shall have the binomial bin$(n, k/n)$ distribution.

  (i)  Show that $\mathbb{E}(Y) = \frac{1}{2}n$, and $\text{var}(Y) = \frac{1}{12}(n^2 + 4n - 2)$.
  (ii)  Find $\mathbb{E}(X + Y)$ and $\text{var}(X + Y)$.

**14.** Let $X$, $Y$ be discrete integer-valued random variables with the joint mass function

$$f(x, y) = \frac{\lambda^y e^{-2\lambda}}{x!\,(y-x)!}, \qquad 0 \le x \le y < \infty.$$

Show that $X$ and $Y$ are each Poisson-distributed, and that the conditional distribution of $X$ given $Y$ is binomial.

---

## 3.8  Sums of random variables

Much of the classical theory of probability concerns sums of random variables. We have seen already many such sums; the number of heads in $n$ tosses of a coin is one of the simplest such examples, but we shall encounter many situations which are more complicated than this. One particular complication is when the summands are dependent. The first stage in developing a systematic technique is to find a formula for describing the mass function of the sum $Z = X + Y$ of two variables having joint mass function $f(x, y)$.

**(1) Theorem.**  *We have that* $\mathbb{P}(X + Y = z) = \sum_x f(x, z - x)$.

**Proof.**  The union

$$\{X + Y = z\} = \bigcup_x (\{X = x\} \cap \{Y = z - x\})$$

is disjoint, and at most countably many of its contributions have non-zero probability. Therefore

$$\mathbb{P}(X + Y = z) = \sum_x \mathbb{P}(X = x,\ Y = z - x) = \sum_x f(x, z - x). \qquad \blacksquare$$

If $X$ and $Y$ are independent, then

$$\mathbb{P}(X + Y = z) = f_{X+Y}(z) = \sum_x f_X(x) f_Y(z - x) = \sum_y f_X(z - y) f_Y(y).$$

  The mass function of $X + Y$ is called the *convolution* of the mass functions of $X$ and $Y$, and is written

**(2)** $$f_{X+Y} = f_X * f_Y.$$

**(3) Example (3.5.6) revisited.**  Let $X_1$ and $X_2$ be independent geometric variables with common mass function

$$f(k) = p(1 - p)^{k-1}, \qquad k = 1, 2, \dots.$$

By (2), $Z = X_1 + X_2$ has mass function

$$\mathbb{P}(Z = z) = \sum_k \mathbb{P}(X_1 = k)\mathbb{P}(X_2 = z - k)$$

$$= \sum_{k=1}^{z-1} p(1 - p)^{k-1} p(1 - p)^{z-k-1}$$

$$= (z - 1)p^2(1 - p)^{z-2}, \qquad z = 2, 3, \ldots$$

in agreement with Example (3.5.6). The general formula for the sum of a number, $r$ say, of geometric variables can easily be verified by induction.                                              ●

---

## Exercises for Section 3.8

**1.** Let $X$ and $Y$ be independent variables, $X$ being equally likely to take any value in $\{0, 1, \ldots, m\}$, and $Y$ similarly in $\{0, 1, \ldots, n\}$. Find the mass function of $Z = X + Y$. The random variable $Z$ is said to have the *trapezoidal distribution*.

**2.** Let $X$ and $Y$ have the joint mass function

$$f(x, y) = \frac{C}{(x + y - 1)(x + y)(x + y + 1)}, \qquad x, y = 1, 2, 3, \ldots.$$

Find the mass functions of $U = X + Y$ and $V = X - Y$.

**3.** Let $X$ and $Y$ be independent geometric random variables with respective parameters $\alpha$ and $\beta$. Show that

$$\mathbb{P}(X + Y = z) = \frac{\alpha\beta}{\alpha - \beta}\left\{(1 - \beta)^{z-1} - (1 - \alpha)^{z-1}\right\}.$$

**4.** Let $\{X_r : 1 \le r \le n\}$ be independent geometric random variables with parameter $p$. Show that $Z = \sum_{r=1}^n X_r$ has a negative binomial distribution. [Hint: No calculations are necessary.]

**5. Pepys's problem†.** Sam rolls $6n$ dice once; he needs at least $n$ sixes. Isaac rolls $6(n + 1)$ dice; he needs at least $n + 1$ sixes. Who is more likely to obtain the number of sixes he needs?

**6. Stein–Chen equation.** Let $N$ be Poisson distributed with parameter $\lambda$. Show that, for any function $g$ such that the expectations exist, $\mathbb{E}(Ng(N)) = \lambda\mathbb{E}g(N+1)$. More generally, if $S = \sum_{r=1}^N X_r$, where $\{X_r : r \ge 0\}$ are independent identically distributed non-negative integer-valued random variables, show that

$$\mathbb{E}\big(Sg(S)\big) = \lambda\mathbb{E}\big(g(S + X_0)X_0\big).$$

**7. Random sum.** Let $S = \sum_{i=1}^N X_i$, where the $X_i$, $i \ge 1$, are independent, identically distributed random variables with mean $\mu$ and variance $\sigma^2$, and $N$ is positive, integer-valued, and independent of the $X_i$. Show that $\mathbb{E}(S) = \mu\mathbb{E}(N)$, and

$$\text{var}(S) = \sigma^2\mathbb{E}(N) + \mu^2\,\text{var}(N).$$

**8.** Let $X$ and $Y$ be independent random variables with the geometric distributions $f_X(k) = pq^{k-1}$, $f_Y(k) = \lambda\mu^{k-1}$, for $k \ge 1$, where $p + q = \lambda + \mu = 1$ and $q \ne \mu$. Write down $\mathbb{P}(X + Y = n + 1, X = k)$, and hence find the distribution of $X + Y$, and the conditional distribution of $X$ given that $X + Y = n + 1$. Does anything special occur when $q = \mu$?

---

†Pepys put a simple version of this problem to Newton in 1693, but was reluctant to accept the correct reply he received.

## 3.9  Simple random walk

Until now we have dealt largely with general theory; the final two sections of this chapter may provide some lighter relief. One of the simplest random processes is so-called 'simple random walk'†; this process arises in many ways, of which the following is traditional. A gambler G plays the following game at the casino. The croupier tosses a (possibly biased) coin repeatedly; each time heads appears, he gives G one franc, and each time tails appears he takes one franc from G. Writing $S_n$ for G's fortune after $n$ tosses of the coin, we have that $S_{n+1} = S_n + X_{n+1}$ where $X_{n+1}$ is a random variable taking the value 1 with some fixed probability $p$ and $-1$ otherwise; furthermore, $X_{n+1}$ is assumed independent of the results of all previous tosses. Thus

$$(1) \qquad\qquad S_n = S_0 + \sum_{i=1}^{n} X_i,$$

so that $S_n$ is obtained from the initial fortune $S_0$ by the addition of $n$ independent random variables. We are assuming here that there are no constraints on G's fortune imposed externally, such as that the game is terminated if his fortune is reduced to zero.

An alternative picture of 'simple random walk' involves the motion of a particle—a particle which inhabits the set of integers and which moves at each step either one step to the right, with probability $p$, or one step to the left, the directions of different steps being independent of each other. More complicated random walks arise when the steps of the particle are allowed to have some general distribution on the integers, or the reals, so that the position $S_n$ at time $n$ is given by (1) where the $X_i$ are independent and identically distributed random variables having some specified distribution function. Even greater generality is obtained by assuming that the $X_i$ take values in $\mathbb{R}^d$ for some $d \geq 1$, or even some vector space over the real numbers. Random walks may be used with some success in modelling various practical situations, such as the numbers of cars in a toll queue at 5 minute intervals, the position of a pollen grain suspended in fluid at 1 second intervals, or the value of the Dow Jones index each Monday morning. In each case, it may not be too bad a guess that the $(n+1)$th reading differs from the $n$th by a random quantity which is independent of previous jumps but has the same probability distribution‡. The theory of random walks is a basic tool in the probabilist's kit, and we shall concern ourselves here with 'simple random walk' only.

At any instant of time a particle inhabits one of the integer points of the real line. At time 0 it starts from some specified point, and at each subsequent epoch of time $1, 2, \ldots$ it moves from its current position to a new position according to the following law. With probability $p$ it moves one step to the right, and with probability $q = 1 - p$ it moves one step to the left; moves are independent of each other. The walk is called *symmetric* if $p = q = \frac{1}{2}$. Example (1.7.4) concerned a symmetric random walk with 'absorbing' barriers at the points 0 and $N$. In general, let $S_n$ denote the position of the particle after $n$ moves, and set $S_0 = a$. Then

$$(2) \qquad\qquad S_n = a + \sum_{i=1}^{n} X_i$$

---

†Karl Pearson coined the term 'random walk' in 1906, and (using a result of Rayleigh) demonstrated the theorem that "the most likely place to find a drunken walker is somewhere near his starting point", empirical verification of which is not hard to find.

‡It is in fact more common to assume that the *logarithm* of the Dow Jones index is a random walk. Do you see why?

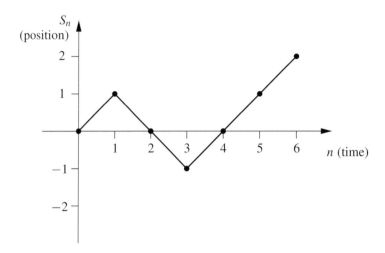

Figure 3.2. A random walk $S_n$.

where $X_1, X_2, \ldots$ is a sequence of independent Bernoulli variables taking values $+1$ and $-1$ (rather than $+1$ and $0$ as before) with probabilities $p$ and $q$.

We record the motion of the particle as the sequence $\{(n, S_n) : n \geq 0\}$ of Cartesian coordinates of points in the plane. This collection of points, joined by solid lines between neighbours, is called the *path* of the particle. In the example shown in Figure 3.2, the particle has visited the points $0, 1, 0, -1, 0, 1, 2$ in succession. This representation has a confusing aspect in that the direction of the particle's steps is parallel to the $y$-axis, whereas we have previously been specifying the movement in the traditional way as to the right or to the left. In future, any reference to the $x$-axis or the $y$-axis will pertain to a diagram of its path as exemplified by Figure 3.2.

The sequence (2) of partial sums has three important properties.

**(3) Lemma.** *The simple random walk is* spatially homogeneous; *that is*

$$\mathbb{P}(S_n = j \mid S_0 = a) = \mathbb{P}(S_n = j + b \mid S_0 = a + b).$$

**Proof.** Both sides equal $\mathbb{P}\left(\sum_1^n X_i = j - a\right)$.                                   ∎

**(4) Lemma.** *The simple random walk is* temporally homogeneous; *that is*

$$\mathbb{P}(S_n = j \mid S_0 = a) = \mathbb{P}(S_{m+n} = j \mid S_m = a).$$

**Proof.** The left- and right-hand sides satisfy

$$\text{LHS} = \mathbb{P}\left(\sum_1^n X_i = j - a\right) = \mathbb{P}\left(\sum_{m+1}^{m+n} X_i = j - a\right) = \text{RHS}.$$                                   ∎

**(5) Lemma.** *The simple random walk has the* Markov property; *that is*

$$\mathbb{P}(S_{m+n} = j \mid S_0, S_1, \ldots, S_m) = \mathbb{P}(S_{m+n} = j \mid S_m), \qquad n \geq 0.$$

Statements such as $\mathbb{P}(S = j \mid X, Y) = \mathbb{P}(S = j \mid X)$ are to be interpreted in the obvious way as meaning that

$$\mathbb{P}(S = j \mid X = x, \ Y = y) = \mathbb{P}(S = j \mid X = x) \quad \text{for all } x \text{ and } y;$$

this is a slight abuse of notation.

**Proof.** If one knows the value of $S_m$, then the distribution of $S_{m+n}$ depends only on the jumps $X_{m+1}, \ldots, X_{m+n}$, and cannot depend on further information concerning the values of $S_0, S_1, \ldots, S_{m-1}$. ∎

This 'Markov property' is often expressed informally by saying that, conditional upon knowing the value of the process at the $m$th step, its values after the $m$th step do not depend on its values before the $m$th step. More colloquially: conditional upon the present, the future does not depend on the past. We shall meet this property again later.

**(6) Example. Absorbing barriers.** Let us revisit Example (1.7.4) for general values of $p$. Equation (1.7.5) gives us the following difference equation for the probabilities $\{p_k\}$ where $p_k$ is the probability of ultimate ruin starting from $k$:

(7)                          $$p_k = p \cdot p_{k+1} + q \cdot p_{k-1} \quad \text{if} \ \ 1 \le k \le N - 1$$

with boundary conditions $p_0 = 1$, $p_N = 0$. The solution of such a difference equation proceeds as follows. Look for a solution of the form $p_k = \theta^k$. Substitute this into (7) and cancel out the power $\theta^{k-1}$ to obtain $p\theta^2 - \theta + q = 0$, which has roots $\theta_1 = 1$, $\theta_2 = q/p$. If $p \ne \frac{1}{2}$ then these roots are distinct and the general solution of (7) is $p_k = A_1\theta_1^k + A_2\theta_2^k$ for arbitrary constants $A_1$ and $A_2$. Use the boundary conditions to obtain

$$p_k = \frac{(q/p)^k - (q/p)^N}{1 - (q/p)^N}.$$

If $p = \frac{1}{2}$ then $\theta_1 = \theta_2 = 1$ and the general solution to (7) is $p_k = A_1 + A_2 k$. Use the boundary conditions to obtain $p_k = 1 - (k/N)$.

A more complicated equation is obtained for the mean number $D_k$ of steps before the particle hits one of the absorbing barriers, starting from $k$. In this case we use conditional expectations and (3.7.4) to find that

(8)               $$D_k = p(1 + D_{k+1}) + q(1 + D_{k-1}) \quad \text{if} \ \ 1 \le k \le N - 1$$

with the boundary conditions $D_0 = D_N = 0$. Try solving this; you need to find a general solution and a particular solution, as in the solution of second-order linear differential equations. This answer is

(9)          $$D_k = \begin{cases} \dfrac{1}{q-p}\left[k - N\left(\dfrac{1 - (q/p)^k}{1 - (q/p)^N}\right)\right] & \text{if } p \ne \frac{1}{2}, \\[4mm] k(N-k) & \text{if } p = \frac{1}{2}. \end{cases}$$   ●

**(10) Example. Retaining barriers.** In Example (1.7.4), suppose that the Jaguar buyer has a rich uncle who will guarantee all his losses. Then the random walk does not end when the particle hits zero, although it cannot visit a negative integer. Instead $\mathbb{P}(S_{n+1} = 0 \mid S_n = 0) = q$ and $\mathbb{P}(S_{n+1} = 1 \mid S_n = 0) = p$. The origin is said to have a 'retaining' barrier (sometimes called 'reflecting').

What now is the expected duration of the game? The mean duration $F_k$, starting from $k$, satisfies the same difference equation (8) as before but subject to different boundary conditions. We leave it as an *exercise* to show that the boundary conditions are $F_N = 0$, $pF_0 = 1 + pF_1$, and hence to find $F_k$. ●

In such examples the techniques of 'conditioning' are supremely useful. The idea is that in order to calculate a probability $\mathbb{P}(A)$ or expectation $\mathbb{E}(Y)$ we condition either on some partition of $\Omega$ (and use Lemma (1.4.4)) or on the outcome of some random variable (and use Theorem (3.7.4) or the forthcoming Theorem (4.6.5)). In this section this technique yielded the difference equations (7) and (8). In later sections the same idea will yield differential equations, integral equations, and functional equations, some of which can be solved.

## Exercises for Section 3.9

**1.** Let $T$ be the time which elapses before a simple random walk is absorbed at either of the absorbing barriers at 0 and $N$, having started at $k$ where $0 \le k \le N$. Show that $\mathbb{P}(T < \infty) = 1$ and $\mathbb{E}(T^k) < \infty$ for all $k \ge 1$.

**2.** For simple random walk $S$ with absorbing barriers at 0 and $N$, let $W$ be the event that the particle is absorbed at 0 rather than at $N$, and let $p_k = \mathbb{P}(W \mid S_0 = k)$. Show that, if the particle starts at $k$ where $0 < k < N$, the conditional probability that the first step is rightwards, given $W$, equals $pp_{k+1}/p_k$. Deduce that the mean duration $J_k$ of the walk, conditional on $W$, satisfies the equation

$$pp_{k+1}J_{k+1} - p_k J_k + (p_k - pp_{k+1})J_{k-1} = -p_k, \quad \text{for } 0 < k < N,$$

subject to the convention that $p_N J_N = 0$. Show that we may take as boundary condition $J_0 = 0$. Find $J_k$ in the symmetric case, when $p = \frac{1}{2}$.

**3.** With the notation of Exercise (3.9.2), suppose further that at any step the particle may remain where it is with probability $r$ where $p + q + r = 1$. Show that $J_k$ satisfies

$$pp_{k+1}J_{k+1} - (1 - r)p_k J_k + qp_{k-1}J_{k-1} = -p_k$$

and that, when $\rho = q/p \ne 1$,

$$J_k = \frac{1}{p - q} \cdot \frac{1}{\rho^k - \rho^N} \left\{ k(\rho^k + \rho^N) - \frac{2N\rho^N(1 - \rho^k)}{1 - \rho^N} \right\}.$$

**4.** **Problem of the points.** A coin is tossed repeatedly, heads turning up with probability $p$ on each toss. Player A wins the game if $m$ heads appear before $n$ tails have appeared, and player B wins otherwise. Let $p_{mn}$ be the probability that A wins the game. Set up a difference equation for the $p_{mn}$. What are the boundary conditions?

**5.** Consider a simple random walk on the set $\{0, 1, 2, \ldots, N\}$ in which each step is to the right with probability $p$ or to the left with probability $q = 1 - p$. Absorbing barriers are placed at 0 and $N$. Show that the number $X$ of positive steps of the walk before absorption satisfies

$$\mathbb{E}(X) = \frac{1}{2}\{D_k - k + N(1 - p_k)\}$$

where $D_k$ is the mean number of steps until absorption having started at $k$, and $p_k$ is the probability of absorption at 0.

**6.   Gambler's ruin revisited.** Let $D_k$ be the duration of a random walk on $\{0, 1, 2, \ldots, a\}$ with absorbing barriers at 0 and $a$, and started at $k$, with steps $X_i$ satisfying

$$\mathbb{P}(X_i = 1) = \mathbb{P}(X_i = -1) = p, \quad \mathbb{P}(X_i = 0) = 1 - 2p.$$

(a) When $p = \frac{1}{2}$, show that

$$\mathrm{var}(D_k) = \tfrac{1}{3}k(a - k)\{(a - k)^2 + k^2 - 2\}.$$

(b) Deduce (without lengthy calculation) that, for $p < \frac{1}{2}$,

$$\mathrm{var}(D_k) = \frac{k(a - k)}{(2p)^2}\left[1 - 2p + \tfrac{1}{3}\{(a - k)^2 + k^2 - 2\}\right].$$

**7.   Returns and visits by random walk.** Consider a simple symmetric random walk on the set $\{0, 1, 2, \ldots, a\}$ with absorbing barriers at 0 and $a$, and starting at $k$ where $0 < k < a$. Let $r_k$ be the probability the walk ever returns to $k$, and let $v_x$ be the mean number of visits to point $x$ before absorption. Find $r_k$, and hence show that

$$v_x = \begin{cases} 2x(a - k)/a & \text{for } 0 < x < k, \\ 2k(a - x)/a & \text{for } k < x < a. \end{cases}$$

**8.**   (a) "Millionaires should always gamble, poor men never" [J. M. Keynes].
(b) "If I wanted to gamble, I would buy a casino" [P. Getty].
(c) "That the chance of gain is naturally overvalued, we may learn from the universal success of lotteries" [Adam Smith, 1776].
Discuss.

---

## 3.10  Random walk: counting sample paths

In the previous section, our principal technique was to condition on the first step of the walk and then solve the ensuing difference equation. Another primitive but useful technique is to count. Let $X_1, X_2, \ldots$ be independent variables, each taking the values $-1$ and $1$ with probabilities $q = 1 - p$ and $p$, as before, and let

**(1)**
$$S_n = a + \sum_{i=1}^{n} X_i$$

be the position of the corresponding random walker after $n$ steps, having started at $S_0 = a$. The set of realizations of the walk is the set of vectors $\mathbf{s} = (s_0, s_1, \ldots)$ with $s_0 = a$ and $s_{i+1} - s_i = \pm 1$, and any such vector may be thought of as a 'sample path' of the walk, drawn in the manner of Figure 3.2. The probability that the first $n$ steps of the walk follow a given path $\mathbf{s} = (s_0, s_1, \ldots, s_n)$ is $p^r q^l$ where $r$ is the number of steps of $s$ to the right and $l$ is the number to the left†; that is to say, $r = |\{i : s_{i+1} - s_i = 1\}|$ and $l = |\{i : s_{i+1} - s_i = -1\}|$. Any event

---

†The words 'right'and 'left'are to be interpreted as meaning in the positive and negative directions respectively, plotted along the $y$-axis as in Figure 3.2.

may be expressed in terms of an appropriate set of paths, and the probability of the event is the sum of the component probabilities. For example, $\mathbb{P}(S_n = b) = \sum_r M_n^r(a, b) p^r q^{n-r}$ where $M_n^r(a, b)$ is the number of paths $(s_0, s_1, \ldots, s_n)$ with $s_0 = a$, $s_n = b$, and having exactly $r$ rightward steps. It is easy to see that $r + l = n$, the total number of steps, and $r - l = b - a$, the aggregate rightward displacement, so that $r = \frac{1}{2}(n + b - a)$ and $l = \frac{1}{2}(n - b + a)$. Thus

$$(2) \qquad\qquad \mathbb{P}(S_n = b) = \binom{n}{\frac{1}{2}(n + b - a)} p^{\frac{1}{2}(n+b-a)} q^{\frac{1}{2}(n-b+a)},$$

since there are exactly $\binom{n}{r}$ paths with length $n$ having $r$ rightward steps and $n - r$ leftward steps. Formula (2) is useful only if $\frac{1}{2}(n + b - a)$ is an integer lying in the range $0, 1, \ldots, n$; otherwise, the probability in question equals 0.

Natural equations of interest for the walk include:

(a) when does the first visit of the random walk to a given point occur; and

(b) what is the furthest rightward point visited by the random walk by time $n$?

Such questions may be answered with the aid of certain elegant results and techniques for counting paths. The first of these is the 'reflection principle'. Here is some basic notation. As in Figure 3.2, we keep a record of the random walk $S$ through its path $\{(n, S_n) : n \geq 0\}$.

Suppose we know that $S_0 = a$ and $S_n = b$. The random walk may or may not have visited the origin between times 0 and $n$. Let $N_n(a, b)$ be the number of possible paths from $(0, a)$ to $(n, b)$, and let $N_n^0(a, b)$ be the number of such paths which contain some point $(k, 0)$ on the $x$-axis.

**(3) Theorem. The reflection principle.** *If $a, b > 0$ then $N_n^0(a, b) = N_n(-a, b)$.*

**Proof.** Each path from $(0, -a)$ to $(n, b)$ intersects the $x$-axis at some earliest point $(k, 0)$. Reflect the segment of the path with $0 \leq x \leq k$ in the $x$-axis to obtain a path joining $(0, a)$ to $(n, b)$ which intersects the $x$-axis (see Figure 3.3). This operation gives a one–one correspondence between the collections of such paths, and the theorem is proved.  ∎

We have, as before, a formula for $N_n(a, b)$.

**(4) Lemma.** $N_n(a, b) = \binom{n}{\frac{1}{2}(n + b - a)}$.

**Proof.** Choose a path from $(0, a)$ to $(n, b)$ and let $\alpha$ and $\beta$ be the numbers of positive and negative steps, respectively, in this path. Then $\alpha + \beta = n$ and $\alpha - \beta = b - a$, so that $\alpha = \frac{1}{2}(n + b - a)$. The number of such paths is the number of ways of picking $\alpha$ positive steps from the $n$ available. That is

$$(5) \qquad\qquad N_n(a, b) = \binom{n}{\alpha} = \binom{n}{\frac{1}{2}(n + b - a)}. \qquad\qquad ∎$$

The famous 'ballot theorem' is a consequence of these elementary results; it was proved first by W. A. Whitworth in 1878.

**(6) Corollary†. Ballot theorem.** *If $b > 0$ then the number of paths from $(0, 0)$ to $(n, b)$ which do not revisit the $x$-axis equals $(b/n)N_n(0, b)$.*

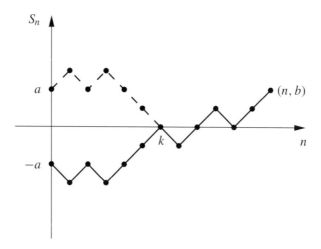

Figure 3.3. A random walk; the dashed line is the reflection of the first segment of the walk.

**Proof.** The first step of all such paths is to $(1, 1)$, and so the number of such path is

$$N_{n-1}(1, b) - N_{n-1}^0(1, b) = N_{n-1}(1, b) - N_{n-1}(-1, b)$$

by the reflection principle. We now use (4) and an elementary calculation to obtain the required result. ∎

As an application, and an explanation of the title of the theorem, we may easily answer the following amusing question. Suppose that, in a ballot, candidate $A$ scores $\alpha$ votes and candidate $B$ scores $\beta$ votes where $\alpha > \beta$. What is the probability that, during the ballot, $A$ was always ahead of $B$? Let $X_i$ equal 1 if the $i$th vote was cast for $A$, and $-1$ otherwise. Assuming that each possible combination of $\alpha$ votes for $A$ and $\beta$ votes for $B$ is equally likely, we have that the probability is question is the proportion of paths from $(0, 0)$ to $(\alpha + \beta, \alpha - \beta)$ which do not revisit the $x$-axis. Using the ballot theorem, we obtain the answer $(\alpha - \beta)/(\alpha + \beta)$.

Here are some applications of the reflection principle to random walks. First, what is the probability that the walk does not revisit its starting point in the first $n$ steps? We may as well assume that $S_0 = 0$, so that $S_1 \neq 0, \ldots, S_n \neq 0$ if and only if $S_1 S_2 \cdots S_n \neq 0$.

**(7) Theorem.** *If $S_0 = 0$ then, for $n \geq 1$,*

**(8)**
$$\mathbb{P}(S_1 S_2 \cdots S_n \neq 0, \ S_n = b) = \frac{|b|}{n} \mathbb{P}(S_n = b),$$

*and therefore*

**(9)**
$$\mathbb{P}(S_1 S_2 \cdots S_n \neq 0) = \frac{1}{n} \mathbb{E}|S_n|.$$

**Proof.** Suppose that $S_0 = 0$ and $S_n = b \, (> 0)$. The event in question occurs if and only if the path of the random walk does not visit the $x$-axis in the time interval $[1, n]$. The number of

---

†Derived from the Latin word *corollarium* meaning 'money paid for a garland' or 'tip'.

such paths is, by the ballot theorem, $(b/n)N_n(0, b)$, and each such path has $\frac{1}{2}(n+b)$ rightward steps and $\frac{1}{2}(n-b)$ leftward steps. Therefore

$$\mathbb{P}(S_1 S_2 \cdots S_n \neq 0, \ S_n = b) = \frac{b}{n} N_n(0, b) p^{\frac{1}{2}(n+b)} q^{\frac{1}{2}(n-b)} = \frac{b}{n} \mathbb{P}(S_n = b)$$

as required. A similar calculation is valid if $b < 0$.  ∎

Another feature of interest is the maximum value attained by the random walk. We write $M_n = \max\{S_i : 0 \le i \le n\}$ for the maximum value up to time $n$, and shall suppose that $S_0 = 0$, so that $M_n \ge 0$. Clearly $M_n \ge S_n$, and the first part of the next theorem is therefore trivial.

**(10) Theorem.** *Suppose that $S_0 = 0$. Then, for $r \ge 1$,*

**(11)**
$$\mathbb{P}(M_n \ge r, \ S_n = b) = \begin{cases} \mathbb{P}(S_n = b) & \text{if } b \ge r, \\ (q/p)^{r-b}\mathbb{P}(S_n = 2r - b) & \text{if } b < r. \end{cases}$$

It follows that, for $r \ge 1$,

**(12)**
$$\mathbb{P}(M_n \ge r) = \mathbb{P}(S_n \ge r) + \sum_{b=-\infty}^{r-1} (q/p)^{r-b}\mathbb{P}(S_n = 2r - b)$$
$$= \mathbb{P}(S_n = r) + \sum_{c=r+1}^{\infty} \left[1 + (q/p)^{c-r}\right]\mathbb{P}(S_n = c),$$

yielding in the symmetric case when $p = q = \frac{1}{2}$ that

**(13)**
$$\mathbb{P}(M_n \ge r) = 2\mathbb{P}(S_n \ge r + 1) + \mathbb{P}(S_n = r),$$

which is easily expressed in terms of the binomial distribution.

**Proof of (10).** We may assume that $r \ge 1$ and $b < r$. Let $N_n^r(0, b)$ be the number of paths from $(0, 0)$ to $(n, b)$ which include some point having height $r$, which is to say some point $(i, r)$ with $0 < i < n$; for such a path $\pi$, let $(i_\pi, r)$ be the earliest such point. We may reflect the segment of the path with $i_\pi \le x \le n$ in the line $y = r$ to obtain a path $\pi'$ joining $(0, 0)$ to $(n, 2r - b)$. Any such path $\pi'$ is obtained thus from a unique path $\pi$, and therefore $N_n^r(0, b) = N_n(0, 2r - b)$. It follows as required that

$$\mathbb{P}(M_n \ge r, \ S_n = b) = N_n^r(0, b) p^{\frac{1}{2}(n+b)} q^{\frac{1}{2}(n-b)}$$
$$= (q/p)^{r-b} N_n(0, 2r - b) p^{\frac{1}{2}(n+2r-b)} q^{\frac{1}{2}(n-2r+b)}$$
$$= (q/p)^{r-b}\mathbb{P}(S_n = 2r - b).$$  ∎

What is the chance that the walk reaches a new maximum at a particular time? More precisely, what is the probability that the walk, starting from 0, reaches the point $b$ ($> 0$) for the first time at the $n$th step? Writing $f_b(n)$ for this probability, we have that

$$f_b(n) = \mathbb{P}(M_{n-1} = S_{n-1} = b - 1, \ S_n = b)$$
$$= p\left[\mathbb{P}(M_{n-1} \geq b - 1, \ S_{n-1} = b - 1) - \mathbb{P}(M_{n-1} \geq b, \ S_{n-1} = b - 1)\right]$$
$$= p\left[\mathbb{P}(S_{n-1} = b - 1) - (q/p)\mathbb{P}(S_{n-1} = b + 1)\right] \quad \text{by (11)}$$
$$= \frac{b}{n}\mathbb{P}(S_n = b)$$

by a simple calculation using (2). A similar conclusion may be reached if $b < 0$, and we arrive at the following.

**(14) Hitting time theorem.** *The probability $f_b(n)$ that a random walk $S$ hits the point $b$ for the first time at the $n$th step, having started from 0, satisfies*

**(15)**
$$f_b(n) = \frac{|b|}{n}\mathbb{P}(S_n = b) \quad \text{if } \ n \geq 1.$$

The conclusion here has a close resemblance to that of the ballot theorem, and particularly Theorem (7). This is no coincidence: a closer examination of the two results leads to another technique for random walks, the technique of 'reversal'. If the first $n$ steps of the original random walk are

$$\{0, S_1, S_2, \ldots, S_n\} = \left\{0, X_1, X_1 + X_2, \ldots, \sum_1^n X_i\right\}$$

then the steps of the *reversed* walk, denoted by $0, T_1, \ldots, T_n$, are given by

$$\{0, T_1, T_2, \ldots, T_n\} = \left\{0, X_n, X_n + X_{n-1}, \ldots, \sum_1^n X_i\right\}.$$

Draw a diagram to see how the two walks correspond to each other. The $X_i$ are independent and identically distributed, and it follows that the two walks have identical distributions even if $p \neq \frac{1}{2}$. Notice that the addition of an extra step to the original walk may change *every* step of the reversed walk.

Now, the original walk satisfies $S_n = b \, (> 0)$ and $S_1 S_2 \cdots S_n \neq 0$ if and only if the reversed walk satisfied $T_n = b$ and $T_n - T_{n-i} = X_1 + \cdots + X_i > 0$ for all $i \geq 1$, which is to say that the first visit of the reversed walk to the point $b$ takes place at time $n$. Therefore

**(16)**
$$\mathbb{P}(S_1 S_2 \cdots S_n \neq 0, \ S_n = b) = f_b(n) \quad \text{if } \ b > 0.$$

This is the 'coincidence' remarked above; a similar argument is valid if $b < 0$. The technique of reversal has other applications. For example, let $\mu_b$ be the mean number of visits of the walk to the point $b$ before it returns to its starting point. If $S_0 = 0$ then, by (16),

**(17)** $$\mu_b = \sum_{n=1}^{\infty} \mathbb{P}(S_1 S_2 \cdots S_n \neq 0, \ S_n = b) = \sum_{n=1}^{\infty} f_b(n) = \mathbb{P}(S_n = b \text{ for some } n),$$

the probability of ultimately visiting $b$. This leads to the following result.

**(18) Theorem.** *If $p = \frac{1}{2}$ and $S_0 = 0$, for any $b$ ($\neq 0$) the mean number $\mu_b$ of visits of the walk to the point $b$ before returning to the origin equals $1$.*

**Proof.** Let $f_b = \mathbb{P}(S_n = b$ for some $n \geq 0)$. We have, by conditioning on the value of $S_1$, that $f_b = \frac{1}{2}(f_{b+1} + f_{b-1})$ for $b > 0$, with boundary condition $f_0 = 1$. The solution of this difference equation is $f_b = Ab + B$ for constants $A$ and $B$. The unique such solution lying in $[0, 1]$ with $f_0 = 1$ is given by $f_b = 1$ for all $b \geq 0$. By symmetry, $f_b = 1$ for $b \leq 0$. However, $f_b = \mu_b$ for $b \neq 0$, and the claim follows.                                    ∎

'The truly amazing implications of this result appear best in the language of fair games. A perfect coin is tossed until the first equalization of the accumulated numbers of heads and tails. The gambler receives one penny for every time that the accumulated number of heads exceeds the accumulated number of tails by $m$. The *"fair entrance fee"* equals $1$ independently of $m$.' [Feller 1968, p. 367]

We conclude with two celebrated properties of the symmetric random walk.

**(19) Theorem. Arc sine law for last visit to the origin.** *Suppose that $p = \frac{1}{2}$ and $S_0 = 0$. The probability that the last visit to $0$ up to time $2n$ occurred at time $2k$ is $\mathbb{P}(S_{2k} = 0)\mathbb{P}(S_{2n-2k} = 0)$.*

In advance of proving this, we note some consequences. Writing $\alpha_{2n}(2k)$ for the probability referred to in the theorem, it follows from the theorem that $\alpha_{2n}(2k) = u_{2k}u_{2n-2k}$ where

$$u_{2k} = \mathbb{P}(S_{2k} = 0) = \binom{2k}{k}2^{-2k}.$$

In order to understand the behaviour of $u_{2k}$ for large values of $k$, we use Stirling's formula:

**(20)**                        $n! \sim n^n e^{-n}\sqrt{2\pi n}$   as   $n \to \infty$,

which is to say that the ratio of the left-hand side to the right-hand side tends to $1$ as $n \to \infty$. Applying this formula, we obtain that $u_{2k} \sim 1/\sqrt{\pi k}$ as $k \to \infty$. This gives rise to the approximation

$$\alpha_{2n}(2k) \simeq \frac{1}{\pi\sqrt{k(n-k)}},$$

valid for values of $k$ which are close to neither $0$ nor $n$. With $T_{2n}$ denoting the time of the last visit to $0$ up to time $2n$, it follows that

$$\mathbb{P}(T_{2n} \leq 2xn) \simeq \sum_{k \leq xn} \frac{1}{\pi\sqrt{k(n-k)}} \sim \int_{u=0}^{xn} \frac{1}{\pi\sqrt{u(n-u)}}\,du = \frac{2}{\pi}\sin^{-1}\sqrt{x},$$

which is to say that $T_{2n}/(2n)$ has a distribution function which is approximately $(2/\pi)\sin^{-1}\sqrt{x}$ when $n$ is sufficiently large. We have proved a limit theorem.

The arc sine law is rather surprising. One may think that, in a long run of $2n$ tosses of a fair coin, the epochs of time at which there have appeared equal numbers of heads and tails should appear rather frequently. On the contrary, there is for example probability $\frac{1}{2}$ that no such epoch arrived in the final $n$ tosses, and indeed probability approximately $\frac{1}{5}$ that no such epoch occurred after the first $\frac{1}{5}n$ tosses. One may think that, in a long run of $2n$ tosses of a fair coin, the last time at which the numbers of heads and tails were equal tends to be close to the end. On the contrary, the distribution of this time is symmetric around the midpoint.

How much time does a symmetric random walk spend to the right of the origin? More precisely, for how many values of $k$ satisfying $0 \le k \le 2n$ is it the case that $S_k > 0$? Intuitively, one might expect the answer to be around $n$ with large probability, but the truth is quite different. With large probability, the proportion of time spent to the right (or to the left) of the origin is near to 0 or to 1, but not near to $\frac{1}{2}$. That is to say, in a long sequence of tosses of a fair coin, there is large probability that one face (either heads or tails) will lead the other for a disproportionate amount of time.

**(21) Theorem. Arc sine law for sojourn times.** *Suppose that* $p = \frac{1}{2}$ *and* $S_0 = 0$. *The probability that the walk spends exactly $2k$ intervals of time, up to time $2n$, to the right of the origin equals* $\mathbb{P}(S_{2k} = 0)\mathbb{P}(S_{2n-2k} = 0)$.

We say that the interval $(k, k + 1)$ is spent to the right of the origin if either $S_k > 0$ or $S_{k+1} > 0$. It is clear that the number of such intervals is even if the total number of steps is even. The conclusion of this theorem is most striking. First, the answer is the same as that of Theorem (19). Secondly, by the calculations following (19) we have that the probability that the walk spends $2xn$ units of time or less to the right of the origin is approximately $(2/\pi) \sin^{-1} \sqrt{x}$.

**Proof of (19).** The probability in question is

$$\alpha_{2n}(2k) = \mathbb{P}(S_{2k} = 0)\mathbb{P}(S_{2k+1}S_{2k+2} \cdots S_{2n} \ne 0 \mid S_{2k} = 0)$$
$$= \mathbb{P}(S_{2k} = 0)\mathbb{P}(S_1 S_2 \cdots S_{2n-2k} \ne 0).$$

Now, setting $m = n - k$, we have by (8) that

$$\textbf{(22)} \qquad \mathbb{P}(S_1 S_2 \cdots S_{2m} \ne 0) = 2\sum_{k=1}^{m} \frac{2k}{2m}\mathbb{P}(S_{2m} = 2k) = 2\sum_{k=1}^{m} \frac{2k}{2m}\binom{2m}{m+k}\left(\frac{1}{2}\right)^{2m}$$

$$= 2\left(\frac{1}{2}\right)^{2m} \sum_{k=1}^{m}\left[\binom{2m-1}{m+k-1} - \binom{2m-1}{m+k}\right]$$

$$= 2\left(\frac{1}{2}\right)^{2m}\binom{2m-1}{m}$$

$$= \binom{2m}{m}\left(\frac{1}{2}\right)^{2m} = \mathbb{P}(S_{2m} = 0). \qquad \blacksquare$$

In passing, note the proof in (22) that

$$\textbf{(23)} \qquad\qquad \mathbb{P}(S_1 S_2 \cdots S_{2m} \ne 0) = \mathbb{P}(S_{2m} = 0)$$

for the simple symmetric random walk.

**Proof of (21).** Let $\beta_{2n}(2k)$ be the probability in question, and write $u_{2m} = \mathbb{P}(S_{2m} = 0)$ as before. We are claiming that, for all $m \ge 1$,

$$\textbf{(24)} \qquad\qquad \beta_{2m}(2k) = u_{2k}u_{2m-2k} \quad \text{if} \ \ 0 \le k \le m.$$

First,

$$\mathbb{P}(S_1 S_2 \cdots S_{2m} > 0) = \mathbb{P}(S_1 = 1, \ S_2 \ge 1, \ldots, \ S_{2m} \ge 1)$$
$$= \tfrac{1}{2}\mathbb{P}(S_1 \ge 0, \ S_2 \ge 0, \ldots, \ S_{2m-1} \ge 0),$$

where the second line follows by considering the walk $S_1 - 1, S_2 - 1, \ldots, S_{2m} - 1$. Now $S_{2m-1}$ is an odd number, so that $S_{2m-1} \ge 0$ implies that $S_{2m} \ge 0$ also. Thus

$$\mathbb{P}(S_1 S_2 \cdots S_{2m} > 0) = \tfrac{1}{2}\mathbb{P}(S_1 \ge 0, S_2 \ge 0, \ldots, S_{2m} \ge 0),$$

yielding by (23) that

$$\tfrac{1}{2}u_{2m} = \mathbb{P}(S_1 S_2 \cdots S_{2m} > 0) = \tfrac{1}{2}\beta_{2m}(2m),$$

and (24) follows for $k = m$, and therefore for $k = 0$ also by symmetry.

Let $n$ be a positive integer, and let $T$ be the time of the first return of the walk to the origin. If $S_{2n} = 0$ then $T \le 2n$; the probability mass function $f_{2r} = \mathbb{P}(T = 2r)$ satisfies

$$\mathbb{P}(S_{2n} = 0) = \sum_{r=1}^{n} \mathbb{P}(S_{2n} = 0 \mid T = 2r)\mathbb{P}(T = 2r) = \sum_{r=1}^{n} \mathbb{P}(S_{2n-2r} = 0)\mathbb{P}(T = 2r),$$

which is to say that

$$(25) \qquad\qquad\qquad u_{2n} = \sum_{r=1}^{n} u_{2n-2r} f_{2r}.$$

Let $1 \le k \le n - 1$, and consider $\beta_{2n}(2k)$. The corresponding event entails that $T = 2r$ for some $r$ satisfying $1 \le r < n$. The time interval $(0, T)$ is spent entirely either to the right or the left of the origin, and each possibility has probability $\tfrac{1}{2}$. Therefore,

$$(26) \qquad \beta_{2n}(2k) = \sum_{r=1}^{k} \tfrac{1}{2}\mathbb{P}(T = 2r)\beta_{2n-2r}(2k - 2r) + \sum_{r=1}^{n-k} \tfrac{1}{2}\mathbb{P}(T = 2r)\beta_{2n-2r}(2k).$$

We conclude the proof by using induction. Certainly (24) is valid for all $k$ if $m = 1$. Assume (24) is valid for all $k$ and all $m < n$.

From (26),

$$\beta_{2n}(2k) = \tfrac{1}{2}\sum_{r=1}^{k} f_{2r} u_{2k-2r} u_{2n-2k} + \tfrac{1}{2}\sum_{r=1}^{n-k} f_{2r} u_{2k} u_{2n-2k-2r}$$
$$= \tfrac{1}{2}u_{2n-2k}u_{2k} + \tfrac{1}{2}u_{2k}u_{2n-2k} = u_{2k}u_{2n-2k}$$

by (25), as required.                                                                                            ■

## Exercises for Section 3.10

**1.**    Consider a symmetric simple random walk $S$ with $S_0 = 0$. Let $T = \min\{n \geq 1 : S_n = 0\}$ be the time of the first return of the walk to its starting point. Show that

$$\mathbb{P}(T = 2n) = \frac{1}{2n-1}\binom{2n}{n}2^{-2n},$$

and deduce that $\mathbb{E}(T^\alpha) < \infty$ if and only if $\alpha < \frac{1}{2}$. You may need Stirling's formula: $n! \sim n^{n+\frac{1}{2}}e^{-n}\sqrt{2\pi}$.

**2.**    For a symmetric simple random walk starting at 0, show that the mass function of the maximum satisfies $\mathbb{P}(M_n = r) = \mathbb{P}(S_n = r) + \mathbb{P}(S_n = r+1)$ for $r \geq 0$.

**3.**    For a symmetric simple random walk starting at 0, show that the probability that the first visit to $S_{2n}$ takes place at time $2k$ equals the product $\mathbb{P}(S_{2k} = 0)\mathbb{P}(S_{2n-2k} = 0)$, for $0 \leq k \leq n$.

**4.    Samuels' theorem.**  A simple random walk on the integers $\mathbb{Z}$ moves one step rightwards with probability $p$ and otherwise one step leftwards, where $p \in (0, 1)$. Suppose it starts at 0 and has absorbing barriers at $\pm a$. Show that the time and place of absorption are independent.

**5.    Hitting time theorem.**  Let $\{X_m : m \geq 1\}$ be independent, identically distributed random variables taking integer values such that $\mathbb{P}(X_1 \geq -1) = 1$. Let $S_n$ be the (generalized) random walk given by $S_n = k + X_1 + X_2 + \cdots + X_n$, where $k \geq 0$ is given, and let $T = \inf\{n \geq 0 : S_n = 0\}$ be the hitting time of 0.
  Show by induction that $\mathbb{P}(T = n) = (k/n)\mathbb{P}(S_n = 0)$ when $n \geq 1$, $k \geq 0$.

## 3.11  Problems

**1.**    (a) Let $X$ and $Y$ be independent discrete random variables, and let $g, h : \mathbb{R} \to \mathbb{R}$. Show that $g(X)$ and $h(Y)$ are independent.
   (b) Show that two discrete random variables $X$ and $Y$ are independent if and only if $f_{X,Y}(x, y) = f_X(x)f_Y(y)$ for all $x, y \in \mathbb{R}$.
   (c) More generally, show that $X$ and $Y$ are independent if and only if $f_{X,Y}(x, y)$ can be factorized as the product $g(x)h(y)$ of a function of $x$ alone and a function of $y$ alone.

**2.**    Show that if $\text{var}(X) = 0$ then $X$ is almost surely constant; that is, there exists $a \in \mathbb{R}$ such that $\mathbb{P}(X = a) = 1$. (First show that if $\mathbb{E}(X^2) = 0$ then $\mathbb{P}(X = 0) = 1$.)

**3.**    (a) Let $X$ be a discrete random variable and let $g : \mathbb{R} \to \mathbb{R}$. Show that, when the sum is absolutely convergent,

$$\mathbb{E}(g(X)) = \sum_x g(x)\mathbb{P}(X = x).$$

   (b) If $X$ and $Y$ are independent and $g, h : \mathbb{R} \to \mathbb{R}$, show that $\mathbb{E}(g(X)h(Y)) = \mathbb{E}(g(X))\mathbb{E}(h(Y))$ whenever these expectations exist.

**4.**    Let $\Omega = \{\omega_1, \omega_2, \omega_3\}$, with $\mathbb{P}(\omega_1) = \mathbb{P}(\omega_2) = \mathbb{P}(\omega_3) = \frac{1}{3}$. Define $X, Y, Z : \Omega \to \mathbb{R}$ by

$$X(\omega_1) = 1, \quad X(\omega_2) = 2, \quad X(\omega_3) = 3,$$
$$Y(\omega_1) = 2, \quad Y(\omega_2) = 3, \quad Y(\omega_3) = 1,$$
$$Z(\omega_1) = 2, \quad Z(\omega_2) = 2, \quad Z(\omega_3) = 1.$$

Show that $X$ and $Y$ have the same mass functions. Find the mass functions of $X + Y$, $XY$, and $X/Y$. Find the conditional mass functions $f_{Y|Z}$ and $f_{Z|Y}$.

**5.** For what values of $k$ and $\alpha$ is $f$ a mass function, where:
(a) $f(n) = k/\{n(n + 1)\}$, $n = 1, 2, \ldots$,
(b) $f(n) = kn^\alpha$, $n = 1, 2, \ldots$ (*zeta* or *Zipf distribution*)?

**6.** Let $X$ and $Y$ be independent Poisson variables with respective parameters $\lambda$ and $\mu$. Show that:
(a) $X + Y$ is Poisson, parameter $\lambda + \mu$,
(b) the conditional distribution of $X$, given $X + Y = n$, is binomial, and find its parameters.

**7.** If $X$ is geometric, show that $\mathbb{P}(X = n + k \mid X > n) = \mathbb{P}(X = k)$ for $k, n \geq 1$. Why do you think that this is called the 'lack of memory' property? Does any other distribution on the positive integers have this property?

**8.** Show that the sum of two independent binomial variables, $\text{bin}(m, p)$ and $\text{bin}(n, p)$ respectively, is $\text{bin}(m + n, p)$.

**9.** Let $N$ be the number of heads occurring in $n$ tosses of a biased coin. Write down the mass function of $N$ in terms of the probability $p$ of heads turning up on each toss. Prove and utilize the identity

$$\sum_i \binom{n}{2i} x^{2i} y^{n-2i} = \tfrac{1}{2}\{(x + y)^n + (y - x)^n\}$$

in order to calculate the probability $p_n$ that $N$ is even. Compare with Problem (1.8.20).

**10.** An urn contains $N$ balls, $b$ of which are blue and $r$ ($= N - b$) of which are red. A random sample of $n$ balls is withdrawn without replacement from the urn. Show that the number $B$ of blue balls in this sample has the mass function

$$\mathbb{P}(B = k) = \binom{b}{k}\binom{N - b}{n - k} \Big/ \binom{N}{n}.$$

This is called the *hypergeometric distribution* with parameters $N$, $b$, and $n$. Show further that if $N$, $b$, and $r$ approach $\infty$ in such a way that $b/N \to p$ and $r/N \to 1 - p$, then

$$\mathbb{P}(B = k) \to \binom{n}{k} p^k (1 - p)^{n-k}.$$

You have shown that, for small $n$ and large $N$, the distribution of $B$ barely depends on whether or not the balls are replaced in the urn immediately after their withdrawal.

**11.** Let $X$ and $Y$ be independent $\text{bin}(n, p)$ variables, and let $Z = X + Y$. Show that the conditional distribution of $X$ given $Z = N$ is the hypergeometric distribution of Problem (3.11.10).

**12.** Suppose $X$ and $Y$ take values in $\{0, 1\}$, with joint mass function $f(x, y)$. Write $f(0, 0) = a$, $f(0, 1) = b$, $f(1, 0) = c$, $f(1, 1) = d$, and find necessary and sufficient conditions for $X$ and $Y$ to be: (a) uncorrelated, (b) independent.

**13. Tail sum for expectation.**
(a) If $X$ takes non-negative integer values show that

$$\mathbb{E}(X) = \sum_{n=0}^{\infty} \mathbb{P}(X > n).$$

(b) An urn contains $b$ blue and $r$ red balls. Balls are removed at random until the first blue ball is drawn. Show that the expected number drawn is $(b + r + 1)/(b + 1)$.

(c) The balls are replaced and then removed at random until all the remaining balls are of the same colour. Find the expected number remaining in the urn.

(d) Let $X$ and $Y$ be independent random variables taking values in the non-negative integers, with finite means. Let $U = \min\{X, Y\}$ and $V = \max\{X, Y\}$. Show that

$$\mathbb{E}(U) = \sum_{r=1}^{\infty} \mathbb{P}(X \geq r)\mathbb{P}(Y \geq r),$$

$$\mathbb{E}(V) = \sum_{r=1}^{\infty} \left[\mathbb{P}(X \geq r) + \mathbb{P}(Y \geq r) - \mathbb{P}(X \geq r)\mathbb{P}(Y \geq r)\right],$$

$$\mathbb{E}(UV) = \sum_{r,s=1}^{\infty} \mathbb{P}(X \geq r)\mathbb{P}(Y \geq s).$$

(e) Let $X$ take values in the non-negative integers. Show that

$$\mathbb{E}(X^2) = \mathbb{E}(X) + 2\sum_{r=0}^{\infty} r\mathbb{P}(X > r) = \sum_{r=0}^{\infty}(2r + 1)\mathbb{P}(X > r),$$

and find a similar formula for $\mathbb{E}(X^3)$.

**14.** Let $X_1, X_2, \ldots, X_n$ be independent random variables, and suppose that $X_k$ is Bernoulli with parameter $p_k$. Show that $Y = X_1 + X_2 + \cdots + X_n$ has mean and variance given by

$$\mathbb{E}(Y) = \sum_1^n p_k, \quad \text{var}(Y) = \sum_1^n p_k(1 - p_k).$$

Show that, for $\mathbb{E}(Y)$ fixed, var$(Y)$ is a maximum when $p_1 = p_2 = \cdots = p_n$. That is to say, the variation in the sum is greatest when individuals are most alike. Is this contrary to intuition?

**15.** Let $\mathbf{X} = (X_1, X_2, \ldots, X_n)$ be a vector of random variables. The *covariance matrix* $\mathbf{V}(\mathbf{X})$ of $\mathbf{X}$ is defined to be the symmetric $n$ by $n$ matrix with entries $(v_{ij} : 1 \leq i, j \leq n)$ given by $v_{ij} = \text{cov}(X_i, X_j)$. Show that $|\mathbf{V}(\mathbf{X})| = 0$ if and only if the $X_i$ are linearly dependent with probability one, in that $\mathbb{P}(a_1X_1 + a_2X_2 + \cdots + a_nX_n = b) = 1$ for some $\mathbf{a}$ and $b$. ($|\mathbf{V}|$ denotes the determinant of $\mathbf{V}$.)

**16.** Let $X$ and $Y$ be independent Bernoulli random variables with parameter $\frac{1}{2}$. Show that $X + Y$ and $|X - Y|$ are dependent though uncorrelated.

**17.** A secretary drops $n$ matching pairs of letters and envelopes down the stairs, and then places the letters into the envelopes in a random order. Use indicators to show that the number $X$ of correctly matched pairs has mean and variance 1 for all $n \geq 2$. Show that the mass function of $X$ converges to a Poisson mass function as $n \to \infty$.

**18.** Let $\mathbf{X} = (X_1, X_2, \ldots, X_n)$ be a vector of independent random variables each having the Bernoulli distribution with parameter $p$. Let $f : \{0, 1\}^n \to \mathbb{R}$ be *increasing*, which is to say that $f(\mathbf{x}) \leq f(\mathbf{y})$ whenever $x_i \leq y_i$ for each $i$.

(a) Let $e(p) = \mathbb{E}(f(\mathbf{X}))$. Show that $e(p_1) \leq e(p_2)$ if $p_1 \leq p_2$.

(b) **FKG inequality**†. Let $f$ and $g$ be increasing functions from $\{0, 1\}^n$ into $\mathbb{R}$. Show by induction on $n$ that $\text{cov}(f(\mathbf{X}), g(\mathbf{X})) \geq 0$.

**19.** Let $R(p)$ be the reliability function of a network $G$ with a given source and sink, each edge of which is working with probability $p$, and let $A$ be the event that there exists a working connection from source to sink. Show that

$$R(p) = \sum_{\omega} I_A(\omega) p^{N(\omega)} (1 - p)^{m-N(\omega)}$$

†Named after C. Fortuin, P. Kasteleyn, and J. Ginibre 1971, but due in this form to T. E. Harris 1960.

where $\omega$ is a typical realization (i.e. outcome) of the network, $N(\omega)$ is the number of working edges of $\omega$, and $m$ is the total number of edges of $G$.

Deduce that $R'(p) = \text{cov}(I_A, N)/\{p(1-p)\}$, and hence that

$$\frac{R(p)(1-R(p))}{p(1-p)} \le R'(p) \le \sqrt{\frac{mR(p)(1-R(p))}{p(1-p)}}.$$

**20.** Let $R(p)$ be the reliability function of a network $G$, each edge of which is working with probability $p$.

(a) Show that $R(p_1 p_2) \le R(p_1)R(p_2)$ if $0 \le p_1, p_2 \le 1$.

(b) Show that $R(p^\gamma) \le R(p)^\gamma$ for all $0 \le p \le 1$ and $\gamma \ge 1$.

**21. DNA fingerprinting.** In a certain style of detective fiction, the sleuth is required to declare "the criminal has the unusual characteristics ... ; find this person and you have your man". Assume that any given individual has these unusual characteristics with probability $10^{-7}$ independently of all other individuals, and that the city in question contains $10^7$ inhabitants. Calculate the expected number of such people in the city.

(a) Given that the police inspector finds such a person, what is the probability that there is at least one other?

(b) If the inspector finds two such people, what is the probability that there is at least one more?

(c) How many such people need be found before the inspector can be reasonably confident that he has found them all?

(d) For the given population, how improbable should the characteristics of the criminal be, in order that he (or she) be specified uniquely?

**22.** In 1710, J. Arbuthnot observed that male births had exceeded female births in London for 82 successive years. Arguing that this showed the two sexes cannot be equally likely, since $2^{-82}$ is very small, he attributed this run of masculinity to Divine Providence. Let us assume that each birth results in a girl with probability $p = 0.485$, and that the outcomes of different confinements are independent of each other. Ignoring the possibility of twins (and so on), show that the probability that girls outnumber boys in $2n$ live births is no greater than $\binom{2n}{n}p^n q^n \{q/(q-p)\}$, where $q = 1 - p$. Suppose that 20,000 children are born in each of 82 successive years. Show that the probability that boys outnumber girls every year is at least 0.99. You may need Stirling's formula.

**23.** Consider a symmetric random walk with an absorbing barrier at $N$ and a reflecting barrier at $0$ (so that, when the particle is at $0$, it moves to $1$ at the next step). Let $\alpha_k(j)$ be the probability that the particle, having started at $k$, visits $0$ exactly $j$ times before being absorbed at $N$. We make the convention that, if $k = 0$, then the starting point counts as one visit. Show that

$$\alpha_k(j) = \frac{N-k}{N^2}\left(1 - \frac{1}{N}\right)^{j-1}, \qquad j \ge 1,\ 0 \le k \le N.$$

**24. Problem of the points (3.9.4).** A coin is tossed repeatedly, heads turning up with probability $p$ on each toss. Player A wins the game if heads appears at least $m$ times before tails has appeared $n$ times; otherwise player B wins the game. Find the probability that A wins the game.

**25.** A coin is tossed repeatedly, heads appearing on each toss with probability $p$. A gambler starts with initial fortune $k$ (where $0 < k < N$); he wins one point for each head and loses one point for each tail. If his fortune is ever $0$ he is bankrupted, whilst if it ever reaches $N$ he stops gambling to buy a Jaguar. Suppose that $p < \frac{1}{2}$. Show that the gambler can increase his chance of winning by doubling the stakes. You may assume that $k$ and $N$ are even.

What is the corresponding strategy if $p \ge \frac{1}{2}$?

**26.** A compulsive gambler is never satisfied. At each stage he wins £1 with probability $p$ and loses £1 otherwise. Find the probability that he is ultimately bankrupted, having started with an initial fortune of £$k$.

**27. Range of random walk.** Let $\{X_n : n \geq 1\}$ be independent, identically distributed random variables taking integer values. Let $S_0 = 0$, $S_n = \sum_{i=1}^{n} X_i$. The *range* $R_n$ of $S_0, S_1, \ldots, S_n$ is the number of distinct values taken by the sequence. Show that $\mathbb{P}(R_n = R_{n-1}+1) = \mathbb{P}(S_1 S_2 \cdots S_n \neq 0)$, and deduce that, as $n \to \infty$,

$$\frac{1}{n}\mathbb{E}(R_n) \to \mathbb{P}(S_k \neq 0 \text{ for all } k \geq 1).$$

Hence show that, for the simple random walk, $n^{-1}\mathbb{E}(R_n) \to |p - q|$ as $n \to \infty$.

**28. Arc sine law for maxima.** Consider a symmetric random walk $S$ starting from the origin, and let $M_n = \max\{S_i : 0 \leq i \leq n\}$. Show that, for $i = 2k$, $2k + 1$, the probability that the walk reaches $M_{2n}$ for the first time at time $i$ equals $\frac{1}{2}\mathbb{P}(S_{2k} = 0)\mathbb{P}(S_{2n-2k} = 0)$.

**29.** Let $S$ be a symmetric random walk with $S_0 = 0$, and let $N_n$ be the number of points that have been visited by $S$ exactly once up to time $n$. Show that $\mathbb{E}(N_n) = 2$.

**30. Family planning.** Consider the following fragment of verse entitled 'Note for the scientist'.

> People who have three daughters try for more,
> And then its fifty–fifty they'll have four,
> Those with a son or sons will let things be,
> Hence all these surplus women, QED.

(a) What do you think of the argument?

(b) Show that the mean number of children of either sex in a family whose fertile parents have followed this policy equals 1. (You should assume that each delivery yields exactly one child whose sex is equally likely to be male or female.) Discuss.

**31. Dirichlet distribution.** Let $\beta > 1$, let $p_1, p_2, \ldots$ denote the prime numbers, and let $N(1), N(2)$, $\ldots$ be independent random variables, $N(i)$ having mass function $\mathbb{P}(N(i) = k) = (1 - \gamma_i)\gamma_i^k$ for $k \geq 0$, where $\gamma_i = p_i^{-\beta}$ for all $i$. Show that $M = \prod_{i=1}^{\infty} p_i^{N(i)}$ is a random integer with mass function $\mathbb{P}(M = m) = Cm^{-\beta}$ for $m \geq 1$ (this may be called the *Dirichlet distribution*), where $C$ is a constant satisfying

$$C = \prod_{i=1}^{\infty}\left(1 - \frac{1}{p_i^{\beta}}\right) = \left(\sum_{m=1}^{\infty}\frac{1}{m^{\beta}}\right)^{-1}.$$

**32.** $N + 1$ plates are laid out around a circular dining table, and a hot cake is passed between them in the manner of a symmetric random walk: each time it arrives on a plate, it is tossed to one of the two neighbouring plates, each possibility having probability $\frac{1}{2}$. The game stops at the moment when the cake has visited every plate at least once. Show that, with the exception of the plate where the cake began, each plate has probability $1/N$ of being the last plate visited by the cake.

**33. Simplex algorithm†.** There are $\binom{n}{m}$ points ranked in order of merit with no matches. You seek to reach the best, $B$. If you are at the $j$th best, you step to any one of the $j - 1$ better points, with equal probability of stepping to each. Let $r_j$ be the expected number of steps to reach $B$ from the $j$th best vertex. Show that $r_j = \sum_{k=1}^{j-1} k^{-1}$. Give an asymptotic expression for the expected time to reach $B$ from the worst vertex, for large $m, n$.

**34. Dimer problem.** There are $n$ unstable molecules in a row, $m_1, m_2, \ldots, m_n$. One of the $n - 1$ pairs of neighbours, chosen at random, combines to form a stable dimer; this process continues until there remain $U_n$ isolated molecules no two of which are adjacent. Show that the probability that $m_1$ remains isolated is $\sum_{r=0}^{n-1}(-1)^r/r! \to e^{-1}$ as $n \to \infty$. Deduce that $\lim_{n\to\infty} n^{-1}\mathbb{E}U_n = e^{-2}$.

---

†Due to George Dantzig (1914–2005), not to be confused with David van Dantzig.

**35. Poisson approximation.** Let $\{I_r : 1 \le r \le n\}$ be independent Bernoulli random variables with respective parameters $\{p_r : 1 \le r \le n\}$ satisfying $p_r \le c < 1$ for all $r$ and some $c$. Let $\lambda = \sum_{r=1}^{n} p_r$ and $X = \sum_{r=1}^{n} X_r$. Show that

$$\mathbb{P}(X = k) = \frac{\lambda^k e^{-\lambda}}{k!} \left\{ 1 + O\left( \lambda \max_r p_r + \frac{k^2}{\lambda} \max_r p_r \right) \right\}.$$

**36. Sampling.** The length of the tail of the $r$th member of a troop of $N$ chimeras is $x_r$. A random sample of $n$ chimeras is taken (without replacement) and their tails measured. Let $I_r$ be the indicator of the event that the $r$th chimera is in the sample. Set

$$X_r = x_r I_r, \quad \overline{Y} = \frac{1}{n} \sum_{r=1}^{N} X_r, \quad \mu = \frac{1}{N} \sum_{r=1}^{N} x_r, \quad \sigma^2 = \frac{1}{N} \sum_{r=1}^{N} (x_r - \overline{x})^2.$$

Show that $\mathbb{E}(\overline{Y}) = \mu$, and $\text{var}(\overline{Y}) = (N - n)\sigma^2/\{n(N - 1)\}$.

**37. Berkson's fallacy.** Any individual in a group $G$ contracts a certain disease $C$ with probability $\gamma$; such individuals are hospitalized with probability $c$. Independently of this, anyone in $G$ may be in hospital with probability $a$, for some other reason. Let $X$ be the number in hospital, and $Y$ the number in hospital who have $C$ (including those with $C$ admitted for any other reason). Show that the correlation between $X$ and $Y$ is

$$\rho(X, Y) = \sqrt{\frac{\gamma p}{1 - \gamma p} \cdot \frac{(1 - a)(1 - \gamma c)}{a + \gamma c - a\gamma c}},$$

where $p = a + c - ac$.

It has been stated erroneously that, when $\rho(X, Y)$ is near unity, this is evidence for a causal relation between being in $G$ and contracting $C$.

**38.** A telephone sales company attempts repeatedly to sell new kitchens to each of the $N$ families in a village. Family $i$ agrees to buy a new kitchen after it has been solicited $K_i$ times, where the $K_i$ are independent identically distributed random variables with mass function $f(n) = \mathbb{P}(K_i = n)$. The value $\infty$ is allowed, so that $f(\infty) \ge 0$. Let $X_n$ be the number of kitchens sold at the $n$th round of solicitations, so that $X_n = \sum_{i=1}^{N} I_{\{K_i=n\}}$. Suppose that $N$ is a random variable with the Poisson distribution with parameter $\nu$.

(a) Show that the $X_n$ are independent random variables, $X_r$ having the Poisson distribution with parameter $\nu f(r)$.
(b) The company loses heart after the $T$th round of calls, where $T = \inf\{n : X_n = 0\}$. Let $S = X_1 + X_2 + \cdots + X_T$ be the number of solicitations made up to time $T$. Show further that $\mathbb{E}(S) = \nu \mathbb{E}(F(T))$ where $F(k) = f(1) + f(2) + \cdots + f(k)$.

**39.** A particle performs a random walk on the non-negative integers as follows. When at the point $n$ ($> 0$) its next position is uniformly distributed on the set $\{0, 1, 2, \ldots, n + 1\}$. When it hits 0 for the first time, it is absorbed. Suppose it starts at the point $a$.

(a) Find the probability that its position never exceeds $a$, and prove that, with probability 1, it is absorbed ultimately.
(b) Find the probability that the final step of the walk is from 1 to 0 when $a = 1$.
(c) Find the expected number of steps taken before absorption when $a = 1$.

**40.** Let $G$ be a finite graph with neither loops nor multiple edges, and write $d_v$ for the degree of the vertex $v$. An *independent set* is a set of vertices no pair of which is joined by an edge. Let $\alpha(G)$ be the size of the largest independent set of $G$. Use the probabilistic method to show that $\alpha(G) \ge \sum_v 1/(d_v + 1)$. [This conclusion is sometimes referred to as *Turán's theorem*.]

**41. Kelly betting, or proportional investment.** A gambler (or 'investor') makes a sequence of bets of the following type: on each bet, for a given stake $S$, the return is either the loss of the stake with probability $q$ ($= 1 - p$) or a win totalling $(1 + r)S$ with probability $p$. (Assume the usual independence.) The entry fee is $cS$ where $c < r$. Show that the mean gain per play for stake $S$ is $gS$, where $g = pr - q - c$ (with a negative value indicating a loss).

The gambler decides to bet a fixed fraction $f$ of her current fortune at each stage. That is, given a current fortune $F$, she bets $fF$ for some fixed $f$. Show that her resulting fortune is

$$F' = F\{1 + f[(1 + r)I - (1 + c)]\},$$

where $I$ is the indicator function of a win.

Suppose $p > (1 + c)/(1 + r)$. The gambler considers two policies for choosing $f$: for given $F$,
(a) maximize $\mathbb{E}(F')$,
(b) maximize $\mathbb{E}(\log F')$.
Find an expression for $f$ in each case. Show that $f_a > f_b$, and explain why a cautious gambler may prefer (b) to (a) even though this entails a slower rate of growth of her expected fortune.

**42. Random adding sequence.** Let $x_1, x_2, \ldots, x_r$ be given reals with $r \geq 2$, and let the sequence $\{X_n : n \geq 1\}$ of random variables be given as follows. First, $X_n = x_n$ for $n \leq r$. For $n \geq r$, we set $X_{n+1} = X_{U_n} + X_{V_n}$ where $U_n$, $V_n$ are uniformly distributed on $\{1, 2, \ldots, n\}$, and the family $\{U_n, V_n : n \geq r\}$ are independent. Show that

$$\mathbb{E}(X_n) = \frac{2n}{r(r+1)} \sum_{k=1}^{r} x_k.$$

Enthusiasts may care to show that, when $r = 1 = x_1$ and $n \to \infty$,

$$\frac{1}{n^2} \mathbb{E}(X_n^2) \to \frac{1}{2\pi} \sinh \pi.$$

**43. Random subtracting sequence.** Let $X_1 = 1$. For $n \geq 1$, let $X_{n+1} = X_{U_n} - X_{V_n}$ where $U_n$, $V_n$ are uniformly distributed on $\{1, 2, \ldots, n\}$, and the family $\{U_n, V_n : n \geq 1\}$ are independent. Show that

$$\frac{1}{n} \operatorname{var}(X_n) \to -\frac{\sin(\pi \sqrt{3})}{\pi \sqrt{3}} \qquad \text{as } n \to \infty.$$

You may find it useful to recall Euler's sine formula:

$$\sin(\pi x) = \pi x \prod_{n=1}^{\infty} \left(1 - \frac{x^2}{n^2}\right).$$

**44. Bilbo baffled.** Gollum has concealed the ring of power in a box chosen randomly from a row of $n \geq 1$ such boxes. Bilbo opens a box at random. If the ring is not there, his occult powers are sufficient for him to learn whether the ring lies to the right or the left, and he opens further boxes accordingly. Find an expression for the mean number $b_n$ of boxes opened before finding the ring, and deduce that $b_n \sim 2 \log n$ as $n \to \infty$.

**45. Bilbo reloaded.** In a variant of the previous problem, we have $n = 2^r - 1$, and Bilbo invariably chooses the middle box. Find the mean number $m_r$ of boxes inspected, and the asymptotics of $m_r$ as $r \to \infty$.

**46. Fairies.** A bad fairy has cursed you. A good fairy has concealed the magic word that cancels the curse in one of $n$ numbered boxes, and she has told you that it is in box $i$ with probability $p_i$, for $i = 1, 2, \ldots, n$. Each day you are permitted to look in one box.

(a) Assume that, each day, you inspect a box chosen at random, box $i$ being chosen with probability $c_i$, and, furthermore, boxes chosen on different days are independent. Find the mean number of days that elapse before your release from the curse, and find the mass function $c$ that minimizes this mean value.

(b) Suppose now that you remember the results of your previous failed searches. What now is your optimal policy, and what is the mean number of days that elapse?

(c) After each search, the bad fairy removes the magic word, which is immediately replaced by the good fairy in an independently chosen box, with the same distribution at each replacement. What now is your optimal policy? Find the mean number of elapsed days.

**47. Duration of play.** Gwen and John play 'best of $2n + 1$ games', and play stops as soon as either has won $n + 1$ games. Gwen wins each game with probability $\gamma \in (0, 1)$ and John otherwise (with probability $\delta = 1 - \gamma$). Different games have independent winners. Write down an expression for the probability $f_r$ that Gwen wins $r$ games altogether, given that John has won the match, and deduce that $r f_r = (n + r)\gamma f_{r-1}$ for $0 < r \le n$.

Hence or otherwise, prove that the mean total number $T_n$ of games in the match is

$$T_n = (n + 1) \left( \frac{\gamma(1 - P_n)}{\delta} + \frac{\delta P_n}{\gamma} + 1 \right) - (2n + 1) \binom{2n}{n} (\gamma\delta)^n,$$

where $P_n$ is the probability that Gwen wins the match.

When $p = \frac{1}{2}$, show that

$$T_n = 2n - 2\sqrt{n/\pi} + 2 + O(n^{-1/2}), \qquad \text{as } n \to \infty.$$

**48. Stirling numbers, Bell numbers, Dobinski's formula.** Let $S(n, k)$ be the number of ways to partition $N = \{1, 2, \dots, n\}$ into $k$ non-empty parts. Suppose each element of $N$ is coloured with one of $c$ distinct colours. Prove that

$$c^n = \sum_{k=1}^{n} S(n, k)c(c - 1)\cdots(c - k + 1).$$

Deduce that the $n$th moment of the Poisson distribution with parameter 1 equals the number $b_n$ of ways to partition $N$, that is, $b_n = \sum_{k=1}^{n} S(n, k)$.

**49. Server advantage?** Let $\beta, \gamma \in [0, 1]$. Bertha and Harold play a game of rackets. Bertha wins the point with probability $\beta$ when she serves, and Harold wins with probability $\gamma$ when he serves. The first player to win $n$ points wins the game, and Bertha serves first. Consider the following rules for changing server.

(a) Service alternates between players.

(b) Service is retained until a point is lost, and then passes to the other player.

(c) Service is retained until a point is won, and then passes to the other player.

(d) Bertha serves the first $n$ points, and then Harold serves any further points required to decide the outcome.

Show that $\mathbb{P}(\text{Bertha wins})$ is the same for all four rules.

**50. Entropy.** Let $X, Y$ be discrete random variables with respective mass functions $f_X, f_Y$, and joint mass function $f_{X,Y}$. Define the:

$$\text{entropy of } X \; : \; H(X) = -\mathbb{E}\big(\log f_X(X)\big),$$
$$\text{joint entropy of } X \text{ given } Y \; : \; H(X, Y) = -\mathbb{E}\big(\log f_{X,Y}(X, Y)\big),$$
$$\text{conditional entropy of } X \text{ given } Y \; : \; H(X \mid Y) = -\mathbb{E}\big(\log f_{X\mid Y}(X \mid Y)\big).$$

[It is normal to use logarithms to base 2 in information theory, but we use natural logarithms here.]
(a) Show that $H(X + a) = H(X)$ for $a \in \mathbb{R}$.
(b) Show that $H(X) - H(X \mid Y) = I(X; Y)$, where $I$ is the mutual information of Exercise (3.6.5).
(c) Show when $X$ and $Y$ are independent that $H(X, Y) = H(X) + H(Y)$.
(d) Show that the entropy of the binomial distribution $\text{bin}(n, p)$ is non-decreasing in $n$.
(e) Find the entropy of the geometric distribution, parameter $p$, and show it is decreasing in $p$.

**51.** (a) Show that the entropy $H(\lambda)$, as defined in Problem (3.11.50), of the Poisson distribution with parameter $\lambda$ (using natural logarithms) is given by

$$H(\lambda) = \lambda - \lambda \log \lambda + e^{-\lambda} \sum_{m=0}^{\infty} \frac{\lambda^m \log(m!)}{m!}.$$

(b) Recalling Exercise (3.6.5) and Example (3.7.5), show that the mutual information of the number $N$ of hens and the number $K$ of chicks is $I(N; K) = H(\lambda) - H(\lambda(1 - p))$. Deduce that $H(\lambda)$ is increasing in $\lambda$.
(c) Spend a short time seeking to show the last statement directly from the expression in part (a).

**52. Dirichlet distribution revisited.** Let $\beta > 1$, and let $X$, $Y$ be independent random variables with the Dirichlet distribution with parameter $\beta$ of Problem (3.11.31).
(a) Show that the events $E_p = \{X \text{ is divisible by } p\}$ are independent for $p$ prime.
(b) Deduce Euler's formula

$$\prod_{p \text{ prime}} \left(1 - \frac{1}{p^\beta}\right) = \frac{1}{\zeta(\beta)},$$

where $\zeta(\beta)$ is the Riemann zeta function, $\zeta(\beta) = \sum_{m=1}^{\infty} m^{-\beta}$, and $\beta > 1$.
(c) Show that the probability that $X$ is 'square-free' (that is, indivisible by any perfect square other than 1) equals $1/\zeta(2\beta)$.
(d) Let $H$ be the highest common factor of $X$ and $Y$. Prove that

$$\mathbb{P}(H = m) = \frac{m^{-2\beta}}{\zeta(2\beta)}, \qquad m = 1, 2, \ldots.$$

**53. Strict Oxford secrets.** A *strict Oxford secret* is a secret which you may tell to no more than one other person†. Of a group of $n + 1$ Oxonians, one learns a strict secret. In accordance with the rules, she tells it to one other person selected uniformly at random from the rest of the group. Each confidante tells the secret to exactly one person picked at random from the rest of the group, excluding the person from whom they heard the secret. When someone who already knows the secret hears it repeated, the entire process ceases.

Let $S$ be the total number of people who eventually know the secret. Find the distribution of $S$, and show that

$$\frac{1}{\sqrt{n}} \mathbb{E}(S) \to \sqrt{\pi/2}, \qquad \frac{1}{n} \text{var}(S) \to \tfrac{1}{2}(4 - \pi), \qquad \text{as } n \to \infty.$$

With $\underline{r}$ denoting $n!/(n - r)!$, you may find it useful that

$$\sum_{r=1}^{\infty} \frac{n^{\underline{r}}}{n^r} \sim \sqrt{\pi n/2}, \qquad \sum_{r=1}^{\infty} r \frac{n^{\underline{r}}}{n^r} \sim n.$$

---

†Oliver Franks (1905–1992) defined a secret in the Oxford sense as one that you can tell to no more than one person *at a time*.

**54. Transposition shuffle.** From a pack of $n$ cards, two different cards are selected randomly, and they are transposed. Let $p_r$ be the probability that any given card (say, the top card) is in its original position after $r > 0$ such independent transpositions.

(a) Show that

$$p_r = \frac{1}{n} + \frac{n-1}{n}\left(\frac{n-3}{n-1}\right)^r.$$

(b) Find $\mathbb{E}(C_r)$, where $C_r$ is the number of cards in their original place after $r$ random transpositions.

(c) Show that, for large $n$, the number $r$ of transpositions needed for $\mathbb{E}(C_r) \approx 2$ is approximately $\frac{1}{2}n\log n$.

**55. Random walk on the $d$-cube.** The $d$-cube $C_d$ is the graph with vertex-set $\{0, 1\}^d$, and an edge between two vertices $\mathbf{x} = (x_1, x_2, \ldots, x_d)$ and $\mathbf{y} = (y_1, y_2, \ldots, y_d)$ if and only if $\sum_i |x_i - y_i| = 1$. A particle pursues a random walk on $C_d$. At each epoch of time, it moves from its current position to a neighbour chosen uniformly at random, with the usual independence. Two vertices are called 'antipodal' if the graph-distance between them equals $d$.

Show that the mean first passage time $m_d$ of the walker between two antipodal vertices satisfies $\mu_d \sim 2^d$ as $d \to \infty$.

**56. The lost boarding pass, Problem (1.8.39) revisited.**

(a) A particle performs a type of random walk on the set $S = \{1, 2, \ldots, n\}$. Let $X_r$ be the particle's position at time $r$. Given that $X_r = x$, $X_{r+1}$ is chosen uniformly at random from the set $\{1\} \cup \{x + 1, x + 2, \ldots, n\}$. The particle stops moving at the first instant it arrives at either 1 or $n$ (so that 1 and $n$ are 'absorbing', but absorption does not occur at time 0 even if $X_0 \in \{1, n\}$). The point $m \in \{1, 2, \ldots, n\}$ is said to be *hit* if $X_r = m$ for some $r \geq 1$. Show that

$$\mathbb{P}(m \text{ is hit} \mid X_0 = 1) = \begin{cases} \dfrac{1}{2} & \text{if } m = 1, \\[2mm] \dfrac{1}{n-m+2} & \text{if } m \geq 2. \end{cases}$$

(b) The $n$ passengers on a flight in an airplane with $n$ seats have been told their seat numbers. They board the plane one by one. The first person sits in a seat chosen uniformly at random from the $n$ seats then available. Subsequent passengers sit in their assigned seats whenever they find them available, or otherwise in a randomly chosen empty seat. For $m \geq 2$, what is the probability that the $m$th passenger finds his or her assigned seat to be already occupied?

**57. Paley–Zygmund 'second moment' inequalities.**

(a) Let $X$ be a random variable with $\mathbb{E}(X) > 0$ and $0 < \mathbb{E}(X^2) < \infty$. Show that

$$\mathbb{P}(X > a\mathbb{E}(X)) \geq \frac{(1-a)^2 \mathbb{E}(X)^2}{\mathbb{E}(X^2)} \qquad \text{for } 0 \leq a \leq 1.$$

(b) Deduce that, if $\mathbb{P}(X \geq 0) = 1$,

$$\mathbb{P}(X = 0) \leq \frac{\text{var}(X)}{\mathbb{E}(X^2)} \leq \frac{\text{var}(X)}{\mathbb{E}(X)^2}.$$

(c) Let $A_1, A_2, \ldots, A_n$ be events, and let $X = \sum_{r=1}^{n} I_{A_r}$ be the sum of their indicator functions. Show that

$$\mathbb{P}(X = 0) \leq \frac{1}{\mathbb{E}(X)} + \frac{1}{\mathbb{E}(X)^2} \sum{}^{*} \mathbb{P}(A_r \cap A_s),$$

where the summation $\sum^{*}$ is over all distinct unordered pairs $r$, $s$ such that $A_r$ and $A_s$ are not independent.

# 4

# Continuous random variables

*Summary.* The distribution of a continuous random variable may be specified via its probability density function. The key notion of independence is explored for continuous random variables. The concept of expectation and its consequent theory are discussed in depth. Conditional distributions and densities are studied, leading to the notion of conditional expectation. Certain specific distributions are introduced, including the exponential and normal distributions, and the multivariate normal distribution. The density function following a change of variables is derived by the Jacobian formula. The study of sums of random variables leads to the convolution formula for density functions. Methods for sampling from given distributions are presented. The method of coupling is discussed with examples, and the Stein–Chen approximation to the Poisson distribution is proved. The final section is devoted to questions of geometrical probability.

## 4.1 Probability density functions

Recall that a random variable $X$ is *continuous* if its distribution function $F(x) = \mathbb{P}(X \leq x)$ can be written as†

$$(1) \qquad\qquad F(x) = \int_{-\infty}^{x} f(u)\, du$$

for some integrable $f : \mathbb{R} \to [0, \infty)$.

**(2) Definition.** The function $f$ is called the **(probability) density function** of the continuous random variable $X$.

The density function of $F$ is not prescribed uniquely by (1) since two integrable functions which take identical values except at some specific point have the same integrals. However, if $F$ is differentiable at $u$ then we shall normally set $f(u) = F'(u)$. We may write $f_X(u)$ to stress the role of $X$.

---

†Never mind what type of integral this is, at this stage.

**(3) Example (2.3.4) revisited.** The random variables $X$ and $Y$ have density functions

$$f_X(x) = \begin{cases} (2\pi)^{-1} & \text{if } 0 \le x \le 2\pi, \\ 0 & \text{otherwise,} \end{cases} \qquad f_Y(y) = \begin{cases} y^{-\frac{1}{2}}/(4\pi) & \text{if } 0 \le y \le 4\pi^2, \\ 0 & \text{otherwise.} \end{cases}$$

The random variable $X$ is said to have the uniform distribution on the interval $[0, 2\pi]$.     ●

These density functions are non-zero if and only if $x \in [0, 2\pi]$ and $y \in [0, 4\pi^2]$. In such cases in the future, we shall write simply $f_X(x) = (2\pi)^{-1}$ for $0 \le x \le 2\pi$, and similarly for $f_Y$, with the implicit implication that the functions in question equal zero elsewhere.

Continuous variables contrast starkly with discrete variables in that they satisfy $\mathbb{P}(X = x) = 0$ for all $x \in \mathbb{R}$; this may seem paradoxical since $X$ necessarily takes *some* value. Very roughly speaking, the resolution of this paradox lies in the observation that there are *uncountably* many possible values for $X$; this number is so large that the probability of $X$ taking any particular value cannot exceed zero.

The numerical value $f(x)$ is *not* a probability. However, we can think of $f(x)\,dx$ as the element of probability $\mathbb{P}(x < X \le x + dx)$, since

$$\mathbb{P}(x < X \le x + dx) = F(x + dx) - F(x) \simeq f(x)\,dx.$$

From equation (1), the probability that $X$ takes a value in the interval $[a, b]$ is

$$\mathbb{P}(a \le X \le b) = \int_a^b f(x)\,dx.$$

Intuitively speaking, in order to calculate this probability, we simply add up all the small elements of probability which contribute. More generally, if $B$ is a sufficiently nice subset of $\mathbb{R}$ (such as an interval, or a countable union of intervals, and so on), then it is reasonable to expect that

**(4)** $$\mathbb{P}(X \in B) = \int_B f(x)\,dx,$$

and indeed this turns out to be the case.

We have deliberately used the same letter $f$ for mass functions and density functions† since these functions perform exactly analogous tasks for the appropriate classes of random variables. In many cases proofs of results for discrete variables can be rewritten for continuous variables by replacing any summation sign by an integral sign, and any probability mass $f(x)$ by the corresponding element of probability $f(x)\,dx$.

**(5) Lemma.** *If $X$ has density function $f$ then*
(a) $\int_{-\infty}^{\infty} f(x)\,dx = 1$,
(b) $\mathbb{P}(X = x) = 0$ *for all* $x \in \mathbb{R}$,
(c) $\mathbb{P}(a \le X \le b) = \int_a^b f(x)\,dx$.

**Proof.** Exercise.                                                                              ∎

---

†Some writers prefer to use the letter $p$ to denote a mass function, the better to distinguish mass functions from density functions.

Part (a) of the lemma characterizes those non-negative integrable functions which are density functions of some random variable.

We conclude this section with a technical note for the more critical reader. For what sets $B$ is (4) meaningful, and why does (5a) characterize density functions? Let $\mathcal{J}$ be the collection of all open intervals in $\mathbb{R}$. By the discussion in Section 1.6, $\mathcal{J}$ can be extended to a unique smallest $\sigma$-field $\mathcal{B} = \sigma(\mathcal{J})$ which contains $\mathcal{J}$; $\mathcal{B}$ is called the *Borel $\sigma$-field* and contains *Borel sets*. Equation (4) holds for all $B \in \mathcal{B}$. Setting $\mathbb{P}_X(B) = \mathbb{P}(X \in B)$, we can check that $(\mathbb{R}, \mathcal{B}, \mathbb{P}_X)$ is a probability space. Secondly, suppose that $f : \mathbb{R} \to [0, \infty)$ is integrable and $\int_{-\infty}^{\infty} f(x)\, dx = 1$. For any $B \in \mathcal{B}$, we define

$$\mathbb{P}(B) = \int_B f(x)\, dx.$$

Then $(\mathbb{R}, \mathcal{B}, \mathbb{P})$ is a probability space and $f$ is the density function of the identity random variable $X : \mathbb{R} \to \mathbb{R}$ given by $X(x) = x$ for any $x \in \mathbb{R}$. Assiduous readers will verify the steps of this argument for their own satisfaction (or see Clarke 1975, p. 53).

---

## Exercises for Section 4.1

**1.**   For what values of the parameters are the following functions probability density functions?

(a) $f(x) = C\{x(1-x)\}^{-\frac{1}{2}}$, $0 < x < 1$, the density function of the 'arc sine law'.
(b) $f(x) = C \exp(-x - e^{-x})$, $x \in \mathbb{R}$, the density function of the 'extreme-value distribution'.
(c) $f(x) = C(1 + x^2)^{-m}$, $x \in \mathbb{R}$.

**2.**   Find the density function of $Y = aX$, where $a > 0$, in terms of the density function of $X$. Show that the continuous random variables $X$ and $-X$ have the same distribution function if and only if $f_X(x) = f_X(-x)$ for all $x \in \mathbb{R}$.

**3.**   If $f$ and $g$ are density functions of random variables $X$ and $Y$, show that $\alpha f + (1 - \alpha)g$ is a density function for $0 \le \alpha \le 1$, and describe a random variable of which it is the density function.

**4.**   **Survival.** Let $X$ be a positive random variable with density function $f$ and distribution function $F$. Define the *hazard function* $H(x) = -\log[1 - F(x)]$ and the *hazard rate*

$$r(x) = \lim_{h \downarrow 0} \frac{1}{h} \mathbb{P}(X \le x + h \mid X > x), \qquad x \ge 0.$$

Show that:
(a) $r(x) = H'(x) = f(x)/\{1 - F(x)\}$,
(b) If $r(x)$ increases with $x$ then $H(x)/x$ increases with $x$,
(c) $H(x)/x$ increases with $x$ if and only if $[1 - F(x)]^{\alpha} \le 1 - F(\alpha x)$ for all $0 \le \alpha \le 1$,
(d) If $H(x)/x$ increases with $x$, then $H(x + y) \ge H(x) + H(y)$ for all $x, y \ge 0$.

---

## 4.2  Independence

This section contains the counterpart of Section 3.2 for continuous variables, though it contains a definition and theorem which hold for any pair of variables, regardless of their types (continuous, discrete, and so on). We cannot continue to define the independence of $X$ and $Y$ in terms of events such as $\{X = x\}$ and $\{Y = y\}$, since these events have zero probability and are trivially independent.

**(1) Definition.** Random variables $X$ and $Y$ are called **independent** if

**(2)**            $\{X \leq x\}$   and   $\{Y \leq y\}$   are independent events for all $x, y \in \mathbb{R}$.

The reader should verify that discrete variables satisfy (2) if and only if they are independent in the sense of Section 3.2. Definition (1) is the general definition of the independence of any two variables $X$ and $Y$, regardless of their types. The following general result holds for the independence of functions of random variables. Let $X$ and $Y$ be random variables, and let $g, h : \mathbb{R} \to \mathbb{R}$. Then $g(X)$ and $h(Y)$ are functions which map $\Omega$ into $\mathbb{R}$ by

$$g(X)(\omega) = g(X(\omega)), \qquad h(Y)(\omega) = h(Y(\omega))$$

as in Theorem (3.2.3). Let us suppose that $g(X)$ and $h(Y)$ are random variables. (This holds if they are $\mathcal{F}$-measurable; it is valid for instance if $g$ and $h$ are sufficiently smooth or regular by being, say, continuous or monotonic. The correct condition on $g$ and $h$ is actually that, for all Borel subsets $B$ of $\mathbb{R}$, $g^{-1}(B)$ and $h^{-1}(B)$ are Borel sets also.) In the rest of this book, *we assume that any expression of the form '$g(X)$', where $g$ is a function and $X$ is a random variable, is itself a random variable.*

**(3) Theorem.** *If $X$ and $Y$ are independent, then so are $g(X)$ and $h(Y)$.*

Move immediately to the next section unless you want to prove this.

**Proof.** Some readers may like to try and prove this on their second reading. The proof does not rely on any property such as continuity. The key lies in the requirement of Definition (2.1.3) that random variables be $\mathcal{F}$-measurable, and in the observation that $g(X)$ is $\mathcal{F}$-measurable if $g : \mathbb{R} \to \mathbb{R}$ is *Borel measurable*, which is to say that $g^{-1}(B) \in \mathcal{B}$, the Borel $\sigma$-field, for all $B \in \mathcal{B}$. Complete the proof yourself (*exercise*).                                ∎

---

## Exercises for Section 4.2

**1.**    I am selling my house, and have decided to accept the first offer exceeding £$K$. Assuming that offers are independent random variables with common distribution function $F$, find the expected number of offers received before I sell the house.

**2.**    Let $X$ and $Y$ be independent random variables with common distribution function $F$ and density function $f$. Show that $V = \max\{X, Y\}$ has distribution function $\mathbb{P}(V \leq x) = F(x)^2$ and density function $f_V(x) = 2f(x)F(x)$, $x \in \mathbb{R}$. Find the density function of $U = \min\{X, Y\}$.

**3.**    The annual rainfall figures in Bandrika are independent identically distributed continuous random variables $\{X_r : r \geq 1\}$. Find the probability that:
(a) $X_1 < X_2 < X_3 < X_4$,
(b) $X_1 > X_2 < X_3 < X_4$.

**4.**    Let $\{X_r : r \geq 1\}$ be independent and identically distributed with distribution function $F$ satisfying $F(y) < 1$ for all $y$, and let $Y(y) = \min\{k : X_k > y\}$. Show that

$$\lim_{y \to \infty} \mathbb{P}\big(Y(y) \leq \mathbb{E}Y(y)\big) = 1 - e^{-1}.$$

**5.**    **Peripheral points.** Let $P_i = (X_i, Y_i)$, $1 \leq i \leq n$, be independent, uniformly distributed points in the unit square $[0, 1]^2$. A point $P_i$ is called *peripheral* if, for all $r = 1, 2, \ldots, n$, either $X_r \leq X_i$ or $Y_r \leq Y_i$, or both. Show that the mean number of peripheral points is $n\left(\frac{3}{4}\right)^{n-1}$.

**6.** Let $U$ and $V$ be independent and uniformly distributed on the interval $[0, 1]$.

(a) Show that

$$\mathbb{P}(x < V < U^2) = \tfrac{1}{3} - x + \tfrac{2}{3}x^{3/2}, \qquad x \in [0, 1).$$

(b) Find the conditional density function of $V$ given $U^2 > V$.

(c) Find the probability that the equation $x^2 + 2Ux + V = 0$ has two distinct real roots.

(d) Given that the two roots $R_1$ and $R_2$ are real and distinct, find the probability that both roots have absolute value less than 1.

**7.   Random hemispheres.**

(a) We select $n$ points independently and uniformly at random on the perimeter of a circle. What is the probability that they all lie within some semicircle?

(b) This time we place our $n$ points uniformly on the surface of a sphere in $\mathbb{R}^3$. Show that they all lie within some hemisphere with probability $(n^2 - n + 2)2^{-n}$.

# 4.3  Expectation

The expectation of a discrete variable $X$ is $\mathbb{E}X = \sum_x x \mathbb{P}(X = x)$. This is an average of the possible values of $X$, each value being weighted by its probability. For continuous variables, expectations are defined as integrals.

**(1) Definition.** The **expectation** (or **mean**) of a continuous random variable $X$ with density function $f$ is given by

$$\mathbb{E}X = \int_{-\infty}^{\infty} xf(x)\, dx$$

whenever this integral exists.

There are various ways of defining the integral of a function $g : \mathbb{R} \to \mathbb{R}$, but it is not appropriate to explore this here. Note that usually we shall allow the existence of $\int g(x)\, dx$ only if $\int |g(x)|\, dx < \infty$.

The expectation of a discrete random variable is a sum, and of a continuous random variable an integral. The notation for expectation may be unified by use of the notation $\mathbb{E}X = \int x\, dF_X(x)$, where this is interpreted as $\sum x f(x)$ for a discrete random variable with mass function $f$, and as $\int xf(x)\, dx$ for a continuous random variable with density function $f$. This notation is suitable also for the expectations of random variables that are neither discrete nor continuous. This is made formal in the forthcoming Section 5.6.

**(2) Examples (2.3.4) and (4.1.3) revisited.** The random variables $X$ and $Y$ of these examples have mean values

$$\mathbb{E}(X) = \int_0^{2\pi} \frac{x}{2\pi}\, dx = \pi, \quad \mathbb{E}(Y) = \int_0^{4\pi^2} \frac{\sqrt{y}}{4\pi}\, dy = \tfrac{4}{3}\pi^2. \qquad \bullet$$

Roughly speaking, the expectation operator $\mathbb{E}$ has the same properties for continuous variables as it has for discrete variables.

**(3) Theorem. Change of variable formula.** *If $X$ and $g(X)$ are continuous random variables then*

$$\mathbb{E}(g(X)) = \int_{-\infty}^{\infty} g(x)f_X(x)\, dx.$$

We give a simple proof for the case when $g$ takes only non-negative values, and we leave it to the reader to extend this to the general case. Our proof is a corollary of the next lemma.

**(4) Lemma.** *If $X$ has density function $f$ with $f(x) = 0$ when $x < 0$, and distribution function $F$, then*

$$\mathbb{E}X = \int_0^\infty [1 - F(x)]\, dx.$$

**Proof.**

$$\int_0^\infty [1 - F(x)]\, dx = \int_0^\infty \mathbb{P}(X > x)\, dx = \int_0^\infty \int_{y=x}^\infty f(y)\, dy\, dx.$$

Now change the order of integration in the last term.                              ∎

**Proof of (3).** Assume first that $g \geq 0$. By (4),

$$\mathbb{E}(g(X)) = \int_0^\infty \mathbb{P}\big(g(X) > x\big)\, dx = \int_0^\infty \left( \int_B f_X(y)\, dy \right) dx$$

where $B = \{y : g(y) > x\}$. We interchange the order of integration here to obtain

$$\mathbb{E}(g(X)) = \int_{-\infty}^\infty \int_0^{g(y)} dx\, f_X(y)\, dy = \int_{-\infty}^\infty g(y) f_X(y)\, dy.$$

For a general function $g : \mathbb{R} \to \mathbb{R}$, we write $g = g^+ - g^-$ where $g^+(x) = \max\{0, g(x)\}$ and $g^-(x) = -\min\{0, g(x)\}$. Since $g^+, g^- \geq 0$, we have that

$$\mathbb{E}(g(X)) = \int_{-\infty}^\infty g^+(y) f_X(y)\, dy - \int_{-\infty}^\infty g^-(y) f_X(y)\, dy = \int_{-\infty}^\infty g(y) f_X(y)\, dy,$$

as required.                                                                          ∎

**(5) Example (2) continued.** Lemma (4) enables us to find $\mathbb{E}(Y)$ without calculating $f_Y$, for

$$\mathbb{E}(Y) = \mathbb{E}(X^2) = \int_0^{2\pi} x^2 f_X(x)\, dx = \int_0^{2\pi} \frac{x^2}{2\pi}\, dx = \tfrac{4}{3}\pi^2.$$     ●

We were careful to describe many characteristics of discrete variables—such as moments, covariance, correlation, and linearity of $\mathbb{E}$ (see Sections 3.3 and 3.6)—in terms of the operator $\mathbb{E}$ itself. Exactly analogous discussion holds for continuous variables. We do not spell out the details here but only indicate some of the less obvious emendations required to establish these results. For example, Definition (3.3.5) defines the $k$th moment of the discrete variable $X$ to be

**(6)**                                      $m_k = \mathbb{E}(X^k);$

we define the $k$th moment of a continuous variable $X$ by the same equation. Of course, the moments of $X$ may not exist since the integral

$$\mathbb{E}(X^k) = \int x^k f(x)\, dx$$

may not converge (see Example (4.4.7) for an instance of this).

## Exercises for Section 4.3

**1.**   For what values of $\alpha$ is $\mathbb{E}(|X|^\alpha)$ finite, if the density function of $X$ is:
(a) $f(x) = e^{-x}$ for $x \geq 0$,
(b) $f(x) = C(1 + x^2)^{-m}$ for $x \in \mathbb{R}$?
If $\alpha$ is not integral, then $\mathbb{E}(|X|^\alpha)$ is called the *fractional moment of order* $\alpha$ of $X$, whenever the expectation is well defined; see Exercise (3.3.5).

**2.**   Let $X_1, X_2, \ldots, X_n$ be independent identically distributed random variables, and let $S_m = X_1 + X_2 + \cdots + X_m$. Assume that $\mathbb{P}(S_n = 0) = 0$. Show that, if $m \leq n$, then $\mathbb{E}(S_m/S_n) = m/n$.

**3.**   Let $X$ be a non-negative random variable with density function $f$. Show that

$$\mathbb{E}(X^r) = \int_0^\infty r x^{r-1} \mathbb{P}(X > x) \, dx$$

for any $r \geq 1$ for which the expectation is finite.

**4.**   **Mallows's inequality.** Show that the mean $\mu$, median $m$, and variance $\sigma^2$ of the continuous random variable $X$ satisfy $(\mu - m)^2 \leq \sigma^2$. [It can be shown that $|\mu - m| \leq \sigma\sqrt{0.6}$. See Basu and Dasgupta 1997.]

**5.**   Let $X$ be a random variable with mean $\mu$ and continuous distribution function $F$.
(a) Show that

$$\int_{-\infty}^a F(x) \, dx = \int_a^\infty [1 - F(x)] \, dx,$$

if and only if $a = \mu$.
(b) Show that the set of medians of $X$ is the set of all $a$ for which $\mathbb{E}|X - a|$ is a minimum.

**6.**   Let $X_1, X_2, \ldots, X_n$ be non-negative random variables with finite means. Show that

$$\mathbb{E}(\max_j X_j) = \sum_j \mathbb{E}(X_j) - \sum_{i<j} \mathbb{E}(\min\{X_i, X_j\}) + \cdots + (-1)^{n+1}\mathbb{E}(\min_j X_j).$$

**7.**   **Tails and moments.** If $X$ is a continuous random variable and $\mathbb{E}(X^r)$ exists, where $r \geq 1$ is an integer, show that

$$\int_0^\infty x^{r-1}\mathbb{P}(|X| > x) \, dx < \infty, \quad \text{and} \quad x^r \mathbb{P}(|X| > x) \to 0 \quad \text{as } x \to \infty.$$

**8.**   **Integral inequality.** Let $X$ be a random variable taking values in $[0, a]$, with density function $f$ satisfying $f(x) \leq b$ for all $x$. Show that $\mathbb{E}(X) \geq (2b)^{-1}$. More generally, show for $n \geq 1$ that $\mathbb{E}(X^n) \geq 1/[(n+1)b^n]$.
[Hint: You may find a form of Steffensen's integral inequality to be useful, namely: for integrable functions $g$ and $h$ satisfying $0 \leq h(x) \leq 1$ and $g$ increasing on $[0, a]$, we have

$$\int_0^s g(x) \, dx \leq \int_0^a g(x)h(x) \, dx,$$

where $s = \int_0^a h(x) \, dx$.]

**9.**   Let $X$ be continuous with finite variance. Show that $g(a) = \mathbb{E}((X - a)^2)$ is a minimum when $a = \mathbb{E}(X)$.

**10. Archery.** A target comprising a disc with centre O and unit radius is hit by $n$ arrows. They strike independent spots on the target, and for each the chance of hitting within distance $a \in (0, 1)$ of O is $a^2$. Let $R$ be the radius of the smallest circle centred at O that covers all the arrow strikes. Show that $R$ has density $f(r) = 2nr^{2n-1}, 0 \leq r \leq 1$, and find the mean area of this circle.

   The arrow furthest from O falls out. Show that the area of the smallest circle centred at O and covering the remaining arrows is $(n - 1)\pi/(n + 1)$.

**11. Johnson–Rogers inequality.** Let $X$ be continuous with variance $\sigma^2$, and also unimodal about a mode $M$. Show that $|M - \mathbb{E}X| \leq \sigma\sqrt{3}$, with equality if and only if $X$ is uniformly distributed. [You may use without proof the fact that $X$, with distribution function $F$, is unimodal about 0 if and only if there exist independent random variables $U$ and $Y$, where $U$ is uniformly distributed on $(0, 1)$, such that $UY$ has distribution function $F$; this is called Khinchin's representation.]

   What can you say about $|M - m|$ where $m$ is a median of $X$?

**12.** A football is placed at a uniformly distributed position $W$ in the unit interval $[0, 1]$, and kicked with a non-zero random velocity $V$ having a distribution that is symmetric about 0. Assuming that $V$ and $W$ are independent, show (neglecting air resistance) that the time $T$ at which the football hits an endpoint of the interval is unimodal.

**13. Continuation.** Here are two further questions concerning the football of the previous exercise.

(a)  Suppose that the football's speed $|V|$ has density function $g(u) = u^{-3}e^{-1/u}$ for $u > 0$. Show that the hitting time $T$ is exponentially distributed with parameter 1.

(b)  Find the density function of $T$ when $V$ has the so-called Laplace density function $g(v) = \frac{1}{2}e^{-|v|}$.

---

## 4.4 Examples of continuous variables

**(1) Uniform distribution.** The random variable $X$ is *uniform* on $[a, b]$ (or, equivalently, on the open interval $(a, b)$) if it has density function $f$ and distribution function $F$ given by

$$f(x) = \begin{cases} \dfrac{1}{b-a} & \text{if } a < x < b, \\ 0 & \text{otherwise,} \end{cases} \qquad F(x) = \begin{cases} 0 & \text{if } x \leq a, \\ \dfrac{x-a}{b-a} & \text{if } a < x \leq b, \\ 1 & \text{if } x > b. \end{cases}$$

Very roughly speaking, $X$ takes any value between $a$ and $b$ with equal probability. See Example (2.3.4) for an example of a uniform random variable $X$.                    ●

**(2) Exponential distribution.** The random variable $X$ is *exponential* with parameter $\lambda\ (> 0)$ if it has density function $f$ and distribution function $F$ given by

**(3)**                 $$f(x) = \lambda e^{-\lambda x}, \quad F(x) = 1 - e^{-\lambda x}, \qquad \text{for } x \geq 0.$$

This arises as the 'continuous limit' of the waiting time distribution of Example (3.5.5) and very often occurs in practice as a description of the time elapsing between unpredictable events (such as telephone calls, earthquakes, emissions of radioactive particles, and arrivals of buses, girls, and so on). Suppose, as in Example (3.5.5), that a sequence of Bernoulli trials is performed at time epochs $\delta, 2\delta, 3\delta, \dots$ and let $W$ be the waiting time for the first success. Then

$$\mathbb{P}(W > k\delta) = (1 - p)^k \qquad \text{and} \qquad \mathbb{E}W = \delta/p.$$

Now fix a time $t$. By this time, roughly $k = t/\delta$ trials have been made. We shall let $\delta \downarrow 0$. In order that the limiting distribution $\lim_{\delta \downarrow 0} \mathbb{P}(W > t)$ be non-trivial, we shall need to assume that $p \downarrow 0$ also and that $p/\delta$ approaches some positive constant $\lambda$. Then

$$\mathbb{P}(W > t) = \mathbb{P}\left(W > \left(\frac{t}{\delta}\right)\delta\right) \simeq (1 - \lambda\delta)^{t/\delta} \to e^{-\lambda t}$$

which yields (3).

The exponential distribution (3) has mean

$$\mathbb{E}X = \int_0^\infty [1 - F(x)] \, dx = \frac{1}{\lambda}.$$

Further properties of the exponential distribution will be discussed in Section 4.7 and Problem (4.14.5); this distribution proves to be the cornerstone of the theory of Markov processes in continuous time, to be discussed later.                                                      ●

**(4) Normal distribution.**   Arguably the most important continuous distribution is the *normal*†
(or *Gaussian*) distribution, which has two parameters $\mu$ and $\sigma^2$ and density function

$$f(x) = \frac{1}{\sqrt{2\pi\sigma^2}} \exp\left(-\frac{(x-\mu)^2}{2\sigma^2}\right), \qquad -\infty < x < \infty.$$

It is denoted by $N(\mu, \sigma^2)$. If $\mu = 0$ and $\sigma^2 = 1$ then

$$f(x) = \frac{1}{\sqrt{2\pi}} e^{-\frac{1}{2}x^2}, \qquad -\infty < x < \infty,$$

is the density of the *standard* normal distribution. It is an *exercise* in analysis (Problem (4.14.1)) to show that $f$ satisfies $\int_{-\infty}^\infty f(x) \, dx = 1$, and is indeed therefore a density function.

The normal distribution arises in many ways. In particular it can be obtained as a continuous limit of the binomial distribution $\text{bin}(n, p)$ as $n \to \infty$ (this is the 'de Moivre–Laplace limit theorem'). This result is a special case of the central limit theorem to be discussed in Chapter 5; it transpires that in many cases the sum of a large number of independent (or at least not too dependent) random variables is approximately normally distributed. The binomial random variable has this property because it is the sum of Bernoulli variables (see Example (3.5.2)).

Let $X$ be $N(\mu, \sigma^2)$, where $\sigma > 0$, and let

(5)                              $$Y = \frac{X - \mu}{\sigma}.$$

For the distribution of $Y$,

$$\mathbb{P}(Y \le y) = \mathbb{P}\big((X - \mu)/\sigma \le y\big) = \mathbb{P}(X \le y\sigma + \mu)$$

$$= \frac{1}{\sigma\sqrt{2\pi}} \int_{-\infty}^{y\sigma+\mu} \exp\left(-\frac{(x-\mu)^2}{2\sigma^2}\right) dx$$

$$= \frac{1}{\sqrt{2\pi}} \int_{-\infty}^{y} e^{-\frac{1}{2}v^2} \, dv \quad \text{by substituting } x = v\sigma + \mu.$$

---

†Probably first named 'normal' by Francis Galton before 1885, though some attribute the name to C. S. Peirce, who is famous for his erroneous remark "Probability is the only branch of mathematics in which good mathematicians frequently get results which are entirely wrong".

Thus $Y$ is $N(0, 1)$. Routine integrations (see Problem (4.14.1)) show that $\mathbb{E}Y = 0$, $\text{var}(Y) = 1$, and it follows immediately from (5) and Theorems (3.3.8), (3.3.11) that the mean and variance of the $N(\mu, \sigma^2)$ distribution are $\mu$ and $\sigma^2$ respectively, thus explaining the notation.

Traditionally we denote the density and distribution functions of $Y$ by $\phi$ and $\Phi$:

$$\phi(v) = \frac{1}{\sqrt{2\pi}} e^{-\frac{1}{2}v^2}, \qquad \Phi(y) = \mathbb{P}(Y \le y) = \int_{-\infty}^{y} \phi(v) \, dv. \qquad \bullet$$

**(6) Gamma distribution.** The random variable $X$ has the *gamma* distribution with parameters $\lambda, t > 0$, denoted† $\Gamma(\lambda, t)$, if it has density

$$f(x) = \frac{1}{\Gamma(t)} \lambda^t x^{t-1} e^{-\lambda x}, \qquad x \ge 0.$$

Here, $\Gamma(t)$ is the *gamma function*

$$\Gamma(t) = \int_0^\infty x^{t-1} e^{-x} \, dx.$$

If $t = 1$ then $X$ is exponentially distributed with parameter $\lambda$. We remark that if $\lambda = \frac{1}{2}$, $t = \frac{1}{2}d$, for some integer $d$, then $X$ is said to have the *chi-squared distribution* $\chi^2(d)$ with $d$ degrees of freedom (see Problem (4.14.12)). $\qquad \bullet$

**(7) Cauchy distribution.** The random variable $X$ has the *Cauchy* distribution‡ if it has density function

$$f(x) = \frac{1}{\pi(1 + x^2)}, \qquad -\infty < x < \infty.$$

This distribution is notable for having no moments and for its frequent appearances in counter-examples (but see Problem (4.14.4)). $\qquad \bullet$

**(8) Beta distribution.** The random variable $X$ is *beta*, parameters $a, b > 0$, if it has density function

$$f(x) = \frac{1}{B(a, b)} x^{a-1}(1 - x)^{b-1}, \qquad 0 \le x \le 1.$$

We denote this distribution by $\beta(a, b)$. The 'beta function'

$$B(a, b) = \int_0^1 x^{a-1}(1 - x)^{b-1} \, dx$$

is chosen so that $f$ has total integral equal to one. You may care to prove that $B(a, b) = \Gamma(a)\Gamma(b)/\Gamma(a + b)$. If $a = b = 1$ then $X$ is uniform on $[0, 1]$. $\qquad \bullet$

**(9) Weibull distribution.** The random variable $X$ is *Weibull*, parameters $\alpha, \beta > 0$, if it has distribution function

$$F(x) = 1 - \exp(-\alpha x^\beta), \qquad x \ge 0.$$

Differentiate to find that

$$f(x) = \alpha\beta x^{\beta-1} \exp(-\alpha x^\beta), \qquad x \ge 0.$$

Set $\beta = 1$ to obtain the exponential distribution. $\qquad \bullet$

---

†Do not confuse the order of the parameters. Some authors denote this distribution $\Gamma(t, \lambda)$.

‡This distribution was considered earlier by Poisson, and the name is another example of Stigler's law of eponymy.

## Exercises for Section 4.4

**1.** Prove that the gamma function satisfies $\Gamma(t) = (t-1)\Gamma(t-1)$ for $t > 1$, and deduce that $\Gamma(n) = (n-1)!$ for $n = 1, 2, \ldots$. Show that $\Gamma(\frac{1}{2}) = \sqrt{\pi}$ and deduce a closed form for $\Gamma(n + \frac{1}{2})$ for $n = 0, 1, 2, \ldots$.

**2.** Show, as in paragraph (4.4.8), that the beta function satisfies $B(a, b) = \Gamma(a)\Gamma(b)/\Gamma(a + b)$.

**3.** Let $X$ have the uniform distribution on $[0, 1]$. For what function $g$ does $Y = g(X)$ have the exponential distribution with parameter 1?

**4.** Find the distribution function of a random variable $X$ with the Cauchy distribution. For what values of $\alpha$ does $|X|$ have a finite (possibly fractional) moment of order $\alpha$?

**5.** **Log-normal distribution.** Let $Y = e^X$ where $X$ has the $N(0, 1)$ distribution. Find the density function of $Y$.

**6.** **Stein's identity.** Let $X$ be $N(\mu, \sigma^2)$. Show that $\mathbb{E}\{(X - \mu)g(X)\} = \sigma^2\mathbb{E}(g'(X))$ when both sides exist.

**7.** With the terminology of Exercise (4.1.4), find the hazard rate when:
(a) $X$ has the Weibull distribution, $\mathbb{P}(X > x) = \exp(-\alpha x^{\beta-1})$, $x \geq 0$,
(b) $X$ has the exponential distribution with parameter $\lambda$,
(c) $X$ has density function $\alpha f + (1 - \alpha)g$, where $0 < \alpha < 1$ and $f$ and $g$ are the densities of exponential variables with respective parameters $\lambda$ and $\mu$. What happens to this last hazard rate $r(x)$ in the limit as $x \to \infty$?

**8.** **Mills's ratio.** (a) For the standard normal density function $\phi(x)$, show that $\phi'(x) + x\phi(x) = 0$. Hence show that *Mills's ratio* $M(x) = (1 - \Phi(x))/\phi(x)$ satisfies

$$\frac{1}{x} - \frac{1}{x^3} < M(x) < \frac{1}{x} - \frac{1}{x^3} + \frac{3}{x^5}, \qquad x > 0.$$

Here, $\Phi$ denotes the $N(0, 1)$ distribution function.
(b) Let $X$ have the $N(\mu, \sigma^2)$ distribution where $\sigma^2 > 0$, and show that

$$\mathbb{E}(X \mid X > c) = \mu + \frac{\sigma}{M((c - \mu)/\sigma)},$$

$$\mathbb{E}(X \mid X < c) = \mu - \frac{\sigma}{M((c - \mu)/\sigma)}.$$

**9.** **Ordered exponentials.** Let $U, V, W$ be independent, exponentially distributed random variables with respective parameters $\lambda, \mu, \nu$. Show that

$$\mathbb{P}(U \leq V \leq W) = \frac{\lambda\mu\nu}{\nu(\nu + \mu)(\nu + \mu + \lambda)}.$$

**10.** Let $U$ and $X$ be independent, where $U$ is uniform on $(0, 1)$ and $X$ is exponentially distributed with parameter $\lambda$. Show that $\mathbb{E}(\min\{U, X\}) = \lambda^{-1}e^{-\lambda} - \lambda^{-2}(1 - e^{-\lambda})$.

**11.** **Pareto distribution.** (a) Let $X$ be uniformly distributed on $[0, 1]$, and let $a > 0$. Find the distribution of $Y = aX/(1 - X)$.
(b) Let $Y_1, Y_2, \ldots, Y_n$ be independent random variables with the same distribution as $Y$. Show that $S = \min\{Y_1, Y_2, \ldots, Y_n\}$ has the Pareto density function

$$f(s) = \frac{n}{a}\left(\frac{a}{a + s}\right)^{n+1}, \qquad s > 0,$$

and find its mean value.

(c) Find the distribution function of $T = \max\{Y_1, Y_2, \ldots, Y_n\}$.

**12. Arc sine distribution.** Let $X$ have the Cauchy distribution, and show that $Y = 1/(1 + X^2)$ has the arc sine density function

$$f_Y(y) = \frac{2}{\pi \sqrt{1 - y^2}}, \qquad 0 \le y \le 1.$$

**13.** Let $X_1, X_2, \ldots$ be independent and uniformly distributed on $[-c, c]$.
(a) Find expressions for the probability that
    (i)  $X_i \ge b$ for $1 \le i \le n$,
    (ii) $X_i \le b$ for $1 \le i \le n$,
where $b \in [-c, c]$.
(b) Show that the median $Z$ of $X_1, X_2, \ldots, X_{2n+1}$ (that is, the middle value) has density function

$$f_Z(z) = \frac{(2n + 1)!}{(n!)^2} \cdot \frac{(c^2 - z^2)^n}{(2c)^{2n+1}}, \qquad z \in [-c, c].$$

(c) Deduce the value of the integral $\int_{-c}^{c} (c^2 - z^2) \, dz$.
(d) Calculate the mean and variance of $Z$.

**14. Beta distribution of the second kind.** Let $X$ have the beta distribution $\beta(a, b)$. Show that $Y = X/(1 - X)$ has density function

$$f(y) = \frac{\Gamma(a + b)}{\Gamma(a)\Gamma(b)} \cdot \frac{y^{a-1}}{(1 + y)^{a+b}}, \qquad y > 0.$$

This is called the *beta distribution of the second kind*. Find $\mathbb{E}(Y^n)$ where $1 \le n < b$.

**15.** Let $Z$ have the $\Gamma(1, t)$ distribution of paragraph (4.4.6). Show that

$$\mathbb{E}\big(|Z - t|/\sqrt{t}\big) = \frac{2e^{-t}t^{t-\frac{1}{2}}}{\Gamma(t)}.$$

# 4.5  Dependence

Many interesting probabilistic statements about a pair $X, Y$ of variables concern the way $X$ and $Y$ vary together as functions on the same domain $\Omega$.

**(1) Definition.** The **joint distribution function** of $X$ and $Y$ is the function $F : \mathbb{R}^2 \to [0, 1]$ given by
$$F(x, y) = \mathbb{P}(X \le x, \ Y \le y).$$

If $X$ and $Y$ are continuous then we cannot talk of their joint mass function (see Definition (3.6.2)) since this is identically zero. Instead we need another density function.

**(2) Definition.** The random variables $X$ and $Y$ are **(jointly) continuous** with **joint (proba-bility) density function** $f : \mathbb{R}^2 \to [0, \infty)$ if
$$F(x, y) = \int_{v=-\infty}^{y} \int_{u=-\infty}^{x} f(u, v) \, du \, dv \qquad \text{for each } x, y \in \mathbb{R}.$$

If $F$ is sufficiently differentiable at the point $(x, y)$, then we usually specify

$$f(x, y) = \frac{\partial^2}{\partial x \partial y} F(x, y).$$

The properties of joint distribution and density functions are very much the same as those of the corresponding functions of a single variable, and the reader is left to find them. We note the following facts. Let $X$ and $Y$ have joint distribution function $F$ and joint density function $f$. (Sometimes we write $F_{X,Y}$ and $f_{X,Y}$ to stress the roles of $X$ and $Y$.)

**(3) Probabilities.**

$$\mathbb{P}(a \le X \le b, \ c \le Y \le d) = F(b, d) - F(a, d) - F(b, c) + F(a, c)$$

$$= \int_{y=c}^{d} \int_{x=a}^{b} f(x, y) \, dx \, dy.$$

Think of $f(x, y) \, dx dy$ as the element of probability $\mathbb{P}(x < X \le x+dx, \ y < Y \le y+dy)$, so that if $B$ is a sufficiently nice subset of $\mathbb{R}^2$ (such as a rectangle or a union of rectangles and so on) then

**(4)**
$$\mathbb{P}\big((X, Y) \in B\big) = \iint_B f(x, y) \, dx \, dy.$$

We can think of $(X, Y)$ as a point chosen randomly from the plane; then $\mathbb{P}\big((X, Y) \in B\big)$ is the probability that the outcome of this random choice lies in the subset $B$.

**(5) Marginal distributions.** The *marginal distribution functions* of $X$ and $Y$ are

$$F_X(x) = \mathbb{P}(X \le x) = F(x, \infty), \qquad F_Y(y) = \mathbb{P}(Y \le y) = F(\infty, y),$$

where $F(x, \infty)$ is shorthand for $\lim_{y \to \infty} F(x, y)$; now,

$$F_X(x) = \int_{-\infty}^{x} \left( \int_{-\infty}^{\infty} f(u, y) \, dy \right) du$$

and it follows that the *marginal density function* of $X$ is

$$f_X(x) = \int_{-\infty}^{\infty} f(x, y) \, dy.$$

Similarly, the *marginal density function*† of $Y$ is

$$f_Y(y) = \int_{-\infty}^{\infty} f(x, y) \, dx.$$

---

†If $X$ and $Y$ are jointly continuous, then each is continuous. It does not follow that the marginal density functions are themselves continuous functions. See Exercise (4.5.1).

**(6) Expectation.** If $g : \mathbb{R}^2 \to \mathbb{R}$ is a sufficiently nice function (see the proof of Theorem (4.2.3) for an idea of what this means) then

$$\mathbb{E}(g(X, Y)) = \int_{-\infty}^{\infty} \int_{-\infty}^{\infty} g(x, y) f(x, y) \, dx \, dy;$$

in particular, setting $g(x, y) = ax + by$,

$$\mathbb{E}(aX + bY) = a\mathbb{E}X + b\mathbb{E}Y.$$

**(7) Independence.** The random variables $X$ and $Y$ are *independent* if and only if

$$F(x, y) = F_X(x) F_Y(y) \qquad \text{for all} \quad x, y \in \mathbb{R},$$

which, for continuous random variables, is equivalent to requiring that

$$f(x, y) = f_X(x) f_Y(y)$$

whenever $F$ is differentiable at $(x, y)$ (see Problem (4.14.6) also) where $f$, $f_X$, $f_Y$ are taken to be the appropriate derivatives of $F$, $F_X$ and $F_Y$. Continuous random variables are uncorrelated whenever they are independent (see Definitions (3.3.10) and (3.6.7)).

**(8) Example. Buffon's needle.** A plane is ruled by the lines $y = n$ $(n = 0, \pm1, \pm2, \dots)$ and a needle of unit length is cast randomly on to the plane. What is the probability that it intersects some line? We suppose that the needle shows no preference for position or direction.

**Solution.** Let $(X, Y)$ be the coordinates of the centre of the needle and let $\Theta$ be the angle, modulo $\pi$, made by the needle and the $x$-axis. Denote the distance from the needle's centre and the nearest line beneath it by $Z = Y - \lfloor Y \rfloor$, where $\lfloor Y \rfloor$ is the greatest integer not greater than $Y$. We need to interpret the statement 'a needle is cast randomly', and do this by assuming that:

(a) $Z$ is uniformly distributed on $[0, 1]$, so that $f_Z(z) = 1$ if $0 \le z \le 1$,
(b) $\Theta$ is uniformly distributed on $[0, \pi]$, so that $f_\Theta(\theta) = 1/\pi$ if $0 \le \theta \le \pi$,
(c) $Z$ and $\Theta$ are independent, so that $f_{Z,\Theta}(z, \theta) = f_Z(z) f_\Theta(\theta)$.

Thus the pair $Z$, $\Theta$ has joint density function $f(z, \theta) = 1/\pi$ for $0 \le z \le 1, 0 \le \theta \le \pi$. Draw a diagram to see that an intersection occurs if and only if $(Z, \Theta) \in B$ where $B \subseteq [0, 1] \times [0, \pi]$ is given by

$$B = \left\{ (z, \theta) : z \le \tfrac{1}{2} \sin\theta \text{ or } 1 - z \le \tfrac{1}{2} \sin\theta \right\}.$$

Hence

$$\mathbb{P}(\text{intersection}) = \iint_B f(z, \theta) \, dz \, d\theta = \frac{1}{\pi} \int_0^\pi \left( \int_0^{\frac{1}{2}\sin\theta} dz + \int_{1-\frac{1}{2}\sin\theta}^1 dz \right) d\theta = \frac{2}{\pi}.$$

Buffon† designed the experiment in order to estimate the numerical value of $\pi$. Try it if you have time.                                                                  ●

---

†Georges LeClerc, Comte de Buffon. In 1777 he investigated the St Petersburg problem by flipping a coin 2084 times, perhaps the first recorded example of a Monte Carlo method in use.

**(9) Example. Bivariate normal distribution.** Let $f : \mathbb{R}^2 \to \mathbb{R}$ be given by

**(10)** $$f(x, y) = \frac{1}{2\pi\sqrt{1-\rho^2}} \exp\left(-\frac{1}{2(1-\rho^2)}(x^2 - 2\rho xy + y^2)\right)$$

where $\rho$ is a constant satisfying $-1 < \rho < 1$. Check that $f$ is a joint density function by verifying that

$$f(x, y) \geq 0, \qquad \int_{-\infty}^{\infty}\int_{-\infty}^{\infty} f(x, y)\,dx\,dy = 1;$$

$f$ is called the *standard bivariate normal* density function of some pair $X$ and $Y$. Calculation of its marginals shows that $X$ and $Y$ are $N(0, 1)$ variables (*exercise*). Furthermore, the covariance

$$\text{cov}(X, Y) = \mathbb{E}(XY) - \mathbb{E}(X)\mathbb{E}(Y)$$

is given by

$$\text{cov}(X, Y) = \int_{-\infty}^{\infty}\int_{-\infty}^{\infty} xyf(x, y)\,dx\,dy = \rho;$$

you should check this. Remember that independent variables are uncorrelated, but the converse is not true in general. In this case, however, if $\rho = 0$ then

$$f(x, y) = \left(\frac{1}{\sqrt{2\pi}}e^{-\frac{1}{2}x^2}\right)\left(\frac{1}{\sqrt{2\pi}}e^{-\frac{1}{2}y^2}\right) = f_X(x)f_Y(y)$$

and so $X$ and $Y$ are independent. We reach the following important conclusion. *Standard bivariate normal variables are independent if and only if they are uncorrelated.*

The general bivariate normal distribution is more complicated. We say that the pair $X$, $Y$ has the bivariate normal distribution with means $\mu_1$ and $\mu_2$, variances $\sigma_1^2$ and $\sigma_2^2$, and correlation $\rho$ if their joint density function is

$$f(x, y) = \frac{1}{2\pi\sigma_1\sigma_2\sqrt{1-\rho^2}} \exp\left[-\tfrac{1}{2}Q(x, y)\right]$$

where $\sigma_1, \sigma_2 > 0$ and $Q$ is the following quadratic form

$$Q(x, y) = \frac{1}{(1-\rho^2)}\left[\left(\frac{x-\mu_1}{\sigma_1}\right)^2 - 2\rho\left(\frac{x-\mu_1}{\sigma_1}\right)\left(\frac{y-\mu_2}{\sigma_2}\right) + \left(\frac{y-\mu_2}{\sigma_2}\right)^2\right].$$

Routine integrations (*exercise*) show that:
(a) $X$ is $N(\mu_1, \sigma_1^2)$ and $Y$ is $N(\mu_2, \sigma_2^2)$,
(b) the correlation between $X$ and $Y$ is $\rho$,
(c) *$X$ and $Y$ are independent if and only if $\rho = 0$.*
Finally, here is a hint about calculating integrals associated with normal density functions. It is an analytical exercise (Problem (4.14.1)) to show that

$$\int_{-\infty}^{\infty} e^{-\frac{1}{2}x^2}\,dx = \sqrt{2\pi}$$

and hence that

$$f(x) = \frac{1}{\sqrt{2\pi}} e^{-\frac{1}{2}x^2}$$

is indeed a density function. Similarly, a change of variables in the integral shows that the more general function

$$f(x) = \frac{1}{\sigma\sqrt{2\pi}} \exp\left[-\frac{1}{2}\left(\frac{x-\mu}{\sigma}\right)^2\right]$$

is itself a density function. This knowledge can often be used to shorten calculations. For example, let $X$ and $Y$ have joint density function given by (10). By completing the square in the exponent of the integrand, we see that

$$\mathrm{cov}(X, Y) = \iint xyf(x, y)\, dx\, dy$$

$$= \int y \frac{1}{\sqrt{2\pi}} e^{-\frac{1}{2}y^2}\left(\int xg(x, y)\, dx\right) dy$$

where

$$g(x, y) = \frac{1}{\sqrt{2\pi(1-\rho^2)}} \exp\left(-\frac{1}{2}\frac{(x-\rho y)^2}{(1-\rho^2)}\right)$$

is the density function of the $N(\rho y, 1-\rho^2)$ distribution. Therefore $\int xg(x, y)\, dx$ is the mean, $\rho y$, of this distribution, giving

$$\mathrm{cov}(X, Y) = \rho \int y^2 \frac{1}{\sqrt{2\pi}} e^{-\frac{1}{2}y^2}\, dy.$$

However, the integral here is, in turn, the variance of the $N(0, 1)$ distribution, and we deduce that $\mathrm{cov}(X, Y) = \rho$, as was asserted previously.    ●

**(11) Example.** Here is another example of how to manipulate density functions. Let $X$ and $Y$ have joint density function

$$f(x, y) = \frac{1}{y} \exp\left(-y - \frac{x}{y}\right), \qquad 0 < x, y < \infty.$$

Find the marginal density function of $Y$.

**Solution.** We have that

$$f_Y(y) = \int_{-\infty}^{\infty} f(x, y)\, dx = \int_0^{\infty} \frac{1}{y} \exp\left(-y - \frac{x}{y}\right) dx = e^{-y}, \qquad y > 0,$$

and hence $Y$ is exponentially distributed.    ●

Following the final paragraph of Section 4.3, we should note that the expectation operator $\mathbb{E}$ has similar properties when applied to a family of continuous variables as when applied to discrete variables. Consider just one example of this.

**(12) Theorem. Cauchy–Schwarz inequality.** *For any pair $X$, $Y$ of jointly continuous variables, we have that*

$$\{\mathbb{E}(XY)\}^2 \le \mathbb{E}(X^2)\mathbb{E}(Y^2),$$

*with equality if and only if $\mathbb{P}(aX = bY) = 1$ for some real $a$ and $b$, at least one of which is non-zero.*

**Proof.** Exactly as for Theorem (3.6.9).    ■

## Exercises for Section 4.5

**1. Clarke's example.** Let

$$f(x, y) = \frac{|x|}{\sqrt{8\pi}} \exp\{-|x| - \tfrac{1}{2}x^2 y^2\}, \qquad x, y \in \mathbb{R}.$$

Show that $f$ is a continuous joint density function, but that the (first) marginal density function $g(x) = \int_{-\infty}^{\infty} f(x, y)\, dy$ is not continuous. Let $Q = \{q_n : n \geq 1\}$ be a set of real numbers, and define

$$f_Q(x, y) = \sum_{n=1}^{\infty} (\tfrac{1}{2})^n f(x - q_n, y).$$

Show that $f_Q$ is a continuous joint density function whose first marginal density function is discontinuous at the points in $Q$. Can you construct a continuous joint density function whose first marginal density function is continuous nowhere?

**2. Buffon's needle revisited.** Two grids of parallel lines are superimposed: the first grid contains lines distance $a$ apart, and the second contains lines distance $b$ apart which are perpendicular to those of the first set. A needle of length $r$ ($< \min\{a, b\}$) is dropped at random. Show that the probability it intersects a line equals $r(2a + 2b - r)/(\pi ab)$.

**3. Buffon's cross.** The plane is ruled by the lines $y = n$, for $n = 0, \pm 1, \ldots$, and onto this plane we drop a cross formed by welding together two unit needles perpendicularly at their midpoints.

(a) Let $Z$ be the number of intersections of the cross with the grid of parallel lines. Show that $\mathbb{E}(Z/2) = 2/\pi$ and that

$$\mathrm{var}(Z/2) = \frac{3 - \sqrt{2}}{\pi} - \frac{4}{\pi^2}.$$

(b) If you had the choice of using either a needle of unit length, or the cross, in estimating $2/\pi$, which would you use?

(c) Would it be preferable to use a unit needle on the grid of Exercise (4.5.2) with $a = b = 1$?

**4.** Let $X$ and $Y$ be independent random variables each having the uniform distribution on $[0, 1]$. Let $U = \min\{X, Y\}$ and $V = \max\{X, Y\}$. Find $\mathbb{E}(U)$, and hence calculate $\mathrm{cov}(U, V)$.

**5.** (a) Let $X$ and $Y$ be independent continuous random variables. Show that

$$\mathbb{E}(g(X)h(Y)) = \mathbb{E}(g(X))\mathbb{E}(h(Y)),$$

whenever these expectations exist. If $X$ and $Y$ have the exponential distribution with parameter 1, find $\mathbb{E}\{\exp(\tfrac{1}{2}(X + Y))\}$.

(b) Let $X$ have finite variance. Show that $2\,\mathrm{var}(X) = \mathbb{E}((X - Y)^2)$ where $Y$ is independent and distributed as $X$.

**6.** Three points A, B, C are chosen independently at random on the circumference of a circle. Let $b(x)$ be the probability that at least one of the angles of the triangle ABC exceeds $x\pi$. Show that

$$b(x) = \begin{cases} 1 - (3x - 1)^2 & \text{if } \tfrac{1}{3} \leq x \leq \tfrac{1}{2}, \\ 3(1 - x)^2 & \text{if } \tfrac{1}{2} \leq x \leq 1. \end{cases}$$

Hence find the density and expectation of the largest angle in the triangle.

**7.** Let $\{X_r : 1 \leq r \leq n\}$ be independent and identically distributed with finite variance, and define $\overline{X} = n^{-1} \sum_{r=1}^{n} X_r$. Show that $\mathrm{cov}(\overline{X}, X_r - \overline{X}) = 0$.

**8.** Let $X$ and $Y$ be independent random variables with finite variances, and let $U = X + Y$ and $V = XY$. Under what condition are $U$ and $V$ uncorrelated?

**9.** Let $X$ and $Y$ be independent continuous random variables, and let $U$ be independent of $X$ and $Y$ taking the values $\pm 1$ with probability $\frac{1}{2}$. Define $S = UX$ and $T = UY$. Show that $S$ and $T$ are in general dependent, but $S^2$ and $T^2$ are independent.

**10.** Let $X, Y, Z$ be independent and identically distributed continuous random variables. Show that $\mathbb{P}(X > Y) = \mathbb{P}(Z > Y) = \frac{1}{2}$. What is $\mathbb{P}(Z > Y \mid X > Y)$?

**11. Hoeffding's identity.** Let $(X, Y)$ and $(U, V)$ be independent random vectors with common distribution function $F(x, y)$ and marginal distribution functions $F_X(x)$ and $F_Y(y)$. Show that, when $|\mathrm{cov}(X, Y)| < \infty$, we have that $\mathbb{E}\{(X - U)(Y - V)\} = 2\,\mathrm{cov}(X, Y)$ and

$$\mathrm{cov}(X, Y) = \iint_{\mathbb{R}^2} \left[ F(x, y) - F_X(x) F_Y(y) \right] dx\, dy.$$

**12.** Let the pair $X, Y$ have the bivariate normal distribution with means 0, variances 1, and correlation $\rho$. Show that

$$\mathbb{E}\left( \max\{X, Y\} \right) = \sqrt{\frac{1 - \rho}{\pi}}.$$

**13.** Let $X$ be exponentially distributed with parameter $\lambda$. Let $N$ be the greatest integer not greater than $X$, and set $M = X - N$. Show that $M$ and $N$ are independent. Find the density function of $M$ and the distribution of $N$.

**14.** Let $X$ and $Y$ be independent $N(0, 1)$ random variables. Show, if $2a < 1$, $2b < 1$, and $4b^2 < (1 - 2a)(1 - 2c)$, that

$$\mathbb{E}\left( \exp\{aX^2 + 2bXY + cY^2\} \right) = \frac{1}{\sqrt{(1 - 2a)(1 - 2c) - 4b^2}}.$$

**15. Contingency coefficient.** Let $X, Y$ be random variables with joint density function $f(x, y)$ and marginal densities $g, h$. Their *contingency coefficient* is given as

$$\phi^2 := \mathbb{E}\left( \frac{f(X, Y)}{g(X)h(Y)} \right) - 1.$$

If $X, Y$ are bivariate normal with correlation $\rho \in (-1, 1)$, show that $1 + \phi^2 = 1/(1 - \rho^2)$. Show in addition that their mutual information (see Exercise (3.6.5)) is $I = -\frac{1}{2} \log(1 - \rho^2)$.

**16.** Let $X$ be uniformly distributed on $[-1, 1]$. Are the random variables $Z_n = \cos(n\pi X)$, $n = 1, 2, \ldots$, correlated? Are they independent? Explain your answers.

**17. Mean absolute difference.** The mean absolute difference, or MAD, of two independent random variables $X, Y$, with common distribution function $F$, is given by $\mathrm{MAD} = \mathbb{E}|X - Y|$. Show that, for non-negative random variables $X, Y$,

$$\mathrm{MAD} = 2\left\{ \mathbb{E}(X) - \int_0^\infty (1 - F(x))^2\, dx \right\}.$$

More generally, for $X$ and $Y$ taking values in $\mathbb{R}$, with a common continuous distribution function with unique inverse $Q$, show that

$$\mathrm{MAD} = \iint_{(0,1)^2} |Q(u) - Q(v)|\, du\, dv.$$

Find MAD for the following distributions:

(a) uniform on $[0, 1]$,

(b) exponential with parameter $\lambda$,

(c) Pareto with $F(x) = 1 - x^{-a}$, $x > 1$,

(d) normal $N(0, 1)$.

**18. Absolute normals.** Let $X, Y$ be independent $N(0, 1)$ random variables, and $U = \min\{|X|, |Y|\}$, $V = \max\{|X|, |Y|\}$. Show that $\mathbb{E}(U/V) = (4/\pi) \log \sqrt{2}$.

**19. Another random triangle.** A point $P = (X, Y)$ is selected uniformly at random inside a triangle $\Delta$ with corners $(1, 0)$, $(1, 1)$, $(0, 1)$. Show that a triangle ABC with $BC = X$, $CA = Y$, $AB = 2 - X - Y$ can always be constructed, and prove that the angle $\widehat{ABC}$ is obtuse with probability $3 - 4 \log 2$.

**20.** Random variables $X, Y$ have joint density function $f(x, y) = 3\{(x + y) - (x^2 + y^2)\}$ for $x, y \in [0, 1]$. Show that $X$ and $Y$ are uncorrelated but dependent.

---

## 4.6 Conditional distributions and conditional expectation

Suppose that $X$ and $Y$ have joint density function $f$. We wish to discuss the conditional distribution of $Y$ given that $X$ takes the value $x$. However, the probability $\mathbb{P}(Y \leq y \mid X = x)$ is undefined since (see Definition (1.4.1)) we may only condition on events which have strictly positive probability. We proceed as follows. If $f_X(x) > 0$ then, by equation (4.5.4),

$$\mathbb{P}\big(Y \leq y \mid x \leq X \leq x + dx\big) = \frac{\mathbb{P}(Y \leq y, \ x \leq X \leq x + dx)}{\mathbb{P}(x \leq X \leq x + dx)}$$

$$\simeq \frac{\int_{v=-\infty}^{y} f(x, v)\, dx\, dv}{f_X(x)\, dx}$$

$$= \int_{v=-\infty}^{y} \frac{f(x, v)}{f_X(x)}\, dv.$$

As $dx \downarrow 0$, the left-hand side of this equation approaches our intuitive notion of the probability that $Y \leq y$ given that $X = x$, and it is appropriate to make the following definition.

**(1) Definition.** The **conditional distribution function** of $Y$ given $X = x$ is the function $F_{Y|X}(\cdot \mid x)$ given by

$$F_{Y|X}(y \mid x) = \int_{-\infty}^{y} \frac{f(x, v)}{f_X(x)}\, dv$$

for any $x$ such that $f_X(x) > 0$. It is sometimes denoted $\mathbb{P}(Y \leq y \mid X = x)$.

Remembering that distribution functions are integrals of density functions, we are led to the following definition.

**(2) Definition.** The **conditional density function** of $F_{Y|X}$, written $f_{Y|X}$, is given by

$$f_{Y|X}(y \mid x) = \frac{f(x, y)}{f_X(x)}$$

for any $x$ such that $f_X(x) > 0$.

Of course, $f_X(x) = \int_{-\infty}^{\infty} f(x, y) \, dy$, and therefore

$$f_{Y|X}(y \mid x) = \frac{f(x, y)}{\int_{-\infty}^{\infty} f(x, y) \, dy}.$$

Definition (2) is easily remembered as $f_{Y|X} = f_{X,Y}/f_X$. Here is an example of a conditional density function in action.

**(3) Example.** Let $X$ and $Y$ have joint density function

$$f_{X,Y}(x, y) = \frac{1}{x}, \qquad 0 \le y \le x \le 1.$$

Show for yourself (*exercise*) that

$$f_X(x) = 1 \quad \text{if} \quad 0 \le x \le 1, \qquad f_{Y|X}(y \mid x) = \frac{1}{x} \quad \text{if} \quad 0 \le y \le x \le 1,$$

which is to say that $X$ is uniformly distributed on $[0, 1]$ and, conditional on the event $\{X = x\}$, $Y$ is uniform on $[0, x]$. In order to calculate probabilities such as $\mathbb{P}(X^2 + Y^2 \le 1 \mid X = x)$, say, we proceed as follows. If $x > 0$, define

$$A(x) = \{y \in \mathbb{R} : 0 \le y \le x, \; x^2 + y^2 \le 1\};$$

clearly $A(x) = \left[0, \min\{x, \sqrt{1 - x^2}\}\right]$. Also,

$$\mathbb{P}(X^2 + Y^2 \le 1 \mid X = x) = \int_{A(x)} f_{Y|X}(y \mid x) \, dy$$

$$= \frac{1}{x} \min\{x, \sqrt{1 - x^2}\} = \min\{1, \sqrt{x^{-2} - 1}\}.$$

Next, let us calculate $\mathbb{P}(X^2 + Y^2 \le 1)$. Let $A = \{(x, y) : 0 \le y \le x \le 1, \; x^2 + y^2 \le 1\}$. Then

(4) $$\mathbb{P}(X^2 + Y^2 \le 1) = \iint_A f_{X,Y}(x, y) \, dx \, dy$$

$$= \int_{x=0}^1 f_X(x) \int_{y \in A(x)} f_{Y|X}(y \mid x) \, dy \, dx$$

$$= \int_0^1 \min\{1, \sqrt{x^{-2} - 1}\} \, dx = \log(1 + \sqrt{2}). \qquad \bullet$$

From Definitions (1) and (2) it is easy to see that the *conditional expectation* of $Y$ given $X$ can be defined as in Section 3.7 by $\mathbb{E}(Y \mid X) = \psi(X)$ where

$$\psi(x) = \mathbb{E}(Y \mid X = x) = \int_{-\infty}^{\infty} y f_{Y|X}(y \mid x) \, dy;$$

once again, $\mathbb{E}(Y \mid X)$ has the following important property

**(5) Theorem.** *The conditional expectation $\psi(X) = \mathbb{E}(Y \mid X)$ satisfies*

$$\mathbb{E}(\psi(X)) = \mathbb{E}(Y).$$

We shall use this result repeatedly; it is normally written as $\mathbb{E}(\mathbb{E}(Y \mid X)) = \mathbb{E}(Y)$, and it provides a useful method for calculating $\mathbb{E}(Y)$ since it asserts that

$$\mathbb{E}(Y) = \int_{-\infty}^{\infty} \mathbb{E}(Y \mid X = x) f_X(x) \, dx.$$

The proof of (5) proceeds exactly as for discrete variables (see Theorem (3.7.4)); indeed the theorem holds for all pairs of random variables, regardless of their types. For example, in the special case when $X$ is continuous and $Y$ is the discrete random variable $I_B$, the indicator function of an event $B$, the theorem asserts that

**(6)**
$$\mathbb{P}(B) = \mathbb{E}(\psi(X)) = \int_{-\infty}^{\infty} \mathbb{P}(B \mid X = x) f_X(x) \, dx,$$

of which equation (4) may be seen as an application.

**(7) Example.** Let $X$ and $Y$ have the standard bivariate normal distribution of Example (4.5.9). Then

$$f_{Y \mid X}(y \mid x) = f_{X,Y}(x, y)/f_X(x) = \frac{1}{\sqrt{2\pi(1 - \rho^2)}} \exp\left(-\frac{(y - \rho x)^2}{2(1 - \rho^2)}\right)$$

is the density function of the $N(\rho x, 1 - \rho^2)$ distribution. Thus $\mathbb{E}(Y \mid X = x) = \rho x$, giving that $\mathbb{E}(Y \mid X) = \rho X$. ●

**(8) Example.** Continuous and discrete variables have mean values, but what can we say about variables which are neither continuous nor discrete, such as $X$ in Example (2.3.5)? In that example, let $A$ be the event that a tail turns up. Then

$$\begin{aligned}
\mathbb{E}(X) &= \mathbb{E}\big(\mathbb{E}(X \mid I_A)\big) \\
&= \mathbb{E}(X \mid I_A = 1)\mathbb{P}(I_A = 1) + \mathbb{E}(X \mid I_A = 0)\mathbb{P}(I_A = 0) \\
&= \mathbb{E}(X \mid \text{tail})\mathbb{P}(\text{tail}) + \mathbb{E}(X \mid \text{head})\mathbb{P}(\text{head}) \\
&= -1 \cdot q + \pi \cdot p = \pi p - q
\end{aligned}$$

since $X$ is uniformly distributed on $[0, 2\pi]$ if a head turns up. ●

**(9) Example (3) revisited.** Suppose, in the notation of Example (3), that we wish to calculate $\mathbb{E}(Y)$. By Theorem (5),

$$\mathbb{E}(Y) = \int_0^1 \mathbb{E}(Y \mid X = x) f_X(x) \, dx = \int_0^1 \tfrac{1}{2} x \, dx = \tfrac{1}{4}$$

since, conditional on $\{X = x\}$, $Y$ is uniformly distributed on $[0, x]$. ●

There is a more general version of Theorem (5) which will be of interest later.

**(10) Theorem.** *The conditional expectation $\psi(X) = \mathbb{E}(Y \mid X)$ satisfies*

**(11)** $$\mathbb{E}\big(\psi(X)g(X)\big) = \mathbb{E}(Yg(X))$$

*for any function g for which both expectations exist.*

As in Section 3.7, we recapture Theorem (5) by setting $g(x) = 1$ for all $x$. We omit the proof, which is an elementary *exercise*. Conclusion (11) may be taken as a definition of the conditional expectation of $Y$ given $X$, that is as a function $\psi(X)$ such that (11) holds for all appropriate functions $g$. We shall return to this discussion in later chapters.

## Exercises for Section 4.6

**1.**  A point is picked uniformly at random on the surface of a unit sphere. Writing $\Theta$ and $\Phi$ for its longitude and latitude, find the conditional density functions of $\Theta$ given $\Phi$, and of $\Phi$ given $\Theta$.

**2.**  Show that the conditional expectation $\psi(X) = \mathbb{E}(Y \mid X)$ satisfies $\mathbb{E}(\psi(X)g(X)) = \mathbb{E}(Yg(X))$, for any function $g$ for which both expectations exist.

**3.**  Construct an example of two random variables $X$ and $Y$ for which $\mathbb{E}(Y) = \infty$ but such that $\mathbb{E}(Y \mid X) < \infty$ almost surely.

**4.**  Find the conditional density function and expectation of $Y$ given $X$ when they have joint density function:
(a) $f(x, y) = \lambda^2 e^{-\lambda y}$ for $0 \le x \le y < \infty$,
(b) $f(x, y) = xe^{-x(y+1)}$ for $x, y \ge 0$.

**5.**  Let $Y$ be distributed as $\text{bin}(n, X)$, where $X$ is a random variable having a beta distribution on $[0, 1]$ with parameters $a$ and $b$. Describe the distribution of $Y$, and find its mean and variance. What is the distribution of $Y$ in the special case when $X$ is uniform?

**6.**  Let $\{X_r : r \ge 1\}$ be independent and uniformly distributed on $[0, 1]$. Let $0 < x < 1$ and define

$$N = \min\{n \ge 1 : X_1 + X_2 + \cdots + X_n > x\}.$$

Show that $\mathbb{P}(N > n) = x^n/n!$, and hence find the mean and variance of $N$.

**7.**  Let $X$ and $Y$ be random variables with correlation $\rho$. Show that $\mathbb{E}(\text{var}(Y \mid X)) \le (1 - \rho^2) \, \text{var} \, Y$.

**8.**  Let $X$, $Y$, $Z$ be independent and exponential random variables with respective parameters $\lambda$, $\mu$, $\nu$. Find $\mathbb{P}(X < Y < Z)$.

**9.**  Let $X$ and $Y$ have the joint density $f(x, y) = cx(y - x)e^{-y}$, $0 \le x \le y < \infty$.
(a) Find $c$.
(b) Show that:
$$f_{X \mid Y}(x \mid y) = 6x(y - x)y^{-3}, \qquad 0 \le x \le y,$$
$$f_{Y \mid X}(y \mid x) = (y - x)e^{x-y}, \qquad 0 \le x \le y < \infty.$$

(c) Deduce that $\mathbb{E}(X \mid Y) = \frac{1}{2}Y$ and $\mathbb{E}(Y \mid X) = X + 2$.

**10.** Let $\{X_r : r \ge 0\}$ be independent and identically distributed random variables with density function $f$ and distribution function $F$. Let $N = \min\{n \ge 1 : X_n > X_0\}$ and $M = \min\{n \ge 1 : X_0 \ge X_1 \ge \cdots \ge X_{n-1} < X_n\}$. Show that $X_N$ has distribution function $F + (1 - F)\log(1 - F)$, and find $\mathbb{P}(M = m)$.

**11.** Let the point $P = (X, Y)$ be uniformly distributed in the square $S = [0, 1]^2$, and denote by $f(x \mid D)$ the density function of $X$ conditional on the event $D$ that $P$ lies on the diagonal of $S$ through the origin.

(a) Let $U = X - Y$. Find $f(x \mid D)$ by conditioning on $U = 0$.
(b) Let $V = Y/X$. Find $f(x \mid D)$ by conditioning on $V = 1$.
(c) Explain.

12. **Threshold game.** (a) Let $X$ and $Y$ be independent with density functions $f_X$, $f_Y$ and distribution functions $F_X$, $F_Y$. Show that

$$\mathbb{P}(X < Y) = \int_{-\infty}^{\infty} F_X(y) f_Y(y) \, dy.$$

(b) $A$ and $B$ play a game as follows. Each creates a random variable, denoted respectively $U_A$ and $U_B$, which is uniformly distributed on $[0, 1]$; assume $U_A$ and $U_B$ are independent, and neither player knows the opponent's value. Each player has the option to (once) discard their number and to resample from the uniform distribution. After any such choices, the two numbers are compared and the larger wins. Show that the strategy 'replace your number if and only if it is less than $\frac{1}{2}(\sqrt{5} - 1)$' ensures a win with probability at least $\frac{1}{2}$.

13. **Record times.** Let $X_1, X_2, \ldots$ be independent, each with density function $f : \mathbb{R} \to [0, \infty)$. The index $r > 1$ is called a *record time* if $X_r > \max\{X_1, X_2, \ldots, X_{r-1}\}$, and $r = 1$ is called a record time by convention. Let $A_r$ be the event that $r$ is a record time.
(a) Show that the $A_r$ are independent events with $\mathbb{P}(A_r) = 1/r$.
(b) Show that the number $R_n$ of record times up to time $n$ has variance

$$\text{var}(R_n) = \sum_{r=1}^{n} (r^{-1} - r^{-2}).$$

(c) Show that the first record time $T$ after $r = 1$ has mean $\mathbb{E}(T) = \infty$.

14. **Stick breaking.** A unit stick is broken uniformly at random, and the larger piece is broken again uniformly at random. Show that the probability the three pieces may be used to form a triangle is $2 \log 2 - 1$.

15. Let $U_1, U_2, \ldots, U_{n+1}$ be independent random variables with the uniform distribution on $[0, 1]$, and let
$$U_{(1)} = \min_{1 \le r \le n} U_r, \qquad U_{(n)} = \max_{1 \le r \le n} U_r.$$
Show that $\mathbb{E}(U_{(1)}) = 1/(n + 1)$ and deduce that

$$\mathbb{P}\big(U_{(n)} - U_{(1)} \le U_{n+1}\big) = \frac{2}{n + 1}.$$

## 4.7 Functions of random variables

Let $X$ be a random variable with density function $f$, and let $g : \mathbb{R} \to \mathbb{R}$ be a sufficiently nice function (in the sense of the discussion after Theorem (4.2.3)). Then $Y = g(X)$ is a random variable also. In order to calculate the distribution of $Y$, we proceed thus†:

$$\mathbb{P}(Y \le y) = \mathbb{P}\big(g(X) \le y\big) = \mathbb{P}\big(g(X) \in (-\infty, y]\big)$$
$$= \mathbb{P}\big(X \in g^{-1}(-\infty, y]\big) = \int_{g^{-1}(-\infty, y]} f(x) \, dx.$$

†If $A \subseteq \mathbb{R}$ then $g^{-1}A = \{x \in \mathbb{R} : g(x) \in A\}$.

Example (2.3.4) contains an instance of this calculation, when $g(x) = x^2$.

**(1) Example.** Let $X$ be $N(0, 1)$ and let $g(x) = x^2$. Then $Y = g(X) = X^2$ has distribution function

$$\mathbb{P}(Y \leq y) = \mathbb{P}(X^2 \leq y) = \mathbb{P}\left(-\sqrt{y} \leq X \leq \sqrt{y}\right)$$
$$= \Phi(\sqrt{y}) - \Phi(-\sqrt{y}) = 2\Phi(\sqrt{y}) - 1 \quad \text{if } y \geq 0,$$

by the fact that $\Phi(x) = 1 - \Phi(-x)$. Differentiate to obtain

$$f_Y(y) = 2\frac{d}{dy}\Phi(\sqrt{y}) = \frac{1}{\sqrt{y}}\Phi'(\sqrt{y}) = \frac{1}{\sqrt{2\pi y}}e^{-\frac{1}{2}y}$$

for $y \geq 0$. Compare with Example (4.4.6) to see that $X^2$ is $\Gamma(\frac{1}{2}, \frac{1}{2})$, or chi-squared with one degree of freedom†. See Problem (4.14.12) also.                                          ●

**(2) Example.** Let $g(x) = ax + b$ for fixed $a, b \in \mathbb{R}$. Then $Y = g(X) = aX + b$ has distribution function

$$\mathbb{P}(Y \leq y) = \mathbb{P}(aX + b \leq y) = \begin{cases} \mathbb{P}\left(X \leq (y - b)/a\right) & \text{if } a > 0, \\ \mathbb{P}\left(X \geq (y - b)/a\right) & \text{if } a < 0. \end{cases}$$

Differentiate to obtain $f_Y(y) = |a|^{-1}f_X((y - b)/a)$.                                          ●

More generally, if $X_1$ and $X_2$ have joint density function $f$, and $g, h$ are functions mapping $\mathbb{R}^2$ to $\mathbb{R}$, then what is the joint density function of the pair $Y_1 = g(X_1, X_2)$, $Y_2 = h(X_1, X_2)$? Recall how to change variables within an integral. Let $y_1 = y_1(x_1, x_2)$, $y_2 = y_2(x_1, x_2)$ be a one–one mapping $T : (x_1, x_2) \mapsto (y_1, y_2)$ taking some domain $D \subseteq \mathbb{R}^2$ onto some range $R \subseteq \mathbb{R}^2$. The transformation can be inverted as $x_1 = x_1(y_1, y_2)$, $x_2 = x_2(y_1, y_2)$; the *Jacobian*‡ of this inverse is defined to be the determinant

$$J = \begin{vmatrix} \dfrac{\partial x_1}{\partial y_1} & \dfrac{\partial x_2}{\partial y_1} \\[2mm] \dfrac{\partial x_1}{\partial y_2} & \dfrac{\partial x_2}{\partial y_2} \end{vmatrix} = \frac{\partial x_1}{\partial y_1}\frac{\partial x_2}{\partial y_2} - \frac{\partial x_1}{\partial y_2}\frac{\partial x_2}{\partial y_1}$$

which we express as a function $J = J(y_1, y_2)$. We assume that these partial derivatives are continuous.

**(3) Theorem.** *If $g : \mathbb{R}^2 \to \mathbb{R}$, and $T$ maps the set $A \subseteq D$ onto the set $B \subseteq R$ then*

$$\iint_A g(x_1, x_2)\, dx_1\, dx_2 = \iint_B g\left(x_1(y_1, y_2), x_2(y_1, y_2)\right)|J(y_1, y_2)|\, dy_1\, dy_2.$$

**(4) Corollary.** *If $X_1, X_2$ have joint density function $f$, then the pair $Y_1, Y_2$ given by $(Y_1, Y_2) = T(X_1, X_2)$ has joint density function*

$$f_{Y_1, Y_2}(y_1, y_2) = \begin{cases} f\left(x_1(y_1, y_2), x_2(y_1, y_2)\right)|J(y_1, y_2)| & \text{if } (y_1, y_2) \text{ is in the range of } T, \\ 0 & \text{otherwise.} \end{cases}$$

---

†This implies that $\Gamma(\frac{1}{2}) = \sqrt{\pi}$.

‡Introduced by Cauchy (1815) ahead of Jacobi (1841), the nomenclature conforming to Stigler's law.

A similar result holds for mappings of $\mathbb{R}^n$ into $\mathbb{R}^n$. This technique is sometimes referred to as the method of *change of variables*.

**Proof of Corollary.** Let $A \subseteq D$, $B \subseteq R$ be typical sets such that $T(A) = B$. Then $(X_1, X_2) \in A$ if and only if $(Y_1, Y_2) \in B$. Thus

$$
\mathbb{P}\big((Y_1, Y_2) \in B\big) = \mathbb{P}\big((X_1, X_2) \in A\big) = \iint_A f(x_1, x_2)\, dx_1\, dx_2
$$

$$
= \iint_B f\big(x_1(y_1, y_2), x_2(y_1, y_2)\big)|J(y_1, y_2)|\, dy_1\, dy_2
$$

by Example (4.5.4) and Theorem (3). Compare this with the definition of the joint density function of $Y_1$ and $Y_2$,

$$
\mathbb{P}\big((Y_1, Y_2) \in B\big) = \iint_B f_{Y_1, Y_2}(y_1, y_2)\, dy_1\, dy_2 \quad \text{for suitable sets } B \subseteq \mathbb{R}^2,
$$

to obtain the result. ∎

**(5) Example.** Suppose that

$$
X_1 = aY_1 + bY_2, \quad X_2 = cY_1 + dY_2,
$$

where $ad - bc \neq 0$. Check that

$$
f_{Y_1, Y_2}(y_1, y_2) = |ad - bc| f_{X_1, X_2}(ay_1 + by_2, cy_1 + dy_2). \qquad \bullet
$$

**(6) Example.** If $X$ and $Y$ have joint density function $f$, show that the density function of $U = XY$ is

$$
f_U(u) = \int_{-\infty}^{\infty} f(x, u/x)|x|^{-1}\, dx.
$$

**Solution.** Let $T$ map $(x, y)$ to $(u, v)$ by

$$
u = xy, \qquad v = x.
$$

The inverse $T^{-1}$ maps $(u, v)$ to $(x, y)$ by $x = v$, $y = u/v$, and the Jacobian is

$$
J(u, v) = \begin{vmatrix} \dfrac{\partial x}{\partial u} & \dfrac{\partial y}{\partial u} \\[2mm] \dfrac{\partial x}{\partial v} & \dfrac{\partial y}{\partial v} \end{vmatrix} = -\frac{1}{v}.
$$

Thus $f_{U,V}(u, v) = f(v, u/v)|v|^{-1}$. Integrate over $v$ to obtain the result. $\bullet$

**(7) Example.** Let $X_1$ and $X_2$ be independent exponential variables, parameter $\lambda$. Find the joint density function of $Y_1 = X_1 + X_2$ and $Y_2 = X_1/X_2$, and show that they are independent.

**Solution.** Let $T$ map $(x_1, x_2)$ to $(y_1, y_2)$ by

$$y_1 = x_1 + x_2, \qquad y_2 = x_1/x_2, \qquad x_1, x_2, y_1, y_2 \geq 0.$$

The inverse $T^{-1}$ maps $(y_1, y_2)$ to $(x_1, x_2)$ by

$$x_1 = y_1 y_2/(1 + y_2), \qquad x_2 = y_1/(1 + y_2)$$

and the Jacobian is

$$J(y_1, y_2) = -y_1/(1 + y_2)^2,$$

giving

$$f_{Y_1, Y_2}(y_1, y_2) = f_{X_1, X_2}\big(y_1 y_2/(1 + y_2), y_1/(1 + y_2)\big) \frac{|y_1|}{(1 + y_2)^2}.$$

However, $X_1$ and $X_2$ are independent and exponential, so that

$$f_{X_1, X_2}(x_1, x_2) = f_{X_1}(x_1) f_{X_2}(x_2) = \lambda^2 e^{-\lambda(x_1 + x_2)} \quad \text{if} \quad x_1, x_2 \geq 0,$$

whence

$$f_{Y_1, Y_2}(y_1, y_2) = \frac{\lambda^2 e^{-\lambda y_1} y_1}{(1 + y_2)^2} \quad \text{if} \quad y_1, y_2 \geq 0$$

is the joint density function of $Y_1$ and $Y_2$. However,

$$f_{Y_1, Y_2}(y_1, y_2) = [\lambda^2 y_1 e^{-\lambda y_1}] \frac{1}{(1 + y_2)^2}$$

factorizes as the product of a function of $y_1$ and a function of $y_2$; therefore, by Problem (4.14.6), they are independent. Suitable normalization of the functions in this product gives

$$f_{Y_1}(y_1) = \lambda^2 y_1 e^{-\lambda y_1}, \qquad f_{Y_2}(y_2) = \frac{1}{(1 + y_2)^2}. \qquad \bullet$$

**(8) Example.** Let $X_1$ and $X_2$ be given by the previous example and let

$$X = X_1, \qquad S = X_1 + X_2.$$

By Corollary (4), $X$ and $S$ have joint density function

$$f(x, s) = \lambda^2 e^{-\lambda s} \quad \text{if} \quad 0 \leq x \leq s.$$

This may look like the product of a function of $x$ with a function of $s$, implying that $X$ and $S$ are independent; a glance at the domain of $f$ shows this to be false. Suppose we know that $S = s$. What now is the conditional distribution of $X$, given $S = s$?
**Solution.**

$$\mathbb{P}(X \leq x \mid S = s) = \int_{-\infty}^{x} f(u, s)\, du \Big/ \int_{-\infty}^{\infty} f(u, s)\, du$$

$$= \frac{x\lambda^2 e^{-\lambda s}}{s\lambda^2 e^{-\lambda s}} = \frac{x}{s} \quad \text{if} \quad 0 \leq x \leq s.$$

Therefore, conditional on $S = s$, the random variable $X$ is uniformly distributed on $[0, s]$. This result, and its later generalization, is of great interest to statisticians.                                    ●

**(9) Example. A warning.** Let $X_1$ and $X_2$ be independent exponential variables (as in Examples (7) and (8)). What is the conditional density function of $X_1 + X_2$ given $X_1 = X_2$? '*Solution*' 1. Let $Y_1 = X_1 + X_2$ and $Y_2 = X_1/X_2$. Now $X_1 = X_2$ if and only if $Y_2 = 1$. We have from (7) that $Y_1$ and $Y_2$ are independent, and it follows that the conditional density function of $Y_1$ is its marginal density function

(10)                       $$f_{Y_1}(y_1) = \lambda^2 y_1 e^{-\lambda y_1} \quad \text{for} \quad y_1 \geq 0.$$

'*Solution*' 2. Let $Y_1 = X_1 + X_2$ and $Y_3 = X_1 - X_2$. It is an *exercise* to show that $f_{Y_1, Y_3}(y_1, y_3) = \frac{1}{2}\lambda^2 e^{-\lambda y_1}$ for $|y_3| \leq y_1$, and therefore the conditional density function of $Y_1$ given $Y_3$ is

$$f_{Y_1|Y_3}(y_1 \mid y_3) = \lambda e^{-\lambda(y_1 - |y_3|)} \quad \text{for} \quad |y_3| \leq y_1.$$

Now $X_1 = X_2$ if and only if $Y_3 = 0$, and the required conditional density function is therefore

(11)                       $$f_{Y_1|Y_3}(y_1 \mid 0) = \lambda e^{-\lambda y_1} \quad \text{for} \quad y_1 \geq 0.$$

Something is wrong: (10) and (11) are different. The error derives from the original question: what does it mean to condition on the event $\{X_1 = X_2\}$, an event having probability 0? As we have seen, the answer depends upon how we do the conditioning—one cannot condition on such events quite so blithely as one may on events having strictly positive probability. In Solution 1, we are essentially conditioning on the event $\{X_1 \leq X_2 \leq (1+h)X_1\}$ for small $h$, whereas in Solution 2 we are conditioning on $\{X_1 \leq X_2 \leq X_1 + h\}$; these two events contain different sets of information. Recall Exercise (4.6.11) for another example of this.                                    ●

**(12) Example. Bivariate normal distribution.** Let $X$ and $Y$ be independent random variables each having the normal distribution wth mean 0 and variance 1. Define

(13)                       $$U = \sigma_1 X,$$

(14)                       $$V = \sigma_2 \rho X + \sigma_2 \sqrt{1 - \rho^2}\, Y,$$

where $\sigma_1, \sigma_2 > 0$ and $|\rho| < 1$. By Corollary (4), the pair $U, V$ has joint density function

(15)                       $$f(u, v) = \frac{1}{2\pi \sigma_1 \sigma_2 \sqrt{1 - \rho^2}} \exp\left[-\tfrac{1}{2}Q(u, v)\right]$$

where

$$Q(u, v) = \frac{1}{(1 - \rho^2)}\left[\left(\frac{u}{\sigma_1}\right)^2 - 2\rho \left(\frac{u}{\sigma_1}\right)\left(\frac{v}{\sigma_2}\right) + \left(\frac{v}{\sigma_2}\right)^2\right].$$

We deduce that the pair $U, V$ has a bivariate normal distribution.

This fact may be used to derive many properties of the bivariate normal distribution without having recourse to unpleasant integrations. For example, we have that

$$\mathbb{E}(UV) = \sigma_1 \sigma_2 \left\{ \rho \mathbb{E}(X^2) + \sqrt{1 - \rho^2}\, \mathbb{E}(XY) \right\} = \sigma_1 \sigma_2 \rho,$$

whence the correlation coefficient of $U$ and $V$ equals $\rho$.

Here is a second example. Conditional on the event $\{U = u\}$, we have that

$$V = \frac{\sigma_2 \rho}{\sigma_1} u + \sigma_2 Y \sqrt{1 - \rho^2}.$$

Hence $\mathbb{E}(V \mid U) = (\sigma_2 \rho / \sigma_1) U$, and $\text{var}(V \mid U) = \sigma_2^2 (1 - \rho^2)$.   ●

The technology above is satisfactory when the change of variables is one–one, but a problem can arise if the transformation is many–one. The simplest examples arise of course for one-dimensional transformations. For example, if $y = x^2$ then the associated transformation $T : x \mapsto x^2$ is not one–one, since it loses the sign of $x$. It is easy to deal with this complication for transformations which are piecewise one–one (and sufficiently smooth). For example, the above transformation $T$ maps $(-\infty, 0)$ smoothly onto $(0, \infty)$ and similarly for $[0, \infty)$: there are two contributions to the density function of $Y = X^2$, one from each of the intervals $(-\infty, 0)$ and $[0, \infty)$. Arguing similarly but more generally, one arrives at the following conclusion, the proof of which is left as an exercise.

Let $I_1, I_2, \dots, I_n$ be intervals which partition $\mathbb{R}$ (it is not important whether or not these intervals contain their endpoints), and suppose that $Y = g(x)$ where $g$ is strictly monotone and continuously differentiable on every $I_i$. For each $i$, the function $g : I_i \to \mathbb{R}$ is invertible on $g(I_i)$, and we write $h_i$ for the inverse function. Then

**(16)**
$$f_Y(y) = \sum_{i=1}^{n} f_X(h_i(y)) |h_i'(y)|$$

with the convention that the $i$th summand is $0$ if $h_i$ is not defined at $y$. There is a natural extension of this formula to transformations in two and more dimensions.

---

## Exercises for Section 4.7

**1.** Let $X, Y$, and $Z$ be independent and uniformly distributed on $[0, 1]$. Find the joint density function of $XY$ and $Z^2$, and show that $\mathbb{P}(XY < Z^2) = \frac{5}{9}$.

**2.** Let $X$ and $Y$ be independent exponential random variables with parameter 1. Find the joint density function of $U = X + Y$ and $V = X/(X + Y)$, and deduce that $V$ is uniformly distributed on $[0, 1]$.

**3.** Let $X$ be uniformly distributed on $[0, \frac{1}{2}\pi]$. Find the density function of $Y = \sin X$.

**4.** Find the density function of $Y = \sin^{-1} X$ when:
(a) $X$ is uniformly distributed on $[0, 1]$,
(b) $X$ is uniformly distributed on $[-1, 1]$.

**5.** **Normal orthant probability.** Let $X$ and $Y$ have the bivariate normal density function

$$f(x, y) = \frac{1}{2\pi \sqrt{1 - \rho^2}} \exp\left\{ -\frac{1}{2(1 - \rho^2)}(x^2 - 2\rho xy + y^2) \right\}.$$

Show that $X$ and $Z = (Y - \rho X)/\sqrt{1 - \rho^2}$ are independent $N(0, 1)$ variables, and deduce that

$$\mathbb{P}(X > 0, \ Y > 0) = \frac{1}{4} + \frac{1}{2\pi} \sin^{-1} \rho.$$

**6.** Let $X$ and $Y$ have the standard bivariate normal density function of Exercise (4.7.5), and define $Z = \max\{X, Y\}$. Show that $\mathbb{E}(Z) = \sqrt{(1-\rho)/\pi}$, and $\mathbb{E}(Z^2) = 1$.

**7.** Let $X$ and $Y$ be independent exponential random variables with parameters $\lambda$ and $\mu$. Show that $Z = \min\{X, Y\}$ is independent of the event $\{X < Y\}$. Find:
(a) $\mathbb{P}(X = Z)$,
(b) the distributions of $U = \max\{X - Y, 0\}$, denoted $(X - Y)^+$, and $V = \max\{X, Y\} - \min\{X, Y\}$,
(c) $\mathbb{P}(X \leq t < X + Y)$ where $t > 0$.

**8.** A point $(X, Y)$ is picked at random uniformly in the unit circle. Find the joint density of $R$ and $X$, where $R^2 = X^2 + Y^2$.

**9.** A point $(X, Y, Z)$ is picked uniformly at random inside the unit ball of $\mathbb{R}^3$. Find the joint density of $Z$ and $R$, where $R^2 = X^2 + Y^2 + Z^2$.

**10.** Let $X$ and $Y$ be independent and exponentially distributed with parameters $\lambda$ and $\mu$. Find the joint distribution of $S = X + Y$ and $R = X/(X + Y)$. What is the density of $R$?

**11.** Find the density of $Y = a/(1 + X^2)$, where $X$ has the Cauchy distribution.

**12.** Let $(X, Y)$ have the bivariate normal density of Exercise (4.7.5) with $0 \leq \rho < 1$. Show that

$$[1 - \Phi(a)][1 - \Phi(c)] \leq \mathbb{P}(X > a, \ Y > b) \leq [1 - \Phi(a)][1 - \Phi(c)] + \frac{\rho\phi(b)[1 - \Phi(d)]}{\phi(a)},$$

where $c = (b - \rho a)/\sqrt{1 - \rho^2}, d = (a - \rho b)/\sqrt{1 - \rho^2}$, and $\phi$ and $\Phi$ are the density and distribution function of the $N(0, 1)$ distribution.

**13.** Let $X$ have the Cauchy distribution. Show that $Y = X^{-1}$ has the Cauchy distribution also. Find another non-trivial distribution with this property of invariance.

**14.** Let $X$ and $Y$ be independent and gamma distributed as $\Gamma(\lambda, \alpha)$, $\Gamma(\lambda, \beta)$ respectively.
(a) Show that $W = X + Y$ and $Z = X/(X + Y)$ are independent, and that $Z$ has the beta distribution with parameters $\alpha, \beta$.
(b) Show that $R = X/Y$ has a beta distribution of the second kind (see Exercise (4.4.14)).

**15. Frailty.** Let $X, Y$ be independent, positive, continuous random variables, such that $1 - F_Y(y) = (1 - F_X(y))^\lambda$ for $y > 0$. The positive parameter $\lambda$ may be called the *frailty* or *proportional hazard*. In more general models, $\lambda$ may itself be a random variable.
Show that

$$\mathbb{P}(X > Y) = \lambda\mathbb{P}(Y > X) = \frac{\lambda}{1 + \lambda}.$$

**16. Rayleigh distribution.** Let $X$ and $Y$ be independent random variables, where $X$ has an arc sine distribution and $Y$ a Rayleigh distribution:

$$f_X(x) = \frac{1}{\pi\sqrt{1 - x^2}}, \quad |x| < 1, \qquad f_Y(y) = ye^{-\frac{1}{2}y^2}, \quad y > 0.$$

Write down the joint density function of the pair $(Y, XY)$, and deduce that $XY$ has the standard normal distribution.

**17. Binary expansions.** Let $U$ be uniformly distributed on the interval $(0, 1)$.
(a) Let $S$ be a (measurable) subset of $(0, 1)$ with strictly positive measure (length). Show that the conditional distribution of $U$, given that $U \in S$, is uniform on $S$.
(b) Let $V = \sqrt{U}$, and write the binary expansions of $U$ and $V$ as $U = \sum_{r=1}^{\infty} U_r 2^{-r}$ and $V = \sum_{r=1}^{\infty} V_r 2^{-r}$. Show that $U_r$ and $U_s$ are independent for $r \neq s$, while $\text{cov}(V_1, V_2) = -\frac{1}{32}$. Prove that $\lim_{n \to \infty} \mathbb{P}(V_r = 1) = \frac{1}{2}$.

**18.** Let $(X, Y)$ have the standard bivariate normal distribution with correlation $\rho$.
(a) Let $\rho = 0$. By changing from Cartesian to polar coordinates, or otherwise, show that $Z = Y/X$ has the Cauchy distribution.
(b) Let $\rho > 0$. Show that $Z = Y/X$ has density function

$$f(z) = \frac{\sqrt{1 - \rho^2}}{\pi(1 - 2\rho z + z^2)}, \qquad z \in \mathbb{R}.$$

**19. Inverse Mills's ratio.** Let $(X, Y)$ have the standard bivariate normal distribution with correlation $\rho$. Show that

$$\mathbb{E}(Y \mid X > x) = \rho \frac{\phi(x)}{\Phi(-x)},$$

where $\phi$ and $\Phi$ are the $N(0, 1)$ density and distribution functions.

**20.** Let $(X, Y)$ have the standard bivariate normal distribution with density function $f$ and correlation $\rho$. Show that the probability that $(X, Y)$ lies in the interior of the ellipse $f(x, y) = k$ is $1 - \exp\{-A/(2\pi\sqrt{1 - \rho^2})\}$, where $A$ is the area of the ellipse.

**21.** Let $R = X/Y$ where $X, Y$ are independent and uniformly distributed on $(0, 1)$. Find the probability that the integer closest to $R$ is odd.

**22.** Let $(X, Y)$ have the standard bivariate normal distribution with correlation $\rho$ and density function $f$. Show that $f(X, Y)$ is uniformly distributed on the interval $(0, \zeta)$ where $\zeta = 1/\{2\pi\sqrt{1 - \rho^2}\}$.

**23.** A number $n$ of friends each visit Old Slaughter's coffee house independently and uniformly at random during their lunch break from noon to 1pm. Each leaves after $\delta$ hours (or at 1pm if that is sooner), where $\delta < 1/(n-1)$. Show that the probability that none of them meet inside is $(1 - (n-1)\delta)^n$.

**24.** Let $X$ and $Y$ be independent random variables with the uniform distribution on $(0, 1)$. Find the joint density function of $W = XY$ and $Z = Y/X$, and deduce their marginal density functions.

**25. Stein's identity.** (a) Let $X$ have the $N(\mu, \sigma^2)$ distribution with $\sigma^2 > 0$. Show that, for suitable $g : \mathbb{R} \to \mathbb{R}$,

$$\mathbb{E}\{g(X)(X - \mathbb{E}X)\} = \text{var}(X)\mathbb{E}(g'(X)),$$

when both sides exist.
(b) More generally, if $X$ and $Y$ have a bivariate normal distribution, show that

$$\text{cov}(g(X), Y) = \mathbb{E}(g'(X))\text{cov}(X, Y).$$

(c) Let $X$ be a random variable such that $\mathbb{E}(Xg(X)) = \mathbb{E}(g'(X))$ for all appropriate smooth functions $g$ satisfying $g(x)e^{-\frac{1}{2}x^2} \to 0$ as $x \to -\infty$. Show that $X$ has the $N(0, 1)$ distribution.

**26. Chernoff–Cacoullos inequalities.** Let $X$ have the $N(0, 1)$ distribution, and let $G$ be a function with derivative $g$. Show that

$$\{\mathbb{E}g(X)\}^2 \le \text{var}\, G(X) \le \mathbb{E}(g(X)^2).$$

**27.** The joint density function of the pair $(X, Y)$ is

$$f(x, y) = \tfrac{2}{3}(x + y)e^{-x}, \qquad x \in (0, \infty), \; y \in (0, 1).$$

Find the joint density function of the pair $U = X$, and $V = X + Y$, and deduce the density function of $V$.

## 4.8  Sums of random variables

This section contains an important result which is a very simple application of the change of variable technique.

**(1) Theorem.**  *If $X$ and $Y$ have joint density function $f$ then $X + Y$ has density function*

$$f_{X+Y}(z) = \int_{-\infty}^{\infty} f(x, z - x)\, dx.$$

**Proof.** Let $A = \{(x, y) : x + y \le z\}$. Then

$$\mathbb{P}(X + Y \le z) = \iint_A f(u, v)\, du\, dv = \int_{u=-\infty}^{\infty} \int_{v=-\infty}^{z-u} f(u, v)\, dv\, du$$

$$= \int_{x=-\infty}^{\infty} \int_{y=-\infty}^{z} f(x, y - x)\, dy\, dx$$

by the substitution $x = u$, $y = v + u$. Reverse the order of integration to obtain the result. ∎

If $X$ and $Y$ are independent, the result becomes

$$f_{X+Y}(z) = \int_{-\infty}^{\infty} f_X(x) f_Y(z - x)\, dx = \int_{-\infty}^{\infty} f_X(z - y) f_Y(y)\, dy.$$

The function $f_{X+Y}$ is called the *convolution* of $f_X$ and $f_Y$, and is written

**(2)**                                     $$f_{X+Y} = f_X * f_Y.$$

**(3) Example.**  Let $X$ and $Y$ be independent $N(0, 1)$ variables. Then $Z = X + Y$ has density function

$$f_Z(z) = \frac{1}{2\pi} \int_{-\infty}^{\infty} \exp\left[-\tfrac{1}{2}x^2 - \tfrac{1}{2}(z - x)^2\right] dx$$

$$= \frac{1}{2\sqrt{\pi}} e^{-\frac{1}{4}z^2} \int_{-\infty}^{\infty} \frac{1}{\sqrt{2\pi}} e^{-\frac{1}{2}v^2}\, dv$$

by the substitution $v = (x - \tfrac{1}{2}z)\sqrt{2}$. Therefore,

$$f_Z(z) = \frac{1}{2\sqrt{\pi}} e^{-\frac{1}{4}z^2},$$

showing that $Z$ is $N(0, 2)$. More generally, if $X$ is $N(\mu_1, \sigma_1^2)$ and $Y$ is $N(\mu_2, \sigma_2^2)$, and $X$ and $Y$ are independent, then $Z = X + Y$ is $N(\mu_1 + \mu_2, \sigma_1^2 + \sigma_2^2)$. You should check this.    ●

**(4) Example (4.6.3) revisited.**  You must take great care in applying (1) when the domain of $f$ depends on $x$ and $y$. For example, in the notation of Example (4.6.3),

$$f_{X+Y}(z) = \int_A \frac{1}{x}\, dx, \qquad 0 \le z \le 2,$$

where $A = \{x : 0 \le z - x \le x \le 1\} = \left[\frac{1}{2}z, \min\{z, 1\}\right]$. Thus

$$f_{X+Y}(z) = \begin{cases} \log 2 & 0 \le z \le 1, \\ \log(2/z) & 1 \le z \le 2. \end{cases}$$

●

**(5) Example. Bivariate normal distribution.** It is required to calculate the distribution of the linear combination $Z = aU' + bV'$ where the pair $U'$, $V'$ has the bivariate normal density function of equation (4.7.15). Let $X$ and $Y$ be independent random variables, each having the normal distribution with mean 0 and variance 1, and let $U$ and $V$ be given by equations (4.7.13) and (4.7.14). It follows from the result of that example that the pairs $(U, V)$ and $(U', V')$ have the same joint distribution. Therefore $Z$ has the same distribution as $aU + bV$, which equals $(a\sigma_1 + b\sigma_2\rho)X + b\sigma_2 Y\sqrt{1 - \rho^2}$. The distribution of the last sum is easily found by the method of Example (3) to be $N(0, a^2\sigma_1^2 + 2ab\sigma_1\sigma_2\rho + b^2\sigma_2^2)$.          ●

---

## Exercises for Section 4.8

**1.** Let $X$ and $Y$ be independent variables having the exponential distribution with parameters $\lambda$ and $\mu$ respectively. Find the density function of $X + Y$.

**2.** Let $X$ and $Y$ be independent variables with the Cauchy distribution. Find the density function of $\alpha X + \beta Y$ where $\alpha\beta \ne 0$. (Do you know about contour integration?)

**3.** Find the density function of $Z = X + Y$ when $X$ and $Y$ have joint density function $f(x, y) = \frac{1}{2}(x + y)e^{-(x+y)}$, $x, y \ge 0$.

**4. Hypoexponential distribution.** Let $\{X_r : r \ge 1\}$ be independent exponential random variables with respective parameters $\{\lambda_r : r \ge 1\}$ no two of which are equal. Find the density function of $S_n = \sum_{r=1}^n X_r$. [Hint: Use induction.]

**5.** (a) Let $X, Y, Z$ be independent and uniformly distributed on $[0, 1]$. Find the density function of $X + Y + Z$.

(b) If $\{X_r : r \ge 1\}$ are independent and uniformly distributed on $[0, 1]$, show that the density of $S_n = \sum_{r=1}^n X_r$ at any point $x \in (0, n)$ is a polynomial in $x$ of degree $n - 1$. Show in particular that the density function $f_n$ of $S_n$ satisfies $f_n(x) = x^{n-1}/(n - 1)!$ for $x \in [0, 1]$.

(c) Let $n \ge 3$. What is the probability that the $X_1, X_2, \ldots, X_n$ of part (b) can be the lengths of the edges of an $n$-gon?

**6.** For independent identically distributed random variables $X$ and $Y$, show that $U = X + Y$ and $V = X - Y$ are uncorrelated but not necessarily independent. Show that $U$ and $V$ are independent if $X$ and $Y$ are $N(0, 1)$.

**7.** Let $X$ and $Y$ have a bivariate normal density with zero means, variances $\sigma^2$, $\tau^2$, and correlation $\rho$. Show that:

(a) $\mathbb{E}(X \mid Y) = \dfrac{\rho\sigma}{\tau} Y$,

(b) $\text{var}(X \mid Y) = \sigma^2(1 - \rho^2)$,

(c) $\mathbb{E}(X \mid X + Y = z) = \dfrac{(\sigma^2 + \rho\sigma\tau)z}{\sigma^2 + 2\rho\sigma\tau + \tau^2}$,

(d) $\text{var}(X \mid X + Y = z) = \dfrac{\sigma^2\tau^2(1 - \rho^2)}{\tau^2 + 2\rho\sigma\tau + \sigma^2}$.

**8.** Let $X$ and $Y$ be independent $N(0, 1)$ random variables, and let $Z = X + Y$. Find the distribution and density of $Z$ given that $X > 0$ and $Y > 0$. Show that

$$\mathbb{E}(Z \mid X > 0, \ Y > 0) = 2\sqrt{2/\pi}.$$

**9.**   Let $X$ and $Y$ be independent $N(0, 1)$ random variables. Find the joint density function of $U = aX + bY$ and $V = bX - aY$, and hence show that $U$ is $N(0, a^2 + b^2)$. Deduce a proof of the last part of Example (4.8.3).

**10.** Let $X$ and $Y$ have joint distribution function $F$, with marginals $F_X$ and $F_Y$. Show that, if $\mu = \mathbb{E}(X + Y)$ is well defined, then its value is determined by knowledge of $F_X$ and $F_Y$.

**11.** Let $S = U + V + W$ be the sum of three independent random variables with the uniform distribution on $(0, 1)$.

(a) Show that $\mathbb{E}\{\text{var}(U \mid S)\} = \frac{1}{18}$.
(b) Find $v(s) = \text{var}(U \mid S = s)$ for $0 < s < 3$.

---

## 4.9 Multivariate normal distribution

The centrepiece of the normal density function is the function $\exp(-x^2)$, and of the bivariate normal density function the function $\exp(-x^2 - bxy - y^2)$ for suitable $b$. Both cases feature a quadratic in the exponent, and there is a natural generalization to functions of $n$ variables which is of great value in statistics. Roughly speaking, we say that $X_1, X_2, \ldots, X_n$ have the multivariate normal distribution if their joint density function is obtained by 'rescaling' the function $\exp\left(-\sum_i x_i^2 - 2\sum_{i<j} b_{ij} x_i x_j\right)$ of the $n$ real variables $x_1, x_2, \ldots, x_n$. The exponent here is a 'quadratic form', but not all quadratic forms give rise to density function. A *quadratic form* is a function $Q : \mathbb{R}^n \to \mathbb{R}$ of the form

**(1)** $$Q(\mathbf{x}) = \sum_{1 \le i, j \le n} a_{ij} x_i x_j = \mathbf{x}\mathbf{A}\mathbf{x}'$$

where $\mathbf{x} = (x_1, x_2, \ldots, x_n)$, $\mathbf{x}'$ is the transpose of $\mathbf{x}$, and $\mathbf{A} = (a_{ij})$ is a real symmetric matrix with non-zero determinant. A well-known theorem about diagonalizing matrices states that there exists an orthogonal matrix $\mathbf{B}$ such that

**(2)** $$\mathbf{A} = \mathbf{B}\boldsymbol{\Lambda}\mathbf{B}'$$

where $\boldsymbol{\Lambda}$ is the diagonal matrix with the eigenvalues $\lambda_1, \lambda_2, \ldots, \lambda_n$ of $\mathbf{A}$ on its diagonal. Substitute (2) into (1) to obtain

**(3)** $$Q(\mathbf{x}) = \mathbf{y}\boldsymbol{\Lambda}\mathbf{y}' = \sum_i \lambda_i y_i^2$$

where $\mathbf{y} = \mathbf{x}\mathbf{B}$. The function $Q$ (respectively the matrix $\mathbf{A}$) is called a *positive definite quadratic form* (respectively *matrix*) if $Q(\mathbf{x}) > 0$ for all vectors $\mathbf{x}$ having some non-zero coordinate, and we write $Q > 0$ (respectively $\mathbf{A} > 0$) if this holds. From (3), $Q > 0$ if and only if $\lambda_i > 0$ for all $i$. This is all elementary matrix theory. We are concerned with the following question: when is the function $f : \mathbb{R}^n \to \mathbb{R}$ given by

$$f(\mathbf{x}) = K \exp\left(-\tfrac{1}{2} Q(\mathbf{x})\right), \qquad \mathbf{x} \in \mathbb{R}^n,$$

the joint density function of some collection of $n$ random variables? It is necessary and sufficient that:

(a) $f(\mathbf{x}) \geq 0$ for all $\mathbf{x} \in \mathbb{R}^n$,

(b) $\int_{\mathbb{R}^n} f(\mathbf{x})\,d\mathbf{x} = 1$,

(this integral is shorthand for $\int \cdots \int f(x_1, \ldots, x_n)\,dx_1 \cdots dx_n$).

It is clear that (a) holds whenever $K > 0$. Next we investigate (b). First note that $Q$ must be positive definite, since otherwise $f$ has an infinite integral. If $Q > 0$,

$$
\int_{\mathbb{R}^n} f(\mathbf{x})\,d\mathbf{x} = \int_{\mathbb{R}^n} K \exp\left(-\tfrac{1}{2}Q(\mathbf{x})\right) d\mathbf{x}
$$

$$
= \int_{\mathbb{R}^n} K \exp\left(-\tfrac{1}{2}\sum_i \lambda_i y_i^2\right) d\mathbf{y}
$$

by (4.7.3) and (3), since $|J| = 1$ for orthogonal transformations

$$
= K \prod_i \int_{-\infty}^{\infty} \exp(-\tfrac{1}{2}\lambda_i y_i^2)\,dy_i
$$

$$
= K\sqrt{(2\pi)^n/(\lambda_1\lambda_2\cdots\lambda_n)} = K\sqrt{(2\pi)^n/|\mathbf{A}|}
$$

where $|\mathbf{A}|$ denotes the determinant of $\mathbf{A}$. Hence (b) holds whenever $K = \sqrt{(2\pi)^{-n}|\mathbf{A}|}$.

We have seen that

$$
f(\mathbf{x}) = \sqrt{\frac{|\mathbf{A}|}{(2\pi)^n}} \exp(-\tfrac{1}{2}\mathbf{x}\mathbf{A}\mathbf{x}'), \qquad \mathbf{x} \in \mathbb{R}^n,
$$

is a joint density function if and only if $\mathbf{A}$ is positive definite. Suppose that $\mathbf{A} > 0$ and that $\mathbf{X} = (X_1, X_2, \ldots, X_n)$ is a sequence of variables with joint density function $f$. It is easy to see that each $X_i$ has zero mean; just note that $f(\mathbf{x}) = f(-\mathbf{x})$, and so $(X_1, \ldots, X_n)$ and $(-X_1, \ldots, -X_n)$ are identically distributed random vectors; however, $\mathbb{E}|X_i| < \infty$ and so $\mathbb{E}(X_i) = \mathbb{E}(-X_i)$, giving $\mathbb{E}(X_i) = 0$. The vector $\mathbf{X}$ is said to have the *multivariate normal distribution* with zero means. More generally, if $\mathbf{Y} = (Y_1, Y_2, \ldots, Y_n)$ is given by

$$
\mathbf{Y} = \mathbf{X} + \boldsymbol{\mu}
$$

for some vector $\boldsymbol{\mu} = (\mu_1, \mu_2, \ldots, \mu_n)$ of constants, then $\mathbf{Y}$ is said to have the *multivariate normal distribution*.

**(4) Definition.** The vector $\mathbf{X} = (X_1, X_2, \ldots, X_n)$ has the **multivariate normal distribution** (or **multinormal distribution**), written $N(\boldsymbol{\mu}, \mathbf{V})$, if its joint density function is

$$
f(\mathbf{x}) = \frac{1}{\sqrt{(2\pi)^n|\mathbf{V}|}} \exp\left[-\tfrac{1}{2}(\mathbf{x} - \boldsymbol{\mu})\mathbf{V}^{-1}(\mathbf{x} - \boldsymbol{\mu})'\right], \qquad \mathbf{x} \in \mathbb{R}^n,
$$

where $\mathbf{V}$ is a positive definite symmetric matrix.

We have replaced $\mathbf{A}$ by $\mathbf{V}^{-1}$ in this definition. The reason for this lies in part (b) of the following theorem.

**(5) Theorem.** *If* $\mathbf{X}$ *is* $N(\boldsymbol{\mu}, \mathbf{V})$ *then*

(a) $\mathbb{E}(\mathbf{X}) = \boldsymbol{\mu}$, *which is to say that* $\mathbb{E}(X_i) = \mu_i$ *for all* $i$,

(b) $\mathbf{V} = (v_{ij})$ *is called the* covariance matrix, *because* $v_{ij} = \mathrm{cov}(X_i, X_j)$.

**Proof.** Part (a) follows by the argument before (4). Part (b) may be proved by performing an elementary integration, and more elegantly by the forthcoming method of characteristic functions; see Example (5.8.6). ■

We often write
$$V = \mathbb{E}\big((X - \mu)'(X - \mu)\big)$$

since $(X - \mu)'(X - \mu)$ is a matrix with $(i, j)$th entry $(X_i - \mu_i)(X_j - \mu_j)$.

A very important property of this distribution is its invariance of type under linear changes of variables.

**(6) Theorem.** *If* $X = (X_1, X_2, \ldots, X_n)$ *is* $N(0, V)$ *and* $Y = (Y_1, Y_2, \ldots, Y_m)$ *is given by* $Y = XD$ *for some matrix* $D$ *of rank* $m \le n$, *then* $Y$ *is* $N(0, D'VD)$.

**Proof when m = n.** The mapping $T : x \mapsto y = xD$ is a non-singular and can be inverted as $T^{-1} : y \mapsto x = yD^{-1}$. Use this change of variables in Theorem (4.7.3) to show that, if $A$, $B \subseteq \mathbb{R}^n$ and $B = T(A)$, then

$$\mathbb{P}(Y \in B) = \int_A f(x)\, dx = \int_A \frac{1}{\sqrt{(2\pi)^n |V|}} \exp(-\tfrac{1}{2} x V^{-1} x')\, dx$$
$$= \int_B \frac{1}{\sqrt{(2\pi)^n |W|}} \exp(-\tfrac{1}{2} y W^{-1} y')\, dy$$

where $W = D'VD$ as required. The proof for values of $m$ strictly smaller than $n$ is more difficult and is omitted (but see Kingman and Taylor 1966, p. 372). ■

A similar result holds for linear transformations of $N(\mu, V)$ variables.

There are various (essentially equivalent) ways of defining the multivariate normal distribution, of which the above way is perhaps neither the neatest nor the most useful. Here is another.

**(7) Definition.** The vector $X = (X_1, X_2, \ldots, X_n)$ of random variables is said to have the **multivariate normal distribution** whenever, for all $a \in \mathbb{R}^n$, the linear combination $Xa' = a_1 X_1 + a_2 X_2 + \cdots + a_n X_n$ has a normal distribution.

That is to say, $X$ is multivariate normal if and only if every linear combination of the $X_i$ is univariate normal. It often easier to work with this definition, which differs in one important respect from the earlier one. Using (6), it is easy to see that vectors $X$ satisfying (4) also satisfy (7). Definition (7) is, however, slightly more general than (4) as the following indicates. Suppose that $X$ satisfies (7), and in addition there exists $a \in \mathbb{R}^n$ and $b \in \mathbb{R}$ such that $a \ne 0$ and $\mathbb{P}(Xa' = b) = 1$, which is to say that the $X_i$ are linearly related; in this case there are strictly fewer than $n$ 'degrees of freedom' in the vector $X$, and we say that $X$ has a *singular* multivariate normal distribution. It may be shown (see Exercise (5.8.6)) that, if $X$ satisfies (7) and in addition its distribution is non-singular, then $X$ satisfies (4) for appropriate $\mu$ and $V$. The singular case is, however, not covered by (4). If (8) holds, then $0 = \text{var}(Xa') = aVa'$, where $V$ is the covariance matrix of $X$. Hence $V$ is a singular matrix, and therefore possesses no inverse. In particular, Definition (4) cannot apply.

## Exercises for Section 4.9

**1.** A symmetric matrix is called *non-negative* (respectively *positive*) *definite* if its eigenvalues are non-negative (respectively strictly positive). Show that a non-negative definite symmetric matrix $\mathbf{V}$ has a square root, in that there exists a symmetric matrix $\mathbf{W}$ satisfying $\mathbf{W}^2 = \mathbf{V}$. Show further that $\mathbf{W}$ is non-singular if and only if $\mathbf{V}$ is positive definite.

**2.** If $\mathbf{X}$ is a random vector with the $N(\boldsymbol{\mu}, \mathbf{V})$ distribution where $\mathbf{V}$ is non-singular, show that $\mathbf{Y} = (\mathbf{X} - \boldsymbol{\mu})\mathbf{W}^{-1}$ has the $N(\mathbf{0}, \mathbf{I})$ distribution, where $\mathbf{I}$ is the identity matrix and $\mathbf{W}$ is a symmetric matrix satisfying $\mathbf{W}^2 = \mathbf{V}$. The random vector $\mathbf{Y}$ is said to have the *standard* multivariate normal distribution.

**3.** Let $\mathbf{X} = (X_1, X_2, \dots, X_n)$ have the $N(\boldsymbol{\mu}, \mathbf{V})$ distribution, and show that $Y = a_1 X_1 + a_2 X_2 + \cdots + a_n X_n$ has the (univariate) $N(\mu, \sigma^2)$ distribution where

$$\mu = \sum_{i=1}^{n} a_i \mathbb{E}(X_i), \qquad \sigma^2 = \sum_{i=1}^{n} a_i^2 \operatorname{var}(X_i) + 2 \sum_{i<j} a_i a_j \operatorname{cov}(X_i, X_j).$$

**4.** Let $X$ and $Y$ have the bivariate normal distribution with zero means, unit variances, and correlation $\rho$. Find the joint density function of $X + Y$ and $X - Y$, and their marginal density functions.

**5.** Let $X$ have the $N(0, 1)$ distribution and let $a > 0$. Show that the random variable $Y$ given by

$$Y = \begin{cases} X & \text{if } |X| < a \\ -X & \text{if } |X| \geq a \end{cases}$$

has the $N(0, 1)$ distribution, and find an expression for $\rho(a) = \operatorname{cov}(X, Y)$ in terms of the density function $\phi$ of $X$. Does the pair $(X, Y)$ have a bivariate normal distribution?

**6.** Let $\{Y_r : 1 \leq r \leq n\}$ be independent $N(0, 1)$ random variables, and define $X_j = \sum_{r=1}^{n} c_{jr} Y_r$, $1 \leq r \leq n$, for constants $c_{jr}$. Show that

$$\mathbb{E}(X_j \mid X_k) = \left( \frac{\sum_r c_{jr} c_{kr}}{\sum_r c_{kr}^2} \right) X_k.$$

What is $\operatorname{var}(X_j \mid X_k)$?

**7.** Let the vector $(X_r : 1 \leq r \leq n)$ have a multivariate normal distribution with covariance matrix $\mathbf{V} = (v_{ij})$. Show that, conditional on the event $\sum_1^n X_r = x$, $X_1$ has the $N(a, b)$ distribution where $a = (\rho s/t)x$, $b = s^2(1 - \rho^2)$, and $s^2 = v_{11}$, $t^2 = \sum_{ij} v_{ij}$, $\rho = \sum_i v_{i1}/(st)$.

**8.** Let $X$, $Y$, and $Z$ have a standard trivariate normal distribution centred at the origin, with zero means, unit variances, and correlation coefficients $\rho_1$, $\rho_2$, and $\rho_3$. Show that

$$\mathbb{P}(X > 0, Y > 0, Z > 0) = \frac{1}{8} + \frac{1}{4\pi} \{\sin^{-1} \rho_1 + \sin^{-1} \rho_2 + \sin^{-1} \rho_3\}.$$

**9.** Let $X, Y, Z$ have the standard trivariate normal density of Exercise (4.9.8), with $\rho_1 = \rho(X, Y)$. Show that

$$\mathbb{E}(Z \mid X, Y) = \{(\rho_3 - \rho_1 \rho_2)X + (\rho_2 - \rho_1 \rho_3)Y\}/(1 - \rho_1^2),$$
$$\operatorname{var}(Z \mid X, Y) = \{1 - \rho_1^2 - \rho_2^2 - \rho_3^2 + 2\rho_1 \rho_2 \rho_3\}/(1 - \rho_1^2).$$

**10. Rotated normals.** Let $X$ and $Y$ have the standard bivariate normal distribution with correlation $\rho \neq 0$, and let $U = X \cos\theta + Y \sin\theta$ and $V = -X \sin\theta + Y \cos\theta$, where $\theta \in [0, \pi)$. Is there a value of $\theta$ such that $U$ and $V$ are independent?

**11.** Let $X$ have the $N(0, 1)$ distribution, and let $Y = BX$ where $B$ is independent of $X$ with mass function $\mathbb{P}(B = 1) = \mathbb{P}(B = -1) = \frac{1}{2}$. Show that the pair $(X, Y)$ has a singular distribution (in the sense that the support of the joint density function has zero (Lebesgue) area in the plane), but is not singular bivariate normal.

## 4.10  Distributions arising from the normal distribution

This section contains some distributional results which have applications in statistics. The reader may omit it without prejudicing his or her understanding of the rest of the book.

Statisticians are frequently faced with a collection $X_1, X_2, \ldots, X_n$ of random variables arising from a sequence of experiments. They might be prepared to make a general assumption about the unknown distribution of these variables without specifying the numerical values of certain parameters. Commonly they might suppose that $X_1, X_2, \ldots, X_n$ is a collection of independent $N(\mu, \sigma^2)$ variables for some fixed but unknown values of $\mu$ and $\sigma^2$; this assumption is sometimes a very close approximation to reality. They might then proceed to estimate the values of $\mu$ and $\sigma^2$ by using functions of $X_1, X_2, \ldots, X_n$. For reasons which are explained in statistics textbooks, they will commonly use the *sample mean*

$$\overline{X} = \frac{1}{n} \sum_1^n X_i$$

as a guess at the value of $\mu$, and the *sample variance*†

$$S^2 = \frac{1}{n-1} \sum_1^n (X_i - \overline{X})^2$$

as a guess at the value of $\sigma^2$; these at least have the property of being 'unbiased' in that $\mathbb{E}(\overline{X}) = \mu$ and $\mathbb{E}(S^2) = \sigma^2$. The two quantities $\overline{X}$ and $S^2$ are related in a striking and important way.

**(1) Theorem.** *If $X_1, X_2, \ldots$ are independent $N(\mu, \sigma^2)$ variables then $\overline{X}$ and $S^2$ are independent. We have that $\overline{X}$ is $N(\mu, \sigma^2/n)$ and $(n-1)S^2/\sigma^2$ is $\chi^2(n-1)$.*

Remember from Example (4.4.6) that $\chi^2(d)$ denotes the chi-squared distribution with $d$ degrees of freedom.

**Proof.** Define $Y_i = (X_i - \mu)/\sigma$, and

$$\overline{Y} = \frac{1}{n} \sum_1^n Y_i = \frac{\overline{X} - \mu}{\sigma}.$$

From Example (4.4.5), $Y_i$ is $N(0, 1)$, and clearly

$$\sum_1^n (Y_i - \overline{Y})^2 = \frac{(n-1)S^2}{\sigma^2}.$$

---

†In some texts the sample variance is defined with $n$ in place of $(n-1)$.

The joint density function of $Y_1, Y_2, \ldots, Y_n$ is

$$f(\mathbf{y}) = \frac{1}{\sqrt{(2\pi)^n}} \exp\left(-\tfrac{1}{2} \sum_1^n y_i^2\right).$$

This function $f$ has spherical symmetry in the sense that, if $\mathbf{A} = (a_{ij})$ is an orthogonal rotation of $\mathbb{R}^n$ and

(2) $$Y_i = \sum_{j=1}^n Z_j a_{ji} \quad \text{and} \quad \sum_1^n Y_i^2 = \sum_1^n Z_i^2,$$

then $Z_1, Z_2, \ldots, Z_n$ are independent $N(0, 1)$ variables also. Now choose

(3) $$Z_1 = \frac{1}{\sqrt{n}} \sum_1^n Y_i = \sqrt{n}\,\overline{Y}.$$

It is left to the reader to check that $Z_1$ is $N(0, 1)$. Then let $Z_2, Z_3, \ldots, Z_n$ be any collection of variables such that (2) holds, where $\mathbf{A}$ is orthogonal. From (2) and (3),

(4) $$\sum_2^n Z_i^2 = \sum_1^n Y_i^2 - \frac{1}{n}\left(\sum_1^n Y_i\right)^2$$

$$= \sum_1^n Y_i^2 - \frac{2}{n}\sum_{i=1}^n\sum_{j=1}^n Y_i Y_j + \frac{1}{n^2}\sum_{i=1}^n\left(\sum_{j=1}^n Y_j\right)^2$$

$$= \sum_{i=1}^n\left(Y_i - \frac{1}{n}\sum_1^n Y_j\right)^2 = \frac{(n-1)S^2}{\sigma^2}.$$

Now, $Z_1$ is independent of $Z_2, Z_3, \ldots, Z_n$, and so by (3) and (4), $\overline{Y}$ is independent of the random variable $(n-1)S^2/\sigma^2$. By (3) and Example (4.4.4), $\overline{Y}$ is $N(0, 1/n)$ and so $\overline{X}$ is $N(\mu, \sigma^2/n)$. Finally, $(n-1)S^2/\sigma^2$ is the sum of the squares of $n-1$ independent $N(0, 1)$ variables, and the result of Problem (4.14.12) completes the proof. ■

We may observe that $\sigma$ is only a scaling factor for $\overline{X}$ and $S\ (= \sqrt{S^2})$. That is to say,

$$U = \frac{n-1}{\sigma^2}S^2 \quad \text{is} \quad \chi^2(n-1)$$

which does not depend on $\sigma$, and

$$V = \frac{\sqrt{n}}{\sigma}(\overline{X} - \mu) \quad \text{is} \quad N(0, 1)$$

which does not depend on $\sigma$. Hence the random variable

$$T = \frac{V}{\sqrt{U/(n-1)}}$$

has a distribution which does not depend on $\sigma$. The random variable $T$ is the ratio of two independent random variables, the numerator being $N(0, 1)$ and the denominator the square root of $(n-1)^{-1}$ times a $\chi^2(n-1)$ variable; $T$ is said to have the $t$ *distribution* with $n-1$ degrees of freedom, written $t(n-1)$. It is sometimes called 'Student's $t$ distribution' in honour of a famous experimenter at the Guinness factory in Dublin. Let us calculate its density function. The joint density of $U$ and $V$ is

$$f(u, v) = \frac{(\frac{1}{2})^{\frac{1}{2}r} e^{-\frac{1}{2}u} u^{\frac{1}{2}r-1}}{\Gamma(\frac{1}{2}r)} \cdot \frac{1}{\sqrt{2\pi}} \exp(-\tfrac{1}{2}v^2)$$

where $r = n - 1$. Then map $(u, v)$ to $(s, t)$ by $s = u$, $t = v\sqrt{r/u}$. Use Corollary (4.7.4) to obtain

$$f_{U,T}(s, t) = \sqrt{s/r} f\left(s, t\sqrt{s/r}\right)$$

and integrate over $s$ to obtain

$$f_T(t) = \frac{\Gamma(\frac{1}{2}(r+1))}{\sqrt{\pi r}\,\Gamma(\frac{1}{2}r)} \left(1 + \frac{t^2}{r}\right)^{-\frac{1}{2}(r+1)}, \qquad -\infty < t < \infty,$$

as the density function of the $t(r)$ distribution.

Another important distribution in statistics is the $F$ distribution which arises as follows. Let $U$ and $V$ be independent variables with the $\chi^2(r)$ and $\chi^2(s)$ distributions respectively. Then

$$F = \frac{U/r}{V/s}$$

is said to have the $F$ *distribution* with $r$ and $s$ degrees of freedom, written $F(r, s)$. The following properties are obvious:

(a) $F^{-1}$ is $F(s, r)$,
(b) $T^2$ is $F(1, r)$ if $T$ is $t(r)$.

As an *exercise* in the techniques of Section 4.7, show that the density function of the $F(r, s)$ distribution is

$$f(x) = \frac{r\Gamma(\frac{1}{2}(r+s))}{s\Gamma(\frac{1}{2}r)\Gamma(\frac{1}{2}s)} \cdot \frac{(rx/s)^{\frac{1}{2}r-1}}{[1 + (rx/s)]^{\frac{1}{2}(r+s)}}, \qquad x > 0.$$

In Exercises (5.7.7, 8) we shall encounter more general forms of the $\chi^2$, $t$, and $F$ distributions; these are the (so-called) 'non-central' versions of these distributions.

## Exercises for Section 4.10

**1.** Let $X_1$ and $X_2$ be independent variables with the $\chi^2(m)$ and $\chi^2(n)$ distributions respectively. Show that $X_1 + X_2$ has the $\chi^2(m+n)$ distribution.

**2.** Show that the mean of the $t(r)$ distribution is 0, and that the mean of the $F(r, s)$ distribution is $s/(s-2)$ if $s > 2$. What happens if $s \leq 2$?

**3.** Show that the $t(1)$ distribution and the Cauchy distribution are the same.

**4.** Let $X$ and $Y$ be independent variables having the exponential distribution with parameter 1. Show that $X/Y$ has an $F$ distribution. Which?

**5.**   Show the independence of the sample mean and sample variance of an independent sample from the $N(\mu, \sigma^2)$ distribution. This may be done by either or both of: (i) the result of Exercise (4.5.7), (ii) induction on $n$.

**6.**   Let $\{X_r : 1 \leq r \leq n\}$ be independent $N(0, 1)$ variables. Let $\Psi \in [0, \pi]$ be the angle between the vector $(X_1, X_2, \ldots, X_n)$ and some fixed vector in $\mathbb{R}^n$. Show that $\Psi$ has density $f(\psi) = (\sin \psi)^{n-2}/B(\frac{1}{2}, \frac{1}{2}n - \frac{1}{2})$, $0 \leq \psi < \pi$, where $B$ is the beta function.

**7.**   Let $X_1, X_2, \ldots, X_n$ be independent $N(0, 1)$ random variables, and let $V_n \geq 0$ be given by $V_n^2 = \sum_{i=1}^{n} X_i^2$. Show that the random vector $Y = (X_1, X_2, \ldots, X_n)/V_n$ is uniformly distributed on the unit sphere. Deduce the result of Exercise (4.10.6).

---

# 4.11 Sampling from a distribution

It is frequently necessary to conduct numerical experiments involving random variables with a given distribution†. Such experiments are useful in a wide variety of settings, ranging from the evaluation of integrals (see Section 2.6) to the statistical theory of image reconstruction. The target of the current section is to describe a portfolio of techniques for sampling from a given distribution. The range of available techniques has grown enormously over recent years, and we give no more than an introduction here. The fundamental question is as follows. Let $F$ be a distribution function. How may we find a numerical value for a random variable having distribution function $F$?

Various interesting questions arise. What does it mean to say that a real number has a non-trivial distribution function? In a universe whose fundamental rules may be deterministic, how can one simulate randomness? In practice, one makes use of deterministic sequences of real numbers produced by what are called 'congruential generators'. Such sequences are sprinkled uniformly over their domain, and statistical tests indicate acceptance of the hypothesis that they are independent and uniformly distributed. Strictly speaking, these numbers are called 'pseudo-random' but the prefix is often omitted. They are commonly produced by a suitable computer program called a 'random number generator'. With a little cleverness, such a program may be used to generate a sequence $U_1, U_2, \ldots$ of (pseudo-)random numbers which may be assumed to be independent and uniformly distributed on the interval $[0, 1]$. Henceforth in this section we will denote by $U$ a random variable with this distribution.

A basic way of generating a random variable with given distribution function is to use the following theorem.

**(1) Theorem. Inverse transform technique.** *Let $F$ be a distribution function, and let $U$ be uniformly distributed on the interval $[0, 1]$.*

  (a) *If $F$ is a continuous function, the random variable $X = F^{-1}(U)$ has distribution function $F$.*

  (b) *Let $F$ be the distribution function of a random variable taking non-negative integer values. The random variable $X$ given by*

$$X = k \quad \text{if and only if} \quad F(k-1) < U \leq F(k)$$

  *has distribution function $F$.*

---

The inverse function $F^{-1}$ of part (a) is called the *quantile function* of $X$.

**Proof.**  Part (a) is Problem (4.14.4a). Part (b) is a straightforward exercise, on noting that

$$\mathbb{P}\big(F(k-1) < U \le F(k)\big) = F(k) - F(k-1).$$

This part of the theorem is easily extended to more general discrete distributions.    ∎

The inverse transform technique is conceptually easy but has practical drawbacks. In the continuous case, it is required to know or calculate the inverse function $F^{-1}$; in the discrete case, a large number of comparisons may be necessary. Despite the speed of modern computers, such issues remain problematic for extensive simulations.

Here are three examples of the inverse transform technique in practice. Further examples may be found in the exercises at the end of this section.

**(2) Example. Binomial sampling.** Let $U_1, U_2, \ldots, U_n, \ldots$ be independent random variables with the uniform distribution on $[0, 1]$. The sequence $X_k = I_{\{U_k \le p\}}$ of indicator variables contains random variables having the Bernoulli distribution with parameter $p$. The sum $S = \sum_{k=1}^{n} X_k$ has the $\mathrm{bin}(n, p)$ distribution.    ●

**(3) Example. Negative binomial sampling.** With the $X_k$ as in the previous example, let $W_r$ be given by

$$W_r = \min\left\{ n : \sum_{k=1}^{n} X_k = r \right\},$$

the 'time of the $r$th success'. Then $W_r$ has the negative binomial distribution; see Example (3.5.6).    ●

**(4) Example. Gamma sampling.** With the $U_k$ as in Example (2), let

$$X_k = -\frac{1}{\lambda} \log U_k.$$

It is an easy calculation (or use Problem (4.14.4a)) to see that the $X_k$ are independent exponential random variables with parameter $\lambda$. It follows that $S = \sum_{k=1}^{n} X_k$ has the $\Gamma(\lambda, n)$ distribution; see Problem (4.14.10).    ●

Here are two further methods of sampling from a given distribution.

**(5) Example. The rejection method.** It is required to sample from the distribution having density function $f$. Let us suppose that we are provided with a pair $(U, Z)$ of random variables such that:

  (i) $U$ and $Z$ are independent,
  (ii) $U$ is uniformly distribution on $[0, 1]$, and
  (iii) $Z$ has density function $f_Z$, and there exists $a \in \mathbb{R}$ such that $f(z) \le a f_Z(z)$ for all $z$.
  We note the following calculation:

$$\mathbb{P}\big(Z \le x \,\big|\, aUf_Z(Z) \le f(Z)\big) = \frac{\int_{-\infty}^{x} \mathbb{P}\big(aUf_Z(Z) \le f(Z) \,\big|\, Z = z\big) f_Z(z)\, dz}{\int_{-\infty}^{\infty} \mathbb{P}\big(aUf_Z(Z) \le f(Z) \,\big|\, Z = z\big) f_Z(z)\, dz}.$$

Now,

$$\mathbb{P}\big(aU f_Z(Z) \le f(Z) \,\big|\, Z = z\big) = \mathbb{P}\big(U \le f(z)/\{af_Z(z)\}\big) = \frac{f(z)}{af_Z(z)}$$

whence

$$\mathbb{P}\big(Z \le x \,\big|\, aU f_Z(Z) \le f(Z)\big) = \int_{-\infty}^{x} f(z)\,dz.$$

That is to say, conditional on the event $E = \{aU f_Z(Z) \le f(Z)\}$, the random variable $Z$ has the required density function $f$.

We use this fact in the following way. Let us assume that one may use a random number generator to obtain a pair $(U, Z)$ as above. We then check whether or not the event $E$ occurs. If $E$ occurs, then $Z$ has the required density function. If $E$ does not occur, we *reject* the pair $(U, Z)$, and use the random number generator to find another pair $(U', Z')$ with the properties (i)–(iii) above. This process is iterated until the event corresponding to $E$ occurs, and this results in a sample from the given density function.

Each sample pair $(U, Z)$ satisfies the condition of $E$ with probability $a$. It follows by the independence of repeated sampling that the mean number of samples before $E$ is first satisfied is $a^{-1}$.

A similar technique exists for sampling from a discrete distribution.                    ●

**(6) Example. Ratio of uniforms.** There are other 'rejection methods' than that described in the above example, and here is a further example. Once again, let $f$ be a density function from which a sample is required. For a reason which will become clear soon, we shall assume that $f$ satisfies $f(x) = 0$ if $x \le 0$, and $f(x) \le \min\{1, x^{-2}\}$ if $x > 0$. The latter inequality may be relaxed in the following, but at the expense of a complication.

Suppose that $U_1$ and $U_2$ are independent and uniform on $[0, 1]$, and define $R = U_2/U_1$. We claim that, conditional on the event $E = \{U_1 \le \sqrt{f(U_2/U_1)}\}$, the random variable $R$ has density function $f$. This provides the basis for a rejection method using uniform random variables only. We argue as follows in order to show the claim. We have that

$$\mathbb{P}\big(E \cap \{R \le x\}\big) = \iint_{T \cap [0,1]^2} du_1\,du_2$$

where $T = \big\{(u_1, u_2) : u_1 \le \sqrt{f(u_2/u_1)},\ u_2 \le xu_1\big\}$. We make the change of variables $s = u_2/u_1,\ t = u_1$, to obtain that

$$\mathbb{P}\big(E \cap \{R \le x\}\big) = \int_{s=0}^{x} \int_{t=0}^{\sqrt{f(s)}} t\,dt\,ds = \tfrac{1}{2} \int_0^{x} f(s)\,ds,$$

from which it follows as required that

$$\mathbb{P}\big(R \le x \,\big|\, E\big) = \int_0^{x} f(s)\,ds.$$                          ●

In sampling from a distribution function $F$, the structure of $F$ may itself propose a workable approach.

**(7) Example. Mixtures.** Let $F_1$ and $F_2$ be distribution functions and let $0 \le \alpha \le 1$. It is required to sample from the 'mixed' distribution function $G = \alpha F_1 + (1 - \alpha) F_2$. This may be done in a process of two stages:

   (i) first toss a coin which comes up heads with probability $\alpha$ (or, more precisely, utilize the random variable $I_{\{U \le \alpha\}}$ where $U$ has the usual uniform distribution),

  (ii) if the coin shows heads (respectively, tails) sample from $F_1$ (respectively, $F_2$).

   As an example of this approach in action, consider the density function

$$g(x) = \frac{1}{\pi \sqrt{1 - x^2}} + 3x(1 - x), \qquad 0 \le x \le 1,$$

and refer to Theorem (1) and Exercises (4.11.5) and (4.11.13).      ●

   This example leads naturally to the following more general formulation. Assume that the distribution function $G$ may be expressed in the form

$$G(x) = \mathbb{E}(F(x, Y)), \qquad x \in \mathbb{R},$$

where $Y$ is a random variable, and where $F(\cdot, y)$ is a distribution function for each possible value $y$ of $Y$. Then $G$ may be sampled by:

   (i) sampling from the distribution of $Y$, obtaining the value $y$, say,

  (ii) sampling from the distribution function $F(\cdot, y)$.

**(8) Example. Compound distributions.** Here is a further illustrative example. Let $Z$ have the beta distribution with parameters $a$ and $b$, and let

$$p_k = \mathbb{E}\left( \binom{n}{k} Z^k (1 - Z)^{n-k} \right), \qquad k = 0, 1, 2, \dots, n.$$

It is an *exercise* to show that

$$p_k \propto \binom{n}{k} \Gamma(a + k) \Gamma(n + b - k), \qquad k = 0, 1, 2, \dots, n,$$

where $\Gamma$ denotes the gamma function; this distribution is termed a *negative hypergeometric distribution*. In sampling from the mass function $(p_k : k = 0, 1, 2, \dots, n)$ it is convenient to sample first from the beta distribution of $Z$ and then from the binomial distribution $\mathrm{bin}(n, Z)$; see Exercise (4.11.4) and Example (2).      ●

---

## Exercises for Section 4.11

**1. Uniform distribution.** If $U$ is uniformly distributed on $[0, 1]$, what is the distribution of $X = \lfloor nU \rfloor + 1$?

**2. Random permutation.** Given the first $n$ integers in any sequence $S_0$, proceed thus:

(a) pick any position $P_0$ from $\{1, 2, \dots, n\}$ at random, and swap the integer in that place of $S_0$ with the integer in the $n$th place of $S_0$, yielding $S_1$.

(b) pick any position $P_1$ from $\{1, 2, \dots, n - 1\}$ at random, and swap the integer in that place of $S_1$ with the integer in the $(n - 1)$th place of $S_1$, yielding $S_2$,

(c)  at the $(r-1)$th stage the integer in position $P_{r-1}$, chosen randomly from $\{1, 2, \ldots, n-r+1\}$, is swapped with the integer at the $(n-r+1)$th place of the sequence $S_{r-1}$.

Show that $S_{n-1}$ is equally likely to be any of the $n!$ permutations of $\{1, 2, \ldots, n\}$.

**3.    Gamma distribution.** Use the rejection method to sample from the gamma density $\Gamma(\lambda, t)$ where $t\ (\geq 1)$ may not be assumed integral. [Hint: You might want to start with an exponential random variable with parameter $1/t$.]

**4.    Beta distribution.** Show how to sample from the beta density $\beta(\alpha, \beta)$ where $\alpha, \beta \geq 1$. [Hint: Use Exercise (4.11.3).]

**5.    Describe** three distinct methods of sampling from the density $f(x) = 6x(1-x)$, $0 \leq x \leq 1$.

**6.    Aliasing method.** A finite real vector is called a *probability vector* if it has non-negative entries with sum 1. Show that a probability vector $\mathbf{p}$ of length $n$ may be written in the form

$$\mathbf{p} = \frac{1}{n-1} \sum_{r=1}^{n} \mathbf{v}_r,$$

where each $\mathbf{v}_r$ is a probability vector with at most two non-zero entries. Describe a method, based on this observation, for sampling from $\mathbf{p}$ viewed as a probability mass function.

**7.    Box–Muller normals.** Let $U_1$ and $U_2$ be independent and uniformly distributed on $[0, 1]$, and let $T_i = 2U_i - 1$. Show that, conditional on the event that $R = \sqrt{T_1^2 + T_2^2} \leq 1$,

$$X = \frac{T_1}{R}\sqrt{-2\log R^2}, \quad Y = \frac{T_2}{R}\sqrt{-2\log R^2},$$

are independent standard normal random variables.

**8.    Let** $U$ be uniform on $[0, 1]$ and $0 < q < 1$. Show that $X = 1 + \lfloor \log U / \log q \rfloor$ has a geometric distribution.

**9.    A point** $(X, Y)$ is picked uniformly at random in the semicircle $x^2 + y^2 \leq 1$, $x \geq 0$. What is the distribution of $Z = Y/X$?

**10.  Hazard-rate technique.** Let $X$ be a non-negative integer-valued random variable with $h(r) = \mathbb{P}(X = r \mid X \geq r)$. If $\{U_i : i \geq 0\}$ are independent and uniform on $[0, 1]$, show that $Z = \min\{n : U_n \leq h(n)\}$ has the same distribution as $X$.

**11.  Antithetic variables†.** Let $g(x_1, x_2, \ldots, x_n)$ be an increasing function in all its variables, and let $\{U_r : r \geq 1\}$ be independent and identically distributed random variables having the uniform distribution on $[0, 1]$. Show that

$$\text{cov}\big\{g(U_1, U_2, \ldots, U_n), g(1-U_1, 1-U_2, \ldots, 1-U_n)\big\} \leq 0.$$

[Hint: Use the FKG inequality of Problem (3.11.18).] Explain how this can help in the efficient estimation of $I = \int_0^1 g(\mathbf{x})\, d\mathbf{x}$.

**12.  Importance sampling.** We wish to estimate $I = \int g(x) f_X(x)\, dx = \mathbb{E}(g(X))$, where either it is difficult to sample from the density $f_X$, or $g(X)$ has a very large variance. Let $f_Y$ be equivalent to $f_X$, which is to say that, for all $x$, $f_X(x) = 0$ if and only if $f_Y(x) = 0$. Let $\{Y_i : 0 \leq i \leq n\}$ be independent random variables with density function $f_Y$, and define

$$J = \frac{1}{n} \sum_{r=1}^{n} \frac{g(Y_r) f_X(Y_r)}{f_Y(Y_r)}.$$

†A technique invented by J. M. Hammersley and K. W. Morton in 1956.

Show that:

(a) $\mathbb{E}(J) = I = \mathbb{E}\left[\dfrac{g(Y)f_X(Y)}{f_Y(Y)}\right]$,

(b) $\mathrm{var}(J) = \dfrac{1}{n}\left[\mathbb{E}\left(\dfrac{g(Y)^2 f_X(Y)^2}{f_Y(Y)^2}\right) - I^2\right]$,

(c) $J \xrightarrow{\text{a.s.}} I$ as $n \to \infty$. (See Chapter 7 for an account of convergence.)

The idea here is that $f_Y$ should be easy to sample from, and chosen if possible so that var $J$ is much smaller than $n^{-1}[\mathbb{E}(g(X)^2) - I^2]$. The function $f_Y$ is called the *importance density*.

**13.** Construct two distinct methods of sampling from the arc sine density function

$$f(x) = \frac{2}{\pi\sqrt{1-x^2}}, \qquad 0 \le x \le 1.$$

**14. Marsaglia's method.** Let $(X, Y)$ be a random point chosen uniformly on the unit disk. Show that

$$(U, V, W) = \left(2X\sqrt{1-X^2-Y^2},\, 2Y\sqrt{1-X^2-Y^2},\, 1-2(X^2+Y^2)\right)$$

is uniformly distributed on the unit sphere.

**15. Acceptance–complement method.** It is desired to sample from a density function $f_X$ satisfying $f_X = f_1 + f_2$ where $f_1, f_2 : \mathbb{R} \to [0, \infty)$, and $f_1 \le f_Y$ for some random variable $Y$. Let $U$ be uniformly distributed on $[0, 1]$, and independent of $Y$. From $Y$ and $U$, we generate another random variable $Z$ as follows. If $U > f_1(Y)/f_Y(Y)$, let $Z$ have density $f_2(z)/\int_{-\infty}^{\infty} f_2(u)\,du$; otherwise set $Z = Y$. Show that $Z$ has density function $f_X$.

**16. Hardy–Littlewood transform.** The function $F^{-1}$ defined in Theorem (4.11.1) may be called the *quantile function* of the random variable in question.

(a) For a random variable $U$ with the uniform distribution on $(0, 1)$, and a continuously increasing function $G : (0, 1) \to \mathbb{R}$, show that the quantile function of $G(U)$ is $G$.

(b) For a random variable $X$ with distribution function $F$ and density function $f$, define $H_X(v) = \mathbb{E}(F^{-1}(U) \mid U > v)$, where $U$ is as in part (a). The *Hardy–Littlewood transform* of $X$ is defined as $H_X(V)$, where $V$ is uniformly distributed on $(0, 1)$ and independent of $U$. Show that the quantile function of $H_X(V)$ is $(1-v)^{-1}\int_v^{\infty} F^{-1}(w)\,dw$, for $v \in (0, 1)$.

**17. Generalized inverse transform†.** For an arbitrary (not necessarily continuous) distribution function $F = F_X$, define

$$G(x, u) = \mathbb{P}(X < x) + u\mathbb{P}(X = x)$$
$$= F(x-) + u(F(x) - F(x-)), \qquad u \in [0, 1].$$

Let $U$ be uniformly distributed on $(0, 1)$ and independent of $X$. We call $G(X, U)$ the *generalized distributional transform* of $X$. Show that $W = G(X, U)$ is uniformly distributed on $(0, 1)$, and that the random variable $F^{-1}(W)$ has distribution function $F$.

**18. Copulas.** Let $(X_1, X_2, \ldots, X_n)$ be a random vector with joint distribution function $F$ and continuous marginal distribution functions $F_1, F_2, \ldots, F_n$. Show that there exists a 'copula-form' distribution function $G(u_1, u_2, \ldots, u_n)$ such that $F(x_1, x_2, \ldots, x_n) = G(F_1(x_1), F_2(x_2), \ldots, F_n(x_n))$. [A 'copula-form' distribution function is one with the property that all its marginals are uniform on $[0, 1]$.]

---

†Also known in this context as the 'quantile function'.

## 4.12  Coupling and Poisson approximation

It is frequently necessary to compare the distributions of two random variables $X$ and $Y$. Since $X$ and $Y$ may not be defined on the same sample space $\Omega$, it is in general impossible to compare $X$ and $Y$ themselves. An extremely useful and modern technique is to construct copies $X'$ and $Y'$ (of $X$ and $Y$) on the same sample space $\Omega$, and then to compare $X'$ and $Y'$. This approach is known as *coupling*†, and it has many important applications. There is more than one possible coupling of a pair $X$ and $Y$, and the secret of success in coupling is to find the coupling which is well suited to the particular application.

Note that any two distributions may be coupled in a trivial way, since one may always find *independent* random variables $X$ and $Y$ with the required distributions; this may be done via the construction of a product space as in Section 1.6. This coupling has little interest, precisely because the value of $X$ does not influence the value of $Y$.

**(1) Example. Stochastic ordering.** Let $X$ and $Y$ be random variables whose distribution functions satisfy

**(2)**                              $F_X(x) \leq F_Y(x)$        for all $x \in \mathbb{R}$.

In this case, we say that $X$ *dominates* $Y$ *stochastically* and we write $X \geq_{\text{st}} Y$. Note that $X$ and $Y$ need not be defined on the same probability space.

The following theorem asserts in effect that $X \geq_{\text{st}} Y$ if and only if there exist copies of $X$ and $Y$ which are 'pointwise ordered'.

**(3) Theorem.** *Suppose that $X \geq_{\text{st}} Y$. There exists a probability space $(\Omega, \mathcal{F}, \mathbb{P})$ and two random variable $X'$ and $Y'$ on this space such that*:
(a)  *$X'$ and $X$ have the same distribution,*
(b)  *$Y'$ and $Y$ have the same distribution,*
(c)  *$\mathbb{P}(X' \geq Y') = 1$.*

**Proof.** Take $\Omega = [0, 1]$, $\mathcal{F}$ the Borel $\sigma$-field of $\Omega$, and let $\mathbb{P}$ be Lebesgue measure, which is to say that, for any sub-interval $I$ of $\Omega$, $\mathbb{P}(I)$ is defined to be the length of $I$.

For any distribution function $F$, we may define a random variable $Z_F$ on $(\Omega, \mathcal{F}, \mathbb{P})$ by

$$Z_F(\omega) = \inf\{z : \omega \leq F(z)\}, \qquad \omega \in \Omega.$$

Note that

**(4)**                          $\omega \leq F(z)$   if and only if   $Z_F(\omega) \leq z$.

It follows that

$$\mathbb{P}(Z_F \leq z) = \mathbb{P}\big([0, F(z)]\big) = F(z),$$

whence $Z_F$ has distribution function $F$.

Suppose now that $X \geq_{\text{st}} Y$ and write $G$ and $H$ for the distribution functions of $X$ and $Y$. Since $G(x) \leq H(x)$ for all $x$, we have from (4) that $Z_H \leq Z_G$. We set $X' = Z_G$ and $Y' = Z_H$. ∎ ●

---

†The term 'coupling' was introduced by Frank Spitzer around 1970. The coupling method was developed by W. Doeblin in 1938 to study Markov chains. See Lindvall 2002 for details of the history and the mathematics of coupling.

Here is a more physical coupling.

**(5) Example. Buffon's weldings.**  Suppose we cast at random two of Buffon's needles (introduced in Example (4.5.8)), labelled $N_1$ and $N_2$. Let $X$ (respectively, $Y$) be the indicator function of a line-crossing by $N_1$ (respectively, $N_2$). Whatever the relationship between $N_1$ and $N_2$, we have that $\mathbb{P}(X = 1) = \mathbb{P}(Y = 1) = 2/\pi$. The needles may however be coupled in various ways.

   (a)  The needles are linked by a frictionless universal joint at one end.
   (b)  The needles are welded at their ends to form a straight needle with length 2.
   (c)  The needles are welded perpendicularly at their midpoints, yielding the Buffon cross of
        Exercise (4.5.3).

We leave it as an *exercise* to calculate for each of these weldings (or 'couplings') the probability that both needles intersect a line.                                                    ●

**(6) Poisson convergence.**  Consider a large number of independent events each having small probability. In a sense to be made more specific, the number of such events which actually occur has a distribution which is close to a Poisson distribution. An instance of this remarkable observation was the proof in Example (3.5.4) that the bin$(n, \lambda/n)$ distribution approaches the Poisson distribution with parameter $\lambda$, in the limit as $n \to \infty$. Here is a more general result, proved using coupling.

The better to state the result, we introduce first a metric on the space of distribution functions. Let $F$ and $G$ be the distribution functions of discrete distributions which place masses $f_n$ and $g_n$ at the points $x_n$, for $n \geq 1$, and define

**(7)**
$$d_{\mathrm{TV}}(F, G) = \sum_{k \geq 1} |f_k - g_k|.$$

The definition of $d_{\mathrm{TV}}(F, G)$ may be extended to arbitrary distribution functions as in Problem (7.11.16); the quantity $d_{\mathrm{TV}}(F, G)$ is called the *total variation distance*† between $F$ and $G$. For random variables $X$ and $Y$, we define $d_{\mathrm{TV}}(X, Y) = d_{\mathrm{TV}}(F_X, F_Y)$. We note from Exercise (4.12.3) (see also Problem (2.7.13)) that

**(8)**
$$d_{\mathrm{TV}}(X, Y) = 2 \sup_{A \subseteq \mathbb{R}} \left| \mathbb{P}(X \in A) - \mathbb{P}(Y \in A) \right|$$

for discrete random variables $X, Y$.

**(9) Theorem‡.**  *Let* $\{X_r : 1 \leq r \leq n\}$ *be independent Bernoulli random variables with respective parameters* $\{p_r : 1 \leq r \leq n\}$, *and let* $S = \sum_{r=1}^{n} X_r$. *Then*

$$d_{\mathrm{TV}}(S, P) \leq 2 \sum_{r=1}^{n} p_r^2$$

*where* $P$ *is a random variable having the Poisson distribution with parameter* $\lambda = \sum_{r=1}^{n} p_r$.

---

†Some authors define the total variation distance to be one half of that given in (7).

‡Proved by Lucien Le Cam in 1960.

**Proof.** The trick is to find a suitable coupling of $S$ and $P$, and we do this as follows. Let $(X_r, Y_r), 1 \leq r \leq n$, be a sequence of independent pairs, where the pair $(X_r, Y_r)$ takes values in the set $\{0, 1\} \times \{0, 1, 2, \ldots\}$ with mass function

$$\mathbb{P}(X_r = x, \ Y_r = y) = \begin{cases} 1 - p_r & \text{if } x = y = 0, \\ e^{-p_r} - 1 + p_r & \text{if } x = 1, \ y = 0, \\ \dfrac{p_r^y}{y!} e^{-p_r} & \text{if } x = 1, \ y \geq 1. \end{cases}$$

It is easy to check that $X_r$ is Bernoulli with parameter $p_r$, and $Y_r$ has the Poisson distribution with parameter $p_r$.

We set

$$S = \sum_{r=1}^{n} X_r, \qquad P = \sum_{r=1}^{n} Y_r,$$

noting that $P$ has the Poisson distribution with parameter $\lambda = \sum_{r=1}^{n} p_r$; cf. Problem (3.11.6a).

Now,

$$\begin{aligned} \left| \mathbb{P}(S = k) - \mathbb{P}(P = k) \right| &= \left| \mathbb{P}(S = k, \ P \neq k) - P(S \neq k, \ P = k) \right| \\ &\leq \mathbb{P}(S = k, \ S \neq P) + \mathbb{P}(P = k, \ S \neq P), \end{aligned}$$

whence

$$d_{\mathrm{TV}}(S, P) = \sum_{k} \left| \mathbb{P}(S = k) - \mathbb{P}(P = k) \right| \leq 2\mathbb{P}(S \neq P).$$

We have as required that

$$\begin{aligned} \mathbb{P}(S \neq P) \leq \mathbb{P}(X_r \neq Y_r \text{ for some } r) &\leq \sum_{r=1}^{n} \mathbb{P}(X_r \neq Y_r) \\ &= \sum_{r=1}^{n} \left\{ e^{-p_r} - 1 + p_r + \mathbb{P}(Y_r \geq 2) \right\} \\ &= \sum_{r=1}^{n} p_r (1 - e^{-p_r}) \leq \sum_{r=1}^{n} p_r^2. \end{aligned}$$ ∎

**(10) Example.** Set $p_r = \lambda/n$ for $1 \leq r \leq n$ to obtain the inequality $d_{\mathrm{TV}}(S, P) \leq 2\lambda^2/n$, which provides a rate of convergence in the binomial–Poisson limit theorem of Example (3.5.4). ●

In many applications of interest, the Bernoulli trials $X_r$ are not independent. Nevertheless one may prove a Poisson limit theorem so long as they are not 'too dependent'. A beautiful way of doing this is to use the so-called 'Stein–Chen method', as follows.

As before, we suppose that $\{X_r : 1 \leq r \leq n\}$ are Bernoulli random variables with respective parameters $p_r$, but we make no assumption concerning their independence. With $S = \sum_{r=1}^{n} X_r$, we assume that there exists a sequence $V_1, V_2, \ldots, V_n$ of random variables with the property that

**(11)**                $\mathbb{P}(V_r = k - 1) = \mathbb{P}(S = k \mid X_r = 1), \qquad 1 \leq k \leq n.$

[We may assume that $p_r \neq 0$ for all $r$, whence $\mathbb{P}(X_r = 1) > 0$.] We shall see in the forthcoming Example (14) how such $V_r$ may sometimes be constructed in a natural way.

**(12) Theorem. Stein–Chen approximation.** *Let P be a random variable having the Poisson distribution with parameter $\lambda = \sum_{r=1}^n p_r$. The total variation distance between S and P satisfies*

$$d_{\mathrm{TV}}(S, P) \leq 2(1 \wedge \lambda^{-1}) \sum_{r=1}^n p_r \mathbb{E}|S - V_r|.$$

Recall that $x \wedge y = \min\{x, y\}$. The bound for $d_{\mathrm{TV}}(X, Y)$ takes a simple form in a situation where $\mathbb{P}(S \geq V_r) = 1$ for every $r$. If this holds,

$$\sum_{r=1}^n p_r \mathbb{E}|S - V_r| = \sum_{r=1}^n p_r\big(\mathbb{E}(S) - \mathbb{E}(V_r)\big) = \lambda^2 - \sum_{r=1}^n p_r \mathbb{E}(V_r).$$

By (11),

$$p_r \mathbb{E}(V_r) = p_r \sum_{k=1}^n (k-1)\mathbb{P}(S = k \mid X_r = 1) = \sum_{k=1}^n (k-1)\mathbb{P}(X_r = 1 \mid S = k)\mathbb{P}(S = k)$$

$$= \sum_{k=1}^n (k-1)\mathbb{E}(X_r \mid S = k)\mathbb{P}(S = k),$$

whence

$$\sum_{r=1}^n p_r \mathbb{E}(V_r) = \sum_{k=1}^n (k-1)k\mathbb{P}(S = k) = \mathbb{E}(S^2) - \mathbb{E}(S).$$

It follows by Theorem (12) that, in such a situation,

**(13)**                   $$d_{\mathrm{TV}}(S, P) \leq 2(1 \wedge \lambda^{-1})\big(\lambda - \mathrm{var}(S)\big).$$

Before proving Theorem (12), we give an example of its use.

**(14) Example. Balls in boxes.** There are $m$ balls and $n$ boxes. Each ball is placed in a box chosen uniformly at random, different balls being allocated to boxes independently of one another. The number $S$ of empty boxes may be written as $S = \sum_{r=1}^n X_r$ where $X_r$ is the indicator function of the event that the $r$th box is empty. It is easy to see that

$$p_r = \mathbb{P}(X_r = 1) = \left(\frac{n-1}{n}\right)^m,$$

whence $\lambda = np_r = n(1 - n^{-1})^m$. Note that the $X_r$ are not independent.

We now show how to generate a random sequence $V_r$ satisfying (11) in such a way that $\sum_r p_r \mathbb{E}|S - V_r|$ is small. If the $r$th box is empty, we set $V_r = S - 1$. If the $r$th box is not empty, we take the balls therein and distribute them randomly around the other $n-1$ boxes; we let $V_r$ be the number of these $n-1$ boxes which are empty at the end of this further allocation. It becomes evident after a little thought that (11) holds, and furthermore $V_r \le S$. Now,

$$\mathbb{E}(S^2) = \sum_{i,j} \mathbb{E}(X_i X_j) = \sum_i \mathbb{E}(X_i^2) + 2\sum_{i<j} \mathbb{E}(X_i X_j)$$

$$= \mathbb{E}(S) + n(n-1)\mathbb{E}(X_1 X_2),$$

where we have used the facts that $X_i^2 = X_i$ and $\mathbb{E}(X_i X_j) = \mathbb{E}(X_1 X_2)$ for $i \ne j$. Furthermore,

$$\mathbb{E}(X_1 X_2) = \mathbb{P}(\text{boxes 1 and 2 are empty}) = \left(\frac{n-2}{n}\right)^m,$$

whence, by (13),

$$d_{\mathrm{TV}}(S, P) \le 2(1 \wedge \lambda^{-1})\left\{\lambda^2 - n(n-1)\left(1 - \frac{2}{n}\right)^m\right\}.$$    ●

**Proof of Theorem (12).** Let $g : \{0, 1, 2, \dots\} \to \mathbb{R}$ be bounded, and define

$$\Delta g = \sup_r \{|g(r+1) - g(r)|\},$$

so that

**(15)**    $$|g(l) - g(k)| \le |l - k| \cdot \Delta g.$$

We have that

**(16)**    $$\left|\mathbb{E}\{\lambda g(S+1) - Sg(S)\}\right| = \left|\sum_{r=1}^n \{p_r \mathbb{E}g(S+1) - \mathbb{E}(X_r g(S))\}\right|$$

$$= \left|\sum_{r=1}^n p_r \mathbb{E}\{g(S+1) - g(V_r+1)\}\right|    \text{by (11)}$$

$$\le \Delta g \sum_{r=1}^n p_r \mathbb{E}|S - V_r|    \text{by (15)}.$$

Let $A$ be a set of non-negative integers. We choose the function $g = g_A$ in a special way so that $g_A(0) = 0$ and

**(17)**    $$\lambda g_A(r+1) - r g_A(r) = I_A(r) - \mathbb{P}(P \in A),    r \ge 0.$$

One may check that $g_A$ is given explicitly by

**(18)**    $$g_A(r+1) = \frac{r! e^\lambda}{\lambda^{r+1}}\left\{\mathbb{P}(\{P \le r\} \cap \{P \in A\}) - \mathbb{P}(P \le r)\mathbb{P}(P \in A)\right\},    r \ge 0.$$

A bound for $\Delta g_A$ appears in the next lemma, the proof of which is given later.

**(19) Lemma.** *We have that $\Delta g_A \leq 1 \wedge \lambda^{-1}$.*

We now substitute $r = S$ in (17) and take expectations, to obtain by (16), Lemma (19), and (8), that

$$d_{\mathrm{TV}}(S, P) = 2 \sup_A \left| \mathbb{P}(S \in A) - \mathbb{P}(P \in A) \right| \leq 2(1 \wedge \lambda^{-1}) \sum_{r=1}^{n} p_r \mathbb{E}|S - V_r|. \qquad \blacksquare$$

**Proof of Lemma (19).** Let $g_j = g_{\{j\}}$ for $j \geq 0$. From (18),

$$g_j(r+1) = \begin{cases} -\dfrac{r! \, e^\lambda}{\lambda^{r+1}} \mathbb{P}(P = j) \displaystyle\sum_{k=0}^{r} \dfrac{\lambda^k e^{-\lambda}}{k!} & \text{if } r < j, \\[4mm] \dfrac{r! \, e^\lambda}{\lambda^{r+1}} \mathbb{P}(P = j) \displaystyle\sum_{k=r+1}^{\infty} \dfrac{\lambda^k e^{-\lambda}}{k!} & \text{if } r \geq j, \end{cases}$$

implying that $g_j(r+1)$ is negative and decreasing when $r < j$, and is positive and decreasing when $r \geq j$. Therefore the only positive value of $g_j(r+1) - g_j(r)$ is when $r = j$, for which

$$g_j(j+1) - g_j(j) = \frac{e^{-\lambda}}{\lambda} \left\{ \sum_{k=j+1}^{\infty} \frac{\lambda^k}{k!} + \sum_{k=1}^{j} \frac{\lambda^k}{k!} \cdot \frac{k}{j} \right\}$$

$$\leq \frac{e^{-\lambda}}{\lambda}(e^\lambda - 1) = \frac{1 - e^{-\lambda}}{\lambda}$$

when $j \geq 1$. If $j = 0$, we have that $g_j(r+1) - g_j(r) \leq 0$ for all $r$.

Since $g_A(r+1) = \sum_{j \in A} g_j(r+1)$, it follows from the above remarks that

$$g_A(r+1) - g_A(r) \leq \frac{1 - e^{-\lambda}}{\lambda} \qquad \text{for all} \quad r \geq 1.$$

Finally, $-g_A = g_{A^c}$, and therefore $\Delta g_A \leq \lambda^{-1}(1 - e^{-\lambda})$. The claim of the lemma follows on noting that $\lambda^{-1}(1 - e^{-\lambda}) \leq 1 \wedge \lambda^{-1}$. $\qquad \blacksquare$

---

## Exercises for Section 4.12

**1.** Show that $X$ is stochastically larger than $Y$ if and only if $\mathbb{E}(u(X)) \geq \mathbb{E}(u(Y))$ for any non-decreasing function $u$ for which the expectations exist.

**2.** Let $X$ and $Y$ be Poisson distributed with respective parameters $\lambda$ and $\mu$. Show that $X$ is stochastically larger than $Y$ if $\lambda \geq \mu$.

**3.** Show that the total variation distance between two discrete variables $X, Y$ satisfies

$$d_{\mathrm{TV}}(X, Y) = 2 \sup_{A \subseteq \mathbb{R}} \left| \mathbb{P}(X \in A) - \mathbb{P}(Y \in A) \right|.$$

**4. Maximal coupling.** Show for discrete random variables $X, Y$ that $\mathbb{P}(X = Y) \leq 1 - \frac{1}{2} d_{\mathrm{TV}}(X, Y)$, where $d_{\mathrm{TV}}$ denotes total variation distance.

**5.  Maximal coupling continued.** Show that equality is possible in the inequality of Exercise (4.12.4) in the following sense. For any pair $X$, $Y$ of discrete random variables, there exists a pair $X'$, $Y'$ having the same marginal distributions as $X$, $Y$ such that $\mathbb{P}(X' = Y') = 1 - \frac{1}{2}d_{\mathrm{TV}}(X, Y)$.

**6.**  Let $X$ and $Y$ be indicator variables with $\mathbb{E}X = p$, $\mathbb{E}Y = q$. What is the maximum possible value of $\mathbb{P}(X = Y)$, as a function of $p, q$? Explain how $X, Y$ need to be distributed in order that $\mathbb{P}(X = Y)$ be:   (a) maximized,   (b) minimized.

**7.  Stop-loss ordering.** The random variable $X$ is said to be smaller than $Y$ in *stop-loss order* if $\mathbb{E}((X - a)^+) \le \mathbb{E}((Y - a)^+)$ for all $a \in \mathbb{R}$, in which case we write $X \le_{\mathrm{sl}} Y$. Show that $X \le_{\mathrm{sl}} Y$ is equivalent to: $\mathbb{E}(X^+)$, $\mathbb{E}(Y^+) < \infty$ and $\mathbb{E}(c(X)) \le \mathbb{E}(c(Y))$ for all increasing convex functions $c$.

**8.  Convex ordering.** The random variable $X$ is said to be smaller than $Y$ in *convex order* if $\mathbb{E}(u(X)) \le \mathbb{E}(u(Y))$ for all convex functions $u$ such that the expectations exist, in which case we write $X \le_{\mathrm{cx}} Y$.
(a)  Show that, if $X \le_{\mathrm{cx}} Y$, then $\mathbb{E}X = \mathbb{E}Y$ and $\mathrm{var}(X) \le \mathrm{var}(Y)$.
(b)  Show that $X \le_{\mathrm{cx}} Y$ if and only if $\mathbb{E}X = \mathbb{E}Y$ and $X \le_{\mathrm{sl}} Y$.

---

# 4.13  Geometrical probability

In many practical situations, one encounters pictures of apparently random shapes. For example, in a frozen section of some animal tissue, you will see a display of shapes; to undertake any serious statistical inference about such displays requires an appropriate probability model. Radio telescopes observe a display of microwave radiation emanating from the hypothetical 'Big Bang'. If you look at a forest floor, or at the microscopic structure of materials, or at photographs of a cloud chamber or of a foreign country seen from outer space, you will see apparently random patterns of lines, curves, and shapes.

Two problems arise in making precise the idea of a line or shape 'chosen at random'. The first is that, whereas a point in $\mathbb{R}^n$ is parametrized by its $n$ coordinates, the parametrizations of more complicated geometrical objects usually have much greater complexity. As a consequence, the most appropriate choice of density function is rarely obvious. Secondly, the appropriate sample space is often too large to allow an interpretation of 'choose an element uniformly at random'. For example, there is no 'uniform' probability measure on the line, or even on the set of integers. The usual way out of the latter difficulty is to work with the uniform probability measure on a large bounded subset of the state space.

The first difficulty referred to above may be illustrated by an example.

**(1) Example. Bertrand's paradox.**  What is the probability that an equilateral triangle, based on a random chord of a circle, is contained within the circle? This ill-posed question leads us to explore methods of interpreting the concept of a 'random chord'. Let $C$ be a circle with centre O and unit radius. Let $X$ denote the length of such a chord, and consider three cases.

(i)  A point P is picked at random in the interior of $C$, and taken as the midpoint of AB. Clearly $X > \sqrt{3}$ if and only if OP $< \frac{1}{2}$. Hence $\mathbb{P}(X > \sqrt{3}) = (\frac{1}{2})^2 = \frac{1}{4}$.

(ii)  Pick a point P at random on a randomly chosen radius of $C$, and take P as the midpoint of AB. Then $X > \sqrt{3}$ if and only if OP $< \frac{1}{2}$. Hence $\mathbb{P}(X > \sqrt{3}) = \frac{1}{2}$.

(iii)  A and B are picked independently at random on the circumference of $C$. Then $X > \sqrt{3}$ if and only if B lies in the third of the circumference most distant from A. Hence $\mathbb{P}(X > \sqrt{3}) = \frac{1}{3}$.                              ●

The different answers of this example arise because of the differing methods of interpreting 'pick a chord at random'. Do we have any reason to prefer any one of these methods above the others? It is easy to show that if the chord $L$ is determined by $\Pi$ and $\Theta$, where $\Pi$ is the length of the perpendicular from O to $L$, and $\Theta$ is the angle $L$ makes with a given direction, then the three choices given above correspond to the joint density function for the pair $(\Pi, \Theta)$ given respectively by:

   (i)   $f_1(p, \theta) = 2p/\pi$,

   (ii)  $f_2(p, \theta) = 1/\pi$,

   (iii) $f_3(p, \theta) = 2/\{\pi^2 \sqrt{1 - p^2}\}$,

for $0 \le p \le 1, 0 \le \theta \le \pi$.

It was shown by Poincaré that the uniform density of case (ii) may be used as a basis for the construction of a system of many random lines in the plane, whose probabilities are invariant under translation, rotation, and reflection. Since these properties seem desirable for the distribution of a single 'random line', the density function $f_2$ is commonly used†. With these preliminaries out of the way, we return to Buffon's needle.

**(2) Example. Buffon's needle: Example (4.5.8) revisited.** A needle of length $L$ is cast 'at random' onto a plane which is ruled by parallel straight lines, distance $d$ ($> L$) apart. It is not difficult to extend the argument of Example (4.5.8) to obtain that the probability that the needle is intersected by some line is $2L/(\pi d)$. See Problem (4.14.31).

Suppose we change our viewpoint; consider the needle to be fixed, and drop the grid of lines at random. For definiteness, we take the needle to be the line interval with centre at O, length $L$, and lying along the $x$-axis of $\mathbb{R}^2$. 'Casting the plane at random' is taken to mean the following. Draw a circle with centre O and diameter $d$. Pick a random chord of $C$ according to case (ii) above (re-scaled to take into account the fact that $C$ does not have unit radius), and draw the grid in the unique way such that it contains this random chord. It is easy to show that the probability that a line of the grid crosses the needle is $2L/(\pi d)$; see Problem (4.14.31b).

If we replace the needle by a curve $S$ having finite length $L(S)$, lying inside $C$, then the mean number of intersections between $S$ and the random chord is $2L(S)/(\pi d)$. See Problem (4.14.31c).

An interesting consequence is the following. Suppose that the curve $S$ is the boundary of a convex region. Then the number $I$ of intersections between the random chord and $S$ takes values in the set $\{0, 1, 2, \infty\}$, but only the values 0 and 2 have strictly positive probabilities. We deduce that

$$\mathbb{P}\big(\text{the random chord intersects } S\big) = \tfrac{1}{2}\mathbb{E}(I) = \frac{L(S)}{\pi d}.$$

Suppose further that $S'$ is the boundary of a convex subset of the inside of $S$, with length $L(S')$. If the random chord intersects $S'$ then it must surely intersect $S$, whence the conditional probability that it intersects $S'$ given that it intersects $S$ is $L(S')/L(S)$. This conclusion may be extended to include the case of two convex figures which are either disjoint or overlapping. See Exercise (4.13.2).     ●

We conclude with a few simple illustrative and amusing examples. In a classical branch of geometrical probability, one seeks to study geometrical relationships between points dropped

---

†See the discussion of the so-called uniform Poisson line process in Example (6.13.25).

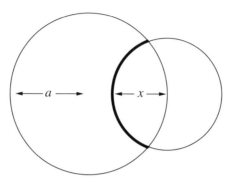

Figure 4.1. Two intersecting circles with radii $a$ and $x$ . The centre of the second circle lies on the first circle. The length of the emboldened arc is $2x \cos^{-1}(x/2a)$.

at random, where 'at random' is intended to imply a uniform density. An early example was recorded by Lewis Carroll: in order to combat insomnia, he solved mathematical problems in his head (that is to say, without writing anything down). On the night of 20th January 1884 he concluded that, if points A, B, C are picked at random in the plane, the probability that ABC is an obtuse triangle is $\frac{1}{8}\pi/\{\frac{1}{3}\pi - \frac{1}{4}\sqrt{3}\}$. This problem is not well posed as stated. We have no way of choosing a point uniformly at random in the plane. One interpretation is to choose the points at random within some convex figure of diameter $d$, to obtain the answer as a function of $d$, and then take the limit as $d \to \infty$. Unfortunately, this can yield different answers depending on the choice of figure (see Exercise (4.13.5)).

Furthermore, Carroll's solution proceeded by constructing axes depending on the largest side of the triangle ABC, and this conditioning affects the distribution of the position of the remaining point. It is possible to formulate a different problem to which Carroll's answer is correct. Other examples of this type may be found in the exercises.

A useful method for tackling a class of problems in geometrical probability is a technique called *Crofton's method*. The basic idea, developed by Crofton to obtain many striking results, is to identify a real-valued parameter of the problem in question, and to establish a differential equation for the probability or expectation in question, in terms of this parameter. This vague synopsis may be made more precise by an example.

**(3) Example.** Two arrows $A$ and $B$ strike at random a circular target of unit radius. What is the density function of the distance $X$ between the points struck by the arrows?

**Solution.** Let us take as target the disk of radius $a$ given in polar coordinates as $\{(r, \theta) : r \le a\}$. We shall establish a differential equation in the variable $a$. Let $f(\cdot, a)$ denote the density function of $X$.

We have by conditional probability that

$$
\textbf{(4)} \qquad f(x, a + \delta a) = f_0(x, a + \delta a)\mathbb{P}_{a+\delta a}(R_0) + f_1(x, a + \delta a)\mathbb{P}_{a+\delta a}(R_1) \\
+ f_2(x, a + \delta a)\mathbb{P}_{a+\delta a}(R_2),
$$

where $R_i$ is the event that exactly $i$ arrows strike the annulus $\{(r, \theta) : a \le r \le a + \delta a\}$, $f_i(x, a + \delta a)$ is the density function of $X$ given the event $R_i$, and $\mathbb{P}_y$ is the probability measure appropriate for a disk of radius $y$.

Conditional on $R_0$, the arrows are uniformly distributed on the disk of radius $a$, whence $f_0(x, a + \delta a) = f(x, a)$. By considering Figure 4.1, we have that†

$$f_1(x, a + \delta a) = \frac{2x}{\pi a^2} \cos^{-1}\left(\frac{x}{2a}\right) + o(1), \quad \text{as } \delta a \to 0,$$

and by the independence of the arrows,

$$\mathbb{P}_{a+\delta a}(R_0) = \left(\frac{a}{a + \delta a}\right)^4 = 1 - \frac{4\,\delta a}{a} + o(\delta a),$$

$$\mathbb{P}_{a+\delta a}(R_1) = \frac{4\,\delta a}{a} + o(\delta a), \quad \mathbb{P}_{a+\delta a}(R_2) = o(\delta a).$$

Taking the limit as $\delta a \to 0$, we obtain the differential equation

(5) $$\frac{\partial f}{\partial a}(x, a) = -\frac{4}{a} f(x, a) + \frac{8x}{\pi a^3} \cos^{-1}\left(\frac{x}{2a}\right).$$

Subject to a suitable boundary condition, it follows that

$$a^4 f(x, a) = \int_0^a \frac{8xu}{\pi} \cos^{-1}\left(\frac{x}{2u}\right) du$$

$$= \frac{2xa^2}{\pi} \left\{ 2\cos^{-1}\left(\frac{x}{2a}\right) - \frac{x}{a}\sqrt{1 - \left(\frac{x}{2a}\right)^2} \right\}, \quad 0 \le x \le 2a.$$

The last integral may be verified by use of a symbolic algebra package, or by looking it up elsewhere, or by using the fundamental theorem of calculus. Fans of unarmed combat may use the substitution $\theta = \cos^{-1}\{x/(2u)\}$. The required density function is $f(x, 1)$.   ●

We conclude with some amusing and classic results concerning areas of random triangles. Triangles have the useful property that, given any two triangles $T$ and $T'$, there exists an affine transformation (that is, an orthogonal projection together with a change of scale) which transforms $T$ into $T'$. Such transformations multiply areas by a constant factor, leaving many probabilities and expectations of interest unchanged. In the following, we denote by $|ABC|$ the area of the triangle with vertices A, B, C.

**(6) Example. Area of a random triangle.** Three points P, Q, R are picked independently at random in the triangle ABC. Show that

(7) $$\mathbb{E}|PQR| = \tfrac{1}{12}|ABC|.$$

**Solution.** We proceed via a sequence of lemmas which you may illustrate with diagrams.

**(8) Lemma.** *Let $G_1$ and $G_2$ be the centres of gravity of ABM and AMC, where M is the midpoint of BC. Choose P at random in the triangle ABM, and Q at random (independently of P) in the triangle AMC. Then*

(9) $$\mathbb{E}|APQ| = \mathbb{E}|AG_1G_2| = \tfrac{2}{9}|ABC|.$$

---

†See Subsection (10) of Appendix I for a reminder about Landau's O/o notation.

**Proof.** Elementary; this is Exercise (4.13.7).                                        ∎

**(10) Lemma.** *Choose* P *and* Q *independently at random in the triangle* ABC*. Then*

(11)                                $$\mathbb{E}|APQ| = \tfrac{4}{27}|ABC|.$$

**Proof.** By the property of affine transformations discussed above, there exists a real number $\alpha$, independent of the choice of ABC, such that

(12)                                $$\mathbb{E}|APQ| = \alpha|ABC|.$$

Denote ABM by $T_1$ and AMC by $T_2$, and let $C_{ij}$ be the event that $\{P \in T_i, Q \in T_j\}$, for $i, j \in \{1, 2\}$. Using conditional expectation and the fact that $\mathbb{P}(C_{ij}) = \tfrac{1}{4}$ for each pair $i, j$,

$$\mathbb{E}|APQ| = \sum_{i,j} \mathbb{E}(|APQ| \mid C_{ij})\mathbb{P}(C_{ij})$$

$$= \alpha|ABM|\mathbb{P}(C_{11}) + \alpha|AMC|\mathbb{P}(C_{22}) + \tfrac{2}{9}|ABC|\big(\mathbb{P}(C_{12}) + \mathbb{P}(C_{21})\big) \quad \text{by (9)}$$

$$= \tfrac{1}{4}\alpha|ABC| + \tfrac{1}{2} \cdot \tfrac{2}{9}|ABC|.$$

We use (12) and divide by $|ABC|$ to obtain $\alpha = \tfrac{4}{27}$, as required.                    ∎

**(13) Lemma.** *Let* P *and* Q *be chosen independently at random in the triangle* ABC*, and* R *be chosen independently of* P *and* Q *at random on the side* BC*. Then*

$$\mathbb{E}|PQR| = \tfrac{1}{9}|ABC|.$$

**Proof.** If the length of BC is $a$, then $|BR|$ is uniformly distributed on the interval $(0, a)$. Denote the triangles ABR and ARC by $S_1$ and $S_2$, and let $D_{ij} = \{P \in S_i, Q \in S_j\}$ for $i, j \in \{1, 2\}$. Let $x \geq 0$, and let $\mathbb{P}_x$ and $\mathbb{E}_x$ denote probability and expectation conditional on the event $\{|BR| = x\}$. We have that

$$\mathbb{P}_x(D_{11}) = \frac{x^2}{a^2}, \quad \mathbb{P}_x(D_{22}) = \left(\frac{a-x}{a}\right)^2, \quad \mathbb{P}_x(D_{12}) = \mathbb{P}_x(D_{21}) = \frac{x(a-x)}{a^2}.$$

By conditional expectation,

$$\mathbb{E}_x|PQR| = \sum_{i,j} \mathbb{E}_x(|PQR| \mid D_{ij})\mathbb{P}(D_{ij}).$$

By Lemma (10),

$$\mathbb{E}_x(|PQR| \mid D_{11}) = \frac{4}{27}\mathbb{E}_x|ABR| = \frac{4}{27} \cdot \frac{x}{a}|ABC|,$$

and so on, whence

$$\mathbb{E}_x|PQR| = \left\{\frac{4}{27}\left(\frac{x}{a}\right)^3 + \frac{4}{27}\left(\frac{a-x}{a}\right)^3 + \frac{2}{9}\frac{x(a-x)}{a^2}\right\}|ABC|.$$

Averaging over |BR| we deduce that

$$\mathbb{E}|PQR| = \frac{1}{a}\int_0^a \mathbb{E}_x|PQR|\,dx = \frac{1}{9}|ABC|. \qquad \blacksquare$$

We may now complete the proof of (7).

**Proof of (7).** By the property of affine transformations mentioned above, it is sufficient to show that $\mathbb{E}|PQR| = \frac{1}{12}|ABC|$ for any single given triangle ABC. Consider the special choice $A = (0,0)$, $B = (x,0)$, $C = (0,x)$, and denote by $\mathbb{P}_x$ the appropriate probability measure when three points P, Q, R are picked from ABC. We write $A(x)$ for the mean area $\mathbb{E}_x|PQR|$. We shall use Crofton's method, with $x$ as the parameter to be varied. Let $\Delta$ be the trapezium with vertices $(0,x)$, $(0, x+\delta x)$, $(x+\delta x, 0)$, $(x,0)$. Then

$$\mathbb{P}_{x+\delta x}(P, Q, R \in ABC) = \left\{\frac{x^2}{(x+\delta x)^2}\right\}^3 = 1 - \frac{6\,\delta x}{x} + o(\delta x)$$

and

$$\mathbb{P}_{x+\delta x}\big(\{P, Q \in ABC\} \cap \{R \in \Delta\}\big) = \frac{2\,\delta x}{x} + o(\delta x).$$

Hence, by conditional expectation and Lemma (13),

$$A(x+\delta x) = A(x)\left(1 - \frac{6\,\delta x}{x}\right) + \frac{1}{9}\cdot\frac{1}{2}x^2\cdot\frac{6\,\delta x}{x} + o(\delta x),$$

leading, in the limit as $\delta x \to 0$, to the equation

$$\frac{dA}{dx} = -\frac{6A}{x} + \frac{1}{3}x,$$

with boundary condition $A(0) = 0$. The solution is $A(x) = \frac{1}{24}x^2$. Since $|ABC| = \frac{1}{2}x^2$, the proof is complete.   $\blacksquare\bullet$

---

## Exercises for Section 4.13

With apologies to those who prefer their exercises better posed ...

**1.**   Pick two points A and B independently at random on the circumference of a circle $C$ with centre O and unit radius. Let $\Pi$ be the length of the perpendicular from O to the line AB, and let $\Theta$ be the angle AB makes with the horizontal. Show that $(\Pi, \Theta)$ has joint density

$$f(p, \theta) = \frac{1}{\pi^2\sqrt{1-p^2}}, \qquad 0 \le p \le 1,\ 0 \le \theta < 2\pi.$$

**2.**   Let $S_1$ and $S_2$ be disjoint convex shapes with boundaries of length $b(S_1)$, $b(S_2)$, as illustrated in the figure beneath. Let $b(H)$ be the length of the boundary of the convex hull of $S_1$ and $S_2$, incorporating their exterior tangents, and $b(X)$ the length of the crossing curve using the interior tangents to loop round $S_1$ and $S_2$. Show that the probability that a random line crossing $S_1$ also crosses $S_2$ is $\{b(X) - b(H)\}/b(S_1)$. (See Example (4.13.2) for an explanation of the term 'random line'.) How is this altered if $S_1$ and $S_2$ are not disjoint?

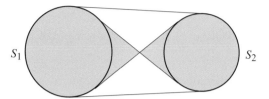

The circles are the shapes $S_1$ and $S_2$. The shaded regions are denoted $A$ and $B$, and $b(X)$ is
the sum of the perimeter lengths of $A$ and $B$.

**3.** Let $S_1$ and $S_2$ be convex figures such that $S_2 \subseteq S_1$. Show that the probability that two independent
random lines $\lambda_1$ and $\lambda_2$, crossing $S_1$, meet within $S_2$ is $2\pi |S_2|/b(S_1)^2$, where $|S_2|$ is the area of $S_2$
and $b(S_1)$ is the length of the boundary of $S_1$. (See Example (4.13.2) for an explanation of the term
'random line'.)

**4.** Let $Z$ be the distance between two points picked independently at random in a disk of radius $a$.
Show that $\mathbb{E}(Z) = 128a/(45\pi)$, and $\mathbb{E}(Z^2) = a^2$.

**5.** Pick two points A and B independently at random in a ball with centre O. Show that the probability
that the angle $\widehat{AOB}$ is obtuse is $\frac{5}{8}$. Compare this with the corresponding result for two points picked
at random in a circle.

**6.** A triangle is formed by A, B, and a point P picked at random in a set $S$ with centre of gravity G.
Show that $\mathbb{E}|ABP| = |ABG|$.

**7.** A point D is fixed on the side BC of the triangle ABC. Two points P and Q are picked independently
at random in ABD and ADC respectively. Show that $\mathbb{E}|APQ| = |AG_1 G_2| = \frac{2}{9}|ABC|$, where $G_1$ and
$G_2$ are the centres of gravity of ABD and ADC.

**8.** From the set of all triangles that are similar to the triangle ABC, similarly oriented, and inside
ABC, one is selected uniformly at random. Show that its mean area is $\frac{1}{10}|ABC|$.

**9.** Two points $X$ and $Y$ are picked independently at random in the interval $(0, a)$. By varying $a$,
show that $F(z, a) = \mathbb{P}(|X - Y| \leq z)$ satisfies

$$\frac{\partial F}{\partial a} + \frac{2}{a}F = \frac{2z}{a^2}, \qquad 0 \leq z \leq a,$$

and hence find $F(z, a)$. Let $r \geq 1$, and show that $m_r(a) = \mathbb{E}(|X - Y|^r)$ satisfies

$$a\frac{dm_r}{da} = 2\left\{\frac{a^r}{r+1} - m_r\right\}.$$

Hence find $m_r(a)$.

**10.** Lines are laid down independently at random on the plane, dividing it into polygons. Show that
the average number of sides of this set of polygons is 4. [Hint: Consider $n$ random great circles of a
sphere of radius $R$; then let $R$ and $n$ increase.]

**11.** A point P is picked at random in the triangle ABC. The lines AP, BP, CP, produced, meet BC,
AC, AB respectively at L, M, N. Show that $\mathbb{E}|LMN| = (10 - \pi^2)|ABC|$.

**12. Sylvester's problem.** If four points are picked independently at random inside the triangle ABC,
show that the probability that no one of them lies inside the triangle formed by the other three is $\frac{2}{3}$.

**13.** If three points P, Q, R are picked independently at random in a disk of radius $a$, show that $\mathbb{E}|PQR| =
35a^2/(48\pi)$. [You may find it useful that $\int_0^\pi \int_0^\pi \sin^3 x \sin^3 y \sin|x - y| \, dx \, dy = 35\pi/128$.]

**14.** Two points A and B are picked independently at random inside a disk $C$. Show that the probability that the circle having centre A and radius |AB| lies inside $C$ is $\frac{1}{6}$.

**15.** Two points A and B are picked independently at random inside a ball $S$. Show that the probability that the sphere having centre A and radius |AB| lies inside $S$ is $\frac{1}{20}$.

**16.** Pick two points independently and uniformly at random on the surface of the sphere of $\mathbb{R}^3$ with radius 1. Show that the density of the Euclidean distance $D$ between them is $f(d) = \frac{1}{2}d$ for $d \in (0, 2)$.

**17.** Two points are chosen independently and uniformly at random on the perimeter (including the diameter) of a semicircle with unit radius. What is the probability of the event $D$ that exactly one of them lies on the diameter, and of the event $N$ that neither lies on the diameter?

Let $A$ be the area of the triangle formed by the two points and the midpoint of the diameter. Show that $\mathbb{E}(A \mid D) = 1/(2\pi)$, and $\mathbb{E}(A \mid N) = 1/\pi$. Hence or otherwise show that $\mathbb{E}(A) = 1/(2 + \pi)$.

**18. Stevens's solution of Jeffreys's bicycle wheel problem.** After passing over a stretch of road strewn with tacks, a cyclist looks repeatedly at the front wheel to check whether the tyre has picked up a tack. One inspection of the wheel covers a fraction $x$ of the tyre. After $n$ independent such inspections, each uniformly positioned on the wheel, show that the probability of having inspected the entire tyre is $1 - \binom{n}{1}(1-x)^{n-1} + \binom{n}{2}(1-2x)^{n-1} - \cdots$, where the series terminates at the term in $1 - kx$ where $k = \lfloor 1/x \rfloor$ is the integer part of $1/x$.

## 4.14 Problems

**1.** (a) Show that $\int_{-\infty}^{\infty} e^{-x^2} dx = \sqrt{\pi}$, and deduce that

$$f(x) = \frac{1}{\sigma\sqrt{2\pi}} \exp\left\{-\frac{(x-\mu)^2}{2\sigma^2}\right\}, \qquad -\infty < x < \infty,$$

is a density function if $\sigma > 0$.
(b) Calculate the mean and variance of a standard normal variable.
(c) Show that the $N(0, 1)$ distribution function $\Phi$ satisfies

$$(x^{-1} - x^{-3})e^{-\frac{1}{2}x^2} < \sqrt{2\pi}[1 - \Phi(x)] < x^{-1}e^{-\frac{1}{2}x^2}, \qquad x > 0.$$

These bounds are of interest because $\Phi$ has no closed form.
(d) Let $X$ be $N(0, 1)$, and $a > 0$. Show that $\mathbb{P}(X > x + a/x \mid X > x) \to e^{-a}$ as $x \to 0$.

**2.** Let $X$ be continuous with density function $f(x) = C(x - x^2)$, where $\alpha < x < \beta$ and $C > 0$.
(a) What are the possible values of $\alpha$ and $\beta$?
(b) What is $C$?

**3.** Let $X$ be a random variable which takes non-negative values only. Show that

$$\sum_{i=1}^{\infty} (i - 1)I_{A_i} \leq X < \sum_{i=1}^{\infty} iI_{A_i},$$

where $A_i = \{i - 1 \leq X < i\}$. Deduce that

$$\sum_{i=1}^{\infty} \mathbb{P}(X \geq i) \leq \mathbb{E}(X) < 1 + \sum_{i=1}^{\infty} \mathbb{P}(X \geq i).$$

4.   (a) Let $X$ have a continuous distribution function $F$. Show that
     (i) $F(X)$ is uniformly distributed on $[0, 1]$,
     (ii) $-\log F(X)$ is exponentially distributed.

(b) A straight line $l$ touches a circle with unit diameter at the point P which is diametrically opposed on the circle to another point Q. A straight line QR joins Q to some point R on $l$. If the angle $\widehat{PQR}$ between the lines PQ and QR is a random variable with the uniform distribution on $[-\frac{1}{2}\pi, \frac{1}{2}\pi]$, show that the length of PR has the Cauchy distribution (this length is measured positive or negative depending upon which side of P the point R lies).

(c) Let the net scores in the two halves of a game between teams $A$ and $B$ be independent and identically distributed random variables $X$, $Y$ that are symmetric about 0 and have a continuous distribution function $F$. Team $A$ wins (respectively, loses) if $X+Y > 0$ (respectively, $X+Y \le 0$). Find the probability that $A$ wins conditional on the half-time score $X$.

5.   **Lack of memory property.** (a) Let $g : [0, \infty) \to (0, \infty)$ be such that $g(s+t) = g(s)g(t)$ for $s, t \ge 0$. If $g$ is monotone, show that $g(s) = e^{\mu s}$ for some $\mu \in \mathbb{R}$.

(b) Let $X$ have an exponential distribution. Show that $\mathbb{P}(X > s+x \mid X > s) = \mathbb{P}(X > x)$, for $x, s \ge 0$. This is the 'lack of memory' property again. Show that the exponential distribution is the only continuous distribution with this property.

(c) The height $M$ of the largest tidal surge in a certain estuary has the exponential distribution with parameter 1, and it costs $h$ to build a barrier with height $h$ in order to protect an upstream city. If $M \le h$, the surge costs nothing; if $M > h$, the surge entails an extra cost of $a + b(M - h)$. Show that, for $a + b \ge 1$, the expected total cost is minimal when $h = \log(a+b)$. What if $a + b < 1$?

6.   Show that $X$ and $Y$ are independent continuous variables if and only if their joint density function $f$ factorizes as the product $f(x, y) = g(x)h(y)$ of functions of the single variables $x$ and $y$ alone.

     Explain why the density function $f(x, y) = Ce^{-x-y}$ (for $x, y \ge 0, x+y > 1$) does not provide a counterexample to this assertion.

7.   Let $X$ and $Y$ have joint density function $f(x, y) = 2e^{-x-y}$, $0 < x < y < \infty$. Are they independent? Find their marginal density functions and their covariance.

8.   **Bertrand's paradox extended.** A chord of the unit circle is picked at random. What is the probability that an equilateral triangle with the chord as base can fit inside the circle if:

(a) the chord passes through a point P picked uniformly in the disk, and the angle it makes with a fixed direction is uniformly distributed on $[0, 2\pi)$,

(b) the chord passes through a point P picked uniformly at random on a randomly chosen radius, and the angle it makes with the radius is uniformly distributed on $[0, 2\pi)$.

9.   **Monte Carlo.** It is required to estimate $J = \int_0^1 g(x)\,dx$ where $0 \le g(x) \le 1$ for all $x$, as in Example (2.6.3). Let $X$ and $Y$ be independent random variables with common density function $f(x) = 1$ if $0 < x < 1$, $f(x) = 0$ otherwise. Let $U = I_{\{Y \le g(X)\}}$, the indicator function of the event that $Y \le g(X)$, and let $V = g(X)$, $W = \frac{1}{2}\{g(X)+g(1-X)\}$. Show that $\mathbb{E}(U) = \mathbb{E}(V) = \mathbb{E}(W) = J$, and that $\text{var}(W) \le \text{var}(V) \le \text{var}(U)$, so that, of the three, $W$ is the most 'efficient' estimator of $J$.

10.  Let $X_1, X_2, \ldots, X_n$ be independent exponential variables, parameter $\lambda$. Show by induction that $S = X_1 + X_2 + \cdots + X_n$ has the $\Gamma(\lambda, n)$ distribution.

11.  Let $X$ and $Y$ be independent variables, $\Gamma(\lambda, m)$ and $\Gamma(\lambda, n)$ respectively.
(a) Use the result of Problem (4.14.10) to show that $X + Y$ is $\Gamma(\lambda, m+n)$ when $m$ and $n$ are integral (the same conclusion is actually valid for non-integral $m$ and $n$).
(b) Find the joint density function of $X + Y$ and $X/(X + Y)$, and deduce that they are independent.
(c) If $Z$ is Poisson with parameter $\lambda t$, and $m$ is integral, show that $\mathbb{P}(Z < m) = \mathbb{P}(X > t)$.
(d) If $0 < m < n$ and $B$ is independent of $Y$ with the beta distribution with parameters $m$ and $n - m$, show that $YB$ has the same distribution as $X$.

**12.** Let $X_1, X_2, \ldots, X_n$ be independent $N(0, 1)$ variables.

(a) Show that $X_1^2$ is $\chi^2(1)$.

(b) Show that $X_1^2 + X_2^2$ is $\chi^2(2)$ by expressing its distribution function as an integral and changing to polar coordinates.

(c) More generally, show that $X_1^2 + X_2^2 + \cdots + X_n^2$ is $\chi^2(n)$.

**13.** Let $X$ and $Y$ have the bivariate normal distribution with means $\mu_1, \mu_2$, variances $\sigma_1^2, \sigma_2^2$, and correlation $\rho$. Show that

(a) $\mathbb{E}(X \mid Y) = \mu_1 + \rho\sigma_1(Y - \mu_2)/\sigma_2$,

(b) the variance of the conditional density function $f_{X|Y}$ is $\text{var}(X \mid Y) = \sigma_1^2(1 - \rho^2)$.

**14.** Let $X$ and $Y$ have joint density function $f$. Find the density function of $Y/X$.

**15.** Let $X$ and $Y$ be independent variables with common density function $f$. Show that $\tan^{-1}(Y/X)$ has the uniform distribution on $(-\frac{1}{2}\pi, \frac{1}{2}\pi)$ if and only if

$$\int_{-\infty}^{\infty} f(x)f(xy)|x|\,dx = \frac{1}{\pi(1 + y^2)}, \qquad y \in \mathbb{R}.$$

Verify that this is valid if either $f$ is the $N(0, 1)$ density function or $f(x) = a(1 + x^4)^{-1}$ for some constant $a$.

**16. Rayleigh distribution.** Let $X$ and $Y$ be independent $N(0, 1)$ variables, and think of $(X, Y)$ as a random point in the plane. Change to polar coordinates $(R, \Theta)$ given by $R^2 = X^2 + Y^2$, $\tan \Theta = Y/X$; show that $R^2$ is $\chi^2(2)$, $\tan \Theta$ has the Cauchy distribution, and $R$ and $\Theta$ are independent. Find the density of $R$.

Find $\mathbb{E}(X^2/R^2)$ and

$$\mathbb{E}\left\{ \frac{\min\{|X|, |Y|\}}{\max\{|X|, |Y|\}} \right\}.$$

**17.** If $X$ and $Y$ are independent random variables, show that $U = \min\{X, Y\}$ and $V = \max\{X, Y\}$ have distribution functions

$$F_U(u) = 1 - \{1 - F_X(u)\}\{1 - F_Y(u)\}, \qquad F_V(v) = F_X(v)F_Y(v).$$

Let $X$ and $Y$ be independent exponential variables, parameter 1. Show that

(a) $U$ is exponential, parameter 2,

(b) $V$ has the same distribution as $X + \frac{1}{2}Y$. Hence find the mean and variance of $V$.

**18.** Let $X$ and $Y$ be independent variables having the exponential distribution with parameters $\lambda$ and $\mu$ respectively. Let $U = \min\{X, Y\}$, $V = \max\{X, Y\}$, and $W = V - U$.

(a) Find $\mathbb{P}(U = X) = \mathbb{P}(X \leq Y)$.

(b) Show that $U$ and $W$ are independent.

**19.** Let $X$ and $Y$ be independent non-negative random variables with continuous density functions on $(0, \infty)$.

(a) If, given $X + Y = u$, $X$ is uniformly distributed on $[0, u]$ whatever the value of $u$, show that $X$ and $Y$ have the exponential distribution.

(b) If, given that $X + Y = u$, $X/u$ has a given beta distribution (parameters $\alpha$ and $\beta$, say) whatever the value of $u$, show that $X$ and $Y$ have gamma distributions.

You may need the fact that the only non-negative continuous solutions of the functional equation $g(s + t) = g(s)g(t)$ for $s, t \geq 0$, with $g(0) = 1$, are of the form $g(s) = e^{\mu s}$. Remember Problem (4.14.5).

**20.** Show that it cannot be the case that $U = X + Y$ where $U$ is uniformly distributed on $[0, 1]$ and $X$ and $Y$ are independent and identically distributed. You should not assume that $X$ and $Y$ are continuous variables.

**21. Order statistics.** Let $X_1, X_2, \ldots, X_n$ be independent identically distributed variables with a common density function $f$. Such a collection is called a *random sample*. For each $\omega \in \Omega$, arrange the sample values $X_1(\omega), \ldots, X_n(\omega)$ in non-decreasing order $X_{(1)}(\omega) \le X_{(2)}(\omega) \le \cdots \le X_{(n)}(\omega)$, where $(1), (2), \ldots, (n)$ is a (random) permutation of $1, 2, \ldots, n$. The new variables $X_{(1)}, X_{(2)}, \ldots, X_{(n)}$ are called the *order statistics*. Show, by a symmetry argument, that the joint distribution function of the order statistics satisfies

$$\mathbb{P}(X_{(1)} \le y_1, \ldots, X_{(n)} \le y_n) = n! \, \mathbb{P}(X_1 \le y_1, \ldots, X_n \le y_n, \; X_1 < X_2 < \cdots < X_n)$$

$$= \int \cdots \int_{\substack{x_1 \le y_1 \\ x_2 \le y_2 \\ \vdots \\ x_n \le y_n}} L(x_1, \ldots, x_n) n! \, f(x_1) \cdots f(x_n) \, dx_1 \cdots dx_n$$

where $L$ is given by

$$L(\mathbf{x}) = \begin{cases} 1 & \text{if } x_1 < x_2 < \cdots < x_n, \\ 0 & \text{otherwise,} \end{cases}$$

and $\mathbf{x} = (x_1, x_2, \ldots, x_n)$. Deduce that the joint density function of $X_{(1)}, \ldots, X_{(n)}$ is $g(\mathbf{y}) = n! \, L(\mathbf{y}) f(y_1) \cdots f(y_n)$.

**22.** Find the marginal density function of the $k$th order statistic $X_{(k)}$ of a sample with size $n$:
(a) by integrating the result of Problem (4.14.21),
(b) directly.

**23.** Find the joint density function of the order statistics of $n$ independent uniform variables on $[0, T]$.

**24.** Let $X_1, X_2, \ldots, X_n$ be independent and uniformly distributed on $[0, 1]$, with order statistics $X_{(1)}, X_{(2)}, \ldots, X_{(n)}$.
(a) Show that, for fixed $k$, the density function of $n X_{(k)}$ converges as $n \to \infty$, and find and identify the limit function.
(b) Show that $\log X_{(k)}$ has the same distribution as $-\sum_{i=k}^{n} i^{-1} Y_i$, where the $Y_i$ are independent random variables having the exponential distribution with parameter 1.
(c) Show that $Z_1, Z_2, \ldots, Z_n$, defined by $Z_k = (X_{(k)}/X_{(k+1)})^k$ for $k < n$ and $Z_n = (X_{(n)})^n$, are independent random variables with the uniform distribution on $[0, 1]$.

**25.** Let $X_1, X_2, X_3$ be independent variables with the uniform distribution on $[0, 1]$. What is the probability that rods of lengths $X_1, X_2$, and $X_3$ may be used to make a triangle? Generalize your answer to $n$ rods used to form a polygon.

**26. Stick breaking.** Let $X_1$ and $X_2$ be independent variables with the uniform distribution on $[0, 1]$. A stick of unit length is broken at points distance $X_1$ and $X_2$ from one of the ends. What is the probability that the three pieces may be used to make a triangle? Generalize your answer to a stick broken in $n$ places.

**27.** Let $X, Y$ be a pair of jointly continuous variables.
(a) **Hölder's inequality.** Show that if $p, q > 1$ and $p^{-1} + q^{-1} = 1$ then

$$\mathbb{E}|XY| \le \{\mathbb{E}|X^p|\}^{1/p} \{\mathbb{E}|Y^q|\}^{1/q}.$$

Set $p = q = 2$ to deduce the Cauchy–Schwarz inequality $\mathbb{E}(XY)^2 \le \mathbb{E}(X^2)\mathbb{E}(Y^2)$.
(b) **Minkowski's inequality.** Show that, if $p \ge 1$, then

$$\{\mathbb{E}(|X + Y|^p)\}^{1/p} \le \{\mathbb{E}|X^p|\}^{1/p} + \{\mathbb{E}|Y^p|\}^{1/p}.$$

Note that in both cases your proof need not depend on the continuity of $X$ and $Y$; deduce that the same inequalities hold for discrete variables.

**28.** Let $Z$ be a random variable. Choose $X$ and $Y$ appropriately in the Cauchy–Schwarz (or Hölder) inequality to show that $g(p) = \log \mathbb{E}|Z^p|$ is a convex function of $p$ on the interval of values of $p$ such that $\mathbb{E}|Z^p| < \infty$. Deduce **Lyapunov's inequality**:

$$\{\mathbb{E}|Z^r|\}^{1/r} \geq \{\mathbb{E}|Z^s|\}^{1/s} \quad \text{whenever } r \geq s > 0.$$

You have shown in particular that, if $Z$ has finite $r$th moment, then $Z$ has finite $s$th moment for all positive $s \leq r$.

Show more generally that, if $r_1, r_2, \ldots, r_n$ are real numbers such that $\mathbb{E}|X^{r_k}| < \infty$ and $s = n^{-1}\sum_{k=1}^n r_k$, then

$$\left\{ \prod_{k=1}^n \mathbb{E}|X^{r_k}| \right\} - \{\mathbb{E}(|X|^s)\} \geq 0.$$

**29.** Show that, using the obvious notation, $\mathbb{E}\{\mathbb{E}(X \mid Y, Z) \mid Y\} = \mathbb{E}(X \mid Y)$.

**30. Rényi's parking problem.** Motor cars of unit length park randomly in a street in such a way that the centre of each car, in turn, is positioned uniformly at random in the space available to it. Let $m(x)$ be the expected number of cars which are able to park in a street of length $x$. Show that

$$m(x+1) = \frac{1}{x} \int_0^x \{m(y) + m(x-y) + 1\} \, dy.$$

It is possible to deduce that $m(x)$ is about as big as $\frac{3}{4}x$ when $x$ is large.

**31. Buffon's needle revisited: Buffon's noodle.**
(a) A plane is ruled by the lines $y = nd$ ($n = 0, \pm1, \ldots$). A needle with length $L$ ($< d$) is cast randomly onto the plane. Show that the probability that the needle intersects a line is $2L/(\pi d)$.
(b) Now fix the needle and let $C$ be a circle diameter $d$ centred at the midpoint of the needle. Let $\lambda$ be a line whose direction and distance from the centre of $C$ are independent and uniformly distributed on $[0, 2\pi]$ and $[0, \frac{1}{2}d]$ respectively. This is equivalent to 'casting the ruled plane at random'. Show that the probability of an intersection between the needle and $\lambda$ is $2L/(\pi d)$.
(c) Let $S$ be a curve within $C$ having finite length $L(S)$. Use indicators to show that the expected number of intersections between $S$ and $\lambda$ is $2L(S)/(\pi d)$.
This type of result is used in stereology, which seeks knowledge of the contents of a cell by studying its cross sections.

**32. Buffon's needle ingested.** In the excitement of calculating $\pi$, Mr Buffon (no relation) inadvertently swallows the needle and is X-rayed. If the needle exhibits no preference for direction in the gut, what is the distribution of the length of its image on the X-ray plate? If he swallowed Buffon's cross (see Exercise (4.5.3)) also, what would be the joint distribution of the lengths of the images of the two arms of the cross?

**33.** Let $X_1, X_2, \ldots, X_n$ be independent exponential variables with parameter $\lambda$, and let $X_{(1)} \leq X_{(2)} \leq \cdots \leq X_{(n)}$ be their order statistics. Show that

$$Y_1 = nX_{(1)}, \quad Y_r = (n+1-r)(X_{(r)} - X_{(r-1)}), \quad 1 < r \leq n$$

are also independent and have the same joint distribution as the $X_i$.

**34.** Let $X_{(1)}, X_{(2)}, \ldots, X_{(n)}$ be the order statistics of a family of independent variables with common continuous distribution function $F$. Show that

$$Y_n = \{F(X_{(n)})\}^n, \quad Y_r = \left\{ \frac{F(X_{(r)})}{F(X_{(r+1)})} \right\}^r, \quad 1 \leq r < n,$$

are independent and uniformly distributed on $[0, 1]$. This is equivalent to Problem (4.14.33). Why?

**35. Secretary/marriage problem.** You are permitted to inspect the $n$ prizes at a fête in a given order, at each stage either rejecting or accepting the prize under consideration. There is no recall, in the sense that no rejected prize may be accepted later. It may be assumed that, given complete information, the prizes may be ranked in a strict order of preference, and that the order of presentation is independent of this ranking. Find the strategy which maximizes the probability of accepting the best prize, and describe its behaviour when $n$ is large.

**36. Fisher's spherical distribution.** Let $R^2 = X^2 + Y^2 + Z^2$ where $X$, $Y$, $Z$ are independent normal random variables with means $\lambda$, $\mu$, $\nu$, and common variance $\sigma^2$, where $(\lambda, \mu, \nu) \neq (0, 0, 0)$. Show that the conditional density of the point $(X, Y, Z)$ given $R = r$, when expressed in spherical polar coordinates relative to an axis in the direction $\mathbf{e} = (\lambda, \mu, \nu)$, is of the form

$$ f(\theta, \phi) = \frac{a}{4\pi \sinh a} e^{a \cos \theta} \sin \theta, \quad 0 \le \theta < \pi, \ 0 \le \phi < 2\pi, $$

where $a = r|\mathbf{e}|$.

**37.** Let $\phi$ be the $N(0, 1)$ density function, and define the functions $H_n$, $n \ge 0$, by $H_0 = 1$, and $(-1)^n H_n \phi = \phi^{(n)}$, the $n$th derivative of $\phi$. Show that:
(a) $H_n(x)$ is a polynomial of degree $n$ having leading term $x^n$, and

$$ \int_{-\infty}^{\infty} H_m(x) H_n(x) \phi(x) \, dx = \begin{cases} 0 & \text{if } m \neq n, \\ n! & \text{if } m = n. \end{cases} $$

(b) $\displaystyle\sum_{n=0}^{\infty} H_n(x) \frac{t^n}{n!} = \exp(tx - \tfrac{1}{2}t^2).$

**38. Lancaster's theorem.** Let $X$ and $Y$ have a standard bivariate normal distribution with zero means, unit variances, and correlation coefficient $\rho$, and suppose $U = u(X)$ and $V = v(Y)$ have finite variances. Show that $|\rho(U, V)| \le |\rho|$. [Hint: Use Problem (4.14.37) to expand the functions $u$ and $v$. You may assume that $u$ and $v$ lie in the linear span of the $H_n$.]

**39.** Let $X_{(1)}, X_{(2)}, \ldots, X_{(n)}$ be the order statistics of $n$ independent random variables, uniform on $[0, 1]$. Show that:
(a) $\mathbb{E}(X_{(r)}) = \dfrac{r}{n+1}$, (b) $\mathrm{cov}(X_{(r)}, X_{(s)}) = \dfrac{r(n-s+1)}{(n+1)^2(n+2)}$ for $r \le s$.

**40.** (a) Let $X$, $Y$, $Z$ be independent $N(0, 1)$ variables, and set $R = \sqrt{X^2 + Y^2 + Z^2}$. Show that $X^2/R^2$ has a beta distribution with parameters $\tfrac{1}{2}$ and $1$, and is independent of $R^2$.

(b) Let $X$, $Y$, $Z$ be independent and uniform on $[-1, 1]$ and set $R = \sqrt{X^2 + Y^2 + Z^2}$. Find the density of $X^2/R^2$ given that $R^2 \le 1$.

**41.** (a) **Skew normal distribution.** Let $\phi$ and $\Phi$ be the density and distribution functions of the random variable $X$ with the standard normal distribution. Show, for $\lambda \in \mathbb{R}$, that $g(x) = 2\phi(x)\Phi(\lambda x)$, $x \in \mathbb{R}$, is the density function of some random variable (denoted by $Y$). Show that $|X|$ and $|Y|$ have the same distributions.
Let $X$ have the $N(0, 1)$ distribution and be independent of $Y$, and define $U = (X + \lambda|Y|)/\sqrt{1 + \lambda^2}$. Write down the joint density of $U$ and $V = |Y|$, and deduce that $U$ has density function $g$.
(b) **General skew distributions.** For $i = 1, 2$, let $X_i$ have a density function $f_i$, which is symmetric about $0$ with distribution function $F_i$.
  (i) Let $\lambda \in \mathbb{R}$, and show that $g_1(x) = 2f_1(x)F_2(\lambda x)$ and $g_2(x) = 2f_2(x)F_1(\lambda x)$ are density functions.
  (ii) Let $X_1$ and $X_2$ be independent. Find the conditional density function of $X_2$ given that $X_1 < \lambda X_2$.

**42.** The six coordinates $(X_i, Y_i)$, $1 \le i \le 3$, of three points A, B, C in the plane are independent $N(0, 1)$. Show that the the probability that C lies inside the circle with diameter AB is $\tfrac{1}{4}$.

**43.** The coordinates $(X_i, Y_i, Z_i)$, $1 \le i \le 3$, of three points A, B, C are independent $N(0, 1)$. Show that the probability that C lies inside the sphere with diameter AB is $\dfrac{1}{3} - \dfrac{\sqrt{3}}{4\pi}$.

**44. Skewness.** Let $X$ have variance $\sigma^2$ and write $m_k = \mathbb{E}(X^k)$. Define the *skewness* of $X$ by $\mathrm{skw}(X) = \mathbb{E}[(X - m_1)^3]/\sigma^3$. Show that:

(a) $\mathrm{skw}(X) = (m_3 - 3m_1 m_2 + 2m_1^3)/\sigma^3$,

(b) $\mathrm{skw}(S_n) = \mathrm{skw}(X_1)/\sqrt{n}$, where $S_n = \sum_{r=1}^n X_r$ is a sum of independent identically distributed random variables,

(c) $\mathrm{skw}(X) = (1 - 2p)/\sqrt{npq}$, when $X$ is bin$(n, p)$ where $p + q = 1$,

(d) $\mathrm{skw}(X) = 1/\sqrt{\lambda}$, when $X$ is Poisson with parameter $\lambda$,

(e) $\mathrm{skw}(X) = 2/\sqrt{t}$, when $X$ is gamma $\Gamma(\lambda, t)$, and $t$ is integral.

**45. Kurtosis.** Let $X$ have variance $\sigma^2$ and $\mathbb{E}(X^k) = m_k$. Define the *kurtosis* of $X$ by $\mathrm{kur}(X) = \mathbb{E}[(X - m_1)^4]/\sigma^4$. Show that:

(a) $\mathrm{kur}(X) = 3$, when $X$ is $N(\mu, \sigma^2)$,

(b) $\mathrm{kur}(X) = 9$, when $X$ is exponential with parameter $\lambda$,

(c) $\mathrm{kur}(X) = 3 + \lambda^{-1}$, when $X$ is Poisson with parameter $\lambda$,

(d) $\mathrm{kur}(S_n) = 3 + \{\mathrm{kur}(X_1) - 3\}/n$, where $S_n = \sum_{r=1}^n X_r$ is a sum of independent identically distributed random variables.

**46. Extreme value. Fisher–Gumbel–Tippett distribution.** Let $X_r$, $1 \le r \le n$, be independent and exponentially distributed with parameter 1. Show that $X_{(n)} = \max\{X_r : 1 \le r \le n\}$ satisfies

$$\lim_{n \to \infty} \mathbb{P}(X_{(n)} - \log n \le x) = \exp(-e^{-x}), \qquad -\infty < x < \infty.$$

Hence show that $\int_0^\infty \{1 - \exp(-e^{-x}) - \exp(-e^x)\}\, dx = \gamma$ where $\gamma$ is Euler's constant.

**47. Squeezing.** Let $S$ and $X$ have density functions satisfying $b(x) \le f_S(x) \le a(x)$ and $f_S(x) \le f_X(x)$. Let $U$ be uniformly distributed on $[0, 1]$ and independent of $X$. Given the value $X$, we implement the following algorithm:

$$\begin{aligned}
&\text{if } Uf_X(X) > a(X), && \text{reject } X; \\
&\text{otherwise: if } Uf_X(X) < b(X), && \text{accept } X; \\
&\text{otherwise: if } Uf_X(X) \le f_S(X), && \text{accept } X; \\
&\text{otherwise: reject } X.
\end{aligned}$$

Show that, conditional on ultimate acceptance, $X$ is distributed as $S$. Explain when you might use this method of sampling.

**48.** Let $X$, $Y$, and $\{U_r : r \ge 1\}$ be independent random variables, where:

$$\mathbb{P}(X = x) = (e - 1)e^{-x}, \quad \mathbb{P}(Y = y) = \frac{1}{(e - 1)y!} \quad \text{for } x, y = 1, 2, \ldots,$$

and the $U_r$ are uniform on $[0, 1]$. Let $M = \max\{U_1, U_2, \ldots, U_Y\}$, and show that $Z = X - M$ is exponentially distributed.

**49.** Let $U$ and $V$ be independent and uniform on $[0, 1]$. Set $X = -\alpha^{-1} \log U$ and $Y = -\log V$ where $\alpha > 0$.

(a) Show that, conditional on the event $Y \ge \frac{1}{2}(X - \alpha)^2$, $X$ has density function $f(x) = \sqrt{2/\pi}\, e^{-\frac{1}{2}x^2}$ for $x > 0$.

(b) In sampling from the density function $f$, it is decided to use a rejection method: for given $\alpha > 0$, we sample $U$ and $V$ repeatedly, and we accept $X$ the first time that $Y \ge \frac{1}{2}(X - \alpha)^2$. What is the optimal value of $\alpha$?

(c) Describe how to use these facts in sampling from the $N(0, 1)$ distribution.

**50.** Let $S$ be a semicircle of unit radius on a diameter $D$.

(a) A point P is picked at random on $D$. If $X$ is the distance from P to $S$ along the perpendicular to $D$, show $\mathbb{E}(X) = \pi/4$.

(b) A point Q is picked at random on $S$. If $Y$ is the perpendicular distance from Q to $D$, show $\mathbb{E}(Y) = 2/\pi$.

**51.** (Set for the Fellowship examination of St John's College, Cambridge in 1858.) 'A large quantity of pebbles lies scattered uniformly over a circular field; compare the labour of collecting them one by one:

(i) at the centre O of the field,

(ii) at a point A on the circumference.'

To be precise, if $L_O$ and $L_A$ are the respective labours per stone, show that $\mathbb{E}(L_O) = \frac{2}{3}a$ and $\mathbb{E}(L_A) = 32a/(9\pi)$ for some constant $a$.

(iii) Suppose you take each pebble to the nearer of two points A or B at the ends of a diameter. Show in this case that the labour per stone satisfies

$$\mathbb{E}(L_{AB}) = \frac{4a}{3\pi}\left\{\frac{16}{3} - \frac{17}{6}\sqrt{2} + \frac{1}{2}\log(1 + \sqrt{2})\right\} \simeq 1.13 \times \frac{2}{3}a.$$

(iv) Finally suppose you take each pebble to the nearest vertex of an equilateral triangle ABC inscribed in the circle. Why is it obvious that the labour per stone now satisfies $\mathbb{E}(L_{ABC}) < \mathbb{E}(L_O)$? Enthusiasts are invited to calculate $\mathbb{E}(L_{ABC})$.

**52.** The lines $L$, $M$, and $N$ are parallel, and P lies on $L$. A line picked at random through P meets $M$ at Q. A line picked at random through Q meets $N$ at R. What is the density function of the angle $\Theta$ that RP makes with $L$? [Hint: Recall Exercise (4.8.2) and Problem (4.14.4).]

**53.** Let $\Delta$ denote the event that you can form a triangle with three given parts of a rod $R$.

(a) $R$ is broken at two points chosen independently and uniformly. Show that $\mathbb{P}(\Delta) = \frac{1}{4}$.

(b) $R$ is broken in two uniformly at random, the longer part is broken in two uniformly at random. Show that $\mathbb{P}(\Delta) = \log(4/e)$.

(c) $R$ is broken in two uniformly at random, a randomly chosen part is broken into two equal parts. Show that $\mathbb{P}(\Delta) = \frac{1}{2}$.

(d) In case (c) show that, given $\Delta$, the triangle is obtuse with probability $3 - 2\sqrt{2}$.

**54.** You break a rod at random into two pieces. Let $R$ be the ratio of the lengths of the shorter to the longer piece. Find the density function $f_R$, together with the mean and variance of $R$.

**55.** Let $R$ be the distance between two points picked at random inside a square of side $a$. Show that $\mathbb{E}(R^2) = \frac{1}{3}a^2$, and that $R^2/a^2$ has density function

$$f(r) = \begin{cases} r - 4\sqrt{r} + \pi & \text{if } 0 \le r \le 1, \\ 4\sqrt{r-1} - 2 - r + 2\sin^{-1}\sqrt{r-1} - 2\sin^{-1}\sqrt{1-r^{-1}} & \text{if } 1 \le r \le 2. \end{cases}$$

**56.** Show that a sheet of paper of area $A$ cm$^2$ can be placed on the square lattice with period 1 cm in such a way that at least $\lceil A \rceil$ points are covered.

**57.** Show that it is possible to position a convex rock of surface area $S$ in sunlight in such a way that its shadow has area at least $\frac{1}{4}S$.

**58. Dirichlet distribution.** Let $\{X_r : 1 \le r \le k+1\}$ be independent $\Gamma(\lambda, \beta_r)$ random variables (respectively).

(a) Show that $Y_r = X_r/(X_1 + \cdots + X_r)$, $2 \le r \le k+1$, are independent random variables.

(b) Show that $Z_r = X_r/(X_1 + \cdots + X_{k+1})$, $1 \le r \le k$, have the joint *Dirichlet density*

$$\frac{\Gamma(\beta_1 + \cdots + \beta_{k+1})}{\Gamma(\beta_1) \cdots \Gamma(\beta_{k+1})} z_1^{\beta_1-1} z_2^{\beta_2-1} \cdots z_k^{\beta_k-1} (1 - z_1 - z_2 - \cdots - z_k)^{\beta_{k+1}-1}.$$

**59. Hotelling's theorem.** Let $\mathbf{X}_r = (X_{1r}, X_{2r}, \ldots, X_{mr})$, $1 \le r \le n$, be independent multivariate normal random vectors having zero means and the same covariance matrix $\mathbf{V} = (v_{ij})$. Show that the two random variables

$$S_{ij} = \sum_{r=1}^{n} X_{ir} X_{jr} - \frac{1}{n} \sum_{r=1}^{n} X_{ir} \sum_{r=1}^{n} X_{jr}, \quad T_{ij} = \sum_{r=1}^{n-1} X_{ir} X_{jr},$$

are identically distributed.

**60.** Choose P, Q, and R independently at random in the square $S(a)$ of side $a$. Show that $\mathbb{E}|PQR| = 11a^2/144$. Deduce that four points picked at random in a parallelogram form a convex quadrilateral with probability $(\frac{5}{6})^2$.

**61.** Choose P, Q, and R uniformly at random within the convex region $C$ illustrated beneath. By considering the event that the convex hull of four randomly chosen points is a triangle, or otherwise, show that the mean area of the shaded region is three times the mean area of the triangle PQR.

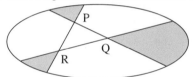

**62. Multivariate normal sampling.** Let $\mathbf{V}$ be a positive-definite symmetric $n \times n$ matrix, and $\mathbf{L}$ a lower-triangular matrix such that $\mathbf{V} = \mathbf{L}'\mathbf{L}$; this is called the *Cholesky decomposition* of $\mathbf{V}$. Let $\mathbf{X} = (X_1, X_2, \ldots, X_n)$ be a vector of independent random variables distributed as $N(0, 1)$. Show that the vector $\mathbf{Z} = \boldsymbol{\mu} + \mathbf{XL}$ has the multivariate normal distribution with mean vector $\boldsymbol{\mu}$ and covariance matrix $\mathbf{V}$.

**63. Verifying matrix multiplications.** We need to decide whether or not $\mathbf{AB} = \mathbf{C}$ where $\mathbf{A}, \mathbf{B}, \mathbf{C}$ are given $n \times n$ matrices, and we adopt the following random algorithm. Let $\mathbf{x}$ be a random $\{0, 1\}^n$-valued vector, each of the $2^n$ possibilities being equally likely. If $(\mathbf{AB} - \mathbf{C})\mathbf{x} = \mathbf{0}$, we decide that $\mathbf{AB} = \mathbf{C}$, and otherwise we decide that $\mathbf{AB} \ne \mathbf{C}$. Show that

$$\mathbb{P}(\text{the decision is correct}) \begin{cases} = 1 & \text{if } \mathbf{AB} = \mathbf{C}, \\ \ge \frac{1}{2} & \text{if } \mathbf{AB} \ne \mathbf{C}. \end{cases}$$

Describe a similar procedure which results in an error probability which may be made as small as desired.

**64. Coupon collecting, Exercise (3.3.2), revisited.** Each box of cereal contains a worthless and inedible object. The objects in different boxes are independent and equally likely to be any of the $n$ available types. Let $T_n$ be the number of boxes opened before collection of a full set.
(a) Use the result of Exercise (4.3.6) to show that

$$\frac{1}{n}\mathbb{E}(T_n) = \sum_{r=1}^{n} \frac{(-1)^{r+1}}{r} \binom{n}{r} = \sum_{r=1}^{n} \frac{1}{r}.$$

(b) Prove the above combinatorial identity directly.

(c) By considering $\mathbb{E}(T_n^2)$, show that

$$2n \sum_{r=1}^{n} \frac{(-1)^{r+1}}{r^2} \binom{n}{r} = \sum_{r=1}^{n} \frac{1}{r^2} + \frac{n-1}{n} \sum_{r=1}^{n} \frac{1}{r} + \left( \sum_{r=1}^{n} \frac{1}{r} \right)^2.$$

**65.** Points P and Q are picked independently and uniformly at random in the triangle ABC, in such a way that the straight line passing through P and Q divides ABC into a triangle $T$ and a quadrilateral $R$. Show that the ratio of the mean area of $T$ to that of $R$ is 4 : 5.

**66.** Let $X$ and $Y$ be independent, identically distributed random variables with finite means. Show that

$$\mathbb{E}|X - Y| \leq \mathbb{E}|X + Y|,$$

with equality if and only if their common distribution is symmetric about 0.

**67. Malmquist's theorem. Simulated order statistics.** Let $U_1, U_2, \ldots, U_n$ be independent and identically distributed on $[0, 1]$. Define $X_1, X_2, \ldots, X_n$ recursively as follows:

$$X_1 = U_1^{1/n}, \ X_2 = X_1 U_2^{1/(n-1)}, \ldots, \ X_j = X_{j-1} U_j^{1/(n-j+1)}, \ldots, \ X_n = X_{n-1} U_n.$$

Show that the random variables $X_n < X_{n-1} < \cdots < X_1$ have the same joint distribution as the order statistics $U_{(1)}, U_{(2)}, \ldots, U_{(n)}$. You may do this in two different ways: first by construction, and second by using the change of variables formula.

**68. Semi-moment.** The value $X$ of a financial index in one year's time has the normal distribution with mean $\mu$ and standard deviation $\sigma > 0$. You possess a derivative contract that will pay you $\max\{X - a, 0\}$, where $a$ is a predetermined constant. Show that the expected payout $V$ of the contract is given by $V = \sigma\{\phi(y) - y(1 - \Phi(y))\}$, where $\phi$ and $\Phi$ are the $N(0, 1)$ density and distribution functions, and $y = (a - \mu)/\sigma$. Deduce that, for large positive $y$,

$$V \approx \frac{\sigma^3}{(a - \mu)^2} \phi((a - \mu)/\sigma).$$

**69. Gauss's inequality.** Let $g$ be the density function of a positive, unimodal random variable $X$ with finite variance, with a unique mode at 0 (so that $g$ is non-increasing on $[0, \infty)$, recall the definitions of Problem (2.7.23)). For any $x$ such that $g(x) > 0$, let $y = x + g(x)^{-1} \int_x^\infty g(v) \, dv$.

(a) Prove that, for $0 < x < y < \infty$, we have that $(y - x)x^2 \leq \frac{4}{9} \int_0^y v^2 \, dv$.

(b) Deduce that

$$\mathbb{P}(|X| > x) \leq \frac{4}{9x^2} \mathbb{E}(X^2), \qquad x > 0.$$

(c) Prove the inequality of (b) for a continuous random variable $X$ with a unique mode at 0.

**70. Bruss's odds rule.** Let $I_i$ be the indicator function of success in the $i$th of $n$ independent trials. Let $p_i = \mathbb{E}I_i = 1 - q_i > 0$, and let $r_i = p_i/q_i$ be the $i$th 'odds ratio'. Let $R_k = \sum_{i=k}^n r_i$, and for the moment assume that $p_i < 1$ for all $i$.

(a) Show that the probability of exactly one success after the $k$th trial is $\sigma_k = R_{k+1} \prod_{i=k+1}^n q_i$.

(b) Prove that $\sigma_k$ is a unimodal function of $k$, in that there is a unique $m$ such that $\sigma_k$ is greatest either for $k = m$ or for $k \in \{m, m + 1\}$.

(c) It is desired to stop the process at the last success. Show that the optimal rule for achieving this is to stop at the first success at time $\tau$ or later, where $\tau = \max\{1, \max\{k : R_k \geq 1\}\}$.

(d) What can be said if $p_i = 1$ for some $i$?

(e) Show that the probability of stopping at the final success is $\sigma_\tau = R_\tau \prod_{i=\tau}^n q_i$.

(f) Use this result to solve the marriage problem (4.14.35).

**71. Prophet inequality.** Let $X_0, X_1, \ldots, X_n$ be non-negative, independent random variables. Their values are revealed to you in order, and you are required to stop the process at some time $T$, and exit with $X_T$. Your target is to maximize $\mathbb{E}(X_T)$, and you are permitted to choose any 'stopping strategy' $T$ that depends only on the past and present values $X_0, X_1, \ldots, X_T$. (Such a random variable is called a 'stopping time', see Exercise (6.1.6).)

Show that $\mathbb{E}(\max_r X_r) \leq 2 \sup_T \mathbb{E}(X_T)$, where $\sup_T$ is the supremum over all stopping strategies.

**72. Stick breaking.** A stick of length $s$ is broken at $n$ points chosen independently and uniformly in the interval $[0, s]$. Let the order statistics of the resulting lengths of sticks be $S_1 < S_2 < \cdots < S_{n+1}$. Fix $y > 0$ and write $p_n(s, y) = \mathbb{P}(S_1 > y)$. Show that

$$p_{n+1}(s, y) = \frac{n+1}{s^{n+1}} \int_0^{s-y} x^n p_n(x, y)\, dx.$$

Deduce that

(i) $p_n(s, y) = s^{-n}\{(s - (n+1)y)^+\}^n$ for $n \geq 1$, where $x^+ = \max\{0, x\}$.

(ii) $\mathbb{E}S_1 = s/(n+1)^2$.

(iii) $\mathbb{E}S_r = s(H_{n+1} - H_{n-r+1})/(n+1)$, where $H_k = \sum_{r=1}^k r^{-1}$.

**73.** Let $\alpha > -1$, and let $X$ and $Y$ have joint density function $f(x, y) = cx^\alpha$ for $x, y \in (0, 1)$, $x + y > 1$. We set $f(x, y) = 0$ for other pairs $x, y$.

(a) Find the value of $c$ and the joint distribution function of $X, Y$.

(b) Show that it is possible to construct a triangle with side-lengths $X, Y, 2 - X - Y$, with probability one.

(c) Show that the angle opposite the side with length $Y$ is obtuse with probability

$$\pi = c \int_0^1 \frac{x^{\alpha+1} - x^{\alpha+2}}{2 - x}\, dx,$$

and find $\pi$ when $\alpha = 0$.

**74. Size-biasing on the freeway.** You are driving at constant speed $v$ along a vast multi-lane highway, and other traffic in the same direction is moving at independent random speeds having the same distribution as a random variable $X$ with density function $f$ and finite variance.

(a) Show that the expected value of the speed of vehicles passing you or being passed by you is

$$m(v) = \frac{\mathbb{E}\{X|X - v|\}}{\mathbb{E}|X - v|}.$$

(b) Suppose that $f$ is symmetric about its mean $\mu$, and that $v < \mu$. Show that the difference $m(v) - \mu$ is positive, and that it has a maximum value for some value $v = v_{\max} < \mu$.

(c) What can be said if $v > \mu$? Discuss.

(d) Suppose $X$ is uniformly distributed on the interval $\{60, 61, \ldots, 80\}$. Show for $v = 63$ that $m(v) \approx 74$, while for $v = 77$ we have $m(v) \approx 66$.

# 5

## Generating functions and their applications

*Summary.* A key method for studying distributions is via transforms such as the probability generating function of a discrete random variable, or the moment generating function and characteristic function of a general random variable. Such transforms are particularly suited to the study of sums of independent random variables, and their areas of application include renewal theory, random walks, and branching processes. The inversion theorem tells how to obtain the distribution function from knowledge of its characteristic function. The continuity theorem allows us to use characteristic functions in studying limits of random variables. Two principal applications are to the law of large numbers and the central limit theorem. The theory of large deviations concerns the estimation of probabilities of 'exponentially unlikely' events.

### 5.1 Generating functions

A sequence $a = \{a_i : i = 0, 1, 2, \ldots\}$ of real numbers may contain a lot of information. One concise way of storing this information is to wrap up the numbers together in a 'generating function'. For example, the (ordinary) *generating function* of the sequence $a$ is the function $G_a$ defined by

$$(1) \qquad G_a(s) = \sum_{i=0}^{\infty} a_i s^i \quad \text{for } s \in \mathbb{R} \text{ for which the sum converges†.}$$

The sequence $a$ may in principle be reconstructed from the function $G_a$ by setting $a_i = G_a^{(i)}(0)/i!$, where $f^{(i)}$ denotes the $i$th derivative of the function $f$. In many circumstances it is easier to work with the generating function $G_a$ than with the original sequence $a$.

**(2) Example. De Moivre's theorem.** The sequence $a_n = (\cos\theta + i\sin\theta)^n$ has generating function

$$G_a(s) = \sum_{n=0}^{\infty} [s(\cos\theta + i\sin\theta)]^n = \frac{1}{1 - s(\cos\theta + i\sin\theta)}$$

---

†More generally, the character $s$ may be indeterminate (not necessarily a real number), and $G_a$ is then viewed as a formal power series.

if $|s| < 1$; here $i = \sqrt{-1}$. It is easily checked by examining the coefficient of $s^n$ that

$$\left[1 - s(\cos\theta + i\sin\theta)\right] \sum_{n=0}^{\infty} s^n\left[\cos(n\theta) + i\sin(n\theta)\right] = 1$$

when $|s| < 1$. Thus

$$\sum_{n=0}^{\infty} s^n\left[\cos(n\theta) + i\sin(n\theta)\right] = \frac{1}{1 - s(\cos\theta + i\sin\theta)}$$

if $|s| < 1$. Equating the coefficients of $s^n$ we obtain the well-known fact that $\cos(n\theta) + i\sin(n\theta) = (\cos\theta + i\sin\theta)^n$. ●

There are several different types of generating function, of which $G_a$ is perhaps the simplest. Another is the *exponential generating function* $E_a$ given by

$$(3) \qquad\qquad E_a(s) = \sum_{i=0}^{\infty} \frac{a_i s^i}{i!} \qquad \text{for } s \in \mathbb{R} \text{ for which the sum converges.}$$

Whilst such generating functions have many uses in mathematics, the ordinary generating function (1) is of greater value when the $a_i$ are probabilities. This is because 'convolutions' are common in probability theory, and (ordinary) generating functions provide an invaluable tool for studying them.

**(4) Convolution.** The *convolution* of the real sequences $a = \{a_i : i \geq 0\}$ and $b = \{b_i : i \geq 0\}$ is the sequence $c = \{c_i : i \geq 0\}$ defined by

$$(5) \qquad\qquad c_n = a_0 b_n + a_1 b_{n-1} + \cdots + a_n b_0;$$

we write $c = a * b$. If $a$ and $b$ have generating functions $G_a$ and $G_b$, then the generating function of $c$ is

$$(6) \qquad\qquad G_c(s) = \sum_{n=0}^{\infty} c_n s^n = \sum_{n=0}^{\infty}\left(\sum_{i=0}^{n} a_i b_{n-i}\right) s^n$$

$$= \sum_{i=0}^{\infty} a_i s^i \sum_{n=i}^{\infty} b_{n-i} s^{n-i} = G_a(s) G_b(s).$$

Thus we learn that, if $c = a * b$, then $G_c(s) = G_a(s) G_b(s)$; convolutions are numerically complicated operations, and it is often easier to work with generating functions.

**(7) Example.** The combinatorial identity

$$\sum_i \binom{n}{i}^2 = \binom{2n}{n}$$

may be obtained as follows. The left-hand side is the convolution of the sequence $a_i = \binom{n}{i}$, $i = 0, 1, 2, \ldots$, with itself. However, $G_a(s) = \sum_i \binom{n}{i} s^i = (1+s)^n$, so that

$$G_{a*a}(s) = G_a(s)^2 = (1+s)^{2n} = \sum_i \binom{2n}{i} s^i.$$

Equating the coefficients of $s^n$ yields the required identity. ●

**(8) Example.** Let $X$ and $Y$ be independent random variables having the Poisson distribution with parameters $\lambda$ and $\mu$ respectively. What is the distribution of $Z = X + Y$?

**Solution.** We have from equation (3.8.2) that the mass function of $Z$ is the convolution of the mass functions of $X$ and $Y$, $f_Z = f_X * f_Y$. The generating function of the sequence $\{f_X(i) : i \geq 0\}$ is

$$
(9) \qquad\qquad G_X(s) = \sum_{i=0}^{\infty} \frac{\lambda^i e^{-\lambda}}{i!} s^i = e^{\lambda(s-1)},
$$

and similarly $G_Y(s) = e^{\mu(s-1)}$. Hence the generating function $G_Z$ of $\{f_Z(i) : i \geq 0\}$ satisfies $G_Z(s) = G_X(s)G_Y(s) = \exp[(\lambda+\mu)(s-1)]$, which we recognize from (9) as the generating function of the Poisson mass function with parameter $\lambda + \mu$.                                        ●

The last example is canonical: generating functions provide a basic technique for dealing with sums of independent random variables. With this example in mind, we make an important definition. Suppose that $X$ is a discrete random variable taking values in the non-negative integers $\{0, 1, 2, \ldots\}$; its distribution is specified by the sequence of probabilities $f(i) = \mathbb{P}(X = i)$.

**(10) Definition.** The **(probability) generating function** of the random variable $X$ is defined to be the generating function $G(s) = \mathbb{E}(s^X)$ of its probability mass function.

Note that $G$ does indeed generate the sequence $\{f(i) : i \geq 0\}$ since

$$
\mathbb{E}(s^X) = \sum_i s^i \mathbb{P}(X = i) = \sum_i s^i f(i)
$$

by Lemma (3.3.3). We write $G_X$ when we wish to stress the role of $X$. If $X$ takes values in the non-negative integers, its generating function $G_X$ converges at least when $|s| \leq 1$ and sometimes in a larger interval. Generating functions can be defined for random variables taking negative as well as positive integer values. Such generating functions generally converge for values of $s$ satisfying $\alpha < |s| < \beta$ for some $\alpha, \beta$ such that $\alpha \leq 1 \leq \beta$. We shall make occasional use of such generating functions, but we do not develop their theory systematically.

In advance of giving examples and applications of the method of generating functions, we recall some basic properties of power series. Let $G(s) = \sum_0^{\infty} a_i s^i$ where $a = \{a_i : i \geq 0\}$ is a real sequence.

**(11) Convergence.** There exists a *radius of convergence* $R$ ($\geq 0$) such that the sum converges absolutely if $|s| < R$ and diverges if $|s| > R$. The sum is uniformly convergent on sets of the form $\{s : |s| \leq R'\}$ for any $R' < R$.

**(12) Differentiation.** $G_a(s)$ may be differentiated or integrated term by term any number of times at points $s$ satisfying $|s| < R$.

**(13) Uniqueness.** If $G_a(s) = G_b(s)$ for $|s| < R'$ where $0 < R' \leq R$ then $a_n = b_n$ for all $n$. Furthermore

$$
(14) \qquad\qquad a_n = \frac{1}{n!} G_a^{(n)}(0).
$$

**(15) Abel's theorem.** If $a_i \geq 0$ for all $i$ and $G_a(s)$ is finite for $|s| < 1$, then $\lim_{s\uparrow 1} G_a(s) = \sum_{i=0}^{\infty} a_i$, whether the sum is finite or equals $+\infty$. This standard result is useful when the radius of convergence $R$ satisfies $R = 1$, since then one has no *a priori* right to take the limit as $s \uparrow 1$.

Returning to the discrete random variable $X$ taking values in $\{0, 1, 2, \ldots\}$ we have that $G(s) = \sum_{0}^{\infty} s^i \mathbb{P}(X = i)$, so that

**(16)** $$G(0) = \mathbb{P}(X = 0), \quad G(1) = 1.$$

In particular, the radius of convergence of a probability generating function is at least 1. Here are some examples of probability generating functions.

**(17) Examples.**
 (a) **Constant variables.** If $\mathbb{P}(X = c) = 1$ then $G(s) = \mathbb{E}(s^X) = s^c$.
 (b) **Bernoulli variables.** If $\mathbb{P}(X = 1) = p$ and $\mathbb{P}(X = 0) = 1 - p$ then

$$G(s) = \mathbb{E}(s^X) = (1 - p) + ps.$$

 (c) **Geometric distribution.** If $X$ is geometrically distributed with parameter $p$, so that $\mathbb{P}(X = k) = p(1 - p)^{k-1}$ for $k \geq 1$, then

$$G(s) = \mathbb{E}(s^X) = \sum_{k=1}^{\infty} s^k p(1 - p)^{k-1} = \frac{ps}{1 - s(1 - p)}.$$

 (d) **Poisson distribution.** If $X$ is Poisson distributed with parameter $\lambda$ then

$$G(s) = \mathbb{E}(s^X) = \sum_{k=0}^{\infty} s^k \frac{\lambda^k}{k!} e^{-\lambda} = e^{\lambda(s-1)}. \qquad \bullet$$

Generating functions are useful when working with integer-valued random variables. Problems arise when random variables take negative or non-integer values. Later in this chapter we shall see how to construct another function, called a 'characteristic function', which is very closely related to $G_X$ but which exists for all random variables regardless of their types.
There are two major applications of probability generating functions: in calculating moments, and in calculating the distributions of *sums* of independent random variables. We begin with moments.

**(18) Theorem.** *If $X$ has generating function $G(s)$ then*
 (a) $\mathbb{E}(X) = G'(1)$,
 (b) *more generally,* $\mathbb{E}\big[X(X - 1) \cdots (X - k + 1)\big] = G^{(k)}(1)$.

Of course, $G^{(k)}(1)$ is shorthand for $\lim_{s\uparrow 1} G^{(k)}(s)$ whenever the radius of convergence of $G$ is 1. The quantity $\mathbb{E}[X(X - 1) \cdots (X - k + 1)]$ is known as the *kth factorial moment* of $X$.

**Proof of (b).** Take $s < 1$ and calculate the $k$th derivative of $G$ to obtain

$$G^{(k)}(s) = \sum_{i} s^{i-k} i(i - 1) \cdots (i - k + 1) f(i) = \mathbb{E}\big[s^{X-k} X(X - 1) \cdots (X - k + 1)\big].$$

Let $s \uparrow 1$ and use Abel's theorem (15) to obtain

$$G^{(k)}(s) \to \sum_i i(i-1)\cdots(i-k+1)f(i) = \mathbb{E}\big[X(X-1)\cdots(X-k+1)\big]. \qquad \blacksquare$$

In order to calculate the variance of $X$ in terms of $G$, we proceed as follows:

$$\textbf{(19)} \qquad \mathrm{var}(X) = \mathbb{E}(X^2) - \mathbb{E}(X)^2 = \mathbb{E}\big(X(X-1) + X\big) - \mathbb{E}(X)^2$$
$$= \mathbb{E}\big(X(X-1)\big) + \mathbb{E}(X) - \mathbb{E}(X)^2 = G''(1) + G'(1) - G'(1)^2.$$

**Exercise.** Find the means and variances of the distributions in (17) by this method.

**(20) Example.** Recall the hypergeometric distribution (3.11.10) with mass function

$$f(k) = \binom{b}{k}\binom{N-b}{n-k} \Big/ \binom{N}{n}.$$

Then $G(s) = \sum_k s^k f(k)$, which can be recognized as the coefficient of $x^n$ in

$$Q(s, x) = (1 + sx)^b (1 + x)^{N-b} \Big/ \binom{N}{n}.$$

Hence the mean $G'(1)$ is the coefficient of $x^n$ in

$$\frac{\partial Q}{\partial s}(1, x) = xb(1 + x)^{N-1} \Big/ \binom{N}{n}$$

and so $G'(1) = bn/N$. Now calculate the variance yourself (*exercise*).   $\bullet$

If you are more interested in the moments of $X$ than in its mass function, you may prefer to work not with $G_X$ but with the function $M_X$ defined by $M_X(t) = G_X(e^t)$. This change of variable is convenient for the following reason. Expanding $M_X(t)$ as a power series in $t$, we obtain

$$\textbf{(21)} \qquad M_X(t) = \sum_{k=0}^{\infty} e^{tk}\mathbb{P}(X = k) = \sum_{k=0}^{\infty}\sum_{n=0}^{\infty} \frac{(tk)^n}{n!}\mathbb{P}(X = k)$$
$$= \sum_{n=0}^{\infty} \frac{t^n}{n!}\left(\sum_{k=0}^{\infty} k^n \mathbb{P}(X = k)\right) = \sum_{n=0}^{\infty} \frac{t^n}{n!}\mathbb{E}(X^n),$$

the exponential generating function of the moments $\mathbb{E}(X^0), \mathbb{E}(X^1), \ldots$ of $X$. The function $M_X$ is called the *moment generating function* of the random variable $X$. We have assumed in (21) that the series in question converge. Some complications can arise in using moment generating functions unless the series $\sum_n t^n \mathbb{E}(X^n)/n!$ has a strictly positive radius of convergence.

**(22) Example.** We have from (9) that the moment generating function of the Poisson distribution with parameter $\lambda$ is $M(t) = \exp[\lambda(e^t - 1)]$.   $\bullet$

We turn next to sums and convolutions. Much of probability theory is concerned with sums of random variables. To study such a sum we need a useful way of describing its distribution in terms of the distributions of its summands, and generating functions prove to be an invaluable asset in this respect. The formula in Theorem (3.8.1) for the mass function of the sum of two independent discrete variables, $\mathbb{P}(X + Y = z) = \sum_x \mathbb{P}(X = x)\mathbb{P}(Y = z - x)$, involves a complicated calculation; the corresponding generating functions provide a more economical way of specifying the distribution of this sum.

**(23) Theorem.** *If X and Y are independent then $G_{X+Y}(s) = G_X(s)G_Y(s)$.*

**Proof.** The direct way of doing this is to use equation (3.8.2) to find that $f_Z = f_X * f_Y$, so that the generating function of $\{f_Z(i) : i \geq 0\}$ is the product of the generating functions of $\{f_X(i) : i \geq 0\}$ and $\{f_Y(i) : i \geq 0\}$, by (4). Alternatively, $g(X) = s^X$ and $h(Y) = s^Y$ are independent, by Theorem (3.2.3), and so $\mathbb{E}(g(X)h(Y)) = \mathbb{E}(g(X))\mathbb{E}(h(Y))$, as required. ∎

**(24) Example. Binomial distribution.** Let $X_1, X_2, \ldots, X_n$ be independent Bernoulli variables, parameter $p$, with sum $S = X_1 + X_2 + \cdots + X_n$. Each $X_i$ has generating function $G(s) = qs^0 + ps^1 = q + ps$, where $q = 1 - p$. Apply (23) repeatedly to find that the $\text{bin}(n, p)$ variable $S$ has generating function

$$G_S(s) = [G(s)]^n = (q + ps)^n.$$

The sum $S_1 + S_2$ of two independent variables, $\text{bin}(n, p)$ and $\text{bin}(m, p)$ respectively, has generating function

$$G_{S_1+S_2}(s) = G_{S_1}(s)G_{S_2}(s) = (q + ps)^{m+n}$$

and is thus $\text{bin}(m + n, p)$. This was Problem (3.11.8).                                        ●

Theorem (23) tells us that the sum $S = X_1 + X_2 + \cdots + X_n$ of independent variables taking values in the non-negative integers has generating function given by

$$G_S = G_{X_1}G_{X_2}\cdots G_{X_n}.$$

If $n$ is itself the outcome of a random experiment then the answer is not quite so simple.

**(25) Theorem. Random sum formula.** *If $X_1, X_2, \ldots$ is a sequence of independent identically distributed random variables with common generating function $G_X$, and N $(\geq 0)$ is a random variable which is independent of the $X_i$ and has generating function $G_N$, then $S = X_1 + X_2 + \cdots + X_N$ has generating function given by*

**(26)**                                $$G_S(s) = G_N(G_X(s)).$$

This has many important applications, one of which we shall meet in Section 5.4. It is an example of a process known as *compounding* with respect to a parameter. Formula (26) is easily remembered; possible confusion about the order in which the functions $G_N$ and $G_X$ are compounded is avoided by remembering that if $\mathbb{P}(N = n) = 1$ then $G_N(s) = s^n$ and $G_S(s) = G_X(s)^n$. Incidentally, we adopt the usual convention that, in the case when $N = 0$, the sum $X_1 + X_2 + \cdots + X_N$ is the 'empty' sum, and equals 0 also.

**Proof.** Use conditional expectation and Theorem (3.7.4) to find that

$$G_S(s) = \mathbb{E}(s^S) = \mathbb{E}\big(\mathbb{E}(s^S \mid N)\big) = \sum_n \mathbb{E}(s^S \mid N = n)\mathbb{P}(N = n)$$

$$= \sum_n \mathbb{E}(s^{X_1 + \cdots + X_n})\mathbb{P}(N = n)$$

$$= \sum_n \mathbb{E}(s^{X_1}) \cdots \mathbb{E}(s^{X_n})\mathbb{P}(N = n) \quad \text{by independence}$$

$$= \sum_n G_X(s)^n \mathbb{P}(N = n) = G_N(G_X(s)). \qquad \blacksquare$$

**(27) Example (3.7.5) revisited.** A hen lays $N$ eggs, where $N$ is Poisson distributed with parameter $\lambda$. Each egg hatches with probability $p$, independently of all other eggs. Let $K$ be the number of chicks. Then $K = X_1 + X_2 + \cdots + X_N$ where $X_1, X_2, \ldots$ are independent Bernoulli variables with parameter $p$. How is $K$ distributed? Clearly

$$G_N(s) = \sum_{n=0}^{\infty} s^n \frac{\lambda^n}{n!} e^{-\lambda} = e^{\lambda(s-1)}, \qquad G_X(s) = q + ps,$$

and so $G_K(s) = G_N(G_X(s)) = e^{\lambda p(s-1)}$, which, by comparison with $G_N$, we see to be the generating function of a Poisson variable with parameter $\lambda p$. ●

Just as information about a mass function can be encapsulated in a generating function, so may joint mass functions be similarly described.

**(28) Definition.** The **joint (probability) generating function** of variables $X_1$ and $X_2$ taking values in the non-negative integers is defined by

$$G_{X_1, X_2}(s_1, s_2) = \mathbb{E}(s_1^{X_1} s_2^{X_2}).$$

There is a similar definition for the joint generating function of an arbitrary family of random variables. Joint generating functions have important uses also, one of which is the following characterization of independence.

**(29) Theorem.** *Random variables $X_1$ and $X_2$ are independent if and only if*

$$G_{X_1, X_2}(s_1, s_2) = G_{X_1}(s_1)G_{X_2}(s_2) \quad \text{for all } s_1 \text{ and } s_2.$$

**Proof.** If $X_1$ and $X_2$ are independent then so are $g(X_1) = s_1^{X_1}$ and $h(X_2) = s_2^{X_2}$; then proceed as in the proof of (23). To prove the converse, equate the coefficients of terms such as $s_1^i s_2^j$ to deduce after some manipulation that $\mathbb{P}(X_1 = i, \ X_2 = j) = \mathbb{P}(X_1 = i)\mathbb{P}(X_2 = j)$. $\blacksquare$

So far we have only considered random variables $X$ which take finite values only, and consequently their generating functions $G_X$ satisfy $G_X(1) = 1$. In the near future we shall encounter variables which can take the value $+\infty$ (see the first passage time $T_0$ of Section 5.3 for example). For such variables $X$ we note that $G_X(s) = \mathbb{E}(s^X)$ converges so long as $|s| < 1$, and furthermore

**(30)** $$\lim_{s \uparrow 1} G_X(s) = \sum_k \mathbb{P}(X = k) = 1 - \mathbb{P}(X = \infty).$$

We can no longer find the moments of $X$ in terms of $G_X$; of course, they all equal $+\infty$. If $\mathbb{P}(X = \infty) > 0$ then we say that $X$ is *defective* with defective distribution function $F_X$.

## Exercises for Section 5.1

**1.**   Find the generating functions of the following mass functions, and state where they converge. Hence calculate their means and variances.

(a)  $f(m) = \binom{n+m-1}{m} p^n (1-p)^m$, for $m \geq 0$.

(b)  $f(m) = \{m(m+1)\}^{-1}$, for $m \geq 1$.

(c)  $f(m) = (1-p)p^{|m|}/(1+p)$, for $m = \ldots, -1, 0, 1, \ldots$.

The constant $p$ satisfies $0 < p < 1$.

**2.**   Let $X$ ($\geq 0$) have probability generating function $G$ and write $t(n) = \mathbb{P}(X > n)$ for the 'tail' probabilities of $X$. Show that the generating function of the sequence $\{t(n) : n \geq 0\}$ is $T(s) = (1 - G(s))/(1-s)$. Show that $\mathbb{E}(X) = T(1)$ and $\text{var}(X) = 2T'(1) + T(1) - T(1)^2$.

**3.**   Let $G_{X,Y}(s, t)$ be the joint probability generating function of $X$ and $Y$. Show that $G_X(s) = G_{X,Y}(s, 1)$ and $G_Y(t) = G_{X,Y}(1, t)$. Show that

$$\mathbb{E}(XY) = \frac{\partial^2}{\partial s\, \partial t} G_{X,Y}(s, t) \bigg|_{s=t=1}.$$

**4.**   Find the joint generating functions of the following joint mass functions, and state for what values of the variables the series converge.

(a)  $f(j, k) = (1-\alpha)(\beta-\alpha)\alpha^j \beta^{k-j-1}$, for $0 \leq k \leq j$, where $0 < \alpha < 1$, $\alpha < \beta$.

(b)  $f(j, k) = (e-1)e^{-(2k+1)}k^j/j!$, for $j, k \geq 0$.

(c)  $f(j, k) = \binom{k}{j}p^{j+k}(1-p)^{k-j}/[k\log\{1/(1-p)\}]$, for $0 \leq j \leq k$, $k \geq 1$, where $0 < p < 1$.

Deduce the marginal probability generating functions and the covariances.

**5.**   A coin is tossed $n$ times, and heads turns up with probability $p$ on each toss. Assuming the usual independence, show that the joint probability generating function of the numbers $H$ and $T$ of heads and tails is $G_{H,T}(x, y) = \{px + (1-p)y\}^n$. Generalize this conclusion to find the joint probability generating function of the multinomial distribution of Exercise (3.5.1).

**6.**   Let $X$ have the binomial distribution $\text{bin}(n, U)$, where $U$ is uniform on $(0, 1)$. Show that $X$ is uniformly distributed on $\{0, 1, 2, \ldots, n\}$.

**7.**   Show that

$$G(x, y, z, w) = \tfrac{1}{8}(xyzw + xy + yz + zw + zx + yw + xz + 1)$$

is the joint generating function of four variables that are pairwise and triplewise independent, but are nevertheless *not* independent.

**8.**   Let $p_r > 0$ and $a_r \in \mathbb{R}$ for $1 \leq r \leq n$. Which of the following is a moment generating function, and for what random variable?

(a)  $M(t) = 1 + \sum_{r=1}^{n} p_r t^r$,    (b)  $M(t) = \sum_{r=1}^{n} p_r e^{a_r t}$.

**9.**   Let $G_1$ and $G_2$ be probability generating functions, and suppose that $0 \leq \alpha \leq 1$. Show that $G_1 G_2$, and $\alpha G_1 + (1-\alpha)G_2$ are probability generating functions. Is $G(\alpha s)/G(\alpha)$ necessarily a probability generating function?

**10.**   Let $X_1, X_2, \ldots$ be independent, continuous random variables with common distribution function $F$, that are independent of the positive integer-valued random variable $Z$. Define the maximum $M = \max\{X_1, X_2, \ldots, X_Z\}$. Show that

$$\mathbb{E}(Z \mid M = m) = 1 + \frac{F(m)G''(F(m))}{G'(F(m))},$$

where $G$ is the probability generating function of $Z$.

**11. Truncated geometric distribution.** Let $0 < p = 1 - q < 1$, and let $X$ have the geometric mass function $f(x) = q^{x-1}p$ for $x = 1, 2, \ldots$. Find the probability generating function of $Y = \min\{n, X\}$ for fixed $n \geq 1$, and show that $\mathbb{E}(Y) = (1 - q^n)/p$.

**12. Van Dantzig's collective marks.** Let $X_i$ be the number of balls in the $i$th of a sequence of bins, and assume the $X_i$ are independent with common probability generating function $G_X$. There are $N$ such bins, where $N$ is independent of the $X_i$ with probability generating function $G_N$. Each ball is 'unmarked' with probability $u$, and marked otherwise, with marks appearing independently.

(a) Show that the probability $\pi$, that all the balls in the $i$th bin are unmarked, satisfies $\pi = G_X(u)$.
(b) Deduce that the probability that all the balls are unmarked equals $G_N(G_X(u))$, and deduce the random sum formula of Theorem (5.1.25).
(c) Find the mean and variance of the total number $T$ of balls in terms of the moments of $X$ and $N$, and compare the argument with the method of Exercise (3.7.4).
(d) Find the probability generating function of the total number $U$ of unmarked balls, and deduce the mean and variance of $U$.

## 5.2 Some applications

Generating functions provide a powerful tool, particularly in the presence of difference equations and convolutions. This section contains a variety of examples of this tool in action.

**(1) Example. Problem of the points†.** A coin is tossed repeatedly and heads turns up with probability $p$ on each toss. Player A wins if $m$ heads appear before $n$ tails, and player B wins otherwise. We have seen, in Exercise (3.9.4) and Problem (3.11.24), two approaches to the problem of determining the probability that A wins. It is elementary, by conditioning on the outcome of the first toss, that the probability $p_{mn}$, that A wins, satisfies

$$(2) \qquad p_{mn} = p p_{m-1,n} + q p_{m,n-1}, \quad \text{for} \quad m, n \geq 1,$$

where $p + q = 1$. The boundary conditions are $p_{m0} = 0$, $p_{0n} = 1$ for $m, n > 0$. We may solve equation (2) by introducing the generating function

$$G(x, y) = \sum_{m=0}^{\infty} \sum_{n=0}^{\infty} p_{mn} x^m y^n$$

subject to the convention that $p_{00} = 0$. Multiplying throughout (2) by $x^m y^n$ and summing over $m, n \geq 1$, we obtain

$$(3) \qquad G(x, y) - \sum_{m=1}^{\infty} p_{m0} x^m - \sum_{n=1}^{\infty} p_{0n} y^n$$

$$= px \sum_{m,n=1}^{\infty} p_{m-1,n} x^{m-1} y^n + qy \sum_{m,n=1}^{\infty} p_{m,n-1} x^m y^{n-1},$$

---

†First recorded by Pacioli in 1494, and eventually solved by Pascal in 1654. Our method is due to Laplace.

and hence, using the boundary conditions,

$$G(x, y) - \frac{y}{1 - y} = px\,G(x, y) + qy\left(G(x, y) - \frac{y}{1 - y}\right), \qquad |y| < 1.$$

Therefore,

(4) $$G(x, y) = \frac{y(1 - qy)}{(1 - y)(1 - px - qy)},$$

from which one may derive the required information by expanding in powers of $x$ and $y$ and finding the coefficient of $x^m y^n$. A cautionary note: in passing from (2) to (3), one should be very careful with the limits of the summations. ●

**(5) Example. Matching revisited.** The famous (mis)matching problem of Example (3.4.3) involves the random placing of $n$ different letters into $n$ differently addressed envelopes. What is the probability $p_n$ that no letter is placed in the correct envelope? Let $M$ be the event that the first letter is put into its correct envelope, and let $N$ be the event that no match occurs. Then

(6) $$p_n = \mathbb{P}(N) = \mathbb{P}(N \mid M^c)\mathbb{P}(M^c),$$

where $\mathbb{P}(M^c) = 1 - n^{-1}$. It is convenient to think of $\alpha_n = \mathbb{P}(N \mid M^c)$ in the following way. It is the probability that, given $n - 2$ pairs of matching white letters and envelopes together with a non-matching red letter and blue envelope, there are no colour matches when the letters are inserted randomly into the envelopes. Either the red letter is placed into the blue envelope or it is not, and a consideration of these two cases gives that

(7) $$\alpha_n = \frac{1}{n - 1} p_{n-2} + \left(1 - \frac{1}{n - 1}\right)\alpha_{n-1}.$$

Combining (6) and (7) we obtain, for $n \geq 3$,

(8) $$p_n = \left(1 - \frac{1}{n}\right)\alpha_n = \left(1 - \frac{1}{n}\right)\left[\frac{1}{n - 1}p_{n-2} + \left(1 - \frac{1}{n - 1}\right)\alpha_{n-1}\right]$$
$$= \left(1 - \frac{1}{n}\right)\left(\frac{1}{n - 1}p_{n-2} + p_{n-1}\right) = \frac{1}{n}p_{n-2} + \left(1 - \frac{1}{n}\right)p_{n-1},$$

a difference relation subject to the boundary conditions $p_1 = 0$, $p_2 = \frac{1}{2}$. We may solve this difference relation by using the generating function

(9) $$G(s) = \sum_{n=1}^{\infty} p_n s^n.$$

We multiply throughout (8) by $ns^{n-1}$ and sum over all suitable values of $n$ to obtain

$$\sum_{n=3}^{\infty} ns^{n-1}p_n = s\sum_{n=3}^{\infty} s^{n-2}p_{n-2} + s\sum_{n=3}^{\infty}(n - 1)s^{n-2}p_{n-1}$$

which we recognize as yielding

$$G'(s) - p_1 - 2p_2s = sG(s) + s[G'(s) - p_1]$$

or $(1 - s)G'(s) = sG(s) + s$, since $p_1 = 0$ and $p_2 = \frac{1}{2}$. This differential equation is easily solved subject to the boundary condition $G(0) = 0$ to obtain $G(s) = (1 - s)^{-1}e^{-s} - 1$. Expanding as a power series in $s$ and comparing with (9), we arrive at the conclusion

$$(10) \qquad p_n = \frac{(-1)^n}{n!} + \frac{(-1)^{n-1}}{(n-1)!} + \cdots + \frac{(-1)}{1!} + 1, \quad \text{for} \quad n \geq 1,$$

as in the conclusion of Example (3.4.3) with $r = 0$.                                    ●

**(11) Example. Matching and occupancy.** The matching problem above is one of the simplest of a class of problems involving putting objects randomly into containers. In a general approach to such questions, we suppose that we are given a collection $\mathcal{A} = \{A_i : 1 \leq i \leq n\}$ of events, and we ask for properties of the random number $X$ of these events which occur (in the previous example, $A_i$ is the event that the $i$th letter is in the correct envelope, and $X$ is the number of correctly placed letters). The problem is to express the mass function of $X$ in terms of probabilities of the form $\mathbb{P}(A_{i_1} \cap A_{i_2} \cap \cdots \cap A_{i_m})$. We introduce the notation

$$S_m = \sum_{i_1 < i_2 < \cdots < i_m} \mathbb{P}(A_{i_1} \cap A_{i_2} \cap \cdots \cap A_{i_m}),$$

the sum of the probabilities of the intersections of exactly $m$ of the events in question. We make the convention that $S_0 = 1$. It is easily seen as follows that

$$(12) \qquad S_m = \mathbb{E}\binom{X}{m},$$

the mean value of the (random) binomial coefficient $\binom{X}{m}$: writing $N_m$ for the number of sub-families of $\mathcal{A}$ having size $m$, all of whose component events occur, we have that

$$S_m = \sum_{i_1 < \cdots < i_m} \mathbb{E}(I_{A_{i_1}} I_{A_{i_2}} \cdots I_{A_{i_m}}) = \mathbb{E}(N_m),$$

whereas $N_m = \binom{X}{m}$. It follows from (12) that

$$(13) \qquad S_m = \sum_{i=0}^{n} \binom{i}{m} \mathbb{P}(X = i).$$

We introduce the generating functions

$$G_S(x) = \sum_{m=0}^{n} x^m S_m, \quad G_X(x) = \sum_{i=0}^{n} x^i \mathbb{P}(X = i),$$

and we then multiply throughout (13) by $x^m$ and sum over $m$, obtaining

$$G_S(x) = \sum_i \mathbb{P}(X = i) \sum_m x^m \binom{i}{m} = \sum_i (1 + x)^i \mathbb{P}(X = i) = G_X(1 + x).$$

Hence $G_X(x) = G_S(x - 1)$, and equating coefficients of $x^i$ yields

**(14)**
$$\mathbb{P}(X = i) = \sum_{j=i}^{n} (-1)^{j-i} \binom{j}{i} S_j \quad \text{for} \quad 0 \le i \le n.$$

This formula, sometimes known as 'Waring's theorem'†, is a complete generalization of certain earlier results, including (10). It may be derived without using generating functions, but at considerable personal cost.                                                   ●

**(15) Example. Recurrent events.** Meteorites fall from the sky, your car runs out of fuel, there is a power failure, you fall ill. Each such event recurs at regular or irregular intervals; one cannot generally predict just when such an event will happen next, but one may be prepared to hazard guesses. A simplistic mathematical model is the following. We call the happening in question $H$, and suppose that, at each time point $1, 2, \ldots$, either $H$ occurs or $H$ does not occur. We write $X_1$ for the first time at which $H$ occurs, $X_1 = \min\{n : H \text{ occurs at time } n\}$, and $X_m$ for the time which elapses between the $(m - 1)$th and $m$th occurrence of $H$. Thus the $m$th occurrence of $H$ takes place at time

**(16)**
$$T_m = X_1 + X_2 + \cdots + X_m.$$

Here are our main assumptions. We assume that the 'inter-occurrence' times $X_1, X_2, \ldots$ are independent random variables taking values in $\{1, 2, \ldots\}$, and furthermore that $X_2, X_3, \ldots$ are identically distributed. That is to say, whilst we assume that *inter*-occurrence times are independent and identically distributed, we allow the time to the *first* occurrence to have a special distribution.

Given the distributions of the $X_i$, how may we calculate the probability that $H$ occurs at some given time? Define $u_n = \mathbb{P}(H \text{ occurs at time } n)$. We have by conditioning on $X_1$ that

**(17)**
$$u_n = \sum_{i=1}^{n} \mathbb{P}(H_n \mid X_1 = i)\mathbb{P}(X_1 = i),$$

where $H_n$ is the event that $H$ occurs at time $n$. Now

$$\mathbb{P}(H_n \mid X_1 = i) = \mathbb{P}(H_{n-i+1} \mid X_1 = 1) = \mathbb{P}(H_{n-i+1} \mid H_1),$$

using the 'translation invariance' entailed by the assumption that the $X_i, i \ge 2$, are independent and identically distributed. A similar conditioning on $X_2$ yields

**(18)**
$$\mathbb{P}(H_m \mid H_1) = \sum_{j=1}^{m-1} \mathbb{P}(H_m \mid H_1, X_2 = j)\mathbb{P}(X_2 = j)$$

$$= \sum_{j=1}^{m-1} \mathbb{P}(H_{m-j} \mid H_1)\mathbb{P}(X_2 = j)$$

---

†This is another instance of Stigler's law of eponymy.

for $m \geq 2$, by translation invariance once again. Multiplying through (18) by $x^{m-1}$ and summing over $m$, we obtain

$$
\text{(19)}\qquad \sum_{m=2}^{\infty} x^{m-1}\mathbb{P}(H_m \mid H_1) = \mathbb{E}(x^{X_2}) \sum_{n=1}^{\infty} x^{n-1}\mathbb{P}(H_n \mid H_1),
$$

so that $G_H(x) = \sum_{m=1}^{\infty} x^{m-1}\mathbb{P}(H_m \mid H_1)$ satisfies $G_H(x) - 1 = F(x)G_H(x)$, where $F(x)$ is the common probability generating function of the inter-occurrence times, and hence

$$
\text{(20)}\qquad G_H(x) = \frac{1}{1 - F(x)}.
$$

Returning to (17), we obtain similarly that $U(x) = \sum_{n=1}^{\infty} x^n u_n$ satisfies

$$
\text{(21)}\qquad U(x) = D(x)G_H(x) = \frac{D(x)}{1 - F(x)}
$$

where $D(x)$ is the probability generating function of $X_1$. Equation (21) contains much of the information relevant to the process, since it relates the occurrences of $H$ to the generating functions of the elements of the sequence $X_1, X_2, \ldots$. We should like to extract information out of (21) about $u_n = \mathbb{P}(H_n)$, the coefficient of $x^n$ in $U(x)$, particularly for large values of $n$.

In principle, one may expand $D(x)/[1 - F(x)]$ as a polynomial in $x$ in order to find $u_n$, but this is difficult in practice. There is one special situation in which this may be done with ease, and this is the situation when $D(x)$ is the function $D = D^*$ given by

$$
\text{(22)}\qquad D^*(x) = \frac{1 - F(x)}{\mu(1 - x)} \quad \text{for} \quad |x| < 1,
$$

and $\mu = \mathbb{E}(X_2)$ is the mean inter-occurrence time. Let us first check that $D^*$ is indeed a suitable probability generating function. The coefficient of $x^n$ in $D^*$ is easily seen to be $(1 - f_1 - f_2 - \cdots - f_n)/\mu$, where $f_i = \mathbb{P}(X_2 = i)$. This coefficient is non-negative since the $f_i$ form a mass function; furthermore, by L'Hôpital's rule,

$$
D^*(1) = \lim_{x \uparrow 1} \frac{1 - F(x)}{\mu(1 - x)} = \lim_{x \uparrow 1} \frac{-F'(x)}{-\mu} = 1
$$

since $F'(1) = \mu$, the mean inter-occurrence time. Hence $D^*(x)$ is indeed a probability generating function, and with this choice for $D$ we obtain that $U = U^*$ where

$$
\text{(23)}\qquad U^*(x) = \frac{1}{\mu(1 - x)}
$$

from (21). Writing $U^*(x) = \sum_n u_n^* x^n$ we find that $u_n^* = \mu^{-1}$ for all $n$. That is to say, for the special choice of $D^*$, the corresponding sequence of the $u_n^*$ is *constant*, so that the density of occurrences of $H$ is constant as time passes. This special process is called a *stationary recurrent-event process*.

How relevant is the choice of $D$ to the behaviour of $u_n$ for large $n$? Intuitively speaking, the choice of distribution of $X_1$ should not affect greatly the behaviour of the process over long time periods, and so one might expect that $u_n \to \mu^{-1}$ as $n \to \infty$, irrespective of the choice of $D$. This is indeed the case, so long as we rule out the possibility that there is 'periodicity' in the process. We call the process *non-arithmetic* if $\gcd\{n : \mathbb{P}(X_2 = n) > 0\} = 1$; certainly the process is non-arithmetic if, for example, $\mathbb{P}(X_2 = 1) > 0$. Note that gcd stands for greatest common divisor.

**(24) Renewal theorem.** *If the mean inter-occurrence time $\mu$ is finite and the process is non-arithmetic, then $u_n = \mathbb{P}(H_n)$ satisfies $u_n \to \mu^{-1}$ as $n \to \infty$.*

**Sketch proof.** The classical proof of this theorem is a purely analytical approach to the equation (21) (see Feller 1968, pp. 335–8). There is a much neater probabilistic proof using the technique of 'coupling'. We do not give a complete proof at this stage, but merely a sketch. The main idea is to introduce a second recurrent-event process, which is stationary and independent of the first. Let $X = \{X_i : i \geq 1\}$ be the first and inter- occurrence times of the original process, and let $X^* = \{X_i^* : i \geq 1\}$ be another sequence of independent random variables, independent of $X$, such that $X_2^*, X_3^*, \ldots$ have the common distribution of $X_2, X_3, \ldots,$ and $X_1^*$ has probability generating function $D^*$. Let $H_n$ and $H_n^*$ be the events that $H$ occurs at time $n$ in the first and second process (respectively), and let $T = \min\{n : H_n \cap H_n^*$ occurs$\}$ be the earliest time at which $H$ occurs simultaneously in both processes. It may be shown that $T < \infty$ with probability 1, using the assumptions that $\mu < \infty$ and that the processes are non-arithmetic; it is intuitively natural that a coincidence occurs sooner or later, but this is not quite so easy to prove, and we omit a rigorous proof at this point, returning to complete the job in Example (5.10.21). The point is that, once the time $T$ has passed, the non-stationary and stationary recurrent-event processes are indistinguishable from each other, since they have had simultaneous occurrences of $H$. That is to say, we have that

$$u_n = \mathbb{P}(H_n \mid T \leq n)\mathbb{P}(T \leq n) + \mathbb{P}(H_n \mid T > n)\mathbb{P}(T > n)$$
$$= \mathbb{P}(H_n^* \mid T \leq n)\mathbb{P}(T \leq n) + \mathbb{P}(H_n \mid T > n)\mathbb{P}(T > n)$$

since, if $T \leq n$, then the two processes have already coincided and the (conditional) probability of $H_n$ equals that of $H_n^*$. Similarly

$$u_n^* = \mathbb{P}(H_n^* \mid T \leq n)\mathbb{P}(T \leq n) + \mathbb{P}(H_n^* \mid T > n)\mathbb{P}(T > n),$$

so that $|u_n - u_n^*| \leq \mathbb{P}(T > n) \to 0$ as $n \to \infty$. However, $u_n^* = \mu^{-1}$ for all $n$, so that $u_n \to \mu^{-1}$ as $n \to \infty$. ∎

---

### Exercises for Section 5.2

**1.**   Let $X$ be the number of events in the sequence $A_1, A_2, \ldots, A_n$ which occur. Let $S_m = \mathbb{E}\binom{X}{m}$, the mean value of the random binomial coefficient $\binom{X}{m}$, and show that

$$\mathbb{P}(X \geq i) = \sum_{j=i}^{n} (-1)^{j-i} \binom{j-1}{i-1} S_j, \qquad \text{for } 1 \leq i \leq n,$$

$$\text{where} \quad S_m = \sum_{j=m}^{n} \binom{j-1}{m-1} \mathbb{P}(X \geq j), \qquad \text{for } 1 \leq m \leq n.$$

**2.** Each person in a group of $n$ people chooses another at random. Find the probability:
(a) that exactly $k$ people are chosen by nobody,
(b) that at least $k$ people are chosen by nobody.

**3. Compounding.**
(a) Let $X$ have the Poisson distribution with parameter $Y$, where $Y$ has the Poisson distribution with parameter $\mu$. Show that $G_{X+Y}(x) = \exp\{\mu(xe^{x-1} - 1)\}$.
(b) Let $X_1, X_2, \ldots$ be independent identically distributed random variables with the *logarithmic* mass function

$$f(k) = \frac{(1-p)^k}{k \log(1/p)}, \qquad k \geq 1,$$

where $0 < p < 1$. If $N$ is independent of the $X_i$ and has the Poisson distribution with parameter $\mu$, show that $Y = \sum_{i=1}^{N} X_i$ has a type of negative binomial distribution.

**4.** Let $X$ have the binomial distribution with parameters $n$ and $p$, and show that

$$\mathbb{E}\left(\frac{1}{1+X}\right) = \frac{1 - (1-p)^{n+1}}{(n+1)p}.$$

Find the limit of this expression as $n \to \infty$ and $p \to 0$, the limit being taken in such a way that $np \to \lambda$ where $0 < \lambda < \infty$. Comment.

**5.** A coin is tossed repeatedly, and heads turns up with probability $p$ on each toss. Let $h_n$ be the probability of an even number of heads in the first $n$ tosses, with the convention that 0 is an even number. Find a difference equation for the $h_n$ and deduce that they have generating function $\frac{1}{2}\{(1 + 2ps - s)^{-1} + (1-s)^{-1}\}$.

**6.** An unfair coin is flipped repeatedly, where $\mathbb{P}(H) = p = 1 - q$. Let $X$ be the number of flips until HTH first appears, and $Y$ the number of flips until either HTH or THT appears. Show that $\mathbb{E}(s^X) = (p^2qs^3)/(1 - s + pqs^2 - pq^2s^3)$ and find $\mathbb{E}(s^Y)$.

**7. Matching again.** The pile of (by now dog-eared) letters is dropped again and enveloped at random, yielding $X_n$ matches. Show that $\mathbb{P}(X_n = j) = (j+1)\mathbb{P}(X_{n+1} = j+1)$. Deduce that the derivatives of the $G_n(s) = \mathbb{E}(s^{X_n})$ satisfy $G'_{n+1} = G_n$, and hence derive the conclusion of Example (3.4.3), namely:

$$\mathbb{P}(X_n = r) = \frac{1}{r!}\left(\frac{1}{2!} - \frac{1}{3!} + \cdots + \frac{(-1)^{n-r}}{(n-r)!}\right).$$

**8.** Let $X$ have a Poisson distribution with parameter $\Lambda$, where $\Lambda$ is exponential with parameter $\mu$. Show that $X$ has a geometric distribution.

**9. Coupons.** Recall from Exercise (3.3.2) that each packet of an overpriced commodity contains a worthless plastic object. There are four types of object, and each packet is equally likely to contain any of the four. Let $T$ be the number of packets you open until you first have the complete set. Find $\mathbb{E}(s^T)$ and $\mathbb{P}(T = k)$.

**10. Library books.** A library permits a reader to hold at most $m$ books at any one time. Your holding before a visit is $B$ books, having the binomial bin$(m, p)$ distribution. On each visit, a held book is retained with probability $r$, and returned otherwise; given the number $R$ of retained books, you borrow a further $N$ books with the bin$(m - R, \alpha)$ distribution. Assuming the usual independence, find the distribution of the number $A = R + N$ of books held after a visit. Show that, after a large number of visits, the number of books held has approximately the bin$(m, s)$ distribution where $s = \alpha/[1 - r(1 - \alpha)]$. You may assume that $0 < \alpha, p, r < 1$.

**11. Matching yet again, Exercise (5.2.7) revisited.** Each envelope is addressed correctly with probability $t$, and incorrectly otherwise (assume the usual independence). It is decided that an incorrectly addressed envelope cannot count as a match even when containing the correct letter. Show that the number $Y_n$ of correctly addressed matches satisfies

$$\mathbb{P}(Y_n = y) = \frac{t^y}{y!} \sum_{k=0}^{n-y} \frac{(-1)^k t^k}{k!} \to \frac{e^{-t} t^y}{y!} \qquad \text{as } n \to \infty, \text{ for } y = 0, 1, 2, \dots.$$

Note the Poisson limit distribution.

**12. Sampling discrete random variables.** A random variable is called *simple* if it may take only finitely many values. Show that a simple random variable $X$ may be simulated using a finite sequence of Bernoulli random variables, in the sense that there exist independent Bernoulli random variables $\{B_j : j = 1, 2, \dots, m\}$ with respective parameters $b_j$, and real numbers $\{r_j\}$, such that $Y = \sum_j r_j B_j$ has the same distribution as $X$.

**13. General Bonferroni inequalities.** (a) For a generating function $A(z) = \sum_{i=0}^{\infty} a_i z^i$, show that $a_0 + a_1 + \dots + a_m$ is the coefficient of $z^m$ in $B(z) := A(z)(1 - z^{m+1})/(1 - z)$.
(b) With the definitions and notation of Example (5.2.11), show that, for $r$ an even integer,

$$S_i - \binom{i+1}{i} S_{i+1} + \dots - \binom{i+r+1}{i} S_{i+r+1}$$

$$\leq \mathbb{P}(X = i) \leq S_i - \binom{i+1}{i} S_{i+1} + \dots + \binom{i+r}{i} S_{i+r}.$$

---

# 5.3  Random walk

Generating functions are particularly valuable when studying random walks. As before, we suppose that $X_1, X_2, \dots$ are independent random variables, each taking the value 1 with probability $p$, and $-1$ otherwise, and we write $S_n = \sum_{i=1}^{n} X_i$; the sequence $S = \{S_i : i \geq 0\}$ is a simple random walk starting at the origin. Natural questions of interest concern the sequence of random times at which the particle subsequently returns to the origin. To describe this sequence we need only find the distribution of the time until the particle returns for the first time, since subsequent times between consecutive visits to the origin are independent copies of this.

Let $p_0(n) = \mathbb{P}(S_n = 0)$ be the probability of being at the origin after $n$ steps, and let $f_0(n) = \mathbb{P}(S_1 \neq 0, \dots, S_{n-1} \neq 0, S_n = 0)$ be the probability that the first return occurs after $n$ steps. Denote the generating functions of these sequences by

$$P_0(s) = \sum_{n=0}^{\infty} p_0(n) s^n, \qquad F_0(s) = \sum_{n=1}^{\infty} f_0(n) s^n.$$

$F_0$ is the probability generating function of the random time $T_0$ until the particle makes its first return to the origin. That is $F_0(s) = \mathbb{E}(s^{T_0})$. Take care here: $T_0$ may be defective, and so it may be the case that $F_0(1) = \mathbb{P}(T_0 < \infty)$ satisfies $F_0(1) < 1$.

**(1) Theorem.** *We have that*:
- (a) $P_0(s) = 1 + P_0(s)F_0(s)$,
- (b) $P_0(s) = (1 - 4pqs^2)^{-\frac{1}{2}}$,
- (c) $F_0(s) = 1 - (1 - 4pqs^2)^{\frac{1}{2}}$.

**Proof.** (a) Let $A$ be the event that $S_n = 0$, and let $B_k$ be the event that the first return to the origin happens at the $k$th step. Clearly the $B_k$ are disjoint and so, by Lemma (1.4.4),

$$\mathbb{P}(A) = \sum_{k=1}^{n} \mathbb{P}(A \mid B_k)\mathbb{P}(B_k).$$

However, $\mathbb{P}(B_k) = f_0(k)$ and $\mathbb{P}(A \mid B_k) = p_0(n-k)$ by temporal homogeneity, giving

$$(2) \qquad\qquad p_0(n) = \sum_{k=1}^{n} p_0(n-k)f_0(k) \quad \text{if} \quad n \geq 1.$$

Multiply (2) by $s^n$, sum over $n$ remembering that $p_0(0) = 1$, and use the convolution property of generating functions to obtain $P_0(s) = 1 + P_0(s)F_0(s)$.

(b) $S_n = 0$ if and only if the particle takes equal numbers of steps to the left and to the right during its first $n$ steps. The number of ways in which it can do this is $\binom{n}{\frac{1}{2}n}$ and each such way occurs with probability $(pq)^{n/2}$, giving

$$(3) \qquad\qquad p_0(n) = \binom{n}{\frac{1}{2}n}(pq)^{n/2}.$$

We have that $p_0(n) = 0$ if $n$ is odd. This sequence (3) has the required generating function $P_0(s)$.

(c) This follows immediately from (a) and (b). ■

**(4) Corollary.**
(a) *The probability that the particle ever returns to the origin is*

$$\sum_{n=1}^{\infty} f_0(n) = F_0(1) = 1 - |p - q|.$$

(b) *If eventual return is certain, that is $F_0(1) = 1$ and $p = \frac{1}{2}$, then the expected time to the first return is*

$$\sum_{n=1}^{\infty} nf_0(n) = F_0'(1) = \infty.$$

We call the process *recurrent* (or *persistent*) if eventual return to the origin is (almost) certain; otherwise it is called *transient*. It is immediately obvious from (4a) that the process is recurrent if and only if $p = \frac{1}{2}$. This is consistent with our intuition, which suggests that if $p > \frac{1}{2}$ or $p < \frac{1}{2}$, then the particle tends to stray a long way to the right or to the left of the origin respectively. Even when $p = \frac{1}{2}$ the time until first return has infinite mean.

**Proof.** (a) Let $s \uparrow 1$ in (1c), and remember equation (5.1.30).

(b) Eventual return is certain if and only if $p = \frac{1}{2}$. But then the generating function of the time $T_0$ to the first return is $F_0(s) = 1 - (1 - s^2)^{\frac{1}{2}}$ and $\mathbb{E}(T_0) = \lim_{s \uparrow 1} F_0'(s) = \infty$. ∎

Now let us consider the times of visits to the point $r$. Define

$$f_r(n) = \mathbb{P}(S_1 \neq r, \ldots, \; S_{n-1} \neq r, \; S_n = r)$$

to be the probability that the first such visit occurs at the $n$th step, with generating function $F_r(s) = \sum_{n=1}^{\infty} f_r(n)s^n$.

**(5) Theorem.** *We have that*:

(a) $F_r(s) = [F_1(s)]^r$ *for* $r \geq 1$,

(b) $F_1(s) = \left[1 - (1 - 4pqs^2)^{\frac{1}{2}}\right]/(2qs)$.

**Proof.** (a) The same argument which yields (2) also shows that

$$f_r(n) = \sum_{k=1}^{n-1} f_{r-1}(n-k)f_1(k) \quad \text{if} \quad r > 1.$$

Multiply by $s^n$ and sum over $n$ to obtain

$$F_r(s) = F_{r-1}(s)F_1(s) = F_1(s)^r.$$

We could have written this out in terms of random variables instead of probabilities, and then used Theorem (5.1.23). To see this, let $T_r = \min\{n \geq 0 : S_n = r\}$ be the number of steps taken before the particle reaches $r$ for the first time ($T_r$ may equal $+\infty$ if $r > 0$ and $p < \frac{1}{2}$ or if $r < 0$ and $p > \frac{1}{2}$). [Until further notice, we have $T_0 = 0$.] In order to visit $r$, the particle must first visit the point 1; this requires $T_1$ steps. After visiting 1 the particle requires a further number, $T_{1,r}$ say, of steps to reach $r$; $T_{1,r}$ is distributed in the manner of $T_{r-1}$ by 'spatial homogeneity'. Thus

$$T_r = \begin{cases} \infty & \text{if } T_1 = \infty, \\ T_1 + T_{1,r} & \text{if } T_1 < \infty, \end{cases}$$

and the result follows from (5.1.23). Some difficulties arise from the possibility that $T_1 = \infty$, but these are resolved fairly easily (*exercise*).

(b) Condition on $X_1$ to obtain, for $n > 1$,

$$\mathbb{P}(T_1 = n) = \mathbb{P}(T_1 = n \mid X_1 = 1)p + \mathbb{P}(T_1 = n \mid X_1 = -1)q$$

$$= 0 \cdot p + \mathbb{P}(\text{first visit to 1 takes } n - 1 \text{ steps} \mid S_0 = -1) \cdot q$$

$$\qquad\qquad\qquad\qquad\qquad\qquad\qquad \text{by temporal homogeneity}$$

$$= \mathbb{P}(T_2 = n - 1)q \qquad\qquad\qquad \text{by spatial homogeneity}$$

$$= qf_2(n - 1).$$

Therefore $f_1(n) = qf_2(n - 1)$ if $n > 1$, and $f_1(1) = p$. Multiply by $s^n$ and sum to obtain

$$F_1(s) = ps + sq\,F_2(s) = ps + qs\,F_1(s)^2$$

by (a). Solve this quadratic to find its two roots. Only one can be a probability generating function; why? [Hint: $F_1(0) = 0$.] ∎

**(6) Corollary.** *The probability that the walk ever visits the positive part of the real axis is*

$$F_1(1) = \frac{1 - |p - q|}{2q} = \min\{1, p/q\}.$$

Knowledge of Theorem (5) enables us to calculate $F_0(s)$ directly without recourse to (1). The method of doing this relies upon a symmetry within the collection of paths which may be followed by a random walk. Condition on the value of $X_1$ as usual to obtain

$$f_0(n) = q f_1(n - 1) + p f_{-1}(n - 1)$$

and thus

$$F_0(s) = qs F_1(s) + ps F_{-1}(s).$$

We need to find $F_{-1}(s)$. Consider any possible path $\pi$ that the particle may have taken to arrive at the point $-1$ and replace each step in the path by its mirror image, positive steps becoming negative and negative becoming positive, to obtain a path $\pi^*$ which ends at $+1$. This operation of reflection provides a one–one correspondence between the collection of paths ending at $-1$ and the collection of paths ending at $+1$. If $\mathbb{P}(\pi; p, q)$ is the probability that the particle follows $\pi$ when each step is to the right with probability $p$, then $\mathbb{P}(\pi; p, q) = \mathbb{P}(\pi^*; q, p)$; thus

$$F_{-1}(s) = \frac{1 - (1 - 4pqs^2)^{\frac{1}{2}}}{2ps},$$

giving that $F_0(s) = 1 - (1 - 4pqs^2)^{\frac{1}{2}}$ as before.

We made use in the last paragraph of a version of the reflection principle discussed in Section 3.10. Generally speaking, results obtained using the reflection principle may also be obtained using generating functions, sometimes in greater generality than before. Consider for example the hitting time theorem (3.10.14): the mass function of the time $T_b$ of the first visit of $S$ to the point $b$ is given by

$$\mathbb{P}(T_b = n) = \frac{|b|}{n} \mathbb{P}(S_n = b) \quad \text{if} \quad n \geq 1.$$

We shall state and prove a version of this for random walks of a more general nature. Consider a sequence $X_1, X_2, \ldots$ of independent identically distributed random variables taking values in the integers (positive and negative). We may think of $S_n = X_1 + X_2 + \cdots + X_n$ as being the $n$th position of a random walk which takes steps $X_i$; for the simple random walk, each $X_i$ is required to take the values $\pm 1$ only. We call a random walk *right-continuous* (respectively *left-continuous*) if $\mathbb{P}(X_i \leq 1) = 1$ (respectively $\mathbb{P}(X_i \geq -1) = 1$), which is to say that the maximum rightward (respectively leftward) step is no greater than 1. In order to avoid certain situations of no interest, we shall consider only right-continuous walks (respectively left-continuous walks) for which $\mathbb{P}(X_i = 1) > 0$ (respectively $\mathbb{P}(X_i = -1) > 0$).

**(7) Hitting time theorem.** *Assume that $S$ is a right-continuous random walk, and let $T_b$ be the first hitting time of the point $b$. Then*

$$\mathbb{P}(T_b = n) = \frac{b}{n} \mathbb{P}(S_n = b) \quad \text{for} \quad b, n \geq 1.$$

For left-continuous walks, the conclusion becomes

(8) $$\mathbb{P}(T_{-b} = n) = \frac{b}{n}\mathbb{P}(S_n = -b) \quad \text{for} \quad b, n \geq 1.$$

**Proof.** This is the classical proof; see Exercise (3.10.5) for a more elementary proof. We introduce the functions

$$G(z) = \mathbb{E}(z^{-X_1}) = \sum_{n=-\infty}^{1} z^{-n}\mathbb{P}(X_1 = n), \quad F_b(z) = \mathbb{E}(z^{T_b}) = \sum_{n=0}^{\infty} z^n\mathbb{P}(T_b = n).$$

These are functions of the complex variable $z$. The function $G(z)$ has a simple pole at the origin, and the sum defining $F_b(z)$ converges for $|z| < 1$.

Since the walk is assumed to be right-continuous, in order to reach $b$ (where $b > 0$) it must pass through the points $1, 2, \ldots, b - 1$. The argument leading to (5a) may therefore be applied, and we find that

(9) $$F_b(z) = F_1(z)^b \quad \text{for} \quad b \geq 0.$$

The argument leading to (5b) may be expressed as

$$F_1(z) = \mathbb{E}(z^{T_1}) = \mathbb{E}\big(\mathbb{E}(z^{T_1} \mid X_1)\big) = \mathbb{E}(z^{1+T_J}) \quad \text{where } J = 1 - X_1$$

since, conditional on $X_1$, the further time required to reach 1 has the same distribution as $T_{1-X_1}$. Now $1 - X_1 \geq 0$, and therefore

$$F_1(z) = z\mathbb{E}\big(F_{1-X_1}(z)\big) = z\mathbb{E}\big(F_1(z)^{1-X_1}\big) = zF_1(z)G(F_1(z)),$$

yielding

(10) $$z = \frac{1}{G(w)}$$

where

(11) $$w = w(z) = F_1(z).$$

Inverting (10) to find $F_1(z)$, and hence $F_b(z) = F_1(z)^b$, is a standard exercise in complex analysis using what is called Lagrange's inversion formula.

**(12) Theorem. Lagrange's inversion formula.** *Let $z = w/f(w)$ where $w/f(w)$ is an analytic function of $w$ on a neighbourhood of the origin. If $g$ is infinitely differentiable, then*

(13) $$g(w(z)) = g(0) + \sum_{n=1}^{\infty} \frac{1}{n!}z^n \left[\frac{d^{n-1}}{du^{n-1}}[g'(u)f(u)^n]\right]_{u=0}.$$

We apply this as follows. Define $w = F_1(z)$ and $f(w) = wG(w)$, so that (10) becomes $z = w/f(w)$. Note that $f(w) = \mathbb{E}(w^{1-X_1})$ which, by the right-continuity of the walk, is a power series in $w$ which converges for $|w| < 1$. Also $f(0) = \mathbb{P}(X_1 = 1) > 0$, and hence $w/f(w)$ is analytic on a neighbourhood of the origin. We set $g(w) = w^b (= F_1(z)^b = F_b(z)$, by (9)). The inversion formula now yields

$$(14) \qquad\qquad F_b(z) = g(w(z)) = g(0) + \sum_{n=1}^{\infty} \frac{1}{n!} z^n D_n$$

where

$$D_n = \frac{d^{n-1}}{du^{n-1}} \left[ bu^{b-1} u^n G(u)^n \right]\Big|_{u=0} .$$

We pick out the coefficient of $z^n$ in (14) to obtain

$$(15) \qquad\qquad \mathbb{P}(T_b = n) = \frac{1}{n!} D_n \quad \text{for} \quad n \geq 1.$$

Now $G(u)^n = \sum_{i=-\infty}^{n} u^{-i} \mathbb{P}(S_n = i)$, so that

$$D_n = \frac{d^{n-1}}{du^{n-1}} \left( b \sum_{i=-\infty}^{n} u^{b+n-1-i} \mathbb{P}(S_n = i) \right)\Big|_{u=0} = b(n-1)! \mathbb{P}(S_n = b),$$

which may be combined with (15) as required. ∎

Once we have the hitting time theorem, we are in a position to derive a magical result called the Pollaczek–Spitzer identity, relating the distributions of the maxima of a random walk to those of the walk itself. This identity is valid in considerable generality; the proof given here uses the hitting time theorem, and is therefore valid only for right-continuous walks (and *mutatis mutandis* for left-continuous walks and their minima).

**(16) Theorem. Pollaczek–Spitzer identity.** *Assume that S is a right-continuous random walk, and let* $M_n = \max\{S_i : 0 \leq i \leq n\}$ *be the maximum of the walk up to time n. Then, for* $|s|, |t| < 1$,

$$(17) \qquad\qquad \log\left( \sum_{n=0}^{\infty} t^n \mathbb{E}(s^{M_n}) \right) = \sum_{n=1}^{\infty} \frac{1}{n} t^n \mathbb{E}(s^{S_n^+})$$

*where* $S_n^+ = \max\{0, S_n\}$ *as usual.*

This curious and remarkable identity relates the generating function of the probability generating functions of the maxima $M_n$ to the corresponding object for $S_n^+$. It contains full information about the distributions of the maxima.

**Proof.** Writing $f_j(n) = \mathbb{P}(T_j = n)$ as in Section 3.10, we have that

$$(18) \qquad\qquad \mathbb{P}(M_n = k) = \sum_{j=0}^{n} f_k(j) \mathbb{P}(T_1 > n - j) \quad \text{for} \quad k \geq 0,$$

since $M_n = k$ if the passage to $k$ occurs at some time $j$ ($\leq n$), and in addition the walk does not rise above $k$ during the next $n - j$ steps; remember that $T_1 = \infty$ if no visit to 1 takes place. Multiply throughout (18) by $s^k t^n$ (where $|s|, |t| \leq 1$) and sum over $k, n \geq 0$ to obtain

$$\sum_{n=0}^{\infty} t^n \mathbb{E}(s^{M_n}) = \sum_{k=0}^{\infty} s^k \left( \sum_{n=0}^{\infty} t^n \mathbb{P}(M_n = k) \right) = \sum_{k=0}^{\infty} s^k F_k(t) \left( \frac{1 - F_1(t)}{1 - t} \right),$$

by the convolution formula for generating functions. We have used the result of Exercise (5.1.2) here; as usual, $F_k(t) = \mathbb{E}(t^{T_k})$. Now $F_k(t) = F_1(t)^k$, by (9), and therefore

(19)
$$\sum_{n=0}^{\infty} t^n \mathbb{E}(s^{M_n}) = D(s, t)$$

where

(20)
$$D(s, t) = \frac{1 - F_1(t)}{(1 - t)(1 - s F_1(t))}.$$

We shall find $D(s, t)$ by finding an expression for $\partial D / \partial t$ and integrating with respect to $t$.
   By the hitting time theorem, for $n \geq 0$,

(21)
$$n\mathbb{P}(T_1 = n) = \mathbb{P}(S_n = 1) = \sum_{j=0}^{n} \mathbb{P}(T_1 = j)\mathbb{P}(S_{n-j} = 0),$$

as usual; multiply throughout by $t^n$ and sum over $n$ to obtain that $t F_1'(t) = F_1(t) P_0(t)$. Hence

(22)
$$\frac{\partial}{\partial t} \log[1 - s F_1(t)] = \frac{-s F_1'(t)}{1 - s F_1(t)} = -\frac{s}{t} F_1(t) P_0(t) \sum_{k=0}^{\infty} s^k F_1(t)^k$$

$$= -\sum_{k=1}^{\infty} \frac{s^k}{t} F_k(t) P_0(t)$$

by (9). Now $F_k(t) P_0(t)$ is the generating function of the sequence

$$\sum_{j=0}^{n} \mathbb{P}(T_k = j)\mathbb{P}(S_{n-j} = 0) = \mathbb{P}(S_n = k)$$

as in (21), which implies that

$$\frac{\partial}{\partial t} \log[1 - s F_1(t)] = -\sum_{n=1}^{\infty} t^{n-1} \sum_{k=1}^{\infty} s^k \mathbb{P}(S_n = k).$$

Hence

$$\frac{\partial}{\partial t} \log D(s, t) = -\frac{\partial}{\partial t} \log(1 - t) + \frac{\partial}{\partial t} \log[1 - F_1(t)] - \frac{\partial}{\partial t} \log[1 - s F_1(t)]$$

$$= \sum_{n=1}^{\infty} t^{n-1} \left( 1 - \sum_{k=1}^{\infty} \mathbb{P}(S_n = k) + \sum_{k=1}^{\infty} s^k \mathbb{P}(S_n = k) \right)$$

$$= \sum_{n=1}^{\infty} t^{n-1} \left( \mathbb{P}(S_n \leq 0) + \sum_{k=1}^{\infty} s^k \mathbb{P}(S_n = k) \right) = \sum_{n=1}^{\infty} t^{n-1} \mathbb{E}(s^{S_n^+}).$$

Integrate over $t$, noting that both sides of (19) equal 1 when $t = 0$, to obtain (17).          ∎

For our final example of the use of generating functions, we return to simple random walk, for which each jump equals 1 or $-1$ with probabilities $p$ and $q = 1 - p$. Suppose that we are told that $S_{2n} = 0$, so that the walk is 'tied down', and we ask for the number of steps of the walk which were not within the negative half-line. In the language of gambling, $L_{2n}$ is the amount of time that the gambler was ahead of the bank. In the arc sine law for sojourn times, Theorem (3.10.21), we explored the distribution of $L_{2n}$ without imposing the condition that $S_{2n} = 0$. Given that $S_{2n} = 0$, we might think that $L_{2n}$ would be about $n$, but, as can often happen, the contrary turns out to be the case.

**(23) Theorem. Leads for tied-down random walk.** *For the simple random walk S,*

$$\mathbb{P}(L_{2n} = 2k \mid S_{2n} = 0) = \frac{1}{n+1}, \quad k = 0, 1, 2, \ldots, n.$$

Thus each possible value of $L_{2n}$ is equally likely. Unlike the related results of Section 3.10, we prove this using generating functions. Note that the distribution of $L_{2n}$ does not depend on the value of $p$. This is not surprising since, conditional on $\{S_{2n} = 0\}$, the joint distribution of $S_0, S_1, \ldots, S_{2n}$ does not depend on $p$ *(exercise)*.

**Proof.** Assume $|s|, |t| < 1$, and define $G_{2n}(s) = \mathbb{E}(s^{L_{2n}} \mid S_{2n} = 0)$, $F_0(s) = \mathbb{E}(s^{T_0})$ where $T_0$ is re-set to $T_0 = \min\{n \geq 1 : S_n = 0\}$, and the bivariate generating function

$$H(s, t) = \sum_{n=0}^{\infty} t^{2n} \mathbb{P}(S_{2n} = 0) G_{2n}(s).$$

By conditioning on the time of the first return to the origin,

(24)          $$G_{2n}(s) = \sum_{r=1}^{n} \mathbb{E}(s^{L_{2n}} \mid S_{2n} = 0, \, T_0 = 2r) \mathbb{P}(T_0 = 2r \mid S_{2n} = 0).$$

We may assume without loss of generality that $p = q = \frac{1}{2}$, so that

$$\mathbb{E}(s^{L_{2n}} \mid S_{2n} = 0, \, T_0 = 2r) = G_{2n-2r}(s)\left(\tfrac{1}{2} + \tfrac{1}{2} s^{2r}\right),$$

since, under these conditions, $L_{2r}$ has (conditional) probability $\frac{1}{2}$ of being equal to either 0 or $2r$. Also

$$\mathbb{P}(T_0 = 2r \mid S_{2n} = 0) = \frac{\mathbb{P}(T_0 = 2r)\mathbb{P}(S_{2n-2r} = 0)}{\mathbb{P}(S_{2n} = 0)},$$

so that (24) becomes

$$G_{2n}(s) = \sum_{r=1}^{n} \frac{\left[G_{2n-2r}(s)\mathbb{P}(S_{2n-2r} = 0)\right]\left[\tfrac{1}{2}(1 + s^{2r})\mathbb{P}(T_0 = 2r)\right]}{\mathbb{P}(S_{2n} = 0)}.$$

Multiply throughout by $t^{2n}\mathbb{P}(S_{2n} = 0)$ and sum over $n \geq 1$, to find that

$$H(s, t) - 1 = \tfrac{1}{2} H(s, t)[F_0(t) + F_0(st)].$$

Hence

$$H(s, t) = \frac{2}{\sqrt{1 - t^2} + \sqrt{1 - s^2 t^2}} = \frac{2\left[\sqrt{1 - s^2 t^2} - \sqrt{1 - t^2}\right]}{t^2(1 - s^2)}$$

$$= \sum_{n=0}^{\infty} t^{2n} \mathbb{P}(S_{2n} = 0) \left(\frac{1 - s^{2n+2}}{(n + 1)(1 - s^2)}\right)$$

after a little work using (1b). We deduce that $G_{2n}(s) = \sum_{k=0}^{n}(n + 1)^{-1} s^{2k}$, and the proof is finished.    ∎

---

## Exercises for Section 5.3

**1.**  For a simple random walk $S$ with $S_0 = 0$ and $p = 1 - q < \frac{1}{2}$, show that the maximum $M = \max\{S_n : n \geq 0\}$ satisfies $\mathbb{P}(M \geq r) = (p/q)^r$ for $r \geq 0$.

**2.**  Use generating functions to show that, for a symmetric random walk,
(a) $2k f_0(2k) = \mathbb{P}(S_{2k-2} = 0)$ for $k \geq 1$, and
(b) $\mathbb{P}(S_1 S_2 \cdots S_{2n} \neq 0) = \mathbb{P}(S_{2n} = 0)$ for $n \geq 1$.

**3.**  A particle performs a random walk on the corners of the square ABCD. At each step, the probability of moving from corner $c$ to corner $d$ equals $\rho_{cd}$, where

$$\rho_{AB} = \rho_{BA} = \rho_{CD} = \rho_{DC} = \alpha, \qquad \rho_{AD} = \rho_{DA} = \rho_{BC} = \rho_{CB} = \beta,$$

and $\alpha, \beta > 0$, $\alpha + \beta = 1$. Let $G_A(s)$ be the generating function of the sequence $(\rho_{AA}(n) : n \geq 0)$, where $\rho_{AA}(n)$ is the probability that the particle is at A after $n$ steps, having started at A. Show that

$$G_A(s) = \frac{1}{2}\left\{\frac{1}{1 - s^2} + \frac{1}{1 - |\beta - \alpha|^2 s^2}\right\}.$$

Hence find the probability generating function of the time of the first return to A.

**4.**  A particle performs a symmetric random walk in two dimensions starting at the origin: each step is of unit length and has equal probability $\frac{1}{4}$ of being northwards, southwards, eastwards, or westwards. The particle first reaches the line $x + y = m$ at the point $(X, Y)$ and at the time $T$. Find the probability generating functions of $T$ and $X - Y$, and state where they converge.

**5.**  Derive the arc sine law for sojourn times, Theorem (3.10.21), using generating functions. That is to say, let $L_{2n}$ be the length of time spent (up to time $2n$) by a simple symmetric random walk to the right of its starting point. Show that

$$\mathbb{P}(L_{2n} = 2k) = \mathbb{P}(S_{2k} = 0)\mathbb{P}(S_{2n-2k} = 0) \qquad \text{for } 0 \leq k \leq n.$$

**6.**  Let $\{S_n : n \geq 0\}$ be a simple symmetric random walk with $S_0 = 0$, and let $T = \min\{n > 0 : S_n = 0\}$. Show that

$$\mathbb{E}\left(\min\{T, 2m\}\right) = 2\mathbb{E}|S_{2m}| = 4m\mathbb{P}(S_{2m} = 0) \qquad \text{for } m \geq 0.$$

**7.**  Let $S_n = \sum_{r=0}^{n} X_r$ be a left-continuous random walk on the integers with a retaining barrier at zero. More specifically, we assume that the $X_r$ are identically distributed integer-valued random variables with $X_1 \geq -1$, $\mathbb{P}(X_1 = 0) \neq 0$, and

$$S_{n+1} = \begin{cases} S_n + X_{n+1} & \text{if } S_n > 0, \\ S_n + X_{n+1} + 1 & \text{if } S_n = 0. \end{cases}$$

Show that the distribution of $S_0$ may be chosen in such a way that $\mathbb{E}(z^{S_n}) = \mathbb{E}(z^{S_0})$ for all $n$, if and only if $\mathbb{E}(X_1) < 0$, and in this case

$$\mathbb{E}(z^{S_n}) = \frac{(1-z)\mathbb{E}(X_1)\mathbb{E}(z^{X_1})}{1-\mathbb{E}(z^{X_1})}.$$

**8.**   Consider a simple random walk starting at 0 in which each step is to the right with probability $p\ (= 1 - q)$. Let $T_b$ be the number of steps until the walk first reaches $b$ where $b > 0$. Show that $\mathbb{E}(T_b \mid T_b < \infty) = b/|p - q|$.

**9.   Conditioned random walk.** Let $S = \{S_k : k = 0, 1, 2, \dots\}$ be a simple random walk on the non-negative integers, with $S_0 = i$ and an absorbing barrier at 0. A typical jump $X$ has mass function $\mathbb{P}(X = 1) = p$ and $\mathbb{P}(X = -1) = q = 1 - p$, where $p \in (\frac{1}{2}, 1)$. Let $H$ be the event that the walk is ultimately absorbed at 0. Show that, conditional on $H$, the walk has the same distribution as a simple random walk $W = \{W_k : k = 0, 1, 2, \dots\}$ for which a typical jump $Y$ satisfies $\mathbb{P}(Y = 1) = q$, $\mathbb{P}(Y = -1) = p$.

---

## 5.4  Branching processes

Besides gambling, many probabilists have been interested in reproduction. Accurate models for the evolution of a population are notoriously difficult to handle, but there are simpler non-trivial models which are both tractable and mathematically interesting. The branching process is such a model. Suppose that a population evolves in generations, and let $Z_n$ be the number of members of the $n$th generation. Each member of the $n$th generation gives birth to a family, possibly empty, of members of the $(n + 1)$th generation; the size of this family is a random variable. We shall make the following assumptions about these family sizes:

(a)  the family sizes of the individuals of the branching process form a collection of independent random variables;

(b)  all family sizes have the same probability mass function $f$ and generating function $G$.

These assumptions, together with information about the distribution of the number $Z_0$ of founding members, specify the random evolution of the process. We assume here that $Z_0 = 1$. There is nothing notably human about this model, which may be just as suitable a description for the growth of a population of cells, or for the increase of neutrons in a reactor, or for the spread of an epidemic in some population. See Figure 5.1 for a picture of a branching process.

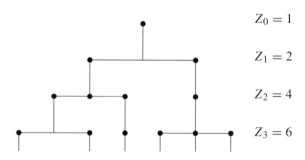

Figure 5.1. The family tree of a branching process.

We are interested in the random sequence $Z_0, Z_1, \dots$ of generation sizes. Let $G_n(s) = \mathbb{E}(s^{Z_n})$ be the generating function of $Z_n$.

**(1) Theorem.** *It is the case that $G_{m+n}(s) = G_m(G_n(s)) = G_n(G_m(s))$, and thus $G_n(s) = G(G(\dots(G(s))\dots))$ is the $n$-fold iterate of $G$.*

**Proof.** Each member of the $(m+n)$th generation has a unique ancestor in the $m$th generation. Thus

$$Z_{m+n} = X_1 + X_2 + \cdots + X_{Z_m}$$

where $X_i$ is the number of members of the $(m+n)$th generation which stem from the $i$th member of the $m$th generation. This is the sum of a random number $Z_m$ of variables. These variables are independent by assumption (a); furthermore, by assumption (b) they are identically distributed with the same distribution as the number $Z_n$ of the $n$th-generation offspring of the first individual in the process. Now use Theorem (5.1.25) to obtain $G_{m+n}(s) = G_m(G_{X_1}(s))$ where $G_{X_1}(s) = G_n(s)$. Iterate this relation to obtain

$$G_n(s) = G_1(G_{n-1}(s)) = G_1(G_1(G_{n-2}(s))) = G_1(G_1(\dots(G_1(s))\dots))$$

and notice that $G_1(s)$ is what we called $G(s)$.                        ∎

In principle, Theorem (1) tells us all about $Z_n$ and its distribution, but in practice $G_n(s)$ may be hard to evaluate. The moments of $Z_n$, at least, may be routinely computed in terms of the moments of a typical family size $Z_1$. For example:

**(2) Lemma.** *Let $\mu = \mathbb{E}(Z_1)$ and $\sigma^2 = \text{var}(Z_1)$. Then*

$$\mathbb{E}(Z_n) = \mu^n, \qquad \text{var}(Z_n) = \begin{cases} n\sigma^2 & \text{if } \mu = 1, \\ \dfrac{\sigma^2(\mu^n - 1)\mu^{n-1}}{\mu - 1} & \text{if } \mu \neq 1. \end{cases}$$

**Proof.** Differentiate $G_n(s) = G(G_{n-1}(s))$ once at $s = 1$ to obtain $\mathbb{E}(Z_n) = \mu\mathbb{E}(Z_{n-1})$; by iteration, $\mathbb{E}(Z_n) = \mu^n$. Differentiate twice to obtain

$$G_n''(1) = G''(1)G_{n-1}'(1)^2 + G'(1)G_{n-1}''(1)$$

and use equation (5.1.19) to obtain the second result.                   ∎

**(3) Example. Geometric branching.** Suppose that each family size has the mass function $f(k) = qp^k$, for $k \geq 0$, where $q = 1 - p$. Then $G(s) = q(1 - ps)^{-1}$, and each family size is one member less than a geometric variable. We can show by induction that

$$G_n(s) = \begin{cases} \dfrac{n - (n-1)s}{n + 1 - ns} & \text{if } p = q = \tfrac{1}{2}, \\[2ex] \dfrac{q\left[p^n - q^n - ps(p^{n-1} - q^{n-1})\right]}{p^{n+1} - q^{n+1} - ps(p^n - q^n)} & \text{if } p \neq q. \end{cases}$$

This result can be useful in providing inequalities for more general distributions. What can we say about the behaviour of this process after many generations? In particular, does it eventually become extinct, or, conversely, do all generations have non-zero size? For this example, we can answer this question from a position of strength since we know $G_n(s)$ in closed form. In fact

$$\mathbb{P}(Z_n = 0) = G_n(0) = \begin{cases} \dfrac{n}{n+1} & \text{if } p = q, \\[2mm] \dfrac{q(p^n - q^n)}{p^{n+1} - q^{n+1}} & \text{if } p \neq q. \end{cases}$$

Let $n \to \infty$ to obtain

$$\mathbb{P}(Z_n = 0) \to \mathbb{P}(\text{ultimate extinction}) = \begin{cases} 1 & \text{if } p \leq q, \\ q/p & \text{if } p > q. \end{cases}$$

We have used Theorem (1.3.5) here surreptitiously, since

**(4)**                    $$\{\text{ultimate extinction}\} = \bigcup_n \{Z_n = 0\}$$

and $A_n = \{Z_n = 0\}$ satisfies $A_n \subseteq A_{n+1}$.                                             ●

We saw in this example that extinction occurs almost surely if and only if $\mu = \mathbb{E}(Z_1) = p/q$ satisfies $\mathbb{E}(Z_1) \leq 1$. This is a very natural condition; it seems reasonable that if $\mathbb{E}(Z_n) = \mathbb{E}(Z_1)^n \leq 1$ then $Z_n = 0$ sooner or later. Actually this result holds in general.

**(5) Theorem.** *As $n \to \infty$, $\mathbb{P}(Z_n = 0) \to \mathbb{P}(\text{ultimate extinction}) = \eta$, say, where $\eta$ is the smallest non-negative root of the equation $s = G(s)$. Also, $\eta = 1$ if $\mu < 1$, and $\eta < 1$ if $\mu > 1$. If $\mu = 1$ then $\eta = 1$ so long as the family-size distribution has strictly positive variance.*

**Proof†.** Let $\eta_n = \mathbb{P}(Z_n = 0)$. Then, by (1),

$$\eta_n = G_n(0) = G(G_{n-1}(0)) = G(\eta_{n-1}).$$

In the light of the remarks about equation (4) we know that $\eta_n \uparrow \eta$, and the continuity of $G$ guarantees that $\eta = G(\eta)$. We show next that if $\psi$ is any non-negative root of the equation $s = G(s)$ then $\eta \leq \psi$. Note that $G$ is non-decreasing on $[0, 1]$ and so

$$\eta_1 = G(0) \leq G(\psi) = \psi.$$

Similarly

$$\eta_2 = G(\eta_1) \leq G(\psi) = \psi$$

and hence, by induction, $\eta_n \leq \psi$ for all $n$, giving $\eta \leq \psi$. Thus $\eta$ is the smallest non-negative root of the equation $s = G(s)$.

---

†This method of solution was first attempted by H. W. Watson in 1873 in response to a challenge posed by F. Galton in the April 1 edition of the Educational Times. For this reason, a branching process is sometimes termed a 'Galton–Watson process'. The correct solution in modern format was supplied independently by J. F. Steffensen and C. M. Christensen in 1930; I. J. Bienaymé and J. B. S. Haldane had earlier realized what the extinction probability should be, but failed to provide the required reasoning. See Albertsen 1995.

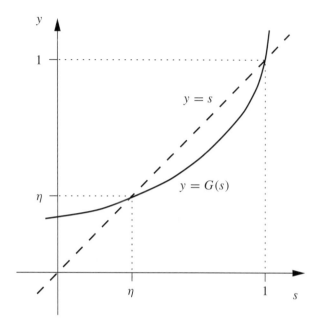

Figure 5.2. A sketch of $G(s)$ showing the roots of the equation $G(s) = s$.

To verify the second assertion of the theorem, we need the fact that $G$ is convex on $[0, 1]$. This holds because

$$G''(s) = \mathbb{E}\left[Z_1(Z_1 - 1)s^{Z_1-2}\right] \geq 0 \quad \text{if} \quad s \geq 0.$$

So $G$ is convex and non-decreasing on $[0, 1]$ with $G(1) = 1$. We can verify that the two curves $y = G(s)$ and $y = s$ generally have two intersections in $[0, 1]$, and these occur at $s = \eta$ and $s = 1$. A glance at Figure 5.2 (and a more analytical verification) tells us that these intersections are coincident if $\mu = G'(1) < 1$. On the other hand, if $\mu > 1$ then these two intersections are not coincident. In the special case when $\mu = 1$ we need to distinguish between the non-random case in which $\sigma^2 = 0$, $G(s) = s$, and $\eta = 0$, and the random case in which $\sigma^2 > 0$, $G(s) > s$ for $0 \leq s < 1$, and $\eta = 1$. ∎

We have seen that, for large $n$, the $n$th generation is empty with probability approaching $\eta$. However, what if the process does *not* die out? If $\mathbb{E}(Z_1) > 1$ then $\eta < 1$ and extinction is not certain. Indeed $\mathbb{E}(Z_n)$ grows geometrically as $n \to \infty$, and it can be shown that

$$\mathbb{P}(Z_n \to \infty \mid \text{non-extinction}) = 1$$

when this conditional probability is suitably interpreted. To see just how fast $Z_n$ grows, we define $W_n = Z_n/\mathbb{E}(Z_n)$ where $\mathbb{E}(Z_n) = \mu^n$, and we suppose that $\mu > 1$. Easy calculations show that

$$\mathbb{E}(W_n) = 1, \quad \text{var}(W_n) = \frac{\sigma^2(1 - \mu^{-n})}{\mu^2 - \mu} \to \frac{\sigma^2}{\mu^2 - \mu} \quad \text{as } n \to \infty,$$

and it seems that $W_n$ may have some non-trivial limit†, called $W$ say. In order to study $W$, define $g_n(s) = \mathbb{E}(s^{W_n})$. Then

$$g_n(s) = \mathbb{E}(s^{Z_n\mu^{-n}}) = G_n(s^{\mu^{-n}})$$

and (1) shows that $g_n$ satisfies the functional recurrence relation

$$g_n(s) = G\big(g_{n-1}(s^{1/\mu})\big).$$

Now, as $n \to \infty$, we have that $W_n \to W$ and $g_n(s) \to g(s) = \mathbb{E}(s^W)$, and we obtain

**(6)**    $$g(s) = G\big(g(s^{1/\mu})\big)$$

by abandoning some of our current notions of mathematical rigour. This functional equation can be established rigorously (see Example (7.8.5)) and has various uses. For example, although we cannot solve it for $g$, we can reach such conclusions as 'if $\mathbb{E}(Z_1^2) < \infty$ then $W$ is continuous, apart from a point mass of size $\eta$ at zero'.

We have made considerable progress with the theory of branching processes. They are reasonably tractable because they have the Markov property (see Example (3.9.5)). Can you formulate and prove this property?

---

## Exercises for Section 5.4

**1.**  Let $Z_n$ be the size of the $n$th generation in an ordinary branching process with $Z_0 = 1, \mathbb{E}(Z_1) = \mu$, and var$(Z_1) > 0$. Show that $\mathbb{E}(Z_n Z_m) = \mu^{n-m}\mathbb{E}(Z_m^2)$ for $m \leq n$. Hence find the correlation coefficient $\rho(Z_m, Z_n)$ in terms of $\mu$.

**2.**  Consider a branching process with generation sizes $Z_n$ satisfying $Z_0 = 1$ and $\mathbb{P}(Z_1 = 0) = 0$. Pick two individuals at random (with replacement) from the $n$th generation and let $L$ be the index of the generation which contains their most recent common ancestor. Show that $\mathbb{P}(L \geq r) \geq \mathbb{E}(Z_r^{-1})$ for $0 \leq r < n$. Show when $r \neq 0$ that equality holds if and only if $Z_1$ is a.s. constant. What can be said if $\mathbb{P}(Z_1 = 0) > 0$?

**3.**  Consider a branching process whose family sizes have the geometric mass function $f(k) = qp^k$, $k \geq 0$, where $p + q = 1$, and let $Z_n$ be the size of the $n$th generation. Let $T = \min\{n : Z_n = 0\}$ be the extinction time, and suppose that $Z_0 = 1$. Find $\mathbb{P}(T = n)$. For what values of $p$ is it the case that $\mathbb{E}(T) < \infty$?

**4.**  Let $Z_n$ be the size of the $n$th generation of a branching process, and assume $Z_0 = 1$. Find an expression for the generating function $G_n$ of $Z_n$, in the cases when $Z_1$ has generating function:
(a) $G(s) = 1 - \alpha(1-s)^\beta$, $0 < \alpha, \beta < 1$.
(b) $G(s) = f^{-1}\{P(f(s))\}$, where $P$ is a probability generating function, and $f$ is a suitable function satisfying $f(1) = 1$.
(c) Suppose in the latter case that $f(x) = x^m$ and $P(s) = s\{\gamma - (\gamma-1)s\}^{-1}$ where $\gamma > 1$. Calculate the answer explicitly.

**5.  Branching with immigration.** Each generation of a branching process (with a single progenitor) is augmented by a random number of immigrants who are indistinguishable from the other members of the population. Suppose that the numbers of immigrants in different generations are independent of each other and of the past history of the branching process, each such number having probability

---

†We are asserting that the sequence $\{W_n\}$ of variables converges to a limit variable $W$. The convergence of random variables is a complicated topic described in Chapter 7. We overlook the details for the moment.

generating function $H(s)$. Show that the probability generating function $G_n$ of the size of the $n$th generation satisfies $G_{n+1}(s) = G_n(G(s))H(s)$, where $G$ is the probability generating function of a typical family of offspring.

**6.**   Let $Z_n$ be the size of the $n$th generation in a branching process with $\mathbb{E}(s^{Z_1}) = (2 - s)^{-1}$ and $Z_0 = 1$. Let $V_r$ be the total number of generations of size $r$. Show that $\mathbb{E}(V_1) = \frac{1}{6}\pi^2$, and $\mathbb{E}(2V_2 - V_3) = \frac{1}{6}\pi^2 - \frac{1}{90}\pi^4$.

**7.**   Let $T$ be the total number of individuals in a branching process with family-size distribution $\text{bin}(2, p)$, where $p \neq \frac{1}{2}$. Show that

$$\mathbb{E}(T \mid T < \infty) = \frac{1}{|2p - 1|}.$$

What is the family-size distribution conditional on $T < \infty$?

**8.**   Let $Z$ be a branching process with $Z_0 = 1$, $\mathbb{E}(Z_1) = \mu > 1$, and $\text{var}(Z_1) = \sigma^2$. Use the Paley–Zygmund inequality to show that the extinction probability $\eta_n = \mathbb{P}(Z_n = 0)$ satisfies

$$\eta_n \leq \frac{\sigma^2}{\mu(\mu - 1)}(1 - \mu^{-n}).$$

---

## 5.5 Age-dependent branching processes

Here is a more general model for the growth of a population. It incorporates the observation that generations are not contemporaneous in most populations; in fact, individuals in the same generation give birth to families at different times. To model this we attach another random variable, called 'age', to each individual; we shall suppose that the collection of all ages is a set of variables which are independent of each other and of all family sizes, and which are continuous, positive, and have the common density function $f_T$. Each individual lives for a period of time, equal to its 'age', before it gives birth to its family of next-generation descendants as before. See Figure 5.3 for a picture of an age-dependent branching process.

Let $Z(t)$ denote the size of the population at time $t$; we shall assume that $Z(0) = 1$. The population-size generating function $G_t(s) = \mathbb{E}(s^{Z(t)})$ is now a function of $t$ as well. As usual, we hope to find an equation involving $G_t$ by conditioning on some suitable event. In this case we condition on the age of the initial individual in the population.

**(1) Theorem.**   $G_t(s) = \displaystyle\int_0^t G\big(G_{t-u}(s)\big) f_T(u)\, du + \int_t^\infty s f_T(u)\, du.$

**Proof.**   Let $T$ be the age of the initial individual. By the use of conditional expectation,

$$(2) \qquad G_t(s) = \mathbb{E}(s^{Z(t)}) = \mathbb{E}\big(\mathbb{E}(s^{Z(t)} \mid T)\big) = \int_0^\infty \mathbb{E}(s^{Z(t)} \mid T = u) f_T(u)\, du.$$

If $T = u$, then at time $u$ the initial individual dies and is replaced by a random number $N$ of offspring, where $N$ has generating function $G$. Each of these offspring behaves in the future as their ancestor did in the past, and the effect of their ancestor's death is to replace the process by the sum of $N$ independent copies of the process displaced in time by an amount $u$. Now if

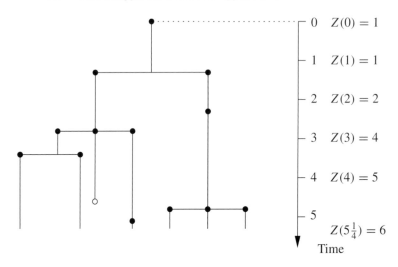

Figure 5.3. The family tree of an age-dependent branching process; • indicates the birth of an individual, and ∘ indicates the death of an individual which has no descendants.

$u > t$ then $Z(t) = 1$ and $\mathbb{E}(s^{Z(t)} \mid T = u) = s$, whilst if $u < t$ then $Z(t) = Y_1 + Y_2 + \cdots + Y_N$ is the sum of $N$ independent copies of $Z(t-u)$ and so $\mathbb{E}(s^{Z(t)} \mid T = u) = G(G_{t-u}(s))$ by Theorem (5.1.25). Substitute into (2) to obtain the result. ∎

Unfortunately we cannot solve equation (1) except in certain special cases. Possibly the most significant case with which we can make some progress arises when the ages are exponentially distributed. In this case, $f_T(t) = \lambda e^{-\lambda t}$ for $t \geq 0$, and the reader may show (*exercise*) that

$$
(3) \qquad \frac{\partial}{\partial t} G_t(s) = \lambda \big[ G(G_t(s)) - G_t(s) \big].
$$

It is no mere coincidence that this case is more tractable. In this very special instance, and in no other, $Z(t)$ has a Markov property; it is called a Markov process, and we shall return to the general theory of such processes in Chapter 6.

Some information about the moments of $Z(t)$ is fairly readily available from (1). For example,

$$
m(t) = \mathbb{E}(Z(t)) = \lim_{s \uparrow 1} \frac{\partial}{\partial s} G_t(s)
$$

satisfies the integral equation

$$
(4) \qquad m(t) = \mu \int_0^t m(t-u) f_T(u)\, du + \int_t^\infty f_T(u)\, du \quad \text{where} \quad \mu = G'(1).
$$

We can find the general solution to this equation only by numerical or series methods. It is reasonably amenable to Laplace transform methods and produces a closed expression for the Laplace transform of $m$. Later we shall use renewal theory arguments (see Example (10.4.22)) to show that there exist $\delta > 0$ and $\beta > 0$ such that $m(t) \sim \delta e^{\beta t}$ as $t \to \infty$ whenever $\mu > 1$.

Finally observe that, in some sense, the age-dependent process $Z(t)$ contains the old process $Z_n$. We say that $Z_n$ is *imbedded* in $Z(t)$ in that we can recapture $Z_n$ by aggregating the generation sizes of $Z(t)$. This imbedding enables us to use properties of $Z_n$ to derive corresponding properties of the less tractable $Z(t)$. For instance, $Z(t)$ dies out if and only if $Z_n$ dies out, and so Theorem (5.4.5) provides us immediately with the extinction probability of the age-dependent process. This technique has uses elsewhere as well. With any non-Markov process we can try to find an imbedded Markov process which provides information about the original process. We consider examples of this later.

---

## Exercises for Section 5.5

**1.**   Let $Z(t)$ be the population-size at time $t$ in an age-dependent branching process, the lifetime distribution of which is exponential with parameter $\lambda$. If $Z(0) = 1$, show that the probability generating function $G_t(s)$ of $Z(t)$ satisfies

$$\frac{\partial}{\partial t} G_t(s) = \lambda \{ G(G_t(s)) - G_t(s) \},$$

where $G$ is the probability generating function of a typical family-size. Show in the case of 'exponential binary fission', when $G(s) = s^2$, that

$$G_t(s) = \frac{se^{-\lambda t}}{1 - s(1 - e^{-\lambda t})}$$

and hence derive the probability mass function of the population size $Z(t)$ at time $t$.

**2.**   Solve the differential equation of Exercise (5.5.1) when $\lambda = 1$ and $G(s) = \frac{1}{2}(1 + s^2)$, to obtain

$$G_t(s) = \frac{2s + t(1 - s)}{2 + t(1 - s)}.$$

Hence find $\mathbb{P}(Z(t) \geq k)$, and deduce that

$$\mathbb{P}\big( Z(t)/t \geq x \mid Z(t) > 0 \big) \to e^{-2x} \quad \text{as } t \to \infty.$$

---

# 5.6 Expectation revisited

This section is divided into parts A and B. All readers must read part A before they proceed to the next section; part B is for people with a keener appreciation of detailed technique. We are about to extend the definition of probability generating functions to more general types of variables than those concentrated on the non-negative integers, and it is a suitable moment to insert some discussion of the expectation of an arbitrary random variable regardless of its type (discrete, continuous, and so on). Up to now we have made only guarded remarks about such variables.

## (A) Notation
Remember that the expectations of discrete and continuous variables are given respectively by

**(1)**                    $\mathbb{E}X = \sum xf(x)$       if $X$ has mass function $f$,

**(2)**                    $\mathbb{E}X = \int xf(x)\,dx$    if $X$ has density function $f$.

We require a single piece of notation which incorporates both these cases. Suppose $X$ has distribution function $F$. Subject to a trivial and unimportant condition, (1) and (2) can be rewritten as

$$\text{(3)} \qquad \mathbb{E}X = \sum x \, dF(x) \quad \text{where } dF(x) = F(x) - \lim_{y \uparrow x} F(y) = f(x),$$

$$\text{(4)} \qquad \mathbb{E}X = \int x \, dF(x) \quad \text{where } dF(x) = \frac{dF}{dx} dx = f(x) \, dx.$$

This suggests that we denote $\mathbb{E}X$ by

$$\text{(5)} \qquad \qquad \mathbb{E}X = \int x \, dF \quad \text{or} \quad \int x \, dF(x)$$

whatever the type of $X$, where (5) is interpreted as (3) for discrete variables and as (4) for continuous variables. We adopt this notation forthwith. Those readers who fail to conquer an aversion to this notation should read $dF$ as $f(x) \, dx$. Previous properties of expectation received two statements and proofs which can now be unified. For instance, (3.3.3) and (4.3.3) become

$$\text{(6)} \qquad \qquad \text{if } g : \mathbb{R} \to \mathbb{R} \quad \text{then} \quad \mathbb{E}(g(X)) = \int g(x) \, dF.$$

**(B) Abstract integration**
The expectation of a random variable $X$ is specified by its distribution function $F$. But $F$ itself is describable in terms of $X$ and the underlying probability space, and it follows that $\mathbb{E}X$ can be thus described also. This part contains a brief sketch of how to integrate on a probability space $(\Omega, \mathcal{F}, \mathbb{P})$. It contains no details, and the reader is left to check up on his or her intuition elsewhere (see Clarke 1975 or Williams 1991 for example). Let $(\Omega, \mathcal{F}, \mathbb{P})$ be some probability space.

**(7)** The random variable $X : \Omega \to \mathbb{R}$ is called *simple* if it takes only finitely many distinct values. Simple variables can be written in the form

$$X = \sum_{i=1}^{n} x_i I_{A_i}$$

for some partition $A_1, A_2, \dots, A_n$ of $\Omega$ and some real numbers $x_1, x_2, \dots, x_n$; we define the *integral* of $X$, written $\mathbb{E}X$ or $\mathbb{E}(X)$, to be

$$\mathbb{E}(X) = \sum_{i=1}^{n} x_i \mathbb{P}(A_i).$$

**(8)** Any non-negative random variable $X : \Omega \to [0, \infty)$ is the limit of some increasing sequence $\{X_n\}$ of simple variables. That is, $X_n(\omega) \uparrow X(\omega)$ for all $\omega \in \Omega$. We define the *integral* of $X$, written $\mathbb{E}(X)$, to be

$$\mathbb{E}(X) = \lim_{n \to \infty} \mathbb{E}(X_n).$$

This is well defined in the sense that two increasing sequences of simple functions, both converging to $X$, have the same limit for their sequences of integrals. The limit $\mathbb{E}(X)$ can be $+\infty$.

**(9)** Any random variable $X : \Omega \to \mathbb{R}$ can be written as the difference $X = X^+ - X^-$ of non-negative random variables

$$X^+(\omega) = \max\{X(\omega), 0\}, \quad X^-(\omega) = -\min\{X(\omega), 0\}.$$

If at least one of $\mathbb{E}(X^+)$ and $\mathbb{E}(X^-)$ is finite, then we define the *integral* of $X$, written $\mathbb{E}(X)$, to be

$$\mathbb{E}(X) = \mathbb{E}(X^+) - \mathbb{E}(X^-).$$

**(10)** Thus, $\mathbb{E}(X)$ is well defined, at least for any variable $X$ such that

$$\mathbb{E}|X| = \mathbb{E}(X^+ + X^-) < \infty.$$

**(11)** In the language of measure theory $\mathbb{E}(X)$ is denoted by

$$\mathbb{E}(X) = \int_{\Omega} X(\omega)\, d\mathbb{P} \quad \text{or} \quad \mathbb{E}(X) = \int_{\Omega} X(\omega)\mathbb{P}(d\omega).$$

The *expectation operator* $\mathbb{E}$ defined in this way has all the properties which were described in detail for discrete and continuous variables.

**(12) Continuity of $\mathbb{E}$.** Important further properties are the following. If $\{X_n\}$ is a sequence of variables with $X_n(\omega) \to X(\omega)$ for all $\omega \in \Omega$ then

(a) (*monotone convergence*) if $X_n(\omega) \geq 0$ and $X_n(\omega) \leq X_{n+1}(\omega)$ for all $n$ and $\omega$, then $\mathbb{E}(X_n) \to \mathbb{E}(X)$,

(b) (*dominated convergence*) if $|X_n(\omega)| \leq Y(\omega)$ for all $n$ and $\omega$, and $\mathbb{E}|Y| < \infty$, then $\mathbb{E}(X_n) \to \mathbb{E}(X)$,

(c) (*bounded convergence*, a special case of dominated convergence) if $|X_n(\omega)| \leq c$ for some constant $c$ and all $n$ and $\omega$ then $\mathbb{E}(X_n) \to \mathbb{E}(X)$.

Rather more is true. Events having zero probability (that is, null events) make no contributions to expectations, and may therefore be ignored. Consequently, it suffices to assume above that $X_n(\omega) \to X(\omega)$ for all $\omega$ *except possibly on some null event*, with a similar weakening of the hypotheses of (a), (b), and (c). For example, the bounded convergence theorem is normally stated as follows: if $\{X_n\}$ is a sequence of random variables satisfying $X_n \to X$ a.s. and $|X_n| \leq c$ a.s. for some constant $c$, then $\mathbb{E}(X_n) \to \mathbb{E}(X)$. The expression 'a.s.' is an abbreviation for 'almost surely', and means 'except possibly on an event of zero probability'.

Here is a useful consequence of monotone convergence. Let $Z_1, Z_2, \ldots$ be non-negative random variables with finite expectations, and let $X = \sum_{i=1}^{\infty} Z_i$. We have by monotone convergence applied to the partial sums of the $Z_i$ that

**(13)** 
$$\mathbb{E}(X) = \sum_{i=1}^{\infty} \mathbb{E}(Z_i),$$

whether or not the summation is finite.

One further property of expectation is called *Fatou's lemma*: if $\{X_n\}$ is a sequence of random variables such that $X_n \geq Y$ a.s. for all $n$ and some $Y$ with $\mathbb{E}|Y| < \infty$, then

**(14)**
$$\mathbb{E}\left(\liminf_{n \to \infty} X_n\right) \leq \liminf_{n \to \infty} \mathbb{E}(X_n).$$

This inequality is often applied in practice with $Y = 0$.

**(15) Lebesgue–Stieltjes integral.** Let $X$ have distribution function $F$. The function $F$ gives rise to a probability measure $\mu_F$ on the Borel sets of $\mathbb{R}$ as follows:

(a) define $\mu_F\big((a, b]\big) = F(b) - F(a)$,

(b) as in the discussion after (4.1.5), the domain of $\mu_F$ can be extended to include the Borel $\sigma$-field $\mathcal{B}$, being the smallest $\sigma$-field containing all half-open intervals $(a, b]$.

So $(\mathbb{R}, \mathcal{B}, \mu_F)$ is a probability space; its completion (see Section 1.6) is denoted by the triple $(\mathbb{R}, \mathcal{L}_F, \mu_F)$, where $\mathcal{L}_F$ is the smallest $\sigma$-field containing $\mathcal{B}$ and all subsets of $\mu_F$-null sets. If $g : \mathbb{R} \to \mathbb{R}$ (is $\mathcal{L}_F$-measurable) then the abstract integral $\int g \, d\mu_F$ is called the *Lebesgue–Stieltjes integral* of $g$ with respect to $\mu_F$, and we normally denote it by $\int g(x) \, dF$ or $\int g(x) \, dF(x)$. Think of it as a special case of the abstract integral (11). The purpose of this discussion is the assertion that if $g : \mathbb{R} \to \mathbb{R}$ (and $g$ is suitably measurable) then $g(X)$ is random variable and

$$\mathbb{E}(g(X)) = \int g(x) \, dF,$$

and we adopt this forthwith as the official notation for expectation. Here is a final word of caution. If $g(x) = I_B(x) h(x)$ where $I_B$ is the indicator function of some $B \subseteq \mathbb{R}$ then

$$\int g(x) \, dF = \int_B h(x) \, dF.$$

We do not in general obtain the same result when we integrate over $B_1 = [a, b]$ and $B_2 = (a, b)$ unless $F$ is continuous at $a$ and $b$, and so we do not use the notation $\int_a^b h(x) \, dF$ unless there is no danger of ambiguity.

## Exercises for Section 5.6

1. (a) **Jensen's inequality.** A function $u : \mathbb{R} \to \mathbb{R}$ is called *convex* if, for $a \in \mathbb{R}$, there exists $\lambda = \lambda(a)$ such that $u(x) \geq u(a) + \lambda(x - a)$ for all $x$. Draw a diagram to illustrate this definition†. The convex function $u$ is called *strictly convex* if $\lambda(a)$ is strictly increasing in $a$.
   (i) Show that, if $u$ is convex and $X$ is a random variable with finite mean, then $\mathbb{E}(u(X)) \geq u(\mathbb{E}X)$.
   (ii) Show further that, if $u$ is strictly convex and $\mathbb{E}(u(X)) = u(\mathbb{E}X)$, then $X$ is a.s. constant.
   (b) The *entropy* of a probability density function $f$ is defined by $H(f) = -\int_{\mathbb{R}} f(x) \log f(x) \, dx$, and the *support* of $f$ is $S(f) = \{x \in \mathbb{R} : f(x) > 0\}$. Show that, among density functions with support $\mathbb{R}$, and with finite mean $\mu$ and variance $\sigma^2 > 0$, the normal $N(\mu, \sigma^2)$ density function, and no other, has maximal entropy.

2. Let $X_1, X_2, \ldots$ be random variables satisfying $\mathbb{E}\left(\sum_{i=1}^{\infty} |X_i|\right) < \infty$. Show that

$$\mathbb{E}\left(\sum_{i=1}^{\infty} X_i\right) = \sum_{i=1}^{\infty} \mathbb{E}(X_i).$$

---

†There is room for debate about the 'right' definition of a convex function. We adopt the above definition since it is convenient for our uses, and is equivalent to the more usual one.

**3.** Let $\{X_n\}$ be a sequence of random variables satisfying $X_n \leq Y$ a.s. for some $Y$ with $\mathbb{E}|Y| < \infty$. Show that

$$\mathbb{E}\left(\limsup_{n\to\infty} X_n\right) \geq \limsup_{n\to\infty} \mathbb{E}(X_n).$$

**4.** Suppose that $\mathbb{E}|X^r| < \infty$ where $r > 0$. Deduce that $x^r \mathbb{P}(|X| \geq x) \to 0$ as $x \to \infty$. Conversely, suppose that $x^r \mathbb{P}(|X| \geq x) \to 0$ as $x \to \infty$ where $r \geq 0$, and show that $\mathbb{E}|X^s| < \infty$ for $0 \leq s < r$.

**5.** Show that $\mathbb{E}|X| < \infty$ if and only if the following holds: for all $\epsilon > 0$, there exists $\delta > 0$, such that $\mathbb{E}(|X|I_A) < \epsilon$ for all $A$ such that $\mathbb{P}(A) < \delta$.

**6.** Let $M = \max\{X, Y\}$ where $X$, $Y$ have some joint distribution. Show that $\mathrm{var}(M) \leq \mathrm{var}(X) + \mathrm{var}(Y)$.

**7.** Let $A_1, A_2, \ldots, A_n$ be events, and let $S$ be the number of them which occur. Show that

$$\mathbb{P}(S > 0) \geq \sum_{r=1}^{n} \frac{\mathbb{P}(I_r = 1)}{\mathbb{E}(S \mid I_r = 1)},$$

where $I_r$ is the indicator function of $A_r$.

## 5.7  Characteristic functions

Probability generating functions proved to be very useful in handling non-negative integral random variables. For more general variables $X$ it is natural to make the substitution $s = e^t$ in the quantity $G_X(s) = \mathbb{E}(s^X)$.

**(1) Definition.** The **moment generating function** of a variable $X$ is the function $M : \mathbb{R} \to [0, \infty)$ given by $M(t) = \mathbb{E}(e^{tX})$.

Moment generating functions are related to Laplace transforms† since

$$M(t) = \int e^{tx}\, dF(x) = \int e^{tx} f(x)\, dx$$

if $X$ is continuous with density function $f$. They have properties similar to those of probability generating functions. For example, if $M(t) < \infty$ on some open interval containing the origin then:

(a) $\mathbb{E}X = M'(0)$, $\mathbb{E}(X^k) = M^{(k)}(0)$;

(b) the function $M$ may be expanded via Taylor's theorem within its circle of convergence,

$$M(t) = \sum_{k=0}^{\infty} \frac{\mathbb{E}(X^k)}{k!} t^k,$$

which is to say that $M$ is the 'exponential generating function' of the sequence of moments of $X$;

(c) if $X$ and $Y$ are independent then‡ $M_{X+Y}(t) = M_X(t)M_Y(t)$.

---

†Note the change of sign from the usual Laplace transform of $f$, namely $\widehat{f}(t) = \int e^{-tx} f(x)\, dx$.

‡This is essentially the assertion that the Laplace transform of a convolution (see equation (4.8.2)) is the product of the Laplace transforms.

Moment generating functions provide a very useful technique but suffer the disadvantage that the integrals which define them may not always be finite. Rather than explore their properties in detail we move on immediately to another class of functions that are equally useful and whose finiteness is guaranteed.

**(2) Definition.** The **characteristic function** of $X$ is the function $\phi : \mathbb{R} \to \mathbb{C}$ defined by

$$\phi(t) = \mathbb{E}(e^{itX}) \quad \text{where} \quad i = \sqrt{-1}.$$

We often write $\phi_X$ for the characteristic function of the random variable $X$. Characteristic functions are related to Fourier transforms, since $\phi(t) = \int e^{itx} \, dF(x)$. In the notation of Section 5.6, $\phi$ is the abstract integral of a complex-valued random variable. It is well defined in the terms of Section 5.6 by $\phi(t) = \mathbb{E}(\cos tX) + i\mathbb{E}(\sin tX)$. Furthermore, $\phi$ is better behaved than the moment generating function $M$.

**(3) Theorem.** *The characteristic function $\phi$ satisfies*:
  (a) $\phi(0) = 1$, $|\phi(t)| \leq 1$ *for all $t$*,
  (b) $\phi$ *is uniformly continuous on* $\mathbb{R}$,
  (c) $\phi$ *is non-negative definite, which is to say that* $\sum_{j,k} \phi(t_j - t_k) z_j \bar{z}_k \geq 0$ *for all real*
    $t_1, t_2, \dots, t_n$ *and complex* $z_1, z_2, \dots, z_n$.

**Proof.** (a) Clearly $\phi(0) = \mathbb{E}(1) = 1$. Furthermore

$$|\phi(t)| \leq \int |e^{itx}| \, dF = \int dF = 1.$$

  (b) We have that

$$|\phi(t+h) - \phi(t)| = \left| \mathbb{E}(e^{i(t+h)X} - e^{itX}) \right| \leq \mathbb{E}\left| e^{itX}(e^{ihX} - 1) \right| \leq \mathbb{E}(Y(h))$$

where $Y(h) = |e^{ihX} - 1|$. However, $|Y(h)| \leq 2$ and $Y(h) \to 0$ as $h \to 0$, and so $\mathbb{E}(Y(h)) \to 0$ by bounded convergence (5.6.12).
  (c) We have that

$$\sum_{j,k} \phi(t_j - t_k) z_j \bar{z}_k = \sum_{j,k} \int [z_j \exp(it_j x)][\bar{z}_k \exp(-it_k x)] \, dF$$

$$= \mathbb{E}\left( \left| \sum_j z_j \exp(it_j X) \right|^2 \right) \geq 0. \qquad \blacksquare$$

Theorem (3) characterizes characteristic functions in the sense that $\phi$ is a characteristic function if and only if it satisfies (3a), (3b), and (3c). This result is called Bochner's theorem, for which we offer no proof. Many of the properties of characteristic functions rely for their proofs on a knowledge of complex analysis. This is a textbook on probability theory, and will not include such proofs unless they indicate some essential technique. We have asserted that the method of characteristic functions is very useful; however, we warn the reader that we shall not make use of them until Section 5.10. In the meantime we shall establish some of their properties.

First and foremost, from a knowledge of $\phi_X$ we can recapture the distribution of $X$. The full power of this statement is deferred until the next section; here we concern ourselves only with the moments of $X$. Several of the interesting characteristic functions are not very well behaved, and we must move carefully.

**(4) Theorem.**

(a) *If $\phi^{(k)}(0)$ exists then*
$$\begin{cases} \mathbb{E}|X^k| < \infty & \text{if } k \text{ is even}, \\ \mathbb{E}|X^{k-1}| < \infty & \text{if } k \text{ is odd.} \end{cases}$$

(b) *If $\mathbb{E}|X^k| < \infty$ then†*
$$\phi(t) = \sum_{j=0}^{k} \frac{\mathbb{E}(X^j)}{j!}(it)^j + o(t^k),$$

*and so $\phi^{(k)}(0) = i^k \mathbb{E}(X^k)$.*

**Proof.** This is essentially Taylor's theorem for a function of a complex variable. For the proof, see Moran 1968 or Kingman and Taylor 1966. ∎

One of the useful properties of characteristic functions is that they enable us to handle sums of independent variables with the minimum of fuss.

**(5) Theorem.** *If $X$ and $Y$ are independent then $\phi_{X+Y}(t) = \phi_X(t)\phi_Y(t)$.*

**Proof.** We have that
$$\phi_{X+Y}(t) = \mathbb{E}(e^{it(X+Y)}) = \mathbb{E}(e^{itX}e^{itY}).$$

Expand each exponential term into cosines and sines, multiply out, use independence, and put back together to obtain the result. ∎

**(6) Theorem.** *If $a, b \in \mathbb{R}$ and $Y = aX + b$ then $\phi_Y(t) = e^{itb}\phi_X(at)$.*

**Proof.** We have that
$$\phi_Y(t) = \mathbb{E}(e^{it(aX+b)}) = \mathbb{E}(e^{itb}e^{i(at)X})$$
$$= e^{itb}\mathbb{E}(e^{i(at)X}) = e^{itb}\phi_X(at). \qquad \blacksquare$$

We shall make repeated use of these last two theorems. We sometimes need to study collections of variables which may be dependent.

**(7) Definition.** The **joint characteristic function** of $X$ and $Y$ is the function $\phi_{X,Y} : \mathbb{R}^2 \to \mathbb{R}$ given by $\phi_{X,Y}(s, t) = \mathbb{E}(e^{isX}e^{itY})$.

Notice that $\phi_{X,Y}(s, t) = \phi_{sX+tY}(1)$. As usual we shall be interested mostly in independent variables.

**(8) Theorem.** *Random variables $X$ and $Y$ are independent if and only if*
$$\phi_{X,Y}(s, t) = \phi_X(s)\phi_Y(t) \quad \text{for all } s \text{ and } t.$$

---

†See Subsection (10) of Appendix I for a reminder about Landau's O/o notation.

**Proof.** If $X$ and $Y$ are independent then the conclusion follows by the argument of (5). The converse is proved by extending the inversion theorem of the next section to deal with joint distributions and showing that the joint distribution function factorizes. ∎

Note particularly that for $X$ and $Y$ to be independent it is not sufficient that

(9) $$\phi_{X,Y}(t, t) = \phi_X(t)\phi_Y(t) \quad \text{for all } t.$$

**Exercise.** Can you find an example of dependent variables which satisfy (9)?

We have seen in Theorem (4) that it is an easy calculation to find the moments of $X$ by differentiating its characteristic function $\phi_X(t)$ at $t = 0$. A similar calculation gives the 'joint moments' $\mathbb{E}(X^j Y^k)$ of two variables from a knowledge of their joint characteristic function $\phi_{X,Y}(s, t)$ (see Problem (5.12.30) for details).

The properties of moment generating functions are closely related to those of characteristic functions. In the rest of the text we shall use the latter whenever possible, but it will be appropriate to use the former for any topic whose analysis employs Laplace transforms; for example, this is the case for the queueing theory of Chapter 11.

**(10) Remark. Moment problem.** If I am given a distribution function $F$, then I can calculate the corresponding moments $m_k(F) = \int_{-\infty}^{\infty} x^k \, dF(x), k = 1, 2, \dots$, whenever these integrals exist. Is the converse true: does the collection of moments $(m_k(F) : k = 1, 2, \dots)$ specify $F$ uniquely? The answer is *no*: there exist distribution functions $F$ and $G$, all of whose moments exist, such that $F \neq G$ but $m_k(F) = m_k(G)$ for all $k$. The usual example is obtained by using the log-normal distribution (see Problem (5.12.43)).

Under what conditions on $F$ is it the case that no such $G$ exists? Various sets of conditions are known which guarantee that $F$ is specified by its moments, but no necessary and sufficient condition is known which is easy to apply to a general distribution. Perhaps the simplest sufficient condition is that the moment generating function of $F$, $M(t) = \int_{-\infty}^{\infty} e^{tx} \, dF(x)$, be finite in some neighbourhood of the point $t = 0$. Those familiar with the theory of Laplace transforms will understand why this is sufficient.

**(11) Remark. Moment generating function.** The characteristic function of a distribution is closely related to its moment generating function, in a manner made rigorous in the following theorem, the proof of which is omitted. [See Lukacs 1970, pp. 197–198.]

**(12) Theorem.** *Let* $M(t) = \mathbb{E}(e^{tX})$, $t \in \mathbb{R}$, *and* $\phi(t) = \mathbb{E}(e^{itX})$, $t \in \mathbb{C}$, *be the moment generating function and characteristic function, respectively, of a random variable* $X$. *For any* $a > 0$, *the following three statements are equivalent:*
  (a) $|M(t)| < \infty$ *for* $|t| < a$,
  (b) $\phi$ *is analytic on the strip* $|\mathrm{Im}(z)| < a$,
  (c) *the moments* $m_k = \mathbb{E}(X^k)$ *exist for* $k = 1, 2, \dots$ *and satisfy*

$$\limsup_{k\to\infty}\{|m_k|/k!\}^{1/k} \le \frac{1}{a}.$$

If any of these conditions hold for $a > 0$, the power series expansion for $M(t)$ may be extended analytically to the strip $|\mathrm{Im}(t)| < a$, resulting in a function $M$ with the property that $\phi(t) = M(it)$. [See Moran 1968, p. 260.]

## Exercises for Section 5.7

1.  Find two dependent random variables $X$ and $Y$ such that $\phi_{X+Y}(t) = \phi_X(t)\phi_Y(t)$ for all $t$.

2.  If $\phi$ is a characteristic function, show that $\mathrm{Re}\{1 - \phi(t)\} \ge \frac{1}{4}\mathrm{Re}\{1 - \phi(2t)\}$, and deduce that $1 - |\phi(2t)| \le 8\{1 - |\phi(t)|\}$.

3.  The **cumulant generating function** $K_X(\theta)$ of the random variable $X$ is defined by $K_X(\theta) = \log \mathbb{E}(e^{\theta X})$, the logarithm of the moment generating function of $X$. If the latter is finite in a neighbourhood of the origin, then $K_X$ has a convergent Taylor expansion:

$$K_X(\theta) = \sum_{n=1}^{\infty} \frac{1}{n!} k_n(X)\theta^n$$

and $k_n(X)$ is called the $n$th *cumulant* (or *semi-invariant*) of $X$.
(a) Express $k_1(X)$, $k_2(X)$, and $k_3(X)$ in terms of the moments of $X$.
(b) If $X$ and $Y$ are independent random variables, show that $k_n(X + Y) = k_n(X) + k_n(Y)$.

4.  Let $X$ be $N(0, 1)$, and show that the cumulants of $X$ are $k_2(X) = 1$, $k_m(X) = 0$ for $m \ne 2$.

5.  The random variable $X$ is said to have a *lattice distribution* if there exist $a$ and $b$ such that $X$ takes values in the set $L(a, b) = \{a + bm : m = 0, \pm 1, \dots\}$. The *span* of such a variable $X$ is the maximal value of $b$ for which there exists $a$ such that $X$ takes values in $L(a, b)$.
(a) Suppose that $X$ has a lattice distribution with span $b$. Show that $|\phi_X(2\pi/b)| = 1$, and that $|\phi_X(t)| < 1$ for $0 < t < 2\pi/b$.
(b) Suppose that $|\phi_X(\theta)| = 1$ for some $\theta \ne 0$. Show that $X$ has a lattice distribution with span $2\pi k/\theta$ for some integer $k$.

6.  Let $X$ be a random variable with density function $f$. Show that $|\phi_X(t)| \to 0$ as $t \to \pm\infty$.

7.  Let $X_1, X_2, \dots, X_n$ be independent variables, $X_i$ being $N(\mu_i, 1)$, and let $Y = X_1^2 + X_2^2 + \dots + X_n^2$. Show that the characteristic function of $Y$ is

$$\phi_Y(t) = \frac{1}{(1 - 2it)^{n/2}} \exp\left(\frac{it\theta}{1 - 2it}\right)$$

where $\theta = \mu_1^2 + \mu_2^2 + \dots + \mu_n^2$. The random variables $Y$ is said to have the *non-central chi-squared distribution* with $n$ degrees of freedom and non-centrality parameter $\theta$, written $\chi^2(n; \theta)$.

8.  Let $X$ be $N(\mu, 1)$ and let $Y$ be $\chi^2(n)$, and suppose that $X$ and $Y$ are independent. The random variable $T = X/\sqrt{Y/n}$ is said to have the *non-central t-distribution* with $n$ degrees of freedom and non-centrality parameter $\mu$. If $U$ and $V$ are independent, $U$ being $\chi^2(m; \theta)$ and $V$ being $\chi^2(n)$, then $F = (U/m)/(V/n)$ is said to have the *non-central F-distribution* with $m$ and $n$ degrees of freedom and non-centrality parameter $\theta$, written $F(m, n; \theta)$.
(a) Show that $T^2$ is $F(1, n; \mu^2)$.
(b) Show that

$$\mathbb{E}(F) = \frac{n(m + \theta)}{m(n - 2)} \quad \text{if } n > 2.$$

9.  Let $X$ be a random variable with density function $f$ and characteristic function $\phi$. Show, subject to an appropriate condition on $f$, that

$$\int_{-\infty}^{\infty} f(x)^2 \, dx = \frac{1}{2\pi} \int_{-\infty}^{\infty} |\phi(t)|^2 \, dt.$$

**10.** If $X$ and $Y$ are continuous random variables, show that

$$\int_{-\infty}^{\infty} \phi_X(y) f_Y(y) e^{-ity} \, dy = \int_{-\infty}^{\infty} \phi_Y(x-t) f_X(x) \, dx.$$

**11. Tilted distributions.** (a) Let $X$ have distribution function $F$ and let $\tau$ be such that $M(\tau) = \mathbb{E}(e^{\tau X}) < \infty$. Show that $F_\tau(x) = M(\tau)^{-1} \int_{-\infty}^{x} e^{\tau y} \, dF(y)$ is a distribution function, called a 'tilted distribution' of $X$, and find its moment generating function.
(b) Suppose $X$ and $Y$ are independent and $\mathbb{E}(e^{\tau X}), \mathbb{E}(e^{\tau Y}) < \infty$. Find the moment generating function of the tilted distribution of $X + Y$ in terms of those of $X$ and $Y$.

**12.** Let $X$ and $Y$ be independent with the distributions $N(\mu, \sigma^2)$ and $N(0, \sigma^2)$, where $\sigma^2 > 0$. Show that $R = \sqrt{X^2 + Y^2}$ has density function

$$f(r) = \frac{r}{\pi \sigma^2} \exp\left\{-\frac{\mu^2 + r^2}{2\sigma^2}\right\} \int_0^\pi \exp\left\{\frac{r\mu \cos\theta}{\sigma^2}\right\} d\theta, \qquad r > 0.$$

The integral may be expressed in terms of a modified Bessel function.

**13. Joint moment generating function.** For each of the following joint density functions of the pair $(X, Y)$, find the joint moment generating function $M(s, t) = \mathbb{E}(e^{sX + tY})$, and hence find $\operatorname{cov}(X, Y)$.
(a) We have that $f(x, y) = 2e^{-x-y}$ for $0 < x < y < \infty$.
(b) We have that

$$f(x, y) = \frac{x^2 + y^2 + c^2}{2\pi(2 + c^2)} \exp\{-\tfrac{1}{2}(x^2 + y^2)\}, \qquad x, y \in \mathbb{R}.$$

---

## 5.8 Examples of characteristic functions

Those who feel daunted by $i = \sqrt{-1}$ should find it a useful exercise to work through this section using $M(t) = \mathbb{E}(e^{tX})$ in place of $\phi(t) = \mathbb{E}(e^{itX})$. Many calculations here are left as *exercises*.

**(1) Example. Bernoulli distribution.** If $X$ is Bernoulli with parameter $p$ then

$$\phi(t) = \mathbb{E}(e^{itX}) = e^{it0} \cdot q + e^{it1} \cdot p = q + pe^{it}. \qquad \bullet$$

**(2) Example. Binomial distribution.** If $X$ is $\operatorname{bin}(n, p)$ then $X$ has the same distribution as the sum of $n$ independent Bernoulli variables $Y_1, Y_2, \ldots, Y_n$. Thus

$$\phi_X(t) = \phi_{Y_1}(t)\phi_{Y_2}(t) \cdots \phi_{Y_n}(t) = (q + pe^{it})^n. \qquad \bullet$$

**(3) Example. Exponential distribution.** If $f(x) = \lambda e^{-\lambda x}$ for $x \geq 0$ then

$$\phi(t) = \int_0^\infty e^{itx} \lambda e^{-\lambda x} \, dx.$$

This is a complex integral and its solution relies on a knowledge of how to integrate around contours in $\mathbb{R}^2$ (the appropriate contour is a sector). Alternatively, the integral may be evaluated by writing $e^{itx} = \cos(tx) + i\sin(tx)$, and integrating the real and imaginary part separately. Do not fall into the trap of treating $i$ as if its were a real number, even though this malpractice yields the correct answer in this case:

$$\phi(t) = \frac{\lambda}{\lambda - it}.$$

●

**(4) Example. Cauchy distribution.** If $f(x) = 1/\{\pi(1+x^2)\}$ then

$$\phi(t) = \frac{1}{\pi} \int_{-\infty}^{\infty} \frac{e^{itx}}{1+x^2}\, dx.$$

Treating $i$ as a real number will not help you to avoid the contour integral this time. Those who are interested should try integrating around a semicircle with diameter $[-R, R]$ on the real axis, thereby obtaining the required characteristic function $\phi(t) = e^{-|t|}$. Alternatively, you might work backwards from the answer thus: you can calculate the Fourier transform of the function $e^{-|t|}$, and then use the Fourier inversion theorem.                       ●

**(5) Example. Normal distribution.** If $X$ is $N(0, 1)$ then

$$\phi(t) = \mathbb{E}(e^{itX}) = \int_{-\infty}^{\infty} \frac{1}{\sqrt{2\pi}} \exp(itx - \tfrac{1}{2}x^2)\, dx.$$

Again, do not treat $i$ as a real number. Consider instead the moment generating function of $X$

$$M(s) = \mathbb{E}(e^{sX}) = \int_{-\infty}^{\infty} \frac{1}{\sqrt{2\pi}} \exp(sx - \tfrac{1}{2}x^2)\, dx.$$

Complete the square in the integrand and use the hint at the end of Example (4.5.9) to obtain $M(s) = e^{\frac{1}{2}s^2}$. We may not substitute $s = it$ without justification. In this particular instance the theory of analytic continuation of functions of a complex variable provides this justification, see Remark (5.7.11)†, and we deduce that

$$\phi(t) = e^{-\frac{1}{2}t^2}.$$

By Theorem (5.7.6), the characteristic function of the $N(\mu, \sigma^2)$ variable $Y = \sigma X + \mu$ is

$$\phi_Y(t) = e^{it\mu}\phi_X(\sigma t) = \exp(i\mu t - \tfrac{1}{2}\sigma^2 t^2).$$

●

**(6) Example. Multivariate normal distribution.** If $X_1, X_2, \ldots, X_n$ has the multivariate normal distribution $N(\mathbf{0}, \mathbf{V})$ then its joint density function is

$$f(\mathbf{x}) = \frac{1}{\sqrt{(2\pi)^n |\mathbf{V}|}} \exp(-\tfrac{1}{2}\mathbf{x}\mathbf{V}^{-1}\mathbf{x}').$$

---

†See Exercise (5.8.17) for a derivation that avoids functions of a complex variable.

The joint characteristic function of $X_1, X_2, \ldots , X_n$ is the function $\phi(\mathbf{t}) = \mathbb{E}(e^{i\mathbf{tX}'})$ where $\mathbf{t} = (t_1, t_2, \ldots , t_n)$ and $\mathbf{X} = (X_1, X_2, \ldots , X_n)$. One way to proceed is to use the fact that $\mathbf{tX}'$ is univariate normal. Alternatively,

$$(7) \qquad \phi(\mathbf{t}) = \int_{\mathbb{R}^n} \frac{1}{\sqrt{(2\pi)^n |\mathbf{V}|}} \exp\left(i\mathbf{tx}' - \tfrac{1}{2}\mathbf{xV}^{-1}\mathbf{x}'\right) d\mathbf{x}.$$

As in the discussion of Section 4.9, there is a linear transformation $\mathbf{y} = \mathbf{xB}$ such that

$$\mathbf{xV}^{-1}\mathbf{x}' = \sum_j \lambda_j y_j^2$$

just as in equation (4.9.3). Make this transformation in (7) to see that the integrand factorizes into the product of functions of the single variables $y_1, y_2, \ldots , y_n$. Then use (5) to obtain

$$\phi(t) = \exp(-\tfrac{1}{2}\mathbf{tVt}').$$

It is now an easy *exercise* to prove Theorem (4.9.5), that $\mathbf{V}$ is the covariance matrix of $\mathbf{X}$, by using the result of Problem (5.12.30). ●

**(8) Example. Gamma distribution.** If $X$ is $\Gamma(\lambda, s)$ then

$$\phi(t) = \int_0^\infty \frac{1}{\Gamma(s)} \lambda^s x^{s-1} \exp(itx - \lambda x)\, dx.$$

As in the case of the exponential distribution (3), routine methods of complex analysis give

$$\phi(t) = \left(\frac{\lambda}{\lambda - it}\right)^s.$$

Why is this similar to the result of (3)? This example includes the chi-squared distribution because a $\chi^2(d)$ variable is $\Gamma(\tfrac{1}{2}, \tfrac{1}{2}d)$ and thus has characteristic function

$$\phi(t) = (1 - 2it)^{-d/2}.$$

You may try to prove this from the result of Problem (4.14.12). ●

---

### Exercises for Section 5.8

**1.** If $\phi$ is a characteristic function, show that $\bar\phi$, $\phi^2$, $|\phi|^2$, $\mathrm{Re}(\phi)$ are characteristic functions. Show that $|\phi|$ is not necessarily a characteristic function.

**2.** Show that

$$\mathbb{P}(X \geq x) \leq \inf_{t \geq 0}\{e^{-tx} M_X(t)\},$$

where $M_X$ is the moment generating function of $X$. Deduce that, if $X$ has the $N(0, 1)$ distribution,

$$\mathbb{P}(X \geq x) \leq e^{-\frac{1}{2}x^2}, \qquad x > 0.$$

**3.** Let $X$ have the $\Gamma(\lambda, m)$ distribution and let $Y$ be independent of $X$ with the beta distribution with parameters $n$ and $m - n$, where $m$ and $n$ are non-negative integers satisfying $n \leq m$. Show that $Z = XY$ has the $\Gamma(\lambda, n)$ distribution.

**4.** Find the characteristic function of $X^2$ when $X$ has the $N(\mu, \sigma^2)$ distribution.

**5.** Let $X_1, X_2, \ldots$ be independent $N(0, 1)$ variables. Use characteristic functions to find the distribution of: (a) $X_1^2$, (b) $\sum_{i=1}^n X_i^2$, (c) $X_1/X_2$, (d) $X_1 X_2$, (e) $X_1 X_2 + X_3 X_4$.

**6.** Let $X_1, X_2, \ldots, X_n$ be such that, for all $a_1, a_2, \ldots, a_n \in \mathbb{R}$, the linear combination $a_1 X_1 + a_2 X_2 + \cdots + a_n X_n$ has a normal distribution. Show that the joint characteristic function of the $X_m$ is $\exp(i\mathbf{t}\boldsymbol{\mu}' - \frac{1}{2}\mathbf{t}\mathbf{V}\mathbf{t}')$, for an appropriate vector $\boldsymbol{\mu}$ and matrix $\mathbf{V}$. Deduce that the vector $(X_1, X_2, \ldots, X_n)$ has a multivariate normal *density function* so long as $\mathbf{V}$ is invertible.

**7.** Let $X$ and $Y$ be independent $N(0, 1)$ variables, and let $U$ and $V$ be independent of $X$ and $Y$. Show that $Z = (UX + VY)/\sqrt{U^2 + V^2}$ has the $N(0, 1)$ distribution. Formulate an extension of this result to cover the case when $X$ and $Y$ have a bivariate normal distribution with zero means, unit variances, and correlation $\rho$.

**8.** Let $X$ be exponentially distributed with parameter $\lambda$. Show by elementary integration that $\mathbb{E}(e^{itX}) = \lambda/(\lambda - it)$.

**9.** Find the characteristic functions of the following density functions:

(a) $f(x) = \frac{1}{2}e^{-|x|}$ for $x \in \mathbb{R}$,

(b) $f(x) = \frac{1}{2}|x|e^{-|x|}$ for $x \in \mathbb{R}$.

**10. Ulam's redistribution of energy.** Is it possible for $X$, $Y$, and $Z$ to have the same distribution and satisfy $X = U(Y + Z)$, where $U$ is uniform on $[0, 1]$, and $Y$, $Z$ are independent of $U$ and of one another? (This question arises in modelling energy redistribution among physical particles.)

**11.** Find the joint characteristic function of two random variables having a bivariate normal distribution with zero means. (No integration is needed.)

**12. Sampling from the normal distribution.** Let $X_1, X_2, \ldots, X_n$ be independent random variables with the $N(\mu, \sigma^2)$ distribution, where $\sigma > 0$. Let

$$\overline{X} = \frac{1}{n}\sum_{i=1}^n X_i, \qquad S^2 = \frac{1}{n-1}\sum_{i=1}^n \left(\frac{X_i - \overline{X}}{\sigma}\right)^2,$$

be the sample mean and variance. Show that $\mathrm{cov}(\overline{X}, X_i - \overline{X}) = 0$, and deduce that $\overline{X}$ and $S^2$ are independent.

Show that

$$\frac{(n-1)S^2}{\sigma^2} + \frac{n}{\sigma^2}(\overline{X} - \mu)^2 = \sum_{i=1}^n \left(\frac{X_i - \mu}{\sigma}\right)^2,$$

and use characteristic functions to prove that $(n - 1)S^2/\sigma^2$ has the $\chi^2(n - 1)$ distribution.

**13. Isserlis's theorem.** Let the vector $(X_1, X_2, \ldots, X_n)$ have a multivariate normal distribution with zero means. Show that, for $n$ odd, $\mathbb{E}(X_1 X_2 \cdots X_n) = 0$, while for $n = 2m$,

$$\mathbb{E}(X_1 X_2 \cdots X_n) = \sum_r \prod_{i_r < j_r} \mathbb{E}(X_{i_r} X_{j_r}),$$

that is, the sum over all products of expectations of $m$ distinct pairs of variables.

**14. Hypoexponential distribution, Exercise (4.8.4) revisited.** Let $X_1, X_2, \ldots, X_n$ be independent random variables with, respectively, the exponential distribution with parameter $\lambda_r$ for $r = 1, 2, \ldots, n$.

Use moment generating functions to find the density function of the sum $S = X_1 + X_2 + \cdots + X_n$. Deduce that, for any set $\{\lambda_r : 1 \le r \le n\}$ of distinct positive numbers,

$$\sum_{r=1}^{n} \prod_{\substack{s=1 \\ s \ne r}}^{n} \frac{\lambda_s}{\lambda_s - \lambda_r} = 1, \qquad \sum_{r=1}^{n} \frac{1}{\lambda_r} \prod_{\substack{s=1 \\ s \ne r}}^{n} \frac{\lambda_s}{\lambda_s - \lambda_r} = \sum_{r=1}^{n} \frac{1}{\lambda_r}.$$

**15.** Let $X$ have the $N(0, 1)$ distribution, and let $f : \mathbb{R} \to \mathbb{R}$ be sufficiently smooth. Show that

$$\mathbb{E}(e^{\theta X} f(X)) = e^{\frac{1}{2}\theta^2} \mathbb{E}(f(X + \theta)),$$

and deduce that $\mathbb{E}(Xf(X)) = \mathbb{E}(f'(X))$.

**16. Normal characteristic function.** Find the characteristic function of the $N(0, 1)$ distribution without using the methods of complex analysis. [Hint: Consider the derivative of the characteristic function, and use an appropriate theorem to differentiate through the integral.]

**17. Size-biased distribution.** Let $X$ be a non-negative random variable with $0 < \mu = \mathbb{E}(X) < \infty$. The random variable $Y$ is said to have the *size-biased* $X$ distribution if $dF_Y(x) = (x/\mu)dF_X(x)$ for all $x$. (You may think of this as saying that $f_Y(x) \propto xf_X(x)$ if $X$ is either discrete with mass function $f_X$, or continuous with density function $f_X$.)

Show that:
(a) the characteristic functions satisfy $\phi_Y(t) = \phi_X'(t)/(i\mu)$,
(b) when $X$ has the $\Gamma(\lambda, r)$ distribution, $Y$ has the $\Gamma(\lambda, r + 1)$ distribution,
(c) the non-negative integer-valued random variable $X$ has the Poisson distribution with parameter $\lambda$ if and only if $Y$ is distributed as $X + 1$.
What is the size-biased $X$ distribution when $X$ has the binomial $\text{bin}(n, p)$ distribution?

---

## 5.9 Inversion and continuity theorems

This section contains accounts of two major ways in which characteristic functions are useful. The first of these states that the distribution of a random variable is specified by its characteristic function. That is to say, if $X$ and $Y$ have the same characteristic function then they have the same distribution†. Furthermore, there is a formula which tells us how to recapture the distribution function $F$ corresponding to the characteristic function $\phi$. Here is a special case first.

**(1) Theorem.** *If $X$ is continuous with density function $f$ and characteristic function $\phi$ then*

$$f(x) = \frac{1}{2\pi} \int_{-\infty}^{\infty} e^{-itx} \phi(t) \, dt$$

*at every point $x$ at which $f$ is differentiable.*

**Proof.** This is the Fourier inversion theorem and can be found in any introduction to Fourier transforms. If the integral fails to converge absolutely then we interpret it as its principal value (see Apostol 1974, p. 277). ∎

---

†For $F_X = F_Y$, it is necessary that $\phi_X(t) = \phi_Y(t)$ *for all* $t$. See Problem (5.12.27(ii)) for an example in which $\phi_X$ and $\phi_Y$ agree on a bounded interval, and yet $F_X \ne F_Y$.

A sufficient, but not necessary condition that a characteristic function $\phi$ be the characteristic function of a continuous variable is that

$$\int_{-\infty}^{\infty} |\phi(t)| \, dt < \infty.$$

The general case is more complicated, and is contained in the next theorem.

**(2) Inversion theorem.** *Let $X$ have distribution function $F$ and characteristic function $\phi$. Define $\overline{F} : \mathbb{R} \to [0, 1]$ by*

$$\overline{F}(x) = \frac{1}{2}\left\{ F(x) + \lim_{y\uparrow x} F(y) \right\}.$$

*Then*

$$\overline{F}(b) - \overline{F}(a) = \lim_{N\to\infty} \int_{-N}^{N} \frac{e^{-iat} - e^{-ibt}}{2\pi i t} \phi(t) \, dt.$$

**Proof.** See Kingman and Taylor 1966.  ∎

**(3) Corollary.** *Random variables $X$ and $Y$ have the same characteristic function if and only if they have the same distribution function.*

**Proof.** If $\phi_X = \phi_Y$ then, by (2),

$$\overline{F}_X(b) - \overline{F}_X(a) = \overline{F}_Y(b) - \overline{F}_Y(a).$$

Let $a \to -\infty$ to obtain $\overline{F}_X(b) = \overline{F}_Y(b)$; now, for any fixed $x \in \mathbb{R}$, let $b \downarrow x$ and use right-continuity and Lemma (2.1.6c) to obtain $F_X(x) = F_Y(x)$.  ∎

Exactly similar results hold for jointly distributed random variables. For example, if $X$ and $Y$ have joint density function $f$ and joint characteristic function $\phi$ then whenever $f$ is differentiable at $(x, y)$

$$f(x, y) = \frac{1}{4\pi^2} \iint_{\mathbb{R}^2} e^{-isx} e^{-ity} \phi(s, t) \, ds \, dt$$

and Theorem (5.7.8) follows straightaway for this special case.

The second result of this section deals with a sequence $X_1, X_2, \ldots$ of random variables. Roughly speaking it asserts that if the distribution functions $F_1, F_2, \ldots$ of the sequence approach some limit $F$ then the characteristic functions $\phi_1, \phi_2, \ldots$ of the sequence approach the characteristic function of the distribution function $F$.

**(4) Definition.** We say that the sequence $F_1, F_2, \ldots$ of distribution functions **converges to** the distribution function $F$, written $F_n \to F$, if $F(x) = \lim_{n\to\infty} F_n(x)$ at each point $x$ where $F$ is continuous.

The reason for the condition of continuity of $F$ at $x$ is indicated by the following example. Define the distribution functions $F_n$ and $G_n$ by

$$F_n(x) = \begin{cases} 0 & \text{if } x < n^{-1}, \\ 1 & \text{if } x \geq n^{-1}, \end{cases} \qquad G_n(x) = \begin{cases} 0 & \text{if } x < -n^{-1}, \\ 1 & \text{if } x \geq -n^{-1}. \end{cases}$$

We have as $n \to \infty$ that

$$F_n(x) \to F(x) \qquad \text{if } x \neq 0, \quad F_n(0) \to 0,$$
$$G_n(x) \to F(x) \qquad \text{for all } x,$$

where $F$ is the distribution function of a random variable which is constantly zero. Indeed $\lim_{n\to\infty} F_n(x)$ is not even a distribution function since it is not right-continuous at zero. It is intuitively reasonable to demand that the sequences $\{F_n\}$ and $\{G_n\}$ have the same limit, and so we drop the requirement that $F_n(x) \to F(x)$ at the point of discontinuity of $F$.

**(5) Continuity theorem.** *Suppose that $F_1, F_2, \ldots$ is a sequence of distribution functions with corresponding characteristic functions $\phi_1, \phi_2, \ldots$.*
   (a) *If $F_n \to F$ for some distribution function $F$ with characteristic function $\phi$, then $\phi_n(t) \to \phi(t)$ for all $t$.*
   (b) *Conversely, if $\phi(t) = \lim_{n\to\infty} \phi_n(t)$ exists and is continuous at $t = 0$, then $\phi$ is the characteristic function of some distribution function $F$, and $F_n \to F$.*

**Proof.** As for (2). See also Problem (5.12.35). ∎

**(6) Example. Stirling's formula.** This well-known formula† states that $n! \sim n^n e^{-n}\sqrt{2\pi n}$ as $n \to \infty$, which is to say that

$$\frac{n!}{n^n e^{-n}\sqrt{2\pi n}} \to 1 \quad \text{as} \quad n \to \infty.$$

A more general form of this relation states that

(7)
$$\frac{\Gamma(t)}{t^{t-1}e^{-t}\sqrt{2\pi t}} \to 1 \quad \text{as} \quad t \to \infty$$

where $\Gamma$ is the gamma function, $\Gamma(t) = \int_0^\infty x^{t-1}e^{-x}\, dx$. Remember that $\Gamma(t) = (t-1)!$ if $t$ is a positive integer; see Example (4.4.6) and Exercise (4.4.1). To prove (7) is an 'elementary' exercise in analysis, (see Exercise (5.9.6)), but it is perhaps amusing to see how simply (7) follows from the Fourier inversion theorem (1).

Let $Y$ be a random variable with the $\Gamma(1, t)$ distribution. Then $X = (Y-t)/\sqrt{t}$ has density function

(8)
$$f_t(x) = \frac{1}{\Gamma(t)}\sqrt{t}\left(x\sqrt{t}+t\right)^{t-1}\exp\left[-\left(x\sqrt{t}+t\right)\right], \quad -\sqrt{t} \le x < \infty,$$

and characteristic function

$$\phi_t(u) = \mathbb{E}(e^{iuX}) = \exp\left(-iu\sqrt{t}\right)\left(1 - \frac{iu}{\sqrt{t}}\right)^{-t}.$$

Now $f_t(x)$ is differentiable with respect to $x$ on $(-\sqrt{t}, \infty)$. We apply Theorem (1) at $x = 0$ to obtain

(9)
$$f_t(0) = \frac{1}{2\pi}\int_{-\infty}^{\infty} \phi_t(u)\, du.$$

---

†Due to de Moivre.

However, $f_t(0) = t^{t-\frac{1}{2}} e^{-t}/\Gamma(t)$ from (8); also

$$\phi_t(u) = \exp\left[-iu\sqrt{t} - t\log\left(1 - \frac{iu}{\sqrt{t}}\right)\right]$$

$$= \exp\left[-iu\sqrt{t} - t\left(-\frac{iu}{\sqrt{t}} + \frac{u^2}{2t} + O(u^3 t^{-\frac{3}{2}})\right)\right]$$

$$= \exp\left[-\tfrac{1}{2}u^2 + O(u^3 t^{-\frac{1}{2}})\right] \to e^{-\frac{1}{2}u^2} \quad \text{as} \quad t \to \infty.$$

Taking the limit in (9) as $t \to \infty$, we find that

$$\lim_{t\to\infty}\left(\frac{1}{\Gamma(t)}t^{t-\frac{1}{2}}e^{-t}\right) = \lim_{t\to\infty}\frac{1}{2\pi}\int_{-\infty}^{\infty}\phi_t(u)\,du$$

$$= \frac{1}{2\pi}\int_{-\infty}^{\infty}\left(\lim_{t\to\infty}\phi_t(u)\right)du$$

$$= \frac{1}{2\pi}\int_{-\infty}^{\infty}e^{-\frac{1}{2}u^2}\,du = \frac{1}{\sqrt{2\pi}}$$

as required for (7). A spot of rigour is needed to justify the interchange of the limit and the integral sign above, and this may be provided by the dominated convergence theorem.   ●

---

## Exercises for Section 5.9

**1.** Let $X_n$ be a discrete random variable taking values in $\{1, 2, \ldots, n\}$, each possible value having probability $n^{-1}$. Show that, as $n \to \infty$, $\mathbb{P}(n^{-1}X_n \le y) \to y$, for $0 \le y \le 1$.

**2.** Let $X_n$ have distribution function

$$F_n(x) = x - \frac{\sin(2n\pi x)}{2n\pi}, \qquad 0 \le x \le 1.$$

(a) Show that $F_n$ is indeed a distribution function, and that $X_n$ has a density function.
(b) Show that, as $n \to \infty$, $F_n$ converges to the uniform distribution function, but that the density function of $F_n$ does not converge to the uniform density function.

**3.** A coin is tossed repeatedly, with heads turning up with probability $p$ on each toss. Let $N$ be the minimum number of tosses required to obtain $k$ heads. Show that, as $p \downarrow 0$, the distribution function of $2Np$ converges to that of a gamma distribution.

**4.** If $X$ is an integer-valued random variable with characteristic function $\phi$, show that

$$\mathbb{P}(X = k) = \frac{1}{2\pi}\int_{-\pi}^{\pi}e^{-itk}\phi(t)\,dt.$$

What is the corresponding result for a random variable whose distribution is arithmetic with span $\lambda$ (that is, there is probability one that $X$ is a multiple of $\lambda$, and $\lambda$ is the largest positive number with this property)?

**5.** Use the inversion theorem to show that

$$\int_{-\infty}^{\infty}\frac{\sin(at)\sin(bt)}{t^2}\,dt = \pi\min\{a, b\}.$$

**6.   Stirling's formula.** Let $f_n(x)$ be a differentiable function on $\mathbb{R}$ with a a global maximum at $a > 0$, and such that $\int_0^\infty \exp\{f_n(x)\}\, dx < \infty$. Laplace's method of steepest descent (related to Watson's lemma and saddlepoint methods) asserts under mild conditions that

$$\int_0^\infty \exp\{f_n(x)\}\, dx \sim \int_0^\infty \exp\{f_n(a) + \tfrac{1}{2}(x - a)^2 f_n''(a)\}\, dx \quad \text{as } n \to \infty.$$

By setting $f_n(x) = n \log x - x$, prove Stirling's formula: $n! \sim n^n e^{-n} \sqrt{2\pi n}$.

**7.**   Let $\mathbf{X} = (X_1, X_2, \ldots, X_n)$ have the multivariate normal distribution with zero means, and covariance matrix $\mathbf{V} = (v_{ij})$ satisfying $|\mathbf{V}| > 0$ and $v_{ij} > 0$ for all $i, j$. Show that

$$\frac{\partial f}{\partial v_{ij}} = \begin{cases} \dfrac{\partial^2 f}{\partial x_i \partial x_j} & \text{if } i \neq j, \\[2mm] \dfrac{1}{2}\dfrac{\partial^2 f}{\partial x_i^2} & \text{if } i = j, \end{cases}$$

and deduce that $\mathbb{P}(\max_{k \leq n} X_k \leq u) \geq \prod_{k=1}^n \mathbb{P}(X_k \leq u)$.

**8.**   Let $X_1, X_2$ have a bivariate normal distribution with zero means, unit variances, and correlation $\rho$. Use the inversion theorem to show that

$$\frac{\partial}{\partial \rho} \mathbb{P}(X_1 > 0, \ X_2 > 0) = \frac{1}{2\pi \sqrt{1 - \rho^2}}.$$

Hence find $\mathbb{P}(X_1 > 0, \ X_2 > 0)$.

**9.**   (a) Let $X_1, X_2, \ldots$ be independent, identically distributed random variables with characteristic function satisfying $\phi(t) = 1 - c|t| + o(t)$ as $t \to 0$, where $c > 0$. Show that, as $n \to \infty$, the distribution of $Y_n = (X_1 + X_2 + \cdots + X_n)/(cn)$ converges to the Cauchy distribution.

(b) Let $U$ have the uniform distribution on $[-1, 1]$, and show that

$$\phi_{1/U}(t) = 1 - |t| \int_{|t|}^\infty \frac{1 - \cos x}{x^2}\, dx.$$

When $U_1, U_2, \ldots, U_n$ are independent and distributed as $U$, write down the limiting distribution as $n \to \infty$ of $Y_n = (2/(n\pi)) \sum_{r=1}^n U_r^{-1}$.

# 5.10  Two limit theorems

We are now in a position to prove two very celebrated theorems in probability theory, the 'law of large numbers' and the 'central limit theorem'. The first of these explains the remarks of Sections 1.1 and 1.3, where we discussed a heuristic foundation of probability theory. Part of our intuition about chance is that if we perform many repetitions of an experiment which has numerical outcomes then the average of all the outcomes settles down to some fixed number. This observation deals in the convergence of sequences of random variables, the general theory of which is dealt with later. Here it suffices to introduce only one new definition.

**(1) Definition.** If $X, X_1, X_2, \ldots$ is a sequence of random variables with respective distribution functions $F, F_1, F_2, \ldots$, we say that $X_n$ **converges in distribution**† to $X$, written $X_n \xrightarrow{\text{D}} X$, if $F_n \to F$ as $n \to \infty$.

This is just Definition (5.9.4) rewritten in terms of random variables.

**(2) Theorem. Law of large numbers.** *Let* $X_1, X_2, \ldots$ *be a sequence of independent identically distributed random variables with finite means* $\mu$. *Their partial sums* $S_n = X_1 + X_2 + \cdots + X_n$ *satisfy*

$$\frac{1}{n} S_n \xrightarrow{\mathrm{D}} \mu \quad as \quad n \to \infty.$$

**Proof.** The theorem asserts that, as $n \to \infty$,

$$\mathbb{P}(n^{-1} S_n \leq x) \to \begin{cases} 0 & \text{if } x < \mu, \\ 1 & \text{if } x > \mu. \end{cases}$$

The method of proof is clear. By the continuity theorem (5.9.5) we need to show that the characteristic function of $n^{-1} S_n$ approaches the characteristic function of the constant random variable $\mu$. Let $\phi$ be the common characteristic function of the $X_i$, and let $\phi_n$ be the characteristic function of $n^{-1} S_n$. By Theorems (5.7.5) and (5.7.6),

(3) $$\phi_n(t) = \left\{ \phi_X(t/n) \right\}^n.$$

The behaviour of $\phi_X(t/n)$ for large $n$ is given by Theorem (5.7.4) as $\phi_X(t) = 1 + it\mu + o(t)$. Substitute into (3) to obtain

$$\phi_n(t) = \left\{ 1 + \frac{i\mu t}{n} + o\left(\frac{t}{n}\right) \right\}^n \to e^{it\mu} \quad as \quad n \to \infty.$$

However, this limit is the characteristic function of the constant $\mu$, and the result follows. ∎

So, for large $n$, the sum $S_n$ is approximately as big as $n\mu$. What can we say about the difference $S_n - n\mu$? There is an extraordinary answer to this question, valid whenever the $X_i$ have finite variance:

(a)  $S_n - n\mu$ is about as big as $\sqrt{n}$,
(b)  the distribution of $(S_n - n\mu)/\sqrt{n}$ approaches the normal distribution as $n \to \infty$ *irrespective* of the distribution of the $X_i$.

**(4) Central limit theorem.** *Let* $X_1, X_2, \ldots$ *be a sequence of independent identically distributed random variables with finite mean* $\mu$ *and finite non-zero variance* $\sigma^2$, *and let* $S_n = X_1 + X_2 + \cdots + X_n$. *Then*

$$\frac{S_n - n\mu}{\sqrt{n\sigma^2}} \xrightarrow{\mathrm{D}} N(0, 1) \quad as \quad n \to \infty.$$

Note that the assertion of the theorem is an abuse of notation, since $N(0, 1)$ is a distribution and not a random variable; the above is admissible because convergence in distribution involves only the corresponding distribution functions. The method of proof is the same as for the law of large numbers.

---

†Also termed *weak convergence* or *convergence in law*. See Section 7.2.

**Proof.** First, write $Y_i = (X_i - \mu)/\sigma$, and let $\phi_Y$ be the characteristic function of the $Y_i$. We have by Theorem (5.7.4) that $\phi_Y(t) = 1 - \frac{1}{2}t^2 + o(t^2)$. Also, the characteristic function $\psi_n$ of

$$U_n = \frac{S_n - n\mu}{\sqrt{n\sigma^2}} = \frac{1}{\sqrt{n}} \sum_{i=1}^{n} Y_i$$

satisfies, by Theorems (5.7.5) and (5.7.6),

$$\psi_n(t) = \{\phi_Y(t/\sqrt{n})\}^n = \left\{ 1 - \frac{t^2}{2n} + o\left(\frac{t^2}{n}\right) \right\}^n \rightarrow e^{-\frac{1}{2}t^2} \quad \text{as} \quad n \to \infty.$$

The last function is the characteristic function of the $N(0, 1)$ distribution, and an application of the continuity theorem (5.9.5) completes the proof. ∎

Numerous generalizations of the law of large numbers and the central limit theorem are available. For example, in Chapter 7 we shall meet two stronger versions of (2), involving weaker assumptions on the $X_i$ and more powerful conclusions. The central limit theorem can be generalized in several directions, two of which deal with dependent variables and differently distributed variables respectively. Some of these are within the reader's grasp. Here is an example of such a central limit theorem, together with the Berry–Esseen estimate for the rate of convergence.

**(5) Theorem.** *Let* $X_1, X_2, \ldots$ *be independent random variables satisfying*

$$\mathbb{E}X_j = 0, \quad \text{var}(X_j) = \sigma_j^2, \quad \mathbb{E}|X_j^3| = t_j,$$

*where* $0 < \sigma_j^2, t_j < \infty$. *Write* $S_n = \sum_{j=1}^{n} X_j$ *and* $\sigma(n)^2 = \text{var}(S_n) = \sum_{j=1}^{n} \sigma_j^2$.
  (a) **Berry–Esseen bound.** *There exists an absolute constant* $C \in (0.4906, 0.5600)$ *such that*

$$\sup_{z \in \mathbb{R}} \left| \mathbb{P}\left( \frac{S_n}{\sigma(n)} \leq z \right) - \Phi(z) \right| \leq \frac{C}{\sigma(n)^3} \sum_{j=1}^{n} t_j,$$

*where* $\Phi$ *is the distribution function of the* $N(0, 1)$ *distribution.*
  (b) *If*

$$\frac{1}{\sigma(n)^3} \sum_{j=1}^{n} t_j \rightarrow 0 \quad \text{as} \quad n \to \infty,$$

*then*

$$\frac{S_n}{\sigma(n)} \xrightarrow{D} N(0, 1).$$

**Proof.** See Feller (1971, p. 542), Loève (1977, p. 287), and also Problem (5.12.40). ∎

The roots of central limit theory are little short of 300 years old. The first proof of (4) was found by de Moivre around 1733 for the special case of Bernoulli variables with $p = \frac{1}{2}$. General values of $p$ were treated later by Laplace. Their methods involved the direct estimation of sums of the form

$$\sum_{\substack{k: \\ k \leq np + x\sqrt{npq}}} \binom{n}{k} p^k q^{n-k} \quad \text{where} \quad p + q = 1.$$

The first rigorous proof of (4) was discovered by Lyapunov around 1901, thereby confirming a less rigorous proof of Laplace. A glance at these old proofs confirms that the method of characteristic functions is outstanding in its elegance and brevity.

The central limit theorem (4) asserts that the *distribution function* of $S_n$, suitably normalized to have mean 0 and variance 1, converges to the distribution function of the $N(0, 1)$ distribution. Is the corresponding result valid at the level of density functions and mass functions? Broadly speaking the answer is yes, but some condition of smoothness is necessary; after all, if $F_n(x) \to F(x)$ as $n \to \infty$ for all $x$, it is not necessarily the case that the derivatives satisfy $F_n'(x) \to F'(x)$. [See Exercise (5.9.2).] The result which follows is called a 'local central limit theorem' since it deals in the local rather than in the cumulative behaviour of the random variables in question. In order to simplify the statement of the theorem, we shall assume that the $X_i$ have zero mean and unit variance.

**(6) Local central limit theorem.** *Let $X_1, X_2, \ldots$ be independent identically distributed random variables with zero mean and unit variance, and suppose further that their common characteristic function $\phi$ satisfies*

$$ (7) \qquad \int_{-\infty}^{\infty} |\phi(t)|^r \, dt < \infty $$

*for some integer $r \geq 1$. The density function $g_n$ of $U_n = (X_1 + X_2 + \cdots + X_n)/\sqrt{n}$ exists for $n \geq r$, and furthermore*

$$ (8) \qquad g_n(x) \to \frac{1}{\sqrt{2\pi}} e^{-\frac{1}{2}x^2} \qquad \text{as } n \to \infty, \text{ uniformly in } x \in \mathbb{R}. $$

A similar result† is valid for sums of lattice-valued random variables, suitably adjusted to have zero mean and unit variance. We state this here, leaving its proof as an *exercise*. In place of (7) we assume that the $X_i$ are restricted to take the values $a, a \pm h, a \pm 2h, \ldots$, where $h$ is the largest positive number for which such a restriction holds. Then $U_n$ is restricted to values of the form $x = (na + kh)/\sqrt{n}$ for $k = 0, \pm 1, \ldots$. For such a number $x$, we write $g_n(x) = \mathbb{P}(U_n = x)$ and leave $g_n(y)$ undefined for other values of $y$. It is the case that

$$ (9) \qquad \frac{\sqrt{n}}{h} g_n(x) \to \frac{1}{\sqrt{2\pi}} e^{-\frac{1}{2}x^2} \qquad \text{as } n \to \infty, \text{ uniformly in appropriate } x. $$

**Proof of (6).** A certain amount of analysis is inevitable here. First, the assumption that $|\phi|^r$ is integrable for some $r \geq 1$ implies that $|\phi|^n$ is integrable for $n \geq r$, since $|\phi(t)| \leq 1$; hence $g_n$ exists and is given by the Fourier inversion formula

$$ (10) \qquad g_n(x) = \frac{1}{2\pi} \int_{-\infty}^{\infty} e^{-itx} \psi_n(t) \, dt, $$

where $\psi_n(t) = \phi(t/\sqrt{n})^n$ is the characteristic function of $U_n$. The Fourier inversion theorem is valid for the normal distribution, and therefore

$$ (11) \qquad \left| g_n(x) - \frac{1}{\sqrt{2\pi}} e^{-\frac{1}{2}x^2} \right| \leq \frac{1}{2\pi} \left| \int_{-\infty}^{\infty} e^{-itx} \left[ \phi(t/\sqrt{n})^n - e^{-\frac{1}{2}t^2} \right] dt \right| \leq I_n $$

---

†Due to B. V. Gnedenko.

where

$$I_n = \frac{1}{2\pi} \int_{-\infty}^{\infty} \left| \phi(t\sqrt{n})^n - e^{-\frac{1}{2}t^2} \right| dt.$$

It suffices to show that $I_n \to 0$ as $n \to \infty$. We have from Theorem (5.7.4) that $\phi(t) = 1 - \frac{1}{2}t^2 + o(t^2)$ as $t \to 0$, and therefore there exists $\delta$ ($> 0$) such that

(12)                          $|\phi(t)| \le e^{-\frac{1}{4}t^2}$    if    $|t| \le \delta$.

Now, for any $a > 0$, $\phi(t/\sqrt{n})^n \to e^{-\frac{1}{2}t^2}$ as $n \to \infty$ uniformly in $t \in [-a, a]$ (to see this, investigate the proof of (4) slightly more carefully), so that

(13)                  $\int_{-a}^{a} \left| \phi(t/\sqrt{n})^n - e^{-\frac{1}{2}t^2} \right| dt \to 0$   as   $n \to \infty$,

for any $a$. Also, by (12),

(14)          $\int_{a < |t| \le \delta\sqrt{n}} \left| \phi(t/\sqrt{n})^n - e^{-\frac{1}{2}t^2} \right| dt \le 2 \int_{a}^{\infty} 2e^{-\frac{1}{4}t^2} dt$

which tends to zero as $a \to \infty$.

It remains to deal with the contribution to $I_n$ arising from $|t| > \delta\sqrt{n}$. From the fact that $g_n$ exists for $n \ge r$, we have from Exercises (5.7.5) and (5.7.6) that $|\phi(t)^r| < 1$ for $t \ne 0$, and $|\phi(t)^r| \to 0$ as $t \to \pm\infty$. Hence $|\phi(t)| < 1$ for $t \ne 0$, and $|\phi(t)| \to 0$ as $t \to \pm\infty$, and therefore $\eta = \sup\{|\phi(t)| : |t| \ge \delta\}$ satisfies $\eta < 1$. Now, for $n \ge r$,

(15)  $\int_{|t| > \delta\sqrt{n}} \left| \phi(t/\sqrt{n})^n - e^{-\frac{1}{2}t^2} \right| dt \le \eta^{n-r} \int_{-\infty}^{\infty} |\phi(t/\sqrt{n})|^r dt + 2 \int_{\delta\sqrt{n}}^{\infty} e^{-\frac{1}{2}t^2} dt$

$$= \eta^{n-r} \sqrt{n} \int_{-\infty}^{\infty} |\phi(u)|^r du + 2 \int_{\delta\sqrt{n}}^{\infty} e^{-\frac{1}{2}t^2} dt$$

$$\to 0 \quad \text{as} \quad n \to \infty.$$

Combining (13)–(15), we deduce that

$$\lim_{n\to\infty} I_n \le 4 \int_{a}^{\infty} e^{-\frac{1}{4}t^2} dt \to 0 \quad \text{as} \quad a \to \infty,$$

so that $I_n \to 0$ as $n \to \infty$ as required.   ∎

(16) **Example. Random walks.** Here is an application of the law of large numbers to the recurrence of random walks. A simple random walk performs steps of size 1, to the right or left with probability $p$ and $1 - p$. We saw in Section 5.3 that a simple random walk is recurrent (that is, returns to its starting point with probability 1) if and only if it is symmetric (which is to say that $p = 1 - p = \frac{1}{2}$). Think of this as saying that the walk is recurrent if and only if the mean value of a typical step $X$ satisfies $\mathbb{E}(X) = 0$, that is, each step is 'unbiased'. This conclusion is valid in much greater generality.

Let $X_1, X_2, \ldots$ be independent identically distributed integer-valued random variables, and let $S_n = X_1 + X_2 + \cdots + X_n$. We think of $X_i$ as being the $i$th jump of a random walk, so that $S_n$ is the position of the random walker after $n$ jumps, having started at $S_0 = 0$. We call the walk *recurrent* (or *persistent*) if $\mathbb{P}(S_n = 0$ for some $n \geq 1) = 1$ and *transient* otherwise.

**(17) Theorem.** *The random walk is recurrent if the mean size of jumps is 0.*

The converse is valid also: the walk is transient if the mean size of jumps is non-zero (Problem (5.12.44)).

**Proof.** Suppose that $\mathbb{E}(X_1) = 0$. For $i \in \mathbb{Z}$, let $V_i$ denote the mean number of visits of the walk to the point $i$,

$$V_i = \mathbb{E}\big|\{n \geq 0 : S_n = i\}\big| = \mathbb{E}\left(\sum_{n=0}^{\infty} I_{\{S_n=i\}}\right) = \sum_{n=0}^{\infty} \mathbb{P}(S_n = i),$$

where $I_A$ is the indicator function of the event $A$. We shall prove first that $V_0 = \infty$, and from this we shall deduce the recurrence of the walk.

For $i \neq 0$, let $T_i = \inf\{n \geq 1 : S_n = i\}$ be the time of the first visit of the walk to $i$, with the convention that $T_i = \infty$ if $i$ is never visited. Then

$$V_i = \sum_{n=0}^{\infty} \mathbb{P}(S_n = i) = \sum_{n=0}^{\infty}\sum_{t=0}^{\infty} \mathbb{P}(S_n = i \mid T_i = t)\mathbb{P}(T_i = t)$$

$$= \sum_{t=0}^{\infty}\sum_{n=t}^{\infty} \mathbb{P}(S_n = i \mid T_i = t)\mathbb{P}(T_i = t)$$

since $S_n \neq i$ for $n < T_i$. By the spatial homogeneity of the walk,

**(18)**
$$V_i = \sum_{t=0}^{\infty} V_0 \mathbb{P}(T_i = t) = V_0 \mathbb{P}(T_i < \infty) \leq V_0.$$

The mean number of time points $n$ for which $|S_n| \leq K$ satisfies

$$\sum_{n=0}^{\infty} \mathbb{P}(|S_n| \leq K) = \sum_{i=-K}^{K} V_i \leq (2K + 1)V_0$$

by (18), and hence

**(19)**
$$V_0 \geq \frac{1}{2K + 1} \sum_{n=0}^{\infty} \mathbb{P}(|S_n| \leq K).$$

Now we use the law of large numbers. For $\epsilon > 0$, it is the case that $\mathbb{P}(|S_n| \leq n\epsilon) \to 1$ as $n \to \infty$, so that there exists $m$ such that $\mathbb{P}(|S_n| \leq n\epsilon) > \frac{1}{2}$ for $n \geq m$. If $n\epsilon \leq K$ then $\mathbb{P}(|S_n| \leq n\epsilon) \leq \mathbb{P}(|S_n| \leq K)$, so that

**(20)**
$$\mathbb{P}(|S_n| \leq K) > \tfrac{1}{2} \quad \text{for} \quad m \leq n \leq K/\epsilon.$$

Substituting (20) into (19), we obtain

$$V_0 \geq \frac{1}{2K+1} \sum_{m \leq n \leq K/\epsilon} \mathbb{P}(|S_n| \leq K) > \frac{1}{2(2K+1)} \left( \frac{K}{\epsilon} - m - 1 \right).$$

This is valid for all large $K$, and we may therefore let $K \to \infty$ and $\epsilon \downarrow 0$ in that order, finding that $V_0 = \infty$ as claimed.

It is now fairly straightforward to deduce that the walk is recurrent. Let $T(1)$ be the time of the first return to 0, with the convention that $T(1) = \infty$ if this never occurs. If $T(1) < \infty$, we write $T(2)$ for the subsequent time which elapses until the next visit to 0. It is clear from the homogeneity of the process that, conditional on $\{T(1) < \infty\}$, the random variable $T(2)$ has the same distribution as $T(1)$. Continuing likewise, we see that the times of returns to 0 are distributed in the same way as the sequence $U_1, U_1 + U_2, \ldots$, where $U_1, U_2, \ldots$ are independent identically distributed random variables having the same distribution as $T(1)$. We wish to exclude the possibility that $\mathbb{P}(T(1) = \infty) > 0$. There are several ways of doing this, one of which is to make use of the recurrent-event analysis of Example (5.2.15). We shall take a slightly more direct route here. Suppose that $\beta = \mathbb{P}(T(1) = \infty)$ satisfies $\beta > 0$, and let $I = \min\{i : U_i = \infty\}$. The event $\{I = i\}$ corresponds to exactly $i - 1$ returns to the origin. Thus, the mean number of returns is $\sum_{i=1}^{\infty} (i-1)\mathbb{P}(I = i)$. However, $I = i$ if and only if $U_j < \infty$ for $1 \leq j < i$ and $U_i = \infty$, an event with probability $(1 - \beta)^{i-1}\beta$. Hence the mean number of returns to 0 is $\sum_{i=1}^{\infty} (i-1)(1 - \beta)^{i-1}\beta = (1 - \beta)/\beta$, which is finite. This contradicts the infiniteness of $V_0$, and hence $\beta = 0$.  ∎

We have proved that a walk whose jumps have zero mean must (with probability 1) return to its starting point. It follows that it must return *infinitely often*, since otherwise there exists some $T_i$ which equals infinity, an event having zero probability.  ●

**(21) Example. Recurrent events.** The renewal theorem of Example (5.2.15) is one of the basic results of applied probability, and it will recur in various forms through this book. Our 'elementary' proof in Example (5.2.15) was incomplete, but we may now complete it with the aid of the last theorem (17) concerning the recurrence of random walks.

Suppose that we are provided with two sequences $X_1, X_2, \ldots$ and $X_1^*, X_2^*, \ldots$ of independent identically distributed random variables taking values in the positive integers $\{1, 2, \ldots\}$. Let $Y_n = X_n - X_n^*$ and $S_n = \sum_{i=1}^{n} Y_i = \sum_{i=1}^{n} X_i - \sum_{i=1}^{n} X_i^*$. Then $S = \{S_n : n \geq 0\}$ may be thought of as a random walk on the integers with steps $Y_1, Y_2, \ldots$; the mean step size satisfies $\mathbb{E}(Y_1) = \mathbb{E}(X_1) - \mathbb{E}(X_1^*) = 0$, and therefore this walk is recurrent, by Theorem (17). Furthermore, the walk must revisit its starting point *infinitely often* (with probability 1), which is to say that $\sum_{i=1}^{n} X_i = \sum_{i=1}^{n} X_i^*$ for infinitely many values of $n$.

What have we proved about recurrent-event processes? Consider two independent recurrent-event processes for which the first occurrence times, $X_1$ and $X_1^*$, have the same distribution as the inter-occurrence times. Not only does there exist some finite time $T$ at which the event $H$ occurs simultaneously in both processes, but also: (i) there exist infinitely many such times $T$, and (ii) there exist infinitely many such times $T$ even if one insists that, by time $T$, the event $H$ has occurred the *same number of times* in the two processes.

We need to relax the assumption that $X_1$ and $X_1^*$ have the same distribution as the inter-occurrence times, and it is here that we require that the process be non-arithmetic. Suppose that $X_1 = u$ and $X_1^* = v$. Now $S_n = S_1 + \sum_{i=2}^{n} Y_i$ is a random walk with mean jump size

0 and starting point $S_1 = u - v$. By the foregoing argument, there exist (with probability 1) infinitely many values of $n$ such that $S_n = u - v$, which is to say that

$$(22) \qquad \sum_{i=2}^{n} X_i = \sum_{i=2}^{n} X_i^*;$$

we denote these (random) times by the increasing sequence $N_1, N_2, \ldots$.

The process is non-arithmetic, and it follows that, for any integer $x$, there exist integers $r$ and $s$ such that

$$(23) \qquad \gamma(r, s; x) = \mathbb{P}\big((X_2 + X_3 + \cdots + X_r) - (X_2^* + X_3^* + \cdots + X_s^*) = x\big) > 0.$$

To check this is an elementary *exercise* (5.10.4) in number theory. The reader may be satisfied with the following proof for the special case when $\beta = \mathbb{P}(X_2 = 1)$ satisfies $\beta > 0$. Then

$$\mathbb{P}\big(X_2 + X_3 + \cdots + X_{x+1} = x\big) \geq \mathbb{P}\big(X_i = 1 \text{ for } 2 \leq i \leq x + 1\big) = \beta^x > 0$$

if $x \geq 0$, and

$$\mathbb{P}\big(-X_2^* - X_3^* - \cdots - X_{|x|+1}^* = x\big) \geq \mathbb{P}\big(X_i^* = 1 \text{ for } 2 \leq i \leq |x| + 1\big) = \beta^{|x|} > 0$$

if $x < 0$, so that (23) is valid with $r = x + 1$, $s = 1$ and $r = 1$, $s = |x| + 1$ in these two respective cases. Without more ado we shall accept that such $r$, $s$ exist under the assumption that the process is non-arithmetic. We set $x = -(u - v)$, choose $r$ and $s$ accordingly, and write $\gamma = \gamma(r, s; x)$.

Suppose now that (22) occurs for some value of $n$. Then

$$\sum_{i=1}^{n+r-1} X_i - \sum_{i=1}^{n+s-1} X_i^* = (X_1 - X_1^*) + \left( \sum_{i=n+1}^{n+r-1} X_i - \sum_{i=n+1}^{n+s-1} X_i^* \right)$$

which equals $(u - v) - (u - v) = 0$ with strictly positive probability (since the contents of the final parentheses have, by (23), strictly positive probability of equalling $-(u - v)$). Therefore, for each $n$ satisfying (22), there is a strictly positive probability $\gamma$ that the $(n + r - 1)$th recurrence of the first process coincides with the $(n + s - 1)$th recurrence of the second. There are infinitely many such values $N_i$ for $n$, and one of infinitely many shots at a target must succeed! More rigorously, define $M_1 = N_1$, and $M_{i+1} = \min\{N_j : N_j > M_i + \max\{r, s\}\}$; the sequence of the $M_i$ is an infinite subsequence of the $N_j$ satisfying $M_{i+1} - M_i > \max\{r, s\}$. Call $M_i$ a *failure* if the $(M_i + r - 1)$th recurrence of the first process does not coincide with the $(M_i + s - 1)$th of the second. Then the events $F_I = \{M_i \text{ is a failure for } 1 \leq i \leq I\}$ satisfy

$$\mathbb{P}(F_{I+1}) = \mathbb{P}(M_{I+1} \text{ is a failure} \mid F_I)\mathbb{P}(F_I) = (1 - \gamma)\mathbb{P}(F_I),$$

so that $\mathbb{P}(F_I) = (1 - \gamma)^I \to 0$ as $I \to \infty$. However, $\{F_I : I \geq 1\}$ is a decreasing sequence of events with limit $\{M_i \text{ is a failure for all } i\}$, which event therefore has zero probability. Thus one of the $M_i$ is *not* a failure, with probability 1, implying that some recurrence of the first process coincides with some recurrence of the second, as required.

The above argument is valid for all 'initial values' $u$ and $v$ for $X_1$ and $X_1^*$, and therefore for all choices of the distribution of $X_1$ and $X_1^*$:

$$
\begin{aligned}
\mathbb{P}(\text{coincident recurrences}) &= \sum_{u,v} \mathbb{P}\big(\text{coincident recurrences}\,\big|\, X_1 = u,\ X_1^* = v\big) \\
&\qquad\qquad \times \mathbb{P}(X_1 = u)\mathbb{P}(X_1^* = v) \\
&= \sum_{u,v} 1 \cdot \mathbb{P}(X_1 = u)\mathbb{P}(X_1^* = v) = 1.
\end{aligned}
$$

In particular, the conclusion is valid when $X_1^*$ has probability generating function $D^*$ given by equation (5.2.22); the proof of the renewal theorem is thereby completed.   ●

---

## Exercises for Section 5.10

**1.** Prove that, for $x \geq 0$, as $n \to \infty$,

(a)
$$
\sum_{\substack{k: \\ |k-\frac{1}{2}n| \leq \frac{1}{2}x\sqrt{n}}} \binom{n}{k} \sim 2^n \int_{-x}^{x} \frac{1}{\sqrt{2\pi}} e^{-\frac{1}{2}u^2}\, du,
$$

(b)
$$
\sum_{\substack{k: \\ |k-n| \leq x\sqrt{n}}} \frac{n^k}{k!} \sim e^n \int_{-x}^{x} \frac{1}{\sqrt{2\pi}} e^{-\frac{1}{2}u^2}\, du.
$$

**2.** It is well known that infants born to mothers who smoke tend to be small and prone to a range of ailments. It is conjectured that also they look abnormal. Nurses were shown selections of photographs of babies, one half of whom had smokers as mothers; the nurses were asked to judge from a baby's appearance whether or not the mother smoked. In 1500 trials the correct answer was given 910 times. Is the conjecture plausible? If so, why?

**3.** Let $X$ have the $\Gamma(1, s)$ distribution; given that $X = x$, let $Y$ have the Poisson distribution with parameter $x$. Find the characteristic function of $Y$, and show that

$$
\frac{Y - \mathbb{E}(Y)}{\sqrt{\operatorname{var}(Y)}} \xrightarrow{\text{D}} N(0, 1) \qquad \text{as } s \to \infty.
$$

Explain the connection with the central limit theorem.

**4.** Let $X_1, X_2, \ldots$ be independent random variables taking values in the positive integers, whose common distribution is non-arithmetic, in that $\gcd\{n : \mathbb{P}(X_1 = n) > 0\} = 1$. Prove that, for all integers $x$, there exist non-negative integers $r = r(x)$, $s = s(x)$, such that

$$
\mathbb{P}\big(X_1 + \cdots + X_r - X_{r+1} - \cdots - X_{r+s} = x\big) > 0.
$$

**5.** Prove the local central limit theorem for sums of random variables taking integer values. You may assume for simplicity that the summands have span 1, in that $\gcd\{|x| : \mathbb{P}(X = x) > 0\} = 1$.

**6.** Let $X_1, X_2, \ldots$ be independent random variables having common density function $f(x) = 1/\{2|x|(\log|x|)^2\}$ for $|x| < e^{-1}$. Show that the $X_i$ have zero mean and finite variance, and that the density function $f_n$ of $X_1 + X_2 + \cdots + X_n$ satisfies $f_n(x) \to \infty$ as $x \to 0$. Deduce that the $X_i$ do not satisfy the local limit theorem.

7.  **First-passage density.** Let $X$ have the density function $f(x) = \sqrt{2\pi x^{-3}} \exp(-\{2x\}^{-1})$, $x > 0$. Show that $\phi(is) = \mathbb{E}(e^{-sX}) = e^{-\sqrt{2s}}$, $s > 0$, and deduce that $X$ has characteristic function

$$\phi(t) = \begin{cases} \exp\{-(1-i)\sqrt{t}\} & \text{if } t \geq 0, \\ \exp\{-(1+i)\sqrt{|t|}\} & \text{if } t \leq 0. \end{cases}$$

[Hint: Use the result of Problem (5.12.18). This distribution is called the *Lévy distribution*.]

8.  Let $\{X_r : r \geq 1\}$ be independent with the distribution of the preceding Exercise (5.10.7). Let $U_n = n^{-1}\sum_{r=1}^{n} X_r$, and $T_n = n^{-1}U_n$. Show that:
(a) $\mathbb{P}(U_n < c) \to 0$ for any $c < \infty$,
(b) $T_n$ has the same distribution as $X_1$.

9.  A sequence of biased coins is flipped; the chance that the $r$th coin shows a head is $\Theta_r$, where $\Theta_r$ is a random variable taking values in $(0, 1)$. Let $X_n$ be the number of heads after $n$ flips. Does $X_n$ obey the central limit theorem when:
(a) the $\Theta_r$ are independent and identically distributed?
(b) $\Theta_r = \Theta$ for all $r$, where $\Theta$ is a random variable taking values in $(0, 1)$?

10.  **Elections.** In an election with two candidates, each of the $v$ voters is equally likely to vote for either candidate, and they vote independently of one another. Show that, when $v$ is large, the probability that the winner was ahead when $\lambda v$ votes had been counted (where $\lambda \in (0, 1)$) is approximately $\frac{1}{2} + (1/\pi)\sin^{-1}\sqrt{\lambda}$.

11.  **Stirling's formula again.** By considering the central limit theorem for the sum of independent Poisson-distributed random variables, show that

$$\frac{\sqrt{n}e^{-n}n^n}{n!} \to \frac{1}{\sqrt{2\pi}} \qquad \text{as } n \to \infty.$$

12.  **Size-biased distribution.** Let $f$ be a mass function on $\{1, 2, \dots\}$ with finite mean $\mu$. Consider a large population of households with independent sizes distributed as $f$. An individual A is selected uniformly at random from the entire population. Show that the probability A belongs to a household of size $x$ is approximately $xf(x)/\mu$. Find the probability generating function of this distribution in terms of that of $f$.

13.  **More size-biasing.** Let $X_1, X_2, \dots$ be independent, strictly positive continuous random variables with common density function $f_X$ and finite mean, and let $Z_r = X_r/S_n$ where $S_k = \sum_{r=1}^{k} X_r$. The $Z_r$ give rise to a partition of $(0, 1]$ into the intervals $I_r = (S_{r-1}, S_r]/S_n$ for $r = 1, 2, \dots, n$. Let $U$ be uniform on $(0, 1]$ and independent of the $X_r$, and let $L$ be the length of the interval into which $U$ falls. Show that $L$ has density function $f_L(z) = (z/\mathbb{E}Z)f_Z(z)$, where $Z = Z_1$. Find the characteristic function of $L$ in terms of that of $Z$.

14.  **Area process for random walk.** Let $S_n = \sum_{r=1}^{n} X_r$ be a continuous, symmetric random walk on $\mathbb{R}$ started at 0, whose jumps are independent $N(0, 1)$ random variables $X_i$. Define the *area process* $A_n = \sum_{m=1}^{n} S_m$. Let $I_m = \{-1 < A_m < 1\}$, and show that, with probability 1, only finitely many of the $I_m$ occur.

15.  **Outguessing machines.** At Bell Labs in the early 1950s, David Hagelbarger and Claude Shannon built machines to predict whether a human coin-flipper would call heads or tails. Hagelbarger's machine (a "sequence extrapolating robot", or "SEER") was correct on 5218 trials of 9795, and Shannon's machine (a "mind-reading (?) machine") was correct on 5010 trials of 8517. In each case, what is the probability of doing at least as well by chance?
[When playing against his own machine, Shannon could beat it in around 60% of trials, in the long run.]

## 5.11 Large deviations

The law of large numbers asserts that, in a certain sense, the sum $S_n$ of $n$ independent identically distributed variables is approximately $n\mu$, where $\mu$ is a typical mean. The central limit theorem implies that the deviations of $S_n$ from $n\mu$ are typically of the order $\sqrt{n}$, that is, small compared with the mean. Now, $S_n$ may deviate from $n\mu$ by quantities of greater order than $\sqrt{n}$, say $n^\alpha$ where $\alpha > \frac{1}{2}$, but such 'large deviations' have probabilities which tend to zero as $n \to \infty$. It is often necessary in practice to estimate such probabilities. The theory of large deviations studies the asymptotic behaviour of $\mathbb{P}(|S_n - n\mu| > n^\alpha)$ as $n \to \infty$, for values of $\alpha$ satisfying $\alpha > \frac{1}{2}$; of particular interest is the case when $\alpha = 1$, corresponding to deviations of $S_n$ from its mean $n\mu$ having the same order as the mean. The behaviour of such quantities is somewhat delicate, depending on rather more than the mean and variance of a typical summand.

Let $X_1, X_2, \ldots$ be a sequence of independent identically distributed random variables with mean $\mu$ and partial sums $S_n = X_1 + X_2 + \cdots + X_n$. It is our target to estimate $\mathbb{P}(S_n > na)$ where $a > \mu$. The quantity central to the required estimate is the moment generating function $M(t) = \mathbb{E}(e^{tX})$ of a typical $X_i$, or more exactly its logarithm $\Lambda(t) = \log M(t)$. The function $\Lambda$ is also known as the *cumulant generating function* of the $X_i$ (recall Exercise (5.7.3)).

Before proceeding, we note some properties of $\Lambda$. First,

**(1)**    $$\Lambda(0) = \log M(0) = 0, \quad \Lambda'(0) = \frac{M'(0)}{M(0)} = \mu \quad \text{if } M'(0) \text{ exists.}$$

Secondly, $\Lambda(t)$ is convex wherever it is finite, since

**(2)**    $$\Lambda''(t) = \frac{M(t)M''(t) - M'(t)^2}{M(t)^2} = \frac{\mathbb{E}(e^{tX})\mathbb{E}(X^2 e^{tX}) - \mathbb{E}(Xe^{tX})^2}{M(t)^2}$$

which is non-negative, by the Cauchy–Schwarz inequality (4.5.12) applied to the random variables $Xe^{\frac{1}{2}tX}$ and $e^{\frac{1}{2}tX}$. We define the *Fenchel–Legendre transform* of $\Lambda(t)$ to be the function $\Lambda^*(a)$ given by

**(3)**    $$\Lambda^*(a) = \sup_{t \in \mathbb{R}}\{at - \Lambda(t)\}, \quad a \in \mathbb{R}.$$

The relationship between $\Lambda$ and $\Lambda^*$ is illustrated in Figure 5.4.

**(4) Theorem. Large deviations†.** *Let* $X_1, X_2, \ldots$ *be independent identically distributed random variables with mean* $\mu$, *and suppose that their moment generating function* $M(t) = \mathbb{E}(e^{tX})$ *is finite in some neighbourhood of the origin* $t = 0$. *Let* $a$ *be such that* $a > \mu$ *and* $\mathbb{P}(X > a) > 0$. *Then* $\Lambda^*(a) > 0$ *and*

**(5)**    $$\frac{1}{n}\log \mathbb{P}(S_n > na) \to -\Lambda^*(a) \quad as \quad n \to \infty.$$

Thus, under the conditions of the theorem, $\mathbb{P}(S_n > na)$ decays exponentially in the manner of $e^{-n\Lambda^*(a)}$. We note that $\mathbb{P}(S_n > na) = 0$ if $\mathbb{P}(X > a) = 0$. The theorem may appear to

---

†A version of this theorem was first published by Cramér in 1938 using different methods. Such theorems and their ramifications have had a very substantial impact on modern probability theory and its applications.

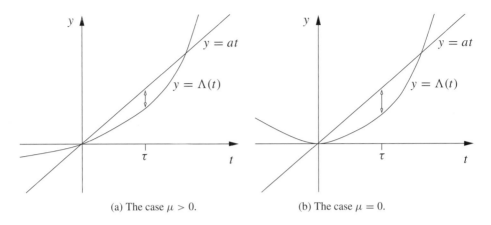

(a) The case $\mu > 0$.     (b) The case $\mu = 0$.

Figure 5.4. A sketch of the function $\Lambda(t) = \log M(t)$ in the two cases when $\Lambda'(0) = \mu > 0$ and when $\Lambda'(0) = \mu = 0$. The value of $\Lambda^*(a)$ is found by maximizing the function $g_a(t) = at - \Lambda(t)$, as indicated by the arrows. In the regular case, the supremum is achieved within the domain of convergence of $M$.

deal only with deviations of $S_n$ in *excess* of its mean; the corresponding result for deviations of $S_n$ below the mean is obtained by replacing $X_i$ by $-X_i$.

**Proof.** We may assume without loss of generality that $\mu = 0$; if $\mu \neq 0$, we replace $X_i$ by $X_i - \mu$, noting in the obvious notation that $\Lambda_X(t) = \Lambda_{X-\mu}(t) + \mu t$ and $\Lambda_X^*(a) = \Lambda_{X-\mu}^*(a - \mu)$. Assume henceforth that $\mu = 0$.

We prove first that $\Lambda^*(a) > 0$ under the assumptions of the theorem. By the remarks after Definition (5.7.1),

$$at - \Lambda(t) = \log\left(\frac{e^{at}}{M(t)}\right) = \log\left(\frac{1 + at + o(t)}{1 + \frac{1}{2}\sigma^2 t^2 + o(t^2)}\right)$$

for small positive $t$, where $\sigma^2 = \mathrm{var}(X)$; we have used here the assumption that $M(t) < \infty$ near the origin. For sufficiently small positive $t$, $1 + at + o(t) > 1 + \frac{1}{2}\sigma^2 t^2 + o(t^2)$, whence $\Lambda^*(a) > 0$ by (3).

We make two notes for future use. First, since $\Lambda$ is convex with $\Lambda'(0) = \mathbb{E}(X) = 0$, and since $a > 0$, the supremum of $at - \Lambda(t)$ over $t \in \mathbb{R}$ is unchanged by the restriction $t > 0$, which is to say that

**(6)** $$\Lambda^*(a) = \sup_{t>0}\{at - \Lambda(t)\}, \qquad a > 0.$$

(See Figure 5.4.) Secondly,

**(7)** $\Lambda$ is strictly convex wherever the second derivative $\Lambda''$ exists.

To see this, note that $\mathrm{var}(X) > 0$ under the hypotheses of the theorem, implying by (2) and Theorem (4.5.12) that $\Lambda''(t) > 0$.

The upper bound for $\mathbb{P}(S_n > na)$ is derived in much the same way as was Bernstein's inequality (2.2.4). For $t > 0$, we have that $e^{tS_n} > e^{nat} I_{\{S_n > na\}}$, so that

$$\mathbb{P}(S_n > na) \leq e^{-nat}\mathbb{E}(e^{tS_n}) = \{e^{-at} M(t)\}^n = e^{-n(at - \Lambda(t))}.$$

This is valid for all $t > 0$, whence, by (6),

(8)
$$\frac{1}{n} \log \mathbb{P}(S_n > na) \leq - \sup_{t>0}\{at - \Lambda(t)\} = -\Lambda^*(a).$$

More work is needed for the lower bound, and there are two cases which we term the *regular* and *non-regular* cases. The regular case covers most cases of practical interest, and concerns the situation when the supremum defining $\Lambda^*(a)$ in (6) is achieved *strictly* within the domain of convergence of the moment generating function $M$. Under this condition, the required argument is interesting but fairly straightforward. Let $T = \sup\{t : M(t) < \infty\}$, noting that $0 < T \leq \infty$. Assume that we are in the regular case, which is to say that there exists $\tau \in (0, T)$ such that the supremum in (6) is achieved at $\tau$; that is,

(9)
$$\Lambda^*(a) = a\tau - \Lambda(\tau),$$

as sketched in Figure 5.4. Since $at - \Lambda(t)$ has a maximum at $\tau$, and since $\Lambda$ is infinitely differentiable on $(0, T)$, the derivative of $at - \Lambda(t)$ equals 0 at $t = \tau$, and therefore

(10)
$$\Lambda'(\tau) = a.$$

Let $F$ be the common distribution function of the $X_i$. We introduce an ancillary distribution function $\widetilde{F}$, sometimes called an 'exponential change of distribution' or a 'tilted distribution' (recall Exercise (5.7.11)), by

(11)
$$d\widetilde{F}(u) = \frac{e^{\tau u}}{M(\tau)} dF(u)$$

which some may prefer to interpret as

$$\widetilde{F}(y) = \frac{1}{M(\tau)} \int_{-\infty}^{y} e^{\tau u} dF(u).$$

Let $\widetilde{X}_1, \widetilde{X}_2, \ldots$ be independent random variables having distribution function $\widetilde{F}$, and write $\widetilde{S}_n = \widetilde{X}_1 + \widetilde{X}_2 + \cdots + \widetilde{X}_n$. We note the following properties of the $\widetilde{X}_i$. The moment generating function of the $\widetilde{X}_i$ is

(12)
$$\widetilde{M}(t) = \int_{-\infty}^{\infty} e^{tu} d\widetilde{F}(u) = \int_{-\infty}^{\infty} \frac{e^{(t+\tau)u}}{M(\tau)} dF(u) = \frac{M(t+\tau)}{M(\tau)}.$$

The first two moments of the $\widetilde{X}_i$ satisfy

(13)
$$\mathbb{E}(\widetilde{X}_i) = \widetilde{M}'(0) = \frac{M'(\tau)}{M(\tau)} = \Lambda'(\tau) = a \qquad \text{by (10)},$$

$$\text{var}(\widetilde{X}_i) = \mathbb{E}(\widetilde{X}_i^2) - \mathbb{E}(\widetilde{X}_i)^2 = \widetilde{M}''(0) - \widetilde{M}'(0)^2$$

$$= \Lambda''(\tau) \in (0, \infty) \qquad \text{by (2) and (7)}.$$

Since $\widetilde{S}_n$ is the sum of $n$ independent variables, it has moment generating function

$$\left(\frac{M(t+\tau)}{M(\tau)}\right)^n = \frac{\mathbb{E}(e^{(t+\tau)S_n})}{M(\tau)^n} = \frac{1}{M(\tau)^n} \int_{-\infty}^{\infty} e^{(t+\tau)u} dF_n(u)$$

where $F_n$ is the distribution function of $S_n$. Therefore, the distribution function $\widetilde{F}_n$ of $\widetilde{S}_n$ satisfies

(14)
$$d\widetilde{F}_n(u) = \frac{e^{\tau u}}{M(\tau)^n} \, dF_n(u).$$

Let $b > a$. We have that

$$
\begin{aligned}
\mathbb{P}(S_n > na) &= \int_{na}^{\infty} dF_n(u) \\
&= \int_{na}^{\infty} M(\tau)^n e^{-\tau u} \, d\widetilde{F}_n(u) \qquad \text{by (14)} \\
&\geq M(\tau)^n e^{-\tau nb} \int_{na}^{nb} d\widetilde{F}_n(u) \\
&\geq e^{-n(\tau b - \Lambda(\tau))} \mathbb{P}(na < \widetilde{S}_n < nb).
\end{aligned}
$$

Since the $\widetilde{X}_i$ have mean $a$ and non-zero variance, we have by the central limit theorem applied to the $\widetilde{X}_i$ that $\mathbb{P}(\widetilde{S}_n > na) \to \frac{1}{2}$ as $n \to \infty$, and by the law of large numbers that $\mathbb{P}(\widetilde{S}_n < nb) \to 1$. Therefore,

$$
\begin{aligned}
\frac{1}{n} \log \mathbb{P}(S_n > na) &\geq -(\tau b - \Lambda(\tau)) + \frac{1}{n} \log \mathbb{P}(na < \widetilde{S}_n < nb) \\
&\to -(\tau b - \Lambda(\tau)) && \text{as } n \to \infty \\
&\to -(\tau a - \Lambda(\tau)) = -\Lambda^*(a) && \text{as } b \downarrow a, \text{ by (9)}.
\end{aligned}
$$

This completes the proof in the regular case.

Finally, we consider the non-regular case. Let $c$ be a real number satisfing $c > a$, and write $Z^c = \min\{Z, c\}$, the truncation of the random variable $Z$ at level $c$. Since $\mathbb{P}(X^c \leq c) = 1$, we have that $M^c(t) = \mathbb{E}(e^{tX^c}) \leq e^{tc}$ for $t > 0$, and therefore $M^c(t) < \infty$ for all $t > 0$. Note that $\mathbb{E}(X^c) \leq \mathbb{E}(X) = 0$, and $\mathbb{E}(X^c) \to 0$ as $c \to \infty$, by the monotone convergence theorem.

Since $\mathbb{P}(X > a) > 0$, there exists $b \in (a, c)$ such that $\mathbb{P}(X > b) > 0$. It follows that $\Lambda^c(t) = \log M^c(t)$ satisfies

$$at - \Lambda^c(t) \leq at - \log\{e^{tb}\mathbb{P}(X > b)\} \to -\infty \quad \text{as } t \to \infty.$$

We deduce that the supremum of $at - \Lambda^c(t)$ over values $t > 0$ is attained at some point $\tau = \tau^c \in (0, \infty)$. The random sequence $X_1^c, X_2^c, \ldots$ is therefore a regular case of the large deviation problem, and $a > \mathbb{E}(X^c)$, whence

(15)
$$\frac{1}{n} \log \mathbb{P}\left( \sum_{i=1}^{n} X_i^c > na \right) \to -\Lambda^{c*}(a) \quad \text{as } n \to \infty,$$

by the previous part of this proof, where

(16)
$$\Lambda^{c*}(a) = \sup_{t>0}\{at - \Lambda^c(t)\} = a\tau - \Lambda^c(\tau).$$

Now $\Lambda^c(t) = \mathbb{E}(e^{tX^c})$ is non-decreasing in $c$ when $t > 0$, implying that $\Lambda^{c*}$ is non-increasing. Therefore there exists a real number $\Lambda^{\infty*}$ such that

(17)                                $\Lambda^{c*}(a) \downarrow \Lambda^{\infty*}$    as $c \uparrow \infty$.

Since $\Lambda^{c*}(a) < \infty$ and $\Lambda^{c*}(a) \geq -\Lambda^c(0) = 0$, we have that $0 \leq \Lambda^{\infty*} < \infty$.
   Evidently $S_n \geq \sum_{i=1}^n X_i^c$, whence

$$\frac{1}{n} \log \mathbb{P}(S_n > na) \geq \frac{1}{n} \log \mathbb{P}\left( \sum_{i=1}^n X_i^c > na \right),$$

and it therefore suffices by (15)–(17) to prove that

(18)                                $\Lambda^{\infty*} \leq \Lambda^*(a).$

Since $\Lambda^{\infty*} \leq \Lambda^{c*}(a)$, the set $I_c = \{t \geq 0 : at - \Lambda^c(t) \geq \Lambda^{\infty*}\}$ is non-empty. Using the smoothness of $\Lambda^c$, and aided by a glance at Figure 5.4, we see that $I_c$ is a non-empty closed interval. Since $\Lambda^c(t)$ is non-decreasing in $c$, the sets $I_c$ are non-increasing. Since the intersection of nested compact sets is non-empty, the intersection $\bigcap_{c>a} I_c$ contains at least one real number $\zeta$. By the monotone convergence theorem, $\Lambda^c(\zeta) \to \Lambda(\zeta)$ as $c \to \infty$, whence

$$a\zeta - \Lambda(\zeta) = \lim_{c \to \infty} \{a\zeta - \Lambda^c(\zeta)\} \geq \Lambda^{\infty*}$$

so that

$$\Lambda^*(a) = \sup_{t>0} \{at - \Lambda(t)\} \geq \Lambda^{\infty*}$$

as required in (18).                                                                          ∎

---

## Exercises for Section 5.11

**1.** A fair coin is tossed $n$ times, showing heads $H_n$ times and tails $T_n$ times. Let $S_n = H_n - T_n$. Show that

$$\mathbb{P}(S_n > an)^{1/n} \to \frac{1}{\sqrt{(1+a)^{1+a}(1-a)^{1-a}}} \qquad \text{if } 0 < a < 1.$$

What happens if $a \geq 1$?

**2.** Show that

$$T_n^{1/n} \to \frac{4}{\sqrt{(1+a)^{1+a}(1-a)^{1-a}}}$$

as $n \to \infty$, where $0 < a < 1$ and

$$T_n = \sum_{\substack{k: \\ |k - \frac{1}{2}n| > \frac{1}{2}an}} \binom{n}{k}.$$

Find the asymptotic behaviour of $T_n^{1/n}$ where

$$T_n = \sum_{\substack{k: \\ k > n(1+a)}} \frac{n^k}{k!}, \qquad \text{where } a > 0.$$

**3.**    Show that the moment generating function of $X$ is finite in a neighbourhood of the origin if and only if $X$ has exponentially decaying tails, in the sense that there exist positive constants $\lambda$ and $\mu$ such that $\mathbb{P}(|X| \geq a) \leq \mu e^{-\lambda a}$ for $a > 0$. [Seen in the light of this observation, the condition of the large deviation theorem (5.11.4) is very natural].

**4.**    Let $X_1, X_2, \ldots$ be independent random variables having the Cauchy distribution, and let $S_n = X_1 + X_2 + \cdots + X_n$. Find $\mathbb{P}(S_n > an)$.

**5.    Chernoff inequality for Bernoulli trials.** Let $S = \sum_{r=1}^n X_r$ be a sum of independent Bernoulli random variables $X_r$ taking values in $\{0, 1\}$, where $\mathbb{E}(X_r) = p_r$ and $\mathbb{E}(S) = \mu > 0$. Show that

$$\mathbb{P}(S > (1 + \epsilon)\mu) \leq \exp\{-\mu[(1 + \epsilon)\log(1 + \epsilon) - \epsilon]\}, \qquad \epsilon > 0.$$

---

# 5.12 Problems

**1.**    A die is thrown ten times. What is the probability that the sum of the scores is 27?

**2.**    A coin is tossed repeatedly, heads appearing with probability $p$ on each toss.

(a) Let $X$ be the number of tosses until the first occasion by which three heads have appeared successively. Write down a difference equation for $f(k) = \mathbb{P}(X = k)$ and solve it. Now write down an equation for $\mathbb{E}(X)$ using conditional expectation. (Try the same thing for the first occurrence of HTH).

(b) Let $N$ be the number of heads in $n$ tosses of the coin. Write down $G_N(s)$. Hence find the probability that:    (i) $N$ is divisible by 2,    (ii) $N$ is divisible by 3.

**3.**    A coin is tossed repeatedly, heads occurring on each toss with probability $p$. Find the probability generating function of the number $T$ of tosses before a run of $n$ heads has appeared for the first time.

**4.**    Find the generating function of the negative binomial mass function

$$f(k) = \binom{k-1}{r-1} p^r (1 - p)^{k-r}, \qquad k = r, r+1, \ldots,$$

where $0 < p < 1$ and $r$ is a positive integer. Deduce the mean and variance.

**5.**    For the simple random walk, show that the probability $p_0(2n)$ that the particle returns to the origin at the $(2n)$th step satisfies $p_0(2n) \sim (4pq)^n / \sqrt{\pi n}$, and use this to prove that the walk is recurrent if and only if $p = \frac{1}{2}$. You will need Stirling's formula: $n! \sim n^{n+\frac{1}{2}} e^{-n} \sqrt{2\pi}$.

**6.**    A symmetric random walk in two dimensions is defined to be a sequence of points $\{(X_n, Y_n) : n \geq 0\}$ which evolves in the following way: if $(X_n, Y_n) = (x, y)$ then $(X_{n+1}, Y_{n+1})$ is one of the four points $(x \pm 1, y), (x, y \pm 1)$, each being picked with equal probability $\frac{1}{4}$. If $(X_0, Y_0) = (0, 0)$:

(a) show that $\mathbb{E}(X_n^2 + Y_n^2) = n$,

(b) find the probability $p_0(2n)$ that the particle is at the origin after the $(2n)$th step, and deduce that the probability of ever returning to the origin is 1.

**7.**    Consider the one-dimensional random walk $\{S_n\}$ given by

$$S_{n+1} = \begin{cases} S_n + 2 & \text{with probability } p, \\ S_n - 1 & \text{with probability } q = 1 - p, \end{cases}$$

where $0 < p < 1$.

(a) What is the probability of ever reaching the origin starting from $S_0 = a$ where $a > 0$?

(b) Let $A_n$ be the mean number of points in $\{0, 1, \ldots, n\}$ that the walk never visits. Find the limit $a = \lim_{n \to \infty} A_n/n$ when $p > \frac{1}{3}$, and verify for $p = \frac{1}{2}$ that $a = \frac{1}{2}(7 - 3\sqrt{5})$.

**8.** Let $X$ and $Y$ be independent variables taking values in the positive integers such that

$$\mathbb{P}(X = k \mid X + Y = n) = \binom{n}{k} p^k (1-p)^{n-k}$$

for some $p$ and all $0 \leq k \leq n$. Show that $X$ and $Y$ have Poisson distributions.

**9.** In a branching process whose family sizes have mean $\mu$ and variance $\sigma^2$, find the variance of $Z_n$, the size of the $n$th generation, given that $Z_0 = 1$.

**10. Waldegrave's problem.** A group $\{A_1, A_2, \ldots, A_r\}$ of $r$ ($> 2$) people play the following game. $A_1$ and $A_2$ wager on the toss of a fair coin. The loser puts £1 in the pool, the winner goes on to play $A_3$. In the next wager, the loser puts £1 in the pool, the winner goes on to play $A_4$, and so on. The winner of the $(r-1)$th wager goes on to play $A_1$, and the cycle recommences. The first person to beat all the others in sequence takes the pool.

(a) Find the probability generating function of the duration of the game.

(b) Find an expression for the probability that $A_k$ wins.

(c) Find an expression for the expected size of the pool at the end of the game, given that $A_k$ wins.

(d) Find an expression for the probability that the pool is intact after the $n$th spin of the coin.

This problem was discussed by Montmort, Bernoulli, de Moivre, Laplace, and others.

**11.** A branching process has a single progenitor.

(a) Show that the generating function $H_n$ of the *total* number of individuals in the first $n$ generations satisfies $H_n(s) = sG(H_{n-1}(s))$.

(b) Let $T$ be the total number of individuals who ever exist, with $Q(s) = \mathbb{E}(s^T)$. Show, for $s \in [0, 1)$, that $Q(s) = sG(Q(s))$. Writing $\mu < 1$ for the mean of the family-size distribution, and $\sigma^2 > 0$ for its variance, show that:

  (i) $Q(1) := \lim_{s \uparrow 1} Q(s) = 1$,

  (ii) $\mathbb{E}(T) = 1/(1 - \mu)$,

  (iii) $\mathrm{var}(T) = \sigma^2/(1 - \mu)^3$.

(c) Find $Q(s)$ when $G(s) = p/(1 - qs)$ where $0 < p = 1 - q < 1$. Discuss the properties of $Q$ in the two cases $p < q$ and $p \geq q$.

(d) Let $G$ be as in part (c), and write $H_n(s) = y_n(s)/x_n(s)$ for appropriate polynomials $x_n$, $y_n$. Show that $x_n$ satisfies $x_n(s) = x_{n-1}(s) - spqx_{n-2}(s)$, with $x_0 = 1$ and $x_1(s) = 1 - qs$. Deduce the form of $Q(s)$.

**12.** Show that the number $Z_n$ of individuals in the $n$th generation of a branching process satisfies $\mathbb{P}(Z_n > N \mid Z_m = 0) \leq G_m(0)^N$ for $n < m$.

**13.** (a) A hen lays $N$ eggs where $N$ is Poisson with parameter $\lambda$. The weight of the $n$th egg is $W_n$, where $W_1, W_2, \ldots$ are independent identically distributed variables with common probability generating function $G(s)$. Show that the generating function $G_W$ of the total weight $W = \sum_{i=1}^N W_i$ is given by $G_W(s) = \exp\{-\lambda + \lambda G(s)\}$. The quantity $W$ is said to have a *compound Poisson distribution*. Show further that, for any positive integral value of $n$, $G_W(s)^{1/n}$ is the probability generating function of some random variable; $W$ (or its distribution) is said to be *infinitely divisible* in this regard.

(b) Show that if $H(s)$ is the probability generating function of some infinitely divisible distribution on the non-negative integers then $H(s) = \exp\{-\lambda + \lambda G(s)\}$ for some $\lambda$ ($> 0$) and some probability generating function $G(s)$.

(c) Can the compound Poisson distribution of $W$ in part (a) be a Poisson distribution for any choice of $G$?

**14.** The distribution of a random variable $X$ is called *infinitely divisible* if, for all positive integers $n$, there exists a sequence $Y_1^{(n)}, Y_2^{(n)}, \ldots, Y_n^{(n)}$ of independent identically distributed random variables such that $X$ and $Y_1^{(n)} + Y_2^{(n)} + \cdots + Y_n^{(n)}$ have the same distribution.

(a) Show that the normal, Poisson, and gamma distributions are infinitely divisible.

(b) Show that the characteristic function $\phi$ of an infinitely divisible distribution has no real zeros, in that $\phi(t) \neq 0$ for all real $t$.

15. Let $X_1, X_2, \ldots$ be independent variables each taking the values 0 or 1 with probabilities $1-p$ and $p$, where $0 < p < 1$. Let $N$ be a random variable taking values in the positive integers, independent of the $X_i$, and write $S = X_1 + X_2 + \cdots + X_N$. Write down the conditional generating function of $N$ given that $S = N$, in terms of the probability generating function $G$ of $N$. Show that $N$ has a Poisson distribution if and only if $\mathbb{E}(x^N)^p = \mathbb{E}(x^N \mid S = N)$ for all $p$ and $x$.

16. If $X$ and $Y$ have joint probability generating function

$$G_{X,Y}(s,t) = \mathbb{E}(s^X t^Y) = \frac{\{1 - (p_1 + p_2)\}^n}{\{1 - (p_1 s + p_2 t)\}^n} \qquad \text{where } p_1 + p_2 \leq 1,$$

find the marginal mass functions of $X$ and $Y$, and the mass function of $X+Y$. Find also the conditional probability generating function $G_{X \mid Y}(s \mid y) = \mathbb{E}(s^X \mid Y = y)$ of $X$ given that $Y = y$. The pair $X, Y$ is said to have the *bivariate negative binomial distribution*.

17. If $X$ and $Y$ have joint probability generating function

$$G_{X,Y}(s,t) = \exp\{\alpha(s-1) + \beta(t-1) + \gamma(st-1)\}$$

find the marginal distributions of $X$, $Y$, and the distribution of $X+Y$, showing that $X$ and $Y$ have the Poisson distribution, but that $X+Y$ does not unless $\gamma = 0$.

18. Define

$$I(a,b) = \int_0^\infty \exp(-a^2 u^2 - b^2 u^{-2}) \, du$$

for $a, b > 0$. Show that
(a) $I(a,b) = a^{-1} I(1, ab)$,     (b) $\partial I / \partial b = -2I(1, ab)$,
(c) $I(a,b) = \sqrt{\pi} e^{-2ab} / (2a)$.
(d) If $X$ has density function $(d/\sqrt{x}) e^{-c/x - gx}$ for $x > 0$, then

$$\mathbb{E}(e^{-tX}) = d \sqrt{\frac{\pi}{g+t}} \exp\left(-2\sqrt{c(g+t)}\right), \qquad t > -g.$$

(e) If $X$ has density function $(2\pi x^3)^{-\frac{1}{2}} e^{-1/(2x)}$ for $x > 0$, then $X$ has moment generating function given by $\mathbb{E}(e^{-tX}) = \exp\{-\sqrt{2t}\}$, $t \geq 0$. [Note that $\mathbb{E}(X^n) = \infty$ for $n \geq 1$.]

19. Let $X$, $Y$, $Z$ be independent $N(0, 1)$ variables. Use characteristic functions and moment generating functions (Laplace transforms) to find the distributions of
(a) $U = X/Y$,
(b) $V = X^{-2}$,
(c) $W = XYZ/\sqrt{X^2 Y^2 + Y^2 Z^2 + Z^2 X^2}$.

20. Let $X$ have density function $f$ and characteristic function $\phi$, and suppose that $\int_{-\infty}^\infty |\phi(t)| \, dt < \infty$. Deduce that

$$f(x) = \frac{1}{2\pi} \int_{-\infty}^\infty e^{-itx} \phi(t) \, dt.$$

21. **Conditioned branching process.** Consider a branching process whose family sizes have the geometric mass function $f(k) = qp^k$, $k \geq 0$, where $\mu = p/q > 1$. Let $Z_n$ be the size of the $n$th generation, and assume $Z_0 = 1$. Show that the conditional distribution of $Z_n / \mu^n$, given that $Z_n > 0$, converges as $n \to \infty$ to the exponential distribution with parameter $1 - \mu^{-1}$.

**22.** A random variable $X$ is called *symmetric* if $X$ and $-X$ are identically distributed. Show that $X$ is symmetric if and only if the imaginary part of its characteristic function is identically zero.

**23.** Let $X$ and $Y$ be independent identically distributed variables with means 0 and variances 1. Let $\phi(t)$ be their common characteristic function, and suppose that $X + Y$ and $X - Y$ are independent. Show that $\phi(2t) = \phi(t)^3 \phi(-t)$, and deduce that $X$ and $Y$ are $N(0, 1)$ variables.

More generally, suppose that $X$ and $Y$ are independent and identically distributed with means 0 and variances 1, and furthermore that $\mathbb{E}(X - Y \mid X + Y) = 0$ and $\text{var}(X - Y \mid X + Y) = 2$. Deduce that $\phi(s)^2 = \phi'(s)^2 - \phi(s)\phi''(s)$, and hence that $X$ and $Y$ are independent $N(0, 1)$ variables.

**24.** Show that the average $Z = n^{-1} \sum_{i=1}^{n} X_i$ of $n$ independent Cauchy variables has the Cauchy distribution too. Why does this not violate the law of large numbers?

**25.** Let $X$ and $Y$ be independent random variables each having the Cauchy density function $f(x) = \{\pi(1 + x^2)\}^{-1}$, and let $Z = \frac{1}{2}(X + Y)$.
(a) Show by using characteristic functions that $Z$ has the Cauchy distribution also.
(b) Show by the convolution formula that $Z$ has the Cauchy density function. You may find it helpful to check first that

$$f(x)f(y-x) = \frac{f(x) + f(y-x)}{\pi(4 + y^2)} + g(y)\{xf(x) + (y-x)f(y-x)\}$$

where $g(y) = 2/\{\pi y(4 + y^2)\}$.

**26.** Let $X_1, X_2, \ldots, X_n$ be independent variables with characteristic functions $\phi_1, \phi_2, \ldots, \phi_n$. Describe random variables which have the following characteristic functions:
  (a) $\phi_1(t)\phi_2(t)\cdots\phi_n(t)$,                    (b) $|\phi_1(t)|^2$,
  (c) $\sum_1^n p_j\phi_j(t)$ where $p_j \geq 0$ and $\sum_1^n p_j = 1$,    (d) $(2 - \phi_1(t))^{-1}$,
  (e) $\int_0^\infty \phi_1(ut)e^{-u} \, du$.

**27.** (a) Find the characteristic functions corresponding to the following density functions on $\mathbb{R}$:
  (i) $1/\cosh(\pi x)$,              (ii) $(1 - \cos x)/(\pi x^2)$,
  (iii) $\exp(-x - e^{-x})$,         (iv) $\frac{1}{2}e^{-|x|}$.
  Show that the mean of the 'extreme-value distribution' in part (iii) is Euler's constant $\gamma$.
(b) **Characteristic functions coincident on an interval.** Write down the density function with characteristic function $\phi(t) = \max\{0, 1 - |t|/\pi\}$ for $t \in \mathbb{R}$.
  Show that the periodic function $\psi(t)$ with period $2\pi$ given by $\psi(t) = \max\{0, 1 - |t|/\pi\}$ for $|t| \leq \pi$, is the characteristic function of the discrete random variable with mass function

$$f(0) = \frac{1}{2}, \quad f(2k+1) = f(-2k-1) = \frac{2}{\pi^2(2k+1)^2}, \quad k = 0, 1, 2, \ldots.$$

**28.** Which of the following are characteristic functions:
  (a) $\phi(t) = 1 - |t|$ if $|t| \leq 1$, $\phi(t) = 0$ otherwise,
  (b) $\phi(t) = (1 + t^4)^{-1}$,    (c) $\phi(t) = \exp(-t^4)$,
  (d) $\phi(t) = \cos t$,            (e) $\phi(t) = 2(1 - \cos t)/t^2$.

**29.** Show that the characteristic function $\phi$ of a random variable $X$ satisfies $|1 - \phi(t)| \leq \mathbb{E}|tX|$.

**30.** Suppose $X$ and $Y$ have joint characteristic function $\phi(s, t)$. Show that, subject to the appropriate conditions of differentiability,

$$i^{m+n}\mathbb{E}(X^m Y^n) = \left.\frac{\partial^{m+n}\phi}{\partial s^m \partial t^n}\right|_{s=t=0}$$

for any positive integers $m$ and $n$.

**31.** If $X$ has distribution function $F$ and characteristic function $\phi$, show that for $t > 0$

(a)
$$\int_{[-t^{-1},t^{-1}]} x^2\, dF \le \frac{3}{t^2}[1 - \operatorname{Re} \phi(t)],$$

(b)
$$\mathbb{P}\left(|X| \ge \frac{1}{t}\right) \le \frac{7}{t}\int_0^t [1 - \operatorname{Re} \phi(v)]\, dv.$$

**32.** Let $X_1, X_2, \dots$ be independent variables which are uniformly distributed on $[0, 1]$. Let $M_n = \max\{X_1, X_2, \dots, X_n\}$ and show that $n(1 - M_n) \overset{D}{\to} X$ where $X$ is exponentially distributed with parameter 1. You need not use characteristic functions.

**33.** If $X$ is either (a) Poisson with parameter $\lambda$, or (b) $\Gamma(1, \lambda)$, show that the distribution of $Y_\lambda = (X - \mathbb{E}X)/\sqrt{\operatorname{var} X}$ approaches the $N(0, 1)$ distribution as $\lambda \to \infty$.
(c) Show that
$$e^{-n}\left(1 + n + \frac{n^2}{2!} + \dots + \frac{n^n}{n!}\right) \to \frac{1}{2} \qquad \text{as } n \to \infty.$$

**34. Coupon collecting.** Recall that you regularly buy quantities of some ineffably dull commodity. To attract your attention, the manufacturers add to each packet a small object which is also dull, and in addition useless, but there are $n$ different types. Assume that each packet is equally likely to contain any one of the different types, as usual. Let $T_n$ be the number of packets bought before you acquire a complete set of $n$ objects. Show that $n^{-1}(T_n - n\log n) \overset{D}{\to} T$, where $T$ is a random variable with distribution function $\mathbb{P}(T \le x) = \exp(-e^{-x})$, $-\infty < x < \infty$.

**35.** Find a sequence $(\phi_n)$ of characteristic functions with the property that the limit given by $\phi(t) = \lim_{n\to\infty} \phi_n(t)$ exists for all $t$, but such that $\phi$ is not itself a characteristic function.

**36.** (a) Use generating functions to show that it is not possible to load two dice in such a way that the sum of the values which they show is equally likely to take any value between 2 and 12. Compare with your method for Problem (2.7.12).
(b) **Sicherman dice.** Show that it is possible to number the faces of two fair dice, in a different manner from two standard dice, such that the sum of the scores when rolled has the same distribution as the sum of the scores of two standard dice.

**37.** A biased coin is tossed $N$ times, where $N$ is a random variable which is Poisson distributed with parameter $\lambda$. Prove that the total number of heads shown is independent of the total number of tails. Show conversely that if the numbers of heads and tails are independent, then $N$ has the Poisson distribution.

**38.** A *binary tree* is a tree (as in the section on branching processes) in which each node has exactly two descendants. Suppose that each node of the tree is coloured black with probability $p$, and white otherwise, independently of all other nodes. For any path $\pi$ containing $n$ nodes beginning at the root of the tree, let $B(\pi)$ be the number of black nodes in $\pi$, and let $X_n(k)$ be the number of such paths $\pi$ for which $B(\pi) \ge k$. Show that there exists $\beta_c$ such that

$$\mathbb{E}\{X_n(\beta n)\} \to \begin{cases} 0 & \text{if } \beta > \beta_c, \\ \infty & \text{if } \beta < \beta_c, \end{cases}$$

and show how to determine the value $\beta_c$.
    Prove that

$$\mathbb{P}(X_n(\beta n) \ge 1) \to \begin{cases} 0 & \text{if } \beta > \beta_c, \\ 1 & \text{if } \beta < \beta_c. \end{cases}$$

**39.** Use the continuity theorem (5.9.5) to show that, as $n \to \infty$,
(a) if $X_n$ is bin$(n, \lambda/n)$ then the distribution of $X_n$ converges to a Poisson distribution,

(b) if $Y_n$ is geometric with parameter $p = \lambda/n$ then the distribution of $Y_n/n$ converges to an exponential distribution.

**40.** Let $X_1, X_2, \ldots$ be independent random variables with zero means and such that $\mathbb{E}|X_j^3| < \infty$ for all $j$. Show that $S_n = X_1 + X_2 + \cdots + X_n$ satisfies $S_n/\sqrt{\text{var}(S_n)} \xrightarrow{D} N(0, 1)$ as $n \to \infty$ if

$$\sum_{j=1}^{n} \mathbb{E}|X_j^3| = o\left(\{\text{var}(S_n)\}^{-\frac{3}{2}}\right).$$

The following steps may be useful. Let $\sigma_j^2 = \text{var}(X_j)$, $\sigma(n)^2 = \text{var}(S_n)$, $\rho_j = \mathbb{E}|X_j^3|$, and $\phi_j$ and $\psi_n$ be the characteristic functions of $X_j$ and $S_n/\sigma(n)$ respectively.

(i) Use Taylor's theorem to show that $|\phi_j(t) - 1| \le 2t^2\sigma_j^2$ and $|\phi_j(t) - 1 + \frac{1}{2}\sigma_j^2 t^2| \le |t|^3\rho_j$ for $j \ge 1$.

(ii) Show that $|\log(1 + z) - z| \le |z|^2$ if $|z| \le \frac{1}{2}$, where the logarithm has its principal value.

(iii) Show that $\sigma_j^3 \le \rho_j$, and deduce from the hypothesis that $\max_{1 \le j \le n} \sigma_j/\sigma(n) \to 0$ as $n \to \infty$, implying that $\max_{1 \le j \le n} |\phi_j(t/\sigma(n)) - 1| \to 0$.

(iv) Deduce an upper bound for $\left|\log \phi_j(t/\sigma(n)) - \frac{1}{2}t^2\sigma_j^2/\sigma(n)^2\right|$, and sum to obtain that $\log \psi_n(t) \to -\frac{1}{2}t^2$.

**41.** Let $X_1, X_2, \ldots$ be independent variables each taking values $+1$ or $-1$ with probabilities $\frac{1}{2}$ and $\frac{1}{2}$. Show that

$$\sqrt{\frac{3}{n^3}} \sum_{k=1}^{n} kX_k \xrightarrow{D} N(0, 1) \qquad \text{as } n \to \infty.$$

**42. Normal sample.** Let $X_1, X_2, \ldots, X_n$ be independent $N(\mu, \sigma^2)$ random variables. Define $\overline{X} = n^{-1}\sum_1^n X_i$ and $Z_i = X_i - \overline{X}$. Find the joint characteristic function of $\overline{X}, Z_1, Z_2, \ldots, Z_n$, and hence prove that $\overline{X}$ and $S^2 = (n-1)^{-1}\sum_1^n (X_i - \overline{X})^2$ are independent.

**43. Log-normal distribution.** Let $X$ be $N(0, 1)$, and let $Y = e^X$; $Y$ is said to have the *log-normal* distribution. Show that the density function of $Y$ is

$$f(x) = \frac{1}{x\sqrt{2\pi}} \exp\left\{-\frac{1}{2}(\log x)^2\right\}, \qquad x > 0.$$

For $|a| \le 1$, define $f_a(x) = \{1 + a\sin(2\pi \log x)\}f(x)$. Show that $f_a$ is a density function with finite moments of all (positive) orders, none of which depends on the value of $a$. The family $\{f_a : |a| \le 1\}$ contains density functions which are not specified by their moments.

**44.** Consider a random walk whose steps are independent and identically distributed integer-valued random variables with non-zero mean. Prove that the walk is transient.

**45. Recurrent events.** Let $\{X_r : r \ge 1\}$ be the integer-valued identically distributed intervals between the times of a recurrent event process. Let $L$ be the earliest time by which there has been an interval of length $a$ containing no occurrence time. Show that, for integral $a$,

$$\mathbb{E}(s^L) = \frac{s^a \mathbb{P}(X_1 > a)}{1 - \sum_{r=1}^{a} s^r \mathbb{P}(X_1 = r)}.$$

**46.** A biased coin shows heads with probability $p \ (= 1 - q)$. It is flipped repeatedly until the first time $W_n$ by which it has shown $n$ consecutive heads. Let $\mathbb{E}(s^{W_n}) = G_n(s)$. Show that $G_n = psG_{n-1}/(1 - qsG_{n-1})$, and deduce that

$$G_n(s) = \frac{(1 - ps)p^n s^n}{1 - s + qp^n s^{n+1}}.$$

**47.** In $n$ flips of a biased coin which shows heads with probability $p (= 1 - q)$, let $L_n$ be the length of the longest run of heads. Show that, for $r \geq 1$,

$$1 + \sum_{n=1}^{\infty} s^n \mathbb{P}(L_n < r) = \frac{1 - p^r s^r}{1 - s + q p^r s^{r+1}}.$$

**48.** The random process $\{X_n : n \geq 1\}$ decays geometrically fast in that, in the absence of external input, $X_{n+1} = \frac{1}{2} X_n$. However, at any time $n$ the process is also increased by $Y_n$ with probability $\frac{1}{2}$, where $\{Y_n : n \geq 1\}$ is a sequence of independent exponential random variables with parameter $\lambda$. Find the limiting distribution of $X_n$ as $n \to \infty$.

**49.** Let $G(s) = \mathbb{E}(s^X)$ where $X \geq 0$. Show that $\mathbb{E}\{(X + 1)^{-1}\} = \int_0^1 G(s)\,ds$, and evaluate this when $X$ is (a) Poisson with parameter $\lambda$, (b) geometric with parameter $p$, (c) binomial bin$(n, p)$, (d) logarithmic with parameter $p$ (see Exercise (5.2.3)). Is there a non-trivial choice for the distribution of $X$ such that $\mathbb{E}\{(X + 1)^{-1}\} = \{\mathbb{E}(X + 1)\}^{-1}$?

**50.** Find the density function of $\sum_{r=1}^{N} X_r$, where $\{X_r : r \geq 1\}$ are independent and exponentially distributed with parameter $\lambda$, and $N$ is geometric with parameter $p$ and independent of the $X_r$.

**51.** Let $X$ have finite non-zero variance and characteristic function $\phi(t)$. Show that

$$\psi(t) = -\frac{1}{\mathbb{E}(X^2)} \frac{d^2 \phi}{dt^2}$$

is a characteristic function, and find the corresponding distribution.

**52. Triangular distribution.** Let $X$ and $Y$ have joint density function

$$f(x, y) = \tfrac{1}{4}\{1 + xy(x^2 - y^2)\}, \qquad |x| < 1, \ |y| < 1.$$

Show that $\phi_X(t)\phi_Y(t) = \phi_{X+Y}(t)$, and that $X$ and $Y$ are dependent. Find the probability density function of $X + Y$.

**53. Exercise (4.6.6) revisited.** Let $X_1, X_2, \ldots$ be independent and uniformly distributed on $(0, 1)$, and let $m(x) = \mathbb{E}(N)$ where $N = \min\{n : \sum_{r=1}^{n} X_r > x\}$ for $x > 0$. Show that $m'(x) = m(x) - m(x - 1)$, and deduce that the Laplace transform $m^*(s) = \int_0^{\infty} m(x)e^{-sx}\,dx$ is given by $m^*(s) = 1/(e^{-s} + s - 1)$ for $s \neq 0$. Hence prove that

$$m(x) = \sum_{r=0}^{\lfloor x \rfloor} \frac{(-1)^r}{r!}(x - r)^r e^{x-r}, \qquad x > 0.$$

**54. Rounding error.** Let $S_n = \sum_{r=1}^{n} X_r$ be a partial sum of a sequence of independent random variables with the uniform distribution on $(0, 1)$. For $x \in \mathbb{R}$, let $\{x\}$ denote the nearest integer to $x$, and write $R_n = \sum_{r=1}^{n} \{X_r\}$.
(a) Show that $X_r - \{X_r\}$ is uniformly distributed on $(-\frac{1}{2}, \frac{1}{2})$.
(b) Show that

$$\mathbb{P}(\{S_n\} = R_n) = 2 f_{n+1}(0) = \int_{-\infty}^{\infty} \frac{1}{\pi}\left(\frac{\sin t}{t}\right)^{n+1} dt,$$

where $f_n(y)$ is the density function of the sum of $n$ independent random variables with the uniform distribution on $(-1, 1)$.
(c) Find a similar expression for $\mathbb{P}(\{S_n\} - R_n = k)$.

**55. Maxwell's molecules.**

(a) Let $V = (V_1, V_2, V_3)$ be the velocity in $\mathbb{R}^3$ of a molecule $M$ of a perfect gas, and assume that in any Cartesian coordinate system the coordinates of $V$ are independent random variables with mean 0 and finite variance $\sigma^2$. Show that the $V_i$ are independent with the $N(0, \sigma^2)$ distribution.

   When $\sigma^2 = 1$, show that $|V|$ has the *Maxwell density* $f(v) = \sqrt{2/\pi}\, v^2 e^{-\frac{1}{2}v^2}$ for $v > 0$.

(b) A physicist assumes that, initially, $M$ is equally likely to be anywhere in the region $R$ between two parallel planes with distance 1 apart, independently of its velocity. Assuming $\sigma^2 = 1$, show that the probability $p(t)$ that $M$ is in $R$ at time $t$ is

$$p(t) = \frac{1}{\sqrt{2\pi}} \left[ \int_{-1}^{1} \frac{1}{t} \exp\left\{ -\frac{x^2}{2t^2} \right\} dx - 2t \left( 1 - \exp\left\{ -\frac{1}{2t^2} \right\} \right) \right].$$

(c) Deduce that the density function of the time $T$ at which $M$ exits $R$ is

$$f_T(t) = \sqrt{2/\pi} \left[ 1 - \exp\left\{ -\frac{1}{2t^2} \right\} \right], \qquad t > 0.$$

**56.** Let $X, Y$ be independent random variables with a joint distribution with circular symmetry about 0 in the $x/y$-plane, and with finite variances. Show that the distribution of $R = X \cos\theta + Y \sin\theta$ does not depend on the value of $\theta$.

   With the usual notation for characteristic functions, show that:

(a) $\phi_X(t \cos\theta)\phi_Y(t \sin\theta) = \phi_R(t)$,

(b) $\phi_X(t) = \phi_X(-t) = \phi_Y(t) = \phi_Y(-t)$,

(c) $\phi_X(t) = \psi(t^2)$ for some continuous real-valued function $\psi$,

(d) $X$ and $Y$ are $N(0, \sigma^2)$ for some $\sigma^2 \geq 0$.

# 6

# Markov chains

*Summary.* A Markov chain is a random process with the property that, conditional on its present value, the future is independent of the past. The Chapman–Kolmogorov equations are derived, and used to explore the recurrence and transience of states. Stationary distributions are studied at length, and the limit theorem for irreducible chains is proved using coupling. The reversibility of Markov chains is discussed. After a section devoted to branching processes, the theory of Poisson processes and birth–death processes is considered in depth, and the theory of continuous-time chains is presented. The technique of imbedding a discrete-time chain inside a continuous-time chain is exploited in different settings. The basic properties of spatial Poisson processes are described, and the chapter ends with an account of the technique of Markov chain Monte Carlo.

## 6.1 Markov processes

The simple random walk (5.3) and the branching process (5.4) are two examples of sequences of random variables that evolve in some random but prescribed manner. Such collections are called† 'random processes'. A typical random process $X$ is a family $\{X_t : t \in T\}$ of random variables indexed by some set $T$. In the above examples $T = \{0, 1, 2, \dots\}$ and we call the process a 'discrete-time' process; in other important examples $T = \mathbb{R}$ or $T = [0, \infty)$ and we call it a 'continuous-time' process. In either case we think of a random process as a family of variables that evolve as time passes. These variables may even be independent of each other, but then the evolution is not very surprising and this very special case is of little interest to us in this chapter. Rather, we are concerned with more general, and we hope realistic, models for random evolution. Simple random walks and branching processes shared the following property: conditional on their values at the $n$th step, their future values did not depend on their previous values. This property proved to be very useful in their analysis, and it is to the general theory of processes with this property that we turn our attention now.

Until further notice we shall be interested in discrete-time processes. Let $\{X_0, X_1, \dots\}$ be a sequence of random variables which take values in some countable set $S$, called the *state*

---

†Such collections are often called 'stochastic' processes. The Greek verb στοχάζομαι means 'to shoot at, aim at, guess at', and the adjective στοχαστικός was used, for example by Plato, to mean 'proceeding by guesswork'.

*space*†. Each $X_n$ is a discrete random variable that takes one of $N$ possible values, where $N = |S|$; it may be the case that $N = \infty$.

**(1) Definition.** The process $X$ is a **Markov chain**‡ if it has the **Markov property**:

$$\mathbb{P}(X_n = s \mid X_0 = x_0, \ X_1 = x_1, \dots, \ X_{n-1} = x_{n-1}) = \mathbb{P}(X_n = s \mid X_{n-1} = x_{n-1})$$

for all $n \geq 1$ and all $s, x_1, \dots, x_{n-1} \in S$.

A proof that the random walk is a Markov chain was given in Lemma (3.9.5). The reader can check that the Markov property is equivalent to each of the stipulations (2) and (3) below: for each $s \in S$ and for every sequence $\{x_i : i \geq 0\}$ in $S$,

(2)     $$\mathbb{P}(X_{n+1} = s \mid X_{n_1} = x_{n_1}, \ X_{n_2} = x_{n_2}, \ X_{n_k} = x_{n_k}) = \mathbb{P}(X_{n+1} = s \mid X_{n_k} = x_{n_k})$$
$$\text{for all } n_1 < n_2 < \cdots < n_k \leq n,$$

(3)     $$\mathbb{P}(X_{m+n} = s \mid X_0 = x_0, \ X_1 = x_1, \dots, \ X_m = x_m) = \mathbb{P}(X_{m+n} = s \mid X_m = x_m)$$
$$\text{for any } m, n \geq 0.$$

We have assumed that $X$ takes values in some *countable set $S$*. The reason for this is essentially the same as the reason for treating discrete and continuous variables separately. Since $S$ is assumed countable, it can be put in one–one correspondence with some subset $S'$ of the integers, and without loss of generality we can assume that $S$ *is* this set $S'$ of integers. If $X_n = i$, then we say that the chain is 'in state $i$'; we can also talk of the chain as 'having the value $i$', or 'visiting $i$', depending upon the context of the remark.

The evolution of a chain is described by its 'transition probabilities' $\mathbb{P}(X_{n+1} = j \mid X_n = i)$; it can be quite complicated in general since these probabilities depend upon the three quantities $n$, $i$, and $j$. We shall restrict our attention to the case when they do not depend on $n$ but only upon $i$ and $j$.

**(4) Definition.** The Markov chain $X$ is called **homogeneous** if

$$\mathbb{P}(X_{n+1} = j \mid X_n = i) = \mathbb{P}(X_1 = j \mid X_0 = i)$$

for all $n, i, j$. The **transition matrix** $\mathbf{P} = (p_{ij})$ is the $|S| \times |S|$ matrix of **transition probabilities**

$$p_{ij} = \mathbb{P}(X_{n+1} = j \mid X_n = i).$$

In former times some authors wrote $p_{ji}$ in place of $p_{ij}$ here, so beware; sometimes we write $p_{i,j}$ for $p_{ij}$. Henceforth, *all Markov chains are assumed homogeneous* unless otherwise specified; we assume that the process $X$ is a Markov chain, and we denote the transition matrix of such a chain by $\mathbf{P}$.

---

†There is, of course, an underlying probability space $(\Omega, \mathcal{F}, \mathbb{P})$, and each $X_n$ is an $\mathcal{F}$-measurable function which maps $\Omega$ into $S$.

‡The expression 'stochastically determined process' was in use until around 1930, when Khinchin suggested this more functional label. We require the Markov property whenever the conditional probabilities are well defined.

**(5) Theorem.** *The transition matrix* **P** *is a* stochastic matrix, *which is to say that*:
  (a) **P** *has non-negative entries, or* $p_{ij} \geq 0$ *for all* $i, j$,
  (b) **P** *has row sums equal to one, or* $\sum_j p_{ij} = 1$ *for all* $i$.

**Proof.** An easy *exercise*.  ■

We can easily see that (5) characterizes transition matrices.

Broadly speaking, we are interested in the evolution of $X$ over two different time scales, the 'short term' and the 'long term'. In the short term the random evolution of $X$ is described by **P**, whilst long-term changes are described in the following way.

**(6) Definition.** The *$n$-step transition matrix* $\mathbf{P}(m, m + n) = (p_{ij}(m, m + n))$ is the matrix of *$n$-step transition probabilities* $p_{ij}(m, m + n) = \mathbb{P}(X_{m+n} = j \mid X_m = i)$.

By the assumption of homogeneity, $\mathbf{P}(m, m + 1) = \mathbf{P}$. That $\mathbf{P}(m, m + n)$ does not depend on $m$ is a consequence of the following important fact.

**(7) Theorem. Chapman–Kolmogorov equations.**

$$p_{ij}(m, m + n + r) = \sum_k p_{ik}(m, m + n) p_{kj}(m + n, m + n + r).$$

*Therefore,* $\mathbf{P}(m, m + n + r) = \mathbf{P}(m, m + n)\mathbf{P}(m + n, m + n + r)$, *and* $\mathbf{P}(m, m + n) = \mathbf{P}^n$, *the nth power of* **P**.

**Proof.** We have as required that

$$p_{ij}(m, m + n + r) = \mathbb{P}(X_{m+n+r} = j \mid X_m = i)$$
$$= \sum_k \mathbb{P}(X_{m+n+r} = j, \ X_{m+n} = k \mid X_m = i)$$
$$= \sum_k \mathbb{P}(X_{m+n+r} = j \mid X_{m+n} = k, \ X_m = i)\mathbb{P}(X_{m+n} = k \mid X_m = i)$$
$$= \sum_k \mathbb{P}(X_{m+n+r} = j \mid X_{m+n} = k)\mathbb{P}(X_{m+n} = k \mid X_m = i),$$

where we have used the fact that $\mathbb{P}(A \cap B \mid C) = \mathbb{P}(A \mid B \cap C)\mathbb{P}(B \mid C)$, proved in Exercise (1.4.2), together with the Markov property (2). The established equation may be written in matrix form as $\mathbf{P}(m, m + n + r) = \mathbf{P}(m, m + n)\mathbf{P}(m + n, m + n + r)$, and it follows by iteration that $\mathbf{P}(m, m + n) = \mathbf{P}^n$.  ■

It is a consequence of Theorem (7) that $\mathbf{P}(m, m + n) = \mathbf{P}(0, n)$, and we write henceforth $\mathbf{P}_n$ for $\mathbf{P}(m, m + n)$, and $p_{ij}(n)$ for $p_{ij}(m, m + n)$. This theorem relates long-term development to short-term development, and tells us how $X_n$ depends on the initial variable $X_0$. Let $\mu_i^{(n)} = \mathbb{P}(X_n = i)$ be the mass function of $X_n$, and write $\boldsymbol{\mu}^{(n)}$ for the row vector with entries $(\mu_i^{(n)} : i \in S)$.

**(8) Lemma.** $\boldsymbol{\mu}^{(m+n)} = \boldsymbol{\mu}^{(m)}\mathbf{P}_n$, *and hence* $\boldsymbol{\mu}^{(n)} = \boldsymbol{\mu}^{(0)}\mathbf{P}^n$.

**Proof.** We have that

$$\mu_j^{(m+n)} = \mathbb{P}(X_{m+n} = j) = \sum_i \mathbb{P}(X_{m+n} = j \mid X_m = i)\mathbb{P}(X_m = i)$$

$$= \sum_i \mu_i^{(m)} p_{ij}(n) = (\mu^{(m)}\mathbf{P}_n)_j$$

and the result follows from Theorem (7). ∎

Thus we reach the important conclusion that the random evolution of the chain is determined by the transition matrix $\mathbf{P}$ and the initial mass function $\mu^{(0)}$. Many questions about the chain can be expressed in terms of these quantities, and the study of the chain is thus largely reducible to the study of algebraic properties of matrices.

**(9) Example. Simple random walk.** The simple random walk on the integers has state space $S = \{0, \pm 1, \pm 2, \ldots\}$ and transition probabilities

$$p_{ij} = \begin{cases} p & \text{if } j = i + 1, \\ q = 1 - p & \text{if } j = i - 1, \\ 0 & \text{otherwise.} \end{cases}$$

The argument leading to equation (3.10.2) shows that

$$p_{ij}(n) = \begin{cases} \binom{n}{\frac{1}{2}(n+j-i)} p^{\frac{1}{2}(n+j-i)} q^{\frac{1}{2}(n-j+i)} & \text{if } n + j - i \text{ is even,} \\ 0 & \text{otherwise.} \end{cases} \quad \bullet$$

**(10) Example.  Branching process.** As in Section 5.4, $S = \{0, 1, 2, \ldots\}$ and $p_{ij}$ is the coefficient of $s^j$ in $G(s)^i$. Also, $p_{ij}(n)$ is the coefficient of $s^j$ in $G_n(s)^i$. $\bullet$

**(11) Example. Gene frequencies.** One of the most interesting and extensive applications of probability theory is to genetics, and particularly to the study of gene frequencies. The problem may be inadequately and superficially described as follows. For definiteness suppose the population is human. Genetic information is (mostly) contained in chromosomes, which are strands of chemicals grouped in cell nuclei. In humans ordinary cells carry 46 chromosomes, 44 of which are homologous pairs. For our purposes a chromosome can be regarded as an ordered set of $n$ sites, the states of which can be thought of as a sequence of random variables $C_1, C_2, \ldots, C_n$. The possible values of each $C_i$ are certain combinations of chemicals, and these values influence (or determine) some characteristic of the owner such as hair colour or leg length.

Now, suppose that $A$ is a possible value of $C_1$, say, and let $X_n$ be the number of individuals in the $n$th generation for which $C_1$ has the value $A$. What is the behaviour of the sequence $X_1, X_2, \ldots, X_n, \ldots$? The first important (and obvious) point is that the sequence is random, because of the following factors.

(a) The value $A$ for $C_1$ may affect the owner's chances of contributing to the next generation. If $A$ gives you short legs, you stand a better chance of being caught by a sabre-toothed tiger. The breeding population is randomly selected from those born, but there may be bias for or against the gene $A$.

(b) The breeding population is randomly combined into pairs to produce offspring. Each parent contributes 23 chromosomes to its offspring, but here again, if $A$ gives you short legs you may have a smaller (or larger) chance of catching a mate.

(c) Sex cells having half the normal complement of chromosomes are produced by a special and complicated process called 'meiosis'. We shall not go into details, but essentially the homologous pairs of the parent are shuffled to produce new and different chromosomes for offspring. The sex cells from each parent (with 23 chromosomes) are then combined to give a new cell (with 46 chromosomes).

(d) Since meiosis involves a large number of complex chemical operations it is hardly surprising that things go wrong occasionally, producing a new value for $C_1$, $\widehat{A}$ say. This is a 'mutation'.

The reader can now see that if generations are segregated (in a laboratory, say), then we can suppose that $X_1, X_2, \ldots$ is a Markov chain with a finite state space. If generations are not segregated and $X(t)$ is the frequency of $A$ in the population at time $t$, then $X(t)$ may be a continuous-time Markov chain.

For a simple example, suppose that the population size is $N$, a constant. If $X_n = i$, it may seem reasonable that any member of the $(n+1)$th generation carries $A$ with probability $i/N$, independently of the others. Then we have a *Wright–Fisher model*

$$p_{ij} = \mathbb{P}(X_{n+1} = j \mid X_n = i) = \binom{N}{j}\left(\frac{i}{N}\right)^j\left(1 - \frac{i}{N}\right)^{N-j}.$$

Even more simply, suppose that at each stage exactly one individual dies and is replaced by a new individual; each individual is picked for death with probability $1/N$. If $X_n = i$, we assume that the probability that the replacement carries $A$ is $i/N$. Then we have *Moran's model*,

$$p_{ij} = \begin{cases} \dfrac{i(N-i)}{N^2} & \text{if } j = i \pm 1, \\[2ex] 1 - 2\dfrac{i(N-i)}{N^2} & \text{if } j = i, \\[2ex] 0 & \text{otherwise.} \end{cases}$$
●

**(12) Example. Recurrent events.** Suppose that $X$ is a Markov chain on $S$, with $X_0 = i$. Let $T(1)$ be the time of the first return of the chain to $i$: that is, $T(1) = \min\{n \geq 1 : X_n = i\}$, with the convention that $T(1) = \infty$ if $X_n \neq i$ for all $n \geq 1$. Suppose that you tell me that $T(1) = 3$, say, which is to say that $X_n \neq i$ for $n = 1, 2$, and $X_3 = i$. The future evolution of the chain $\{X_3, X_4, \ldots\}$ depends, by the Markov property, only on the fact that the new starting point $X_3$ equals $i$, and does not depend further on the values of $X_0, X_1, X_2$. Thus the future process $\{X_3, X_4, \ldots\}$ has the same distribution as had the original process $\{X_0, X_1, \ldots\}$ starting from state $i$. The same argument is valid for any given value of $T(1)$, and we are therefore led to the following observation. Having returned to its starting point for the first time, the future of the chain has the same distribution as had the original chain. Let $T(2)$ be the time which elapses between the first and second return of the chain to its starting point. Then $T(1)$ and $T(2)$ must be independent and identically distributed random variables. Arguing similarly for future returns, we deduce that the time of the $n$th return of the chain to its starting point may be represented as $T(1) + T(2) + \cdots + T(n)$, where $T(1), T(2), \ldots$ are independent identically distributed random variables. That is to say, the return times of the chain form a 'recurrent event process'; see Example (5.2.15). Some care is needed in order to make this argument fully rigorous, and this is the challenge of Exercise (6.1.5).

A problem arises with the above argument if $T(1)$ takes the value $\infty$ with strictly positive probability, which is to say that the chain is not (almost) certain to return to its starting point. For the moment we overlook this difficulty, and suppose not only that $\mathbb{P}(T(1) < \infty) = 1$, but also that $\mu = \mathbb{E}(T(1))$ satisfies $\mu < \infty$. It is now an immediate consequence of the renewal theorem (5.2.24) that

$$p_{ii}(n) = \mathbb{P}(X_n = i \mid X_0 = i) \to \frac{1}{\mu} \quad \text{as } n \to \infty$$

so long as the distribution of $T(1)$ is non-arithmetic; the latter condition is certainly satisfied if, say, $p_{ii} > 0$. ●

**(13) Example. Bernoulli process.** Let $S = \{0, 1, 2, \ldots\}$ and define the Markov chain $Y$ by $Y_0 = 0$ and

$$\mathbb{P}(Y_{n+1} = s + 1 \mid Y_n = s) = p, \quad \mathbb{P}(Y_{n+1} = s \mid Y_n = s) = 1 - p,$$

for all $n \geq 0$, where $0 < p < 1$. You may think of $Y_n$ as the number of heads thrown in $n$ tosses of a coin. It is easy to see that

$$\mathbb{P}(Y_{m+n} = j \mid Y_m = i) = \binom{n}{j - i} p^{j-i}(1 - p)^{n-j+i}, \quad 0 \leq j - i \leq n.$$

Viewed as a Markov chain, $Y$ is not a very interesting process. Suppose, however, that the value of $Y_n$ is counted using a conventional digital decimal meter, and let $X_n$ be the final digit of the reading, $X_n = Y_n$ modulo 10. It may be checked that $X = \{X_n : n \geq 0\}$ is a Markov chain on the state space $S' = \{0, 1, 2, \ldots, 9\}$ with transition matrix

$$\mathbf{P} = \begin{pmatrix} 1 - p & p & 0 & \cdots & 0 \\ 0 & 1 - p & p & \cdots & 0 \\ \vdots & \vdots & \vdots & \ddots & \vdots \\ p & 0 & 0 & \cdots & 1 - p \end{pmatrix}.$$

There are various ways of studying the behaviour of $X$. If we are prepared to use the renewal theorem (5.2.24), then we might argue as follows. The process $X$ passes through the values $0, 1, 2, \ldots, 9, 0, 1, \ldots$ sequentially. Consider the times at which $X$ takes the value $i$, say. These times form a recurrent event process for which a typical inter-occurrence time $T$ satisfies

$$T = \begin{cases} 1 & \text{with probability } 1 - p, \\ 1 + Z & \text{with probability } p, \end{cases}$$

where $Z$ has the negative binomial distribution with parameters 9 and $p$. Therefore $\mathbb{E}(T) = 1 + p\mathbb{E}(Z) = 1 + p(9/p) = 10$. It is now an immediate consequence of the renewal theorem that $\mathbb{P}(X_n = i) \to \frac{1}{10}$ for $i = 0, 1, \ldots, 9$, as $n \to \infty$. ●

**(14) Example. Markov's other chain (1910).** Let $Y_1, Y_3, Y_5, \ldots$ be a sequence of independent identically distributed random variables such that

**(15)**                $$\mathbb{P}(Y_{2k+1} = -1) = \mathbb{P}(Y_{2k+1} = 1) = \tfrac{1}{2}, \quad k = 0, 1, 2, \ldots,$$

and define $Y_{2k} = Y_{2k-1}Y_{2k+1}$, for $k = 1, 2, \ldots$. You may check that $Y_2, Y_4, \ldots$ is a sequence of independent identically distributed variables with the same distribution (15). Now $\mathbb{E}(Y_{2k}Y_{2k+1}) = \mathbb{E}(Y_{2k-1}Y_{2k+1}^2) = \mathbb{E}(Y_{2k-1}) = 0$, and so (by the result of Problem (3.11.12)) the sequence $Y_1, Y_2, \ldots$ is pairwise independent. Hence $p_{ij}(n) = \mathbb{P}(Y_{m+n} = j \mid Y_m = i)$ satisfies $p_{ij}(n) = \frac{1}{2}$ for all $n$ and $i, j = \pm 1$, and it follows easily that the Chapman–Kolmogorov equations are satisfied.

Is $Y$ a Markov chain? *No*, because $\mathbb{P}(Y_{2k+1} = 1 \mid Y_{2k} = -1) = \frac{1}{2}$, whereas

$$\mathbb{P}(Y_{2k+1} = 1 \mid Y_{2k} = -1, \ Y_{2k-1} = 1) = 0.$$

Thus, whilst the Chapman–Kolmogorov equations are *necessary* for the Markov property, they are not *sufficient*; this is for much the same reason that pairwise independence is weaker than independence.

Although $Y$ is not a Markov chain, we can find a Markov chain by enlarging the state space. Let $Z_n = (Y_n, Y_{n+1})$, taking values in $S = \{-1, +1\}^2$. It is an *exercise* to check that $Z$ is a (non-homogeneous) Markov chain with, for example,

$$\mathbb{P}\big(Z_{n+1} = (1, 1) \mid Z_n = (1, 1)\big) = \begin{cases} \frac{1}{2} & \text{if } n \text{ even}, \\ 1 & \text{if } n \text{ odd}. \end{cases}$$

This technique of 'imbedding' $Y$ in a Markov chain on a larger state space turns out to be useful in many contexts of interest.    ●

---

## Exercises for Section 6.1

**1.**    Show that any sequence of independent random variables taking values in the countable set $S$ is a Markov chain. Under what condition is this chain homogeneous?

**2.**    A die is rolled repeatedly. Which of the following are Markov chains? For those that are, supply the transition matrix.
(a) The largest number $X_n$ shown up to the $n$th roll.
(b) The number $N_n$ of sixes in $n$ rolls.
(c) At time $r$, the time $C_r$ since the most recent six.
(d) At time $r$, the time $B_r$ until the next six.

**3.**    Let $\{S_n : n \geq 0\}$ be a simple random walk with $S_0 = 0$, and show that $X_n = |S_n|$ defines a Markov chain; find the transition probabilities of this chain. Let $M_n = \max\{S_k : 0 \leq k \leq n\}$, and show that $Y_n = M_n - S_n$ defines a Markov chain. What happens if $S_0 \neq 0$?

**4.**    Let $X$ be a Markov chain and let $\{n_r : r \geq 0\}$ be an unbounded increasing sequence of positive integers. Show that $Y_r = X_{n_r}$ constitutes a (possibly non-homogeneous) Markov chain. Find the transition matrix of $Y$ when $n_r = 2r$ and $X$ is: (a) simple random walk, and (b) a branching process.

**5.**    Let $X$ be a Markov chain on $S$, and let $I : S^n \to \{0, 1\}$. Show that the distribution of $X_n, X_{n+1}, \ldots$, conditional on $\{I(X_1, \ldots, X_n) = 1\} \cap \{X_n = i\}$, is identical to the distribution of $X_n, X_{n+1}, \ldots$ conditional on $\{X_n = i\}$.

**6.**    **Strong Markov property.** Let $X$ be a Markov chain on $S$, and let $T$ be a random variable taking values in $\{0, 1, 2, \ldots\}$ with the property that the indicator function $I_{\{T=n\}}$, of the event that $T = n$, is a function of the variables $X_1, X_2, \ldots, X_n$. Such a random variable $T$ is called a *stopping time*, and the above definition requires that it is decidable whether or not $T = n$ with a knowledge only of the past and present, $X_0, X_1, \ldots, X_n$, and with no further information about the future.

Show that

$$\mathbb{P}\left(X_{T+m} = j \mid X_k = x_k \text{ for } 0 \le k < T, \; X_T = i\right) = \mathbb{P}(X_{T+m} = j \mid X_T = i)$$

for $m \ge 0$, $i, j \in S$, and all sequences $(x_k)$ of states.

**7.** Let $X$ be a Markov chain with state space $S$, and suppose that $h : S \to T$ is one–one. Show that $Y_n = h(X_n)$ defines a Markov chain on $T$. Must this be so if $h$ is not one–one?

**8.** Let $X$ and $Y$ be Markov chains on the set $\mathbb{Z}$ of integers.
(a) Is the sequence $Z_n = X_n + Y_n$ necessarily a Markov chain?
(b) Is $Z$ a Markov chain if $X$ and $Y$ are independent chains? Give a proof or a counterexample.
(c) Show that $Z$ is a Markov chain if $X$ and $Y$ are independent of one another and have independent increments.

**9.** Let $X$ be a Markov chain. Which of the following are Markov chains?
(a) $X_{m+r}$ for $r \ge 0$.
(b) $X_{2m}$ for $m \ge 0$.
(c) The sequence of pairs $(X_n, X_{n+1})$ for $n \ge 0$.

**10. Two-sided Markov property.** Let $X$ be a Markov chain. Show that, for $1 < r < n$,

$$\mathbb{P}(X_r = k \mid X_i = x_i \text{ for } i = 1, 2, \dots, r-1, r+1 \dots, n)$$
$$= \mathbb{P}(X_r = k \mid X_{r-1} = x_{r-1}, \; X_{r+1} = x_{r+1}).$$

**11.** Let $\{X_n : n \ge 1\}$ be independent identically distributed integer-valued random variables. Let $S_n = \sum_{r=1}^{n} X_r$, with $S_0 = 0$, $Y_n = X_n + X_{n-1}$ with $X_0 = 0$, and $Z_n = \sum_{r=0}^{n} S_r$. Which of the following constitute Markov chains: (a) $S_n$, (b) $Y_n$, (c) $Z_n$, (d) the sequence of pairs $(S_n, Z_n)$?

**12.** A stochastic matrix $\mathbf{P}$ is called *doubly stochastic* if $\sum_i p_{ij} = 1$ for all $j$. It is called *sub-stochastic* if $\sum_i p_{ij} \le 1$ for all $j$. Show that, if $\mathbf{P}$ is stochastic (respectively, doubly stochastic, sub-stochastic), then $\mathbf{P}^n$ is stochastic (respectively, doubly stochastic, sub-stochastic) for all $n$.

**13. Lumping.** Let $X$ be a Markov chain on the finite state space $S$ with transition matrix $\mathbf{P}$, and let $\mathcal{C} = \{C_j : j \in J\}$ be a partition of $S$. Let $Y_n = j$ if $X_n \in C_j$. The chain $X$ is called $\mathcal{C}$-*lumpable* if $Y$ is a Markov chain.

Show that $X$ is $\mathcal{C}$-lumpable if and only if, for $a, b \in J$, $\mathbb{P}(X_{n+1} \in C_b \mid X_n = i)$ is constant for $i \in C_a$.

# 6.2 Classification of states

We can think of the development of a Markov chain as the motion of a notional particle which jumps between the states of the state space $S$ at each epoch of time. As in Section 5.3, we may be interested in the (possibly infinite) time which elapses before the particle returns to its starting point. We saw there that it sufficed to find the distribution of the length of time until the particle returns for the first time, since other interarrival times are merely independent copies of this. However, need the particle ever return to its starting point? With this question in mind we make the following Definition (1).

In advance of stating the definition, we introduce some useful notation. Let $\mathbb{P}_i$ denote the probability measure $\mathbb{P}$ conditional on $X(0) = i$ (and similarly the expectation $\mathbb{E}_i$). That is,

$$\mathbb{P}_i(A) = \mathbb{P}(A \mid X_0 = i), \quad \mathbb{E}_i(X) = \mathbb{E}(X \mid X_0 = i).$$

**(1) Definition.** State $i$ is called **recurrent** (or **persistent**) if

$$\mathbb{P}_i(X_n = i \text{ for some } n \geq 1) = 1,$$

which is to say that the probability of eventual return to $i$, having started from $i$, is 1. If this probability is strictly less than 1, the state $i$ is called **transient**.

As in Section 5.3, we are interested in the *first passage times* of the chain. Let

$$f_{ij}(n) = \mathbb{P}_i(X_1 \neq j, \ X_2 \neq j, \ldots, \ X_{n-1} \neq j, X_n = j)$$

be the probability that the first visit to state $j$, starting from $i$, takes place at the $n$th step. Define

$$(2) \qquad\qquad f_{ij} = \sum_{n=1}^{\infty} f_{ij}(n)$$

to be the probability that the chain ever visits $j$, starting from $i$. Of course, $j$ is recurrent if and only if $f_{jj} = 1$. We seek a criterion for recurrence in terms of the $n$-step transition probabilities. Following our random walk experience, we define the generating functions

$$P_{ij}(s) = \sum_{n=0}^{\infty} s^n p_{ij}(n), \qquad F_{ij}(s) = \sum_{n=0}^{\infty} s^n f_{ij}(n),$$

with the conventions that $p_{ij}(0) = \delta_{ij}$, the Kronecker delta, and $f_{ij}(0) = 0$ for all $i$ and $j$. Clearly $f_{ij} = F_{ij}(1)$. We usually assume that $|s| < 1$, since $P_{ij}(s)$ is then guaranteed to converge. On occasions when we require properties of $P_{ij}(s)$ as $s \uparrow 1$, we shall appeal to Abel's theorem (5.1.15).

**(3) Theorem.**
(a) $P_{ii}(s) = 1 + F_{ii}(s)P_{ii}(s)$.
(b) $P_{ij}(s) = F_{ij}(s)P_{jj}(s)$ if $i \neq j$.

**Proof.** The proof is exactly as that of Theorem (5.3.1). Fix $i, j \in S$ and let $A_m = \{X_m = j\}$ and $B_m$ be the event that the first visit to $j$ (after time 0) takes place at time $m$; that is, $B_m = \{X_r \neq j \text{ for } 1 \leq r < m, \ X_m = j\}$. The $B_m$ are disjoint, so that

$$\mathbb{P}_i(A_m) = \sum_{r=1}^{m} \mathbb{P}_i(A_m \cap B_r).$$

Now, using the Markov property (as found in Exercises (6.1.5) or (6.1.6)),

$$\mathbb{P}_i(A_m \cap B_r) = \mathbb{P}(A_m \mid B_r, \ X_0 = i)\mathbb{P}_i(B_r)$$
$$= \mathbb{P}(A_m \mid X_r = j)\mathbb{P}_i(B_r), \qquad r = 1, 2, \ldots, m.$$

Hence

$$p_{ij}(m) = \sum_{r=1}^{m} f_{ij}(r)p_{jj}(m - r), \qquad m = 1, 2, \ldots.$$

Multiply throughout by $s^m$, where $|s| < 1$, and sum over $m$ ($\geq 1$) to find that $P_{ij}(s) - \delta_{ij} = F_{ij}(s) P_{jj}(s)$ as required. ∎

**(4) Corollary.**
   (a) *State $j$ is recurrent if $\sum_n p_{jj}(n) = \infty$, and if this holds then $\sum_n p_{ij}(n) = \infty$ for all $i$ such that $f_{ij} > 0$.*
   (b) *State $j$ is transient if $\sum_n p_{jj}(n) < \infty$, and if this holds then $\sum_n p_{ij}(n) < \infty$ for all $i$.*

   See Exercise (6.2.3) for a direct proof of this corollary.

**Proof.** First we show that $j$ is recurrent if and only if $\sum_n p_{jj}(n) = \infty$. From (3a),

$$P_{jj}(s) = \frac{1}{1 - F_{jj}(s)} \qquad \text{if } |s| < 1.$$

Hence, as $s \uparrow 1$, $P_{jj}(s) \to \infty$ if and only if $f_{jj} = F_{jj}(1) = 1$. Now use Abel's theorem (5.1.15) to obtain $\lim_{s \uparrow 1} P_{jj}(s) = \sum_n p_{jj}(n)$ and our claim is shown. Use (3b) to complete the proof. ∎

**(5) Corollary.** *If $j$ is transient then $p_{ij}(n) \to 0$ as $n \to \infty$ for all $i$.*

**Proof.** This is immediate from (4). ∎

   An important application of Theorem (4) is to the recurrence of symmetric random walk; see Problem (5.12.5).

   Each state is either recurrent or transient. It is intuitively clear that the number $N(i)$ of times which the chain visits its starting point $i$ satisfies

**(6)** $$\mathbb{P}\big(N(i) = \infty\big) = \begin{cases} 1 & \text{if } i \text{ is recurrent,} \\ 0 & \text{if } i \text{ is transient,} \end{cases}$$

since after each such visit, subsequent return is assured if and only if $f_{ii} = 1$ (see Problem (6.15.5) for a more detailed argument).

   Here is another important classification of states. Let

$$T_j = \min\{n \geq 1 : X_n = j\}$$

be the time of the first visit to $j$, with the convention that $T_j = \infty$ if this visit never occurs; $\mathbb{P}_i(T_i = \infty) > 0$ if and only if $i$ is transient, and in this case $\mathbb{E}_i(T_i) = \infty$.

**(7) Definition.** The **mean recurrence time** $\mu_i$ of a state $i$ is defined as

$$\mu_i = \mathbb{E}_i(T_i) = \begin{cases} \sum_n n f_{ii}(n) & \text{if } i \text{ is recurrent,} \\ \infty & \text{if } i \text{ is transient.} \end{cases}$$

   Note that $\mu_i$ may be infinite even if $i$ is recurrent.

**(8) Definition.** For a recurrent state $i$,

$$i \text{ is called } \begin{cases} \textbf{null} & \text{if } \mu_i = \infty, \\ \textbf{positive (or non-null)} & \text{if } \mu_i < \infty. \end{cases}$$

There is a simple criterion for nullity in terms of the transition probabilities.

**(9) Theorem.** *A recurrent state $i$ is null if and only if $p_{ii}(n) \to 0$ as $n \to \infty$; if this holds then $p_{ji}(n) \to 0$ for all $j$.*

**Proof.** We defer this until note (a) after Theorem (6.4.20).     ∎

Finally, for technical reasons we shall sometimes be interested in the epochs of time at which return to the starting point is possible.

**(10) Definition.** The **period** $d(i)$ of a state $i$ is defined by $d(i) = \gcd\{n : p_{ii}(n) > 0\}$, the greatest common divisor of the epochs at which return is possible. We call $i$ **periodic** if $d(i) > 1$ and **aperiodic** if $d(i) = 1$.

That is to say, $p_{ii}(n) = 0$ unless $n$ is a multiple of $d(i)$, and $d(i)$ is maximal with this property.

**(11) Definition.** A state is called **ergodic** if it is recurrent, positive, and aperiodic.

**(12) Example. Random walk.** Corollary (5.3.4) and Problem (5.12.5) show that the states of the simple random walk are all periodic with period 2, and

(a)  transient, if $p \neq \frac{1}{2}$,

(b)  null recurrent, if $p = \frac{1}{2}$.                          ●

**(13) Example. Branching process.** Consider the branching process of Section 5.4 and suppose that $\mathbb{P}(Z_1 = 0) > 0$. Then 0 is called an *absorbing* state, because the chain never leaves it once it has visited it; all other states are transient.         ●

---

## Exercises for Section 6.2

**1.   Last exits.** Let $l_{ij}(n) = \mathbb{P}(X_n = j, X_k \neq i \text{ for } 1 \leq k < n \mid X_0 = i)$, the probability that the chain passes from $i$ to $j$ in $n$ steps without revisiting $i$. Writing

$$L_{ij}(s) = \sum_{n=1}^{\infty} s^n l_{ij}(n),$$

show that $P_{ij}(s) = P_{ii}(s)L_{ij}(s)$ if $i \neq j$. Deduce that the first passage times and last exit times have the same distribution for any Markov chain for which $P_{ii}(s) = P_{jj}(s)$ for all $i$ and $j$. Give an example of such a chain.

**2.   ** Let $X$ be a Markov chain containing an absorbing state $s$ with which all other states $i$ communicate, in the sense that $p_{is}(n) > 0$ for some $n = n(i)$. Show that all states other than $s$ are transient.

**3.   ** Show that a state $i$ is recurrent if and only if the mean number of visits of the chain to $i$, having started at $i$, is infinite. That is to say, $i$ is recurrent if and only if $\sum_n p_{ii}(n) = \infty$.

**4.   Visits.** Let $V_j = |\{n \geq 1 : X_n = j\}|$ be the number of visits of the Markov chain $X$ to $j$, and define $\eta_{ij} = \mathbb{P}_i(V_j = \infty)$. Show that:

(a)  $\eta_{ii} = \begin{cases} 1 & \text{if } i \text{ is recurrent,} \\ 0 & \text{if } i \text{ is transient,} \end{cases}$

(b)  $\eta_{ij} = \begin{cases} \mathbb{P}_i(T_j < \infty) & \text{if } j \text{ is recurrent,} \\ 0 & \text{if } j \text{ is transient,} \end{cases}$  where $T_j = \min\{n \geq 1 : X_n = j\}$.

**5.  Symmetry.** The distinct pair $i$, $j$ of states of a Markov chain is called *symmetric* if

$$\mathbb{P}_i(T_j < T_i) = \mathbb{P}_j(T_i < T_j),$$

where $T_i = \min\{n \geq 1 : X_n = i\}$. Show that, if $X_0 = i$ and $i$, $j$ is symmetric, the expected number of visits to $j$ before the chain revisits $i$ is 1. [Cf. the quotation following Theorem (3.10.18).]

**6.  Van Dantzig's collective marks.** Let $X$ be a Markov chain, and let $T$ be a geometric random variable with $\mathbb{P}(T > n) = s^n$ for $n \geq 0$, which is independent of $X$. By considering the expected number of visits by $X$ to a given state before time $T$, prove Theorem (6.2.3).

**7.  Constrained first-passage time.** Let $X$ be an ergodic Markov chain started from $a$, and assume $X$ is irreducible (in that, for states $i$, $j$, there exists $m \geq 0$ such that $p_{ij}(m) > 0$). Let $a, b, c$ be distinct states, and let $T(a, b, \neg c)$ be the time until $X$ first visits $b$, with no intermediate visit to $c$. That is, if $X$ visits $b$ before $c$, then $T(a, b, \neg c)$ is that time, and if $X$ visits $c$ before $b$ we set $T(a, b, \neg c) = \infty$. Let

$$G(a, b, \neg c; s) = \sum_{n=1}^{\infty} s^n \mathbb{P}_a\big(T(a, b, \neg c) = n\big).$$

Show that

$$G(a, b, \neg c; s) = \frac{F_{ab} - F_{ac}F_{cb}}{1 - F_{bc}F_{cb}},$$

where, for example, $F_{ab}(s)$ is the probability generating function of the first passage time $T_{ab}$ from $a$ to $b$ irrespective of intermediate visits to $c$. Show further that

$$\mathbb{P}_a\big(T(a, b, \neg c) < \infty\big) = \frac{\mu_{ac} + \mu_{cb} - \mu_{ab}}{\mu_{bc} + \mu_{cb}},$$

where, for example, $\mu_{ab} = \mathbb{E}_a(T_{ab})$.

---

# 6.3  Classification of chains

We consider next the ways in which the states of a Markov chain are related to one other. This investigation will help us to achieve a full classification of the states in the language of the previous section.

**(1) Definition.** We say $i$ **communicates with** $j$, written $i \rightarrow j$, if the chain visits state $j$ with strictly positive probability, having started from $i$. That is, $i \rightarrow j$ if $p_{ij}(m) > 0$ for some $m \geq 0$. We say $i$ and $j$ **intercommunicate** if $i \rightarrow j$ and $j \rightarrow i$, in which case we write $i \leftrightarrow j$.†

If $i \neq j$, then $i \rightarrow j$ if and only if $f_{ij} > 0$. Clearly $i \rightarrow i$ since $p_{ii}(0) = 1$, and it follows that $\leftrightarrow$ is an equivalence relation (*exercise*: if $i \leftrightarrow j$ and $j \leftrightarrow k$, show that $i \leftrightarrow k$). The state space $S$ can be partitioned into the equivalence classes of $\leftrightarrow$. Within each equivalence class all states are of the same type.

**(2) Theorem.** *If $i \leftrightarrow j$ then:*

  (a) *$i$ and $j$ have the same period,*

---

†Some authors say that state $j$ is 'reachable' or 'accessible' from $i$ if $i \rightarrow j$, while $i$ 'communicates' with $j$ if $i \leftrightarrow j$.

(b) $i$ is transient if and only if $j$ is transient,

(c) $i$ is null recurrent if and only if $j$ is null recurrent.

**Proof.** (a) For $k \in S$, let $D_k = \{m \geq 1 : p_{kk}(m) > 0\}$, so that the period of $k$ is $d(k) = \gcd(D_k)$. Let $i \leftrightarrow j$ and find $m, n \geq 1$ such that $\alpha := p_{ij}(m)p_{ji}(n) > 0$. By the Chapman–Kolmogorov equations (6.1.7),

$$p_{ii}(m + r + n) \geq \alpha p_{jj}(r), \qquad r \geq 0.$$

Thus $d(i) \mid m + r + n$ for $r \in \{0\} \cup D_j$. In particular, $d(i) \mid m + n$, and hence $d(i) \mid r$ for $r \in D_j$. Therefore, $d(i) \mid d(j)$. Similarly $d(j) \mid d(i)$, and therefore $d(i) = d(j)$.

(b) If $i \leftrightarrow j$ then there exist $m, n \geq 0$ such that $\alpha = p_{ij}(m)p_{ji}(n) > 0$. By the Chapman–Kolmogorov equations,

$$p_{ii}(m + r + n) \geq p_{ij}(m)p_{jj}(r)p_{ji}(n) = \alpha p_{jj}(r),$$

for any non-negative integer $r$. Now sum over $r$ to obtain

$$\sum_r p_{jj}(r) < \infty \quad \text{if} \quad \sum_r p_{ii}(r) < \infty.$$

Thus, by Corollary (6.2.4), $j$ is transient if $i$ is transient. The converse holds similarly and (b) is shown.

(c) We defer this until the next section. A possible route is by way of Theorem (6.2.9), but we prefer to proceed differently in order to avoid the danger of using a circular argument.    ∎

**(3) Definition.** A set $C$ of states is called:

  (a) **closed** if $p_{ij} = 0$ for all $i \in C$, $j \notin C$,

  (b) **irreducible** if $i \leftrightarrow j$ for all $i, j \in C$.

Once the chain arrives in a closed set $C$ of states then it never leaves $C$ subsequently. A closed set containing exactly one state is called *absorbing*; for example, the state 0 is absorbing for the branching process. It is clear that the equivalence classes of $\leftrightarrow$ are irreducible. We call an irreducible set $C$ aperiodic (or *recurrent, null*, and so on) if all the states in $C$ have this property; Theorem (2) ensures that this is meaningful. If the whole state space $S$ is irreducible, then we say that the chain, and its transition matrix $\mathbf{P}$, are themselves irreducible.

**(4) Decomposition theorem.** *The state space $S$ can be partitioned uniquely as*

$$S = T \cup C_1 \cup C_2 \cup \cdots$$

*where $T$ is the set of transient states, and the $C_i$ are irreducible closed sets of recurrent states.*

**Proof.** Let $C_1, C_2, \ldots$ be the recurrent equivalence classes of $\leftrightarrow$. We need only show that each $C_r$ is closed. Suppose on the contrary that there exist $i \in C_r$, $j \notin C_r$, such that $p_{ij} > 0$. Now $j \nrightarrow i$, and therefore

$$\mathbb{P}_i(X_n \neq i \text{ for all } n \geq 1) \geq \mathbb{P}_i(X_1 = j) > 0,$$

in contradiction of the assumption that $i$ is recurrent.    ∎

The decomposition theorem clears the air a little. On the one hand, if $X_0 \in C_r$, say, the chain never leaves $C_r$ and we might as well take $C_r$ to be the whole state space. On the other hand, if $X_0 \in T$, then the chain either stays in $T$ forever or moves eventually to one of the $C_k$ where it subsequently remains. Thus, either the chain always takes values in the set of transient states or it lies eventually in some irreducible closed set of recurrent states. For the special case when $S$ is finite the first of these possibilities cannot occur.

**(5) Lemma.** *If $S$ is finite, then at least one state is recurrent and all recurrent states are positive.*

**Proof.** If all states are transient, then take the limit through the summation sign to obtain the contradiction

$$1 = \lim_{n \to \infty} \sum_j p_{ij}(n) = 0$$

by Corollary (6.2.5). The same contradiction arises by Theorem (6.2.9) for the closed set of all null recurrent states, should this set be non-empty.  ∎

**(6) Example.** Let $S = \{1, 2, 3, 4, 5, 6\}$ and

$$\mathbf{P} = \begin{pmatrix} \frac{1}{2} & \frac{1}{2} & 0 & 0 & 0 & 0 \\ \frac{1}{4} & \frac{3}{4} & 0 & 0 & 0 & 0 \\ \frac{1}{4} & \frac{1}{4} & \frac{1}{4} & \frac{1}{4} & 0 & 0 \\ \frac{1}{4} & 0 & \frac{1}{4} & \frac{1}{4} & 0 & \frac{1}{4} \\ 0 & 0 & 0 & 0 & \frac{1}{2} & \frac{1}{2} \\ 0 & 0 & 0 & 0 & \frac{1}{2} & \frac{1}{2} \end{pmatrix}.$$

The sets $\{1, 2\}$ and $\{5, 6\}$ are irreducible and closed, and therefore contain recurrent positive states. States 3 and 4 are transient because $3 \to 4 \to 6$ but return from 6 is impossible. All states have period 1 because $p_{ii}(1) > 0$ for all $i$. Hence, 3 and 4 are transient, and 1, 2, 5, and 6 are ergodic. Easy calculations give

$$f_{11}(n) = \begin{cases} p_{11} = \frac{1}{2} & \text{if } n = 1, \\ p_{12}(p_{22})^{n-2} p_{21} = \frac{1}{2}(\frac{3}{4})^{n-2}\frac{1}{4} & \text{if } n \ge 2, \end{cases}$$

and hence $\mu_1 = \sum_n n f_{11}(n) = 3$. Other mean recurrence times can be found similarly. The next section gives another way of finding the $\mu_i$ which usually requires less computation.  ●

---

## Exercises for Section 6.3

**1.** Let $X$ be a Markov chain on $\{0, 1, 2, \ldots\}$ with transition matrix given by $p_{0j} = a_j$ for $j \ge 0$, $p_{ii} = r$ and $p_{i,i-1} = 1 - r$ for $i \ge 1$. Classify the states of the chain, and find their mean recurrence times.

**2.** Determine whether or not the random walk on the integers having transition probabilities $p_{i,i+2} = p$, $p_{i,i-1} = 1 - p$, for all $i$, is recurrent.

**3.**   Classify the states of the Markov chains with transition matrices

(a)
$$\begin{pmatrix} 1-2p & 2p & 0 \\ p & 1-2p & p \\ 0 & 2p & 1-2p \end{pmatrix},$$

(b)
$$\begin{pmatrix} 0 & p & 0 & 1-p \\ 1-p & 0 & p & 0 \\ 0 & 1-p & 0 & p \\ p & 0 & 1-p & 0 \end{pmatrix}.$$

In each case, calculate $p_{ij}(n)$ and the mean recurrence times of the states.

**4.**   A particle performs a random walk on the vertices of a cube. At each step it remains where it is with probability $\frac{1}{4}$, or moves to one of its neighbouring vertices each having probability $\frac{1}{4}$. Let $v$ and $w$ be two diametrically opposite vertices. If the walk starts at $v$, find:
(a) the mean number of steps until its first return to $v$,
(b) the mean number of steps until its first visit to $w$,
(c) the mean number of visits to $w$ before its first return to $v$.

**5.   Visits.** With the notation of Exercise (6.2.4), show that
(a) if $i \to j$ and $i$ is recurrent, then $\eta_{ij} = \eta_{ji} = 1$,
(b) $\eta_{ij} = 1$ if and only if $\mathbb{P}_i(T_j < \infty) = \mathbb{P}_j(T_j < \infty) = 1$.

**6.   Hitting probabilities.** Let $T_A = \min\{n \geq 0 : X_n \in A\}$, where $X$ is a Markov chain and $A$ is a subset of the state space $S$, and let $\eta_j = \mathbb{P}_j(T_A < \infty)$. Show that

$$\eta_j = \begin{cases} 1 & \text{if } j \in A, \\ \sum_{k \in S} p_{jk}\eta_k & \text{if } j \notin A. \end{cases}$$

Show further that if $\mathbf{x} = (x_j : j \in S)$ is any non-negative solution of these equations then $x_j \geq \eta_j$ for all $j$.

**7.   Mean hitting times.** In the notation of Exercise (6.3.6), let $\rho_j = \mathbb{E}_j(T_A)$. Show that

$$\rho_j = \begin{cases} 0 & \text{if } j \in A, \\ 1 + \sum_{k \in S} p_{jk}\rho_k & \text{if } j \notin A, \end{cases}$$

and that if $\mathbf{x} = (x_j : j \in S)$ is any non-negative solution of these equations then $x_j \geq \rho_j$ for all $j$.

**8.**   Let $X$ be an irreducible Markov chain and let $A$ be a subset of the state space. Let $S_r$ and $T_r$ be the successive times at which the chain enters $A$ and visits $A$ respectively. Are the sequences $\{X_{S_r} : r \geq 1\}$, $\{X_{T_r} : r \geq 1\}$ Markov chains? What can be said about the times at which the chain exits $A$?

**9.**   (a) Show that for each pair $i$, $j$ of states of an irreducible aperiodic chain, there exists $N = N(i, j)$ such that $p_{ij}(r) > 0$ for all $r \geq N$.
(b) Show that there exists a function $f$ such that, if $\mathbf{P}$ is the transition matrix of an irreducible aperiodic Markov chain with $n$ states, then $p_{ij}(r) > 0$ for all states $i$, $j$, and all $r \geq f(n)$.
(c) Show further that $f(4) \geq 6$ and $f(n) \geq (n-1)(n-2)$.
[Hint: The postage stamp lemma asserts that, for $a$, $b$ coprime, the smallest $n$ such that all integers strictly exceeding $n$ have the form $\alpha a + \beta b$ for some integers $\alpha, \beta \geq 0$ is $(a-1)(b-1)$.]

**10.**   An urn initially contains $n$ green balls and $n + 2$ red balls. A ball is picked at random: if it is green then a red ball is also removed and both are discarded; if it is red then it is replaced together with an extra red and an extra green ball. This is repeated until there are no green balls in the urn. Show that the probability the process terminates is $1/(n+1)$.

Now reverse the rules: if the ball is green, it is replaced together with an extra green and an extra red ball; if it is red it is discarded along with a green ball. Show that the expected number of iterations until no green balls remain is $\sum_{j=1}^{n}(2j+1) = n(n+2)$. [Thus, a minor perturbation of a simple symmetric random walk can be positive recurrent, whereas the original is null recurrent.]

---

## 6.4 Stationary distributions and the limit theorem

How does a Markov chain $X$ behave after a long time $n$ has elapsed? The sequence $\{X_n\}$ cannot generally converge to some particular state $s$ since it enjoys the inherent random fluctuation which is specified by the transition matrix. However, we might hold out some hope that the *distribution* of $X_n$ settles down. Indeed, subject to certain conditions this turns out to be the case. The classical study of limiting distributions proceeds by algebraic manipulation of the generating functions of Theorem (6.2.3); we shall avoid this here, contenting ourselves for the moment with results which are not quite the best possible but which have attractive probabilistic proofs. This section is in two parts, dealing respectively with stationary distributions and limit theorems.

**(A) Stationary distributions.** We shall see that the existence of a limiting distribution for $X_n$, as $n \to \infty$, is closely bound up with the existence of so-called 'stationary distributions'.

**(1) Definition.** The vector $\pi$ is called a **stationary distribution** of the chain if $\pi$ has entries $(\pi_j : j \in S)$ such that:
  (a) $\pi_j \geq 0$ for all $j$, and $\sum_j \pi_j = 1$,
  (b) $\pi = \pi\mathbf{P}$, which is to say that $\pi_j = \sum_i \pi_i p_{ij}$ for all $j$.

Such a distribution is called stationary† for the following reason. Iterate (1b) to obtain $\pi\mathbf{P}^2 = (\pi\mathbf{P})\mathbf{P} = \pi\mathbf{P} = \pi$, and so

$$(2) \qquad\qquad \pi\mathbf{P}^n = \pi \qquad \text{for all } n \geq 0.$$

Now use Lemma (6.1.8) to see that if $X_0$ has distribution $\pi$ then $X_n$ has distribution $\pi$ for all $n$, showing that the distribution of $X_n$ is 'stationary' (or 'invariant') as time passes; in such a case, of course, $\pi$ is also the limiting distribution of $X_n$ as $n \to \infty$.

Following the discussion after the decomposition theorem (6.3.4), we shall assume henceforth that the chain is irreducible and shall investigate the existence of stationary distributions. No assumption of aperiodicity is required at this stage.

**(3) Theorem.** *An irreducible chain has a stationary distribution $\pi$ if and only if all the states are positive recurrent. In this case, $\pi$ is the unique stationary distribution and is given by $\pi_i = \mu_i^{-1}$ for each $i \in S$, where $\mu_i$ is the mean recurrence time of $i$.*

Stationary distributions $\pi$ satisfy $\pi = \pi\mathbf{P}$. A vector $\mathbf{x} = (x_j : j \in S)$ is called a *measure* if $\mathbf{x} \neq \mathbf{0}$ and $x_j \geq 0$ for all $j$. Furthermore, the measure $\mathbf{x}$ is called *stationary* if $\mathbf{x}\mathbf{P} = \mathbf{x}$. We

---

†The term 'invariant distribution' is frequently used instead.

may display a stationary measure **x** explicitly as follows, whenever the chain is irreducible and recurrent.

Suppose the chain is irreducible. Fix $k \in S$ and let $\rho_i(k)$ be the mean number of visits to the state $i$ prior to a return to the starting state $k$. More precisely, $\rho_i(k) = \mathbb{E}_k(N_i)$ where

$$ N_i = \sum_{n=0}^{\infty} I(X_n = i, \ T_k > n) \tag{4} $$

and $T_k = \min\{r \geq 1 : X_r = k\}$ is the first passage time to state $k$ (note that, for the moment, we allow the possibility that $\mathbb{P}_k(T_k = \infty) > 0$). Clearly, $N_k = 1$ so that $\rho_k(k) = 1$, and

$$ \rho_i(k) = \sum_{n=0}^{\infty} \mathbb{P}_k(X_n = i, \ T_k > n). \tag{5} $$

Let $\rho(k) = (\rho_i(k) : i \in S)$. Regardless of whether or not $T_k < \infty$, we have that $T_k = \sum_{i \in S} N_i$, since the time prior to return to $k$ must be spent somewhere.

If $k$ is recurrent then $\mathbb{P}_k(T_k < \infty) = 1$. In this case, we find on taking expectations that

$$ \mu_k = \sum_{i \in S} \rho_i(k). \tag{6} $$

**(7) Lemma.** *Consider an irreducible chain with transition matrix* **P**, *and fix $k \in S$.*
(a) *Suppose $k$ is recurrent, so that (6) holds. The vector $\rho(k)$ satisfies $\rho_k(k) = 1$, and $\rho_i(k) < \infty$ for all $i$. Furthermore $\rho(k) = \rho(k)$**P**, and so $\rho(k)$ is a stationary measure.*
(b) *Let* **x** *be a stationary measure such that $x_k = 1$. Then* **x** $\geq \rho(k)$. *If, in addition, the chain is recurrent, then* **x** $= \rho(k)$.
(c) *Let* **x** *be a stationary measure. Then $x_j > 0$ for $j \in S$.*

Part (b) will be useful in the treatment of continuous-time chains in Section 6.10.

**Proof.** (a) Write
$$ l_{ki}(n) = \mathbb{P}_k(X_n = i, \ T_k \geq n), $$

the probability that the chain reaches $i$ in $n$ steps but with no intermediate return to $k$.

Let $k$ be recurrent, so that $\mathbb{P}_k(T_k < \infty) = 1$. In this case, (5) may be expressed in the form

$$ \rho_i(k) = \sum_{n=1}^{\infty} l_{ki}(n). \tag{8} $$

We show first that $\rho_i(k) < \infty$ when $i \neq k$. Clearly, $f_{kk}(n+r) \geq l_{ki}(n)f_{ik}(r)$; this holds since the first return time to $k$ equals $n+r$ if: (a) $X_n = i$, (b) there is no return to $k$ up to time $n$, and (c) the next subsequent visit to $k$ takes place after another $r$ steps. By the irreducibility of the chain, there exists $r \geq 1$ such that $f_{ik}(r) > 0$. With this choice of $r$, we have that $l_{ki}(n) \leq f_{kk}(n+r)/f_{ik}(r)$, and so

$$ \rho_i(k) = \sum_{n=1}^{\infty} l_{ki}(n) \leq \frac{1}{f_{ik}(r)} \sum_{n=1}^{\infty} f_{kk}(n+r) \leq \frac{1}{f_{ik}(r)} < \infty. $$

For the stationarity, we argue as follows. We have that $l_{ki}(1) = p_{ki}$. By conditioning on the value of $X_{n-1}$,

$$l_{ki}(n) = \sum_{j:j\neq k} \mathbb{P}_k(X_n = i,\ X_{n-1} = j,\ T_k \geq n) = \sum_{j:j\neq k} l_{kj}(n-1)p_{ji} \qquad \text{for } n \geq 2.$$

Summing over $n$, we obtain

$$\rho_i(k) = p_{ki} + \sum_{j:j\neq k}\left(\sum_{n\geq 2} l_{kj}(n-1)\right)p_{ji} = \rho_k(k)p_{ki} + \sum_{j:j\neq k}\rho_j(k)p_{ji}$$

since $\rho_k(k) = 1$. Part (a) is proved.

(b) This is deferred to Exercise (6.4.17).

(c) Let $\mathbf{x}$ be a stationary measure, and suppose $j$ is such that $x_j = 0$. Since $\mathbf{x} = \mathbf{x}\mathbf{P}^n$ for $n \geq 1$, we have that

$$0 = x_j = \sum_i x_i p_{ij}(n) \geq x_i p_{ij}(n) \qquad \text{for all } i \text{ and } n,$$

yielding that $x_i = 0$ whenever $i \to j$. The chain is assumed irreducible, so that $x_i = 0$ for all $i$, in contradiction of the fact that the $x_i$ have sum 1.  ∎

By equation (6) and Lemma (7a), for an irreducible recurrent chain, the vector $\rho(k)$ satisfies $\rho(k) = \rho(k)\mathbf{P}$, and furthermore the components of $\rho(k)$ are non-negative with sum $\mu_k$. Hence, if $\mu_k < \infty$, the vector $\pi$ with entries $\pi_i = \rho_i(k)/\mu_k$ satisfies $\pi = \pi\mathbf{P}$ and furthermore has non-negative entries which sum to 1; that is to say, $\pi$ is a stationary distribution. We have proved that every positive recurrent irreducible chain has a stationary distribution, an important step towards the proof of the main theorem (3).

Before continuing with the rest of the proof of (3), we note a consequence of the results so far. Lemma (7a, c) implies the existence of a measure $\mathbf{x}$ satisfying the equation $\mathbf{x} = \mathbf{x}\mathbf{P}$ whenever the chain is irreducible and recurrent. By Lemma (7b) this root is unique up to scalar multiplication, and we arrive therefore at the following useful conclusion.

**(9) Theorem.** *If the chain is irreducible and recurrent, there exists a measure $\mathbf{x}$ satisfying the equation $\mathbf{x} = \mathbf{x}\mathbf{P}$ which is unique up to a multiplicative constant, and has strictly positive entries. The chain is positive if $\sum_i x_i < \infty$ and null if $\sum_i x_i = \infty$.*

**Proof of Theorem (3).** Suppose that $\pi$ is a stationary distribution of the chain. If all states are transient then $p_{ij}(n) \to 0$, as $n \to \infty$, for all $i$ and $j$ by Corollary (6.2.5). From (2),

$$(10) \qquad \pi_j = \sum_i \pi_i p_{ij}(n) \to 0 \quad \text{as} \quad n \to \infty, \qquad \text{for all } i \text{ and } j,$$

which contradicts (1a). Thus all states are recurrent. To see the validity of the limit in (10)†, let $F$ be a finite subset of $S$ and write

$$\sum_i \pi_i p_{ij}(n) \leq \sum_{i\in F} \pi_i p_{ij}(n) + \sum_{i\notin F} \pi_i$$

$$\to \sum_{i\notin F} \pi_i \quad \text{as } n \to \infty, \quad \text{since } F \text{ is finite}$$

$$\to 0 \quad \text{as} \quad F \uparrow S.$$

---

†This argument is a form of the bounded convergence theorem (5.6.12) applied to sums instead of to integrals. We shall make repeated use of this technique.

We show next that the existence of $\pi$ implies that all states are positive and that $\pi_i = \mu_i^{-1}$ for each $i$. Suppose that $X_0$ has distribution $\pi$, so that $\mathbb{P}(X_0 = i) = \pi_i$ for each $i$. By Problem (3.11.13a),

$$\pi_j \mu_j = \sum_{n=1}^{\infty} \mathbb{P}(X_0 = j)\mathbb{P}_j(T_j \geq n) = \sum_{n=1}^{\infty} \mathbb{P}(T_j \geq n, \ X_0 = j).$$

However, $\mathbb{P}(T_j \geq 1, \ X_0 = j) = \mathbb{P}(X_0 = j)$, and for $n \geq 2$

$$\begin{aligned}
\mathbb{P}(T_j \geq n, X_0 = j) &= \mathbb{P}(X_0 = j, \ X_m \neq j \text{ for } 1 \leq m \leq n-1)\\
&= \mathbb{P}(X_m \neq j \text{ for } 1 \leq m \leq n-1) - \mathbb{P}(X_m \neq j \text{ for } 0 \leq m \leq n-1)\\
&= \mathbb{P}(X_m \neq j \text{ for } 0 \leq m \leq n-2) - \mathbb{P}(X_m \neq j \text{ for } 0 \leq m \leq n-1)\\
&\qquad \text{by homogeneity}\\
&= a_{n-2} - a_{n-1}
\end{aligned}$$

where $a_n = \mathbb{P}(X_m \neq j \text{ for } 0 \leq m \leq n)$. Sum over $n$ to obtain

$$\pi_j \mu_j = \mathbb{P}(X_0 = j) + \mathbb{P}(X_0 \neq j) - \lim_{n\to\infty} a_n = 1 - \lim_{n\to\infty} a_n.$$

However, $a_n \to \mathbb{P}(X_m \neq j \text{ for all } m) = 0$ as $n \to \infty$, by the recurrence of $j$ (and surreptitious use of Problem (6.15.6)). We have shown that

**(11)**                                           $\pi_j \mu_j = 1,$

so that $\mu_j = \pi_j^{-1} < \infty$ by Lemma (7c). In conclusion, all states of the chain are positive, and (11) specifies $\pi_j$ uniquely as $\mu_j^{-1}$.

Thus, if $\pi$ exists then it is unique and all the states of the chain are positive recurrent. Conversely, if the states of the chain are positive recurrent then the chain has a stationary distribution given by (11). ∎

We may now complete the proof of Theorem (6.3.2c).

**Proof of (6.3.2c).** Let $C(i)$ be the irreducible closed equivalence class of states which contains the positive recurrent state $i$. Suppose that $X_0 \in C(i)$. Then $X_n \in C(i)$ for all $n$, and (7) and (3) combine to tell us that all states in $C(i)$ are positive. ∎

**(12) Example (6.3.6) revisited.** To find $\mu_1$ and $\mu_2$ consider the irreducible closed set $C = \{1, 2\}$. If $X_0 \in C$, then solve the equation $\pi = \pi \mathbf{P}_C$ for $\pi = (\pi_1, \pi_2)$ in terms of

$$\mathbf{P}_C = \begin{pmatrix} \frac{1}{2} & \frac{1}{2} \\ \frac{1}{4} & \frac{3}{4} \end{pmatrix}$$

to find the unique stationary distribution $\pi = (\frac{1}{3}, \frac{2}{3})$, giving that $\mu_1 = \pi_1^{-1} = 3$ and $\mu_2 = \pi_2^{-1} = \frac{3}{2}$. Now find the other mean recurrence times yourself (*exercise*). ●

Theorem (3) provides a useful criterion for deciding whether or not an irreducible chain is positive recurrent: just look for a stationary distribution†. There is a similar criterion for the transience of irreducible chains.

**(13) Theorem.** *Let* $s \in S$ *be any state of an irreducible chain. The chain is transient if and only if there exists a non-zero solution* $\{y_j : j \neq s\}$, *satisfying* $|y_j| \leq 1$ *for all* $j$, *to the equations*

**(14)**
$$y_i = \sum_{j:j\neq s} p_{ij} y_j, \qquad i \neq s.$$

**Proof.** The chain is transient if and only if $s$ is transient. First suppose $s$ is transient and define

**(15)**
$$\tau_i(n) = \mathbb{P}_i(\text{no visit to } s \text{ in first } n \text{ steps})$$
$$= \mathbb{P}_i(X_m \neq s \text{ for } 1 \leq m \leq n).$$

Then
$$\tau_i(1) = \sum_{j:j\neq s} p_{ij}, \qquad \tau_i(n+1) = \sum_{j:j\neq s} p_{ij} \tau_j(n).$$

Furthermore, $\tau_i(n) \geq \tau_i(n+1)$, and so

$$\tau_i = \lim_{n\to\infty} \tau_i(n) = \mathbb{P}_i(\text{no visit to } s \text{ ever}) = 1 - f_{is}$$

satisfies (14). (Can *you* prove this? Use the method of proof of (10).) Also $\tau_i > 0$ for some $i$, since otherwise $f_{is} = 1$ for all $i \neq s$, and therefore

$$f_{ss} = p_{ss} + \sum_{i:i\neq s} p_{si} f_{is} = \sum_i p_{si} = 1$$

by conditioning on $X_1$. This contradicts the transience of $s$.

Conversely, let $\mathbf{y} \neq \mathbf{0}$ satisfy (14) with $|y_i| \leq 1$. Then

$$|y_i| \leq \sum_{j:j\neq s} p_{ij}|y_j| \leq \sum_{j:j\neq s} p_{ij} = \tau_i(1),$$
$$|y_i| \leq \sum_{j:j\neq s} p_{ij} \tau_j(1) = \tau_i(2),$$

and so on, where the $\tau_i(n)$ are given by (15). Thus $|y_i| \leq \tau_i(n)$ for all $n$. Let $n \to \infty$ to show that

$$\tau_i = \lim_{n\to\infty} \tau_i(n) \geq |y_i| > 0$$

for some $i$, which implies that $s$ is transient by the result of Problem (6.15.6).  ∎

---

†We emphasize that a stationary distribution is a *left* eigenvector of the transition matrix, not a *right* eigenvector.

This theorem provides a necessary and sufficient condition for recurrence: an irreducible chain is recurrent if and only if the only bounded solution to (14) is the zero solution. This combines with (3) to give a condition for null recurrence. Another condition is the following (see Exercise (6.4.10)); a corresponding result holds for any countably infinite state space $S$.

**(16) Theorem.** *Let $s \in S$ be any state of an irreducible chain on $S = \{0, 1, 2, \dots\}$. The chain is recurrent if there exists a solution $\{y_j : j \neq s\}$ to the inequalities*

(17)
$$y_i \geq \sum_{j:j\neq s} p_{ij} y_j, \qquad i \neq s,$$

*such that $y_i \to \infty$ as $i \to \infty$.*

**(18) Example. Random walk with retaining barrier.** A particle performs a random walk on the non-negative integers with a retaining barrier at 0. The transition probabilities are

$$p_{0,0} = q, \qquad p_{i,i+1} = p \quad \text{for } i \geq 0, \qquad p_{i,i-1} = q \quad \text{for } i \geq 1,$$

where $p = 1 - q \in (0, 1)$. Let $\rho = p/q$.
  (a) If $q < p$, take $s = 0$ to see that $y_j = 1 - \rho^{-j}$ satisfies (14), and so the chain is transient.
  (b) Solve the equation $\pi = \pi P$ to find that there exists a stationary distribution, with $\pi_j = \rho^j (1 - \rho)$, if and only if $q > p$. Thus the chain is positive recurrent if and only if $q > p$.
  (c) If $q = p = \frac{1}{2}$, take $s = 0$ in (16) and check that $y_j = j$, $j \geq 1$, solves (14). Thus the chain is null recurrent. Alternatively, argue as follows. The chain is recurrent since symmetric random walk is recurrent (just reflect negative excursions of a symmetric random walk into the positive half-line). Solve the equation $\mathbf{x} = \mathbf{x}P$ to find that $x_i = 1$ for all $i$ provides a root, unique up to a multiplicative constant. However, $\sum_i x_i = \infty$ so that the chain is null, by Theorem (9).
These conclusions match our intuitions well.                                           ●

**(B) Limit theorems.** Next we explore the link between the existence of a stationary distribution and the limiting behaviour of the probabilities $p_{ij}(n)$ as $n \to \infty$. The following example indicates a difficulty which arises from periodicity.

**(19) Example.** If $S = \{1, 2\}$ and $p_{12} = p_{21} = 1$, then

$$p_{11}(n) = p_{22}(n) = \begin{cases} 0 & \text{if } n \text{ is odd,} \\ 1 & \text{if } n \text{ is even.} \end{cases}$$

Clearly $p_{ii}(n)$ does not converge as $n \to \infty$; the reason is that both states are periodic with period 2.                                                                              ●

Until further notice we shall deal only with irreducible *aperiodic* chains. The principal result is the following theorem.

**(20) Markov chain limit theorem.** *For an irreducible aperiodic chain, we have that*

$$p_{ij}(n) \to \frac{1}{\mu_j} \quad \text{as} \quad n \to \infty, \qquad \text{for all } i \text{ and } j.$$

We make the following remarks.

(a) If the chain is *transient* or *null recurrent* then $p_{ij}(n) \to 0$ for all $i$ and $j$, since $\mu_j = \infty$. We are now in a position to prove Theorem (6.2.9). Let $C(i)$ be the irreducible closed set of states which contains the recurrent state $i$. If $C(i)$ is aperiodic then the result is an immediate consequence of (20); the periodic case can be treated similarly, but with slightly more difficulty (see note (d) following).

(b) If the chain is *positive recurrent* then $p_{ij}(n) \to \pi_j = \mu_j^{-1}$, where $\pi$ is the unique stationary distribution by (3).

(c) It follows from (20) that the limit probability, $\lim_{n \to \infty} p_{ij}(n)$, does not depend on the starting point $X_0 = i$; that is, the chain forgets its origin. It is now easy to check that

$$\mathbb{P}(X_n = j) = \sum_i \mathbb{P}(X_0 = i) p_{ij}(n) \to \frac{1}{\mu_j} \quad \text{as} \quad n \to \infty$$

by Lemma (6.1.8), irrespective of the distribution of $X_0$.

(d) If $X = \{X_n\}$ is an irreducible chain with period $d$, then $Y = \{Y_n = X_{nd} : n \geq 0\}$ is an aperiodic chain, and it follows that

$$p_{jj}(nd) = \mathbb{P}_j(Y_n = j) \to \frac{d}{\mu_j} \quad \text{as} \quad n \to \infty.$$

**Proof of (20).** If the chain is transient then the result holds from Corollary (6.2.5). The recurrent case is treated by an important technique known as 'coupling' which we met first in Section 4.12. Construct a 'coupled chain' $Z = (X, Y)$, being an ordered pair $X = \{X_n : n \geq 0\}$, $Y = \{Y_n : n \geq 0\}$ of *independent* Markov chains, each having state space $S$ and transition matrix $\mathbf{P}$. Then $Z = \{Z_n = (X_n, Y_n) : n \geq 0\}$ takes values in $S \times S$, and it is easy to check that $Z$ is a Markov chain with transition probabilities

$$p_{ij,kl} = \mathbb{P}(Z_{n+1} = (k, l) \mid Z_n = (i, j))$$
$$= \mathbb{P}(X_{n+1} = k \mid X_n = i)\mathbb{P}(Y_{n+1} = l \mid Y_n = j) \quad \text{by independence}$$
$$= p_{ik} p_{jl}.$$

Since $X$ is irreducible and aperiodic, for any states $i, j, k, l$ there exists $N = N(i, j, k, l)$ such that $p_{ik}(n) p_{jl}(n) > 0$ for all $n \geq N$; thus $Z$ also is irreducible (see Exercise (6.3.9) or Problem (6.15.4); *only here* do we require that $X$ be aperiodic).

Suppose that $X$ is positive recurrent. Then $X$ has a unique stationary distribution $\pi$, by (3), and it is easy to see that $Z$ has a stationary distribution $\nu = (\nu_{ij} : i, j \in S)$ given by $\nu_{ij} = \pi_i \pi_j$; thus $Z$ is also positive recurrent, by (3). Now, suppose that $X_0 = i$ and $Y_0 = j$, so that $Z_0 = (i, j)$. Choose any state $s \in S$ and let

$$T = \min\{n \geq 1 : Z_n = (s, s)\}$$

denote the time of the first passage of $Z$ to $(s, s)$; from Problem (6.15.6) and the recurrence of $Z$, $\mathbb{P}(T < \infty) = 1$. The central idea of the proof is the following observation. If $m \leq n$ and $X_m = Y_m = s$, then the conditional distributions of $X_n$ and $Y_n$ are identical, since the distributions of $X_n$ and $Y_n$ depend only upon the shared transition matrix $\mathbf{P}$ and upon the shared value of the chains at the $m$th stage. We shall use this fact, together with the finiteness of $T$, to show that the ultimate distributions of $X$ and $Y$ are independent of their starting points. More precisely, starting from $Z_0 = (X_0, Y_0) = (i, j)$,

$$p_{ik}(n) = \mathbb{P}(X_n = k)$$
$$= \mathbb{P}(X_n = k, \ T \le n) + \mathbb{P}(X_n = k, \ T > n)$$
$$= \mathbb{P}(Y_n = k, \ T \le n) + \mathbb{P}(X_n = k, \ T > n)$$

because, given that $T \le n$, $X_n$ and $Y_n$ are identically distributed

$$\le \mathbb{P}(Y_n = k) + \mathbb{P}(T > n)$$
$$= p_{jk}(n) + \mathbb{P}(T > n).$$

This, and the related inequality with $i$ and $j$ interchanged, yields the 'coupling inequality'

$$|p_{ik}(n) - p_{jk}(n)| \le \mathbb{P}(T > n) \to 0 \quad \text{as} \quad n \to \infty$$

because $\mathbb{P}(T < \infty) = 1$; therefore,

(21)                    $p_{ik}(n) - p_{jk}(n) \to 0 \quad \text{as} \quad n \to \infty \quad$ for all $i$, $j$, and $k$.

Thus, if the limit $\lim_{n\to\infty} p_{ik}(n)$ exists, then it does not depend on $i$. To show that it exists, write

(22)                    $\pi_k - p_{jk}(n) = \sum_i \pi_i \left( p_{ik}(n) - p_{jk}(n) \right) \to 0 \quad \text{as} \quad n \to \infty,$

giving the result. To see that the limit in (22) follows from (21), use the bounded convergence argument in the proof of (10); for any finite subset $F$ of $S$,

$$\sum_i \pi_i |p_{ik}(n) - p_{jk}(n)| \le \sum_{i \in F} |p_{ik}(n) - p_{jk}(n)| + 2 \sum_{i \notin F} \pi_i$$
$$\to 2 \sum_{i \notin F} \pi_i \quad \text{as} \quad n \to \infty$$

which in turn tends to zero as $F \uparrow S$.

Finally, suppose that $X$ is null recurrent; the argument is a little trickier in this case. If $Z$ is transient, then from Corollary (6.2.5) applied to $Z$,

$$\mathbb{P}(Z_n = (j, j) \mid Z_0 = (i, i)) = p_{ij}(n)^2 \to 0 \quad \text{as} \quad n \to \infty$$

and the result holds. If $Z$ is positive recurrent then, starting from $Z_0 = (i, i)$, the epoch $T_{ii}^Z$ of the first return of $Z$ to $(i, i)$ is no smaller than the epoch $T_i$ of the first return of $X$ to $i$; however, $\mathbb{E}(T_i) = \infty$ and $\mathbb{E}(T_{ii}^Z) < \infty$ which is a contradiction. Lastly, suppose that $Z$ is null recurrent. The argument which leads to (21) still holds, and we wish to deduce that

$$p_{ij}(n) \to 0 \quad \text{as} \quad n \to \infty \quad \text{for all } i \text{ and } j.$$

If this does not hold then there exists a subsequence $n_1, n_2, \ldots$ along which

(23)                    $p_{ij}(n_r) \to \alpha_j \quad \text{as} \quad r \to \infty \quad$ for all $i$ and $j$,

for some $\alpha$, where the $\alpha_j$ are not all zero and are independent of $i$ by (21); this is an application of the principle of 'diagonal selection' (see Billingsley 1995, Feller (1968, p. 336), or Exercise (6.4.5)). Equation (23) implies that, for any finite set $F$ of states,

$$\sum_{j \in F} \alpha_j = \lim_{r \to \infty} \sum_{j \in F} p_{ij}(n_r) \le 1$$

and so $\alpha = \sum_j \alpha_j$ satisfies $0 < \alpha \le 1$. Furthermore

$$\sum_{k \in F} p_{ik}(n_r) p_{kj} \le p_{ij}(n_r + 1) = \sum_k p_{ik} p_{kj}(n_r);$$

let $r \to \infty$ here to deduce from (23) and bounded convergence (as used in the proof of (22)) that

$$\sum_{k \in F} \alpha_k p_{kj} \le \sum_k p_{ik} \alpha_j = \alpha_j,$$

and so, letting $F \uparrow S$, we obtain $\sum_k \alpha_k p_{kj} \le \alpha_j$ for each $j \in S$. However, equality must hold here, since if strict inequality holds for some $j$ then

$$\sum_k \alpha_k = \sum_{k,j} \alpha_k p_{kj} < \sum_j \alpha_j,$$

which is a contradiction. Therefore,

$$\sum_k \alpha_k p_{kj} = \alpha_j \qquad \text{for each } j \in S,$$

giving that $\pi = \{\alpha_j / \alpha : j \in S\}$ is a stationary distribution for $X$; this contradicts the nullity of $X$ by (3).                                                                                          ∎

The original and more general version of the limit theorem (20) for Markov chains does *not* assume that the chain is irreducible. We state it here; for a proof see Theorem (5.2.24) or Example (10.4.20).

**(24) Theorem.** *For any aperiodic state $j$ of a Markov chain, $p_{jj}(n) \to \mu_j^{-1}$ as $n \to \infty$. Furthermore, if $i$ is any other state then $p_{ij}(n) \to f_{ij}/\mu_j$ as $n \to \infty$.*

**(25) Corollary.** *Let*

$$\tau_{ij}(n) = \frac{1}{n} \sum_{m=1}^{n} p_{ij}(m)$$

*be the mean proportion of elapsed time up to the nth step during which the chain was in state $j$, starting from $i$. If $j$ is aperiodic, $\tau_{ij}(n) \to f_{ij}/\mu_j$ as $n \to \infty$.*

**Proof.** *Exercise*: prove and use the fact that, as $n \to \infty$, $n^{-1} \sum_1^n x_i \to x$ if $x_n \to x$.                ∎

**(26) Example. The coupling game.**  You may be able to amaze your friends and break the ice at parties with the following card 'trick'. A pack of cards is shuffled, and you deal the cards (face up) one by one. You instruct the audience as follows. Each person is to select some card, secretly, chosen from the first six or seven cards, say. If the face value of this card is $m$ (aces count 1 and court cards count 10), let the next $m - 1$ cards pass and note the face value of the $m$th. Continuing according to this rule, there will arrive a last card in this sequence, face value $X$ say, with fewer than $X$ cards remaining. Call $X$ the 'score'. Each person's score is known to that person but not to you, and can generally be any number between 1 and 10. At the end of the game, using an apparently fiendishly clever method you announce to the audience a number between 1 and 10. If few errors have been made, the majority of the audience will find that your number agrees with their score. Your popularity will then be assured, for a short while at least.

This is the 'trick'. You follow the same rules as the audience, beginning for the sake of simplicity with the first card. You will obtain a 'score' of $Y$, say, and it happens that there is a large probability that any given person obtains the score $Y$ also; therefore you announce the score $Y$.

Why does the game often work? Suppose that someone picks the $m_1$th card, $m_2$th card, and so on, and you pick the $n_1$ ($= 1$)th, $n_2$th, etc. If $n_i = m_j$ for some $i$ and $j$, then the two of you are 'stuck together' forever after, since the rules of the game require you to follow the same pattern henceforth; when this happens first, we say that 'coupling' has occurred. Prior to coupling, each time you read the value of a card, there is a positive probability that you will arrive at the next stage on exactly the same card as the other person. If the pack of cards were infinitely large, then coupling would certainly take place sooner or later, and it turns out that there is a good chance that coupling takes place before the last card of a regular pack has been dealt.

You may recognize the argument above as being closely related to that used near the beginning of the proof of Theorem (20).                                                          ●

---

## Exercises for Section 6.4

**1.**   The proof copy of a book is read by an infinite sequence of editors checking for mistakes. Each mistake is detected with probability $p$ at each reading; between readings the printer corrects the detected mistakes but introduces a random number of new errors (errors may be introduced even if no mistakes were detected). Assuming as much independence as usual, and that the numbers of new errors after different readings are identically distributed, find an expression for the probability generating function of the stationary distribution of the number $X_n$ of errors after the $n$th editor–printer cycle, whenever this exists. Find it explicitly when the printer introduces a Poisson-distributed number of errors at each stage.

**2.**   Do the appropriate parts of Exercises (6.3.1)–(6.3.4) again, making use of the new techniques at your disposal.

**3.**   **Dams.** Let $X_n$ be the amount of water in a reservoir at noon on day $n$. During the 24 hour period beginning at this time, a quantity $Y_n$ of water flows into the reservoir, and just before noon on each day exactly one unit of water is removed (if this amount can be found). The maximum capacity of the reservoir is $K$, and excessive inflows are spilled and lost. Assume that the $Y_n$ are independent and identically distributed random variables and that, by rounding off to some laughably small unit of volume, all numbers in this exercise are non-negative integers. Show that $(X_n)$ is a Markov chain, and find its transition matrix and an expression for its stationary distribution in terms of the probability generating function $G$ of the $Y_n$.

Find the stationary distribution when $Y$ has probability generating function $G(s) = p(1-qs)^{-1}$.

**4.**   Show by example that chains which are not irreducible may have many different stationary distributions.

**5.    Diagonal selection.** Let $(x_i(n) : i, n \geq 1)$ be a bounded collection of real numbers. Show that there exists an increasing sequence $n_1, n_2, \ldots$ of positive integers such that $\lim_{r \to \infty} x_i(n_r)$ exists for all $i$. Use this result to prove that, for an irreducible Markov chain, if it is not the case that $p_{ij}(n) \to 0$ as $n \to \infty$ for all $i$ and $j$, then there exists a sequence $(n_r : r \geq 1)$ and a vector $\boldsymbol{\alpha} \, (\neq \mathbf{0})$ such that $p_{ij}(n_r) \to \alpha_j$ as $r \to \infty$ for all $i$ and $j$.

**6.    Random walk on a graph.** A particle performs a random walk on the vertex set of a connected graph $G$, which for simplicity we assume to have neither loops nor multiple edges. At each stage it moves to a neighbour of its current position, each such neighbour being chosen with equal probability. If $G$ has $\eta \, (< \infty)$ edges, show that the stationary distribution is given by $\pi_v = d_v/(2\eta)$, where $d_v$ is the degree of vertex $v$.

**7.    **Show that a random walk on the infinite binary tree is transient.

**8.    **At each time $n = 0, 1, 2, \ldots$ a number $Y_n$ of particles enters a chamber, where $\{Y_n : n \geq 0\}$ are independent and Poisson distributed with parameter $\lambda$. Lifetimes of particles are independent and geometrically distributed with parameter $p$. Let $X_n$ be the number of particles in the chamber at time $n$. Show that $X$ is a Markov chain, and find its stationary distribution.

**9.    **A random sequence of convex polygons is generated by picking two edges of the current polygon at random, joining their midpoints, and picking one of the two resulting smaller polygons at random to be the next in the sequence. Let $X_n + 3$ be the number of edges of the $n$th polygon thus constructed. Find $\mathbb{E}(X_n)$ in terms of $X_0$, and find the stationary distribution of the Markov chain $X$.

**10.    **Let $s$ be a state of an irreducible Markov chain on the non-negative integers. Show that the chain is recurrent if there exists a solution $\mathbf{y}$ to the equations $y_i \geq \sum_{j : j \neq s} p_{ij} y_j$, $i \neq s$, satisfying $y_i \to \infty$.

**11.    Bow ties.** A particle performs a random walk on a bow tie ABCDE drawn beneath on the left, where C is the knot. From any vertex its next step is equally likely to be to any neighbouring vertex. Initially it is at A. Find the expected value of:

(a) the time of first return to A,
(b) the number of visits to D before returning to A,
(c) the number of visits to C before returning to A,
(d) the time of first return to A, given no prior visit by the particle to E,
(e) the number of visits to D before returning to A, given no prior visit by the particle to E.

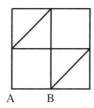

**12.    **A particle starts at A and executes a symmetric random walk on the graph drawn above on the right. Find the expected number of visits to B before it returns to A.

**13.    Top-to-random shuffling.** A pack contains 52 cards labelled $1, 2, \ldots, 52$, and initially they are in increasing order from top to bottom. At each stage of the shuffling process, the top card is moved to one of the 52 available places determined by the other 51 cards, this place being chosen uniformly at random, independently of all previous stages. Find the mean number of stages until card 52 is first on top.

Show that, after the moment at which the card labelled 52 is inserted at random from the top, the order of the pack is uniformly distributed over the 52! possibilities.

**14. Random-to-top shuffling.** A pack contains 52 cards labelled 1, 2, . . . , 52, and initially they are in increasing order from top to bottom. At each stage, a card is picked uniformly at random from the pack and placed on top, independently of all previous stages. Find the mean number of stages until every card has been selected at least once.

Show that, after the moment at which the final card to be selected at random is placed on top, the order of the pack is uniformly distributed over the 52! possibilities.

**15. Quality control.** Dick and Jim are writing exercises in sequence for inclusion in a textbook. Dick writes them and Jim checks them. Each exercise is faulty with probability $p$, independently of other exercises. Jim has two modes of operation. In Mode A, he inspects every exercise as it is produced. In Mode B, he inspects each exercise with probability $1/r$ where $r > 1$, independently of all other events.

Let $N \geq 1$. Jim operates in Mode A until he has found $N$ consecutive non-defective exercises, at which point he changes to Mode B. He operates in Mode B until the first defective exercise is found, and then he reverts to Mode A.

Let $X$ be the Markov chain which is in state $i$ if Jim is operating in Mode A and the last $i$ consecutive exercises since entering Mode A have been found to be non-defective, and is in state $N$ if Jim is in Mode B.

(a) Write down the transition probabilities of $X$ and find its stationary distribution.
(b) Show that the long run proportion of exercises that are inspected is $1/[1 + (r - 1)(1 - p^N)]$.
(c) Find an expression for the long run proportion of defective exercises which are not detected by Jim.

**16. Exercise (3.11.39) revisited.** A particle performs a random walk on the non-negative integers as follows. When at position $k \geq 0$, its next position is uniformly distributed on the set $\{0, 1, \ldots, k, k+1\}$. Show that the sequence of positions forms an aperiodic, positive recurrent Markov chain, and find its stationary distribution.

Find the mean number $\mu$ of steps required to reach position 0 for the first time from position 1.

**17.** Let $\{X_n : n \geq 0\}$ be an irreducible Markov chain with state space $S$ and transition matrix $\mathbf{P}$ (the chain may be either transient or recurrent). Let $k \in S$ and let $\mathbf{x}$ be a stationary measure such that $x_k = 1$. Prove that $\mathbf{x} \geq \boldsymbol{\rho}(k)$ where $\boldsymbol{\rho}(k)$ is given in equation (6.4.5). If the chain is recurrent, show that $\mathbf{x} = \boldsymbol{\rho}(k)$.

---

## 6.5  Reversibility

Most laws of physics have the property that they would make the same assertions if the universal clock were reversed and time were made to run backwards. It may be implausible that nature works in such ways (have *you* ever seen the fragments of a shattered teacup re-assemble themselves on the table from which it fell?), and so one may be led to postulate a non-decreasing quantity called 'entropy'. However, never mind such objections; let us think about the reversal of the time scale of a Markov chain.

Suppose that $\{X_n : 0 \leq n \leq N\}$ is an irreducible positive recurrent Markov chain, with transition matrix $\mathbf{P}$ and stationary distribution $\pi$. Suppose further that $X_0$ has distribution $\pi$, so that $X_n$ has distribution $\pi$ for every $n$. Define the 'reversed chain' $Y$ by $Y_n = X_{N-n}$ for $0 \leq n \leq N$. We first check as follows that $Y$ is a Markov chain.

**(1) Theorem.**  *The sequence $Y$ is a Markov chain with* $\mathbb{P}(Y_{n+1} = j \mid Y_n = i) = (\pi_j/\pi_i)p_{ji}$.

**Proof.** We have as required that

$$\mathbb{P}\big(Y_{n+1} = i_{n+1} \mid Y_n = i_n, \, Y_{n-1} = i_{n-1}, \ldots, \, Y_0 = i_0\big)$$
$$= \frac{\mathbb{P}(Y_k = i_k, \, 0 \le k \le n+1)}{\mathbb{P}(Y_k = i_k, \, 0 \le k \le n)}$$
$$= \frac{\mathbb{P}(X_{N-n-1} = i_{n+1}, \, X_{N-n} = i_n, \ldots, \, X_N = i_0)}{\mathbb{P}(X_{N-n} = i_n, \ldots, \, X_N = i_0)}$$
$$= \frac{\pi_{i_{n+1}} p_{i_{n+1},i_n} p_{i_n,i_{n-1}} \cdots p_{i_1,i_0}}{\pi_{i_n} p_{i_n,i_{n-1}} \cdots p_{i_1,i_0}} = \frac{\pi_{i_{n+1}} p_{i_{n+1},i_n}}{\pi_{i_n}}. \qquad \blacksquare$$

We call the chain $Y$ the *time-reversal* of the chain $X$, and we say that $X$ is *reversible* if $X$ and $Y$ have the same transition probabilities.

**(2) Definition.** Let $X = \{X_n : 0 \le n \le N\}$ be an irreducible Markov chain such that $X_n$ has the stationary distribution $\pi$ for all $n$. The chain is called **reversible** if the transition matrices of $X$ and its time-reversal $Y$ are the same, which is to say that

**(3)** $$\pi_i p_{ij} = \pi_j p_{ji} \qquad \text{for all } i, j.$$

Equations (3) are called the *detailed balance* equations, and they are pivotal to the study of reversible chains. More generally we say that a transition matrix $\mathbf{P}$ and a distribution $\lambda$ are *in detailed balance* if $\lambda_i p_{ij} = \lambda_j p_{ji}$ for all $i, j \in S$. An irreducible chain $X$ having a stationary distribution $\pi$ is called *reversible in equilibrium* if its transition matrix $\mathbf{P}$ is in detailed balance with $\pi$. It may be noted that a chain having a tridiagonal transition matrix is reversible in equilibrium; see Exercise (6.5.1) and Problem (6.15.16d).

The following theorem provides a useful way of finding the stationary distribution of an irreducible chain whose transition matrix $\mathbf{P}$ is in detailed balance with some distribution $\lambda$.

**(4) Theorem.** *Let $\mathbf{P}$ be the transition matrix of an irreducible chain $X$, and suppose that there exists a distribution $\pi$ such that $\pi_i p_{ij} = \pi_j p_{ji}$ for all $i, j \in S$. Then $\pi$ is a stationary distribution of the chain. Furthermore, $X$ is reversible in equilibrium.*

**Proof.** Suppose that $\pi$ satisfies the conditions of the theorem. Then

$$\sum_i \pi_i p_{ij} = \sum_i \pi_j p_{ji} = \pi_j \sum_i p_{ji} = \pi_j$$

and so $\pi = \pi\mathbf{P}$, whence $\pi$ is stationary. The reversibility in equilibrium of $X$ follows from the definition (2). $\qquad \blacksquare$

Although the above definition of reversibility applies to a Markov chain defined on only finitely many time points $0, 1, 2, \ldots, N$, it is easily seen to apply to the infinite time set $0, 1, 2, \ldots$. It may be extended also to the doubly-infinite time set $\ldots, -2, -1, 0, 1, 2, \ldots$. In the last case it is necessary to note the following fact. Let $X = \{X_n : -\infty < n < \infty\}$ be a Markov chain with stationary distribution $\pi$. In order that $X_n$ have distribution $\pi$ for all $n$, it is not generally sufficient that $X_0$ has distribution $\pi$.

**(5) Example. Ehrenfest model of diffusion†.**   Two containers A and B are placed adjacent
to each other and gas is allowed to pass through a small aperture joining them. A total of $m$
gas molecules is distributed between the containers. We assume that at each epoch of time
one molecule, picked uniformly at random from the $m$ available, passes through this aperture.
Let $X_n$ be the number of molecules in container A after $n$ units of time has passed. Clearly
$\{X_n\}$ is a Markov chain with transition matrix

$$p_{i,i+1} = 1 - \frac{i}{m}, \quad p_{i,i-1} = \frac{i}{m} \quad \text{if} \quad 0 \le i \le m.$$

Rather than solve the equation $\boldsymbol{\pi} = \boldsymbol{\pi}\mathbf{P}$ to find the stationary distribution, we note that such
a reasonable diffusion model might be reversible in equilibrium. Look for solutions of the
detailed balance equations $\pi_i p_{ij} = \pi_j p_{ji}$ to obtain $\pi_i = \binom{m}{i}(\frac{1}{2})^m$.                    ●

Here is a way of thinking about reversibility and the equations $\pi_i p_{ij} = \pi_j p_{ji}$. Suppose
we are provided with a Markov chain with state space $S$ and stationary distribution $\boldsymbol{\pi}$. To this
chain there corresponds a 'network' as follows. The nodes of the network are the states in $S$,
and arrows are added between certain pairs of nodes; an arrow is added pointing from state
$i$ to state $j$ whenever $p_{ij} > 0$. We are provided with one unit of material (disease, water,
or sewage, perhaps) which is distributed about the nodes of the network and allowed to flow
along the arrows. The transportation rule is as follows: at each epoch of time a proportion $p_{ij}$
of the amount of material at node $i$ is transported to node $j$. Initially the material is distributed
in such a way that exactly $\pi_i$ of it is at node $i$, for each $i$. It is a simple calculation that
the amount at node $i$ after one epoch of time is $\sum_j \pi_j p_{ji}$, which equals $\pi_i$, since $\boldsymbol{\pi} = \boldsymbol{\pi}\mathbf{P}$.
Therefore the system is in equilibrium: there is a 'global balance' in the sense that the total
quantity leaving each node equals the total quantity arriving there. There may or may not be a
'local balance', in the sense that, for all $i$, $j$, the amount flowing from $i$ to $j$ equals the amount
flowing from $j$ to $i$. Local balance occurs if and only if $\pi_i p_{ij} = \pi_j p_{ji}$ for all $i$, $j$, which is to
say that $\mathbf{P}$ and $\boldsymbol{\pi}$ are in detailed balance.

## Exercises for Section 6.5

**1.**   A random walk on the set $\{0, 1, 2, \ldots, b\}$ has transition matrix given by $p_{00} = 1 - \lambda_0$, $p_{bb} = 1 - \mu_b$, $p_{i,i+1} = \lambda_i$ and $p_{i+1,i} = \mu_{i+1}$ for $0 \le i < b$, where $0 < \lambda_i, \mu_i < 1$ for all $i$, and
$\lambda_i + \mu_i = 1$ for $1 \le i < b$. Show that this process is reversible in equilibrium.

**2.**   Let $X$ be an irreducible, positive recurrent, aperiodic Markov chain on the state space $S$.
(a) **Kolmogorov's reversibility criterion.** Show that $X$ is reversible in equilibrium if and only if

$$p_{j_1,j_2} p_{j_2,j_3} \cdots p_{j_{n-1},j_n} p_{j_n,j_1} = p_{j_1,j_n} p_{j_n,j_{n-1}} \cdots p_{j_2,j_1}$$

for all $n$ and all finite sequences $j_1, j_2, \ldots, j_n$ of states.
(b) **Kelly's reversibility condition.** Show that $X$ is reversible in equilibrium if, for all distinct triples
$i, j, k \in S$,

$$p_{ij} p_{jk} p_{ki} = p_{ik} p_{kj} p_{ji},$$

and in addition there exists $c \in S$ such that $p_{ic} > 0$ for all $i \ne c$.
(c) Consider a chain with $n \ge 3$ states. Show that Kolmogorov's criterion, as expressed above, may
require the verification of up to $\frac{1}{2}\sum_{r=3}^n \binom{n}{r}(r-1)!$ equations, whereas Kelly's condition, if
appropriate, requires no more than $\binom{n-1}{3}$.

---

†Originally introduced by Paul and Tatiana Ehrenfest as the 'dog–flea model', and solved by Mark Kac.

(d) Show that a random walk on a finite tree is reversible in equilibrium.

**3.** Let $X$ be a reversible Markov chain, and let $C$ be a non-empty subset of the state space $S$. Define the Markov chain $Y$ on $S$ by the transition matrix $\mathbf{Q} = (q_{ij})$ where

$$q_{ij} = \begin{cases} \beta p_{ij} & \text{if } i \in C \text{ and } j \notin C, \\ p_{ij} & \text{otherwise,} \end{cases}$$

for $i \neq j$, and where $\beta$ is a constant satisfying $0 < \beta < 1$. The diagonal terms $q_{ii}$ are arranged so that $\mathbf{Q}$ is a stochastic matrix. Show that $Y$ is reversible in equilibrium, and find its stationary distribution. Describe the situation in the limit as $\beta \downarrow 0$.

**4.** Can a reversible chain be periodic?

**5. Ehrenfest dog–flea model.** The dog–flea model of Example (6.5.5) is a Markov chain $X$ on the state space $\{0, 1, \ldots, m\}$ with transition probabilities

$$p_{i,i+1} = 1 - \frac{i}{m}, \quad p_{i,i-1} = \frac{i}{m}, \quad \text{for } 0 \le i \le m.$$

Show that, if $X_0 = i$,

$$\mathbb{E}\left( X_n - \frac{m}{2} \right) = \left( i - \frac{m}{2} \right) \left( 1 - \frac{2}{m} \right)^n \to 0 \quad \text{as } n \to \infty.$$

**6.** Which of the following (when stationary) are reversible Markov chains?

(a) The chain $X = \{X_n\}$ having transition matrix $\mathbf{P} = \begin{pmatrix} 1 - \alpha & \alpha \\ \beta & 1 - \beta \end{pmatrix}$ where $\alpha + \beta > 0$.

(b) The chain $Y = \{Y_n\}$ having transition matrix $\mathbf{P} = \begin{pmatrix} 0 & p & 1-p \\ 1-p & 0 & p \\ p & 1-p & 0 \end{pmatrix}$ where $0 < p < 1$.

(c) $Z_n = (X_n, Y_n)$, where $X_n$ and $Y_n$ are independent and satisfy (a) and (b).

**7.** Let $X_n, Y_n$ be independent simple random walks. Let $Z_n$ be $(X_n, Y_n)$ truncated to lie in the region $X_n \ge 0$, $Y_n \ge 0$, $X_n + Y_n \le a$ where $a$ is integral. Find the stationary distribution of $Z_n$.

**8.** Show that an irreducible Markov chain with a finite state space and transition matrix $\mathbf{P}$ is reversible in equilibrium if and only if $\mathbf{P} = \mathbf{DS}$ for some symmetric matrix $\mathbf{S}$ and diagonal matrix $\mathbf{D}$ with strictly positive diagonal entries. Show further that for reversibility in equilibrium to hold, it is necessary but not sufficient that $\mathbf{P}$ has real eigenvalues.

**9. Random walk on a graph.** Let $G$ be a finite connected graph with neither loops nor multiple edges, and let $X$ be a random walk on $G$ as in Exercise (6.4.6). Show that $X$ is reversible in equilibrium.

**10.** Consider a random walk on the strictly positive integers with transition probabilities

$$p_{i,i-1} = \frac{1}{2} \cdot \frac{i+2}{i+1}, \quad p_{i,i+1} = \frac{1}{2} \cdot \frac{i}{i+1}, \quad i \ge 2,$$

and $p_{11} = \frac{3}{4}$, $p_{12} = \frac{1}{4}$. Show that the walk is positive recurrent and find the mean recurrence time of the state $i$.

**11.** An aleatory beetle performs a random walk on five vertices comprising the principal points (labelled $n, e, s, w$) and centre (labelled $c$) of a compass, with transition probabilities

$$p_{en} = p_{ws} = \tfrac{1}{4},$$

$$p_{cn} = p_{ne} = p_{ec} = p_{ce} = p_{sw} = p_{wn} = \tfrac{1}{8},$$

$$p_{nc} = p_{es} = p_{sc} = p_{cs} = p_{se} = p_{wc} = \tfrac{1}{16},$$

$$p_{cw} = p_{nw} = \tfrac{1}{32}.$$

Other moves have probability 0. Show that the mean recurrence time of the centre is $\mu_c = \frac{11}{2}$.

**12. Lazy Markov chain.** Let $X$ be an irreducible (but not necessarily aperiodic) Markov chain on the countable state space $S$ with transition matrix $\mathbf{P}$, with invariant distribution $\pi$. Let $a \in (0, 1)$ and let $\mathbf{L} = a\mathbf{P} + (1 - a)\mathbf{I}$ where $\mathbf{I}$ is the identity matrix.

(a) Show that $\mathbf{L}$ is the transition matrix of an irreducible, aperiodic Markov chain $Y$ with invariant distribution $\pi$.

(b) Show that, if $X$ is reversible in equilibrium, then so is $Y$.

---

## 6.6  Chains with finitely many states

The theory of Markov chains is much simplified by the condition that $S$ be finite. By Lemma (6.3.5), if $S$ is finite and irreducible then it is necessarily positive recurrent. It may even be possible to calculate the $n$-step transition probabilities explicitly. Of central importance here is the following algebraic theorem, in which $i = \sqrt{-1}$. Let $N$ denote the cardinality of $S$.

**(1) Theorem (Perron–Frobenius).**  *Let* $\mathbf{P}$ *be the transition matrix of a finite irreducible chain with period $d \geq 1$. Then*:

(a) $\lambda_1 = 1$ *is an eigenvalue of* $\mathbf{P}$,

(b) *the $d$ complex roots of unity*

$$\lambda_1 = \omega^0, \ \lambda_2 = \omega^1, \dots, \ \lambda_d = \omega^{d-1} \quad \text{where } \omega = e^{2\pi i/d},$$

 *are eigenvalues of* $\mathbf{P}$,

(c) *the remaining eigenvalues $\lambda_{d+1}, \dots, \lambda_N$ satisfy $|\lambda_r| < 1$,*

(d) *there exists a unique distribution $\pi$ satisfying $\pi\mathbf{P} = \pi$; furthermore, all components of $\pi$ are strictly positive.*

Suppose the eigenvalues $\lambda_1, \dots, \lambda_N$ of the theorem are distinct, and let $\pi_r$ be a left eigenvector of $\mathbf{P}$ with eigenvalue $\lambda_r$ (in particular, $\pi_1 = \pi$ as in part (d)). Let $\mathbf{B}$ be the matrix whose $r$th row is $\pi_r$. Then $\mathbf{P} = \mathbf{B}^{-1}\boldsymbol{\Lambda}\mathbf{B}$ where $\boldsymbol{\Lambda}$ is the diagonal matrix with entries $\lambda_1, \dots, \lambda_N$. Therefore, $\mathbf{P} = \mathbf{B}^{-1}\boldsymbol{\Lambda}\mathbf{B}$, implying that

$$\mathbf{P}^n = \mathbf{B}^{-1}\boldsymbol{\Lambda}^n\mathbf{B} = \mathbf{B}^{-1}\begin{pmatrix} \lambda_1^n & 0 & \cdots & 0 \\ 0 & \lambda_2^n & \cdots & 0 \\ \vdots & \vdots & \ddots & \vdots \\ 0 & 0 & \cdots & \lambda_N^n \end{pmatrix}\mathbf{B}.$$

We can use the Perron–Frobenius theorem to explore the properties of $\mathbf{P}^n$ for large $n$. For example, if the chain is aperiodic then $d = 1$ and

$$\mathbf{P}^n \to \mathbf{B}^{-1}\begin{pmatrix} 1 & 0 & \cdots & 0 \\ 0 & 0 & \cdots & 0 \\ \vdots & \vdots & \ddots & \vdots \\ 0 & 0 & \cdots & 0 \end{pmatrix}\mathbf{B} \quad \text{as} \quad n \to \infty.$$

It may be checked that the limiting matrix has identical rows $\pi$, whence the $(i, j)$th entry $p_{ij}(n)$ of $\mathbf{P}^n$ satisfies $p_{ij}(n) \to \pi_j$ as $n \to \infty$. The speed of convergence is controlled† by

---

†The constant $\mu$ is sometimes called the SLEM, the *second largest eigenvalue modulus*.

$\mu := \max\{|\lambda_r| : r = 2, 3, \ldots, N\}$, and there exists $c < \infty$ such that

$$|p_{ij}(n) - \pi_j| \le c\mu^n, \qquad i, j = 1, 2, \ldots, N.$$

When the eigenvalues of the matrix $\mathbf{P}$ are not distinct, then $\mathbf{P}$ cannot always be reduced to the diagonal canonical form in this way. The best that we may be able to do is to rewrite $\mathbf{P}$ in its 'Jordan canonical form' $\mathbf{P} = \mathbf{B}^{-1}\mathbf{MB}$ where

$$\mathbf{M} = \begin{pmatrix} \mathbf{J}_1 & \mathbf{0} & \mathbf{0} & \cdots \\ \mathbf{0} & \mathbf{J}_2 & \mathbf{0} & \cdots \\ \mathbf{0} & \mathbf{0} & \mathbf{J}_3 & \cdots \\ \vdots & \vdots & \vdots & \ddots \end{pmatrix}$$

and $\mathbf{J}_1, \mathbf{J}_2, \ldots$ are square matrices given as follows. Let $\lambda_1, \lambda_2, \ldots, \lambda_m$ be the distinct eigenvalues of $\mathbf{P}$ and let $k_r$ be the multiplicity of $\lambda_r$. Then

$$\mathbf{J}_r = \begin{pmatrix} \lambda_r & 1 & 0 & 0 & \cdots \\ 0 & \lambda_r & 1 & 0 & \cdots \\ 0 & 0 & \lambda_r & 1 & \cdots \\ \vdots & \vdots & \vdots & \vdots & \ddots \end{pmatrix}$$

is a $k_r \times k_r$ matrix with each diagonal term $\lambda_r$, each superdiagonal term 1, and all other terms 0. Once again we have that $\mathbf{P}^n = \mathbf{B}^{-1}\mathbf{M}^n\mathbf{B}$, where $\mathbf{M}^n$ has quite a simple form (see Cox and Miller (1965, p. 118 *et seq.*) for more details).

**(2) Example. Inbreeding.** Consider the genetic model described in Example (6.1.11c) and suppose that $C_1$ can take the values $A$ or $a$ on each of two homologous chromosomes. Then the possible types of individuals can be denoted by

$$AA, \ Aa \ (\equiv aA), \ aa,$$

and mating between types is denoted by

$$AA \times AA, \ AA \times Aa, \quad \text{and so on.}$$

As described in Example (6.1.11c), meiosis causes the offspring's chromosomes to be selected randomly from each parent; in the simplest case (since there are two choices for each of two places) each outcome has probability $\frac{1}{4}$. Thus for the offspring of $AA \times Aa$ the four possible outcomes are

$$AA, \ Aa, \ AA, \ Aa$$

and $\mathbb{P}(AA) = \mathbb{P}(Aa) = \frac{1}{2}$. For the cross $Aa \times Aa$,

$$\mathbb{P}(AA) = \mathbb{P}(aa) = \tfrac{1}{2}\mathbb{P}(Aa) = \tfrac{1}{4}.$$

Clearly the offspring of $AA \times AA$ can only be $AA$, and those of $aa \times aa$ can only be $aa$.

We now construct a Markov chain by mating an individual with itself, then crossing a single resulting offspring with itself, and so on. (This scheme is possible with plants.) The genetic types of this sequence of individuals constitute a Markov chain with three states, $AA, Aa, aa$. In view of the above discussion, the transition matrix is

$$\mathbf{P} = \begin{pmatrix} 1 & 0 & 0 \\ \frac{1}{4} & \frac{1}{2} & \frac{1}{4} \\ 0 & 0 & 1 \end{pmatrix}$$

and the reader may verify that

$$\mathbf{P}^n = \begin{pmatrix} 1 & 0 & 0 \\ \frac{1}{2} - (\frac{1}{2})^{n+1} & (\frac{1}{2})^n & \frac{1}{2} - (\frac{1}{2})^{n+1} \\ 0 & 0 & 1 \end{pmatrix} \to \begin{pmatrix} 1 & 0 & 0 \\ \frac{1}{2} & 0 & \frac{1}{2} \\ 0 & 0 & 1 \end{pmatrix} \quad \text{as} \quad n \to \infty.$$

Thus, ultimately, inbreeding produces a pure ($AA$ or $aa$) line for which all subsequent offspring have the same type. In like manner one can consider the progress of many different breeding schemes which include breeding with rejection of unfavourable genes, back-crossing to encourage desirable genes, and so on.                                           ●

---

## Exercises for Section 6.6

The first two exercises provide proofs that a Markov chain with finitely many states has a stationary distribution.

**1.**   The Markov–Kakutani theorem asserts that, for any convex compact subset $C$ of $\mathbb{R}^n$ and any linear continuous mapping $T$ of $C$ into $C$, $T$ has a fixed point (in the sense that $T(x) = x$ for some $x \in C$). Use this to prove that a finite stochastic matrix has a non-negative non-zero left eigenvector corresponding to the eigenvalue 1.

**2.**   Let $\mathbf{T}$ be a $m \times n$ matrix and let $\mathbf{v} \in \mathbb{R}^n$. Farkas's theorem asserts that exactly one of the following holds:
  (i) there exists $\mathbf{x} \in \mathbb{R}^m$ such that $\mathbf{x} \geq \mathbf{0}$ and $\mathbf{xT} = \mathbf{v}$,
  (ii) there exists $\mathbf{y} \in \mathbb{R}^n$ such that $\mathbf{yv}' < 0$ and $\mathbf{Ty}' \geq \mathbf{0}$.
Use this to prove that a finite stochastic matrix has a non-negative non-zero left eigenvector corresponding to the eigenvalue 1.

**3.**   **Arbitrage.** Suppose you are betting on a race with $m$ possible outcomes. There are $n$ bookmakers, and a unit stake with the $i$th bookmaker yields $t_{ij}$ if the $j$th outcome of the race occurs. A vector $\mathbf{x} = (x_1, x_2, \dots, x_n)$, where $x_r \in (-\infty, \infty)$ is your stake with the $r$th bookmaker, is called a *betting scheme*. Show that exactly one of (a) and (b) holds:
  (a) there exists a probability mass function $\mathbf{p} = (p_1, p_2, \dots, p_m)$ such that $\sum_{j=1}^m t_{ij} p_j = 0$ for all values of $i$,
  (b) there exists a betting scheme $\mathbf{x}$ for which you surely win, that is, $\sum_{i=1}^n x_i t_{ij} > 0$ for all $j$.

**4.**   Let $X$ be a Markov chain with state space $S = \{1, 2, 3\}$ and transition matrix

$$\mathbf{P} = \begin{pmatrix} 1-p & p & 0 \\ 0 & 1-p & p \\ p & 0 & 1-p \end{pmatrix}$$

where $0 < p < 1$. Prove that

$$\mathbf{P}^n = \begin{pmatrix} a_{1n} & a_{2n} & a_{3n} \\ a_{3n} & a_{1n} & a_{2n} \\ a_{2n} & a_{3n} & a_{1n} \end{pmatrix}$$

where $a_{1n} + \omega a_{2n} + \omega^2 a_{3n} = (1 - p + p\omega)^n$, $\omega$ being a complex cube root of 1.

**5.** Let $\mathbf{P}$ be the transition matrix of a Markov chain with finite state space. Let $\mathbf{I}$ be the identity matrix, $\mathbf{U}$ the $|S| \times |S|$ matrix with all entries unity, and $\mathbf{1}$ the row $|S|$-vector with all entries unity. Let $\boldsymbol{\pi}$ be a non-negative vector with $\sum_i \pi_i = 1$. Show that $\boldsymbol{\pi}\mathbf{P} = \boldsymbol{\pi}$ if and only if $\boldsymbol{\pi}(\mathbf{I} - \mathbf{P} + \mathbf{U}) = \mathbf{1}$. Deduce that if $\mathbf{P}$ is irreducible then $\boldsymbol{\pi} = \mathbf{1}(\mathbf{I} - \mathbf{P} + \mathbf{U})^{-1}$.

**6.    Chess.** A chess piece performs a random walk on a chessboard; at each step it is equally likely to make any one of the available moves. What is the mean recurrence time of a corner square if the piece is a: (a) king?   (b) queen?   (c) bishop?   (d) knight?   (e) rook?

**7.    Chess continued.** A rook and a bishop perform independent symmetric random walks with synchronous steps on a $4 \times 4$ chessboard (16 squares). If they start together at a corner, show that the expected number of steps until they meet again at the same corner is 448/3.

**8.** Find the $n$-step transition probabilities $p_{ij}(n)$ for the chain $X$ having transition matrix

$$
\mathbf{P} = \begin{pmatrix} 0 & \frac{1}{2} & \frac{1}{2} \\ \frac{1}{3} & \frac{1}{4} & \frac{5}{12} \\ \frac{2}{3} & \frac{1}{4} & \frac{1}{12} \end{pmatrix}.
$$

**9.    The 'PageRank' Markov chain.** The pages on the worldwide web form a directed graph $W$ with $n$ vertices (representing pages) joined by directed edges (representing links). The existence of a link from $i$ to $j$ is denoted by $i \to j$, and the graph is specified by its adjacency matrix $L = (l_{ij})$ where $l_{ij} = 1$ if $i \to j$ and $l_{ij} = 0$ otherwise. The *out-degree* $d_i$ (respectively, *in-degree* $c_i$) of vertex $i$ is the number of links pointing away from $i$ (respectively, towards $i$). Vertex $i$ is said to *dangle* if $d_i = 0$.

The behaviour of a swiftly bored web surfer is modelled by a random walk on $W$. Let $b \in (0, 1)$. From any dangling vertex, the random walk moves to a randomly chosen vertex of $W$, each vertex having probability $1/n$. When at a non-dangling vertex $i$, with probability $b < 1$ the walk moves to a random linked vertex (each having probability $1/d_i$), while with probability $1 - b$ it moves to a random vertex of $W$ (each having probability $1/n$).

(a) Show that the transition matrix $\mathbf{P}$ may be written in the form $\mathbf{P} = b\mathbf{Q} + \mathbf{v}'\mathbf{e}$, where $\mathbf{Q} = (q_{ij})$ with

$$
q_{ij} = \begin{cases} 1/d_i & \text{if } i \to j, \\ 1/n & \text{if } i \text{ dangles}, \\ 0 & \text{otherwise,} \end{cases}
$$

and $\mathbf{v} = (v_i)$ is a row vector with $v_i = (1 - b)/n$, and $\mathbf{e}$ is a row vector with entries 1.
(b) Deduce that the stationary distribution $\boldsymbol{\pi}$ is given by $\boldsymbol{\pi} = \{(1 - b)/n\}\mathbf{e}(\mathbf{I} - b\mathbf{Q})^{-1}$ where $\mathbf{I}$ is the identity matrix.
(c) Explain why the elements of $\boldsymbol{\pi}$, when rearranged in decreasing order, supply a description of the relative popularities of web pages (called 'PageRank' by Google, that being their trademark for the patented algorithm).

**10.** Let $\mathbf{P}$ be the transition matrix of an irreducible Markov chain on a finite state space, and let $\boldsymbol{\pi}$ be a left eigenvector of $\mathbf{P}$ corresponding to the eigenvalue 1. Show from the equation $\boldsymbol{\pi} = \boldsymbol{\pi}\mathbf{P}$ directly that the entries of $\boldsymbol{\pi}$ are either all positive or all negative, and hence prove Theorem (6.6.1d): there exists a unique distribution $\boldsymbol{\pi}$ satisfying $\boldsymbol{\pi}\mathbf{P} = \boldsymbol{\pi}$, and, furthermore, all components of $\boldsymbol{\pi}$ are strictly positive.

## 6.7  Branching processes revisited

The foregoing general theory is an attractive and concise account of the evolution through time of a Markov chain. Unfortunately, it is an inadequate description of many specific Markov chains. Consider for example a branching process $\{Z_0, Z_1, \ldots\}$ where $Z_0 = 1$. If there is strictly positive probability $\mathbb{P}(Z_1 = 0)$ that each family is empty then 0 is an absorbing state. Hence 0 is positive recurrent, and all other states are transient. The chain is not irreducible but there exists a unique stationary distribution $\boldsymbol{\pi}$ given by $\pi_0 = 1, \pi_i = 0$ if $i > 0$. These facts tell us next to nothing about the behaviour of the process, and we must look elsewhere for detailed information. The difficulty is that the process may behave in one of various qualitatively different ways depending, for instance, on whether or not it ultimately becomes extinct. One way of approaching the problem is to study the behaviour of the process *conditional* upon the occurrence of some event, such as extinction, or on the value of some random variable, such as the total number $\sum_i Z_i$ of progeny. This section contains an outline of such a method.

Let $f$ and $G$ be the mass function and generating function of a typical family size $Z_1$:

$$f(k) = \mathbb{P}(Z_1 = k), \qquad G(s) = \mathbb{E}(s^{Z_1}).$$

Let $T = \inf\{n : Z_n = 0\}$ be the time of extinction, with the convention that the infimum of the empty set is $+\infty$. Roughly speaking, if $T = \infty$ then the process will grow beyond all possible bounds, whilst if $T < \infty$ then the size of the process never becomes very large and subsequently reduces to zero. Think of $\{Z_n\}$ as a fluctuating sequence which either becomes so large that it escapes to $\infty$ or is absorbed at 0 during one of its fluctuations. From the results of Section 5.4, the probability $\mathbb{P}(T < \infty)$ of ultimate extinction is the smallest non-negative root of the equation $s = G(s)$. Now let

$$E_n = \{n < T < \infty\}$$

be the event that extinction occurs at some time after $n$. We shall study the distribution of $Z_n$ conditional upon the occurrence of $E_n$. We introduce the traditional but problematic notation

$$_0 p_j(n) = \mathbb{P}(Z_n = j \mid E_n)$$

for the conditional probability that $Z_n = j$ given the future extinction of $Z$. We are interested in the limiting value

$$_0\pi_j = \lim_{n \to \infty} {_0 p_j(n)},$$

if this limit exists. To avoid certain trivial cases we assume throughout this section that

$$0 < f(0) + f(1) < 1, \qquad f(0) > 0;$$

these conditions imply that $0 < \mathbb{P}(E_n) < 1$, that the probability $\eta$ of ultimate extinction satisfies $0 < \eta \le 1$, and that $\mathrm{var}(Z_1) \neq 0$.

**(1) Lemma.** *If* $\mathbb{E}(Z_1) < \infty$ *then* $\lim_{n \to \infty} {_0 p_j(n)} = {_0\pi_j}$ *exists. The generating function*

$$G^\pi(s) = \sum_j {_0\pi_j s^j}$$

*satisfies the functional equation*

(2) $$G^{\pi}\left(\eta^{-1}G(s\eta)\right) = mG^{\pi}(s) + 1 - m$$

*where $\eta$ is the probability of ultimate extinction and $m = G'(\eta)$.*

Note that if $\mu = \mathbb{E}Z_1 \leq 1$ then $\eta = 1$ and $m = \mu$. Thus (2) reduces to

$$G^{\pi}(G(s)) = \mu G^{\pi}(s) + 1 - \mu.$$

Whatever the value of $\mu$, we have that $G'(\eta) \leq 1$, with equality if and only if $\mu = 1$.

**Proof.** For $s \in [0, 1)$, let

$$G_n^{\pi}(s) = \mathbb{E}(s^{Z_n} \mid E_n) = \sum_j {}_0 p_j(n) s^j$$

$$= \sum_{j=1}^{\infty} s^j \frac{\mathbb{P}(Z_n = j, E_n)}{\mathbb{P}(E_n)} = \frac{G_n(s\eta) - G_n(0)}{\eta - G_n(0)}$$

where $G_n(s) = \mathbb{E}(s^{Z_n})$ as before, since

$$\mathbb{P}(Z_n = j, E_n) = \mathbb{P}(Z_n = j \text{ and all subsequent lines die out})$$
$$= \mathbb{P}(Z_n = j)\eta^j \quad \text{if} \quad j \geq 1,$$

and $\mathbb{P}(E_n) = \mathbb{P}(T < \infty) - \mathbb{P}(T \leq n) = \eta - G_n(0)$. Let

$$H_n(s) = \frac{\eta - G_n(s)}{\eta - G_n(0)}, \quad h(s) = \frac{\eta - G(s)}{\eta - s}, \quad 0 \leq s < \eta,$$

so that

(3) $$G_n^{\pi}(s) = 1 - H_n(s\eta).$$

Note that $H_n$ has domain $[0, \eta)$ and $G_n^{\pi}$ has domain $[0, 1)$. By Theorem (5.4.1),

$$\frac{H_n(s)}{H_{n-1}(s)} = \frac{h(G_{n-1}(s))}{h(G_{n-1}(0))}.$$

However, $G_{n-1}$ is non-decreasing, and $h$ is non-decreasing because $G$ is convex on $[0, \eta)$, giving that $H_n(s) \geq H_{n-1}(s)$ for $s < \eta$. Hence, by (3), the limits

$$\lim_{n\to\infty} G_n^{\pi}(s) = G^{\pi}(s) \quad \text{and} \quad \lim_{n\to\infty} H_n(s\eta) = H(s\eta)$$

exist for $s \in [0, 1)$ and satisfy

(4) $$G^{\pi}(s) = 1 - H(s\eta) \quad \text{if} \quad 0 \leq s < 1.$$

Thus the coefficient $_0\pi_j$ of $s^j$ in $G^\pi(s)$ exists for all $j$ as required. Furthermore, if $0 \le s < \eta$,

(5)
$$H_n(G(s)) = \frac{\eta - G_n(G(s))}{\eta - G_n(0)} = \frac{\eta - G(G_n(0))}{\eta - G_n(0)} \cdot \frac{\eta - G_{n+1}(s)}{\eta - G_{n+1}(0)}$$
$$= h(G_n(0))H_{n+1}(s).$$

As $n \to \infty$, $G_n(0) \uparrow \eta$ and so

$$h(G_n(0)) \to \lim_{s \uparrow \eta} \frac{\eta - G(s)}{\eta - s} = G'(\eta).$$

Let $n \to \infty$ in (5) to obtain

(6)
$$H(G(s)) = G'(\eta)H(s) \quad \text{if} \quad 0 \le s < \eta$$

and (2) follows from (4).                                                         ∎

**(7) Corollary.** *If $\mu \ne 1$, then $\sum_j {}_0\pi_j = 1$.*
*If $\mu = 1$, then $_0\pi_j = 0$ for all $j$.*

**Proof.** We have that $\mu = 1$ if and only if $G'(\eta) = 1$. If $\mu \ne 1$ then $G'(\eta) \ne 1$ and letting $s$ increase to $\eta$ in (6) gives $\lim_{s \uparrow \eta} H(s) = 0$; therefore, from (4), $\lim_{s \uparrow 1} G^\pi(s) = 1$, or

$$\sum_j {}_0\pi_j = 1.$$

If $\mu = 1$ then $G'(\eta) = 1$, and (2) becomes $G^\pi(G(s)) = G^\pi(s)$. However, $G(s) > s$ for all $s < 1$ and so $G^\pi(s) = G^\pi(0) = 0$ for all $s < 1$. Thus $_0\pi_j = 0$ for all $j$.        ∎

So long as $\mu \ne 1$, the distribution of $Z_n$, conditional on future extinction, converges as $n \to \infty$ to some limit $\{_0\pi_j\}$ which is a proper distribution, and may be called a *quasi-stationary distribution*. The so-called 'critical' branching process with $\mu = 1$ is more difficult to study in that, for $j \ge 1$,

$$\mathbb{P}(Z_n = j) \to 0 \quad \text{because extinction is certain,}$$
$$\mathbb{P}(Z_n = j \mid E_n) \to 0 \quad \text{because } Z_n \to \infty, \text{ conditional on } E_n.$$

However, it is possible to show, in the spirit of the discussion at the end of Section 5.4, that the distribution of

$$Y_n = \frac{Z_n}{n\sigma^2} \quad \text{where} \quad \sigma^2 = \text{var } Z_1,$$

conditional on $E_n$, converges as $n \to \infty$.

**(8) Theorem.** *If $\mu = 1$ and $G''(1) < \infty$ then $Y_n = Z_n/(n\sigma^2)$ satisfies*

$$\mathbb{P}(Y_n \le y \mid E_n) \to 1 - e^{-2y} \quad \text{as} \quad n \to \infty.$$

**Proof.** See Athreya and Ney (1972, p. 20).                                     ∎

If $\mu = 1$, the distribution of $Y_n$, given $E_n$, is asymptotically exponential with parameter 2. In this case, the branching process is called *critical*; the cases $\mu < 1$ and $\mu > 1$ are called *subcritical* and *supercritical* respectively. See Athreya and Ney 1972 for further details.

## Exercises for Section 6.7

**1.** Let $Z_n$ be the size of the $n$th generation of a branching process with $Z_0 = 1$ and $\mathbb{P}(Z_1 = k) = 2^{-k}$ for $k \geq 0$. Show directly that, as $n \to \infty$, $\mathbb{P}(Z_n \leq 2yn \mid Z_n > 0) \to 1 - e^{-2y}$, $y > 0$, in agreement with Theorem (6.7.8).

**2.** Let $Z$ be a supercritical branching process with $Z_0 = 1$ and family-size generating function $G$. Assume that the probability $\eta$ of extinction satisfies $0 < \eta < 1$. Find a way of describing the process $Z$, *conditioned on its ultimate extinction*.

**3.** Let $Z_n$ be the size of the $n$th generation of a branching process with $Z_0 = 1$ and $\mathbb{P}(Z_1 = k) = qp^k$ for $k \geq 0$, where $p + q = 1$ and $p > \frac{1}{2}$. Use your answer to Exercise (6.7.2) to show that, if we condition on the ultimate extinction of $Z$, then the process grows in the manner of a branching process with generation sizes $\tilde{Z}_n$ satisfying $\tilde{Z}_0 = 1$ and $\mathbb{P}(\tilde{Z}_1 = k) = pq^k$ for $k \geq 0$.

**4.** (a) Show that $\mathbb{E}(X \mid X > 0) \leq \mathbb{E}(X^2)/\mathbb{E}(X)$ for any random variable $X$ taking non-negative values.

(b) Let $Z_n$ be the size of the $n$th generation of a branching process with $Z_0 = 1$ and $\mathbb{P}(Z_1 = k) = qp^k$ for $k \geq 0$, where $p > \frac{1}{2}$. Use part (a) to show that $\mathbb{E}(Z_n/\mu^n \mid Z_n > 0) \leq 2p/(p - q)$, where $\mu = p/q$.

(c) Show that, in the notation of part (b), $\mathbb{E}(Z_n/\mu^n \mid Z_n > 0) \to p/(p - q)$ as $n \to \infty$.

## 6.8  Birth processes and the Poisson process

Many processes in nature may change their values at any instant of time rather than at certain specified epochs only. Such a process is a family $\{X(t) : t \geq 0\}$ of random variables indexed by the half-line $[0, \infty)$ and taking values in a state space $S$. Depending on the underlying random mechanism, $X$ may or may not be a Markov process. Before attempting to study any general theory of continuous-time processes, we explore some simple but non-trivial examples in detail: first, the Poisson process in Subsection (A), followed by the birth process in (B). These inform our discussion of some basic properties of Markov processes in Subsection (C).

**(A) The Poisson process.** Given the right equipment, we should have no difficulty in observing that the process of emission of particles from a radioactive source seems to behave in a manner which is not totally predictable. If we switch on our Geiger counter at time zero, then the reading $N(t)$ which it shows at a later time $t$ is the outcome of some random process. This process $\{N(t) : t \geq 0\}$ has certain natural properties, such as:

(a)  $N(0) = 0$, and $N(t) \in \{0, 1, 2, \dots\}$,

(b)  if $s < t$ then $N(s) \leq N(t)$,

but it is not so easy to specify more detailed properties. We might use the following description. In the time interval $(t, t + h)$ there may or may not be some emissions. If $h$ is small then the likelihood of an emission is roughly proportional to $h$; it is not very likely that two or more emissions will occur in a small interval. More formally, we make the following definition of a Poisson process†.

---

†Developed separately but contemporaneously by Erlang, Bateman, and Campbell in 1909, and named after Poisson by Feller before 1940.

**(1) Definition.** A **Poisson process with intensity** (or **rate**) $\lambda$ is a process $N = \{N(t) : t \geq 0\}$ taking values in $S = \{0, 1, 2, \ldots\}$ such that:

(a) $N(0) = 0$, and $N(s) \leq N(t)$ for $s < t$,

(b) $\mathbb{P}\big(N(t + h) = n + m \,\big|\, N(t) = n\big) = \begin{cases} \lambda h + o(h) & \text{if } m = 1, \\ o(h) & \text{if } m > 1, \\ 1 - \lambda h + o(h) & \text{if } m = 0, \end{cases}$

(c) if $s < t$, the number $N(t) - N(s)$ of emissions in the interval $(s, t]$ is independent of the times of emissions during $[0, s]$.

We speak of $N(t)$ as the number of 'arrivals' or 'occurrences' or 'events', or in this example 'emissions', of the process by time $t$. The process $N$ is called a 'counting process' and is one of the simplest and most fundamental examples of a continuous-time Markov chain. We shall consider the general theory of such processes in the next section; here we study special properties of Poisson processes and their generalizations.

There are in fact several ways of defining a Poisson process, including: (i) the above 'infinitesimal' definition, (ii) via so-called 'interarrival times' (see Theorem (10)), and (iii) via general distributional properties (see Example (28)). We concentrate on the first two of these in this section.

We are interested first in the distribution of the random variable $N(t)$.

**(2) Theorem.** *The random variable $N(t)$ has the Poisson distribution with parameter $\lambda t$. That is to say,*

$$\mathbb{P}\big(N(t) = j\big) = \frac{(\lambda t)^j}{j!} e^{-\lambda t}, \qquad j = 0, 1, 2, \ldots .$$

**Proof.** Condition $N(t + h)$ on $N(t)$ to obtain

$$\mathbb{P}\big(N(t + h) = j\big) = \sum_i \mathbb{P}\big(N(t) = i\big)\mathbb{P}\big(N(t + h) = j \,\big|\, N(t) = i\big)$$

$$= \sum_i \mathbb{P}\big(N(t) = i\big)\mathbb{P}\big((j - i) \text{ arrivals in } (t, t + h]\big)$$

$$= \mathbb{P}\big(N(t) = j - 1\big)\mathbb{P}(\text{one arrival}) + \mathbb{P}\big(N(t) = j\big)\mathbb{P}(\text{no arrivals}) + o(h).$$

Thus $p_j(t) = \mathbb{P}(N(t) = j)$ satisfies

$$p_j(t + h) = \lambda h p_{j-1}(t) + (1 - \lambda h) p_j(t) + o(h) \quad \text{if} \quad j \neq 0,$$
$$p_0(t + h) = (1 - \lambda h) p_0(t) + o(h).$$

Subtract $p_j(t)$ from each side of the first of these equations, divide by $h$, and let $h \downarrow 0$ to obtain

(3)    $$p_j'(t) = \lambda p_{j-1}(t) - \lambda p_j(t) \quad \text{if} \quad j \neq 0,$$

and likewise

(4)    $$p_0'(t) = -\lambda p_0(t).$$

The boundary condition is

**(5)**
$$p_j(0) = \delta_{j0} = \begin{cases} 1 & \text{if } j = 0, \\ 0 & \text{if } j \neq 0. \end{cases}$$

Equations (3) and (4) form a collection of differential–difference equations. Here are two methods of solution, both of which have applications elsewhere.

*Method A. Induction.* Solve (4) subject to the condition $p_0(0) = 1$ to obtain $p_0(t) = e^{-\lambda t}$. Substitute this into (3) with $j = 1$ to obtain $p_1(t) = \lambda t e^{-\lambda t}$ and iterate, to obtain by induction that
$$p_j(t) = \frac{(\lambda t)^j}{j!} e^{-\lambda t}.$$

*Method B. Generating functions.* Define the generating function
$$G(s, t) = \sum_{j=0}^{\infty} p_j(t) s^j = \mathbb{E}(s^{N(t)}).$$

Multiply (3) by $s^j$ and sum over $j$ to obtain
$$\frac{\partial G}{\partial t} = \lambda(s - 1)G$$

with the boundary condition $G(s, 0) = 1$. The solution is, as required,

**(6)**
$$G(s, t) = e^{\lambda(s-1)t} = e^{-\lambda t} \sum_{j=0}^{\infty} \frac{(\lambda t)^j}{j!} s^j. \qquad \blacksquare$$

This result seems very like the account in Example (3.5.4) that the binomial $\text{bin}(n, p)$ distribution approaches the Poisson distribution if $n \to \infty$ and $np \to \lambda$. Why is this no coincidence?

There is an important alternative and equivalent formulation of a Poisson process which provides much insight into its behaviour. Let $T_0, T_1, \ldots$ be the *arrival times* given by

**(7)**
$$T_0 = 0, \qquad T_n = \inf\{t : N(t) = n\}.$$

Then $T_n$ is the time of the $n$th arrival. The *interarrival times* are the random variables $U_0, U_1, \ldots$ given by

**(8)**
$$U_n = T_{n+1} - T_n, \qquad n \geq 0.$$

From knowledge of $N$, we can find the values of $U_0, U_1, \ldots$ by (7) and (8). Conversely, we can reconstruct $N$ from a knowledge of the $U_i$ by

**(9)**
$$T_n = \sum_{i=0}^{n-1} U_i, \qquad N(t) = \max\{n : T_n \leq t\}.$$

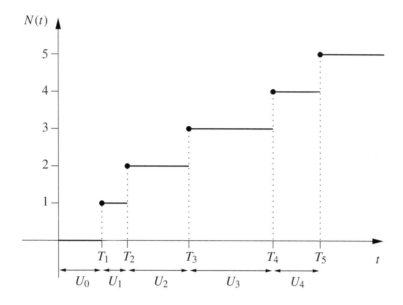

Figure 6.1. A typical realization of a Poisson process $N(t)$.

Figure 6.1 is an illustration of this.

**(10) Theorem.** *The random variables $U_0, U_1, \ldots$ are independent, each having the exponential distribution with parameter $\lambda$.*

There is an important generalization of this result to arbitrary continuous-time Markov chains with countable state space (see Theorem (6.9.6)). We shall investigate this in the next section.

**Proof.** First consider $U_0$:

$$\mathbb{P}(U_0 > t) = \mathbb{P}(N(t) = 0) = e^{-\lambda t}$$

and so $U_0$ is exponentially distributed with parameter $\lambda$. Now, conditional on $U_0$,

$$\mathbb{P}(U_1 > t \mid U_0 = t_0) = \mathbb{P}(\text{no arrival in } (t_0, t_0 + t] \mid U_0 = t_0).$$

The event $\{U_0 = t_0\}$ relates to arrivals during the time interval $[0, t_0]$, whereas the event $\{$no arrival in $(t_0, t_0 + t]\}$ relates to arrivals after time $t_0$. These events are independent, by (1c), and therefore

$$\mathbb{P}(U_1 > t \mid U_0 = t_0) = \mathbb{P}(\text{no arrival in } (t_0, t_0 + t]) = e^{-\lambda t}.$$

Thus $U_1$ is independent of $U_0$, and has the same distribution. Similarly,

$$\mathbb{P}(U_n > t \mid U_0 = t_0, \ldots, \ U_{n-1} = t_{n-1}) = \mathbb{P}(\text{no arrival in } (T, T + t])$$

where $T = t_0 + t_1 + \cdots + t_{n-1}$, and the claim of the theorem follows by induction on $n$. ∎

It is not difficult to see that the process $N$, constructed by (9) from a sequence $U_0, U_1, \ldots$, is a Poisson process if and only if the $U_i$ are independent identically distributed exponential variables (*exercise*: use the lack-of-memory property of Problem (4.14.5)). If the $U_i$ form such a sequence, it is a simple matter to deduce the distribution of $N(t)$ directly, as follows. In this case, $T_n = \sum_{i=0}^{n-1} U_i$ is $\Gamma(\lambda, n)$ and $N(t)$ is specified by the useful remark that

$$N(t) \geq j \quad \text{if and only if} \quad T_j \leq t.$$

Therefore

$$\mathbb{P}\big(N(t) = j\big) = \mathbb{P}(T_j \leq t < T_{j+1}) = \mathbb{P}(T_j \leq t) - \mathbb{P}(T_{j+1} \leq t)$$

$$= \frac{(\lambda t)^j}{j!} e^{-\lambda t}$$

using the properties of gamma variables and integration by parts (see Problem (4.14.11c)).

Finally in this subsection, we consider the following question. Given that $N(t) = n$, what can be said about the times of the first $n$ arrivals?

**(11) Theorem.** *The joint distribution of the arrival times $T_1, T_2, \ldots, T_n$, conditional on the event $\{N(t) = n\}$, is the same as the joint distribution of the order statistics of a family of $n$ independent variables which are uniformly distributed on $[0, t]$.*

This is something of a mouthful, and asserts that, if we know that there are $n$ arrivals by time $t$, then the set of arrival times is indistinguishable from a set of $n$ points chosen uniformly at random in the interval $[0, t]$. Recall the *order statistics* of Problem (4.14.21).

**Proof.** We seek the conditional density function of $\mathbf{T} = (T_1, T_2, \ldots, T_n)$ given that $N(t) = n$. First note that the interarrival times $U_0, U_1, \ldots, U_{n-1}$ are independent exponential variables with parameter $\lambda$ so that $\mathbf{U} = (U_0, U_1, \ldots, U_{n-1})$ has joint density function

$$f_{\mathbf{U}}(\mathbf{u}) = \lambda^n \exp\left(-\lambda \sum_{i=0}^{n-1} u_i\right).$$

Make the transformation $\mathbf{U} \mapsto \mathbf{T}$ and use the change of variable formula (4.7.4) to find that

$$f_{\mathbf{T}}(\mathbf{t}) = \lambda^n e^{-\lambda t_n} \quad \text{if} \quad t_1 < t_2 < \cdots < t_n.$$

Let $C \subset \mathbb{R}^n$. Then

(12)  $$\mathbb{P}\big(\mathbf{T} \in C \mid N(t) = n\big) = \frac{\mathbb{P}(N(t) = n \text{ and } \mathbf{T} \in C)}{\mathbb{P}(N(t) = n)},$$

where

(13)  $$\mathbb{P}\big(N(t) = n \text{ and } \mathbf{T} \in C\big) = \int_C \mathbb{P}\big(N(t) = n \mid \mathbf{T} = \mathbf{t}\big) f_{\mathbf{T}}(\mathbf{t})\, d\mathbf{t}$$

$$= \int_C \mathbb{P}\big(N(t) = n \mid T_n = t_n\big) f_{\mathbf{T}}(\mathbf{t})\, d\mathbf{t}, \quad \text{for} \quad t_n \leq t,$$

and

**(14)**     $\mathbb{P}\big(N(t) = n \mid T_n = t_n\big) = \mathbb{P}(U_{n+1} > t - t_n) = e^{-\lambda(t-t_n)}, \quad$ for $\quad t_n \leq t.$

Substitute (14) into (13) and (13) into (12) to obtain

$$\mathbb{P}\big(\mathbf{T} \in C \mid N(t) = n\big) = \int_C L(\mathbf{t}) n! \, t^{-n} \, d\mathbf{t}$$

where

$$L(\mathbf{t}) = \begin{cases} 1 & \text{if } t_1 < t_2 < \cdots < t_n, \\ 0 & \text{otherwise.} \end{cases}$$

We recognize $g(\mathbf{t}) = L(\mathbf{t}) n! \, t^{-n}$ from the result of Problem (4.14.23) as the joint density function of the order statistics of $n$ independent uniform variables on $[0, t]$.  ∎

**(B) Birth processes and explosion.** The Poisson process is a very satisfactory model for radioactive emissions from a sample of uranium-235 since this isotope has a half-life of $7 \times 10^8$ years and decays fairly slowly. However, for a newly produced sample of strontium-92, which has a half-life of 2.7 hours, we need a more sophisticated process which takes into account the retardation in decay rate over short time intervals. We might suppose that the rate $\lambda$ at which emissions are detected depends on the number detected already.

**(15) Definition.** A **birth process with birth rates** (or **intensities**) $\lambda_0, \lambda_1, \ldots$ is a process $\{N(t) : t \geq 0\}$ taking values in $S = \{0, 1, 2, \ldots\}$ such that:
  (a) $N(0) \geq 0$, and $N(s) \leq N(t)$ for $s < t$,
  (b) $\mathbb{P}\big(N(t+h) = n + m \mid N(t) = n\big) = \begin{cases} \lambda_n h + o(h) & \text{if } m = 1, \\ o(h) & \text{if } m > 1, \\ 1 - \lambda_n h + o(h) & \text{if } m = 0, \end{cases}$
  (c) if $s < t$ then, conditional on the value of $N(s)$, the increment $N(t) - N(s)$ is independent of all arrivals prior to $s$.

Here are some interesting special cases.

(a) **Poisson process.** $\lambda_n = \lambda$ for all $n$.  ●

(b) **Simple birth.** $\lambda_n = n\lambda$. This models the growth of a population in which living individuals give birth independently of one another, each giving birth to a new individual with probability $\lambda h + o(h)$ in the interval $(t, t + h)$. No individuals may die. The number $M$ of births in the interval $(t, t + h)$ satisfies:

$$\mathbb{P}\big(M = m \mid N(t) = n\big) = \binom{n}{m} (\lambda h)^m (1 - \lambda h)^{n-m} + o(h)$$

$$= \begin{cases} 1 - n\lambda h + o(h) & \text{if } m = 0, \\ n\lambda h + o(h) & \text{if } m = 1, \\ o(h) & \text{if } m > 1. \end{cases}$$

Note the effective birth rate $\lambda n$ when in state $n$.  ●

**(c) Simple birth with immigration.** $\lambda_n = n\lambda + \nu$. This models a simple birth process which experiences immigration at constant rate $\nu$ from elsewhere.    ●

A significant complication arises when the birth rates $\lambda_n$ increase quickly with $n$, namely that the process $N$ may pass through all (finite) states in *bounded* time. We say that *explosion* occurs if this happens with a strictly positive probability. With $T_n$ the time of the $n$th arrival, let $T_\infty = \lim_{n \to \infty} T_n$.

**(16) Definition.** We call the process $N$ **non-explosive** (or **honest**) if $\mathbb{P}(T_\infty = \infty) = 1$, and **explosive** otherwise.

**(17) Theorem.** *The birth process $N$ with strictly positive birth rates $\lambda_n$ is non-explosive if and only if $\sum_n \lambda_n^{-1} = \infty$.*

This beautiful theorem asserts that if the birth rates are not too large then $N(t)$ is almost surely finite, but if they are sufficiently large that $\sum \lambda_n^{-1}$ converges then births occur so frequently that there is positive probability of infinitely many births occurring in a finite interval of time; thus $N(t)$ may take the value $+\infty$ instead of a non-negative integer. When this happens, we have that $\sum_j p_{ij}(t) < 1$, and the deficit $1 - \sum_j p_{ij}(t)$ is the probability $\mathbb{P}(T_\infty \le t)$ of escaping to infinity by time $t$, starting from $i$.

Theorem (17) is a consequence of the following lemma.

**(18) Lemma.** *Let $U_0, U_1, \ldots$ be independent random variables, $U_n$ having the exponential distribution with parameter $\lambda_n$, and let $T_\infty = \sum_n U_n$. We have that*

$$\mathbb{P}(T_\infty < \infty) = \begin{cases} 0 & \text{if } \sum_n \lambda_n^{-1} = \infty, \\ 1 & \text{if } \sum_n \lambda_n^{-1} < \infty. \end{cases}$$

**Proof.** By equation (5.6.13),

$$\mathbb{E}(T_\infty) = \mathbb{E}\left(\sum_{n=0}^{\infty} U_n\right) = \sum_{n=0}^{\infty} \frac{1}{\lambda_n}.$$

If $\sum_n \lambda_n^{-1} < \infty$ then $\mathbb{E}(T_\infty) < \infty$, whence $\mathbb{P}(T_\infty = \infty) = 0$.

In order to study the atom of $T_\infty$ at $\infty$ we work with the bounded random variable $e^{-T_\infty}$, defined as the limit as $n \to \infty$ of $e^{-T_n}$. By monotone convergence (5.6.12),

$$\mathbb{E}(e^{-T_\infty}) = \mathbb{E}\left(\prod_{n=0}^{\infty} e^{-U_n}\right) = \lim_{N \to \infty} \mathbb{E}\left(\prod_{n=0}^{N} e^{-U_n}\right)$$

$$= \lim_{N \to \infty} \prod_{n=0}^{N} \mathbb{E}(e^{-U_n}) \qquad \text{by independence}$$

$$= \lim_{N \to \infty} \prod_{n=0}^{N} \frac{1}{1 + \lambda_n^{-1}} = \left\{\prod_{n=0}^{\infty} (1 + \lambda_n^{-1})\right\}^{-1}.$$

The last product† equals $\infty$ if $\sum_n \lambda_n^{-1} = \infty$, implying in turn that $\mathbb{E}(e^{-T_\infty}) = 0$. However, $e^{-T_\infty} \ge 0$, and therefore $\mathbb{P}(T_\infty = \infty) = \mathbb{P}(e^{-T_\infty} = 0) = 1$ as required.    ■

---

†See Subsection (8) of Appendix I for some notes about infinite products.

We now turn to a pair of differential equations that govern the transition probabilities

$$p_{ij}(t) = \mathbb{P}\big(N(s+t) = j \mid N(s) = i\big) = \mathbb{P}\big(N(t) = j \mid N(0) = i\big)$$

of a birth process $N$. The phenomenon of explosion complicates the derivation and analysis of such equations, and we return in Section 6.10 (see Theorem (6.10.7)) for a rigorous general account, confining ourselves here to an account tailored to the birth process.

Let $N$ be a birth process with strictly positive intensities $\lambda_0, \lambda_1, \ldots$. Let us proceed as for the Poisson process. We condition $N(t+h)$ on $N(t)$ and let $h \downarrow 0$ as we did for (3) and (4), to obtain the so-called

**(19) Forward system of equations:** $p'_{ij}(t) = \lambda_{j-1} p_{i,j-1}(t) - \lambda_j p_{ij}(t)$ for $j \geq i$,

with the convention that $\lambda_{-1} = 0$, and the boundary condition $p_{ij}(0) = \delta_{ij}$. Alternatively we might condition $N(t+h)$ on $N(h)$ and let $h \downarrow 0$ to obtain the so-called

**(20) Backward system of equations:** $p'_{ij}(t) = \lambda_i p_{i+1,j}(t) - \lambda_i p_{ij}(t)$ for $j \geq i$,

with the boundary condition $p_{ij}(0) = \delta_{ij}$. These derivations are not rigorous since they involve interchanges of limiting operations.

Can we solve these equations as we did for the Poisson process?

**(21) Theorem.** *The forward system has a unique solution, which satisfies the backward system.*

**Proof.** Note first that $p_{ij}(t) = 0$ if $j < i$. Solve the forward equation with $j = i$ to obtain $p_{ii}(t) = e^{-\lambda_i t}$. Substitute into the forward equation with $j = i+1$ to find $p_{i,i+1}(t)$. Continue this operation to deduce that the forward system has a unique solution. To obtain more information about this solution, define the Laplace transforms‡

$$\widehat{p}_{ij}(\theta) = \int_0^\infty e^{-\theta t} p_{ij}(t)\, dt.$$

Transform the forward system to obtain

$$(\theta + \lambda_j)\widehat{p}_{ij}(\theta) = \delta_{ij} + \lambda_{j-1}\widehat{p}_{i,j-1}(\theta).$$

This is a difference equation which is readily solved to obtain

$$\text{(22)} \qquad \widehat{p}_{ij}(\theta) = \frac{1}{\lambda_j} \frac{\lambda_i}{\theta + \lambda_i} \frac{\lambda_{i+1}}{\theta + \lambda_{i+1}} \cdots \frac{\lambda_j}{\theta + \lambda_j} \quad \text{for} \quad j \geq i.$$

This determines $p_{ij}(t)$ uniquely by the inversion theorem for Laplace transforms.

To see that this solution satisfies the backward system, transform this system similarly to obtain that any solution $\pi_{ij}(t)$ to the backward equation, with Laplace transform

$$\widehat{\pi}_{ij}(\theta) = \int_0^\infty e^{-\theta t} \pi_{ij}(t)\, dt,$$

‡See Section F of Appendix I for some properties of Laplace transforms.

satisfies

$$(\theta + \lambda_i)\widehat{\pi}_{ij}(\theta) = \delta_{ij} + \lambda_i\widehat{\pi}_{i+1,j}(\theta).$$

The $\widehat{p}_{ij}$, given by (22), satisfy this equation, and so the $p_{ij}$ satisfy the backward system.    ■

We have not been able to show that the backward system has a unique solution, for the very good reason that this may not be true. All we can show is that it has a minimal solution.

**(23) Theorem.** *If* $\{p_{ij}(t)\}$ *is the unique solution of the forward system, then any solution* $\{\pi_{ij}(t)\}$ *of the backward system satisfies* $p_{ij}(t) \le \pi_{ij}(t)$ *for all* $i, j, t$.

**Proof.** See Feller (1968, pp. 475–477) for the proof. A more general result is presented at Theorem (6.10.7).    ■

There may seem something wrong here, because the condition

**(24)**                                $$\sum_j p_{ij}(t) = 1$$

in conjunction with the result of (23) would constrain $\{p_{ij}(t)\}$ to be the *unique* solution of the backward system. The point is that (24) may fail to hold, since it is equivalent to $\mathbb{P}(T_\infty > t) = 1$. That is to say, (24) holds for all $t$ if and only if $N$ is honest.

The forward and backward equations (19)–(20) are sometimes called the Kolmogorov equations, and they are studied rigorously, and in greater generality, in Section 6.10. We shall see there that the 'unique' solution of Theorem (23) corresponds to the so-called 'minimal chain' introduced around the forthcoming (6.9.12). We shall conclude that the backward equation has a unique solution (subject to the boundary condition) if and only if the process is non-explosive.

In summary, we have considered several random processes, indexed by continuous time, which model phenomena occurring in nature. However, certain dangers arise unless we take care in the construction of such processes. They may even find a way to the so-called 'boundary' of the state space by exploding in finite time.

**(C) The strong Markov property and right-continuity.** We terminate this section with brief discussions of the Markov property for birth processes, and of the continuity of random processes.

Recall that a sequence $X = \{X_n : n \ge 0\}$ is said to satisfy the Markov property if, conditional on the event $\{X_n = i\}$, events relating to the collection $\{X_m : m > n\}$ are independent of events relating to $\{X_m : m < n\}$. Birth processes have a similar property. Let $N$ be a birth process and let $T$ be a fixed time. Conditional on the event $\{N(T) = i\}$, the evolution of the process subsequent to time $T$ is independent of that prior to $T$; this is an immediate consequence of (15c), and is called the 'weak Markov property'. It is often desirable to make use of a stronger property, in which $T$ is allowed to be a *random variable* rather than merely a constant. On the other hand, such a conclusion cannot be valid for all random $T$, since if $T$ 'looks into the future' as well as the past, then information about the past may generally be relevant to the future (*exercise*: find a random variable $T$ for which the desired conclusion is false). A useful class of random times are those whose values depend only on the past, and here is a formal definition. We call the random time $T$ a *stopping time* for the process $N$ if, for all $t \ge 0$, the indicator function of the event $\{T \le t\}$ is a function of the values $\{N(s) : s \le t\}$ of the process up to time $t$; that is to say, we require that it be decidable whether or not $T$ has occurred by time $t$ knowing only the values of the process up to time $t$. Examples of stopping times are the times $T_1, T_2, \ldots$ of arrivals; examples of times which are not stopping times are $T_4 - 2$, $\frac{1}{2}(T_1 + T_2)$, and other random variables which 'look into the future'. Recall the limit $T_\infty = \lim_{n \to \infty} T_n$.

**(25) Theorem. Strong Markov property.** *Let $N$ be a birth process and let $T$ be a stopping time for $N$. Conditional on the event $I = \{T < T_\infty\} \cap \{N(T) = i\}$, the process $N^* = \{N^*(u) = N(T + u) : u \geq 0\}$ is a birth process with initial size $i$, and is independent of the past $\{N(s) : s \leq T\}$.*

**Outline proof.** The following argument may be made rigorous. Let $B$ denote a complete description of the past $\{N(s) : s \leq T\}$ (problems of measurability arise here, but these are not serious since birth processes have only countably many arrivals). Knowledge of $B$ carries with it knowledge of the value of the stopping time $T$, which we write as $T = T(B)$.

Let $A$ be an event defined in terms of the 'future' $\{N(T + u) : u \geq 0\}$, and consider the conditional probability $\mathbb{P}(A \mid I \cap B)$. The event $I \cap B$ specifies: (i) the value of $T$, (ii) the value of $N(T)$, and (iii) the history of the process up to time $T$; it is by virtue of the fact that $T$ is a stopping time that this event is defined in terms of $\{N(s) : s \leq T(B)\}$. By the *weak* Markov property, since $T$ is constant on this event, we may discount information in (iii), so that

$$(26) \qquad\qquad \mathbb{P}(A \mid I \cap B) = \mathbb{P}\big(A \,\big|\, N(T) = i,\ T = T(B) < T_\infty\big).$$

The process $N$ is temporally homogeneous, and $A$ is defined in terms of $\{N(s) : s > T\}$. Therefore, the (conditional) probability of $A$ depends only on the value of $N(T)$, which is to say that

$$\mathbb{P}\big(A \,\big|\, N(T) = i,\ T = T(B) < T_\infty\big) = \mathbb{P}\big(A \,\big|\, T < T_\infty,\ N(T) = i\big)$$

as required.                                                                        ∎

We used two properties of birth processes in our outline proof of the strong Markov property: temporal homogeneity and the weak Markov property. The strong Markov property plays an important role in the study of continuous-time Markov chains and processes, and we shall encounter it in a more general form later. When applied to a birth process $N$, it implies that the new process $N'$, defined by $N'(t) = N(t + T) - N(T)$, $t \geq 0$, conditional on $\{T < \infty,\ N(T) = i\}$ is also a birth process, whenever $T$ is a stopping time for $N$; it is easily seen that this new birth process has intensities $\lambda_i, \lambda_{i+1}, \ldots$. In the case of the Poisson process, we have that $N'(t) = N(t + T) - N(T)$ is a Poisson process also.

We turn now to the issue of continuity of sample paths.

**(27) Right-continuity of random processes.** Let $M = \{M(t) : t \geq 0\}$ be a random process taking values in some countable set $S$. There is an underlying sample space $\Omega$, and we may write $M(\cdot, \omega)$ for the 'sample path' of $M$ corresponding to $\omega \in \Omega$. The sample path $M(\cdot, \omega)$ is called *right-continuous* if, for $t \in [0, \infty)$, there exists $\epsilon_{t,\omega} > 0$ such that

$$M(t, \omega) = M(t + u, \omega) \quad \text{for} \quad 0 \leq u < \epsilon_{t,\omega},$$

which is to say that $M(\cdot, \omega)$ is constant on the interval $[t, t + \epsilon)$. The process $M$ is called *right-continuous* if all its sample paths are right-continuous†. The sample paths of a right-continuous process are step functions.                                                   ●

---

†With $S$ endowed with the discrete topology, this is the usual notion of right-continuity.

The processes considered here will normally be assumed to be right-continuous. There are two reasons for this. First, there is a technical reason to do with measurability which arises since the underlying probability measure is *countably* additive, whereas the index set of $X$ is the *uncountable* set $\mathbb{R}$ (see, for example, Norris 1997). Secondly, as we will see in the next section, the complications of the general theory of continuous-time Markov chains are partly neutralized by the assumption of right-continuity.

**(28) Example.** A Poisson process $N$ is said to have 'stationary independent increments', since: (a) the distribution of $N(t) - N(s)$ depends only on $t - s$, and (b) the increments $\{N(t_i) - N(s_i) : i = 1, 2, \ldots, n\}$ are independent if $s_1 \le t_1 \le s_2 \le t_2 \le \cdots \le t_n$. This property is nearly a characterization of the Poisson process.

Suppose that $M = \{M(t) : t \ge 0\}$ is a non-decreasing right-continuous integer-valued process with $M(0) = 0$, having stationary independent increments, and with the extra property that $M$ has only jump discontinuities of size 1. Note first that, for $u, v \ge 0$,

$$\mathbb{E}M(u + v) = \mathbb{E}M(u) + \mathbb{E}\big[M(u + v) - M(u)\big] = \mathbb{E}M(u) + \mathbb{E}M(v)$$

by the assumption of stationary increments. Now $\mathbb{E}M(u)$ is non-decreasing in $u$, so that there exists $\lambda$ such that

**(29)**                                    $\mathbb{E}M(u) = \lambda u, \qquad u \ge 0.$

Let $T = \sup\{t : M(t) = 0\}$ be the time of the first jump of $M$. We have from the right-continuity of $M$ that $M(T) = 1$ (almost surely), so that $T$ is a stopping time for $M$. Now

**(30)**                                    $\mathbb{E}M(s) = \mathbb{E}\big\{\mathbb{E}(M(s) \mid T)\big\}.$

Certainly $\mathbb{E}(M(s) \mid T) = 0$ if $s < T$, and for $s \ge t$

$$
\begin{aligned}
\mathbb{E}\big(M(s) \mid T = t\big) &= \mathbb{E}\big(M(t) \mid T = t\big) + \mathbb{E}\big(M(s) - M(t) \mid T = t\big) \\
&= 1 + \mathbb{E}\big(M(s) - M(t) \mid M(t) = 1, \ M(u) = 0 \text{ for } u < t\big) \\
&= 1 + \mathbb{E}M(s - t)
\end{aligned}
$$

by the assumption of stationary independent increments. We substitute this into (30) to obtain

$$\mathbb{E}M(s) = \int_0^s \big[1 + \mathbb{E}M(s - t)\big] \, dF(t)$$

where $F$ is the distribution function of $T$. Now $\mathbb{E}M(s) = \lambda s$ for all $s$, so that

**(31)**                                    $\lambda s = F(s) + \lambda \int_0^s (s - t) \, dF(t),$

an integral equation for the unknown function $F$. One of the standard ways of solving such an equation is to use Laplace transforms. We leave it as an *exercise* to deduce from (31) that $F(t) = 1 - e^{-\lambda t}, t \ge 0$, so that $T$ has the exponential distribution. An argument similar to that used for Theorem (10) now shows that the 'inter-jump' times of $M$ are independent and have the exponential distribution. Hence $M$ is a Poisson process with intensity $\lambda$.    ●

## Exercises for Section 6.8

**1.   Superposition.** Flies and wasps land on your dinner plate in the manner of independent Poisson processes with respective intensities $\lambda$ and $\mu$. Show that the arrivals of flying objects form a Poisson process with intensity $\lambda + \mu$.

**2.   Thinning.** Insects land in the soup in the manner of a Poisson process with intensity $\lambda$, and each such insect is green with probability $p$, independently of the colours of all other insects. Show that the arrivals of green insects form a Poisson process with intensity $\lambda p$.

**3.** Let $T_n$ be the time of the $n$th arrival in a Poisson process $N$ with intensity $\lambda$, and define the excess lifetime process $E(t) = T_{N(t)+1} - t$, being the time one must wait subsequent to $t$ before the next arrival. Show by conditioning on $T_1$ that

$$\mathbb{P}\big(E(t) > x\big) = e^{-\lambda(t+x)} + \int_0^t \mathbb{P}\big(E(t-u) > x\big)\lambda e^{-\lambda u}\, du.$$

Solve this integral equation in order to find the distribution function of $E(t)$. Explain your conclusion.

**4.** Let $B$ be a simple birth process of paragraph (6.8.15b) with $B(0) = I$; the birth rates are $\lambda_n = n\lambda$. Write down the forward system of equations for the process and deduce that

$$\mathbb{P}\big(B(t) = k\big) = \binom{k-1}{I-1} e^{-I\lambda t}\big(1 - e^{-\lambda t}\big)^{k-I}, \qquad k \geq I.$$

Show also that $\mathbb{E}(B(t)) = Ie^{\lambda t}$ and $\operatorname{var}(B(t)) = Ie^{2\lambda t}(1 - e^{-\lambda t})$.

**5.** Let $B$ be a process of simple birth with immigration (6.8.11c) with parameters $\lambda$ and $\nu$, and with $B(0) = 0$; the birth rates are $\lambda_n = n\lambda + \nu$. Write down the sequence of differential–difference equations for $p_n(t) = \mathbb{P}(B(t) = n)$. Without solving these equations, use them to show that $m(t) = \mathbb{E}(B(t))$ satisfies $m'(t) = \lambda m(t) + \nu$, and solve for $m(t)$.

**6.** Let $N$ be a birth process with intensities $\lambda_0, \lambda_1, \ldots$, and let $N(0) = 0$. Show that $p_n(t) = \mathbb{P}(N(t) = n)$ is given by

$$p_n(t) = \frac{1}{\lambda_n} \sum_{i=0}^{n} \lambda_i e^{-\lambda_i t} \prod_{\substack{j=0 \\ j \neq i}}^{n} \frac{\lambda_j}{\lambda_j - \lambda_i}$$

provided that $\lambda_i \neq \lambda_j$ whenever $i \neq j$.

**7.** Suppose that the general birth process of the previous exercise is such that $\sum_n \lambda_n^{-1} < \infty$. Show that $\lambda_n p_n(t) \to f(t)$ as $n \to \infty$ where $f$ is the density function of the random variable $T = \sup\{t : N(t) < \infty\}$. Deduce that $\mathbb{E}(N(t) \mid N(t) < \infty)$ is finite or infinite depending on the convergence or divergence of $\sum_n n\lambda_n^{-1}$.

Find the Laplace transform of $f$ in closed form for the case when $\lambda_n = (n + \frac{1}{2})^2$, and deduce an expression for $f$.

**8.   Traffic lights.** A traffic light is green at time 0, and subsequently alternates between green and red at the instants of a Poisson process with intensity $\lambda$. Starting from time $x > 0$, let $W(x)$ be the waiting time until the light is green for the first time. Find the distribution of $W(x)$.

**9.   Conditional property of simple birth.** Let $X = \{X(t) : t \geq 0\}$ be a simple birth process with rate $\lambda$, and let $X(0) = 1$. Let $b \geq 1$. Show that, conditional on the event $\{X(t) = b + 1\}$, the times of the $b$ births have the same distribution as the order statistics of a random sample of size $b$ from the density function

$$f(x) = \frac{\lambda e^{-\lambda(t-x)}}{1 - e^{-\lambda t}}, \qquad 0 \leq x \leq t.$$

**10. Coincidences.** Boulders fall down a chute (or couloir) at the instants of a Poisson process with intensity $\lambda$, and mountaineers ascend the chute at the instants of a Poisson process with intensity $\mu$ (the two processes are independent of one another). If a fall and an ascent occur during any interval of length $c$ or less, it is said that a *coincidence* has occurred. Show that the time $T$ until the first coincidence has mean

$$\mathbb{E}T = \frac{1}{\lambda+\mu}\left\{1+\frac{\lambda^2+\mu^2}{\lambda\mu}+2e^{-(\lambda+\mu)c}-e^{-2(\lambda+\mu)c}\right\}\Big/\left\{1-e^{-2(\lambda+\mu)c}\right\}.$$

Show that $c\mathbb{E}(T) \to (2\lambda\mu)^{-1}$ as $c \downarrow 0$. Can you prove the last result directly?

**11.** Let $S_1$, $S_2$ be the times of the first two arrivals in a Poisson process with intensity $\lambda$, in order of arrival. Show that

$$\mathbb{P}(s < S_1 \leq t < S_2) = \lambda(t-s)e^{-\lambda t}, \qquad 0 < s < t < \infty,$$

and deduce the joint density function of $S_1$ and $S_2$.

**12. Gig economy.** A freelance salesman is paid $R$ units for each sale, and commissions arrive at the instants of a Poisson process of rate $\lambda$. Living costs consume his resources at unit rate. If his initial wealth is $S > 0$, show that the probability that he ever becomes bankrupt is $a^{S/R}$, where $a$ is the smallest $x > 0$ such that $x = e^{\lambda R(x-1)}$.

Prove that $a \in (0, 1)$ if $\lambda R > 1$, while $a = 1$ if $\lambda R < 1$.

---

# 6.9 Continuous-time Markov chains

Let $X = \{X(t) : t \geq 0\}$ be a family of random variables taking values in some countable state space $S$ and indexed by the half-line $[0, \infty)$. We shall generally assume that $S$ is a subset of the integers.

## (A) Markov property.

**(1) Definition.** The process $X$ satisfies the **Markov property** if

$$\mathbb{P}\big(X(t_n) = j \mid X(t_1) = i_1, \ldots, X(t_{n-1}) = i_{n-1}\big) = \mathbb{P}\big(X(t_n) = j \mid X(t_{n-1}) = i_{n-1}\big)$$

for all $j, i_1, \ldots, i_{n-1} \in S$ and any increasing sequence $t_1 < t_2 < \cdots < t_n$ of times.

There is some complexity in the general theory of processes with the Markov property when $S$ is infinite (the finite case is more straightforward). The way to handle this in generality is too difficult to describe in detail here, and the reader should look elsewhere (see, e.g., Chung 1960, Freedman 1971, or Liggett 2010). We follow a different route by requiring a degree of smoothness of the sample paths of $X$, namely right-continuity (see (6.8.27) for the definition of right-continuity in this context).

**(2) Definition.** The process $X$ is a **(continuous-time) Markov chain** if it is right-continuous and satisfies the Markov property.

If $X$ is a continuous-time Markov chain, any homogeneous discretization of time yields a discrete-time Markov chain, as follows. Let $h > 0$, and define $Z_n = X(nh)$. It is an immediate consequence of the Markov property (1) that $Z = \{Z_n : n \geq 0\}$ is a discrete-time Markov chain, which we may call a *skeleton* of $X$.

First we address certain basics.

**(3) Definition.** The **transition probability** $p_{ij}(s,t)$ of the Markov chain $X$ is defined to be

$$p_{ij}(s,t) = \mathbb{P}\big(X(t) = j \mid X(s) = i\big) \quad \text{for} \quad s \leq t.$$

The chain is called **homogeneous** if $p_{ij}(s,t) = p_{ij}(0, t-s)$ for all $i, j, s, t$, and we write $p_{ij}(t-s)$ for $p_{ij}(s,t)$.

*Henceforth we suppose that $X$ is a homogeneous Markov chain*, and we write $\mathbf{P}_t$ for the $|S| \times |S|$ matrix with entries $p_{ij}(t)$. The family $\{\mathbf{P}_t : t \geq 0\}$ is called the *transition semigroup* of the chain.

**(4) Theorem.** *The family $\{\mathbf{P}_t : t \geq 0\}$ is a* stochastic semigroup, *in that it has the following properties.*
  (a) $\mathbf{P}_0 = \mathbf{I}$, *the identity matrix.*
  (b) $\mathbf{P}_t$ *is stochastic, that is $\mathbf{P}_t$ has non-negative entries and row sums* 1.
  (c) *The* Chapman–Kolmogorov equations *hold:* $\mathbf{P}_{s+t} = \mathbf{P}_s \mathbf{P}_t$ *if $s, t \geq 0$.*

The Chapman–Kolmogorov equations of part (c) are also known as the *semigroup property*.

**Proof.** Part (a) is obvious.
(b) With $\mathbf{1}$ a row vector of ones, we have that

$$(\mathbf{P}_t \mathbf{1}')_i = \sum_j p_{ij}(t) = \mathbb{P}\bigg(\bigcup_j \{X(t) = j\} \,\bigg|\, X(0) = i\bigg) = 1.$$

(c) Using the Markov property,

$$\begin{aligned}
p_{ij}(s+t) &= \mathbb{P}\big(X(s+t) = j \mid X(0) = i\big) \\
&= \sum_k \mathbb{P}\big(X(s+t) = j \mid X(s) = k, \ X(0) = i\big)\mathbb{P}\big(X(s) = k \mid X(0) = i\big) \\
&= \sum_k p_{ik}(s)p_{kj}(t),
\end{aligned}$$

as in the discrete case of Theorem (6.1.7). ∎

The evolution of $X$ is specified by the stochastic semigroup $\{\mathbf{P}_t\}$ and the distribution of $X(0)$. Most questions about $X$ can be rephrased in terms of these matrices and their properties.

Let $X$ be a continuous-time Markov chain. We write $\mathbb{P}_i$ for the probability measure conditional on starting in the state $X(0) = i$, and $\mathbb{E}_i$ for the corresponding expectation. A certain amount of measure theory is needed in the rigorous study of continuous-time Markov chains. We shall avoid that here by presenting arguments that are occasionally intuitive. A greater degree of rigour may be found in Liggett 2010 and Norris 1997.

There is a more general form of the Markov property than (1) above. For $t \geq 0$, we call an event $A$ *t-historical* if it is given in terms of the past history $\{X(s) : s < t\}$, and *t-future* if it is given in terms of the future $\{X(s) : s > t\}$. We call the random time $T$ a *stopping time* for the process $X$ if, for all $t \geq 0$, the indicator function of the event $\{T \leq t\}$ is a function of the family $\{X(s) : s \leq t\}$.

In order to handle the potential issue of explosion, we let $T_n$ be the time of the $n$th change in value of $X$, and write $T_\infty = \lim_{n \to \infty} T_n$. The times $T_n$ are well defined by the assumption of right-continuity.

**(5) Theorem.**

   (a) **Extended Markov property.** *Let $t > 0$, and let $H$ be a $t$-historical event and $F$ a $t$-future event. Then*

$$\mathbb{P}\big(F \mid X(t) = j, \ H\big) = \mathbb{P}\big(F \mid X(t) = j\big), \qquad j \in S, \ t \geq 0.$$

   (b) **Strong Markov property.** *Let $T$ be a stopping time for $X$. Conditional on the event $I = \{T < T_\infty\} \cap \{X(T) = i\}$, the process $X^* = \{X^*(u) = X(T + u) : u \geq 0\}$ is a continuous-time Markov chain with the same transition probabilities as $X$, that starts in state $i$ and is independent of the past $\{X(s) : s < T\}$.*

The proof is omitted (see Norris 1997, p. 227), though diligent readers may adapt the ideas of the proof of Theorem (6.8.25).

We are ready to describe the architecture of the chain $X$. Let $X(0) = i$, and let $U_0 = \inf\{t : X(t) \neq i\}$ be the time of the first change of value. Since the chain is right-continuous, we have $U_0 > 0$. The chain evolves in time in the manner of the next two theorems.

**(6) Theorem.** *The random variable $U_0$ has the exponential distribution with a parameter denoted $g_i$ and satisfying $g_i \in [0, \infty)$. Furthermore, $U_0$ is a stopping time.*

We can have $g_i = 0$ but not $g_i = \infty$. More specifically, the case $g_i = 0$ corresponds to $\mathbb{P}_i(U_0 = \infty) = 1$, which is to say that $i$ is absorbing. Secondly, the case $g_i = \infty$ corresponds to $\mathbb{P}_i(U_0 = 0) = 1$, which is impossible by the assumption of right-continuity.

**Proof.** By the extended Markov property at time $s$, and the homogeneity of the chain.

$$\mathbb{P}_i(U_0 > s + t \mid U_0 > s) = \mathbb{P}_i(U_0 > s + t \mid X(s) = i) = \mathbb{P}_i(U_0 > t),$$

so that $Q(t) := \mathbb{P}_i(U_0 > t)$ satisfies $Q(s + t) = Q(s)Q(t)$ for $s, t \geq 0$. By the lack-of-memory property (see Problem (4.14.15)), $U_0$ has the exponential distribution with some parameter $g_i$.

By right-continuity, we have $\mathbb{P}_i(U_0 > 0) = 1$, whence $g_i = \infty$ is impossible. To see that $U_0$ is a stopping time, write $\{U_0 > t\} = \{X(u) = 0 \text{ for } u \leq t\}$. ∎

Let the (Markov) transition matrix $\mathbf{Y} = (y_{ij} : i, j \in S)$ be given as follows. If $g_i = 0$, we set $y_{ij} = \delta_{ij}$, the Kronecker delta. If $g_i > 0$, we set

$$y_{ij} = \mathbb{P}_i(X(U_0) = j), \qquad j \in S.$$

Key to the analysis of the Markov chain $X$ is the matrix $\mathbf{G} = (g_{ij} : i, j \in S)$, called the *generator* of the chain†, and defined by

**(7)**
$$g_{ij} = \begin{cases} g_i y_{i,j} & \text{if } j \neq i, \\ -g_i & \text{if } j = i. \end{cases}$$

---

†Some writers use the historic notation $q_{ij}$ in place of $g_{ij}$, and call the resulting matrix $\mathbf{Q}$ the '$Q$-matrix' of the process.

Since **Y** has row sums 1, **G** has row sums 0, and so

(8)
$$\sum_{j:\, j\neq i} g_{ij} = g_i, \qquad i \in S.$$

The element $g_{ij}$ may be interpreted as the 'intensity' of jumps from $i$ to $j$, that is,

(9)
$$\mathbb{P}\big(X(t+h) = j \mid X(t) = i\big) = g_{ij}h + o(h), \qquad j \neq i.$$

This is easily proved when $S$ is finite†, but needs be treated with care when $S$ is infinite. We note for later convenience that

(10)
$$g_{ij} = g_i(y_{ij} - \delta_{ij}), \qquad i, j \in S.$$

**(11) Theorem.** *Let $X(0) = i$ where $g_i > 0$.*
  (a) *The random variable $X(U_0)$ is independent of $U_0$.*
  (b) *Write $X^*(s) = X(U_0 + s)$ for $s \geq 0$. Conditional on $X(U_0) = j$, the process $X^*$ is a continuous-time Markov chain with the same transition probabilities as $X$, and with initial state $X^*(0) = j$. In particular, given $X(U_0) = j$, $X^*$ is independent of $U_0$ and the initial state $i$.*

**Proof.** (a) By the extended Markov property at time $u$,

$$\mathbb{P}_i\big(X(U_0) = j \mid U_0 > u\big) = \mathbb{P}_i\big(X(U_0) = j \mid U_0 > 0\big) = \mathbb{P}_i(X(U_0) = j),$$

so that $X(U_0)$ and $U_0$ are independent.
  (b) This follows similarly from the strong Markov property at the stopping time $U_0$.  ∎

In summary, having started in state $i$, the chain $X$ remains in $i$ for an exponentially distributed time $U_0$, with parameter $g_i$. It then jumps to a new state, $j$ say, chosen with probabilities as in the transition matrix **Y**. The chain then proceeds in a similar manner, starting in $j$, and otherwise independent of the history of $X$ prior to time $U_0$. Let $T_m$ be the time of the $m$th change in value of $X$ (with $T_0 = 0$), and let $U_m = T_{m+1} - T_m$. After the $m$th jump of $X$, the chain is constant for the exponentially-distributed time $U_m$ called the $m$th 'holding time', before jumping to a new state according to **Y**. A diagram of the evolution of $X$ appears in Figure 6.2.

The continuous-time chain $X$ may be described as follows‡ in terms of its so-called holding times $(U_m)$ and jump chain $Y$. Let $S$ be a countable state space, and let $\mathbf{Y} = (y_{ij} : i, j \in S)$ be the transition matrix of a discrete-time Markov chain $Y = \{Y_n : n \geq 0\}$ taking values in $S$. We shall assume that $y_{ii} = 0$ for all $i \in S$; this is not an essential assumption, but recognizes the fact that jumps from any state $i$ to itself will be invisible in continuous time. Let $g_i, i \in S$, be non-negative constants. We now construct a continuous-time chain $X$ as follows. First,

---

†See Exercise (6.9.11).

‡Above, we started with the continuous-time chain $X$, and we constructed the holding times and the jump chain. This process is now reversed: start with the parameters $(g_i)$ and the jump chain $Y$, and patch them together to create a continuous-time chain $X$. William Feller (1957) remarked that we can visualize $Y$ as a road map giving possibilities and probabilities for movement, while the parameters $g_i$ give the speeds of the moves.

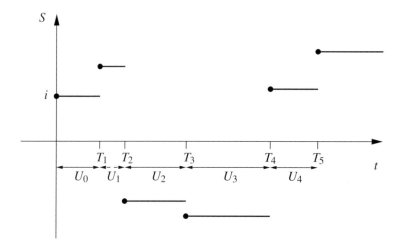

Figure 6.2. An illustration of the evolution of a Markov chain starting in state $i$.

let $X(0) = Y_0$. After a holding time $U_0$ having the exponential distribution with parameter $g_{Y_0}$, the process jumps to the state $Y_1$. After a further holding time $U_1$ having the exponential distribution with parameter $g_{Y_1}$, the chain jumps to $Y_2$, and so on. The $U_m$ are the *holding times* and the discrete-time chain $Y$ is called the *jump chain*.

We argue slightly more fully as follows. Conditional on the values $(Y_n)$ of the chain $Y$, let $U_0, U_1, \ldots$ be independent random variables having the respective exponential distributions with parameters $g_{Y_0}, g_{Y_1}, \ldots$, and set $T_n = U_0 + U_1 + \cdots + U_n$. We now define

**(12)**
$$X(t) = \begin{cases} Y_n & \text{if } T_n \le t < T_{n+1} \text{ for some } n, \\ \infty & \text{otherwise.} \end{cases}$$

Equation (12) indicates a potential complication, namely the issue of 'explosion'. Let $T_\infty = \lim_{n \to \infty} T_n$. It can be the case that $T_\infty$ is finite with strictly positive probability, in which case $X(t)$ is not defined for $t \ge T_\infty$. If this holds, then $X$ fails to satisfy the definition (2) of a Markov chain. This gap is filled by introducing a new state, which we choose to label $\infty$, so that the new state space is $S' = S \cup \{\infty\}$. We define

**(13)**
$$X(t) = \infty \quad \text{for } t \ge T_\infty.$$

There are other ways than this of extending the time-domain of the process beyond the time interval $[0, T_\infty)$, and (13) defines the so-called *minimal* process, since this process survives for the minimal time $T_\infty$. Recall the case of birth processes and Lemma (6.8.18). The process $X$ is said to *explode* from state $i$ if $\mathbb{P}_i(T_\infty < \infty) > 0$; *explosion* is said to occur if $X$ explodes from some state $i$. Sufficient conditions for non-explosion are given in Theorem (17).

**(14) Proposition.** *Let $X$ be a minimal process with transition probabilities $\{\mathbf{P}_t\}$. Then $\{\mathbf{P}_t\}$ satisfies the semigroup property $\mathbf{P}_{s+t} = \mathbf{P}_s \mathbf{P}_t$ for $s, t \ge 0$.*

**Proof.** The point is that an explosive process $X$ may attain the additional state labelled $\infty$, but it cannot return. Thus, for $i, j \in S$,

$$p_{ij}(s+t) = \sum_{k \in S} p_{ik}(s) p_{kj}(t), \qquad i, j \in S,$$

as required. ∎

Recall the generator $\mathbf{G} = (g_{ij} : i, j \in S)$ of the chain. There is a sense in which the generator plays the role of the transition matrix $\mathbf{P}$ for discrete-time chains. We have that

(15)
$$\sum_j g_{ij} = 0 \quad \text{for all } i, \quad \text{or} \quad \mathbf{G1}' = \mathbf{0}',$$

where $\mathbf{1}$ and $\mathbf{0}$ are row vectors of ones and zeros. A $S \times S$ matrix $\mathbf{K} = (k_{ij} : i, j \in S)$ is called a *generator* if $\mathbf{K}$ has non-negative off-diagonal elements and row sums 0.

(16) **Example. Birth process (6.8.15).** From the definition of this process, it is clear that

$$g_{ii} = -\lambda_i, \quad g_{i,i+1} = \lambda_i, \quad g_{ij} = 0 \quad \text{if} \quad j < i \quad \text{or} \quad j > i + 1.$$

The generator is

$$\mathbf{G} = \begin{pmatrix} -\lambda_0 & \lambda_0 & 0 & 0 & 0 & \cdots \\ 0 & -\lambda_1 & \lambda_1 & 0 & 0 & \cdots \\ 0 & 0 & -\lambda_2 & \lambda_2 & 0 & \cdots \\ \vdots & \vdots & \vdots & \vdots & \vdots & \ddots \end{pmatrix}.$$

Recall from Theorem (6.8.17) that the chain explodes if and only if $\sum_n \lambda_n^{-1} < \infty$. ●

It is useful to have verifiable conditions under which a chain does not explode.

(17) **Theorem.** *Let $i \in S$. The chain $X$ is non-explosive from state $i$ if any of the following three conditions holds*:
  (a) *$S$ is finite,*
  (b) *$\sup_j g_j < \infty$,*
  (c) *$i$ is a recurrent state for the jump chain $Y$.*

The conditions of the above theorem are far from being necessary for non-explosion, as revealed by consideration of a Poisson or birth process.

**Proof.** First we prove that (b) suffices, noting in advance that (a) implies (b). We shall use the fact that, if $Z$ has the exponential distribution with parameter $\mu > 0$, then $\mu Z$ is exponential with parameter 1 (*exercise*). Suppose that $g_j < \gamma < \infty$ for all $j$. The holding time $U_n$ of the chain has the exponential distribution with parameter $g_{Y_n}$, where $Y$ is the jump chain. If $g_{Y_n} > 0$, then $V_n = g_{Y_n} U_n$ has the exponential distribution with parameter 1. If $g_{Y_n} = 0$, then $U_n = \infty$ almost surely. Therefore,

$$\gamma T_\infty = \begin{cases} \infty & \text{if } g_{Y_n} = 0 \text{ for some } n, \\ \sum_{n=1}^\infty \gamma U_n \geq \sum_{n=1}^\infty V_n & \text{otherwise.} \end{cases}$$

It follows by Lemma (6.8.18) that the last sum is almost surely infinite; therefore, explosion does not occur.

Suppose now that (c) holds. If $g_i = 0$, then $X(t) = i$ for all $t$, and there is nothing to prove. Suppose that $g_i > 0$. Since $Y_0 = i$ and $i$ is recurrent for $Y$, there exists almost surely an infinity of times $N_0 < N_1 < \cdots$ at which $Y$ takes the value $i$. Now,

$$g_i T_\infty \geq \sum_{j=0}^\infty g_i U_{N_j},$$

and the $U_{N_j}$ are independent and exponentially distributed with parameter $g_i$. We may once again appeal to Lemma (6.8.18). ∎

**(18) Example.** Here is an example in which the transition probabilities $p_{ij}(t)$ may be given explicitly in terms of those of the jump chain. Let $Y$ be a discrete-time chain with transition matrix $\mathbf{Y} = (y_{ij})$ satisfying $y_{ii} = 0$ for all $i \in S$, and let $N$ be a Poisson process with intensity $\lambda$. We define $X$ by $X(t) = Y_n$ if $T_n \le t < T_{n+1}$ where $T_n$ is the time of the $n$th arrival in the Poisson process (and $T_0 = 0$). The process $X$ has transition semigroup $\mathbf{P}_t = (p_{ij}(t))$ given by

$$p_{ij}(t) = \mathbb{P}_i(X(t) = j)$$

$$= \sum_{n=0}^{\infty} \mathbb{P}_i\big(X(t) = j, \ N(t) = n\big)$$

$$= \sum_{n=0}^{\infty} \frac{(\lambda t)^n}{n!} e^{-\lambda t} \mathbb{P}_i(Y_n = j) = \sum_{n=0}^{\infty} e^{-\lambda t} \frac{(\lambda t)^n (\mathbf{Y}^n)_{ij}}{n!}.$$

We note that

$$e^{-\lambda t \mathbf{I}} = \sum_{n=0}^{\infty} \frac{(-\lambda t)^n}{n!} \mathbf{I}^n = \mathbf{I} e^{-\lambda t},$$

whence $\mathbf{P}_t = e^{\lambda t (\mathbf{Y}-\mathbf{I})}$, where the matrix exponential $e^{\mathbf{A}}$ is given as $e^{\mathbf{A}} = \sum_{n=0}^{\infty} \mathbf{A}^n / n!$.  ●

**(B) Irreducibility and recurrence.** We move on to the classification of states; this is slightly less of a chore than it was for discrete-time chains.

**(19) Definition.** The chain is called **irreducible** if, for any pair $i$, $j$ of states, there exists $t > 0$ such that $p_{ij}(t) > 0$.

If the state space $S$ is a singleton, it is trivial that the chain is irreducible. If $|S| \ge 2$, then the irreducibility of $X$ implies that $g_i > 0$ for $i \in S$.

**(20) Theorem.**
  (a) *If the chain $X$ is irreducible, then $p_{ij}(t) > 0$ for all $i$, $j \in S$ and $t > 0$.*
  (b) *If $X$ is irreducible, then so is the jump chain $Y$. If $Y$ is irreducible and $g_i > 0$ for $i \in S$, then $X$ is irreducible.*

Part (a) is essentially the so-called *Lévy dichotomy*: for transition probabilities $p_{ij}(t)$ and states $i$, $j$, either $p_{ij}(t) = 0$ for all $t > 0$, or $p_{ij}(t) > 0$ for all $t > 0$.

**Proof.** (a) Let $X$ be irreducible, and assume that $|S| \ge 2$, so that $g_i > 0$ for $i \in S$. First, we have

$$p_{ii}(t) \ge \mathbb{P}_i(U_0 > t) = 1 - e^{-g_i t} > 0, \qquad t \ge 0.$$

Secondly, let $i$, $j \in S$ be distinct. Since $p_{ij}(t) > 0$ for some $t > 0$, it is necessarily the case that $i \to j$ for the jump chain $Y$. Therefore, there exist distinct states $i_0, i_1, \ldots, i_n$ with $i = i_0$, $j = i_n$ such that $y_{i_0,i_1} y_{i_1,i_2} \cdots y_{i_{n-1},i_n} > 0$. If $u$, $v \in S$ are such that $y_{uv} > 0$, then

$$p_{uv}(s) \ge \mathbb{P}_u\big(U_0 \le s, \ Y_1 = v, \ U_1 > s\big)$$

$$= (1 - e^{-g_u s}) y_{u,v} e^{-g_v s} > 0, \qquad s > 0.$$

Therefore,

$$p_{ij}(t) \geq p_{i_0,i_1}(t/n)\,p_{i_1,i_2}(t/n) \cdots p_{i_{n-1},i_n}(t/n) > 0,$$

as required.

(b) As in the proof of part (a), if $Y$ is irreducible and $g_i > 0$ for $i \in S$, then $X$ is irreducible. Conversely, if $Y$ is not irreducible, there exist $i, j \in S$ such that $Y$ can never reach $j$ having started in $i$. It is then clear that $p_{ij}(t) = 0$ for all $t > 0$. ∎

The birth process is *not* irreducible, since it is non-decreasing. See Problem (6.15.15) for a condition for irreducibility in terms of the generator $\mathbf{G}$ of the chain.

The interplay between a continuous-time chain $X$ and its jump chain $Y$ provides a basic tool for the study of the former. We present just one example of this statement; others may be found in the exercises. We call the state $i$

(21)

$$\text{\textit{recurrent} for } X \text{ if } \mathbb{P}_i\big(\text{the set } \{t : X(t) = i\} \text{ is unbounded}\big) = 1,$$

$$\text{\textit{transient} for } X \text{ if } \mathbb{P}_i\big(\text{the set } \{t : X(t) = i\} \text{ is unbounded}\big) = 0.$$

We shall see in Theorem (22b) that a given state $i$ is either recurrent or transient for $X$.†

A recurrent state $i$ is called *positive* (or *non-null*‡) if either $S = \{i\}$, or the earliest return time to $i$ has finite mean, which is to say that the *mean return time* $m_i := \mathbb{E}_i(R_i)$ satisfies $m_i < \infty$, where $R_i := \inf\{t > U_0 : X(t) = i\}$. Positive recurrence is important in the context of stationary distributions, to which we shall turn in the next section.

**(22) Theorem.** *Consider the continuous-time chain $X$ with jump chain $Y$.*

(a) *If $g_i = 0$, the state $i$ is recurrent for $X$.*

(b) *Assume $g_i > 0$. State $i$ is recurrent (respectively, transient) for $X$ if and only if it is recurrent (respectively, transient) for the jump chain $Y$. Furthermore, $i$ is recurrent if the transition probabilities $p_{ii}(t) = \mathbb{P}_i(X(t) = i)$ satisfy $\int_0^\infty p_{ii}(t)\,dt = \infty$, and is transient otherwise.*

(c) *Assume $X$ is irreducible. Either every state is recurrent or every state is transient.*

**Proof.** It is trivial that $i$ is recurrent if $g_i = 0$, since then the chain $X$ remains forever in the state $i$ once it has first visited it.

Assume $g_i > 0$ and $X(0) = i$. If $i$ is transient for the jump chain $Y$, there exists almost surely a last visit of $Y$ to $i$; this implies the almost sure boundedness of the set $\{t : X(t) = i\}$, whence $i$ is transient for $X$. Suppose $i$ is recurrent for $Y$. By Theorem (17c), the chain $X$ does not explode. By the recurrence of $i$, there exists almost surely an infinity of values $n$ with $Y_n = i$. Since there is no explosion, the times $T_n$ of these visits are almost surely unbounded, whence $i$ is recurrent for $X$.

Now, the integrands being positive, we may interchange limits to obtain

$$\int_0^\infty p_{ii}(t)\,dt = \int_0^\infty \mathbb{E}_i\big(I_{\{X(t)=i\}}\big)\,dt$$

$$= \mathbb{E}_i\left[\int_0^\infty I_{\{X(t)=i\}}\,dt\right] = \mathbb{E}_i\left[\sum_{n=0}^\infty U_n I_{\{Y_n=i\}}\right]$$

---

†The word 'persistent' is frequently used in place of recurrent. An irreducible chain is called recurrent if every state is recurrent, and otherwise transient; see Theorem (22c).

‡A recurrent state $k$ is *null* if $m_k = \infty$. The mean return time $m_k$ is the equivalent in continuous-time of the mean recurrence time $\mu_k$ of (6.2.7).

where $\{U_n : n \geq 0\}$ are the holding times of $X$. The right side equals

$$\sum_{n=0}^{\infty} \mathbb{E}_i(U_0) y_{ii}(n) = \frac{1}{g_i} \sum_{n=0}^{\infty} y_{ii}(n)$$

where $y_{ii}(n)$ is the appropriate $n$-step transition probability of $Y$. By Corollary (6.2.4), the last sum diverges if and only if $i$ is recurrent for $Y$.

The final claim is trivial if $|S| = 1$, and we assume $|S| \geq 2$. Since $X$ is assumed irreducible, we have that $g_i > 0$ for $i \in S$, and also the jump chain $Y$ is irreducible. The claim now follows by the corresponding fact for irreducible discrete-time chains, namely Theorem (6.3.2).    ∎

---

## Exercises for Section 6.9

**1.** Let $\lambda\mu > 0$ and let $X$ be a Markov chain on $\{1, 2\}$ with generator

$$\mathbf{G} = \begin{pmatrix} -\mu & \mu \\ \lambda & -\lambda \end{pmatrix}.$$

(a) Write down the forward equations and solve them for the transition probabilities $p_{ij}(t)$, $i, j = 1, 2$.

(b) Calculate $\mathbf{G}^n$ and hence find $\sum_{n=0}^{\infty}(t^n/n!)\mathbf{G}^n$. Compare your answer with that to part (a).

(c) Solve the equation $\boldsymbol{\pi}\mathbf{G} = \mathbf{0}$ in order to find the stationary distribution. Verify that $p_{ij}(t) \to \pi_j$ as $t \to \infty$.

**2.** As a continuation of the previous exercise, find:

(a) $\mathbb{P}(X(t) = 2 \mid X(0) = 1, X(3t) = 1)$,

(b) $\mathbb{P}(X(t) = 2 \mid X(0) = 1, X(3t) = 1, X(4t) = 1)$.

**3.** Jobs arrive in a computer queue in the manner of a Poisson process with intensity $\lambda$. The central processor handles them one by one in the order of their arrival, and each has an exponentially distributed runtime with parameter $\mu$, the runtimes of different jobs being independent of each other and of the arrival process. Let $X(t)$ be the number of jobs in the system (either running or waiting) at time $t$, where $X(0) = 0$. Explain why $X$ is a Markov chain, and write down its generator. Show that a stationary distribution exists if and only if $\lambda < \mu$, and find it in this case.

**4.  Pasta property.** Let $X = \{X(t) : t \geq 0\}$ be a Markov chain having stationary distribution $\boldsymbol{\pi}$. We may sample $X$ at the times of a Poisson process: let $N$ be a Poisson process with intensity $\lambda$, independent of $X$, and define $Y_n = X(T_n)$. Show that $Y = \{Y_n : n \geq 0\}$ is a discrete-time Markov chain with the same stationary distribution as $X$. (This exemplifies the 'Pasta' property: Poisson arrivals see time averages.)

[The full assumption of the independence of $N$ and $X$ is not necessary for the conclusion. It suffices that $\{N(s) : s \geq t\}$ be independent of $\{X(s) : s \leq t\}$, a property known as 'lack of anticipation'. It is not even necessary that $X$ be Markov; the Pasta property holds for many suitable ergodic processes.]

**5.  Hitting probabilities.** Let $X$ be a continuous-time Markov chain with generator $\mathbf{G}$ satisfying $g_i = -g_{ii} > 0$ for all $i$. Let $H_A = \inf\{t \geq 0 : X(t) \in A\}$ be the hitting time of the set $A$ of states, and let $\eta_j = \mathbb{P}_j(H_A < \infty)$ be the chance of ever reaching $A$ from $j$. By using properties of the jump chain, which you may assume to be well behaved, show that $\sum_k g_{jk}\eta_k = 0$ for $j \notin A$.

**6.  Mean hitting times.** In continuation of the preceding exercise, let $\mu_j = \mathbb{E}_j(H_A)$. Show that the vector $\boldsymbol{\mu}$ is the minimal non-negative solution of the equations

$$\mu_j = 0 \quad \text{if } j \in A, \qquad 1 + \sum_{k \in S} g_{jk}\mu_k = 0 \quad \text{if } j \notin A.$$

**7.** Let $X$ be a continuous-time Markov chain with transition probabilities $p_{ij}(t)$ and define $F_i = \inf\{t > T_1 : X(t) = i\}$ where $T_1$ is the time of the first jump of $X$. Show that, if $g_{ii} \neq 0$, then $\mathbb{P}_i(F_i < \infty) = 1$ if and only if $i$ is recurrent.

**8.** Let $X$ be the simple symmetric random walk on the integers in continuous time, so that

$$p_{i,i+1}(h) = p_{i,i-1}(h) = \tfrac{1}{2}\lambda h + o(h).$$

Show that the walk is recurrent. Let $T$ be the time spent visiting $m$ during an excursion from 0. Find the distribution of $T$.

**9.** Let $i$ be a transient state of a continuous-time Markov chain $X$ with $X(0) = i$. Show that the total time spent in state $i$ has an exponential distribution.

**10.** Let $X$ be an asymmetric simple random walk in continuous time on the non-negative integers with retention at 0, so that

$$p_{ij}(h) = \begin{cases} \lambda h + o(h) & \text{if } j = i+1, \ i \geq 0, \\ \mu h + o(h) & \text{if } j = i-1, \ i \geq 1. \end{cases}$$

Suppose that $X(0) = 0$ and $\lambda > \mu$. Show that the total time $V_r$ spent in state $r$ is exponentially distributed with parameter $\lambda - \mu$.

Assume now that $X(0)$ has some general distribution with probability generating function $G$. Find the expected amount of time spent at 0 in terms of $G$.

**11.** Let $X$ be the continuous-time Markov chain on the finite state space $S$ with generator $\mathbf{G} = (g_{ij})$. Show from first principles that the transition probabilities satisfy

$$p_{ij}(h) = \begin{cases} 1 + g_{ii}h + o(h) & \text{if } i = j, \\ g_{ij}h + o(h) & \text{if } i \neq j. \end{cases}$$

**12.** Let $X$ be a Markov chain on the integers $\mathbb{Z}$ with generator satisfying $g_{i,i-1} = g_{i,i+1} = 2^i$ for $i \in \mathbb{Z}$, and $g_{i,j} = 0$ for other pairs $(i, j)$ with $i \neq j$. Does $X$ explode?

**13. Growth with annihilation.** A population grows subject to the threat of total annihilation. It is modelled as a Markov chain with generator $\mathbf{G} = (g_{ij})$ satisfying

$$g_{i,i+1} = \frac{1}{i+2}, \qquad\qquad i \geq 0,$$

$$g_{i,0} = \frac{1}{(i+1)(i+2)}, \qquad i \geq 1,$$

the other off-diagonal elements of $\mathbf{G}$ being 0. Show that the chain is null recurrent.

**14. Skeletons.** Let $Z_n = X(nh)$ where $X$ is a Markov chain and $h > 0$. Show that $i$ is recurrent for $Z$ if and only if it is recurrent for $X$. Show that $Z$ is irreducible if and only if $X$ is irreducible.

## 6.10 Kolmogorov equations and the limit theorem

As in the last section, $X$ is a continuous-time Markov chain on the countable state space $S$, with transition semigroup $\{\mathbf{P}_t\}$ and generator $\mathbf{G}$. This section begins with an account of the relationship between $\{\mathbf{P}_t\}$ and $\mathbf{G}$.

**(A) The forward and backward equations.** The transition semigroup satisfies two differential equations, called the (Kolmogorov) forward and backward equations, which we introduce next.

Suppose for the moment that *the state space S is finite*, and let $X(0) = i$. By conditioning $X(t + h)$ on $X(t)$, we obtain

$$p_{ij}(t + h) = \sum_k p_{ik}(t)p_{kj}(h)$$

$$= p_{ij}(t)(1 + g_{jj}h) + \sum_{k: k \neq j} p_{ik}(t)g_{kj}h + o(h) \quad \text{by (6.9.9)}$$

$$= p_{ij}(t) + h \sum_{k \in S} p_{ik}(t)g_{kj} + o(h),$$

giving that

**(1)**     $\dfrac{1}{h}\left[p_{ij}(t + h) - p_{ij}(t)\right] \to \sum_{k \in S} p_{ik}(t)g_{kj} = (\mathbf{P}_t\mathbf{G})_{ij}$     as $h \downarrow 0$.

The above argument involves the passing of a limit through a summation. This is proper when $S$ is finite, but requires justification when $S$ is infinite.

With $\mathbf{P}'_t$ the matrix with entries $p'_{ij}(t)$, we write the *forward equations* (1) in the form

**(2)**                                   $\mathbf{P}'_t = \mathbf{P}_t\mathbf{G},$     $t \geq 0$.

A similar argument, by conditioning $X(t + h)$ on $X(h)$, yields the *backward equation*

**(3)**                                   $\mathbf{P}'_t = \mathbf{G}\mathbf{P}_t,$     $t \geq 0$.

The common boundary condition for the forward and backward equations is $\mathbf{P}_0 = \mathbf{I}$.

Since $S$ is assumed finite, the forward and backward systems each forms a finite set of first-order linear differential equations which, subject to the boundary condition, has a unique solution. This solution may be expressed in the form

**(4)**                                   $\mathbf{P}_t = \sum_{n=0}^{\infty} \dfrac{t^n}{n!} \mathbf{G}^n,$

the sum of powers of matrices (remember that $\mathbf{G}^0 = \mathbf{I}$). Equation (4) is deducible from (2) or (3) in very much the same way as we might show that the function of the single variable $p(t) = e^{gt}$ solves the differential equation $p'(t) = gp(t)$. The representation (4) for $\mathbf{P}_t$ is very useful and is usually written as

**(5)**                                   $\mathbf{P}_t = e^{t\mathbf{G}}$   or   $\mathbf{P}_t = \exp(t\mathbf{G}),$

where $e^{\mathbf{A}}$ is the natural abbreviation for $\sum_{n=0}^{\infty}(1/n!)\mathbf{A}^n$ whenever $\mathbf{A}$ is a square matrix. Some properties of matrix exponentials may be found in Exercise (6.10.1).

**(6) Example.** Consider a two-state chain $X$ with $S = \{1, 2\}$; $X$ jumps between 1 and 2 as time passes. There are two equivalent ways of describing the chain, depending on whether we specify $\mathbf{G}$ or we specify the holding times:

(a) $X$ has generator $\mathbf{G} = \begin{pmatrix} -\alpha & \alpha \\ \beta & -\beta \end{pmatrix}$,

(b) if the chain is in state 1 (respectively, 2), then it stays in this state for a length of time which is exponentially distributed with parameter $\alpha$ (respectively, $\beta$) before jumping to 2 (respectively, 1).

The equations of the forward system (2) take the form

$$p'_{11}(t) = -\alpha p_{11}(t) + \beta p_{12}(t),$$

and are easily solved to find the transition probabilities of the chain (*exercise*).   ●

We now lift the assumption that $S$ be finite. The above derivation requires an interchange of limits when $S$ is infinite, and the story becomes more complicated. It turns out that the forward and backward equations may have more than one solution. Recall the term 'minimal' introduced in the discussion of explosion at (6.9.13). The $S \times S$ matrix $\mathbf{H} = (h_{ij})$ is called a *generator* on $S$ if $h_{ij} \geq 0$ for $i \neq j$ and $\mathbf{H}$ has row sums 0.

**(7) Theorem.** *Let $S$ be a countable set, and $\mathbf{G}$ a generator on $S$. Let $X$ be a minimal Markov chain with generator $\mathbf{G}$.*
   (a) *The transition semigroup $\{\mathbf{P}_t\}$ of $X$ is the minimal non-negative solution of the backward equation $\mathbf{P}'_t = \mathbf{G}\mathbf{P}_t$, subject to the boundary condition $\mathbf{P}_0 = \mathbf{I}$.*
   (b) *The semigroup $\{\mathbf{P}_t\}$ is also the minimal non-negative solution of the forward equation.*

In saying that $\{\mathbf{P}_t\}$ is the 'minimal' non-negative solution, we mean that $p_{ij}(t) \leq \pi_{ij}(t)$ for any other non-negative solution $\{\mathbf{\Pi}_t\}$, subject to the boundary condition. If there is no explosion, then $\sum_j p_{ij}(t) = 1$ for all $i \in S$ and $t > 0$. Furthermore, it is an *exercise* that any solution to either the forward or backward equation (when subject to the boundary condition) is sub-stochastic in that $\sum_j p_{ij}(t) \leq 1$ for $i \in S$ and $t > 0$. It follows that, in the absence of explosion, the transition semigroup $\{\mathbf{P}_t\}$ is the unique solution to both the forward and backward equations.

**(8) Example. Birth process.** Recall the birth process of Definition (6.8.15), and suppose that $\lambda_i > 0$ for all $i$. We saw in Section 6.8 that the condition $\sum_i \lambda_i^{-1} = \infty$ is necessary and sufficient for the forward and backward equations to have a unique solution.   ●

**Proof of Theorem (7).** We prove part (a), and refer the reader to Norris (1997, p. 100), for the proof of part (b). Write $p_{ij}(t)$ for the transition probabilities of the minimal chain $X$. By considering whether or not $T_1 > t$,

(9)  $$p_{ij}(t) = \mathbb{P}_i(T_1 > t, X(t) = j) + \sum_{k:\, k \neq i} \mathbb{P}_i\big(T_1 \leq t, X(T_1) = k, X(t) = j\big)$$

$$= \delta_{ij} e^{-g_i t} + \sum_{k:\, k \neq i} \int_0^t g_i e^{-g_i s} y_{ik} p_{kj}(t - s)\, ds,$$

where $\delta_{ij}$ is the Kronecker delta. We change variables $u = t - s$ in the integral, and interchange the summation and integral (using a version of Fubini's theorem), to obtain

$$e^{g_i t} p_{ij}(t) = \delta_{ij} + \int_0^t \sum_{k:\, k \neq i} g_i e^{g_i u} y_{ik} p_{kj}(u)\, du.$$

Now differentiate this equation (having verified the formalities), and substitute for the $y_{ik}$, to obtain the backward equation

(10)
$$p'_{ij}(t) = \sum_{k \in S} g_{ik} p_{kj}(t), \qquad i \in S.$$

We show next that $\{P_t\}$ is the minimal solution of the backward equation (10). Let $\{\widehat{P}_t\}$ be a non-negative solution to (10) subject to the boundary condition. By reversing the steps from (9) to (10), we find that

(11)
$$\widehat{p}_{ij}(t) = \delta_{ij} e^{-g_i t} + \sum_{k:\, k \neq i} \int_0^t g_i e^{-g_i s} y_{ik} \widehat{p}_{kj}(t - s)\, ds.$$

We claim that

(12)
$$\widehat{p}_{ij}(t) \geq \mathbb{P}_i(X(t) = j, T_n > t), \qquad i, j \in S,\ t > 0,\ n \geq 1,$$

and we prove this by induction on $n$. Consider first the case $n = 1$. By (11),

$$\widehat{p}_{ij}(t) \geq \delta_{ij} e^{-g_i t} = \mathbb{P}_i(X(t) = j, T_1 > t), \qquad i, j \in S,\ t > 0.$$

Suppose (12) holds for $1 \leq n \leq N$ and consider the case $n = N + 1$. By the argument leading to (9), we may obtain

$$\mathbb{P}_i(X(t) = j, T_{N+1} > t)$$
$$= \delta_{ij} e^{-g_i t} + \sum_{k:\, k \neq i} \int_0^t g_i e^{-g_i s} y_{ik} \mathbb{P}_k\big(X(t - s) = j, T_N > t - s\big)\, ds.$$

On comparison with (11), we deduce (12) with $n = N + 1$. The induction step is complete.

The events on the right side of (12) are increasing in $n$, and furthermore $T_n \to T_\infty$. By the monotonicity of probability measures and the minimality of $X$,

$$\widehat{p}_{ij}(t) \geq \lim_{n \to \infty} \mathbb{P}_i(X(t) = j, T_n > t)$$
$$= \mathbb{P}_i(X(t) = j, T_\infty > t) = p_{ij}(t),$$

as required.                                                                    ∎

**(B) Stationary measures.** Next we consider the limiting distribution of $X(t)$ as $t \to \infty$, in the non-explosive case. As in the discrete-time case, the large-$t$ behaviour of $X(t)$ is closely bound up with the existence of stationary distributions and measures†. This definition may be compared with Definition (6.4.1). Recall that a *measure* $\mathbf{x}$ on $S$ is a collection $\mathbf{x} = (x_i : i \in S)$ of non-negative reals satisfying $\mathbf{x} \neq \mathbf{0}$.

A measure $\pi$ with a given property $\Pi$ is said to be *unique* if, given any other measure $\mathbf{v}$ with $\Pi$, there exists $\alpha > 0$ such that $\mathbf{v} = \alpha \pi$. That is, uniqueness is subject to scalar multiplication.

---

†A stationary measure is also known as an *invariant measure*.

**(13) Definition.**  Suppose $X$ is irreducible and non-explosive with transition semigroup $\{\mathbf{P}_t\}$. The vector $\boldsymbol{\pi}$ is a **stationary measure** for $X$ if $\boldsymbol{\pi}$ is a measure satisfying $\boldsymbol{\pi} = \boldsymbol{\pi}\mathbf{P}_t$ for $t \geq 0$. It is called a **stationary distribution** if, in addition, $\sum_j \pi_j = 1$.

By definition, if $\boldsymbol{\pi}$ is a stationary distribution for $X$, then it is stationary for any skeleton $Z$ of $X$ (see Exercise (6.9.14)); by Theorem (6.4.3) applied to $Z$, $\boldsymbol{\pi}$ is therefore the unique solution to the equations $\boldsymbol{\pi} = \boldsymbol{\pi}\mathbf{P}_t$.

If $X(0)$ has distribution $\boldsymbol{\mu}^{(0)}$ then the distribution $\boldsymbol{\mu}^{(t)}$ of $X(t)$ is given by

**(14)** $$\boldsymbol{\mu}^{(t)} = \boldsymbol{\mu}^{(0)}\mathbf{P}_t.$$

If $\boldsymbol{\mu}^{(0)} = \boldsymbol{\mu}$, a stationary distribution, then $X(t)$ has distribution $\boldsymbol{\mu}$ for all $t$. For discrete-time chains we found stationary distributions by solving the equations $\boldsymbol{\pi} = \boldsymbol{\pi}\mathbf{P}$; the corresponding equations $\boldsymbol{\pi} = \boldsymbol{\pi}\mathbf{P}_t$ for continuous-time chains may seem complicated but they amount to a simple condition relating $\boldsymbol{\pi}$ and $\mathbf{G}$.

Here is the main theorem concerning the existence of stationary distributions. The reader is reminded of the first return time $R_i := \inf\{t > U_0 : X(t) = i\}$, with mean $m_i = \mathbb{E}_i(R_i)$.

**(15) Theorem.**  *Let $X$ be an irreducible Markov chain with state space $S$, where $|S| \geq 2$.*
  (a) *If some $k \in S$ is positive recurrent, there exists a unique stationary distribution $\boldsymbol{\pi}$. Furthermore, $\boldsymbol{\pi}$ is the unique distribution that satisfies $\boldsymbol{\pi}\mathbf{G} = \mathbf{0}$, and all states are positive recurrent.*
  (b) *If $X$ is non-explosive and there exists a distribution $\boldsymbol{\pi}$ satisfying $\boldsymbol{\pi}\mathbf{G} = \mathbf{0}$, then: (i) all states are positive recurrent, (ii) $\boldsymbol{\pi}$ is stationary, and (iii) $\pi_k = 1/(m_k g_k)$ for $k \in S$.*

The proof is preceded by a lemma. Recall the transition matrix $\mathbf{Y}$ of the jump chain $Y$.

**(16) Lemma.**  *Let $X$ be an irreducible Markov chain with state space $S$, where $|S| \geq 2$.*
  (a) *A measure $\mathbf{x}$ satisfies $\mathbf{x}\mathbf{G} = \mathbf{0}$ if and only if the measure $\boldsymbol{v} = (v_i : i \in S)$ given by $v_i = x_i g_i$ satisfies $\boldsymbol{v}\mathbf{Y} = \boldsymbol{v}$. If $X$ is recurrent, such $\mathbf{x}$ is the unique measure (up to scalar multiplication) satisfying $\mathbf{x}\mathbf{G} = \mathbf{0}$.*
  (b) *If $\mathbf{x}$ is a measure satisfying $\mathbf{x}\mathbf{G} = \mathbf{0}$, then $x_j > 0$ for all $j \in S$.*
  (c) *Suppose $X$ is recurrent, and let $k \in S$. The measure $\boldsymbol{\mu}(k) = (\mu_j(k) : j \in S)$ given by*

**(17)** $$\mu_j(k) = \mathbb{E}_k\left(\int_0^{R_k} I_{\{X(s)=j\}}\,ds\right)$$

*satisfies $\boldsymbol{\mu}(k)\mathbf{G} = \mathbf{0}$ and is stationary for $X$.*

Here as usual, $I_A$ denotes the indicator function of the event $A$. The quantity $\mu_j(k)$ should be viewed as the mean time spent in state $j$ between two consecutive visits to $k$, and it plays the role of the vector $\boldsymbol{\rho}(k)$ in the theory of discrete-time chains (see Lemma (6.4.7)).

**Proof of Lemma (16).**  By irreducibility, $g_i > 0$ for $i \in S$.
(a) By (6.9.7) and (6.9.10),

$$(-\boldsymbol{v} + \boldsymbol{v}\mathbf{Y})_j = -v_j + \sum_{i \in S} v_i y_{ij}$$

$$= -x_j g_j + \sum_{i \in S} x_i(g_{ij} + g_i \delta_{ij}) = \mathbf{x}\mathbf{G},$$

as required. If $X$ is recurrent, by Lemma (6.4.7b), $\boldsymbol{\nu}$ is the unique stationary measure of $\mathbf{Y}$ (up to scalar multiplication), so that $\mathbf{x}$ is the unique measure satisfying $\mathbf{x}\mathbf{G} = \mathbf{0}$.

(b) By part (a), $\mathbf{x}\mathbf{G} = \mathbf{0}$ if and only if $\boldsymbol{\nu}\mathbf{Y} = \boldsymbol{\nu}$. In this case, by Lemma (6.4.7c), $\nu_j = x_j g_j > 0$ for $j \in S$, and the claim follows.

(c) Since $X$ is irreducible and recurrent, by (6.9.17c) it is non-explosive. Let $\mu_j(k)$ be the mean time spent in state $j$ before the first return time $R_k$ to $k$, as in (17). Such time is the sum of exponentially distributed random variables with parameter $g_j$, indexed by visits of the jump chain $Y$ to $j$ prior to $R_k$. Therefore,

$$(18) \qquad \qquad \mu_j(k) = \frac{1}{g_j} \rho_j(k),$$

where $\rho_j(k)$ is the mean number of visits by $Y$ to $j$ before its first return to its starting state $k$ (recall (6.4.5)). By Lemma (6.4.7), $\boldsymbol{\rho}(k) = (\rho_j(k) : j \in S)$ is a stationary measure for $Y$. By part (a) above, $\boldsymbol{\mu}(k)$ satisfies $\boldsymbol{\mu}(k)\mathbf{G} = \mathbf{0}$.

Next, we prove that $\boldsymbol{\mu}(k)$ is a stationary measure for $X$, in that $\boldsymbol{\mu}(k)\mathbf{P}_t = \boldsymbol{\mu}(k)$. By the strong Markov property (6.9.5b) at time $R_k$,

$$\mathbb{E}_k \left( \int_0^t I_{\{X(s)=j\}} \, ds \right) = \mathbb{E}_k \left( \int_{R_k}^{R_k+t} I_{\{X(s)=j\}} \, ds \right), \qquad t \geq 0.$$

By expressing $\int_0^{R_k}$ as $\int_0^t + \int_t^{R_k}$,

$$\mu_j(k) = \mathbb{E}_k \left( \int_t^{t+R_k} I_{\{X(s)=j\}} \, ds \right) = \mathbb{E}_k \left( \int_t^{\infty} I_{\{X(s)=j\}} I_{\{s-t<R_k\}} \, ds \right)$$

$$= \int_0^{\infty} \mathbb{P}_k \big( X(u+t) = j, u < R_k \big) \, du \qquad \text{where } u = s - t$$

$$= \int_0^{\infty} \sum_{i \in S} \mathbb{P}_k \big( X(u) = i, u < R_k \big) p_{ij}(t) \, du$$

$$= \sum_{i \in S} \mu_i(k) p_{ij}(t),$$

as claimed.                                                                                                ■

**Proof of Theorem (15).** By irreducibility, $g_i > 0$ for $i \in S$.

(a) Let $k$ be a positive recurrent state. By irreducibility and Theorem (6.9.22c), all states are recurrent, and hence by Theorem (6.9.17c) $X$ is non-explosive. We have that

$$(19) \qquad \qquad R_k = \sum_j \int_0^{R_k} I_{\{X(s)=j\}} \, ds.$$

(We note for future use that, if $X$ explodes, the left side of (19) can be infinite while the right side is finite.)

Let $\boldsymbol{\mu}(k)$ be given as in (17). By (19),

$$\sum_j \mu_j(k) = \mathbb{E}_k(R_k) = m_k < \infty.$$

By Lemma (16a, c) and the comment after (13), $\pi := \mu(k)/m_k$ is the unique stationary distribution and the unique root of $\pi\mathbf{G} = 0$. Note, in particular, that $\pi_k = \mu_k(k)/m_k$. By (17), $\mu_k(k) = 1/g_k$, so that

$$(20) \qquad\qquad\qquad\qquad \pi_k = \frac{1}{m_k g_k}.$$

Let $i \in S$. We apply Lemma (16c) with $k$ replaced by $i$ to find that $\mu(i)$ satisfies $\mu(i)\mathbf{G} = 0$. By the uniqueness of (16a), there exists $C_i \in (0, \infty)$ such that $\mu(i) = C_i\mu(k)$, whence

$$m_i = \sum_j \mu_j(i) = C_i \sum_j \mu_j(k) = C_i m_k.$$

Therefore, $m_i < \infty$, so that $i$ is positive.

(b) Suppose $X$ is non-explosive and $\pi$ is a distribution satisfying $\pi\mathbf{G} = 0$. By Lemma (16a), $v = (\pi_i g_i : i \in S)$ satisfies $v\mathbf{Y} = v$. By (16b), $\pi_j > 0$ for $j \in S$.

Let $k \in S$ and $v' = ((\pi_i g_i)/(\pi_k g_k) : i \in S)$, so that $v'\mathbf{Y} = v'$ and $v'_k = 1$. By Lemma (6.4.7b), with $\rho(k)$ as in (18), we have $\rho(k) \leq v'$, which is to say that

$$\rho_i(k) \leq \frac{\pi_i g_i}{\pi_k g_k}, \qquad i \in S.$$

By non-explosivity, equation (19) holds. On taking expectations, we find as in the proof of part (a) that

$$m_k = \sum_i \mu_i(k) = \sum_i \frac{\rho_i(k)}{g_i} \leq \frac{1}{\pi_k g_k} < \infty,$$

proving that every $k$ is positive recurrent. Stationarity holds by part (a).

Equation (20) was proved under the assumption that $k$ is recurrent. It holds, therefore, for every $k \in S$.    ∎

(21) **Example.**  Here is an example of an irreducible Markov chain for which the equation $\pi\mathbf{G} = 0$ is satisfied by two distinct distributions. Take $S = \mathbb{Z}$, the integers, and define $\mathbf{G}$ by

$$g_{m,m+1} = \begin{cases} 4^m & \text{if } m \geq 0, \\ \frac{1}{2} \cdot 4^{-m} & \text{if } m \leq -1, \end{cases} \qquad g_{m,m-1} = \begin{cases} \frac{1}{2} \cdot 4^m & \text{if } m \geq 1, \\ 4^{-m} & \text{if } m \leq 0, \end{cases}$$

with $g_{m,n} = 0$ for other $m \neq n$. It is easily checked that $\pi(i)\mathbf{G} = 0$ for $i = 1, 2$, where

$$\pi_m(1) = \frac{1}{3} \cdot \frac{1}{2^{|m|}}, \qquad \pi_m(2) = \begin{cases} \dfrac{1}{3} \cdot \dfrac{1}{4^m} & \text{if } m \geq 0, \\[2ex] \dfrac{1}{3} \cdot \dfrac{2^{-m+1} - 1}{4^{-m}} & \text{if } m \leq -1, \end{cases}$$

By Theorem (15b), the chain must be explosive.    ●

(**C) The limit theorem.**  We turn to the limit theorem for Markov chains, sometimes known as the ergodic theorem. What can be said about the limiting distribution of $X(t)$ as $t \to \infty$?

**(22) Markov chain limit theorem.** *Let X be irreducible and non-explosive.*
  (a) *If there exists a stationary distribution* $\pi$, *then it is unique and*

$$p_{ij}(t) \to \pi_j \quad as \quad t \to \infty, \quad for\ all\ i\ and\ j.$$

  (b) *If there is no stationary distribution, then* $p_{ij}(t) \to 0$ *as* $t \to \infty$, *for all i and j.*

The proof is preceded by a lemma concerning the uniform continuity of transition probabilities.

**(23) Lemma.** *Let X be a minimal Markov chain with state space S, transition semigroup* $\{\mathbf{P}_t\}$, *and generator* $\mathbf{G}$. *Then*

$$|p_{ij}(t+u) - p_{ij}(t)| \le 1 - e^{-g_i u}, \qquad i, j \in S, \quad t, u \ge 0.$$

**Proof.** Since $\{\mathbf{P}_t\}$ has the semigroup property (6.9.14),

$$|p_{ij}(t+u) - p_{ij}(t)| \le \left| \sum_{k \in S} p_{ik}(u) p_{kj}(t) - p_{ij}(t) \right|$$

$$= \left| \left( \sum_{k:\, k \neq i} p_{ik}(u) p_{kj}(t) \right) - p_{ij}(t)[1 - p_{ii}(u)] \right|$$

$$\le 1 - p_{ii}(u)$$

$$\le \mathbb{P}_i(U_0 \le u) = 1 - e^{-g_i u},$$

as required.                                                                          ∎

**Proof of Theorem (22).** We may assume that $|S| \ge 2$, whence, by irreducibility, $g_i > 0$ for $i \in S$. Fix $h > 0$ and let $Z_n = X(nh)$. Then $Z = \{Z_n\}$ is an irreducible, aperiodic, discrete-time Markov chain, called a *skeleton* of $X$ (see Exercise (6.9.14)). By the comments following Theorem (6.4.20),

**(24)**                                   $$p_{ij}(nh) \to l_j^h \qquad as\ n \to \infty,$$

where

**(25)**                          $$l_j^h = \begin{cases} \pi_j^h & \text{if } Z \text{ is positive recurrent,} \\ 0 & \text{otherwise,} \end{cases}$$

and, in the positive recurrent case, $\pi^h = (\pi_j^h : j \in S)$ is the unique stationary distribution of $Z$. We now fill in the gaps via the continuity of Lemma (23).

Let $i, j \in S$, $\epsilon > 0$, and pick $h > 0$ such that

$$1 - e^{-g_i h} \le \tfrac{1}{2}\epsilon.$$

By (24), we may find $N$ such that

$$|p_{ij}(nh) - l_j^h| \le \tfrac{1}{2}\epsilon, \qquad n \ge N.$$

For $t > Nh$, let $n$ be such that $nh \le t < (n+1)h$. By Lemma (23),

$$|p_{ij}(t) - l_j^h| \le |p_{ij}(t) - p_{ij}(nh)| + |p_{ij}(nh) - l_j^h| \le \epsilon.$$

The limits follow.

In the positive recurrent case, the uniqueness of $\pi$ holds since any stationary distribution for $X$ is also stationary for $Z$. ∎

**(26) Example (6) revisited.** This section closes with a fairly elementary calculation. If $\alpha, \beta > 0$ and $S = \{1, 2\}$, then

$$\mathbf{G} = \begin{pmatrix} -\alpha & \alpha \\ \beta & -\beta \end{pmatrix}$$

is the generator of a stochastic semigroup $\{\mathbf{P}_t\}$ given by the following calculation. Diagonalize $\mathbf{G}$ to obtain $\mathbf{G} = \mathbf{B}\Lambda\mathbf{B}^{-1}$ where

$$\mathbf{B} = \begin{pmatrix} \alpha & 1 \\ -\beta & 1 \end{pmatrix}, \qquad \Lambda = \begin{pmatrix} -(\alpha + \beta) & 0 \\ 0 & 0 \end{pmatrix}.$$

Therefore

$$\mathbf{P}_t = \sum_{n=0}^{\infty} \frac{t^n}{n!} \mathbf{G}^n = \mathbf{B}\left(\sum_{n=0}^{\infty} \frac{t^n}{n!} \Lambda^n\right) \mathbf{B}^{-1}$$

$$= \mathbf{B} \begin{pmatrix} h(t) & 0 \\ 0 & 1 \end{pmatrix} \mathbf{B}^{-1} \qquad \text{since} \quad \Lambda^0 = \mathbf{I}$$

$$= \frac{1}{\alpha + \beta} \begin{pmatrix} \alpha h(t) + \beta & \alpha[1 - h(t)] \\ \beta[1 - h(t)] & \alpha + \beta h(t) \end{pmatrix}$$

where $h(t) = e^{-t(\alpha+\beta)}$. Let $t \to \infty$ to obtain

$$\mathbf{P}_t \to \begin{pmatrix} 1-\rho & \rho \\ 1-\rho & \rho \end{pmatrix} \qquad \text{where} \quad \rho = \frac{\alpha}{\alpha + \beta},$$

and so

$$\mathbb{P}(X(t) = i) \to \begin{cases} 1-\rho & \text{if } i = 1, \\ \rho & \text{if } i = 2, \end{cases}$$

irrespective of the initial distribution of $X(0)$. This shows that $\pi = (1 - \rho, \rho)$ is the limiting distribution. Check that $\pi\mathbf{G} = \mathbf{0}$. The method of Example (6) provides an alternative and easier route to these results. ●

---

## Exercises for Section 6.10

1.  Let $N < \infty$, and let $\mathcal{Q}$ be the space of $N \times N$ matrices with real entries, with norm

$$|Q| = \sup_{\mathbf{x} \ne \mathbf{0}} \frac{|Q\mathbf{x}|}{|\mathbf{x}|}, \qquad Q \in \mathcal{Q},$$

where $|\mathbf{y}|$ is the Euclidean norm of the vector $\mathbf{y}$, and the supremum is over all non-zero column vectors.

(a) Show for $Q_1, Q_2 \in \mathcal{Q}$ that

$$|Q_1 + Q_2| \le |Q_1| + |Q_2|, \quad |Q_1 Q_2| \le |Q_1| \cdot |Q_2|.$$

(b) Show for $Q \in \mathcal{Q}$ that $E_n := \sum_{k=0}^{n} Q^k/k!$ converges with respect to $|\cdot|$ to a limit which we denote $E = E(Q)$.
(c) Show that $E(Q_1 + Q_2) = E(Q_1)E(Q_2)$ if $Q_1$ and $Q_2$ are commuting elements of $\mathcal{Q}$.

**2.**   Let $Y$ be an irreducible discrete-time Markov chain on a countably infinite state space $S$, having transition matrix $\mathbf{Y} = (y_{ij})$ satisfying $y_{ii} = 0$ for all states $i$, and with stationary distribution $\boldsymbol{v}$. Construct a continuous-time process $X$ on $S$ for which $Y$ is the jump chain, such that $X$ has no stationary distribution.

**3.**   Let $X = (X(t) : t \ge 0)$ be a Markov chain on $\mathbb{Z}$ with generator $\mathbf{G} = (g_{ij})$ given by

$$g_{i,i-1} = i^2 + 1, \quad g_{i,i} = -2(i^2+1), \quad g_{i,i+1} = i^2 + 1, \quad i \in \mathbb{Z}.$$

Show that $X$ is recurrent. Is $X$ positive recurrent?

**4.**   Let $X$ be a Markov chain on $\mathbb{Z}$ with generator $G = (g_{ij})$ satisfying

$$g_{i,i-1} = 3^{|i|}, \quad g_{i,i} = -3^{|i|+1}, \quad g_{i,i+1} = 2 \cdot 3^{|i|}, \quad i \in \mathbb{Z}.$$

Show that $X$ is transient, but has an invariant distribution. Explain.

**5.**   Let $X = (X_t : t \ge 0)$ be a Markov chain with generator $G = (g_{ij})$ on the finite state space $S$, and let $f : S \to \mathbb{R}$ be a function, which we identify with the vector $f = (f(i) : i \in S)$. Show that

$$\mathbf{G}f(i) = \sum_{j \in S} g_{ij}\left(f(j) - f(i)\right),$$

where $\mathbf{G}f$ denotes the standard matrix multiplication.
    Show that

$$\mathbf{G}f(i) = \lim_{t \to 0} \frac{1}{t}\left[\mathbb{E}_i f(X_t) - f(i)\right], \quad i \in S,$$

and deduce that

$$\mathbb{E}_i f(X_t) = f(i) + \int_0^t \mathbb{E}_i(\mathbf{G}f(X_s))\, ds.$$

## 6.11  Birth–death processes and imbedding

A birth process is a continuous-time Markov chain with generator $\mathbf{G} = (g_{ij})$ given by

$$g_{ij} = \begin{cases} \lambda_i & \text{if } j = i+1, \\ 0 & \text{if } j \ne i, i+1, \end{cases}$$

where $\lambda_i$ is the intensity of birth when in state $i$. More realistic continuous-time models for population growth incorporate death also. Suppose that the number $X(t)$ of individuals alive in some population at time $t$ evolves in the following way:
    (a) $X$ is a Markov chain taking values in $\{0, 1, 2, \dots\}$,

(b) $X$ has generator $\mathbf{G} = (g_{ij} : i, j \geq 0)$ given by

$$
\mathbf{G} = \begin{pmatrix}
-\lambda_0 & \lambda_0 & 0 & 0 & 0 & \cdots \\
\mu_1 & -(\lambda_1 + \mu_1) & \lambda_1 & 0 & 0 & \cdots \\
0 & \mu_2 & -(\lambda_2 + \mu_2) & \lambda_2 & 0 & \cdots \\
0 & 0 & \mu_3 & -(\lambda_3 + \mu_3) & \lambda_3 & \cdots \\
\vdots & \vdots & \vdots & \vdots & \vdots & \ddots
\end{pmatrix},
$$

where the 'birth rates' $\lambda_0, \lambda_1, \ldots$ and the 'death rates' $\mu_0, \mu_1, \ldots$ satisfy $\lambda_i \geq 0$, $\mu_i \geq 0$, $\mu_0 = 0$.

Such a process $X$ is called a *birth–death process*. When in state $i$, the intensity of birth is $\lambda_i$ and the intensity of death is $\mu_i$. In particular cases, we may have that $\lambda_0 = 0$, and then $0$ is an absorbing state and the chain is not irreducible.

The transition probabilities $p_{ij}(t) = \mathbb{P}_i(X(t) = j)$ may in principle be calculated from a knowledge of the birth and death rates, although in practice these functions rarely have nice forms. It is an easier matter to determine the asymptotic behaviour of the process as $t \to \infty$.

Suppose that

(2) $\qquad \lambda_i > 0 \quad \text{for} \quad i \geq 0, \qquad \mu_i > 0 \quad \text{for} \quad i \geq 1, \qquad \mu_0 = 0,$

whence, in particular, the chain is irreducible. Following Theorem (6.10.15), we look for measures $\mathbf{x}$ that satisfy the equation $\mathbf{x}\mathbf{G} = \mathbf{0}$, which is to say that

$$-\lambda_0 x_0 + \mu_1 x_1 = 0,$$
$$\lambda_{n-1} x_{n-1} - (\lambda_n + \mu_n) x_n + \mu_{n+1} x_{n+1} = 0 \qquad \text{for } n \geq 1.$$

A simple induction† yields that

(3) $$x_n = \frac{\lambda_0 \lambda_1 \cdots \lambda_{n-1}}{\mu_1 \mu_2 \cdots \mu_n} x_0, \qquad n \geq 1.$$

Such a measure $\mathbf{x}$ is a distribution if and only if $\sum_n x_n = 1$. This may happen if and only if

(4) $$\sum_{n=0}^{\infty} \frac{\lambda_0 \lambda_1 \cdots \lambda_{n-1}}{\mu_1 \mu_2 \cdots \mu_n} < \infty,$$

where the term $n = 0$ is interpreted as 1. If the last holds, then

(5) $$x_0 = \left( \sum_{n=0}^{\infty} \frac{\lambda_0 \lambda_1 \cdots \lambda_{n-1}}{\mu_1 \mu_2 \cdots \mu_n} \right)^{-1} > 0.$$

In summary, there exists a distribution $\boldsymbol{\pi}$ satisfying $\boldsymbol{\pi}\mathbf{G} = \mathbf{0}$ if and only if (4) holds.

---

†Alternatively, note that the matrix $\mathbf{G}$ is tridiagonal, whence the chain is reversible in equilibrium (see Problem (6.15.16)). Now seek a solution to the detailed balance equations.

Equation (5) is insufficient to deduce that $\pi$ is a stationary distribution for $X$. By Theorem (6.10.15b), we require in addition that $X$ is non-explosive. By Theorem (6.9.17c), $X$ is non-explosive if it is recurrent, which is in turn equivalent to requiring that the jump chain $Y$ is recurrent (by Theorem (6.9.22)). By Lemma (6.10.16a), it suffices for the recurrence of $Y$ that $v = (v_n = x_n g_n : n \geq 0)$ is a stationary measure for $Y$ with $\sum_n v_n < \infty$. The last holds if and only if

$$\text{(6)} \qquad \sum_{n=0}^{\infty} \frac{\lambda_0 \lambda_1 \cdots \lambda_{n-1}}{\mu_1 \mu_2 \cdots \mu_n} (\lambda_n + \mu_n) < \infty.$$

We summarize the above discussion in a theorem.

**(7) Theorem.** *Let $X$ be a birth–death process with generator $\mathbf{G}$ satisfying (2).*
(a) *The measure $\mathbf{x}$ of (3) satisfies $\mathbf{xG} = \mathbf{0}$.*
(b) *There is a distribution $\pi$ satisfying $\pi \mathbf{G} = \mathbf{0}$ if and only if (4) holds.*
(c) *The $\pi$ of part (b) is a stationary distribution of $X$ if, in addition, (6) holds.*

**(8) Example.** Consider the birth–death chain with birth rates $\lambda_i = \lambda g_i$ and death rates $\mu_i = \mu g_i$, where $\lambda, \mu > 0$, $\lambda + \mu = 1$, and $g_i > 0$ for all $i$. By (3), the measure

$$x_n = (\lambda/\mu)^n \frac{1}{g_n}, \qquad n \geq 0,$$

satisfies $\mathbf{xG} = \mathbf{0}$. As in (4), $\mathbf{x}$ is summable if and only if

$$\text{(9)} \qquad \sum_{n=0}^{\infty} (\lambda/\mu)^n \frac{1}{g_n} < \infty.$$

On the other hand, $v = (x_n g_n : n \geq 0)$ is summable (as in (6)) if and only if

$$\text{(10)} \qquad \sum_{n=0}^{\infty} (\lambda/\mu)^n < \infty.$$

If $g_n \equiv 1$, conditions (9) and (10) are equivalent, and they hold if and only if $\lambda/\mu < 1$. Suppose, in contrast, that $g_n = 2^n$ and $\lambda/\mu \in (1, 2)$. Then (9) holds but not (10). In this case, the jump chain is transient, implying that $X$ is transient also (by Theorem (6.9.22b)). However, $\mathbf{x}$ is summable, so it may be normalized to a distribution $\pi$. That is, $X$ is transient, whereas there exists a distribution $\pi$ satisfying $\pi \mathbf{G} = \mathbf{0}$. By Theorem (6.10.15b), the only possibility is that $X$ is explosive. ●

By Theorem (6.10.22), the process $X$ settles into equilibrium (with stationary distribution given by (3) and (5)) if the summations in (4) and (6) are finite, a condition requiring that the birth rates are not too large relative to the death rates.

Here are some examples of birth–death processes.

**(11) Example. Pure birth.** The death rates satisfy $\mu_n = 0$ for all $n$. ●

**(12) Example. Simple death with immigration.** Let us model a population which evolves as follows. At time zero the size $X(0)$ of the population equals $I$. Individuals do not reproduce, but new individuals immigrate into the population at the arrival times of a Poisson process with intensity $\lambda > 0$. Each individual dies at rate $\mu$. We recognize $X$ as an irreducible birth–death process with parameters

**(13)**
$$\lambda_n = \lambda, \quad \mu_n = n\mu.$$

We may ask for the distribution of $X(t)$ and for the limiting distribution of the chain as $t \to \infty$. The former question is answered by solving the forward equations; this is Problem (6.15.18). The latter question is answered by the following.

**(14) Theorem.** *In the limit as $t \to \infty$, $X(t)$ is asymptotically Poisson distributed with parameter $\rho = \lambda/\mu$. That is,*

$$\mathbb{P}\big(X(t) = n\big) \to \frac{\rho^n}{n!}e^{-\rho}, \qquad n = 0, 1, 2, \dots .$$

**Proof.** Inequalities (4) and (6) are valid. By the earlier discussion, $X$ is non-explosive with stationary distribution $\pi_n \propto \rho^n/n!$. The claim follows by Theorem (6.10.22).  ∎●

**(15) Example. Simple birth–death.** This is the birth–death process with birth/death rates

$$\lambda_n = n\lambda, \quad \mu_n = n\mu,$$

and we explore this model in some detail. Assume that $X(0) = I > 0$, and note that the state 0 is absorbing. We cannot deduce from Theorem (6.9.17) that $X$ is non-explosive for general $\lambda$, $\mu$, and so we assume for the moment that $X$ is the minimal process. The distribution of $X(t)$ may be found using its generating function.

**(16) Theorem.** *The generating function of $X(t)$ is*

$$G(s, t) = \mathbb{E}(s^{X(t)}) = \begin{cases} \left(\dfrac{\lambda t(1 - s) + s}{\lambda t(1 - s) + 1}\right)^I & \text{if } \mu = \lambda, \\[4mm] \left(\dfrac{\mu(1 - s) - (\mu - \lambda s)e^{-t(\lambda - \mu)}}{\lambda(1 - s) - (\mu - \lambda s)e^{-t(\lambda - \mu)}}\right)^I & \text{if } \mu \neq \lambda. \end{cases}$$

**Proof.** This resembles Proof B of Theorem (6.8.2)†. By Theorem (6.10.7), the probabilities $p_j(t) = \mathbb{P}_I(X(t) = j)$, satisfy the forward equations:

$$p_j'(t) = \lambda(j - 1)p_{j-1}(t) - (\lambda + \mu)jp_j(t) + \mu(j + 1)p_{j+1}(t) \quad \text{if } j \geq 1,$$
$$p_0'(t) = \mu p_1(t).$$

Multiply the $j$th equation by $s^j$ and sum to obtain

$$\sum_{j=0}^{\infty} s^j p_j'(t) = \lambda s^2 \sum_{j=1}^{\infty}(j - 1)s^{j-2}p_{j-1}(t) - (\lambda + \mu)s \sum_{j=0}^{\infty} js^{j-1}p_j(t)$$
$$+ \mu \sum_{j=0}^{\infty}(j + 1)s^j p_{j+1}(t).$$

---

†It appears that Conrad Palm was the first to use generating functions in this way, some time before Maurice Bartlett's independent use in 1945.

Put $G(s, t) = \sum_0^\infty s^j p_j(t) = \mathbb{E}(s^{X(t)})$ to obtain

(17)
$$\frac{\partial G}{\partial t} = \lambda s^2 \frac{\partial G}{\partial s} - (\lambda + \mu)s \frac{\partial G}{\partial s} + \mu \frac{\partial G}{\partial s}$$

$$= (\lambda s - \mu)(s - 1)\frac{\partial G}{\partial s}$$

with boundary condition $G(s, 0) = s^I$. The solution to this partial differential equation is given by (16); to see this either solve (17) by standard methods, or substitute the conclusion of (16) into (17). ∎

Note that $G(1, t) = 1$ for all $t$, whence $X$ is non-explosive for all $\lambda$ and $\mu$. To find the mean and variance of $X(t)$, differentiate $G$:

$$\mathbb{E}_I(X(t)) = Ie^{(\lambda - \mu)t}, \qquad \mathrm{var}_I(X(t)) = \begin{cases} 2I\lambda t & \text{if } \lambda = \mu, \\ I\dfrac{\lambda + \mu}{\lambda - \mu}e^{(\lambda - \mu)t}[e^{(\lambda - \mu)t} - 1] & \text{if } \lambda \neq \mu. \end{cases}$$

Write $\rho = \lambda/\mu$ and notice that

$$\mathbb{E}_I(X(t)) \to \begin{cases} 0 & \text{if } \rho < 1, \\ \infty & \text{if } \rho > 1. \end{cases}$$

**(18) Corollary.** *The extinction probabilities* $\eta(t) = \mathbb{P}_I(X(t) = 0)$ *satisfy, as* $t \to \infty$,

$$\eta(t) \to \begin{cases} 1 & \text{if } \rho \leq 1, \\ \rho^{-I} & \text{if } \rho > 1. \end{cases}$$

**Proof.** We have that $\eta(t) = G(0, t)$. Substitute $s = 0$ in $G(s, t)$ to find $\eta(t)$ explicitly. ∎

The observant reader will have noticed that these results are almost identical to those obtained for the branching process, except in that they pertain to a process in continuous time. There are (at least) two discrete Markov chains imbedded in $X$.

*(A) Imbedded random walk.* The jump chain $Y$ is a simple random walk on $\{0, 1, 2, \dots\}$ with parameter $p = \lambda/(\lambda + \mu)$, initial position $I$, and with an absorbing barrier at 0. As in Example (3.9.6), the probability of ultimate absorption at 0 is given by (18). Other properties of random walks (see Sections 3.9 and 5.3) are applicable also.

*(B) Imbedded branching process.* We can think of the birth–death process $X$ in the following way. Let $X(0) = 1$. After birth an individual lives for a certain length of time which is exponentially distributed with parameter $\lambda + \mu$. When this period is complete it dies, leaving behind it either no individuals, with probability $\mu/(\lambda+\mu)$, or two individuals, with probability $\lambda/(\lambda + \mu)$. This process is an age-dependent branching process with age density function

(19)
$$f_T(u) = (\lambda + \mu)e^{-(\lambda+\mu)u}, \qquad u \geq 0,$$

and family-size generating function

(20)
$$G(s) = \frac{\mu + \lambda s^2}{\mu + \lambda},$$

in the notation of Section 5.5 (do not confuse $G$ in (20) with $G(s, t) = \mathbb{E}(s^{X(t)})$). Thus, if $I = 1$, the generating function $G(s, t) = \mathbb{E}(s^{X(t)})$ satisfies the differential equation

$$(21) \qquad \frac{\partial G}{\partial t} = \lambda G^2 - (\lambda + \mu)G + \mu.$$

After (17), this is the *second* differential equation for $G(s, t)$. Needless to say, (21) is really just the backward equation of the process; the reader should check this and verify that it has the same solution as the forward equation (17). Suppose we lump together the members of each generation of this age-dependent branching process. Then we obtain an ordinary branching process with family-size generating function $G(s)$ given by (20). From the general theory, the extinction probability of the process is the smallest non-negative root of the equation $s = G(s)$, and we can verify that this is given by (18) with $I = 1$.                  ●

**(22) Example. A more general branching process.** Finally, we consider a more general type of age-dependent branching process than that above, and we investigate its explosivity. Suppose that each individual in a population lives for an exponentially distributed time with parameter $\lambda$ say. After death it leaves behind it a (possibly empty) family of offspring: the size $N$ of this family has mass function $f(k) = \mathbb{P}(N = k)$ and generating function $G_N$. Let $X(t)$ be the size of the population at time $t$; we assume that $X(0) = 1$. From Section 5.5 the backward equation for $G(s, t) = \mathbb{E}(s^{X(t)})$ is

$$\frac{\partial G}{\partial t} = \lambda\big(G_N(G) - G\big)$$

with boundary condition $G(s, 0) = s$. The solution is given by

$$(23) \qquad \int_s^{G(s,t)} \frac{du}{G_N(u) - u} = \lambda t$$

provided that $G_N(u) - u$ has no zeros within the domain of the integral. There are many interesting questions about this process; for example, is it non-explosive?

**(24) Theorem.** *The process $X$ is non-explosive if and only if*

$$(25) \qquad \int_{1-\epsilon}^1 \frac{du}{G_N(u) - u} \qquad \textit{diverges for all } \epsilon > 0.$$

**Proof.** See Harris (1963, p. 107).                                         ■

If condition (25) fails then the population size may explode to $+\infty$ in finite time.

**(26) Corollary.** *The chain $X$ is non-explosive if $\mathbb{E}(N) < \infty$.*

**Proof.** Expand $G_N(u) - u$ about $u = 1$ to find that

$$G_N(u) - u = [\mathbb{E}(N) - 1](u - 1) + o(u - 1) \quad \text{as} \quad u \uparrow 1.$$                  ■●

## Exercises for Section 6.11

**1.** Describe the jump chain for a birth–death process with rates $\lambda_n$ and $\mu_n$.

**2.** Consider an immigration–death process $X$, being a birth–death process with birth rates $\lambda_n = \lambda$ and death rates $\mu_n = n\mu$. Find the transition matrix of the jump chain $Y$, and show that it has as stationary distribution

$$\pi_n = \frac{1}{2(n!)}\left(1 + \frac{n}{\rho}\right)\rho^n e^{-\rho}$$

where $\rho = \lambda/\mu$. Explain why this differs from the stationary distribution of $X$.

**3.** Consider the birth–death process $X$ with $\lambda_n = n\lambda$ and $\mu_n = n\mu$ for all $n \geq 0$. Suppose $X(0) = 1$ and let $\eta(t) = \mathbb{P}_1(X(t) = 0)$. Show that $\eta$ satisfies the differential equation

$$\eta'(t) + (\lambda + \mu)\eta(t) = \mu + \lambda\eta(t)^2.$$

Hence find $\eta(t)$, and calculate $\mathbb{P}_1(X(t) = 0 \mid X(u) = 0)$ for $0 < t < u$.

**4.** For the birth–death process of the previous exercise with $\lambda < \mu$, show that the distribution of $X(t)$, conditional on the event $\{X(t) > 0\}$, converges as $t \to \infty$ to a geometric distribution.

**5.** Let $X$ be a birth–death process with $\lambda_n = n\lambda$ and $\mu_n = n\mu$, and suppose $X(0) = 1$. Show that the time $T$ at which $X(t)$ first takes the value $0$ satisfies

$$\mathbb{E}(T \mid T < \infty) = \begin{cases} \dfrac{1}{\lambda}\log\left(\dfrac{\mu}{\mu - \lambda}\right) & \text{if } \lambda < \mu, \\[2mm] \dfrac{1}{\mu}\log\left(\dfrac{\lambda}{\lambda - \mu}\right) & \text{if } \lambda > \mu. \end{cases}$$

What happens when $\lambda = \mu$?

**6.** Let $X$ be the birth–death process of Exercise (6.11.5) with $\lambda \neq \mu$, and let $V_r(t)$ be the total amount of time the process has spent in state $r \geq 0$, up to time $t$. Find the distribution of $V_1(\infty)$ and the generating function $\sum_r s^r \mathbb{E}(V_r(t))$. Hence show in two ways that $\mathbb{E}(V_1(\infty)) = [\max\{\lambda, \mu\}]^{-1}$. Show further that $\mathbb{E}(V_r(\infty)) = \lambda^{r-1} r^{-1}[\max\{\lambda, \mu\}]^{-r}$.

**7.** Repeat the calculations of Exercise (6.11.6) in the case $\lambda = \mu$.

**8.** Consider a birth–death process $X$ with birth rates $\lambda_n > 0$ for $n \geq 0$, death rates $\mu_n > 0$ for $n > 0$, and $\mu_0 = 0$. Let $X(0) = n > 0$, and let $D_n$ be the time until the process first takes the value $n - 1$.
(a) Show that $d_n = \mathbb{E}(D_n)$ satisfies

$$\lambda_n d_{n+1} = \mu_n d_n - 1, \qquad n \geq 1.$$

(b) Show that the moment generating function $M_n(\theta) = \mathbb{E}(e^{\theta D_n})$ satisfies

$$(\lambda_n + \mu_n - \theta)M_n(\theta) = \mu_n + \lambda_n M_n(\theta)M_{n+1}(\theta), \qquad n \geq 1.$$

**9.** **Biofilms.** In a model for a biofilm population, we assume there are $n$ colonizable 'niches' (or 'food sources'). Let $X(t)$ be the number of occupied niches at time $t$, and assume $X$ is a Markov chain that evolves as follows. The lifetime of any colony is exponentially distributed with parameter $\mu$; if $X(t) = i$, the rate of establishment of a new colony in an empty niche is $\lambda i(n - i)$. The usual independence may be assumed.

For $X(0) = 1, 2, \ldots$, find the mean time until the population is extinct, which is to say that no niches are occupied. Discuss the implications when $n$ is large.

## 6.12 Special processes

There are many more general formulations of the processes which we modelled in Sections 6.8 and 6.11. Here is a very small selection of some of them, with some details of the areas in which they have been found useful.

**(1) Non-homogeneous chains.** We may relax the assumption that the transition probabilities $p_{ij}(s, t) = \mathbb{P}(X(t) = j \mid X(s) = i)$ satisfy the homogeneity condition $p_{ij}(s, t) = p_{ij}(0, t - s)$. This leads to some difficult problems. Progress can be made in the special case when $X$ is the simple birth–death process of the previous section, for which $\lambda_n = n\lambda$ and $\mu_n = n\mu$. The parameters $\lambda$ and $\mu$ are now assumed to be non-constant functions of $t$. (After all, most populations have birth and death rates which vary from season to season.) It is easy to check that the forward equation (6.11.17) remains unchanged:

$$\frac{\partial G}{\partial t} = [\lambda(t)s - \mu(t)](s - 1)\frac{\partial G}{\partial s}.$$

The solution is

$$G(s, t) = \left[ 1 + \left( \frac{e^{r(t)}}{s - 1} - \int_0^t \lambda(u)e^{r(u)}\, du \right)^{-1} \right]^t$$

where $I = X(0)$ and

$$r(t) = \int_0^t \left[ \mu(u) - \lambda(u) \right] du.$$

The extinction probability $\mathbb{P}_I(X(t) = 0)$ is the coefficient of $s^0$ in $G(s, t)$, and it is left as an *exercise* for the reader to prove the next result.

**(2) Theorem.** *We have that $\mathbb{P}_I(X(t) = 0) \to 1$ if and only if*

$$\int_0^\infty \mu(u)e^{r(u)}\, du = \infty.$$   ●

**(3) A bivariate branching process.** We advertised the branching process as a feasible model for the growth of cell populations; we should also note one of its inadequacies in this role. Even the age-dependent process cannot meet the main objection, which is that the time of division of a cell may depend rather more on the *size* of the cell than on its *age*. So here is a model for the growth and degradation of long-chain polymers†.

A population comprises *particles*. Let $N(t)$ be the number of particles present at time $t$, and suppose that $N(0) = 1$. We suppose that the $N(t)$ particles are partitioned into $W(t)$ groups of size $N_1, N_2, \ldots, N_W$ such that the particles in each group are aggregated into a *unit cell*. Think of the cells as a collection of $W(t)$ long-chain (linear) polymers, containing $N_1, N_2, \ldots, N_W$ particles respectively. As time progresses each cell grows and divides. We suppose that each cell accumulates one particle from outside the system at rate $\lambda$. As cells become larger they are more likely to divide. Given a cell of size $N$, and a value $M \in \{1, 2, \ldots, N - 1\}$, we

---

†In physical chemistry, a *long-chain polymer* is a chain of molecules, neighbouring pairs of which are joined by bonds.

assume that the cell divides into two cells of sizes $M$ and $N - M$ at rate $\mu$. The total rate of division of an $N$-cell is thus $\mu(N - 1)$. The linearity of the rate of division seems reasonable for polymer degradation since the particles are strung together in a line and any of the $N - 1$ 'links' between pairs of neighbouring particles may sever.

At time $t$ there are $N(t)$ particles and $W(t)$ cells, and the process is said to be in state $X(t) = (N(t), W(t))$. During the interval $(t, t + h)$ various transitions for $X$ are possible. Either some cell grows or some cell divides, or more than one such event occurs. The total rate of cell growth is $\lambda W$ since there are $W$ chances of this happening; the total rate of division is $\mu(N_1 + \cdots + N_W - W) = \mu(N - W)$, since there are $N - W$ links in all. Putting this information together results in a bivariate Markov chain $X(t) = (N(t), W(t))$ with state space $S = \{1, 2, \ldots\}^2$ and generator $\mathbf{G} = (g_{(n,w),(n',w')} : (n, w), (n', w') \in S)$ given by

$$g_{(n,w),(n,w)+\epsilon} = \begin{cases} \lambda w & \text{if } \epsilon = (1, 0), \\ \mu(n - w) & \text{if } \epsilon = (0, 1), \\ 0 & \text{if } \epsilon \neq (0, 0), (1, 0), (0, 1). \end{cases}$$

Write down the forward equations as usual to obtain that the joint generating function

$$G(x, y; t) = \mathbb{E}(x^{N(t)} y^{W(t)})$$

satisfies the partial differential equation

$$\frac{\partial G}{\partial t} = \mu x (y - 1) \frac{\partial G}{\partial x} + y \big[ \lambda(x - 1) - \mu(y - 1) \big] \frac{\partial G}{\partial y}$$

with $G(x, y; 0) = xy$. The joint moments of $N$ and $W$ are easily derived from this equation. More sophisticated techniques show that $N(t) \to \infty$, $W(t) \to \infty$, and $N(t)/W(t)$ approaches some constant as $t \to \infty$.

Unfortunately, most cells in nature are irritatingly non-Markovian!    ●

**(4) A non-linear epidemic.** Consider a population of constant size $N + 1$, and watch the spread of a disease about its members. Let $X(t)$ be the number of healthy individuals at time $t$ and suppose that $X(0) = N$. We assume that if $X(t) = n$ then the rate of new infections is proportional to the number of possible encounters between ill folk and healthy folk. That is, $X$ has generator $\mathbf{G}$ with

$$g_{ij} = \begin{cases} \lambda i (N + 1 - i) & \text{if } j = i - 1 \geq 0, \\ 0 & \text{if } j \neq i - 1, i. \end{cases}$$

Nobody recovers from this disease. In the usual way, the reader can show that

$$G(s, t) = \mathbb{E}(s^{X(t)}) = \sum_{n=0}^{N} s^n \mathbb{P}(X(t) = n)$$

satisfies

$$\frac{\partial G}{\partial t} = \lambda(1 - s) \left( N \frac{\partial G}{\partial s} - s \frac{\partial^2 G}{\partial s^2} \right)$$

with $G(s, 0) = s^N$. There is no simple way of solving this equation, though a lot of information is available about approximate solutions.                                                                          ●

**(5) Birth–death with immigration.** We saw in Example (6.11.12) that populations are not always closed and that there is sometimes a chance that a new process will be started by an arrival from outside. This may be due to mutation (if we are counting genes), or leakage (if we are counting neutrons), or irresponsibility (if we are counting cases of rabies).

Suppose that there is one individual in the population at time zero; this individual is the founding member of some birth–death process $N$ with fixed but unspecified parameters. Suppose further that other individuals immigrate into the population in the manner of a Poisson process $I$ with intensity $\nu$. Each immigrant starts a new birth–death process which is an independent identically distributed copy of the original process $N$ but displaced in time according to its time of arrival. Let $T_0 \, (= 0), T_1, T_2, \ldots$ be the times at which immigrants arrive, and let $U_0, U_1, \ldots$ be the interarrival times $U_n = T_{n+1} - T_n$. The total population at time $t$ is the aggregate of the processes generated by the $I(t) + 1$ immigrants up to time $t$. Call this total $Y(t)$ to obtain

$$
\textbf{(6)} \qquad\qquad\qquad Y(t) = \sum_{i=0}^{I(t)} N_i(t - T_i)
$$

where $N_1, N_2, \ldots$ are independent copies of $N = N_0$. The problem is to find how the distribution of $Y$ depends on the typical process $N$ and the immigration rate $\nu$; this is an example of the problem of compounding discussed in Theorem (5.1.25).

We are now ready to describe $Y(t)$ in terms of the constituent processes $N_i$.

**(7) Theorem.** *If $N(t)$ has generating function $G_N(s, t) = \mathbb{E}(s^{N(t)})$ then the generating function $G(s, t) = \mathbb{E}(s^{Y(t)})$ satisfies*

$$
G(s, t) = G_N(s, t) \exp\left( \nu \int_0^t [G_N(s, u) - 1] \, du \right).
$$

**Proof.** Let $U_0, U_1, \ldots$ be a sequence of independent uniform variables on $[0, t]$. By (6),

$$
\mathbb{E}(s^{Y(t)}) = \mathbb{E}\left( s^{N_0(t) + N_1(t - T_1) + \cdots + N_I(t - T_I)} \right)
$$

where $I = I(t)$. By independence, conditional expectation, and Theorem (6.8.11) applied to the Poisson process $I$,

$$
\textbf{(8)} \qquad \mathbb{E}(s^{Y(t)}) = \mathbb{E}(s^{N_0(t)}) \mathbb{E}\left\{ \mathbb{E}\left( s^{N_1(t - T_1) + \cdots + N_I(t - T_I)} \mid I \right) \right\}
$$

$$
= G_N(s, t) \mathbb{E}\left\{ \mathbb{E}\left( s^{N_1(t - U_1) + \cdots + N_I(t - U_I)} \mid I \right) \right\}
$$

$$
= G_N(s, t) \mathbb{E}\left\{ \mathbb{E}\left( s^{N_1(t - U_1)} \right)^I \right\}.
$$

However,

$$
\textbf{(9)} \qquad \mathbb{E}(s^{N_1(t - U_0)}) = \mathbb{E}\left\{ \mathbb{E}(s^{N_1(t - U_0)} \mid U_0) \right\}
$$

$$
= \int_0^t \frac{1}{t} G_N(s, t - u) \, du = H(s, t),
$$

and

**(10)**
$$\mathbb{E}(H^I) = \sum_{k=0}^{\infty} H^k \frac{(vt)^k}{k!} e^{-vt} = e^{vt(H-1)}.$$

Substitute (9) and (10) into (8) to obtain the result.                          ∎●

**(11) Branching random walk.** Another characteristic of many interesting populations is their distribution about the space that they inhabit. We introduce this spatial aspect gently, by assuming that each individual lives at some point on the real line. (This may seem a fair description of a sewer, river, or hedge.) Let us suppose that the evolution proceeds as follows. After its birth, a typical individual inhabits a randomly determined spot $X$ in $\mathbb{R}$ for a random time $T$. After this time has elapsed it dies, leaving behind a family containing $N$ offspring which it distributes at points $X + Y_1, X + Y_2, \ldots, X + Y_N$ where $Y_1, Y_2, \ldots$ are independent and identically distributed. These individuals then behave as their ancestor did, producing the next generation offspring after random times at points $X + Y_i + Y_{ij}$, where $Y_{ij}$ is the displacement of the $j$th offspring of the $i$th individual, and the $Y_{ij}$ are independent and identically distributed. We shall be interested in the way that living individuals are distributed about $\mathbb{R}$ at some time $t$.

Suppose that the process begins with a single newborn individual at the point 0. We require some notation. Write $G_N(s)$ for the generating function of a typical family size $N$ and let $F$ be the distribution function of a typical $Y$. Let $Z(x, t)$ be the number of living individuals at points in the interval $(-\infty, x]$ at time $t$. We shall study the generating function

$$G(s; x, t) = \mathbb{E}(s^{Z(x,t)}).$$

Let $T$ be the lifetime of the initial individual, $N$ its family size, and $Y_1, Y_2, \ldots, Y_N$ the positions of its offspring. We shall condition $Z$ on all these variables to obtain a type of backward equation. We must be careful about the order in which we do this conditioning, because the length of the sequence $Y_1, Y_2, \ldots$ depends on $N$. Hold your breath, and note from Problem (4.14.29) that

$$G(s; x, t) = \mathbb{E}\left\{\mathbb{E}\left[\mathbb{E}\langle\mathbb{E}(s^Z \mid T, N, \mathbf{Y}) \mid T, N\rangle \mid T\right]\right\}.$$

Clearly
$$Z(x, t) = \begin{cases} Z(x, 0) & \text{if } T > t, \\ \sum_{i=1}^N Z_i(x - Y_i, t - T) & \text{if } T \le t, \end{cases}$$
where the processes $Z_1, Z_2, \ldots$ are independent copies of $Z$. Hence

$$\mathbb{E}(s^Z \mid T, N, \mathbf{Y}) = \begin{cases} G(s; x, 0) & \text{if } T > t, \\ \prod_{i=1}^N G(s; x - Y_i, t - T) & \text{if } T \le t. \end{cases}$$

Thus, if $T \le t$ then

$$\mathbb{E}\left[\mathbb{E}\langle\mathbb{E}(s^Z \mid T, N, \mathbf{Y}) \mid T, N\rangle \mid T\right] = \mathbb{E}\left[\left(\int_{-\infty}^{\infty} G(s; x - y, t - T)\, dF(y)\right)^N \middle| T\right]$$

$$= G_N\left(\int_{-\infty}^{\infty} G(s; x - y, t - T)\, dF(y)\right).$$

Now breathe again. We consider here only the Markovian case when $T$ is exponentially distributed with some parameter $\mu$. Then

$$G(s; x, t) = \int_0^t \mu e^{-\mu u} G_N\left(\int_{-\infty}^{\infty} G(s; x - y, t - u) \, dF(y)\right) du + e^{-\mu t} G(s; x, 0).$$

Substitute $v = t - u$ inside the integral and differentiate with respect to $t$ to obtain

$$\frac{\partial G}{\partial t} + \mu G = \mu G_N\left(\int_{-\infty}^{\infty} G(s; x - y, t) \, dF(y)\right).$$

It is not immediately clear that this is useful. However, differentiate with respect to $s$ at $s = 1$ to find that $m(x, t) = \mathbb{E}(Z(x, t))$ satisfies

$$\frac{\partial m}{\partial t} + \mu m = \mu \mathbb{E}(N) \int_{-\infty}^{\infty} m(x - y, t) \, dF(y)$$

which equation is approachable by Laplace transform techniques. Such results can be generalized to higher dimensions.                                                        ●

**(12) Spatial growth.**  Here is a simple model for skin cancer. Suppose that each point $(x, y)$ of the two-dimensional square lattice $\mathbb{Z}^2 = \{(x, y) : x, y = 0, \pm 1, \pm 2, \ldots\}$ is a skin cell. There are two types of cell, called $b$-cells (*benign* cells) and $m$-cells (*malignant* cells). Each cell lives for an exponentially distributed period of time, parameter $\beta$ for $b$-cells and parameter $\mu$ for $m$-cells, after which it splits into two similar cells, one of which remains at the point of division and the other displaces one of the four nearest neighbours, each chosen at random with probability $\frac{1}{4}$. The displaced cell moves out of the system. Thus there are two competing types of cell. We assume that $m$-cells divide at least as fast as $b$-cells; the ratio $\kappa = \mu/\beta \geq 1$ is called the 'carcinogenic advantage'.

Suppose that there is only one $m$-cell initially and that all other cells are benign. What happens to the resulting tumour of malignant cells?

**(13) Theorem.**  *If $\kappa = 1$, the m-cells die out with probability 1, but the mean time until extinction is infinite. If $\kappa > 1$, there is probability $\kappa^{-1}$ that the m-cells die out, and probability $1 - \kappa^{-1}$ that their number grows beyond all bounds.*

Thus there is strictly positive probability of the malignant cells becoming significant if and only if the carcinogenic advantage exceeds one.

**Proof.**  Let $X(t)$ be the number of $m$-cells at time $t$, and let $T_0 (= 0), T_1, T_2, \ldots$ be the sequence of times at which $X$ changes its value. Consider the imbedded discrete-time process $X = \{X_n\}$, where $X_n = X(T_n+)$ is the number of $m$-cells just after the $n$th transition; $X$ is a Markov chain taking values in $\{0, 1, 2, \ldots\}$. Remember the imbedded random walk of the birth–death process, Example (6.11.15); in the case under consideration a little thought shows that $X$ has transition probabilities

$$p_{i,i+1} = \frac{\mu}{\mu + \beta} = \frac{\kappa}{\kappa + 1}, \quad p_{i,i-1} = \frac{1}{\kappa + 1} \quad \text{if} \ \ i \neq 0, \qquad p_{0,0} = 1.$$

Therefore $X_n$ is simply a random walk with parameter $p = \kappa/(\kappa + 1)$ and with an absorbing barrier at 0. The probability of ultimate extinction from the starting point $X(0) = 1$ is $\kappa^{-1}$.

The walk is symmetric and null recurrent if $\kappa = 1$ and all non-zero states are transient if $\kappa > 1$.                                                                                    ∎

If $\kappa = 1$, the same argument shows that the $m$-cells certainly die out whenever there is a finite number of them to start with. However, suppose that they are distributed initially at the points of some (possibly infinite) set. It is possible to decide what happens after a long length of time; roughly speaking this depends on the relative densities of benign and malignant cells over large distances. One striking result is the following.

**(14) Theorem.** *If $\kappa = 1$, the probability that a specified finite collection of points contains only one type of cell approaches one as $t \to \infty$.*

**Sketch proof.** If two cells have a common ancestor then they are of the same type. Since offspring displace any neighbour with equal probability, the line of ancestors of any cell performs a symmetric random walk in two dimensions stretching backwards in time. Therefore, given any two cells at time $t$, the probability that they have a common ancestor is the probability that two symmetric and independent random walks $S_1$ and $S_2$ which originate at these points have met by time $t$. The difference $S_1 - S_2$ is also a type of symmetric random walk, and, as in Theorem (5.10.17), $S_1 - S_2$ almost certainly visits the origin sooner or later, implying that $\mathbb{P}(S_1(t) = S_2(t)$ for some $t) = 1$.                                                     ∎ ●

**(15) Example. Simple queue.** Here is a simple model for a queueing system. Customers enter a shop in the manner of a Poisson process, parameter $\lambda$. They are served in the order of their arrival by a single assistant; each service period is a random variable which we assume to be exponentially distributed with parameter $\mu$ and which is independent of all other considerations. Let $X(t)$ be the length of the waiting line at time $t$ (including any person being served). It is easy to see that $X$ is a birth–death process with parameters $\lambda_n = \lambda$ for $n \geq 0$, $\mu_n = \mu$ for $n \geq 1$. The server would be very unhappy indeed if the queue length $X(t)$ were to tend to infinity as $t \to \infty$, since then he or she would have very few tea breaks. It is not difficult to see that the distribution of $X(t)$ settles down to a limit distribution, as $t \to \infty$, if and only if $\lambda < \mu$, which is to say that arrivals occur more slowly than departures on average (see conditions (6.11.4,6)). We shall consider this process in detail in Chapter 11, together with other more complicated queueing models.                                                     ●

---

### Exercises for Section 6.12

**1.** Customers entering a shop are served in the order of their arrival by the single server. They arrive in the manner of a Poisson process with intensity $\lambda$, and their service times are independent exponentially distributed random variables with parameter $\mu$. By considering the jump chain, show that the expected duration of a busy period $B$ of the server is $(\mu - \lambda)^{-1}$ when $\lambda < \mu$. (The busy period runs from the moment a customer arrives to find the server free until the earliest subsequent time when the server is again free.)

**2.    Disasters.** Immigrants arrive at the instants of a Poisson process of rate $\nu$, and each independently founds a simple birth process of rate $\lambda$. At the instants of an independent Poisson process of rate $\delta$, the population is annihilated. Find the probability generating function of the population $X(t)$, given that $X(0) = 0$.

**3.    More disasters.** In the framework of Exercise (6.12.2), suppose that each immigrant gives rise to a simple birth–death process of rates $\lambda$ and $\mu$. Show that the mean population size stays bounded if and only if $\delta > \lambda - \mu$.

**4.   The queue M/G/∞.** An ftp server receives clients at the times of a Poisson process with parameter $\lambda$, beginning at time 0. The $i$th client remains connected for a length $S_i$ of time, where the $S_i$ are independent identically distributed random variables, independent of the process of arrivals. Assuming that the server has an infinite capacity, show that the number of clients being serviced at time $t$ has the Poisson distribution with parameter $\lambda \int_0^t [1 - G(x)]\,dx$, where $G$ is the common distribution function of the $S_i$. Show that the mean of this distribution converges to $\lambda\mathbb{E}(S)$ as $t \to \infty$.

## 6.13  Spatial Poisson processes

The Poisson process of Section 6.8 is a cornerstone of the theory of continuous-time Markov chains. It is also a beautiful process in its own right, with rich theory and many applications. While the process of Section 6.8 was restricted to the time axis $\mathbb{R}_+ = [0, \infty)$, there is an important generalization to the Euclidean space $\mathbb{R}^d$ where $d \geq 1$.

We begin with a technicality. Recall that the essence of the Poisson process of Section 6.8 was the set of arrival times, a random countable subset of $\mathbb{R}_+$. Similarly, a realization of a Poisson process on $\mathbb{R}^d$ will be a countable subset $\Pi$ of $\mathbb{R}^d$. We shall study the distribution of $\Pi$ through the number $|\Pi \cap A|$ of its points lying in a typical subset $A$ of $\mathbb{R}^d$. Some regularity will be assumed about such sets $A$, namely that there is a well defined notion of the 'volume' of $A$. Specifically, we shall assume that $A \in \mathcal{B}^d$, where $\mathcal{B}^d$ denotes the Borel $\sigma$-field of $\mathbb{R}^d$, being the smallest $\sigma$-field containing all *boxes* of the form $\prod_{i=1}^d (a_i, b_i]$. Members of $\mathcal{B}^d$ are called *Borel sets*, and we write $|A|$ for the volume (or Lebesgue measure) of the Borel set $A$.

**(1) Definition.** The random countable subset $\Pi$ of $\mathbb{R}^d$ is called a **Poisson process with (constant) intensity** $\lambda$ if, for all $A \in \mathcal{B}^d$, the random variables $N(A) = |\Pi \cap A|$ satisfy:
  (a) $N(A)$ has the Poisson distribution with parameter $\lambda|A|$, and
  (b) if $A_1, A_2, \ldots, A_n$ are disjoint sets in $\mathcal{B}^d$, then $N(A_1), N(A_2), \ldots, N(A_n)$ are independent random variables.

We often refer to the counting process $N$ as being itself a Poisson process if it satisfies (a) and (b) above. In the case when $\lambda > 0$ and $|A| = \infty$, the number $|\Pi \cap A|$ has the Poisson distribution with parameter $\infty$, a statement to be interpreted as $\mathbb{P}(|\Pi \cap A| = \infty) = 1$.

It is not difficult to see the equivalence of (1) and Definition (6.8.1) when $d = 1$. That is, if $d = 1$ and $N$ satisfies (1), then

**(2)**
$$M(t) = N([0, t]), \quad t \geq 0,$$

satisfies (6.8.1). Conversely, if $M$ satisfies (6.8.1), one may find a process $N$ satisfying (1) such that (2) holds. Attractive features of the above definition include the facts that the origin plays no special role, and that the definition may be extended to sub-regions of $\mathbb{R}^d$ as well as to $\sigma$-fields of subsets of general measure spaces.

There are many stochastic models based on the Poisson process. One-dimensional processes were used by Bateman, Erlang, Geiger, and Rutherford around 1909 in their investigations of practical situations involving radioactive particles and telephone calls. Examples in two and higher dimensions include the positions of animals in their habitat, the distribution of stars in a galaxy or of galaxies in the universe, the locations of active sites in a chemical reaction or of the weeds in your lawn, and the incidence of thunderstorms and tornadoes. Even when a Poisson process is not a perfect description of such a system, it can provide a relatively simple yardstick against which to measure the improvements which may be offered by more sophisticated but often less tractable models.

Definition (1) utilizes as reference measure the Lebesgue measure on $\mathbb{R}^d$, in the sense that the volume of a set $A$ is its Euclidean volume. It is useful to have a definition of a Poisson process with other measures than Lebesgue measure, and such processes are termed 'non-homogeneous'. Replacing the Euclidean element $\lambda \, d\mathbf{x}$ with the element $\lambda(\mathbf{x}) \, d\mathbf{x}$, we obtain the following, in which $\Lambda(A)$ is given by

$$(3) \qquad\qquad \Lambda(A) = \int_A \lambda(\mathbf{x}) \, d\mathbf{x}, \quad A \in \mathcal{B}^d.$$

**(4) Definition.** Let $d \geq 1$ and let $\lambda : \mathbb{R}^d \to \mathbb{R}$ be a non-negative measurable function such that $\Lambda(A) < \infty$ for all bounded $A$. The random countable subset $\Pi$ of $\mathbb{R}^d$ is called a **non-homogeneous Poisson process with intensity function** $\lambda$ if, for all $A \in \mathcal{B}^d$, the random variables $N(A) = |\Pi \cap A|$ satisfy:

(a) $N(A)$ has the Poisson distribution with parameter $\Lambda(A)$, and

(b) if $A_1, A_2, \ldots, A_n$ are disjoint sets in $\mathcal{B}^d$, then $N(A_1), N(A_2), \ldots, N(A_n)$ are independent random variables.

We call the function $\Lambda(A)$, $A \in \mathcal{B}^d$, the *mean measure* of the process $\Pi$. We have constructed $\Lambda$ as the integral (3) of the intensity function $\lambda$; one may in fact dispense altogether with the function $\lambda$, working instead with measures $\Lambda$ which 'have no atoms' in the sense that $\Lambda(\{\mathbf{x}\}) = 0$ for all $\mathbf{x} \in \mathbb{R}^d$. The condition that $\Lambda(A) < \infty$ for bounded $A$ is optional, but will be convenient later.

Our first theorem states that the union of two independent Poisson processes is also a Poisson process.

**(5) Superposition theorem.** *Let $\Pi'$ and $\Pi''$ be independent Poisson processes on $\mathbb{R}^d$ with respective intensity functions $\lambda'$ and $\lambda''$. The set $\Pi = \Pi' \cup \Pi''$ is a Poisson process with intensity function $\lambda = \lambda' + \lambda''$.*

The same proof may be used to show the stronger fact that the superposition of *countably* many independent Poisson processes with mean measures $\Lambda_i$ is a Poisson process, whenever $\sum_i \Lambda_i(A) < \infty$ for bounded sets $A$.

**Proof.** Let $N'(A) = |\Pi' \cap A|$ and $N''(A) = |\Pi'' \cap A|$. Then $N'(A)$ and $N''(A)$ are independent Poisson-distributed random variables with respective parameters $\Lambda'(A)$ and $\Lambda''(A)$, the integrals (3) of $\lambda'$ and $\lambda''$. It follows that the sum $S(A) = N'(A) + N''(A)$ has the Poisson distribution with parameter $\Lambda'(A) + \Lambda''(A)$. Furthermore, if $A_1, A_2, \ldots$ are disjoint, the random variables $S(A_1), S(A_2), \ldots$ are independent. It remains to show that, almost surely, $S(A) = |\Pi \cap A|$ for all $A$, which is to say that no point of $\Pi'$ coincides with a point of $\Pi''$. This is a rather technical step, and the proof may be omitted on a first reading.

Since $\mathbb{R}^d$ is a countable union of bounded sets, it is enough to show that, for every bounded $A \subseteq \mathbb{R}^d$, $A$ contains almost surely no point common to $\Pi'$ and $\Pi''$. Let $n \geq 0$ and, for $\mathbf{k} = (k_1, k_2, \ldots, k_d) \in \mathbb{Z}^d$, let $B_{\mathbf{k}}(n) = \prod_{i=1}^d (k_i 2^{-n}, (k_i + 1)2^{-n}]$; cubes of this form are termed *n-cubes* or *n-boxes*. Let $A$ be a bounded subset of $\mathbb{R}^d$, and let $\bar{A}$ be the (bounded) union of all $B_{\mathbf{k}}(0)$ which intersect $A$. The probability that $A$ contains a point common to $\Pi'$ and $\Pi''$ is bounded for all $n$ by the probability that some $B_{\mathbf{k}}(n)$ lying in $\bar{A}$ contains a common point. This is no greater than the mean number of such boxes, whence

$$\mathbb{P}(\Pi' \cap \Pi'' \cap A \neq \varnothing) \leq \sum_{\mathbf{k}: B_{\mathbf{k}}(n) \subseteq \overline{A}} \mathbb{P}(N'(B_{\mathbf{k}}(n)) \geq 1, \ N''(B_{\mathbf{k}}(n)) \geq 1)$$

$$= \sum_{\mathbf{k}: B_{\mathbf{k}}(n) \subseteq \overline{A}} \left(1 - e^{-\Lambda'(B_{\mathbf{k}}(n))}\right)\left(1 - e^{-\Lambda''(B_{\mathbf{k}}(n))}\right)$$

$$\leq \sum_{\mathbf{k}: B_{\mathbf{k}}(n) \subseteq \overline{A}} \Lambda'(B_{\mathbf{k}}(n)) \Lambda''(B_{\mathbf{k}}(n)) \quad \text{since } 1 - e^{-x} \leq x \text{ for } x \geq 0$$

$$\leq \max_{\mathbf{k}: B_{\mathbf{k}}(n) \subseteq \overline{A}} \{\Lambda'(B_{\mathbf{k}}(n))\} \sum_{\mathbf{k}: B_{\mathbf{k}}(n) \subseteq \overline{A}} \Lambda''(B_{\mathbf{k}}(n))$$

$$= M_n(A) \Lambda''(\overline{A})$$

where

$$M_n(A) = \max_{\mathbf{k}: B_{\mathbf{k}}(n) \subseteq \overline{A}} \Lambda'(B_{\mathbf{k}}(n)).$$

It is the case that $M_n(A) \to 0$ as $n \to \infty$. This is easy to prove when $\lambda'$ is a constant function, since then $M_n(A) \propto |B_{\mathbf{k}}(n)| = 2^{-nd}$. It is not quite so easy to prove for general $\lambda'$. Since we shall need a slightly more general argument later, and we state next the required result.

**(6) Lemma.** *Let $\mu$ be a measure on the pair $(\mathbb{R}^d, \mathcal{B}^d)$ which has no atoms, which is to say that $\mu(\{\mathbf{y}\}) = 0$ for all $\mathbf{y} \in \mathbb{R}^d$. Let $n \geq 0$ and $\mathbf{k} \in \mathbb{Z}^d$, and let $B_{\mathbf{k}}(n) = \prod_{i=1}^d (k_i 2^{-n}, (k_i+1)2^{-n}]$. For any bounded set $A$, we have that*

$$\max_{\mathbf{k}: B_{\mathbf{k}}(n) \subseteq A} \mu(B_{\mathbf{k}}(n)) \to 0 \qquad \text{as } n \to \infty.$$

Returning to the proof of the theorem, it follows by Lemma (6) applied to the set $\overline{A}$ that $M_n(A) \to 0$ as $n \to \infty$, and the proof is complete.  ∎

**Proof of Lemma (6).** We may assume without loss of generality that $A$ is a finite union of 0-cubes. Let

$$M_n(A) = \max_{\mathbf{k}: B_{\mathbf{k}}(n) \subseteq A} \mu(B_{\mathbf{k}}(n)),$$

and note that $M_n \geq M_{n+1}$. Suppose that $M_n(A) \not\to 0$. There exists $\delta > 0$ such that $M_n(A) > \delta$ for all $n$, and therefore, for every $n \geq 0$, there exists an $n$-cube $B_{\mathbf{k}}(n) \subseteq A$ with $\mu(B_{\mathbf{k}}(n)) > \delta$. We colour an $m$-cube $C$ *black* if, for all $n \geq m$, there exists an $n$-cube $C' \subseteq C$ such that $\mu(C') > \delta$. Now $A$ is the union of finitely many translates of $(0, 1]^d$, and for at least one of these, $B_0$ say, there exist infinitely many $n$ such that $B_0$ contains some $n$-cube $B'$ with $\mu(B') > \delta$. Since $\mu(\cdot)$ is monotonic, the 0-cube $B_0$ is black. By a similar argument, $B_0$ contains some black 1-cube $B_1$. Continuing similarly, we obtain an infinite sequence $B_0 \supseteq B_1 \supseteq B_2 \supseteq \cdots$ such that each $B_r$ is a black $r$-cube. In particular, $\mu(B_r) > \delta$ for all $r$, whereas†

$$\lim_{r \to \infty} \mu(B_r) = \mu\left(\bigcap_r B_r\right) = 0,$$

---

†We use here a property of continuity of general measures, proved in the manner of Theorem (1.3.5).

by assumption, since $\bigcap_r B_r$ is either a singleton or the empty set. The conclusion of the lemma follows from this contradiction. ∎

It is possible to avoid the complication of Lemma (6) at this stage, but we introduce the lemma here since it will be useful in the forthcoming proof of Rényi's theorem (17). In an alternative and more general approach to Poisson processes, instead of the 'random set' $\Pi$ one studies the 'random measure' $N$. This leads to substantially easier proofs of results corresponding to (5) and the forthcoming (8), but at the expense of extra abstraction.

The following 'mapping theorem' enables us to study the image of a Poisson process $\Pi$ under a (measurable) mapping $f : \mathbb{R}^d \to \mathbb{R}^s$. Suppose that $\Pi$ is a non-homogeneous Poisson process on $\mathbb{R}^d$ with intensity function $\lambda$, and consider the set $f(\Pi)$ of images of $\Pi$ under $f$. We shall need that $f(\Pi)$ contains (with probability 1) no multiple points, and this imposes a constraint on the pair $\lambda$, $f$. The subset $B \subseteq \mathbb{R}^s$ contains the images of points of $\Pi$ lying in $f^{-1}B$, whose cardinality is a random variable having the Poisson distribution with parameter $\Lambda(f^{-1}B)$. The key assumption on the pair $\lambda$, $f$ will therefore be that

(7)                              $\Lambda(f^{-1}\{\mathbf{y}\}) = 0 \quad \text{for all } \mathbf{y} \in \mathbb{R}^s,$

where $\Lambda$ is the integral (3) of $\lambda$.

**(8) Mapping theorem.** *Let $\Pi$ be a non-homogeneous Poisson process on $\mathbb{R}^d$ with intensity function $\lambda$, and let $f : \mathbb{R}^d \to \mathbb{R}^s$ satisfy (7). Assume further that*

(9)                 $$\mu(B) = \Lambda(f^{-1}B) = \int_{f^{-1}B} \lambda(\mathbf{x})\, d\mathbf{x}, \quad B \in \mathcal{B}^s,$$

*satisfies $\mu(B) < \infty$ for all bounded sets $B$. Then $f(\Pi)$ is a non-homogeneous Poisson process on $\mathbb{R}^s$ with mean measure $\mu$.*

**Proof.** Assume for the moment that the points in $f(\Pi)$ are almost surely distinct. The number of points of $f(\Pi)$ lying in the set $B \subseteq \mathbb{R}^s$ is $|\Pi \cap f^{-1}B|$, which has the Poisson distribution with parameter $\Lambda(f^{-1}B) = \mu(B)$, as required. If $B_1, B_2, \ldots$ are disjoint, their pre-images $f^{-1}B_1, f^{-1}B_2, \ldots$ are disjoint also, whence the numbers of points in the $B_i$ are independent. It follows that $f(\Pi)$ is a Poisson process with the given mean measure, and it remains only to show the assumed distinctness of $f(\Pi)$. The proof of this is similar to that of Theorem (5), and may be omitted on first read.

Let $U = [0, 1)^s \subseteq \mathbb{R}^s$. We shall show that the points in $f(\Pi) \cap U$ are almost surely distinct. Since $\mathbb{R}^s$ is the union of countably many translates of $U$, to each of which the above conclusion applies, we deduce as required that $\mathbb{R}^s$ contains almost surely no repeated points of $f(\Pi)$.

Let $n \geq 0$; later we shall let $n \to \infty$. The number $N_\mathbf{k} = |\Pi \cap f^{-1}B_\mathbf{k}|$ of 'strikes' inside the box $B_\mathbf{k} = \prod_{i=1}^s (k_i 2^{-n}, (k_i + 1)2^{-n}]$ of $\mathbb{R}^s$ has the Poisson distribution with parameter $\mu_\mathbf{k} = \mu(B_\mathbf{k})$. Therefore,

$$\mathbb{P}(N_\mathbf{k} \geq 2) = 1 - e^{-\mu_\mathbf{k}} - \mu_\mathbf{k} e^{-\mu_\mathbf{k}}$$
$$\leq 1 - (1 + \mu_\mathbf{k})(1 - \mu_\mathbf{k}) = \mu_\mathbf{k}^2,$$

where we have used the fact that $e^{-x} \geq 1 - x$ for $x \geq 0$. The mean number of such $n$-boxes within the unit cube $U$ is no greater than

$$\sum_\mathbf{k} \mu_\mathbf{k}^2 \leq M_n \sum_\mathbf{k} \mu_\mathbf{k} = M_n \mu(U),$$

where

$$M_n = \max_{\mathbf{k}} \mu_{\mathbf{k}} \to 0 \qquad \text{as } n \to \infty$$

by hypothesis (7) and Lemma (6). Now $\mu(U) < \infty$, and we deduce as in the proof of Theorem (5) that $U$ contains almost surely no repeated points. The proof is complete. ∎

**(10) Example. Polar coordinates.** Let $\Pi$ be a Poisson process on $\mathbb{R}^2$ with constant rate $\lambda$, and let $f : \mathbb{R}^2 \to \mathbb{R}^2$ be the polar coordinate function $f(x, y) = (r, \theta)$ where

$$r = \sqrt{x^2 + y^2}, \qquad \theta = \tan^{-1}(y/x).$$

It is straightforward to check that (7) holds, and we deduce that $f(\Pi)$ is a Poisson process on $\mathbb{R}^2$ with mean measure

$$\mu(B) = \int_{f^{-1}B} \lambda \, dx \, dy = \int_{B \cap S} \lambda r \, dr \, d\theta,$$

where $S = f(\mathbb{R}^2)$ is the strip $\{(r, \theta) : r \geq 0, \ 0 \leq \theta < 2\pi\}$. We may think of $f(\Pi)$ as a Poisson process on the strip $S$ having intensity function $\lambda r$. ●

We turn now to one of the most important attributes of the Poisson process, which unlocks the door to many other useful results. This is the so-called 'conditional property', of which we saw a simple version in Theorem (6.8.11).

**(11) Theorem. Conditional property.** *Let $\Pi$ be a non-homogeneous Poisson process on $\mathbb{R}^d$ with intensity function $\lambda$, and let $A$ be a subset of $\mathbb{R}^d$ such that $0 < \Lambda(A) < \infty$. Conditional on the event that $|\Pi \cap A| = n$, the $n$ points of the process lying in $A$ have the same distribution as $n$ points chosen independently at random in $A$ according to the common probability measure*

$$\mathbb{Q}(B) = \frac{\Lambda(B)}{\Lambda(A)}, \qquad B \subseteq A.$$

Since

**(12)**
$$\mathbb{Q}(B) = \int_B \frac{\lambda(\mathbf{x})}{\Lambda(A)} \, d\mathbf{x},$$

the relevant density function is $\lambda(\mathbf{x})/\Lambda(A)$ for $\mathbf{x} \in A$. When $\Pi$ has constant intensity $\lambda$, the theorem implies that, given $|\Pi \cap A| = n$, the $n$ points in question are distributed uniformly and independently at random in $A$.

**Proof.** Write $N(B) = |\Pi \cap B|$, and let $A_1, A_2, \ldots, A_k$ be a partition of $A$. It is an elementary calculation that, if $n_1 + n_2 + \cdots + n_k = n$,

**(13)**
$$\mathbb{P}\big(N(A_1) = n_1, N(A_2) = n_2, \ldots, N(A_k) = n_k \,\big|\, N(A) = n\big)$$

$$= \frac{\prod_i \mathbb{P}(N(A_i) = n_i)}{\mathbb{P}(N(A) = n)} \qquad \text{by independence}$$

$$= \frac{\prod_i \Lambda(A_i)^{n_i} e^{-\Lambda(A_i)}/n_i!}{\Lambda(A)^n e^{-\Lambda(A)}/n!}$$

$$= \frac{n!}{n_1! n_2! \cdots n_k!} \mathbb{Q}(A_1)^{n_1} \mathbb{Q}(A_2)^{n_2} \cdots \mathbb{Q}(A_k)^{n_k}.$$

The conditional distribution of the positions of the $n$ points is specified by this function of $A_1, A_2, \ldots, A_n$.

We recognize the multinomial distribution of (13) as the joint distribution of $n$ points selected independently from $A$ according to the probability measure $\mathbb{Q}$. It follows that the joint distribution of the points in $\Pi \cap A$, conditional on there being exactly $n$ of them, is the same as that of the independent sample.                                                                      ∎

The conditional property enables a proof of the existence of Poisson processes, and it aids the simulation thereof. Let $\lambda > 0$ and let $A_1, A_2, \ldots$ be a partition of $\mathbb{R}^d$ into Borel sets of finite Lebesgue measure. For each $i$, we simulate a random variable $N_i$ having the Poisson distribution with parameter $\lambda |A_i|$. Then we sample $n$ independently chosen points in $A_i$, each being uniformly distributed on $A_i$. The union over $i$ of all such sets of points is a Poisson process with constant intensity $\lambda$. A similar construction is valid for a non-homogeneous process. The method may be facilitated by a careful choice of the $A_i$, perhaps as unit cubes of $\mathbb{R}^d$.

The following colouring theorem may be viewed as complementary to the superposition theorem (5). As in the latter case, there is a version of the theorem in which points are marked with one of countably many colours rather than just two.

**(14) Colouring theorem.** *Let $\Pi$ be a non-homogeneous Poisson process on $\mathbb{R}^d$ with intensity function $\lambda$. We colour the points of $\Pi$ in the following way. A point of $\Pi$ at position $\mathbf{x}$ is coloured green with probability $\gamma(\mathbf{x})$; otherwise it is coloured scarlet (with probability $\sigma(\mathbf{x}) = 1 - \gamma(\mathbf{x})$). Points are coloured independently of one another. Let $\Gamma$ and $\Sigma$ be the sets of points coloured green and scarlet, respectively. Then $\Gamma$ and $\Sigma$ are independent Poisson processes with respective intensity functions $\gamma(\mathbf{x})\lambda(\mathbf{x})$ and $\sigma(\mathbf{x})\lambda(\mathbf{x})$.*

**Proof.** Let $A \subseteq \mathbb{R}^d$ with $\Lambda(A) < \infty$. By the conditional property (11), if $|\Pi \cap A| = n$, these points have the same distribution as $n$ points chosen independently at random from $A$ according to the probability measure $\mathbb{Q}(B) = \Lambda(B)/\Lambda(A)$. We may therefore consider $n$ points chosen in this way. By the independence of the points, their colours are independent of one another. The chance that a given point is coloured green is $\overline{\gamma} = \int_A \gamma(\mathbf{x}) \, d\mathbb{Q}$, the corresponding probability for the colour scarlet being $\overline{\sigma} = 1 - \overline{\gamma} = \int_A \sigma(\mathbf{x}) \, d\mathbb{Q}$. It follows that, conditional on $|\Pi \cap A| = n$, the numbers $N_g$ and $N_s$ of green and scarlet points in $A$ have, jointly, the binomial distribution

$$\mathbb{P}\big(N_g = g, \ N_s = s \,\big|\, N(A) = n\big) = \frac{n!}{g! \, s!} \overline{\gamma}^g \overline{\sigma}^s, \qquad \text{where } g + s = n.$$

The unconditional probability is therefore

$$\mathbb{P}(N_g = g, \ N_s = s) = \frac{(g+s)!}{g! \, s!} \overline{\gamma}^g \overline{\sigma}^s \frac{\Lambda(A)^{g+s} e^{-\Lambda(A)}}{(g+s)!}$$

$$= \frac{(\overline{\gamma}\Lambda(A))^g e^{-\overline{\gamma}\Lambda(A)}}{g!} \cdot \frac{(\overline{\sigma}\Lambda(A))^s e^{-\overline{\sigma}\Lambda(A)}}{s!}$$

which is to say that the numbers of green and scarlet points in $A$ are independent. Furthermore they have, by (12), Poisson distributions with parameters

$$\overline{\gamma}\Lambda(A) = \int_A \gamma(\mathbf{x})\Lambda(A) \, d\mathbb{Q} = \int_A \gamma(\mathbf{x})\lambda(\mathbf{x}) \, d\mathbf{x},$$

$$\overline{\sigma}\Lambda(A) = \int_A \sigma(\mathbf{x})\Lambda(A) \, d\mathbb{Q} = \int_A \sigma(\mathbf{x})\lambda(\mathbf{x}) \, d\mathbf{x}.$$

Independence of the counts of points in disjoint regions follows trivially from the fact that $\Pi$ has this property.                                                                                       ∎

**(15) Example.** The Alternative Millennium Dome contains $n$ zones, and visitors are required to view them all in sequence. Visitors arrive at the instants of a Poisson process on $\mathbb{R}_+$ with constant intensity $\lambda$, and the $r$th visitor spends time $X_{r,s}$ in the $s$th zone, where the random variables $X_{r,s}$, $r \geq 1$, $1 \leq s \leq n$, are independent, and furthermore, for given $s$, the distribution of $X_{r,s}$ does not depend on $r$.

Let $t \geq 0$, and let $V_s(t)$ be the number of visitors in zone $s$ at time $t$. Show that, for fixed $t$, the random variables $V_s(t)$, $1 \leq s \leq n$, are independent, each with a Poisson distribution.
**Solution.** Let $T_1 < T_2 < \cdots$ be the times of arrivals of visitors, and let $c_1, c_2, \ldots, c_n, \delta$ be distinct colours. A point of the Poisson process at time $x$ is coloured $c_s$ if and only if

$$
\textbf{(16)} \qquad\qquad x + \sum_{v=1}^{s-1} X_v \leq t < x + \sum_{v=1}^{s} X_v
$$

where $X_1, X_2, \ldots, X_n$ are the times to be spent in the zones by a visitor arriving at time $x$. If (16) holds for no $s$, we colour the point at $x$ with the colour $\delta$; at time $t$, such a visitor has either not yet arrived or has already departed. Note that the colours of different points of the Poisson process are independent, and that a visitor arriving at time $x$ is coloured $c_s$ if and only if this individual is in zone $s$ at time $t$.

The required independence follows by a version of the colouring theorem with $n + 1$ available colours instead of just two.                                                        ●

Before moving to other things, we note yet another characterization of the Poisson process. It turns out that one needs only check that the probability of a given region being empty is given by the Poisson formula. Recall from the proof of (5) that a *box* is a region of $\mathbb{R}^d$ of the form $B_{\mathbf{k}}(n) = \prod_{i=1}^{d} (k_i 2^{-n}, (k_i + 1)2^{-n}]$ for some $\mathbf{k} \in \mathbb{Z}^d$ and $n \geq 0$.

**(17) Rényi's theorem.** *Let $\Pi$ be a random countable subset of $\mathbb{R}^d$, and let $\lambda : \mathbb{R}^d \to \mathbb{R}$ be a non-negative measurable function satisfying $\Lambda(A) = \int_A \lambda(\mathbf{x})\,d\mathbf{x} < \infty$ for all bounded $A$. If*

$$
\textbf{(18)} \qquad\qquad \mathbb{P}(\Pi \cap A = \varnothing) = e^{-\Lambda(A)}
$$

*for any finite union $A$ of boxes, then $\Pi$ is a Poisson process with intensity function $\lambda$.*

**Proof.** Let $n \geq 0$, and denote by $I_{\mathbf{k}}(n)$ the indicator function of the event that $B_{\mathbf{k}}(n)$ is non-empty. By (18), the events $I_{\mathbf{k}}(n)$, $\mathbf{k} \in \mathbb{Z}^d$, are independent.

Let $A$ be a bounded open set in $\mathbb{R}^d$, and let $\mathcal{K}_n(A)$ be the set of all $\mathbf{k}$ such that $B_{\mathbf{k}}(n) \subseteq A$. Since $A$ is open, we have that

$$
\textbf{(19)} \qquad N(A) = |\Pi \cap A| = \lim_{n \to \infty} T_n(A) \quad \text{where} \quad T_n(A) = \sum_{\mathbf{k} \in \mathcal{K}_n(A)} I_{\mathbf{k}}(n);
$$

note that, by the nesting of the boxes $B_{\mathbf{k}}(n)$, this is a monotone increasing limit. We have also that

$$
\textbf{(20)} \qquad\qquad \Lambda(A) = \lim_{n \to \infty} \sum_{\mathbf{k} \in \mathcal{K}_n(A)} \Lambda(B_{\mathbf{k}}(n)).
$$

The quantity $T_n(A)$ is the sum of independent variables, and has probability generating function

(21) $$\mathbb{E}(s^{T_n(A)}) = \prod_{\mathbf{k} \in \mathcal{K}_n(A)} \left\{ s + (1-s)e^{-\Lambda(B_{\mathbf{k}}(n))} \right\}.$$

We have by Lemma (6) that $\Lambda(B_{\mathbf{k}}(n)) \to 0$ uniformly in $\mathbf{k} \in \mathcal{K}_n(A)$, as $n \to \infty$. Also, for fixed $s \in [0, 1)$, there exists $\phi(\delta)$ satisfying $\phi(\delta) \uparrow 1$ as $\delta \downarrow 0$ such that

(22) $$e^{-(1-s)\alpha} \le s + (1-s)e^{-\alpha} \le e^{-(1-s)\phi(\delta)\alpha} \quad \text{if } 0 \le \alpha \le \delta.$$

[The left inequality holds by the convexity of $e^{-x}$, and the right inequality by Taylor's theorem.] It follows by (19), (21), and monotone convergence, that

$$\mathbb{E}(s^{N(A)}) = \lim_{n \to \infty} \prod_{\mathbf{k} \in \mathcal{K}_n(A)} \left\{ s + (1-s)e^{-\Lambda(B_{\mathbf{k}}(n))} \right\} \qquad \text{for } 0 \le s < 1,$$

and by (20) and (22) that, for fixed $s \in [0, 1)$,

$$e^{-(1-s)\Lambda(A)} \le \mathbb{E}(s^{N(A)}) \le e^{-(1-s)\phi(\delta)\Lambda(A)} \qquad \text{for all } \delta > 0.$$

We take the limit as $\delta \downarrow 0$ to obtain the Poisson distribution of $N(A)$.

The variables $N(A_1), N(A_2), \ldots$ are independent for disjoint open sets $A_1, A_2, \ldots$. This holds since $T_n(A_1), T_n(A_2), \ldots$ are independent, and $T_n(A_i) \to N(A_i)$ as $n \to \infty$. It is now an exercise (omitted) in measure theory to extend the above from open subsets to general measurable subsets of $\mathbb{R}^d$.                                                                              ∎

There are many applications of the theory of Poisson processes in which the points of a process have an effect elsewhere in the space. A well-known practical example concerns the fortune of someone who plays a lottery. The player wins prizes at the times of a Poisson process $\Pi$ on $\mathbb{R}_+$, and the amounts won are independent identically distributed random variables. Gains are discounted at rate $\alpha \ge 0$. The total gain $G(t)$ by time $t$ may be expressed in the form

$$G(t) = \sum_{x \in \Pi \cap [0,t]} e^{-\alpha(t-x)} W_x,$$

where $W_x$ is the amount won at time $x \in \Pi$. We may write

$$G(t) = \sum_{x \in \Pi} r(t-x) W_x$$

where

$$r(u) = \begin{cases} 0 & \text{if } u < 0, \\ e^{-\alpha u} & \text{if } u \ge 0. \end{cases}$$

Such sums may be studied by way of the next theorem. We state this in the special case of a homogeneous Poisson process on the line $\mathbb{R}$, but it is easily generalized. The one-dimensional problem is sometimes termed *shot noise*, since one may think of the sum as the cumulative effect of pulses which arrive in a system, and whose amplitudes decay exponentially.

**(23) Theorem†.**    *Let* Π *be a Poisson process on* $\mathbb{R}$ *with constant intensity* $\lambda$, *let* $r : \mathbb{R} \to \mathbb{R}$ *be integrable over bounded intervals, and let* $\{W_x : x \in \Pi\}$ *be independent identically distributed random variables, independent of* Π. *The sum*

$$G(t) = \sum_{x \in \Pi \cap [0,t]} r(t - x) W_x$$

*has characteristic function*

$$\mathbb{E}(e^{i\theta G(t)}) = \exp\left\{\lambda \int_0^t \left(\mathbb{E}(e^{i\theta W r(s)}) - 1\right) ds\right\},$$

*where* $W$ *has the common distribution of the* $W_x$. *If* $\mathbb{E}|W| < \infty$, *we have that*

$$\mathbb{E}(G(t)) = \lambda \mathbb{E}(W) \int_0^t r(s) \, ds.$$

**Proof.** This runs just like that of Theorem (6.12.7), which is in fact a special case. It is left as an *exercise* to check the details. The mean of $G(t)$ is calculated from its characteristic function by differentiating the latter at $\theta = 0$.    ∎

A similar idea works in higher dimensions, as the following demonstrates.

**(24) Example. Olbers's paradox‡.**   Suppose that stars occur in $\mathbb{R}^3$ at the points $\{\mathbf{R}_i : i \geq 1\}$ of a Poisson process with constant intensity $\lambda$. The star at $\mathbf{R}_i$ has brightness $B_i$, where the $B_i$ are independent and identically distributed with mean $\beta$. The intensity of the light striking an observer at the origin O from a star of brightness $B$, distance $r$ away, is (in the absence of intervening clouds of dust) equal to $cB/r^2$, for some absolute constant $c$. Hence the total illumination at O from stars within a large ball $S$ with radius $a$ is

$$I_a = \sum_{i: |\mathbf{R}_i| \leq a} \frac{cB_i}{|\mathbf{R}_i|^2}.$$

Conditional on the event that the number $N_a$ of such stars satisfies $N_a = n$, we have from the conditional property (11) that these $n$ stars are uniformly and independently distributed over $S$. Hence

$$\mathbb{E}(I_a \mid N_a) = N_a c \beta \frac{1}{|S|} \int_S \frac{1}{|\mathbf{r}|^2} \, dV.$$

Now $\mathbb{E}(N_a) = \lambda |S|$, whence

$$\mathbb{E}(I_a) = \lambda c \beta \int_S \frac{1}{|\mathbf{r}|^2} \, dV = \lambda c \beta (4\pi a).$$

---

†This theorem is sometimes called the Campbell–Hardy theorem. See Exercise (6.13.2) for a more general version.

‡Put to Newton and Halley by William Stukeley in 1721, more than a century before Olbers's work, and earlier used by Johannes Kepler in 1606 as an argument for the universe being finite.

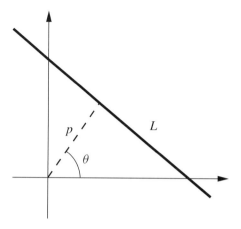

Figure 6.3. The line $L$ is parametrized as the pair $(p, \theta)$.

The fact that this is unbounded as $a \to \infty$ is called 'Olbers's paradox', and suggests that the celestial sphere should be uniformly bright at night. The fact that it is not is a problem whose resolution is still a matter for debate. One plausible explanation relies on a sufficiently fast rate of expansion of the Universe, and another on the limited age of its galaxies.    ●

**(25) Example. Poisson line process.** Let $S$ be the set of doubly-infinite, undirected straight lines in $\mathbb{R}^2$. A *line process* is a random countable subset of $S$. In order to explain the concept of a 'Poisson' line process, we need a coordinate system on $S$, and a natural candidate for this is as follows.

Let $L$ be a line in $\mathbb{R}^2$, and drop a perpendicular $L^\perp$ from the origin $\mathbf{0}$ onto $L$. Write $\theta \in [0, \pi)$ for the inclination of $L^\perp$, and $p \in (-\infty, \infty)$ for its signed length. The mapping $f : L \mapsto (p, \theta)$ is a bijection between $S$ and its image $S^\perp = \mathbb{R} \times [0, \pi)$. See Figure 6.3.

Let $\lambda : S^\perp \to [0, \infty)$ satisfy $\Lambda(A) = \int_A \lambda(\mathbf{x})\, d\mathbf{x} < \infty$ for bounded sets $A$, and let $\Pi$ be a Poisson process on $S^\perp$ with intensity function $\lambda$. The inverse image $f^{-1}\Pi$ is a *Poisson line process* with intensity function $\lambda$. The Poisson line process is called *uniform* if $\lambda$ is constant. Such processes arose in the discussion of Bertrand's paradox in Example (4.13.1ii).

The uniform Poisson line processes have some useful properties, namely that their distributions are invariant under mappings of the plane that are compositions of translations, rotations, and reflections. They are in fact the only Poisson processes with these properties.    ●

## Exercises for Section 6.13

**1.** In a certain town at time $t = 0$ there are no bears. Brown bears and grizzly bears arrive as independent Poisson processes $B$ and $G$ with respective intensities $\beta$ and $\gamma$.
(a) Show that the first bear is brown with probability $\beta/(\beta + \gamma)$.
(b) Find the probability that between two consecutive brown bears, there arrive exactly $r$ grizzly bears.
(c) Given that $B(1) = 1$, find the expected value of the time at which the first bear arrived.

**2.    Campbell–Hardy theorem.** Let $\Pi$ be the points of a non-homogeneous Poisson process on $\mathbb{R}^d$ with intensity function $\lambda$. Let $S = \sum_{\mathbf{x} \in \Pi} g(\mathbf{x})$ where $g$ is a (measurable) function which we assume for convenience to be non-negative.

(a)  Show directly that $\mathbb{E}(S) = \int_{\mathbb{R}^d} g(\mathbf{x})\lambda(\mathbf{x})\,d\mathbf{x}$ and $\mathrm{var}(S) = \int_{\mathbb{R}^d} g(\mathbf{x})^2\lambda(\mathbf{x})\,d\mathbf{x}$, provided these integrals converge.

(b)  Show that

$$\mathbb{E}(e^{-tS}) = \exp\left\{-\int_{\mathbb{R}^d}(1 - e^{-tg(\mathbf{x})})\lambda(\mathbf{x})\,d\mathbf{x}\right\}, \qquad t > 0,$$

and deduce that $\mathbb{P}(S < \infty) = 1$ if $\int_{\mathbb{R}^d} \min\{1, g(\mathbf{x})\}\lambda(\mathbf{x})\,d\mathbf{x} < \infty$.

(c)  If the integral condition of part (b) holds, show that the characteristic function $\phi$ of $S$ satisfies

$$\phi(t) = \exp\left\{-\int_{\mathbb{R}^d}(1 - e^{itg(\mathbf{x})})\lambda(\mathbf{x})\,d\mathbf{x}\right\}, \qquad t \in \mathbb{R},$$

**3.**   Let $\Pi$ be a Poisson process with constant intensity $\lambda$ on the surface of the sphere of $\mathbb{R}^3$ with radius 1. Let $P$ be the process given by the $(X, Y)$ coordinates of the points projected on a plane passing through the centre of the sphere. Show that $P$ is a Poisson process, and find its intensity function.

**4.**   Repeat Exercise (6.13.3), when $\Pi$ is a homogeneous Poisson process on the ball $\{(x_1, x_2, x_3) : x_1^2 + x_2^2 + x_3^2 \leq 1\}$.

**5.**   You stick pins in a Mercator projection of the Earth in the manner of a Poisson process with constant intensity $\lambda$. What is the intensity function of the corresponding process on the globe? What would be the intensity function on the map if you formed a Poisson process of constant intensity $\lambda$ of meteorite strikes on the surface of the Earth?

**6.   Shocks.** The $r$th point $T_r$ of a Poisson process $N$ of constant intensity $\lambda$ on $\mathbb{R}_+$ gives rise to an effect $X_r e^{-\alpha(t-T_r)}$ at time $t \geq T_r$, where the $X_r$ are independent and identically distributed with finite variance. Find the mean and variance of the total effect $S(t) = \sum_{r=1}^{N(t)} X_r e^{-\alpha(t-T_r)}$ in terms of the first two moments of the $X_r$, and calculate $\mathrm{cov}(S(s), S(t))$.

What is the behaviour of the correlation $\rho(S(s), S(t))$ as $s \to \infty$ with $t - s$ fixed?

**7.**   Let $N$ be a non-homogeneous Poisson process on $\mathbb{R}_+$ with intensity function $\lambda$. Find the joint density of the first two inter-event times, and deduce that they are not in general independent.

**8.   Competition lemma.** Let $\{N_r(t) : r \geq 1\}$ be a collection of independent Poisson processes on $\mathbb{R}_+$ with respective constant intensities $\{\lambda_r : r \geq 1\}$, such that $\sum_r \lambda_r = \lambda < \infty$. Set $N(t) = \sum_r N_r(t)$, and let $I$ denote the index of the process supplying the first point in $N$, occurring at time $T$. Show that

$$\mathbb{P}(I = i, \ T \geq t) = \mathbb{P}(I = i)\mathbb{P}(T \geq t) = \frac{\lambda_i}{\lambda}e^{-\lambda t}, \qquad i \geq 1.$$

**9.   Poisson line process.** Let $\Pi$ be a Poisson process on $\mathbb{R}^2 \setminus \mathbf{0}$ with intensity function $\lambda(u, v) = (u^2 + v^2)^{-3/2}$. Each point $(U, V) \in \Pi$ gives rise to a line $Ux + Vy = 1$ in the $x/y$-plane $\mathbb{L}$.

(a)  Let $(p, \theta)$ be the polar coordinates of the foot of the perpendicular from the origin onto the line $ux + vy = 1$ in the $x/y$-plane $\mathbb{L}$. Express $(p, \theta)$ in terms of $(u, v)$.

(b)  Show that the Poisson line process is mapped by the map of part (a) to a uniform Poisson process on the strip $S^\perp = \mathbb{R} \times [0, \pi)$.

(c)  Show that the line process in $\mathbb{L}$ is invariant under translations and rotations of $\mathbb{L}$.

**10.   Attracted by traffic.** A large continent is traversed by a doubly-infinite straight freeway on which lorries are parked at the points of a Poisson process with constant intensity 1. The masses of the lorries are independent, identically distributed random variables that are independent of the parking places. Let $G$ be the gravitational attraction due to the lorries on a pedestrian of unit mass standing beside the freeway. You may take the gravitational constant to be 1.

Show that $G$ has characteristic function of the form $\phi(t) = \exp(-c|t|^{1/2})$ where $c > 0$. Express $c$ in terms of the mean of a typical mass $M$.

## 6.14 Markov chain Monte Carlo

In applications of probability and statistics, we are frequently required to compute quantities of the form $\int_\Theta g(\theta)\pi(\theta)\,d\theta$ or $\sum_{\theta \in \Theta} g(\theta)\pi(\theta)$, where $g : \Theta \to \mathbb{R}$ and $\pi$ is a density or mass function, as appropriate. When the domain $\Theta$ is large and $\pi$ is complicated, it can be beyond the ability of modern computers to perform such a computation, and we may resort to 'Monte Carlo' methods (recall Section 2.6). Such situations arise surprisingly frequently in areas as disparate as statistical inference and physics. Monte Carlo techniques do not invariably yield exact answers, but can instead give a sequence of approximations to the required quantity.

**(1) Example. Bayesian inference.** A prior mass function $\pi(\theta)$ is postulated on the discrete set $\Theta$ of possible values of $\theta$, and data $x$ is collected. The posterior mass function $\pi(\theta \mid x)$ is given by

$$\pi(\theta \mid x) = \frac{f(x \mid \theta)\pi(\theta)}{\sum_{\psi \in \Theta} f(x \mid \psi)\pi(\psi)}.$$

It is required to compute some characteristic of the posterior, of the form

$$\mathbb{E}(g(\theta) \mid x) = \sum_\theta g(\theta)\pi(\theta \mid x).$$

Depending on the circumstances, such a quantity can be hard to compute. This problem arises commonly in statistical applications including the theory of image analysis, spatial statistics, and more generally in the analysis of large structured data sets.                    ●

**(2) Example. Ising model†.**   We are given a finite connected graph $G = (V, E)$ with vertex set $V$ and edge set $E$. Each vertex may be in either of two states, $-1$ or $1$, and a *configuration* is a vector $\theta = \{\theta_v : v \in V\}$ lying in the state space $\Theta = \{-1, 1\}^V$. The configuration $\theta$ is assigned the probability

$$\pi(\theta) = \frac{1}{Z}\exp\left\{\sum_{v \sim w} \theta_v\theta_w\right\}$$

where the sum is over all pairs $v, w$ of distinct neighbours in the graph $G$ (the relation $\sim$ denoting adjacency), and $Z$ is the appropriate normalizing constant, or 'partition function',

$$Z = \sum_{\theta \in \Theta}\exp\left\{\sum_{v \sim w} \theta_v\theta_w\right\}.$$

For $u, v \in V$, the chance that $u$ and $v$ have the same state is

$$\sum_{\theta : \theta_u = \theta_v} \pi(\theta) = \sum_\theta \tfrac{1}{2}(\theta_u\theta_v + 1)\pi(\theta).$$

The calculation of such probabilities can be strangely complicated.                    ●

It can be difficult to calculate the sums in such examples, even with the assistance of ordinary Monte Carlo methods. For example, the elementary Monte Carlo method of Section

---

†This famous model of ferromagnetism was proposed by Lenz in 1920, and was studied by Ising around 1924.

2.6 relied upon having a supply of independent random variables with mass function $\pi$. In practice, $\Theta$ is often large and highly structured, and $\pi$ may have complicated form, with the result that it may be hard to simulate directly from $\pi$.

We present two methods for approaching such problems: the first yields approximations, and the second exact solutions. Each has its merits and its demerits.

**(A) Sampling forwards in time.** The 'Markov chain Monte Carlo' (McMC) approach is to construct a Markov chain having the following properties:

(a) the chain has $\pi$ as unique stationary distribution,

(b) the transition probabilities of the chain have a simple form.

Property (b) ensures the easy simulation of the chain, and property (a) ensures that the distribution thereof approaches the required distribution as time passes. Let $X = \{X_n : n \geq 0\}$ be such a chain. Subject to weak conditions, the averages of $g(X_r)$ satisfy

$$\frac{1}{n}\sum_{r=0}^{n-1} g(X_r) \to \sum_{\theta} g(\theta)\pi(\theta).$$

The convergence is usually in mean square and almost surely (see Problem (6.15.44) and Chapter 7), and thus the averages provide the required approximations.

Although the methods of this chapter may be adapted to *continuous* spaces $\Theta$, we consider here only the case when $\Theta$ is finite. Suppose that we are given a finite set $\Theta$ and a mass function $\pi = (\pi_i : i \in \Theta)$, termed the 'target distribution'. Our task is to discuss how to construct an ergodic discrete-time Markov chain $X$ on $\Theta$ with transition matrix $\mathbf{P} = (p_{ij})$, having given stationary distribution $\pi$, and with the property that realizations of the $X$ may be readily simulated.

There is a wide choice of such Markov chains. Computation and simulation is easier for reversible chains, and we shall therefore restrict out attention to chains whose transition probabilities $p_{ij}$ satisfy the detailed balance equations

**(3)**                         $\pi_k p_{kj} = \pi_j p_{jk}, \quad j, k \in \Theta.$

Recall Definition (6.5.2). Producing a suitable chain $X$ turns out to be remarkably straightforward. There are two steps in the following simple algorithm. Suppose that $X_n = i$, and it is required to construct $X_{n+1}$.

(i) Let $\mathbf{H} = (h_{ij} : i, j \in \Theta)$ be an arbitrary stochastic matrix, called the 'proposal matrix'. We pick $Y \in \Theta$ according to the probabilities $\mathbb{P}(Y = j \mid X_n = i) = h_{ij}$.

(ii) Let $\mathbf{A} = (a_{ij} : i, j \in \Theta)$ be a matrix with entries satisfying $0 \leq a_{ij} \leq 1$; the $a_{ij}$ are called 'acceptance probabilities'. Given that $Y = j$, we set

$$X_{n+1} = \begin{cases} j & \text{with probability } a_{ij}, \\ X_n & \text{with probability } 1 - a_{ij}. \end{cases}$$

How do we determine the matrices $\mathbf{H}, \mathbf{A}$? The proposal matrix $\mathbf{H}$ is chosen in such a way that it is easy and cheap to simulate according to it. The acceptance matrix $\mathbf{A}$ is chosen in such a way that the detailed balance equations (3) hold. Since $p_{ij}$ is given by

**(4)**                 $p_{ij} = \begin{cases} h_{ij} a_{ij} & \text{if } i \neq j, \\ 1 - \sum_{k: k \neq i} h_{ik} a_{ik} & \text{if } i = j, \end{cases}$

the detailed balance equations (3) will be satisfied if we choose

$$
(5) \qquad\qquad a_{ij} = 1 \wedge \left( \frac{\pi_j h_{ji}}{\pi_i h_{ij}} \right)
$$

where $x \wedge y = \min\{x, y\}$ as usual. This choice of **A** leads to an algorithm called the *Hastings algorithm*†. It may be considered desirable to accept as many proposals as possible, and this may be achieved as follows. Let $(t_{ij})$ be a symmetric matrix with non-negative entries satisfying $a_{ij} t_{ij} \leq 1$ for all $i, j \in \Theta$, and let $a_{ij}$ be given by (5). It is easy to see that one may choose any acceptance probabilities $a'_{ij}$ given by $a'_{ij} = a_{ij} t_{ij}$. Such a generalization is termed *Hastings's general algorithm*.

While the above provides a general approach to McMC, further ramifications are relevant in practice. It is often the case in applications that the space $\Theta$ is a product space. For example, it was the case in (2) that $\Theta = \{-1, 1\}^V$ where $V$ is the vertex set of a certain graph; in the statistical analysis of images, one may take $\Theta = S^V$ where $S$ is the set of possible states of a given pixel and $V$ is the set of all pixels. It is natural to exploit this product structure in devising the required Markov chain, and this may be done as follows.

Suppose that $S$ is a finite set of 'local states', that $V$ is a finite index set, and set $\Theta = S^V$. For a given target distribution $\pi$ on $\Theta$, we seek to construct an approximating Markov chain $X$. One way to proceed is to restrict ourselves to transitions which flip the value of the current state at only one coordinate $v \in V$; this is called 'updating at $v$'. That is, given that $X_n = i = (i_w : w \in V)$, we decide that $X_{n+1}$ takes a value in the set of all $j = (j_w : w \in V)$ such that $j_w = i_w$ whenever $w \neq v$. This may be achieved by following the above recipe in a way specific to the choice of the index $v$.

How do we decide on the choice of $v$? Several ways present themselves, of which the following two are obvious examples. One way is to select $v$ uniformly at random from $V$ at each step of the chain $X$. Another is to cycle through the elements of $V$ is some deterministic manner.

**(6) Example. Gibbs sampler, or heat bath algorithm.** As in Example (2), take $\Theta = S^V$ where the 'local state space' $S$ and the index set $V$ are finite. For $i = (i_w : w \in V) \in \Theta$ and $v \in V$, let $\Theta_{i,v} = \{j \in \Theta : j_w = i_w \text{ for } w \neq v\}$. Suppose that $X_n = i$ and that we have decided to update at $v$. We take

$$
(7) \qquad\qquad h_{ij} = \frac{\pi_j}{\sum_{k \in \Theta_{i,v}} \pi_k}, \qquad j \in \Theta_{i,v},
$$

which is to say that the proposal $Y$ is chosen from $\Theta_{i,v}$ according to the conditional distribution given the other components $i_w$, $w \neq v$.

We have from (5) that $a_{ij} = 1$ for all $j \in \Theta_{i,v}$, on noting that $\Theta_{i,v} = \Theta_{j,v}$ if $j \in \Theta_{i,v}$. Therefore $\mathbb{P}_v(X_{n+1} = j \mid X_n = i) = h_{ij}$ for $j \in \Theta_{i,v}$, where $\mathbb{P}_v$ denotes the probability measure associated with updating at $v$.

We may choose the value of $v$ either by flipping coins or by cycling through $V$ in some pre-determined manner.       ●

**(8) Example. Metropolis algorithm.** If the matrix **H** is symmetric, equation (5) gives $a_{ij} = 1 \wedge (\pi_j/\pi_i)$, whence $p_{ij} = h_{ij}\{1 \wedge (\pi_j/\pi_i)\}$ for $i \neq j$.

---

†Or the *Metropolis–Hastings algorithm*; see Example (8).

A simple choice for the proposal probabilities $h_{ij}$ would be to sample the proposal 'uniformly at random' from the set of available changes. In the notation of Example (6), we might take

$$
h_{ij} = \begin{cases} \dfrac{1}{|\Theta_{i,v}| - 1} & \text{if } j \neq i,\ j \in \Theta_{i,v}, \\ 0 & \text{if } j = i. \end{cases}
$$
●

The accuracy of McMC hinges on the rate at which the Markov chain $X$ approaches its stationary distribution $\pi$. In practical cases, it is notoriously difficult to decide whether or not $X_n$ is close to its equilibrium, although certain theoretical results are available. The choice of distribution $\alpha$ of $X_0$ is relevant, and it is worthwhile choosing $\alpha$ in such a way that $X_0$ has strictly positive probability of lying in any part of the set $\Theta$ where $\pi$ has positive weight. One might choose to estimate $\sum_\theta g(\theta)\pi(\theta)$ by $n^{-1}\sum_{r=M}^{M+n-1} g(X_r)$ for some large 'mixing time' $M$. We do not pursue here the determination of suitable $M$.

This section closes with a precise mathematical statement concerning the rate of convergence of the distribution $\alpha \mathbf{P}^n$ to the stationary distribution $\pi$. We assume for simplicity that $X$ is aperiodic and irreducible. Recall from the Perron–Frobenius theorem (6.6.1) that $\mathbf{P}$ has $T = |\Theta|$ eigenvalues $\lambda_1, \lambda_2, \ldots, \lambda_T$ such that $\lambda_1 = 1$ and $|\lambda_j| < 1$ for $j \neq 1$. We write $\lambda_2$ for the eigenvalue with second largest modulus. It may be shown in some generality that

$$
\mathbf{P}^n = \mathbf{I}\pi' + \mathrm{O}(n^{m-1}|\lambda_2|^n),
$$

where $\mathbf{I}$ is the identity matrix, $\pi'$ is the column vector $(\pi_i : i \in \Theta)$, and $m$ is the multiplicity of $\lambda_2$. Here is a concrete result in the reversible case.

**(9) Theorem.** *Let $X$ be an aperiodic irreducible reversible Markov chain on the finite state space $\Theta$, with transition matrix $\mathbf{P}$ and stationary distribution $\pi$. Then*

$$
(10) \qquad \sum_{k \in \Theta} |p_{ik}(n) - \pi_k| \leq |\Theta| \cdot |\lambda_2|^n \sup\{|v_r(i)| : r \in \Theta\}, \qquad i \in \Theta,\ n \geq 1,
$$

*where $v_r(i)$ is the $i$th term of the $r$th right-eigenvector $\mathbf{v}_r$ of $\mathbf{P}$.*

We note that the left side of (10) is the total variation distance (see equation (4.12.7)) between the mass functions $p_{i\cdot}(n)$ and $\pi$.

**Proof.** Let $T = |\Theta|$ and number the states in $\Theta$ as $1, 2, \ldots, T$. Using the notation and result of Exercise (6.14.1), we have that $\mathbf{P}$ is self-adjoint. Therefore the right eigenvectors $\mathbf{v}_1, \mathbf{v}_2, \ldots, \mathbf{v}_T$ corresponding to the eigenvalues $\lambda_1, \lambda_2, \ldots, \lambda_T$, are real. We may take $\mathbf{v}_1, \mathbf{v}_2, \ldots, \mathbf{v}_T$ to be an orthonormal basis of $\mathbb{R}^T$ with respect to the given scalar product. The unit vector $\mathbf{e}_k$, having 1 in its $k$th place and 0 elsewhere, may be written

$$
(11) \qquad \mathbf{e}_k = \sum_{r=1}^{T} \langle \mathbf{e}_k, \mathbf{v}_r \rangle \mathbf{v}_r = \sum_{r=1}^{T} v_r(k)\pi_k \mathbf{v}_r.
$$

Now $\mathbf{P}^n \mathbf{e}_k = (p_{1k}(n), p_{2k}(n), \ldots, p_{Tk}(n))'$, and $\mathbf{P}^n \mathbf{v}_r = \lambda_r^n \mathbf{v}_r$. We pre-multiply (11) by $\mathbf{P}^n$ and deduce that

$$
p_{ik}(n) = \sum_{r=1}^{T} v_r(k)\pi_k \lambda_r^n v_r(i).
$$

Now $v_1 = 1$ and $\lambda_1 = 1$, so that the term of the sum corresponding to $r = 1$ is simply $\pi_k$. It follows that

$$\sum_k |p_{ik}(n) - \pi_k| \le \sum_{r=2}^{T} |\lambda_r|^n |v_r(i)| \sum_k \pi_k |v_r(k)|.$$

By the Cauchy–Schwarz inequality,

$$\sum_k \pi_k |v_r(k)| \le \sqrt{\sum_k \pi_k |v_r(k)|^2} = 1,$$

and (10) follows.                                                                       ∎

Despite the theoretical appeal of such results, they are not always useful when $\mathbf{P}$ is large, because of the effort required to compute the right side of (10). It is thus important to establish readily computed bounds for $|\lambda_2|$, and bounds on $|p_{ik}(n) - \pi_k|$, which do not depend on the $v_j$. We give a representative bound without proof.

**(12) Theorem. Conductance bound.** *We have under the assumptions of Theorem* (9) *that* $1 - 2\Psi \le \lambda_2 \le 1 - \frac{1}{2}\Psi^2$ *where*

$$\Psi = \inf\left\{ \sum_{i \in B} \pi_i p_{ij} \Big/ \sum_{i \in B} \pi_i : B \subseteq \Theta,\ 0 < \sum_{i \in B} \pi_i \le \tfrac{1}{2} \right\}.$$

**(B) Coupling from the past.** The above McMC methods result in an *approximation* to the target distribution $\pi$. There is a beautiful method, called *coupling from the past* (cftp)†, for sampling *exactly* from $\pi$. Whereas the cost of McMC lies in the accuracy of the approximation, the cost of cftp lies in the fact that the time required is itself a random variable.

Let $\pi = (\pi_i : i \in \Theta)$ be a mass function on a finite set $\Theta$. We call $\pi$ *positive* if $\pi_i > 0$ for $i \in \Theta$, and we assume henceforth that $\pi$ is positive. First, we find an irreducible transition matrix $\mathbf{P} = (p_{ij})$ on $\Theta \times \Theta$ such that $\pi$ is stationary for $\mathbf{P}$, in that $\pi \mathbf{P} = \pi$. There are many choices for $\mathbf{P}$, and the actual choice is immaterial for the moment. Next, let $W = (W(i) : i \in \Theta)$ be a vector of random variables such that

**(13)**                             $\mathbb{P}(W(i) = j) = p_{ij}, \qquad j \in \Theta.$

Thus, if $X$ is a Markov chain with transition matrix $\mathbf{P}$ starting at $X_0 = i$, then $X_1$ has the same distribution as $W(i)$. No assumption is made at this stage about the *joint* distribution of the $W(i)$, and we return to this later in the section.

Let $W_{-1}, W_{-2}, \dots$ be independent random vectors distributed as $W$, that is, each $W_{-m}$ is distributed as (13). We construct a sequence $\{Y_{-n} : n = 0, 1, 2, \dots\}$ of random maps from $\Theta$ to $\Theta$ by the following inductive procedure. First, we define $Y_0 : \Theta \to \Theta$ to be the identity mapping. Having found $Y_0, Y_{-1}, Y_{-2}, \dots, Y_{-m}$ for $m \ge 0$, we define

**(14)**                       $Y_{-m-1}(i) = Y_{-m}(W_{-m-1}(i)), \qquad i \in \Theta.$

---

†Or the *Propp–Wilson algorithm*.

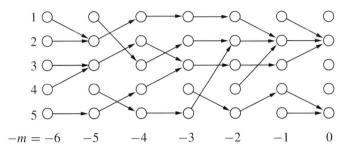

Figure 6.4. An illustration of 'coupling from the past' with $\Theta = \{1, 2, 3, 4, 5\}$. For $m = 1, 2, \ldots, 6$ and each $i \in \Theta$, we pick $j \in \Theta$ with probability $p_{ij}$, as indicated by the arrows. For $i \in \Theta$, the value $Y_{-m}(i)$ is found by following the arrows from element $i$ in the column labelled $-m$ to the column labelled $0$. We see that $Y_{-6}$ is the constant function taking the value 2, and in fact $M = 6$ in this case.

That is, $Y_{-m-1}(i)$ is obtained from $i$ by passing in one time step to the value $W_{-m-1}(i)$, and then applying $Y_{-m}$ to this new state. The exact dependence structure of this scheme is an important ingredient of its analysis.

The *coalescence time $M$* is given by

$$(15) \qquad\qquad M = \inf\{m : Y_{-m}(\cdot) \text{ is a constant function}\},$$

noting that $M$ may be finite or infinite. It follows by the definition of coalescence time that, if $M < \infty$, then the value $Y_{-M} = Y_{-M}(i)$ does not depend on the choice of $i$. The process of coalescence is illustrated in Figure 6.4. We prove next that $Y_{-M}$ has law $\pi$, so long as $\mathbb{P}(M < \infty) = 1$.

**(16) Theorem.** *Let $\pi$ be a positive mass function on the finite set $\Theta$, and let $\mathbf{P}$ be irreducible with stationary distribution $\pi$.*

(a) *Either $\mathbb{P}(M = \infty) = 1$ or $\mathbb{P}(M < \infty) = 1$, and in the latter case there exists $\beta > 0$ such that*

$$(17) \qquad\qquad \mathbb{P}(M \geq m) \leq e^{-\beta m}, \qquad m \geq 0.$$

(b) *Suppose $\mathbb{P}(M < \infty) = 1$. Then $Y_{-M}$ has distribution $\pi$.*

Whether or not $M$ is a.s. finite depends on the joint distribution of the $W(i)$. If the $W(i)$ are independent and $\mathbf{P}$ is aperiodic, then $M < \infty$ a.s. On the other hand, if $\pi$ is uniformly distributed on $\Theta$, and the vector $(W(i) : i \in \Theta)$ is a random permutation of $\Theta$, then $M = \infty$ a.s. See Exercise (6.14.5).

In classical Monte Carlo experiments, the time-$n$ measure converges to the target measure as $n \to \infty$. An estimate of the rate of convergence is necessary in order to know when to cease the process. Such estimates are not central to 'coupling from the past', since this method results, after a finite (random) time, in a sample having exactly the target distribution. That said, the method of proof implies a geometric rate of convergence, in that there exists $\beta > 0$ such that (17) holds.

**Proof of Theorem (16).** (a) For $L \geq 1$, let

$$(18) \qquad\qquad \alpha_L = \mathbb{P}(M \leq L) = \mathbb{P}(Y_{-L} \text{ is a constant function}).$$

If $\alpha_L = 0$ for all $L$, then

$$\mathbb{P}(M < \infty) = \lim_{L \to \infty} \alpha_L = 0.$$

Suppose, conversely, that $\alpha_L > 0$ for some $L$, and pick $L$ accordingly. We extend the notation prior to (14) as follows. Let $(Y_{-s,-t} : 0 \le t \le s)$ be functions mapping $\Theta$ to $\Theta$ given by:

(i) $Y_{-t,-t}$ is the identity map, for $t = 0, 1, 2, \ldots$,

(ii) $Y_{-s,-t}(i) = Y_{-s+1,-t}(W_{-s}(i))$, for $t = 0, 1, \ldots, s - 1$.

By the independence of the $W_{-m}$, the maps $Y_{-kL,-(k-1)L}, k = 1, 2, \ldots$, are independent and identically distributed. By (18), each is a constant function with probability $\alpha_L > 0$, and hence there exists almost surely a (random) integer $K$ such that $Y_{-KL,-(K-1)L}$ is a constant function. It follows that $M \le KL$, whence $\mathbb{P}(M < \infty) = 1$. Since $K$ has a geometric distribution, (17) holds for appropriate $\beta > 0$.

(b) Suppose that $\mathbb{P}(M < \infty) = 1$. Let $X$ be chosen randomly from $\Theta$ with law $\pi$, and write $X_m = Y_{-m}(X)$. Since $\pi$ is a stationary distribution of the transition matrix $\mathbf{P}$, $X_m$ has distribution $\pi$ for all $m = 0, 1, 2, \ldots$. By the definition of $M$,

$$Y_{-M} = X_m \quad \text{on the event} \quad \{M \le m\}.$$

For $j \in \Theta$ and $m = 0, 1, 2, \ldots$,

$$\begin{aligned}
\mathbb{P}(Y_{-M} = j) &= \mathbb{P}(Y_{-M} = j, \ M \le m) + \mathbb{P}(Y_{-M} = j, \ M > m) \\
&= \mathbb{P}(X_m = j, \ M \le m) + \mathbb{P}(Y_{-M} = j, \ M > m) \\
&\le \pi_j + \mathbb{P}(M > m),
\end{aligned}$$

and similarly,

$$\pi_j = \mathbb{P}(X_m = j) \le \mathbb{P}(Y_{-M} = j) + \mathbb{P}(M > m).$$

We combine these two inequalities to obtain that

$$\left| \mathbb{P}(Y_{-M} = j) - \pi_j \right| \le \mathbb{P}(M > m), \qquad j \in \Theta,$$

and we let $m \to \infty$ to obtain the result.  ∎

The above procedure may seem unwieldy in practice, since $\Theta$ will often be large, and it is necessary to keep track in (14) of the $Y_{-m}(i)$ *for every* $i \in \Theta$. The reality can be simpler in certain situations of interest including the Ising model of Example (2). The additional property required for this amounts to the state space $\Theta$ being partially ordered (with a least element and a greatest element), and there being a matrix $\mathbf{P}$ and a corresponding random vector $W$ that preserves order in the sense that

(19)                        $W(i) \le W(j) \quad \text{whenever} \quad i \le j.$

This is the point where assumptions are needed about the *joint* distribution of the $W(i)$.

We amplify this for the special case in which $\Theta = \{0, 1\}^V$, which we assume henceforth. Each $i \in \Theta$ may be written as a 0/1-vector $i = (i_v : v \in V)$. For $i, j \in \Theta$, we write $i \le j$ if $i_v \le j_v$ for $v \in V$. Note that $\Theta$ has a minimal element $\mathbf{0} = (0, 0, \ldots, 0)$ and a maximal element $\mathbf{1} = (1, 1, \ldots, 1)$. The state space of the Ising model of (2) may be written in this form by relabelling the vertex-state $-1$ as 0, and adjusting the Ising measure accordingly.

The positive distribution $\pi$ is said to satisfy the 'FKG lattice condition'†

**(20)** $$\pi_{i \vee j} \pi_{i \wedge j} \geq \pi_i \pi_j, \qquad i, j \in \Theta,$$

where

$$[i \vee j](v) = \max\{i_v, j_v\}, \qquad [i \wedge j](v) = \min\{i_v, j_v\}.$$

**(21) Lemma.** *Let* $\Theta = \{0, 1\}^V$ *where* $V$ *is finite, and let* $\pi$ *be a positive distribution on* $\Theta$ *satisfying the FKG lattice condition. There exists an irreducible transition matrix* $\mathbf{P}$ *and a corresponding random vector* $W$ *such that* (19) *holds. With this choice, the coalescence time* $M$ *satisfies* $\mathbb{P}(M < \infty) = 1$.

**Proof.** Let $\pi$ be a positive distribution on $\Theta$. The matrix $\mathbf{P}$ is constructed in the manner of the Gibbs sampler (6). That is, we choose $v \in V$ uniformly at random, and we set $\mathbf{P} = \mathbf{P}_v$ where $\mathbf{P}_v = \mathbf{H}$ as in (7). Having chosen $\mathbf{P}$, the marginal distribution of each $W(i)$ is determined, and we explain next how to sample them simultaneously.

Fix $v \in V$, and let $i \in \Theta$. In the notation of (6), we have that $|\Theta_{i,v}| = 2$. We write $i(1)$ (respectively, $i(0)$) for the vector that agrees with $i$ at vertices $u \neq v$ and takes the value 1 (respectively, 0) at $v$. By (7),

**(22)** $$h_{i,i(1)} = 1 - h_{i,i(0)} = \frac{\pi_{i(1)}}{\pi_{i(0)} + \pi_{i(1)}}.$$

We claim that

**(23)** $$h_{i,i(1)} \leq h_{j,j(1)}, \qquad \text{for } i \leq j.$$

This is proved by writing each side as in (22), then multiplying up, and finally observing that the required inequality $\pi_{j(1)}\pi_{i(0)} \geq \pi_{i(1)}\pi_{j(0)}$ holds for $i \leq j$, by (20).

Let $U$ be uniformly distributed on $[0, 1]$, and let $v$ be fixed as above. For $i \in \Theta$, let $W(i)$ be given by

**(24)** $$W(i) = \begin{cases} i(1) & \text{if } U \leq h_{i,i(1)}, \\ i(0) & \text{otherwise.} \end{cases}$$

Equation (19) holds by (23).

In summary, $W$ is constructed as follows:
1. pick $v \in V$ uniformly at random,
2. sample a uniform random variable $U$,
3. for each $i \in \Theta$, sample $W(i)$ according to (24).

The resulting $W$ has the required property.

Suppose the above recipe is adopted, and let $M$ be the coalescence time. Let $S$ be the smallest number of iterations until every $v$ has been picked at least once, and write $U_1, U_2, \ldots, U_S$ for the corresponding uniform random variables, as in (24). Conditional on $S$, there is a strictly positive probability that $U_i < h_{\min}$ for every $i$, where $h_{\min} = \min\{h_{i,i(1)} : i \in \Theta\} > 0$ by

---

†The FKG lattice condition is sufficient for the FKG inequality of Exercise (3.12.18b). See, e.g., Grimmett (2018, Thm 4.11).

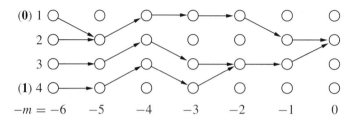

Figure 6.5. Suppose $V = \{u, v\}$ and $\Theta = \{0, 1\}^V$. Label the elements of $\Theta$ as 1, 2, 3, 4, with 1 the minimal element and 4 the maximal element. It suffices to follow the trajectories originating at 1 and 4, and wait for them to coalesce. All other trajectories are trapped between the minimal and maximal trajectories.

the positivity of $\pi$. On this event, we have that coalescence has occurred in that $M \leq S$. Therefore, $\mathbb{P}(M < \infty) > 0$, and the claim follows by Theorem (16a). ∎

Let $\pi$ be a positive distribution satisfying the FKG lattice condition. We choose $W$ according to Lemma (21). By (14) and (19),

$$Y_{-m}(\mathbf{0}) \leq Y_{-m}(i) \leq Y_{-m}(\mathbf{1}), \qquad i \in \Theta,$$

and thus we may rewrite (15) as

$$M = \inf\{m : Y_{-m}(\mathbf{0}) = Y_{-m}(\mathbf{1})\},$$

since the $Y_{-m}(i)$, $i \neq \mathbf{0}, \mathbf{1}$, are sandwiched between the two extremes. In this situation, it suffices to keep track only of $Y_{-m}(\mathbf{0})$ and $Y_{-m}(\mathbf{1})$. See the illustration of Figure 6.5.

**(25) Example. Ising model revisited.** It is an *exercise*† to show that the Ising measure of Example (2) satisfies the FKG lattice condition (20). In order to sample exactly from the Ising measure, one may couple from the past using the recipe of Lemma (21) until the minimal and maximal trajectories are in the same state. This common value is an exact sample. ●

---

## Exercises for Section 6.14

**1.** Let $\mathbf{P}$ be a stochastic matrix on the finite set $\Theta$ with stationary distribution $\pi$. Define the inner product $\langle \mathbf{x}, \mathbf{y} \rangle = \sum_{k \in \Theta} x_k y_k \pi_k$, and let $l^2(\pi) = \{\mathbf{x} \in \mathbb{R}^\Theta : \langle \mathbf{x}, \mathbf{x} \rangle < \infty\}$. Show, in the obvious notation, that $\mathbf{P}$ is reversible with respect to $\pi$ if and only if $\langle \mathbf{x}, \mathbf{P}\mathbf{y} \rangle = \langle \mathbf{P}\mathbf{x}, \mathbf{y} \rangle$ for all $\mathbf{x}, \mathbf{y} \in l^2(\pi)$.

**2. Barker's algorithm.** Show that a possible choice for the acceptance probabilities in Hastings's general algorithm is

$$b_{ij} = \frac{\pi_j g_{ji}}{\pi_i g_{ij} + \pi_j g_{ji}},$$

where $\mathbf{G} = (g_{ij})$ is the proposal matrix.

**3.** Let $S$ be a countable set. For each $j \in S$, the sets $A_{jk}$, $k \in S$, form a partition of the interval $[0, 1]$. Let $g : S \times [0, 1] \to S$ be given by $g(j, u) = k$ if $u \in A_{jk}$. The sequence $\{X_n : n \geq 0\}$ of random variables is generated recursively by $X_{n+1} = g(X_n, U_{n+1})$, $n \geq 0$, where $\{U_n : n \geq 1\}$ are independent random variables with the uniform distribution on $[0, 1]$. Show that $X$ is a Markov chain, and find its transition matrix.

---

†See Exercise (6.14.6).

**4.   Dobrushin's bound.** Let $\mathbf{U} = (u_{st})$ be a finite $|S| \times |T|$ stochastic matrix. *Dobrushin's ergodic coefficient* is defined to be

$$d(\mathbf{U}) = \tfrac{1}{2} \sup_{i,j \in S} \sum_{t \in T} |u_{it} - u_{jt}|.$$

(a) Show that, if $\mathbf{V}$ is a finite $|T| \times |U|$ stochastic matrix, then $d(\mathbf{UV}) \le d(\mathbf{U})d(\mathbf{V})$.
(b) Let $X$ and $Y$ be discrete-time Markov chains with the same transition matrix $\mathbf{P}$, and show that

$$\sum_k \big| \mathbb{P}(X_n = k) - \mathbb{P}(Y_n = k) \big| \le d(\mathbf{P})^n \sum_k \big| \mathbb{P}(X_0 = k) - \mathbb{P}(Y_0 = k) \big|.$$

**5.**   Let $\pi$ be a positive mass function on the finite set $\Theta$, and let $\mathbf{P}$ be the transition matrix of an irreducible, aperiodic Markov chain with stationary distribution $\pi$. Let $W = (W(i) : i \in \Theta)$ be a vector of random variables such that $\mathbb{P}(W(i) = j) = p_{ij}$ for $i, j \in \Theta$, and use $W$ as an update rule in the coupling-from-the-past algorithm for sampling from $\pi$.
(a) If the $W(i)$, $i \in \Theta$, are independent, show that the coalescence time is a.s. finite.
(b) Give two examples of situations in which the coalescence time is a.s. infinite.

**6.   Ising model.** Show that the Ising distribution of Example (6.14.2) satisfies the FKG lattice condition (6.14.20).

---

# 6.15 Problems

**1.**   Classify the states of the discrete-time Markov chains with state space $S = \{1, 2, 3, 4\}$ and transition matrices

(a) $$\begin{pmatrix} \frac{1}{3} & \frac{2}{3} & 0 & 0 \\ \frac{1}{2} & \frac{1}{2} & 0 & 0 \\ \frac{1}{4} & 0 & \frac{1}{4} & \frac{1}{2} \\ 0 & 0 & 0 & 1 \end{pmatrix}$$
(b) $$\begin{pmatrix} 0 & \frac{1}{2} & \frac{1}{2} & 0 \\ \frac{1}{3} & 0 & 0 & \frac{2}{3} \\ 1 & 0 & 0 & 0 \\ 0 & 0 & 1 & 0 \end{pmatrix}.$$

In case (a), calculate $f_{34}(n)$, and deduce that the probability of ultimate absorption in state 4, starting from 3, equals $\frac{2}{3}$. Find the mean recurrence times of the states in case (b).

**2.**   A transition matrix is called *doubly stochastic* if all its column sums equal 1, that is, if $\sum_i p_{ij} = 1$ for all $j \in S$.
(a) Show that if a finite chain has a doubly stochastic transition matrix, then all its states are positive recurrent, and that if it is, in addition, irreducible and aperiodic then $p_{ij}(n) \to N^{-1}$ as $n \to \infty$, where $N$ is the number of states.
(b) Show that, if an infinite irreducible chain has a doubly stochastic transition matrix, then its states are either all null recurrent or all transient.

**3.**   Prove that intercommunicating states of a Markov chain have the same period.

**4.**   (a) Show that for each pair $i$, $j$ of states of an irreducible aperiodic chain, there exists $N = N(i, j)$ such that $p_{ij}(n) > 0$ for all $n \ge N$.
(b) Let $X$ and $Y$ be independent irreducible aperiodic chains with the same state space $S$ and transition matrix $\mathbf{P}$. Show that the bivariate chain $Z_n = (X_n, Y_n)$, $n \ge 0$, is irreducible and aperiodic.
(c) Show that the bivariate chain $Z$ may be reducible if $X$ and $Y$ are periodic.

**5.** Suppose $\{X_n : n \geq 0\}$ is a discrete-time Markov chain with $X_0 = i$. Let $N$ be the total number of visits made subsequently by the chain to the state $j$. Show that

$$\mathbb{P}(N = n) = \begin{cases} 1 - f_{ij} & \text{if } n = 0, \\ f_{ij}(f_{jj})^{n-1}(1 - f_{jj}) & \text{if } n \geq 1, \end{cases}$$

and deduce that $\mathbb{P}(N = \infty) = 1$ if and only if $f_{ij} = f_{jj} = 1$.

**6.** Let $i$ and $j$ be two states of a discrete-time Markov chain. Show that if $i$ communicates with $j$, then there is positive probability of reaching $j$ from $i$ without revisiting $i$ in the meantime. Deduce that, if the chain is irreducible and recurrent, then the probability $f_{ij}$ of ever reaching $j$ from $i$ equals 1 for all $i$ and $j$.

**7.** Let $\{X_n : n \geq 0\}$ be a recurrent irreducible Markov chain on the state space $S$ with transition matrix $\mathbf{P}$, and let $\mathbf{x}$ be a positive solution of the equation $\mathbf{x} = \mathbf{x}\mathbf{P}$.
(a) Show that

$$q_{ij}(n) = \frac{x_j}{x_i} p_{ji}(n), \qquad i, j \in S, \ n \geq 1,$$

defines the $n$-step transition probabilities of a recurrent irreducible Markov chain on $S$ whose first-passage probabilities are given by

$$g_{ij}(n) = \frac{x_j}{x_i} l_{ji}(n), \qquad i \neq j, \ n \geq 1,$$

where $l_{ji}(n) = \mathbb{P}_j(X_n = i, T > n)$ and $T = \min\{m > 0 : X_m = j\}$.
(b) Show that $\mathbf{x}$ is unique up to a multiplicative constant.
(c) Let $T_j = \min\{n \geq 1 : X_n = j\}$ and define $h_{ij} = \mathbb{P}_i(T_j \leq T_i)$. Show that $x_i h_{ij} = x_j h_{ji}$ for all $i, j \in S$.

**8.** **Renewal sequences.** The sequence $u = \{u_n : n \geq 0\}$ is called a 'renewal sequence' if

$$u_0 = 1, \quad u_n = \sum_{i=1}^{n} f_i u_{n-i} \qquad \text{for } n \geq 1,$$

for some collection $f = \{f_n : n \geq 1\}$ of non-negative numbers summing to 1.
(a) Show that $u$ is a renewal sequence if and only if there exists a Markov chain $X$ on a countable state space $S$ such that $u_n = \mathbb{P}(X_n = s \mid X_0 = s)$, for some recurrent $s \in S$ and all $n \geq 1$.
(b) Show that if $u$ and $v$ are renewal sequences then so is $\{u_n v_n : n \geq 0\}$.

**9.** Consider the symmetric random walk in three dimensions on the set of points $\{(x, y, z) : x, y, z = 0, \pm 1, \pm 2, \dots\}$; this process is a sequence $\{X_n : n \geq 0\}$ of points such that $\mathbb{P}(X_{n+1} = X_n + \epsilon) = \frac{1}{6}$ for $\epsilon = (\pm 1, 0, 0), (0, \pm 1, 0), (0, 0, \pm 1)$. Suppose that $X_0 = (0, 0, 0)$. Show that

$$\mathbb{P}(X_{2n} = (0, 0, 0)) = \left(\frac{1}{6}\right)^{2n} \sum_{i+j+k=n} \frac{(2n)!}{(i! \, j! \, k!)^2} = \left(\frac{1}{2}\right)^{2n} \binom{2n}{n} \sum_{i+j+k=n} \left(\frac{n!}{3^n i! \, j! \, k!}\right)^2$$

and deduce by Stirling's formula that the origin is a transient state.

**10.** Consider the three-dimensional version of the cancer model (6.12.12). If $\kappa = 1$, are the empires of Theorem (6.12.14) inevitable in this case?

**11.** Let $X$ be a discrete-time Markov chain with state space $S = \{1, 2\}$, and transition matrix

$$\mathbf{P} = \begin{pmatrix} 1 - \alpha & \alpha \\ \beta & 1 - \beta \end{pmatrix}.$$

Classify the states of the chain. Suppose that $\alpha\beta > 0$ and $\alpha\beta \neq 1$. Find the $n$-step transition probabilities and show directly that they converge to the unique stationary distribution as $n \to \infty$. For what values of $\alpha$ and $\beta$ is the chain reversible in equilibrium?

**12. Another diffusion model.** $N$ black balls and $N$ white balls are placed in two urns so that each contains $N$ balls. After each unit of time one ball is selected at random from each urn, and the two balls thus selected are interchanged. Let the number of black balls in the first urn denote the state of the system. Write down the transition matrix of this Markov chain and find the unique stationary distribution. Is the chain reversible in equilibrium?

**13.** Consider a Markov chain on the set $S = \{0, 1, 2, \dots\}$ with transition probabilities $p_{i,i+1} = a_i$, $p_{i,0} = 1 - a_i$, $i \geq 0$, where $(a_i : i \geq 0)$ is a sequence of constants which satisfy $0 < a_i < 1$ for all $i$. Let $b_0 = 1$, $b_i = a_0 a_1 \cdots a_{i-1}$ for $i \geq 1$. Show that the chain is
(a) recurrent if and only if $b_i \to 0$ as $i \to \infty$,
(b) positive recurrent if and only if $\sum_i b_i < \infty$,
and write down the stationary distribution if the latter condition holds.

Let $A$ and $\beta$ be positive constants and suppose that $a_i = 1 - Ai^{-\beta}$ for all large $i$. Show that the chain is
(c) transient if $\beta > 1$,
(d) positive recurrent if $\beta < 1$.
Finally, if $\beta = 1$ show that the chain is
(e) positive recurrent if $A > 1$,
(f) null recurrent if $A \leq 1$.

**14.** Let $X$ be a continuous-time Markov chain with countable state space $S$ and semigroup $\{\mathbf{P}_t\}$. Show that $p_{ij}(t)$ is a continuous function of $t$. Let $g(t) = -\log p_{ii}(t)$; show that $g$ is a continuous function, $g(0) = 0$, and $g(s+t) \leq g(s) + g(t)$. We say that $g$ is 'subadditive', and a well known theorem gives the result that

$$\lim_{t \downarrow 0} \frac{g(t)}{t} = \lambda \quad \text{exists and} \quad \lambda = \sup_{t > 0} \frac{g(t)}{t} \leq \infty.$$

Deduce that the limit $g_{ii} = \lim_{t \downarrow 0} t^{-1}\{p_{ii}(t) - 1\}$ exists.

**15.** Let $X$ be a continuous-time Markov chain with generator $\mathbf{G} = (g_{ij})$. Show that $X$ is irreducible if and only if for any pair $i$, $j$ of distinct states there exists a sequence $i, k_1, k_2, \dots, k_n, j$ of distinct states such that $g_{i,k_1} g_{k_1,k_2} \cdots g_{k_n,j} > 0$.

**16. Reversibility.**
(a) Let $T > 0$ and let $X = \{X(t) : 0 \leq t \leq T\}$ be an irreducible, non-explosive Markov chain with stationary distribution $\boldsymbol{\pi}$, and suppose that $X(0)$ has distribution $\boldsymbol{\pi}$. Let $Y(t) = X(T - t)$ for $0 \leq t \leq T$. We call $X$ *reversible (in equilibrium)* if $X$ and $Y$ have the same joint distributions.
   (i) Show that $Y$ is a (left-continuous) Markov chain with transition probabilities $\hat{p}_{ij}(t) = (\pi_j/\pi_i)p_{ji}(t)$ and generator $\hat{\mathbf{G}}$ satisfying $\pi_j \hat{g}_{ji} = \pi_i g_{ij}$, where the $p_{ji}(t)$ and $\mathbf{G} = (g_{ij})$ are those of $X$. Show that $Y$ is irreducible and non-explosive with stationary distribution $\boldsymbol{\pi}$.
   (ii) Show that $X$ is reversible in equilibrium if and only if the *detailed balance equations* $\pi_i g_{ij} = \pi_j g_{ji}$ (for all $i$ and $j$) hold.
   (iii) Show that a measure $\boldsymbol{\nu}$ satisfies $\boldsymbol{\nu}\mathbf{G} = \mathbf{0}$ if it satisfies the detailed balance equations.
(b) Let $X$ be irreducible and non-explosive with stationary distribution $\boldsymbol{\pi}$, and assume $X(0)$ has distribution $\boldsymbol{\pi}$.
   (i) **Kolmogorov's criterion.** Show that $X$ is reversible if and only if, for all $n$ and all finite sequences $k_1, k_2, \dots, k_n$ of states,

$$g_{k_1,k_2} g_{k_2,k_3} \cdots g_{k_{n-1},k_n} g_{k_n,k_1} = g_{k_1,k_n} g_{k_n,k_{n-1}} \cdots g_{k_2,k_1}.$$

   (ii) **Kelly's criterion.** Show that $X$ is reversible if, for all distinct triples $i, j, k \in S$, we have $g_{ij} g_{jk} g_{ki} = g_{ik} g_{kj} g_{ji}$, and in addition there exists $c \in S$ such that $g_{ic} > 0$ for all $i \neq c$.

(c) Show that every irreducible chain $X$ with exactly two states is reversible in equilibrium.

(d) Show that every non-explosive birth–death process $X$ having a stationary distribution is reversible in equilibrium.

**17. Elfving's imbedding problem.** Show that not every discrete-time Markov chain can be imbedded in a continuous-time chain. More precisely, let

$$\mathbf{P} = \begin{pmatrix} \alpha & 1-\alpha \\ 1-\alpha & \alpha \end{pmatrix} \qquad \text{for some } 0 < \alpha < 1$$

be a transition matrix. Show that there exists a semigroup $\{\mathbf{P}_t\}$ of transition probabilities in continuous time such that $\mathbf{P}_1 = \mathbf{P}$, if and only if $\frac{1}{2} < \alpha < 1$. In this case show that $\{\mathbf{P}_t\}$ is unique and calculate it in terms of $\alpha$.

**18.** Consider an immigration–death process $X(t)$, being a birth–death process with rates $\lambda_n = \lambda$, $\mu_n = n\mu$. Show that its generating function $G(s, t) = \mathbb{E}(s^{X(t)})$ is given by

$$G(s, t) = \big\{ 1 + (s-1)e^{-\mu t} \big\}^I \exp\{\rho(s-1)(1 - e^{-\mu t})\}$$

where $\rho = \lambda/\mu$ and $X(0) = I$. Deduce the limiting distribution of $X(t)$ as $t \to \infty$.

**19.** Let $N$ be a non-homogeneous Poisson process on $\mathbb{R}_+ = [0, \infty)$ with intensity function $\lambda$. Write down the forward and backward equations for $N$, and solve them.

Let $N(0) = 0$, and find the density function of the time $T$ until the first arrival in the process. If $\lambda(t) = c/(1+t)$, show that $\mathbb{E}(T) < \infty$ if and only if $c > 1$.

**20.** Successive offers for my house are independent identically distributed random variables $X_1$, $X_2, \ldots$, having density function $f$ and distribution function $F$. Let $Y_1 = X_1$, let $Y_2$ be the first offer exceeding $Y_1$, and generally let $Y_{n+1}$ be the first offer exceeding $Y_n$. Show that $Y_1, Y_2, \ldots$ are the times of arrivals in a non-homogeneous Poisson process with intensity function $\lambda(t) = f(t)/(1 - F(t))$. The $Y_i$ are called 'record values'.

Now let $Z_1$ be the first offer received which is the second largest to date, and let $Z_2$ be the second such offer, and so on. Show that the $Z_i$ are the arrival times of a non-homogeneous Poisson process with intensity function $\lambda$.

**21.** Let $N$ be a Poisson process with constant intensity $\lambda$, and let $Y_1, Y_2, \ldots$ be independent random variables with common characteristic function $\phi$ and density function $f$. The process $N^*(t) = Y_1 + Y_2 + \cdots + Y_{N(t)}$ is called a *compound* Poisson process. $Y_n$ is the change in the value of $N^*$ at the $n$th arrival of the Poisson process $N$. Think of it like this. A 'random alarm clock' rings at the arrival times of a Poisson process. At the $n$th ring the process $N^*$ accumulates an extra quantity $Y_n$. Write down a forward equation for $N^*$ and hence find the characteristic function of $N^*(t)$. Can you see directly why it has the form which you have found?

**22.** If the intensity function $\lambda$ of a non-homogeneous Poisson process $N$ is itself a random process, then $N$ is called a *doubly stochastic* Poisson process (or *Cox process*).

(a) Consider the case when $\lambda(t) = \Lambda$ for all $t$, and $\Lambda$ is a random variable taking either of two values $\lambda_1$ or $\lambda_2$, each being picked with equal probability $\frac{1}{2}$. Find the probability generating function of $N(t)$, and deduce its mean and variance.

(b) For a doubly stochastic Poisson process $N$, show that $\text{var}(N(t)) \geq \mathbb{E}(N(t))$.

(c) Let $M$ be an ordinary Poisson process on the time-interval $[0, \infty)$ with constant rate 1. Let $M^*$ be obtained from $M$ by deleting the $k$th arrival for every odd value of $k$. Is $M^*$ either: (i) a Poisson process, or (ii) a doubly stochastic Poisson process?

**23.** Show that a simple birth process $X$ with parameter $\lambda$ is a doubly stochastic Poisson process with intensity function $\lambda(t) = \lambda X(t)$.

**24. Pólya's process.** The Markov chain $X = \{X(t) : t \geq 0\}$ is a birth process whose intensities $\lambda_k(t)$ depend also on the time $t$ and are given by

$$\mathbb{P}\big(X(t+h) = k+1 \,\big|\, X(t) = k\big) = \frac{1 + \mu k}{1 + \mu t}h + o(h)$$

as $h \downarrow 0$. Show that the probability generating function $G(s, t) = \mathbb{E}(s^{X(t)})$ satisfies

$$\frac{\partial G}{\partial t} = \frac{s-1}{1+\mu t}\left\{ G + \mu s \frac{\partial G}{\partial s} \right\}, \qquad 0 < s < 1.$$

Hence find the mean and variance of $X(t)$ when $X(0) = I$.

**25.** (a) Let $X$ be a birth–death process with strictly positive birth rates $\lambda_0, \lambda_1, \ldots$ and death rates $\mu_1, \mu_2, \ldots$ Let $\eta_i$ be the probability that $X(t)$ ever takes the value 0 starting from $X(0) = i$. Show that

$$\lambda_j \eta_{j+1} - (\lambda_j + \mu_j)\eta_j + \mu_j \eta_{j-1} = 0, \qquad j \geq 1,$$

and deduce that $\eta_i = 1$ for all $i$ so long as $\sum_1^\infty e_j = \infty$ where $e_j = \mu_1 \mu_2 \cdots \mu_j/(\lambda_1 \lambda_2 \cdots \lambda_j)$.
(b) For the discrete-time chain on the non-negative integers with

$$p_{j,j+1} = \frac{(j+1)^2}{j^2 + (j+1)^2} \quad \text{and} \quad p_{j,j-1} = \frac{j^2}{j^2 + (j+1)^2},$$

find the probability that the chain ever visits 0, starting from 1.

**26.** Find a good necessary condition and a good sufficient condition for the birth–death process $X$ of Problem (6.15.25a) to be honest.

**27.** Let $X$ be a simple symmetric birth–death process with $\lambda_n = \mu_n = n\lambda$, and let $T$ be the time until extinction. Show that

$$\mathbb{P}(T \leq x \mid X(0) = I) = \left( \frac{\lambda x}{1 + \lambda x} \right)^I,$$

and deduce that extinction is certain if $\mathbb{P}(X(0) < \infty) = 1$.
Show that $\mathbb{P}(\lambda T/I \leq x \mid X(0) = I) \to e^{-1/x}$ as $I \to \infty$.

**28. Immigration–death with disasters.** Let $X$ be an immigration–death–disaster process, that is, a birth–death process with parameters $\lambda_i = \lambda$, $\mu_i = i\mu$, and with the additional possibility of 'disasters' which reduce the population to 0. Disasters occur at the times of a Poisson process with intensity $\delta$, independently of all previous births and deaths.
(a) Show that $X$ has a stationary distribution, and find an expression for the generating function of this distribution.
(b) Show that, in equilibrium, the mean of $X(t)$ is $\lambda/(\delta + \mu)$.

**29.** With any sufficiently nice (Lebesgue measurable, say) subset $B$ of the real line $\mathbb{R}$ is associated a random variable $X(B)$ such that
(a) $X(B)$ takes values in $\{0, 1, 2, \ldots\}$,
(b) if $B_1, B_2, \ldots, B_n$ are disjoint then $X(B_1), X(B_2), \ldots, X(B_n)$ are independent, and furthermore $X(B_1 \cup B_2) = X(B_1) + X(B_2)$,
(c) the distribution of $X(B)$ depends only on $B$ through its Lebesgue measure ('length') $|B|$, and

$$\frac{\mathbb{P}(X(B) \geq 1)}{\mathbb{P}(X(B) = 1)} \to 1 \quad \text{as } |B| \to 0.$$

Show that $X$ is a Poisson process.

**30. Poisson forest.** Let $N$ be a Poisson process in $\mathbb{R}^2$ with constant intensity $\lambda$, and let $R_{(1)} < R_{(2)} < \cdots$ be the ordered distances from the origin of the points of the process.
(a) Show that $R_{(1)}^2, R_{(2)}^2, \ldots$ are the points of a Poisson process on $\mathbb{R}_+ = [0, \infty)$ with intensity $\lambda \pi$.
(b) Show that $R_{(k)}$ has density function

$$f(r) = \frac{2\pi \lambda r (\lambda \pi r^2)^{k-1} e^{-\lambda \pi r^2}}{(k-1)!}, \qquad r > 0.$$

**31.** Let $X$ be a $n$-dimensional Poisson process with constant intensity $\lambda$. Show that the volume of the largest ($n$-dimensional) sphere centred at the origin which contains no point of $X$ is exponentially distributed. Deduce the density function of the distance $R$ from the origin to the nearest point of $X$. Show that $\mathbb{E}(R) = \Gamma(1/n)/\{n(\lambda c)^{1/n}\}$ where $c$ is the volume of the unit ball of $\mathbb{R}^n$ and $\Gamma$ is the gamma function.

**32.** A village of $N + 1$ people suffers an epidemic. Let $X(t)$ be the number of ill people at time $t$, and suppose that $X(0) = 1$ and $X$ is a birth process with rates $\lambda_i = \lambda i(N + 1 - i)$. Let $T$ be the length of time required until every member of the population has succumbed to the illness. Show that

$$\mathbb{E}(T) = \frac{1}{\lambda} \sum_{k=1}^{N} \frac{1}{k(N + 1 - k)}$$

and deduce that

$$\mathbb{E}(T) = \frac{2(\log N + \gamma)}{\lambda(N + 1)} + O(N^{-2})$$

where $\gamma$ is Euler's constant. It is striking that $\mathbb{E}(T)$ decreases with $N$, for large $N$.

**33.** A particle has velocity $V(t)$ at time $t$, where $V(t)$ is assumed to take values in $\{n + \frac{1}{2} : n \geq 0\}$. Transitions during $(t, t + h)$ are possible as follows:

$$\mathbb{P}\big(V(t + h) = w \mid V(t) = v\big) = \begin{cases} (v + \frac{1}{2})h + o(h) & \text{if } w = v + 1, \\ 1 - 2vh + o(h) & \text{if } w = v, \\ (v - \frac{1}{2})h + o(h) & \text{if } w = v - 1. \end{cases}$$

Initially $V(0) = \frac{1}{2}$. Let

$$G(s, t) = \sum_{n=0}^{\infty} s^n \mathbb{P}\big(V(t) = n + \tfrac{1}{2}\big).$$

(a) Show that

$$\frac{\partial G}{\partial t} = (1 - s)^2 \frac{\partial G}{\partial s} - (1 - s)G$$

and deduce that $G(s, t) = \{1 + (1 - s)t\}^{-1}$.

(b) Show that the expected length $m_n(T)$ of time for which $V = n + \frac{1}{2}$ during the time interval $[0, T]$ is given by

$$m_n(T) = \int_0^T \mathbb{P}\big(V(t) = n + \tfrac{1}{2}\big) dt$$

and that, for fixed $k$, $m_k(T) - \log T \to -\sum_{i=1}^{k} i^{-1}$ as $T \to \infty$.

(c) What is the expected velocity of the particle at time $t$?

**34.** A sequence $X_0, X_1, \ldots$ of random integers is generated as follows. First, $X_0 = 0$, $X_1 = 1$. For $n \geq 1$, conditional on $X_0, X_1, \ldots, X_n$, the next value $X_{n+1}$ is equally likely to be either $X_n + X_{n-1}$ or $|X_n - X_{n-1}|$.

(a) Is $X$ a Markov chain?

(b) Use the Markov chain $Y_n = (X_{n-1}, X_n)$ to find the probability that $X$ hits the value 3 before it revisits 0.

(c) Show that the probability that $Y$ ever reaches the state $(1, 1)$, having started at $(1, 2)$, is $\frac{1}{2}(3 - \sqrt{5})$.

**35.** Take a regular hexagon and join opposite corners by straight lines meeting at the point C. A particle performs a symmetric random walk on these 7 vertices, starting at A ($\neq$ C). Find:

(a) the probability of return to A without hitting C,

(b) the expected time to return to A,

(c) the expected number of visits to C before returning to A,

(d) the expected time to return to A, given that there is no prior visit to C.

**36. Diffusion, osmosis.** Markov chains are defined by the following procedures at any time $n$:

(a) **Bernoulli model.** Two adjacent containers A and B each contain $m$ particles; $m$ are of type I and $m$ are of type II. A particle is selected at random in each container. If they are of opposite types they are exchanged with probability $\alpha$ if the type I is in A, or with probability $\beta$ if the type I is in B. Let $X_n$ be the number of type I particles in A at time $n$.

(b) **Ehrenfest dog–flea model.** Two adjacent containers contain $m$ particles in all. A particle is selected at random. If it is in A it is moved to B with probability $\alpha$, if it is in B it is moved to A with probability $\beta$. Let $Y_n$ be the number of particles in A at time $n$.

In each case find the transition matrix and stationary distribution of the chain.

**37.** Let $X$ be an irreducible continuous-time Markov chain on the state space $S$ with transition probabilities $p_{jk}(t)$ and unique stationary distribution $\boldsymbol{\pi}$, and write $\mathbb{P}(X(t) = j) = a_j(t)$. If $c(x)$ is a concave function, show that the function $d(t) = \sum_{j \in S} \pi_j c(a_j(t)/\pi_j)$ increases to $c(1)$ as $t \to \infty$.

The *relative entropy* (or *Kullback–Leibler divergence*) of two strictly positive probability mass functions $f, g$ on a subset $S$ of integers is defined as

$$D(f; g) = \sum_{i \in S} f(i) \log\big(f(i)/g(i)\big).$$

Prove that, if $X$ has a finite state space and stationary distribution $\boldsymbol{\pi}$, the relative entropy $D(a(t); \boldsymbol{\pi})$ decreases monotonely to 0 as $t \to \infty$.

**38.** With the notation of the preceding problem, let $u_k(t) = \mathbb{P}(X(t) = k \mid X(0) = 0)$, and suppose the chain is reversible in equilibrium (see Problem (6.15.16)). Show that $u_0(2t) = \sum_j (\pi_0/\pi_j) u_j(t)^2$, and deduce that $u_0(t)$ decreases to $\pi_0$ as $t \to \infty$.

**39. Perturbing a Poisson process.** Let $\Pi$ be the set of points in a Poisson process on $\mathbb{R}^d$ with constant intensity $\lambda$. Each point is displaced, where the displacements are independent and identically distributed. Show that the resulting point process is a Poisson process with intensity $\lambda$.

**40. Perturbations continued.** Suppose for convenience in Problem (6.15.39) that the displacements have a continuous distribution function and finite mean, and that $d = 1$. Suppose also that you are at the origin originally, and you move to $a$ in the perturbed process. Let $L_R$ be the number of points formerly on your left that are now on your right, and $R_L$ the number of points formerly on your right that are now on your left. Show that $\mathbb{E}(L_R) = \mathbb{E}(R_L)$ if and only if $a = \mu$ where $\mu$ is the mean displacement of a particle.

Deduce that if cars enter the start of a long road at the instants of a Poisson process, having independent identically distributed velocities, then, if you travel at the average speed, in the long run the rate at which you are overtaken by other cars equals the rate at which you overtake other cars.

**41.** Ants enter a kitchen at the instants of a Poisson process $N$ of rate $\lambda$; they each visit the pantry and then the sink, and leave. The $r$th ant spends time $X_r$ in the pantry and $Y_r$ in the sink (and $X_r + Y_r$ in the kitchen altogether), where the vectors $V_r = (X_r, Y_r)$ and $V_s$ are independent for $r \neq s$. At time $t = 0$ the kitchen is free of ants. Find the joint distribution of the numbers $A(t)$ of ants in the pantry and $B(t)$ of ants in the sink at time $t$.

Show that, as $t \to \infty$, the number of ants in the kitchen converges in distribution, provided $\mathbb{E}(X_r + Y_r) < \infty$.

Now suppose the ants arrive in pairs at the times of the Poisson process, but then separate to behave independently as above. Find the joint distribution of the numbers of ants in the two locations.

**42.** Let $\{X_r : r \geq 1\}$ be independent exponential random variables with parameter $\lambda$, and set $S_n = \sum_{r=1}^{n} X_r$. Show that:

(a) $Y_k = S_k/S_n$, $1 \leq k \leq n - 1$, have the same distribution as the order statistics of independent variables $\{U_k : 1 \leq k \leq n - 1\}$ which are uniformly distributed on $(0, 1)$,

(b) $Z_k = X_k/S_n$, $1 \leq k \leq n$, have the same joint distribution as the coordinates of a point $(U_1, \ldots, U_n)$ chosen uniformly at random on the simplex $\sum_{r=1}^n u_r = 1$, $u_r \geq 0$ for all $r$.

**43.** Let $X$ be a discrete-time Markov chain with a finite number of states and transition matrix $\mathbf{P} = (p_{ij})$ where $p_{ij} > 0$ for all $i$, $j$. Show that there exists $\lambda \in (0, 1)$ such that $|p_{ij}(n) - \pi_j| < \lambda^n$, where $\pi$ is the stationary distribution.

**44.** Under the conditions of Problem (6.15.43), let $V_i(n) = \sum_{r=0}^{n-1} I_{\{X_r=i\}}$ be the number of visits of the chain to $i$ before time $n$. Show that

$$\mathbb{E}\left(\left|\frac{1}{n}V_i(n) - \pi_i\right|^2\right) \to 0 \quad \text{as } n \to \infty.$$

Show further that, if $f$ is any bounded function on the state space, then

$$\mathbb{E}\left(\left|\frac{1}{n}\sum_{r=0}^{n-1} f(X_r) - \sum_{i \in S} f(i)\pi_i\right|^2\right) \to 0.$$

**45. Conditional entropy.** Let $A$ and $\mathbf{B} = (B_0, B_1, \ldots, B_n)$ be a discrete random variable and vector, respectively. The *conditional entropy* of $A$ with respect to $\mathbf{B}$ is defined as $H(A \mid \mathbf{B}) = \mathbb{E}(\mathbb{E}\{-\log f(A \mid \mathbf{B}) \mid \mathbf{B}\})$ where $f(a \mid \mathbf{b}) = \mathbb{P}(A = a \mid \mathbf{B} = \mathbf{b})$. Let $X$ be an aperiodic Markov chain on a finite state space. Show that

$$H(X_{n+1} \mid X_0, X_1 \ldots, X_n) = H(X_{n+1} \mid X_n),$$

and that

$$H(X_{n+1} \mid X_n) \to -\sum_i \pi_i \sum_j p_{ij} \log p_{ij} \qquad \text{as } n \to \infty,$$

if $X$ is aperiodic with a unique stationary distribution $\pi$.

**46. Coupling.** Let $X$ and $Y$ be independent recurrent birth–death processes with the same parameters (and no explosions). It is not assumed that $X_0 = Y_0$. Show that:
(a) for any $A \subseteq \mathbb{R}$, $|\mathbb{P}(X_t \in A) - \mathbb{P}(Y_t \in A)| \to 0$ as $t \to \infty$,
(b) if $\mathbb{P}(X_0 \leq Y_0) = 1$, then $\mathbb{E}[g(X_t)] \leq \mathbb{E}[g(Y_t)]$ for any increasing function $g$.

**47. Resources.** The number of birds in a wood at time $t$ is a continuous-time Markov process $X$. Food resources impose the constraint $0 \leq X(t) \leq n$. Competition entails that the transition probabilities obey

$$p_{k,k+1}(h) = \lambda(n - k)h + o(h), \qquad p_{k,k-1}(h) = \mu k h + o(h).$$

Find $\mathbb{E}(s^{X(t)})$, together with the mean and variance of $X(t)$, when $X(0) = r$. What happens as $t \to \infty$?

**48. Parrondo's paradox.** A counter performs an irreducible random walk on the vertices 0, 1, 2 of the triangle in the figure beneath, with transition matrix

$$\mathbf{P} = \begin{pmatrix} 0 & p_0 & q_0 \\ q_1 & 0 & p_1 \\ p_2 & q_2 & 0 \end{pmatrix}$$

where $p_i + q_i = 1$ for all $i$. Show that the stationary distribution $\pi$ has

$$\pi_0 = \frac{1 - q_2 p_1}{3 - q_1 p_0 - q_2 p_1 - q_0 p_2},$$

with corresponding formulae for $\pi_1, \pi_2$.

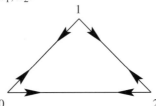

Suppose that you gain one peseta for each clockwise step of the walk, and you lose one peseta for each anticlockwise step. Show that, in equilibrium, the mean yield per step is

$$\gamma = \sum_i (2p_i - 1)\pi_i = \frac{3(2p_0 p_1 p_2 - p_0 p_1 - p_1 p_2 - p_2 p_0 + p_0 + p_1 + p_2 - 1)}{3 - q_1 p_0 - q_2 p_1 - q_0 p_2}.$$

Consider now three cases of this process:

A. We have $p_i = \frac{1}{2} - a$ for each $i$, where $a > 0$. Show that the mean yield per step satisfies $\gamma_A < 0$.

B. We have that $p_0 = \frac{1}{10} - a$, $p_1 = p_2 = \frac{3}{4} - a$, where $a > 0$. Show that $\gamma_B < 0$ for sufficiently small $a$.

C. At each step the counter is equally likely to move according to the transition probabilities of case A or case B, the choice being made independently at every step. Show that, in this case, $p_0 = \frac{3}{10} - a$, $p_1 = p_2 = \frac{5}{8} - a$. Show that $\gamma_C > 0$ for sufficiently small $a$.

The fact that two systematically unfavourable games may be combined to make a favourable game is called Parrondo's paradox. Such bets are not available in casinos.

**49.** Cars arrive at the beginning of a long road in a Poisson stream of rate $\lambda$ from time $t = 0$ onwards. A car has a fixed velocity $V > 0$ which is a random variable. The velocities of cars are independent and identically distributed, and independent of the arrival process. Cars can overtake each other freely. Show that the number of cars on the first $x$ miles of the road at time $t$ has the Poisson distribution with parameter $\lambda \mathbb{E}[V^{-1} \min\{x, Vt\}]$.

**50.** Events occur at the times of a Poisson process with intensity $\lambda$, and you are offered a bet based on the process. Let $t > 0$. You are required to say the word 'now' immediately after the event which you think will be the last to occur prior to time $t$. You win if you succeed, otherwise you lose. If no events occur before $t$ you lose. If you have not selected an event before time $t$ you lose.

Consider the strategy in which you choose the first event to occur after a specified time $s$, where $0 < s < t$.

(a) Calculate an expression for the probability that you win using this strategy.

(b) Which value of $s$ maximizes this probability?

(c) If $\lambda t \geq 1$, show that the probability that you win using this value of $s$ is $e^{-1}$.

**51.** A new Oxbridge professor wishes to buy a house, and can afford to spend up to one million pounds. Declining the services of conventional estate agents, she consults her favourite internet property page on which houses are announced at the times of a Poisson process with intensity $\lambda$ per day. House prices may be assumed to be independent random variables which are uniformly distributed over the interval $(800{,}000, 2{,}000{,}000)$. She decides to view every affordable property announced during the next 30 days. The time spent viewing any given property is uniformly distributed over the range $(1, 2)$ hours. What is the moment generating function of the total time spent viewing houses?

**52. Kemeny's constant.** Let $X = \{X_n : n \geq 0\}$ be an irreducible, aperiodic Markov chain on the finite state space $S$, and let $h_{ij} = \mathbb{E}_i(\min\{n \geq 0 : X_n = j\})$ denote the mean hitting time of $j$ (noting that $h_{ii} = 0$). Let $K_i = \sum_j h_{ij}\pi_j$ be the mean time to hit a state $Z$ chosen at random according to the stationary distribution $\pi$. Show that $K_i$ is independent of the choice of $i$.

**53.** A professor travels between home and work on foot. She possesses a total of $r$ umbrellas, being those at home and at work. If it is raining when she leaves either home or work, she takes an umbrella with her (if there is one available). Assume that it is raining at the start of any given walk with probability $p$ (subject to the usual independence). Let $X_n$ be the number of umbrellas available to her at the start of her $n$th walk.

(a) Explain why $X$ is a Markov chain, and write down its transition matrix.

(b) Show that the chain has stationary distribution $\pi$ given by

$$
\pi_i = \begin{cases} \dfrac{1-p}{r+1-p} & \text{if } i = 0, \\[2ex] \dfrac{1}{r+1-p} & \text{if } i = 1, 2, \dots, r. \end{cases}
$$

What proportion of walks result in her getting wet, in the long run?

(c) Let $r = 1$ and $X_1 = 1$. Calculate the mean number of walks made before she gets wet.

**54. Spiders.** A spider climbs a vertical spout of height $h$ at speed 1. At the instants of a Poisson process of constant rate $\lambda$, the spider is flushed back to the bottom of the spout. It then recommences its climb. Let $T$ be the time to reach the top, and $N$ the number of intermediate flushes. Show that

$$
\mathbb{E}(e^{-\theta T} s^N) = \frac{(\lambda+\theta)e^{-(\lambda+\theta)h}}{\lambda+\theta - \lambda s(1 - e^{-(\lambda+\theta)h})}, \qquad \theta, s \in \mathbb{R}.
$$

By calculating $\mathbb{E}(e^{-\theta T} \mid N = n)$ or otherwise, determine $\mathbb{E}(T \mid N = n)$ for $n \ge 0$.

**55. Probability flows.** Let $X$ be an ergodic Markov chain with transition matrix $\mathbf{P}$ and stationary distribution $\pi$. Show for any set $A$ of states that

$$
\sum_{\substack{i \in A \\ j \notin A}} \pi_i p_{ij} = \sum_{\substack{i \notin A \\ j \in A}} \pi_i p_{ij}.
$$

**56. Attractive hiking.** A hiker of unit mass stands at the origin of the plain $\mathbb{R}^2$. Boulders with independent, identically distributed masses $M_1, M_2, \dots$ lie scattered on the plain at the points of a Poisson process with intensity 1. Let $G_R$ be the $x$-component of the gravitational attraction on the hiker of the boulders within distance $R$ of the hiker. You may take the gravitational constant to be 1.

Show that, as $R \to \infty$, $G_R$ converges in distribution to a Cauchy distribution with characteristic function of the form $\phi(t) = e^{-c|t|}$, and express $c$ in terms of a typical mass $M$.

**57. Holtsmark distribution for stellar gravity.** Let stars of common mass $m$ be positioned at the points of a Poisson process with intensity 1 in $\mathbb{R}^3$. Let $G_R$ be the $x$-component of the gravitational attraction due to the stars within distance $R$ of the origin, upon a hitchhiker of unit mass at the origin. You may take the gravitational constant to be 1.

(a) Show that, as $R \to \infty$, $G_R$ converges in distribution to a symmetric distribution with characteristic function $\phi(t) = \exp\{-c|t|^{3/2}\}$ where $c > 0$.

(b) What is the answer if the stars have independent, identically distributed random masses $M_i$?

# 7

## Convergence of random variables

*Summary.* The many modes of convergence of a sequence of random variables are discussed and placed in context, and criteria are developed for proving convergence. These include standard inequalities, Skorokhod's theorem, the Borel–Cantelli lemmas, and the zero–one law. Laws of large numbers, including the strong law, are proved using elementary arguments. Martingales are defined, and the martingale convergence theorem proved, with applications. The relationship between prediction and conditional expectation is explored, and the condition of uniform integrability described.

## 7.1 Introduction

Expressions such as 'in the long run' and 'on the average' are commonplace in everyday usage, and express our faith that the averages of the results of repeated experimentation show less and less random fluctuation as they settle down to some limit.

**(1) Example. Buffon's needle (4.5.8).** In order to estimate the numerical value of $\pi$, Buffon devised the following experiment. Fling a needle a large number $n$ of times onto a ruled plane and count the number $S_n$ of times that the needle intersects a line. In accordance with the result of Example (4.5.8), the proportion $S_n/n$ of intersections is found to be near to the probability $2/\pi$. Thus $X_n = 2n/S_n$ is a plausible estimate for $\pi$; this estimate converges as $n \to \infty$, and it seems reasonable to write $X_n \to \pi$ as $n \to \infty$. ●

**(2) Example. Decimal expansion.** Any number $y$ satisfying $0 \le y < 1$ has a decimal expansion

$$y = 0 \cdot y_1 y_2 \cdots = \sum_{j=1}^{\infty} y_j 10^{-j},$$

where each $y_j$ takes some value in the set $\{0, 1, 2, \ldots, 9\}$. Now think of $y_j$ as the outcome of a random variable $Y_j$ where $\{Y_j\}$ is a family of independent variables each of which may take any value in $\{0, 1, 2, \ldots, 9\}$ with equal probability $\frac{1}{10}$. The quantity

$$Y = \sum_{j=1}^{\infty} Y_j 10^{-j}$$

is a random variable taking values in $[0, 1]$. It seems likely that $Y$ is uniformly distributed on $[0, 1]$, and this turns out to be the case (see Problem (7.11.4)). More rigorously, this amounts to asserting that the sequence $\{X_n\}$ given by

$$X_n = \sum_{j=1}^{n} Y_j 10^{-j}$$

converges in some sense as $n \to \infty$ to a limit $Y$, and that this limit random variable is uniformly distributed on $[0, 1]$.      ●

In both these examples we encountered a sequence $\{X_n\}$ of random variables together with the assertion that

(3)                           $X_n \to X$    as    $n \to \infty$

for some other random variable $X$. However, random variables are real-valued functions on some sample space, and so (3) is a statement about the convergence of a sequence of *functions*. It is not immediately clear how such convergence is related to our experience of the theory of convergence of sequences $\{x_n\}$ of real numbers, and so we digress briefly to discuss sequences of functions.

Suppose for example that $f_1(\cdot), f_2(\cdot), \ldots$ is a sequence of functions mapping $[0, 1]$ into $\mathbb{R}$. In what manner may they converge to some limit function $f$?

**(4) Convergence pointwise.** If, for all $x \in [0, 1]$, the sequence $\{f_n(x)\}$ of real numbers satisfies $f_n(x) \to f(x)$ as $n \to \infty$ then we say that $f_n \to f$ *pointwise*.      ●

**(5) Norm convergence.** Let $V$ be a collection of functions mapping $[0, 1]$ into $\mathbb{R}$, and assume $V$ is endowed with a function $\| \cdot \| : V \to \mathbb{R}$ satisfying:
- (a) $\|f\| \geq 0$ for all $f \in V$,
- (b) $\|f\| = 0$ if and only if $f$ is the zero function (or equivalent to it, in some sense to be specified),
- (c) $\|af\| = |a| \cdot \|f\|$ for all $a \in \mathbb{R}$, $f \in V$,
- (d) $\|f + g\| \leq \|f\| + \|g\|$ (this is called the *triangle inequality*).

The function $\| \cdot \|$ is called a *norm*. If $\{f_n\}$ is a sequence of members of $V$ then we say that $f_n \to f$ *with respect to the norm* $\| \cdot \|$ if

$$\|f_n - f\| \to 0 \quad \text{as} \quad n \to \infty.$$

Certain special and important norms are given by the $L_p$ norm

$$\|g\|_p = \left( \int_0^1 |g(x)|^p \, dx \right)^{1/p}$$

for $p \geq 1$ and any function $g$ satisfying $\|g\|_p < \infty$.      ●

**(6) Convergence in measure.** Let $\epsilon > 0$ be prescribed, and define the 'distance' between two functions $g, h : [0, 1] \to \mathbb{R}$ by

$$d_\epsilon(g, h) = \int_E dx$$

where $E = \{u \in [0, 1] : |g(u) - h(u)| > \epsilon\}$. We say that $f_n \to f$ *in measure* if

$$d_\epsilon(f_n, f) \to 0 \quad \text{as} \quad n \to \infty \quad \text{for all } \epsilon > 0. \qquad \bullet$$

The convergence of $\{f_n\}$ according to one definition does not necessarily imply its convergence according to another. For example, we shall see later that:

(a) if $f_n \to f$ pointwise then $f_n \to f$ in measure, but the converse is not generally true,
(b) there exist sequences which converge pointwise but not with respect to $\| \cdot \|_1$, and vice versa.

In this chapter we shall see how to adapt these modes of convergence to suit families of *random variables*. Major applications of the ensuing theory include the study of the sequence

$$(7) \qquad\qquad\qquad S_n = X_1 + X_2 + \cdots + X_n$$

of partial sums of an independent identically distributed sequence $\{X_i\}$; the law of large numbers of Section 5.10 will appear as a special case.

It will be clear, from our discussion and the reader's experience, that probability theory is indispensable in descriptions of many processes which occur naturally in the world. Often in such cases we are interested in the future values of the process, and thus in the long-term behaviour within the mathematical model; this is why we need to prove limit theorems for sequences of random variables. Many of these sequences are generated by less tractable operations than, say, the partial sums in (7), and general results such as the law of large numbers may not be enough. It turns out that many other types of sequence are guaranteed to converge; in particular we shall consider later the remarkable theory of 'martingales' which has important applications throughout theoretical and applied probability. This chapter continues in Sections 7.7 and 7.8 with a simple account of the convergence theorem for martingales, together with some examples of its use; these include the asymptotic behaviour of the branching process and provide rigorous derivations of certain earlier remarks (such as (5.4.6)). Conditional expectation is put on a firm footing in Section 7.9.

All readers should follow the chapter up to and including Section 7.4. The subsequent material may be omitted at the first reading.

---

## Exercises for Section 7.1

**1.** Let $r \geq 1$, and define $\|X\|_r = \{\mathbb{E}|X^r|\}^{1/r}$. Show that:
(a) $\|cX\|_r = |c| \cdot \|X\|_r$ for $c \in \mathbb{R}$,
(b) $\|X + Y\|_r \leq \|X\|_r + \|Y\|_r$,
(c) $\|X\|_r = 0$ if and only if $\mathbb{P}(X = 0) = 1$.
This amounts to saying that $\| \cdot \|_r$ is a norm on the set of equivalence classes of random variables on a given probability space with finite $r$th moment, the equivalence relation being given by $X \sim Y$ if and only if $\mathbb{P}(X = Y) = 1$.

**2.** Define $\langle X, Y \rangle = \mathbb{E}(XY)$ for random variables $X$ and $Y$ having finite variance, and define $\|X\| = \sqrt{\langle X, X \rangle}$. Show that:
(a) $\langle aX + bY, Z \rangle = a\langle X, Z \rangle + b\langle Y, Z \rangle$,
(b) $\|X + Y\|^2 + \|X - Y\|^2 = 2(\|X\|^2 + \|Y\|^2)$, the *parallelogram property*,
(c) if $\langle X_i, X_j \rangle = 0$ for all $i \neq j$ then

$$\left\| \sum_{i=1}^{n} X_i \right\|^2 = \sum_{i=1}^{n} \|X_i\|^2.$$

**3.** Let $\epsilon > 0$. Let $g, h : [0, 1] \to \mathbb{R}$, and define $d_\epsilon(g, h) = \int_E dx$ where $E = \{u \in [0, 1] : |g(u) - h(u)| > \epsilon\}$. Show that $d_\epsilon$ does not satisfy the triangle inequality.

**4.   Lévy metric.** For two distribution functions $F$ and $G$, let

$$d(F, G) = \inf\{\delta > 0 : F(x - \delta) - \delta \leq G(x) \leq F(x + \delta) + \delta \text{ for all } x \in \mathbb{R}\}.$$

Show that $d$ is a metric on the space of distribution functions.

**5.**   Find random variables $X, X_1, X_2, \ldots$ such that $\mathbb{E}(|X_n - X|^2) \to 0$ as $n \to \infty$, but $\mathbb{E}|X_n| = \infty$ for all $n$.

## 7.2  Modes of convergence

There are four principal ways of interpreting the statement '$X_n \to X$ as $n \to \infty$'. Three of these are related to (7.1.4), (7.1.5), and (7.1.6), and the fourth is already familiar to us.

**(1) Definition.** Let $X, X_1, X_2, \ldots$ be random variables on some probability space $(\Omega, \mathcal{F}, \mathbb{P})$. We say:

(a) $X_n \to X$ **almost surely**, written $X_n \xrightarrow{\text{a.s.}} X$, if $\{\omega \in \Omega : X_n(\omega) \to X(\omega) \text{ as } n \to \infty\}$ is an event whose probability is 1,

(b) $X_n \to X$ **in $r$th mean**, where $r \geq 1$, written $X_n \xrightarrow{r} X$, if $\mathbb{E}|X_n^r| < \infty$ for all $n$ and

$$\mathbb{E}(|X_n - X|^r) \to 0 \quad \text{as} \quad n \to \infty,$$

(c) $X_n \to X$ **in probability**, written $X_n \xrightarrow{\text{P}} X$, if

$$\mathbb{P}(|X_n - X| > \epsilon) \to 0 \quad \text{as} \quad n \to \infty \quad \text{for all } \epsilon > 0,$$

(d) $X_n \to X$ **in distribution**, written† $X_n \xrightarrow{\text{D}} X$, if

$$\mathbb{P}(X_n \leq x) \to \mathbb{P}(X \leq x) \quad \text{as} \quad n \to \infty$$

for all points $x$ at which the function $F_X(x) = \mathbb{P}(X \leq x)$ is continuous.

It is appropriate to make some remarks about the four sections of this potentially bewildering definition.

(a) The natural adaptation of Definition (7.1.4) is to say that $X_n \to X$ *pointwise* if the set $A = \{\omega \in \Omega : X_n(\omega) \to X(\omega) \text{ as } n \to \infty\}$ satisfies $A = \Omega$. Such a condition is of little interest to probabilists since it contains no reference to probabilities. In part (a) of (1) we do not require that $A$ is the whole of $\Omega$, but rather that its complement $A^c$ is a null set. There are several notations for this mode of convergence, and we shall use these later. They include

$$X_n \to X \text{ almost everywhere, or } X_n \xrightarrow{\text{a.e.}} X,$$
$$X_n \to X \text{ with probability 1, or } X_n \to X \text{ w.p.1.}$$

---

†Many authors avoid this notation since convergence in distribution pertains only to the *distribution function* of $X$ and not to the variable $X$ itself. We use it here for the sake of uniformity of notation, but refer the reader to note (d) below.

(b) It is easy to check by Minkowski's inequality (4.14.27) that

$$\|Y\|_r = (\mathbb{E}|Y^r|)^{1/r} = \left( \int |y|^r \, dF_Y \right)^{1/r}$$

defines a norm on the collection of random variables with finite $r$th moment, for any value of $r \geq 1$. Rewrite Definition (7.1.5) with this norm to obtain Definition (1b). Here we shall only consider positive integral values of $r$, though the subsequent theory can be extended without difficulty to deal with any real $r$ not smaller than 1. Of most use are the values $r = 1$ and $r = 2$, in which cases we write respectively

$$X_n \xrightarrow{1} X, \text{ or } X_n \to X \text{ in mean, or l.i.m. } X_n = X,$$

$$X_n \xrightarrow{2} X, \text{ or } X_n \to X \text{ in mean square, or } X_n \xrightarrow{\text{m.s.}} X.$$

(c) The functions of Definition (7.1.6) had a common domain $[0, 1]$; the $X_n$ have a common domain $\Omega$, and the distance function $d_\epsilon$ is naturally adapted to become

$$d_\epsilon(Y, Z) = \mathbb{P}(|Y - Z| > \epsilon) = \int_E d\mathbb{P}$$

where $E = \{\omega \in \Omega : |Y(\omega) - Z(\omega)| > \epsilon\}$. This notation will be familiar to those readers with knowledge of the abstract integral of Section 5.6.

(d) We have seen this already in Section 5.9 where we discussed the continuity condition. Further examples of convergence in distribution are to be found in Chapter 6, where we saw, for example, that an irreducible ergodic Markov chain converges in distribution to its unique stationary distribution. Convergence in distribution is also termed *weak convergence* or *convergence in law*. Note that if $X_n \xrightarrow{\text{D}} X$ then $X_n \xrightarrow{\text{D}} X'$ for any $X'$ which has the same distribution as $X$.

It is no surprise to learn that the four modes of convergence are not equivalent to each other. You may guess after some reflection that convergence in distribution is the weakest, since it is a condition only on the *distribution functions* of the $X_n$; it contains no reference to the *sample space* $\Omega$ and no information about, say, the dependence or independence of the $X_n$. The following example is a partial confirmation of this.

(2) **Example.** Let $X$ be a Bernoulli variable taking values 0 and 1 with equal probability $\frac{1}{2}$. Let $X_1, X_2, \ldots$ be identical random variables given by $X_n = X$ for all $n$. The $X_n$ are certainly not independent, but $X_n \xrightarrow{\text{D}} X$. Let $Y = 1 - X$. Clearly $X_n \xrightarrow{\text{D}} Y$ also, since $X$ and $Y$ have the same distribution. However, $X_n$ cannot converge to $Y$ in any other mode because $|X_n - Y| = 1$ always.                                                                                    ●

**Cauchy convergence.** As in the case of sequences of real numbers, it is often convenient to work with a definition of convergence which does not make explicit reference to the limit. For example, we say that the sequence $\{X_n : n \geq 1\}$ of random variables on the probability space $(\Omega, \mathcal{F}, \mathbb{P})$ is *almost surely Cauchy convergent* if the set of points $\omega$ of the sample space for which the real sequence $\{X_n(\omega) : n \geq 1\}$ is Cauchy convergent is an event having probability 1, which is to say that

$$\mathbb{P}\Big( \{\omega \in \Omega : X_m(\omega) - X_n(\omega) \to 0 \text{ as } m, n \to \infty\} \Big) = 1.$$

(See Appendix I for a brief discussion of the Cauchy convergence of a sequence of real numbers.) Now, a sequence of reals converges if and only if it is Cauchy convergent. Thus, for any $\omega \in \Omega$, the real sequence $\{X_n(\omega) : n \geq 1\}$ converges if and only if it is Cauchy convergent, implying that $\{X_n : n \geq 1\}$ converges almost surely if and only if it is almost surely Cauchy convergent. Other modes of Cauchy convergence appear in Exercise (7.3.1) and Problem (7.11.11).

Here is the chart of implications between the modes of convergence. Learn it well. Statements such as

$$(X_n \xrightarrow{P} X) \Rightarrow (X_n \xrightarrow{D} X)$$

mean that any sequence which converges in probability also converges in distribution to the same limit.

**(3) Theorem.** *The following implications hold*:

$$(X_n \xrightarrow{a.s.} X) \searrow$$
$$\qquad\qquad (X_n \xrightarrow{P} X) \Rightarrow (X_n \xrightarrow{D} X)$$
$$(X_n \xrightarrow{r} X) \nearrow$$

*for any $r \geq 1$. Also, if $r > s \geq 1$ then*

$$(X_n \xrightarrow{r} X) \Rightarrow (X_n \xrightarrow{s} X).$$

*No other implications hold in general†.*

The four basic implications of this theorem are of the general form 'if $A$ holds, then $B$ holds'. The converse implications are false in general, but become true if certain extra conditions are imposed; such partial converses take the form 'if $B$ holds together with $C$, then $A$ holds'. These two types of statement are sometimes said to be of the 'Abelian' and 'Tauberian' types, respectively; these titles are derived from the celebrated theory of the summability of series. Usually, there are many possible choices for appropriate sets $C$ of extra conditions, and it is often difficult to establish attractive 'corrected converses'.

**(4) Theorem.**
(a) If $X_n \xrightarrow{D} c$, where $c$ is constant, then $X_n \xrightarrow{P} c$.
(b) If $X_n \xrightarrow{P} X$ and $\mathbb{P}(|X_n| \leq k) = 1$ for all $n$ and some $k$, then $X_n \xrightarrow{r} X$ for all $r \geq 1$.
(c) If $P_n(\epsilon) = \mathbb{P}(|X_n - X| > \epsilon)$ satisfies $\sum_n P_n(\epsilon) < \infty$ for all $\epsilon > 0$, then $X_n \xrightarrow{a.s.} X$.

You should become well acquainted with Theorems (3) and (4). The proofs follow as a series of lemmas. These lemmas contain some other relevant and useful results.

Consider briefly the first and principal part of Theorem (3). We may already anticipate some way of showing that convergence in probability implies convergence in distribution, since both modes involve probabilities of the form $\mathbb{P}(Y \leq y)$ for some random variable $Y$ and real $y$. The other two implications require intermediate steps. Specifically, the relation between convergence in $r$th mean and convergence in probability requires a link between expectations and distributions. We have to move very carefully in this context; even apparently 'natural'

†But see (14).

statements may be false. For example, if $X_n \xrightarrow{\text{a.s.}} X$ (and therefore $X_n \xrightarrow{\text{P}} X$ also) then it does *not* necessarily follow that $\mathbb{E}X_n \to \mathbb{E}X$ (see (9) for an instance of this); this matter is explored fully in Section 7.10. The proof of the appropriate stage of Theorem (3) requires Markov's inequality (7).

**(5) Lemma.** *If $X_n \xrightarrow{\text{P}} X$ then $X_n \xrightarrow{\text{D}} X$. The converse assertion fails in general†.*

**Proof.** Suppose $X_n \xrightarrow{\text{P}} X$ and write

$$F_n(x) = \mathbb{P}(X_n \leq x), \quad F(x) = \mathbb{P}(X \leq x),$$

for the distribution functions of $X_n$ and $X$ respectively. If $\epsilon > 0$,

$$F_n(x) = \mathbb{P}(X_n \leq x) = \mathbb{P}(X_n \leq x, \ X \leq x + \epsilon) + \mathbb{P}(X_n \leq x, \ X > x + \epsilon)$$
$$\leq F(x + \epsilon) + \mathbb{P}(|X_n - X| > \epsilon).$$

Similarly,

$$F(x - \epsilon) = \mathbb{P}(X \leq x - \epsilon) = \mathbb{P}(X \leq x - \epsilon, \ X_n \leq x) + \mathbb{P}(X \leq x - \epsilon, \ X_n > x)$$
$$\leq F_n(x) + \mathbb{P}(|X_n - X| > \epsilon).$$

Thus

$$F(x - \epsilon) - \mathbb{P}(|X_n - X| > \epsilon) \leq F_n(x) \leq F(x + \epsilon) + \mathbb{P}(|X_n - X| > \epsilon).$$

Let $n \to \infty$ to obtain

$$F(x - \epsilon) \leq \liminf_{n\to\infty} F_n(x) \leq \limsup_{n\to\infty} F_n(x) \leq F(x + \epsilon)$$

for all $\epsilon > 0$. If $F$ is continuous at $x$ then

$$F(x - \epsilon) \uparrow F(x) \quad \text{and} \quad F(x + \epsilon) \downarrow F(x) \quad \text{as} \quad \epsilon \downarrow 0,$$

and the result is proved. Example (2) shows that the converse is false. ∎

**(6) Lemma.**

(a) *If $r > s \geq 1$ and $X_n \xrightarrow{r} X$ then $X_n \xrightarrow{s} X$.*

(b) *If $X_n \xrightarrow{1} X$ then $X_n \xrightarrow{\text{P}} X$.*

*The converse assertions fail in general.*

This includes the fact that convergence in $r$th mean implies convergence in probability. Here is a useful inequality which we shall use in the proof of this lemma.

**(7) Lemma. Markov's inequality.** *If $X$ is any random variable with finite mean then*

$$\mathbb{P}(|X| \geq a) \leq \frac{\mathbb{E}|X|}{a} \quad \text{for any } a > 0.$$

---

†But see (14).

**Proof.** Let $A = \{|X| \geq a\}$. Then $|X| \geq aI_A$ where $I_A$ is the indicator function of $A$. Take expectations to obtain the result. ∎

**Proof of Lemma (6).**

(a) By the result of Problem (4.14.28),

$$\left[\mathbb{E}(|X_n - X|^s)\right]^{1/s} \leq \left[\mathbb{E}(|X_n - X|^r)\right]^{1/r}$$

and the result follows immediately. To see that the converse fails, define an independent sequence $\{X_n\}$ by

**(8)**
$$X_n = \begin{cases} n & \text{with probability } n^{-\frac{1}{2}(r+s)}, \\ 0 & \text{with probability } 1 - n^{-\frac{1}{2}(r+s)}. \end{cases}$$

It is an easy *exercise* to check that

$$\mathbb{E}|X_n^s| = n^{\frac{1}{2}(s-r)} \to 0, \qquad \mathbb{E}|X_n^r| = n^{\frac{1}{2}(r-s)} \to \infty.$$

(b) By Markov's inequality (7),

$$\mathbb{P}(|X_n - X| > \epsilon) \leq \frac{\mathbb{E}|X_n - X|}{\epsilon} \qquad \text{for all } \epsilon > 0$$

and the result follows immediately. To see that the converse fails, define an independent sequence $\{X_n\}$ by

**(9)**
$$X_n = \begin{cases} n^3 & \text{with probability } n^{-2}, \\ 0 & \text{with probability } 1 - n^{-2}. \end{cases}$$

Then $\mathbb{P}(|X| > \epsilon) = n^{-2}$ for all large $n$, and so $X_n \xrightarrow{\text{P}} 0$. However, $\mathbb{E}|X_n| = n \to \infty$. ∎

**(10) Lemma.** *Let $A_n(\epsilon) = \{|X_n - X| > \epsilon\}$ and $B_m(\epsilon) = \bigcup_{n \geq m} A_n(\epsilon)$. Then:*

(a) $X_n \xrightarrow{\text{a.s.}} X$ *if and only if* $\mathbb{P}(B_m(\epsilon)) \to 0$ *as* $m \to \infty$, *for all* $\epsilon > 0$,

(b) $X_n \xrightarrow{\text{a.s.}} X$ *if* $\sum_n \mathbb{P}(A_n(\epsilon)) < \infty$ *for all* $\epsilon > 0$,

(c) *if* $X_n \xrightarrow{\text{a.s.}} X$ *then* $X_n \xrightarrow{\text{P}} X$, *but the converse fails in general.*

**Proof.**

(a) Let $C = \{\omega \in \Omega : X_n(\omega) \to X(\omega) \text{ as } n \to \infty\}$ and let

$$A(\epsilon) = \left\{\omega \in \Omega : \omega \in A_n(\epsilon) \text{ for infinitely many values of } n\right\} = \bigcap_m \bigcup_{n=m}^{\infty} A_n(\epsilon).$$

Now $X_n(\omega) \to X(\omega)$ if and only if $\omega \notin A(\epsilon)$ for all $\epsilon > 0$. Hence $\mathbb{P}(C) = 1$ implies $\mathbb{P}(A(\epsilon)) = 0$ for all $\epsilon > 0$. On the other hand, if $\mathbb{P}(A(\epsilon)) = 0$ for all $\epsilon > 0$, then

$$\mathbb{P}(C^c) = \mathbb{P}\left(\bigcup_{\epsilon > 0} A(\epsilon)\right) = \mathbb{P}\left(\bigcup_{m=1}^{\infty} A(m^{-1})\right) \qquad \text{since } A(\epsilon) \subseteq A(\epsilon') \text{ if } \epsilon \geq \epsilon'$$

$$\leq \sum_{m=1}^{\infty} \mathbb{P}(A(m^{-1})) = 0.$$

It follows that $\mathbb{P}(C) = 1$ if and only if $\mathbb{P}(A(\epsilon)) = 0$ for all $\epsilon > 0$.

In addition, $\{B_m(\epsilon) : m \geq 1\}$ is a decreasing sequence of events with limit $A(\epsilon)$ (see Problem (1.8.16)), and therefore $\mathbb{P}(A(\epsilon)) = 0$ if and only if $\mathbb{P}(B_m(\epsilon)) \to 0$ as $m \to \infty$.

(b) From the definition of $B_m(\epsilon)$,

$$\mathbb{P}(B_m(\epsilon)) \leq \sum_{n=m}^{\infty} \mathbb{P}(A_n(\epsilon))$$

and so $\mathbb{P}(B_m(\epsilon)) \to 0$ as $m \to \infty$ whenever $\sum_n \mathbb{P}(A_n(\epsilon)) < \infty$.

(c) We have that $A_n(\epsilon) \subseteq B_n(\epsilon)$, and therefore $\mathbb{P}(|X_n - X| > \epsilon) = \mathbb{P}(A_n(\epsilon)) \to 0$ whenever $\mathbb{P}(B_n(\epsilon)) \to 0$. To see that the converse fails, define an independent sequence $\{X_n\}$ by

$$(11) \qquad\qquad X_n = \begin{cases} 1 & \text{with probability } n^{-1}, \\ 0 & \text{with probability } 1 - n^{-1}. \end{cases}$$

Clearly $X_n \xrightarrow{P} 0$. However, if $0 < \epsilon < 1$,

$$\mathbb{P}(B_m(\epsilon)) = 1 - \lim_{r \to \infty} \mathbb{P}\big(X_n = 0 \text{ for all } n \text{ such that } m \leq n \leq r\big) \quad \text{by Theorem (1.3.5)}$$

$$= 1 - \left(1 - \frac{1}{m}\right)\left(1 - \frac{1}{m+1}\right)\cdots \qquad\qquad\qquad \text{by independence}$$

$$= 1 - \lim_{M \to \infty}\left(\frac{m-1}{m} \frac{m}{m+1} \frac{m+1}{m+2} \cdots \frac{M}{M+1}\right)$$

$$= 1 - \lim_{M \to \infty} \frac{m-1}{M+1} = 1 \qquad\qquad\qquad\qquad \text{for all } m,$$

and so $\{X_n\}$ does not converge almost surely.                                                               ∎

**(12) Lemma.** *There exist sequences which*:
  (a) *converge almost surely but not in mean*,
  (b) *converge in mean but not almost surely*.

**Proof.**
  (a) Consider Example (9). Use (10b) to show that $X_n \xrightarrow{\text{a.s.}} 0$.
  (b) Consider Example (11).                                                                                      ∎

This completes the proof of Theorem (3), and we move to Theorem (4).

**Proof of Theorem (4).**
  (a) We have that

$$\mathbb{P}(|X_n - c| > \epsilon) = \mathbb{P}(X_n < c - \epsilon) + \mathbb{P}(X_n > c + \epsilon) \to 0 \quad \text{if} \quad X_n \xrightarrow{D} c.$$

  (b) If $X_n \xrightarrow{P} X$ and $\mathbb{P}(|X_n| \leq k) = 1$ then $\mathbb{P}(|X| \leq k) = 1$ also, since

$$\mathbb{P}(|X| \leq k + \epsilon) = \lim_{n \to \infty} \mathbb{P}(|X_n| \leq k + \epsilon) = 1$$

for all $\epsilon > 0$. Now, let $A_n(\epsilon) = \{|X_n - X| > \epsilon\}$, with complement $A_n(\epsilon)^c$. Then

$$|X_n - X|^r \le \epsilon^r I_{A_n(\epsilon)^c} + (2k)^r I_{A_n(\epsilon)}$$

with probability 1. Take expectations to obtain

$$\mathbb{E}(|X_n - X|^r) \le \epsilon^r + [(2k)^r - \epsilon^r]\mathbb{P}(A_n(\epsilon)) \to \epsilon^r \quad \text{as} \quad n \to \infty.$$

Let $\epsilon \downarrow 0$ to obtain that $X_n \overset{r}{\to} X$.
 (c) This is just (10b).                                                              ∎

Note that any sequence $\{X_n\}$ which satisfies $X_n \overset{P}{\to} X$ necessarily contains a subsequence $\{X_{n_i} : 1 \le i < \infty\}$ which converges almost surely.

**(13) Theorem.** *If* $X_n \overset{P}{\to} X$, *there exists a non-random increasing sequence of integers* $n_1, n_2, \ldots$ *such that* $X_{n_i} \overset{\text{a.s.}}{\longrightarrow} X$ *as* $i \to \infty$.

**Proof.** Since $X_n \overset{P}{\to} X$, we have that

$$\mathbb{P}(|X_n - X| > \epsilon) \to 0 \quad \text{as} \quad n \to \infty, \quad \text{for all } \epsilon > 0.$$

Pick an increasing sequence $n_1, n_2, \ldots$ of positive integers such that

$$\mathbb{P}(|X_{n_i} - X| > i^{-1}) \le i^{-2}.$$

For any $\epsilon > 0$,

$$\sum_{i > \epsilon^{-1}} \mathbb{P}(|X_{n_i} - X| > \epsilon) \le \sum_{i > \epsilon^{-1}} \mathbb{P}(|X_{n_i} - X| > i^{-1}) < \infty$$

and the result follows from (10b).                                                    ∎

We have seen that convergence in distribution is the weakest mode of convergence since it involves distribution functions only and makes no reference to an underlying probability space (see Theorem (5.9.4) for an equivalent formulation of convergence in distribution which involves distribution functions alone). However, assertions of the form '$X_n \overset{D}{\to} X$' (or equivalently '$F_n \to F$', where $F_n$ and $F$ are the distribution functions of $X_n$ and $X$) have important and useful representations in terms of almost sure convergence.

**(14) Skorokhod's representation theorem.** *If* $\{X_n\}$ *and* $X$, *with distribution functions* $\{F_n\}$ *and* $F$, *are such that*

$$X_n \overset{D}{\to} X \ (\text{or, equivalently, } F_n \to F) \text{ as } n \to \infty$$

*then there exists a probability space* $(\Omega', \mathcal{F}', \mathbb{P}')$ *and random variables* $\{Y_n\}$ *and* $Y$, *mapping* $\Omega'$ *into* $\mathbb{R}$, *such that:*
 (a) $\{Y_n\}$ *and* $Y$ *have distribution functions* $\{F_n\}$ *and* $F$,
 (b) $Y_n \overset{\text{a.s.}}{\longrightarrow} Y$ *as* $n \to \infty$.

Therefore, although $X_n$ may fail to converge to $X$ in any mode other than in distribution, there exists a sequence $\{Y_n\}$ such that $Y_n$ is distributed identically to $X_n$ for every $n$, which converges almost surely to a copy of $X$. The proof is elementary.

**Proof.** Let $\Omega' = (0, 1)$, let $\mathcal{F}'$ be the Borel $\sigma$-field generated by the intervals of $\Omega'$ (see the discussion at the end of Section 4.1), and let $\mathbb{P}'$ be the probability measure induced on $\mathcal{F}'$ by the requirement that, for any interval $I = (a, b) \subseteq \Omega'$, $\mathbb{P}'(I) = (b - a)$; $\mathbb{P}'$ is called *Lebesgue measure*. For $\omega \in \Omega'$, define

$$Y_n(\omega) = \inf\{x : \omega \leq F_n(x)\},$$
$$Y(\omega) = \inf\{x : \omega \leq F(x)\}.$$

Note that $Y_n$ and $Y$ are essentially the inverse functions of $F_n$ and $F$ since

**(15)**
$$\omega \leq F_n(x) \Leftrightarrow Y_n(\omega) \leq x,$$
$$\omega \leq F(x) \Leftrightarrow Y(\omega) \leq x.$$

It follows immediately that $Y_n$ and $Y$ satisfy (14a) since, for example, from (15)

$$\mathbb{P}'(Y \leq y) = \mathbb{P}'\big((0, F(y)]\big) = F(y).$$

To show (14b), proceed as follows. Given $\epsilon > 0$ and $\omega \in \Omega'$, pick a point $x$ of continuity of $F$ such that

$$Y(\omega) - \epsilon < x < Y(\omega).$$

We have by (15) that $F(x) < \omega$. However, $F_n(x) \to F(x)$ as $n \to \infty$ and so $F_n(x) < \omega$ for all large $n$, giving that

$$Y(\omega) - \epsilon < x < Y_n(\omega) \quad \text{for all large } n;$$

now let $n \to \infty$ and $\epsilon \downarrow 0$ to obtain

**(16)**
$$\liminf_{n \to \infty} Y_n(\omega) \geq Y(\omega) \quad \text{for all } \omega.$$

Finally, if $\omega < \omega' < 1$, pick a point $x$ of continuity of $F$ such that

$$Y(\omega') < x < Y(\omega') + \epsilon.$$

We have by (15) that $\omega < \omega' \leq F(x)$, and so $\omega < F_n(x)$ for all large $n$, giving that

$$Y_n(\omega) \leq x < Y(\omega') + \epsilon \quad \text{for all large } n;$$

now let $n \to \infty$ and $\epsilon \downarrow 0$ to obtain

**(17)**
$$\limsup_{n \to \infty} Y_n(\omega) \leq Y(\omega') \quad \text{whenever } \omega < \omega'.$$

Combine this with (16) to see that $Y_n(\omega) \to Y(\omega)$ for all points $\omega$ of continuity of $Y$. However, $Y$ is monotone non-decreasing and so that set $D$ of discontinuities of $Y$ is countable; thus $\mathbb{P}'(D) = 0$ and the proof is complete. ∎

We complete this section with two elementary applications of the representation theorem (14). The results in question are standard, but the usual classical proofs are tedious.

**(18) Theorem.** *If* $X_n \xrightarrow{D} X$ *and* $g : \mathbb{R} \to \mathbb{R}$ *is continuous then* $g(X_n) \xrightarrow{D} g(X)$.

**Proof.** Let $\{Y_n\}$ and $Y$ be given as in (14). By the continuity of $g$,

$$\{\omega : g(Y_n(\omega)) \to g(Y(\omega))\} \supseteq \{\omega : Y_n(\omega) \to Y(\omega)\},$$

and so $g(Y_n) \xrightarrow{a.s.} g(Y)$ as $n \to \infty$. Therefore $g(Y_n) \xrightarrow{D} g(Y)$; however, $g(Y_n)$ and $g(Y)$ have the same distributions as $g(X_n)$ and $g(X)$.  ∎

**(19) Theorem.** *The following three statements are equivalent.*

(a) $X_n \xrightarrow{D} X$.

(b) $\mathbb{E}(g(X_n)) \to \mathbb{E}(g(X))$ *for all bounded continuous functions* $g$.

(c) $\mathbb{E}(g(X_n)) \to \mathbb{E}(g(X))$ *for all functions* $g$ *of the form* $g(x) = f(x)I_{[a,b]}(x)$ *where* $f$ *is continuous on* $[a, b]$ *and* $a$ *and* $b$ *are points of continuity of the distribution function of the random variable* $X$.

Condition (b) is usually taken as the definition of what is called *weak convergence*. It is not important in (c) that $g$ be continuous on the *closed* interval $[a, b]$. The same proof is valid if $g$ in part (c) is of the form $g(x) = f(x)I_{(a,b)}(x)$ where $f$ is bounded and continuous on the open interval $(a, b)$. Theorem (19) is part of the so-called 'portmanteau theorem'; the full theorem supplies several further conditions that are equivalent to convergence in distribution.

**Proof.** First we prove that (a) implies (b). Suppose that $X_n \xrightarrow{D} X$ and $g$ is bounded and continuous. By the Skorokhod representation theorem (14), there exist random variables $Y, Y_1, Y_2, \ldots$ having the same distributions as $X, X_1, X_2, \ldots$ and such that $Y_n \xrightarrow{a.s.} Y$. Therefore $g(Y_n) \xrightarrow{a.s.} g(Y)$ by the continuity of $g$, and furthermore the $g(Y_n)$ are uniformly bounded random variables. We apply the bounded convergence theorem (5.6.12) to deduce that $\mathbb{E}(g(Y_n)) \to \mathbb{E}(g(Y))$, and (b) follows since $\mathbb{E}(g(Y_n)) = \mathbb{E}(g(X_n))$ and $\mathbb{E}(g(Y)) = \mathbb{E}(g(X))$.

We write $C$ for the set of points of continuity of $F_X$. Now $F_X$ is monotone and has therefore at most countably many points of discontinuity; hence $C^c$ is countable.

Suppose now that (b) holds. For (c), it suffices to prove that $\mathbb{E}(h(X_n)) \to \mathbb{E}(h(X))$ for all functions $h$ of the form $h(x) = f(x)I_{(-\infty,b]}(x)$, where $f$ is bounded and continuous, and $b \in C$; the general result follows by an exactly analogous argument. Suppose then that $h(x) = f(x)I_{(-\infty,b]}(x)$ as prescribed. The idea is to approximate to $h$ by a continuous function. For $\delta > 0$, define the continuous functions $h'$ and $h''$ by

$$h'(x) = \begin{cases} h(x) & \text{if } x \notin (b, b+\delta), \\ \left(1 + \dfrac{b-x}{\delta}\right)h(b) & \text{if } x \in (b, b+\delta), \end{cases}$$

$$h''(x) = \begin{cases} \left(1 + \dfrac{x-b}{\delta}\right)h(b) & \text{if } x \in (b-\delta, b), \\ \left(1 + \dfrac{b-x}{\delta}\right)h(b) & \text{if } x \in [b, b+\delta), \\ 0 & \text{otherwise.} \end{cases}$$

It may be helpful to draw a picture. Now

$$\left|\mathbb{E}\big(h(X_n) - h'(X_n)\big)\right| \leq \left|\mathbb{E}(h''(X_n))\right|, \qquad \left|\mathbb{E}\big(h(X) - h'(X)\big)\right| \leq \left|\mathbb{E}(h''(X))\right|$$

so that

$$\left|\mathbb{E}(h(X_n)) - \mathbb{E}(h(X))\right| \leq \left|\mathbb{E}(h''(X_n))\right| + \left|\mathbb{E}(h''(X))\right| + \left|\mathbb{E}(h'(X_n)) - \mathbb{E}(h'(X))\right|$$
$$\rightarrow 2\left|\mathbb{E}(h''(X))\right| \quad \text{as} \quad n \rightarrow \infty$$

by assumption (b). We now observe that

$$\left|\mathbb{E}(h''(X))\right| \leq |h(b)|\mathbb{P}(b - \delta < X < b + \delta) \rightarrow 0 \quad \text{as} \quad \delta \downarrow 0,$$

by the assumption that $\mathbb{P}(X = b) = 0$. Hence (c) holds.

Suppose finally that (c) holds, and that $b$ is such that $\mathbb{P}(X = b) = 0$. By considering the function $f(x) = 1$ for all $x$, we have that, if $a \in C$,

**(20)**    $\mathbb{P}(X_n \leq b) \geq \mathbb{P}(a \leq X_n \leq b) \rightarrow \mathbb{P}(a \leq X \leq b) \quad \text{as } n \rightarrow \infty$
$$\rightarrow \mathbb{P}(X \leq b) \qquad \text{as } a \rightarrow -\infty \text{ through } C.$$

A similar argument, but taking the limit in the other direction, yields for $b' \in C$

**(21)**            $\mathbb{P}(X_n \geq b') \geq \mathbb{P}(b' \leq X_n \leq c) \qquad \text{if } c \geq b'$
$$\rightarrow \mathbb{P}(b' \leq X \leq c) \qquad \text{as } n \rightarrow \infty, \text{ if } c \in C$$
$$\rightarrow \mathbb{P}(X \geq b') \qquad \text{as } c \rightarrow \infty \text{ through } C.$$

It follows from (20) and (21) that, if $b, b' \in C$ and $b < b'$, then for any $\epsilon > 0$ there exists $N$ such that

$$\mathbb{P}(X \leq b) - \epsilon \leq \mathbb{P}(X_n \leq b) \leq \mathbb{P}(X_n < b') \leq \mathbb{P}(X < b') + \epsilon$$

for all $n \geq N$. Take the limits as $n \rightarrow \infty$ and $\epsilon \downarrow 0$, and $b' \downarrow b$ through $C$, in that order, to obtain that $\mathbb{P}(X_n \leq b) \rightarrow \mathbb{P}(X \leq b)$ as $n \rightarrow \infty$ if $b \in C$, the required result.  ∎

---

## Exercises for Section 7.2

**1.**   (a) Suppose $X_n \xrightarrow{r} X$ where $r \geq 1$. Show that $\mathbb{E}|X_n^r| \rightarrow \mathbb{E}|X^r|$.

(b) Suppose $X_n \xrightarrow{1} X$. Show that $\mathbb{E}(X_n) \rightarrow \mathbb{E}(X)$. Is the converse true?

(c) Suppose $X_n \xrightarrow{2} X$. Show that $\text{var}(X_n) \rightarrow \text{var}(X)$.

**2.   Dominated convergence.** Suppose $|X_n| \leq Z$ for all $n$, where $\mathbb{E}(Z) < \infty$. Prove that if $X_n \xrightarrow{P} X$ then $X_n \xrightarrow{1} X$.

**3.**   (a) Give a rigorous proof that $\mathbb{E}(XY) = \mathbb{E}(X)\mathbb{E}(Y)$ for any pair $X, Y$ of independent non-negative random variables on $(\Omega, \mathcal{F}, \mathbb{P})$ with finite means. [Hint: For $k \geq 0$, $n \geq 1$, define $X_n = k/n$ if $k/n \leq X < (k+1)/n$, and similarly for $Y_n$. Show that $X_n$ and $Y_n$ are independent, and $X_n \leq X$, and $Y_n \leq Y$. Deduce that $\mathbb{E}X_n \rightarrow \mathbb{E}X$ and $\mathbb{E}Y_n \rightarrow \mathbb{E}Y$, and also $\mathbb{E}(X_n Y_n) \rightarrow \mathbb{E}(XY)$.]

(b) Give an example to show that the product $XY$ of dependent random variables $X$, $Y$ with finite means may have $\mathbb{E}(XY) = \infty$.

**4.**    Show that convergence in distribution is equivalent to convergence with respect to the Lévy metric of Exercise (7.1.4).

**5.**    (a) Suppose that $X_n \xrightarrow{D} X$ and $Y_n \xrightarrow{P} c$, where $c$ is a constant. Show that $X_n Y_n \xrightarrow{D} cX$, and that $X_n/Y_n \xrightarrow{D} X/c$ if $c \neq 0$.

(b) Suppose that $X_n \xrightarrow{D} 0$ and $Y_n \xrightarrow{P} Y$, and let $g : \mathbb{R}^2 \to \mathbb{R}$ be continuous. Show that $g(X_n, Y_n) \xrightarrow{P} g(0, Y)$.

[These results are sometimes referred to as 'Slutsky's theorem(s)'.]

**6.**    Let $X_1, X_2, \ldots$ be random variables on the probability space $(\Omega, \mathcal{F}, \mathbb{P})$. Show that the set $A = \{\omega \in \Omega : \text{the sequence } X_n(\omega) \text{ converges}\}$ is an event (that is, lies in $\mathcal{F}$), and that there exists a random variable $X$ (that is, an $\mathcal{F}$-measurable function $X : \Omega \to \mathbb{R}$) such that $X_n(\omega) \to X(\omega)$ for $\omega \in A$.

**7.**    Let $\{X_n\}$ be a sequence of random variables, and let $\{c_n\}$ be a sequence of reals converging to the limit $c$. For convergence almost surely, in $r$th mean, in probability, and in distribution, show that the convergence of $X_n$ to $X$ entails the convergence of $c_n X_n$ to $cX$.

**8.**    Let $\{X_n\}$ be a sequence of independent random variables which converges in probability to the limit $X$. Show that $X$ is almost surely constant.

**9.**    **Convergence in total variation.**  The sequence of discrete random variables $X_n$, with mass functions $f_n$, is said to *converge in total variation* to $X$ with mass function $f$ if

$$\sum_x |f_n(x) - f(x)| \to 0 \quad \text{as} \quad n \to \infty.$$

Suppose $X_n \to X$ in total variation, and $u : \mathbb{R} \to \mathbb{R}$ is bounded. Show that $\mathbb{E}(u(X_n)) \to \mathbb{E}(u(X))$.

**10.** Let $\{X_r : r \geq 1\}$ be independent Poisson variables with respective parameters $\{\lambda_r : r \geq 1\}$. Show that $\sum_{r=1}^{\infty} X_r$ converges or diverges almost surely according as $\sum_{r=1}^{\infty} \lambda_r$ converges or diverges.

**11. Waiting for a coincidence.**  Let $X_1, X_2, \ldots$ be independent random variables, uniformly distributed on $\{1, 2, \ldots, m\}$, and let $I_m = \min\{r \geq 2 : X_r = X_s \text{ for some } s < r\}$ be the earliest index of a coincidence of values. Show that $I_m/\sqrt{m}$ converges in distribution as $m \to \infty$, with as limit the Rayleigh distribution with density function $f(x) = xe^{-\frac{1}{2}x^2}$, $x > 0$.

**12. Moments.**  The random variable $X$ has finite moments of all orders, and satisfies $\mathbb{P}(X > 0) = 1$, $\mathbb{P}(X > x) > 0$ for $x > 0$.

(a) Show that $c = \sum_{n=0}^{\infty} 1/\mathbb{E}(X^n)$ satisfies $c < \infty$.

(b) Let $M$ take values in $\{1, 2, \ldots\}$ with mass function $f(m) = 1/(c\mathbb{E}(X^m))$ for $m \geq 1$. Show that $\mathbb{E}(x^M) < \infty$ for all $x \geq 0$, while $\mathbb{E}(X^M) = \infty$.

---

# 7.3  Some ancillary results

Next we shall develop some refinements of the methods of the last section; these will prove to be of great value later. There are two areas of interest. The first deals with inequalities and generalizes Markov's inequality, Lemma (7.2.7). The second deals with infinite families of events and the Borel–Cantelli lemmas; it is related to the result of Theorem (7.2.4c).

Markov's inequality is easily generalized.

**(1) Theorem.** *Let $h : \mathbb{R} \to [0, \infty)$ be a non-negative function. Then*

$$\mathbb{P}(h(X) \geq a) \leq \frac{\mathbb{E}(h(X))}{a} \quad \text{for all } a > 0.$$

**Proof.** Denote by $A$ the event $\{h(X) \geq a\}$, so that $h(X) \geq aI_A$. Take expectations to obtain the result. ∎

We note some special cases of this.

**(2) Example. Markov's inequality.** Set $h(x) = |x|$. ●

**(3) Example†. Chebyshov's inequality.** Set $h(x) = x^2$ to obtain

$$\mathbb{P}(|X| \geq a) \leq \frac{\mathbb{E}(X^2)}{a^2} \quad \text{if } a > 0.$$

This inequality was also discovered by Bienaymé and others. ●

**(4) Example.** More generally, let $g : [0, \infty) \to [0, \infty)$ be a strictly increasing non-negative function, and set $h(x) = g(|x|)$ to obtain

$$\mathbb{P}(|X| \geq a) \leq \frac{\mathbb{E}(g(|X|))}{g(a)} \quad \text{if } a > 0.$$ ●

Theorem (1) provides an upper bound for the probability $\mathbb{P}(h(X) \geq a)$. Lower bounds are harder to find in general, but pose no difficulty in the case when $h$ is a uniformly bounded function.

**(5) Theorem.** *If $h : \mathbb{R} \to [0, M]$ is a non-negative function taking values bounded by some number $M$, then*

$$\mathbb{P}(h(X) \geq a) \geq \frac{\mathbb{E}(h(X)) - a}{M - a} \quad \text{whenever } 0 \leq a < M.$$

**Proof.** Let $A = \{h(X) \geq a\}$ as before and note that $h(X) \leq MI_A + aI_{A^c}$. ●

The reader is left to apply this result to the special cases (2), (3), and (4). This is an appropriate moment to note three other important inequalities. Let $X$ and $Y$ be random variables.

**(6) Theorem. Hölder's inequality.** *If $p, q > 1$ and $p^{-1} + q^{-1} = 1$ then*

$$\mathbb{E}|XY| \leq (\mathbb{E}|X^p|)^{1/p}(\mathbb{E}|Y^q|)^{1/q}.$$

---

†Our transliteration of Чебышёв (Chebyshov) is at odds with common practice, but dispenses with the need for clairvoyance in pronunciation. It was explained by Eugene Dynkin to the authors around 1980 that the dots over the letter ё are included in Russian works only if they are intended for children or foreigners.

**(7) Theorem. Minkowski's inequality.** *If $p \geq 1$ then*

$$\left[\mathbb{E}(|X+Y|^p)\right]^{1/p} \leq (\mathbb{E}|X^p|)^{1/p} + (\mathbb{E}|Y^p|)^{1/p}.$$

**Proof of (6) and (7).** You did these for Problem (4.14.27).    ∎

**(8) Theorem. $C_p$ inequality.** $\mathbb{E}(|X+Y|^p) \leq C_p\left[\mathbb{E}|X^p| + \mathbb{E}|Y^p|\right]$ *where $p > 0$ and*

$$C_p = \begin{cases} 1 & \text{if } 0 < p \leq 1, \\ 2^{p-1} & \text{if } p > 1. \end{cases}$$

**Proof.** It is not difficult to show that $|x+y|^p \leq C_p[|x|^p + |y|^p]$ for all $x, y \in \mathbb{R}$ and $p > 0$. Now complete the details.    ∎

Inequalities (6) and (7) assert that

$$\|XY\|_1 \leq \|X\|_p \|Y\|_q \qquad \text{if } p^{-1} + q^{-1} = 1,$$
$$\|X+Y\|_p \leq \|X\|_p + \|Y\|_p \qquad \text{if } p \geq 1,$$

where $\| \cdot \|_p$ is the $L_p$ norm $\|X\|_p = (\mathbb{E}|X^p|)^{1/p}$.

Here is an application of these inequalities. It is related to the fact that if $x_n \to x$ and $y_n \to y$ then $x_n + y_n \to x + y$.

**(9) Theorem.**
(a) *If $X_n \overset{\text{a.s.}}{\longrightarrow} X$ and $Y_n \overset{\text{a.s.}}{\longrightarrow} Y$ then $X_n + Y_n \overset{\text{a.s.}}{\longrightarrow} X + Y$.*
(b) *If $X_n \overset{r}{\to} X$ and $Y_n \overset{r}{\to} Y$ then $X_n + Y_n \overset{r}{\to} X + Y$.*
(c) *If $X_n \overset{P}{\to} X$ and $Y_n \overset{P}{\to} Y$ then $X_n + Y_n \overset{P}{\to} X + Y$.*
(d) *It is not in general true that $X_n + Y_n \overset{D}{\to} X + Y$ whenever $X_n \overset{D}{\to} X$ and $Y_n \overset{D}{\to} Y$.*

**Proof.** *You* do it. You will need either (7) or (8) to prove part (b).    ∎

Theorem (7.2.4) contains a criterion for a sequence to converge almost surely. It is a special case of two very useful results called the 'Borel–Cantelli lemmas'. Let $A_1, A_2, \ldots$ be an infinite sequence of events from some probability space $(\Omega, \mathcal{F}, \mathbb{P})$. We shall often be interested in finding out how many of the $A_n$ occur. Recall (Problem (1.8.16)) that the event that infinitely many of the $A_n$ occur, sometimes written $\{A_n$ infinitely often$\}$ or $\{A_n$ i.o.$\}$, satisfies

$$\{A_n \text{ i.o.}\} = \limsup_{n \to \infty} A_n = \bigcap_n \bigcup_{m=n}^{\infty} A_m.$$

**(10) Theorem. Borel–Cantelli lemmas.** *Let $A = \bigcap_n \bigcup_{m=n}^{\infty} A_m$ be the event that infinitely many of the $A_n$ occur. Then:*
(a) $\mathbb{P}(A) = 0$ *if* $\sum_n \mathbb{P}(A_n) < \infty$,
(b) $\mathbb{P}(A) = 1$ *if* $\sum_n \mathbb{P}(A_n) = \infty$ *and* $A_1, A_2, \ldots$ *are independent events.*

It is easy to see that statement (b) is false if the assumption of independence is dropped. Just consider some event $E$ with $0 < \mathbb{P}(E) < 1$ and define $A_n = E$ for all $n$. Then $A = E$ and $\mathbb{P}(A) = \mathbb{P}(E)$.

**Proof.**
   (a) We have that $A \subseteq \bigcup_{m=n}^{\infty} A_m$ for all $n$, and so

$$\mathbb{P}(A) \leq \sum_{m=n}^{\infty} \mathbb{P}(A_m) \rightarrow 0 \quad \text{as} \quad n \rightarrow \infty,$$

whenever $\sum_n \mathbb{P}(A_n) < \infty$.
   (b) It is an easy *exercise* in set theory to check that

$$A^{\mathrm{c}} = \bigcup_n \bigcap_{m=n}^{\infty} A_m^{\mathrm{c}}.$$

However,

$$\mathbb{P}\left( \bigcap_{m=n}^{\infty} A_m^{\mathrm{c}} \right) = \lim_{r \to \infty} \mathbb{P}\left( \bigcap_{m=n}^{r} A_m^{\mathrm{c}} \right) \qquad \text{by Theorem (1.3.5)}$$

$$= \prod_{m=n}^{\infty} \left[1 - \mathbb{P}(A_m)\right] \qquad \text{by independence}$$

$$\leq \prod_{m=n}^{\infty} \exp[-\mathbb{P}(A_m)] \qquad \text{since } 1 - x \leq e^{-x} \text{ if } x \geq 0$$

$$= \exp\left( -\sum_{m=n}^{\infty} \mathbb{P}(A_m) \right) = 0$$

whenever $\sum_n \mathbb{P}(A_n) = \infty$. Thus

$$\mathbb{P}(A^{\mathrm{c}}) = \lim_{n \to \infty} \mathbb{P}\left( \bigcap_{m=n}^{\infty} A_m^{\mathrm{c}} \right) = 0,$$

giving $\mathbb{P}(A) = 1$ as required.                                                ∎

**(11) Example. Markov chains.** Let $\{X_n\}$ be a Markov chain with $X_0 = i$ for some state $i$. Let $A_n = \{X_n = i\}$ be the event that the chain returns to $i$ after $n$ steps. State $i$ is recurrent if and only if $\mathbb{P}(A_n \text{ i.o.}) = 1$. By the first Borel–Cantelli lemma,

$$\mathbb{P}(A_n \text{ i.o.}) = 0 \quad \text{if} \quad \sum_n \mathbb{P}(A_n) < \infty$$

and it follows that $i$ is transient if $\sum_n p_{ii}(n) < \infty$, which is part of an earlier result, Corollary (6.2.4). We cannot establish the converse by this method since the $A_n$ are not independent. ●

If the events $A_1, A_2, \ldots$ of Theorem (10) are independent then $\mathbb{P}(A)$ equals either 0 or 1 depending on whether or not $\sum \mathbb{P}(A_n)$ converges. This is an example of a general theorem called a 'zero–one law'. There are many such results, of which the following is a simple example.

**(12) Theorem. Zero–one law.** *Let $A_1, A_2, \ldots$ be a collection of events, and let $\mathcal{A}$ be the smallest $\sigma$-field of subsets of $\Omega$ which contains all of them. If $A \in \mathcal{A}$ is an event which is independent of the finite collection $A_1, A_2, \ldots, A_n$ for each value of $n$, then*

$$\text{either} \quad \mathbb{P}(A) = 0 \quad \text{or} \quad \mathbb{P}(A) = 1.$$

**Proof.** Roughly speaking, the assertion that $A$ belongs to $\mathcal{A}$ means that $A$ is definable in terms of $A_1, A_2, \ldots$. Examples of such events include $B_1, B_2$, and $B_3$ defined by

$$B_1 = A_7 \setminus A_9, \quad B_2 = A_3 \cup A_6 \cup A_9 \cup \cdots, \quad B_3 = \bigcup_n \bigcap_{m=n}^{\infty} A_m.$$

A standard result of measure theory asserts that if $A \in \mathcal{A}$ then there exists a sequence of events $\{C_n\}$ such that

**(13)** $$C_n \in \mathcal{A}_n \quad \text{and} \quad \mathbb{P}(A \bigtriangleup C_n) \to 0 \quad \text{as } n \to \infty$$

where $\mathcal{A}_n$ is the smallest $\sigma$-field which contains the finite collection $A_1, A_2, \ldots, A_n$. But $A$ is assumed independent of this collection, and so is independent of $C_n$ for all $n$. From (13),

**(14)** $$\mathbb{P}(A \cap C_n) \to \mathbb{P}(A).$$

However, by independence,

$$\mathbb{P}(A \cap C_n) = \mathbb{P}(A)\mathbb{P}(C_n) \to \mathbb{P}(A)^2$$

which may be combined with (14) to give $\mathbb{P}(A) = \mathbb{P}(A)^2$, and so $\mathbb{P}(A)$ is 0 or 1. ∎

Read on for another zero–one law. Let $X_1, X_2, \ldots$ be a collection of random variables on the probability space $(\Omega, \mathcal{F}, \mathbb{P})$. For any subcollection $\{X_i : i \in I\}$, write $\sigma(X_i : i \in I)$ for the smallest $\sigma$-field with respect to which each of the variables $X_i$ ($i \in I$) is measurable. This $\sigma$-field exists by the argument of Section 1.6. It contains events which are 'defined in terms of $\{X_i : i \in I\}$'. Let $\mathcal{H}_n = \sigma(X_{n+1}, X_{n+2}, \ldots)$. Then $\mathcal{H}_n \supseteq \mathcal{H}_{n+1} \supseteq \cdots$; write

$$\mathcal{H}_\infty = \bigcap_n \mathcal{H}_n.$$

$\mathcal{H}_\infty$ is called the *tail $\sigma$-field* of the $X_n$ and contains events such as

$$\{X_n > 0 \text{ i.o.}\}, \quad \left\{ \limsup_{n \to \infty} X_n = \infty \right\}, \quad \left\{ \sum_n X_n \text{ converges} \right\},$$

the definitions of which need never refer to any finite subcollection $\{X_1, X_2, \ldots, X_n\}$. Events in $\mathcal{H}_\infty$ are called *tail events*.

**(15) Theorem. Kolmogorov's zero–one law.** *If $X_1, X_2, \ldots$ are independent variables then all events $H \in \mathcal{H}_\infty$ satisfy either $\mathbb{P}(H) = 0$ or $\mathbb{P}(H) = 1$.*

Such a $\sigma$-field $\mathcal{H}_\infty$ is called *trivial* since it contains only null events and their complements. You may try to prove this theorem using the techniques in the proof of (12); it is not difficult.

**(16) Example.** Let $X_1, X_2, \ldots$ be independent random variables and let

$$
H_1 = \left\{ \omega \in \Omega : \sum_n X_n(\omega) \text{ converges} \right\},
$$

$$
H_2 = \left\{ \omega \in \Omega : \limsup_{n \to \infty} X_n(\omega) = \infty \right\}.
$$

Each $H_i$ has either probability 0 or probability 1. ●

We can associate many other random variables with the sequence $X_1, X_2, \ldots$; these include

$$
Y_1 = \tfrac{1}{2}(X_3 + X_6), \quad Y_2 = \limsup_{n \to \infty} X_n, \quad Y_3 = Y_1 + Y_2.
$$

We call such a variable $Y$ a *tail function* if it is $\mathcal{H}_\infty$-measurable, where $\mathcal{H}_\infty$ is the tail $\sigma$-field of the $X_n$. Roughly speaking, $Y$ is a tail function if its definition includes no essential reference to any finite subsequence $X_1, X_2, \ldots, X_n$. The random variables $Y_1$ and $Y_3$ are *not* tail functions; can you see why $Y_2$ *is* a tail function? More rigorously (see the discussion after Definition (2.1.3)) $Y$ is a tail function if and only if

$$
\{\omega \in \Omega : Y(\omega) \le y\} \in \mathcal{H}_\infty \quad \text{for all } y \in \mathbb{R}.
$$

Thus, if $\mathcal{H}_\infty$ is trivial then the distribution function $F_Y(y) = \mathbb{P}(Y \le y)$ takes the values 0 and 1 only. Such a function is the distribution function of a random variable which is constant (see Example (2.1.7)), and we have shown the following useful result.

**(17) Theorem.** *Let $Y$ be a tail function of the independent sequence $X_1, X_2, \ldots$. There exists $k$ satisfying $-\infty \le k \le \infty$ such that $\mathbb{P}(Y = k) = 1$.*

**Proof.** Let $k = \inf\{y : \mathbb{P}(Y \le y) = 1\}$, with the convention that the infimum of an empty set is $+\infty$. Then

$$
\mathbb{P}(Y \le y) = \begin{cases} 0 & \text{if } y < k, \\ 1 & \text{if } y \ge k. \end{cases} \quad ∎
$$

**(18) Example.** Let $X_1, X_2, \ldots$ be independent variables with partial sums $S_n = \sum_{i=1}^n X_i$. Then

$$
Z_1 = \liminf_{n \to \infty} \frac{1}{n} S_n, \quad Z_2 = \limsup_{n \to \infty} \frac{1}{n} S_n
$$

are almost surely constant (but possibly infinite). To see this, note that if $m \le n$ then

$$
\frac{1}{n} S_n = \frac{1}{n} \sum_{i=1}^m X_i + \frac{1}{n} \sum_{i=m+1}^n X_i = S_n(1) + S_n(2), \text{ say.}
$$

However, $S_n(1) \to 0$ pointwise as $n \to \infty$, and so $Z_1$ and $Z_2$ depend in no way upon the values of $X_1, X_2, \dots , X_m$. It follows that the event

$$\left\{ \frac{1}{n} S_n \text{ converges} \right\} = \{Z_1 = Z_2\}$$

has either probability 1 or probability 0. That is, $n^{-1} S_n$ converges either almost everywhere or almost nowhere; this was, of course, deducible from (15) since $\{Z_1 = Z_2\} \in \mathcal{H}_\infty$.   ●

---

## Exercises for Section 7.3

**1.**  (a) Suppose that $X_n \xrightarrow{P} X$. Show that $\{X_n\}$ is *Cauchy convergent in probability* in that, for all $\epsilon > 0$, $\mathbb{P}(|X_n - X_m| > \epsilon) \to 0$ as $n, m \to \infty$. In what sense is the converse true?
(b) Let $\{X_n\}$ and $\{Y_n\}$ be sequences of random variables such that the pairs $(X_i, X_j)$ and $(Y_i, Y_j)$ have the same distributions for all $i, j$. If $X_n \xrightarrow{P} X$, show that $Y_n$ converges in probability to some limit $Y$ having the same distribution as $X$.

**2.**  Show that the probability that infinitely many of the events $\{A_n : n \geq 1\}$ occur satisfies $\mathbb{P}(A_n \text{ i.o.}) \geq \lim \sup_{n \to \infty} \mathbb{P}(A_n)$.

**3.**  Let $\{S_n : n \geq 0\}$ be a simple random walk which moves to the right with probability $p$ at each step, and suppose that $S_0 = 0$. Write $X_n = S_n - S_{n-1}$.
(a) Show that $\{S_n = 0 \text{ i.o.}\}$ is not a tail event of the sequence $\{X_n\}$.
(b) Show that $\mathbb{P}(S_n = 0 \text{ i.o.}) = 0$ if $p \neq \frac{1}{2}$.
(c) Let $T_n = S_n/\sqrt{n}$, and show that

$$\left\{ \liminf_{n \to \infty} T_n \leq -x \right\} \cap \left\{ \limsup_{n \to \infty} T_n \geq x \right\}$$

is a tail event of the sequence $\{X_n\}$, for all $x > 0$, and deduce directly that $\mathbb{P}(S_n = 0 \text{ i.o.}) = 1$ if $p = \frac{1}{2}$.

**4.**  **Hewitt–Savage zero–one law.** Let $X_1, X_2, \dots$ be independent identically distributed random variables. The event $A$, defined in terms of the $X_n$, is called *exchangeable* if $A$ is invariant under finite permutations of the coordinates, which is to say that its indicator function $I_A$ satisfies $I_A(X_1, X_2, \dots, X_n, \dots) = I_A(X_{i_1}, X_{i_2}, \dots, X_{i_n}, X_{n+1}, \dots)$ for all $n \geq 1$ and all permutations $(i_1, i_2, \dots, i_n)$ of $(1, 2, \dots, n)$. Show that all exchangeable events $A$ are such that either $\mathbb{P}(A) = 0$ or $\mathbb{P}(A) = 1$.

**5.**  Returning to the simple random walk $S$ of Exercise (7.3.3), show that $\{S_n = 0 \text{ i.o.}\}$ is an exchangeable event with respect to the steps of the walk, and deduce from the Hewitt–Savage zero–one law that it has probability either 0 or 1.

**6.**  **Weierstrass's approximation theorem.** Let $f : [0, 1] \to \mathbb{R}$ be a continuous function, and let $S_n$ be a random variable having the binomial distribution with parameters $n$ and $x$. Using the formula $\mathbb{E}(Z) = \mathbb{E}(ZI_A) + \mathbb{E}(ZI_{A^c})$ with $Z = f(x) - f(n^{-1} S_n)$ and $A = \{|n^{-1} S_n - x| > \delta\}$, show that

$$\lim_{n \to \infty} \sup_{0 \leq x \leq 1} \left| f(x) - \sum_{k=0}^{n} f(k/n) \binom{n}{k} x^k (1 - x)^{n-k} \right| = 0.$$

You have proved Weierstrass's approximation theorem, which states that every continuous function on $[0, 1]$ may be approximated by a polynomial uniformly over the interval.

**7.**  **Complete convergence.** A sequence $X_1, X_2, \dots$ of random variables is said to be *completely convergent* to $X$ if

$$\sum_n \mathbb{P}(|X_n - X| > \epsilon) < \infty \qquad \text{for all } \epsilon > 0.$$

Show that, for sequences of independent variables, complete convergence is equivalent to a.s. convergence. Find a sequence of (dependent) random variables which converges a.s. but not completely.

**8.**    Let $X_1, X_2, \ldots$ be independent identically distributed random variables with common mean $\mu$ and finite variance. Show that

$$\binom{n}{2}^{-1} \sum_{1 \le i < j \le n} X_i X_j \xrightarrow{\text{P}} \mu^2 \qquad \text{as } n \to \infty.$$

**9.**    Let $\{X_n : n \ge 1\}$ be independent and exponentially distributed with parameter 1. Show that

$$\mathbb{P}\left( \limsup_{n \to \infty} \frac{X_n}{\log n} = 1 \right) = 1.$$

**10.**    Let $\{X_n : n \ge 1\}$ be independent $N(0, 1)$ random variables. Show that:

(a)  $\mathbb{P}\left( \limsup\limits_{n \to \infty} \dfrac{|X_n|}{\sqrt{\log n}} = \sqrt{2} \right) = 1$,

(b)  $\mathbb{P}(X_n > a_n \text{ i.o.}) = \begin{cases} 0 & \text{if } \sum_n \mathbb{P}(X_1 > a_n) < \infty, \\ 1 & \text{if } \sum_n \mathbb{P}(X_1 > a_n) = \infty. \end{cases}$

**11.**    Construct an example to show that the convergence in distribution of $X_n$ to $X$ does not imply the convergence of the unique medians of the sequence $X_n$.

**12.**    (i) Let $\{X_r : r \ge 1\}$ be independent, non-negative and identically distributed with infinite mean. Show that $\limsup_{r \to \infty} X_r / r = \infty$ almost surely.
(ii) Let $\{X_r\}$ be a stationary Markov chain on the positive integers with transition probabilities

$$p_{jk} = \begin{cases} \dfrac{j}{j+2} & \text{if } k = j+1, \\[2mm] \dfrac{2}{j+2} & \text{if } k = 1. \end{cases}$$

(a)  Find the stationary distribution of the chain, and show that it has infinite mean.
(b)  Show that $\limsup_{r \to \infty} X_r / r \le 1$ almost surely.

**13.**    Let $\{X_r : 1 \le r \le n\}$ be independent and identically distributed with mean $\mu$ and finite variance $\sigma^2$. Let $\overline{X} = n^{-1} \sum_{r=1}^{n} X_r$. Show that

$$\sum_{r=1}^{n} (X_r - \mu) \bigg/ \sqrt{\sum_{r=1}^{n} (X_r - \overline{X})^2}$$

converges in distribution to the $N(0, 1)$ distribution as $n \to \infty$.

**14.**    For a random variable $X$ with mean 0, variance $\sigma^2$, and $\mathbb{E}(X^4) < \infty$, show that

$$\mathbb{P}(|X| > t) \le \frac{\mathbb{E}(X^4) - \sigma^4}{\mathbb{E}(X^4) - 2\sigma^2 t^2 + t^4}, \qquad t > 0.$$

## 7.4 Laws of large numbers

Let $\{X_n\}$ be a sequence of random variables with partial sums $S_n = \sum_{i=1}^{n} X_i$. We are interested in the asymptotic behaviour of $S_n$ as $n \to \infty$; this long-term behaviour depends crucially upon the sequence $\{X_i\}$. The general problem may be described as follows. Under what conditions does the following convergence occur?

$$(1) \qquad\qquad \frac{S_n}{b_n} - a_n \to S \quad \text{as} \quad n \to \infty$$

where $a = \{a_n\}$ and $b = \{b_n\}$ are sequences of real numbers, $S$ is a random variable, and the convergence takes place in some mode to be specified.

**(2) Example.** Let $X_1, X_2, \ldots$ be independent identically distributed variables with mean $\mu$ and variance $\sigma^2$. By Theorems (5.10.2) and (5.10.4), we have that

$$\frac{S_n}{n} \xrightarrow{\mathrm{D}} \mu \quad \text{and} \quad \frac{S_n}{\sigma\sqrt{n}} - \frac{\mu\sqrt{n}}{\sigma} \xrightarrow{\mathrm{D}} N(0, 1).$$

There may not be a *unique* collection $a, b, S$ such that (1) occurs.                              ●

The convergence problem (1) can often be simplified by setting $a_n = 0$ for all $n$, whenever the $X_i$ have finite means. Just rewrite the problem in terms of $X_i' = X_i - \mathbb{E}X_i$ and $S_n' = S_n - \mathbb{E}S_n$.

The general theory of relations such as (1) is well established and extensive. We shall restrict our attention here to a small but significant part of the theory when the $X_i$ are independent and identically distributed random variables. Suppose for the moment that this is true. We saw in Example (2) that (at least) two types of convergence may be established for such sequences, so long as they have finite second moments. The law of large numbers admits stronger forms than that given in (2). For example, notice that $n^{-1}S_n$ converges in distribution to a constant limit, and use Theorem (7.2.4) to see that $n^{-1}S_n$ converges in probability also. Perhaps we can strengthen this further to include convergence in $r$th mean, for some $r$, or almost sure convergence. Indeed, this turns out to be possible when suitable conditions are imposed on the common distribution of the $X_i$. We shall not use the method of characteristic functions of Chapter 5, preferring to approach the problem more directly in the spirit of Section 7.2.

We shall say that the sequence $\{X_n\}$ obeys the 'weak law of large numbers' if there exists a constant $\mu$ such that $n^{-1}S_n \xrightarrow{\mathrm{P}} \mu$. If the stronger result $n^{-1}S_n \xrightarrow{\text{a.s.}} \mu$ holds, then we call it the 'strong law of large numbers'. We seek sufficient, and if possible necessary, conditions on the common distribution of the $X_i$ for the weak and strong laws to hold. As the title suggests, the weak law is implied by the strong law, since convergence in probability is implied by almost sure convergence. A sufficient condition for the strong law is given by the following theorem.

**(3) Theorem.** *Let* $X_1, X_2, \ldots$ *be independent identically distributed random variables with* $\mathbb{E}(X_1^2) < \infty$ *and* $\mathbb{E}(X_1) = \mu$. *Then*

$$\frac{1}{n} \sum_{i=1}^{n} X_i \to \mu \quad \text{almost surely and in mean square.}$$

This strong law holds whenever the $X_i$ have finite second moment. The proof of mean square convergence is very easy; almost sure convergence is harder to demonstrate (but see Problem (7.11.6) for an easy proof of almost sure convergence subject to the stronger condition that $\mathbb{E}(X_1^4) < \infty$).

**Proof.** To show mean square convergence, calculate

$$\mathbb{E}\left(\left(\frac{1}{n}S_n - \mu\right)^2\right) = \mathbb{E}\left(\frac{1}{n^2}(S_n - \mathbb{E}S_n)^2\right) = \frac{1}{n^2}\operatorname{var}\left(\sum_1^n X_i\right)$$

$$= \frac{1}{n^2}\sum_1^n \operatorname{var}(X_i) \qquad\qquad \text{by independence and Theorem (3.3.11)}$$

$$= \frac{1}{n}\operatorname{var}(X_1) \to 0 \qquad \text{as}\quad n \to \infty,$$

since $\operatorname{var}(X_1) < \infty$ by virtue of the assumption that $\mathbb{E}(X_1^2) < \infty$.

Next we show almost sure convergence. We saw in Theorem (7.2.13) that there necessarily exists a subsequence $n_1, n_2, \ldots$ along which $n^{-1}S_n$ converges to $\mu$ almost surely; we can find such a subsequence explicitly. Write $n_i = i^2$ and use Chebyshov's inequality, Example (7.3.3), to find that

$$\mathbb{P}\left(\frac{1}{i^2}|S_{i^2} - i^2\mu| > \epsilon\right) \leq \frac{\operatorname{var}(S_{i^2})}{i^4\epsilon^2} = \frac{\operatorname{var}(X_1)}{i^2\epsilon^2}.$$

Sum over $i$ and use Theorem (7.2.4c) to find that

**(4)**
$$\frac{1}{i^2}S_{i^2} \xrightarrow{\text{a.s.}} \mu \quad \text{as}\quad i \to \infty.$$

We need to fill in the gaps in this limit process. Suppose for the moment that the $X_i$ are *non-negative*. Then $\{S_n\}$ is monotonic non-decreasing, and so

$$S_{i^2} \leq S_n \leq S_{(i+1)^2} \quad \text{if}\quad i^2 \leq n \leq (i+1)^2.$$

Divide by $n$ to find that

$$\frac{1}{(i+1)^2}S_{i^2} \leq \frac{1}{n}S_n \leq \frac{1}{i^2}S_{(i+1)^2} \quad \text{if}\quad i^2 \leq n \leq (i+1)^2;$$

now let $n \to \infty$ and use (4), remembering that $i^2/(i+1)^2 \to 1$ as $i \to \infty$, to deduce that

**(5)**
$$\frac{1}{n}S_n \xrightarrow{\text{a.s.}} \mu \quad \text{as}\quad n \to \infty$$

as required, whenever the $X_i$ are non-negative. Finally we lift the non-negativity condition. For general $X_i$, define random variables $X_n^+, X_n^-$ by

$$X_n^+(\omega) = \max\{X_n(\omega), 0\}, \quad X_n^-(\omega) = -\min\{X_n(\omega), 0\};$$

then $X_n^+$ and $X_n^-$ are non-negative and

$$X_n = X_n^+ - X_n^-, \quad \mathbb{E}(X_n) = \mathbb{E}(X_n^+) - \mathbb{E}(X_n^-).$$

Furthermore, $X_n^+ \leq |X_n|$ and $X_n^- \leq |X_n|$, so that $\mathbb{E}((X_1^+)^2) < \infty$ and $\mathbb{E}((X_1^-)^2) < \infty$. Now apply (5) to the sequences $\{X_n^+\}$ and $\{X_n^-\}$ to find, by Theorem (7.3.9a), that

$$\frac{1}{n} S_n = \frac{1}{n}\left(\sum_1^n X_i^+ - \sum_1^n X_i^-\right)$$

$$\xrightarrow{\text{a.s.}} \mathbb{E}(X_1^+) - \mathbb{E}(X_1^-) = \mathbb{E}(X_1) \quad \text{as} \quad n \to \infty. \qquad \blacksquare$$

Is the result of Theorem (3) as sharp as possible? It is not difficult to see that the condition $\mathbb{E}(X_1^2) < \infty$ is both necessary and sufficient for mean square convergence to hold. For almost sure convergence the weaker condition that

**(6)** $$\mathbb{E}|X_1| < \infty$$

will turn out to be necessary and sufficient, but the proof of this is slightly more difficult and is deferred until the next section. There exist sequences which satisfy the weak law but not the strong law. Indeed, the characteristic function technique (see Section 5.10) can be used to prove the following necessary and sufficient condition for the weak law. We offer no proof, but see Laha and Rohatgi (1979, p. 320), Feller (1971, p. 565), and Problem (7.11.15).

**(7) Theorem.** *The independent identically distributed sequence $\{X_n\}$, with common distribution function $F$, satisfies*

$$\frac{1}{n} \sum_{i=1}^n X_i \xrightarrow{\text{P}} \mu$$

*for some constant $\mu$, if and only if one of the following conditions (8) or (9) holds:*

**(8)** $$n\mathbb{P}(|X_1| > n) \to 0 \quad \text{and} \quad \int_{[-n,n]} x \, dF \to \mu \quad \text{as } n \to \infty,$$

**(9)** *the characteristic function $\phi(t)$ of the $X_j$ is differentiable at $t = 0$ and $\phi'(0) = i\mu$.*

Of course, the integral in (8) can be rewritten as

$$\int_{[-n,n]} x \, dF = \mathbb{E}(X_1 \mid |X_1| \leq n)\mathbb{P}(|X_1| \leq n) = \mathbb{E}(X_1 I_{\{|X_1|\leq n\}}).$$

Thus, a sequence satisfies the weak law but not the strong law whenever (8) holds without (6); as an example of this, suppose the $X_j$ are symmetric (in that $X_1$ and $-X_1$ have the same distribution) but their common distribution function $F$ satisfies

$$1 - F(x) \sim \frac{1}{x \log x} \quad \text{as} \quad x \to \infty.$$

Some distributions fail even to satisfy (8).

**(10) Example.** Let the $X_j$ have the Cauchy distribution with density function

$$f(x) = \frac{1}{\pi(1 + x^2)}.$$

Then the first part of (8) is violated. Indeed, the characteristic function of $U_n = n^{-1} S_n$ is

$$\phi_{U_n}(t) = \phi_{X_1}\left(\frac{t}{n}\right) \cdots \phi_{X_n}\left(\frac{t}{n}\right) = \left[\exp\left(-\frac{|t|}{n}\right)\right]^n = e^{-|t|}$$

and so $U_n$ itself has the Cauchy distribution for all values of $n$. In particular, (1) holds with $b_n = n$, $a_n = 0$, where $S$ is Cauchy, and the convergence is in distribution. ●

---

## Exercises for Section 7.4

**1.** Let $X_2, X_3, \ldots$ be independent random variables such that

$$\mathbb{P}(X_n = n) = \mathbb{P}(X_n = -n) = \frac{1}{2n \log n}, \quad \mathbb{P}(X_n = 0) = 1 - \frac{1}{n \log n}.$$

Show that this sequence obeys the weak law but not the strong law, in the sense that $n^{-1} \sum_1^n X_i$ converges to 0 in probability but not almost surely.

**2.** Construct a sequence $\{X_r : r \geq 1\}$ of independent random variables with zero mean such that $n^{-1} \sum_{r=1}^n X_r \to -\infty$ almost surely, as $n \to \infty$.

**3.** Let $N$ be a spatial Poisson process with constant intensity $\lambda$ in $\mathbb{R}^d$, where $d \geq 2$. Let $S$ be the ball of radius $r$ centred at zero. Show that $N(S)/|S| \to \lambda$ almost surely as $r \to \infty$, where $|S|$ is the volume of the ball.

**4. Proportional betting.** In each of a sequence of independent bets, a gambler either wins 30%, or loses 25% of her current fortune, each with probability $\frac{1}{2}$. Denoting her fortune after $n$ bets by $F_n$, show that $\mathbb{E}(F_n) \to \infty$ as $n \to \infty$, while $F_n \to 0$ almost surely.

**5. General weak law.** Let $S_n = X_1 + X_2 + \cdots + X_n$ be the sum of independent, identically distributed random variables. Let $\delta, \epsilon > 0$, and define the truncated variables $Y_j = X_j I_{\{|X_j| \leq \delta n\}}$. Let $A = \{X_j = Y_j \text{ for } j = 1, 2, \ldots, n\}$, and $B = \{|S_n - n\mathbb{E}(Y_1)| \geq \epsilon n\}$. Prove that:
(a) $\mathbb{P}(B) \leq \mathbb{P}(A^c) + \mathbb{P}(B \mid A)$,
(b) $\mathbb{P}(A^c) \leq n\mathbb{P}(|X_1| > \delta n)$,
(c) $\mathbb{P}(B \mid A) \leq \mathbb{E}(Y_1^2)/(n\epsilon^2)$.
Deduce the weak law of large numbers, namely that, if $\mathbb{E}|X_1| < \infty$,

$$\mathbb{P}\big(|S_n/n - \mu| > \epsilon\big) \to 0 \qquad \text{as } n \to \infty,$$

where $\mu = \mathbb{E}(X_1)$.

## 7.5 The strong law

This section is devoted to the proof of the strong law of large numbers.

**(1) Theorem. Strong law of large numbers.** *Let* $X_1, X_2, \dots$ *be independent identically distributed random variables. Then*

$$\frac{1}{n} \sum_{i=1}^{n} X_i \to \mu \text{ almost surely, as } n \to \infty,$$

*for some constant* $\mu$, *if and only if* $\mathbb{E}|X_1| < \infty$. *In this case* $\mu = \mathbb{E}X_1$.

The traditional proof of this theorem is long and difficult, and proceeds by a generalization of Chebyshov's inequality. We avoid that here, and give a relatively elementary proof which is an adaptation of the method used to prove Theorem (7.4.3). We make use of the technique of *truncation*, used earlier in the proof of the large deviation theorem (5.11.4).

**Proof.** Suppose first that the $X_i$ are *non-negative* random variables with $\mathbb{E}|X_1| = \mathbb{E}(X_1) < \infty$, and write $\mu = \mathbb{E}(X_1)$. We 'truncate' the $X_n$ to obtain a new sequence $\{Y_n\}$ given by

$$(2) \qquad\qquad Y_n = X_n I_{\{X_n < n\}} = \begin{cases} X_n & \text{if } X_n < n, \\ 0 & \text{if } X_n \geq n. \end{cases}$$

Note that

$$\sum_n \mathbb{P}(X_n \neq Y_n) = \sum_n \mathbb{P}(X_n \geq n) \leq \mathbb{E}(X_1) < \infty$$

by the result of Problem (4.14.3). Of course, $\mathbb{P}(X_n \geq n) = \mathbb{P}(X_1 \geq n)$ since the $X_i$ are identically distributed. By the first Borel–Cantelli lemma (7.3.10a),

$$\mathbb{P}\big(X_n \neq Y_n \text{ for infinitely many values of } n\big) = 0,$$

and so

$$(3) \qquad\qquad \frac{1}{n} \sum_{i=1}^{n} (X_i - Y_i) \xrightarrow{\text{a.s.}} 0 \quad \text{as} \quad n \to \infty;$$

thus it will suffice to show that

$$(4) \qquad\qquad \frac{1}{n} \sum_{i=1}^{n} Y_i \xrightarrow{\text{a.s.}} \mu \quad \text{as} \quad n \to \infty.$$

We shall need the following elementary observation. If $\alpha > 1$ and $\beta_k = \lfloor \alpha^k \rfloor$, the integer part of $\alpha^k$, then there exists $A > 0$ such that

$$(5) \qquad\qquad \sum_{k=m}^{\infty} \frac{1}{\beta_k^2} \leq \frac{A}{\beta_m^2} \quad \text{for} \quad m \geq 1.$$

This holds because, for large $m$, the convergent series on the left side is 'nearly' geometric with first term $\beta_m^{-2}$. Note also that

$$(6) \qquad\qquad \frac{\beta_{k+1}}{\beta_k} \to \alpha \quad \text{as} \quad k \to \infty.$$

Write $S_n' = \sum_{i=1}^{n} Y_i$. For $\alpha > 1, \epsilon > 0$, use Chebyshov's inequality to find that

$$(7) \qquad \sum_{n=1}^{\infty} \mathbb{P}\left(\frac{1}{\beta_n}|S_{\beta_n}' - \mathbb{E}(S_{\beta_n}')| > \epsilon\right) \leq \frac{1}{\epsilon^2} \sum_{n=1}^{\infty} \frac{1}{\beta_n^2} \operatorname{var}(S_{\beta_n}')$$

$$= \frac{1}{\epsilon^2} \sum_{n=1}^{\infty} \frac{1}{\beta_n^2} \sum_{i=1}^{\beta_n} \operatorname{var}(Y_i) \quad \text{by independence}$$

$$\leq \frac{A}{\epsilon^2} \sum_{i=1}^{\infty} \frac{1}{i^2} \mathbb{E}(Y_i^2)$$

by changing the order of summation and using (5).

Let $B_{ij} = \{j - 1 \leq X_i < j\}$, and note that $\mathbb{P}(B_{ij}) = \mathbb{P}(B_{1j})$. Now

$$(8) \qquad\qquad \sum_{i=1}^{\infty} \frac{1}{i^2}\mathbb{E}(Y_i^2) = \sum_{i=1}^{\infty} \frac{1}{i^2} \sum_{j=1}^{i} \mathbb{E}(Y_i^2 I_{B_{ij}}) \quad \text{by (2)}$$

$$\leq \sum_{i=1}^{\infty} \frac{1}{i^2} \sum_{j=1}^{i} j^2 \mathbb{P}(B_{ij})$$

$$\leq \sum_{j=1}^{\infty} j^2 \mathbb{P}(B_{1j}) \frac{2}{j} \leq 2\big[\mathbb{E}(X_1) + 1\big] < \infty.$$

Combine (7) and (8) and use Theorem (7.2.4c) to deduce that

$$(9) \qquad\qquad \frac{1}{\beta_n}\big[S_{\beta_n}' - \mathbb{E}(S_{\beta_n}')\big] \xrightarrow{\text{a.s.}} 0 \quad \text{as} \quad n \to \infty.$$

Also,

$$\mathbb{E}(Y_n) = \mathbb{E}(X_n I_{\{X_n < n\}}) = \mathbb{E}(X_1 I_{\{X_1 < n\}}) \to \mathbb{E}(X_1) = \mu$$

as $n \to \infty$, by monotone convergence (5.6.12). Thus

$$\frac{1}{\beta_n}\mathbb{E}(S_{\beta_n}') = \frac{1}{\beta_n} \sum_{i=1}^{\beta_n} \mathbb{E}(Y_i) \to \mu \quad \text{as} \quad n \to \infty$$

(remember the hint in the proof of Corollary (6.4.22)), yielding from (9) that

$$(10) \qquad\qquad \frac{1}{\beta_n}S_{\beta_n}' \xrightarrow{\text{a.s.}} \mu \quad \text{as} \quad n \to \infty;$$

this is a partial demonstration of (4). In order to fill in the gaps, use the fact that the $Y_i$ are non-negative, implying that the sequence $\{S'_n\}$ is monotonic non-decreasing, to deduce that

(11) $$\frac{1}{\beta_{n+1}} S'_{\beta_n} \leq \frac{1}{m} S'_m \leq \frac{1}{\beta_n} S'_{\beta_{n+1}} \qquad \text{if} \quad \beta_n \leq m \leq \beta_{n+1}.$$

Let $m \to \infty$ in (11) and remember (6) to find that

(12) $$\alpha^{-1}\mu \leq \liminf_{m\to\infty} \frac{1}{m} S'_m \leq \limsup_{m\to\infty} \frac{1}{m} S'_m \leq \alpha\mu \quad \text{almost surely.}$$

This holds for all $\alpha > 1$; let $\alpha \downarrow 1$ to obtain (4), and deduce by (3) that

(13) $$\frac{1}{n}\sum_{i=1}^{n} X_i \xrightarrow{\text{a.s.}} \mu \quad \text{as} \quad n \to \infty$$

whenever the $X_i$ are non-negative. Now proceed exactly as in the proof of Theorem (7.4.3) in order to lift the non-negativity condition. Note that we have proved the main part of the theorem without using the full strength of the independence assumption; we have used only the fact that the $X_i$ are *pairwise* independent.

In order to prove the converse, suppose that $n^{-1}\sum_{i=1}^{n} X_i \xrightarrow{\text{a.s.}} \mu$. Then $n^{-1}X_n \xrightarrow{\text{a.s.}} 0$ by the theory of convergent real series, and the second Borel–Cantelli lemma (7.3.10b) gives

$$\sum_n \mathbb{P}(|X_n| \geq n) < \infty,$$

since the divergence of this sum would imply that $\mathbb{P}(n^{-1}|X_n| \geq 1 \text{ i.o.}) = 1$ (only here do we use the full assumption of independence). By Problem (4.14.3),

$$\mathbb{E}|X_1| \leq 1 + \sum_{n=1}^{\infty} \mathbb{P}(|X_1| \geq n) = 1 + \sum_{n=1}^{\infty} \mathbb{P}(|X_n| \geq n),$$

and hence $\mathbb{E}|X_1| < \infty$, which completes the proof of the theorem.  ∎

---

## Exercises for Section 7.5

**1.   Entropy.** The interval $[0, 1]$ is partitioned into $n$ disjoint sub-intervals with lengths $p_1, p_2, \ldots, p_n$, and the *entropy* of this partition is defined to be

$$h = -\sum_{i=1}^{n} p_i \log p_i.$$

Let $X_1, X_2, \ldots$ be independent random variables having the uniform distribution on $[0, 1]$, and let $Z_m(i)$ be the number of the $X_1, X_2, \ldots, X_m$ which lie in the $i$th interval of the partition above. Show that

$$R_m = \prod_{i=1}^{n} p_i^{Z_m(i)}$$

satisfies $m^{-1} \log R_m \to -h$ almost surely as $m \to \infty$.

**2.    Recurrent events.** Catastrophes occur at the times $T_1, T_2, \ldots$ where $T_i = X_1 + X_2 + \cdots + X_i$ and the $X_i$ are independent identically distributed positive random variables. Let $N(t) = \max\{n : T_n \leq t\}$ be the number of catastrophes which have occurred by time $t$. Prove that if $\mathbb{E}(X_1) < \infty$ then $N(t) \to \infty$ and $N(t)/t \to 1/\mathbb{E}(X_1)$ as $t \to \infty$, almost surely.

**3.    Random walk.** Let $X_1, X_2, \ldots$ be independent identically distributed random variables taking values in the integers $\mathbb{Z}$ and having a finite mean. Show that the Markov chain $S = \{S_n\}$ given by $S_n = \sum_1^n X_i$ is transient if $\mathbb{E}(X_1) \neq 0$.

## 7.6  The law of the iterated logarithm

Let $S_n = X_1 + X_2 + \cdots + X_n$ be the partial sum of independent identically distributed variables, as usual, and suppose further that $\mathbb{E}(X_i) = 0$ and $\mathrm{var}(X_i) = 1$ for all $i$. To date, we have two results about the growth rate of $\{S_n\}$.

*Law of large numbers:* $\dfrac{1}{n} S_n \to 0$ a.s. and in mean square.

*Central limit theorem:* $\dfrac{1}{\sqrt{n}} S_n \xrightarrow{\mathrm{D}} N(0, 1)$.

Thus the sequence $U_n = S_n/\sqrt{n}$ enjoys a random fluctuation which is asymptotically normally distributed. Apart from this long-term trend towards the normal distribution, the sequence $\{U_n\}$ may suffer some large but rare fluctuations. The law of the iterated logarithm is an extraordinary result which tells us exactly how large these fluctuations are. First note that, in the language of Section 7.3,

$$U = \limsup_{n \to \infty} \frac{U_n}{\sqrt{2 \log \log n}}$$

is a tail function of the sequence of the $X_i$. The zero–one law, Theorem (7.3.17), tells us that there exists a number $k$, possibly infinite, such that $\mathbb{P}(U = k) = 1$. The next theorem asserts that $k = 1$.

**(1) Theorem. Law of the iterated logarithm.** *If $X_1, X_2, \ldots$ are independent identically distributed random variables with mean 0 and variance 1 then*

$$\mathbb{P}\left( \limsup_{n \to \infty} \frac{S_n}{\sqrt{2n \log \log n}} = 1 \right) = 1.$$

The proof is long and difficult and is omitted (but see the discussion in Billingsley 1995 or Laha and Rohatgi 1979). The theorem amounts to the assertion that

$$A_n = \left\{ S_n \geq c\sqrt{2n \log \log n} \right\}$$

occurs for infinitely many values of $n$ if $c < 1$ and for only finitely many values of $n$ if $c > 1$, with probability 1. It is an immediate corollary of (1) that

$$\mathbb{P}\left( \liminf_{n \to \infty} \frac{S_n}{\sqrt{2n \log \log n}} = -1 \right) = 1;$$

just apply (1) to the sequence $-X_1, -X_2, \ldots$.

---

## Exercise for Section 7.6

**1.**    A function $\phi(x)$ is said to belong to the 'upper class' if, in the notation of this section, $\mathbb{P}(S_n > \phi(n)\sqrt{n}$ i.o.) $= 0$. A consequence of the law of the iterated logarithm is that $\sqrt{\alpha \log \log x}$ is in the upper class for all $\alpha > 2$. Use the first Borel–Cantelli lemma to prove the much weaker fact that $\phi(x) = \sqrt{\alpha \log x}$ is in the upper class for all $\alpha > 2$, in the special case when the $X_i$ are independent $N(0, 1)$ variables.

---

## 7.7 Martingales

Many probabilists specialize in limit theorems, and much of applied probability is devoted to finding such results. The accumulated literature is vast and the techniques multifarious. One of the most useful skills for establishing such results is that of martingale divination, because the convergence of martingales is guaranteed.

**(1) Example.** It is appropriate to discuss an example of the use of the word 'martingale' which pertains to gambling, a favourite source of probabilistic illustrations. We are all familiar with the following gambling strategy. A gambler has a large fortune. He wagers £1 on an evens bet. If he loses then he wagers £2 on the next play. If he loses on the $n$th play then he wagers £$2^n$ on the next. Each sum is calculated so that his inevitable ultimate win will cover his lost stakes and profit him by £1. This strategy is called a 'martingale'. Nowadays casinos do not allow its use, and croupiers have instructions to refuse the bets of those who are seen to practise it. Thackeray's advice was to avoid its use at all costs, and his reasoning may have had something to do with the following calculation. Suppose the gambler wins for the first time at the $N$th play. $N$ is a random variable with mass function

$$\mathbb{P}(N = n) = (\tfrac{1}{2})^n$$

and so $\mathbb{P}(N < \infty) = 1$; the gambler is almost surely guaranteed a win in the long run. However, by this time he will have lost an amount £$L$ with mean value

$$\mathbb{E}(L) = \sum_{n=1}^{\infty}(\tfrac{1}{2})^n(1 + 2 + \cdots + 2^{n-2}) = \infty.$$

He must be prepared to lose a lot of money! And so, of course, must the proprietor of the casino.

The perils of playing the martingale are illustrated by the following two excerpts from the memoirs of G. Casanova recalling his stay in Venice in 1754 (see Casanova 1922, ch. 7).

> Playing the martingale, continually doubling my stake, I won every day during the rest of the carnival. I was fortunate enough never to lose the sixth card, and if I had lost it, I should have been without money to play, for I had 2000 sequins on that card. I congratulated myself on having increased the fortune of my dear mistress.

However, some days later:

> I still played the martingale, but with such bad luck that I was soon left without a sequin. As I shared my property with my mistress, I was obliged to tell her of my losses, and at her request sold all her diamonds, losing what I got for them; she had now only 500 sequins. There was no more talk of her escaping from the convent, for we had nothing to live on.

Shortly after these events, Casanova was imprisoned by the authorities, until he escaped to organize a lottery for the benefit of both himself and the French treasury in Paris. Before it became merely a spangle, the sequin was an Italian gold coin.          ●

In the spirit of this diversion, suppose a gambler wagers repeatedly with an initial capital $S_0$, and let $S_n$ be his capital after $n$ plays. We shall think of $S_0, S_1, \ldots$ as a sequence of dependent random variables. Before his $(n + 1)$th wager the gambler knows the numerical values of $S_0, S_1, \ldots, S_n$, but can only guess at the future $S_{n+1}, \ldots$. If the game is fair then, conditional upon the past information, he will expect no change in his present capital on average. That is to say†,

(2)                          $$\mathbb{E}(S_{n+1} \mid S_0, S_1, \dots , S_n) = S_n.$$

Most casinos need to pay at least their overheads, and will find a way of changing this equation to

$$\mathbb{E}(S_{n+1} \mid S_0, S_1, \dots , S_n) \le S_n.$$

The gambler is fortunate indeed if this inequality is reversed. Sequences satisfying (2) are called 'martingales', and they have very special and well-studied properties of convergence. They may be discovered within many probabilistic models, and their general theory may be used to establish limit theorems. We shall now abandon the gambling example, and refer disappointed readers to *How to gamble if you must* by L. Dubins and L. Savage, where they may find an account of the gamblers' ruin theorem.

(3) **Definition.** A sequence $\{S_n : n \ge 1\}$ is a **martingale** with respect to the sequence $\{X_n : n \ge 1\}$ if, for all $n \ge 1$:
   (a) $\mathbb{E}|S_n| < \infty$,
   (b) $\mathbb{E}(S_{n+1} \mid X_1, X_2, \dots , X_n) = S_n.$

   Equation (2) shows that the sequence of gambler's fortunes is a martingale with respect to itself. The extra generality, introduced by the sequence $\{X_n\}$ in (3), is useful for martingales which arise in the following way. A specified sequence $\{X_n\}$ of random variables, such as a Markov chain, may itself *not* be a martingale. However, it is often possible to find some function $\phi$ such that $\{S_n = \phi(X_n) : n \ge 1\}$ *is* a martingale. In this case, the martingale property (2) becomes the assertion that, given the values of $X_1, X_2, \dots , X_n$, the mean value of $S_{n+1} = \phi(X_{n+1})$ is just $S_n = \phi(X_n)$; that is,

(4)                          $$\mathbb{E}(S_{n+1} \mid X_1, \dots , X_n) = S_n.$$

Of course, this condition is without meaning unless $S_n$ is some function, say $\phi_n$, of $X_1, \dots , X_n$ (that is, $S_n = \phi_n(X_1, \dots , X_n)$) since the conditional expectation in (4) is itself a function of $X_1, \dots , X_n$. We shall often omit reference to the underlying sequence $\{X_n\}$, asserting merely that $\{S_n\}$ is a martingale.

(5) **Example. Branching processes, two martingales.** Let $Z_n$ be the size of the $n$th generation of a branching process, with $Z_0 = 1$. Recall that the probability $\eta$ that the process ultimately becomes extinct is the smallest non-negative root of the equation $s = G(s)$, where $G$ is the probability generating function of $Z_1$. There are (at least) two martingales associated with the process. First, conditional on $Z_n = z_n$, $Z_{n+1}$ is the sum of $z_n$ independent family sizes, and so

$$\mathbb{E}(Z_{n+1} \mid Z_n = z_n) = z_n \mu$$

---

†Such conditional expectations appear often in this section. Make sure you understand their meanings. This one is the mean value of $S_{n+1}$, calculated as though $S_0, \dots , S_n$ were already known. Clearly this mean value depends on $S_0, \dots , S_n$; so it is a *function* of $S_0, \dots , S_n$. Assertion (2) is that it has the value $S_n$. Any detailed account of conditional expectations would probe into the guts of measure theory. We shall avoid that here, but describe some important properties at the end of this section and in Section 7.9.

where $\mu = G'(1)$ is the mean family size. Thus, by the Markov property,

$$\mathbb{E}(Z_{n+1} \mid Z_1, Z_2, \dots, Z_n) = Z_n \mu.$$

Now define $W_n = Z_n/\mathbb{E}(Z_n)$ and remember that $\mathbb{E}(Z_n) = \mu^n$ to obtain

$$\mathbb{E}(W_{n+1} \mid Z_1, \dots, Z_n) = W_n,$$

and so $\{W_n\}$ is a martingale (with respect to $\{Z_n\}$). It is not the only martingale which arises from the branching process. Let $V_n = \eta^{Z_n}$ where $\eta$ is the probability of ultimate extinction. Surprisingly perhaps, $\{V_n\}$ is a martingale also, as the following indicates. Write $Z_{n+1} = X_1 + X_2 + \dots + X_{Z_n}$ in terms of the family sizes of the members of the $n$th generation to obtain

$$\mathbb{E}(V_{n+1} \mid Z_1, \dots, Z_n) = \mathbb{E}\left(\eta^{(X_1 + \dots + X_{Z_n})} \mid Z_1, \dots, Z_n\right)$$

$$= \prod_{i=1}^{Z_n} \mathbb{E}(\eta^{X_i} \mid Z_1, \dots, Z_n) \quad \text{by independence}$$

$$= \prod_{i=1}^{Z_n} \mathbb{E}(\eta^{X_i}) = \prod_{i=1}^{Z_n} G(\eta) = \eta^{Z_n} = V_n,$$

since $\eta = G(\eta)$. These facts are very significant in the study of the long-term behaviour of the branching process. ●

**(6) Example.** Let $X_1, X_2, \dots$ be independent variables with zero means. We claim that the sequence of partial sums $S_n = X_1 + X_2 + \dots + X_n$ is a martingale (with respect to $\{X_n\}$). For,

$$\mathbb{E}(S_{n+1} \mid X_1, \dots, X_n) = \mathbb{E}(S_n + X_{n+1} \mid X_1, \dots, X_n)$$

$$= \mathbb{E}(S_n \mid X_1, \dots, X_n) + \mathbb{E}(X_{n+1} \mid X_1, \dots, X_n)$$

$$= S_n + 0, \quad \text{by independence.} \quad ●$$

**(7) Example.  Markov chains.** Let $X_0, X_1, \dots$ be a discrete-time Markov chain taking values in some countable state space $S$ with transition matrix $\mathbf{P}$. Suppose that $\psi : S \to \mathbb{R}$ is a bounded function which satisfies

**(8)** $$\sum_{j \in S} p_{ij} \psi(j) = \psi(i) \quad \text{for all } i \in S.$$

We claim that $S_n = \psi(X_n)$ constitutes a martingale (with respect to $\{X_n\}$). For,

$$\mathbb{E}(S_{n+1} \mid X_1, \dots, X_n) = \mathbb{E}\left(\psi(X_{n+1}) \mid X_1, \dots, X_n\right)$$

$$= \mathbb{E}\left(\psi(X_{n+1}) \mid X_n\right) \quad \text{by the Markov property}$$

$$= \sum_{j \in S} p_{X_n, j} \psi(j)$$

$$= \psi(X_n) = S_n \quad \text{by (8).} \quad ●$$

**(9) Example.** Let $X_1, X_2, \ldots$ be independent variables with zero means, finite variances, and partial sums $S_n = \sum_{i=1}^{n} X_i$. Define

$$T_n = S_n^2 = \left(\sum_{i=1}^{n} X_i\right)^2.$$

Then

$$\mathbb{E}(T_{n+1} \mid X_1, \ldots, X_n) = \mathbb{E}\left(S_n^2 + 2S_n X_{n+1} + X_{n+1}^2 \mid X_1, \ldots, X_n\right)$$
$$= T_n + 2\mathbb{E}(X_{n+1})\mathbb{E}(S_n \mid X_1, \ldots, X_n) + \mathbb{E}(X_{n+1}^2)$$
$$\text{by independence}$$
$$= T_n + \mathbb{E}(X_{n+1}^2) \geq T_n.$$

Thus $\{T_n\}$ is not a martingale, since it only satisfies (4) with $\geq$ in place of $=$; it is called a 'submartingale', and has properties similar to those of a martingale.                    ●

These examples show that martingales are all around us. They are extremely useful because, subject to a condition on their moments, they always converge; this is 'Doob's convergence theorem' and is the main result of the next section. Martingales are explored in considerable detail in Chapter 12.

Finally, here are some properties of conditional expectation. You need not read them now, but may refer back to them when necessary. Recall that the conditional expectation of $X$ given $Y$ is defined by

$$\mathbb{E}(X \mid Y) = \psi(Y) \quad \text{where} \quad \psi(y) = \mathbb{E}(X \mid Y = y)$$

is the mean of the conditional distribution of $X$ given that $Y = y$. Most of the conditional expectations in this chapter take the form $\mathbb{E}(X \mid \mathbf{Y})$, the mean value of $X$ conditional on the values of the variables in the random vector $\mathbf{Y} = (Y_1, Y_2, \ldots, Y_n)$. We stress that $\mathbb{E}(X \mid \mathbf{Y})$ is a function of $\mathbf{Y}$ alone. Expressions such as '$\mathbb{E}(X \mid \mathbf{Y}) = Z$' should sometimes be qualified by 'almost surely'; we generally omit this qualification.

**(10) Lemma.**
  (a) $\mathbb{E}(X_1 + X_2 \mid \mathbf{Y}) = \mathbb{E}(X_1 \mid \mathbf{Y}) + \mathbb{E}(X_2 \mid \mathbf{Y})$.
  (b) $\mathbb{E}(Xg(\mathbf{Y}) \mid \mathbf{Y}) = g(\mathbf{Y})\mathbb{E}(X \mid \mathbf{Y})$ *for (measurable) functions* $g : \mathbb{R}^n \to \mathbb{R}$.
  (c) $\mathbb{E}(X \mid h(\mathbf{Y})) = \mathbb{E}(X \mid \mathbf{Y})$ *if* $h : \mathbb{R}^n \to \mathbb{R}^n$ *is one–one.*

**Sketch proof.**
  (a) This depends on the linearity of expectation only.
  (b) $\mathbb{E}(Xg(\mathbf{Y}) \mid \mathbf{Y} = \mathbf{y}) = g(\mathbf{y})\mathbb{E}(X \mid \mathbf{Y} = \mathbf{y})$.
  (c) Roughly speaking, knowledge of $\mathbf{Y}$ is interchangeable with knowledge of $h(\mathbf{Y})$, in that

$$\mathbf{Y}(\omega) = \mathbf{y} \text{ if and only if } h(\mathbf{Y}(\omega)) = h(\mathbf{y}), \quad \text{for any } \omega \in \Omega. \quad ∎$$

**(11) Lemma. Tower property.** $\mathbb{E}\big[\mathbb{E}(X \mid \mathbf{Y}_1, \mathbf{Y}_2) \big| \mathbf{Y}_1\big] = \mathbb{E}(X \mid \mathbf{Y}_1)$.

**Proof.** Just write down these expectations as integrals involving conditional distributions to see that the result holds. It is a more general version of Problem (4.14.29). ∎

Sometimes we consider the mean value $\mathbb{E}(X \mid A)$ of a random variable $X$ conditional upon the occurrence of some event $A$ having strictly positive probability. This is just the mean of the corresponding distribution function $F_{X\mid A}(x) = \mathbb{P}(X \leq x \mid A)$. We can think of $\mathbb{E}(X \mid A)$ as a constant random variable with domain $A \subseteq \Omega$; it is undefined at points $\omega \in A^c$. The following result is an application of Lemma (1.4.4).

**(12) Lemma.** *If $\{B_i : 1 \leq i \leq n\}$ is a partition of $A$ then*

$$\mathbb{E}(X \mid A)\mathbb{P}(A) = \sum_{i=1}^{n} \mathbb{E}(X \mid B_i)\mathbb{P}(B_i).$$

You may like the following proof:

$$\mathbb{E}(XI_A) = \mathbb{E}\left(X \sum_i I_{B_i}\right) = \sum_i \mathbb{E}(XI_{B_i}).$$

Sometimes we consider mixtures of these two types of conditional expectation. These are of the form $\mathbb{E}(X \mid \mathbf{Y}, A)$ where $X, Y_1, \ldots, Y_n$ are random variables and $A$ is an event. Such quantities are defined in the obvious way and have the usual properties. For example, (11) becomes

**(13)** $$\mathbb{E}(X \mid A) = \mathbb{E}\big[\mathbb{E}(X \mid \mathbf{Y}, A) \,\big|\, A\big].$$

We shall make some use of the following fact soon. If, in (13), $A$ is an event which is defined in terms of the $Y_i$ (such as $A = \{Y_1 \leq 1\}$ or $A = \{|Y_2 Y_3 - Y_4| > 2\}$) then it is not difficult to see that

**(14)** $$\mathbb{E}\big[\mathbb{E}(X \mid \mathbf{Y}) \,\big|\, A\big] = \mathbb{E}\big[\mathbb{E}(X \mid \mathbf{Y}, A) \,\big|\, A\big];$$

just note that evaluating the random variable $\mathbb{E}(X \mid \mathbf{Y}, A)$ at a point $\omega \in \Omega$ yields

$$\mathbb{E}(X \mid \mathbf{Y}, A)(\omega) \begin{cases} = \mathbb{E}(X \mid \mathbf{Y})(\omega) & \text{if } \omega \in A \\ \text{is undefined} & \text{if } \omega \notin A. \end{cases}$$

The sequences $\{S_n\}$ of this section satisfy

**(15)** $$\mathbb{E}|S_n| < \infty, \quad \mathbb{E}(S_{n+1} \mid X_1, \ldots, X_n) = S_n.$$

**(16) Lemma.** *If $\{S_n\}$ satisfies (15) then:*
(a) $\mathbb{E}(S_{m+n} \mid X_1, \ldots, X_m) = S_m$ *for all* $m, n \geq 1$,
(b) $\mathbb{E}(S_n) = \mathbb{E}(S_1)$ *for all* $n$.

**Proof.**
(a) Use (11) with $X = S_{m+n}$, $\mathbf{Y}_1 = (X_1, \ldots, X_m)$, and $\mathbf{Y}_2 = (X_{m+1}, \ldots, X_{m+n-1})$ to obtain

$$\mathbb{E}(S_{m+n} \mid X_1, \ldots, X_m) = \mathbb{E}\big[\mathbb{E}(S_{m+n} \mid X_1, \ldots, X_{m+n-1}) \,\big|\, X_1, \ldots, X_m\big]$$
$$= \mathbb{E}(S_{m+n-1} \mid X_1, \ldots, X_m)$$

and iterate to obtain the result.
(b) $\mathbb{E}(S_n) = \mathbb{E}\big(\mathbb{E}(S_n \mid X_1)\big) = \mathbb{E}(S_1)$ by (a). ∎

For a more satisfactory account of conditional expectation, see Section 7.9.

## Exercises for Section 7.7

**1.** Let $X_1, X_2, \ldots$ be random variables such that the partial sums $S_n = X_1 + X_2 + \cdots + X_n$ determine a martingale. Show that $\mathbb{E}(X_i X_j) = 0$ if $i \neq j$.

**2.** Let $Z_n$ be the size of the $n$th generation of a branching process with immigration, in which the family sizes have mean $\mu \ (\neq 1)$ and the mean number of immigrants in each generation is $m$. Suppose that $\mathbb{E}(Z_0) < \infty$, and show that

$$S_n = \mu^{-n} \left\{ Z_n - m \left( \frac{1 - \mu^n}{1 - \mu} \right) \right\}$$

is a martingale with respect to a suitable sequence of random variables.

**3.** Let $X_0, X_1, X_2, \ldots$ be a sequence of random variables with finite means and satisfying $\mathbb{E}(X_{n+1} \mid X_0, X_1, \ldots, X_n) = aX_n + bX_{n-1}$ for $n \geq 1$, where $0 < a, b < 1$ and $a + b = 1$. Find a value of $\alpha$ for which $S_n = \alpha X_n + X_{n-1}, n \geq 1$, defines a martingale with respect to the sequence $X$.

**4.** Let $X_n$ be the net profit to the gambler of betting a unit stake on the $n$th play in a casino; the $X_n$ may be dependent, but the game is fair in the sense that $\mathbb{E}(X_{n+1} \mid X_1, X_2, \ldots, X_n) = 0$ for all $n$. The gambler stakes $Y$ on the first play, and thereafter stakes $f_n(X_1, X_2, \ldots, X_n)$ on the $(n+1)$th play, where $f_1, f_2, \ldots$ are given functions. Show that her profit after $n$ plays is

$$S_n = \sum_{i=1}^{n} X_i f_{i-1}(X_1, X_2, \ldots, X_{i-1}),$$

where $f_0 = Y$. Show further that the sequence $S = \{S_n\}$ satisfies the martingale condition $\mathbb{E}(S_{n+1} \mid X_1, X_2, \ldots, X_n) = S_n, n \geq 1$, if $Y$ is assumed to be known throughout.

**5.** A *run* in a random permutation $(\pi_1, \pi_2, \ldots, \pi_n)$ of $(1, 2, \ldots, n)$ is a subsequence satisfying $\pi_{r-1} > \pi_r < \pi_{r+1} < \cdots < \pi_s > \pi_{s+1}$. We set $\pi_0 = n + 1$ and $\pi_{n+1} = 0$ by convention. Let $R_n$ be the number of runs. Show that $M_n = nR_n - \frac{1}{2}n(n+1)$ is a martingale. Find $\mathbb{E}(R_n)$ and $\mathbb{E}(R_n^2)$.

## 7.8 Martingale convergence theorem

This section is devoted to the proof and subsequent applications of the following theorem. It receives a section to itself by virtue of its wealth of applications.

**(1) Theorem.** *If $\{S_n\}$ is a martingale with $\mathbb{E}(S_n^2) < M < \infty$ for some $M$ and all $n$, then there exists a random variable $S$ such that $S_n$ converges to $S$ almost surely and in mean square.*

This result has a more general version which, amongst other things,

   (i) deals with submartingales,

   (ii) imposes weaker moment conditions,

   (iii) explores convergence in mean also,

but the proof of this is more difficult. On the other hand, the proof of (1) is within our grasp, and is only slightly more difficult than the proof of the strong law for independent sequences, Theorem (7.5.1); it mimics the traditional proof of the strong law and begins with a generalization of Chebyshov's inequality. We return to the theory of martingales in much greater generality in Chapter 12.

**(2) Theorem. Doob–Kolmogorov inequality.** *If $\{S_n\}$ is a martingale with respect to $\{X_n\}$ then*

$$\mathbb{P}\left(\max_{1\le i\le n}|S_i|\ge\epsilon\right)\le\frac{1}{\epsilon^2}\mathbb{E}(S_n^2)\quad\text{whenever}\quad\epsilon>0.$$

**Proof of (2).** Let $A_0=\Omega$, $A_k=\{|S_i|<\epsilon$ for all $i\le k\}$, and let $B_k=A_{k-1}\cap\{|S_k|\ge\epsilon\}$ be the event that $|S_i|\ge\epsilon$ for the first time when $i=k$. Then

$$A_k\cup\left(\bigcup_{i=1}^{k}B_i\right)=\Omega.$$

Therefore

(3) $$\mathbb{E}(S_n^2)=\sum_{i=1}^{n}\mathbb{E}(S_n^2 I_{B_i})+\mathbb{E}(S_n^2 I_{A_n})\ge\sum_{i=1}^{n}\mathbb{E}(S_n^2 I_{B_i}).$$

However,

$$\mathbb{E}(S_n^2 I_{B_i})=\mathbb{E}\big((S_n-S_i+S_i)^2 I_{B_i}\big)$$
$$=\mathbb{E}\big((S_n-S_i)^2 I_{B_i}\big)+2\mathbb{E}\big((S_n-S_i)S_i I_{B_i}\big)+\mathbb{E}(S_i^2 I_{B_i})$$
$$=\alpha+\beta+\gamma,\text{ say}.$$

Note that $\alpha\ge0$ and $\gamma\ge\epsilon^2\mathbb{P}(B_i)$, because $|S_i|\ge\epsilon$ if $B_i$ occurs. To deal with $\beta$, note that

$$\mathbb{E}\big((S_n-S_i)S_i I_{B_i}\big)=\mathbb{E}\big[S_i I_{B_i}\mathbb{E}(S_n-S_i\mid X_1,\dots,X_i)\big]\quad\text{by Lemma (7.7.10b)}$$
$$=0\qquad\qquad\qquad\qquad\qquad\qquad\text{by Lemma (7.7.16a)},$$

since $B_i$ concerns $X_1,\dots,X_i$ only, by the discussion after (7.7.4). Thus (3) becomes

$$\mathbb{E}(S_n^2)\ge\sum_{i=1}^{n}\epsilon^2\mathbb{P}(B_i)=\epsilon^2\mathbb{P}\left(\max_{1\le i\le n}|S_i|\ge\epsilon\right)$$

and the result is shown.                                                                 ∎

**Proof of (1).** First note that $S_m$ and $(S_{m+n}-S_m)$ are uncorrelated whenever $m,n\ge1$, since

$$\mathbb{E}\big(S_m(S_{m+n}-S_m)\big)=\mathbb{E}\big[S_m\mathbb{E}(S_{m+n}-S_m\mid X_1,\dots,X_m)\big]=0$$

by Lemma (7.7.16). Thus

(4) $$\mathbb{E}(S_{m+n}^2)=\mathbb{E}(S_m^2)+\mathbb{E}\big((S_{m+n}-S_m)^2\big).$$

It follows that $\{\mathbb{E}(S_n^2)\}$ is a non-decreasing sequence, which is bounded above, by the assumption in (1); hence we may suppose that the constant $M$ is chosen such that

$$\mathbb{E}(S_n^2)\uparrow M\quad\text{as}\quad n\to\infty.$$

We shall show that the sequence $\{S_n(\omega) : n \geq 1\}$ is almost-surely Cauchy convergent (see the notes on Cauchy convergence after Example (7.2.2)). Let $C = \{\omega \in \Omega : \{S_n(\omega)\}$ is Cauchy convergent$\}$. For $\omega \in C$, the sequence $S_n(\omega)$ converges as $n \to \infty$ to some limit $S(\omega)$, and we shall show that $\mathbb{P}(C) = 1$. Note that

$$C = \left\{ \forall \epsilon > 0, \; \exists m \text{ such that } |S_{m+i} - S_{m+j}| < \epsilon \text{ for all } i, j \geq 1 \right\}.$$

By the triangle inequality

$$|S_{m+i} - S_{m+j}| \leq |S_{m+i} - S_m| + |S_{m+j} - S_m|,$$

so that

$$C = \left\{ \forall \epsilon > 0, \; \exists m \text{ such that } |S_{m+i} - S_m| < \epsilon \text{ for all } i \geq 1 \right\}$$
$$= \bigcap_{\epsilon > 0} \bigcup_m \{|S_{m+i} - S_m| < \epsilon \text{ for all } i \geq 1\}.$$

The complement of $C$ may be expressed as

$$C^c = \bigcup_{\epsilon > 0} \bigcap_m \{|S_{m+i} - S_m| \geq \epsilon \text{ for some } i \geq 1\} = \bigcup_{\epsilon > 0} \bigcap_m A_m(\epsilon)$$

where $A_m(\epsilon) = \{|S_{m+i} - S_m| \geq \epsilon \text{ for some } i \geq 1\}$. Now $A_m(\epsilon) \subseteq A_m(\epsilon')$ if $\epsilon \geq \epsilon'$, so that

$$\mathbb{P}(C^c) = \lim_{\epsilon \downarrow 0} \mathbb{P}\left( \bigcap_m A_m(\epsilon) \right) \leq \lim_{\epsilon \downarrow 0} \lim_{m \to \infty} \mathbb{P}(A_m(\epsilon)).$$

In order to prove that $\mathbb{P}(C^c) = 0$ as required, it suffices to show that $\mathbb{P}(A_m(\epsilon)) \to 0$ as $m \to \infty$ for all $\epsilon > 0$. To this end we shall use the Doob–Kolmogorov inequality.

For a given choice of $m$, define the sequence $Y = \{Y_n : n \geq 1\}$ by $Y_n = S_{m+n} - S_m$. It may be checked that $Y$ is a martingale with respect to itself:

$$\mathbb{E}(Y_{n+1} \mid Y_1, \dots, Y_n) = \mathbb{E}\big[\mathbb{E}(Y_{n+1} \mid X_1, \dots, X_{m+n}) \,\big|\, Y_1, \dots, Y_n\big]$$
$$= \mathbb{E}(Y_n \mid Y_1, \dots, Y_n) = Y_n$$

by Lemma (7.7.11) and the martingale property. We apply the Doob–Kolmogorov inequality (2) to this martingale to find that

$$\mathbb{P}\big(|S_{m+i} - S_m| \geq \epsilon \text{ for some } 1 \leq i \leq n\big) \leq \frac{1}{\epsilon^2} \mathbb{E}\big((S_{m+n} - S_m)^2\big).$$

Letting $n \to \infty$ and using (4) we obtain

$$\mathbb{P}(A_m(\epsilon)) \leq \frac{1}{\epsilon^2}\big(M - \mathbb{E}(S_m^2)\big),$$

and hence $\mathbb{P}(A_m(\epsilon)) \to 0$ as $m \to \infty$ as required for almost-sure convergence. We have proved that there exists a random variable $S$ such that $S_n \xrightarrow{\text{a.s.}} S$.

It remains only to prove convergence of $S_n$ to $S$ in mean square. For this we need Fatou's lemma (5.6.13). It is the case that

$$\mathbb{E}\big((S_n - S)^2\big) = \mathbb{E}\Big(\liminf_{m\to\infty}(S_n - S_m)^2\Big) \leq \liminf_{m\to\infty}\mathbb{E}\big((S_n - S_m)^2\big)$$

$$= M - \mathbb{E}(S_n^2) \to 0 \quad\text{as}\quad n \to \infty,$$

and the proof is finished.  ∎

Here are some applications of the martingale convergence theorem.

**(5) Example. Branching processes.** Recall Example (7.7.5). By Lemma (5.4.2), $W_n = Z_n/\mathbb{E}(Z_n)$ has second moment

$$\mathbb{E}(W_n^2) = 1 + \frac{\sigma^2(1 - \mu^{-n})}{\mu(\mu - 1)} \quad\text{if}\quad \mu \neq 1$$

where $\sigma^2 = \mathrm{var}(Z_1)$. Thus, if $\mu \neq 1$, there exists a random variable $W$ such that $W_n \xrightarrow{\text{a.s.}} W$, and so $W_n \xrightarrow{\text{D}} W$ also; their characteristic functions satisfy $\phi_{W_n}(t) \to \phi_W(t)$ by Theorem (5.9.5). This makes the discussion at the end of Section 5.4 fully rigorous, and we can rewrite equation (5.4.6) as

$$\phi_W(\mu t) = G(\phi_W(t)). \qquad \bullet$$

**(6) Example. Markov chains.** Suppose that the chain $X_0, X_1, \ldots$ of Example (7.7.7) is irreducible and recurrent, and let $\psi$ be a bounded function mapping $S$ into $\mathbb{R}$ which satisfies equation (7.7.8). Then the sequence $\{S_n\}$, given by $S_n = \psi(X_n)$, is a martingale and satisfies the condition $\mathbb{E}(S_n^2) \leq M$ for some $M$, by the boundedness of $\psi$. For any state $i$, the event $\{X_n = i\}$ occurs for infinitely many values of $n$ with probability 1. However, $\{S_n = \psi(i)\} \supseteq \{X_n = i\}$ and so

$$S_n \xrightarrow{\text{a.s.}} \psi(i) \quad\text{for all } i,$$

which is clearly impossible unless $\psi(i)$ is the same for all $i$. We have shown that any bounded solution of equation (7.7.8) is constant.  $\bullet$

**(7) Example. Genetic model.** Recall Example (6.1.11), which dealt with gene frequencies in the evolution of a population. We encountered there a Markov chain $X_0, X_1, \ldots$ taking values in $\{0, 1, \ldots, N\}$ with transition probabilities given by

**(8)** $$p_{ij} = \mathbb{P}(X_{n+1} = j \mid X_n = i) = \binom{N}{j}\left(\frac{i}{N}\right)^j\left(1 - \frac{i}{N}\right)^{N-j}.$$

Then

$$\mathbb{E}(X_{n+1} \mid X_0, \ldots, X_n) = \mathbb{E}(X_{n+1} \mid X_n) \qquad\text{by the Markov property}$$

$$= \sum_j jp_{X_n, j} = X_n$$

by (8). Thus $X_0, X_1, \ldots$ is a martingale. Also, let $Y_n$ be defined by $Y_n = X_n(N - X_n)$, and suppose that $N > 1$. Then

$$\mathbb{E}(Y_{n+1} \mid X_0, \ldots, X_n) = \mathbb{E}(Y_{n+1} \mid X_n) \quad \text{by the Markov property}$$

and

$$\mathbb{E}(Y_{n+1} \mid X_n = i) = \sum_j j(N - j)p_{ij} = i(N - i)(1 - N^{-1})$$

by (8). Thus

(9) $$\mathbb{E}(Y_{n+1} \mid X_0, \ldots, X_n) = Y_n(1 - N^{-1}),$$

and we see that $\{Y_n\}$ is not itself a martingale. However, set $S_n = Y_n/(1 - N^{-1})^n$ to obtain from (9) that

$$\mathbb{E}(S_{n+1} \mid X_0, \ldots, X_n) = S_n;$$

we deduce that $\{S_n\}$ is a martingale.

The martingale $\{X_n\}$ has uniformly bounded second moments, and so there exists an $X$ such that $X_n \xrightarrow{\text{a.s.}} X$. Unlike the previous example, this chain is not irreducible. In fact, 0 and $N$ are absorbing states, and $X$ takes these values only. Can you find the probability $\mathbb{P}(X = 0)$ that the chain is ultimately absorbed at 0? The results of the next section will help you with this.

Finally, what happens when we apply the convergence theorem to $\{S_n\}$? ●

---

## Exercises for Section 7.8

**1.** **Kolmogorov's inequality.** Let $X_1, X_2, \ldots$ be independent random variables with zero means and finite variances, and let $S_n = X_1 + X_2 + \cdots + X_n$. Use the Doob–Kolmogorov inequality to show that

$$\mathbb{P}\left( \max_{1 \le j \le n} |S_j| > \epsilon \right) \le \frac{1}{\epsilon^2} \sum_{j=1}^n \text{var}(X_j) \qquad \text{for } \epsilon > 0.$$

**2.** Let $X_1, X_2, \ldots$ be independent random variables such that $\sum_n n^{-2} \text{var}(X_n) < \infty$. Use Kolmogorov's inequality to prove that

$$\sum_{i=1}^n \frac{X_i - \mathbb{E}(X_i)}{i} \xrightarrow{\text{a.s.}} Y \qquad \text{as } n \to \infty,$$

for some finite random variable $Y$, and deduce that

$$\frac{1}{n} \sum_{i=1}^n (X_i - \mathbb{E}X_i) \xrightarrow{\text{a.s.}} 0 \qquad \text{as } n \to \infty.$$

(You may find Kronecker's lemma to be useful: if $(a_n)$ and $(b_n)$ are real sequences with $b_n \uparrow \infty$ and $\sum_i a_i/b_i < \infty$, then $b_n^{-1} \sum_{i=1}^n a_i \to 0$ as $n \to \infty$.)

**3.** Let $S$ be a martingale with respect to $X$, such that $\mathbb{E}(S_n^2) < K < \infty$ for some $K \in \mathbb{R}$. Suppose that $\text{var}(S_n) \to 0$ as $n \to \infty$, and prove that $S = \lim_{n \to \infty} S_n$ exists and is constant almost surely.

## 7.9  Prediction and conditional expectation

Probability theory is not merely an intellectual pursuit, but provides also a framework for estimation and prediction. Practical men and women often need to make decisions based on quantities which are not easily measurable, either because they lie in the future or because of some intrinsic inaccessibility; in doing so they usually make use of some current or feasible observation. Economic examples are commonplace (business trends, inflation rates, and so on); other examples include weather prediction, the climate in prehistoric times, the state of the core of a nuclear reactor, the cause of a disease in an individual or a population, or the paths of celestial bodies. This last problem has the distinction of being amongst the first to be tackled by mathematicians using a modern approach to probability.

At its least complicated, a question of prediction or estimation involves an unknown or unobserved random variable $Y$, about which we are provided with the value of some (observable) random variable $X$. The problem is to deduce information about the value of $Y$ from a knowledge of the value of $X$. Thus we seek a function $h(X)$ which is (in some sense) close to $Y$; we write $\widehat{Y} = h(X)$ and call $\widehat{Y}$ an 'estimator' of $Y$. As we saw in Section 7.1, there are many different ways in which two random variables may be said to be close to one another—pointwise, in $r$th mean, in probability, and so on. A particular way of especial convenience is to work with the norm given by

(1) $$\|U\|_2 = \sqrt{\mathbb{E}(U^2)},$$

so that the distance between two random variables $U$ and $V$ is

(2) $$\|U - V\|_2 = \sqrt{\mathbb{E}\{(U - V)^2\}};$$

The norm $\|\cdot\|_2$ is often called the $L_2$ norm, and the corresponding notion of convergence is of course convergence in mean square:

(3) $$\|U_n - U\|_2 \to 0 \quad \text{if and only if} \quad U_n \xrightarrow{\text{m.s.}} U.$$

This norm is a special case of the '$L_p$ norm' given by $\|X\|_p = \{\mathbb{E}|X^p|\}^{1/p}$ where $p \geq 1$.

We recall that $\|\cdot\|_2$ satisfies the *triangle inequality*:

(4) $$\|U + V\|_2 \leq \|U\|_2 + \|V\|_2.$$

With this notation, we make the following definition.

(5) **Definition.** Let $X$ and $Y$ be random variables on $(\Omega, \mathcal{F}, \mathbb{P})$ such that $\mathbb{E}(Y^2) < \infty$. The **minimum mean-squared-error predictor** (or **best predictor**) of $Y$ given $X$ is the function $\widehat{Y} = h(X)$ of $X$ for which $\|Y - \widehat{Y}\|_2$ is a minimum.

We shall commonly use the term 'best predictor' in this context; the word 'best' is only shorthand, and should not be interpreted literally.

Let $H$ be the set of all functions of $X$ having finite second moment:

(6) $$H = \{h(X) : h \text{ maps } \mathbb{R} \text{ to } \mathbb{R}, \ \mathbb{E}(h(X)^2) < \infty\}.$$

The best (or minimum mean-squared-error) predictor of $Y$ is (if it exists) a random variable $\widehat{Y}$ belonging to $H$ such that $\mathbb{E}((Y - \widehat{Y})^2) \leq \mathbb{E}((Y - Z)^2)$ for all $Z \in H$. Does there exist such a $\widehat{Y}$? The answer is yes, and furthermore there is (essentially) a unique such $\widehat{Y}$ in $H$. In proving this we shall make use of two properties of $H$, that it is a linear space, and that it is closed (with respect to the norm $\| \cdot \|_2$); that is to say, for $Z_1, Z_2, \ldots \in H$ and $a_1, a_2, \ldots \in \mathbb{R}$,

(7) $$a_1 Z_1 + a_2 Z_2 + \cdots + a_n Z_n \in H, \quad \text{and}$$

(8) if $\| Z_m - Z_n \|_2 \to 0$ as $m, n \to \infty$, there exists $Z \in H$ such that $Z_n \xrightarrow{\text{m.s.}} Z$.

(See Exercise (7.9.6a).) More generally, we call a set $H$ of random variables a *closed linear space* (with respect to $\| \cdot \|_2$) if $\| X \|_2 < \infty$ for all $X \in H$, and $H$ satisfies (7) and (8).

**(9) Theorem.** *Let $H$ be a closed linear space (with respect to $\| \cdot \|_2$) of random variables. Let $Y$ be a random variable on $(\Omega, \mathcal{F}, \mathbb{P})$ with finite variance. There exists a random variable $\widehat{Y}$ in $H$ such that*

(10) $$\| Y - \widehat{Y} \|_2 \leq \| Y - Z \|_2 \quad \text{for all } Z \in H,$$

*and which is unique in the sense that $\mathbb{P}(\widehat{Y} = \overline{Y}) = 1$ for any $\overline{Y} \in H$ with $\| Y - \overline{Y} \|_2 = \| Y - \widehat{Y} \|_2$.*

**Proof.** Let $d = \inf\{\| Y - Z \|_2 : Z \in H\}$, and find a sequence $Z_1, Z_2, \ldots$ in $H$ such that $\lim_{n \to \infty} \| Y - Z_n \|_2 = d$. Now, for any $A, B \in H$, the 'parallelogram rule' holds†:

(11) $$\| A - B \|_2^2 = 2\left[ \| Y - A \|_2^2 - 2\| Y - \tfrac{1}{2}(A + B) \|_2^2 + \| Y - B \|_2^2 \right];$$

to show this, just expand the right side. Note that $\tfrac{1}{2}(A + B) \in H$ since $H$ is a linear space. Setting $A = Z_n$, $B = Z_m$, we obtain using the definition of $d$ that

$$\| Z_m - Z_n \|_2^2 \leq 2\left( \| Y - Z_m \|_2^2 - 2d + \| Y - Z_n \|_2^2 \right) \to 0 \quad \text{as} \quad m, n \to \infty.$$

Therefore $\| Z_m - Z_n \|_2 \to 0$, so that there exists $\widehat{Y} \in H$ such that $Z_n \xrightarrow{\text{m.s.}} \widehat{Y}$; it is here that we use the fact (8) that $H$ is closed. It follows by the triangle inequality (4) that

$$\| Y - \widehat{Y} \|_2 \leq \| Y - Z_n \|_2 + \| Z_n - \widehat{Y} \|_2 \to d \quad \text{as} \quad n \to \infty,$$

so that $\widehat{Y}$ satisfies (10).

Finally, suppose that $\overline{Y} \in H$ satisfies $\| Y - \overline{Y} \|_2 = d$. Apply (11) with $A = \overline{Y}$, $B = \widehat{Y}$, to obtain

$$\| \overline{Y} - \widehat{Y} \|_2^2 = 4\left[ d^2 - \| Y - \tfrac{1}{2}(\overline{Y} + \widehat{Y}) \|_2^2 \right] \leq 4(d^2 - d^2) = 0.$$

Hence $\mathbb{E}((\overline{Y} - \widehat{Y})^2) = 0$ and so $\mathbb{P}(\widehat{Y} = \overline{Y}) = 1$. ∎

**(12) Example.** Let $Y$ have mean $\mu$ and variance $\sigma^2$. With no information about $Y$, it is appropriate to ask for the real number $h$ which minimizes $\| Y - h \|_2$. Now $\| Y - h \|_2^2 = \mathbb{E}((Y - h)^2) = \sigma^2 + (\mu - h)^2$, so that $\mu$ is the best predictor of $Y$. The set $H$ of possible estimators is the real line $\mathbb{R}$. ●

---

† $\| Z \|_2^2$ denotes $\{\| Z \|_2\}^2$.

**(13) Example.** Let $X_1, X_2, \ldots$ be uncorrelated random variables with zero means and unit variances. It is desired to find the best predictor of $Y$ amongst the class of linear combinations of the $X_i$. Clearly

$$\mathbb{E}\left(\left(Y - \sum_i a_i X_i\right)^2\right) = \mathbb{E}(Y^2) - 2\sum_i a_i \mathbb{E}(X_i Y) + \sum_i a_i^2$$

$$= \mathbb{E}(Y^2) + \sum_i \left[a_i - \mathbb{E}(X_i Y)\right]^2 - \sum_i \mathbb{E}(X_i Y)^2.$$

This is a minimum when $a_i = \mathbb{E}(X_i Y)$ for all $i$, so that $\widehat{Y} = \sum_i X_i \mathbb{E}(X_i Y)$. (*Exercise:* Prove that $\mathbb{E}(\widehat{Y}^2) < \infty$.) This is seen best in the following light. Thinking of the $X_i$ as orthogonal (that is, uncorrelated) unit vectors in the space $H$ of linear combinations of the $X_i$, we have found that $\widehat{Y}$ is the weighted average of the $X_i$, weighted in proportion to the magnitudes of their 'projections' onto $Y$. The geometry of this example is relevant to Theorem (14).    ●

**(14) Projection theorem.** *Let $H$ be a closed linear space (with respect to $\| \cdot \|_2$) of random variables, and let $Y$ satisfy $\mathbb{E}(Y^2) < \infty$. Let $M \in H$. The following two statements are equivalent:*

**(15)**                          $\mathbb{E}\big((Y - M)Z\big) = 0 \quad \text{for all } Z \in H,$

**(16)**                          $\|Y - M\|_2 \le \|Y - Z\|_2 \quad \text{for all } Z \in H.$

Here is the geometrical intuition. Let $L_2(\Omega, \mathcal{F}, \mathbb{P})$ be the set of random variables on $(\Omega, \mathcal{F}, \mathbb{P})$ having finite second moment. Now $H$ is a linear subspace of $L_2(\Omega, \mathcal{F}, \mathbb{P})$; think of $H$ as a hyperplane in a vector space of very large dimension. If $Y \notin H$ then the shortest route from $Y$ to $H$ is along the perpendicular from $Y$ onto $H$. Writing $\widehat{Y}$ for the foot of this perpendicular, we have that $Y - \widehat{Y}$ is perpendicular to any vector in the hyperplane $H$. Translating this geometrical remark back into the language of random variables, we conclude that $\langle Y - \widehat{Y}, Z \rangle = 0$ for all $Z \in H$, where $\langle U, V \rangle$ is the scalar product in $L_2(\Omega, \mathcal{F}, \mathbb{P})$ defined by $\langle U, V \rangle = \mathbb{E}(UV)$. These remarks do not of course constitute a proof of the theorem.

**Proof.** Suppose first that $M \in H$ satisfies (15). Then, for $M' \in H$,

$$\mathbb{E}\big((Y - M')^2\big) = \mathbb{E}\big((Y - M + M - M')^2\big)$$

$$= \mathbb{E}\big((Y - M)^2\big) + \mathbb{E}\big((M - M')^2\big)$$

by (15), since $M - M' \in H$; therefore $\|Y - M\|_2 \le \|Y - M'\|_2$ for all $M' \in H$.

Conversely, suppose that $M$ satisfies (16), but that there exists $Z \in H$ such that

$$\mathbb{E}\big((Y - M)Z\big) = d > 0.$$

We may assume without loss of generality that $\mathbb{E}(Z^2) = 1$; otherwise replace $Z$ by $Z/\sqrt{\mathbb{E}(Z^2)}$, noting that $\mathbb{E}(Z^2) \ne 0$ since $\mathbb{P}(Z = 0) \ne 1$. Writing $M' = M + dZ$, we have that

$$\mathbb{E}\big((Y - M')^2\big) = \mathbb{E}\big((Y - M + M - M')^2\big)$$

$$= \mathbb{E}\big((Y - M)^2\big) - 2d\mathbb{E}\big((Y - M)Z\big) + d^2\mathbb{E}(Z^2)$$

$$= \mathbb{E}\big((Y - M)^2\big) - d^2,$$

in contradiction of the minimality of $\mathbb{E}((Y - M)^2)$.    ∎

It is only a tiny step from the projection theorem (14) to the observation, well known to statisticians, that the best predictor of $Y$ given $X$ is just the conditional expectation $\mathbb{E}(Y \mid X)$. This fact, easily proved directly (*exercise*), follows immediately from the projection theorem.

**(17) Theorem.** *Let $X$ and $Y$ be random variables, and suppose that $\mathbb{E}(Y^2) < \infty$. The best predictor of $Y$ given $X$ is the conditional expectation $\mathbb{E}(Y \mid X)$.*

**Proof.** Let $H$ be the closed linear space of functions of $X$ having finite second moment. Define $\psi(x) = \mathbb{E}(Y \mid X = x)$. Certainly $\psi(X)$ belongs to $H$, since

$$\mathbb{E}(\psi(X)^2) = \mathbb{E}\big(\mathbb{E}(Y \mid X)^2\big) \le \mathbb{E}\big(\mathbb{E}(Y^2 \mid X)\big) = \mathbb{E}(Y^2),$$

where we have used the Cauchy–Schwarz inequality. On the other hand, for $Z = h(X) \in H$,

$$\begin{aligned}
\mathbb{E}\big([Y - \psi(X)]Z\big) &= \mathbb{E}(Yh(X)) - \mathbb{E}\big(\mathbb{E}(Y \mid X)h(X)\big) \\
&= \mathbb{E}(Yh(X)) - \mathbb{E}\big(\mathbb{E}(Yh(X) \mid X)\big) \\
&= \mathbb{E}(Yh(X)) - \mathbb{E}\big(Yh(X)\big) = 0,
\end{aligned}$$

using the elementary fact that $\mathbb{E}(Yh(X) \mid X) = h(X)\mathbb{E}(Y \mid X)$. Applying the projection theorem, we find that $M = \psi(X) (= \mathbb{E}(Y \mid X))$ minimizes $\|Y - M\|_2$ for $M \in H$, which is the claim of the theorem. $\blacksquare$

Here is an important step. We may take the conclusion of (17) as a *definition* of the conditional expectation $\mathbb{E}(Y \mid X)$: if $\mathbb{E}(Y^2) < \infty$, the *conditional expectation* $\mathbb{E}(Y \mid X)$ of $Y$ given $X$ is defined to be the best predictor of $Y$ given $X$.

There are two major advantages of defining conditional expectation in this way. First, it is a definition which is valid for all pairs $X, Y$ such that $\mathbb{E}(Y^2) < \infty$, regardless of their types (discrete, continuous, and so on). Secondly, it provides a route to a much more general notion of conditional expectation which turns out to be particularly relevant to the martingale theory of Chapter 12.

**(18) Example.** Let $X = \{X_i : i \in I\}$ be a family of random variables, and let $H$ be the space of all functions of the $X_i$ with finite second moments. If $\mathbb{E}(Y^2) < \infty$, the *conditional expectation* $\mathbb{E}(Y \mid X_i, i \in I)$ of $Y$ given the $X_i$ is defined to be the function $M = \psi(X) \in H$ which minimizes the mean squared error $\|Y - M\|_2$ over all $M$ in $H$. Note that $\psi(X)$ satisfies

**(19)** $$\mathbb{E}\big([Y - \psi(X)]Z\big) = 0 \quad \text{for all } Z \in H,$$

and $\psi(X)$ is unique in the sense that $\mathbb{P}(\psi(X) = N) = 1$ if $\|Y - \psi(X)\|_2 = \|Y - N\|_2$ for any $N \in H$. We note here that, strictly speaking, conditional expectations are not actually *unique*; this causes no difficulty, and we shall therefore continue to speak in terms of *the* conditional expectation. $\bullet$

We move on to an important generalization of the idea of conditional expectation, involving 'conditioning on a $\sigma$-field'. Let $Y$ be a random variable on $(\Omega, \mathcal{F}, \mathbb{P})$ having finite second moment, and let $\mathcal{G}$ be a sub-$\sigma$-field of $\mathcal{F}$. Let $H$ be the space of random variables which are $\mathcal{G}$-measurable and have finite second moment. That is to say, $H$ contains those random variables $Z$ such that $\mathbb{E}(Z^2) < \infty$ and $\{Z \le z\} \in \mathcal{G}$ for all $z \in \mathbb{R}$. It is not difficult to see that $H$ is a closed linear space with respect to $\|\cdot\|_2$. We have from (9) that there exists an element

$M$ of $H$ such that $\|Y - M\|_2 \le \|Y - Z\|_2$ for all $Z \in H$, and furthermore $M$ is unique (in the usual way) with this property. We call $M$ the 'conditional expectation of $Y$ given the $\sigma$-field $\mathcal{G}$', written $\mathbb{E}(Y \mid \mathcal{G})$.

This is a more general definition of conditional expectation than that obtained by conditioning on a family of random variables (as in the previous example). To see this, take $\mathcal{G}$ to be the smallest $\sigma$-field with respect to which every member of the family $X = \{X_i : i \in I\}$ is measurable. It is clear that $\mathbb{E}(Y \mid \mathcal{G}) = \mathbb{E}(Y \mid X_i, \ i \in I)$, in the sense that they are equal with probability 1.

We arrive at the following definition by use of the projection theorem (14).

**(20) Definition.** Let $(\Omega, \mathcal{F}, \mathbb{P})$ be a probability space, and let $Y$ be a random variable satisfying $\mathbb{E}(Y^2) < \infty$. If $\mathcal{G}$ is a sub-$\sigma$-field of $\mathcal{F}$, the **conditional expectation** $\mathbb{E}(Y \mid \mathcal{G})$ is a $\mathcal{G}$-measurable random variable satisfying

**(21)** $$\mathbb{E}\big([Y - \mathbb{E}(Y \mid \mathcal{G})]Z\big) = 0 \quad \text{for all } Z \in H,$$

where $H$ is the collection of all $\mathcal{G}$-measurable random variables with finite second moment.

There are certain members of $H$ with particularly simple form, being the indicator functions of events in $\mathcal{G}$. It may be shown without great difficulty that condition (21) may be replaced by

**(22)** $$\mathbb{E}\big([Y - \mathbb{E}(Y \mid \mathcal{G})]I_G\big) = 0 \quad \text{for all } G \in \mathcal{G}.$$

Setting $G = \Omega$, we deduce the important fact that

**(23)** $$\mathbb{E}\big(\mathbb{E}(Y \mid \mathcal{G})\big) = \mathbb{E}(Y).$$

**(24) Example. Doob's martingale.** (Though some ascribe this to Lévy.) Let $Y$ have finite second moment, and let $X_1, X_2, \ldots$ be a sequence of random variables. Define

$$Y_n = \mathbb{E}(Y \mid X_1, X_2, \ldots, X_n).$$

Then $\{Y_n\}$ is a martingale with respect to $\{X_n\}$. To show this it is necessary to prove that $\mathbb{E}|Y_n| < \infty$ and $\mathbb{E}(Y_{n+1} \mid X_1, X_2, \ldots, X_n) = Y_n$. Certainly $\mathbb{E}|Y_n| < \infty$, since $\mathbb{E}(Y_n^2) < \infty$. For the other part, let $H_n$ be the space of functions of $X_1, X_2, \ldots, X_n$ having finite second moment. We have by (19) that, for $Z \in H_n$,

$$0 = \mathbb{E}\big((Y - Y_n)Z\big) = \mathbb{E}\big((Y - Y_{n+1} + Y_{n+1} - Y_n)Z\big)$$
$$= \mathbb{E}\big((Y_{n+1} - Y_n)Z\big) \quad \text{since} \quad Z \in H_n \subseteq H_{n+1}.$$

Therefore $Y_n = \mathbb{E}(Y_{n+1} \mid X_1, X_2, \ldots, X_n)$.

Here is a more general formulation. Let $Y$ be a random variable on $(\Omega, \mathcal{F}, \mathbb{P})$ with $\mathbb{E}(Y^2) < \infty$, and let $\{\mathcal{G}_n : n \ge 1\}$ be a sequence of $\sigma$-fields contained in $\mathcal{F}$ and satisfying $\mathcal{G}_n \subseteq \mathcal{G}_{n+1}$ for all $n$. Such a sequence $\{\mathcal{G}_n\}$ is called a *filtration*; in the context of the previous paragraph we might take $\mathcal{G}_n$ to be the smallest $\sigma$-field with respect to which $X_1, X_2, \ldots, X_n$ are each measurable. We define $Y_n = \mathbb{E}(Y \mid \mathcal{G}_n)$. As before $\{Y_n\}$ satisfies $\mathbb{E}(Y_{n+1} \mid \mathcal{G}_n) = Y_n$ and $\mathbb{E}|Y_n| < \infty$; such a sequence is called a 'martingale with respect to the filtration $\{\mathcal{G}_n\}$'.  ●

This new type of conditional expectation has many useful properties. We single out one of these, namely the *pull-through property*, thus named since it involves a random variable being pulled through a paranthesis.

**(25) Theorem.** *Let $Y$ have finite second moment and let $\mathcal{G}$ be a sub-$\sigma$-field of the $\sigma$-field $\mathcal{F}$. Then $\mathbb{E}(XY \mid \mathcal{G}) = X\mathbb{E}(Y \mid \mathcal{G})$ for all $\mathcal{G}$-measurable random variables $X$ with finite second moments.*

**Proof.** Let $X$ be $\mathcal{G}$-measurable with finite second moment. Clearly

$$Z = \mathbb{E}(XY \mid \mathcal{G}) - X\mathbb{E}(Y \mid \mathcal{G})$$

is $\mathcal{G}$-measurable and satisfies

$$Z = X\big[Y - \mathbb{E}(Y \mid \mathcal{G})\big] - \big[XY - \mathbb{E}(XY \mid \mathcal{G})\big]$$

so that, for $G \in \mathcal{G}$,

$$\mathbb{E}(ZI_G) = \mathbb{E}\big(\big[Y - \mathbb{E}(Y \mid \mathcal{G})\big]XI_G\big) - \mathbb{E}\big(\big[XY - \mathbb{E}(XY \mid \mathcal{G})\big]I_G\big) = 0,$$

the first term being zero by the fact that $XI_G$ is $\mathcal{G}$-measurable with finite second moment, and the second by the definition of $\mathbb{E}(XY \mid \mathcal{G})$. Any $\mathcal{G}$-measurable random variable $Z$ satisfying $\mathbb{E}(ZI_G) = 0$ for all $G \in \mathcal{G}$ is such that $\mathbb{P}(Z = 0) = 1$ (just set $G_1 = \{Z > 0\}$, $G_2 = \{Z < 0\}$ in turn), and the result follows. ∎

In all our calculations so far, we have used the norm $\|\cdot\|_2$, leading to a definition of $\mathbb{E}(Y \mid \mathcal{G})$ for random variables $Y$ with $\mathbb{E}(Y^2) < \infty$. This condition of finite second moment is of course too strong in general, and needs to be replaced by the natural weaker condition that $\mathbb{E}|Y| < \infty$. One way of doing this would be to rework the previous arguments using instead the norm $\|\cdot\|_1$. An easier route is to use the technique of 'truncation' as in the following proof.

**(26) Theorem.** *Let $Y$ be a random variable on $(\Omega, \mathcal{F}, \mathbb{P})$ with $\mathbb{E}|Y| < \infty$, and let $\mathcal{G}$ be a sub-$\sigma$-field of $\mathcal{F}$. There exists a random variable $Z$ such that:*
  (a) *$Z$ is $\mathcal{G}$-measurable,*
  (b) *$\mathbb{E}|Z| < \infty$,*
  (c) *$\mathbb{E}((Y - Z)I_G) = 0$ for all $G \in \mathcal{G}$.*
*$Z$ is unique in the sense that, for any $Z'$ satisfying (a), (b), and (c), we have that $\mathbb{P}(Z = Z') = 1$.*

The random variable $Z$ in the theorem is called the 'conditional expectation of $Y$ given $\mathcal{G}$', and is written $\mathbb{E}(Y \mid \mathcal{G})$. It is an *exercise* to prove that

**(27)**                                $$\mathbb{E}(XY \mid \mathcal{G}) = X\mathbb{E}(Y \mid \mathcal{G})$$

for all $\mathcal{G}$-measurable $X$, whenever both sides exist, and also that this definition coincides (almost surely) with the previous one when $Y$ has finite second moment. A meaningful value can be assigned to $\mathbb{E}(Y \mid \mathcal{G})$ under the weaker assumption on $Y$ that either $\mathbb{E}(Y^+) < \infty$ or $\mathbb{E}(Y^-) < \infty$.

**Proof.** Suppose first that $Y \geq 0$ and $\mathbb{E}|Y| < \infty$. Let $Y_n = \min\{Y, n\}$, so that $Y_n \uparrow Y$ as $n \to \infty$. Certainly $\mathbb{E}(Y_n^2) < \infty$, and hence we may use (20) to find the conditional expectation $\mathbb{E}(Y_n \mid \mathcal{G})$, a $\mathcal{G}$-measurable random variable satisfying

$$(28) \qquad\qquad \mathbb{E}\big([Y_n - \mathbb{E}(Y_n \mid \mathcal{G})]I_G\big) = 0 \quad \text{for all } G \in \mathcal{G}.$$

Now $Y_n \leq Y_{n+1}$, and so we may take $\mathbb{E}(Y_n \mid \mathcal{G}) \leq \mathbb{E}(Y_{n+1} \mid \mathcal{G})$; see Exercise (7.9.4iii). Hence $\lim_{n\to\infty} \mathbb{E}(Y_n \mid \mathcal{G})$ exists, and we write $\mathbb{E}(Y \mid \mathcal{G})$ for this limit, a $\mathcal{G}$-measurable random variable. By monotone convergence (5.6.12) and (23), $\mathbb{E}(Y_n I_G) \uparrow \mathbb{E}(Y I_G)$, and

$$\mathbb{E}\big[\mathbb{E}(Y_n \mid \mathcal{G})I_G\big] \uparrow \mathbb{E}\big[\mathbb{E}(Y \mid \mathcal{G})I_G\big] = \mathbb{E}\big[\mathbb{E}(Y I_G \mid \mathcal{G})\big] = \mathbb{E}(Y I_G),$$

so that, by (28), $\mathbb{E}([Y - \mathbb{E}(Y \mid \mathcal{G})]I_G) = 0$ for all $G \in \mathcal{G}$.

Next we lift in the usual way the restriction that $Y$ be non-negative. We express $Y$ as $Y = Y^+ - Y^-$ where $Y^+ = \max\{Y, 0\}$ and $Y^- = -\min\{Y, 0\}$ are non-negative; we define $\mathbb{E}(Y \mid \mathcal{G}) = \mathbb{E}(Y^+ \mid \mathcal{G}) - \mathbb{E}(Y^- \mid \mathcal{G})$. It is easy to check that $\mathbb{E}(Y \mid \mathcal{G})$ satisfies (a), (b), and (c). To see the uniqueness, suppose that there exist two $\mathcal{G}$-measurable random variables $Z_1$ and $Z_2$ satisfying (c). Then $\mathbb{E}((Z_1 - Z_2)I_G) = \mathbb{E}((Y - Y)I_G) = 0$ for all $G \in \mathcal{G}$. Setting $G = \{Z_1 > Z_2\}$ and $G = \{Z_1 < Z_2\}$ in turn, we find that $\mathbb{P}(Z_1 = Z_2) = 1$ as required. ∎

Having defined $\mathbb{E}(Y \mid \mathcal{G})$, we can of course define conditional probabilities also: if $A \in \mathcal{F}$, we define $\mathbb{P}(A \mid \mathcal{G}) = \mathbb{E}(I_A \mid \mathcal{G})$. It may be checked that $\mathbb{P}(\varnothing \mid \mathcal{G}) = 0$, $\mathbb{P}(\Omega \mid \mathcal{G}) = 1$ a.s., and $\mathbb{P}(\bigcup_i A_i \mid \mathcal{G}) = \sum_i \mathbb{P}(A_i \mid \mathcal{G})$ a.s. for any sequence $\{A_i : i \geq 1\}$ of disjoint events in $\mathcal{F}$.

It looks as though there should be a way of defining $\mathbb{P}(\cdot \mid \mathcal{G})$ so that it is a probability measure on $(\Omega, \mathcal{F})$. This turns out to be impossible in general, but the details are beyond the scope of this book.

---

## Exercises for Section 7.9

**1.** Let $Y$ be uniformly distributed on $[-1, 1]$ and let $X = Y^2$.
(a) Find the best predictor of $X$ given $Y$, and of $Y$ given $X$.
(b) Find the best linear predictor of $X$ given $Y$, and of $Y$ given $X$.

**2.** (a) Let the pair $(X, Y)$ have a general bivariate normal distribution. Find $\mathbb{E}(Y \mid X)$.
(b) Let $U_1, U_2, \ldots, U_n$ be independent $N(0, 1)$ random variables, and let $(a_i)$, $(b_i)$ be real non-zero vectors. Show that $X = \sum_i a_i U_i$ and $Y = \sum_i b_i U_i$ satisfy

$$\mathbb{E}(Y \mid X) = X \frac{\sum_i a_i b_i}{\sum_i a_i^2}.$$

**3.** Let $X_1, X_2, \ldots, X_n$ be random variables with zero means and covariance matrix $\mathbf{V} = (v_{ij})$, and let $Y$ have finite second moment. Find the linear function $h$ of the $X_i$ which minimizes the mean squared error $\mathbb{E}\{(Y - h(X_1, \ldots, X_n))^2\}$.

**4.** Verify the following properties of conditional expectation. You may assume that the relevant expectations exist.
(a) $\mathbb{E}\{\mathbb{E}(Y \mid \mathcal{G})\} = \mathbb{E}(Y)$.
(b) $\mathbb{E}(\alpha Y + \beta Z \mid \mathcal{G}) = \alpha \mathbb{E}(Y \mid \mathcal{G}) + \beta \mathbb{E}(Z \mid \mathcal{G})$ for $\alpha, \beta \in \mathbb{R}$.
(c) $\mathbb{E}(Y \mid \mathcal{G}) \geq 0$ if $Y \geq 0$.
(d) $\mathbb{E}(Y \mid \mathcal{G}) = \mathbb{E}\{\mathbb{E}(Y \mid \mathcal{H}) \mid \mathcal{G}\}$ if $\mathcal{G} \subseteq \mathcal{H}$.
(e) $\mathbb{E}(Y \mid \mathcal{G}) = \mathbb{E}(Y)$ if $Y$ is independent of $I_G$ for every $G \in \mathcal{G}$.

(f) **Jensen's inequality.** $g\{\mathbb{E}(Y \mid \mathcal{G})\} \le \mathbb{E}\{g(Y) \mid \mathcal{G}\}$ for all convex functions $g$.

(g) If $Y_n \xrightarrow{\text{a.s.}} Y$ and $|Y_n| \le Z$ a.s. where $\mathbb{E}(Z) < \infty$, then $\mathbb{E}(Y_n \mid \mathcal{G}) \xrightarrow{\text{a.s.}} \mathbb{E}(Y \mid \mathcal{G})$.

Statements (b)–(f) are of course to be interpreted 'almost surely'.

**5.**   Let $X$ and $Y$ have joint mass function $f(x, y) = \{x(x+1)\}^{-1}$ for $x = y = 1, 2, \ldots$. Show that $\mathbb{E}(Y \mid X) < \infty$ while $\mathbb{E}(Y) = \infty$.

**6.**   Let $(\Omega, \mathcal{F}, \mathbb{P})$ be a probability space and let $\mathcal{G}$ be a sub-$\sigma$-field of $\mathcal{F}$. Let $H$ be the space of $\mathcal{G}$-measurable random variables with finite second moment.

(a) Show that $H$ is closed with respect to the norm $\| \cdot \|_2$.

(b) Let $Y$ be a random variable satisfying $\mathbb{E}(Y^2) < \infty$, and show the equivalence of the following two statements for any $M \in H$:

   (i) $\mathbb{E}\{(Y - M)Z\} = 0$ for all $Z \in H$,

   (ii) $\mathbb{E}\{(Y - M)I_G\} = 0$ for all $G \in \mathcal{G}$.

**7.**   **Maximal correlation coefficient.** For possibly dependent random variables $X$ and $Y$, define the *maximal correlation coefficient* $m(X, Y) = \sup \rho(f(X), g(Y))$, where $\rho$ denotes ordinary correlation, and the supremum is over all functions $f$ and $g$ such that $f(X)$ and $g(Y)$ have finite non-zero variances. Show that:

(a) $m(X, Y) = 0$ if and only if $X$ and $Y$ are independent,

(b) $m(X, Y)^2 = \sup_g \text{var}(\mathbb{E}(g(Y) \mid X))$, where the supremum is over all functions $g$ such that $\text{var}(g(Y)) = 1$,

(c) $\widehat{f}(X) m(X, Y) = \mathbb{E}(\widehat{g}(Y) \mid X)$ a.s., where $\widehat{f}$ and $\widehat{g}$ are functions such that $m = \rho(\widehat{f}(X), \widehat{g}(Y))$.

(d) We have, a.s., that

$$\mathbb{E}\big(\mathbb{E}(\widehat{f}(X) \mid Y) \,\big|\, X\big) = m(X, Y)^2 \widehat{f}(X), \quad \mathbb{E}\big(\mathbb{E}(\widehat{g}(Y) \mid X) \,\big|\, Y\big) = m(X, Y)^2 \widehat{g}(Y).$$

(e) If the ordered triple $X, Y, Z$ is a Markov chain, show that $m(X, Z) \le m(X, Y)m(Y, Z)$, with equality if $(X, Y)$ and $(Z, Y)$ are identically distributed.

(f) Deduce that, for a pair $(U, V)$ with the standard bivariate normal distribution, $m(U, V)$ is an increasing function of the modulus $|\rho|$ of correlation. [It can be shown that $m(U, V) = |\rho|$ in this case.]

**8.**   **Monotone correlation.** Let $\rho_{\text{mon}}(X, Y) = \sup \rho(f(X), g(Y))$, where the supremum is over all monotonic functions $f$ and $g$ such that $f(X)$ and $g(Y)$ have finite, non-zero variances. Show that $\rho_{\text{mon}}(X, Y) = 0$ if and only if $X$ and $Y$ are independent.

For random variables $X, Y$ with finite, non-zero variances, show that $\rho(X, Y) \le \rho_{\text{mon}}(X, Y) \le m(X, Y)$, where $m$ is the maximal correlation coefficient.

**9.**   **Prediction.** Let $(X, Y)$ have joint density function $f(x, y) = 2e^{-x-y}$ for $0 < x \le y < \infty$.

(a) Find the minimum mean-squared-error predictor of $X$ given that $Y = y$.

(b) Find the minimum mean-squared-error *linear* predictor of $X$ given that $Y = y$.

(c) Compare these.

---

# 7.10   Uniform integrability

Suppose that we are presented with a sequence $\{X_n : n \ge 1\}$ of random variables, and we are able to prove that $X_n \xrightarrow{\text{P}} X$. Convergence in probability tells us little about the behaviour of $\mathbb{E}(X_n)$, as the trite example

$$Y_n = \begin{cases} n & \text{with probability } n^{-1}, \\ 0 & \text{otherwise}, \end{cases}$$

shows; in this special case, $Y_n \xrightarrow{P} 0$ but $\mathbb{E}(Y_n) = 1$ for all $n$. Should we wish to prove that $\mathbb{E}(X_n) \to \mathbb{E}(X)$, or further that $X_n \xrightarrow{1} X$ (which is to say that $\mathbb{E}|X_n - X| \to 0$), then an additional condition is required.

We encountered in an earlier Exercise (7.2.2) an argument of the kind required. If $X_n \xrightarrow{P} X$ and $|X_n| \le Y$ for some $Y$ such that $\mathbb{E}|Y| < \infty$, then $X_n \xrightarrow{1} X$. This extra condition, that $\{X_n\}$ be dominated *uniformly*, is often too strong an assumption in cases of interest. A weaker condition is provided by the following definition. As usual $I_A$ denotes the indicator function of the event $A$.

**(1) Definition.** A sequence $X_1, X_2, \ldots$ of random variables is said to be **uniformly integrable** if

(2)
$$\sup_n \mathbb{E}(|X_n| I_{\{|X_n| \ge a\}}) \to 0 \quad \text{as} \quad a \to \infty.$$

Let us investigate this condition briefly. A random variable $Y$ is called 'integrable' if $\mathbb{E}|Y| < \infty$, which is to say that

$$\mathbb{E}(|Y| I_{\{|Y| \ge a\}}) = \int_{|y| \ge a} |y| \, dF_Y(y)$$

tends to 0 as $a \to \infty$ (see Exercise (5.6.5)). Therefore, a family $\{X_n : n \ge 1\}$ is 'integrable' if

$$\mathbb{E}(|X_n| I_{\{X_n \ge a\}}) \to 0 \quad \text{as} \quad a \to \infty$$

for all $n$, and is 'uniformly integrable' if the convergence is uniform in $n$. Roughly speaking, the condition of integrability restricts the amount of probability in the tails of the distribution, and uniform integrability restricts such quantities *uniformly* over the family of random variables in question.

The principal use of uniform integrability is demonstrated by the following theorem.

**(3) Theorem.** *Suppose that $X_1, X_2, \ldots$ is a sequence of random variables satisfying $X_n \xrightarrow{P} X$. The following three statements are equivalent to one another.*
  (a) *The family $\{X_n : n \ge 1\}$ is uniformly integrable.*
  (b) $\mathbb{E}|X_n| < \infty$ *for all $n$, $\mathbb{E}|X| < \infty$, and $X_n \xrightarrow{1} X$.*
  (c) $\mathbb{E}|X_n| < \infty$ *for all $n$, and $\mathbb{E}|X_n| \to \mathbb{E}|X| < \infty$.*

In advance of proving this, we note some sufficient conditions for uniform integrability.

**(4) Example.** Suppose $|X_n| \le Y$ for all $n$, where $\mathbb{E}|Y| < \infty$. Then

$$|X_n| I_{\{|X_n| \ge a\}} \le |Y| I_{\{|Y| \ge a\}},$$

so that

$$\sup_n \mathbb{E}(|X_n| I_{\{|X_n| \ge a\}}) \le \mathbb{E}(|Y| I_{\{|Y| \ge a\}})$$

which tends to zero as $a \to \infty$, since $\mathbb{E}|Y| < \infty$.     ●

**(5) Example.** Suppose that there exist $\delta > 0$ and $K < \infty$ such that $\mathbb{E}(|X_n|^{1+\delta}) \leq K$ for all $n$. Then

$$\mathbb{E}(|X_n|I_{\{|X_n|\geq a\}}) \leq \frac{1}{a^\delta}\mathbb{E}(|X_n|^{1+\delta}I_{\{|X_n|\geq a\}})$$

$$\leq \frac{1}{a^\delta}\mathbb{E}(|X_n|^{1+\delta}) \leq \frac{K}{a^\delta} \to 0$$

as $a \to \infty$, so that the family is uniformly integrable. ●

Turning to the proof of Theorem (3), we note first a preliminary lemma which is of value in its own right.

**(6) Lemma.** *A family $\{X_n : n \geq 1\}$ is uniformly integrable if and only if both of the following hold:*

(a) $\sup_n \mathbb{E}|X_n| < \infty$,

(b) *for all $\epsilon > 0$, there exists $\delta > 0$ such that, for all $n$, $\mathbb{E}(|X_n|I_A) < \epsilon$ for any event $A$ such that $\mathbb{P}(A) < \delta$.*

The equivalent statement for a single random variable $X$ is the assertion that $\mathbb{E}|X| < \infty$ if and only if

**(7)** $$\sup_{A:\mathbb{P}(A)<\delta} \mathbb{E}(|X|I_A) \to 0 \quad \text{as} \quad \delta \to 0;$$

see Exercise (5.6.5).

**Proof of (6).** Suppose first that $\{X_n\}$ is uniformly integrable. For any $a > 0$,

$$\mathbb{E}|X_n| = \mathbb{E}(|X_n|I_{\{X_n<a\}}) + \mathbb{E}(|X_n|I_{\{X_n\geq a\}}),$$

and therefore

$$\sup_n \mathbb{E}|X_n| \leq a + \sup_n \mathbb{E}(|X_n|I_{\{X_n\geq a\}}).$$

We use uniform integrability to find that $\sup_n \mathbb{E}|X_n| < \infty$. Next,

**(8)** $$\mathbb{E}(|X_n|I_A) = \mathbb{E}(|X_n|I_{A\cap B_n(a)}) + \mathbb{E}(|X_n|I_{A\cap B_n(a)^c})$$

where $B_n(a) = \{|X_n| \geq a\}$. Now

$$\mathbb{E}(|X_n|I_{A\cap B_n(a)}) \leq \mathbb{E}(|X_n|I_{B_n(a)})$$

and

$$\mathbb{E}(|X_n|I_{A\cap B_n(a)^c}) \leq a\mathbb{E}(I_A) = a\mathbb{P}(A).$$

Let $\epsilon > 0$ and pick $a$ such that $\mathbb{E}(|X_n|I_{B_n(a)}) < \frac{1}{2}\epsilon$ for all $n$. We have from (8) that $\mathbb{E}(|X_n|I_A) \leq \frac{1}{2}\epsilon + a\mathbb{P}(A)$, which is smaller than $\epsilon$ whenever $\mathbb{P}(A) < \epsilon/(2a)$.

Secondly, suppose that (a) and (b) hold; let $\epsilon > 0$ and pick $\delta$ according to (b). We have that

$$\mathbb{E}|X_n| \geq \mathbb{E}(|X_n|I_{B_n(a)}) \geq a\mathbb{P}(B_n(a))$$

(this is Markov's inequality) so that

$$\sup_n \mathbb{P}(B_n(a)) \le \frac{1}{a} \sup_n \mathbb{E}|X_n| < \infty.$$

Pick $a$ such that $a^{-1} \sup_n \mathbb{E}|X_n| < \delta$, implying that $\mathbb{P}(B_n(a)) < \delta$ for all $n$. It follows from (b) that $\mathbb{E}(|X_n|I_{B_n(a)}) < \epsilon$ for all $n$, and hence $\{X_n\}$ is uniformly integrable. ∎

**Proof of Theorem (3).** The main part is the statement that (a) implies (b), and we prove this first. Suppose that the family is uniformly integrable. Certainly each member is integrable, so that $\mathbb{E}|X_n| < \infty$ for all $n$. Since $X_n \xrightarrow{\text{P}} X$, there exists a subsequence $\{X_{n_k} : k \ge 1\}$ such that $X_{n_k} \xrightarrow{\text{a.s.}} X$ (see Theorem (7.2.13)). By Fatou's lemma (5.6.13),

(9) $$\mathbb{E}|X| = \mathbb{E}\left(\liminf_{k\to\infty} |X_{n_k}|\right) \le \liminf_{k\to\infty} \mathbb{E}|X_{n_k}| \le \sup_n \mathbb{E}|X_n|,$$

which is finite as a consequence of Lemma (6).

To prove convergence in mean, we write, for $\epsilon > 0$,

(10) $$\mathbb{E}|X_n - X| = \mathbb{E}\left(|X_n - X|I_{\{|X_n-X|<\epsilon\}} + |X_n - X|I_{\{|X_n-X|\ge\epsilon\}}\right)$$
$$\le \epsilon + \mathbb{E}(|X_n|I_{A_n}) + \mathbb{E}(|X|I_{A_n})$$

where $A_n = \{|X_n - X| > \epsilon\}$. Now $\mathbb{P}(A_n) \to 0$ in the limit as $n \to \infty$, and hence $\mathbb{E}(|X_n|I_{A_n}) \to 0$ as $n \to \infty$, by Lemma (6). Similarly $\mathbb{E}(|X|I_{A_n}) \to 0$ as $n \to \infty$, by (7), so that $\limsup_{n\to\infty} \mathbb{E}|X_n - X| \le \epsilon$. Let $\epsilon \downarrow 0$ to obtain that $X_n \xrightarrow{1} X$.

That (b) implies (c) is immediate from the observation that

$$\big|\mathbb{E}|X_n| - \mathbb{E}|X|\big| \le \mathbb{E}|X_n - X|,$$

and it remains to prove that (c) implies (a). Suppose then that (c) holds. Clearly

(11) $$\mathbb{E}(|X_n|I_{\{|X_n|\ge a\}}) = \mathbb{E}|X_n| - \mathbb{E}(u(X_n))$$

where $u(x) = |x|I_{(-a,a)}(x)$. Now $u$ is a continuous bounded function on $(-a, a)$ and $X_n \xrightarrow{\text{D}} X$; hence

$$\mathbb{E}(u(X_n)) \to \mathbb{E}(u(X)) = \mathbb{E}(|X|I_{\{|X|<a\}})$$

if $a$ and $-a$ are points of continuity of the distribution function $F_X$ of $X$ (see Theorem (7.2.19) and the comment thereafter). The function $F_X$ is monotone, and therefore the set $\Delta$ of discontinuities of $F_X$ is at most countable. It follows from (11) that

(12) $$\mathbb{E}(|X_n|I_{\{|X_n|\ge a\}}) \to \mathbb{E}|X| - \mathbb{E}(|X|I_{\{|X|<a\}}) = \mathbb{E}(|X|I_{\{|X|\ge a\}})$$

if $a \notin \Delta$. For any $\epsilon > 0$, on the one hand there exists $b \notin \Delta$ such that $\mathbb{E}(|X|I_{\{|X|\ge b\}}) < \epsilon$; with this choice of $b$, there exists by (12) an integer $N$ such that $\mathbb{E}(|X_n|I_{\{|X_n|\ge b\}}) < 2\epsilon$ for all $n \ge N$. On the other hand, there exists $c$ such that $\mathbb{E}(|X_k|I_{\{|X_k|\ge c\}}) < 2\epsilon$ for all $k < N$, since only finitely many terms are involved. If $a > \max\{b, c\}$, we have that $\mathbb{E}(|X_n|I_{\{|X_n|\ge a\}}) < 2\epsilon$ for all $n$, and we have proved that $\{X_n\}$ is uniformly integrable. ∎

The concept of uniform integrability will be of particular value when we return in Chapter 12 to the theory of martingales. The following example may be seen as an illustration of this.

**(13) Example.** Let $Y$ be a random variable on $(\Omega, \mathcal{F}, \mathbb{P})$ with $\mathbb{E}|Y| < \infty$, and let $\{\mathcal{G}_n : n \geq 1\}$ be a filtration, which is to say that $\mathcal{G}_n$ is a sub-$\sigma$-field of $\mathcal{F}$, and furthermore $\mathcal{G}_n \subseteq \mathcal{G}_{n+1}$ for all $n$. Let $X_n = \mathbb{E}(Y \mid \mathcal{G}_n)$. The sequence $\{X_n : n \geq 1\}$ is uniformly integrable, as may be seen in the following way.

It is a consequence of Jensen's inequality, Exercise (7.9.4vi), that

$$|X_n| = \left|\mathbb{E}(Y \mid \mathcal{G}_n)\right| \leq \mathbb{E}\left(|Y| \mid \mathcal{G}_n\right)$$

almost surely, so that $\mathbb{E}(|X_n|I_{\{|X_n|\geq a\}}) \leq \mathbb{E}(Z_n I_{\{Z_n \geq a\}})$ where $Z_n = \mathbb{E}(|Y| \mid \mathcal{G}_n)$. By the definition of conditional expectation, $\mathbb{E}\left\{(|Y| - Z_n)I_{\{Z_n \geq a\}}\right\} = 0$, so that

**(14)** $$\mathbb{E}(|X_n|I_{\{|X_n|\geq a\}}) \leq \mathbb{E}(|Y|I_{\{Z_n \geq a\}}).$$

We now repeat an argument used before. By Markov's inequality,

$$\mathbb{P}(Z_n \geq a) \leq a^{-1}\mathbb{E}(Z_n) = a^{-1}\mathbb{E}|Y|,$$

and therefore $\mathbb{P}(Z_n \geq a) \to 0$ as $a \to \infty$, uniformly in $n$. Using (7), we deduce that $\mathbb{E}(|Y|I_{\{Z_n \geq a\}}) \to 0$ as $a \to \infty$, uniformly in $n$, implying that the sequence $\{X_n\}$ is uniformly integrable. ●

We finish this section with an application.

**(15) Example. Convergence of moments.** Suppose that $X_1, X_2, \ldots$ is a sequence satisfying $X_n \xrightarrow{D} X$, and furthermore $\sup_n \mathbb{E}(|X_n|^\alpha) < \infty$ for some $\alpha > 1$. It follows that

**(16)** $$\mathbb{E}(X_n^\beta) \to \mathbb{E}(X^\beta)$$

for any integer $\beta$ satisfying $1 \leq \beta < \alpha$. This may be proved either directly or via Theorem (3). First, if $\beta$ is an integer satisfying $1 \leq \beta < \alpha$, then $\{X_n^\beta : n \geq 1\}$ is uniformly integrable by (5), and furthermore $X_n^\beta \xrightarrow{D} X^\beta$ (easy *exercise*, or use Theorem (7.2.18)). If it were the case that $X_n^\beta \xrightarrow{P} X^\beta$ then Theorem (3) would imply the result. In any case, by the Skorokhod representation theorem (7.2.14), there exist random variables $Y, Y_1, Y_2, \ldots$ having the same distributions as $X, X_1, X_2, \ldots$ such that $Y_n^\beta \xrightarrow{P} Y^\beta$. Thus $\mathbb{E}(Y_n^\beta) \to \mathbb{E}(Y^\beta)$ by Theorem (3). However, $\mathbb{E}(Y_n^\beta) = \mathbb{E}(X_n^\beta)$ and $\mathbb{E}(Y^\beta) = \mathbb{E}(X^\beta)$, and the proof is complete. ●

---

## Exercises for Section 7.10

**1.** Show that the sum $\{X_n + Y_n\}$ of two uniformly integrable sequences $\{X_n\}$ and $\{Y_n\}$ gives a uniformly integrable sequence.

**2.** (a) Suppose that $X_n \xrightarrow{r} X$ where $r \geq 1$. Show that $\{|X_n|^r : n \geq 1\}$ is uniformly integrable, and deduce that $\mathbb{E}(X_n^r) \to \mathbb{E}(X^r)$ if $r$ is an integer.
  (b) Conversely, suppose that $\{|X_n|^r : n \geq 1\}$ is uniformly integrable where $r \geq 1$, and show that $X_n \xrightarrow{r} X$ if $X_n \xrightarrow{P} X$.

**3.** Let $g : [0, \infty) \to [0, \infty)$ be an increasing function satisfying $g(x)/x \to \infty$ as $x \to \infty$. Show that the sequence $\{X_n : n \geq 1\}$ is uniformly integrable if $\sup_n \mathbb{E}\{g(|X_n|)\} < \infty$.

**4.** Let $\{Z_n : n \geq 0\}$ be the generation sizes of a branching process with $Z_0 = 1$, $\mathbb{E}(Z_1) = 1$, $\text{var}(Z_1) \neq 0$. Show that $\{Z_n : n \geq 0\}$ is not uniformly integrable.

**5. Pratt's lemma.** Suppose that $X_n \leq Y_n \leq Z_n$ where $X_n \xrightarrow{P} X$, $Y_n \xrightarrow{P} Y$, and $Z_n \xrightarrow{P} Z$. If $\mathbb{E}(X_n) \to \mathbb{E}(X)$ and $\mathbb{E}(Z_n) \to \mathbb{E}(Z)$, show that $\mathbb{E}(Y_n) \to \mathbb{E}(Y)$.

**6.** Let $\{X_n : n \geq 1\}$ be a sequence of variables satisfying $\mathbb{E}(\sup_n |X_n|) < \infty$. Show that $\{X_n\}$ is uniformly integrable.

**7.** Give an example of a uniformly integrable sequence $\{X_n\}$ of random variables and a $\sigma$-field $\mathcal{G}$ such that $X_n \xrightarrow{\text{a.s.}} X$ as $n \to \infty$, but $\mathbb{E}(X_n \mid \mathcal{G})$ does not converge a.s. to $\mathbb{E}(X \mid \mathcal{G})$.

## 7.11 Problems

**1.** Let $X_n$ have density function

$$f_n(x) = \frac{n}{\pi(1 + n^2 x^2)}, \qquad n \geq 1.$$

With respect to which modes of convergence does $X_n$ converge as $n \to \infty$?

**2.** (i) Suppose that $X_n \xrightarrow{\text{a.s.}} X$ and $Y_n \xrightarrow{\text{a.s.}} Y$, and show that $X_n + Y_n \xrightarrow{\text{a.s.}} X + Y$. Show that the corresponding result holds for convergence in $r$th mean and in probability, but not in distribution.

(ii) Show that if $X_n \xrightarrow{\text{a.s.}} X$ and $Y_n \xrightarrow{\text{a.s.}} Y$ then $X_n Y_n \xrightarrow{\text{a.s.}} XY$. Does the corresponding result hold for the other modes of convergence?

**3.** Let $g : \mathbb{R} \to \mathbb{R}$ be continuous. Show that $g(X_n) \xrightarrow{P} g(X)$ if $X_n \xrightarrow{P} X$.

**4.** Let $Y_1, Y_2, \dots$ be independent identically distributed variables, each of which can take any value in $\{0, 1, \dots, 9\}$ with equal probability $\frac{1}{10}$. Let $X_n = \sum_{i=1}^n Y_i \, 10^{-i}$. Show by the use of characteristic functions that $X_n$ converges in distribution to the uniform distribution on $[0, 1]$. Deduce that $X_n \xrightarrow{\text{a.s.}} Y$ for some $Y$ which is uniformly distributed on $[0, 1]$.

**5.** Let $N(t)$ be a Poisson process with constant intensity on $\mathbb{R}$.
(a) Find the covariance of $N(s)$ and $N(t)$.
(b) Show that $N$ is continuous in mean square, which is to say that $\mathbb{E}(\{N(t + h) - N(t)\}^2) \to 0$ as $h \to 0$.
(c) Prove that $N$ is continuous in probability, which is to say that $\mathbb{P}(|N(t + h) - N(t)| > \epsilon) \to 0$ as $h \to 0$, for all $\epsilon > 0$.
(d) Show that $N$ is differentiable in probability but not in mean square.

**6.** Prove that $n^{-1} \sum_{i=1}^n X_i \xrightarrow{\text{a.s.}} 0$ whenever the $X_i$ are independent identically distributed variables with zero means and such that $\mathbb{E}(X_1^4) < \infty$.

**7.** Show that $X_n \xrightarrow{\text{a.s.}} X$ whenever $\sum_n \mathbb{E}(|X_n - X|^r) < \infty$ for some $r > 0$.

**8.** Show that if $X_n \xrightarrow{D} X$ then $a X_n + b \xrightarrow{D} a X + b$ for any real $a$ and $b$.

**9.** (a) **Cantelli, or one-sided Chebyshov inequality.** If $X$ has zero mean and variance $\sigma^2 > 0$, show that

$$\mathbb{P}(X \geq t) \leq \frac{\sigma^2}{\sigma^2 + t^2} \qquad \text{for } t > 0.$$

(b) Deduce that $|\mu - m| \leq \sigma$, where $\mu$, $m$, and $\sigma$ ($> 0$) are the mean, median, and standard deviation of a given distribution.

(c) Use Jensen's inequality to prove part (b) directly.

**10.** Show that $X_n \xrightarrow{P} 0$ if and only if

$$\mathbb{E}\left(\frac{|X_n|}{1 + |X_n|}\right) \to 0 \quad \text{as } n \to \infty.$$

**11.** The sequence $\{X_n\}$ is said to be *mean-square Cauchy convergent* if $\mathbb{E}\{(X_n - X_m)^2\} \to 0$ as $m, n \to \infty$. Show that $\{X_n\}$ converges in mean square to some limit $X$ if and only if it is mean-square Cauchy convergent. Does the corresponding result hold for the other modes of convergence?

**12.** Suppose that $\{X_n\}$ is a sequence of uncorrelated variables with zero means and uniformly bounded variances. Show that $n^{-1} \sum_{i=1}^{n} X_i \xrightarrow{\text{m.s.}} 0$.

**13.** Let $X_1, X_2, \ldots$ be independent identically distributed random variables with the common distribution function $F$, and suppose that $F(x) < 1$ for all $x$. Let $M_n = \max\{X_1, X_2, \ldots, X_n\}$ and suppose that there exists a strictly increasing unbounded positive sequence $a_1, a_2, \ldots$ such that $\mathbb{P}(M_n/a_n \leq x) \to H(x)$ for some distribution function $H$. Let us assume that $H$ is continuous with $0 < H(1) < 1$; substantially weaker conditions suffice but introduce extra difficulties.

(a) Show that $n[1 - F(a_n x)] \to - \log H(x)$ as $n \to \infty$ and deduce that

$$\frac{1 - F(a_n x)}{1 - F(a_n)} \to \frac{\log H(x)}{\log H(1)} \quad \text{if } x > 0.$$

(b) Deduce that if $x > 0$

$$\frac{1 - F(tx)}{1 - F(t)} \to \frac{\log H(x)}{\log H(1)} \quad \text{as } t \to \infty.$$

(c) Set $x = x_1 x_2$ and make the substitution

$$g(x) = \frac{\log H(e^x)}{\log H(1)}$$

to find that $g(x + y) = g(x)g(y)$, and deduce that

$$H(x) = \begin{cases} \exp(-\alpha x^{-\beta}) & \text{if } x \geq 0, \\ 0 & \text{if } x < 0, \end{cases}$$

for some non-negative constants $\alpha$ and $\beta$.

You have shown that $H$ is the distribution function of $Y^{-1}$, where $Y$ has a Weibull distribution.

**14.** Let $X_1, X_2, \ldots, X_n$ be independent and identically distributed random variables with the Cauchy distribution. Show that $M_n = \max\{X_1, X_2, \ldots, X_n\}$ is such that $\pi M_n/n$ converges in distribution, the limiting distribution function being given by $H(x) = e^{-1/x}$ if $x \geq 0$.

**15.** Let $X_1, X_2, \ldots$ be independent and identically distributed random variables whose common characteristic function $\phi$ satisfies $\phi'(0) = i\mu$. Show that $n^{-1} \sum_{j=1}^{n} X_j \xrightarrow{P} \mu$.

**16. Total variation distance.** The *total variation distance* $d_{\text{TV}}(X, Y)$ between two random variables $X$ and $Y$ is defined by

$$d_{\text{TV}}(X, Y) = \sup_{u:\|u\|_\infty = 1} \left|\mathbb{E}(u(X)) - \mathbb{E}(u(Y))\right|$$

where the supremum is over all (measurable) functions $u : \mathbb{R} \to \mathbb{R}$ such that $\|u\|_\infty = \sup_x |u(x)|$ satisfies $\|u\|_\infty = 1$.

(a) If $X$ and $Y$ are discrete with respective masses $f_n$ and $g_n$ at the points $x_n$, show that

$$d_{TV}(X, Y) = \sum_n |f_n - g_n| = 2 \sup_{A \subseteq \mathbb{R}} \left| \mathbb{P}(X \in A) - \mathbb{P}(Y \in A) \right|.$$

(b) If $X$ and $Y$ are continuous with respective density functions $f$ and $g$, show that

$$d_{TV}(X, Y) = \int_{-\infty}^{\infty} |f(x) - g(x)| \, dx = 2 \sup_{A \subseteq \mathbb{R}} \left| \mathbb{P}(X \in A) - \mathbb{P}(Y \in A) \right|.$$

(c) Show that $d_{TV}(X_n, X) \to 0$ implies that $X_n \to X$ in distribution, but that the converse is false.

(d) **Maximal coupling.** Show that $\mathbb{P}(X \neq Y) \geq \frac{1}{2} d_{TV}(X, Y)$, and that there exists a pair $X'$, $Y'$ having the same marginals for which equality holds.

(e) If $X_i$, $Y_j$ are independent random variables, show that

$$d_{TV}\left( \sum_{i=1}^{n} X_i, \sum_{i=1}^{n} Y_i \right) \leq \sum_{i=1}^{n} d_{TV}(X_i, Y_i).$$

**17.** Let $g : \mathbb{R} \to \mathbb{R}$ be bounded and continuous. Show that

$$\sum_{k=0}^{\infty} g(k/n) \frac{(n\lambda)^k}{k!} e^{-n\lambda} \to g(\lambda) \quad \text{as } n \to \infty.$$

**18.** Let $X_n$ and $Y_m$ be independent random variables having the Poisson distribution with parameters $n$ and $m$, respectively. Show that

$$\frac{(X_n - n) - (Y_m - m)}{\sqrt{X_n + Y_m}} \xrightarrow{\text{D}} N(0, 1) \quad \text{as } m, n \to \infty.$$

**19.** (a) Suppose that $X_1, X_2, \ldots$ is a sequence of random variables, each having a normal distribution, and such that $X_n \xrightarrow{\text{D}} X$. Show that $X$ has a normal distribution, possibly degenerate.

(b) For each $n \geq 1$, let $(X_n, Y_n)$ be a pair of random variables having a bivariate normal distribution. Suppose that $X_n \xrightarrow{\text{P}} X$ and $Y_n \xrightarrow{\text{P}} Y$, and show that the pair $(X, Y)$ has a bivariate normal distribution.

**20.** Let $X_1, X_2, \ldots$ be random variables satisfying $\text{var}(X_n) < c$ for all $n$ and some constant $c$. Show that the sequence obeys the weak law, in the sense that $n^{-1} \sum_1^n (X_i - \mathbb{E}X_i)$ converges in probability to 0, if the correlation coefficients satisfy either of the following:

(i) $\rho(X_i, X_j) \leq 0$ for all $i \neq j$,

(ii) $\rho(X_i, X_j) \to 0$ as $|i - j| \to \infty$.

**21.** Let $X_1, X_2, \ldots$ be independent random variables with common density function

$$f(x) = \begin{cases} 0 & \text{if } |x| \leq 2, \\ \dfrac{c}{x^2 \log |x|} & \text{if } |x| > 2, \end{cases}$$

where $c$ is a constant. Show that the $X_i$ have no mean, but $n^{-1} \sum_{i=1}^{n} X_i \xrightarrow{\text{P}} 0$ as $n \to \infty$. Show that convergence does not take place almost surely.

**22.** Let $X_n$ be the Euclidean distance between two points chosen independently and uniformly from the $n$-dimensional unit cube. Show that $\mathbb{E}(X_n)/\sqrt{n} \to 1/\sqrt{6}$ as $n \to \infty$.

**23.** Let $X_1, X_2, \ldots$ be independent random variables having the uniform distribution on $[-1, 1]$. Show that

$$\mathbb{P}\left( \left| \sum_{i=1}^n X_i^{-1} \right| > \tfrac{1}{2} n\pi \right) \to \tfrac{1}{2} \quad \text{as } n \to \infty.$$

**24.** Let $X_1, X_2, \ldots$ be independent random variables, each $X_k$ having mass function given by

$$\mathbb{P}(X_k = k) = \mathbb{P}(X_k = -k) = \frac{1}{2k^2},$$

$$\mathbb{P}(X_k = 1) = \mathbb{P}(X_k = -1) = \frac{1}{2}\left(1 - \frac{1}{k^2}\right) \qquad \text{if } k > 1.$$

Show that $U_n = \sum_1^n X_i$ satisfies $U_n/\sqrt{n} \overset{D}{\to} N(0, 1)$ but $\text{var}(U_n/\sqrt{n}) \to 2$ as $n \to \infty$.

**25.** Let $X_1, X_2, \ldots$ be random variables, and let $N_1, N_2, \ldots$ be random variables taking values in the positive integers such that $N_k \overset{P}{\to} \infty$ as $k \to \infty$. Show that:

(i) if $X_n \overset{D}{\to} X$ and the $X_n$ are independent of the $N_k$, then $X_{N_k} \overset{D}{\to} X$ as $k \to \infty$,

(ii) if $X_n \overset{a.s.}{\to} X$ then $X_{N_k} \overset{P}{\to} X$ as $k \to \infty$.

**26. Stirling's formula.**

(a) Let $a(k, n) = n^k/(k-1)!$ for $1 \le k \le n+1$. Use the fact that $1 - x \le e^{-x}$ if $x \ge 0$ to show that

$$\frac{a(n-k, n)}{a(n+1, n)} \le e^{-k^2/(2n)} \qquad \text{if } k \ge 0.$$

(b) Let $X_1, X_2, \ldots$ be independent Poisson variables with parameter 1, and let $S_n = X_1 + \cdots + X_n$. Define the function $g : \mathbb{R} \to \mathbb{R}$ by

$$g(x) = \begin{cases} -x & \text{if } 0 \ge x \ge -M, \\ 0 & \text{otherwise,} \end{cases}$$

where $M$ is large and positive. Show that, for large $n$,

$$\mathbb{E}\left( g\left\{ \frac{S_n - n}{\sqrt{n}} \right\} \right) = \frac{e^{-n}}{\sqrt{n}} \{a(n+1, n) - a(n-k, n)\}$$

where $k = \lfloor Mn^{1/2} \rfloor$. Now use the central limit theorem and (a) above, to deduce Stirling's formula:

$$\frac{n! \, e^n}{n^{n+\frac{1}{2}}\sqrt{2\pi}} \to 1 \qquad \text{as } n \to \infty.$$

**27. Pólya's urn.** A bag contains red and green balls. A ball is drawn from the bag, its colour noted, and then it is returned to the bag together with a new ball of the same colour. Initially the bag contained one ball of each colour. If $R_n$ denotes the number of red balls in the bag after $n$ additions, show that $S_n = R_n/(n+2)$ is a martingale. Deduce that the ratio of red to green balls converges almost surely to some limit as $n \to \infty$.

**28. Anscombe's theorem.** Let $\{X_i : i \ge 1\}$ be independent identically distributed random variables with zero mean and finite positive variance $\sigma^2$, and let $S_n = \sum_1^n X_i$. Suppose that the integer-valued random process $M(t)$ satisfies $t^{-1}M(t) \overset{P}{\to} \theta$ as $t \to \infty$, where $\theta$ is a positive constant. Show that

$$\frac{S_{M(t)}}{\sigma\sqrt{\theta t}} \overset{D}{\to} N(0, 1) \quad \text{and} \quad \frac{S_{M(t)}}{\sigma\sqrt{M(t)}} \overset{D}{\to} N(0, 1) \quad \text{as } t \to \infty.$$

You should not assume that the process $M$ is independent of the $X_i$.

**29. Kolmogorov's inequality.** Let $X_1, X_2, \ldots$ be independent random variables with zero means, and $S_n = X_1 + X_2 + \cdots + X_n$. Let $M_n = \max_{1 \le k \le n} |S_k|$ and show that $\mathbb{E}(S_n^2 I_{A_k}) > c^2 \mathbb{P}(A_k)$ where $A_k = \{M_{k-1} \le c < M_k\}$ and $c > 0$. Deduce Kolmogorov's inequality:

$$\mathbb{P}\left( \max_{1 \le k \le n} |S_k| > c \right) \le \frac{\mathbb{E}(S_n^2)}{c^2}, \qquad c > 0.$$

**30.** Let $X_1, X_2, \ldots$ be independent random variables with zero means, and let $S_n = X_1 + X_2 + \cdots + X_n$. Using Kolmogorov's inequality or the martingale convergence theorem, show that:

(i) $\sum_{i=1}^{\infty} X_i$ converges almost surely if $\sum_{k=1}^{\infty} \mathbb{E}(X_k^2) < \infty$,

(ii) if there exists an increasing real sequence $(b_n)$ such that $b_n \to \infty$, and satisfying the inequality $\sum_{k=1}^{\infty} \mathbb{E}(X_k^2)/b_k^2 < \infty$, then $b_n^{-1} \sum_{k=1}^{\infty} X_k \xrightarrow{\text{a.s.}} 0$ as $n \to \infty$.

**31. Estimating the transition matrix.** The Markov chain $X_0, X_1, \ldots, X_n$ has initial distribution $f_i = \mathbb{P}(X_0 = i)$ and transition matrix $\mathbf{P}$. The *log-likelihood* function $\lambda(\mathbf{P})$ is defined as $\lambda(\mathbf{P}) = \log(f_{X_0} p_{X_0, X_1} p_{X_1, X_2} \cdots p_{X_{n-1}, X_n})$. Show that:

(a) $\lambda(\mathbf{P}) = \log f_{X_0} + \sum_{i,j} N_{ij} \log p_{ij}$ where $N_{ij}$ is the number of transitions from $i$ to $j$,

(b) viewed as a function of the $p_{ij}$, $\lambda(\mathbf{P})$ is maximal when $p_{ij} = \hat{p}_{ij}$ where $\hat{p}_{ij} = N_{ij}/\sum_k N_{ik}$,

(c) if $X$ is irreducible and ergodic then $\hat{p}_{ij} \xrightarrow{\text{a.s.}} p_{ij}$ as $n \to \infty$.

**32. Ergodic theorem in discrete time.** Let $X$ be an irreducible discrete-time Markov chain, and let $\mu_i$ be the mean recurrence time of state $i$. Let $V_i(n) = \sum_{r=0}^{n-1} I_{\{X_r = i\}}$ be the number of visits to $i$ up to $n - 1$, and let $f$ be any bounded function on $S$. Show that:

(a) $n^{-1} V_i(n) \xrightarrow{\text{a.s.}} \mu_i^{-1}$ as $n \to \infty$,

(b) if $\mu_i < \infty$ for all $i$, then

$$\frac{1}{n} \sum_{r=0}^{n-1} f(X_r) \to \sum_{i \in S} f(i)/\mu_i \quad \text{as } n \to \infty.$$

**33. Ergodic theorem in continuous time.** Let $X$ be an irreducible recurrent continuous-time Markov chain with generator $\mathbf{G}$ and finite mean return times $m_j$.

(a) Show that $\dfrac{1}{t} \displaystyle\int_0^t I_{\{X(s) = j\}} \, ds \xrightarrow{\text{a.s.}} \dfrac{1}{m_j g_j}$ as $t \to \infty$;

(b) deduce that the stationary distribution $\boldsymbol{\pi}$ satisfies $\pi_j = 1/(m_j g_j)$;

(c) show that, if $f$ is a bounded function on $S$,

$$\frac{1}{t} \int_0^t f(X(s)) \, ds \xrightarrow{\text{a.s.}} \sum_i \pi_i f(i) \quad \text{as } t \to \infty.$$

**34. Tail equivalence.** Suppose that the sequences $\{X_n : n \ge 1\}$ and $\{Y_n : n \ge 1\}$ are *tail equivalent*, which is to say that $\sum_{n=1}^{\infty} \mathbb{P}(X_n \ne Y_n) < \infty$. Show that:

(a) $\sum_{n=1}^{\infty} X_n$ and $\sum_{n=1}^{\infty} Y_n$ converge or diverge together,

(b) $\sum_{n=1}^{\infty} (X_n - Y_n)$ converges almost surely,

(c) if there exist a random variable $X$ and a sequence $a_n$ such that $a_n \uparrow \infty$ and $a_n^{-1} \sum_{r=1}^{n} X_r \xrightarrow{\text{a.s.}} X$, then

$$\frac{1}{a_n} \sum_{r=1}^{n} Y_r \xrightarrow{\text{a.s.}} X.$$

**35. Three series theorem.** Let $\{X_n : n \geq 1\}$ be independent random variables. Show that $\sum_{n=1}^{\infty} X_n$ converges a.s. if, for some $a > 0$, the following three series all converge:

(a) $\sum_n \mathbb{P}(|X_n| > a)$,

(b) $\sum_n \text{var}(X_n I_{\{|X_n| \leq a\}})$,

(c) $\sum_n \mathbb{E}(X_n I_{\{|X_n| \leq a\}})$.

[The converse holds also, but is harder to prove.]

**36.** Let $\{X_n : n \geq 1\}$ be independent random variables with continuous common distribution function $F$. We call $X_k$ a *record value* for the sequence if $X_k > X_r$ for $1 \leq r < k$, and we write $I_k$ for the indicator function of the event that $X_k$ is a record value.

(a) Show that the random variables $I_k$ are independent.

(b) Show that $R_m = \sum_{k=1}^{m} I_r$ satisfies $R_m/\log m \xrightarrow{\text{a.s.}} 1$ as $m \to \infty$.

**37. Random harmonic series.** Let $\{X_n : n \geq 1\}$ be a sequence of independent random variables with $\mathbb{P}(X_n = 1) = \mathbb{P}(X_n = -1) = \frac{1}{2}$. Does the series $\sum_{r=1}^{n} X_r/r$ converge a.s. as $n \to \infty$?

**38. Stirling's formula for the gamma function.** Let $X$ have the gamma distribution $\Gamma(1, s)$. By considering the integral of the density function of $Y = (X - s)/\sqrt{s}$, show that $\Gamma(s) \sim \sqrt{2\pi} s^{s-\frac{1}{2}} e^{-s}$ as $s \to \infty$. [Hint: You may find it useful that

$$\int_a^b e^{-u(x)} \, dx \leq \frac{1}{u'(a)} \int_a^b u'(x) e^{-u(x)} \, dx,$$

if $u'(x)$ is strictly positive and increasing.]

**39. Random series.** Let $c_1, c_2, \ldots$ be reals, let $X_1, X_2, \ldots$ be independent random variables with the mass function $f(1) = f(-1) = \frac{1}{2}$, and let $S_n = \sum_{r=1}^{n} c_r X_r$. Write

$$B_n = \sum_{r=1}^{n} c_r^4, \qquad D_n = \sqrt{\sum_{r=1}^{n} c_r^2}.$$

(a) Use characteristic functions to show that $S_n/D_n$ converges in distribution to the $N(0, 1)$ distribution (as $n \to \infty$) if and only if $B_n/D_n^4 \to 0$. [Hint: You may use the fact that $-\frac{2}{3}\theta^4 \leq \frac{1}{2}\theta^2 + \log \cos \theta \leq -\frac{1}{12}\theta^4$ for $-\frac{1}{4}\pi \leq \theta \leq \frac{1}{4}\pi$.]

(b) Find the limit of $S_n/D_n$ in the special case $c_r = 2^{-r}$.

**40. Berge's inequality.** Let $X$ and $Y$ be random variables with mean 0, variance 1, and correlation $\rho$. Show that, for $\epsilon > 0$,

$$\mathbb{P}(|X| \vee |Y| > \epsilon) \leq \frac{1}{\epsilon^2} \left( 1 + \sqrt{1 - \rho^2} \right),$$

where $x \vee y = \max\{x, y\}$. [Hint: If $|t| \leq 1$, the function $g(x, y) = (x^2 - 2txy + y^2)/(\epsilon^2(1 - t^2))$ is non-negative, and moreover satisfies $g(x, y) \geq 1$ when $|x| \vee |y| \geq \epsilon$.]

**41. Poisson tail, balls in bins.** Let $X$ have the Poisson distribution with parameter 1.

(a) Show that $\mathbb{P}(X \geq t) \leq e^{t-1}/t^t$ for $t \geq 1$.

(b) Deduce that the maximum $M_n$ of $n$ independent random variables, distributed as $X$, satisfies

$$\lim_{n \to \infty} \mathbb{P}\left( M_n \geq \frac{(1+a)\log n}{\log \log n} \right) = \begin{cases} 1 & \text{if } a < 0, \\ 0 & \text{if } a > 0. \end{cases}$$

# 8

# Random processes

*Summary.* This brief introduction to random processes includes elementary previews of stationary processes, renewal processes, queueing processes, and the Wiener process (Brownian motion). There are short introductions to Lévy processes, self-similarity, and time changes, followed by a discussion of the Kolmogorov consistency conditions.

## 8.1 Introduction

Recall that a 'random process' $X$ is a family $\{X_t : t \in T\}$ of random variables which map the sample space $\Omega$ into some set $S$. There are many possible choices for the index set $T$ and the state space $S$, and the characteristics of the process depend strongly upon these choices. For example, in Chapter 6 we studied discrete-time $(T = \{0, 1, 2, \dots\})$ and continuous-time $(T = [0, \infty))$ Markov chains which take values in some countable set $S$. Other possible choices for $T$ include $\mathbb{R}^n$ and $\mathbb{Z}^n$, whilst $S$ might be an uncountable set such as $\mathbb{R}$. The mathematical analysis of a random process varies greatly depending on whether $S$ and $T$ are countable or uncountable, just as discrete random variables are distinguishable from continuous variables. The main differences are indicated by those cases in which

(a) $T = \{0, 1, 2, \dots\}$ or $T = [0, \infty)$,

(b) $S = \mathbb{Z}$ or $S = \mathbb{R}$.

There are two levels at which we can observe the evolution of a random process $X$.

(i) Each $X_t$ is a function which maps $\Omega$ into $S$. For any fixed $\omega \in \Omega$, there is a corresponding collection $\{X_t(\omega) : t \in T\}$ of members of $S$; this is called the *realization* or *sample path* of $X$ at $\omega$. We can study properties of sample paths.

(ii) The $X_t$ are not independent in general. If $S \subseteq \mathbb{R}$ and $\mathbf{t} = (t_1, t_2, \dots, t_n)$ is a vector of members of $T$, then the vector $(X_{t_1}, X_{t_2}, \dots, X_{t_n})$ has joint distribution function $F_{\mathbf{t}} : \mathbb{R}^n \to [0, 1]$ given by $F_{\mathbf{t}}(\mathbf{x}) = \mathbb{P}(X_{t_1} \le x_1, \dots, X_{t_n} \le x_n)$. The collection $\{F_{\mathbf{t}}\}$, as $\mathbf{t}$ ranges over all vectors of members of $T$ of any finite length, is called the collection of *finite-dimensional distributions* (abbreviated to *fdds*) of $X$, and it contains all the information which is available about $X$ from the distributions of its component variables $X_t$. We can study the distributional properties of $X$ by using its fdds.

These two approaches do not generally yield the same information about the process in question, since knowledge of the fdds does not yield complete information about the properties of the sample paths. We shall see an example of this in the final section of this chapter†.

We are not concerned here with the general theory of random processes (which tends to be rather abstract and technical), but prefer to study certain specific types of process that are characterized by one or more special properties. This is not a new approach for us. In Chapter 6 we devoted our attention to processes which satisfy the Markov property, whilst large parts of Chapter 7 were devoted to sequences $\{S_n\}$ which were either martingales or the partial sums of independent sequences. In this short chapter we introduce certain other types of process and their characteristic properties. These can be divided broadly under four headings, covering 'stationary processes', 'renewal processes', 'queues', and 'diffusions'; their detailed analysis is left for Chapters 9, 10, 11, and 13 respectively. We also introduce the reader to Lévy processes, subordinators, self-similarity, stability, and time changes.

We shall only be concerned with the cases when $T$ is one of the sets $\mathbb{Z}$, $\{0, 1, 2, \ldots\}$, $\mathbb{R}$, or $[0, \infty)$. If $T$ is an uncountable subset of $\mathbb{R}$, representing continuous time say, then we shall usually write $X(t)$ rather than $X_t$ for ease of notation. Evaluation of $X(t)$ at some $\omega \in \Omega$ yields a point in $S$, which we shall denote by $X(t; \omega)$.

## 8.2  Stationary processes

Many important processes have the property that their finite-dimensional distributions are invariant under time shifts (or space shifts if $T$ is a subset of some Euclidean space $\mathbb{R}^n$).

**(1) Definition.** The process $X = \{X(t) : t \geq 0\}$, taking values in $\mathbb{R}$, is called **strongly stationary** if the families

$$\{X(t_1), X(t_2), \ldots, X(t_n)\} \quad \text{and} \quad \{X(t_1 + h), X(t_2 + h), \ldots, X(t_n + h)\}$$

have the same joint distribution, for all given $t_1, t_2, \ldots, t_n$ and $h > 0$.

Note that, if $X$ is strongly stationary, then $X(t)$ has the same distribution for all $t$.

We saw in Section 3.6 that the covariance of two random variables $X$ and $Y$ contains some information, albeit incomplete, about their joint distribution. With this in mind we formulate another stationarity property which, for processes with $\text{var}(X(t)) < \infty$, is weaker than strong stationarity.

**(2) Definition.** The process $X = \{X(t) : t \geq 0\}$ is called **weakly** (or **second-order** or **covariance**) **stationary** if, for all $t_1$, $t_2$, and $h > 0$,

$$\mathbb{E}(X(t_1)) = \mathbb{E}(X(t_2)) \quad \text{and} \quad \text{cov}\big(X(t_1), X(t_2)\big) = \text{cov}\big(X(t_1 + h), X(t_2 + h)\big).$$

Thus, $X$ is weakly stationary if and only if it has constant mean, and its *autocovariance function*

**(3)**
$$c(t, t + h) = \text{cov}\big(X(t), X(t + h)\big)$$

satisfies

$$c(t, t + h) = c(0, h) \quad \text{for all } t, h \geq 0.$$

---

†See Example (8.9.4).

We emphasize that the autocovariance function $c(s, t)$ of a weakly stationary process is a function of $t - s$ only. Note that $\rho(h) = c(0, h)/c(0, 0)$ is called the *autocorrelation function* of the weakly stationary process $X$.

Definitions similar to (1) and (2) hold for processes with $T = \mathbb{R}$ and for discrete-time processes $X = \{X_n : n \geq 0\}$; the autocovariance function of a weakly stationary discrete-time process $X$ is just a sequence $\{c(0, m) : m \geq 0\}$ of real numbers.

Weak stationarity interests us more than strong stationarity for two reasons. First, the condition of strong stationarity is often too restrictive for certain applications; secondly, many substantial and useful properties of stationary processes are derivable from weak stationarity alone. Thus, the assertion that $X$ is *stationary* should be interpreted to mean that $X$ is *weakly stationary*. Of course, there exist processes which are stationary but not strongly stationary (see Example (5)), and conversely processes without finite second moments may be strongly stationary but not weakly stationary.

**(4) Example. Markov chains.** Let $X = \{X(t) : t \geq 0\}$ be an irreducible, non-explosive Markov chain taking values in some countable subset $S$ of $\mathbb{R}$ and with a unique stationary distribution $\pi$. Then (see Theorem (6.10.22))

$$\mathbb{P}\big(X(t) = j \mid X(0) = i\big) \to \pi_j \quad \text{as} \quad t \to \infty$$

for all $i, j \in S$. The fdds of $X$ depend on the initial distribution $\mu^{(0)}$ of $X(0)$, and it is not generally true that $X$ is stationary. Suppose, however, that $\mu^{(0)} = \pi$. Then the distribution $\mu^{(t)}$ of $X(t)$ satisfies $\mu^{(t)} = \pi \mathbf{P}_t = \pi$, where $\{\mathbf{P}_t\}$ is the transition semigroup of the chain. Thus $X(t)$ has distribution $\pi$ for all $t$. Furthermore, if $0 < s < s + t$ and $h > 0$, the pairs $(X(s), X(s + t))$ and $(X(s + h), X(s + t + h))$ have the same joint distribution since:
  (a)  $X(s)$ and $X(s + h)$ are identically distributed,
  (b)  the distribution of $X(s + h)$ (respectively $X(s + t + h)$) depends only on the distribution of $X(s)$ (respectively $X(s + t)$) and on the transition matrix $\mathbf{P}_h$.
A similar argument holds for collections of the $X(u)$ which contain more than two elements, and we have shown that $X$ is strongly stationary.                                ●

**(5) Example.** Let $A$ and $B$ be uncorrelated (but not necessarily independent) random variables, each of which has mean 0 and variance 1. Fix a number $\lambda \in [0, \pi]$ and define

**(6)**                         $$X_n = A \cos(\lambda n) + B \sin(\lambda n).$$

Then $\mathbb{E} X_n = 0$ for all $n$ and $X = \{X_n\}$ has autocovariance function

$$
\begin{aligned}
c(m, m + n) &= \mathbb{E}(X_m X_{m+n}) \\
&= \mathbb{E}\big(\big[A \cos(\lambda m) + B \sin(\lambda m)\big]\big[A \cos\{\lambda(m + n)\} + B \sin\{\lambda(m + n)\}\big]\big) \\
&= \mathbb{E}\big(A^2 \cos(\lambda m) \cos\{\lambda(m + n)\} + B^2 \sin(\lambda m) \sin\{\lambda(m + n)\}\big) \\
&= \cos(\lambda n)
\end{aligned}
$$

since $\mathbb{E}(AB) = 0$. Thus $c(m, m + n)$ depends on $n$ alone and so $X$ is stationary. In general $X$ is not strongly stationary unless extra conditions are imposed on the joint distribution of $A$ and $B$; to see this for the case $\lambda = \frac{1}{2}\pi$, simply calculate that

$$\{X_0, X_1, X_2, X_3, \ldots\} = \{A, B, -A, -B, \ldots\}$$

which is strongly stationary if and only if the pairs $(A, B)$, $(B, -A)$, and $(-A, -B)$ have the same joint distributions. It can be shown that $X$ is strongly stationary for any $\lambda$ if $A$ and $B$ are $N(0, 1)$ variables. The reason for this lies in Example (4.5.9), where we saw that normal variables are independent whenever they are uncorrelated.                                           ●

Two major results in the theory of stationary processes are the 'spectral theorem' and the 'ergodic theorem'; we close this section with short discussions of these. First, recall from the theory of Fourier analysis that any function $f : \mathbb{R} \to \mathbb{R}$ which

   (a)  is periodic with period $2\pi$ (that is, $f(x + 2\pi) = f(x)$ for all $x$),
   (b)  is continuous, and
   (c)  has bounded variation,

has a unique Fourier expansion

$$f(x) = \tfrac{1}{2}a_0 + \sum_{n=1}^{\infty}\left[a_n \cos(nx) + b_n \sin(nx)\right]$$

which expresses $f$ as the sum of varying proportions of regular oscillations. In some sense to be specified, a stationary process $X$ is similar to a periodic function since its autocovariances are invariant under time shifts. The spectral theorem asserts that, subject to certain conditions, stationary processes can be decomposed in terms of regular underlying oscillations with random magnitudes. The set of frequencies of oscillations which contribute to this combination is called the 'spectrum' of the process. For example, the process $X$ in (5) is specified precisely in these terms by (6). In spectral theory it is convenient to allow the processes in question to take values in the complex plane. In this case (6) can be rewritten as

(7)                          $X_n = \mathrm{Re}(Y_n)$   where   $Y_n = Ce^{i\lambda n}$;

here $C$ is a complex-valued random variable and $i = \sqrt{-1}$. The sequence $Y = \{Y_n\}$ is stationary also whenever $\mathbb{E}(C) = 0$ and $\mathbb{E}(C\overline{C}) < \infty$, where $\overline{C}$ is the complex conjugate of $C$ (but see Definition (9.1.1)).

The ergodic theorem deals with the partial sums of a stationary sequence $X = \{X_n : n \geq 0\}$. Consider first the following two extreme examples of stationarity.

**(8) Example. Independent sequences.** Let $X = \{X_n : n \geq 0\}$ be a sequence of independent identically distributed variables with zero means and unit variances. Certainly $X$ is stationary, and its autocovariance function is given by

$$c(m, m + n) = \mathbb{E}(X_m X_{m+n}) = \begin{cases} 1 & \text{if } n = 0, \\ 0 & \text{if } n \neq 0. \end{cases}$$

The strong law of large numbers implies that $n^{-1} \sum_{j=1}^{n} X_j \xrightarrow{\text{a.s.}} 0$.                                           ●

**(9) Example. Identical sequences.** Let $Y$ be a random variable with zero mean and unit variance, and let $X = \{X_n : n \geq 0\}$ be the stationary sequence given by $X_n = Y$ for all $n$. Then $X$ has autocovariance function $c(m, m + n) = \mathbb{E}(X_m X_{m+n}) = 1$ for all $n$. It is clear that $n^{-1} \sum_{j=1}^{n} X_j \xrightarrow{\text{a.s.}} Y$ since each term in the sum is $Y$ itself.                                           ●

These two examples are, in some sense, extreme examples of stationarity since the first deals with independent variables and the second deals with identical variables. In both examples, however, the averages $n^{-1}\sum_{j=1}^{n} X_j$ converge as $n \to \infty$. In the first case the limit is constant, whilst in the second the limit is a random variable with a non-trivial distribution. This indicates a shared property of stationary processes, and we shall see that any stationary sequence $X = \{X_n : n \geq 0\}$ with finite means satisfies

$$\frac{1}{n}\sum_{j=1}^{n} X_j \xrightarrow{\text{a.s.}} Y$$

for some random variable $Y$. This result is called the ergodic theorem for stationary sequences. A similar result holds for continuous-time stationary processes.

The theory of stationary processes is important and useful in statistics. Many sequences $\{x_n : 0 \leq n \leq N\}$ of observations, indexed by the time at which they were taken, are suitably modelled by random processes, and statistical problems such as the estimation of unknown parameters and the prediction of the future values of the sequence are often studied in this context. Such sequences are called 'time series' and they include many examples which are well known to us already, such as the successive values of the Financial Times Share Index, or the frequencies of sunspots in successive years. Statisticians and politicians often seek to find some underlying structure in such sequences, and to this end they may study 'moving average' processes $Y$, which are smoothed versions of a stationary sequence $X$,

$$Y_n = \sum_{i=0}^{r} \alpha_i X_{n-i},$$

where $\alpha_0, \alpha_1, \ldots, \alpha_r$ are constants. Alternatively, they may try to fit a model to their observations, and may typically consider 'autoregressive schemes' $Y$, being sequences which satisfy

$$Y_n = \sum_{i=1}^{r} \alpha_i Y_{n-i} + Z_n$$

where $\{Z_n\}$ is a sequence of uncorrelated variables with zero means and constant finite variance.

An introduction to the theory of stationary processes is given in Chapter 9.

---

## Exercises for Section 8.2

1.  **Flip–flop.** Let $\{X_n\}$ be a Markov chain on the state space $S = \{0, 1\}$ with transition matrix

$$\mathbf{P} = \begin{pmatrix} 1 - \alpha & \alpha \\ \beta & 1 - \beta \end{pmatrix},$$

where $\alpha + \beta > 0$. Find:
(a) the correlation $\rho(X_m, X_{m+n})$, and its limit as $m \to \infty$ with $n$ remaining fixed,
(b) $\lim_{n\to\infty} n^{-1}\sum_{r=1}^{n} \mathbb{P}(X_r = 1)$.
Under what condition is the process strongly stationary?

2.  **Random telegraph.** Let $\{N(t) : t \geq 0\}$ be a Poisson process of intensity $\lambda$, and let $T_0$ be an independent random variable such that $\mathbb{P}(T_0 = \pm 1) = \frac{1}{2}$. Define $T(t) = T_0(-1)^{N(t)}$. Show that $\{T(t) : t \geq 0\}$ is stationary and find: (a) $\rho(T(s), T(s+t))$, (b) the mean and variance of $X(t) = \int_0^t T(s)\,ds$.

[The so-called Goldstein–Kac process $X(t)$ denotes the position of a particle moving with unit speed, starting from the origin along the positive $x$-axis, whose direction is reversed at the instants of a Poisson process.]

**3.  Korolyuk–Khinchin theorem.** An integer-valued counting process $\{N(t) : t \geq 0\}$ with $N(0) = 0$ is called *crudely stationary* if $p_k(s, t) = \mathbb{P}(N(s+t) - N(s) = k)$ depends only on the length $t - s$ and not on the location $s$. It is called *simple* if, almost surely, it has jump discontinuities of size 1 only. Show that, for a simple crudely stationary process $N$, $\lim_{t \downarrow 0} t^{-1} \mathbb{P}(N(t) > 0) = \mathbb{E}(N(1))$.

**4.  Kac's ergodic formula.** Let $X = \{X_n : n \geq 0\}$ be an ergodic Markov chain started in its stationary distribution $\boldsymbol{\pi}$. Let $A$ be a subset of states, with stationary probability $\pi(A)$, and let $T_A = \min\{n \geq 1 : X_n \in A\}$. Show that $\mathbb{E}(T_A \mid X_0 \in A) = 1/\pi(A)$.

---

# 8.3  Renewal processes

We are often interested in the successive occurrences of events such as the emission of radioactive particles, the failures of light bulbs, or the incidences of earthquakes.

**(1) Example. Light bulb failures.** This is the archetype of renewal processes. A room is lit by a single light bulb. When this bulb fails it is replaced immediately by an apparently identical copy. Let $X_i$ be the (random) lifetime of the $i$th bulb, and suppose that the first bulb is installed at time $t = 0$. Then $T_n = X_1 + X_2 + \cdots + X_n$ is the time until the $n$th failure (where, by convention, we set $T_0 = 0$), and

$$N(t) = \max\{n : T_n \leq t\}$$

is the number of bulbs which have failed by time $t$. It is natural to assume that the $X_i$ are independent and identically distributed random variables.                                ●

**(2) Example. Markov chains.** Let $\{Y_n : n \geq 0\}$ be a Markov chain, and choose some state $i$. We are interested in the time epochs at which the chain is in the state $i$. The times $0 < T_1 < T_2 < \cdots$ of successive visits to $i$ are given by

$$T_1 = \min\{n \geq 1 : Y_n = i\},$$
$$T_{m+1} = \min\{n > T_m : Y_n = i\} \quad \text{for} \quad m \geq 1;$$

they may have defective distributions unless the chain is irreducible and recurrent. Assume $X$ is irreducible and recurrent, and let $\{X_m : m \geq 1\}$ be given by

$$X_m = T_m - T_{m-1} \quad \text{for} \quad m \geq 1,$$

where we set $T_0 = 0$ by convention. It is clear that the $X_m$ are independent, and that $X_2, X_3, \ldots$ are identically distributed since each is the elapsed time between two successive visits to $i$. On the other hand, $X_1$ does *not* have this shared distribution in general, unless the chain began in the state $Y_0 = i$. The number of visits to $i$ which have occurred by time $t$ is given by $N(t) = \max\{n : T_n \leq t\}$.                                ●

Both examples above contain a continuous-time random process $N = \{N(t) : t \geq 0\}$, where $N(t)$ represents the number of occurrences of some event in the time interval $[0, t)$.

Such a process $N$ is called a 'renewal' or 'counting' process for obvious reasons; the Poisson process of Section 6.8 provides another example of a renewal process.

---

**(3) Definition.** A **renewal process** $N = \{N(t) : t \geq 0\}$ is a process for which

$$N(t) = \max\{n : T_n \leq t\}$$

where

$$T_0 = 0, \quad T_n = X_1 + X_2 + \cdots + X_n \quad \text{for} \quad n \geq 1,$$

and the $X_m$ are independent identically distributed non-negative random variables.

---

This definition describes $N$ in terms of an underlying sequence $\{X_n\}$. In the absence of knowledge about this sequence we can construct it from $N$; just define

(4)                         $T_n = \inf\{t : N(t) = n\}, \quad X_n = T_n - T_{n-1}.$

Note that the finite-dimensional distributions of a renewal process $N$ are specified by the distribution of the $X_m$. For example, if the $X_m$ are exponentially distributed then $N$ is a Poisson process. We shall try to use the notation of (3) consistently in Chapter 10, in the sense that $\{N(t)\}$, $\{T_n\}$, and $\{X_n\}$ will always denote variables satisfying (4).

It is sometimes appropriate to allow $X_1$ to have a different distribution from the shared distribution of $X_2, X_3, \ldots$; in this case $N$ is called a *delayed* (or *modified*) renewal process. The process $N$ in (2) is a delayed renewal process whatever the initial $Y_0$; if $Y_0 = i$ then $N$ is an ordinary renewal process.

A special case arises when the interarrival times $X_i$ have an exponential distribution†.

**(5) Theorem.** *Poisson processes are the only renewal processes that are Markov chains.*

If you like, think of renewal processes as a generalization of Poisson processes in which we have dropped the condition that interarrival times be exponentially distributed.

There are two principal areas of interest concerning renewal processes. First, suppose that we interrupt a renewal process $N$ at some specified time $s$. By this time, $N(s)$ occurrences have already taken place and we are awaiting the $(N(s) + 1)$th. That is, $s$ belongs to the random interval

$$I_s = [T_{N(s)}, T_{N(s)+1}).$$

Here are definitions of three random variables of interest.

**(6)** The *excess* (or *residual*) *lifetime* of $I_s$: $E(s) = T_{N(s)+1} - s$.

**(7)** The *current lifetime* (or *age*) of $I_s$: $C(s) = s - T_{N(s)}$.

**(8)** The *total lifetime* of $I_s$: $D(s) = E(s) + C(s)$.

We shall be interested in the distributions of these random variables, which are illustrated in Figure 8.1.

It will come as no surprise to the reader to learn that the other principal topic concerns the asymptotic behaviour of a renewal process $N(t)$ as $t \to \infty$. Here we turn our attention to the *renewal function* $m(t)$ given by

(9)                                   $m(t) = \mathbb{E}(N(t)).$

---

†See Exercise (8.3.5).

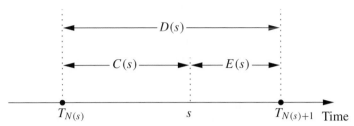

Figure 8.1. Excess, current, and total lifetimes at time $s$.

For a Poisson process $N$ with intensity $\lambda$, Theorem (6.8.2) shows that $m(t) = \lambda t$. In general, $m$ is *not* a linear function of $t$; however, it is not too difficult to show that $m$ is asymptotically linear, in that

$$\frac{1}{t} m(t) \to \frac{1}{\mu} \quad \text{as} \quad t \to \infty, \quad \text{where} \quad \mu = \mathbb{E}(X_1).$$

The 'renewal theorem' is a refinement of this result and asserts that

$$m(t+h) - m(t) \to \frac{h}{\mu} \quad \text{as} \quad t \to \infty,$$

subject to a certain condition on $X_1$.

   An introduction to the theory of renewal processes is given in Chapter 10.

## Exercises for Section 8.3

**1.**  Let $(f_n : n \geq 1)$ be a probability distribution on the positive integers, and define a sequence $(u_n : n \geq 0)$ by $u_0 = 1$ and $u_n = \sum_{r=1}^{n} f_r u_{n-r}$, $n \geq 1$. Explain why such a sequence is called a *renewal sequence*, and show that $u$ is a renewal sequence if and only if there exists a Markov chain $U$ and a state $s$ such that $u_n = \mathbb{P}(U_n = s \mid U_0 = s)$.

**2.**  Let $\{X_i : i \geq 1\}$ be the inter-event times of a discrete renewal process on the integers. Show that the excess lifetime $B_n$ constitutes a Markov chain. Write down the transition probabilities of the sequence $\{B_n\}$ when reversed in equilibrium. Compare these with the transition probabilities of the chain $U$ of your solution to Exercise (8.3.1).

**3.**  Let $(u_n : n \geq 1)$ satisfy $u_0 = 1$ and $u_n = \sum_{r=1}^{n} f_r u_{n-r}$ for $n \geq 1$, where $(f_r : r \geq 1)$ is a non-negative sequence. Show that:
(a) $v_n = \rho^n u_n$ is a renewal sequence if $\rho > 0$ and $\sum_{n=1}^{\infty} \rho^n f_n = 1$,
(b) as $n \to \infty$, $\rho^n u_n$ converges to some constant $c$.

**4.**  Events occur at the times of a discrete-time renewal process $N$ (see Example (5.2.15)). Let $u_n$ be the probability of an event at time $n$, with generating function $U(s)$, and let $F(s)$ be the probability generating function of a typical inter-event time. Show that, if $|s| < 1$:

$$\sum_{r=0}^{\infty} \mathbb{E}(N(r)) s^r = \frac{F(s)U(s)}{1-s} \quad \text{and} \quad \sum_{t=0}^{\infty} \mathbb{E}\left[\binom{N(t)+k}{k}\right] s^t = \frac{U(s)^k}{1-s} \quad \text{for } k \geq 0.$$

**5.**  Prove Theorem (8.3.5): Poisson processes are the only renewal processes that are Markov chains.

## 8.4 Queues

The theory of queues is attractive and popular for two main reasons. First, queueing models are easily described and draw strongly from our intuitions† about activities such as shopping or dialling a telephone operator. Secondly, even the solutions to the simplest models use much of the apparatus which we have developed in this book. Queues are, in general, non-Markovian, non-stationary, and quite difficult to study. Subject to certain conditions, however, their analysis uses ideas related to imbedded Markov chains, convergence of sequences of random variables, martingales, stationary processes, and renewal processes. We present a broad account of their theory in Chapter 11.

Customers arrive at a service point or counter at which a number of servers are stationed. An arriving customer may have to wait until one of these servers becomes available. The customer then moves to the head of the queue and is served, subsequently leaving the system on the completion of service. We must specify a number of details about this queueing system before we are able to model it adequately. For example,

   (a) in what manner do customers enter the system?

   (b) in what order are they served?

   (c) how long are their service times?

For the moment we shall suppose that the answers to these questions are as follows.

   (a) The number $N(t)$ of customers who have entered by time $t$ is a renewal process. That is, if $T_n$ is the time of arrival of the $n$th customer (with the convention that $T_0 = 0$) then the *interarrival times* $X_n = T_n - T_{n-1}$ are independent and identically distributed.

   (b) Arriving customers join the end of a single line of people who receive attention on a 'first come, first served' basis. There are a certain number of servers. On becoming free, a server turns to the customer at the head of the waiting line. We shall usually suppose that the queue has a single server only.

   (c) Service times are independent identically distributed random variables. That is, if $S_n$ is the service time of the $n$th customer to arrive, then $\{S_n\}$ is a sequence of independent identically distributed non-negative random variables which do not depend on the arriving stream $N$ of customers.

It requires only a little imagination to think of various other systems. Here are some examples.

**(1)** *Queues with baulking.* If the line of waiting customers is long then an arriving customer may, with a certain probability, decide not to join it.

**(2)** *Continental queueing.* In the absence of queue discipline, unoccupied servers pick a customer at random from the waiting mêlée.

**(3)** *Airline baggage check-in.* The waiting customers divide into several lines, one for each server. The servers themselves enter and leave the system at random, causing the attendant customers to change lines as necessary.

**(4)** *Last come, first served.* Arriving documents are placed on the top of an in-tray. An available server takes the next document from the top of the pile.

**(5)** *Group service.* Waiting customers are served in batches. This is appropriate for lift queues and bus queues.

**(6)** *Student discipline.* Arriving customers jump the queue, joining it near a friend

---

†and frustrations, including listening to the wrong types of music on a telephone.

We shall consider mostly the single-server queues described by (a), (b), and (c) above. Such queues are specified by the distribution of a typical interarrival time and the distribution of a typical service time; the method of analysis depends partly upon how much information we have about these quantities.

The state of the queue at time $t$ is described by the number $Q(t)$ of waiting customers ($Q(t)$ *includes* customers who are in the process of being served at this time). It would be unfortunate if $Q(t) \to \infty$ as $t \to \infty$, and we devote special attention to finding out when this occurs. We call a queue *stable* if the distribution of $Q(t)$ settles down as $t \to \infty$ in some well-behaved way; otherwise we call it *unstable*. We choose not to define stability more precisely at this stage, wishing only to distinguish between such extremes as

(a) queues which either grow beyond all bounds or enjoy large wild fluctuations in length,
(b) queues whose lengths converge in distribution, as $t \to \infty$, to some 'equilibrium distribution'.

Let $S$ and $X$ be a typical service time and a typical interarrival time, respectively, in a single-server queue; the ratio

$$\rho = \frac{\mathbb{E}(S)}{\mathbb{E}(X)}$$

is called the *traffic intensity*.

**(7) Theorem.** *Let $Q = \{Q(t) : t \geq 0\}$ be a queue with a single server and traffic intensity $\rho$.*
(a) *If $\rho < 1$ then $Q$ is stable.*
(b) *If $\rho > 1$ then $Q$ is unstable.*
(c) *If $\rho = 1$ and at least one of $S$ and $X$ has strictly positive variance then $Q$ is unstable.*

The conclusions of this theorem are intuitively very attractive. Why?

A more satisfactory account of this theorem is given in Section 11.5.

---

## Exercises for Section 8.4

**1.**   The two tellers in a bank each take an exponentially distributed time to deal with any customer; their parameters are $\lambda$ and $\mu$ respectively. You arrive to find exactly two customers present, each occupying a teller.
(a) You take a fancy to a randomly chosen teller, and queue for that teller to be free; no later switching is permitted. Assuming any necessary independence, what is the probability $p$ that you are the last of the three customers to leave the bank?
(b) If you choose to be served by the quicker teller, find $p$.
(c) Suppose you go to the teller who becomes free first. Find $p$.

**2.**   Customers arrive at a desk according to a Poisson process of intensity $\lambda$. There is one clerk, and the service times are independent and exponentially distributed with parameter $\mu$. At time 0 there is exactly one customer, currently in service. Show that the probability that the next customer arrives before time $t$ and finds the clerk busy is

$$\frac{\lambda}{\lambda + \mu}(1 - e^{-(\lambda+\mu)t}).$$

**3.**   Vehicles pass a crossing at the instants of a Poisson process of intensity $\lambda$; you need a gap of length at least $a$ in order to cross. Let $T$ be the first time at which you could succeed in crossing to the other side. Show that $\mathbb{E}(T) = (e^{a\lambda} - 1)/\lambda$, and find $\mathbb{E}(e^{\theta T})$.

Suppose there are two lanes to cross, carrying independent Poissonian traffic with respective rates $\lambda$ and $\mu$. Find the expected time to cross in the two cases when: (a) there is an island or refuge between the two lanes, (b) you must cross both in one go. Which is the greater?

**4.**   Customers arrive at the instants of a Poisson process of intensity $\lambda$, and the single server has exponential service times with parameter $\mu$. An arriving customer who sees $n$ customers present (including anyone in service) will join the queue with probability $(n + 1)/(n + 2)$, otherwise leaving for ever. Under what condition is there a stationary distribution? Find the mean of the time spent in the queue (not including service time) by a customer who joins it when the queue is in equilibrium. What is the probability that an arrival joins the queue when in equilibrium?

**5.**   Customers enter a shop at the instants of a Poisson process of rate 2. At the door, two representatives separately demonstrate a new corkscrew. This typically occupies the time of a customer and the representative for a period which is exponentially distributed with parameter 1, independently of arrivals and other demonstrators. If both representatives are busy, customers pass directly into the shop. No customer passes a free representative without being stopped, and all customers leave by another door. If both representatives are free at time 0, show the probability that both are busy at time $t$ is $\frac{2}{5} - \frac{2}{3}e^{-2t} + \frac{4}{15}e^{-5t}$.

---

# 8.5  The Wiener process

Most of the random processes considered so far are 'discrete' in the sense that they take values in the integers or in some other countable set. Perhaps the simplest example is simple random walk $\{S_n\}$, a process which jumps one unit to the left or to the right at each step. This random walk $\{S_n\}$ has two interesting and basic properties:

(a) *time-homogeneity*, in that, for all non-negative $m$ and $n$, $S_m$ and $S_{m+n} - S_n$ have the same distribution (we assume $S_0 = 0$); and

(b) *independent increments*, in that the increments $S_{n_i} - S_{m_i}$ ($i \geq 1$) are independent whenever the intervals $(m_i, n_i]$ are disjoint†.

What is the 'continuous' analogue of this random walk? It is reasonable to require that such a 'continuous' random process has the two properties above, and these are the defining characteristics of the so-called Lévy processes of Section 8.6. Subject to certain further assumptions about means and variances, there is essentially only one such process, called the *Wiener process*. This is a process $W = \{W(t) : t \geq 0\}$, indexed by continuous time and taking values in the real line $\mathbb{R}$, which is time-homogeneous with independent increments, and with the vital extra property that $W(t)$ has the normal distribution with mean 0 and variance $\sigma^2 t$ for some constant $\sigma^2$. This process is sometimes called *Brownian motion*, and is a cornerstone of the modern theory of random processes. Think about it as a model for a particle which diffuses randomly along a line. There is no difficulty in constructing Wiener processes in higher dimensions, leading to models for such processes as the Dow–Jones index or the diffusion of a gas molecule in a container. Note that $W(0) = 0$; the definition of a Wiener process may be easily extended to allow more general starting points.

The joint distributions of a Wiener process $W$ are thus multivariate normal. As discussed in Section 8.1, this fact is insufficient to answer important questions concerning the sample paths of $W$. It turns out that there are different processes with the given distributions, and of these it is convenient to select one whose sample paths are invariably continuous. This is

---

†Increments may be positive or negative.

made formal in Definition (13.3.1), and the existence of such a process is proved later in that section. Further discussion of the continuity of sample paths in a more general context may be found at the end of Section 8.9†.

What are the finite-dimensional distributions of the Wiener process $W$? These are easily calculated as follows.

**(1) Lemma.** *The vector of random variables $W(t_1)$, $W(t_1)$, ... , $W(t_n)$ has the multivariate normal distribution with zero means and covariance matrix $(v_{ij})$ where $v_{ij} = \sigma^2 \min\{t_i, t_j\}$.*

The Wiener process is an important example of a more general class of processes called 'Gaussian processes', which are characterized by having joint distributions that are multivariate normal. Gaussian processes are the subject of Section 9.6.

**Proof.** By assumption, $W(t_i)$ has the normal distribution with zero mean and variance $\sigma^2 t_i$. It therefore suffices to prove that $\text{cov}(W(s), W(t)) = \sigma^2 \min\{s, t\}$. Now, if $s < t$, then

$$\mathbb{E}(W(s)W(t)) = \mathbb{E}\big(W(s)^2 + W(s)[W(t) - W(s)]\big) = \mathbb{E}(W(s)^2) + 0,$$

since $W$ has independent increments and $\mathbb{E}(W(s)) = 0$. Hence

$$\text{cov}(W(s), W(t)) = \text{var}(W(s)) = \sigma^2 s. \qquad \blacksquare$$

We note from Lemma (1) that

**(2)**                    $$\text{cov}(W(s), W(t)) = \tfrac{1}{2}\big(s + t - |s - t|\big)\sigma^2, \qquad s, t \geq 0.$$

A Wiener process $W$ is called *standard* if $W(0) = 0$ and $\sigma^2 = 1$. A more extended treatment of Wiener processes appears in Chapter 13.

---

## Exercises for Section 8.5

**1.**   For a Wiener process $W$ with $W(0) = 0$, show that

$$\mathbb{P}\big(W(s) > 0, \ W(t) > 0\big) = \frac{1}{4} + \frac{1}{2\pi} \sin^{-1}\sqrt{\frac{s}{t}} \quad \text{for } s < t.$$

Calculate $\mathbb{P}(W(s) > 0, \ W(t) > 0, \ W(u) > 0)$ when $s < t < u$.

**2.**   Let $W$ be a Wiener process. Show that, for $s < t < u$, the conditional distribution of $W(t)$ given $W(s)$ and $W(u)$ is normal

$$N\left(\frac{(u - t)W(s) + (t - s)W(u)}{u - s}, \ \frac{(u - t)(t - s)}{u - s}\right).$$

Deduce that the conditional correlation between $W(t)$ and $W(u)$, given $W(s)$ and $W(v)$, where $s < t < u < v$, is

$$\sqrt{\frac{(v - u)(t - s)}{(v - t)(u - s)}}.$$

---

†Recall also the assumption in Section 6.9 that continuous-time Markov chains have right-continuous sample paths.

**3.**   For what values of $a$ and $b$ is $aW_1 + bW_2$ a standard Wiener process, where $W_1$ and $W_2$ are independent standard Wiener processes?

**4.**   Show that a Wiener process $W$ with variance parameter $\sigma^2$ satisfies

$$\sum_{j=0}^{n-1}\{W((j+1)t/n) - W(jt/n)\}^2 \overset{\text{m.s.}}{\longrightarrow} \sigma^2 t \quad \text{as } n \to \infty.$$

The process $W$ is said to have *finite quadratic variation*.

**5.**   Let $W$ be a Wiener process. Which of the following define Wiener processes?
   (a) $-W(t)$,   (b) $\sqrt{t}W(1)$,   (c) $W(2t) - W(t)$.

**6.**   Find the distribution of $\int_0^t [W(u)/u]\,du$ where $W$ is the Wiener process. [The integral is interpreted as the limit as $\epsilon \downarrow 0$ of the integral from $\epsilon$ to $t$.]

**7.**   **Wiener process in $n$ dimensions.**   An $n$-dimensional Wiener process is a process $W(t) = (W_1(t), W_2(t), \dots, W_n(t))$ taking values in $\mathbb{R}^n$ with independent increments, started at the origin $W(0) = 0$, and such that $W(t) - W(s)$ is multivariate normal with means 0 and covariance matrix $(t - s)\mathbf{I}$ where $\mathbf{I}$ is the $n \times n$ identity matrix. Let $W(t)$ be an $n$-dimensional Wiener process, and let $A$ be an orthonormal $n \times n$ matrix. Show that $AW(t)$ is an $n$-dimensional Wiener process.

**8.**   Let $W$ be a standard Wiener process. Show that, for $p \geq 0$,

$$\mathbb{E}\big(|W(t) - W(s)|^p\big) = c_p|t - s|^{p/2}, \qquad s, t \in \mathbb{R},$$

where

$$c_p = \frac{1}{\sqrt{2\pi}} \int_{-\infty}^{\infty} |y|^p e^{-\frac{1}{2}y^2}\,dy.$$

# 8.6 Lévy processes and subordinators

The Poisson and Wiener processes share the property of having independent, stationary increments†. They are important examples of so-called 'Lévy processes'.

**(1) Definition.**   The process $X = \{X(t) : t \geq 0\}$ is called a **Lévy process** if:
   (a) $X$ takes values in $\mathbb{R}$, and $X(0) = 0$,
   (b) $X$ has *independent increments*, in that the differences $X(t_j) - X(s_j)$ are independent whenever the intervals $(s_j, t_j]$ are disjoint,
   (c) $X$ has *stationary increments*, in that the distribution of $X(t) - X(s)$ depends only on $t - s$,
   (d) $X$ has sample paths that are *continuous in probability*, in that for $t \geq 0$, we have $X(t + h) \overset{\text{P}}{\to} X(t)$ as $h \to 0$.
A Lévy process with non-decreasing sample paths is called a **subordinator**.

Lévy processes are important for a variety of reasons: they are natural extensions of random walks to continuous time, they have the Markov property, they provide a tractable case of the general class (called 'semimartingales') of suitable integrators in stochastic integration

---

†See Example (6.8.23) for the Poisson process, and Section 8.5 for the Wiener process.

(see Section 13.7), and they include special cases of self-similar processes that may exhibit interesting fractal behaviour. In addition, Lévy processes appear frequently in models for real-world phenomena including finance, and the movements of many types of animal, in particular the Lévy flight foraging hypothesis.

It turns out (rather as in the forthcoming Theorem (8.9.6)) that a Lévy process has a version (that is, a process with the same fdds) having sample paths that are right-continuous with left limits, and it is usual to assume that a Lévy process under study has these properties.

A useful technique of probability theory is to change the rate at which the clock runs. One example is the construction of a continuous-time Markov chain $X$ from its jump chain $Y$: $X$ follows $Y$ when it changes its value, but between two changes it holds its current state for a random length of time. *Subordinators* are natural and useful time-change processes†. Given a process $Y$ and a subordinator $T$, we may observe $Y$ against a clock that reads $T(t)$ at time $t$, thus yielding the time-changed process $Z(t) = Y(T(t))$. See Section 8.8.

Lévy processes are intimately linked to so-called 'infinitely divisible distributions' (recall Problem (5.12.14)). A random variable $Y$ (or its distribution) is said to be *infinitely divisible* if, for $n \geq 1$, there exist independent, identically distributed random variables $Y_1^{(n)}, Y_2^{(n)}, \ldots, Y_n^{(n)}$ such that the sum $Y_1^{(n)} + Y_2^{(n)} + \cdots + Y_n^{(n)}$ has the same distribution as $Y$. If $X$ is a Lévy process, it is immediate that $X(t)$ is infinitely divisible, since it may be expressed as the sum

$$X(t) = \sum_{j=1}^{n} \left[ X(jt/n) - X((j-1)t/n) \right]$$

of independent, identically distributed increments. The converse turns out to hold also: given an infinitely divisible distribution, there exists a Lévy process $X$ such that $X(1)$ has the given distribution (see Bertoin 1996). We explore this a little further in the next theorem, concerning characteristic functions. A word of warning: since the symbol $t$ denotes time in the current context, we shall use the symbol $\theta$ to denote the variable of a characteristic function.

**(2) Theorem.** *The characteristic function* $\phi(t, \theta) = \mathbb{E}(e^{i\theta X(t)})$ *of a Lévy process* $X = \{X(t) : t \geq 0\}$ *has the form*

$$\phi(t, \theta) = e^{t\Lambda(\theta)}, \qquad \theta \in \mathbb{R}, \ t \geq 0,$$

*for some function* $\Lambda : \mathbb{R} \to \mathbb{R}$ *called the* Lévy symbol *or* Lévy characteristic exponent *of* $X$.

**Proof.** Since

$$X(s + t) = X(s) + [X(s + t) - X(s)],$$

and $X$ has independent, stationary increments, we have that $\phi$ satisfies

**(3)** $$\phi(s + t, \theta) = \phi(s, \theta)\phi(t, \theta), \qquad s, t \geq 0.$$

Now, $X$ is continuous in probability, and it follows‡ that the only characteristic function solutions to (3) satisfying $\phi(0, \theta) = 1$ are the exponential functions $\phi(t, \theta) = e^{t\Lambda(\theta)}$ for some $\Lambda(\theta)$. ∎

---

†Also termed *directing processes*.

‡See Problem (4.14.5) and Exercise (8.6.1).

The admissible Lévy symbols are precisely those of the infinitely divisible distributions. Lévy and Khinchin proved that the characteristic functions of the infinitely divisible distribution are exactly the functions $\phi(\theta) = e^{\Lambda(\theta)}$ such that $\Lambda$ has the form

$$\Lambda(\theta) = ia\theta + \tfrac{1}{2}\sigma^2\theta^2 + \int_{-\infty}^{\infty} \left[1 - e^{i\theta x} + i\theta x I(|x| < 1)\right] d\Pi(x),$$

for some $a \in \mathbb{R}$, $\sigma^2 \geq 0$, and some measure $\Pi$ on $\mathbb{R} \setminus \{0\}$ satisfying $\int_{\mathbb{R}} \min\{1, |x|^2\} d\Pi < \infty$. See Feller (1971, ch. XVII) for the Lévy–Khinchin formula, and Bertoin 1996 for the above form of the formula.

**(4) Example. Poisson process.** For the Poisson process $N = \{N(t) : t \geq 0\}$ with rate $\lambda$, $N(1)$ has the Poisson distribution with parameter $\lambda$, so that $\Lambda(\theta) = \lambda(e^{i\theta} - 1)$.     ●

**(5) Example. Wiener process.** For the standard Wiener process $W$ of Section 8.5, $W(1)$ has the $N(0, 1)$ distribution, whence $\Lambda(\theta) = -\tfrac{1}{2}\theta^2$.     ●

When considering a subordinator $T$, it is customary to use its Laplace transform

$$M(t, \theta) = \mathbb{E}(e^{-\theta T(t)}), \qquad \theta \geq 0,$$

rather than its characteristic function (note the minus sign in the exponent). As in Theorem (2), $M$ takes the form

$$M(t, \theta) = e^{-t\Lambda(\theta)},$$

for some $\Lambda(\theta)$ called the *Laplace exponent* of $T$.

**(6) Example. Moran gamma process.** This process is a subordinator $T$ whose increments have the gamma distribution $\Gamma(1, u)$, where $u$ is the length of the increment. The Laplace transform of the last distribution is

$$M_u(\theta) = \frac{1}{(1 + \theta)^u} = \exp\left\{-u \int_0^{\infty} (1 - e^{-\theta y}) \frac{1}{y} e^{-y} \, dy\right\},$$

which gives a representation for the Laplace exponent, namely

$$\Lambda(\theta) = \int_0^{\infty} (1 - e^{-\theta y}) \frac{1}{y} e^{-y} \, dy.$$     ●

The relevance of subordinators as time changes is indicated by the following theorem, the proof of which is omitted (see Exercise (8.6.4)).

**(7) Theorem.** *Let $X$ be a Lévy process and $Y$ an independent subordinator defined on the same probability space as $X$. The process $Y$ given by $Y(t) = X(T(t))$ is a Lévy process.*

## Exercises for Section 8.6

**1.**   Prove that the characteristic function $\phi(t, \theta) = \mathbb{E}(e^{i\theta X(t)})$ of a Lévy process $X$ is a continuous function of $t$.

**2.**   Show that a compound Poisson process is a Lévy process, and find its Lévy symbol.

**3.**   Verify the formula of Example (8.6.6) for the Laplace exponent of the Moran gamma process.

**4.**   Prove Theorem (8.6.7): the property of being a Lévy process is preserved under time change by an independent subordinator.

**5.**   Let $N$ be a Poisson process of rate 1, and $T$ an independent random variable with the $\Gamma(1, t)$ distribution. Show that $Y = N(T)$ has probability generating function $\mathbb{E}(s^Y) = (2 - s)^{-t}$ for $s < 2$.

**6.**   Let $X$ be a Lévy process with finite variances. Show that the following are martingales:
(a) $X(t) - \mathbb{E}(X(t))$,
(b) $Z(t)^2 - \mathbb{E}(Z(t)^2)$, where $Z(t) = X(t) - \mathbb{E}(X(t))$,
(c) $e^{i\theta X(t)}/\phi(t, \theta)$, where $\phi(t, \theta) = \mathbb{E}(e^{i\theta X(t)})$.

**7.**   Let $X$ be a continuous-time martingale and $T$ an independent subordinator. Show that, subject to the moment condition $\mathbb{E}|Y(t)| < \infty$, $Y(t) = X(T(t))$ defines a martingale with respect to a suitable filtration. Show that the moment condition is valid whenever $X$ is a positive martingale.

## 8.7  Self-similarity and stability

When examined on a different scale, and suitably dilated, a process may sometimes resemble the original. Such 'self-similarity' is a phenomenon familiar to naturalists, geographers, digital artists, and more widely†. *Random processes* may also have properties of self-similarity, and an important example is provided by the Wiener process: for $a > 0$ and a standard Wiener process $W(t)$, the process $aW(t/a^2)$ is also a standard Wiener process. (See Problems (9.7.18) and (13.12.1).) In general, a random process is called *self-similar* if its distributions are invariant under appropriate scalings of space and time. Self-similar processes may be used to model time-series, and other systems with certain long-range dependence.

For two random variables $Y$, $Z$, we write $Y \overset{\mathrm{D}}{=} Z$ if they have the same distribution. We may also write $Y \overset{\mathrm{D}}{=} F$ to mean that $Y$ has distribution function $F$.

**(1) Definition.**   A random process $X = \{X(t) : t \geq 0\}$ is called **self-similar** if, for $a > 0$ there exists $b > 0$ such that, for all $t \geq 0$, $X(at) \overset{\mathrm{D}}{=} bX(t)$.

A random process $X$ is called *degenerate* if, for all $t \geq 0$, $X(t)$ is almost surely constant.

**(2) Theorem.**   *Let $X$ be a non-degenerate random process that is self-similar, and continuous in probability at $t = 0$. There exists $H \geq 0$, called the* Hurst exponent, *such that, for $a > 0$ and $t \geq 0$, $X(at) \overset{\mathrm{D}}{=} a^H X(t)$. Furthermore, if $H > 0$ then $X(0) = 0$, and if $H = 0$ then, for $t \geq 0$, $X(t) = X(0)$ a.s.*

**Proof.**   By self-similarity, for $a > 0$ there exists $b = b(a) > 0$ such that $X(at) \overset{\mathrm{D}}{=} b(a)X(t)$. Since $X$ is non-degenerate, we may choose $T > 0$ such that $X(T)$ is not a.s. constant (if $X(0)$

---

†Self-similarity and fractals have been popularized by Mandelbrot 1983.

is non-degenerate, then so is $X(t)$ for small positive $t$, since $X(t) \xrightarrow{P} X(0)$ as $t \to 0$). By self-similarity, $X(t) \overset{D}{=} b(t/T)X(T)$ is non-degenerate for all $t > 0$.

We note next that, for given $a > 0$, there is a unique value of $b = b(a)$. Suppose $X(at) \overset{D}{=} b_1 X(t) \overset{D}{=} b_2 X(t)$. Then $b_1 = b_2$ by the non-degeneracy of $X(t)$ for $t \geq 0$. In particular, $b(1) = 1$.

For $a_1, a_2 > 0$, we have that

$$X(a_1 a_2 t) \overset{D}{=} b(a_1)X(a_2 t) \overset{D}{=} b(a_1)b(a_2)X(t).$$

On the other hand, $X(a_1 a_2 t) \overset{D}{=} b(a_1 a_2)X(t)$, so that

$$b(a_1 a_2)X(t) \overset{D}{=} b(a_1)b(a_2)X(t), \qquad t \geq 0.$$

Since $X$ is non-degenerate, this implies

(3) $$b(a_1 a_2) = b(a_1)b(a_2), \qquad a_1, a_2 > 0.$$

In particular, $b(1/a) = 1/b(a)$.

Let $a < 1$. Then $X(a^n) \overset{D}{=} b(a)^n X(1)$. Now $X(a^n) \xrightarrow{P} X(0)$ as $n \to \infty$, and hence $X(a^n) \xrightarrow{D} X(0)$. It must, therefore, be the case that $b(a) \leq 1$. By (3), for $a_1 < a_2$,

$$b(a_1)/b(a_2) = b(a_1/a_2) \leq 1,$$

so that $b$ is a non-decreasing function. By Problem (4.14.5a), the unique non-decreasing solution to (3) is given by $b(a) = a^H$ for some $H \geq 0$.

When $H > 0$, then $X(0) \overset{D}{=} a^H X(0)$ for all $a > 0$, whence $\mathbb{P}(X(0) = 0) = 1$. Conversely, if $H = 0$, then $X(at) \sim X(t)$ for all $a > 0$, $t > 0$. That $X(t) = X(0)$ a.s. follows since $X(t) \xrightarrow{P} X(0)$ as $t \to 0$. ∎

(4) **Example. The Wiener process** $W$. Since $W(at) \overset{D}{=} a^{\frac{1}{2}} W(t)$, the standard Wiener process $W$ has Hurst exponent $H = \frac{1}{2}$. Since $\sigma W$ is a Wiener process with variance-parameter $\sigma^2$, we have $H = \frac{1}{2}$ for a Wiener process with $\sigma^2 > 0$.

It is an *exercise*† to show that, for a non-degenerate, self-similar process $X$ with stationary increments, finite variances, and Hurst exponent $H > 0$,

(5) $$\mathbb{E}(X(s)X(t)) = \tfrac{1}{2}\left(s^{2H} + t^{2H} - |t-s|^{2H}\right)\mathbb{E}(X(1)^2), \qquad s, t \geq 0.$$

This may be compared with equation (8.5.2). ●

Formula (5) leads to an important definition in the Gaussian case.

(6) **Definition.** Let $H \in (0, 1]$. A Gaussian process $X = \{X(t) : t \geq 0\}$ with zero means and covariance function

$$\mathrm{cov}(X(s), X(t)) = \tfrac{1}{2}\left(s^{2H} + t^{2H} - |t-s|^{2H}\right)\mathrm{var}(X(1)), \qquad s, t \geq 0,$$

---

†See Exercise (8.7.2).

is called **fractional Brownian motion**, abbreviated to fBM or fBM$_H$.

It turns out that the process fBM$_H$ is self-similar with Hurst exponent $H$, and has stationary increments. Its increments are not independent except in the Wiener case $H = \frac{1}{2}$. The sample paths of fBM may be taken to be a.s. continuous but nowhere differentiable.

We now drop the assumption of Gaussianity. Self-similar processes with stationary, independent increments (that is, Lévy processes) are intimately connected with an important class of distributions called 'stable laws'.

**(7) Definition.** A random variable $X$ (and its distribution) is called **stable** if, for $a, b > 0$, and for random varables $X_1, X_2$ with the same distribution as $X$, there exist $A \in \mathbb{R}$ and $B > 0$ such that $aX_1 + bX_2 \overset{\text{D}}{=} BX + A$. The distribution is called **strictly stable** if this holds with $A = 0$.

The condition of (7) may be expressed thus in terms of the characteristic function $\phi$ of the distribution:

$$(8) \qquad \phi(a\theta)\phi(b\theta) = e^{iA\theta}\phi(B\theta), \qquad \theta \in \mathbb{R}.$$

It is a simple *exercise*† to show that, if $X_1, X_2, \ldots, X_n$ are an independent sample from a stable distribution, then there exist $B_n$ and $A_n$ such that

$$(9) \qquad S_n := X_1 + X_2 + \cdots + X_n \overset{\text{D}}{=} B_n X_1 + A_n,$$

By splitting a sum of $mk$ random variables into $m$ sums of length $k$, we find that $B_{mk} = B_m B_k$, whence it may be shown that

$$(10) \qquad B_m = m^{1/\alpha}$$

for some $\alpha \in (0, 2]$ (see Feller 1971, p. 170), and in this case $X$ is called $\alpha$-*stable*. The parameter $\alpha$ is an important characteristic of the underlying distribution, and it contains information about the appropriate scaling of the sum $S_n$ as $n \to \infty$. It features in the characterization of stable laws presented in Theorem (13).

Every stable distribution is infinitely divisible, whereas there exist distributions which are infinitely divisible but not stable.

**(11) Examples.** The characteristic function $\phi(\theta) = e^{-\frac{1}{2}\theta^2}$ of the $N(0, 1)$ distribution satisfies

$$\phi(a\theta)\phi(b\theta) = \phi\big(\theta\sqrt{a^2 + b^2}\big),$$

and so the distribution is strictly stable. The Cauchy distribution is strictly stable since its characteristic function $\phi(\theta) = e^{-|\theta|}$ satisfies $\phi(a\theta)\phi(b\theta) = \phi(((a + b)\theta)$. ●

Let $X$ be strictly $\alpha$-stable. By (9), the characteristic function $\phi(\theta) = \mathbb{E}(e^{i\theta X})$ satisfies $\phi(t)^n = \phi(B_n\theta)$. Therefore,

$$\phi(\theta)^{mn} = \phi(B_m\theta)^n = \phi(B_n\theta)^m,$$

---

†See Exercise (8.7.1).

so that $\phi(\theta)^{m/n} = \phi((B_m/B_n)\theta)$ for positive, rational $m/n$. By (10), we have that $B_m/B_n = (m/n)^{1/\alpha}$. Let $d \geq 0$, and find a sequence of rationals converging to $d$. By continuity, the characteristic function of a strictly $\alpha$-stable distribution satisfies

$$(12) \qquad\qquad\qquad \phi(\theta)^d = \phi(d^{1/\alpha}\theta), \qquad d \geq 0.$$

Stable distributions may be characterized by their characteristic functions. By comparing the following with the Lévy–Khinchin formula for infinitely divisible distributions (see Section 8.6), we may see that stable distributions form a strict subclass of infinitely divisible distributions.

**(13) Theorem.** *Let $a \in \mathbb{R}$, $b \in [-1, 1]$, $c > 0$, and $\alpha \in (0, 2]$. A distribution $\mu$ is $\alpha$-stable with parameters $a, b, c$ if and only if only if it has characteristic function given by*

$$(14) \qquad\qquad \phi(\theta) = \exp\left\{ia\theta - c^\alpha|\theta|^\alpha\left(1 - ib\,\mathrm{sign}(\theta)\Phi_\alpha(\theta)\right)\right\},$$

*where*

$$\Phi_\alpha(\theta) = \begin{cases} \tan(\tfrac{1}{2}\alpha\pi) & \text{if } \alpha \neq 1, \\[2mm] -\dfrac{2}{\pi}\log|\theta| & \text{if } \alpha = 1. \end{cases}$$

*In this case $\mu$ is denoted $S_\alpha(c, b, a)$.*

**Proof.** See Feller (1971, pp. 570, 576) and Janson 2011. ∎

A stable distribution is symmetric if and only if $\phi(\theta)$ equals its complex conjugate, which is to say that the distribution is $S_\alpha(c, 0, 0)$ for some $\alpha$, $c$, in which case the characteristic function takes the simple form

$$(15) \qquad\qquad\qquad \phi(\theta) = \exp\left(-c^\alpha|\theta|^\alpha\right), \qquad \theta \in \mathbb{R}.$$

It is an *exercise* to derive this directly from (12).

The parameters $a$ and $\alpha$ are 'location' and 'scaling' parameters. The stable distributions with $\alpha = 2$ and $b = 0$ are the $N(a, 2c^2)$ distributions, while those with $\alpha = 1$ may be said to be of Cauchy-type. The support of the stable distributions is generally the real line $\mathbb{R}$, except when $b = 1, \alpha < 1$ (when it is the half-line $(a, \infty)$) or $b = -1, \alpha < 1$ (when it is the half-line $(-\infty, a)$).

Some stable distributions have convergent Laplace transforms. Following an analysis of the representation (13), we find the next result, the proof of which is omitted.

**(16) Theorem.** *For $\alpha \in (0, 2)$, a random variable $X$ with the stable distribution $S_\alpha(c, b, a)$ has a convergent Laplace transform $\mathbb{E}(e^{-\theta X})$ for $\theta \geq 0$ if and only if $b = 1$. In this case,*

$$\mathbb{E}(e^{-\theta X}) = \begin{cases} \exp\left(-a\theta - \dfrac{(c\theta)^\alpha}{\cos(\alpha\pi/2)}\right) & \text{if } \alpha \neq 1, \\[4mm] \exp\left(-a\theta + \dfrac{2}{\pi}c\theta\log\theta\right) & \text{if } \alpha = 1. \end{cases}$$

**(17) Example.** Let $X$ have the stable distribution $S_\alpha(c, 1, 0)$ where $\alpha \in (0, 1)$. Then $X$ is a strictly stable, positive random variable with Laplace transform

$$\mathbb{E}(e^{-\theta X}) = \exp\left(-D\theta^\alpha\right), \qquad \theta \geq 0,$$

where $c = \{D\cos(\alpha\pi/2)\}^{1/\alpha}$. ●

The connection between stability and Lévy processes is as follows.

**(18) Theorem.** *A Lévy process $X$ is self-similar with Hurst exponent $H \in (0, 2]$ if and only if $X(1)$ has a strictly $H^{-1}$-stable distribution.*

**Proof.** Let $X$ be a self-similar Lévy process with Hurst exponent $H \geq 0$. By Theorem (8.6.2), $X(t)$ has characteristic function $\phi(t, \theta) = e^{t\Lambda(\theta)}$, where $\Lambda$ is the Lévy symbol. By self-similarity, $X(t) \overset{D}{=} t^H X(1)$, which entails $\Lambda(t^H \theta) = t\Lambda(\theta)$ for $t \geq 0$ and $\theta \in \mathbb{C}$. The only continuous solutions of this functional equation have the form $\Lambda(\theta) = a|\theta|^{1/H}$ for $a \in \mathbb{R}$, and the corresponding $\phi$ is a characteristic function if and only if $H \geq \frac{1}{2}$. Therefore, $X(1)$ has characteristic function $\phi(1, \theta) = \exp(-c^\alpha|\theta|^\alpha)$ for some $c > 0$ and $\alpha = H^{-1} \in (0, 2]$, which we recognise from (15) as the characteristic function of a symmetric $H^{-1}$-stable distribution.

Suppose, conversely, that $X$ is a Lévy process such that $X(1)$ is strictly $\alpha$-stable. Let $a > 0$. Since $\phi(t, \theta) = e^{t\Lambda(\theta)}$, we have that

$$\phi(at, \theta) = \phi(1, \theta)^{at}$$
$$= \phi(1, a^{1/\alpha}\theta)^t \qquad \text{by (12)}$$
$$= \phi(t, a^{1/\alpha}\theta)$$

Therefore, $X(at) \overset{D}{=} a^{1/\alpha}X(t)$, so that $X$ is self-similar with Hurst exponent $H = 1/\alpha$. ∎

Finally, we note a useful connection between self-similar processes and strongly stationary processes.

**(19) Theorem. Lamperti transformation.** *Let $X = \{X(t) : t \in \mathbb{R}\}$ be a strongly stationary process. Fix $H > 0$ and define $Y(t) = t^H X(\log t)$ for $t > 0$, with $Y(0) = 0$. The process $Y = \{Y(t) : t \geq 0\}$ is self-similar with Hurst exponent $H$. Conversely, any non-degenerate self-similar process $Y = \{Y(t) : t \geq 0\}$ with $Y(0) = 0$ is thus obtained from the strongly stationary process $X$ given by $X(t) = e^{-tH}Y(e^t)$ for $t \in \mathbb{R}$.*

**Proof.** See Embrechts and Maejima (2002, p. 11). ∎

**(20) Example. Ornstein–Uhlenbeck process.** Let $W$ be a standard Wiener process. Then $U(t) = e^{-\frac{1}{2}t}W(e^t)$ defines a stationary process called the *Ornstein–Uhlenbeck process*. See Exercise (13.7.4).

More generally, by applying the Lamperti transformation of Theorem (19) to the fractional Brownian motion fBM$_H$ of Definition (6), we obtain a so-called *fractional Ornstein–Uhlenbeck process*. ●

## Exercises for Section 8.7

**1.** Let $X$ be stable, and let $X_1, X_2, \ldots, X_n$ be a random sample distributed as $X$. Show that, for $n \geq 1$, there exist $A_n, B_n$ such that $X_1 + X_2 + \cdots + X_n \overset{D}{=} B_n X + A_n$.

**2.** Prove that, for a non-degenerate, self-similar process $X$ with stationary increments, finite variances, and Hurst exponent $H > 0$,

$$\text{var}(X(t)) = t^{2H}\,\text{var}(X(1)), \qquad t \geq 0,$$
$$\mathbb{E}(X(s)X(t)) = \tfrac{1}{2}(s^{2H} + t^{2H} - |t - s|^{2H})\mathbb{E}(X(1)^2), \qquad s, t \geq 0.$$

**3.** Show that a stable distribution is infinitely divisible.

**4.** Show that a self-similar Lévy process either is a Wiener process or has infinite variances.

**5.** Prove that a distribution with characteristic function $\phi(\theta) = \exp(-|\theta|^\alpha)$, where $\alpha \in (0, 2]$, is strictly $\alpha$-stable and symmetric.

**6.** Let $X$ and $Y$ be independent stable random variables, where $X$ is symmetric with exponent $\alpha \in (0, 2]$, and $Y$ is non-negative with Laplace transform $M(\theta) = \exp(-k\theta^\beta)$ for $\theta \in [0, \infty)$, where $1 \neq \beta \in (0, 2]$. Show that $Z = XY^{1/\alpha}$ is symmetric and $\alpha\beta$-stable. Deduce that, for independent $N(0, 1)$ random variables $U$, $V$, the ratio $Z = U/V$ has the Cauchy distribution.

---

# 8.8  Time changes

There are many types of random process in continuous time, a fair number of which turn out to be equivalent to one another through the simple device of changing the rate at which the clock is running. Time changes may also be used to keep track of certain features of potential interest. We gather in this section a few of the properties and connections of random time changes, beginning with two examples.

**(1) Example. Stopping times.** Let $X$ be a process in continuous time (the case of discrete time is similar), and let $S$ be a stopping time for $X$ (see Section 6.9 for the definition). Define the time change process $T = \{T(t) : t \geq 0\}$ by

$$T(t) = \begin{cases} t & \text{if } t \leq S, \\ S & \text{if } t > S. \end{cases}$$

The process $X(T(t))$ is $X$ stopped at time $S$.

More generally, let $0 = S_0 < S_1 < S_2 < \cdots$ be a (possibly infinite) increasing sequence of stopping times. They give rise to the time change process

$$T(t) = S_n \quad \text{if} \quad S_n \leq t < S_{n+1}.$$

This generates the time changed process $X(T(t))$, which may be viewed as the discrete-time process $Y_n = X(S_n)$ subject to random holding times.  ●

**(2) Example. Poissonization.** Let $Y$ be a discrete-time Markov chain on the state space $S$, and let $N$ be an independent Poisson process of constant rate $\lambda$. The time-changed process $Z(t) = Y_{N(t)}$ is a continuous-time Markov chain†.

More generally, let $\{g_i : i \in S\}$ be strictly positive reals, and assume for simplicity that the transition matrix $\mathbf{P} = (p_{ij})$ of $Y$ satisfies $p_{ii} = 0$ for $i \in S$. Let $\{V_m : m \geq 0\}$ be independent random variables having the exponential distribution with parameter 1, and assume the $V_i$ are independent of $Y$. The time change process $T$ is given by

$$T(t) = \max\{n : T_n \leq t\}$$

where $T_0 = 0$ and

$$T_n = \sum_{m=0}^{n-1} g_{Y_m} V_m.$$

---

†Such Poissonization seems to be due to M. Kac.

Then $Z(t) = Y_{T(t)}$ is the continuous-time Markov chain having jump chain $Y$ and holding times with parameters $g_i$.                                                                                                   ●

Any proper discussion of time changes requires some measure-theoretic background, which we introduce next.

**(3) Definition.** Let $(\Omega, \mathcal{F}, \mathbb{P})$ be a probability space.

  (a) A **filtration** is an increasing family $\mathcal{F} = \{\mathcal{F}_t : t \geq 0\}$ of sub-$\sigma$-fields of $\mathcal{F}$. The filtration $\mathcal{F}$ is said to satisfy the **usual conditions** if it is:

   (i) right-continuous, in that $\mathcal{F}_t = \bigcap_{s>t} \mathcal{F}_s$ for $t \geq 0$, and

   (ii) complete, in that $\mathcal{F}_0 \supseteq \mathcal{N}$, where $\mathcal{N} = \{A \in \mathcal{F} : \mathbb{P}(A) = 0\}$ is the set of null events.

  (b) A process $X = \{X(t) : t \geq 0\}$ is said to be **adapted** to the filtration $\mathcal{F}$ if, for $t \geq 0$, $X(t)$ is $\mathcal{F}_t$-measurable.

  (c) The **natural filtration** of a process $X$ is given by $\mathcal{F}_t = \sigma(\{X(u) : u \leq t\})$, the smallest $\sigma$-field with respect to which each $X(u)$, $u \leq t$, is measurable.

  (d) Given a filtration $\mathcal{F}$, a **time change** is a non-decreasing, right-continuous, process $T$ taking values in $[0, \infty]$ that is adapted to $\mathcal{F}$.

The *lifetime* of a time change $T$ is the random variable $D = \inf\{t : T(t) = \infty\}$. Recall that, if a time change $T$ is a Lévy process on $[0, D]$, then it is called a *subordinator*. Here are some important special instances of time changes.

**(4) Occupation times.** Let $B$ be a (measurable) subset of the state space $S$ of a random process $X$. The *occupation time* process $I_B$ is given by

$$I_B(t) = \int_0^t I(X(s) \in B)\, ds.$$

**(5) Quadratic variation.** The *quadratic variation* process $\langle X \rangle$ of a process $X$ is defined as

**(6)** $$\langle X \rangle_t = \lim_{|\mathbf{t}| \to 0} \sum_{r=1}^n \left| X(t_r) - X(t_{r-1}) \right|^2, \qquad t \geq 0,$$

whenever this limit exists, where $\mathbf{t}$ denotes the increasing sequence $0 = t_0 < t_1 < \cdots < t_n = t$ and $|\mathbf{t}| = \max\{t_r - t_{r-1} : r \geq 1\}$ is the mesh size of $\mathbf{t}$. The quadratic variation process $\langle X \rangle$ exists for many processes $X$ of interest, including diffusion processes and continuous martingales. See Exercise (8.5.4) for the case of the Wiener process†.

**(7) Local time.** Let $W$ be a Wiener process, and define the *local time* at $x \in \mathbb{R}$ by

$$L(t, x) = \lim_{\epsilon \downarrow 0} \frac{1}{2\epsilon} \int_0^t I(x - \epsilon < W(s) < x + \epsilon)\, ds.$$

Thus $L(t, x)$ supplies a measure of the amount of time spent by the process at position $x$ up to time $t$. The existence of local time was proved by Lévy, and the reader is referred to Rogers and Williams (2000a, sect. I.14).

---

†Note the condition $|\mathbf{t}| \to 0$ in (6). The supremum, over *all* partitions, of the summation can diverge, as it does in the case of the Wiener process $W$.

**(8) Additive functionals.** Let $X$ be a suitable Markov process, and let $f : \mathbb{R} \to [0, \infty)$ be measurable. We define the process $A$ by

$$A(t) = \int_0^t f(X(s)) \, ds.$$

Such $A$ is an 'additive functional', and its inverse process $C(s) = \inf\{t : A(t) > s\}$ is a time change. See Revuz and Yor (1999, ch. X) for more on additive functionals.

A proper exploration of time changes is beyond our scope here. Instead we restrict ourselves to some highlights

**(9) Example. Markov processes.** A continuous process $X$ is *Markov* if it satisfies a suitably formulated Markov property. A Markov process is called *strong* if it satisfies the strong Markov property, suitably formulated. See Exercise (8.8.4) for a continuous Markov process without the strong Markov property.

Let $X$ be a right-continuous, strong Markov process, and let $T$ be the right-continuous inverse of a continuous additive functional $A$ that is adapted to the natural filtration of $X$. The time-changed process $X(T(t))$ is a strong Markov process.                                      ●

**(10) Example. Martingales.** Similar to (9), we have that continuous martingales, when time-changed by a Lévy process, retain the martingale property subject to a certain optional stopping condition. See, for example, Tankov and Cont (2003, p. 508) and Revuz and Yor (1999, p. 181).                                      ●

**(11) Example. Wiener process.** Let $W$ be the standard Wiener process, and let $A$ be the occupation time process for the non-negative half-line,

$$A(t) = \int_0^t I(W(s) \geq 0) \, ds, \qquad t \geq 0.$$

The inverse process

$$C(s) = \inf\{t \geq 0 : A(t) > s\}, \qquad s \geq 0,$$

is a time change. It turns out that the time-changed process $R(t) = W(C(t))$ is the Wiener process reflected at the origin, that is, $R$ has the same fdds as $|W|$. See Karatzas and Shreve (1991, Thm 6.3.1).                                      ●

**(12) Example. Stopping a Wiener process.** One can obtain almost anything one likes through stopping a Wiener process. More precisely, let $X$ be a random variable with mean 0 and finite variance. There exists a Wiener process $W$ and a stopping time $S$ adapted to a suitable filtration such that $W(S) \overset{\mathrm{D}}{=} X$, and in addition $\mathbb{E}(S) = \mathbb{E}(X^2)$. See Billingsley (1995, Thm 37.6).                                      ●

**(13) Example. Stopping a martingale.** Let $M = \{M(t) : t \geq 0\}$ be a continuous martingale with respect to the filtration $\mathcal{F}$, with $M(0) = 0$. Suppose the quadratic variation process $\langle M \rangle$ exists and is almost surely strictly increasing to $\infty$. Define the time change $T(t) = \inf\{u : \langle M \rangle_u > t\}$. Then $W(t) = M(T(t))$ is a Wiener process with respect to the filtration $\{\mathcal{F}_{T(t)} : t \geq 0\}$, and furthermore, $M(t) = W(\langle M \rangle_t)$.

In summary, there is a time change of $M$ that yields $W$, and conversely there is a time change of $W$ that yields $M$. See Revuz and Yor (1999, Thm V.1.6).                                      ●

For far more on this topic see Barndorff-Nielsen and Shiryaev 2015.

## Exercises for Section 8.8

**1.**   Let $Z$ be a continuous-time Markov chain, and let $T$ be an independent subordinator. Show that $X(t) = Z(T(t))$ defines a Markov chain.

**2.**   Let $Z = \{Z_n : n \geq 0\}$ be a discrete-time Markov chain with $n$-step transition probabilities $z_{ij}(n)$. Let $N$ be a Poisson process of rate $\lambda$ that is independent of $Z$, and set $X(t) = Z_{N(t)}$. Show that $X$ is a continuous-time Markov chain with transition probabilities

$$p_{ij}(t) = \sum_{n=0}^{\infty} e^{-\lambda t} \frac{(\lambda t)^n}{n!} z_{ij}(n).$$

**3.**   Let $W$ be the standard Wiener process, and let $T$ be an independent stable subordinator with $\mathbb{E}(e^{-\theta T(t)}) = \exp(-t \theta^{a/2})$ where $a \in (0, 2)$ and $\theta \geq 0$. Show that the time-changed process $Y(t) = W(T(t))$ is a symmetric Lévy process.

**4.**   **Markov, but not strong Markov.** Let $V$ be exponentially distributed with parameter 1, and define $X = \{X(t) : t \geq 0\}$ by: $X(t) = 0$ for $t \leq V$, and $X(t) = t - V$ for $t \geq V$.
(a)  Show that $X$ is a Markov process.
(b)  Show that $V$ is a stopping time for $X$.
(c)  Show that the two processes $X_1(t) = X(t)$ and $X_2(t) = X(t + V)$ satisfy $X_1(0) = X_2(0) = 0$, but have different fdds.

## 8.9  Existence of processes

In our discussions of the properties of random variables, only scanty reference has been made to the underlying probability space $(\Omega, \mathcal{F}, \mathbb{P})$; some may have felt relief at this omission. We have often made assumptions about hypothetical random variables without even checking that such variables exist. For example, we are in the habit of making statements such as 'let $X_1, X_2, \ldots$ be independent variables with common distribution function $F$', but we have made no effort to show that there exists some probability space on which such variables can be constructed. The foundations of such statements require examination. It is the purpose of this section to indicate that such assumptions are justifiable.

First, suppose that $(\Omega, \mathcal{F}, \mathbb{P})$ is a probability space and that $X = \{X_t : t \in T\}$ is some collection of random variables mapping $\Omega$ into $\mathbb{R}$. We saw in Section 8.1 that to any vector $\mathbf{t} = (t_1, t_2, \ldots, t_n)$ containing members of $T$ and of finite length there corresponds a joint distribution function $F_{\mathbf{t}}$; the collection of such functions $F_{\mathbf{t}}$, as $\mathbf{t}$ ranges over all possible vectors of any length, is called the set of *finite-dimensional distributions*, or *fdds*, of $X$. It is clear that these distribution functions satisfy the two *Kolmogorov consistency conditions*:

**(1)**  $F_{(t_1, \ldots, t_n, t_{n+1})}(x_1, \ldots, x_n, x_{n+1}) \to F_{(t_1, \ldots, t_n)}(x_1, \ldots, x_n)$ as $x_{n+1} \to \infty$,
**(2)**  if $\pi$ is a permutation of $(1, 2, \ldots, n)$ and $\pi \mathbf{y}$ denotes the vector $\pi \mathbf{y} = (y_{\pi(1)}, \ldots, y_{\pi(n)})$ for any $n$-vector $\mathbf{y}$, then $F_{\pi \mathbf{t}}(\pi \mathbf{x}) = F_{\mathbf{t}}(\mathbf{x})$ for all $\mathbf{x}, \mathbf{t}, \pi,$ and $n$.

Condition (1) is just a higher-dimensional form of (2.1.6a), and condition (2) says that the operation of permuting the $X_t$ has the obvious corresponding effect on their joint distributions. So fdds always satisfy (1) and (2); furthermore (1) and (2) characterize fdds, in the manner of the so-called Daniell–Kolmogorov Theorem.

**(3) Theorem.** *Let T be any set, and suppose that to each vector* $\mathbf{t} = (t_1, t_2, \ldots, t_n)$ *containing members of T and of finite length, there corresponds a joint distribution function* $F_{\mathbf{t}}$. *If the collection* $\{F_{\mathbf{t}}\}$ *satisfies the Kolmogorov consistency conditions, there exists a probability space* $(\Omega, \mathcal{F}, \mathbb{P})$ *and a collection* $X = \{X_t : t \in T\}$ *of random variables on this space such that* $\{F_{\mathbf{t}}\}$ *is the set of fdds of X.*

The proof of this result lies in the heart of measure theory, as the following sketch indicates.

**Sketch proof.** Let $\Omega = \mathbb{R}^T$, the product of $T$ copies of $\mathbb{R}$; the points of $\Omega$ are collections $\mathbf{y} = \{y_t : t \in T\}$ of real numbers. Let $\mathcal{F} = \mathcal{B}^T$, the $\sigma$-field generated by subsets of the form $\prod_{t \in T} B_t$ for Borel sets $B_t$ all but finitely many of which equal $\mathbb{R}$. It is a fundamental result in measure theory that there exists a probability measure $\mathbb{P}$ on $(\Omega, \mathcal{F})$ such that

$$\mathbb{P}\left(\{\mathbf{y} \in \Omega : y_{t_1} \leq x_1,\ y_{t_2} \leq x_2, \ldots,\ y_{t_n} \leq x_n\}\right) = F_{\mathbf{t}}(\mathbf{x})$$

for all $\mathbf{t}$ and $\mathbf{x}$; this follows by an extension of the argument of Section 1.6. Then $(\Omega, \mathcal{F}, \mathbb{P})$ is the required space. Define $X_t : \Omega \to \mathbb{R}$ by $X_t(\mathbf{y}) = y_t$ to obtain the required family $\{X_t\}$. ∎

We have seen that the fdds are characterized by the consistency conditions (1) and (2). But how much do they tell us about the sample paths of the corresponding process $X$? A simple example is enough to indicate some of the dangers here.

**(4) Example.** Let $U$ be a random variable which is uniformly distributed on $[0, 1]$. Define two processes $X = \{X(t) : 0 \leq t \leq 1\}$ and $Y = \{Y(t) : 0 \leq t \leq 1\}$ by

$$X(t) = 0 \quad \text{for all } t, \quad Y(t) = \begin{cases} 1 & \text{if } U = t, \\ 0 & \text{otherwise.} \end{cases}$$

Clearly $X$ and $Y$ have the same fdds, since $\mathbb{P}(U = t) = 0$ for all $t$. But $X$ and $Y$ are different processes. In particular $\mathbb{P}(X(t) = 0 \text{ for all } t) = 1$ and $\mathbb{P}(Y(t) = 0 \text{ for all } t) = 0$. ●

One may easily construct less trivial examples of different processes having the same fdds; such processes are called *versions* of one another. This complication should not be overlooked with a casual wave of the hand; it is central to any theory which attempts to study properties of sample paths, such as first-passage times. As the above example illustrates, such properties are not generally specified by the fdds, and their validity may therefore depend on which version of the process is under study.

For the random process $\{X(t) : t \in T\}$, where $T = [0, \infty)$ say, knowledge of the fdds amounts to being given a probability space of the form $(\mathbb{R}^T, \mathcal{B}^T, \mathbb{P})$, as in the sketch proof of (3) above. Many properties of sample paths do not correspond to events in $\mathcal{B}^T$. For example, the subset of $\Omega$ given by $A = \{\omega \in \Omega : X(t) = 0 \text{ for all } t \in T\}$ is an *uncountable* intersection of events $A = \bigcap_{t \in T}\{X(t) = 0\}$, and may not itself be an event. Such difficulties would be avoided if all sample paths of $X$ were continuous, since then $A$ is the intersection of $\{X(t) = 0\}$ over all *rational* $t \in T$; this is a *countable* intersection.

**(5) Example.** Let $W$ be the Wiener process of Section 8.5, and let $T$ be the time of the first passage of $W$ to the point 1, so that $S = \inf\{t : W(t) = 1\}$. Then

$$\{S > t\} = \bigcap_{0 \leq s \leq t} \{W(s) \neq 1\}$$

is a set of configurations which does not belong to the Borel $\sigma$-field $\mathcal{B}^{[0,\infty)}$. If all sample paths of $W$ were continuous, and hence uniformly continuous on the interval $[0, t]$, one might write

$$\{S > t\} = \bigcup_{k \geq 1} \bigcap_{\substack{0 \leq s \leq t \\ s \in \mathbb{Q}}} \{|W(s) - 1| > k^{-1}\},$$

the countable union of countable intersections of events. The symbol $\mathbb{Q}$ denotes the rationals. As the construction of Example (4) indicates, there are versions of the Wiener process which have discontinuous sample paths. One of the central results of Chapter 13 is that there exists a version with continuous sample paths, and it is with this version that one normally works. See Theorem (13.3.19).                                                                                    ●

It is too restrictive to require continuity of sample paths in general; after all, processes such as the Poisson process most definitely do not have continuous sample paths. The most which can be required is continuity from either the left or the right. Following a convention, we go for the latter here. Under what conditions may one assume that there exists a version with right-continuous sample paths? An answer is provided by the next theorem; see Breiman (1968, p. 300) for a proof.

**(6) Theorem.** *Let* $X = \{X(t) : t \geq 0\}$ *be a real-valued random process. Let $D$ be a subset of* $[0, \infty)$ *which is dense in* $[0, \infty)$. *If*:

(i) $X$ *is continuous in probability from the right, that is,* $X(t + h) \overset{P}{\to} X(t)$ *as* $h \downarrow 0$, *for all $t$, and*

(ii) *at any accumulation point $a$ of $D$, $X$ has finite right and left limits with probability 1, that is* $\lim_{h \downarrow 0} X(a + h)$ *and* $\lim_{h \uparrow 0} X(a + h)$ *exist, a.s., where the limits are taken as* $h \to 0$ *through $D$,*

*then there exists a version $Y$ of $X$ such that:*

(a) *the sample paths of $Y$ are right-continuous,*

(b) $Y$ *has left limits, in that* $\lim_{h \uparrow 0} Y(t + h)$ *exists for all $t$.*

In other words, if (i) and (ii) hold, there exists a probability space and a process $Y$ defined on this space, such that $Y$ has the same fdds as $X$ in addition to properties (a) and (b). A process which is right-continuous with left limits is called càdlàg by some (largely French speakers), and a Skorokhod map or R-process by others.

---

## 8.10  Problems

**1.**    Let $\{Z_n\}$ be a sequence of uncorrelated real-valued variables with zero means and unit variances, and define the 'moving average'

$$Y_n = \sum_{i=0}^{r} \alpha_i Z_{n-i},$$

for constants $\alpha_0, \alpha_1, \ldots, \alpha_r$. Show that $Y$ is stationary and find its autocovariance function.

**2.**    Let $\{Z_n\}$ be a sequence of uncorrelated real-valued variables with zero means and unit variances. Suppose that $\{Y_n\}$ is an 'autoregressive' stationary sequence in that it satisfies $Y_n = \alpha Y_{n-1} + Z_n$, $-\infty < n < \infty$, for some real $\alpha$ satisfying $|\alpha| < 1$. Show that $Y$ has autocovariance function $c(m) = \alpha^{|m|}/(1 - \alpha^2)$.

**3.** Let $\{X_n\}$ be a sequence of independent identically distributed Bernoulli variables, each taking values 0 and 1 with probabilities $1 - p$ and $p$ respectively. Find the mass function of the renewal process $N(t)$ with interarrival times $\{X_n\}$.

**4.** Customers arrive in a shop in the manner of a Poisson process with parameter $\lambda$. There are infinitely many servers, and each service time is exponentially distributed with parameter $\mu$. Show that the number $Q(t)$ of waiting customers at time $t$ constitutes a birth–death process. Find its stationary distribution.

**5.** Let $X(t) = Y\cos(\theta t) + Z\sin(\theta t)$ where $Y$ and $Z$ are independent $N(0, 1)$ random variables, and let $\widetilde{X}(t) = R\cos(\theta t + \Psi)$ where $R$ and $\Psi$ are independent. Find distributions for $R$ and $\Psi$ such that the processes $X$ and $\widetilde{X}$ have the same fdds.

**6. Bartlett's theorem.** Customers arrive at the entrance to a queueing system at the instants of an non-homogeneous Poisson process with rate function $\lambda(t)$. Their subsequent service histories are independent of each other, and a customer arriving at time $s$ is in state $A$ at time $s + t$ with probability $p(s, t)$. Show that the number of customers in state $A$ at time $t$ is Poisson with parameter $\int_{-\infty}^{t} \lambda(u) p(u, t - u)\, du$.

**7.** An insurance company receives premiums (net of costs) at unit rate per unit time, and the claims of size $X_1, X_2, \ldots$ are independent random variables with common distribution function $F$, arriving at the instants of a Poisson process of rate 1. Show that the probability $r(y)$ of ruin (that is, assets becoming negative) for an initial capital $y$ satisfies

$$\frac{dr(y)}{dy} = \lambda r(y) - \lambda \mathbb{P}(X_1 > y) - \lambda \int_0^y r(y - x)\, dF(x).$$

**8. Fractional Brownian motion.** Let $X$ be fBM$_H$ with var$(X(1)) = 1$, where $H \in (0, 1)$. Show that:

(a) $X(t) - X(s)$ has the $N(0, |t - s|^{2H})$ distribution,

(b) for $r > 0$,

$$\mathbb{E}\big(|X(t) - X(s)|^r\big) = C|t - s|^{rH},$$

where $C$ depends on $r$.

(c) $X$ is a Wiener process if and only if $H = \frac{1}{2}$.

**9. Holtsmark distribution for stellar gravity, Problem (6.15.57) revisited.** Suppose stars are distributed in $\mathbb{R}^3$ in the manner of a Poisson process with intensity $\lambda$, and let $G_\lambda$ be the $x$-coordinate of their aggregated gravitational force exerted on a unit mass at the origin.

(a) Show that $G_{\lambda+\mu} \stackrel{D}{=} G_\lambda + G_\mu$, the sum of two independent variables.

(b) Show by the inverse square law that $G_\lambda \stackrel{D}{=} \lambda^{2/3} G_1$.

(c) Comment on the relevance of the above in studying gravitational attraction in three dimensions.

**10.** In a Prague teashop (U Myšáka) before the Velvet Revolution of 1989, customers queue at the entrance for a blank bill. In the shop there are separate counters for coffee, sweetcakes, pretzels, milk, drinks, and ice cream, and queues form at each of these. At each service point the customers' bills are marked appropriately. There is a restricted number $N$ of seats, and departing customers have to queue in order to pay their bills. If interarrival times and service times are exponentially distributed and the process is in equilibrium, find how much longer a greedy customer must wait if he insists on sitting down. Answers on a postcard to the authors, please.

# 9

---

# Stationary processes

*Summary.* The theory of stationary processes, with discrete or continuous parameter, is developed. Autocovariances and spectral distributions are introduced. A theory of stochastic integration is developed for functions integrated against a stochastic process, and this theory is used to obtain a representation of a stationary process known as the spectral theorem. A result of major importance is the ergodic theorem, which explains the convergence of successive averages of a stationary process. Ergodic theorems are presented for weakly and strongly stationary processes. The final section is an introduction to Gaussian processes.

## 9.1 Introduction

Recall that a process $X$ is *strongly stationary* whenever its *finite-dimensional distributions* are invariant under time shifts; it is *(weakly) stationary* whenever it has constant means and its *autocovariance function* is invariant under time shifts. Section 8.2 contains various examples of such processes. Next, we shall explore some deeper consequences of stationarity, in particular the spectral theorem and the ergodic theorem†.

A special class of random processes comprises those processes whose joint distributions are multivariate normal; these are called 'Gaussian processes'. Section 9.6 contains a brief account of some of the properties of such processes. In general, a Gaussian process is not stationary, but it is easy to characterize those which are.

We shall be interested mostly in continuous-time processes $X = \{X(t) : -\infty < t < \infty\}$, indexed by the whole real line, and will indicate any necessary variations for processes with other index sets, such as discrete-time processes. It is convenient to suppose that $X$ takes values in the complex plane $\mathbb{C}$. This entails few extra complications and provides the natural setting for the theory. No conceptual difficulty is introduced by this generalization, since any complex-valued process $X$ can be decomposed as $X = X_1 + iX_2$ where $X_1$ and $X_2$ are real-valued processes. However, we must take care when discussing the finite-dimensional distributions (fdds) of $X$ since the distribution function of a complex-valued random variable $C = R + iI$ is no longer a function of a single real variable. Thus, our definition of strong

---

†The word 'ergodic' has several meanings, and probabilists tend to use it rather carelessly. We conform to this custom here.

stationarity requires revision; we leave this to the reader. The concept of weak stationarity concerns covariances; we must note an important amendment to the real-valued theory in this context. As before, the expectation operator $\mathbb{E}$ is defined by $\mathbb{E}(R + iI) = \mathbb{E}(R) + i\mathbb{E}(I)$.

**(1) Definition.** The **covariance** of two complex-valued random variables $C_1$ and $C_2$ is defined to be

$$\text{cov}(C_1, C_2) = \mathbb{E}\big((C_1 - \mathbb{E}C_1)\overline{(C_2 - \mathbb{E}C_2)}\big)$$

where $\bar{z}$ denotes the complex conjugate of $z$.

This reduces to the usual definition (3.6.7) when $C_1$ and $C_2$ are real. Note that the operator 'cov' is not symmetrical in its arguments, since

$$\text{cov}(C_2, C_1) = \overline{\text{cov}(C_1, C_2)}.$$

Variances are defined as follows.

**(2) Definition.** The **variance** of a complex-valued random variable $C$ is defined to be

$$\text{var}(C) = \text{cov}(C, C).$$

Decompose $C$ into its real and imaginary parts, $C = R + iI$, and apply (2) to obtain $\text{var}(C) = \text{var}(R) + \text{var}(I)$. We may write

$$\text{var}(C) = \mathbb{E}(|C - \mathbb{E}C|^2).$$

We do not generally speak of complex random variables as being 'uncorrelated', preferring to use a word which emphasizes the geometrical properties of the complex plane.

**(3) Definition.** Complex-valued random variables $C_1$ and $C_2$ are called **orthogonal** if they satisfy $\text{cov}(C_1, C_2) = 0$.

If $X = X_1 + iX_2$ is a complex-valued process with real part $X_1$ and imaginary part $X_2$ then $\overline{X}$ denotes the complex conjugate process of $X$, that is, $\overline{X} = X_1 - iX_2$.

**(4) Example. Functions of the Poisson process.** Let $N$ be a Poisson process with intensity $\lambda$. Let $\alpha$ be a positive number, and define $X(t) = N(t + \alpha) - N(t)$, for $t \geq 0$. It is easily seen (*exercise*) from the definition of a Poisson process that $X$ is a strongly stationary process with mean $\mathbb{E}(X(t)) = \lambda\alpha$ and autocovariance function

$$c(t, t + h) = \mathbb{E}\big(X(t)X(t + h)\big) - (\lambda\alpha)^2 = \begin{cases} 0 & \text{if } h \geq \alpha, \\ \lambda(\alpha - h) & \text{if } h < \alpha, \end{cases}$$

where $t, h \geq 0$.

Here is a second example based on the Poisson process. Let $\beta = e^{2\pi i/m}$ be a complex $m$th root of unity, where $m \geq 2$, and define $Y(t) = \beta^{Z+N(t)}$ where $Z$ is a random variable that is independent of $N$ with mass function $\mathbb{P}(Z = j) = 1/m$, for $1 \leq j \leq m$. Once again, it is left as an *exercise* to show that $Y$ is a strictly stationary (complex-valued) process with mean $\mathbb{E}(Y(t)) = 0$. Its autocovariance function is given by the following calculation:

$$\mathbb{E}\big(Y(t)\overline{Y(t + h)}\big) = \mathbb{E}\big(\beta^{N(t)}\overline{\beta}^{N(t+h)}\big) = \mathbb{E}\big((\beta\overline{\beta})^{N(t)}\overline{\beta}^{N(t+h)-N(t)}\big)$$

$$= \mathbb{E}(\overline{\beta}^{N(h)}) \qquad \text{since } \beta\overline{\beta} = 1$$

$$= \exp\big[\lambda h(\overline{\beta} - 1)\big] \qquad \text{for } t, h \geq 0,$$

where we have used elementary properties of the Poisson process.  ●

## Exercises for Section 9.1

**1.** Let $\ldots, Z_{-1}, Z_0, Z_1, Z_2, \ldots$ be independent real random variables with means $0$ and variances $1$, and let $\alpha, \beta \in \mathbb{R}$. Show that there exists a (weakly) stationary sequence $\{W_n\}$ satisfying $W_n = \alpha W_{n-1} + \beta W_{n-2} + Z_n$, $n = \ldots, -1, 0, 1, \ldots$, if the (possibly complex) zeros of the quadratic equation $z^2 - \alpha z - \beta = 0$ are smaller than $1$ in absolute value.

**2.** Let $U$ be uniformly distributed on $[0, 1]$ with binary expansion $U = \sum_{i=1}^{\infty} X_i 2^{-i}$. Show that the sequence

$$V_n = \sum_{i=1}^{\infty} X_{i+n} 2^{-i}, \qquad n \geq 0,$$

is strongly stationary, and calculate its autocovariance function.

**3.** Let $\{X_n : n = \ldots, -1, 0, 1, \ldots\}$ be a stationary real sequence with mean $0$ and autocovariance function $c(m)$.

(i) Show that the infinite series $\sum_{n=0}^{\infty} a_n X_n$ converges almost surely, and in mean square, whenever $\sum_{n=0}^{\infty} |a_n| < \infty$.

(ii) Let

$$Y_n = \sum_{k=0}^{\infty} a_k X_{n-k}, \qquad n = \ldots, -1, 0, 1, \ldots$$

where $\sum_{k=0}^{\infty} |a_k| < \infty$. Find an expression for the autocovariance function $c_Y$ of $Y$, and show that

$$\sum_{m=-\infty}^{\infty} |c_Y(m)| < \infty.$$

**4.** Let $X = \{X_n : n \geq 0\}$ be a discrete-time Markov chain with countable state space $S$ and stationary distribution $\boldsymbol{\pi}$, and suppose that $X_0$ has distribution $\boldsymbol{\pi}$. Show that the sequence $\{f(X_n) : n \geq 0\}$ is strongly stationary for any function $f : S \to \mathbb{R}$.

**5.** Let $W, X, Y, Z$ have a multivariate normal distribution with zero means and covariance matrix

$$\begin{pmatrix} 1 & 0 & 0 & -1 \\ 0 & 1 & -1 & 0 \\ 0 & -1 & 1 & 0 \\ -1 & 0 & 0 & 1 \end{pmatrix}.$$

Let $U = W + iX$ and $V = Y + iZ$. Show that $U$ and $V$ are uncorrelated but not independent. Why does this not violate the conclusion of Example (4.5.9) that multivariate normal random variables are independent if and only if they are uncorrelated?

**6.** Let $X_n = \cos(nS + U)$ for $n \in \mathbb{Z}$, where $U$ is uniformly distributed on $(-\pi, \pi)$, and $S$ is independent of $U$ with a density function $g$ that is symmetric on its support $(-\pi, \pi)$. Show that $X = \{X_n : n \in \mathbb{Z}\}$ is weakly stationary, and find its autocorrelation function $\rho_X$.

Let $Y_n = Z_n + aZ_{n-1}$ for $n \in \mathbb{Z}$, where $|a| < 1$ and the $Z_n$ are independent, identically distributed random variables with means $0$ and variances $1$. Is it possible to choose $g$ in such a way that $Y = \{Y_n : n \in \mathbb{Z}\}$ has the same autocorrelation function as $X$? If so, how?

## 9.2  Linear prediction

Statisticians painstakingly observe and record processes which evolve in time, not merely for the benefit of historians but also in the belief that it is an advantage to know the past when attempting to predict the future. Most scientific schemes (and many non-scientific schemes) for prediction are 'model' based, in that they make some specific assumptions about the process, and then use past data to extrapolate into the future. For example, in the statistical theory of 'time series', one often assumes that the process is some combination of general trend, periodic fluctuations, and random noise, and it is common to suppose that the noise component is a stationary process having an autocovariance function of a certain form.

Suppose that we are observing a sequence $\{x_n\}$ of numbers, the number $x_n$ being revealed to us at time $n$, and that we are prepared to accept that these numbers are the outcomes of a stationary sequence $\{X_n\}$ with known mean $\mathbb{E}X_n = \mu$ and autocovariance function $c(m) = \mathrm{cov}(X_n, X_{n+m})$. We may be required to estimate the value of $X_{r+k}$ (where $k \geq 1$), given the values $X_r, X_{r-1}, \ldots, X_{r-s}$. We saw in Section 7.9 that the 'best' (that is, the minimum mean-squared-error) predictor of $X_{r+k}$ given $X_r, X_{r-1}, \ldots, X_{r-s}$ is the conditional mean $M = \mathbb{E}(X_{r+k} \mid X_r, X_{r-1}, \ldots, X_{r-s})$; that is to say, the mean squared error $\mathbb{E}((Y - X_{r+k})^2)$ is minimized over all choices of functions $Y$ of $X_r, X_{r-1}, \ldots, X_{r-s}$ by the choice $Y = M$. The calculation of such quantities requires a knowledge of the finite-dimensional distributions (fdds) of $X$ which we do not generally possess. For various reasons, it is not realistic to attempt to estimate the fdds in order to *estimate* the conditional mean. The problem becomes more tractable, and its solution more elegant, if we restrict our attention to *linear* predictors of $X_{r+k}$, which is to say that we seek the best predictor of $X_{r+k}$ amongst the class of linear functions of $X_r, X_{r-1}, \ldots, X_{r-s}$.

**(1) Theorem.** *Let $X$ be a real stationary sequence with zero mean and autocovariance function $c(m)$. Amongst the class of linear functions of the subsequence $X_r, X_{r-1}, \ldots, X_{r-s}$, the best predictor of $X_{r+k}$ (where $k \geq 1$) is*

$$\text{(2)} \qquad \widehat{X}_{r+k} = \sum_{i=0}^{s} a_i X_{r-i}$$

*where the $a_i$ satisfy the equations*

$$\text{(3)} \qquad \sum_{i=0}^{s} a_i c(|i - j|) = c(k + j) \quad \text{for} \quad 0 \leq j \leq s.$$

**Proof.** Let $H$ be the closed linear space of linear functions of $X_r, X_{r-1}, \ldots, X_{r-s}$. We have from the projection theorem (7.9.14) that the element $M$ of $H$ for which $\mathbb{E}((X_{r+k} - M)^2)$ is a minimum is the (almost surely) unique $M$ such that

$$\text{(4)} \qquad \mathbb{E}((X_{r+k} - M)Z) = 0 \quad \text{for all } Z \in H.$$

Certainly $X_{r-j} \in H$ for $0 \leq j \leq s$. Writing $M = \sum_{i=0}^{s} a_i X_{r-i}$ and substituting $Z = X_{r-j}$ in (4), we obtain

$$\mathbb{E}(X_{r+k} X_{r-j}) = \mathbb{E}(M X_{r-j}) = \sum_{i=0}^{s} a_i \mathbb{E}(X_{r-i} X_{r-j}),$$

whence (3) follows by the assumption of zero mean.                                   ∎

Therefore, if we know the autocovariance function $c$, then equation (3) tells us how to find the best linear predictor of future values of the stationary sequence $X$. In practice we may not know $c$, and may instead have to estimate it. Rather than digress further in this direction, the reader is referred to the time series literature, for example Chatfield 2003.

**(5) Example. Autoregressive scheme.** Let $\{Z_n\}$ be a sequence of independent variables with zero means and unit variances, and let $\{Y_n\}$ satisfy

$$(6) \qquad Y_n = \alpha Y_{n-1} + Z_n, \quad -\infty < n < \infty,$$

where $\alpha$ is a real number satisfying $|\alpha| < 1$. We have from Problem (8.10.2) that $Y$ is stationary with zero mean and autocovariance function $c(m) = \mathbb{E}(Y_n Y_{n+m})$ given by

$$(7) \qquad c(m) = \frac{\alpha^{|m|}}{1 - \alpha^2}, \quad -\infty < m < \infty.$$

Suppose we wish to estimate $Y_{r+k}$ (where $k \geq 1$) from a knowledge of $Y_r, Y_{r-1}, \ldots, Y_{r-s}$. The best linear predictor is $\widehat{Y}_{r+k} = \sum_{i=0}^{s} a_i Y_{r-i}$ where the $a_i$ satisfy equations (3):

$$\sum_{i=0}^{s} a_i \alpha^{|i-j|} = \alpha^{k+j}, \quad 0 \leq j \leq s.$$

A solution is $a_0 = \alpha^k$, $a_i = 0$ for $i \geq 1$, so that the best linear predictor is $\widehat{Y}_{r+k} = \alpha^k Y_r$. The mean squared error of prediction is

$$\begin{aligned}
\mathbb{E}\big((Y_{r+k} - \widehat{Y}_{r+k})^2\big) &= \operatorname{var}(Y_{r+k} - \alpha^k Y_r) \\
&= \operatorname{var}(Y_{r+k}) - 2\alpha^k \operatorname{cov}(Y_{r+k}, Y_r) + \alpha^{2k} \operatorname{var}(Y_r) \\
&= c(0) - 2\alpha^k c(k) + \alpha^{2k} c(0) = \frac{1 - \alpha^{2k}}{1 - \alpha^2}, \quad \text{by (7).} \qquad \bullet
\end{aligned}$$

**(8) Example.** Let $X_n = (-1)^n X_0$ where $X_0$ is equally likely to take each of the values $-1$ and $+1$. It is easily checked in this special case that $X$ is stationary with zero mean and autocovariance function $c(m) = (-1)^m \mathbb{E}(X_0^2) = (-1)^m$, $-\infty < m < \infty$. The best linear predictor of $X_{r+k}$ (where $k \geq 1$) based on $X_r, X_{r-1}, \ldots, X_{r-s}$ is obtained by solving the equations

$$\sum_{i=0}^{s} a_i (-1)^{|i-j|} = (-1)^{k+j}, \quad 0 \leq j \leq s.$$

A solution is $a_0 = (-1)^j$, $a_i = 0$ for $i \geq 1$, so that $\widehat{X}_{r+k} = (-1)^k X_r$, and the mean squared error of prediction is zero. $\qquad \bullet$

## Exercises for Section 9.2

1.  Let $X$ be a (weakly) stationary sequence with zero mean and autocovariance function $c(m)$.
 (i) Find the best linear predictor $\widehat{X}_{n+1}$ of $X_{n+1}$ given $X_n$.
 (ii) Find the best linear predictor $\widetilde{X}_{n+1}$ of $X_{n+1}$ given $X_n$ and $X_{n-1}$.
 (iii) Find an expression for $D = \mathbb{E}\{(X_{n+1} - \widehat{X}_{n+1})^2\} - \mathbb{E}\{(X_{n+1} - \widetilde{X}_{n+1})^2\}$, and evaluate this expression when:
       (a) $X_n = \cos(nU)$ where $U$ is uniform on $[-\pi, \pi]$,
       (b) $X$ is an autoregressive scheme with $c(k) = \alpha^{|k|}$ where $|\alpha| < 1$.

2.  Suppose $|a| < 1$. Does there exist a (weakly) stationary sequence $\{X_n : -\infty < n < \infty\}$ with zero means and autocovariance function

$$c(k) = \begin{cases} 1 & \text{if } k = 0, \\ \dfrac{a}{1+a^2} & \text{if } |k| = 1, \\ 0 & \text{if } |k| > 1. \end{cases}$$

Assuming that such a sequence exists, find the best linear predictor $\widehat{X}_n$ of $X_n$ given $X_{n-1}$, $X_{n-2}, \ldots$, and show that the mean squared error of prediction is $(1+a^2)^{-1}$. Verify that $\{\widehat{X}_n\}$ is (weakly) stationary.

# 9.3 Autocovariances and spectra

Let $X = \{X(t) : -\infty < t < \infty\}$ be a (weakly) stationary process which takes values in the complex plane $\mathbb{C}$. It has autocovariance function $c$ given by

$$c(s, s+t) = \operatorname{cov}\big(X(s), X(s+t)\big) \quad \text{for} \quad s, t \in \mathbb{R}$$

where $c(s, s+t)$ depends on $t$ alone. We think of $c$ as a complex-valued function of the single variable $t$, and abbreviate it to

$$c(t) = c(s, s+t) \quad \text{for any } s.$$

Notice that the variance of $X(t)$ is constant for all $t$ since

**(1)**                             $\operatorname{var}(X(t)) = \operatorname{cov}\big(X(t), X(t)\big) = c(0).$

We shall sometimes assume that the mean value $\mathbb{E}(X(t))$ of $X$ equals zero; if this is not true, then define $X'(t) = X(t) - \mathbb{E}(X(t))$ to obtain another stationary process with zero means and the same autocovariance function.

   Autocovariances have the following properties.

**(2) Theorem.** *We have that:*
  (a) $c(-t) = \overline{c(t)}$,
  (b) *c is a non-negative definite function, which is to say that*

$$\sum_{j,k} c(t_k - t_j)z_j\bar{z}_k \geq 0$$

*for all real $t_1, t_2, \ldots, t_n$ and all complex $z_1, z_2, \ldots, z_n$.*

**Proof.**

(a) $c(-t) = \text{cov}(X(t), X(0)) = \overline{\text{cov}(X(0), X(t))} = \overline{c(t)}$.

(b) This resembles the proof of Theorem (5.7.3c). Just write

$$\sum_{j,k} c(t_k - t_j)z_j\bar{z}_k = \sum_{j,k} \text{cov}\big(z_j X(t_j), z_k X(t_k)\big) = \text{cov}(Z, Z) \geq 0$$

where $Z = \sum_j z_j X(t_j)$.   ∎

Of more interest than the autocovariance function is the 'autocorrelation function' (see Definition (3.6.7)).

**(3) Definition.** The **autocorrelation function** of a weakly stationary process $X$ with autocovariance function $c(t)$ is defined by

$$\rho(t) = \frac{\text{cov}(X(0), X(t))}{\sqrt{\text{var}(X(0))\,\text{var}(X(t))}} = \frac{c(t)}{c(0)}$$

whenever $c(0) = \text{var}(X(t)) > 0$.

Of course, $\rho(t)$ is just the correlation between $X(s)$ and $X(s+t)$, for any $s$.

Following the discussion in Section 8.2, we seek to assess the incidence of certain regular oscillations within the random fluctuation of $X$. For a weakly stationary process this is often a matter of studying regular oscillations in its autocorrelation function. The following theorem is often named after Wiener and Khinchin.

**(4) Theorem. Spectral theorem for autocorrelation functions.** *The autocorrelation function $\rho(t)$ of a weakly stationary process $X$ with strictly positive variance is the characteristic function of some distribution function $F$ whenever $\rho(t)$ is continuous at $t = 0$. That is to say,*

**(5)** $$\rho(t) = \int_{-\infty}^{\infty} e^{it\lambda}\, dF(\lambda).$$

**Proof.** This follows immediately from the discussion after Theorem (5.7.3), and is a simple application of Bochner's theorem. Following (2), we need only show that $\rho$ is uniformly continuous. Without loss of generality we can suppose that $\mathbb{E}(X(t)) = 0$ for all $t$. Let $c(t)$ be the autocovariance function of $X$, and use the Cauchy–Schwarz inequality (3.6.9) to obtain

$$\begin{aligned}
\big|c(t+h) - c(t)\big| &= \big|\mathbb{E}\big(X(0)[X(t+h) - X(t)]\big)\big| \\
&\leq \mathbb{E}\big(|X(0)||X(t+h) - X(t)|\big) \\
&\leq \sqrt{\mathbb{E}\big(|X(0)|^2\big)\mathbb{E}\big(|X(t+h) - X(t)|^2\big)} \\
&= \sqrt{c(0)\big[2c(0) - c(h) - c(-h)\big]}.
\end{aligned}$$

Therefore $c$ is uniformly continuous whenever it is continuous at $h = 0$. Thus $\rho(t) = c(t)/c(0)$ is uniformly continuous as claimed, and the result follows.   ∎

Think of equation (5) as follows. With any real $\lambda$ we may associate a complex-valued oscillating function $g_\lambda$ which has period $2\pi/|\lambda|$ and some non-negative amplitude $f_\lambda$, say:

$$g_\lambda(t) = f_\lambda e^{it\lambda};$$

in the less general real-valued theory we might consider oscillations of the form $g'_\lambda(t) = f_\lambda \cos(t\lambda)$ (see equations (8.2.6) and (8.2.7)). With any collection $\lambda_1, \lambda_2, \ldots$ of frequencies we can associate a mixture

(6) $$g_\lambda(t) = \sum_j f_j e^{it\lambda_j}$$

of pure oscillations, where the $f_j$ indicate the relative strengths of the various components. As the number of component frequencies in (6) grows, the summation may approach an integral

(7) $$g(t) = \int_{-\infty}^{\infty} f(\lambda)e^{it\lambda}\, d\lambda$$

where $f$ is some non-negative function which assigns weights to the $\lambda$. The progression from (6) to (7) is akin to the construction of the abstract integral (see Section 5.6). We have seen many expressions which are similar to (7), but in which $f$ is the density function of some continuous random variable. Just as continuous variables are only a special subclass of the larger family of all random variables, so (7) is not the most general limiting form for (6); the general form is

(8) $$g(t) = \int_{-\infty}^{\infty} e^{it\lambda}\, dF(\lambda)$$

where $F$ is a function which maps $\mathbb{R}$ into $[0, \infty)$ and which is right-continuous, non-decreasing, and such that $F(-\infty) = 0$; we omit the details of this, which are very much the same as in part B of Section 5.6. It is easy to see that $F$ is a distribution function if and only if $g(0) = 1$. Theorem (4) asserts that $\rho$ enjoys a decomposition in the form of (8), as a mixture of pure oscillations.

There is an alternative view of (5) which differs slightly from this. If $\Lambda$ is a random variable with distribution function $F$, then $g_\Lambda(t) = e^{it\Lambda}$ is a pure oscillation with a random frequency. Theorem (4) asserts that $\rho$ is the mean value of this random oscillation for some special distribution $F$. Of course, by the uniqueness theorem (5.9.3) there is a unique distribution function $F$ such that (5) holds.

**(9) Definition.** If the autocorrelation function $\rho$ satisfies

$$\rho(t) = \int_{-\infty}^{\infty} e^{it\lambda}\, dF(\lambda)$$

then $F$ is called the **spectral distribution function** of the process. The **spectral density function** is the density function which corresponds to the distribution function $F$ whenever this density exists.

For a given autocorrelation function $\rho$, we can find the spectral distribution function by the inversion techniques of Section 5.9.

In general, there may be certain frequency bands which make no contribution to (5). For example, if the spectral distribution function $F$ satisfies $F(\lambda) = 0$ for all $\lambda \le 0$, then only positive frequencies make non-trivial contributions. If the frequency band $(\lambda - \epsilon, \lambda + \epsilon)$ makes a non-trivial contribution to (5) for all $\epsilon > 0$, then we say that $\lambda$ belongs to the 'spectrum' of the process.

**(10) Definition.** The **spectrum** of $X$ is the set of all real numbers $\lambda$ with the property that

$$F(\lambda + \epsilon) - F(\lambda - \epsilon) > 0 \quad \text{for all } \epsilon > 0$$

where $F$ is the spectral distribution function.

If $X$ is a discrete-time process then the above account is inadequate, since the autocorrelation function $\rho$ now maps $\mathbb{Z}$ into $\mathbb{C}$ and cannot be a characteristic function unless its domain is extended. Theorem (4) remains broadly true, but asserts now that $\rho$ has a representation

**(11)** $$\rho(n) = \int_{-\infty}^{\infty} e^{in\lambda}\, dF(\lambda)$$

for some distribution function $F$ and all integral $n$. No condition of continuity is appropriate here. This representation (11) is not unique because the integrand $g_\lambda(n) = e^{in\lambda}$ is periodic in $\lambda$, which is to say that $g_{\lambda+2\pi}(n) = g_\lambda(n)$ for all $n$. In this case it is customary to rewrite equation (11) as

$$\rho(n) = \sum_{k=-\infty}^{\infty} \int_{((2k-1)\pi,(2k+1)\pi]} e^{in\lambda}\, dF(\lambda),$$

yielding the usual statement of the spectral theorem for discrete-time processes:

**(12)** $$\rho(n) = \int_{(-\pi,\pi]} e^{in\lambda}\, d\widetilde{F}(\lambda)$$

for some appropriate distribution function $\widetilde{F}$ obtained from $F$ and satisfying $\widetilde{F}(-\pi) = 0$ and $\widetilde{F}(\pi) = 1$. A further simplification is possible if $X$ is real valued, since then $\rho(n) = \rho(-n)$, so that

**(13)** $$\rho(n) = \tfrac{1}{2}[\rho(n) + \rho(-n)] = \int_{(-\pi,\pi]} \tfrac{1}{2}(e^{in\lambda} + e^{-in\lambda})\, d\widetilde{F}(\lambda) \quad \text{by (12)}$$

$$= \int_{(-\pi,\pi]} \cos(n\lambda)\, d\widetilde{F}(\lambda).$$

Furthermore $\cos(n\lambda) = \cos(-n\lambda)$, and it follows that $\rho$ may be expressed as

**(14)** $$\rho(n) = \int_{[-\pi,\pi]} \cos(n\lambda)\, dG(\lambda)$$

for some distribution function $G$ of a symmetric distribution on $[-\pi, \pi]$. We note that the validity of (14) for some such $G$ is both necessary and sufficient for $\rho$ to be the autocorrelation function of a real-valued stationary sequence. The necessity of (14) has been shown. For its sufficiency, we shall see at the beginning of Section 9.6 that all symmetric, non-negative definite functions $\rho$ with $\rho(0) = 1$ are autocorrelation functions of stationary sequences whose fdds are multivariate normal.

Equations (12)–(14) express $\rho$ as the Fourier transform of some distribution function. Fourier transforms may be inverted in the usual way to obtain an expression for the spectral distribution in terms of $\rho$. One such expression is the following.

**(15) Theorem.** *Let $\rho$ be the autocorrelation function of a stationary sequence. If the function $\widetilde{F}$ in (12) is differentiable with derivative $f$, then*

$$(16) \qquad\qquad f(\lambda) = \frac{1}{2\pi} \sum_{n=-\infty}^{\infty} e^{-in\lambda} \rho(n)$$

*at every point $\lambda$ at which $f$ is differentiable.*

For real-valued sequences, (16) may be written as

$$(17) \qquad\qquad f(\lambda) = \frac{1}{2\pi} \sum_{n=-\infty}^{\infty} \rho(n) \cos(n\lambda), \quad \pi \le \lambda \le \pi.$$

As in the discussion after Theorem (5.9.1) of characteristic functions, a sufficient (but not necessary) condition for the existence of the spectral density function $f$ is

$$(18) \qquad\qquad \sum_{n=-\infty}^{\infty} |\rho(n)| < \infty.$$

**(19) Example. Independent sequences.** Let $X = \{X_n : n \ge 0\}$ be a sequence of independent variables with zero means and unit variances. In Example (8.2.8) we found that the autocorrelation function is given by

$$\rho(n) = \begin{cases} 1 & \text{if } n = 0, \\ 0 & \text{if } n \ne 0. \end{cases}$$

In order to find the spectral density function, either use (15) or recognize that

$$\rho(n) = \int_{-\pi}^{\pi} e^{in\lambda} \cdot \frac{1}{2\pi} d\lambda$$

to see that the spectral density function is the uniform density function on $[-\pi, \pi]$. The spectrum of $X$ is the interval $[-\pi, \pi]$. Such a sequence $X$ is sometimes called 'discrete white noise'.                                                                          ●

**(20) Example. Identical sequences.** Let $Y$ be a random variable with zero mean and unit variance, and let $X = \{X_n : n \ge 0\}$ be the stationary sequence given by $X_n = Y$ for all $n$. In Example (8.2.9) we calculated the autocorrelation function as $\rho(n) = 1$ for all $n$, and we recognize this as the characteristic function of a distribution which is concentrated at 0. The spectrum of $X$ is the set $\{0\}$.                                                                    ●

**(21) Example. Two-state Markov chains.** Let $X = \{X(t) : t \ge 0\}$ be a Markov chain with state space $S = \{1, 2\}$. Suppose, as in Example (6.9.15), that the times spent in states 1 and 2 are exponentially distributed with parameters $\alpha$ and $\beta$ respectively where $\alpha\beta > 0$. That is to say, $X$ has generator $\mathbf{G}$ given by

$$\mathbf{G} = \begin{pmatrix} -\alpha & \alpha \\ \beta & -\beta \end{pmatrix}.$$

In our solution to Example (6.9.15) we wrote down the Kolmogorov forward equations and found that the transition probabilities

$$p_{ij}(t) = \mathbb{P}(X(t) = j \mid X(0) = i), \quad 1 \le i, j \le 2,$$

are given by

$$p_{11}(t) = 1 - p_{12}(t) = \frac{\beta}{\alpha + \beta} + \frac{\alpha}{\alpha + \beta} e^{-t(\alpha+\beta)},$$

$$p_{22}(t) = 1 - p_{21}(t) = \frac{\alpha}{\alpha + \beta} + \frac{\beta}{\alpha + \beta} e^{-t(\alpha+\beta)},$$

in agreement with Example (6.10.12). Let $t \to \infty$ to find that the chain has a stationary distribution $\pi$ given by

$$\pi_1 = \frac{\beta}{\alpha + \beta}, \quad \pi_2 = \frac{\alpha}{\alpha + \beta}.$$

Suppose now that $X(0)$ has distribution $\pi$. As in Example (8.2.4), $X$ is a strongly stationary process. We are going to find its spectral representation. First, find the autocovariance function. If $t \ge 0$, then a short calculation yields

$$\mathbb{E}(X(0)X(t)) = \sum_i i\,\mathbb{E}(X(t) \mid X(0) = i)\pi_i = \sum_{i,j} ij\,p_{ij}(t)\pi_i$$

$$= \frac{(2\alpha + \beta)^2}{(\alpha + \beta)^2} + \frac{\alpha\beta}{(\alpha + \beta)^2} e^{-t(\alpha+\beta)},$$

and so the autocovariance function $c(t)$ is given by

$$c(t) = \mathbb{E}(X(0)X(t)) - \mathbb{E}(X(0))\mathbb{E}(X(t)) = \frac{\alpha\beta}{(\alpha + \beta)^2} e^{-t(\alpha+\beta)} \quad \text{if} \quad t \ge 0.$$

Hence $c(0) = \alpha\beta/(\alpha + \beta)^2$ and the autocorrelation function $\rho$ is given by

$$\rho(t) = \frac{c(t)}{c(0)} = e^{-t(\alpha+\beta)} \quad \text{if} \quad t \ge 0.$$

The process $X$ is real valued, and so $\rho$ is symmetric; thus

**(22)**                                        $$\rho(t) = e^{-|t|(\alpha+\beta)}.$$

The spectral theorem asserts that $\rho$ is the characteristic function of some distribution. We may use the inversion theorem (5.9.2) to find this distribution; however, this method is long and complicated and we prefer to rely on our experience. Compare (22) with the result of Example (5.8.4), where we saw that if $Y$ is a random variable with the Cauchy density function

$$f(\lambda) = \frac{1}{\pi(1+\lambda^2)}, \quad -\infty < \lambda < \infty,$$

then $Y$ has characteristic function $\phi(t) = e^{-|t|}$. Thus $\rho(t) = \phi(t(\alpha + \beta))$, and $\rho$ is the characteristic function of $(\alpha + \beta)Y$ (see Theorem (5.7.6)). By Example (4.7.2) the density function of $\Lambda = (\alpha + \beta)Y$ is

$$f_\Lambda(\lambda) = \frac{1}{\alpha + \beta} f_Y\left(\frac{\lambda}{\alpha + \beta}\right) = \frac{\alpha + \beta}{\pi[(\alpha + \beta)^2 + \lambda^2]}, \quad -\infty < \lambda < \infty,$$

and this is the spectral density function of $X$. The spectrum of $X$ is the whole real line $\mathbb{R}$. ●

**(23) Example. Autoregressive scheme.** Let $\{Z_n\}$ be uncorrelated random variables with zero means and unit variances, and suppose that

$$X_n = \alpha X_{n-1} + Z_n, \quad -\infty < n < \infty,$$

where $\alpha$ is real and satisfies $|\alpha| < 1$. We saw in Problem (8.7.2) that $X$ has autocorrelation function

$$\rho(n) = \alpha^{|n|}, \quad -\infty < n < \infty.$$

Use (16) to find the spectral density function $f_X$ of $X$:

$$f_X(\lambda) = \frac{1}{2\pi} \sum_{n=-\infty}^{\infty} e^{-in\lambda} \alpha^{|n|}$$

$$= \frac{1 - \alpha^2}{2\pi|1 - \alpha e^{i\lambda}|^2} = \frac{1 - \alpha^2}{2\pi(1 - 2\alpha\cos\lambda + \alpha^2)}, \quad -\pi \le \lambda \le \pi.$$

More generally, suppose that the process $Y$ satisfies

$$Y_n = \sum_{j=1}^{r} \alpha_j Y_{n-j} + Z_n, \quad -\infty < n < \infty$$

where $\alpha_1, \alpha_2, \dots, \alpha_r$ are constants. The same techniques can be applied, though with some difficulty, to find that $Y$ is stationary if the complex roots $\theta_1, \theta_2, \dots, \theta_r$ of the polynomial

$$A(z) = z^r - \alpha_1 z^{r-1} - \cdots - \alpha_r = 0$$

satisfy $|\theta_j| < 1$. If this holds then the spectral density function $f_Y$ of $Y$ is given by

$$f_Y(\lambda) = \frac{1}{2\pi\sigma^2|A(e^{-i\lambda})|^2}, \quad -\pi \le \lambda \le \pi,$$

where $\sigma^2 = \text{var}(Y_0)$.                                                                        ●

## Exercises for Section 9.3

**1.**   Let $X_n = A\cos(n\lambda) + B\sin(n\lambda)$ where $A$ and $B$ are uncorrelated random variables with zero means and unit variances. Show that $X$ is stationary with a spectrum containing exactly one point.

**2.**   Let $U$ be uniformly distributed on $(-\pi, \pi)$, and let $V$ be independent of $U$ with distribution function $F$. Show that $X_n = e^{i(U-Vn)}$ defines a stationary (complex) sequence with spectral distribution function $F$.

**3.**   Find the autocorrelation function of the stationary process $\{X(t) : -\infty < t < \infty\}$ whose spectral density function is:

(i) $N(0, 1)$,   (ii) $f(x) = \tfrac{1}{2}e^{-|x|}$, $-\infty < x < \infty$.

**4.**   Let $X_1, X_2, \ldots$ be a real-valued stationary sequence with zero means and autocovariance function $c(m)$. Show that

$$\text{var}\left(\frac{1}{n}\sum_{j=1}^{n} X_j\right) = c(0)\int_{(-\pi,\pi]} \left(\frac{\sin(n\lambda/2)}{n\sin(\lambda/2)}\right)^2 dF(\lambda)$$

where $F$ is the spectral distribution function. Deduce that $n^{-1}\sum_{j=1}^{n} X_j \xrightarrow{\text{m.s.}} 0$ if and only if $F(0) - F(0-) = 0$, and show that

$$c(0)\{F(0) - F(0-)\} = \lim_{n\to\infty} \frac{1}{n}\sum_{j=0}^{n-1} c(j).$$

**5.**   Let $Z = \{Z_n : n \in \mathbb{Z}\}$ be a sequence of uncorrelated random variables with means 0 and variances 1, and suppose the sequence $X = \{X_n : n \in \mathbb{Z}\}$ satisfies $X_n = \phi X_{n-1} + Z_n + \theta Z_{n-1}$.
(a) For what values of $\phi, \theta$ is $X$ stationary?
(b) For what values of $\phi, \theta$ does $Z_n$ have a series representation in terms of the $X_m$?
(c) Show that the autocorrelation function $\rho$ of $X$ satisfies

$$\rho(1) = \frac{(\phi + \theta)(1 + \phi\theta)}{1 + 2\phi\theta + \theta^2},$$

and find $\rho(k)$ for $k > 1$. Hence evaluate the spectral density in a finite form.

**6.**   Let $Z = \{Z_n : n \in \mathbb{Z}\}$ be a sequence of uncorrelated random variables with means 0 and variances 1, and suppose $X_n = \phi X_{n-1} + \theta X_{n-2} + Z_n$ for $n \in \mathbb{Z}$, where $\phi^2 + 4\theta > 0$. Suppose $d_k$ is a real sequence such that

$$X_n = \sum_{k=0}^{\infty} d_k Z_{n-k}.$$

Show that $d_k = a_1 r_1^k + a_2 r_2^k$, where $a_1 + a_2 = 1$ and $a_1 r_1 + a_2 r_2 = \phi$. Show further that

$$a_1 = \frac{r_1}{r_1 - r_2}, \quad a_2 = \frac{-r_2}{r_1 - r_2}, \quad r_1 r_2 = -\theta, \quad r_1^2 + r_2^2 = \phi^2 + 2\theta.$$

Hence find $\mathbb{E}(X_n^2)$ in terms of $\phi$ and $\theta$ when $|r_1|, |r_2| < 1$.

**7.**   **Kolmogorov–Szegő formula.** Let $X = \{X_n : n \in \mathbb{Z}\}$ be a stationary process with zero means and variances $\sigma^2$, and let $\widehat{X}_{n+1}$ be the best linear predictor of $X_{n+1}$ given $\{X_r : r \le n\}$. (See Section 9.2.) The Kolmogorov–Szegő formula states that

$$\text{var}(X_{n+1} - \widehat{X}_{n+1}) = \exp\left\{\frac{1}{2\pi}\int_{-\pi}^{\pi} \log(2\pi\sigma^2 f(\lambda))\, d\lambda\right\},$$

where $f$ is the spectral density of $X$. Verify this for the autoregressive scheme of Example (9.3.23).

## 9.4 Stochastic integration and the spectral representation

Let $X = \{X(t) : -\infty < t < \infty\}$ be a stationary process which takes values in $\mathbb{C}$, as before. In the last section we saw that the autocorrelation function $\rho$ enjoys the representation

$$(1) \qquad\qquad \rho(t) = \int_{-\infty}^{\infty} e^{it\lambda}\, dF(\lambda)$$

as the characteristic function of some distribution function $F$ whenever $\rho$ is continuous at $t = 0$. This spectral representation is very useful in many contexts, including for example statistical analyses of sequences of data, but it is not the full story. Equation (1) is an analytical result with limited probabilistic content; of more interest to us is the process $X$, and (1) leads us to ask whether $X$ itself enjoys a similar representation. The answer to this is in the affirmative, but the statement of the result is complicated and draws deeply from abstract theory.

Without much loss of generality we can suppose that $X(t)$ has mean 0 and variance 1 for all $t$. With each such stationary process $X$ we can associate another process $S$ called the 'spectral process' of $X$, in much the same way as the spectral distribution function $F$ is associated with the autocorrelation function $\rho$.

**(2) Spectral theorem.** *If $X$ is a stationary process with zero mean, unit variance, continuous autocorrelation function, and spectral distribution function $F$, there exists a complex-valued process $S = \{S(\lambda) : -\infty < \lambda < \infty\}$ such that*

$$(3) \qquad\qquad X(t) = \int_{-\infty}^{\infty} e^{it\lambda}\, dS(\lambda).$$

*Furthermore $S$ has orthogonal increments in the sense that*

$$\mathbb{E}\big([S(v) - S(u)][\overline{S}(t) - \overline{S}(s)]\big) = 0 \quad \text{if} \quad u \leq v \leq s \leq t,$$

*and in addition* $\mathbb{E}(|S(v) - S(u)|^2) = F(v) - F(u)$ *if* $u \leq v$.

The discrete-time stationary process $X = \{X_n : -\infty < n < \infty\}$ has a spectral representation also. The only significant difference is that the domain of the spectral process may be taken to be $(-\pi, \pi]$.

**(4) Spectral theorem.** *If $X$ is a discrete-time stationary process with zero mean, unit variance, and spectral distribution function $F$, there exists a complex-valued process $S = \{S(\lambda) : -\pi < \lambda \leq \pi\}$ such that*

$$(5) \qquad\qquad X_n = \int_{(-\pi,\pi]} e^{in\lambda}\, dS(\lambda).$$

*Furthermore $S$ has orthogonal increments, and*

$$(6) \qquad\qquad \mathbb{E}\big(|S(v) - S(u)|^2\big) = F(v) - F(u) \quad \text{for} \quad u \leq v.$$

A proof of (4) is presented later in this section. The proof of (2) is very similar, Fourier sums being replaced by Fourier integrals; this proof is therefore omitted. The process $S$ in (3) and (5) is called the *spectral process* of $X$.

Before proving the above spectral representation, we embark upon an exploration of the 'stochastic integral', of which (3) and (5) are examples. The theory of stochastic integration is of major importance in modern probability theory, particularly in the study of diffusion processes.

As amply exemplified by the material in this book, probabilists are very often concerned with partial sums $\sum_{i=1}^{n} X_i$ and weighted sums $\sum_{i=1}^{n} a_i X_i$ of sequences of random variables. If $X$ is a continuous-time process rather than a discrete-time sequence, the corresponding objects are integrals of the form $\int_{\alpha}^{\beta} a(u)\, dX(u)$; how should such an integral be defined? It is not an easy matter to discuss the 'stochastic integral' before an audience some of whom have seen little or nothing beyond the Riemann integral. There follows such an attempt.

Let $S = \{S(t) : t \in \mathbb{R}\}$ be a complex-valued continuous-time random process on the probability space $(\Omega, \mathcal{F}, \mathbb{P})$, and suppose that $S$ has the following properties:

**(7)** $$\mathbb{E}(|S(t)|^2) < \infty \quad \text{for all } t,$$

**(8)** $$\mathbb{E}\big(|S(t+h) - S(t)|^2\big) \to 0 \quad \text{as} \quad h \downarrow 0, \quad \text{for all } t,$$

**(9)** the process $S$ has *orthogonal increments* in that

$$\mathbb{E}\big([S(v) - S(u)][\overline{S}(t) - \overline{S}(s)]\big) = 0 \quad \text{whenever} \quad u \le v \le s \le t.$$

Condition (7) is helpful, since we shall work with random variables with finite second moments, and with mean-square convergence. Condition (8) is a continuity assumption which will be useful for technical reasons. Condition (9) will be of central importance in demonstrating the existence of limits necessary for the definition of the stochastic integral.

Let $G(t)$ be defined by

**(10)** $$G(t) = \begin{cases} \mathbb{E}\big(|S(t) - S(0)|^2\big) & \text{if } t \ge 0, \\ -\mathbb{E}\big(|S(t) - S(0)|^2\big) & \text{if } t < 0. \end{cases}$$

It is an elementary calculation that

**(11)** $$\mathbb{E}\big(|S(t) - S(s)|^2\big) = G(t) - G(s), \quad \text{for} \quad s \le t.$$

To see that this holds when $0 \le s \le t$, for example, we argue as follows:

$$\begin{aligned} G(t) &= \mathbb{E}\big(\big|[S(t) - S(s)] + [S(s) - S(0)]\big|^2\big) \\ &= \mathbb{E}\big(|S(t) - S(s)|^2\big) + \mathbb{E}\big(|S(s) - S(0)|^2\big) \\ &\quad + \mathbb{E}\big([S(t) - S(s)][\overline{S}(s) - \overline{S}(0)] + [\overline{S}(t) - \overline{S}(s)][S(s) - S(0)]\big) \\ &= \mathbb{E}\big(|S(t) - S(s)|^2\big) + G(s) \end{aligned}$$

by the assumption of orthogonal increments. It follows from (11) that $G$ is monotonic non-decreasing, and is right-continuous in that

**(12)** $$G(t+h) \to G(t) \quad \text{as} \quad h \downarrow 0.$$

The function $G$ is central to the analysis which follows.

Let $a_1 < a_2 < \cdots < a_n$, and let $c_1, c_2, \ldots, c_{n-1}$ be complex numbers. Define the step function $\phi$ on $\mathbb{R}$ by

$$\phi(t) = \begin{cases} 0 & \text{if } t < a_1 \text{ or } t \geq a_n, \\ c_j & \text{if } a_j \leq t < a_{j+1}, \end{cases}$$

and define the integral $I(\phi)$ of $\phi$ with respect to $S$ by

**(13)** $$I(\phi) = \int_{-\infty}^{\infty} \phi(t) \, dS(t) = \sum_{j=1}^{n-1} c_j [S(a_{j+1}) - S(a_j)];$$

this is a finite sum, and therefore there is no problem concerning its existence.

Suppose that $\phi_1$ and $\phi_2$ are step functions of the type given above. We may assume, by a suitable 'refinement' argument, that $\phi_1$ and $\phi_2$ are of the form

$$\phi_1(t) = \phi_2(t) = 0 \qquad \text{if } t < a_1 \text{ or } t \geq a_n,$$
$$\phi_1(t) = c_j, \; \phi_2(t) = d_j \quad \text{if } a_j \leq t < a_{j+1},$$

for some $a_1 < a_2 < \cdots < a_n$. Then, using the assumption of orthogonal increments,

$$\mathbb{E}\big(I(\phi_1)\overline{I(\phi_2)}\big) = \sum_{j,k} c_j \overline{d_k} \mathbb{E}\big([S(a_{j+1}) - S(a_j)][\overline{S}(a_{k+1}) - \overline{S}(a_k)]\big)$$

$$= \sum_{j} c_j \overline{d_j} \mathbb{E}\big(|S(a_{j+1}) - S(a_j)|^2\big)$$

$$= \sum_{j} c_j \overline{d_j}[G(a_{j+1}) - G(a_j)] \quad \text{by (11)},$$

which may be written as

**(14)** $$\mathbb{E}\big(I(\phi_1)\overline{I(\phi_2)}\big) = \int_{-\infty}^{\infty} \phi_1(t)\overline{\phi_2(t)} \, dG(t).$$

It is now immediate by expansion of the squares that

**(15)** $$\mathbb{E}\big(|I(\phi_1) - I(\phi_2)|^2\big) = \int_{-\infty}^{\infty} |\phi_1(t) - \phi_2(t)|^2 \, dG(t),$$

which is to say that 'integration is distance preserving' in the sense that

**(16)** $$\|I(\phi_1) - I(\phi_2)\|_2 = \|\phi_1 - \phi_2\|,$$

where the first norm is given by

**(17)** $$\|U - V\|_2 = \sqrt{\mathbb{E}(|U - V|^2)} \quad \text{for random variables } U, V,$$

and the second by

**(18)** $$\|f - g\| = \sqrt{\int_{-\infty}^{\infty} |f(t) - g(t)|^2 \, dG(t)} \quad \text{for suitable } f, g : \mathbb{R} \to \mathbb{C}.$$

We are ready to take limits. Let $\psi : \mathbb{R} \to \mathbb{C}$ and let $\{\phi_n\}$ be a sequence of step functions such that $\|\phi_n - \psi\| \to 0$ as $n \to \infty$. Then

$$\|\phi_n - \phi_m\| \le \|\phi_n - \psi\| + \|\phi_m - \psi\| \to 0 \quad \text{as} \quad m, n \to \infty,$$

whence it follows from (16) that the sequence $\{I(\phi_n)\}$ is mean-square Cauchy convergent, and hence convergent in mean square (see Problem (7.11.11)). That is, there exists a random variable $I(\psi)$ such that $I(\phi_n) \xrightarrow{\text{m.s.}} I(\psi)$; we call $I(\psi)$ the integral of $\psi$ with respect to $S$, writing

**(19)**
$$I(\psi) = \int_{-\infty}^{\infty} \psi(t) \, dS(t).$$

Note that the integral is not defined uniquely, but only as any mean-square limit of $I(\phi_n)$; any two such limits $I_1$ and $I_2$ are such that $\mathbb{P}(I_1 = I_2) = 1$.

For which functions $\psi$ do there exist approximating sequences $\{\phi_n\}$ of step functions? The answer is those (measurable) functions for which

**(20)**
$$\int_{-\infty}^{\infty} |\psi(t)|^2 \, dG(t) < \infty.$$

To recap, for any given function $\psi : \mathbb{R} \to \mathbb{C}$ satisfying (20), there exists a random variable

**(21)**
$$I(\psi) = \int_{-\infty}^{\infty} \psi(t) \, dS(t)$$

defined as above. Such integrals have many of the usual properties of integrals, for example:

(a) the integral of the zero function is zero,

(b) $I(\alpha\psi_1 + \beta\psi_2) = \alpha I(\psi_1) + \beta I(\psi_2)$ for $\alpha, \beta \in \mathbb{C}$,

and so on. Such statements should be qualified by the phrase 'almost surely', since integrals are not defined uniquely; we shall omit this qualification here.

Integrals may be defined on *bounded* intervals just as on the whole of the real line. For example, if $\psi : \mathbb{R} \to \mathbb{C}$ and $(a, b)$ is a bounded interval, we define

$$\int_{(a,b)} \psi(t) \, dS(t) = \int_{-\infty}^{\infty} \psi_{ab}(t) \, dS(t)$$

where $\psi_{ab}(t) = \psi(t) I_{(a,b)}(t)$.

The above exposition is directed at integrals $\int \psi(t) \, dS(t)$ where $\psi$ is a given function from $\mathbb{R}$ to $\mathbb{C}$. It is possible to extend this definition to the situation where $\psi$ is itself a random process. Such an integral may be constructed very much as above, but at the expense of adding certain extra assumptions concerning the pair $(\psi, S)$; see Section 13.8.

**Proof of Theorem (4).** Let $H_X$ be the set of all linear combinations of the $X_j$, so that $H_X$ is the set of all random variables of the form $\sum_{j=1}^{n} a_j X_{m(j)}$ for $a_1, a_2, \ldots, a_n \in \mathbb{C}$ and integers $n, m(1), m(2), \ldots, m(n)$. The space $H_X$ is a vector space over $\mathbb{C}$ with a natural inner product given by

**(22)**
$$\langle U, V \rangle_2 = \mathbb{E}(U\overline{V}).$$

The *closure* $\overline{H}_X$ of $H_X$ is defined to be the space $H_X$ together with all limits of mean-square Cauchy-convergent sequences in $H_X$.

Similarly, we let $H_F$ be the set of all linear combinations of the functions $f_n : \mathbb{R} \to \mathbb{C}$ defined by $f_n(x) = e^{inx}$ for $-\infty < x < \infty$. We impose an inner product on $H_F$ by

$$
\textbf{(23)} \qquad \langle u, v \rangle = \int_{(-\pi,\pi]} u(\lambda)\overline{v(\lambda)}\, dF(\lambda) \quad \text{for} \quad u, v \in H_F,
$$

and we write $\overline{H}_F$ for the closure of $H_F$, being the space $H_F$ together with all Cauchy-convergent sequences in $H_F$ (a sequence $\{u_n\}$ is Cauchy convergent if $\langle u_n - u_m, u_n - u_m \rangle \to 0$ as $m, n \to \infty$).

The two spaces $\overline{H}_X$ and $\overline{H}_F$ are Hilbert spaces, and we place them in one–one correspondence in the following way. Define the linear mapping $\mu : H_F \to H_X$ by $\mu(f_j) = X_j$, so that

$$
\mu\left(\sum_{j=1}^{n} a_j f_j\right) = \sum_{j=1}^{n} a_j X_j;
$$

it is seen easily that $\mu$ is one–one, in a formal sense. Furthermore,

$$
\langle \mu(f_n), \mu(f_m) \rangle_2 = \langle X_n, X_m \rangle_2 = \int_{(-\pi,\pi]} e^{i(n-m)\lambda}\, dF(\lambda) = \langle f_n, f_m \rangle
$$

by equations (9.3.12) and (23); therefore, by linearity, $\langle \mu(u), \mu(v) \rangle_2 = \langle u, v \rangle$ for $u, v \in H_F$, so that $\mu$ is 'distance preserving' on $H_F$. The domain of $\mu$ may be extended to $\overline{H}_F$ in the natural way: if $u \in \overline{H}_F$, $u = \lim_{n\to\infty} u_n$ where $u_n \in H_F$, we define $\mu(u) = \lim_{n\to\infty} \mu(u_n)$ where the latter limit is taken in the usual sense for $\overline{H}_X$. The new mapping $\mu$ from $\overline{H}_F$ to $\overline{H}_X$ is not quite one–one, since mean-square limits are not defined uniquely, but this difficulty is easily avoided ($\mu$ is one–one when viewed as a mapping from equivalence classes of functions to equivalence classes of random variables). Furthermore it may easily be checked that $\mu$ is distance preserving on $\overline{H}_F$, and linear in that

$$
\mu\left(\sum_{j=1}^{n} a_j u_j\right) = \sum_{j=1}^{n} a_j \mu(u_j)
$$

for $a_1, a_2, \ldots, a_n \in \mathbb{C}, u_1, u_2, \ldots, u_n \in \overline{H}_F$.

The mapping $\mu$ is sometimes called an *isometric isomorphism*. We now define the process $S = \{S(\lambda) : -\pi < \lambda \le \pi\}$ by

$$
\textbf{(24)} \qquad S(\lambda) = \mu(I_\lambda) \quad \text{for} \quad -\pi < \lambda \le \pi,
$$

where $I_\lambda : \mathbb{R} \to \{0, 1\}$ is the indicator function of the interval $(-\pi, \lambda]$. It is a standard result of Fourier analysis that $I_\lambda \in \overline{H}_F$, so that $\mu(I_\lambda)$ is well defined. We introduce one more piece of notation, defining $J_{\alpha\beta}$ to be the indicator function of the interval $(\alpha, \beta]$; thus $J_{\alpha\beta} = I_\beta - I_\alpha$.

We need to show that $X$ and $S$ are related (almost surely) by (5). To this end, we check first that $S$ satisfies conditions (7)–(9). Certainly $\mathbb{E}(|S(\lambda)|^2) < \infty$ since $S(\lambda) \in \overline{H}_X$. Secondly,

$$
\mathbb{E}\left(|S(\lambda + h) - S(\lambda)|^2\right) = \langle S(\lambda + h) - S(\lambda), S(\lambda + h) - S(\lambda) \rangle_2
$$
$$
= \langle J_{\lambda,\lambda+h}, J_{\lambda,\lambda+h} \rangle
$$

by linearity and the isometry of $\mu$. Now $\langle J_{\lambda,\lambda+h}, J_{\lambda,\lambda+h} \rangle \to 0$ as $h \downarrow 0$, and (8) has been verified. Thirdly, if $u \leq v \leq s \leq t$, then

$$\langle S(v) - S(u), S(t) - S(s) \rangle_2 = \langle J_{uv}, J_{st} \rangle = 0$$

since $J_{uv}(x)J_{st}(x) = 0$ for all $x$. Thus $S$ has orthogonal increments. Furthermore, by (23),

$$\mathbb{E}\big(|S(v) - S(u)|^2\big) = \langle J_{uv}, J_{uv} \rangle = \int_{(u,v]} dF(\lambda) = F(v) - F(u)$$

since $F$ is right-continuous; this confirms (6), and it remains to check that (5) holds.

The process $S$ satisfies conditions (7)–(9), and it follows that the stochastic integral

$$I(\psi) = \int_{(-\pi,\pi]} \psi(\lambda)\, dS(\lambda)$$

is defined for a broad class of functions $\psi : (-\pi, \pi] \to \mathbb{C}$. We claim that

**(25)**                    $I(\psi) = \mu(\psi)$   (almost surely)   for $\psi \in \overline{H}_F$.

The result of the theorem will follow immediately by the choice $\psi = f_n$, for which (25) implies that (almost surely) $I(f_n) = \mu(f_n) = X_n$, which is to say that

$$\int_{(-\pi,\pi]} e^{in\lambda}\, dS(\lambda) = X_n$$

as required.

It remains to prove (25), which we do by systematic approximation. Suppose first that $\psi$ is a step function,

**(26)**                    $\psi(x) = \begin{cases} 0 & \text{if } x < a_1 \text{ or } x \geq a_n, \\ c_j & \text{if } a_j \leq x < a_{j+1}, \end{cases}$

where $-\pi < a_1 < a_2 < \cdots < a_n \leq \pi$ and $c_1, c_2, \ldots, c_n \in \mathbb{C}$. Then

$$I(\psi) = \sum_{j=1}^{n} c_j[S(a_{j+1}) - S(a_j)] = \sum_{j=1}^{n} c_j \mu(J_{a_j,a_{j+1}}) \quad \text{by (24)}$$

$$= \mu\left(\sum_{j=1}^{n} c_j J_{a_j,a_{j+1}}\right) = \mu(\psi) \qquad\qquad \text{by (26).}$$

Hence $I(\psi) = \mu(\psi)$ for all step functions $\psi$. More generally, if $\psi \in \overline{H}_F$ and $\{\psi_n\}$ is a sequence of step functions converging to $\psi$, then $\mu(\psi_n) \to \mu(\psi)$. By the definition of the stochastic integral, it is the case that $I(\psi_n) \to I(\psi)$, and it follows that $I(\psi) = \mu(\psi)$, which proves (25).                                                                                    ∎

## Exercises for Section 9.4

**1.** Let $S$ be the spectral process of a stationary process $X$ with zero mean and unit variance. Show that the increments of $S$ have zero means.

**2.** **Moving average representation.** Let $X$ be a discrete-time stationary process having zero means, continuous strictly positive spectral density function $f$, and with spectral process $S$. Let

$$Y_n = \int_{(-\pi,\pi]} \frac{e^{in\lambda}}{\sqrt{2\pi f(\lambda)}} \, dS(\lambda).$$

Show that $\ldots, Y_{-1}, Y_0, Y_1, \ldots$ is a sequence of uncorrelated random variables with zero means and unit variances.

Show that $X_n$ may be represented as a moving average $X_n = \sum_{j=-\infty}^{\infty} a_j Y_{n-j}$ where the $a_j$ are constants satisfying

$$\sqrt{2\pi f(\lambda)} = \sum_{j=-\infty}^{\infty} a_j e^{-ij\lambda} \qquad \text{for } \lambda \in (-\pi, \pi].$$

**3.** **Gaussian process.** Let $X$ be a discrete-time stationary sequence with zero mean and unit variance, and whose fdds are of the multivariate-normal type. Show that the spectral process of $X$ has independent increments having normal distributions.

## 9.5  The ergodic theorem

The law of large numbers asserts that

**(1)**
$$\frac{1}{n} \sum_{j=1}^{n} X_j \to \mu$$

whenever $\{X_j\}$ is an independent identically distributed sequence with mean $\mu$; the convergence takes place almost surely. This section is devoted to a complete generalization of the law of large numbers, the assumption that the $X_j$ be independent being replaced by the assumption that they form a stationary process. This generalization is called the 'ergodic theorem' and it has more than one form depending on the type of stationarity—weak or strong—and the required mode of convergence; recall the various corresponding forms of the law of large numbers†.

It is usual to state the ergodic theorem for discrete-time processes, and we conform to this habit here. Similar results hold for continuous-time processes, sums of the form $\sum_{1}^{n} X_j$ being replaced by integrals of the form $\int_0^n X(t) \, dt$. Here is the usual form of the ergodic theorem.

**(2) Theorem. Ergodic theorem for strongly stationary processes.** *Let $X = \{X_n : n \geq 1\}$ be a strongly stationary process such that $\mathbb{E}|X_1| < \infty$. There exists a random variable $Y$ with the same mean as the $X_n$ such that*

$$\frac{1}{n} \sum_{j=1}^{n} X_j \to Y \quad \text{a.s. and in mean.}$$

---

†The original weak ergodic theorem was proved by von Neumann, and the later strong theorem by Birkhoff.

The proof of this is difficult, as befits a complete generalization of the strong law of large numbers (see Problem (9.7.10)). The following result is considerably more elementary.

**(3) Theorem. Ergodic theorem for weakly stationary processes.** *If $X = \{X_n : n \geq 1\}$ is a (weakly) stationary process, there exists a random variable $Y$ such that $\mathbb{E}Y = \mathbb{E}X_1$ and*

$$\frac{1}{n}\sum_{j=1}^{n} X_j \xrightarrow{\text{m.s.}} Y.$$

We prove the latter theorem first. The normal proof of the 'strong ergodic theorem' (2) is considerably more difficult, and makes use of harder ideas than those required for the 'weak ergodic theorem' (3). The second part of this section is devoted to a discussion of the strong ergodic theorem, together with a relatively straightforward proof.

Theorems (2) and (3) generalize the laws of large numbers. There are similar generalizations of the central limit theorem and the law of the iterated logarithm, although such results hold only for stationary processes which satisfy certain extra conditions. We give no details of this here, save for pointing out that these extra conditions take the form '$X_m$ and $X_n$ are "nearly independent" when $|m - n|$ is large'.

We give two proofs of (3). Proof A is conceptually easy but has some technical difficulties; we show that $n^{-1}\sum_1^n X_j$ is a mean-square Cauchy-convergent sequence (see Problem (7.11.11)). Proof B uses the spectral representation of $X$; we sketch this here and show that it yields an explicit form for the limit $Y$ as the contribution made towards $X$ by 'oscillations of zero frequency'.

**Proof A.** Recall from (7.11.11) that a sequence $\{Y_n\}$ converges in mean square to some limit if and only if $\{Y_n\}$ is *mean-square Cauchy convergent*, which is to say that

**(4)**                     $$\mathbb{E}\big(|Y_n - Y_m|^2\big) \to 0 \quad \text{as} \quad m, n \to \infty.$$

A similar result holds for complex-valued sequences. We shall show that the sequence $\{n^{-1}\sum_1^n X_j\}$ satisfies (4) whenever $X$ is stationary. This is easy in concept, since it involves expressions involving the autocovariance function of $X$ alone; the proof of the mean-square version of the law of large numbers was easy for the same reason. Unfortunately, the verification of (4) is not a trivial calculation.

For any complex-valued random variable $Z$, define

$$\|Z\| = \sqrt{\mathbb{E}(|Z|^2)};$$

the function $\| \cdot \|$ is a norm (see Section 7.2) when viewed as a function on the collection of equivalence classes of random variables with finite second moment and with $Y \sim Z$ if $\mathbb{P}(Y = Z) = 1$. We wish to show that†

**(5)**                     $$\|\langle X\rangle_n - \langle X\rangle_m\| \to 0 \quad \text{as} \quad n, m \to \infty$$

where

$$\langle X\rangle_n = \frac{1}{n}\sum_{j=1}^{n} X_j.$$

---

†Physicists often use the notation $\langle \cdot \rangle$ to denote expectation.

Set

$$\mu_N = \inf_\lambda \|\lambda_1 X_1 + \lambda_2 X_2 + \cdots + \lambda_N X_N\|$$

where the infimum is calculated over all vectors $\lambda = (\lambda_1, \lambda_2, \ldots, \lambda_N)$ containing non-negative entries with sum 1. Clearly $\mu_N \geq \mu_{N+1}$ and so

$$\mu = \lim_{N \to \infty} \mu_N = \inf_N \mu_N$$

exists. If $m < n$ then

$$\|\langle X \rangle_n + \langle X \rangle_m\| = 2 \left\| \sum_{j=1}^n \lambda_j X_j \right\|$$

where

$$\lambda_j = \begin{cases} \dfrac{1}{2}\left(\dfrac{1}{m} + \dfrac{1}{n}\right) & \text{if } 1 \leq j \leq m, \\[2mm] \dfrac{1}{2n} & \text{if } m < j \leq n, \end{cases}$$

and so

$$\|\langle X \rangle_n + \langle X \rangle_m\| \geq 2\mu.$$

It is not difficult to deduce (see Exercise (7.1.2b) for the first line here) that

$$\|\langle X \rangle_n - \langle X \rangle_m\|^2 = 2\|\langle X \rangle_n\|^2 + 2\|\langle X \rangle_m\|^2 - \|\langle X \rangle_n + \langle X \rangle_m\|^2$$
$$\leq 2\|\langle X \rangle_n\|^2 + 2\|\langle X \rangle_m\|^2 - 4\mu^2$$
$$= 2\left|\|\langle X \rangle_n\|^2 - \mu^2\right| + 2\left|\|\langle X \rangle_m\|^2 - \mu^2\right|$$

and (5) follows as soon as we can show that

**(6)**                    $\|\langle X \rangle_n\| \to \mu$    as    $n \to \infty$.

The remaining part of the proof is devoted to demonstrating (6).

Choose any $\epsilon > 0$ and pick $N$ and $\lambda$ such that

$$\|\lambda_1 X_1 + \lambda_2 X_2 + \cdots + \lambda_N X_N\| \leq \mu + \epsilon$$

where $\lambda_i \geq 0$ and $\sum_1^N \lambda_i = 1$. Define the moving average

$$Y_k = \lambda_1 X_k + \lambda_2 X_{k+1} + \cdots + \lambda_N X_{k+N-1};$$

it is not difficult to see that $Y = \{Y_k\}$ is a stationary process (see Problem (8.7.1)). We shall show that

**(7)**                    $\|\langle Y \rangle_n - \langle X \rangle_n\| \to 0$    as    $n \to \infty$

where

$$\langle Y \rangle_n = \frac{1}{n} \sum_{j=1}^n Y_j.$$

Note first that, by the triangle inequality (7.1.5),

(8)                               $\|\langle Y \rangle_n\| \le \|Y_1\| \le \mu + \epsilon$    for all $n$

since $\|Y_n\| = \|Y_1\|$ for all $n$. Now

$$\langle Y \rangle_n = \lambda_1 \langle X \rangle_{1,n} + \lambda_2 \langle X \rangle_{2,n} + \cdots + \lambda_N \langle X \rangle_{N,n}$$

where

$$\langle X \rangle_{k,n} = \frac{1}{n} \sum_{j=k}^{k+n-1} X_j;$$

now use the facts that $\langle X \rangle_{1,n} = \langle X \rangle_n$, $1 - \lambda_1 = \lambda_2 + \lambda_3 + \cdots + \lambda_N$, and the triangle inequality to deduce that

$$\|\langle Y \rangle_n - \langle X \rangle_n\| \le \sum_{j=2}^N \lambda_j \|\langle X \rangle_{j,n} - \langle X \rangle_{1,n}\|.$$

However, by the triangle inequality again,

$$\|\langle X \rangle_{j,n} - \langle X \rangle_{1,n}\| = \frac{1}{n} \|(X_j + \cdots + X_{j+n-1}) - (X_1 + \cdots + X_n)\|$$

$$= \frac{1}{n} \|(X_{n+1} + \cdots + X_{j+n-1}) - (X_1 + \cdots + X_{j-1})\|$$

$$\le \frac{2j}{n} \|X_1\|$$

since $\|X_n\| = \|X_1\|$ for all $n$. Therefore,

$$\|\langle Y \rangle_n - \langle X \rangle_n\| \le \sum_{j=2}^N \lambda_j \frac{2j}{n} \|X_1\| \le \frac{2N}{n} \|X_1\|;$$

let $n \to \infty$ to deduce that (7) holds. Use (8) to obtain

$$\mu \le \|\langle X \rangle_n\| \le \|\langle X \rangle_n - \langle Y \rangle_n\| + \|\langle Y \rangle_n\|$$

$$\le \|\langle X \rangle_n - \langle Y \rangle_n\| + \mu + \epsilon \to \mu + \epsilon \quad \text{as} \quad n \to \infty.$$

Now $\epsilon$ was arbitrary, and we let $\epsilon \downarrow 0$ to obtain (6).

Since $\langle X \rangle_n \xrightarrow{\text{m.s.}} Y$, we have that $\langle X \rangle_n \xrightarrow{1} Y$, which implies that $\mathbb{E}\langle X \rangle_n \to \mathbb{E}Y$. However, $\mathbb{E}\langle X \rangle_n = \mathbb{E}X_1$, whence $\mathbb{E}Y = \mathbb{E}X_1$.    ∎

**Sketch proof B.** Suppose that $\mathbb{E}(X_n) = 0$ for all $n$. The process $X$ has a spectral representation

$$X_n = \int_{(-\pi,\pi]} e^{in\lambda} \, dS(\lambda).$$

Now,

(9)          $\langle X \rangle_n = \frac{1}{n} \sum_{j=1}^n X_j = \int_{(-\pi,\pi]} \frac{1}{n} \sum_{j=1}^n e^{ij\lambda} \, dS(\lambda) = \int_{(-\pi,\pi]} g_n(\lambda) \, dS(\lambda)$

where

$$
\text{(10)} \qquad g_n(\lambda) = \begin{cases} 1 & \text{if } \lambda = 0, \\ \dfrac{e^{i\lambda}}{n} \dfrac{1 - e^{in\lambda}}{1 - e^{i\lambda}} & \text{if } \lambda \neq 0. \end{cases}
$$

We have that $|g_n(\lambda)| \leq 1$ for all $n$ and $\lambda$, and, as $n \to \infty$,

$$
\text{(11)} \qquad g_n(\lambda) \to g(\lambda) = \begin{cases} 1 & \text{if } \lambda = 0, \\ 0 & \text{if } \lambda \neq 0. \end{cases}
$$

It can be shown that

$$
\int_{(-\pi,\pi]} g_n(\lambda)\, dS(\lambda) \xrightarrow{\text{m.s.}} \int_{(-\pi,\pi]} g(\lambda)\, dS(\lambda) \quad \text{as} \quad n \to \infty,
$$

implying that

$$
\langle X \rangle_n \xrightarrow{\text{m.s.}} \int_{(-\pi,\pi]} g(\lambda)\, dS(\lambda) = S(0) - S(0-),
$$

by the right-continuity of $S$, where $S(0-) = \lim_{y \uparrow 0} S(y)$. This shows that $\langle X \rangle_n$ converges in mean square to the random magnitude of the discontinuity of $S(\lambda)$ at $\lambda = 0$ (this quantity may be zero); in other words, $\langle X \rangle_n$ converges to the 'zero frequency' or 'infinite wavelength' contribution of the spectrum of $X$. This conclusion is natural and memorable, since the average of any oscillation having non-zero frequency is zero.                                                                 ∎

The second proof of Theorem (3) is particularly useful in that it provides an explicit representation for the limit in terms of the spectral process of $X$. It is easy to calculate the first two moments of this limit.

**(12) Lemma.** *If $X$ is a stationary process with zero means and autocovariance function $c(m)$ then the limit variable $Y = \lim_{n \to \infty} \{ n^{-1} \sum_{j=1}^{n} X_j \}$ satisfies*

$$
\mathbb{E}(Y) = 0, \quad \mathbb{E}(|Y|^2) = \lim_{n \to \infty} \frac{1}{n} \sum_{j=1}^{n} c(j).
$$

A similar result holds for processes with non-zero means.

**Proof.** We have that $\langle X \rangle_n \xrightarrow{\text{m.s.}} Y$, and so $\langle X \rangle_n \xrightarrow{1} Y$ by Theorem (7.2.3). The result of Exercise (7.2.1) implies that $\mathbb{E}(\langle X \rangle_n) \to \mathbb{E}(Y)$ as $n \to \infty$; however, $\mathbb{E}(\langle X \rangle_n) = \mathbb{E}(X_1) = 0$ for all $n$.

In order to prove the second part, either use Exercise (7.2.1) again and expand $\mathbb{E}(\langle X \rangle_n^2)$ in terms of $c$ (see Exercise (9.5.2)), or use the method of Proof B of (3). We use the latter method. The autocovariance function $c(m)$ satisfies

$$
\frac{1}{n} \sum_{j=1}^{n} c(j) = c(0) \int_{(-\pi,\pi]} g_n(\lambda)\, dF(\lambda)
$$

$$
\to c(0) \int_{(-\pi,\pi]} g(\lambda)\, dF(\lambda) \quad \text{as} \quad n \to \infty
$$

$$
= c(0)[F(0) - F(0-)]
$$

where $g_n$ and $g$ are given by (10) and (11), $F$ is the spectral distribution function, and $F(0-) = \lim_{y \uparrow 0} F(y)$ as usual. We can now use (9.4.6) and the continuity properties of $S$ to show that

$$c(0)[F(0) - F(0-)] = \mathbb{E}\big(|S(0) - S(0-)|^2\big) = \mathbb{E}(|Y|^2). \qquad \blacksquare$$

We turn now to the strong ergodic theorem (2), which we shall first rephrase slightly. Here is some terminology and general discussion.

A vector $\mathbf{X} = (X_1, X_2, \dots)$ of real-valued random variables takes values in the set of real vectors of the form $\mathbf{x} = (x_1, x_2, \dots)$. We write $\mathbb{R}^T$ for the set of all such real sequences, where $T$ denotes the set $\{1, 2, \dots\}$ of positive integers. The natural $\sigma$-field for $\mathbb{R}^T$ is the product $\mathcal{B}^T$ of the appropriate number of copies of the Borel $\sigma$-field $\mathcal{B}$ of subsets of $\mathbb{R}$. Let $\mathbb{Q}$ be a probability measure on the pair $(\mathbb{R}^T, \mathcal{B}^T)$. The triple $(\mathbb{R}^T, \mathcal{B}^T, \mathbb{Q})$ is our basic probability space, and we make the following crucial definitions.

There is a natural 'shift operator' $\tau$ mapping $\mathbb{R}^T$ onto itself, defined by $\tau(\mathbf{x}) = \mathbf{x}'$ where $\mathbf{x}' = (x_2, x_3, \dots)$; that is, the vector $\mathbf{x} = (x_1, x_2, \dots)$ is mapped to the vector $(x_2, x_3, \dots)$. The measure $\mathbb{Q}$ is called *stationary* if and only if $\mathbb{Q}(A) = \mathbb{Q}(\tau^{-1}A)$ for all $A \in \mathcal{B}^T$ (remember that $\tau^{-1}A = \{\mathbf{x} \in \mathbb{R}^T : \tau(\mathbf{x}) \in A\}$). If $\mathbb{Q}$ is stationary, we call the shift $\tau$ 'measure preserving'. Stationary measures correspond to strongly stationary sequences of random variables, as the following example indicates.

**(13) Example.** Let $\mathbf{X} = (X_1, X_2, \dots)$ be a strongly stationary sequence on the probability space $(\Omega, \mathcal{F}, \mathbb{P})$. Define the probability measure $\mathbb{Q}$ on $(\mathbb{R}^T, \mathcal{B}^T)$ by $\mathbb{Q}(A) = \mathbb{P}(\mathbf{X} \in A)$ for $A \in \mathcal{B}^T$. Now $\mathbf{X}$ and $\tau(\mathbf{X})$ have the same fdds, and therefore

$$\mathbb{Q}(\tau^{-1}A) = \mathbb{P}\big(\tau(\mathbf{X}) \in A\big) = \mathbb{P}(\mathbf{X} \in A) = \mathbb{Q}(A)$$

for all (measurable) subsets $A$ of $\mathbb{R}^T$.

We have seen that every strongly stationary sequence generates a stationary measure on $(\mathbb{R}^T, \mathcal{B}^T)$. The converse is true also. Let $\mathbb{Q}$ be a stationary measure on $(\mathbb{R}^T, \mathcal{B}^T)$, and define the sequence $\mathbf{Y} = (Y_1, Y_2, \dots)$ of random variables by $Y_n(\mathbf{x}) = x_n$, the $n$th component of the real vector $\mathbf{x}$. We have from the stationarity of $\mathbb{Q}$ that, for $A \in \mathcal{B}^T$,

$$\mathbb{Q}(\mathbf{Y} \in A) = \mathbb{Q}(A) = \mathbb{Q}(\tau^{-1}A) = \mathbb{Q}\big(\tau(\mathbf{Y}) \in A\big)$$

so that $\mathbf{Y}$ and $\tau(\mathbf{Y})$ have the same fdds. Hence $\mathbf{Y}$ is a strongly stationary sequence. $\bullet$

There is a certain special class of events in $\mathcal{B}^T$ called *invariant* events.

**(14) Definition.** An event $A$ in $\mathcal{B}^T$ is called **invariant** if $A = \tau^{-1}A$.

An event $A$ is invariant if

**(15)** $$\mathbf{x} \in A \quad \text{if and only if} \quad \tau(\mathbf{x}) \in A,$$

for any $\mathbf{x} \in \mathbb{R}^T$. Now (15) is equivalent to the statement '$\mathbf{x} \in A$ if and only if $\tau^n(\mathbf{x}) \in A$ for all $n \geq 0$'; remembering that $\tau^n(\mathbf{x}) = (x_{n+1}, x_{n+2}, \dots)$, we see therefore that the membership by $\mathbf{x}$ of an invariant event $A$ does not depend on any finite collection of the components of $\mathbf{x}$. Here are some examples of invariant events:

$$A_1 = \left\{ \mathbf{x} : \limsup_{n \to \infty} x_n \leq 3 \right\},$$

$$A_2 = \{\mathbf{x} : \text{the sequence } n^{-1}x_n \text{ converges}\},$$

$$A_3 = \{\mathbf{x} : x_n = 0 \text{ for all large } n\}.$$

We denote by $\mathscr{I}$ the set of all invariant events. It is not difficult to see (Exercise (9.5.1)) that $\mathscr{I}$ is a $\sigma$-field, and therefore $\mathscr{I}$ is a sub-$\sigma$-field of $\mathscr{B}^T$, called the *invariant $\sigma$-field*.

Finally, we need the idea of conditional expectation. Let $U$ be a random variable on $(\mathbb{R}^T, \mathscr{B}^T, \mathbb{Q})$ with finite mean $\mathbb{E}(U)$; here, $\mathbb{E}$ denotes expectation with respect to the measure $\mathbb{Q}$. We saw in Theorem (7.9.26) that there exists an $\mathscr{I}$-measurable random variable $Z$ such that $\mathbb{E}|Z| < \infty$ and $\mathbb{E}((U - Z)I_G) = 0$ for all $G \in \mathscr{I}$; $Z$ is usually denoted by $Z = \mathbb{E}(U \mid \mathscr{I})$ and is called the conditional expectation of $U$ given $\mathscr{I}$.

We are now ready to restate the strong ergodic theorem (2) in the following way.

**(16) Ergodic theorem.** *Let $\mathbb{Q}$ be a stationary probability measure on $(\mathbb{R}^T, \mathscr{B}^T)$, and let $Y$ be a real-valued random variable on the space $(\mathbb{R}^T, \mathscr{B}^T, \mathbb{Q})$. Let $Y_1, Y_2, \ldots$ be the sequence of random variables defined by*

**(17)**
$$Y_i(\mathbf{x}) = Y(\tau^{i-1}(\mathbf{x})) \quad \text{for} \quad \mathbf{x} \in \mathbb{R}^T.$$

*If $Y$ has finite mean, then*

**(18)**
$$\frac{1}{n} \sum_{i=1}^{n} Y_i \to \mathbb{E}(Y \mid \mathscr{I}) \quad \text{a.s. and in mean.}$$

The sequence $\mathbf{Y} = (Y_1, Y_2, \ldots)$ is of course strongly stationary: since $\mathbb{Q}$ is stationary,

$$\mathbb{Q}\big((Y_2, Y_3, \ldots) \in A\big) = \mathbb{Q}\big(\tau(\mathbf{Y}) \in A\big) = \mathbb{Q}(\mathbf{Y} \in \tau^{-1}A) = \mathbb{Q}(\mathbf{Y} \in A) \quad \text{for } A \in \mathscr{B}^T.$$

The above theorem asserts that the average of the first $n$ values of $\mathbf{Y}$ converges as $n \to \infty$, the limit being the conditional mean of $Y$ given $\mathscr{I}$; this is a conclusion very similar to that of the strong law of large numbers (7.5.1).

To understand the relationship between Theorems (2) and (16), consider the situation treated by (2). Let $X_1, X_2, \ldots$ be a strongly stationary sequence on $(\Omega, \mathscr{F}, \mathbb{P})$, and let $\mathbb{Q}$ be the stationary measure on $(\mathbb{R}^T, \mathscr{B}^T)$ defined by $\mathbb{Q}(A) = \mathbb{P}(\mathbf{X} \in A)$ for $A \in \mathscr{B}^T$. We define $Y : \mathbb{R}^T \to \mathbb{R}$ by $Y(\mathbf{x}) = x_1$ for $\mathbf{x} = (x_1, x_2, \ldots) \in \mathbb{R}^T$, so that $Y_i$ in (17) is given by $Y_i(\mathbf{x}) = x_i$. It is clear that the sequences $\{X_n : n \geq 1\}$ and $\{Y_n : n \geq 1\}$ have the same joint distributions, and it follows that the convergence of $n^{-1} \sum_1^n Y_i$ entails the convergence of $n^{-1} \sum_1^n X_i$.

**(19) Definition.** The stationary measure $\mathbb{Q}$ on $(\mathbb{R}^T, \mathscr{B}^T)$ is called **ergodic** if each invariant event has probability either 0 or 1, which is to say that $\mathbb{Q}(A) = 0$ or 1 for all $A \in \mathscr{I}$.

Ergodic stationary measures are of particular importance. The simplest example of such a measure is product measure.

**(20) Example. Independent sequences.** Let $\mathbb{S}$ be a probability measure on $(\mathbb{R}, \mathscr{B})$, and let $\mathbb{Q} = \mathbb{S}^T$, the appropriate product measure on $(\mathbb{R}^T, \mathscr{B}^T)$. Product measures arise in the context of independent random variables, as follows. Let $X_1, X_2, \ldots$ be a sequence of independent identically distributed random variables on a probability space $(\Omega, \mathscr{F}, \mathbb{P})$, and let $\mathbb{S}(A) = \mathbb{P}(X_1 \in A)$ for $A \in \mathscr{B}$. Then $\mathbb{S}$ is a probability measure on $(\mathbb{R}, \mathscr{B})$. The probability space $(\mathbb{R}^T, \mathscr{B}^T, \mathbb{S}^T)$ is the natural space for the vector $\mathbf{X} = (X_1, X_2, \ldots)$; that is, $\mathbb{S}^T(A) = \mathbb{P}(\mathbf{X} \in A)$ for $A \in \mathscr{B}^T$.

Suppose that $A$ ($\in \mathcal{B}^T$) is invariant. Then, for all $n$, $A$ belongs to the $\sigma$-field generated by the subsequence $(X_n, X_{n+1}, \ldots)$, and hence $A$ belongs to the tail $\sigma$-field of the $X_i$. By Kolmogorov's zero–one law (7.3.15), the latter $\sigma$-field is trivial, in that all events therein have probability either 0 or 1. Hence all invariant events have probability either 0 or 1, and therefore the measure $\mathbb{S}^T$ is ergodic. ●

The conclusion (18) of the ergodic theorem takes on a particularly simple form when the measure $\mathbb{Q}$ is ergodic as well as stationary. In this case, the random variable $\mathbb{E}(Y \mid \mathcal{I})$ is (a.s.) constant, as the following argument demonstrates. The conditional expectation $\mathbb{E}(Y \mid \mathcal{I})$ is $\mathcal{I}$-measurable, and therefore the event $A_y = \{\mathbb{E}(Y \mid \mathcal{I}) \le y\}$ belongs to $\mathcal{I}$ for all $y$. However, $\mathcal{I}$ is trivial, in that it contains only events having probability 0 or 1. Hence $\mathbb{E}(Y \mid \mathcal{I})$ takes almost surely the value $\sup\{y : \mathbb{Q}(A_y) = 0\}$. Taking expectations, we find that this value is $\mathbb{E}(Y)$, so that the conclusion (18) becomes

$$(21) \qquad \frac{1}{n}\sum_{i=1}^{n} Y_i \to \mathbb{E}(Y) \quad \text{a.s. and in mean}$$

in the ergodic case.

**Proof of ergodic theorem (16).** We give full details of this for the case when $\mathbb{Q}$ is ergodic, and finish the proof with brief notes describing how to adapt the argument to the general case.

Assume then that $\mathbb{Q}$ is ergodic, so that $\mathbb{E}(Y \mid \mathcal{I}) = \mathbb{E}(Y)$. First we prove almost-sure convergence, which is to say that

$$(22) \qquad \frac{1}{n}\sum_{i=1}^{n} Y_i \to \mathbb{E}(Y) \quad \text{a.s.}$$

It suffices to prove that

$$(23) \qquad \text{if} \quad \mathbb{E}(Y) < 0 \quad \text{then} \quad \limsup_{n\to\infty}\left\{\frac{1}{n}\sum_{i=1}^{n} Y_i\right\} \le 0 \quad \text{a.s.}$$

To see that (23) suffices, we argue as follows. Suppose that (23) holds, and that $Z$ is a (measurable) function on $(\mathbb{R}^T, \mathcal{B}^T, \mathbb{Q})$ with finite mean, and let $\epsilon > 0$. then $Y' = Z - \mathbb{E}(Z) - \epsilon$ and $Y'' = -Z + \mathbb{E}(Z) - \epsilon$ have negative means. Applying (23) to $Y'$ and $Y''$ we obtain

$$\mathbb{E}(Z) - \epsilon \le \liminf_{n\to\infty}\left\{\frac{1}{n}\sum_{i=1}^{n} Z_i\right\} \le \limsup_{n\to\infty}\left\{\frac{1}{n}\sum_{i=1}^{n} Z_i\right\} \le \mathbb{E}(Z) + \epsilon \quad \text{a.s.,}$$

where $Z_i$ is the random variable given by $Z_i(\mathbf{x}) = Z(\tau^{i-1}(\mathbf{x}))$. These inequalities hold for all $\epsilon > 0$, and therefore

$$\liminf_{n\to\infty}\left\{\frac{1}{n}\sum_{i=1}^{n} Z_i\right\} = \limsup_{n\to\infty}\left\{\frac{1}{n}\sum_{i=1}^{n} Z_i\right\} = \mathbb{E}(Z) \quad \text{a.s.}$$

as required for almost-sure convergence.

Turning to the proof of (23), suppose that $\mathbb{E}(Y) < 0$, and introduce the notation $S_n = \sum_{i=1}^{n} Y_i$. Now $S_n \leq M_n$ where $M_n = \max\{0, S_1, S_2, \dots, S_n\}$ satisfies $M_n \leq M_{n+1}$. Hence $S_n \leq M_\infty$ where $M_\infty = \lim_{n\to\infty} M_n$. Therefore

$$(24) \qquad \qquad \limsup_{n\to\infty}\left\{\frac{1}{n}S_n\right\} \leq \limsup_{n\to\infty}\left\{\frac{1}{n}M_\infty\right\},$$

and (23) will be proved once we know that $M_\infty < \infty$ a.s. It is easily seen that the event $\{M_\infty < \infty\}$ is an invariant event, and hence has probability either 0 or 1; it is here that we use the hypothesis that $\mathbb{Q}$ is ergodic. We must show that $\mathbb{Q}(M_\infty < \infty) = 1$, and to this end we assume the contrary, that $\mathbb{Q}(M_\infty = \infty) = 1$.

Now,

$$(25) \qquad\qquad M_{n+1} = \max\{0, S_1, S_2, \dots, S_{n+1}\}$$
$$= \max\big\{0, S_1 + \max\{0, S_2 - S_1, \dots, S_{n+1} - S_1\}\big\}$$
$$= \max\{0, S_1 + M'_n\}$$

where $M'_n = \max\{0, S'_1, S'_2, \dots, S'_n\}$, and $S'_j = \sum_{i=1}^{j} Y_{i+1}$. It follows from (25) that

$$M_{n+1} = M'_n + \max\{-M'_n, Y\},$$

since $S_1 = Y$. Taking expectations and using the fact that $\mathbb{E}(M'_n) = \mathbb{E}(M_n)$, we find that

$$(26) \qquad\qquad 0 \leq \mathbb{E}(M_{n+1}) - \mathbb{E}(M_n) = \mathbb{E}\big(\max\{-M'_n, Y\}\big).$$

If $M_n \uparrow \infty$ a.s. then $M'_n \uparrow \infty$ a.s., implying that $\max\{-M'_n, Y\} \downarrow Y$ a.s. It follows by (26) (and dominated convergence) that $0 \leq \mathbb{E}(Y)$ in contradiction of the assumption that $\mathbb{E}(Y) < 0$. Our initial hypothesis was therefore false, which is to say that $\mathbb{Q}(M_\infty < \infty) = 1$, and (23) is proved.

Having proved almost-sure convergence, convergence in mean will follow by Theorem (7.10.3) once we have proved that the family $\{n^{-1}S_n : n \geq 1\}$ is uniformly integrable. The random variables $Y_1, Y_2, \dots$ are identically distributed with finite mean; hence (see Exercise (5.6.5)) for any $\epsilon > 0$, there exists $\delta > 0$ such that, for all $i$,

$$(27) \qquad\qquad \mathbb{E}\big(|Y_i|I_A\big) < \epsilon \quad \text{for all } A \text{ satisfying } \mathbb{Q}(A) < \delta.$$

Hence, for all $n$,

$$\mathbb{E}\big(|n^{-1}S_n|I_A\big) \leq \frac{1}{n}\sum_{i=1}^{n}\mathbb{E}\big(|Y_i|I_A\big) < \epsilon$$

whenever $\mathbb{Q}(A) < \delta$. We deduce by an appeal to Lemma (7.10.6) that $\{n^{-1}S_n : n \geq 1\}$ is a uniformly integrable family as required.

This completes the proof in the ergodic case. The proof is only slightly more complicated in the general case, and here is a sketch of the additional steps required.

1. Use the definition of $\mathcal{I}$ to show that $\mathbb{E}(Y \mid \mathcal{I}) = \mathbb{E}(Y_i \mid \mathcal{I})$ for all $i$.

2. Replace (23) by the following statement: on the event $\{\mathbb{E}(Y \mid \mathcal{I}) < 0\}$, we have that

$$\limsup_{n\to\infty}\left\{\frac{1}{n}\sum_{i=1}^{n}Y_i\right\} \le 0$$

except possibly for an event of probability 0. Check that this is sufficient for the required result by applying it to the random variables

$$Y' = Z - \mathbb{E}(Z \mid \mathcal{I}) - \epsilon, \quad Y'' = -Z + \mathbb{E}(Z \mid \mathcal{I}) - \epsilon,$$

where $\epsilon > 0$.

3. Moving to (26), prove that $\mathbb{E}(M_n' \mid \mathcal{I}) = \mathbb{E}(M_n \mid \mathcal{I})$, and deduce the inequality $\mathbb{E}\big(\max\{-M_n', Y\} \mid \mathcal{I}\big) \ge 0$.

4. Continuing from (26), show that $\{M_n \to \infty\} = \{M_n' \to \infty\}$, and deduce the inequality $\mathbb{E}(Y \mid \mathcal{I}) \ge 0$ on the event $\{M_n \to \infty\}$. This leads us to the same contradiction as in the ergodic case, and we conclude the proof as before.     ∎

Here are some applications of the ergodic theorem.

**(28) Example. Markov chains.** Let $X = \{X_n\}$ be an irreducible ergodic Markov chain with countable state space $S$, and let $\pi$ be the unique stationary distribution of the chain. Suppose that $X(0)$ has distribution $\pi$; the argument of Example (8.2.4) shows that $X$ is strongly stationary. Choose some state $k$ and define the collection $I = \{I_n : n \ge 0\}$ of indicator functions by

$$I_n = \begin{cases} 1 & \text{if } X_n = k, \\ 0 & \text{otherwise.} \end{cases}$$

Clearly $I$ is strongly stationary. It has autocovariance function

$$c(n, n+m) = \text{cov}(I_n, I_{n+m}) = \pi_k[p_{kk}(m) - \pi_k], \quad m \ge 0,$$

where $p_{kk}(m) = \mathbb{P}(X_m = k \mid X_0 = k)$. The partial sum $S_n = \sum_{j=0}^{n-1} I_j$ is the number of visits to the state $k$ before the $n$th jump, and a short calculation gives

$$\frac{1}{n}\mathbb{E}(S_n) = \pi_k \quad \text{for all } n.$$

It is a consequence of the ergodic theorem (2) that

$$\frac{1}{n}S_n \xrightarrow{\text{a.s.}} S \quad \text{as} \quad n \to \infty,$$

where $S$ is a random variable with mean $\mathbb{E}(S) = \mathbb{E}(I_0) = \pi_k$. Actually $S$ is constant in that $\mathbb{P}(S = \pi_k) = 1$; just note that $c(n, n+m) \to 0$ as $m \to \infty$ and use the result of Problem (9.7.9).     ●

**(29) Example. Binary expansion.** Let $X$ be uniformly distributed on $[0, 1]$. The random number $X$ has a binary expansion

$$X = 0 \cdot X_1 X_2 \cdots = \sum_{j=1}^{\infty} X_j 2^{-j}$$

where $X_1, X_2, \ldots$ is a sequence of independent identically distributed random variables, each taking one of the values 0 or 1 with probability $\frac{1}{2}$ (see Problem (7.11.4)). Define

(30)                          $$Y_n = 0 \cdot X_n X_{n+1} \cdots \quad \text{for} \quad n \geq 1$$

and check for yourself that $Y = \{Y_n : n \geq 1\}$ is strongly stationary. Use (2) to see that

$$\frac{1}{n} \sum_{j=1}^{n} Y_j \xrightarrow{\text{a.s.}} \frac{1}{2} \quad \text{as} \quad n \to \infty.$$

Generalize this example as follows. Let $g : \mathbb{R} \to \mathbb{R}$ be such that:
(a) $g$ has period 1, so that $g(x + 1) = g(x)$ for all $x$,
(b) $g$ is uniformly continuous and integrable over $[0, 1]$,
and define $Z = \{Z_n : n \geq 1\}$ by $Z_n = g(2^{n-1}X)$ where $X$ is uniform on $[0, 1]$ as before. The process $Y$, above, may be constructed in this way by choosing $g(x) = x$ modulo 1. Check for yourself that $Z$ is strongly stationary, and deduce that

$$\frac{1}{n} \sum_{j=1}^{n} g(2^{j-1}X) \xrightarrow{\text{a.s.}} \int_0^1 g(x)\, dx \quad \text{as} \quad n \to \infty.$$

Can you adapt this example to show that

$$\frac{1}{n} \sum_{j=1}^{n} g(X + (j - 1)\pi) \xrightarrow{\text{a.s.}} \int_0^1 g(x)\, dx \quad \text{as} \quad n \to \infty$$

for any fixed positive irrational number $\pi$?                                                  ●

**(31) Example. Range of random walk.** Let $X_1, X_2, \ldots$ be independent identically distributed random variables taking integer values, and let $S_n = X_1 + X_2 + \cdots + X_n$; think of $S_n$ as being the position of a random walk after $n$ steps. Let $R_n$ be the *range* of the walk up to time $n$, which is to say that $R_n$ is the number of distinct values taken by the sequence $S_1, S_2, \ldots, S_n$. It was proved by elementary means in Problem (3.11.27) that

(32)                          $$\frac{1}{n}\mathbb{E}(R_n) \to \mathbb{P}(\text{no return}) \quad \text{as} \quad n \to \infty$$

where the event $\{\text{no return}\} = \{S_k \neq 0 \text{ for all } k \geq 1\}$ is the event that the walk never revisits its starting point $S_0 = 0$.
   Of more interest than (32) is the fact that

(33)                          $$\frac{1}{n}R_n \xrightarrow{\text{a.s.}} \mathbb{P}(\text{no return}),$$

and we shall prove this with the aid of the ergodic theorem (16).
   First, let $N$ be a positive integer, and let $Z_k$ be the number of distinct points visited by $S_{(k-1)N+1}, S_{(k-1)N+2}, \ldots, S_{kN}$; clearly $Z_1, Z_2, \ldots$ are independent identically distributed

variables. Now, if $KN \le n < (K+1)N$, then $|R_n - R_{KN}| \le N$ and $R_{KN} \le Z_1 + Z_2 + \cdots + Z_K$. Therefore

$$\frac{1}{n}R_n \le \frac{1}{KN}(R_{KN} + N) \le \frac{1}{KN}(Z_1 + Z_2 + \cdots + Z_K) + \frac{1}{K}$$

$$\xrightarrow{\text{a.s.}} \frac{1}{N}\mathbb{E}(Z_1) \quad \text{as} \quad K \to \infty$$

by the strong law of large numbers. It is easily seen that $Z_1 = R_N$, and therefore, almost surely,

(34)                  $$\limsup_{n \to \infty} \left\{ \frac{1}{n}R_n \right\} \le \frac{1}{N}\mathbb{E}(R_N) \to \mathbb{P}(\text{no return})$$

as $N \to \infty$, by (32). This is the required upper bound.

For the lower bound, we must work a little harder. Let $V_k$ be the indicator function of the event that the position of the walk at time $k$ is not revisited subsequently; that is,

$$V_k = \begin{cases} 1 & \text{if } S_j \ne S_k \text{ for all } j > k, \\ 0 & \text{otherwise.} \end{cases}$$

The collection of points $S_k$ for which $V_k = 1$ is a collection of distinct points, and it follows that

(35)                          $$R_n \ge V_1 + V_2 + \cdots + V_n.$$

On the other hand, $V_k$ may be represented as $Y(X_{k+1}, X_{k+2}, \dots)$ where $Y : \mathbb{R}^T \to \{0, 1\}$ is defined by

$$Y(x_1, x_2, \dots) = \begin{cases} 1 & \text{if } x_1 + \cdots + x_l \ne 0 \text{ for all } l \ge 1, \\ 0 & \text{otherwise.} \end{cases}$$

The $X_j$ are independent and identically distributed, and therefore Theorem (16) may be applied to deduce that

$$\frac{1}{n}(V_1 + V_2 + \cdots + V_n) \xrightarrow{\text{a.s.}} \mathbb{E}(V_1).$$

Note that $\mathbb{E}(V_1) = \mathbb{P}(\text{no return})$.

It follows from (35) that

$$\liminf_{n \to \infty} \left\{ \frac{1}{n}R_n \right\} \ge \mathbb{P}(\text{no return}) \quad \text{a.s.},$$

which may be combined with (34) to obtain the claimed result (33).                    ●

## Exercises for Section 9.5

**1.** Let $T = \{1, 2, \dots\}$ and let $\mathcal{I}$ be the set of invariant events of $(\mathbb{R}^T, \mathcal{B}^T)$. Show that $\mathcal{I}$ is a $\sigma$-field.

**2.** Assume that $X_1, X_2, \dots$ is a stationary sequence with autocovariance function $c(m)$. Show that

$$\text{var}\left(\frac{1}{n}\sum_{i=1}^{n} X_i\right) = \frac{2}{n^2}\sum_{j=1}^{n}\sum_{i=0}^{j-1} c(i) - \frac{c(0)}{n}.$$

Assuming that $j^{-1}\sum_{i=0}^{j-1} c(i) \to \sigma^2$ as $j \to \infty$, show that

$$\text{var}\left(\frac{1}{n}\sum_{i=1}^{n} X_i\right) \to \sigma^2 \qquad \text{as } n \to \infty.$$

**3.** Let $X_1, X_2, \dots$ be independent identically distributed random variables with zero mean and unit variance. Let

$$Y_n = \sum_{i=0}^{\infty} \alpha_i X_{n+i} \qquad \text{for } n \geq 1$$

where the $\alpha_i$ are constants satisfying $\sum_i \alpha_i^2 < \infty$. Use the martingale convergence theorem to show that the above summation converges almost surely and in mean square. Prove that $n^{-1}\sum_{i=1}^{n} Y_i \to 0$ a.s. and in mean, as $n \to \infty$.

## 9.6 Gaussian processes

Let $X = \{X(t) : -\infty < t < \infty\}$ be a real-valued stationary process with autocovariance function $c(t)$; in line with Theorem (9.3.2), $c$ is a real-valued function which satisfies:

(a) $c(-t) = c(t)$,

(b) $c$ is a non-negative definite function.

It is not difficult to see that a function $c : \mathbb{R} \to \mathbb{R}$ is the autocovariance function of some real-valued stationary process if and only if $c$ satisfies (a) and (b). Subject to these conditions on $c$, there is an explicit construction of a corresponding stationary process.

**(1) Theorem.** *If $c : \mathbb{R} \to \mathbb{R}$ and $c$ satisfies* (a) *and* (b) *above, there exists a real-valued strongly stationary process $X$ with autocovariance function $c$.*

**Proof.** We shall construct $X$ by defining its finite-dimensional distributions (fdds) and then using the Kolmogorov consistency conditions (8.9.3). For any vector $\mathbf{t} = (t_1, t_2, \dots, t_n)$ of real numbers with some finite length $n$, let $F_{\mathbf{t}}$ be the multivariate normal distribution function with zero means and covariance matrix $\mathbf{V} = (v_{jk})$ with entries $v_{jk} = c(t_k - t_j)$ (see Section 4.9).

The family $\{F_{\mathbf{t}} : \mathbf{t} \in \mathbb{R}^n, n = 1, 2, \dots\}$ satisfies the Kolmogorov consistency conditions (8.9.3) and so there exists a process $X$ with this family of fdds. It is clear that $X$ is strongly stationary with autocovariance function $c$. ∎

A result similar to (1) holds for complex-valued functions $c : \mathbb{R} \to \mathbb{C}$, (a) being replaced by the property that

**(2)** $$c(-t) = \overline{c(t)}.$$

We do not explore this here, but choose to consider real-valued processes only. The process $X$ which we have constructed in the foregoing proof is an example of a (real-valued) 'Gaussian process'.

**(3) Definition.** A real-valued continuous-time process $X$ is called a **Gaussian** process if each finite-dimensional vector $(X(t_1), X(t_2), \ldots, X(t_n))$ has the multivariate normal distribution $N(\boldsymbol{\mu}(\mathbf{t}), \mathbf{V}(\mathbf{t}))$ for some mean vector $\boldsymbol{\mu}$ and some covariance matrix $\mathbf{V}$ which may depend on $\mathbf{t} = (t_1, t_2, \ldots, t_n)$.

The $X(t_j)$ may have a singular multivariate normal distribution. We shall often restrict our attention to Gaussian processes with $\mathbb{E}(X(t)) = 0$ for all $t$; as before, similar results are easily found when this fails to hold.

A Gaussian process is not necessarily stationary.

**(4) Theorem.** *The Gaussian process $X$ is stationary if and only if $\mathbb{E}(X(t))$ is constant for all $t$ and the covariance matrix $\mathbf{V}(\mathbf{t})$ in Definition (3) satisfies $\mathbf{V}(\mathbf{t}) = \mathbf{V}(\mathbf{t} + h)$ for all $\mathbf{t}$ and $h > 0$, where $\mathbf{t} + h = (t_1 + h, t_2 + h, \ldots, t_n + h)$.*

**Proof.** This is an easy *exercise*. ∎

It is clear that a Gaussian process is strongly stationary if and only if it is weakly stationary.

Can a Gaussian process be a Markov process? The answer is in the affirmative. First, we must rephrase the Markov property (6.1.1) to deal with processes which take values in the real line.

**(5) Definition.** The continuous-time process $X$, taking values in $\mathbb{R}$, is called a **Markov process** if the following holds:

**(6)** $\quad \mathbb{P}\big(X(t_n) \leq x \,\big|\, X(t_1) = x_1, \ldots, X(t_{n-1}) = x_{n-1}\big) = \mathbb{P}\big(X(t_n) \leq x \,\big|\, X(t_{n-1}) = x_{n-1}\big)$

for all $x, x_1, x_2, \ldots, x_{n-1}$, and all increasing sequences $t_1 < t_2 < \cdots < t_n$ of times.

**(7) Theorem.** *The Gaussian process $X$ is a Markov process if and only if*

**(8)** $\qquad \mathbb{E}\big(X(t_n) \,\big|\, X(t_1) = x_1, \ldots, X(t_{n-1}) = x_{n-1}\big) = \mathbb{E}\big(X(t_n) \,\big|\, X(t_{n-1}) = x_{n-1}\big)$

*for all $x_1, x_2, \ldots, x_{n-1}$ and all increasing sequences $t_1 < t_2 < \cdots < t_n$ of times.*

**Proof.** It is clear from (5) that (8) holds whenever $X$ is Markov. Conversely, suppose that $X$ is Gaussian and satisfies (8). Both the left- and right-hand sides of (6) are normal distribution functions. Any normal distribution is specified by its mean and variance, and so we need only show that the left- and right-hand sides of (6) have equal first two moments. The equality of the first moments is trivial, since this is simply the assertion of (8). Also, if $1 \leq r < n$, then $\mathbb{E}(YX_r) = 0$ where

**(9)** $\qquad\qquad Y = X_n - \mathbb{E}(X_n \mid X_1, \ldots, X_{n-1}) = X_n - \mathbb{E}(X_n \mid X_{n-1})$

and we have written $X_r = X(t_r)$ for ease of notation; to see this, write

$$\mathbb{E}(YX_r) = \mathbb{E}\big(X_n X_r - \mathbb{E}(X_n X_r \mid X_1, \ldots, X_{n-1})\big)$$
$$= \mathbb{E}(X_n X_r) - \mathbb{E}(X_n X_r) = 0.$$

However, $Y$ and $X$ are normally distributed, and furthermore $\mathbb{E}(Y) = 0$; as in Example (4.5.9), $Y$ and $X_r$ are independent. It follows that $Y$ is independent of the collection $X_1, X_2, \ldots, X_{n-1}$, using properties of the multivariate normal distribution.

Write $A_r = \{X_r = x_r\}$ and $A = A_1 \cap A_2 \cap \cdots \cap A_{n-1}$. By the proven independence, $\mathbb{E}(Y^2 \mid A) = \mathbb{E}(Y^2 \mid A_{n-1})$, which may be written as $\text{var}(X_n \mid A) = \text{var}(X_n \mid A_{n-1})$, by (9). Thus the left- and right-hand sides of (6) have the same second moment also, and the result is proved.                                                                                           ∎

**(10) Example. A stationary Gaussian Markov process.** Suppose $X$ is stationary, Gaussian and Markov, and has zero means. Use the result of Problem (4.14.13) to obtain that

$$c(0)\mathbb{E}[X(s+t) \mid X(s)] = c(t)X(s) \quad \text{whenever} \quad t \geq 0,$$

where $c$ is the autocovariance function of $X$. Thus, if $0 \leq s \leq s + t$ then

$$c(0)\mathbb{E}[X(0)X(s+t)] = c(0)\mathbb{E}\Big[\mathbb{E}\big(X(0)X(s+t) \mid X(0), X(s)\big)\Big]$$
$$= c(0)\mathbb{E}\big[X(0)\mathbb{E}\big(X(s+t) \mid X(s)\big)\big]$$
$$= c(t)\mathbb{E}\big(X(0)X(s)\big)$$

by Lemma (7.7.10). Thus

**(11)**                          $$c(0)c(s+t) = c(s)c(t) \quad \text{for} \quad s, t \geq 0.$$

This is satisfied whenever

**(12)**                                    $$c(t) = c(0)e^{-\alpha|t|}.$$

Following Problem (4.14.5) we can see that (12) is the general solution to (11) subject to some condition of regularity such as that $c$ be continuous. We shall see later (see Problem (13.12.4)) that such a process is called a stationary *Ornstein–Uhlenbeck process*.                    ●

**(13) Example. The Wiener process.** Suppose that $\sigma^2 > 0$ and define

**(14)**                          $$c(s, t) = \sigma^2 \min\{s, t\} \quad \text{whenever} \quad s, t \geq 0.$$

We claim that there exists a Gaussian process $W = \{W(t) : t \geq 0\}$ with zero means such that $W(0) = 0$ and $\text{cov}(W(s), W(t)) = c(s, t)$. By the argument in the proof of (1), it is sufficient to show that the matrix $\mathbf{V(t)}$ with entries $(v_{jk})$, where $v_{jk} = c(t_k, t_j)$, is positive definite for all $\mathbf{t} = (t_1, t_2, \ldots, t_n)$. In order to see that this indeed holds, let $z_1, z_2, \ldots, z_n$ be complex numbers and suppose that $0 = t_0 < t_1 < \cdots < t_n$. It is not difficult to check that

$$\sum_{j,k=1}^{n} c(t_k, t_j)z_j\bar{z}_k = \sigma^2 \sum_{j=1}^{n}(t_j - t_{j-1})\left|\sum_{k=j}^{n} z_k\right|^2 > 0$$

whenever one of the $z_j$ is non-zero; this guarantees the existence of $W$. It is called the *Wiener process*; we explore its properties in more detail in Chapter 13, noting only two facts here.

**(15) Lemma.** *The Wiener process $W$ satisfies $\mathbb{E}(W(t)^2) = \sigma^2 t$ for all $t \geq 0$.*

**Proof.** $\mathbb{E}(W(t)^2) = \text{cov}\big(W(t), W(t)\big) = c(t, t) = \sigma^2 t$.                   ∎

**(16) Lemma.** *The Wiener process $W$ has stationary independent increments, that is:*
   (a) *the distribution of $W(t) - W(s)$ depends on $t - s$ alone,*
   (b) *the variables $W(t_j) - W(s_j)$, $1 \leq j \leq n$, are independent whenever the intervals $(s_j, t_j]$ are disjoint.*

**Proof.** The increments of $W$ are jointly normally distributed; their independence follows as soon as we have shown that they are uncorrelated. However, if $u \leq v \leq s \leq t$,

$$\mathbb{E}\big([W(v) - W(u)][W(t) - W(s)]\big) = c(v, t) - c(v, s) + c(u, s) - c(u, t)$$
$$= \sigma^2(v - v + u - u) = 0$$

by (14).

Finally, $W(t) - W(s)$ is normally distributed with zero mean, and with variance

$$\mathbb{E}\big([W(t) - W(s)]^2\big) = \mathbb{E}(W(t)^2) - 2c(s, t) + \mathbb{E}(W(s)^2)$$
$$= \sigma^2(t - s) \quad \text{if} \quad s \leq t.$$                  ∎ ●

---

## Exercises for Section 9.6

**1.**   Show that the function $c(s, t) = \min\{s, t\}$ is positive definite. That is, show that

$$\sum_{j,k=1}^{n} c(t_k, t_j) z_j \bar{z}_k > 0$$

for all $0 \leq t_1 < t_2 < \cdots < t_n$ and all complex numbers $z_1, z_2, \ldots, z_n$ at least one of which is non-zero.

**2.**   Let $X_1, X_2, \ldots$ be a stationary Gaussian sequence with zero means and unit variances which satisfies the Markov property. Find the spectral density function of the sequence in terms of the constant $\rho = \text{cov}(X_1, X_2)$.

**3.**   Show that a Gaussian process is strongly stationary if and only if it is weakly stationary.

**4.**   Let $X$ be a stationary Gaussian process with zero mean, unit variance, and autocovariance function $c(t)$. Find the autocovariance functions of the processes $X^2 = \{X(t)^2 : -\infty < t < \infty\}$ and $X^3 = \{X(t)^3 : -\infty < t < \infty\}$.

**5.**   (a) Let $W$ be a standard Wiener process, and $T : [0, \infty) \to [0, \infty)$ a non-random, right-continuous, non-decreasing function with $T(0) = 0$. Show that $X(t) := W(T(t))$ defines a Gaussian process. Find the characteristic function of $X(t)$ and the covariance function $c(s, t) = \text{cov}(X(s), X(t))$.
   (b) Let $W$ be a Wiener process with $\text{var}(W(t)) = 2t$, and let $T$ be an independent $\alpha$-stable subordinator, where $0 < \alpha < 1$. Show that $X(t) = W(T(t))$ is a symmetric $(2\alpha)$-stable Lévy process.

**6.**   **Brownian sheet.** Let $X(s, t)$ be a two-parameter, zero-mean, Gaussian process on the positive quadrant $[0, \infty)^2$ of $\mathbb{R}^2$, with covariance function $\text{cov}(X(s, t), X(u, v)) = (s \wedge u)(t \wedge v)$, where $x \wedge y = \min\{x, y\}$. For any rectangle $R$ with corners $(s, t), (u, t), (s, v), (u, v)$ where $0 \leq s < u < \infty$ and $0 \leq t < v < \infty$, define

$$Y(R) = X(s, t) + X(u, v) - X(s, v) - X(u, t).$$

Find $\text{var}(Y(R))$, and show that $Y$ has independent increments.

## 9.7 Problems

**1.** Let $\ldots, X_{-1}, X_0, X_1, \ldots$ be uncorrelated random variables with zero means and unit variances, and define

$$Y_n = X_n + \alpha \sum_{i=1}^{\infty} \beta^{i-1} X_{n-i} \qquad \text{for} \; -\infty < n < \infty,$$

where $\alpha$ and $\beta$ are constants satisfying $|\beta| < 1$, $|\beta - \alpha| < 1$. Find the best linear predictor of $Y_{n+1}$ given the entire past $Y_n, Y_{n-1}, \ldots$.

**2.** Let $\{Y_k : -\infty < k < \infty\}$ be a stationary sequence with variance $\sigma_Y^2$, and let

$$X_n = \sum_{k=0}^{r} a_k Y_{n-k}, \qquad -\infty < n < \infty,$$

where $a_0, a_1, \ldots, a_r$ are constants. Show that $X$ has spectral density function

$$f_X(\lambda) = \frac{\sigma_Y^2}{\sigma_X^2} f_Y(\lambda) |G_a(e^{i\lambda})|^2$$

where $f_Y$ is the spectral density function of $Y$, $\sigma_X^2 = \text{var}(X_1)$, and $G_a(z) = \sum_{k=0}^{r} a_k z^k$.

Calculate this spectral density explicitly in the case of 'exponential smoothing', when $r = \infty$, $a_k = \mu^k(1 - \mu)$, and $0 < \mu < 1$.

**3.** Suppose that $\widehat{Y}_{n+1} = \alpha Y_n + \beta Y_{n-1}$ is the best linear predictor of $Y_{n+1}$ given the entire past $Y_n, Y_{n-1}, \ldots$ of the stationary sequence $\{Y_k : -\infty < k < \infty\}$. Find the spectral density function of the sequence.

**4.** **Recurrent events, Example (5.2.15).** Meteorites fall at integer times $T_1, T_2, \ldots$ where $T_n = X_1 + X_2 + \cdots + X_n$. We assume that the $X_i$ are independent, $X_2, X_3, \ldots$ are identically distributed, and the distribution of $X_1$ is such that the probability that a meteorite falls at time $n$ is constant for all $n$. Let $Y_n$ be the indicator function of the event that a meteorite falls at time $n$. Show that $\{Y_n\}$ is stationary and find its spectral density function in terms of the characteristic function of $X_2$.

**5.** Let $X = \{X_n : n \geq 1\}$ be given by $X_n = \cos(nU)$ where $U$ is uniformly distributed on $[-\pi, \pi]$. Show that $X$ is stationary but not strongly stationary. Find the autocorrelation function of $X$ and its spectral density function.

**6.** (a) Let $N$ be a Poisson process with intensity $\lambda$, and let $\alpha > 0$. Define $X(t) = N(t + \alpha) - N(t)$ for $t \geq 0$. Show that $X$ is strongly stationary, and find its spectral density function.

(b) Let $W$ be a Wiener process and define $X = \{X(t) : t \geq 1\}$ by $X(t) = W(t) - W(t-1)$. Show that $X$ is strongly stationary and find its autocovariance function. Find the spectral density function of $X$.

**7.** Let $Z_1, Z_2, \ldots$ be uncorrelated variables, each with zero mean and unit variance.

(a) Define the moving average process $X$ by $X_n = Z_n + \alpha Z_{n-1}$ where $\alpha$ is a constant. Find the spectral density function of $X$.

(b) More generally, let $Y_n = \sum_{i=0}^{r} \alpha_i Z_{n-i}$, where $\alpha_0 = 1$ and $\alpha_1, \ldots, \alpha_r$ are constants. Find the spectral density function of $Y$.

**8.** Show that the complex-valued stationary process $X = \{X(t) : -\infty < t < \infty\}$ has a spectral density function which is bounded and uniformly continuous whenever its autocorrelation function $\rho$ is continuous and satisfies $\int_0^\infty |\rho(t)| \, dt < \infty$.

**9.** Let $X = \{X_n : n \geq 1\}$ be stationary with constant mean $\mu = \mathbb{E}(X_n)$ for all $n$, and such that $\text{cov}(X_1, X_n) \to 0$ as $n \to \infty$. Show that $n^{-1} \sum_{j=1}^{n} X_j \xrightarrow{\text{m.s.}} \mu$.

**10.** Deduce the strong law of large numbers from an appropriate ergodic theorem.

**11.** Let $\mathbb{Q}$ be a stationary measure on $(\mathbb{R}^T, \mathcal{B}^T)$ where $T = \{1, 2, \ldots\}$. Show that $\mathbb{Q}$ is ergodic if and only if

$$\frac{1}{n} \sum_{i=1}^{n} Y_i \to \mathbb{E}(Y) \qquad \text{a.s. and in mean}$$

for all $Y : \mathbb{R}^T \to \mathbb{R}$ for which $\mathbb{E}(Y)$ exists, where $Y_i : \mathbb{R}^T \to \mathbb{R}$ is given by $Y_i(\mathbf{x}) = Y(\tau^{i-1}(\mathbf{x}))$. As usual, $\tau$ is the natural shift operator on $\mathbb{R}^T$.

**12.** The stationary measure $\mathbb{Q}$ on $(\mathbb{R}^T, \mathcal{B}^T)$ is called *strongly mixing* if $\mathbb{Q}(A \cap \tau^{-n} B) \to \mathbb{Q}(A)\mathbb{Q}(B)$ as $n \to \infty$, for all $A, B \in \mathcal{B}^T$; as usual, $T = \{1, 2, \ldots\}$ and $\tau$ is the shift operator on $\mathbb{R}^T$. Show that every strongly mixing measure is ergodic.

**13. Ergodic theorem.** Let $(\Omega, \mathcal{F}, \mathbb{P})$ be a probability space, and let $T : \Omega \to \Omega$ be measurable and measure preserving (i.e. $\mathbb{P}(T^{-1} A) = \mathbb{P}(A)$ for all $A \in \mathcal{F}$). Let $X : \Omega \to \mathbb{R}$ be a random variable, and let $X_i$ be given by $X_i(\omega) = X(T^{i-1}(\omega))$. Show that

$$\frac{1}{n} \sum_{i=1}^{n} X_i \to \mathbb{E}(X \mid \mathcal{I}) \qquad \text{a.s. and in mean}$$

where $\mathcal{I}$ is the $\sigma$-field of invariant events of $T$.

If $T$ is ergodic (in that $\mathbb{P}(A)$ equals 0 or 1 whenever $A$ is invariant), prove that $\mathbb{E}(X \mid \mathcal{I}) = \mathbb{E}(X)$ almost surely.

**14. Borel's normal number theorem.** Consider the probability space $(\Omega, \mathcal{F}, \mathbb{P})$ where $\Omega = [0, 1)$, $\mathcal{F}$ is the set of Borel subsets, and $\mathbb{P}$ is Lebesgue measure.
(a) Show that the shift $T : \Omega \to \Omega$ defined by $T(x) = 2x \pmod 1$ is measurable, measure preserving, and ergodic (in that $\mathbb{P}(A)$ equals 0 or 1 if $A = T^{-1}A$).
(b) Let $X : \Omega \to \mathbb{R}$ be the random variable given by the identity mapping $X(\omega) = \omega$. Show that the proportion of 1's, in the expansion of $X$ to base 2, equals $\frac{1}{2}$ almost surely. This is sometimes called 'Borel's normal number theorem'.
(c) Deduce that, for any continuous random variable $Y$ taking values in $[0, 1)$, the proportion of 1's in its binary expansion to $n$ places converges a.s. to $\frac{1}{2}$ as $n \to \infty$.
(d) Let $\{X_i : i \geq 1\}$ be a sequence of independent Bernoulli random variables with parameter $p \neq \frac{1}{2}$, and set $Z = \sum_{i=1}^{\infty} X_i 2^{-i}$. Is $Z$ a random variable? Is it continuous, or discrete, or what?

**15.** Let $g : \mathbb{R} \to \mathbb{R}$ be periodic with period 1, and uniformly continuous and integrable over $[0, 1]$. Define $Z_n = g(X + (n-1)\alpha)$, $n \geq 1$, where $X$ is uniform on $[0, 1]$ and $\alpha$ is irrational. Show that, as $n \to \infty$,

$$\frac{1}{n} \sum_{j=1}^{n} Z_j \to \int_0^1 g(u)\, du \qquad \text{a.s.}$$

**16.** Let $X = \{X(t) : t \geq 0\}$ be a non-decreasing random process such that:
(a) $X(0) = 0$, $X$ takes values in the non-negative integers,
(b) $X$ has stationary independent increments,
(c) the sample paths $\{X(t, \omega) : t \geq 0\}$ have only jump discontinuities of unit magnitude.
Show that $X$ is a Poisson process.

**17.** Let $X$ be a continuous-time process. Show that:
(a) if $X$ has stationary increments and $m(t) = \mathbb{E}(X(t))$ is a continuous function of $t$, then there exist $\alpha$ and $\beta$ such that $m(t) = \alpha + \beta t$,
(b) if $X$ has stationary independent increments and $v(t) = \text{var}(X(t) - X(0))$ is a continuous function of $t$ then there exists $\sigma^2$ such that $\text{var}(X(s+t) - X(s)) = \sigma^2 t$ for all $s$.

**18.** A Wiener process $W$ is called *standard* if $W(0) = 0$ and $W(1)$ has unit variance. Let $W$ be a standard Wiener process, and let $\alpha$ be a positive constant. Show that:

(a) $\alpha W(t/\alpha^2)$ is a standard Wiener process,

(b) $W(t + \alpha) - W(\alpha)$ is a standard Wiener process,

(c) the process $V$, given by $V(t) = tW(1/t)$ for $t > 0$, $V(0) = 0$, is a standard Wiener process,

(d) the process $W(1) - W(1 - t)$ is a standard Wiener process on $[0, 1]$.

**19.** Let $W$ be a standard Wiener process. Show that the stochastic integrals

$$X(t) = \int_0^t dW(u), \qquad Y(t) = \int_0^t e^{-(t-u)} \, dW(u), \qquad t \geq 0,$$

are well defined, and prove that $X(t) = W(t)$, and that $Y$ has autocovariance function $\mathrm{cov}(Y(s), Y(t)) = \frac{1}{2}(e^{-|s-t|} - e^{-s-t})$, $s < t$.

**20.** Let $W$ be a standard Wiener process. Find the means of the following three processes, and the autocovariance functions in cases (b) and (c):

(a) $X(t) = |W(t)|$,

(b) $Y(t) = e^{W(t)}$,

(c) $Z(t) = \int_0^t W(u) \, du$.

(d) Which of $X$, $Y$, $Z$ are Gaussian processes? Which of these are Markov processes?

(e) Find $\mathbb{E}(Z(t)^n)$ for $n \geq 0$.

**21.** Let $W$ be a standard Wiener process. Find the conditional joint density function of $W(t_2)$ and $W(t_3)$ given that $W(t_1) = W(t_4) = 0$, where $t_1 < t_2 < t_3 < t_4$.

Show that the conditional correlation of $W(t_2)$ and $W(t_3)$ is

$$\rho = \sqrt{\frac{(t_4 - t_3)(t_2 - t_1)}{(t_4 - t_2)(t_3 - t_1)}}.$$

**22. Empirical distribution function.** Let $U_1, U_2, \ldots$ be independent random variables with the uniform distribution on $[0, 1]$. Let $I_j(x)$ be the indicator function of the event $\{U_j \leq x\}$, and define

$$F_n(x) = \frac{1}{n} \sum_{j=1}^n I_j(x), \qquad 0 \leq x \leq 1.$$

The function $F_n$ is called the 'empirical distribution function' of the $U_j$.

(a) Find the mean and variance of $F_n(x)$, and prove that $\sqrt{n}(F_n(x) - x) \xrightarrow{\mathrm{D}} Y(x)$ as $n \to \infty$, where $Y(x)$ is normally distributed.

(b) What is the (multivariate) limit distribution of a collection of random variables of the form $\{\sqrt{n}(F_n(x_i) - x_i) : 1 \leq i \leq k\}$, where $0 \leq x_1 < x_2 < \cdots < x_k \leq 1$?

(c) Show that the autocovariance function of the asymptotic finite-dimensional distributions of $\sqrt{n}(F_n(x) - x)$, in the limit as $n \to \infty$, is the same as that of the process $Z(t) = W(t) - tW(1)$, $0 \leq t \leq 1$, where $W$ is a standard Wiener process. The process $Z$ is called a 'Brownian bridge' or 'tied-down Brownian motion'.

**23. Pólya's urn revisited.** An urn contains initially one red ball and one green ball. At later stages, a ball is picked from the urn uniformly at random, and is returned to the urn together with a fresh ball of the same colour. Assume the usual independence. Let $X_k$ be the indicator function of the event that the $k$th ball picked is red.

(a) Show that, for $x_i \in \{0, 1\}$,

$$\mathbb{P}(X_1 = x_1, X_2 = x_2, \ldots, X_n = x_n) = \frac{r! \, (n - r)!}{(n + 1)!}, \qquad x_1, x_2, \ldots, x_n \in \{0, 1\}, \ r = \sum_{k=1}^n x_k.$$

(b) Show that $X_1, X_2, \ldots$ is a stationary sequence, and that $n^{-1} \sum_{k=1}^n X_k$ converges a.s. as $n \to \infty$ to some random variable $R$.

(c) Find the distribution of $R$.

# 10

---

# Renewals

*Summary.* A renewal process is a recurrent-event process with independent identically distributed interevent times. The asymptotic behaviour of a renewal process is described by the renewal theorem and the elementary renewal theorem, and the key renewal theorem is often useful. The waiting-time paradox leads to a discussion of excess and current lifetimes, and their asymptotic distributions are found. Other renewal-type processes are studied, including alternating and delayed renewal processes, and the use of renewal is illustrated in applications to Markov chains and age-dependent branching processes. The asymptotic behaviour of renewal–reward processes is studied, and Little's formula is proved.

## 10.1 The renewal equation

We saw in Section 8.3 that renewal processes provide attractive models for many natural phenomena. Recall their definition.

**(1) Definition.** A **renewal process** $N = \{N(t) : t \geq 0\}$ is a process such that

$$N(t) = \max\{n : T_n \leq t\}$$

where $T_0 = 0$, $T_n = X_1 + X_2 + \cdots + X_n$ for $n \geq 1$, and $\{X_i\}$ is a sequence of independent identically distributed non-negative random variables.

We commonly think of a renewal process $N(t)$ as representing the number of occurrences of some event in the time interval $[0, t]$; the event in question might be the arrival of a person or particle, or the failure of a light bulb. With this in mind, we shall speak of $T_n$ as the 'time of the $n$th arrival' and $X_n$ as the '$n$th interarrival time'. We shall try to use the notation of (1) consistently throughout, denoting by $X$ and $T$ a typical interarrival time and a typical arrival time of the process $N$.

When is $N$ an honest process, which is to say that $N(t) < \infty$ almost surely (see Definition (6.8.18))?

**(2) Theorem.** $\mathbb{P}(N(t) < \infty) = 1$ *for all $t$ if and only if $\mathbb{E}(X_1) > 0$.*

This amounts to saying that $N$ is honest if and only if the interarrival times are not concentrated at zero. The proof is simple and relies upon the following important observation:

(3)                                        $N(t) \geq n$   if and only if   $T_n \leq t$.

We shall make repeated use of (3). It provides a link between $N(t)$ and the sum $T_n$ of independent variables; we know a lot about such sums already.

**Proof of (2).** Since the $X_i$ are non-negative, if $\mathbb{E}(X_1) = 0$ then $\mathbb{P}(X_i = 0) = 1$ for all $i$. Therefore

$$\mathbb{P}\big(N(t) = \infty\big) = 1 \quad \text{for all } t > 0.$$

Conversely, suppose that $\mathbb{E}(X_1) > 0$. There exists $\epsilon > 0$ such that $\mathbb{P}(X_1 > \epsilon) = \delta > 0$. Let $A_i = \{X_i > \epsilon\}$, and let $A = \{X_i > \epsilon \text{ i.o.}\} = \limsup A_i$ be the event that infinitely many of the $X_i$ exceed $\epsilon$. We have that

$$\mathbb{P}(A^c) = \mathbb{P}\left(\bigcup_m \bigcap_{n>m} A_n^c\right) \leq \sum_m \lim_{n\to\infty} (1-\delta)^{n-m} = \sum_m 0 = 0.$$

Therefore, by (3),

$$\mathbb{P}\big(N(t) = \infty\big) = \mathbb{P}(T_n \leq t \text{ for all } n) \leq \mathbb{P}(A^c) = 0. \qquad \blacksquare$$

Thus $N$ is honest if and only if $X_1$ is *not* concentrated at 0. Henceforth we shall assume not only that $\mathbb{P}(X_1 = 0) < 1$, but also impose the stronger condition that $\mathbb{P}(X_1 = 0) = 0$. That is, *we consider only the case when the $X_i$ are strictly positive in that* $\mathbb{P}(X_1 > 0) = 1$.

It is easy in principle to find the distribution of $N(t)$ in terms of the distribution of a typical interarrival time. Let $F$ be the distribution function of $X_1$, and let $F_k$ be the distribution function of $T_k$.

**(4) Lemma†.** *We have that $F_1 = F$ and $F_{k+1}(x) = \displaystyle\int_0^x F_k(x - y)\, dF(y)$ for $k \geq 1$.*

**Proof.** Clearly $F_1 = F$. Also $T_{k+1} = T_k + X_{k+1}$, and Theorem (4.8.1) gives the result when suitably rewritten for independent variables of general type. $\qquad \blacksquare$

**(5) Lemma.** *We have that $\mathbb{P}(N(t) = k) = F_k(t) - F_{k+1}(t)$.*

**Proof.** $\{N(t) = k\} = \{N(t) \geq k\} \setminus \{N(t) \geq k + 1\}$. Now use (3). $\qquad \blacksquare$

We shall be interested largely in the expected value of $N(t)$.

**(6) Definition.** The **renewal function** $m$ is given by $m(t) = \mathbb{E}(N(t))$.

Again, it is easy to find $m$ in terms of the $F_k$.

**(7) Lemma.** *We have that $m(t) = \displaystyle\sum_{k=1}^{\infty} F_k(t)$.*

---

†Readers of Section 5.6 may notice that the statement of this lemma violates our notation for the domain of an integral. We adopt the convention that expressions of the form $\int_a^b g(y)\, dF(y)$ denote integrals over the half-open interval $(a, b]$, with the left endpoint excluded.

**Proof.** Define the indicator variables

$$I_k = \begin{cases} 1 & \text{if } T_k \leq t, \\ 0 & \text{otherwise.} \end{cases}$$

Then $N(t) = \sum_{k=1}^{\infty} I_k$ and so

$$m(t) = \mathbb{E}\left(\sum_{k=1}^{\infty} I_k\right) = \sum_{k=1}^{\infty} \mathbb{E}(I_k) = \sum_{k=1}^{\infty} F_k(t). \qquad \blacksquare$$

An alternative approach to the renewal function is by way of conditional expectations and the 'renewal equation'. First note that $m$ is the solution of a certain integral equation.

**(8) Lemma.** *The renewal function $m$ satisfies the* renewal equation,

(9) $$m(t) = F(t) + \int_0^t m(t-x)\,dF(x).$$

**Proof.** Use conditional expectation to obtain

$$m(t) = \mathbb{E}(N(t)) = \mathbb{E}(\mathbb{E}[N(t) \mid X_1]).$$

On the one hand,
$$\mathbb{E}(N(t) \mid X_1 = x) = 0 \quad \text{if} \quad t < x$$

since the first arrival occurs after time $t$, but on the other hand,

$$\mathbb{E}(N(t) \mid X_1 = x) = 1 + \mathbb{E}(N(t-x)) \quad \text{if} \quad t \geq x$$

since the process of arrivals, starting from the epoch of the first arrival, is a copy of $N$ itself. Therefore,

$$m(t) = \int_0^\infty \mathbb{E}(N(t) \mid X_1 = x)\,dF(x) = \int_0^t [1 + m(t-x)]\,dF(x). \qquad \blacksquare$$

We know from (7) that

$$m(t) = \sum_{k=1}^{\infty} F_k(t)$$

is a solution to the renewal equation (9). Actually, it is the unique solution to (9) which is bounded on finite intervals. This is a consequence of the next lemma. We shall encounter a more general form of (9) later, and it is appropriate to anticipate this now. The more general case involves solutions $\mu$ to the *renewal-type equation*

(10) $$\mu(t) = H(t) + \int_0^t \mu(t-x)\,dF(x), \quad t \geq 0,$$

for a suitable function $H$.

**(11) Theorem.** *The function $\mu$, given by*

$$\mu(t) = H(t) + \int_0^t H(t - x)\, dm(x),$$

*is a solution of the renewal-type equation* (10). *If $H$ is bounded on finite intervals then $\mu$ is bounded on finite intervals and is the unique solution of* (10) *with this property†.*

We shall make repeated use of this result, the proof of which is simple.

**Proof.** If $h : [0, \infty) \to \mathbb{R}$, define the functions $h * m$ and $h * F$ by

$$(h * m)(t) = \int_0^t h(t - x)\, dm(x), \quad (h * F)(t) = \int_0^t h(t - x)\, dF(x),$$

whenever these integrals exist. The operation $*$ is a type of convolution; do not confuse it with the related but different convolution operator of Sections 3.8 and 4.8. It can be shown that

$$(h * m) * F = h * (m * F),$$

and so we write $h * m * F$ for this double convolution. Note also that:

**(12)**                                $m = F + m * F$    by (9),

**(13)**                           $F_{k+1} = F_k * F = F * F_k$    by (4).

Using this notation, $\mu$ can be written as $\mu = H + H * m$. Convolve with $F$ and use (12) to find that

$$\mu * F = H * F + H * m * F = H * F + H * (m - F)$$
$$= H * m = \mu - H,$$

and so $\mu$ satisfies (10).

If $H$ is bounded on finite intervals then

$$\sup_{0 \le t \le T} |\mu(t)| \le \sup_{0 \le t \le T} |H(t)| + \sup_{0 \le t \le T} \left| \int_0^t H(t - x)\, dm(x) \right|$$
$$\le [1 + m(T)] \sup_{0 \le t \le T} |H(t)| < \infty,$$

and so $\mu$ is indeed bounded on finite intervals; we have used the finiteness of $m$ here (see Problem (10.6.1b)). To show that $\mu$ is the unique such solution of (10), suppose that $\mu_1$ is another bounded solution and write $\delta(t) = \mu(t) - \mu_1(t)$; $\delta$ is a bounded function. Also $\delta = \delta * F$ by (10). Iterate this equation and use (13) to find that $\delta = \delta * F_k$ for all $k \ge 1$, which implies that

$$|\delta(t)| \le F_k(t) \sup_{0 \le u \le t} |\delta(u)| \quad \text{for all } k \ge 1.$$

---

†Think of the integral in (11) as $\int H(t - x) m'(x)\, dx$ if you are unhappy about its present form.

Let $k \to \infty$ to find that $|\delta(t)| = 0$ for all $t$, since

$$F_k(t) = \mathbb{P}\big(N(t) \geq k\big) \to 0 \quad \text{as} \quad k \to \infty$$

by (2). The proof is complete. $\blacksquare$

The method of Laplace–Stieltjes transforms is often useful in renewal theory (see Definition (16) of Appendix I). For example, we can transform (10) to obtain the formula

$$\mu^*(\theta) = \frac{H^*(\theta)}{1 - F^*(\theta)} \quad \text{for} \quad \theta \neq 0,$$

an equation which links the Laplace–Stieltjes transforms of $\mu$, $H$, and $F$. In particular, setting $H = F$, we find from (8) that

(14) $$m^*(\theta) = \frac{F^*(\theta)}{1 - F^*(\theta)},$$

a formula which is directly derivable from (7) and (13). Hence there is a one–one correspondence between renewal functions $m$ and distribution functions $F$ of the interarrival times.

**(15) Example. Poisson process.** This is the only Markovian renewal process, and has exponentially distributed interarrival times with some parameter $\lambda$. The epoch $T_k$ of the $k$th arrival is distributed as $\Gamma(\lambda, k)$; Lemma (7) gives that

$$m(t) = \sum_{k=1}^{\infty} \int_0^t \frac{\lambda(\lambda s)^{k-1} e^{-\lambda s}}{(k-1)!} \, ds = \int_0^t \lambda \, ds = \lambda t.$$

Alternatively, just remember that $N(t)$ has the Poisson distribution with parameter $\lambda t$ to obtain the same result. $\bullet$

## Exercises for Section 10.1

**1.** Prove that $\mathbb{E}(e^{\theta N(t)}) < \infty$ for some strictly positive $\theta$ whenever $\mathbb{E}(X_1) > 0$. [Hint: Consider the renewal process with interarrival times $X_k' = \epsilon I_{\{X_k \geq \epsilon\}}$ for some suitable $\epsilon$.]

**2.** Let $N$ be a renewal process and let $W$ be the waiting time until the length of some interarrival time has exceeded $s$. That is, $W = \inf\{t : C(t) > s\}$, where $C(t)$ is the time which has elapsed (at time $t$) since the last arrival. Show that

$$F_W(x) = \begin{cases} 0 & \text{if } x < s, \\ 1 - F(s) + \int_0^s F_W(x - u) \, dF(u) & \text{if } x \geq s, \end{cases}$$

where $F$ is the distribution function of an interarrival time. If $N$ is a Poisson process with intensity $\lambda$, show that

$$\mathbb{E}(e^{\theta W}) = \frac{\lambda - \theta}{\lambda - \theta e^{(\lambda - \theta)s}} \quad \text{for } \theta < \lambda,$$

and $\mathbb{E}(W) = (e^{\lambda s} - 1)/\lambda$. You may find it useful to rewrite the above integral equation in the form of a renewal-type equation.

**3.** Find an expression for the mass function of $N(t)$ in a renewal process whose interarrival times are: (a) Poisson distributed with parameter $\lambda$, (b) gamma distributed, $\Gamma(\lambda, b)$.

**4.** Let the times between the events of a renewal process $N$ be uniformly distributed on $(0, 1)$. Find the mean and variance of $N(t)$ for $0 \le t \le 1$.

**5.** Let $N$ be the renewal process with interarrival times $X_1, X_2, \ldots$ . Show that, for $t > 0$, the interarrival time $X_{N(t)+1}$ is stochastically larger than $X_1$.

**6.** Suppose the interarrival times $X_i$ of a renewal process have a density function $f$ with ordinary Laplace transform $\widehat{f}(\theta) = \int_0^\infty e^{-\theta x} f(x)\, dx$. Show that the renewal function $m$ has Laplace transform

$$\widehat{m}(\theta) = \frac{\widehat{f}(\theta)}{\theta - \theta \widehat{f}(\theta)}, \qquad \theta > 0.$$

**7.** Let $r(y)$ be the ruin function of the insurance problem (8.10.7). In the notation of that problem, show that the Laplace–Stieltjes transforms $r^*(\theta)$ and $F^*(\theta)$ are related by $\lambda F^* = r^*(\theta - \lambda + \lambda F^*)$.

## 10.2  Limit theorems

We study next the asymptotic behaviour of $N(t)$ and its renewal function $m(t)$ for large values of $t$. There are four main results here, two for each of $N$ and $m$. For the renewal process $N$ itself there is a law of large numbers and a central limit theorem; these rely upon the relation (10.1.3), which links $N$ to the partial sums of independent variables. The two results for $m$ deal also with first- and second-order properties. The first asserts that $m(t)$ is approximately linear in $t$; the second asserts that the gradient of $m$ is asymptotically constant. The proofs are given later in the section.

How does $N(t)$ behave when $t$ is large? Let $\mu = \mathbb{E}(X_1)$ be the mean of a typical interarrival time. Henceforth we assume that $\mu < \infty$.

**(1) Theorem.**
$$\frac{1}{t} N(t) \xrightarrow{\text{a.s.}} \frac{1}{\mu} \quad as\ t \to \infty.$$

**(2) Theorem.** *If $\sigma^2 = \mathrm{var}(X_1)$ satisfies $0 < \sigma < \infty$, then*

$$\frac{N(t) - t/\mu}{\sqrt{t\sigma^2/\mu^3}} \xrightarrow{\text{D}} N(0, 1) \quad as \quad t \to \infty.$$

It is not quite so easy to find the asymptotic behaviour of the renewal function.

**(3) Elementary renewal theorem.**  $\dfrac{1}{t} m(t) \to \dfrac{1}{\mu}$  *as $t \to \infty$.*

The second-order properties of $m$ are hard to find, and we require a preliminary definition.

**(4) Definition.** Call a random variable $X$ and its distribution $F_X$ **arithmetic with span $\lambda$** ($> 0$) if $X$ takes values in the set $\{m\lambda : m = 0, \pm 1, \ldots\}$ with probability 1, and $\lambda$ is maximal with this property.

If the interarrival times of $N$ are arithmetic, with span $\lambda$ say, then so is $T_k$ for each $k$. In this case $m(t)$ may be discontinuous at values of $t$ which are multiples of $\lambda$, and this affects the second-order properties of $m$.

**(5) Blackwell's renewal theorem.** *If $X_1$ is not arithmetic then*

**(6)**                    $$m(t+h) - m(t) \to \frac{h}{\mu} \quad as \quad t \to \infty \quad for\ all\ h.$$

*If $X_1$ is arithmetic with span $\lambda$, then (6) holds whenever $h$ is a multiple of $\lambda$.*

It is appropriate to make some remarks about these theorems before we set to their proofs. Theorems (1) and (2) are straightforward, and use the law of large numbers and the central limit theorem for partial sums of independent sequences. It is perhaps surprising that (3) is harder to demonstrate than (1) since it concerns only the mean value of $N(t)$; it has a suitably probabilistic proof which uses the method of truncation, a technique which proved useful in the proof of the strong law (7.5.1). On the other hand, the proof of (5) is difficult. The usual method of proof is largely an exercise in solving integral equations, and is not appropriate for inclusion here (see Feller 1971, p. 360). There is an alternative proof which is short, beautiful, and probabilistic, and uses 'coupling' arguments related to those in the proof of the ergodic theorem for discrete-time Markov chains. This method requires some results which appear later in this chapter, and so we defer a sketch of the argument until Example (10.4.21). In the case of arithmetic interarrival times, (5) is essentially the same as Theorem (5.2.24), a result about *integer-valued* random variables. There is an apparently more general form of (5) which is deducible from (5). It is called the 'key renewal theorem' because of its many applications.

In the rest of this chapter we shall commonly assume that the interarrival times are *not* arithmetic. Similar results often hold in the arithmetic case, but they are usually more complicated to state.

**(7) Key renewal theorem.** *If $g : [0, \infty) \to [0, \infty)$ is such that*:
  (a) $g(t) \geq 0$ *for all t,*
  (b) $\int_0^\infty g(t)\, dt < \infty$,
  (c) *g is a non-increasing function,*
*then*

$$\int_0^t g(t-x)\, dm(x) \to \frac{1}{\mu} \int_0^\infty g(x)\, dx \quad as \quad t \to \infty$$

*whenever $X_1$ is not arithmetic.*

In order to deduce this theorem from the renewal theorem (5), first prove it for indicator functions of intervals, then for step functions, and finally for limits of increasing sequences of step functions. We omit the details.

**Proof of (1).** This is easy. Just note that

**(8)**                         $$T_{N(t)} \leq t < T_{N(t)+1} \quad for\ all\ t.$$

Therefore, if $N(t) > 0$,

$$\frac{T_{N(t)}}{N(t)} \leq \frac{t}{N(t)} < \frac{T_{N(t)+1}}{N(t)+1}\left(1 + \frac{1}{N(t)}\right).$$

As $t \to \infty$, $N(t) \xrightarrow{a.s.} \infty$, and the strong law of large numbers gives

$$\mu \leq \lim_{t\to\infty}\left(\frac{t}{N(t)}\right) \leq \mu \quad almost\ surely. \qquad \blacksquare$$

**Proof of (2).** This is Problem (10.6.3).                                                       ∎

In preparation for the proof of (3), we recall an important definition. Let $M$ be a random variable taking values in the set $\{1, 2, \ldots\}$. We call the random variable $M$ a *stopping time* with respect to the sequence $X_i$ of interarrival times if, for all $m \geq 1$, the event $\{M \leq m\}$ belongs to the $\sigma$-field of events generated by $X_1, X_2, \ldots, X_m$. Note that $M = N(t) + 1$ is a stopping time for the $X_i$, since

$$\{M \leq m\} = \{N(t) \leq m - 1\} = \left\{ \sum_{i=1}^{m} X_i > t \right\},$$

which is an event defined in terms of $X_1, X_2, \ldots, X_m$. The random variable $N(t)$ is *not* a stopping time.

**(9) Lemma. Wald's equation.** *Let $X_1, X_2, \ldots$ be independent identically distributed random variables with finite mean, and let $M$ be a stopping time with respect to the $X_i$ satisfying $\mathbb{E}(M) < \infty$. Then*

$$\mathbb{E}\left( \sum_{i=1}^{M} X_i \right) = \mathbb{E}(X_1)\mathbb{E}(M).$$

Applying Wald's equation to the sequence of interarrival times together with the stopping time $M = N(t) + 1$, we obtain

**(10)**                                $\mathbb{E}(T_{N(t)+1}) = \mu[m(t) + 1].$

Wald's equation may seem trite, but this is far from being the case. For example, it is not generally true that $\mathbb{E}(T_{N(t)}) = \mu m(t)$; the forthcoming Example (10.3.2) is an example of some of the dangers here.

**Proof of Wald's equation (9).** The basic calculation is elementary. Just note that

$$\sum_{i=1}^{M} X_i = \sum_{i=1}^{\infty} X_i I_{\{M \geq i\}},$$

so that (using dominated convergence or Exercise (5.6.2))

$$\mathbb{E}\left( \sum_{i=1}^{M} X_i \right) = \sum_{i=1}^{\infty} \mathbb{E}(X_i I_{\{M \geq i\}}) = \sum_{i=1}^{\infty} \mathbb{E}(X_i)\mathbb{P}(M \geq i) \qquad \text{by independence,}$$

since $\{M \geq i\} = \{M \leq i - 1\}^c$, an event definable in terms of $X_1, X_2, \ldots, X_{i-1}$ and therefore independent of $X_i$. The final sum equals

$$\mathbb{E}(X_1) \sum_{i=1}^{\infty} \mathbb{P}(M \geq i) = \mathbb{E}(X_1)\mathbb{E}(M). \qquad \blacksquare$$

**Proof of (3).** Half of this is easy. We have from (8) that $t < T_{N(t)+1}$; take expectations of this and use (10) to obtain

$$\frac{m(t)}{t} > \frac{1}{\mu} - \frac{1}{t}.$$

Letting $t \to \infty$, we obtain

(11)
$$\liminf_{t \to \infty} \frac{1}{t} m(t) \geq \frac{1}{\mu}.$$

We may be tempted to proceed as follows in order to bound $m(t)$ above. We have from (8) that $T_{N(t)} \leq t$, and so

(12)    $t \geq \mathbb{E}(T_{N(t)}) = \mathbb{E}(T_{N(t)+1} - X_{N(t)+1}) = \mu[m(t) + 1] - \mathbb{E}(X_{N(t)+1}).$

The problem is that $X_{N(t)+1}$ depends on $N(t)$, and so $\mathbb{E}(X_{N(t)+1}) \neq \mu$ in general. To cope with this, truncate the $X_i$ at some $a > 0$ to obtain a new sequence

$$X_j^a = \begin{cases} X_j & \text{if } X_j < a, \\ a & \text{if } X_j \geq a. \end{cases}$$

Now consider the renewal process $N^a$ with associated interarrival times $\{X_j^a\}$. Apply (12) to $N^a$, noting that $\mu^a = \mathbb{E}(X_j^a) \leq a$, to obtain

(13)    $t \geq \mu^a[\mathbb{E}(N^a(t)) + 1] - a.$

However, $X_j^a \leq X_j$ for all $j$, and so $N^a(t) \geq N(t)$ for all $t$. Therefore

$$\mathbb{E}(N^a(t)) \geq \mathbb{E}(N(t)) = m(t)$$

and (13) implies

$$\frac{m(t)}{t} \leq \frac{1}{\mu^a} + \frac{a - \mu^a}{\mu^a t}.$$

Let $t \to \infty$ to obtain

$$\limsup_{t \to \infty} \frac{1}{t} m(t) \leq \frac{1}{\mu^a};$$

now let $a \to \infty$ and use monotone convergence (5.6.12) to find that $\mu^a \to \mu$, and therefore

$$\limsup_{t \to \infty} \frac{1}{t} m(t) \leq \frac{1}{\mu}.$$

Combine this with (11) to obtain the result. ∎

---

## Exercises for Section 10.2

**1.** Planes land at Heathrow airport at the times of a renewal process with interarrival time distribution function $F$. Each plane contains a random number of people with a given common distribution and finite mean. Assuming as much independence as usual, find an expression for the rate of arrival of passengers over a long time period.

**2.** Let $Z_1, Z_2, \ldots$ be independent identically distributed random variables with mean 0 and finite variance $\sigma^2$, and let $T_n = \sum_{i=1}^n Z_i$. Let $M$ be a finite stopping time with respect to the $Z_i$ such that $\mathbb{E}(M) < \infty$. Show that $\text{var}(T_M) = \mathbb{E}(M)\sigma^2$.

**3.** Show that $\mathbb{E}(T_{N(t)+k}) = \mu(m(t)+k)$ for all $k \geq 1$, but that it is not generally true that $\mathbb{E}(T_{N(t)}) = \mu m(t)$.

**4.** Show that, using the usual notation, the family $\{N(t)/t : 0 \leq t < \infty\}$ is uniformly integrable. How might one make use of this observation?

**5.** Consider a renewal process $N$ having interarrival times with moment generating function $M$, and let $T$ be a positive random variable which is independent of $N$. Find $\mathbb{E}(s^{N(T)})$ when:
(a) $T$ is exponentially distributed with parameter $\nu$,
(b) $N$ is a Poisson process with intensity $\lambda$, in terms of the moment generating function of $T$. What is the distribution of $N(T)$ in this case, if $T$ has the gamma distribution $\Gamma(\nu, b)$?

## 10.3 Excess life

Suppose that we begin to observe a renewal process $N$ at some epoch $t$ of time. A certain number $N(t)$ of arrivals have occurred by then, and the next arrival will be that numbered $N(t) + 1$. That is to say, we have begun our observation at a point in the random interval $I_t = [T_{N(t)}, T_{N(t)+1})$, the endpoints of which are arrival times.

**(1) Definition.**
(a) The **excess lifetime** at $t$ is $E(t) = T_{N(t)+1} - t$.
(b) The **current lifetime** (or **age**) at $t$ is $C(t) = t - T_{N(t)}$.
(c) The **total lifetime** at $t$ is $D(t) = E(t) + C(t) = X_{N(t)+1}$.

That is, $E(t)$ is the time which elapses before the next arrival, $C(t)$ is the elapsed time since the last arrival (with the convention that the zeroth arrival occurs at time 0), and $D(t)$ is the length of the interarrival time which contains $t$ (see Figure 8.1 for a diagram of these random variables).

**(2) Example. Waiting time paradox.** Suppose that $N$ is a Poisson process with parameter $\lambda$. How large is $\mathbb{E}(E(t))$? Consider the two following lines of reasoning.
(A) $N$ is a Markov chain, and so the distribution of $E(t)$ does not depend on the arrivals prior to time $t$. Thus $E(t)$ has the same mean as $E(0) = X_1$, and so $\mathbb{E}(E(t)) = \lambda^{-1}$.
(B) If $t$ is fairly large, then on average it lies near the midpoint of the interarrival interval $I_t$ which contains it. That is

$$\mathbb{E}(E(t)) \simeq \frac{1}{2}\mathbb{E}(T_{N(t)+1} - T_{N(t)}) = \frac{1}{2}\mathbb{E}(X_{N(t)+1}) = \frac{1}{2\lambda}.$$

These arguments cannot both be correct. The reasoning of (B) is false, in that $X_{N(t)+1}$ does *not* have mean $\lambda^{-1}$; we have already observed this after (10.2.12). In fact, $X_{N(t)+1}$ is a very special interarrival time; longer intervals have a higher chance of catching $t$ in their interiors than small intervals. In Problem (10.6.5) we shall see that $\mathbb{E}(X_{N(t)+1}) = (2 - e^{-\lambda t})/\lambda$. For this process, $E(t)$ and $C(t)$ are independent for any $t$; this property holds for no other renewal process with non-arithmetic interarrival times.                                                                  ●

Now we find the distribution of the excess lifetime $E(t)$ for a general renewal process.

**(3) Theorem.** *The distribution function of the excess life $E(t)$ is given by*

$$\mathbb{P}(E(t) \le y) = F(t + y) - \int_0^t [1 - F(t + y - x)]\, dm(x).$$

**Proof.** Condition on $X_1$ in the usual way to obtain

$$\mathbb{P}(E(t) > y) = \mathbb{E}[\mathbb{P}(E(t) > y \mid X_1)].$$

However, you will see after a little thought that

$$\mathbb{P}(E(t) > y \mid X_1 = x) = \begin{cases} \mathbb{P}(E(t - x) > y) & \text{if } x \le t, \\ 0 & \text{if } t < x \le t + y, \\ 1 & \text{if } x > t + y, \end{cases}$$

since $E(t) > y$ if and only if no arrivals occur in $(t, t + y]$. Thus

$$\mathbb{P}\big(E(t) > y\big) = \int_0^\infty \mathbb{P}\big(E(t) > y \mid X_1 = x\big)\,dF(x)$$

$$= \int_0^t \mathbb{P}\big(E(t - x) > y\big)\,dF(x) + \int_{t+y}^\infty dF(x).$$

So, for fixed $y \geq 0$, the function $\mu(t) = \mathbb{P}(E(t) > y)$ satisfies (10.1.10) with $H(t) = 1 - F(t + y)$; use Theorem (10.1.11) to see that

$$\mu(t) = 1 - F(t + y) + \int_0^t [1 - F(t + y - x)]\,dm(x)$$

as required.                                                                    ∎

**(4) Corollary.** *The distribution of the current life $C(t)$ is given by*

$$\mathbb{P}\big(C(t) \geq y\big) = \begin{cases} 0 & \text{if } y > t, \\ 1 - F(t) + \displaystyle\int_0^{t-y} [1 - F(t - x)]\,dm(x) & \text{if } y \leq t. \end{cases}$$

**Proof.** It is the case that $C(t) \geq y$ if and only if there are no arrivals in $(t - y, t]$. Thus

$$\mathbb{P}\big(C(t) \geq y\big) = \mathbb{P}\big(E(t - y) > y\big) \quad \text{if} \quad y \leq t$$

and the result follows from (3).                                              ∎

   Might the renewal process $N$ have stationary increments, in the sense that the distribution of $N(t + s) - N(t)$ depends on $s$ alone when $s \geq 0$? This is true for the Poisson process but fails in general. The reason is simple: generally speaking, the process of arrivals after time $t$ depends on the age $t$ of the process to date. When $t$ is very large, however, it is plausible that the process may forget the date of its inception, thereby settling down into a stationary existence. Thus turns out to be the case. To show this asymptotic stationarity we need to demonstrate that the distribution of $N(t + s) - N(t)$ converges as $t \to \infty$. It is not difficult to see that this is equivalent to the assertion that the distribution of the excess life $E(t)$ settles down as $t \to \infty$, an easy consequence of the key renewal theorem (10.2.7) and Lemma (4.3.4).
   For simplicity of notation, we write $X = X_1$.

**(5) Theorem.** *If $X$ is not arithmetic and $\mu = \mathbb{E}(X) < \infty$ then*

$$\mathbb{P}\big(E(t) \leq y\big) \to \frac{1}{\mu} \int_0^y [1 - F(x)]\,dx \quad \text{as} \quad t \to \infty.$$

   The situation is slightly different if $X$ is arithmetic. For example, if $X$ is concentrated at the value 1 then, as $n \to \infty$,

$$\mathbb{P}\big(E(n + c) \leq \tfrac{1}{2}\big) \to \begin{cases} 1 & \text{if } c = \tfrac{1}{2}, \\ 0 & \text{if } c = \tfrac{1}{4}. \end{cases}$$

The arithmetic case may be treated using the theory of Markov chains in discrete time.

**(6) Theorem.** *Let $X$ take values in the non-negative integers with span $1$, and assume $\mu =$ $\mathbb{E}(X) < \infty$. Then, for $k \in \{1, 2, \dots\}$,*

$$\mathbb{P}(E(n) = k) \to \frac{1}{\mu}\mathbb{P}(X \geq k) \quad \text{as } n \to \infty.$$

**Proof.** The process $E = \{E(n) : n \geq 0\}$ is a Markov chain with transition matrix $\mathbf{P} = (p_{ij})$ given by

$$p_{k,k-1} = 1 \quad \text{for} \quad k \geq 2, \qquad p_{1,j} = \mathbb{P}(X = j) \quad \text{for} \quad j \geq 1.$$

Any stationary distribution $\boldsymbol{\pi} = (\pi_k : k \geq 1)$ satisfies $\boldsymbol{\pi} = \boldsymbol{\pi}\mathbf{P}$. This equation is satisfied by the unique mass function (*exercise*)

$$(7) \qquad\qquad\qquad \pi_k = \frac{1}{\mu}\mathbb{P}(X \geq k), \qquad k \geq 1.$$

The chain $E$ has state space

$$S = \begin{cases} \{1, 2, \dots, M\} & \text{if } M < \infty, \\ \{1, 2, \dots\} & \text{if } M = \infty, \end{cases}$$

where $M = \sup\{m : \mathbb{P}(X = m) > 0\}$. It is irreducible on $S$. By (7) and Theorem (6.4.3), the chain is positive recurrent on $S$. Since $X$ has span 1, the chain is aperiodic. The claim now follows by the Markov chain limit theorem, Theorem (6.4.20). ∎

---

### Exercises for Section 10.3

**1.** Suppose that the distribution of the excess lifetime $E(t)$ does not depend on $t$. Show that the renewal process is a Poisson process.

**2.** Show that the current and excess lifetime processes, $C(t)$ and $E(t)$, are Markov processes.

**3.** Suppose that $X_1$ is non-arithmetic with finite mean $\mu$.
(a) Show that $E(t)$ converges in distribution as $t \to \infty$, the limit distribution function being

$$H(x) = \int_0^x \frac{1}{\mu}[1 - F(y)]\,dy.$$

(b) Show that the $r$th moment of this limit distribution is given by

$$\int_0^\infty x^r \, dH(x) = \frac{\mathbb{E}(X_1^{r+1})}{\mu(r+1)},$$

assuming that this is finite.
(c) Show that

$$\mathbb{E}(E(t)^r) = \mathbb{E}\big(\{(X_1 - t)^+\}^r\big) + \int_0^t h(t - x)\,dm(x)$$

for some suitable function $h$ to be found, and deduce by the key renewal theorem that $\mathbb{E}(E(t)^r) \to \mathbb{E}(X_1^{r+1})/\{\mu(r+1)\}$ as $t \to \infty$, assuming this limit is finite.

**4.** Find an expression for the mean value of the excess lifetime $E(t)$ conditional on the event that the current lifetime $C(t)$ equals $x$.

**5.** Let $M(t) = N(t) + 1$, and suppose that $X_1$ has finite non-zero variance $\sigma^2$.
(a) Show that $\text{var}(T_{M(t)} - \mu M(t)) = \sigma^2(m(t) + 1)$.
(b) In the non-arithmetic case, show that $\text{var}(M(t))/t \to \sigma^2/\mu^3$ as $t \to \infty$.

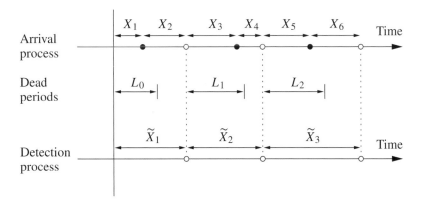

Figure 10.1. Arrivals and detections by a Type I counter; • indicates an undetected arrival, and ○ indicates a detected arrival.

## 10.4 Applications

Here are some examples of the ways in which renewal theory can be applied.

**(1) Example. Counters, and their dead periods.** In Section 6.8 we used an idealized Geiger counter which was able to register radioactive particles, irrespective of the rate of their arrival. In practice, after the detection of a particle such counters require a certain interval of time in order to complete its registration. These intervals are called 'dead periods'; during its dead periods the counter is locked and fails to register arriving particles. There are two common types of counter.

**Type 1.** Each detected arrival locks the counter for a period of time, possibly of random length, during which it ignores all arrivals.

**Type 2.** Each arrival locks the counter for a period of time, possibly of random length, irrespective of whether the counter is already locked or not. The counter registers only those arrivals that occur whilst it is unlocked.

Genuine Geiger counters are of Type 1; this case might also be used to model the process in Example (8.3.1) describing the replacement of light bulbs in rented property when the landlord is either mean or lazy. We consider Type 1 counters briefly; Type 2 counters are harder to analyse, and so are left to the reader.

Suppose that arrivals occur as a renewal process $N$ with renewal function $m$ and interarrival times $X_1, X_2, \ldots$ having distribution function $F$. Let $L_n$ be the length of the dead period induced by the $n$th detected arrival. It is customary and convenient to suppose that an additional dead period, of length $L_0$, begins at time $t = 0$; the reason for this will soon be clear. We suppose that $\{L_n\}$ is a family of independent random variables with the common distribution function $F_L$, where $F_L(0) = 0$. Let $\widetilde{N}(t)$ be the number of arrivals detected by the Type 1 counter by time $t$. Then $\widetilde{N}$ is a stochastic process with interarrival times $\widetilde{X}_1, \widetilde{X}_2, \ldots$ where $\widetilde{X}_{n+1} = L_n + E_n$ and $E_n$ is the excess life of $N$ at the end of the $n$th dead period (see Figure 10.1). The process $\widetilde{N}$ is *not* in general a renewal process, because the $\widetilde{X}_i$ need be neither independent nor identically distributed. In the very special case when $N$ is a Poisson process, the $E_n$ are independent exponential variables and $\widetilde{N}$ is a renewal process; it is easy to construct other examples for which this conclusion fails.

It is not difficult to find the elapsed time $\tilde{X}_1$ until the first detection. Condition on $L_0$ to obtain that

$$\mathbb{P}(\tilde{X}_1 \leq x) = \mathbb{E}\big(\mathbb{P}(\tilde{X}_1 \leq x \mid L_0)\big) = \int_0^x \mathbb{P}(L_0 + E_0 \leq x \mid L_0 = l)\,dF_L(l).$$

However, $E_0 = E(L_0)$, the excess lifetime of $N$ at $L_0$, and so

(2) $$\mathbb{P}(\tilde{X}_1 \leq x) = \int_0^x \mathbb{P}\big(E(l) \leq x - l\big)\,dF_L(l).$$

Now use Theorem (10.3.3) and the integral representation

$$m(t) = F(t) + \int_0^t F(t-x)\,dm(x),$$

which follows from Theorem (10.1.11), to find that

(3) $$\mathbb{P}(\tilde{X}_1 \leq x) = \int_0^x \left(\int_l^x [1 - F(x-y)]\,dm(y)\right) dF_L(l)$$
$$= \int_0^x [1 - F(x-y)]F_L(y)\,dm(y).$$

If $N$ is a Poisson process with intensity $\lambda$, equation (2) becomes

$$\mathbb{P}(\tilde{X}_1 \leq x) = \int_0^x (1 - e^{-\lambda(x-l)})\,dF_L(l).$$

$\tilde{N}$ is now a renewal process, and this equation describes the common distribution of the interarrival times.

If the counter is registering the arrival of radioactive particles, then we may seek an estimate $\lambda$ of the unknown emission rate $\lambda$ of the source based upon our knowledge of the mean length $\mathbb{E}(L)$ of a dead period and the counter reading $\tilde{N}(t)$. Assume that the particles arrive in the manner of a Poisson process, and let $\gamma_t = \tilde{N}(t)/t$ be the density of observed particles. Then

$$\gamma_t \simeq \frac{1}{\mathbb{E}(\tilde{X}_1)} = \frac{1}{\mathbb{E}(L) + \lambda^{-1}} \quad \text{for large } t,$$

and so $\lambda \simeq \hat{\lambda}$ where

$$\hat{\lambda} = \frac{\gamma_t}{1 - \gamma_t \mathbb{E}(L)}. \qquad \bullet$$

**(4) Example. Alternating renewal process.** A machine breaks down repeatedly. After the $n$th breakdown the repairman takes a period of time, length $Y_n$, to repair it; subsequently the machine runs for a period of length $Z_n$ before it breaks down for the next time. We assume that the $Y_m$ and the $Z_n$ are independent of each other, the $Y_m$ having common distribution function $F_Y$ and the $Z_n$ having common distribution function $F_Z$. Suppose that the machine was installed at time $t = 0$. Let $N(t)$ be the number of completed repairs by time $t$ (see

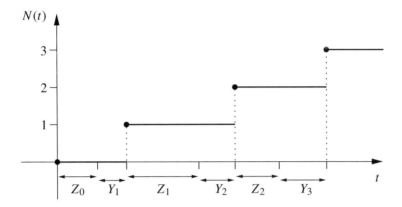

Figure 10.2. An alternating renewal process.

Figure 10.2). Then $N$ is a renewal process with interarrival times $X_1, X_2, \ldots$ given by $X_n = Z_{n-1} + Y_n$ and with distribution function

$$F(x) = \int_0^x F_Y(x - y) \, dF_Z(y).$$

Let $p(t)$ be the probability that the machine is working at time $t$.

**(5) Lemma.** *We have that*

$$p(t) = 1 - F_Z(t) + \int_0^t p(t - x) \, dF(x)$$

*and hence*

$$p(t) = 1 - F_Z(t) + \int_0^t [1 - F_Z(t - x)] \, dm(x)$$

*where $m$ is the renewal function of $N$.*

**Proof.** The probability that the machine is on at time $t$ satisfies

$$
\begin{aligned}
p(t) = \mathbb{P}(\text{on at } t) &= \mathbb{P}(Z_0 > t) + \mathbb{P}(\text{on at } t, \ Z_0 \le t) \\
&= \mathbb{P}(Z_0 > t) + \mathbb{E}\big[\mathbb{P}(\text{on at } t, \ Z_0 \le t \mid X_1)\big] \\
&= \mathbb{P}(Z_0 > t) + \int_0^t \mathbb{P}(\text{on at } t \mid X_1 = x) \, dF(x)
\end{aligned}
$$

$$\text{since } \mathbb{P}(\text{on at } t, \ Z_0 \le t \mid X_1 > t) = 0$$

$$= \mathbb{P}(Z_0 > t) + \int_0^t p(t - x) \, dF(x).$$

Now use Theorem (10.1.11).    ∎

**(6) Corollary.** *If $X_1$ is not arithmetic then $p(t) \to (1 + \rho)^{-1}$ as $t \to \infty$, where $\rho = \mathbb{E}(Y)/\mathbb{E}(Z)$ is the ratio of the mean lengths of a typical repair period and a typical working period.*

**Proof.** Use the key renewal theorem (10.2.7).                                    ■●

**(7) Example. Superposition of renewal processes.** Suppose that a room is illuminated by two lights, the bulbs of which fail independently of each other. On failure, they are replaced immediately. Let $N_1$ and $N_2$ be the renewal processes describing the occurrences of bulb failures in the first and second lights respectively, and suppose that these are independent processes with the same interarrival time distribution function $F$. Let $\widetilde{N}$ be the superposition of these two processes; that is, $\widetilde{N}(t) = N_1(t) + N_2(t)$ is the total number of failures by time $t$. In general $\widetilde{N}$ is not a renewal process. Let us assume for the sake of simplicity that the interarrival times of $N_1$ and $N_2$ are not arithmetic.

**(8) Theorem.** $\widetilde{N}$ *is a renewal process if and only if* $N_1$ *and* $N_2$ *are Poisson processes.*

**Proof.** It is easy to see that $\widetilde{N}$ is a Poisson process with intensity $2\lambda$ whenever $N_1$ and $N_2$ are Poisson processes with intensity $\lambda$. Conversely, suppose that $\widetilde{N}$ is a renewal process, and write $\{X_n(1)\}$, $\{X_n(2)\}$, and $\{\widetilde{X}_n\}$ for the interarrival times of $N_1$, $N_2$, and $\widetilde{N}$ respectively. Clearly $\widetilde{X}_1 = \min\{X_1(1), X_1(2)\}$, and so the distribution function $\widetilde{F}$ of $\widetilde{X}_1$ satisfies

(9) $$1 - \widetilde{F}(y) = [1 - F(y)]^2.$$

Let $E_1(t)$, $E_2(t)$, and $\widetilde{E}(t)$ denote the excess lifetimes of $N_1$, $N_2$, and $\widetilde{N}$ respectively at time $t$. Clearly, $\widetilde{E}(t) = \min\{E_1(t), E_2(t)\}$, and so

$$\mathbb{P}\big(\widetilde{E}(t) > y\big) = \mathbb{P}\big(E_1(t) > y\big)^2.$$

Let $t \to \infty$ and use Theorem (10.3.5) to obtain

(10) $$\frac{1}{\widetilde{\mu}} \int_y^\infty [1 - \widetilde{F}(x)]\, dx = \frac{1}{\mu^2} \left( \int_y^\infty [1 - F(x)]\, dx \right)^2$$

where $\widetilde{\mu} = \mathbb{E}(\widetilde{X}_1)$ and $\mu = \mathbb{E}(X_1(1))$. Differentiate (10) and use (9) to obtain

$$\frac{1}{\widetilde{\mu}}[1 - \widetilde{F}(y)] = \frac{2}{\mu^2}[1 - F(y)] \int_y^\infty [1 - F(x)]\, dx$$

$$= \frac{1}{\widetilde{\mu}}[1 - F(y)]^2$$

(this step needs further justification if $F$ is not continuous). Thus

$$1 - F(y) = \frac{2\widetilde{\mu}}{\mu^2} \int_y^\infty [1 - F(x)]\, dx$$

which is an integral equation with solution

$$F(y) = 1 - \exp\left( -\frac{2\widetilde{\mu}}{\mu^2} y \right).$$                        ■●

**(11) Example. Delayed renewal process.** The Markov chain of Example (8.3.2) indicates that it is sometimes appropriate to allow the first interarrival time $X_1$ to have a distribution which differs from the shared distribution of $X_2, X_3, \ldots$.

**(12) Definition.** Let $X_1, X_2, \ldots$ be independent positive variables such that $X_2, X_3, \ldots$ have the same distribution. Let

$$T_0 = 0, \quad T_n = \sum_1^n X_i, \quad N^{\mathrm{d}}(t) = \max\{n : T_n \leq t\}.$$

Then $N^{\mathrm{d}}$ is called a **delayed** (or **modified**) **renewal process**.

Another example of a delayed renewal process is provided by a variation of the Type 1 counter of (1) with particles arriving in the manner of a Poisson process. It was convenient there to assume that the life of the counter began with a dead period in order that the process $\widetilde{N}$ of detections be a renewal process. In the absence of this assumption $\widetilde{N}$ is a delayed renewal process. The theory of delayed renewal processes is very similar to that of ordinary renewal processes and we do not explore it in detail. The renewal equation (10.1.9) becomes

$$m^{\mathrm{d}}(t) = F^{\mathrm{d}}(t) + \int_0^t m(t - x) \, d F^{\mathrm{d}}(x)$$

where $F^{\mathrm{d}}$ is the distribution function of $X_1$ and $m$ is the renewal function of an ordinary renewal process $N$ whose interarrival times are $X_2, X_3, \ldots$. It is left to the reader to check that

**(13)** $$m^{\mathrm{d}}(t) = F^{\mathrm{d}}(t) + \int_0^t m^{\mathrm{d}}(t - x) \, d F(x)$$

and

**(14)** $$m^{\mathrm{d}}(t) = \sum_{k=1}^\infty F_k^{\mathrm{d}}(t)$$

where $F_k^{\mathrm{d}}$ is the distribution function of $T_k = X_1 + X_2 + \cdots + X_k$ and $F$ is the shared distribution function of $X_2, X_3, \ldots$.

With our knowledge of the properties of $m$, it is not too hard to show that $m^{\mathrm{d}}$ satisfies the renewal theorems. Write $\mu$ for $\mathbb{E}(X_2)$.

**(15) Theorem.** *We have that*:
(a) $\dfrac{1}{t} m^{\mathrm{d}}(t) \to \dfrac{1}{\mu}$ *as* $t \to \infty$.
(b) *If $X_2$ is not arithmetic then*

**(16)** $$m^{\mathrm{d}}(t + h) - m^{\mathrm{d}}(t) \to \frac{h}{\mu} \quad as \quad t \to \infty \quad for\ any\ h.$$

*If $X_2$ is arithmetic with span $\lambda$ then* (16) *remains true whenever $h$ is a multiple of $\lambda$.*

There is an important special case for the distribution function $F^{\mathrm{d}}$.

**(17) Theorem.** *The process $N^{\mathrm{d}}$ has stationary increments if and only if*

**(18)** $$F^{\mathrm{d}}(y) = \frac{1}{\mu} \int_0^y [1 - F(x)] \, dx.$$

If $F^d$ is given by (18), then $N^d$ is called a *stationary* (or *equilibrium*) *renewal process*. We should recognize (18) as the asymptotic distribution (10.3.5) of the excess lifetime of the ordinary renewal process $N$. So the result of (17) is no surprise since $N^d$ starts off with this 'equilibrium' distribution. We shall see that in this case $m^d(t) = t/\mu$ for all $t \geq 0$.

**Proof of (17).** Suppose that $N^d$ has stationary increments. Then

$$m^d(s+t) = \mathbb{E}\big([N^d(s+t) - N^d(s)] + N^d(s)\big)$$
$$= \mathbb{E}(N^d(t)) + \mathbb{E}(N^d(s))$$
$$= m^d(t) + m^d(s).$$

By monotonicity, $m^d(t) = ct$ for some $c > 0$. Substitute into (13) to obtain

$$F^d(t) = c \int_0^t [1 - F(x)]\, dx$$

and let $t \to \infty$ to obtain $c = 1/\mu$.

Conversely, suppose that $F^d$ is given by (18). Substitute (18) into (13) and use the method of Laplace–Stieltjes transforms to deduce that

**(19)**
$$m^d(t) = \frac{t}{\mu}.$$

Now, $N^d$ has stationary increments if and only if the distribution of $E^d(t)$, the excess lifetime of $N^d$ at $t$, does not depend on $t$. But

$$\mathbb{P}\big(E^d(t) > y\big) = \sum_{k=0}^{\infty} \mathbb{P}\big(E^d(t) > y,\ N^d(t) = k\big)$$
$$= \mathbb{P}\big(E^d(t) > y,\ N^d(t) = 0\big)$$
$$+ \sum_{k=1}^{\infty} \int_0^t \mathbb{P}\big(E^d(t) > y,\ N^d(t) = k \,\big|\, T_k = x\big)\, dF_k^d(x)$$
$$= 1 - F^d(t+y) + \int_0^t [1 - F(t+y-x)]\, d\left(\sum_{k=1}^{\infty} F_k^d(x)\right)$$
$$= 1 - F^d(t+y) + \int_0^t [1 - F(t+y-x)]\, dm^d(x)$$

from (14). Now substitute (18) and (19) into this equation to obtain the result. ∎

**(20) Example. Markov chains.** Let $Y = \{Y_n : n \geq 0\}$ be a discrete-time Markov chain with countable state space $S$. At last we are able to prove the ergodic theorem (6.4.21) for $Y$, as a consequence of the renewal theorem (16). Suppose that $Y_0 = i$ and let $j$ be an aperiodic state. We can suppose that $j$ is recurrent, since the result follows from Corollary (6.2.5) if $j$ is transient. Observe the sequence of visits of $Y$ to the state $j$. That is, let

$$T_0 = 0, \quad T_{n+1} = \min\{k > T_n : Y_k = j\} \quad \text{for} \quad n \geq 0.$$

$T_1$ may equal $+\infty$; actually $\mathbb{P}(T_1 < \infty) = f_{ij}$. Conditional on the event $\{T_1 < \infty\}$, the inter-visit times

$$X_n = T_n - T_{n-1} \quad \text{for} \quad n \geq 2$$

are independent and identically distributed; following Example (8.3.2), $N^{\mathrm{d}}(t) = \max\{n :$ $T_n \leq t\}$ defines a delayed renewal process with a renewal function $m^{\mathrm{d}}(t) = \sum_{n=1}^{t} p_{ij}(n)$ for integral $t$. Now, adapt (16) to deal with the possibility that the first interarrival time $X_1 = T_1$ equals infinity, to obtain

$$p_{ij}(n) = m^{\mathrm{d}}(n) - m^{\mathrm{d}}(n-1) \to \frac{f_{ij}}{\mu_j} \quad \text{as} \quad n \to \infty$$

where $\mu_j$ is the mean recurrence time of $j$.                                                        ●

**(21) Example. Sketch proof of the renewal theorem.** There is an elegant proof of the renewal theorem (10.2.5) which proceeds by coupling the renewal process $N$ to an independent delayed renewal process $N^{\mathrm{d}}$; here is a sketch of the method. Let $N$ be a renewal process with interarrival times $\{X_n\}$ and interarrival time distribution function $F$ with mean $\mu$. We suppose that $F$ is non-arithmetic; the proof in the arithmetic case is easier. Let $N^{\mathrm{d}}$ be a stationary renewal process (see (17)) with interarrival times $\{Y_n\}$, where $Y_1$ has distribution function

$$F^{\mathrm{d}}(y) = \frac{1}{\mu} \int_0^y [1 - F(x)] \, dx$$

and $Y_2, Y_3, \ldots$ have distribution function $F$; suppose further that the $X_i$ are independent of the $Y_i$. The idea of the proof is as follows.

(a) For any $\epsilon > 0$, there must exist an arrival time $T_a = \sum_i^a X_i$ of $N$ and an arrival time $T_b^{\mathrm{d}} = \sum_i^b Y_i$ of $N^{\mathrm{d}}$ such that $|T_a - T_b^{\mathrm{d}}| < \epsilon$.

(b) If we replace $X_{a+1}, X_{a+2}, \ldots$ by $Y_{b+1}, Y_{b+2}, \ldots$ in the construction of $N$, then the distributional properties of $N$ are unchanged since all these variables are identically distributed.

(c) But the $Y_i$ are the interarrival times of a stationary renewal process, for which (19) holds; this implies that $m^{\mathrm{d}}(t + h) - m^{\mathrm{d}}(t) = h/\mu$ for all $t, h$. However, $m(t)$ and $m^{\mathrm{d}}(t)$ are nearly the same for large $t$, by the previous remarks, and so $m(t+h) - m(t) \simeq h/\mu$ for large $t$.

The details of the proof are slightly too difficult for inclusion here (see Lindvall 1977).        ●

**(22) Example. Age-dependent branching process.** Consider the branching process $Z(t)$ of Section 5.5 in which each individual lives for a random length of time before splitting into its offspring. We have seen that the expected number $m(t) = \mathbb{E}(Z(t))$ of individuals alive at time $t$ satisfies the integral equation (5.5.4):

**(23)** $$m(t) = \nu \int_0^t m(t - x) \, dF_T(x) + \int_t^\infty dF_T(x)$$

where $F_T$ is the distribution function of a typical lifetime and $\nu$ is the mean number of offspring of an individual; we assume for simplicity that $F_T$ is continuous. We have changed some of the notation of (5.5.4) for obvious reasons. Equation (23) reminds us of the renewal-type equation (10.1.10) but the factor $\nu$ must be assimilated before the solution can be found using the method of Theorem (10.1.11). This presents few difficulties in the supercritical case. If $\nu > 1$, there exists a unique $\beta > 0$ such that

$$F_T^*(\beta) = \int_0^\infty e^{-\beta x}\, dF_T(x) = \frac{1}{\nu};$$

this holds because the Laplace–Stieltjes transform $F_T^*(\theta)$ is a strictly decreasing continuous function of $\theta$ with

$$F_T^*(0) = 1, \qquad F_T^*(\theta) \to 0 \quad \text{as} \quad \theta \to \infty.$$

Now, with this choice of $\beta$, define

$$\widetilde{F}(t) = \nu \int_0^t e^{-\beta x}\, dF_T(x), \qquad g(t) = e^{-\beta t} m(t).$$

Multiply through (23) by $e^{-\beta t}$ to obtain

**(24)** $$g(t) = h(t) + \int_0^t g(t-x)\, d\widetilde{F}(x)$$

where

$$h(t) = e^{-\beta t}[1 - F_T(t)];$$

(24) has the same general form as (10.1.10), since our choice for $\beta$ ensures that $\widetilde{F}$ is the distribution function of a positive random variable. The behaviour of $g(t)$ for large $t$ may be found by applying Theorem (10.1.11) and the key renewal theorem (10.2.7).    ●

---

### Exercises for Section 10.4

**1.** Find the distribution of the excess lifetime for a renewal process each of whose interarrival times is the sum of two independent exponentially distributed random variables having respective parameters $\lambda$ and $\mu$. Show that the excess lifetime has mean

$$\frac{1}{\mu} + \frac{\lambda e^{-(\lambda+\mu)t} + \mu}{\lambda(\lambda+\mu)}.$$

**2. Stationary renewal and size-biasing.** Let $f$ be the density function and $F$ the distribution function of the interarrival times $\{X_i : i \geq 2\}$ of a stationary renewal process, and $\mu$ their common mean. Let $C$, $D$, $E$ be the current, total, and excess lifetimes in equilibrium.
(a) Show that $C$ has density function $h(x) = (1 - F(x))/\mu$ for $x > 0$.
(b) Show that $D$ has the *size-biased* (or *length-biased*) density function $g(y) = (y/\mu)f(y)$.
(c) Let $U$ be uniformly distributed on $(0, 1)$ and independent of $D$. Show that $UD$ has the same distribution as both $C$ and $E$. Explain why this should be so.

**3.** Let $m$ be the renewal function of an ordinary renewal process $N$ whose interarrival times have finite mean.
(a) Show that the mean number of renewals in the interval $(a, b]$ is no larger than $1 + m(b - a)$.
(b) Show that there exists $A > 0$ such that $m(t) \leq A(1 + t)$ for $t \geq 0$.
(c) Suppose the interarrival times $X_r$ are non-arithmetic with finite variance and mean $\mu$. Show that

$$m(t) - \frac{t}{\mu} \to \frac{\sigma^2 - \mu^2}{2\mu^2} \qquad \text{as } t \to \infty.$$

[Hint: Use coupling and Blackwell's renewal theorem (10.2.5).]

## 10.5  Renewal–reward processes

Renewal theory provides models for many situations in real life. In practice, there may be rewards and/or costs associated with such a process, and these may be introduced as follows.

Let $\{(X_i, R_i) : i \geq 1\}$ be independent and identically distributed pairs of random variables such that $X_i > 0$. For a typical pair $(X, R)$, the quantity $X$ is to be interpreted as an interarrival time of a renewal process, and the quantity $R$ as a reward associated with that interarrival time; we do not assume that $X$ and $R$ are independent. Costs count as negative rewards. We now construct the renewal process $N$ by $N(t) = \sup\{n : T_n \leq t\}$ where $T_n = X_1 + X_2 + \cdots + X_n$, and the 'cumulative reward process' $C$ by

$$C(t) = \sum_{i=1}^{N(t)} R_i.$$

The *reward function* is $c(t) = \mathbb{E}C(t)$. The asymptotic properties of $C(t)$ and $c(t)$ are given by the following analogue of Theorems (10.2.1) and (10.2.3).

**(1) Renewal–reward theorem.**  *Suppose that $0 < \mathbb{E}X < \infty$ and $\mathbb{E}|R| < \infty$. Then:*

**(2)**
$$\frac{C(t)}{t} \xrightarrow{\text{a.s.}} \frac{\mathbb{E}R}{\mathbb{E}X} \qquad \text{as } t \to \infty,$$

**(3)**
$$\frac{c(t)}{t} \to \frac{\mathbb{E}R}{\mathbb{E}X} \qquad \text{as } t \to \infty.$$

**Proof.**  We have by the strong law of large numbers and Theorem (10.2.1) that

**(4)**
$$\frac{C(t)}{t} = \frac{\sum_{i=1}^{N(t)} R_i}{N(t)} \cdot \frac{N(t)}{t} \xrightarrow{\text{a.s.}} \frac{\mathbb{E}R}{\mathbb{E}X}.$$

We give two proofs of (3): first, using the renewal equation, and second, using uniform integrability.

**Proof using the renewal equation.**  We saw prior to Lemma (10.2.9) that $N(t) + 1$ is a stopping time for the sequence $\{X_i : i \geq 1\}$, whence it is a stopping time for the sequence of pairs $\{(X_i, R_i) : i \geq 1\}$. By a straightforward generalization of Wald's equation (10.2.9),

$$c(t) = \mathbb{E}\left(\sum_{j=1}^{N(t)+1} R_j\right) - \mathbb{E}(R_{N(t)+1}) = \mathbb{E}\big(N(t) + 1\big)\mathbb{E}(R) - \mathbb{E}(R_{N(t)+1}).$$

The result will follow once we have shown that $t^{-1}\mathbb{E}(R_{N(t)+1}) \to 0$ as $t \to \infty$.

By conditioning on $X_1$, as usual, we find that $r(t) = \mathbb{E}(R_{N(t)+1})$ satisfies the renewal equation

$$r(t) = H(t) + \int_0^t r(t - x)\, dF(x),$$

where $F$ is the distribution function of $X$, $H(t) = \mathbb{E}(RI_{\{X>t\}})$, and $(X, R)$ is a typical interarrival-time/reward pair. We note that

**(5)**
$$H(t) \to 0 \quad \text{as } t \to \infty, \qquad |H(t)| < \mathbb{E}|R| < \infty.$$

By Theorem (10.1.11), the above renewal equation has solution

$$r(t) = H(t) + \int_0^t H(t - x)\,dm(x)$$

where $m(t) = \mathbb{E}(N(t))$. Let $\epsilon > 0$. By (5), there exists $M < \infty$ such that $|H(t)| < \epsilon$ for $t \geq M$. Therefore, when $t \geq M$,

$$\left|\frac{r(t)}{t}\right| \leq \frac{1}{t}\left[|H(t)| + \int_0^{t-M} |H(t-x)|\,dm(x) + \int_{t-M}^t |H(t-x)|\,dm(x)\right]$$

$$\leq \frac{1}{t}\left[\epsilon + \epsilon m(t - M) + \mathbb{E}|R|\big(m(t) - m(t - M)\big)\right]$$

$$\to \frac{\epsilon}{\mathbb{E}X} \qquad \text{as } t \to \infty,$$

by Theorem (10.2.3) and the inequality $m(t) - m(t - M) \leq 1 + m(M)$; see Exercise (10.4.3a). The result follows on letting $\epsilon \downarrow 0$.

**Proof using uniform integrability.** Note that

(6)
$$\frac{|C(t)|}{t} \leq Z(t) \quad \text{where} \quad Z(t) = \frac{1}{t}\sum_{i=1}^{N(t)+1} |R_i|.$$

By Wald's equation as above, and the elementary renewal theorem (10.2.3),

$$\mathbb{E}(Z(t)) = \left(\frac{m(t) + 1}{t}\right)\mathbb{E}|R| \to \frac{\mathbb{E}|R|}{\mathbb{E}(X)} \qquad \text{as } t \to \infty.$$

By Theorem (7.10.3), the family $\{Z(t) : t \geq 0\}$ is uniformly integrable, and by (6), $\{C(t)/t\}$ is uniformly integrable. It follows from (4) that $C(t)/t$ converges in probability and hence, by Theorem (7.10.3) again, in mean. Equation (3) follows by Exercise (7.2.1b). ∎

The reward process $C$ accumulates rewards at the rate of one reward per interarrival time. In practice, rewards may accumulate in a continuous manner, spread over the interval in question, in which case the accumulated reward $\tilde{C}(t)$ at time $t$ is obtained by adding to $C(t)$ that part of $R_{N(t)+1}$ arising from the already elapsed part of the interval in progress at time $t$. This makes no effective difference to the conclusion of the renewal–reward theorem so long as rewards accumulate in a monotone manner. Suppose then that the reward $\tilde{C}(t)$ accumulated at time $t$ necessarily lies between $C(t)$ and $C(t) + R_{N(t)+1}$. We have as in (4) that

$$\frac{1}{t}\big(C(t) + R_{N(t)+1}\big) = \frac{\sum_{i=1}^{N(t)+1} R_i}{N(t) + 1} \cdot \frac{N(t) + 1}{t} \xrightarrow{\text{a.s.}} \frac{\mathbb{E}R}{\mathbb{E}X}.$$

Taken with (4), this implies that

(7)
$$\frac{\tilde{C}(t)}{t} \xrightarrow{\text{a.s.}} \frac{\mathbb{E}R}{\mathbb{E}X} \qquad \text{as } t \to \infty.$$

One has similarly that $\tilde{c}(t) = \mathbb{E}(\tilde{C}(t))$ satisfies $\tilde{c}(t)/t \to \mathbb{E}R/\mathbb{E}X$ as $t \to \infty$.

**(8) Example.** A vital component of an aeroplane is replaced at a cost $b$ whenever it reaches the given age $A$. If it fails earlier, the cost of replacement is $a$. The distribution function of the usual lifetime of a component of this type is $F$, which we assume to have density function $f$. At what rate does the cost of replacing the component accrue?

Let $X_1, X_2, \ldots$ be the runtimes of the component and its replacements. The $X_i$ may be assumed independent with common distribution function

$$H(x) = \begin{cases} F(x) & \text{if } x < A, \\ 1 & \text{if } x \geq A. \end{cases}$$

By Lemma (4.3.4), the mean of the $X_i$ is

$$\mathbb{E}X = \int_0^A [1 - F(x)]\,dx.$$

The cost of replacing a component having runtime $X$ is

$$S(X) = \begin{cases} a & \text{if } X < A, \\ b & \text{if } X \geq A, \end{cases}$$

whence $\mathbb{E}S = aF(A) + b[1 - F(A)]$.

By the renewal–reward theorem (1), the asymptotic cost per unit time is

**(9)**
$$\frac{\mathbb{E}S}{\mathbb{E}X} = \frac{aF(A) + b[1 - F(A)]}{\int_0^A [1 - F(x)]\,dx}.$$

One may choose $A$ to minimize this expression. ●

We give two major applications of the renewal–reward theorem, of which the first is to passage times of Markov chains. Let $X = \{X(t) : t \geq 0\}$ be an irreducible Markov chain in continuous time on the countable state space $S$, with transition semigroup $\{\mathbf{P}_t\}$ and generator $\mathbf{G} = (g_{ij})$. For simplicity we assume that $X$ is the minimal process associated with its jump chain $Y$, as discussed prior to Proposition (6.9.14). Let $U = \inf\{t : X(t) \neq X(0)\}$ be the first 'holding time' of the chain, and define the 'first passage time' of the state $i$ by $F_i = \inf\{t > U : X(t) = i\}$. We define the *mean return time* of $i$ by $m_i = \mathbb{E}_i(F_i)$. In avoid to avoid a triviality, we assume $|S| \geq 2$, implying by the irreducibility of the chain that $g_i = -g_{ii} > 0$ for each $i$.

**(10) Theorem.** *Assume the above conditions, and let $X(0) = i$. If $m_i < \infty$, the proportion of time spent in state $i$, and the expectation of this amount, satisfy, as $t \to \infty$,*

**(11)**
$$\frac{1}{t}\int_0^t I_{\{X(s)=i\}}\,ds \xrightarrow{\text{a.s.}} \frac{1}{m_i g_i},$$

**(12)**
$$\frac{1}{t}\int_0^t p_{ii}(s)\,ds \to \frac{1}{m_i g_i}.$$

We note from Theorem (6.10.15) that the limit in (11) and (12) is the stationary distribution of the chain.

**Proof.** We define the pairs $(P_r, Q_r)$, $r \geq 0$, of times as follows. First, we let $P_0 = 0$ and $Q_0 = \inf\{t : X(t) \neq i\}$, and more generally

$$P_r = \inf\{t > P_{r-1} + Q_{r-1} : X(t) = i\},$$
$$Q_r = \inf\{s > 0 : X(P_r + s) \neq i\}.$$

That is, $P_r$ is the time of the $r$th passage of $X$ into the state $i$, and $Q_r$ is the subsequent holding time in state $i$. The $P_r$ may be viewed as the times of arrivals in a renewal process having interarrival times distributed as $F_i$ conditional on $X(0) = i$, and we write $N(t) = \sup\{r : P_r \leq t\}$ for the associated renewal process. With the interarrival interval $(P_r, P_{r+1})$ we associate the reward $Q_r$.

We have that

(13)
$$\frac{1}{t} \sum_{r=0}^{N(t)-1} Q_r \leq \frac{1}{t} \int_0^t I_{\{X(s)=i\}}\, ds \leq \frac{1}{t} \sum_{r=0}^{N(t)} Q_r.$$

Applying Theorem (1), we identify the limits in (11) and (12) as $\mathbb{E}(Q_0)/\mathbb{E}(P_1)$. Since $Q_0$ has the exponential distribution with parameter $g_i$, and $P_1$ is the first passage time of $i$, we see that $\mathbb{E}(Q_0)/\mathbb{E}(P_1) = (g_i m_i)^{-1}$ as required. ∎

Another important and subtle application of the renewal–reward theorem is to queueing. A striking property of this application is its degree of generality, and only few assumptions are required of the queueing system. Specifically, we shall assume that:

(a) customers arrive one by one, and the $n$th customer spends a 'waiting time' $V_n$ in the system before departing†;

(b) there exists almost surely a finite (random) time $T$ $(> 0)$ such that the process beginning at time $T$ has the same distribution as the process beginning at time 0; the time $T$ is called a 'regeneration point';

(c) the number $Q(t)$ of customers in the system at time $t$ satisfies $Q(0) = Q(T) = 0$.

From (b) follows the almost-sure existence of an infinite sequence of times $0 = T_0 < T_1 < T_2 < \cdots$ each of which is a regeneration point of the process, and whose interarrival times $X_i = T_i - T_{i-1}$ are independent and identically distributed. That is, there exists a renewal process of regeneration points.

Examples of such systems are multifarious, and include various scenarios described in Chapter 11: a stable G/G/1 queue where the $T_i$ are the times at which departing customers leave the queue empty, or alternatively the times at which an arriving customer finds no one waiting; a network of queues, with the regeneration points being stopping times at which the network is empty.

Let us assume that (a), (b), (c) hold. We call the time intervals $[T_{i-1}, T_i)$ *cycles* of the process, and note that the processes $P_i = \{Q(t) : T_{i-1} \leq t < T_i\}$, $i \geq 1$, are independent and identically distributed. We write $N_i$ for the number of arriving customers during the cycle $[T_{i-1}, T_i)$, and $N = N_1$, $X = T_1$. In order to avoid a triviality, we assume also that the regeneration points are chosen in such a way that $N_i > 0$ for all $i$. We shall apply the renewal–reward theorem three times, and shall assume that

(14)
$$\mathbb{E}X < \infty, \quad \mathbb{E}N < \infty, \quad \mathbb{E}(NX) < \infty.$$

†This waiting time includes any service time.

(A) Consider the renewal process with arrival times $T_0, T_1, T_2, \ldots$. The reward associated with the interarrival time $X_i = T_i - T_{i-1}$ is taken to be

$$R_i = \int_{T_{i-1}}^{T_i} Q(u) \, du.$$

The $R_i$ have the same distribution as $R = R_1 = \int_0^X Q(u) \, du$; furthermore $Q(u) \le N$ when $0 \le u \le X$, whence $\mathbb{E}R \le \mathbb{E}(NX) < \infty$ by (14). By the renewal–reward theorem (1) and the discussion before (7),

(15) $$\frac{1}{t} \int_0^t Q(u) \, du \xrightarrow{\text{a.s.}} \frac{\mathbb{E}R}{\mathbb{E}X} \quad \text{as } t \to \infty.$$

The ratio $\mathbb{E}(R)/\mathbb{E}(X)$ is termed the 'long run average queue length' and is denoted by $L$.

(B) Consider now another renewal–reward process with arrival times $T_0, T_1, T_2, \ldots$. The reward associated with the interarrival time $X_i$ is taken to be the number $N_i$ of customers who arrive during the corresponding cycle. By the renewal–reward theorem and the discussion prior to (7), we have from hypothesis (14) that the number $N(t)$ of arrivals by time $t$ satisfies

(16) $$\frac{N(t)}{t} \xrightarrow{\text{a.s.}} \frac{\mathbb{E}N}{\mathbb{E}X} \quad \text{as } t \to \infty.$$

The ratio $\mathbb{E}(N)/\mathbb{E}(X)$ is termed the 'long run rate of arrival' and is denoted by $\lambda$.

(C) Consider now the renewal–reward process with interarrival times $N_1, N_2, \ldots$, the reward $S_i$ associated with the interarrival time $N_i$ being the sum of the waiting times of customers arriving during the $i$th cycle of the queue. The mean reward is $\mathbb{E}S = \mathbb{E}\left(\sum_1^N V_i\right)$; this is no larger than $\mathbb{E}(NX)$ which by (14) is finite. Applying the renewal–reward theorem and the discussion prior to (7), we have that

(17) $$\frac{1}{n} \sum_{i=1}^n V_i \xrightarrow{\text{a.s.}} \frac{\mathbb{E}S}{\mathbb{E}N} \quad \text{as } n \to \infty.$$

The ratio $\mathbb{E}(S)/\mathbb{E}(N)$ is termed the 'long run average waiting time' and is denoted by $W$.

(18) **Little's theorem.** *Under the assumptions above, we have that $L = \lambda W$.*

**Proof.** We have that

$$\frac{L}{\lambda W} = \frac{\mathbb{E}R}{\mathbb{E}X} \cdot \frac{\mathbb{E}X}{\mathbb{E}N} \cdot \frac{\mathbb{E}N}{\mathbb{E}S} = \frac{\mathbb{E}\int_0^X Q(u) \, du}{\mathbb{E}\sum_1^N V_i}$$

so that the result will follow once we have shown that

(19) $$\sum_1^N V_i = \int_0^X Q(u) \, du.$$

Each side of this equation is the mean amount of customer time spent during the first cycle of the system: the left side of (19) counts this by customer, and the right side counts it by unit of time. The required equality follows. ∎

**(20) Example.** Cars arrive at a car wash in the manner of a Poisson process with rate $\nu$. They wait in a line, while the car at the head of the line is washed by the unique car-wash machine. There is space in the line for exactly $K$ cars, including any car currently being washed, and, when the line is full, arriving cars leave and never return. The wash times of cars are independent random variables with distribution function $F$ and mean $\theta$.

Let $p_i$ denote the proportion of time that there are exactly $i$ cars in the line, including any car being washed. Since the queue length is not a Markov chain (unless wash times are exponentially distributed), one should not talk of the system being 'in equilibrium'. Nevertheless, using renewal theory, one may see that there exists an asymptotic proportion $p_i$ of time.

We shall apply Little's theorem to the smaller system comprising the location at the head of the line, that is, the car-wash machine itself. We take as regeneration points the times at which cars depart from the machine leaving the line empty.

The 'long run average queue length' is $L = 1 - p_0$, being the proportion of time that the machine is in use. The 'long run rate of arrival' $\lambda$ is the rate at which cars enter this subsystem, and this equals the rate at which cars join the line. Since an arriving car joins the line with probability $1 - p_K$, and since cars arrive in the manner of a Poisson process with parameter $\nu$, we deduce that $\lambda = \nu(1 - p_K)$. Finally, the 'long run average waiting time' $W$ is the mean time taken by the machine to wash a car, so that $W = \theta$.

We have by Little's theorem (18) that $L = \lambda W$ which is to say that $1 - p_0 = \nu(1 - p_K)\theta$. This equation may be interpreted in terms of the cost of running the machine, which is proportional to $1 - p_0$, and the disappointment of customers who arrive to find the line full, which is proportional to $\nu p_K$.                                                                                   ●

---

## Exercises for Section 10.5

**1.** If $X(t)$ is an irreducible positive recurrent Markov chain, and $u(\cdot)$ is a bounded function on the integers, show that

$$\frac{1}{t} \int_0^t u(X(s))\, ds \overset{\text{a.s.}}{\longrightarrow} \sum_{i \in S} \pi_i u(i),$$

where $\pi$ is the stationary distribution of $X(t)$.

**2.** Let $M(t)$ be an alternating renewal process, with interarrival pairs $\{X_r, Y_r : r \ge 1\}$. Show that

$$\frac{1}{t} \int_0^t I_{\{M(s) \text{ is even}\}}\, ds \overset{\text{a.s.}}{\longrightarrow} \frac{\mathbb{E}X_1}{\mathbb{E}X_1 + \mathbb{E}Y_1} \qquad \text{as } t \to \infty.$$

Is this limit valid for an arbitrary joint distribution of independent pairs $(X_i, Y_i)$?

**3.** Let $C(s)$ be the current lifetime (or age) of a renewal process $N(t)$ with a typical interarrival time $X$. Show that

$$\frac{1}{t} \int_0^t C(s)\, ds \overset{\text{a.s.}}{\longrightarrow} \frac{\mathbb{E}(X^2)}{2\mathbb{E}(X)} \qquad \text{as } t \to \infty.$$

Find the corresponding limit for the excess lifetime.

**4.** Let $j$ and $k$ be distinct states of an irreducible discrete-time Markov chain $X$ with stationary distribution $\pi$. Show that

$$\mathbb{P}(T_j < T_k \mid X_0 = k) = \frac{1/\pi_k}{\mathbb{E}(T_j \mid X_0 = k) + \mathbb{E}(T_k \mid X_0 = j)}$$

where $T_i = \min\{n \ge 1 : X_n = i\}$ is the first passage time to the state $i$. [Hint: Consider the times of return to $j$ having made an intermediate visit to $k$.]

**5.    Total life.** Use the result of Exercise (10.5.2) to show that the limiting distribution function as $t \to \infty$ of the total life $D(t)$ of a renewal process is

$$F_D(y) = \int_0^y \frac{1}{\mu} x f(x) \, dx,$$

where $f$ is the density function of a typical interarrival time $X$, and $\mu = \mathbb{E}(X)$. The integrand is called the *size-biased* (or *length-biased*) density for $X$.

**6.    (a)** Let $X$ be exponentially distributed with parameter $\lambda$. Show that

$$\mathbb{E}(\min\{X, d\}) = \frac{1}{\lambda}(1 - e^{-\lambda d}), \qquad d \ge 0.$$

(b) John's garage offers him the choice of two tyre replacement plans.
   1. The garage undertakes to replace all the tyres of his car at the normal price whenever one of the tyres requires replacing.
   2. The garage undertakes to replace all the tyres of his car at 5% of the normal price two years after they have last been replaced. However, if any tyre needs replacing earlier, the garage will then replace all the tyres at a price that is 5% higher than the normal price.

Assuming that a new tyre has an exponentially distributed lifetime with mean 8 years, determine the long run average cost per year under the two options. Which option should John choose if his car is new?

**7.    Uptime.** A machine $M$ is repaired at time $t = 0$, and its uptime after any repair is exponentially distributed with parameter $\lambda$, at which point it breaks down (assume the usual independence). Following any repair at time $T$, say, it is inspected at times $T, T + m, T + 2m, \ldots$, and instantly repaired if found to be broken (the inspection schedule is then restarted). Show that the long run proportion of time that $M$ is working (the 'uptime ratio') is $m^{-1} \int_0^m e^{-\lambda x} \, dx$.

---

# 10.6  Problems

In the absence of indications to the contrary, $\{X_n : n \ge 1\}$ denotes the sequence of interarrival times of either a renewal process $N$ or a delayed renewal process $N^{\mathrm{d}}$. In either case, $F^{\mathrm{d}}$ and $F$ are the distribution functions of $X_1$ and $X_2$ respectively, though $F^{\mathrm{d}} \ne F$ only if the renewal process is delayed. We write $\mu = \mathbb{E}(X_2)$, and shall usually assume that $0 < \mu < \infty$. The functions $m$ and $m^{\mathrm{d}}$ denote the renewal functions of $N$ and $N^{\mathrm{d}}$. We write $T_n = \sum_{i=1}^n X_i$, the time of the $n$th arrival.

**1.    (a)** Show that $\mathbb{P}(N(t) \to \infty$ as $t \to \infty) = 1$.
(b) Show that $m(t) < \infty$ if $\mu \ne 0$.
(c) More generally show that, for all $k > 0$, $\mathbb{E}(N(t)^k) < \infty$ if $\mu \ne 0$.

**2.**    Let $v(t) = \mathbb{E}(N(t)^2)$. Show that

$$v(t) = m(t) + 2 \int_0^t m(t - s) \, dm(s).$$

Find $v(t)$ when $N$ is a Poisson process.

**3.**    Suppose that $\sigma^2 = \mathrm{var}(X_1) > 0$. Show that the renewal process $N$ satisfies

$$\frac{N(t) - (t/\mu)}{\sqrt{t\sigma^2/\mu^3}} \xrightarrow{\mathrm{D}} N(0, 1), \qquad \text{as } t \to \infty.$$

**4.**    Find the asymptotic distribution of the current life $C(t)$ of $N$ as $t \to \infty$ when $X_1$ is not arithmetic.

**5.** Let $N$ be a Poisson process with intensity $\lambda$. Show that the total life $D(t)$ at time $t$ has distribution function $\mathbb{P}(D(t) \leq x) = 1 - (1 + \lambda \min\{t, x\})e^{-\lambda x}$ for $x \geq 0$. Deduce that $\mathbb{E}(D(t)) = (2 - e^{-\lambda t})/\lambda$.

**6.** A Type 1 counter records the arrivals of radioactive particles. Suppose that the arrival process is Poisson with intensity $\lambda$, and that the counter is locked for a dead period of fixed length $T$ after each detected arrival. Show that the detection process $\widetilde{N}$ is a renewal process with interarrival time distribution $\widetilde{F}(x) = 1 - e^{-\lambda(x-T)}$ if $x \geq T$. Find an expression for $\mathbb{P}(\widetilde{N}(t) \geq k)$.

**7.** Particles arrive at a Type 1 counter in the manner of a renewal process $N$; each detected arrival locks the counter for a dead period of random positive length. Show that

$$\mathbb{P}(\widetilde{X}_1 \leq x) = \int_0^x [1 - F(x - y)]F_L(y)\,dm(y)$$

where $F_L$ is the distribution function of a typical dead period.

**8.** (a) Show that $m(t) = \frac{1}{2}\lambda t - \frac{1}{4}(1 - e^{-2\lambda t})$ if the interarrival times have the gamma distribution $\Gamma(\lambda, 2)$.

(b) Radioactive particles arrive like a Poisson process, intensity $\lambda$, at a counter. The counter fails to register the $n$th arrival whenever $n$ is odd but suffers no dead periods. Find the renewal function $\widetilde{m}$ of the detection process $\widetilde{N}$.

**9.** Show that Poisson processes are the only renewal processes with non-arithmetic interarrival times having the property that the excess lifetime $E(t)$ and the current lifetime $C(t)$ are independent for each choice of $t$.

**10.** Let $N_1$ be a Poisson process, and let $N_2$ be a renewal process which is independent of $N_1$ with non-arithmetic interarrival times having finite mean. Show that $N(t) = N_1(t) + N_2(t)$ is a renewal process if and only if $N_2$ is a Poisson process.

**11.** Let $N$ be a renewal process, and suppose that $F$ is non-arithmetic and that $\sigma^2 = \mathrm{var}(X_1) < \infty$. Use the properties of the moment generating function $F^*(-\theta)$ of $X_1$ to deduce the formal expansion

$$m^*(\theta) = \frac{1}{\theta\mu} + \frac{\sigma^2 - \mu^2}{2\mu^2} + o(1) \quad \text{as } \theta \to 0.$$

Invert this Laplace–Stieltjes transform formally to obtain

$$m(t) = \frac{t}{\mu} + \frac{\sigma^2 - \mu^2}{2\mu^2} + o(1) \quad \text{as } t \to \infty.$$

Prove this rigorously by showing that

$$m(t) = \frac{t}{\mu} - F_E(t) + \int_0^t [1 - F_E(t - x)]\,dm(x),$$

where $F_E$ is the asymptotic distribution function of the excess lifetime (see Exercise (10.3.3)), and applying the key renewal theorem. Compare the result with the renewal theorems.

**12.** Show that the renewal function $m^{\mathrm{d}}$ of a delayed renewal process satisfies

$$m^{\mathrm{d}}(t) = F^{\mathrm{d}}(t) + \int_0^t m^{\mathrm{d}}(t - x)\,dF(x).$$

Show that $v^{\mathrm{d}}(t) = \mathbb{E}(N^{\mathrm{d}}(t)^2)$ satisfies

$$v^{\mathrm{d}}(t) = m^{\mathrm{d}}(t) + 2\int_0^t m^{\mathrm{d}}(t - x)\,dm(x)$$

where $m$ is the renewal function of the renewal process with interarrival times $X_2, X_3, \ldots$.

**13.** Let $m(t)$ be the mean number of living individuals at time $t$ in an age-dependent branching process with exponential lifetimes, parameter $\lambda$, and mean family size $\nu$ ($> 1$). Prove that $m(t) = I e^{(\nu-1)\lambda t}$ where $I$ is the number of initial members.

**14. Alternating renewal process.** The interarrival times of this process are $Z_0, Y_1, Z_1, Y_2, \ldots$, where the $Y_i$ and $Z_j$ are independent with respective common moment generating functions $M_Y$ and $M_Z$. Let $p(t)$ be the probability that the epoch $t$ of time lies in an interval of type $Z$. Show that the Laplace–Stieltjes transform $p^*$ of $p$ satisfies

$$p^*(\theta) = \frac{1 - M_Z(-\theta)}{1 - M_Y(-\theta)M_Z(-\theta)}.$$

**15. Type 2 counters.** Particles are detected by a Type 2 counter of the following sort. The incoming particles constitute a Poisson process with intensity $\lambda$. The $j$th particle locks the counter for a length $Y_j$ of time, and annuls any after-effect of its predecessors. Suppose that $Y_1, Y_2, \ldots$ are independent of each other and of the Poisson process, each having distribution function $G$. The counter is unlocked at time 0.

Let $L$ be the (maximal) length of the first interval of time during which the counter is locked. Show that $H(t) = \mathbb{P}(L > t)$ satisfies

$$H(t) = e^{-\lambda t}[1 - G(t)] + \int_0^t H(t - x)[1 - G(x)]\lambda e^{-\lambda x}\,dx.$$

Solve for $H$ in terms of $G$, and evaluate the ensuing expression in the case $G(x) = 1 - e^{-\mu x}$ where $\mu > 0$.

**16. Thinning.** Consider a renewal process $N$, and suppose that each arrival is 'overlooked' with probability $q$, independently of all other arrivals. Let $M(t)$ be the number of arrivals which are detected up to time $t/p$ where $p = 1 - q$.

(a) Show that $M$ is a renewal process whose interarrival time distribution function $F_p$ is given by $F_p(x) = \sum_{r=1}^{\infty} pq^{r-1} F_r(x/p)$, where $F_n$ is the distribution function of the time of the $n$th arrival in the original process $N$.

(b) Find the characteristic function of $F_p$ in terms of that of $F$, and use the continuity theorem to show that, as $p \downarrow 0$, $F_p(s) \to 1 - e^{-s/\mu}$ for $s > 0$, so long as the interarrival times in the original process have finite mean $\mu$. Interpret!

(c) Suppose that $p < 1$, and $M$ and $N$ are processes with the same fdds. Show that $N$ is a Poisson process.

**17.** (a) A PC keyboard has 100 different keys and a monkey is tapping them (uniformly) at random. Assuming no power failure, use the elementary renewal theorem to find the expected number of keys tapped until the first appearance of the sequence of fourteen characters 'W. Shakespeare'. Answer the same question for the sequence 'omo'.

(b) A coin comes up heads with probability $p$ on each toss. Find the mean number of tosses until the first appearances of the sequences (i) HHH, and (ii) HTH.

**18.** Let $N$ be a stationary renewal process. Let $s$ be a fixed positive real number, and define $X(t) = N(s + t) - N(t)$ for $t \geq 0$. Show that $X$ is a strongly stationary process.

**19.** Bears arrive in a village at the instants of a renewal process; they are captured and confined at a cost of $\$c$ per unit time per bear. When a given number $B$ bears have been captured, an expedition (costing $\$d$) is organized to remove and release them a long way away. What is the long-run average cost of this policy?

**20.** Let $X = \{X_n : n \geq 0\}$ be an ergodic Markov chain on a finite state space $S$, with stationary distribution $\pi$. Let $k \in S$, and let $T$ be a strictly positive stopping time with finite mean such that $X_T = k$. For $j \in S$, let $V_j(k)$ be the number of visits to $j$ by the chain started in $k$ and stopped at time $T$ (set $V_k(k) = 1$). Use the renewal–reward theorem to show that $\mathbb{E}(V_j(k)) = \pi_j \mathbb{E}_k(T)$.

**21. Car trading.** For a given type of car, costing $c$ new, the number of years between its manufacture and being broken up for scrap is $X$ ($\geq 1$), where $X$ is a random variable with distribution function $F$. In its $k$th year after manufacture (with $k \geq 1$), depreciation has reduced its resale value to $c\lambda^k$, and in that year it costs $r\mu^{k-1}$ in repairs and maintenance, where $\mu \neq 1$. (Usually, $\lambda < 1$ and $\mu > 1$.)

You buy a new car immediately your old car either is scrapped or reaches the age of $m$ years, whichever is the sooner. Show that, if you continue the policy indefinitely, in the long run you minimize your expected average cost by choosing $m$ such that

$$\frac{1}{\mathbb{E}(Y)} \left\{ c + \frac{r}{1-\mu} - cG(\lambda) - \frac{r}{1-\mu}G(\mu) \right\}$$

is as small as possible, where $G(s) = \mathbb{E}(s^Y)$ and $Y = \min\{m, X\}$.

# 11

# Queues

*Summary.* A queue may be specified by its arrival process and its queueing and service disciplines. Queues with exponentially distributed interarrival and service times are the easiest to study, since such processes are Markov chains. Imbedded Markov chains allow an analysis when either the interarrival or the service times are exponentially distributed. The general case may be studied using the theory of random walks via Lindley's equation. Open and closed networks of Markovian queues may be studied via their stationary distributions.

## 11.1 Single-server queues

As summarized in Section 8.4, with each queue we can associate two sequences $\{X_n : n \geq 1\}$ and $\{S_n : n \geq 1\}$ of independent positive random variables, the $X_n$ being interarrival times with common distribution function $F_X$ and the $S_n$ being service times with common distribution function $F_S$. We assume that customers arrive in the manner of a renewal process with interarrival times $\{X_n\}$, the $n$th customer arriving at time $T_n = X_1 + X_2 + \cdots + X_n$. Each arriving customer joins the line of customers who are waiting for the attention of the *single* server. When the $n$th customer reaches the head of this line he is served for a period of length $S_n$, after which he leaves the system. Let $Q(t)$ be the number of waiting customers at time $t$ (including any customer whose service is in progress at $t$); clearly $Q(0) = 0$. Thus $Q = \{Q(t) : t \geq 0\}$ is a random process whose finite-dimensional distributions (fdds) are specified by the distribution functions $F_X$ and $F_S$. We seek information about $Q$. For example, we may ask:

(a) When is $Q$ a Markov chain, or when does $Q$ contain an imbedded Markov chain?
(b) When is $Q$ asymptotically stationary, in the sense that the distribution of $Q(t)$ settles down as $t \to \infty$?
(c) When does the queue length grow beyond all bounds, in that the server is not able to cope with the high rate of arrivals?

The answers to these and other similar questions take the form of applying conditions to the distribution functions $F_X$ and $F_S$; the style of the analysis may depend on the types of these distributions functions. With this in mind, it is convenient to use a notation for the queueing system which incorporates information about $F_X$ and $F_S$. The most common notation scheme†

---

†Due to D. G. Kendall.

describes each system by a triple A/B/*s*, where A describes $F_X$, B describes $F_S$, and *s* is the number of servers. Typically, A and B may each be one of the following:

$D(d) \equiv$ almost surely concentrated at the value *d* (D for 'deterministic'),

$M(\lambda) \equiv$ exponential, parameter $\lambda$ (M for 'Markovian'),

$\Gamma(\lambda, k) \equiv$ gamma, parameters $\lambda$ and $k$,

$G \equiv$ some general distribution, fixed but unspecified.

**(1) Example. M($\lambda$)/M($\mu$)/1.** Interarrival times are exponential with parameter $\lambda$ and service times are exponential with parameter $\mu$. Thus customers arrive in the manner of a Poisson process with intensity $\lambda$. The process $Q = \{Q(t)\}$ is a continuous-time Markov chain with state space $\{0, 1, 2, \dots\}$; this follows from the lack-of-memory property of the exponential distribution. Furthermore, such systems are the *only* systems whose queue lengths are homogeneous Markov chains. Why is this?                                                                             ●

**(2) Example. M($\lambda$)/D(1)/1.** Customers arrive in the manner of a Poisson process, and each requires a service time of constant length 1. The process $Q$ is not a Markov chain, but we shall see later that there exists an imbedded discrete-time Markov chain $\{Q_n : n \geq 0\}$ whose properties provide information about $Q$.                                                                             ●

**(3) Example. G/G/1.** In this case we have no special information about $F_X$ or $F_S$. Some authors denote this system by GI/G/1, reserving the title G/G/1 to denote a more complicated system in which the interarrival times may not be independent.                                                                             ●

The notation M($\lambda$) is sometimes abbreviated to M alone. Thus Example (1) becomes M/M/1; this slightly unfortunate abbreviation does *not* imply that $F_X$ and $F_S$ are the same. A similar remark holds for systems described as G/G/1.

Broadly speaking, there are two types of statement to be made about the queue $Q$:

(a) 'time-dependent' statements, which contain information about the queue for finite values of *t*;

(b) 'limiting' results, which discuss the asymptotic properties of the queue as $t \to \infty$. These include conditions for the queue length to grow beyond all bounds.

Statements of the former type are most easily made about M($\lambda$)/M($\mu$)/1, since this is the only Markovian system; such conclusions are more elusive for more general systems, and we shall generally content ourselves with the asymptotic properties of such queues.

In the subsequent sections we explore the systems M/M/1, M/G/1, G/M/1, and G/G/1, in that order. For the reader's convenience, we present these cases roughly in order of increasing difficulty. This is not really satisfactory, since we are progressing from the specific to the general, so we should like to stress that queues with Markovian characteristics are very special systems and that their properties do not always indicate features of more general systems.

Here is a final piece of notation.

**(4) Definition.** The **traffic intensity** $\rho$ of a queue is defined as $\rho = \mathbb{E}(S)/\mathbb{E}(X)$, the ratio of the mean of a typical service time to the mean of a typical interarrival time.

We assume throughout that neither $\mathbb{E}(S)$ nor $\mathbb{E}(X)$ takes the value zero or infinity.

We shall see that queues behave in qualitatively different manners depending on whether $\rho < 1$ or $\rho > 1$. In the latter case, service times exceed interarrival times, on average, and the queue length grows beyond all bounds with probability 1; in the former case, the queue attains an equilibrium as $t \to \infty$. It is a noteworthy conclusion that the threshold between instability and stability depends on the mean values of $F_X$ and $F_S$ alone.

## 11.2 M/M/1

The queue $M(\lambda)/M(\mu)/1$ is very special in that $Q$ is a continuous-time Markov chain. Furthermore, reference to (6.11.1) reminds us that $Q$ is a birth–death process with birth and death rates given by

$$\lambda_n = \lambda \quad \text{for all } n, \qquad \mu_n = \begin{cases} \mu & \text{if } n \geq 1, \\ 0 & \text{if } n = 0. \end{cases}$$

By Theorem (6.9.17b) $Q$ is non-explosive, and by Theorem (6.10.7) the probabilities $p_n(t) = \mathbb{P}(Q(t) = n)$ satisfy the Kolmogorov forward equations:

(1)
$$\frac{dp_n}{dt} = \lambda p_{n-1}(t) - (\lambda + \mu) p_n(t) + \mu p_{n+1}(t) \quad \text{for} \quad n \geq 1,$$

(2)
$$\frac{dp_0}{dt} = -\lambda p_0(t) + \mu p_1(t),$$

subject to the boundary conditions $p_n(0) = \delta_{0n}$, the Kronecker delta. It is slightly tricky to solve these equations, but routine methods provide the answer after some manipulation. There are at least two possible routes: either use generating functions or use Laplace transforms with respect to $t$. We proceed in the latter way here, and define the Laplace transform† of $p_n$ by

$$\widehat{p}_n(\theta) = \int_0^\infty e^{-\theta t} p_n(t) \, dt.$$

(3) **Theorem.** *We have that* $\widehat{p}_n(\theta) = \theta^{-1}[1 - \alpha(\theta)]\alpha(\theta)^n$ *where*

(4)
$$\alpha(\theta) = \frac{(\lambda + \mu + \theta) - \sqrt{(\lambda + \mu + \theta)^2 - 4\lambda\mu}}{2\mu}.$$

The actual probabilities $p_n(t)$ can be deduced in terms of Bessel functions. It turns out that $p_n(t) = K_n(t) - K_{n+1}(t)$ where

$$K_n(t) = \int_0^t (\lambda/\mu)^{\frac{1}{2}n} n s^{-1} e^{-s(\lambda+\mu)} I_n(2s\sqrt{\lambda\mu}) \, ds$$

and $I_n(x)$ is a modified Bessel function (see Feller 1971, p. 482), defined to be the coefficient of $z^n$ in the power series expansion of $\exp[\frac{1}{2}x(z + z^{-1})]$. See Exercise (11.2.5) for another representation of $p_n(t)$.

---

†Do not confuse $\widehat{p}_n$ with the Laplace–Stieltjes transform $p_n^*(\theta) = \int_0^\infty e^{-\theta t} \, dp_n(t)$.

**Proof.** Transform (1) and (2) to obtain

(5) $$\mu \widehat{p}_{n+1} - (\lambda + \mu + \theta)\widehat{p}_n + \lambda \widehat{p}_{n-1} = 0 \quad \text{for} \quad n \geq 1,$$
(6) $$\mu \widehat{p}_1 - (\lambda + \theta)\widehat{p}_0 = -1,$$

where we have used the fact (see equation (14) of Appendix I) that

$$\int_0^\infty e^{-\theta t} \frac{dp_n}{dt}\, dt = \theta \widehat{p}_n - \delta_{0n}, \quad \text{for all } n.$$

Equation (5) is an ordinary difference equation, and standard techniques (see Appendix I) show that it has a unique solution which is bounded as $\theta \to \infty$ and which is given by

(7) $$\widehat{p}_n(\theta) = \widehat{p}_0(\theta)\alpha(\theta)^n$$

where $\alpha$ is given by (4). Substitute (7) into (6) to deduce that $\widehat{p}_0(\theta) = [1 - \alpha(\theta)]/\theta$ and the proof is complete. Alternatively, $\widehat{p}_0(\theta)$ may be calculated from the fact that $\sum_n p_n(t) = 1$, implying that $\sum_n \widehat{p}_n(\theta) = \theta^{-1}$. ∎

The asymptotic behaviour of $Q(t)$ as $t \to \infty$ is deducible from (3), but more direct methods yield the answer more quickly. Remember that $Q$ is a Markov chain.

**(8) Theorem.** *Let $\rho = \lambda/\mu$ be the traffic intensity.*
(a) *If $\rho < 1$, then $\mathbb{P}(Q(t) = n) \to (1 - \rho)\rho^n = \pi_n$ for $n \geq 0$, where $\pi$ is the unique stationary distribution.*
(b) *If $\rho \geq 1$, there is no stationary distribution, and $\mathbb{P}(Q(t) = n) \to 0$ for all $n$.*

The result is very natural. It asserts that the queue settles down into equilibrium if and only if interarrival times exceed service times on average. We shall see later that if $\rho > 1$ then $\mathbb{P}(Q(t) \to \infty$ as $t \to \infty) = 1$, whilst if $\rho = 1$ then the queue length experiences wild oscillations with no reasonable bound on their magnitudes.

**Proof.** The process $Q$ is an irreducible chain. Let us try to find a stationary distribution, as done in greater generality in Section 11.2 for birth–death chains. The mass function $\pi$ is a stationary distribution if and only if $\pi \mathbf{G} = \mathbf{0}$, which is to say that

$$\pi_{n+1} - (1 + \rho)\pi_n + \rho\pi_{n-1} = 0 \quad \text{for} \quad n \geq 1,$$
(9)
$$\pi_1 - \rho\pi_0 = 0.$$

The general solution to (9) is

$$\pi_n = \begin{cases} A + B\rho^n & \text{if } \rho \neq 1, \\ A + Bn & \text{if } \rho = 1, \end{cases}$$

where $A$ and $B$ are arbitrary constants. Thus the only bounded solution to (9) with bounded sum is

$$\pi_n = \begin{cases} B\rho^n & \text{if } \rho < 1, \\ 0 & \text{if } \rho \geq 1. \end{cases}$$

Hence, if $\rho < 1$, $\pi_n = (1 - \rho)\rho^n$ is a stationary distribution, whilst if $\rho \geq 1$ then there exists no stationary distribution. By Theorem (6.10.22), the proof is complete. ∎

The asymptotic behaviour of $Q$ may also be observed via its jump chain. Let $V_n$ be the epoch of time at which the $n$th change in $Q$ occurs. That is to say

$$V_0 = 0, \quad V_{n+1} = \inf\{t > V_n : Q(t) \neq Q(V_n+)\}.$$

Now let $Q_n = Q(V_n+)$ be the number of waiting customers immediately after the $n$th change in $Q$. Clearly $\{Q_n : n \geq 0\}$ is a random walk on the non-negative integers, with

$$Q_{n+1} = \begin{cases} Q_n + 1 & \text{with probability } \dfrac{\lambda}{\lambda + \mu} = \dfrac{\rho}{1+\rho}, \\[2mm] Q_n - 1 & \text{with probability } \dfrac{\mu}{\lambda + \mu} = \dfrac{1}{1+\rho}, \end{cases}$$

whenever $Q_n \geq 1$ (see paragraph A after (6.11.12) for a similar result for another birth–death process). When $Q_n = 0$ we have that

$$\mathbb{P}(Q_{n+1} = 1 \mid Q_n = 0) = 1,$$

so that the walk leaves 0 immediately after arriving there; it is only in this regard that the walk differs from the random walk (6.4.18) with a retaining barrier. Look for stationary distributions of the walk in the usual way to find (*exercise*) that there exists such a distribution if and only if $\rho < 1$, and it is given by

**(10)**            $\pi_0 = \tfrac{1}{2}(1 - \rho), \quad \pi_n = \tfrac{1}{2}(1 - \rho^2)\rho^{n-1} \quad \text{for} \quad n \geq 1.$

By Theorem (6.4.9),

$$\{Q_n\} \text{ is } \begin{cases} \text{positive recurrent} & \text{if } \rho < 1, \\ \text{null recurrent} & \text{if } \rho = 1, \\ \text{transient} & \text{if } \rho > 1. \end{cases}$$

Equation (10) differs from the result of (8) because the walk $\{Q_n\}$ and the process $Q$ behave differently at the state 0. It is possible to deduce (8) from (10) by taking account of the times which elapse between the jumps of the walk (see Exercise (11.2.6) for details). It is clear now that $Q_n \to \infty$ almost surely as $n \to \infty$ if $\rho > 1$, whilst $\{Q_n\}$ experiences large fluctuations in the symmetric case $\rho = 1$.

---

## Exercises for Section 11.2

**1.**   Consider a random walk on the non-negative integers with a reflecting barrier at 0, and which moves rightwards or leftwards with respective probabilities $\rho/(1 + \rho)$ and $1/(1 + \rho)$; when at 0, the particle moves to 1 at the next step. Show that the walk has a stationary distribution if and only if $\rho < 1$, and in this case the unique such distribution $\pi$ is given by $\pi_0 = \tfrac{1}{2}(1-\rho), \pi_n = \tfrac{1}{2}(1-\rho^2)\rho^{n-1}$ for $n \geq 1$.

**2.**   Suppose now that the random walker of Exercise (11.2.1) delays its steps in the following way. When at the point $n$, it waits a random length of time having the exponential distribution with parameter $\theta_n$ before moving to its next position; different 'holding times' are independent of each other and of further information concerning the steps of the walk. Show that, subject to reasonable assumptions on the $\theta_n$, the ensuing continuous-time process settles into an equilibrium distribution $\nu$ given by $\nu_n = C\pi_n/\theta_n$ for some appropriate constant $C$.

By applying this result to the case when $\theta_0 = \lambda, \theta_n = \lambda + \mu$ for $n \geq 1$, deduce that the equilibrium distribution of the $M(\lambda)/M(\mu)/1$ queue is $v_n = (1 - \rho)\rho^n$, $n \geq 0$, where $\rho = \lambda/\mu < 1$.

**3.    Waiting time.** Consider a $M(\lambda)/M(\mu)/1$ queue with $\rho = \lambda/\mu$ satisfying $\rho < 1$, and suppose that the number $Q(0)$ of people in the queue at time 0 has the stationary distribution $\pi_n = (1 - \rho)\rho^n$, $n \geq 0$. Let $W$ be the time spent by a typical new arrival before he begins his service. Show that the distribution of $W$ is given by $\mathbb{P}(W \leq x) = 1 - \rho e^{-x(\mu - \lambda)}$ for $x \geq 0$, and note that $\mathbb{P}(W = 0) = 1 - \rho$.

**4.**    A box contains $i$ red balls and $j$ lemon balls, and they are drawn at random without replacement. Each time a red (respectively lemon) ball is drawn, a particle doing a walk on $\{0, 1, 2, \ldots\}$ moves one step to the right (respectively left); the origin is a retaining barrier, so that leftwards steps from the origin are suppressed. Let $\pi(n; i, j)$ be the probability that the particle ends at position $n$, having started at the origin. Write down a set of difference equations for the $\pi(n; i, j)$, and deduce that

$$\pi(n; i, j) = A(n; i, j) - A(n + 1; i, j) \quad \text{for } i \leq j + n$$

where $A(n; i, j) = \binom{i}{n} / \binom{j+n}{n}$.

**5.**    Let $Q$ be a $M(\lambda)/M(\mu)/1$ queue with $Q(0) = 0$. Show that $p_n(t) = \mathbb{P}(Q(t) = n)$ satisfies

$$p_n(t) = \sum_{i,j \geq 0} \pi(n; i, j) \left( \frac{(\lambda t)^i e^{-\lambda t}}{i!} \right) \left( \frac{(\mu t)^j e^{-\mu t}}{j!} \right)$$

where the $\pi(n; i, j)$ are given in Exercise (11.2.4).

**6.**    Let $Q(t)$ be the length of an $M(\lambda)/M(\mu)/1$ queue at time $t$, and let $Z = \{Z_n\}$ be the jump chain of $Q$. Explain how the stationary distribution of $Q$ may be derived from that of $Z$, and vice versa.

**7.    Tandem queues.** Two queues have one server each, and all service times are independent and exponentially distributed, with parameter $\mu_i$ for queue $i$. Customers arrive at the first queue at the instants of a Poisson process of rate $\lambda$ ($< \min\{\mu_1, \mu_2\}$), and on completing service immediately enter the second queue. The queues are in equilibrium. Show that:
(a) the output of the first queue is a Poisson process with intensity $\lambda$, and that its departures before time $t$ are independent of the length of this queue at time $t$ (this is known as *Burke's theorem*),
(b) the waiting times of a given customer in the two queues are not independent.

---

# 11.3  M/G/1

M/M/1 is the only queue which is a Markov chain; the analysis of other queueing systems requires greater ingenuity. If either interarrival times or service times are exponentially distributed then the general theory of Markov chains still provides a method for studying the queue. The reason is that, for each of these two cases, we may find a discrete-time Markov chain which is imbedded in the continuous-time process $Q$. We consider M/G/1 in this section, which is divided into three parts dealing with equilibrium theory, the 'waiting time' of a typical customer, and the length of a typical 'busy period' during which the server is continuously occupied.

**(A) Asymptotic queue length.** Consider $M(\lambda)/G/1$. Customers arrive in the manner of a Poisson process with intensity $\lambda$. Let $D_n$ be the time of departure of the $n$th customer from the system, and let $Q(D_n)$ be the number of customers which he leaves behind him in the system on his departure (really, we should write $Q(D_n+)$ instead of $Q(D_n)$ to make clear that the departing customer is not included). Then $Q(D) = \{Q(D_n) : n \geq 1\}$ is a sequence of random variables. What can we say about a typical increment $Q(D_{n+1}) - Q(D_n)$? If $Q(D_n) > 0$, the $(n + 1)$th customer begins his service time immediately at time $D_n$; during this service time of length $S_{n+1}$, a random number, $U_n$ say, of customers arrive and join the waiting line. Therefore the $(n + 1)$th customer leaves $U_n + Q(D_n) - 1$ customers behind him as he departs. That is,

**(1)** $$Q(D_{n+1}) = U_n + Q(D_n) - 1 \quad \text{if} \quad Q(D_n) > 0.$$

If $Q(D_n) = 0$, the server must wait for the $(n + 1)$th arrival before she sets to work again. When this service is complete, the $(n + 1)$th customer leaves exactly $U_n$ customers behind him where $U_n$ is the number of arrivals during his service time, as before. That is,

**(2)** $$Q(D_{n+1}) = U_n \quad \text{if} \quad Q(D_n) = 0.$$

Combine (1) and (2) to obtain

**(3)** $$Q(D_{n+1}) = U_n + Q(D_n) - h(Q(D_n))$$

where $h$ is defined by

$$h(x) = \begin{cases} 1 & \text{if } x > 0, \\ 0 & \text{if } x \leq 0. \end{cases}$$

Equation (3) holds for any queue. However, in the case of $M(\lambda)/G/1$ the random variable $U_n$ depends *only* on the length of time $S_{n+1}$, and is independent of $Q(D_n)$, because of the special properties of the Poisson process of arrivals. We conclude from (3) that $Q(D)$ is a Markov chain.

**(4) Theorem.** *The sequence $Q(D)$ is a Markov chain with transition matrix*

$$\mathbf{P}_D = \begin{pmatrix} \delta_0 & \delta_1 & \delta_2 & \cdots \\ \delta_0 & \delta_1 & \delta_2 & \cdots \\ 0 & \delta_0 & \delta_1 & \cdots \\ 0 & 0 & \delta_0 & \cdots \\ \vdots & \vdots & \vdots & \ddots \end{pmatrix}$$

*where*

$$\delta_j = \mathbb{E}\left( \frac{(\lambda S)^j}{j!} e^{-\lambda S} \right)$$

*and S is a typical service time.*

The quantity $\delta_j$ is simply the probability that exactly $j$ customers join the queue during a typical service time.

**Proof.** We need to show that $\mathbf{P}_D$ is the correct transition matrix. In the notation of Chapter 6,

$$p_{0j} = \mathbb{P}\big(Q(D_{n+1}) = j \mid Q(D_n) = 0\big) = \mathbb{E}\big(\mathbb{P}(U_n = j \mid S)\big)$$

where $S = S_{n+1}$ is the service time of the $(n + 1)$th customer. Thus

$$p_{0j} = \mathbb{E}\left( \frac{(\lambda S)^j}{j!} e^{-\lambda S} \right) = \delta_j$$

as required, since, conditional on $S$, $U_n$ has the Poisson distribution with parameter $\lambda S$. Likewise, if $i \geq 1$ then

$$p_{ij} = \mathbb{E}\big(\mathbb{P}(U_n = j - i + 1 \mid S)\big) = \begin{cases} \delta_{j-i+1} & \text{if } j - i + 1 \geq 0, \\ 0 & \text{if } j - i + 1 < 0. \end{cases}$$

∎

This result enables us to observe the behaviour of the process $Q = \{Q(t)\}$ by evaluating it at the time epochs $D_1, D_2, \ldots$ and using the theory of Markov chains. It is important to note that this course of action provides reliable information about the asymptotic behaviour of $Q$ only because $D_n \to \infty$ almost surely as $n \to \infty$. The asymptotic behaviour of $Q(D)$ is described by the next theorem.

**(5) Theorem.** *Let $\rho = \lambda \mathbb{E}(S)$ be the traffic intensity.*
  (a) *If $\rho < 1$, then $Q(D)$ is ergodic with a unique stationary distribution $\pi$, having generating function*

$$G(s) = \sum_j \pi_j s^j = (1 - \rho)(s - 1)\frac{M_S(\lambda(s - 1))}{s - M_S(\lambda(s - 1))},$$

  *where $M_S$ is the moment generating function of a typical service time.*
  (b) *If $\rho > 1$, then $Q(D)$ is transient.*
  (c) *If $\rho = 1$, then $Q(D)$ is null recurrent.*

Here are some consequences of this theorem.

**(6) Busy period.** A *busy period* is a period of time during which the server is continuously occupied. The length $B$ of a typical busy period behaves similarly to the time $B'$ between successive visits of the chain $Q(D)$ to the state 0. Thus

$$
\begin{aligned}
&\text{if} \quad \rho < 1 \quad \text{then} \quad \mathbb{E}(B) < \infty, \\
&\text{if} \quad \rho = 1 \quad \text{then} \quad \mathbb{E}(B) = \infty, \qquad \mathbb{P}(B = \infty) = 0, \\
&\text{if} \quad \rho > 1 \quad \text{then} \quad \mathbb{P}(B = \infty) > 0.
\end{aligned}
$$

See the forthcoming Theorems (17) and (18) for more details about $B$.

**(7) Stationarity of $Q$.** It is an immediate consequence of (5) and Theorem (6.4.17) that $Q(D)$ is asymptotically stationary whenever $\rho < 1$. In this case it can be shown that $Q$ is asymptotically stationary also, in that $\mathbb{P}(Q(t) = n) \to \pi_n$ as $t \to \infty$. Roughly speaking, this is because $Q(t)$ forgets more and more about its origins as $t$ becomes larger.

**Proof of (5).** The sequence $Q(D)$ is irreducible and aperiodic. We proceed by applying Theorems (6.4.3), (6.4.13), and (6.4.16).
  (a) Look for a root of the equation $\pi = \pi \mathbf{P}_D$. Any such $\pi$ satisfies

**(8)**
$$\pi_j = \pi_0 \delta_j + \sum_{i=1}^{j+1} \pi_i \delta_{j-i+1}, \quad \text{for} \quad j \geq 0.$$

First, note that if $\pi_0 \; (\geq 0)$ is given, then (8) has a unique solution $\pi$. Furthermore, this solution has non-negative entries. To see this, add equations (8) for $j = 0, 1, \dots , n$ and solve for $\pi_{n+1}$ to obtain

$$
\text{(9)} \qquad\qquad \pi_{n+1}\delta_0 = \pi_0\epsilon_n + \sum_{i=1}^{n} \pi_i\epsilon_{n-i+1} \quad \text{for} \quad n \geq 0
$$

where

$$
\epsilon_n = 1 - \delta_0 - \delta_1 - \cdots - \delta_n > 0 \quad \text{because} \quad \sum_j \delta_j = 1.
$$

From (9), $\pi_{n+1} \geq 0$ whenever $\pi_i \geq 0$ for all $i \leq n$, and so

$$
\text{(10)} \qquad\qquad \pi_n \geq 0 \quad \text{for all } n
$$

if $\pi_0 \geq 0$, by induction. Return to (8) to see that the generating functions

$$
G(s) = \sum_j \pi_j s^j, \quad \Delta(s) = \sum_j \delta_j s^j,
$$

satisfy

$$
G(s) = \pi_0\Delta(s) + \frac{1}{s}[G(s) - \pi_0]\Delta(s)
$$

and therefore

$$
\text{(11)} \qquad\qquad G(s) = \frac{\pi_0(s-1)\Delta(s)}{s - \Delta(s)}.
$$

The vector $\pi$ is a stationary distribution if and only if $\pi_0 > 0$ and $\lim_{s \uparrow 1} G(s) = 1$. Apply L'Hôpital's rule to (11) to discover that

$$
\pi_0 = 1 - \Delta'(1) > 0
$$

is a necessary and sufficient condition for this to occur, and thus there exists a stationary distribution if and only if

$$
\text{(12)} \qquad\qquad \Delta'(1) < 1.
$$

However,

$$
\Delta(s) = \sum_j s^j \mathbb{E}\left(\frac{(\lambda S)^j}{j!}e^{-\lambda S}\right) = \mathbb{E}\left(e^{-\lambda S}\sum_j \frac{(\lambda s S)^j}{j!}\right)
$$
$$
= \mathbb{E}(e^{\lambda S(s-1)}) = M_S(\lambda(s-1))
$$

where $M_S$ is the moment generating function of $S$. Thus

$$
\text{(13)} \qquad\qquad \Delta'(1) = \lambda M_S'(0) = \lambda\mathbb{E}(S) = \rho
$$

and condition (12) becomes $\rho < 1$. Thus $Q(D)$ is positive recurrent if and only if $\rho < 1$. In this case, $G(s)$ takes the form given in (5a).

(b) Recall from Theorem (6.4.13) that $Q(D)$ is transient if and only if there is a bounded non-zero solution $\{y_j : j \geq 1\}$ to the equation

$$
\textbf{(14)} \qquad\qquad y_1 = \sum_{i=1}^{\infty} \delta_i y_i,
$$

$$
\textbf{(15)} \qquad\qquad y_j = \sum_{i=0}^{\infty} \delta_i y_{j+i-1} \quad \text{for} \quad j \geq 2.
$$

If $\rho > 1$ then $\Delta(s)$ satisfies

$$
0 < \Delta(0) < 1, \quad \Delta(1) = 1, \quad \Delta'(1) > 1,
$$

from (13). Draw a picture (or see Figure 5.1) to see that there exists a number $b \in (0, 1)$ such that $\Delta(b) = b$. By inspection, $y_j = 1 - b^j$ solves (14) and (15), and (b) is shown.

(c) $Q(D)$ is transient if $\rho > 1$ and positive recurrent if and only if $\rho < 1$. We need only show that $Q(D)$ is recurrent if $\rho = 1$. But it is not difficult to see that $\{y_j : j \neq 0\}$ solves equation (6.4.14), when $y_j$ is given by $y_j = j$ for $j \geq 1$, and the result follows.   ∎

**(B) Waiting time.** When $\rho < 1$ the imbedded queue length settles down into a stationary distribution $\pi$. Suppose that a customer joins the queue after some large time has elapsed. He will wait a period $W$ of time before his service begins; $W$ is called his *waiting time* (this definition is at odds with that used by some authors who include the customer's service time in $W$). The distribution of $W$ should not vary much with the time of the customer's arrival since the system is 'nearly' in equilibrium.

**(16) Theorem. Pollaczek–Khinchin formula.** *The waiting time $W$ has moment generating function*

$$
M_W(s) = \frac{(1 - \rho)s}{\lambda + s - \lambda M_S(s)}
$$

*when the imbedded queue is in equilibrium.*

**Proof.** The condition that the imbedded queue be in equilibrium amounts to the supposition that the length $Q(D)$ of the queue on the departure of a customer is distributed according to the stationary distribution $\pi$. Suppose that a customer waits for a period of length $W$ and then is served for a period of length $S$. On departure he leaves behind him all those customers who have arrived during the period, length $W + S$, during which he was in the system. The number $Q$ of such customers is Poisson distributed with parameter $\lambda(W + S)$, and so

$$
\begin{aligned}
\mathbb{E}(s^Q) &= \mathbb{E}\big(\mathbb{E}(s^Q \mid W, S)\big) \\
&= \mathbb{E}(e^{\lambda(W+S)(s-1)}) \\
&= \mathbb{E}(e^{\lambda W(s-1)})\mathbb{E}(e^{\lambda S(s-1)}) \quad \text{by independence} \\
&= M_W\big(\lambda(s - 1)\big)M_S\big(\lambda(s - 1)\big).
\end{aligned}
$$

However, $Q$ has distribution $\pi$ given by (5a) and the result follows.   ∎

**(C) Busy period: a branching process.** Finally, put yourself in the server's shoes. She may not be as interested in the waiting times of her customers as she is in the frequency of her tea breaks. Recall from (6) that a *busy period* is a period of time during which she is continuously occupied, and let $B$ be the length of a typical busy period. That is, if the first customer arrives at time $T_1$ then

$$B = \inf\{t > 0 : Q(t + T_1) = 0\};$$

The quantity $B$ is well defined whether or not $Q(D)$ is ergodic, though it may equal $+\infty$.

**(17) Theorem.** *The moment generating function $M_B$ of $B$ satisfies the functional equation*

$$M_B(s) = M_S\big(s - \lambda + \lambda M_B(s)\big).$$

It can be shown that this functional equation has a unique solution which is the moment generating function of a (possibly infinite) random variable (see Feller 1971, pp. 441, 473). The server may wish to calculate the probability

$$\mathbb{P}(B < \infty) = \lim_{x \to \infty} \mathbb{P}(B \le x)$$

that she is eventually free. It is no surprise to find the following, in agreement with (6).

**(18) Theorem.** *We have that*

$$\mathbb{P}(B < \infty) \begin{cases} = 1 & \text{if } \rho \le 1, \\ < 1 & \text{if } \rho > 1. \end{cases}$$

This may remind you of a similar result for the extinction probability of a branching process. This is no coincidence; we prove (17) and (18) by methods first encountered in the study of branching processes.

**Proof of (17) and (18).** Here is an imbedded branching process. Call customer $C_2$ an 'offspring' of customer $C_1$ if $C_2$ joins the queue while $C_1$ is being served. Since customers arrive in the manner of a Poisson process, the numbers of offspring of different customers are independent random variables. Therefore, the family tree of the 'offspring process' is that of a branching process. The mean number of offspring of a given customer is given in the notation of the proof of (5) as $\mu = \Delta'(1)$, whence $\mu = \rho$ by (13). The offspring process is ultimately extinct if and only if the queue is empty at some time later than the first arrival. That is,

$$\mathbb{P}(B < \infty) = \mathbb{P}(Z_n = 0 \text{ for some } n),$$

where $Z_n$ is the number of customers in the $n$th generation of the process. We have by Theorem (5.4.5) that $\eta = \mathbb{P}(Z_n = 0 \text{ for some } n)$ satisfies

$$\eta = 1 \quad \text{if and only if} \quad \mu \le 1.$$

Therefore $\mathbb{P}(B < \infty) = 1$ if and only if $\rho \le 1$, as required for (18).

Each individual in this branching process has a service time; $B$ is the sum of these service times. Thus

**(19)** 
$$B = S + \sum_{j=1}^{Z} B_j$$

where $S$ is the service time of the first customer, $Z$ is the number of offspring of this customer, and $B_j$ is the sum of the service times of the $j$th such offspring together with all his descendants in the offspring process (this is similar to the argument of Problem (5.12.11)). The two terms on the right side of (19) are *not* independent of each other; after all, if $S$ is large then $Z$ is likely to be large as well. However, condition on $S$ to obtain

$$M_B(s) = \mathbb{E}\left(\mathbb{E}\left\{\exp\left[s\left(S + \sum_{j=1}^{Z} B_j\right)\right]\,\middle|\, S\right\}\right)$$

and remember that, conditional on $Z$, the random variables $B_1, B_2, \ldots, B_Z$ are independent with the same distribution as $B$ to obtain

$$M_B(s) = \mathbb{E}\left(e^{sS} G_{\mathrm{Po}(\lambda S)}\{M_B(s)\}\right)$$

where $G_{\mathrm{Po}(\mu)}$ is the probability generating function of the Poisson distribution with parameter $\mu$. Therefore

$$M_B(s) = \mathbb{E}(e^{S(s-\lambda+\lambda M_B(s))})$$

as required.                                                                                      ∎

---

### Exercises for Section 11.3

**1.** Consider M$(\lambda)$/D$(d)$/1 where $\rho = \lambda d < 1$. Show that the mean queue length at moments of departure in equilibrium is $\frac{1}{2}\rho(2 - \rho)/(1 - \rho)$.

**2.** Consider M$(\lambda)$/M$(\mu)$/1, and show that the moment generating function of a typical busy period is given by

$$M_B(s) = \frac{(\lambda + \mu - s) - \sqrt{(\lambda + \mu - s)^2 - 4\lambda\mu}}{2\lambda}$$

for all sufficiently small but positive values of $s$.

**3.** (a) Show that, for a M/G/1 queue, the sequence of times at which the server passes from being busy to being free constitutes a renewal process.

(b) When the above queue is in equilibrium, what is the moment generating function of the total time that an arriving customer spends in the queue including service?

**4.** **Loss system.** Consider the M$(\lambda)$/G/1 queue with no waiting room. Customers who arrive while the server is busy are lost. Show that the long run proportion of arrivals lost is $1/(1 + \rho^{-1})$ where $\rho = \lambda\mathbb{E}(S)$ is the traffic intensity.

---

## 11.4  G/M/1

The system G/M$(\mu)$/1 contains an imbedded discrete-time Markov chain also, and this chain provides information about the properties of $Q(t)$ for large $t$. This section is divided into two parts, dealing with the asymptotic behaviour of $Q(t)$ and the waiting time distribution.

**(A) Asymptotic queue length.** This time, consider the epoch of time at which the $n$th customer *joins* the queue, and let $Q(A_n)$ be the number of individuals who are ahead of him in the system at the moment of his arrival. The quantity $Q(A_n)$ includes any customer whose service is in progress; more specifically, $Q(A_n) = Q(T_n-)$ where $T_n$ is the instant of the $n$th arrival. The argument of the last section shows that

(1) $$Q(A_{n+1}) = Q(A_n) + 1 - V_n$$

where $V_n$ is the number of departures from the system during the interval $[T_n, T_{n+1})$ between the $n$th and $(n+1)$th arrival. This time, $V_n$ depends on $Q(A_n)$ since not more than $Q(A_n) + 1$ individuals may depart during this interval. However, service times are exponentially distributed, and so, conditional upon $Q(A_n)$ and $X_{n+1} = T_{n+1} - T_n$, the random variable $V_n$ has a truncated Poisson distribution

(2) $$\mathbb{P}(V_n = d \mid Q(A_n) = q, \; X_{n+1} = x) = \begin{cases} \dfrac{(\mu x)^d}{d!} e^{-\mu x} & \text{if } d \le q, \\[3mm] \displaystyle\sum_{m>q} \dfrac{(\mu x)^m}{m!} e^{-\mu x} & \text{if } d = q + 1. \end{cases}$$

Anyway, given $Q(A_n)$, the random variable $V_n$ is independent of the sequence $Q(A_1)$, $Q(A_2)$, ..., $Q(A_{n-1})$, and so $Q(A) = \{Q(A_n) : n \ge 1\}$ is a Markov chain.

(3) **Theorem.** *The sequence $Q(A)$ is a Markov chain with transition matrix*

$$\mathbf{P}_A = \begin{pmatrix} 1 - \alpha_0 & \alpha_0 & 0 & 0 & \cdots \\ 1 - \alpha_0 - \alpha_1 & \alpha_1 & \alpha_0 & 0 & \cdots \\ 1 - \alpha_0 - \alpha_1 - \alpha_2 & \alpha_2 & \alpha_1 & \alpha_0 & \cdots \\ \vdots & \vdots & \vdots & \vdots & \ddots \end{pmatrix}$$

*where*

$$\alpha_j = \mathbb{E}\left( \frac{(\mu X)^j}{j!} e^{-\mu X} \right)$$

*and $X$ is a typical interarrival time.*

The quantity $\alpha_j$ is simply the probability that exactly $j$ events of a Poisson process occur during a typical interarrival time.

**Proof.** This proceeds as for Theorem (11.3.4).                                          ∎

(4) **Theorem.** *Let $\rho = \{\mu \mathbb{E}(X)\}^{-1}$ be the traffic intensity.*
   (a) *If $\rho < 1$, then $Q(A)$ is ergodic with a unique stationary distribution $\pi$ given by*

$$\pi_j = (1 - \eta)\eta^j \quad \text{for} \quad j \ge 0$$

   *where $\eta$ is the smallest positive root of $\eta = M_X(\mu(\eta - 1))$ and $M_X$ is the moment generating function of $X$.*
   (b) *If $\rho > 1$, then $Q(A)$ is transient.*
   (c) *If $\rho = 1$, then $Q(A)$ is null recurrent.*

If $\rho < 1$ then $Q(A)$ is asymptotically stationary. Unlike the case of M/G/1, however, the stationary distribution $\pi$ given by (4a) need *not* be the limiting distribution of $Q$ itself; to see an example of this, just consider D(1)/M/1.

**Proof.** Let $Q_d$ be an $M(\mu)/G/1$ queue whose service times have the same distribution as the interarrival times of $Q$ (the queue $Q_d$ is called the *dual* of $Q$, but more about that later). The traffic intensity $\rho_d$ of $Q_d$ satisfies

(5)                                         $$\rho\rho_d = 1.$$

From the results of Section 11.3, $Q_d$ has an imbedded Markov chain $Q_d(D)$, obtained from the values of $Q_d$ at the epochs of time at which customers depart. We shall see that $Q(A)$ is positive recurrent (respectively transient) if and only if the imbedded chain $Q_d(D)$ of $Q_d$ is transient (respectively positive recurrent) and the results will follow immediately from Theorem (11.3.5) and its proof.

(a) Look for non-negative solutions $\pi$ to the equation

(6)                                         $$\pi = \pi P_A$$

which have sum $\pi \mathbf{1}' = 1$. Expand (6), set

(7)                     $$y_j = \pi_0 + \pi_1 + \cdots + \pi_{j-1} \quad \text{for} \quad j \geq 1,$$

and remember that $\sum_j \alpha_j = 1$ to obtain

(8)                                         $$y_1 = \sum_{i=1}^{\infty} \alpha_i y_i,$$

(9)                     $$y_j = \sum_{i=0}^{\infty} \alpha_i y_{j+i-1} \quad \text{for} \quad j \geq 2.$$

These are the same equations as (11.3.14) and (11.3.15) for $Q_d$. As in the proof of Theorem (11.3.5), it is easy to check that

(10)                                         $$y_j = 1 - \eta^j$$

solves (8) and (9) whenever

$$A(s) = \sum_{j=0}^{\infty} \alpha_j s^j$$

satisfies $A'(1) > 1$, where $\eta$ is the unique root in the interval $(0, 1)$ of the equation $A(s) = s$. However, write $A$ in terms of $M_X$, as before, to find that $A(s) = M_X(\mu(s - 1))$, giving

$$A'(1) = \rho_d = \frac{1}{\rho}.$$

Combine (7) and (10) to find the stationary distribution for the case $\rho < 1$. If $\rho \geq 1$ then $\rho_d \leq 1$ by (5), and so (8) and (9) have no bounded non-zero solution since otherwise $Q_d(D)$ would be transient, contradicting Theorem (11.3.5). Thus $Q(A)$ is positive recurrent if and only if $\rho < 1$.

(b) To prove transience, we seek bounded non-zero solutions $\{y_j : j \geq 1\}$ to the equations

**(11)**
$$y_j = \sum_{i=1}^{j+1} y_i \alpha_{j-i+1} \quad \text{for} \quad j \geq 1.$$

Suppose that $\{y_j\}$ satisfies (11), and that $y_1 \geq 0$. Define $\pi = \{\pi_j : j \geq 0\}$ as follows:

$$\pi_0 = y_1 \alpha_0, \quad \pi_1 = y_1(1 - \alpha_0), \quad \pi_j = y_j - y_{j-1} \quad \text{for} \quad j \geq 2.$$

It is an easy exercise to show that $\pi$ satisfies equation (11.3.8) with the $\delta_j$ replaced by the $\alpha_j$ throughout. But (11.3.8) possesses a non-zero solution with bounded sum if and only if $\rho_d < 1$, which is to say that $Q(A)$ is transient if and only if $\rho = 1/\rho_d > 1$.
(c) $Q(A)$ is transient if and only if $\rho > 1$, and is positive recurrent if and only if $\rho < 1$. If $\rho = 1$ then $Q(A)$ has no choice but null recurrence. ∎

**(B) Waiting time.** An arriving customer waits for just as long as the server needs to complete the current service period and to serve the other waiting customers. That is, the $n$th customer waits for a length $W_n$ of time:

$$W_n = Z_1^* + Z_2 + Z_3 + \cdots + Z_{Q(A_n)} \quad \text{if} \quad Q(A_n) > 0$$

where $Z_1^*$ is the *excess* (or *residual*) *service time* of the customer at the head of the queue, and $Z_2, Z_3, \ldots, Z_{Q(A_n)}$ are the service times of the others. Given $Q(A_n)$, the $Z_i$ are independent, but $Z_1^*$ does not in general have the same distribution as $Z_2, Z_3, \ldots$ . In the case of G/M($\mu$)/1, however, the lack-of-memory property helps us around this difficulty.

**(12) Theorem.** *The waiting time $W$ of an arriving customer has distribution*

$$\mathbb{P}(W \leq x) = \begin{cases} 0 & \text{if } x < 0, \\ 1 - \eta e^{-\mu(1-\eta)x} & \text{if } x \geq 0, \end{cases}$$

*where $\eta$ is given in (4a), when the imbedded queue is in equilibrium.*

Note that $W$ has an atom of size $1 - \eta$ at the origin.

**Proof.** By the lack-of-memory property, $W_n$ is the sum of $Q(A_n)$ independent exponential variables. Use the stationary distribution of $Q(A)$ to find that

$$M_W(s) = (1 - \eta) + \eta \frac{\mu(1 - \eta)}{\mu(1 - \eta) - s}$$

which we recognize as the moment generating function of a random variable which either equals zero (with probability $1 - \eta$) or is exponentially distributed with parameter $\mu(1 - \eta)$ (with probability $\eta$). ∎

Finally, here is a word of caution. There is another quantity called *virtual* waiting time, which must not be confused with *actual* waiting time. The latter is the actual time spent by a customer after his arrival; the former is the time which a customer *would* spend if he were to arrive at some particular instant. The stationary distributions of these waiting times may differ whenever the stationary distribution of $Q$ differs from the stationary distribution of the imbedded Markov chain $Q(A)$.

## Exercises for Section 11.4

**1.** Consider G/M($\mu$)/1, and let $\alpha_j = \mathbb{E}((\mu X)^j e^{-\mu X}/j!)$ where $X$ is a typical interarrival time. Suppose the traffic intensity $\rho$ is less than 1. Show that the equilibrium distribution $\pi$ of the imbedded chain at moments of arrivals satisfies

$$\pi_n = \sum_{i=0}^{\infty} \alpha_i \pi_{n+i-1} \qquad \text{for } n \geq 1.$$

Look for a solution of the form $\pi_n = \theta^n$ for some $\theta$, and deduce that the unique stationary distribution is given by $\pi_j = (1-\eta)\eta^j$ for $j \geq 0$, where $\eta$ is the smallest positive root of the equation $s = M_X(\mu(s-1))$.

**2.** Consider a G/M($\mu$)/1 queue in equilibrium. Let $\eta$ be the smallest positive root of the equation $x = M_X(\mu(x-1))$ where $M_X$ is the moment generating function of an interarrival time. Show that the mean number of customers ahead of a new arrival is $\eta(1-\eta)^{-1}$, and the mean waiting time is $\eta\{\mu(1-\eta)\}^{-1}$.

**3.** Consider D(1)/M($\mu$)/1 where $\mu > 1$. Show that the continuous-time queue length $Q(t)$ does not converge in distribution as $t \to \infty$, even though the imbedded chain at the times of arrivals is ergodic.

## 11.5  G/G/1

If neither interarrival times nor service times are exponentially distributed then the methods of the last three sections fail. This apparent setback leads us to the remarkable discovery that queueing problems are intimately related to random walk problems. This section is divided into two parts, one dealing with the equilibrium theory of G/G/1 and the other dealing with the imbedded random walk.

**(A) Asymptotic waiting time.** Let $W_n$ be the waiting time of the $n$th customer. There is a useful relationship between $W_n$ and $W_{n+1}$ in terms of the service time $S_n$ of the $n$th customer and the length $X_{n+1}$ of time between the $n$th and the $(n+1)$th arrivals.

**(1) Theorem. Lindley's equation.** *We have that*

$$W_{n+1} = \max\{0, W_n + S_n - X_{n+1}\}.$$

**Proof.** The $n$th customer is in the system for a length $W_n + S_n$ of time. If $X_{n+1} > W_n + S_n$ then the queue is empty at the $(n+1)$th arrival, and so $W_{n+1} = 0$. If $X_{n+1} \leq W_n + S_n$ then the $(n+1)$th customer arrives while the $n$th is still present, but only waits for a period of length $W_n + S_n - X_{n+1}$ before the previous customer leaves. ∎

We shall see that Lindley's equation implies that the distribution functions

$$F_n(x) = \mathbb{P}(W_n \leq x)$$

of the $W_n$ converge as $n \to \infty$ to some limit function $F(x)$. Of course, $F$ need not be a proper distribution function; indeed, it is intuitively clear that the queue settles down into equilibrium if and only if $F$ is a distribution function which is not defective.

**(2) Theorem.** *Let $F_n(x) = \mathbb{P}(W_n \leq x)$. Then*

$$F_{n+1}(x) = \begin{cases} 0 & \text{if } x < 0, \\ \displaystyle\int_{-\infty}^{x} F_n(x - y) \, dG(y) & \text{if } x \geq 0, \end{cases}$$

*where $G$ is the distribution function of $U_n = S_n - X_{n+1}$. Thus the limit $F(x) = \lim_{n\to\infty} F_n(x)$ exists.*

Note that $\{U_n : n \geq 1\}$ is a collection of independent identically distributed random variables.

**Proof.** If $x \geq 0$ then

$$\mathbb{P}(W_{n+1} \leq x) = \int_{-\infty}^{\infty} \mathbb{P}(W_n + U_n \leq x \mid U_n = y) \, dG(y)$$

$$= \int_{-\infty}^{x} \mathbb{P}(W_n \leq x - y) \, dG(y) \quad \text{by independence,}$$

and the first part is proved. We claim that

**(3)**                                 $F_{n+1}(x) \leq F_n(x)$    for all $x$ and $n$.

If (3) holds then the second result follows immediately; we prove (3) by induction. Trivially, $F_2(x) \leq F_1(x)$ because $F_1(x) = 1$ for all $x \geq 0$. Suppose that (3) holds for $n = k - 1$, say. Then, for $x \geq 0$,

$$F_{k+1}(x) - F_k(x) = \int_{-\infty}^{x} \left[ F_k(x - y) - F_{k-1}(x - y) \right] dG(y) \leq 0$$

by the induction hypothesis. The proof is complete.                                         ∎

It follows that the distribution functions of $\{W_n\}$ converge as $n \to \infty$. It is clear, by monotone convergence, that the limit $F(x)$ satisfies the Wiener–Hopf equation

$$F(x) = \int_{-\infty}^{x} F(x - y) \, dG(y) \quad \text{for} \quad x \geq 0;$$

this is not easily solved for $F$ in terms of $G$. However, it is not too difficult to find a criterion for $F$ to be a proper distribution function.

**(4) Theorem.** *Let $\rho = \mathbb{E}(S)/\mathbb{E}(X)$ be the traffic intensity*
(a) *If $\rho < 1$, then $F$ is a non-defective distribution function.*
(b) *If $\rho > 1$, then $F(x) = 0$ for all $x$.*
(c) *If $\rho = 1$ and $\mathrm{var}(U) > 0$, then $F(x) = 0$ for all $x$.*

An explicit formula for the moment generating function of $F$ when $\rho < 1$ is given in Theorem (14) below. Theorem (4) classifies the stability of G/G/1 in terms of the sign of $1 - \rho$; note that this information is obtainable from the distribution function $G$ since

**(5)**              $\rho < 1 \quad \Leftrightarrow \quad \mathbb{E}(S) < \mathbb{E}(X) \quad \Leftrightarrow \quad \mathbb{E}(U) = \displaystyle\int_{-\infty}^{\infty} u \, dG(u) < 0$

where $U$ is a typical member of the $U_i$. We call the process *stable* when $\rho < 1$.

The crucial step in the proof of (4) is important in its own right. Use Lindley's equation (1) to see that:

$$W_1 = 0,$$
$$W_2 = \max\{0, W_1 + U_1\} = \max\{0, U_1\},$$
$$W_3 = \max\{0, W_2 + U_2\} = \max\{0, U_2, U_2 + U_1\},$$

and in general

$$(6) \qquad W_{n+1} = \max\{0, U_n, U_n + U_{n-1}, \dots, U_n + U_{n-1} + \dots + U_1\}$$

which expresses $W_{n+1}$ in terms of the partial sums of a sequence of independent identically distributed variables. It is difficult to derive asymptotic properties of $W_{n+1}$ directly from (6) since every non-zero term changes its value as $n$ increases from the value $k$, say, to the value $k + 1$. The following theorem is the crucial observation.

**(7) Theorem.** *The random variable $W_{n+1}$ has the same distribution as*

$$W'_{n+1} = \max\{0, U_1, U_1 + U_2, \dots, U_1 + U_2 + \dots + U_n\}.$$

**Proof.** The vectors $(U_1, U_2, \dots, U_n)$ and $(U_n, U_{n-1}, \dots, U_1)$ are sequences with the same joint distribution. Replace each $U_i$ in (6) by $U_{n+1-i}$. ∎

That is to say, $W_{n+1}$ and $W'_{n+1}$ are *different* random variables but they have the *same* distribution. Thus

$$F(x) = \lim_{n \to \infty} \mathbb{P}(W_n \leq x) = \lim_{n \to \infty} \mathbb{P}(W'_n \leq x).$$

Furthermore,

$$(8) \qquad W'_n \leq W'_{n+1} \quad \text{for all } n \geq 1,$$

a monotonicity property which is not shared by $\{W_n\}$. This property provides another method for deriving the existence of $F$ in (2).

**Proof of (4).** From (8), the limit $W' = \lim_{n \to \infty} W'_n$ exists almost surely (and, in fact, pointwise) but may be $+\infty$. Furthermore,

$$(9) \qquad W' = \max\{0, \Sigma_1, \Sigma_2, \dots\}$$

where

$$\Sigma_n = \sum_{j=1}^{n} U_j$$

and $F(x) = \mathbb{P}(W' \leq x)$. Thus

$$F(x) = \mathbb{P}(\Sigma_n \leq x \text{ for all } n) \quad \text{if} \quad x \geq 0,$$

and the proof proceeds by using properties of the sequence $\{\Sigma_n\}$ of partial sums, such as the strong law (7.5.1):

**(10)**
$$\frac{1}{n}\Sigma_n \xrightarrow{\text{a.s.}} \mathbb{E}(U) \quad \text{as} \quad n \to \infty.$$

Suppose first that $\mathbb{E}(U) < 0$. Then

$$\mathbb{P}(\Sigma_n > 0 \text{ for infinitely many } n) = \mathbb{P}\left(\frac{1}{n}\Sigma_n - \mathbb{E}(U) > |\mathbb{E}(U)| \text{ i.o.}\right) = 0$$

by (10). Thus, from (9), $W'$ is almost surely the maximum of only finitely many terms, and so $\mathbb{P}(W' < \infty) = 1$, implying that $F$ is a non-defective distribution function.

Next suppose that $\mathbb{E}(U) > 0$. Pick any $x > 0$ and choose $N$ such that

$$N \geq \frac{2x}{\mathbb{E}(U)}.$$

For $n \geq N$,

$$\mathbb{P}(\Sigma_n \geq x) = \mathbb{P}\left(\frac{1}{n}\Sigma_n - \mathbb{E}(U) \geq \frac{x}{n} - \mathbb{E}(U)\right)$$

$$\geq \mathbb{P}\left(\frac{1}{n}\Sigma_n - \mathbb{E}(U) \geq -\tfrac{1}{2}\mathbb{E}(U)\right).$$

Let $n \to \infty$ and use the weak law to find that

$$\mathbb{P}(W' \geq x) \geq \mathbb{P}(\Sigma_n \geq x) \to 1 \quad \text{for all } x.$$

Therefore $W'$ almost surely exceeds any finite number, and so $\mathbb{P}(W' < \infty) = 0$ as required.

In the case when $\mathbb{E}(U) = 0$ these crude arguments do not work and we need a more precise measure of the fluctuations of $\Sigma_n$; one way of doing this is by way of the law of the iterated logarithm (7.6.1). If $\text{var}(U) > 0$ and $\mathbb{E}(U_1^2) < \infty$, then $\{\Sigma_n\}$ enjoys fluctuations of order $O(\sqrt{n \log \log n})$ in both positive and negative directions with probability 1, and so

$$\mathbb{P}(\Sigma_n \geq x \text{ for some } n) = 1 \quad \text{for all } x.$$

There are other arguments which yield the same result.  ∎

**(B) Imbedded random walk.** The sequence $\Sigma = \{\Sigma_n : n \geq 0\}$ given by

**(11)**
$$\Sigma_0 = 0, \quad \Sigma_n = \sum_{j=1}^{n} U_j \quad \text{for} \quad n \geq 1,$$

describes the path of a particle which performs a random walk on $\mathbb{R}$, jumping by an amount $U_n$ at the $n$th step. This simple observation leads to a wealth of conclusions about queueing systems. For example, we have just seen that the waiting time $W_n$ of the $n$th customer has the same distribution as the maximum $W_n'$ of the first $n$ positions of the walking particle. If

$\mathbb{E}(U) < 0$ then the waiting time distributions converge as $n \to \infty$, which is to say that the maximum displacement $W' = \lim W'_n$ is almost surely finite. Other properties also can be expressed in terms of this random walk, and the techniques of reflection and reversal which we discussed in Section 3.10 are useful here.

The limiting waiting time distribution is the same as the distribution of the maximum

$$W' = \max\{0, \Sigma_1, \Sigma_2, \dots\},$$

and so it is appropriate to study the so-called 'ladder points' of $\Sigma$. Define an increasing sequence $L(0)$, $L(1)$, ... of random variables by

$$L(0) = 0, \quad L(n+1) = \min\{m > L(n) : \Sigma_m > \Sigma_{L(n)}\};$$

that is, $L(n+1)$ is the earliest epoch $m$ of time at which $\Sigma_m$ exceeds the walk's previous maximum $\Sigma_{L(n)}$. The $L(n)$ are called *ladder points*; *negative ladder points* of $\Sigma$ are defined similarly as the epochs at which $\Sigma$ attains new minimum values. The result of (4) amounts to the assertion that

$$\mathbb{P}(\text{there exist infinitely many ladder points}) = \begin{cases} 0 & \text{if } \mathbb{E}(U) < 0, \\ 1 & \text{if } \mathbb{E}(U) > 0. \end{cases}$$

The total number of ladder points is given by the next lemma.

**(12) Lemma.** *Let $\eta = \mathbb{P}(\Sigma_n > 0 \text{ for some } n \geq 1)$ be the probability that at least one ladder point exists. The total number $\Lambda$ of ladder points has mass function*

$$\mathbb{P}(\Lambda = l) = (1 - \eta)\eta^l \quad \text{for} \quad l \geq 0.$$

**Proof.** The process $\Sigma$ is a discrete-time Markov chain. Thus

$$\mathbb{P}(\Lambda \geq l + 1 \mid \Lambda \geq l) = \eta$$

since the path of the walk after the $l$th ladder point is a copy of $\Sigma$ itself.  ∎

Thus the queue is stable if $\eta < 1$, in which case the maximum $W'$ of $\Sigma$ is related to the height of a typical ladder point. Let

$$Y_j = \Sigma_{L(j)} - \Sigma_{L(j-1)}$$

be the difference in the displacements of the walk at the $(j-1)$th and $j$th ladder points. Conditional on the value of $\Lambda$, $\{Y_j : 1 \leq j \leq \Lambda\}$ is a collection of independent identically distributed variables, by the Markov property. Furthermore,

**(13)** $$W' = \Sigma_{L(\Lambda)} = \sum_{j=1}^{\Lambda} Y_j;$$

this leads to the next lemma, relating the waiting time distribution to the distribution of a typical $Y_j$.

**(14) Lemma.** *If the traffic intensity $\rho$ satisfies $\rho < 1$, the equilibrium waiting time distribution has moment generating function*

$$M_W(s) = \frac{1 - \eta}{1 - \eta M_Y(s)}$$

*where $M_Y$ is the moment generating function of $Y$.*

**Proof.** We have that $\rho < 1$ if and only if $\eta < 1$. Use (13) and (5.1.25) to find that

$$M_W(s) = G_\Lambda(M_Y(s)).$$

Now use the result of Lemma (12).                                    ∎

Lemma (14) describes the waiting time distribution in terms of the distribution of $Y$. Analytical properties of $Y$ are a little tricky to obtain, and we restrict ourselves here to an elegant description of $Y$ which provides a curious link between pairs of 'dual' queueing systems.

The server of the queue enjoys busy periods during which she works continuously; in between busy periods she has *idle periods* during which she drinks tea. Let $I$ be the length of her first idle period.

**(15) Lemma.** *Let $L = \min\{m > 0; \Sigma_m < 0\}$ be the first negative ladder point of $\Sigma$. Then $I = -\Sigma_L$.*

That is, $I$ equals the absolute value of the depth of the first negative ladder point. It is of course possible that $\Sigma$ has *no* negative ladder points.

**Proof.** Call a customer *lucky* if he finds the queue empty as he arrives (customers who arrive at exactly the same time as the previous customer departs are deemed to be unlucky). We claim that the $(L+1)$th customer is the first lucky customer after the very first arrival. If this holds then (15) follows immediately since $I$ is the elapsed time between the $L$th departure and the $(L+1)$th arrival:

$$I = \sum_{j=1}^{L} X_{j+1} - \sum_{j=1}^{L} S_j = -\Sigma_L.$$

To verify the claim remember that

**(16)**     $W_n = \max\{0, V_n\}$   where   $V_n = \max\{U_{n-1}, U_{n-1} + U_{n-2}, \dots, \Sigma_{n-1}\}$

and note that the $n$th customer is lucky if and only if $V_n < 0$. Now

$$V_n \geq \Sigma_{n-1} \geq 0 \quad \text{for} \quad 2 \leq n \leq L,$$

and it remains to show that $V_{L+1} < 0$. To see this, note that

$$U_L + U_{L-1} + \cdots + U_{L-k} = \Sigma_L - \Sigma_{L-k-1} \leq \Sigma_L < 0$$

whenever $0 \leq k < L$. Now use (16) to obtain the result.                   ∎

Now we are ready to extract a remarkable identity which relates 'dual pairs' of queueing systems.

**(17) Definition.** If $Q$ is a queueing process with interarrival time distribution $F_X$ and service time distribution $F_S$, its **dual process** $Q_d$ is a queueing process with interarrival time distribution $F_S$ and service time distribution $F_X$.

For example, the dual of $M(\lambda)/G/1$ is $G/M(\lambda)/1$, and vice versa; we made use of this fact in the proof of (11.4.4). The traffic intensities $\rho$ and $\rho_d$ of $Q$ and $Q_d$ satisfy $\rho \rho_d = 1$; the processes $Q$ and $Q_d$ cannot both be stable except in pathological instances when all their interarrival and service times almost surely take the same constant value.

**(18) Theorem.** *Let $\Sigma$ and $\Sigma_d$ be the random walks associated with the queue $Q$ and its dual $Q_d$. Then $-\Sigma$ and $\Sigma_d$ are identically distributed random walks.*

**Proof.** Let $Q$ have interarrival times $\{X_n\}$ and service times $\{S_n\}$; $\Sigma$ has jumps of size $U_n = S_n - X_{n+1}$ ($n \geq 1$). The reflected walk $-\Sigma$, which is obtained by reflecting $\Sigma$ in the $x$-axis, has jumps of size $-U_n = X_{n+1} - S_n$ ($n \geq 1$) (see Section 3.10 for more details of the reflection principle). Write $\{S_n'\}$ and $\{X_n'\}$ for the interarrival and service times of $Q_d$; $\Sigma_d$ has jumps of size $U_n' = X_n' - S_{n+1}'$ ($n \geq 1$), which have the same distribution as the jumps of $-\Sigma$. ∎

This leads to a corollary.

**(19) Theorem.** *The height $Y$ of the first ladder point of $\Sigma$ has the same distribution as the length $I_d$ of a typical idle period in the dual queue.*

**Proof.** From (15), $-I_d$ is the height of the first ladder point of $-\Sigma_d$, which by (18) is distributed as the height $Y$ of the first ladder point of $\Sigma$. ∎

Here is an example of an application of these facts.

**(20) Theorem.** *Let $Q$ be a stable queueing process with dual process $Q_d$. Let $W$ be a typical equilibrium waiting time of $Q$ and $I_d$ a typical idle period of $Q_d$. Their moment generating functions are related by*

$$M_W(s) = \frac{1 - \eta}{1 - \eta M_{I_d}(s)}$$

*where $\eta = \mathbb{P}(W > 0)$.*

**Proof.** Use (14) and (19). ∎

An application of this result is given in Exercise (11.5.2). Another application is a second derivation of the equilibrium waiting time distribution (11.4.12) of $G/M/1$; just remark that the dual of $G/M/1$ is $M/G/1$, and that idle periods of $M/G/1$ are exponentially distributed (though, of course, the server does not have many such periods if the queue is unstable).

---

## Exercises for Section 11.5

**1.** Show that, for a $G/G/1$ queue, the starting times of the busy periods of the server constitute a renewal process.

**2.** Consider a $G/M(\mu)/1$ queue in equilibrium, together with the dual (unstable) $M(\mu)/G/1$ queue. Show that the idle periods of the latter queue are exponentially distributed. Use the theory of duality of queues to deduce for the former queue that: (a) the waiting-time distribution is a mixture of an exponential distribution and an atom at zero, and (b) the equilibrium queue length is geometric.

**3.**   Consider G/M($\mu$)/1, and let $G$ be the distribution function of $S - X$ where $S$ and $X$ are typical (independent) service and interarrival times. Show that the *Wiener–Hopf equation*

$$F(x) = \int_{-\infty}^{x} F(x - y) \, dG(y), \qquad x \geq 0,$$

for the limiting waiting-time distribution $F$ is satisfied by $F(x) = 1 - \eta e^{-\mu(1-\eta)x}$, $x \geq 0$. Here, $\eta$ is the smallest positive root of the equation $x = M_X(\mu(x - 1))$, where $M_X$ is the moment generating function of $X$.

## 11.6 Heavy traffic

A queue settles into equilibrium if its traffic intensity $\rho$ is less than 1; it is unstable if $\rho > 1$. It is our shared personal experience that many queues (such as in doctors' waiting rooms and at airport check-in desks) have a tendency to become unstable. The reason is simple: employers do not like to see their employees idle, and so they provide only just as many servers as are necessary to cope with the arriving customers. That is, they design the queueing system so that $\rho$ is only slightly smaller than 1; the ensuing queue is long but stable, and the server experiences 'heavy traffic'. As $\rho \uparrow 1$ the equilibrium queue length $Q_\rho$ becomes longer and longer, and it is interesting to ask for the rate at which $Q_\rho$ approaches infinity. Often it turns out that a suitably scaled form of $Q_\rho$ is asymptotically exponentially distributed. We describe this here for the M/D/1 system, leaving it to the readers to amuse themselves by finding corresponding results for other queues. In this special case, $Q_\rho \simeq Z/(1 - \rho)$ as $\rho \uparrow 1$ where $Z$ is an exponential variable.

**(1) Theorem.**   *Let $\rho = \lambda d$ be the traffic intensity of the $M(\lambda)/D(d)/1$ queue, and let $Q_\rho$ be a random variable with the stationary queue length distribution. Then $(1 - \rho)Q_\rho$ converges in distribution as $\rho \uparrow 1$ to the exponential distribution with parameter 2.*

**Proof.**   Use (11.3.5) to see that $Q_\rho$ has moment generating function

$$(2) \qquad\qquad M_\rho(s) = \frac{(1 - \rho)(e^s - 1)}{\exp[s - \rho(e^s - 1)] - 1} \qquad \text{if} \quad \rho < 1.$$

The moment generating function of $(1 - \rho)Q_\rho$ is $M_\rho((1 - \rho)s)$, and we make the appropriate substitution in equation (2). Now let $\rho \uparrow 1$ and use L'Hôpital's rule to deduce that $M_\rho((1 - \rho)s) \to 2/(2 - s)$. ∎

### Exercise for Section 11.6

**1.**   Consider the M($\lambda$)/M($\mu$)/1 queue with $\rho = \lambda/\mu < 1$. Let $Q_\rho$ be a random variable with the equilibrium queue distribution, and show that $(1 - \rho)Q_\rho$ converges in distribution as $\rho \uparrow 1$, the limit distribution being exponential with parameter 1.

## 11.7 Networks of queues

A customer departing from one queue may well be required to enter another. Networks of queues provide natural models for many situations in real life, ranging from the manufacture of components in complex industrial processes to an application for a visa for travel to another country.

We make concrete our notion of a queueing network as follows. There is a finite set $S$ of 'stations' labelled $s_1, s_2, \ldots, s_c$. At time $t$, station $i$ contains $Q_i(t)$ individuals, so that the state of the system may be represented as the vector $\mathbf{Q}(t) = (Q_1(t), Q_2(t), \ldots, Q_c(t))$. We assume for simplicity that the process $\mathbf{Q}$ is Markovian, taking values in the set

$$\mathcal{N} = \left\{ \mathbf{n} = (n_1, n_2, \ldots, n_c) : n_i = 0, 1, 2, \ldots \text{ for } 1 \leq i \leq c \right\}$$

of sequences of non-negative integers. The migration of customers between stations will be described in a rather general way in order to enable a breadth of applications.

Let $\mathbf{G} = (g_{\mathbf{m},\mathbf{n}})$ denote the generator of $\mathbf{Q}$, and write $\mathbf{e}_k = (0, 0, \ldots, 0, 1, 0, \ldots, 0)$ for the row vector of length $c$ with 1 in the $k$th position and 0 elsewhere. Assume that $\mathbf{Q}(t) = \mathbf{m}$. Three types of event may occur, at rates given by the following. For $i \neq j$, we set

$$g_{\mathbf{m},\mathbf{n}} = \begin{cases} \lambda_{ij}\phi_i(m_i) & \text{if } \mathbf{n} = \mathbf{m} - \mathbf{e}_i + \mathbf{e}_j, \\ \nu_j & \text{if } \mathbf{n} = \mathbf{m} + \mathbf{e}_j, \\ \mu_i\phi_i(m_i) & \text{if } \mathbf{n} = \mathbf{m} - \mathbf{e}_i, \\ 0 & \text{for other } \mathbf{n} \neq \mathbf{m}. \end{cases}$$

where $\lambda_{ij}, \nu_j, \mu_i$ are constants and the $\phi_i$ are functions such that $\phi_i(0) = 0$ and $\phi_i(m) > 0$ for $m \geq 1$. We assume for later use that $\lambda_{ii} = 0$ for all $i$. Thus a single customer transfers from station $i$ to station $j$ at rate $\lambda_{ij}\phi_i(m_i)$, a new arrival occurs at station $j$ at rate $\nu_j$, and a single customer departs the system from station $i$ at rate $\mu_i\phi_i(m_i)$. Note that transfers and departures may occur at rates which depend on the number of customers at that station; the functions $\phi_i$ are included to allow for general queue types (see the examples below).

Queueing networks defined in this rather general manner are sometimes termed 'migration processes' or 'Jackson networks'[†]. Here are some concrete instances. First, the network is termed a 'closed migration process' if $\nu_j = \mu_j = 0$ for all $j$, since in this case no arrival from or departure to the outside world is permitted. If some $\nu_j$ or $\mu_j$ is strictly positive, the network is termed an 'open migration process'. Closed migration processes are special inasmuch as they are restricted to a subset of $\mathcal{N}$ containing vectors $\mathbf{n}$ having constant sum.

**(3) Example.** Suppose that each station $i$ has $r$ servers, and that each customer at that station requires a service time having the exponential distribution with parameter $\gamma_i$. On departing station $i$, a customer proceeds to station $j$ ($\neq i$) with probability $p_{ij}$, or departs the system entirely with probability $q_i = 1 - \sum_{j:j\neq i} p_{ij}$. Assuming the usual independence, the corresponding migration process has parameters given by $\phi_i(n) = \min\{n, r\}$, $\lambda_{ij} = \gamma_i p_{ij}$, $\mu_i = \gamma_i q_i$, $\nu_j = 0$. ●

**(4) Example.** Suppose that customers are invisible to one another, in the sense that each proceeds around the network at rates which do not depend on the positions of other customers. In this case, we have $\phi_i(n) = n$. ●

---

[†]Named after J. R. Jackson and R. R. P. Jackson.

**(A) Closed migration processes.** We shall explore the equilibrium behaviour of closed processes, and we assume therefore that $v_j = \mu_j = 0$ for all $j$. The number of customers is constant, and we denote this number by $N$.

Consider first the case $N = 1$, and suppose for convenience that $\phi_j(1) = 1$ for all $j$. When at station $i$, the single customer moves to station $j$ at rate $\lambda_{ij}$. The customer's position is a continuous-time Markov chain with generator $\mathbf{H} = (h_{ij})$ given by

$$
(5) \qquad h_{ij} = \begin{cases} \lambda_{ij} & \text{if } i \neq j, \\ -\sum_k \lambda_{ik} & \text{if } i = j, \end{cases}
$$

and we assume henceforth that the chain is irreducible. It has a stationary distribution $\boldsymbol{\alpha} = (\alpha_i : i \in S)$ satisfying $\boldsymbol{\alpha}\mathbf{H} = \mathbf{0}$, which is to say that

$$
(6) \qquad \sum_j \alpha_j \lambda_{ji} = \alpha_i \sum_j \lambda_{ij} \quad \text{for } i \in S,
$$

and henceforth we let $\boldsymbol{\alpha}$ denote the solution to these equations†. We have by irreducibility that $\alpha_i > 0$ for all $i$. It is the case that $\sum_i \alpha_i = 1$, but this will be irrelevant in the following.

The stationary distribution in the case of a general closed migration process is given in the following theorem. We write $\mathcal{N}_N$ for the set of all vectors in $\mathcal{N}$ having sum $N$. Any empty product is to be interpreted as 1.

**(7) Theorem.** *The unique stationary distribution of an irreducible closed migration process with $N$ customers is given by*

$$
(8) \qquad \pi(\mathbf{n}) = B_N \prod_{i=1}^{c} \left\{ \frac{\alpha_i^{n_i}}{\prod_{r=1}^{n_i} \phi_i(r)} \right\}, \quad \mathbf{n} \in \mathcal{N}_N,
$$

*where $B_N$ is the appropriate normalizing constant.*

Note the product form of the stationary distribution in (8). This does not imply the independence of queue lengths in equilibrium since they are constrained to have constant sum $N$ and must therefore in general be dependent.

**Proof.** By Theorem (6.9.17a) the chain is non-explosive, and by Theorem (6.10.15) it has a unique stationary distribution $\boldsymbol{\gamma} = (\gamma(\mathbf{n}) : \mathbf{n} \in \mathcal{N}_N)$, which is the unique distribution satisfying the equations

$$
(9) \qquad \sum_{i,j} \gamma(\mathbf{n} - \mathbf{e}_i + \mathbf{e}_j)\lambda_{ji}\phi_j(n_j + 1) = \gamma(\mathbf{n}) \sum_{i,j} \lambda_{ij}\phi_i(n_i), \quad \mathbf{n} \in \mathcal{N}_N.
$$

This is a complicated system of equations. If we may solve the equation 'for each $i$', then we obtain a solution to (9) by summing over $i$. Removing from (9) the summation over $i$, we obtain the 'partial balance equations'

$$
(10) \qquad \sum_j \gamma(\mathbf{n} - \mathbf{e}_i + \mathbf{e}_j)\lambda_{ji}\phi_j(n_j + 1) = \gamma(\mathbf{n}) \sum_j \lambda_{ij}\phi_i(n_i), \quad \mathbf{n} \in \mathcal{N}_N^i, \ i \in S,
$$

---

†These equations may be viewed, after a change of variables, as the closed-migration version of the *traffic equations* (16).

where $\mathcal{N}_n^i$ is the subset of $\mathcal{N}$ containing all vectors $\mathbf{n}$ with $n_i \geq 1$. It suffices to check that (8) satisfies (10). With $\pi$ given by (8), we have that

$$\pi(\mathbf{n} - \mathbf{e}_i + \mathbf{e}_j) = \pi(\mathbf{n}) \frac{\alpha_j \phi_i(n_i)}{\alpha_i \phi_j(n_j + 1)}$$

whence $\pi$ satisfies (10) if and only if

$$\sum_j \frac{\alpha_j \phi_i(n_i)}{\alpha_i \phi_j(n_j + 1)} \cdot \lambda_{ji} \phi_j(n_j + 1) = \sum_j \lambda_{ij} \phi_i(n_i), \quad \mathbf{n} \in \mathcal{N}_N^i, \ i \in S.$$

The latter equation is simply (6), and the proof is complete. $\blacksquare$

**(11) Example.** A company has an antiquated telephone exchange with exactly $K$ incoming telephone lines and an ample number of operators. Calls arrive in the manner of a Poisson process with rate $\nu$. Each call occupies an operator for a time which is exponentially distributed with parameter $\lambda$, and then lasts a further period of time having the exponential distribution with parameter $\mu$. At the end of this time, the call ceases and the line becomes available for another incoming call. Arriving calls are lost if all $K$ channels are already in use.

Although the system of calls is a type of *open* queueing system, one may instead consider the *lines* as customers in a network having three stations. That is, at any time $t$ the state vector of the system may be taken to be $\mathbf{n} = (n_1, n_2, n_3)$ where $n_1$ is the number of free lines, $n_2$ is the number of calls being actively serviced by an operator, and $n_3$ is the number of calls still in operation but no longer utilizing an operator. This leads to a closed migration process with transition rates given by

$$\begin{aligned}
\lambda_{12} &= \nu, & \phi_1(n) &= I_{\{n \geq 1\}}, \\
\lambda_{23} &= \lambda, & \phi_2(n) &= n, \\
\lambda_{31} &= \mu, & \phi_3(n) &= n,
\end{aligned}$$

where $I_{\{n \geq 1\}}$ is the indicator function that $n \geq 1$. It is an easy exercise to show from (6) that the relative sizes of $\alpha_1, \alpha_2, \alpha_3$ satisfy $\alpha_1 : \alpha_2 : \alpha_3 = \nu^{-1} : \lambda^{-1} : \mu^{-1}$, whence the stationary distribution is given by

$$(12) \qquad \pi(n_1, n_2, n_3) = B \cdot \frac{1}{\nu^{n_1}} \cdot \frac{1}{\lambda^{n_2} n_2!} \cdot \frac{1}{\mu^{n_3} n_3!}, \qquad n_1 + n_2 + n_3 = K,$$

for an appropriate constant $B$. $\bullet$

**(B) Open migration processes.** We turn now to the general situation in which customers may enter or leave the system. As in the case of a closed migration process, it is valuable to consider first an auxiliary process containing exactly one customer. We attach another station, labelled $\infty$, to those already in existence, and we consider the following closed migration process on the augmented set $S \cup \{\infty\}$ of stations. There is a unique customer who, when at station $i$, moves to station $j$ ($\neq i$) at rate:

$$\begin{cases} \lambda_{ij} & \text{if } 1 \leq i, j \leq c, \\ \mu_i & \text{if } j = \infty, \\ \nu_j & \text{if } i = \infty. \end{cases}$$

*We assume henceforth that this auxiliary process is irreducible.* Let $\mathbf{J}$ be its generator. The chain has a unique stationary distribution $\boldsymbol{\beta} = (\beta_1, \beta_2, \ldots, \beta_c, \beta_\infty)$ which is specified by the equation $\boldsymbol{\beta}\mathbf{J} = \mathbf{0}$. By Exercise (11.7.7), this is equivalent to the equations

$$\beta_\infty v_i + \sum_{j \in S} \beta_j \lambda_{ji} = \beta_i \left( \mu_i + \sum_{j \in S} \lambda_{ij} \right) \quad \text{for } i \in S.$$

Note that $\beta_i > 0$ for $i \in S \cup \{\infty\}$. We set $\alpha_i = \beta_i/\beta_\infty$, to obtain a vector $\boldsymbol{\alpha} = (\alpha_i : i \in S)$ with strictly positive entries such that

(13)
$$v_i + \sum_{j \in S} \alpha_j \lambda_{ji} = \alpha_i \left( \mu_i + \sum_{j \in S} \lambda_{ij} \right) \quad \text{for } i \in S.$$

Since $\boldsymbol{\beta}$ is unique, $\boldsymbol{\alpha}$ is the unique positive solution to (13). We shall make use of this vector $\boldsymbol{\alpha}$ in very much the same way as we used equation (6) for closed migration processes. We let

$$D_i = \sum_{n=0}^{\infty} \frac{\alpha_i^n}{\prod_{r=1}^{n} \phi_i(r)}, \qquad i \in S.$$

(14) **Theorem.** *Assume that the above auxiliary process is irreducible, and that $D_i < \infty$ for all $i \in S$. The open migration process has unique stationary distribution*

(15)
$$\pi(\mathbf{n}) = \prod_{i=1}^{c} \pi_i(n_i), \quad \mathbf{n} \in \mathcal{N},$$

*where*
$$\pi_i(n_i) = D_i^{-1} \frac{\alpha_i^{n_i}}{\prod_{r=1}^{n_i} \phi_i(r)}.$$

Let us substitute $\gamma_i = \alpha_i g_i$ in (13), or equivalenly $\gamma_i = \beta_i g_i/\beta_\infty$, where $g_i = \mu_i + \sum_{j \neq i} \lambda_{ij}$ is the total rate of departure of a customer at station $i$. This leads to the so-called *traffic equations*

(16)
$$\gamma_i = v_i + \sum_{j \in S} \gamma_j y_{ji}, \qquad i \in S,$$

where $\mathbf{Y} = (y_{ij})$ is the transition matrix of the jump chain $Y$ of the single-customer system. These equations describe a notional system in equilibrium in which $\gamma_i$ is the aggregate arrival rate at $i$. Since $\boldsymbol{\alpha}$ is the unique solution to (13), $\boldsymbol{\gamma}$ is the unique solution to the traffic equations†.

**Proof of Theorem (14).** We note first that the chain is irreducible by assumption, and in addition non-explosive (see Exercise (11.7.6)). The latter follows from the fact that the aggregate arrival process is a Poisson process with finite rate $\sum_i v_i$.

---

†Indeed $\boldsymbol{\gamma} = \boldsymbol{\rho}/\rho_\infty$ where $\boldsymbol{\rho}$ is the unique stationary distribution of the jump chain $Y$ of the single-customer system.

Let $\mathbf{G} = (g(\mathbf{n}, \mathbf{n}') : \mathbf{n}, \mathbf{n}' \in \mathcal{N})$ be the generator of the Markov chain $\mathbf{Q}$. By Theorem (6.10.15), a distribution $\gamma = (\gamma(\mathbf{n}) : \mathbf{n} \in \mathcal{N})$ is the unique stationary distribution if only if $\gamma\mathbf{G} = \mathbf{0}$, which is to say that

**(17)**

$$\sum_i \gamma(\mathbf{n} - \mathbf{e}_i)g(\mathbf{n} - \mathbf{e}_i, \mathbf{n}) + \sum_{i,j} \gamma(\mathbf{n} - \mathbf{e}_i + \mathbf{e}_j)g(\mathbf{n} - \mathbf{e}_i + \mathbf{e}_j, \mathbf{n}) + \sum_j \gamma(\mathbf{n} + \mathbf{e}_j)g(\mathbf{n} + \mathbf{e}_j, \mathbf{n})$$

$$= \gamma(\mathbf{n}) \left( \sum_i g(\mathbf{n}, \mathbf{n} - \mathbf{e}_i) + \sum_{i,j} g(\mathbf{n}, \mathbf{n} - \mathbf{e}_i + \mathbf{e}_j) + \sum_j g(\mathbf{n}, \mathbf{n} + \mathbf{e}_j) \right).$$

These are solved by $\gamma$ if it satisfies the 'partial balance equations'

**(18)**   $\gamma(\mathbf{n} - \mathbf{e}_i)g(\mathbf{n} - \mathbf{e}_i, \mathbf{n}) + \sum_j \gamma(\mathbf{n} - \mathbf{e}_i + \mathbf{e}_j)g(\mathbf{n} - \mathbf{e}_i + \mathbf{e}_j, \mathbf{n})$

$$= \gamma(\mathbf{n}) \left( g(\mathbf{n}, \mathbf{n} - \mathbf{e}_i) + \sum_j g(\mathbf{n}, \mathbf{n} - \mathbf{e}_i + \mathbf{e}_j) \right), \quad i \in S,$$

together with the equation

**(19)**                $\sum_j \gamma(\mathbf{n} + \mathbf{e}_j)g(\mathbf{n} + \mathbf{e}_j, \mathbf{n}) = \gamma(\mathbf{n}) \sum_j g(\mathbf{n}, \mathbf{n} + \mathbf{e}_j).$

Equation (18) is concerned with moves between stations, and equation (19) with arrivals and departures from the network.

We now substitute (15) into (18) and divide through by $\pi(\mathbf{n})$ to find that $\pi$ satisfies (18) by reason of (13). Substituting $\pi$ into (19), we obtain the equation

$$\sum_j \alpha_j \mu_j = \sum_j \nu_j,$$

whose validity follows by summing (13) over $i$.                                              ∎

Equation (15) has the important and striking consequence that, in equilibrium, the queue lengths at the different stations are independent random variables. Note, in contrast, that the queue *processes* $Q_i(\cdot)$, $i \in S$, are highly dependent on one another.

It is a remarkable fact that the reversal in time of an open migration process yields another open migration process.

**(20) Theorem.** *Let* $\mathbf{Q} = \{Q(t) : -\infty < t < \infty\}$ *be an irreducible open migration process. Assume that the condition of Theorem (14) holds, and that* $\mathbf{Q}(t)$ *has distribution* $\pi$ *given by* (15) *for* $t \geq 0$. *Then* $\mathbf{Q}'(t) = \mathbf{Q}(-t)$ *is an open migration network with parameters*

$$\lambda'_{ij} = \frac{\alpha_j \lambda_{ji}}{\alpha_i}, \quad \nu'_j = \alpha_j \mu_j, \quad \mu'_i = \frac{\nu_i}{\alpha_i}, \quad \phi'_i(\cdot) = \phi_i(\cdot),$$

*where* $\alpha$ *satisfies* (13).

**Proof.** It is a straightforward exercise in conditional probabilities (see Problem (6.15.16)) to show that $\mathbf{Q}'$ is a Markov chain with generator $\mathbf{G}' = (g'(\mathbf{m}, \mathbf{n}) : \mathbf{m}, \mathbf{n} \in \mathcal{N})$ given by

$$g'(\mathbf{n}, \mathbf{n} - \mathbf{e}_i + \mathbf{e}_j) = \frac{\pi(\mathbf{n} - \mathbf{e}_i + \mathbf{e}_j)g(\mathbf{n} - \mathbf{e}_i + \mathbf{e}_j, \mathbf{n})}{\pi(\mathbf{n})} = \lambda'_{ij}\phi_i(n_i),$$

$$g'(\mathbf{n}, \mathbf{n} - \mathbf{e}_i) = \frac{\pi(\mathbf{n} - \mathbf{e}_i)g(\mathbf{n} - \mathbf{e}_i, \mathbf{n})}{\pi(\mathbf{n})} = \mu'_i\phi_i(n_i),$$

$$g'(\mathbf{n}, \mathbf{n} + \mathbf{e}_j) = \frac{\pi(\mathbf{n} + \mathbf{e}_j)g(\mathbf{n} + \mathbf{e}_j, \mathbf{n})}{\pi(\mathbf{n})} = v'_j,$$

as required. ∎

Here is one noteworthy consequence of this useful theorem. Let $\mathbf{Q}$ be an open migration network in equilibrium, and consider the processes of departures from the network from the various stations. That is, let $D_i(t)$ be the number of customers who depart the system from station $i$ during the time interval $[0, t]$. These departures correspond to arrivals at station $i$ in the reversed process $\mathbf{Q}'$. However, such arrival processes are independent Poisson processes, with respective parameters $v'_i = \alpha_i \mu_i$. It follows that the departure processes are independent Poisson processes with these parameters.

---

## Exercises for Section 11.7

**1.** Consider an open migration process with $c$ stations, in which individuals arrive at station $j$ at rate $v_j$, individuals move from $i$ to $j$ at rate $\lambda_{ij}\phi_i(n_i)$, and individuals depart from $i$ at rate $\mu_i\phi_i(n_i)$, where $n_i$ denotes the number of individuals currently at station $i$. Show when $\phi_i(n_i) = n_i$ for all $i$ that the system behaves as though the customers move independently through the network. Identify the explicit form of the stationary distribution, subject to an assumption of irreducibility, and explain a connection with the Bartlett theorem of Problem (8.10.6).

**2.** Let $Q$ be an M($\lambda$)/M($\mu$)/$s$ queue where $\lambda < s\mu$, and assume $Q$ is in equilibrium. Show that the process of departures is a Poisson process with intensity $\lambda$, and that departures up to time $t$ are independent of the value of $Q(t)$.

**3.** Customers arrive in the manner of a Poisson process with intensity $\lambda$ in a shop having two servers. The service times of these servers are independent and exponentially distributed with respective parameters $\mu_1$ and $\mu_2$. Arriving customers form a single queue, and the person at the head of the queue moves to the first free server. When both servers are free, the next arrival is allocated a server chosen according to one of the following rules:
(a) each server is equally likely to be chosen,
(b) the server who has been free longer is chosen.
Assume that $\lambda < \mu_1 + \mu_2$, and the process is in equilibrium. Show in each case that the process of departures from the shop is a Poisson process, and that departures prior to time $t$ are independent of the number of people in the shop at time $t$.

**4. Difficult customers.** Consider an M($\lambda$)/M($\mu$)/1 queue modified so that on completion of service the customer leaves with probability $\delta$, or rejoins the queue with probability $1 - \delta$. Find the distribution of the total time a customer spends being served. Hence show that equilibrium is possible if $\lambda < \delta\mu$, and find the stationary distribution. Show that, in equilibrium, the departure process is Poisson, but if the rejoining customer goes to the end of the queue, the composite arrival process is not Poisson.

**5.** Consider an open migration process in equilibrium. If there is no path by which an individual at station $k$ can reach station $j$, show that the stream of individuals moving directly from station $j$ to station $k$ forms a Poisson process.

**6.** Show that an open migration process is non-explosive.

**7.** Let $X$ be an irreducible, continuous-time Markov chain with generator $\mathbf{G}$ on the state space $T = S \cup \{\infty\}$, where $S$ is countable and non-empty. Show that a distribution $\pi$ on $T$ satisfies $\pi \mathbf{G} = \mathbf{0}$ if and only if, for $j \in S$, $\sum_{i \in T} \pi_i g_{ij} = 0$.

# 11.8 Problems

**1. Finite waiting room.** Consider M($\lambda$)/M($\mu$)/$k$ with the constraint that arriving customers who see $N$ customers in the line ahead of them leave and never return. Find the stationary distribution of queue length for the cases $k = 1$ and $k = 2$.

**2. Baulking.** Consider M($\lambda$)/M($\mu$)/1 with the constraint that if an arriving customer sees $n$ customers in the line ahead of him, he joins the queue with probability $p(n)$ and otherwise leaves in disgust.

(a) Find the stationary distribution of queue length if $p(n) = (n + 1)^{-1}$.

(b) Find the stationary distribution $\pi$ of queue length if $p(n) = 2^{-n}$, and show that the probability that an arriving customer joins the queue (in equilibrium) is $\mu(1 - \pi_0)/\lambda$.

**3. Series.** In a Moscow supermarket customers queue at the cash desk to pay for the goods they want; then they proceed to a second line where they wait for the goods in question. If customers arrive in the shop like a Poisson process with parameter $\lambda$ and all service times are independent and exponentially distributed, parameter $\mu_1$ at the first desk and $\mu_2$ at the second, find the stationary distributions of queue lengths, when they exist, and show that, at any given time, the two queue lengths are independent in equilibrium.

**4. Batch (or bulk) service.** Consider M/G/1, with the modification that the server may serve up to $m$ customers simultaneously. If the queue length is less than $m$ at the beginning of a service period then she serves everybody waiting at that time. Find a formula which is satisfied by the probability generating function of the stationary distribution of queue length at the times of departures, and evaluate this generating function explicitly in the case when $m = 2$ and service times are exponentially distributed.

**5.** Consider M($\lambda$)/M($\mu$)/1 where $\lambda < \mu$. Find the moment generating function of the length $B$ of a typical busy period, and show that $\mathbb{E}(B) = (\mu - \lambda)^{-1}$ and $\mathrm{var}(B) = (\lambda + \mu)/(\mu - \lambda)^3$. Show that the density function of $B$ is

$$f_B(x) = \frac{\sqrt{\mu/\lambda}}{x} e^{-(\lambda+\mu)x} I_1\left(2x\sqrt{\lambda\mu}\right) \qquad \text{for } x > 0$$

where $I_1$ is a modified Bessel function.

**6.** Consider M($\lambda$)/G/1 in equilibrium. Obtain an expression for the mean queue length at departure times. Show that the mean waiting time in equilibrium of an arriving customer is $\frac{1}{2}\lambda \mathbb{E}(S^2)/(1 - \rho)$ where $S$ is a typical service time and $\rho = \lambda \mathbb{E}(S)$.

Amongst all possible service-time distributions with given mean, find the one for which the mean waiting time is a minimum.

**7.** Let $W_t$ be the time which a customer would have to wait in a M($\lambda$)/G/1 queue if he were to arrive at time $t$. Show that the distribution function $F(x; t) = \mathbb{P}(W_t \le x)$ satisfies

$$\frac{\partial F}{\partial t} = \frac{\partial F}{\partial x} - \lambda F + \lambda \mathbb{P}(W_t + S \le x)$$

where $S$ is a typical service time, independent of $W_t$.

Suppose that $F(x, t) \to H(x)$ for all $x$ as $t \to \infty$, where $H$ is a distribution function satisfying $0 = h - \lambda H + \lambda \mathbb{P}(U + S \le x)$ for $x > 0$, where $U$ is independent of $S$ with distribution function $H$, and $h$ is the density function of $H$ on $(0, \infty)$. Show that the moment generating function $M_U$ of $U$ satisfies

$$M_U(\theta) = \frac{(1 - \rho)\theta}{\lambda + \theta - \lambda M_S(\theta)}$$

where $\rho$ is the traffic intensity. You may assume that $\mathbb{P}(S = 0) = 0$.

**8.**   Consider a G/G/1 queue in which the service times are constantly equal to 2, whilst the interarrival times take either of the values 1 and 4 with equal probability $\frac{1}{2}$. Find the limiting waiting time distribution.

**9.**   Consider an extremely idealized model of a telephone exchange having infinitely many channels available. Calls arrive in the manner of a Poisson process with intensity $\lambda$, and each requires one channel for a length of time having the exponential distribution with parameter $\mu$, independently of the arrival process and of the duration of other calls. Let $Q(t)$ be the number of calls being handled at time $t$, and suppose that $Q(0) = I$.

Determine the probability generating function of $Q(t)$, and deduce $\mathbb{E}(Q(t))$, $\mathbb{P}(Q(t) = 0)$, and the limiting distribution of $Q(t)$ as $t \to \infty$.

Assuming the queue is in equilibrium, find the proportion of time that no channels are occupied, and the mean length of an idle period. Deduce that the mean length of a busy period is $(e^{\lambda/\mu} - 1)/\lambda$.

**10.**   Customers arrive in a shop in the manner of a Poisson process with intensity $\lambda$, where $0 < \lambda < 1$. They are served one by one in the order of their arrival, and each requires a service time of unit length. Let $Q(t)$ be the number in the queue at time $t$. By comparing $Q(t)$ with $Q(t + 1)$, determine the limiting distribution of $Q(t)$ as $t \to \infty$ (you may assume that the quantities in question converge). Hence show that the mean queue length in equilibrium is $\lambda(1 - \frac{1}{2}\lambda)/(1 - \lambda)$.

Let $W$ be the waiting time of a newly arrived customer when the queue is in equilibrium. Deduce from the results above that $\mathbb{E}(W) = \frac{1}{2}\lambda/(1 - \lambda)$.

**11.**   Consider M($\lambda$)/D(1)/1, and suppose that the queue is empty at time 0. Let $T$ be the earliest time at which a customer departs leaving the queue empty. Show that the moment generating function $M_T$ of $T$ satisfies

$$\log\left(1 - \frac{s}{\lambda}\right) + \log M_T(s) = (s - \lambda)\big(1 - M_T(s)\big),$$

and deduce the mean value of $T$, distinguishing between the cases $\lambda < 1$ and $\lambda \geq 1$.

**12.**   Suppose $\lambda < \mu$, and consider a M($\lambda$)/M($\mu$)/1 queue $Q$ in equilibrium.
(a) Show that $Q$ is a reversible Markov chain.
(b) Deduce the equilibrium distributions of queue length and waiting time.
(c) Show that the times of departures of customers form a Poisson process, and that $Q(t)$ is independent of the times of departures prior to $t$.
(d) Consider a sequence of $K$ single-server queues such that customers arrive at the first in the manner of a Poisson process, and (for each $j$) on completing service in the $j$th queue each customer moves to the $(j + 1)$th. Service times in the $j$th queue are exponentially distributed with parameter $\mu_j$, with as much independence as usual. Determine the (joint) equilibrium distribution of the queue lengths, when $\lambda < \mu_j$ for all $j$.

**13.**   Consider the queue M($\lambda$)/M($\mu$)/k, where $k \geq 1$. Show that a stationary distribution $\pi$ exists if and only if $\lambda < k\mu$, and calculate it in this case.

Suppose that the cost of operating this system in equilibrium is

$$Ak + B \sum_{n=k}^{\infty}(n - k + 1)\pi_n,$$

the positive constants $A$ and $B$ representing respectively the costs of employing a server and of the dissatisfaction of delayed customers.

Show that, for fixed $\mu$, there is a unique value $\lambda^*$ in the interval $(0, \mu)$ such that it is cheaper to have $k = 1$ than $k = 2$ if and only if $\lambda < \lambda^*$.

**14.**   Customers arrive in a shop in the manner of a Poisson process with intensity $\lambda$. They form a single queue. There are two servers, labelled 1 and 2, server $i$ requiring an exponentially distributed time with parameter $\mu_i$ to serve any given customer. The customer at the head of the queue is served by the first idle server; when both are idle, an arriving customer is equally likely to choose either.

(a) Show that the queue length settles into equilibrium if and only if $\lambda < \mu_1 + \mu_2$.

(b) Show that, when in equilibrium, the queue length is a time-reversible Markov chain.

(c) Deduce the equilibrium distribution of queue length.

(d) Generalize your conclusions to queues with many servers.

**15.** Consider the D(1)/M($\mu$)/1 queue where $\mu > 1$, and let $Q_n$ be the number of people in the queue just before the $n$th arrival. Let $Q_\mu$ be a random variable having as distribution the stationary distribution of the Markov chain $\{Q_n\}$. Show that $(1 - \mu^{-1})Q_\mu$ converges in distribution as $\mu \downarrow 1$, the limit distribution being exponential with parameter 2.

**16. Kendall's taxicabs.** Taxis arrive at a stand in the manner of a Poisson process with intensity $\tau$, and passengers arrive in the manner of an (independent) Poisson process with intensity $\pi$. If there are no waiting passengers, the taxis wait until passengers arrive, and then move off with the passengers, one to each taxi. If there is no taxi, passengers wait until they arrive. Suppose that initially there are neither taxis nor passengers at the stand. Show that the probability that $n$ passengers are waiting at time $t$ is $(\pi/\tau)^{\frac{1}{2}n}e^{-(\pi+\tau)t}I_n(2t\sqrt{\pi\tau})$, where $I_n(x)$ is the modified Bessel function, i.e. the coefficient of $z^n$ in the power series expansion of $\exp\{\frac{1}{2}x(z + z^{-1})\}$.

**17.** Machines arrive for repair as a Poisson process with intensity $\lambda$. Each repair involves two stages, the $i$th machine to arrive being under repair for a time $X_i + Y_i$, where the pairs $(X_i, Y_i), i = 1, 2, \ldots$, are independent with a common joint distribution. Let $U(t)$ and $V(t)$ be the numbers of machines in the $X$-stage and $Y$-stage of repair at time $t$. Show that $U(t)$ and $V(t)$ are independent Poisson random variables.

**18. Ruin.** An insurance company pays independent and identically distributed claims $\{K_n : n \geq 1\}$ at the instants of a Poisson process with intensity $\lambda$, where $\lambda\mathbb{E}(K_1) < 1$. Premiums are received at constant rate 1. Show that the maximum deficit $M$ the company will ever accumulate has moment generating function

$$\mathbb{E}(e^{\theta M}) = \frac{(1 - \rho)\theta}{\lambda + \theta - \lambda\mathbb{E}(e^{\theta K})}.$$

**19.** (a) **Erlang's loss formula.** Consider M($\lambda$)/M($\mu$)/$s$ with baulking, in which a customer departs immediately if, on arrival, he sees all the servers occupied ahead of him. Show that, in equilibrium, the probability that all servers are occupied is

$$\pi_s = \frac{\rho^s/s!}{\sum_{j=0}^{s}\rho^j/j!}, \qquad \text{where } \rho = \lambda/\mu.$$

(b) Consider an M($\lambda$)/M($\mu$)/$\infty$ queue with channels (servers) numbered $1, 2, \ldots$. On arrival, a customer will choose the lowest numbered channel that is free, and be served by that channel. Show in the notation of part (a) that the fraction $p_c$ of time that channel $c$ is busy is $p_c = \rho(\pi_{c-1} - \pi_c)$ for $c \geq 2$, and $p_1 = \pi_1$.

**20.** For an M($\lambda$)/M($\mu$)/1 queue with $\lambda < \mu$, when in equilibrium, show that the expected time until the queue is first empty is $\lambda(\mu - \lambda)^{-2}$.

**21.** Consider an M($\lambda$)/G/$\infty$ queue. Use the renewal–reward theorem to show that the expected duration of a busy period is $(e^\rho - 1)/\lambda$ where $\rho = \lambda\mathbb{E}(S)$.

# 12

# Martingales

*Summary.* The general theory of martingales and submartingales has many applications. After an account of the concentration inequality for martingales, the martingale convergence theorem is proved via the upcrossings inequality. Stopping times are studied, and the optional stopping theorem proved. This leads to Wald's identity and the maximal inequality. The chapter ends with a discussion of backward martingales and continuous-time martingales. Many examples of the use of martingale theory are included.

## 12.1 Introduction

Random processes come in many forms, and their analysis depends heavily on the assumptions that one is prepared to make about them. There are certain broad classes of processes whose general properties enable one to build attractive theories. Two such classes are Markov processes and stationary processes. A third is the class of martingales.

**(1) Definition.** A sequence $Y = \{Y_n : n \geq 0\}$ is a **martingale** with respect to the sequence $X = \{X_n : n \geq 0\}$ if, for all $n \geq 0$,

(a) $\mathbb{E}|Y_n| < \infty$,

(b) $\mathbb{E}(Y_{n+1} \mid X_0, X_1, \ldots, X_n) = Y_n$.

A warning note: conditional expectations are ubiquitous in this chapter. Remember that they are random variables, and that formulae of the form $\mathbb{E}(A \mid B) = C$ generally hold only 'almost surely'. We shall omit the term 'almost surely' throughout the chapter.

Here are some examples of martingales; further examples may be found in Section 7.7.

**(2) Example. Simple random walk.** A particle jumps either one step to the right or one step to the left, with corresponding probabilities $p$ and $q$ $(= 1 - p)$. Assuming the usual independence of different moves, it is clear that the position $S_n = X_1 + X_2 + \cdots + X_n$ of the particle after $n$ steps satisfies $\mathbb{E}|S_n| \leq n$ and

$$\mathbb{E}(S_{n+1} \mid X_1, X_2, \ldots, X_n) = S_n + (p - q),$$

whence it is easily seen that $Y_n = S_n - n(p - q)$ defines a martingale with respect to $X$. ●

**(3) Example. The martingale.** The following gambling strategy is called a martingale. A gambler has a large fortune. He wagers £1 on an evens bet. If he loses then he wagers £2 on the next bet. If he loses the first $n$ plays, then he bets $\pounds 2^n$ on the $(n+1)$th. He is bound to win sooner or later, say on the $T$th bet, at which point he ceases to play, and leaves with his profit of $2^T - (1 + 2 + 4 + \cdots + 2^{T-1})$. Thus, following this strategy, he is assured an ultimate profit. This sounds like a good policy.

Writing $Y_n$ for the accumulated gain of the gambler after the $n$th play (losses count negative), we have that $Y_0 = 0$ and $|Y_n| \le 1 + 2 + \cdots + 2^{n-1} = 2^n - 1$. Furthermore, $Y_{n+1} = Y_n$ if the gambler has stopped by time $n + 1$, and

$$
Y_{n+1} = \begin{cases} Y_n - 2^n & \text{with probability } \tfrac{1}{2}, \\ Y_n + 2^n & \text{with probability } \tfrac{1}{2}, \end{cases}
$$

otherwise, implying that $\mathbb{E}(Y_{n+1} \mid Y_1, Y_2, \ldots, Y_n) = Y_n$. Therefore $Y$ is a martingale (with respect to itself).

As remarked in Example (7.7.1), this martingale possesses a particularly disturbing feature. The random time $T$ has a geometric distribution, $\mathbb{P}(T = n) = (\tfrac{1}{2})^n$ for $n \ge 1$, so that the mean loss of the gambler just before his ultimate win is

$$
\sum_{n=1}^{\infty} (\tfrac{1}{2})^n (1 + 2 + \cdots + 2^{n-2})
$$

which equals infinity. Do not follow this strategy unless your initial capital is considerably greater than that of the casino.                                                                                                ●

**(4) Example. De Moivre's martingale.** About a century before the martingale was fashionable amongst Paris gamblers, Abraham de Moivre made use of a (mathematical) martingale to answer the following 'gambler's ruin' question. A simple random walk on the set $\{0, 1, 2, \ldots, N\}$ stops when it first hits either of the absorbing barriers at 0 and at $N$; what is the probability that it stops at the barrier 0?

Write $X_1, X_2, \ldots$ for the steps of the walk, and $S_n$ for the position after $n$ steps, where $S_0 = k$. Define $Y_n = (q/p)^{S_n}$ where $p = \mathbb{P}(X_i = 1)$, $p + q = 1$, and $0 < p < 1$. We claim that

(5)                                    $\mathbb{E}(Y_{n+1} \mid X_1, X_2, \ldots, X_n) = Y_n$    for all $n$.

If $S_n$ equals 0 or $N$ then the process has stopped by time $n$, implying that $S_{n+1} = S_n$ and therefore $Y_{n+1} = Y_n$. If on the other hand $0 < S_n < N$, then

$$
\mathbb{E}(Y_{n+1} \mid X_1, X_2, \ldots, X_n) = \mathbb{E}\big((q/p)^{S_n + X_{n+1}} \mid X_1, X_2, \ldots, X_n\big)
$$
$$
= (q/p)^{S_n} \big[ p(q/p) + q(q/p)^{-1} \big] = Y_n,
$$

and (5) is proved. It follows, by taking expectations of (5), that $\mathbb{E}(Y_{n+1}) = \mathbb{E}(Y_n)$ for all $n$, and hence $\mathbb{E}|Y_n| = \mathbb{E}|Y_0| = (q/p)^k$ for all $n$. In particular $Y$ is a martingale (with respect to the sequence $X$).

Let $T$ be the number of steps before the absorption of the particle at either 0 or $N$. De Moivre argued as follows: $\mathbb{E}(Y_n) = (q/p)^k$ for all $n$, and therefore $\mathbb{E}(Y_T) = (q/p)^k$. If you are willing to accept this remark, then the answer to the original question is a simple consequence, as follows. Expanding $\mathbb{E}(Y_T)$, we have that

$$
\mathbb{E}(Y_T) = (q/p)^0 p_k + (q/p)^N (1 - p_k)
$$

where $p_k = \mathbb{P}(\text{absorbed at 0} \mid S_0 = k)$. However, $\mathbb{E}(Y_T) = (q/p)^k$ by assumption, and therefore

$$p_k = \frac{\rho^k - \rho^N}{1 - \rho^N} \quad \text{where} \quad \rho = q/p$$

(so long as $\rho \neq 1$), in agreement with the calculation of Example (3.9.6).

This is a very attractive method, which relies on the statement that $\mathbb{E}(Y_T) = \mathbb{E}(Y_0)$ for a certain type of random variable $T$. A major part of our investigation of martingales will be to determine conditions on such random variables $T$ which ensure that the desired statements are true.    ●

**(6) Example. Markov chains.** Let $X$ be a discrete-time Markov chain taking values in the countable state space $S$ with transition matrix $\mathbf{P}$. Suppose that $\psi : S \to S$ is bounded and *harmonic*, which is to say that

$$\sum_{j \in S} p_{ij} \psi(j) = \psi(i) \quad \text{for all } i \in S.$$

It is easily seen that $Y = \{\psi(X_n) : n \geq 0\}$ is a martingale with respect to $X$: simply use the Markov property in order to perform the calculation:

$$\mathbb{E}\big(\psi(X_{n+1}) \,\big|\, X_1, X_2, \dots, X_n\big) = \mathbb{E}\big(\psi(X_{n+1}) \,\big|\, X_n\big) = \sum_{j \in S} p_{X_n, j} \psi(j) = \psi(X_n).$$

More generally, suppose that $\psi$ is a right eigenvector of $\mathbf{P}$, which is to say that there exists $\lambda \,(\neq 0)$ such that

$$\sum_{j \in S} p_{ij} \psi(j) = \lambda \psi(i), \quad i \in S.$$

Then

$$\mathbb{E}\big(\psi(X_{n+1}) \,\big|\, X_1, X_2, \dots, X_n\big) = \lambda \psi(X_n),$$

implying that $\lambda^{-n} \psi(X_n)$ defines a martingale so long as $\mathbb{E}|\psi(X_n)| < \infty$ for all $n$.    ●

Central to the definition of a martingale is the idea of conditional expectation, a subject developed to some extent in Chapter 7. As described there, the most general form of conditional expectation is of the following nature. Let $Y$ be a random variable on the probability space $(\Omega, \mathcal{F}, \mathbb{P})$ having finite mean, and let $\mathcal{G}$ be a sub-$\sigma$-field of $\mathcal{F}$. The conditional expectation of $Y$ given $\mathcal{G}$, written $\mathbb{E}(Y \mid \mathcal{G})$, is a $\mathcal{G}$-measurable random variable satisfying

(7)                    $\mathbb{E}\big([Y - \mathbb{E}(Y \mid \mathcal{G})]I_G\big) = 0 \quad \text{for all events } G \in \mathcal{G},$

where $I_G$ is the indicator function of $G$. There is a corresponding general definition of a martingale. In preparation for this, we introduce the following terminology. Suppose that $\mathcal{F} = \{\mathcal{F}_0, \mathcal{F}_1, \dots\}$ is a sequence of sub-$\sigma$-fields of $\mathcal{F}$; we call $\mathcal{F}$ a *filtration* if $\mathcal{F}_n \subseteq \mathcal{F}_{n+1}$ for all $n$. A sequence $Y = \{Y_n : n \geq 0\}$ is said to be *adapted* to the filtration $\mathcal{F}$ if $Y_n$ is $\mathcal{F}_n$-measurable for all $n$. Given a filtration $\mathcal{F}$, we normally write $\mathcal{F}_\infty = \lim_{n\to\infty} \mathcal{F}_n$ for the smallest $\sigma$-field containing $\mathcal{F}_n$ for all $n$.

**(8) Definition.** Let $\mathcal{F}$ be a filtration of the probability space $(\Omega, \mathcal{F}, \mathbb{P})$, and let $Y$ be a sequence of random variables which is adapted to $\mathcal{F}$. We call the pair $(Y, \mathcal{F}) = \{(Y_n, \mathcal{F}_n) : n \geq 0\}$ a **martingale** if, for all $n \geq 0$,

(a) $\mathbb{E}|Y_n| < \infty$,

(b) $\mathbb{E}(Y_{n+1} \mid \mathcal{F}_n) = Y_n$.

The former definition (1) is retrieved by choosing $\mathcal{F}_n = \sigma(X_0, X_1, \ldots, X_n)$, the smallest $\sigma$-field with respect to which each of the variables $X_0, X_1, \ldots, X_n$ is measurable. We shall sometimes suppress reference to the filtration $\mathcal{F}$, speaking only of a martingale $Y$.

Note that, if $Y$ is a martingale with respect to $\mathcal{F}$, then it is also a martingale with respect to $\mathcal{G}$, where $\mathcal{G}_n = \sigma(Y_0, Y_1, \ldots, Y_n)$. A further minor point is that martingales need not be infinite in extent: a finite sequence $\{(Y_n, \mathcal{F}_n) : 0 \leq n \leq N\}$ satisfying the above definition is also termed a martingale.

There are many cases of interest in which the martingale condition $\mathbb{E}(Y_{n+1} \mid \mathcal{F}_n) = Y_n$ does not hold, being replaced instead by an inequality: $\mathbb{E}(Y_{n+1} \mid \mathcal{F}_n) \geq Y_n$ for all $n$, or $\mathbb{E}(Y_{n+1} \mid \mathcal{F}_n) \leq Y_n$ for all $n$. Sequences satisfying such inequalities have many of the properties of martingales, and we have special names for them.

**(9) Definition.** Let $\mathcal{F}$ be a filtration of the probability space $(\Omega, \mathcal{F}, \mathbb{P})$, and let $Y$ be a sequence of random variables which is adapted to $\mathcal{F}$. We call the pair $(Y, \mathcal{F})$ a **submartingale** if, for all $n \geq 0$,

(a) $\mathbb{E}(Y_n^+) < \infty$,

(b) $\mathbb{E}(Y_{n+1} \mid \mathcal{F}_n) \geq Y_n$,

or a **supermartingale** if, for all $n \geq 0$,

(c) $\mathbb{E}(Y_n^-) < \infty$,

(d) $\mathbb{E}(Y_{n+1} \mid \mathcal{F}_n) \leq Y_n$.

Remember that $X^+ = \max\{0, X\}$ and $X^- = -\min\{0, X\}$, so that $X = X^+ - X^-$ and $|X| = X^+ + X^-$. The moment conditions (a) and (c) are weaker than the condition that $\mathbb{E}|Y_n| < \infty$. Note that $Y$ is a martingale if and only if it is both a submartingale and a supermartingale. Also, $Y$ is a submartingale if and only if $-Y$ is a supermartingale.

Sometimes we shall write that $(Y_n, \mathcal{F}_n)$ is a (sub/super)martingale in cases where we mean the corresponding statement for $(Y, \mathcal{F})$.

It can be somewhat tiresome to deal with sub(/super)martingales and martingales separately, keeping track of their various properties. The general picture is somewhat simplified by the following result, which expresses a submartingale as the sum of a martingale and an increasing 'predictable' process. We shall not make use of this decomposition in the rest of the chapter. Here is a piece of notation. We call the pair $(S, \mathcal{F}) = \{(S_n, \mathcal{F}_n) : n \geq 0\}$ *predictable* if $S_n$ is $\mathcal{F}_{n-1}$-measurable for all $n \geq 1$. We call a predictable process $(S, \mathcal{F})$ *increasing* if $S_0 = 0$ and $\mathbb{P}(S_n \leq S_{n+1}) = 1$ for all $n$.

**(10) Theorem. Doob decomposition.** *A submartingale* $(Y, \mathcal{F})$ *with finite means may be expressed in the form*

**(11)**
$$Y_n = M_n + S_n$$

*where* $(M, \mathcal{F})$ *is a martingale, and* $(S, \mathcal{F})$ *is an increasing predictable process. This decomposition is unique.*

The process $(S, \mathcal{F})$ in (11) is called the *compensator* of the submartingale $(Y, \mathcal{F})$. Note that compensators have finite mean, since $0 \le S_n \le Y_n^+ - M_n$, implying that

**(12)**                            $\mathbb{E}|S_n| \le \mathbb{E}(Y_n^+) + \mathbb{E}|M_n|.$

**Proof.** We define $M$ and $S$ explicitly as follows: $M_0 = Y_0$, $S_0 = 0$,

$$M_{n+1} - M_n = Y_{n+1} - \mathbb{E}(Y_{n+1} \mid \mathcal{F}_n), \quad S_{n+1} - S_n = \mathbb{E}(Y_{n+1} \mid \mathcal{F}_n) - Y_n,$$

for $n \ge 0$. It is easy to check (*exercise*) that $(M, \mathcal{F})$ and $(S, \mathcal{F})$ satisfy the statement of the theorem. To see uniqueness, suppose that $Y_n = M_n' + S_n'$ is another such decomposition. Then

$$Y_{n+1} - Y_n = (M_{n+1}' - M_n') + (S_{n+1}' - S_n')$$
$$= (M_{n+1} - M_n) + (S_{n+1} - S_n).$$

Take conditional expectations given $\mathcal{F}_n$ to obtain $S_{n+1}' - S_n' = S_{n+1} - S_n$, $n \ge 0$. However, $S_0' = S_0 = 0$, and therefore $S_n' = S_n$, implying that $M_n' = M_n$. (Most of the last few statements should be qualified by 'almost surely'.) ∎

---

## Exercises for Section 12.1

**1.**    (a) If $(Y, \mathcal{F})$ is a martingale, show that $\mathbb{E}(Y_n) = \mathbb{E}(Y_0)$ for all $n$.
(b) If $(Y, \mathcal{F})$ is a submartingale (respectively supermartingale) with finite means, show that $\mathbb{E}(Y_n) \ge \mathbb{E}(Y_0)$ (respectively $\mathbb{E}(Y_n) \le \mathbb{E}(Y_0)$).

**2.**    Let $(Y, \mathcal{F})$ be a martingale, and show that $\mathbb{E}(Y_{n+m} \mid \mathcal{F}_n) = Y_n$ for all $n, m \ge 0$.

**3.**    Let $Z_n$ be the size of the $n$th generation of a branching process with $Z_0 = 1$, having mean family size $\mu$ and extinction probability $\eta$. Show that $Z_n \mu^{-n}$ and $\eta^{Z_n}$ define martingales.

**4.**    Let $\{S_n : n \ge 0\}$ be a simple symmetric random walk on the integers with $S_0 = k$. Show that $S_n$ and $S_n^2 - n$ are martingales. Making assumptions similar to those of de Moivre (see Example (12.1.4)), find the probability of ruin and the expected duration of the game for the gambler's ruin problem.

**5.**    Let $(Y, \mathcal{F})$ be a martingale with the property that $\mathbb{E}(Y_n^2) < \infty$ for all $n$. Show that, for $i \le j \le k$, $\mathbb{E}\{(Y_k - Y_j)Y_i\} = 0$, and $\mathbb{E}\{(Y_k - Y_j)^2 \mid \mathcal{F}_i\} = \mathbb{E}(Y_k^2 \mid \mathcal{F}_i) - \mathbb{E}(Y_j^2 \mid \mathcal{F}_i)$. Suppose there exists $K$ such that $\mathbb{E}(Y_n^2) \le K$ for all $n$. Show that the sequence $\{Y_n\}$ converges in mean square as $n \to \infty$.

**6.**    Let $Y$ be a martingale and let $u$ be a convex function mapping $\mathbb{R}$ to $\mathbb{R}$. Show that $\{u(Y_n) : n \ge 0\}$ is a submartingale provided that $\mathbb{E}(u(Y_n)^+) < \infty$ for all $n$.

Show that $|Y_n|$, $Y_n^2$, and $Y_n^+$ constitute submartingales whenever the appropriate moment conditions are satisfied.

**7.**    Let $Y$ be a submartingale and let $u$ be a convex non-decreasing function mapping $\mathbb{R}$ to $\mathbb{R}$. Show that $\{u(Y_n) : n \ge 0\}$ is a submartingale provided that $\mathbb{E}(u(Y_n)^+) < \infty$ for all $n$.

Show that (subject to a moment condition) $Y_n^+$ constitutes a submartingale, but that $|Y_n|$ and $Y_n^2$ need not constitute submartingales.

**8.**    Let $X$ be a discrete-time Markov chain with countable state space $S$ and transition matrix $\mathbf{P}$. Suppose that $\psi : S \to \mathbb{R}$ is bounded and satisfies $\sum_{j \in S} p_{ij} \psi(j) \le \lambda \psi(i)$ for some $\lambda > 0$ and all $i \in S$. Show that $\lambda^{-n} \psi(X_n)$ constitutes a supermartingale.

**9.**    Let $G_n(s)$ be the probability generating function of the size $Z_n$ of the $n$th generation of a branching process, where $Z_0 = 1$ and $\text{var}(Z_1) > 0$. Let $H_n$ be the inverse function of the function $G_n$, viewed as a function on the interval $[0, 1]$, and show that $M_n = \{H_n(s)\}^{Z_n}$ defines a martingale with respect to the sequence $Z$.

## 12.2 Martingale differences and Hoeffding's inequality

Much of the theory of martingales is concerned with their behaviour as $n \to \infty$, and particularly with their properties of convergence. Of supreme importance is the martingale convergence theorem, a general result of great power and with many applications. Before giving an account of that theorem (in the next section), we describe a bound on the degree of fluctuation of a martingale. This bound is straightforward to derive and has many important applications.

Let $(Y, \mathcal{F})$ be a martingale. The sequence of *martingale differences* is the sequence $D = \{D_n : n \geq 1\}$ defined by $D_n = Y_n - Y_{n-1}$, so that

$$(1) \qquad\qquad Y_n = Y_0 + \sum_{i=1}^{n} D_i .$$

Note that the sequence $D$ is such that $D_n$ is $\mathcal{F}_n$-measurable, $\mathbb{E}|D_n| < \infty$, and

$$(2) \qquad\qquad \mathbb{E}(D_{n+1} \mid \mathcal{F}_n) = 0 \quad \text{for all } n.$$

**(3) Theorem. Hoeffding's inequality.** *Let $(Y, \mathcal{F})$ be a martingale, and suppose that there exists a sequence $K_1, K_2, \ldots$ of real numbers such that $\mathbb{P}(|Y_n - Y_{n-1}| \leq K_n) = 1$ for all $n$. Then*

$$\mathbb{P}(|Y_n - Y_0| \geq x) \leq 2 \exp\left( -\tfrac{1}{2}x^2 \bigg/ \sum_{i=1}^{n} K_i^2 \right), \quad x > 0.$$

That is to say, if the martingale differences are bounded (almost surely) then there is only a small chance of a large deviation of $Y_n$ from its initial value $Y_0$.

**Proof.** We begin with an elementary inequality. If $\psi > 0$, the function $g(d) = e^{\psi d}$ is convex, whence it follows that

$$(4) \qquad\qquad e^{\psi d} \leq \tfrac{1}{2}(1-d)e^{-\psi} + \tfrac{1}{2}(1+d)e^{\psi} \quad \text{if } |d| \leq 1.$$

Applying this to a random variable $D$ having mean 0 and satisfying $\mathbb{P}(|D| \leq 1) = 1$, we obtain

$$(5) \qquad\qquad \mathbb{E}(e^{\psi D}) \leq \tfrac{1}{2}(e^{-\psi} + e^{\psi}) < e^{\frac{1}{2}\psi^2},$$

by a comparison of the coefficients of $\psi^{2n}$ for $n \geq 0$.

Moving to the proof proper, it is a consequence of Markov's inequality, Theorem (7.3.1), that

$$(6) \qquad\qquad \mathbb{P}(Y_n - Y_0 \geq x) \leq e^{-\theta x} \mathbb{E}(e^{\theta(Y_n - Y_0)})$$

for $\theta > 0$. Writing $D_n = Y_n - Y_{n-1}$, we have that

$$\mathbb{E}(e^{\theta(Y_n - Y_0)}) = \mathbb{E}(e^{\theta(Y_{n-1} - Y_0)} e^{\theta D_n}).$$

By conditioning on $\mathcal{F}_{n-1}$, we obtain

(7) $$\mathbb{E}(e^{\theta(Y_n - Y_0)} \mid \mathcal{F}_{n-1}) = e^{\theta(Y_{n-1} - Y_0)}\mathbb{E}(e^{\theta D_n} \mid \mathcal{F}_{n-1})$$
$$\leq e^{\theta(Y_{n-1} - Y_0)}\exp(\tfrac{1}{2}\theta^2 K_n^2),$$

where we have used the fact that $Y_{n-1} - Y_0$ is $\mathcal{F}_{n-1}$-measurable, in addition to (5) applied to the random variable $D_n/K_n$. We take expectations of (7) and iterate to find that

$$\mathbb{E}(e^{\theta(Y_n - Y_0)}) \leq \mathbb{E}(e^{\theta(Y_{n-1} - Y_0)})\exp(\tfrac{1}{2}\theta^2 K_n^2) \leq \exp\left(\tfrac{1}{2}\theta^2 \sum_{i=1}^{n} K_i^2\right).$$

Therefore, by (6),

$$\mathbb{P}(Y_n - Y_0 \geq x) \leq \exp\left(-\theta x + \tfrac{1}{2}\theta^2 \sum_{i=1}^{n} K_i^2\right)$$

for all $\theta > 0$. Suppose $x > 0$, and set $\theta = x \big/ \sum_{i=1}^{n} K_i^2$ (this is the value which minimizes the exponent); we obtain

$$\mathbb{P}(Y_n - Y_0 \geq x) \leq \exp\left(-\tfrac{1}{2}x^2 \big/ \sum_{i=1}^{n} K_i^2\right), \quad x > 0.$$

The same argument is valid with $Y_n - Y_0$ replaced by $Y_0 - Y_n$, and the claim of the theorem follows by adding the two (identical) bounds together. ∎

**(8) Example. Large deviations.** Let $X_1, X_2, \ldots$ be independent random variables, $X_i$ having the Bernoulli distribution with parameter $p$. We set $S_n = X_1 + X_2 + \cdots + X_n$ and $Y_n = S_n - np$ to obtain a martingale $Y$. It is a consequence of Hoeffding's inequality that

$$\mathbb{P}(|S_n - np| \geq x\sqrt{n}) \leq 2\exp(-\tfrac{1}{2}x^2/\mu^2) \quad \text{for} \quad x > 0,$$

where $\mu = \max\{p, 1 - p\}$. This is an inequality of a type encountered already as Bernstein's inequality (2.2.4), and explored in greater depth in Section 5.11. ●

**(9) Example. Bin packing.** The bin packing problem is a basic problem of operations research. Given $n$ objects with sizes $x_1, x_2, \ldots, x_n$, and an unlimited collection of bins each of size 1, what is the minimum number of bins required in order to pack the objects? In the randomized version of this problem, we suppose that the objects have independent random sizes $X_1, X_2, \ldots$ having some common distribution on $[0, 1]$. Let $B_n$ be the (random) number of bins required in order to pack $X_1, X_2, \ldots, X_n$ efficiently; that is, $B_n$ is the minimum number of bins of unit capacity such that the sum of the sizes of the objects in any given bin does not exceed its capacity. It may be shown that $B_n$ grows approximately linearly in $n$, in that there exists a positive constant $\beta$ such that $n^{-1}B_n \to \beta$ a.s. and in mean square as $n \to \infty$. We shall not prove this here, but note its consequence:

(10) $$\frac{1}{n}\mathbb{E}(B_n) \to \beta \quad \text{as} \quad n \to \infty.$$

The next question might be to ask how close $B_n$ is to its mean value $\mathbb{E}(B_n)$, and Hoeffding's inequality may be brought to bear here. For $i \leq n$, let $Y_i = \mathbb{E}(B_n \mid \mathcal{F}_i)$, where $\mathcal{F}_i$ is the $\sigma$-field generated by $X_1, X_2, \ldots , X_i$. It is easily seen that $(Y, \mathcal{F})$ is a martingale, albeit one of finite length. Furthermore $Y_n = B_n$, and $Y_0 = \mathbb{E}(B_n)$ since $\mathcal{F}_0$ is the trivial $\sigma$-field $\{\varnothing, \Omega\}$.

Now, let $B_n(i)$ be the minimal number of bins required in order to pack all the objects *except* the $i$th. Since the objects are packed efficiently, we must have $B_n(i) \leq B_n \leq B_n(i)+1$. Taking conditional expectations given $\mathcal{F}_{i-1}$ and $\mathcal{F}_i$, we obtain

**(11)**
$$\mathbb{E}\big(B_n(i) \,\big|\, \mathcal{F}_{i-1}\big) \leq Y_{i-1} \leq \mathbb{E}\big(B_n(i) \,\big|\, \mathcal{F}_{i-1}\big) + 1,$$
$$\mathbb{E}\big(B_n(i) \,\big|\, \mathcal{F}_i\big) \leq Y_i \;\; \leq \mathbb{E}\big(B_n(i) \,\big|\, \mathcal{F}_i\big) + 1.$$

However, $\mathbb{E}(B_n(i) \mid \mathcal{F}_{i-1}) = \mathbb{E}(B_n(i) \mid \mathcal{F}_i)$, since we are not required to pack the $i$th object, and hence knowledge of $X_i$ is irrelevant. It follows from (11) that $|Y_i - Y_{i-1}| \leq 1$. We may now apply Hoeffding's inequality (3) to find that

**(12)**
$$\mathbb{P}\big(|B_n - \mathbb{E}(B_n)| \geq x\big) \leq 2\exp(-\tfrac{1}{2}x^2/n), \quad x > 0.$$

For example, setting $x = \epsilon n$, we see that the chance that $B_n$ deviates from its mean by $\epsilon n$ (or more) decays exponentially in $n$ as $n \to \infty$. Using (10) we have also that, as $n \to \infty$,

**(13)**
$$\mathbb{P}(|B_n - \beta n| \geq \epsilon n) \leq 2\exp\big\{-\tfrac{1}{2}\epsilon^2 n[1 + \mathrm{o}(1)]\big\}. \qquad \bullet$$

**(14) Example. Travelling salesman problem.** A travelling salesman is required to visit $n$ towns but may choose his route. How does he find the shortest possible route, and how long is it? Here is a randomized version of the problem. Let $P_1 = (U_1, V_1)$, $P_2 = (U_2, V_2), \ldots , P_n = (U_n, V_n)$ be independent and uniformly distributed points in the unit square $[0, 1]^2$; that is, suppose that $U_1, U_2, \ldots , U_n, V_1, V_2, \ldots , V_n$ are independent random variables each having the uniform distribution on $[0, 1]$. It is required to tour these points using an aeroplane. If we tour them in the order $P_{\pi(1)}, P_{\pi(2)}, \ldots , P_{\pi(n)}$, for some permutation $\pi$ of $\{1, 2, \ldots , n\}$, the total length of the journey is

$$d(\pi) = \sum_{i=1}^{n-1} |P_{\pi(i+1)} - P_{\pi(i)}| + |P_{\pi(n)} - P_{\pi(1)}|$$

where $|\cdot|$ denotes Euclidean distance. The shortest tour has length $D_n = \min_\pi d(\pi)$. It turns out that the asymptotic behaviour of $D_n$ for large $n$ is given as follows: there exists a positive constant $\tau$ such that $D_n/\sqrt{n} \to \tau$ a.s. and in mean square. We shall not prove this, but note the consequence that

**(15)**
$$\frac{1}{\sqrt{n}} \mathbb{E}(D_n) \to \tau \quad \text{as} \quad n \to \infty.$$

How close is $D_n$ to its mean? As in the case of bin packing, this question may be answered in part with the aid of Hoeffding's inequality. Once again, we set $Y_i = \mathbb{E}(D_n \mid \mathcal{F}_i)$ for $i \leq n$, where $\mathcal{F}_i$ is the $\sigma$-field generated by $P_1, P_2, \ldots , P_i$. As before, $(Y, \mathcal{F})$ is a martingale, and $Y_n = D_n$, $Y_0 = \mathbb{E}(D_n)$.

Let $D_n(i)$ be the minimal tour-length through the points $P_1, P_2, \ldots, P_{i-1}, P_{i+1}, \ldots, P_n$, and note that $\mathbb{E}(D_n(i) \mid \mathcal{F}_i) = \mathbb{E}(D_n(i) \mid \mathcal{F}_{i-1})$. The vital inequality is

$$(16) \qquad\qquad D_n(i) \leq D_n \leq D_n(i) + 2Z_i, \quad i \leq n-1,$$

where $Z_i$ is the shortest distance from $P_i$ to one of the points $P_{i+1}, P_{i+2}, \ldots, P_n$. It is obvious that $D_n \geq D_n(i)$ since every tour of all $n$ points includes a tour of the subset $P_1, \ldots, P_{i-1}, P_{i+1}, \ldots, P_n$. To obtain the second inequality of (16), we argue as follows. Suppose that $P_j$ is the closest point to $P_i$ amongst the set $\{P_{i+1}, P_{i+2}, \ldots, P_n\}$. One way of visiting all $n$ points is to follow the optimal tour of $P_1, \ldots, P_{i-1}, P_{i+1}, \ldots, P_n$, and on arriving at $P_j$ we make a return trip to $P_i$. The resulting trajectory is not quite a tour, but it can be turned into a tour by not landing at $P_j$ on the return but going directly to the next point; the resulting tour has length no greater than $D_n(i) + 2Z_i$.

We take conditional expectations of (16) to obtain

$$\mathbb{E}\big(D_n(i) \mid \mathcal{F}_{i-1}\big) \leq Y_{i-1} \leq \mathbb{E}\big(D_n(i) \mid \mathcal{F}_{i-1}\big) + 2\mathbb{E}(Z_i \mid \mathcal{F}_{i-1}),$$
$$\mathbb{E}\big(D_n(i) \mid \mathcal{F}_i\big) \leq Y_i \;\; \leq \mathbb{E}\big(D_n(i) \mid \mathcal{F}_i\big) + 2\mathbb{E}(Z_i \mid \mathcal{F}_i),$$

and hence

$$(17) \qquad\qquad |Y_i - Y_{i-1}| \leq 2 \max\big\{\mathbb{E}(Z_i \mid \mathcal{F}_i), \mathbb{E}(Z_i \mid \mathcal{F}_{i-1})\big\}, \quad i \leq n-1.$$

In order to estimate the right side here, let $Q \in [0, 1]^2$, and let $Z_i(Q)$ be the shortest distance from $Q$ to the closest of a collection of $n - i$ points chosen uniformly at random from the unit square. If $Z_i(Q) > x$ then no point lies within the circle $C(x, Q)$ having radius $x$ and centre at $Q$. Note that $\sqrt{2}$ is the largest possible distance between two points in the square. Now, there exists $c$ such that, for all $x \in (0, \sqrt{2}]$, the intersection of $C(x, Q)$ with the unit square has area at least $cx^2$, uniformly in $Q$. Therefore

$$(18) \qquad\qquad \mathbb{P}\big(Z_i(Q) > x\big) \leq (1 - cx^2)^{n-i}, \quad 0 < x \leq \sqrt{2}.$$

Integrating over $x$, we find that

$$\mathbb{E}(Z_i(Q)) \leq \int_0^{\sqrt{2}} (1 - cx^2)^{n-i} \, dx \leq \int_0^{\sqrt{2}} e^{-cx^2(n-i)} \, dx < \frac{C}{\sqrt{n-i}}$$

for some constant $C$; (*exercise*). Returning to (17), we deduce that the random variables $\mathbb{E}(Z_i \mid \mathcal{F}_i)$ and $\mathbb{E}(Z_i \mid \mathcal{F}_{i-1})$ are smaller than $C/\sqrt{n-i}$, whence $|Y_i - Y_{i-1}| \leq 2C/\sqrt{n-i}$ for $i \leq n-1$. For the case $i = n$, we use the trivial bound $|Y_n - Y_{n-1}| \leq 2\sqrt{2}$, being twice the length of the diagonal of the square.

Applying Hoeffding's inequality, we obtain

$$(19) \qquad\qquad \mathbb{P}(|D_n - \mathbb{E}D_n| \geq x) \leq 2\exp\left(-\frac{x^2}{2(8 + \sum_{i=1}^{n-1} 4C^2/i)}\right)$$
$$\leq 2\exp(-Ax^2/\log n), \quad x > 0,$$

for some positive constant $A$. Combining this with (15), we find that

$$\mathbb{P}(|D_n - \tau\sqrt{n}| \geq \epsilon\sqrt{n}) \leq 2\exp(-B\epsilon^2 n/\log n), \quad \epsilon > 0,$$

for some positive constant $B$ and all large $n$.                                    ●

**(20) Example. Markov chains.** Let $X = \{X_n : n \geq 0\}$ be an irreducible aperiodic Markov chain on the finite state space $S$ with transition matrix $\mathbf{P}$. Denote by $\pi$ the stationary distribution of $X$, and suppose that $X_0$ has distribution $\pi$, so that $X$ is stationary. Fix a state $s \in S$, and let $N(n)$ be the number of visits of $X_1, X_2, \ldots, X_n$ to $s$. The sequence $N$ is a delayed renewal process and therefore $n^{-1} N(n) \xrightarrow{\text{a.s.}} \pi_s$ as $n \to \infty$. The convergence is rather fast, as the following (somewhat overcomplicated) argument indicates.

Let $\mathcal{F}_0 = \{\emptyset, \Omega\}$ and, for $0 < m \leq n$, let $\mathcal{F}_m = \sigma(X_1, X_2, \ldots, X_m)$. Set $Y_m = \mathbb{E}(N(n) \mid \mathcal{F}_m)$ for $m \geq 0$, so that $(Y_m, \mathcal{F}_m)$ is a martingale. Note that $Y_n = N(n)$ and $Y_0 = \mathbb{E}(N(n)) = n\pi_s$ by stationarity.

We write $N(m, n) = N(n) - N(m)$, $0 \leq m \leq n$, the number of visits to $s$ by the subsequence $X_{m+1}, X_{m+2}, \ldots, X_n$. Now

$$Y_m = \mathbb{E}\big(N(m) \mid \mathcal{F}_m\big) + \mathbb{E}\big(N(m, n) \mid \mathcal{F}_m\big) = N(m) + \mathbb{E}\big(N(m, n) \mid X_m\big)$$

by the Markov property. Therefore, if $m \geq 1$,

$$Y_m - Y_{m-1} = [N(m) - N(m-1)] + \big[\mathbb{E}\big(N(m, n) \mid X_m\big) - \mathbb{E}\big(N(m-1, n) \mid X_{m-1}\big)\big]$$
$$= \mathbb{E}\big(N(m-1, n) \mid X_m\big) - \mathbb{E}\big(N(m-1, n) \mid X_{m-1}\big)$$

since $N(m) - N(m-1) = \delta_{X_m, s}$, the Kronecker delta. It follows that

$$|Y_m - Y_{m-1}| \leq \max_{t, u \in S}\Big|\mathbb{E}\big(N(m-1, n) \mid X_m = t\big) - \mathbb{E}\big(N(m-1, n) \mid X_{m-1} = u\big)\Big|$$
$$= \max_{t, u \in S} |D_m(t, u)|$$

where, by the time homogeneity of the process,

**(21)**      $D_m(t, u) = \mathbb{E}\big(N(n - m + 1) \mid X_1 = t\big) - \mathbb{E}\big(N(n - m + 1) \mid X_0 = u\big).$

It is easily seen that

$$\mathbb{E}\big(N(n - m + 1) \mid X_1 = t\big) \leq \delta_{ts} + \mathbb{E}(T_{tu}) + \mathbb{E}\big(N(n - m + 1) \mid X_0 = u\big),$$

where $\mathbb{E}(T_{xy})$ is the mean first-passage time from state $x$ to state $y$; just wait for the first passage to $u$, counting one for each moment which elapses. Similarly

$$\mathbb{E}\big(N(n - m + 1) \mid X_0 = u\big) \leq \mathbb{E}(T_{ut}) + \mathbb{E}\big(N(n - m + 1) \mid X_1 = t\big).$$

Hence, by (21), $|D_m(t, u)| \leq 1 + \max\{\mathbb{E}(T_{tu}), \mathbb{E}(T_{ut})\}$, implying that

**(22)**                                $|Y_m - Y_{m-1}| \leq 1 + \mu$

where $\mu = \max\{\mathbb{E}(T_{xy}) : x, y \in S\}$; note that $\mu < \infty$ since $S$ is finite. Applying Hoeffding's inequality, we deduce that

$$\mathbb{P}\big(|N(n) - n\pi_s| \geq x\big) \leq 2 \exp\left(-\frac{x^2}{2n(\mu + 1)}\right), \qquad x > 0.$$

Setting $x = n\epsilon$, we obtain

$$(23) \qquad \mathbb{P}\left(\left|\frac{1}{n}N(n) - \pi_s\right| \geq \epsilon\right) \leq 2 \exp\left(-\frac{n\epsilon^2}{2(\mu + 1)}\right), \qquad \epsilon > 0,$$

a large-deviation estimate which decays exponentially fast as $n \to \infty$. Similar inequalities may be established by other means, more elementary than those used above. ●

---

## Exercises for Section 12.2

**1.    Knapsack problem.** It is required to pack a knapsack to maximum benefit. Suppose you have $n$ objects, the $i$th object having volume $V_i$ and worth $W_i$, where $V_1, V_2, \ldots, V_n, W_1, W_2, \ldots, W_n$ are independent non-negative random variables with finite means, and $W_i \leq M$ for all $i$ and some fixed $M$. Your knapsack has volume $c$, and you wish to maximize the total worth of the objects packed in it. That is, you wish to find the vector $z_1, z_2, \ldots, z_n$ of 0's and 1's such that $\sum_1^n z_i V_i \leq c$ and which maximizes $\sum_1^n z_i W_i$. Let $Z$ be the maximal possible worth of the knapsack's contents, and show that $\mathbb{P}(|Z - \mathbb{E}Z| \geq x) \leq 2\exp\{-x^2/(2nM^2)\}$ for $x > 0$.

**2.    Graph colouring.** Given $n$ vertices $v_1, v_2, \ldots, v_n$, for each $1 \leq i < j \leq n$ we place an edge between $v_i$ and $v_j$ with probability $p$; different pairs are joined independently of each other. We call $v_i$ and $v_j$ *neighbours* if they are joined by an edge. The *chromatic number* $\chi$ of the ensuing graph is the minimal number of pencils of different colours which are required in order that each vertex may be coloured differently from each of its neighbours. Show that $\mathbb{P}(|\chi - \mathbb{E}\chi| \geq x) \leq 2\exp\{-\frac{1}{2}x^2/n\}$ for $x > 0$.

**3.    Maurer's inequality.**

(a) Let $X$ and $Y$ be random variables such that $X \leq b < \infty$ a.s., $\mathbb{E}(X \mid Y) = 0$, and $\mathbb{E}(X^2 \mid Y) \leq \sigma^2 < \infty$. Show that

$$\mathbb{E}(e^{\theta X} \mid Y) \leq \exp\left\{\tfrac{1}{2}\theta^2(b^2 + \sigma^2)\right\}, \qquad \theta \geq 0.$$

(b) Let $(M, \mathcal{F})$ be a martingale with $M_0 = 0$, having differences $D_r = M_r - M_{r-1}$, and suppose that $D_n \leq b_n < \infty$ a.s., and $\mathbb{E}(D_n^2 \mid M_{n-1}) \leq \sigma_n^2 < \infty$ for $n \geq 1$. Show that

$$\mathbb{P}(M_n \geq t) \leq \exp\left\{-\frac{t^2}{2\sum_{r=1}^n (b_r^2 + \sigma_r^2)}\right\}, \qquad t \geq 0.$$

**4.    Quadratic variation.** Let $(M, \mathcal{F})$ be a martingale with $M_0 = 0$, having differences $D_r = M_r - M_{r-1}$. The process given by $Q_n = \sum_{r=1}^n D_r^2$ is called the *optional quadratic variation* of $M$, while $V_n = \sum_{r=1}^n \mathbb{E}(D_r^2 \mid \mathcal{F}_{r-1})$ is called the *predictable quadratic variation* of $M$. Show that $X_n = M_n^2 - Q_n$ and $Y_n = M_n^2 - V_n$ define martingales with respect to $\mathcal{F}$.

## 12.3  Crossings and convergence

Martingales are of immense value in proving convergence theorems, and the following famous result has many applications.

**(1) Martingale convergence theorem.** *Let $(Y, \mathcal{F})$ be a submartingale and suppose that $\mathbb{E}(Y_n^+) \le M$ for some $M$ and all $n$. There exists a random variable $Y_\infty$ such that $Y_n \overset{\text{a.s.}}{\longrightarrow} Y_\infty$ as $n \to \infty$. We have in addition that:*
   *(i) $Y_\infty$ has finite mean if $\mathbb{E}|Y_0| < \infty$, and*
   *(ii) $Y_n \overset{1}{\to} Y_\infty$ if the sequence $\{Y_n : n \ge 0\}$ is uniformly integrable.*

It follows of course that any submartingale or supermartingale $(Y, \mathcal{F})$ converges almost surely if it satisfies $\mathbb{E}|Y_n| \le M$.

The key step in the classical proof of this theorem is 'Snell's upcrossings inequality'. Suppose that $y = \{y_n : n \ge 0\}$ is a real sequence, and $[a, b]$ is a real interval. An upcrossing of $[a, b]$ is defined to be a crossing by $y$ of $[a, b]$ in the upwards direction. More precisely, we define $T_1 = \min\{n : y_n \le a\}$, the first time that $y$ hits the interval $(-\infty, a]$, and $T_2 = \min\{n > T_1 : y_n \ge b\}$, the first subsequent time when $y$ hits $[b, \infty)$; we call the interval $[T_1, T_2]$ an *upcrossing* of $[a, b]$. In addition, let

$$T_{2k-1} = \min\{n > T_{2k-2} : y_n \le a\}, \quad T_{2k} = \min\{n > T_{2k-1} : y_n \ge b\},$$

for $k \ge 2$, so that the upcrossings of $[a, b]$ are the intervals $[T_{2k-1}, T_{2k}]$ for $k \ge 1$. Let $U_n(a, b; y)$ be the number of upcrossings of $[a, b]$ by the subsequence $y_0, y_1, \ldots, y_n$, and let $U(a, b; y) = \lim_{n \to \infty} U_n(a, b; y)$ be the total number of such upcrossings by $y$.

**(2) Lemma.** *If $U(a, b; y) < \infty$ for all rationals $a$ and $b$ satisfying $a < b$, then $\lim_{n \to \infty} y_n$ exists (but may be infinite).*

**Proof.** If $\lambda = \liminf_{n \to \infty} y_n$ and $\mu = \limsup_{n \to \infty} y_n$ satisfy $\lambda < \mu$ then there exist rationals $a, b$ such that $\lambda < a < b < \mu$. Now $y_n \le a$ for infinitely many $n$, and $y_n \ge b$ similarly, implying that $U(a, b; y) = \infty$, a contradiction. Therefore $\lambda = \mu$. ∎

Suppose now that $(Y, \mathcal{F})$ is a submartingale, and let $U_n(a, b; Y)$ be the number of upcrossings of $[a, b]$ by $Y$ up to time $n$.

**(3) Theorem. Upcrossings inequality.** *If $a < b$ then*

$$\mathbb{E}U_n(a, b; Y) \le \frac{\mathbb{E}((Y_n - a)^+)}{b - a}.$$

**Proof.** Setting $Z_n = (Y_n - a)^+$, we have by Exercise (12.1.7) that $(Z, \mathcal{F})$ is a non-negative submartingale. Upcrossings by $Y$ of $[a, b]$ correspond to upcrossings by $Z$ of $[0, b - a]$, so that $U_n(a, b; Y) = U_n(0, b - a; Z)$.

Let $[T_{2k-1}, T_{2k}]$, $k \ge 1$, be the upcrossings by $Z$ of $[0, b - a]$, and define the indicator functions

$$I_i = \begin{cases} 1 & \text{if } i \in (T_{2k-1}, T_{2k}] \text{ for some } k, \\ 0 & \text{otherwise.} \end{cases}$$

Note that $I_i$ is $\mathcal{F}_{i-1}$-measurable, since

$$\{I_i = 1\} = \bigcup_k \{T_{2k-1} \leq i - 1\} \setminus \{T_{2k} \leq i - 1\},$$

an event which depends on $Y_0, Y_1, \ldots, Y_{i-1}$ only. Now

**(4)**        $$(b - a)U_n(0, b - a; Z) \leq \sum_{i=1}^{n}(Z_i - Z_{i-1})I_i,$$

since each upcrossing of $[0, b - a]$ contributes an amount of at least $b - a$ to the summation. However

**(5)**   $$\mathbb{E}\big((Z_i - Z_{i-1})I_i\big) = \mathbb{E}\big(\mathbb{E}[(Z_i - Z_{i-1})I_i \mid \mathcal{F}_{i-1}]\big) = \mathbb{E}\big(I_i[\mathbb{E}(Z_i \mid \mathcal{F}_{i-1}) - Z_{i-1}]\big)$$
$$\leq \mathbb{E}\big[\mathbb{E}(Z_i \mid \mathcal{F}_{i-1}) - Z_{i-1}\big] = \mathbb{E}(Z_i) - \mathbb{E}(Z_{i-1})$$

where we have used the fact that $Z$ is a submartingale to obtain the inequality. Summing over $i$, we obtain from (4) that

$$(b - a)\mathbb{E}U_n(0, b - a; Z) \leq \mathbb{E}(Z_n) - \mathbb{E}(Z_0) \leq \mathbb{E}(Z_n)$$

and the theorem is proved.   ■

**Proof of Theorem (1).** Suppose $(Y, \mathcal{F})$ is a submartingale and $\mathbb{E}(Y_n^+) \leq M$ for all $n$. We have from the upcrossings inequality that, if $a < b$,

$$\mathbb{E}U_n(a, b; Y) \leq \frac{\mathbb{E}(Y_n^+) + |a|}{b - a}$$

so that $U(a, b; Y) = \lim_{n \to \infty} U_n(a, b; Y)$ satisfies

$$\mathbb{E}U(a, b; Y) = \lim_{n \to \infty} \mathbb{E}U_n(a, b; Y) \leq \frac{M + |a|}{b - a}$$

for all $a < b$. Therefore $U(a, b; Y) < \infty$ a.s. for all $a < b$. Since there are only countably many rationals, it follows that, with probability 1, $U(a, b; Y) < \infty$ for all rational $a$ and $b$. By Lemma (2), the sequence $Y_n$ converges almost surely to some limit $Y_\infty$. We argue as follows to show that $\mathbb{P}(|Y_\infty| < \infty) = 1$. Since $|Y_n| = 2Y_n^+ - Y_n$ and $\mathbb{E}(Y_n \mid \mathcal{F}_0) \geq Y_0$, we have that

$$\mathbb{E}\big(|Y_n| \mid \mathcal{F}_0\big) = 2\mathbb{E}(Y_n^+ \mid \mathcal{F}_0) - \mathbb{E}(Y_n \mid \mathcal{F}_0) \leq 2\mathbb{E}(Y_n^+ \mid \mathcal{F}_0) - Y_0.$$

By Fatou's lemma,

**(6)**        $$\mathbb{E}\big(|Y_\infty| \mid \mathcal{F}_0\big) = \mathbb{E}\left(\liminf_{n \to \infty} |Y_n| \,\Big|\, \mathcal{F}_0\right) \leq \liminf_{n \to \infty} \mathbb{E}\big(|Y_n| \mid \mathcal{F}_0\big) \leq 2Z - Y_0$$

where $Z = \liminf_{n \to \infty} \mathbb{E}(Y_n^+ \mid \mathcal{F}_0)$. However $\mathbb{E}(Z) \leq M$ by Fatou's lemma, so that $Z < \infty$ a.s., implying that $\mathbb{E}\big(|Y_\infty| \mid \mathcal{F}_0\big) < \infty$ a.s. Hence $\mathbb{P}(|Y_\infty| < \infty \mid \mathcal{F}_0) = 1$, and therefore

$$\mathbb{P}(|Y_\infty| < \infty) = \mathbb{E}\big[\mathbb{P}(|Y_\infty| < \infty \mid \mathcal{F}_0)\big] = 1.$$

If $\mathbb{E}|Y_0| < \infty$, we may take expectations of (6) to obtain $\mathbb{E}|Y_\infty| \le 2M - \mathbb{E}(Y_0) < \infty$. That uniform integrability is enough to ensure convergence in mean is a consequence of Theorem (7.10.3).                                                                                              ∎

The following is an immediate corollary of the martingale convergence theorem.

**(7) Theorem.** *If $(Y, \mathscr{F})$ is either a non-negative supermartingale or a non-positive submartingale, then $Y_\infty = \lim_{n\to\infty} Y_n$ exists almost surely.*

**Proof.** If $Y$ is a non-positive submartingale then $\mathbb{E}(Y_n^+) = 0$, whence the result follows from Theorem (1). For a non-negative supermartingale $Y$, apply the same argument to $-Y$.    ∎

**(8) Example. Random walk.** Consider de Moivre's martingale of Example (12.1.4), namely $Y_n = (q/p)^{S_n}$ where $S_n$ is the position after $n$ steps of the usual simple random walk. The sequence $\{Y_n\}$ is a non-negative martingale, and hence converges almost surely to some finite limit $Y$ as $n \to \infty$. This is not of much interest if $p = q$, since $Y_n = 1$ for all $n$ in this case. Suppose then that $p \ne q$. The random variable $Y_n$ takes values in the set $\{\rho^k : k = 0, \pm 1, \dots\}$ where $\rho = q/p$. Certainly $Y_n$ cannot converge to any given (possibly random) member of this set, since this would necessarily entail that $S_n$ converges to a finite limit (which is obviously false). Therefore $Y_n$ converges to a limit point of the set, not lying within the set. The only such limit point which is finite is 0, and therefore $Y_n \to 0$ a.s. Hence, $S_n \to -\infty$ a.s. if $p < q$, and $S_n \to \infty$ a.s. if $p > q$. Note that $Y_n$ does not converge in mean, since $\mathbb{E}(Y_n) = \mathbb{E}(Y_0) \ne 0$ for all $n$.                                                                                            ●

**(9) Example. Doob's martingale.** Let $Z$ be a random variable on $(\Omega, \mathscr{F}, \mathbb{P})$ such that $\mathbb{E}|Z| < \infty$. Suppose that $\mathscr{F} = \{\mathscr{F}_0, \mathscr{F}_1, \dots\}$ is a filtration, and write $\mathscr{F}_\infty = \lim_{n\to\infty} \mathscr{F}_n$ for the smallest $\sigma$-field containing every $\mathscr{F}_n$. Now define $Y_n = \mathbb{E}(Z \mid \mathscr{F}_n)$. It is easily seen that $(Y, \mathscr{F})$ is a martingale. First, by Jensen's inequality,

$$\mathbb{E}|Y_n| = \mathbb{E}\big|\mathbb{E}(Z \mid \mathscr{F}_n)\big| \le \mathbb{E}\big\{\mathbb{E}\big(|Z|\,\big|\,\mathscr{F}_n\big)\big\} = \mathbb{E}|Z| < \infty,$$

and secondly

$$\mathbb{E}(Y_{n+1} \mid \mathscr{F}_n) = \mathbb{E}\big[\mathbb{E}(Z \mid \mathscr{F}_{n+1})\,\big|\,\mathscr{F}_n\big] = \mathbb{E}(Z \mid \mathscr{F}_n)$$

since $\mathscr{F}_n \subseteq \mathscr{F}_{n+1}$. Furthermore $\{Y_n\}$ is a uniformly integrable sequence, as shown in Example (7.10.13). It follows by the martingale convergence theorem that $Y_\infty = \lim_{n\to\infty} Y_n$ exists almost surely and in mean.

It is actually the case that $Y_\infty = \mathbb{E}(Z \mid \mathscr{F}_\infty)$, so that†

**(10)**                     $\mathbb{E}(Z \mid \mathscr{F}_n) \to \mathbb{E}(Z \mid \mathscr{F}_\infty)$    a.s. and in mean.

To see this, one argues as follows. Let $N$ be a positive integer. First, $Y_n I_A \to Y_\infty I_A$ a.s. for all $A \in \mathscr{F}_N$. Now $\{Y_n I_A : n \ge N\}$ is uniformly integrable, and therefore $\mathbb{E}(Y_n I_A) \to \mathbb{E}(Y_\infty I_A)$ for all $A \in \mathscr{F}_N$. On the other hand $\mathbb{E}(Y_n I_A) = \mathbb{E}(Y_N I_A) = \mathbb{E}(Z I_A)$ for all $n \ge N$ and all $A \in \mathscr{F}_N$, by the definition of conditional expectation. Hence $\mathbb{E}(Z I_A) = \mathbb{E}(Y_\infty I_A)$ for all $A \in \mathscr{F}_N$. Letting $N \to \infty$ and using a standard result of measure theory, we find that $\mathbb{E}((Z - Y_\infty) I_A) = 0$ for all $A \in \mathscr{F}_\infty$, whence $Y_\infty = \mathbb{E}(Z \mid \mathscr{F}_\infty)$.

---

†Proved earlier by Lévy when $Z$ is an indicator function, giving the zero–one law of Example (12).

There is an important converse to these results.

> **(11) Lemma.** *Let* $(Y, \mathcal{F})$ *be a martingale. Then* $Y_n$ *converges in mean if and only if there exists a random variable* $Z$ *with finite mean such that* $Y_n = \mathbb{E}(Z \mid \mathcal{F}_n)$. *If* $Y_n \xrightarrow{1} Y_\infty$, *then* $Y_n = \mathbb{E}(Y_\infty \mid \mathcal{F}_n)$.

If such a random variable $Z$ exists, we say that the martingale $(Y, \mathcal{F})$ is *closed*.

**Proof.** In the light of the previous discussion, it suffices to prove that, if $(Y, \mathcal{F})$ is a martingale which converges in mean to $Y_\infty$, then $Y_n = \mathbb{E}(Y_\infty \mid \mathcal{F}_n)$. For any positive integer $N$ and event $A \in \mathcal{F}_N$, it is the case that $\mathbb{E}(Y_n I_A) \to \mathbb{E}(Y_\infty I_A)$; just note that $Y_n I_A \xrightarrow{1} Y_\infty I_A$ since

$$\mathbb{E}|(Y_n - Y_\infty)I_A| \le \mathbb{E}|Y_n - Y_\infty| \to 0 \quad \text{as} \quad n \to \infty.$$

On the other hand, $\mathbb{E}(Y_n I_A) = \mathbb{E}(Y_N I_A)$ for $n \ge N$ and $A \in \mathcal{F}_N$, by the martingale property. It follows that $\mathbb{E}(Y_\infty I_A) = \mathbb{E}(Y_N I_A)$ for all $A \in \mathcal{F}_N$, which is to say that $Y_N = \mathbb{E}(Y_\infty \mid \mathcal{F}_N)$ as required.                                                                                    ∎●

**(12) Example. Zero–one law (7.3.12).** Let $X_0, X_1, \ldots$ be independent random variables, and let $\mathcal{T}$ be their tail $\sigma$-field; that is to say, $\mathcal{T} = \bigcap_n \mathcal{H}_n$ where $\mathcal{H}_n = \sigma(X_n, X_{n+1}, \ldots)$. Here is a proof that, for all $A \in \mathcal{T}$, either $\mathbb{P}(A) = 0$ or $\mathbb{P}(A) = 1$.

Let $A \in \mathcal{T}$ and define $Y_n = \mathbb{E}(I_A \mid \mathcal{F}_n)$ where $\mathcal{F}_n = \sigma(X_1, X_2, \ldots, X_n)$. Now $A \in \mathcal{T} \subseteq \mathcal{F}_\infty = \lim_{n \to \infty} \mathcal{F}_n$, and therefore $Y_n \to \mathbb{E}(I_A \mid \mathcal{F}_\infty) = I_A$ a.s. and in mean, by (11). On the other hand $Y_n = \mathbb{E}(I_A \mid \mathcal{F}_n) = \mathbb{P}(A)$, since $A\ (\in \mathcal{T})$ is independent of all events in $\mathcal{F}_n$. Hence $\mathbb{P}(A) = I_A$ almost surely, which is to say that $I_A$ is almost surely constant. However, $I_A$ takes values 0 and 1 only, and therefore either $\mathbb{P}(A) = 0$ or $\mathbb{P}(A) = 1$.                                        ●

This completes the main contents of this section. We terminate it with one further result of interest, being a bound related to the upcrossings inequality. For a certain type of process, one may obtain rather tight bounds on the tail of the number of upcrossings.

**(13) Theorem. Dubins's inequality.** *Let* $(Y, \mathcal{F})$ *be a non-negative supermartingale. Then*

$$\textbf{(14)} \qquad\qquad \mathbb{P}\{U_n(a, b; Y) \ge j\} \le \left(\frac{a}{b}\right)^j \mathbb{E}\big(\min\{1, Y_0/a\}\big)$$

*for* $0 < a < b$ *and* $j \ge 0$.

Summing (14) over $j$, we find that

$$\textbf{(15)} \qquad\qquad \mathbb{E}U_n(a, b; Y) \le \frac{a}{b - a}\mathbb{E}\big(\min\{1, Y_0/a\}\big),$$

an inequality which may be compared with the upcrossings inequality (3).

**Proof.** This is achieved by an adaptation of the proof of the upcrossings inequality (3), and we use the notation of that proof. Fix a positive integer $j$. We replace the indicator function $I_i$ by the random variable

$$J_i = \begin{cases} a^{-1}(b/a)^{k-1} & \text{if } i \in (T_{2k-1}, T_{2k}] \text{ for some } k \le j, \\ 0 & \text{otherwise.} \end{cases}$$

Next we let $X_0, X_1, \ldots$ be given by $X_0 = \min\{1, Y_0/a\}$,

(16)
$$X_n = X_0 + \sum_{i=1}^{n} J_i(Y_i - Y_{i-1}), \quad n \ge 1.$$

If $T_{2j} \le n$, then

$$X_n \ge X_0 + \sum_{k=1}^{j} a^{-1}(b/a)^{k-1}(Y_{T_{2k}} - Y_{T_{2k-1}}).$$

However, $Y_{T_{2k}} \ge b$ and $Y_{T_{2k+1}} \le a$, so that

(17)
$$Y_{T_{2k}} - \frac{b}{a}Y_{T_{2k+1}} \ge 0,$$

implying that

$$X_n \ge X_0 + a^{-1}(b/a)^{j-1}Y_{T_{2j}} - a^{-1}Y_{T_1}, \quad \text{if} \quad T_{2j} \le n.$$

If $Y_0 \le a$ then $T_1 = 0$ and $X_0 - a^{-1}Y_{T_1} = 0$; on the other hand, if $Y_0 > a$ then $X_0 - a^{-1}Y_{T_1} = 1 - a^{-1}Y_{T_1} > 0$. In either case it follows that

(18)
$$X_n \ge (b/a)^j \quad \text{if} \quad T_{2j} \le n.$$

Now $Y$ is a non-negative sequence, and hence $X_n \ge X_0 - a^{-1}Y_{T_1} \ge 0$ by (16) and (17). Take expectations of (18) to obtain

(19)
$$\mathbb{E}(X_n) \ge (b/a)^j \mathbb{P}(U_n(a, b; Y) \ge j),$$

and it remains to bound $\mathbb{E}(X_n)$ above. Arguing as in (5) and using the supermartingale property, we arrive at

$$\mathbb{E}(X_n) = \mathbb{E}(X_0) + \sum_{i=1}^{n} \mathbb{E}\big(J_i(Y_i - Y_{i-1})\big) \le \mathbb{E}(X_0).$$

The conclusion of the theorem follows from (19).                                   ∎

(20) **Example. Simple random walk.** Consider de Moivre's martingale $Y_n = (q/p)^{S_n}$ of Examples (12.1.4) and (8), with $p < q$. By Theorem (13), $\mathbb{P}(U_n(a, b; Y) \ge j) \le (a/b)^j$. An upcrossing of $[a, b]$ by $Y$ corresponds to an upcrossing of $[\log a, \log b]$ by $S$ (with logarithms to the base $q/p$). Hence

$$\mathbb{P}\big(U_n(0, r; S) \ge j\big) = \mathbb{P}\big\{U_n\big(1, (q/p)^r; Y\big) \ge j\big\} \le (p/q)^{rj}, \quad j \ge 0.$$

Actually equality holds here in the limit as $n \to \infty$: $\mathbb{P}(U(0, r; S) \ge j) = (p/q)^{rj}$ for positive integers $r$; see Exercise (5.3.1).                                   ●

## Exercises for Section 12.3

**1.** Give a reasonable definition of a *downcrossing* of the interval $[a, b]$ by the random sequence $Y_0, Y_1, \ldots$.

(a) Show that the number of downcrossings differs from the number of upcrossings by at most 1.

(b) If $(Y, \mathcal{F})$ is a submartingale, show that the number $D_n(a, b; Y)$ of downcrossings of $[a, b]$ by $Y$ up to time $n$ satisfies

$$\mathbb{E}D_n(a, b; Y) \leq \frac{\mathbb{E}\{(Y_n - b)^+\}}{b - a}.$$

**2.** Let $(Y, \mathcal{F})$ be a supermartingale with finite means, and let $U_n(a, b; Y)$ be the number of upcrossings of the interval $[a, b]$ up to time $n$. Show that

$$\mathbb{E}U_n(a, b; Y) \leq \frac{\mathbb{E}\{(Y_n - a)^-\}}{b - a}.$$

Deduce that $\mathbb{E}U_n(a, b; Y) \leq a/(b - a)$ if $Y$ is non-negative and $a \geq 0$.

**3.** Let $X$ be a Markov chain with countable state space $S$ and transition matrix $\mathbf{P}$. Suppose that $X$ is irreducible and recurrent, and that $\psi : S \to S$ is a bounded function satisfying $\sum_{j \in S} p_{ij} \psi(j) \leq \psi(i)$ for $i \in S$. Show that $\psi$ is a constant function.

**4.** Let $Z_1, Z_2, \ldots$ be independent random variables such that:

$$Z_n = \begin{cases} a_n & \text{with probability } \frac{1}{2}n^{-2}, \\ 0 & \text{with probability } 1 - n^{-2}, \\ -a_n & \text{with probability } \frac{1}{2}n^{-2}, \end{cases}$$

where $a_1 = 2$ and $a_n = 4\sum_{j=1}^{n-1} a_j$. Show that $Y_n = \sum_{j=1}^{n} Z_j$ defines a martingale. Show that $Y = \lim Y_n$ exists almost surely, but that there exists no $M$ such that $\mathbb{E}|Y_n| \leq M$ for all $n$.

**5. Random adding martingale.** Let $x_1, x_2, \ldots, x_r \in \mathbb{R}$, and let the sequence $\{X_n : n \geq 1\}$ of random variables be given as follows. We set

$$X_n = \begin{cases} x_n & \text{if } 1 \leq n \leq r, \\ X_{U(n)} + X_{V(n)} & \text{if } n > r, \end{cases}$$

where $U(n)$ and $V(n)$ are uniformly distributed on $\{1, 2, \ldots, n - 1\}$, and the random variables $\{U(n), V(n) : n > r\}$ are independent.

Show that

$$M_n = \frac{1}{n(n+1)} \sum_{k=1}^{n} X_k, \qquad n = r, r + 1 \ldots,$$

is a martingale with respect to the sequence $\{X_n\}$.

Enthusiasts seeking a challenge are invited to show that the $M$ converges almost surely and in mean square to a non-degenerate limit.

**6. Pólya's urn revisited.** Let $R_n$ and $B_n$ be the numbers of red and blue balls, respectively, in an urn at the $n$th stage, and assume $R_0 = B_0 = 1$. At each stage, a ball is drawn and returned together with a fresh ball of the other colour. Show that $M_n = (B_n - R_n)(B_n + R_n - 1)$ defines a martingale. Does it converge almost surely?

## 12.4 Stopping times

We are all called upon on occasion to take an action whose nature is fixed but whose timing is optional. Commonly occurring examples include getting married or divorced, employing a secretary, having a baby, and buying a house. An important feature of such actions is that they are taken in the light of the past and present, and they may not depend on the future. Other important examples arise in considering money markets. The management of portfolios is affected by such rules as: (a) sell a currency if it weakens to a predetermined threshold, (b) buy government bonds if the exchange index falls below a given level, and so on. (Such rules are often sufficiently simple to be left to computers to implement, with occasionally spectacular consequences†.)

A more mathematical example is provided by the gambling analogy. A gambler pursues a strategy which we may assume to be based upon his experience rather than his clairvoyance. That is to say, his decisions to vary his stake (or to stop gambling altogether) depend on the outcomes of the game up to the time of the decision, and no further. A gambler is able to follow the rule 'stop when ahead' but cannot be expected to follow a rule such as 'stop just before a loss'.

Such actions have the common feature that, at any time, we have sufficient information to decide whether or not to take the action *at that time*. The usual way of expressing this property in mathematical terms is as follows. Let $(\Omega, \mathcal{F}, \mathbb{P})$ be a probability space, and let $\mathcal{F} = \{\mathcal{F}_0, \mathcal{F}_1, \ldots\}$ be a filtration. We think of $\mathcal{F}_n$ as representing the information which is available at time $n$, or more precisely the smallest $\sigma$-field with respect to which all observations up to and including time $n$ are measurable.

**(1) Definition.** A random variable $T$ taking values in $\{0, 1, 2, \ldots\} \cup \{\infty\}$ is called a **stopping time** (with respect to the filtration $\mathcal{F}$) if $\{T = n\} \in \mathcal{F}_n$ for all $n \geq 0$.

Note that stopping times $T$ satisfy

**(2)**                           $$\{T > n\} = \{T \leq n\}^c \in \mathcal{F}_n \quad \text{for all } n,$$

since $\mathcal{F}$ is a filtration. They are not required to be finite, but may take the value $\infty$. Stopping times are sometimes called *Markov times*. They were discussed in Section 6.8 in the context of birth processes.

Given a filtration $\mathcal{F}$ and a stopping time $T$, it is useful to introduce some notation to represent information gained up to the random time $T$. We denote by $\mathcal{F}_T$ the collection of all events $A$ such that $A \cap \{T \leq n\} \in \mathcal{F}_n$ for all $n$. It is easily seen that $\mathcal{F}_T$ is a $\sigma$-field, and we think of $\mathcal{F}_T$ as the set of events whose occurrence or non-occurrence is known by time $T$.

**(3) Example. The martingale (12.1.3).** A fair coin is tossed repeatedly; let $T$ be the time of the first head. Writing $X_i$ for the number of heads on the $i$th toss, we have that

$$\{T = n\} = \{X_n = 1, \ X_j = 0 \text{ for } 1 \leq j < n\} \in \mathcal{F}_n$$

where $\mathcal{F}_n = \sigma(X_1, X_2, \ldots, X_n)$. Therefore $T$ is a stopping time. In this case $T$ is finite almost surely.                                                                                      ●

---

†At least one NYSE crash has been attributed to the use of simple online stock-dealing systems programmed to sell whenever a stock price falls to a given threshold. Such systems can be subject to feedback, and the rules have been changed to inhibit this.

**(4) Example. First passage times.** Let $\mathcal{F}$ be a filtration and let the random sequence $X$ be adapted to $\mathcal{F}$, so that $X_n$ is $\mathcal{F}_n$-measurable. For each (sufficiently nice) subset $B$ of $\mathbb{R}$ define the *first passage time* of $X$ to $B$ by $T_B = \min\{n : X_n \in B\}$ with $T_B = \infty$ if $X_n \notin B$ for all $n$. It is easily seen that $T_B$ is a stopping time. ●

Stopping times play an important role in the theory of martingales, as illustrated in the following examples. First, a martingale which is stopped at a random time $T$ remains a martingale, so long as $T$ is a stopping time.

**(5) Theorem.** *Let $(Y, \mathcal{F})$ be a submartingale and let $T$ be a stopping time (with respect to $\mathcal{F}$). Then $(Z, \mathcal{F})$, defined by $Z_n = Y_{T \wedge n}$, is a submartingale.*

Here, as usual, we use the notation $x \wedge y = \min\{x, y\}$. If $(Y, \mathcal{F})$ is a martingale, then it is both a submartingale and a supermartingale, whence $Y_{T \wedge n}$ constitutes a martingale, by (5).

**Proof.** We may write

**(6)**
$$Z_n = \sum_{t=0}^{n-1} Y_i I_{\{T=t\}} + Y_n I_{\{T \geq n\}},$$

whence $Z_n$ is $\mathcal{F}_n$-measurable (using (2)) and

$$\mathbb{E}(Z_n^+) \leq \sum_{t=0}^{n} \mathbb{E}(Y_t^+) < \infty.$$

Also, from (6), $Z_{n+1} - Z_n = (Y_{n+1} - Y_n)I_{\{T>n\}}$, whence, using (2) and the submartingale property,

$$\mathbb{E}(Z_{n+1} - Z_n \mid \mathcal{F}_n) = \mathbb{E}(Y_{n+1} - Y_n \mid \mathcal{F}_n)I_{\{T>n\}} \geq 0.$$ ∎

One strategy open to a gambler in a casino is to change the game (think of the gambler as an investor in stocks, if you wish). If he is fortunate enough to be playing fair games, then he should not gain or lose (on average) at such a change. More formally, let $(X, \mathcal{F})$ and $(Y, \mathcal{F})$ be two martingales with respect to the filtration $\mathcal{F}$. Let $T$ be a stopping time with respect to $\mathcal{F}$; $T$ is the switching time from $X$ to $Y$, and $X_T$ is the 'capital' which is carried forward.

**(7) Theorem. Optional switching.** *Suppose that $X_T = Y_T$ on the event $\{T < \infty\}$. Then*

$$Z_n = \begin{cases} X_n & \text{if } n < T, \\ Y_n & \text{if } n \geq T, \end{cases}$$

*defines a martingale with respect to $\mathcal{F}$.*

**Proof.** We have that

**(8)**
$$Z_n = X_n I_{\{n<T\}} + Y_n I_{\{n \geq T\}};$$

each summand is $\mathcal{F}_n$-measurable, and hence $Z_n$ is $\mathcal{F}_n$-measurable. Also $\mathbb{E}|Z_n| \leq \mathbb{E}|X_n| + \mathbb{E}|Y_n| < \infty$. By the martingale property of $X$ and $Y$,

**(9)**
$$Z_n = \mathbb{E}(X_{n+1} \mid \mathcal{F}_n)I_{\{n<T\}} + \mathbb{E}(Y_{n+1} \mid \mathcal{F}_n)I_{\{n \geq T\}}$$
$$= \mathbb{E}\big(X_{n+1} I_{\{n<T\}} + Y_{n+1} I_{\{n \geq T\}} \mid \mathcal{F}_n\big),$$

since $T$ is a stopping time. Now

(10)    $$X_{n+1}I_{\{n<T\}} + Y_{n+1}I_{\{n\geq T\}} = Z_{n+1} + X_{n+1}I_{\{n+1=T\}} - Y_{n+1}I_{\{n+1=T\}}$$
$$= Z_{n+1} + (X_T - Y_T)I_{\{n+1=T\}}$$

whence, by (9) and the assumption that $X_T = Y_T$ on the event $\{T < \infty\}$, we have that $Z_n = \mathbb{E}(Z_{n+1} \mid \mathcal{F}_n)$, so that $(Z, \mathcal{F})$ is a martingale. ∎

'Optional switching' does not disturb the martingale property. 'Optional sampling' can be somewhat more problematical. Let $(Y, \mathcal{F})$ be a martingale and let $T_1, T_2, \ldots$ be a sequence of stopping times satisfying $T_1 \leq T_2 \leq \cdots < \infty$. Let $Z_0 = Y_0$ and $Z_n = Y_{T_n}$, so that the sequence $Z$ is obtained by 'sampling' the sequence $Y$ at the stopping times $T_j$. It is natural to set $\mathcal{H}_n = \mathcal{F}_{T_n}$, and to ask whether $(Z, \mathcal{H})$ is a martingale. The answer in general is no. To see this, use the simple example when $Y_n$ is the excess of heads over tails in $n$ tosses of a fair coin, with $T_1 = \min\{n : Y_n = 1\}$; for this example $\mathbb{E}Y_0 = 0$ but $\mathbb{E}Y_{T_1} = 1$. The answer is, however, affirmative if the $T_j$ are bounded.

**(11) Optional sampling theorem.** *Let $(Y, \mathcal{F})$ be a submartingale.*
(a) *If $T$ is a stopping time and there exists a deterministic $N$ $(< \infty)$ such that $\mathbb{P}(T \leq N) = 1$, then $\mathbb{E}(Y_T^+) < \infty$ and $\mathbb{E}(Y_T \mid \mathcal{F}_0) \geq Y_0$.*
(b) *If $T_1 \leq T_2 \leq \cdots$ is a sequence of stopping times such that $\mathbb{P}(T_j \leq N_j) = 1$ for some deterministic real sequence $N_j$, then $(Z, \mathcal{H})$, defined by $(Z_0, \mathcal{H}_0) = (Y_0, \mathcal{F}_0)$, $(Z_j, \mathcal{H}_j) = (Y_{T_j}, \mathcal{F}_{T_j})$, is a submartingale.*

If $(Y, \mathcal{F})$ is a martingale, then it is both a submartingale and a supermartingale; Theorem (11) then implies that $\mathbb{E}(Y_T \mid \mathcal{F}_0) = Y_0$ for any bounded stopping time $T$, and furthermore $(Y_{T_j}, \mathcal{F}_{T_j})$ is a martingale for any increasing sequence $T_1, T_2, \ldots$ of bounded stopping times.

**Proof.** Part (b) may be obtained without great difficulty by repeated application of part (a), and we therefore confine outselves to proving (a). Suppose $\mathbb{P}(T \leq N) = 1$. Let $Z_n = Y_{T \wedge n}$, so that $(Z, \mathcal{F})$ is a submartingale, by (5). Therefore $\mathbb{E}(Z_N^+) < \infty$ and

(12)                                $$\mathbb{E}(Z_N \mid \mathcal{F}_0) \geq Z_0 = Y_0,$$

and the proof is finished by observing that $Z_N = Y_{T \wedge N} = Y_T$ a.s. ∎

Certain inequalities are of great value when studying the asymptotic properties of martingales. The following simple but powerful 'maximal inequality' is an easy consequence of the optional sampling theorem.

**(13) Theorem.** *Let $(Y, \mathcal{F})$ be a martingale. For $x > 0$,*

(14)    $$\mathbb{P}\left(\max_{0\leq m\leq n} Y_m \geq x\right) \leq \frac{\mathbb{E}(Y_n^+)}{x} \quad and \quad \mathbb{P}\left(\max_{0\leq m\leq n} |Y_m| \geq x\right) \leq \frac{\mathbb{E}|Y_n|}{x}.$$

**Proof.** Let $x > 0$, and let $T = \min\{m : Y_m \geq x\}$ be the first passage time of $Y$ above the level $x$. Then $T \wedge n$ is a bounded stopping time, and therefore $\mathbb{E}(Y_0) = \mathbb{E}(Y_{T \wedge n}) = \mathbb{E}(Y_n)$ by Theorem (11a) and the martingale property. Now $\mathbb{E}(Y_{T \wedge n}) = \mathbb{E}(Y_T I_{\{T \leq n\}} + Y_n I_{\{T > n\}})$. However,

$$\mathbb{E}(Y_T I_{\{T \leq n\}}) \geq x\mathbb{E}(I_{\{T \leq n\}}) = x\mathbb{P}(T \leq n)$$

since $Y_T \geq x$, and therefore

**(15)**                    $\mathbb{E}(Y_n) = \mathbb{E}(Y_{T \wedge n}) \geq x\mathbb{P}(T \leq n) + \mathbb{E}(Y_n I_{\{T > n\}})$,

whence

$$x\mathbb{P}(T \leq n) \leq \mathbb{E}(Y_n I_{\{T \leq n\}}) \leq \mathbb{E}(Y_n^+)$$

as required for the first part of (14). As for the second part, just note that $(-Y, \mathcal{F})$ is a martingale, so that

$$\mathbb{P}\left( \max_{0 \leq m \leq n} \{-Y_m\} \geq x \right) \leq \frac{\mathbb{E}(Y_n^-)}{x} \quad \text{for} \quad x > 0,$$

which may be added to the first part.                                                                  ∎

We shall explore maximal inequalities for submartingales and supermartingales in the forthcoming Section 12.6.

---

## Exercises for Section 12.4

**1.**  If $T_1$ and $T_2$ are stopping times with respect to a filtration $\mathcal{F}$, show that $T_1 + T_2$, $\max\{T_1, T_2\}$, and $\min\{T_1, T_2\}$ are stopping times also.

**2.**  Let $X_1, X_2, \ldots$ be a sequence of non-negative independent random variables and let $N(t) = \max\{n : X_1 + X_2 + \cdots + X_n \leq t\}$. Show that $N(t) + 1$ is a stopping time with respect to a suitable filtration to be specified.

**3.**  Let $(Y, \mathcal{F})$ be a submartingale and $x > 0$. Show that

$$\mathbb{P}\left( \max_{0 \leq m \leq n} Y_m \geq x \right) \leq \frac{1}{x}\mathbb{E}(Y_n^+).$$

**4.**  Let $(Y, \mathcal{F})$ be a non-negative supermartingale and $x > 0$. Show that

$$\mathbb{P}\left( \max_{0 \leq m \leq n} Y_m \geq x \right) \leq \frac{1}{x}\mathbb{E}(Y_0).$$

**5.**  Let $(Y, \mathcal{F})$ be a submartingale and let $S$ and $T$ be stopping times satisfying $0 \leq S \leq T \leq N$ for some deterministic $N$. Show that $\mathbb{E}Y_0 \leq \mathbb{E}Y_S \leq \mathbb{E}Y_T \leq \mathbb{E}Y_N$.

**6.**  Let $\{S_n\}$ be a simple random walk with $S_0 = 0$ such that $0 < p = \mathbb{P}(S_1 = 1) < \frac{1}{2}$. Use de Moivre's martingale to show that $\mathbb{E}(\sup_m S_m) \leq p/(1 - 2p)$. Show further that this inequality may be replaced by an equality.

**7.**  Let $\mathcal{F}$ be a filtration. For any stopping time $T$ with respect to $\mathcal{F}$, denote by $\mathcal{F}_T$ the collection of all events $A$ such that, for all $n$, $A \cap \{T \leq n\} \in \mathcal{F}_n$. Let $S$ and $T$ be stopping times.
(a) Show that $\mathcal{F}_T$ is a $\sigma$-field, and that $T$ is measurable with respect to this $\sigma$-field.
(b) If $A \in \mathcal{F}_S$, show that $A \cap \{S \leq T\} \in \mathcal{F}_T$.
(c) Let $S$ and $T$ satisfy $S \leq T$. Show that $\mathcal{F}_S \subseteq \mathcal{F}_T$.

**8.**  **Stopping an exchangeable sequence.**
(a) Let $T$ be a stopping time for an exchangeable sequence $X_1, X_2, \ldots, X_n$. Show that, if $\mathbb{P}(T \leq n - r) = 1$, then the random vector $(X_{T+1}, X_{T+2}, \ldots, X_{T+r})$ has the same distribution as $(X_1, X_2, \ldots, X_r)$.
(b) An urn contains $v$ violet balls and $w$ white balls, which are drawn at random without replacement. Let $m < w$, and let $T$ be the number of the draw at which the $m$th white ball is drawn. What is the probability that the next ball is white?

# 12.5 Optional stopping

If you stop a martingale $(Y, \mathcal{F})$ at a fixed time $n$, the mean value $\mathbb{E}(Y_n)$ satisfies $\mathbb{E}(Y_n) = \mathbb{E}(Y_0)$. Under what conditions is this true if you stop after a *random* time $T$; that is, when is it the case that $\mathbb{E}(Y_T) = \mathbb{E}(Y_0)$? The answer to this question is very valuable in studying first-passage properties of martingales (see (12.1.4) for example). It would be unreasonable to expect such a result to hold generally unless $T$ is required to be a stopping time.

Let $T$ be a stopping time which is finite (in that $\mathbb{P}(T < \infty) = 1$), and let $(Y, \mathcal{F})$ be a martingale. Then $T \wedge n \to T$ as $n \to \infty$, so that $Y_{T \wedge n} \to Y_T$ a.s. It follows (as in Theorem (7.10.3)) that $\mathbb{E}(Y_0) = \mathbb{E}(Y_{T \wedge n}) \to \mathbb{E}(Y_T)$ so long as the family $\{Y_{T \wedge n} : n \geq 0\}$ is uniformly integrable.

The following two theorems provide useful conditions which are sufficient for the conclusion $\mathbb{E}(Y_0) = \mathbb{E}(Y_T)$.

**(1) Optional stopping theorem.** *Let $(Y, \mathcal{F})$ be a martingale and let $T$ be a stopping time. Then $\mathbb{E}(Y_T) = \mathbb{E}(Y_0)$ if:*
  (a) $\mathbb{P}(T < \infty) = 1$,
  (b) $\mathbb{E}|Y_T| < \infty$, *and*
  (c) $\mathbb{E}(Y_n I_{\{T > n\}}) \to 0$ *as $n \to \infty$.*

**(2) Theorem.** *Let $(Y, \mathcal{F})$ be a martingale and let $T$ be a stopping time. If the $Y_n$ are uniformly integrable and $\mathbb{P}(T < \infty) = 1$ then $Y_T = \mathbb{E}(Y_\infty \mid \mathcal{F}_T)$ and $Y_0 = \mathbb{E}(Y_T \mid \mathcal{F}_0)$. In particular $\mathbb{E}(Y_0) = \mathbb{E}(Y_T)$.*

**Proof of (1).** It is easily seen that $Y_T = Y_{T \wedge n} + (Y_T - Y_n)I_{\{T > n\}}$. Taking expectations and using the fact that $\mathbb{E}(Y_{T \wedge n}) = \mathbb{E}(Y_0)$ (see Theorem (12.4.11)), we find that

$$\text{(3)} \qquad \mathbb{E}(Y_T) = \mathbb{E}(Y_0) + \mathbb{E}(Y_T I_{\{T > n\}}) - \mathbb{E}(Y_n I_{\{T > n\}}).$$

The last term tends to zero as $n \to \infty$, by assumption (c). As for the penultimate term,

$$\mathbb{E}(Y_T I_{\{T > n\}}) = \sum_{k = n+1}^{\infty} \mathbb{E}(Y_T I_{\{T = k\}})$$

is, by assumption (b), the tail of the convergent series $\mathbb{E}(Y_T) = \sum_k \mathbb{E}(Y_T I_{\{T=k\}})$; therefore $\mathbb{E}(Y_T I_{\{T > n\}}) \to 0$ as $n \to \infty$, and (3) yields $\mathbb{E}(Y_T) = \mathbb{E}(Y_0)$ in the limit as $n \to \infty$. ∎

**Proof of (2).** Since $(Y, \mathcal{F})$ is uniformly integrable, we have by Theorems (12.3.1) and (12.3.11) that the limit $Y_\infty = \lim_{n \to \infty} Y_n$ exists almost surely, and $Y_n = \mathbb{E}(Y_\infty \mid \mathcal{F}_n)$. It follows from the definition (12.1.7) of conditional expectation that

$$\text{(4)} \qquad \mathbb{E}(Y_n I_A) = \mathbb{E}(Y_\infty I_A) \quad \text{for all } A \in \mathcal{F}_n.$$

Now, if $A \in \mathcal{F}_T$ then $A \cap \{T = n\} \in \mathcal{F}_n$, so that

$$\mathbb{E}(Y_T I_A) = \sum_n \mathbb{E}(Y_n I_{A \cap \{T = n\}}) = \sum_n \mathbb{E}(Y_\infty I_{A \cap \{T = n\}}) = \mathbb{E}(Y_\infty I_A),$$

whence $Y_T = \mathbb{E}(Y_\infty \mid \mathcal{F}_T)$. Secondly, since $\mathcal{F}_0 \subseteq \mathcal{F}_T$,

$$\mathbb{E}(Y_T \mid \mathcal{F}_0) = \mathbb{E}\big(\mathbb{E}(Y_\infty \mid \mathcal{F}_T) \mid \mathcal{F}_0\big) = \mathbb{E}(Y_\infty \mid \mathcal{F}_0) = Y_0. \qquad ∎$$

**(5) Example. Markov chains.**  Let $X$ be an irreducible recurrent Markov chain with countable state space $S$ and transition matrix $\mathbf{P}$, and let $\psi : S \to \mathbb{R}$ be a bounded function satisfying

$$\sum_{j \in S} p_{ij} \psi(j) = \psi(i) \quad \text{for all } i \in S.$$

Then $\psi(X_n)$ constitutes a martingale. Let $T_i$ be the first passage time of $X$ to the state $i$, that is, $T_i = \min\{n : X_n = i\}$; it is easily seen that $T_i$ is a stopping time and is (almost surely) finite. Furthermore, the sequence $\{\psi(X_n)\}$ is bounded and therefore uniformly integrable. Applying Theorem (2), we obtain $\mathbb{E}(\psi(X_T)) = \mathbb{E}(\psi(X_0))$, whence $\mathbb{E}(\psi(X_0)) = \psi(i)$ for all states $i$. Therefore $\psi$ is a constant function.                                              ●

**(6) Example. Symmetric simple random walk.**  Let $S_n$ be the position of the particle after $n$ steps and suppose that $S_0 = 0$. Then $S_n = \sum_{i=1}^{n} X_i$ where $X_1, X_2, \ldots$ are independent and equally likely to take each of the values $+1$ and $-1$. It is easy to see as in Example (12.1.2) that $\{S_n\}$ is a martingale. Let $a$ and $b$ be positive integers and let $T = \min\{n : S_n = -a$ or $S_n = b\}$ be the earliest time at which the walk visits either $-a$ or $b$. Certainly $T$ is a stopping time and satisfies the conditions of Theorem (1). Let $p_a$ be the probability that the particle visits $-a$ before it visits $b$. By the optional stopping theorem,

**(7)**              $$\mathbb{E}(S_T) = (-a)p_a + b(1 - p_a), \quad \mathbb{E}(S_0) = 0;$$

therefore $p_a = b/(a + b)$, which agrees with the earlier result of equation (1.7.7) when the notation is translated suitably. The sequence $\{S_n\}$ is not the only martingale available. Let $\{Y_n\}$ be given by $Y_n = S_n^2 - n$; then $\{Y_n\}$ is a martingale also. Apply Theorem (1) with $T$ given as before to obtain $\mathbb{E}(T) = ab$.                                          ●

**(8) Example. De Moivre's martingale (12.1.4).**  Consider now a simple random walk $\{S_n\}$ with $0 < S_0 < N$, for which each step is rightwards with probability $p$ where $0 < p = 1-q < 1$. We have seen that $Y_n = (q/p)^{S_n}$ defines a martingale, and furthermore the first passage time $T$ of the walk to the set $\{0, N\}$ is a stopping time. It is easily checked that conditions (1a)–(1c) of the optional stopping theorem are satisfied, and hence $\mathbb{E}((q/p)^{S_T}) = \mathbb{E}((q/p)^{S_0})$. Therefore $p_k = \mathbb{P}(S_T = 0 \mid S_0 = k)$ satisfies $p_k + (q/p)^N(1 - p_k) = (q/p)^k$, whence $p_k$ may be calculated as in Example (12.1.4).                                              ●

When applying the optional stopping theorem it is sometimes convenient to use a more restrictive set of conditions.

**(9) Theorem.**  *Let* $(Y, \mathcal{F})$ *be a martingale, and let* $T$ *be a stopping time. Then* $\mathbb{E}(Y_T) = \mathbb{E}(Y_0)$ *if the following hold*:
   (a) $\mathbb{P}(T < \infty) = 1$, $\mathbb{E}T < \infty$, *and*
   (b) *there exists a constant* $c$ *such that* $\mathbb{E}(|Y_{n+1} - Y_n| \mid \mathcal{F}_n) \leq c$ *for all* $n < T$.

**Proof.**  By the discussion prior to (1), it suffices to show that the sequence $\{Y_{T \wedge n} : n \geq 0\}$ is uniformly integrable. Let $Z_n = |Y_n - Y_{n-1}|$ for $n \geq 1$, and $W = Z_1 + Z_2 + \cdots + Z_T$. Certainly $|Y_{T \wedge n}| \leq |Y_0| + W$ for all $n$, and it is enough (by Example (7.10.4)) to show that $\mathbb{E}(W) < \infty$. We have that

**(10)**                          $$W = \sum_{i=1}^{\infty} Z_i I_{\{T \geq i\}}.$$

Now

$$\mathbb{E}\big(Z_i I_{\{T \geq i\}} \mid \mathcal{F}_{i-1}\big) = I_{\{T \geq i\}} \mathbb{E}(Z_i \mid \mathcal{F}_{i-1}) \leq c I_{\{T \geq i\}},$$

since $\{T \geq i\} = \{T \leq i - 1\}^c \in \mathcal{F}_{i-1}$. Therefore $\mathbb{E}(Z_i I_{\{T \geq i\}}) \leq c\mathbb{P}(T \geq i)$, giving by (10) that

$$\textbf{(11)} \qquad\qquad \mathbb{E}(W) \leq c \sum_{i=1}^{\infty} \mathbb{P}(T \geq i) = c\mathbb{E}(T) < \infty. \qquad\qquad \blacksquare$$

**(12) Example. Wald's equation (10.2.9).** Let $X_1, X_2, \ldots$ be independent identically distributed random variables with finite mean $\mu$, and let $S_n = \sum_{i=1}^{n} X_i$. It is easy to see that $Y_n = S_n - n\mu$ constitutes a martingale with respect to the filtration $\{\mathcal{F}_n\}$ where $\mathcal{F}_n = \sigma(Y_1, Y_2, \ldots, Y_n)$. Now

$$\mathbb{E}\big(|Y_{n+1} - Y_n| \mid \mathcal{F}_n\big) = \mathbb{E}|X_{n+1} - \mu| = \mathbb{E}|X_1 - \mu| < \infty.$$

We deduce from (9) that $\mathbb{E}(Y_T) = \mathbb{E}(Y_0) = 0$ for any stopping time $T$ with finite mean, implying that

$$\textbf{(13)} \qquad\qquad\qquad \mathbb{E}(S_T) = \mu\mathbb{E}(T),$$

a result derived earlier in the context of renewal theory as Lemma (10.2.9).

If the $X_i$ have finite variance $\sigma^2$, it is also the case that

$$\textbf{(14)} \qquad\qquad \text{var}(Y_T) = \sigma^2 \mathbb{E}(T) \qquad \text{if } \mathbb{E}(T) < \infty.$$

It is possible to prove this by applying the optional stopping theorem to the martingale $Z_n = Y_n^2 - n\sigma^2$, but this is not a simple application of (9). It may also be proved by exploiting Wald's identity (15), or more simply by the method of Exercise (10.2.2). $\qquad\qquad\bullet$

**(15) Example. Wald's identity.** This time, let $X_1, X_2, \ldots$ be independent identically distributed random variables with common moment generating function $M(t) = \mathbb{E}(e^{tX})$; suppose that there exists at least one value of $t$ ($\neq 0$) such that $1 \leq M(t) < \infty$, and fix $t$ accordingly. Let $S_n = X_1 + X_2 + \cdots + X_n$, define

$$\textbf{(16)} \qquad\qquad Y_0 = 1, \quad Y_n = \frac{e^{tS_n}}{M(t)^n} \quad \text{for} \quad n \geq 1,$$

and let $\mathcal{F}_n = \sigma(X_1, X_2, \ldots, X_n)$. It is clear that $(Y, \mathcal{F})$ is a martingale. When are the conditions of Theorem (9) valid? Let $T$ be a stopping time with finite mean, and note that

$$\textbf{(17)} \quad \mathbb{E}\big(|Y_{n+1} - Y_n| \mid \mathcal{F}_n\big) = Y_n \mathbb{E}\left(\left|\frac{e^{tX}}{M(t)} - 1\right|\right) \leq \frac{Y_n}{M(t)}\mathbb{E}\big(e^{tX} + M(t)\big) = 2Y_n.$$

Suppose that $T$ is such that

$$\textbf{(18)} \qquad\qquad\qquad |S_n| \leq C \quad \text{for} \quad n < T,$$

where $C$ is a constant. Now $M(t) \geq 1$, and

$$Y_n = \frac{e^{tS_n}}{M(t)^n} \leq \frac{e^{|t|C}}{M(t)^n} \leq e^{|t|C} \quad \text{for} \quad n < T,$$

giving by (17) that condition (9b) holds. In summary, if $T$ is a stopping time with finite mean such that (18) holds, then

**(19)**                 $$\mathbb{E}\{e^{tS_T} M(t)^{-T}\} = 1 \quad \text{whenever} \quad M(t) \geq 1,$$

an equation usually called *Wald's identity*.

Here is an application of (19). Suppose the $X_i$ have strictly positive variance, and let $T = \min\{n : S_n \leq -a \text{ or } S_n \geq b\}$ where $a, b > 0$; $T$ is the 'first exit time' from the interval $(-a, b)$. Certainly $|S_n| \leq \max\{a, b\}$ if $n < T$. Furthermore $\mathbb{E}T < \infty$, which may be seen as follows. By the non-degeneracy of the $X_i$, there exist $M$ and $\epsilon > 0$ such that $\mathbb{P}(|S_M| > a+b) > \epsilon$. If any of the quantities $|S_M|, |S_{2M} - S_M|, \ldots, |S_{kM} - S_{(k-1)M}|$ exceed $a+b$ then the process must have exited $(-a, b)$ by time $kM$. Therefore $\mathbb{P}(T \geq kM) \leq (1-\epsilon)^k$, implying that

$$\mathbb{E}(T) = \sum_{i=1}^{\infty} \mathbb{P}(T \geq i) \leq M \sum_{k=0}^{\infty} \mathbb{P}(T \geq kM) < \infty.$$

We conclude that (19) is valid. In many concrete cases of interest, there exists $\theta \ (\neq 0)$ such that $M(\theta) = 1$. Applying (19) with $t = \theta$, we obtain $\mathbb{E}(e^{\theta S_T}) = 1$, or

$$\eta_a \mathbb{P}(S_T \leq -a) + \eta_b \mathbb{P}(S_T \geq b) = 1$$

where

$$\eta_a = \mathbb{E}(e^{\theta S_T} \mid S_T \leq -a), \quad \eta_b = \mathbb{E}(e^{\theta S_T} \mid S_T \geq b),$$

and therefore

**(20)**                 $$\mathbb{P}(S_T \leq -a) = \frac{\eta_b - 1}{\eta_b - \eta_a}, \quad \mathbb{P}(S_T \geq b) = \frac{1 - \eta_a}{\eta_b - \eta_a}.$$

When $a$ and $b$ are large, it is reasonable to suppose that $\eta_a \simeq e^{-\theta a}$ and $\eta_b \simeq e^{\theta b}$, giving the approximations

**(21)**                 $$\mathbb{P}(S_T \leq -a) \simeq \frac{e^{\theta b} - 1}{e^{\theta b} - e^{-\theta a}}, \quad \mathbb{P}(S_T \geq b) \simeq \frac{1 - e^{-\theta a}}{e^{\theta b} - e^{-\theta a}}.$$

These approximations are of course exact if $S$ is a simple random walk and $a$ and $b$ are positive integers.                                                                                     ●

**(22) Example. Simple random walk.** Suppose that $\{S_n\}$ is a simple random walk whose steps $\{X_i\}$ take the values $1$ and $-1$ with respective probabilities $p$ and $q \ (= 1 - p)$. For positive integers $a$ and $b$, we have from Wald's identity (19) that

**(23)**      $$e^{-at} \mathbb{E}\big(M(t)^{-T} I_{\{S_T=-a\}}\big) + e^{bt} \mathbb{E}\big(M(t)^{-T} I_{\{S_T=b\}}\big) = 1 \quad \text{if} \quad M(t) \geq 1$$

where $T$ is the first exit time of $(-a, b)$ as before, and $M(t) = pe^t + qe^{-t}$.

Setting $M(t) = s^{-1}$, we obtain a quadratic for $e^t$, and hence $e^t = \lambda_1(s)$ or $e^t = \lambda_2(s)$ where

$$\lambda_1(s) = \frac{1 + \sqrt{1 - 4pqs^2}}{2ps}, \qquad \lambda_2(s) = \frac{1 - \sqrt{1 - 4pqs^2}}{2ps}.$$

Substituting these into equation (23), we obtain two linear equations in the quantities

**(24)**                $P_1(s) = \mathbb{E}(s^T I_{\{S_T=-a\}}), \qquad P_s(s) = \mathbb{E}(s^T I_{\{S_T=b\}}),$

with solutions

$$P_1(s) = \frac{\lambda_1^a \lambda_2^a (\lambda_1^b - \lambda_2^b)}{\lambda_1^{a+b} - \lambda_2^{a+b}}, \qquad P_2(s) = \frac{\lambda_1^a - \lambda_2^a}{\lambda_1^{a+b} - \lambda_2^{a+b}},$$

which we add to obtain the probability generating function of $T$,

**(25)**                $\mathbb{E}(s^T) = P_1(s) + P_2(s), \quad 0 < s \le 1.$

Suppose we let $a \to \infty$, so that $T$ becomes the time until the first passage to the point $b$. From (24), $P_1(s) \to 0$ as $a \to \infty$ if $0 < s < 1$, and a quick calculation gives $P_2(s) \to F_b(s)$ where

$$F_b(s) = \left( \frac{1 - \sqrt{1 - 4pqs^2}}{2qs} \right)^b$$

in agreement with Theorem (5.3.5). Notice that $F_b(1) = (\min\{1, p/q\})^b$.                              ●

---

## Exercises for Section 12.5

**1.**  Let $(Y, \mathscr{F})$ be a martingale and $T$ a stopping time such that $\mathbb{P}(T < \infty) = 1$. Show that $\mathbb{E}(Y_T) = \mathbb{E}(Y_0)$ if either of the following holds:

(a) $\mathbb{E}(\sup_n |Y_{T \wedge n}|) < \infty$,    (b) $\mathbb{E}(|Y_{T \wedge n}|^{1+\delta}) \le c$ for some $c, \delta > 0$ and all $n$.

**2.**  Let $(Y, \mathscr{F})$ be a martingale. Show that $(Y_{T \wedge n}, \mathscr{F}_n)$ is a uniformly integrable martingale for any finite stopping time $T$ such that either:

(a) $\mathbb{E}|Y_T| < \infty$ and $\mathbb{E}(|Y_n| I_{\{T>n\}}) \to 0$ as $n \to \infty$, or

(b) $\{Y_n\}$ is uniformly integrable.

**3.**  Let $(Y, \mathscr{F})$ be a uniformly integrable martingale, and let $S$ and $T$ be finite stopping times satisfying $S \le T$. Prove that $Y_T = \mathbb{E}(Y_\infty \mid \mathscr{F}_T)$ and that $Y_S = \mathbb{E}(Y_T \mid \mathscr{F}_S)$, where $Y_\infty$ is the almost sure limit as $n \to \infty$ of $Y_n$.

**4.**  Let $\{S_n : n \ge 0\}$ be a simple symmetric random walk with $0 < S_0 < N$ and with absorbing barriers at $0$ and $N$. Use the optional stopping theorem to show that the mean time until absorption is $\mathbb{E}\{S_0(N - S_0)\}$.

**5.**  Let $\{S_n : n \ge 0\}$ be a simple symmetric random walk with $S_0 = 0$. Show that

$$Y_n = \frac{\cos\{\lambda[S_n - \frac{1}{2}(b-a)]\}}{(\cos \lambda)^n}$$

constitutes a martingale if $\cos \lambda \ne 0$.

Let $a$ and $b$ be positive integers. Show that the time $T$ until absorption at one of two absorbing barriers at $-a$ and $b$ satisfies

$$\mathbb{E}\big(\{\cos\lambda\}^{-T}\big) = \frac{\cos\{\tfrac{1}{2}\lambda(b-a)\}}{\cos\{\tfrac{1}{2}\lambda(b+a)\}}, \qquad 0 < \lambda < \frac{\pi}{b+a}.$$

**6.**   Let $\{S_n : n \geq 0\}$ be a simple symmetric random walk on the positive and negative integers, with $S_0 = 0$. For each of the three following random variables, determine whether or not it is a stopping time and find its mean:

$$U = \min\{n \geq 5 : S_n = S_{n-5} + 5\}, \quad V = U - 5, \quad W = \min\{n : S_n = 1\}.$$

**7.**   Let $S_n = a + \sum_{r=1}^n X_r$ be a simple symmetric random walk. The walk stops at the earliest time $T$ when it reaches either of the two positions 0 or $K$ where $0 < a < K$. Show that $M_n = \sum_{r=0}^n S_r - \tfrac{1}{3}S_n^3$ is a martingale and deduce that $\mathbb{E}\big(\sum_{r=0}^T S_r\big) = \tfrac{1}{3}(K^2 - a^2)a + a$.

**8.   Gambler's ruin.** Let $X_i$ be independent random variables each equally likely to take the values $\pm 1$, and let $T = \min\{n : S_n \in \{-a, b\}\}$. Verify the conditions of the optional stopping theorem (12.5.1) for the martingale $S_n^2 - n$ and the stopping time $T$.

**9.   Family planning, Problem (3.11.30) revisited.**   Children are either female or male. Their sexes are independent random variables, being female with probability $q$ or male with probability $p = 1 - q$. A woman ceases childbearing at stage $T$, and we write $G_n$ and $B_n$ for the numbers of girls and boys born to her up to and including stage $n$. Assume that $T$ is a finite stopping time for the sequence $\{(G_n, B_n) : n \geq 1\}$. Show that, no matter the stopping rule that yields $T$, we have $\mathbb{E}(G_T)/\mathbb{E}(B_T) = q/p$. What can be said about $\mathbb{E}(G_T/B_T)$?

**10.**   Let $X = \{X_n : n \geq 0\}$ be a Markov chain with state space $\{0, 1, \ldots, b\}$, such that $i \to 0$, $b$ for $i \in \{1, 2, \ldots, b-1\}$. If $X$ is also a martingale, show that 0 and $b$ are absorbing, and that, given $X_0$, the probability of absorption at $b$ is $X_0/b$.

## 12.6 The maximal inequality

In proving the convergence of a sequence $X_1, X_2, \ldots$ of random variables, it is often useful to establish an inequality of the form

$$\mathbb{P}\big(\max\{X_1, X_2, \ldots, X_n\} \geq x\big) \leq A_n(x),$$

and such an inequality is sometimes called a maximal inequality. The bound $A_n(x)$ usually involves an expectation. Examples of such inequalities include Kolmogorov's inequality in the proof of the strong law of large numbers, and the Doob–Kolmogorov inequality (7.8.2) in the proof of the convergence of martingales with bounded second moments. Both these inequalities are special cases of the following maximal inequality for submartingales. In order to simplify the notation of this section, we shall write $X_n^*$ for the maximum of the first $n+1$ members of a sequence $X_0, X_1, \ldots$, so that $X_n^* = \max\{X_i : 0 \leq i \leq n\}$.

**(1) Theorem. Maximal inequality.**

(a) *If* $(Y, \mathcal{F})$ *is a submartingale, then*

$$\mathbb{P}(Y_n^* \geq x) \leq \frac{\mathbb{E}(Y_n^+)}{x} \quad for \quad x > 0.$$

(b) *If* $(Y, \mathcal{F})$ *is a supermartingale and* $\mathbb{E}|Y_0| < \infty$, *then*

$$\mathbb{P}(Y_n^* \geq x) \leq \frac{\mathbb{E}(Y_0) + \mathbb{E}(Y_n^-)}{x} \quad for \quad x > 0.$$

These inequalities may be improved somewhat. For example, a closer look at the proof in case (a) leads to the inequality

(2) $$\mathbb{P}(Y_n^* \geq x) \leq \frac{1}{x} \mathbb{E}(Y_n^+ I_{\{Y_n^* \geq x\}}) \quad for \quad x > 0.$$

**Proof.** This is very similar to that of Theorem (12.4.13). Let $T = \min\{n : Y_n \geq x\}$ where $x > 0$, and suppose first that $(Y, \mathcal{F})$ is a submartingale. Then $(Y^+, \mathcal{F})$ is a non-negative submartingale with finite means by Exercise (12.1.7), and $T = \min\{n : Y_n^+ \geq x\}$ since $x > 0$. Applying the optional sampling theorem (12.4.11b) with stopping times $T_1 = T \wedge n$, $T_2 = n$, we obtain $\mathbb{E}(Y_{T \wedge n}^+) \leq \mathbb{E}(Y_n^+)$. However,

$$\mathbb{E}(Y_{T \wedge n}^+) = \mathbb{E}(Y_T^+ I_{\{T \leq n\}}) + \mathbb{E}(Y_n^+ I_{\{T > n\}})$$
$$\geq x\mathbb{P}(T \leq n) + \mathbb{E}(Y_n^+ I_{\{T > n\}})$$

whence, as required,

(3) $$x\mathbb{P}(T \leq n) \leq \mathbb{E}\big(Y_n^+ (1 - I_{\{T > n\}})\big)$$
$$= \mathbb{E}(Y_n^+ I_{\{T \leq n\}}) \leq \mathbb{E}(Y_n^+).$$

Suppose next that $(Y, \mathcal{F})$ is a supermartingale. By optional sampling $\mathbb{E}(Y_0) \geq \mathbb{E}(Y_{T \wedge n})$. Now

$$\mathbb{E}(Y_{T \wedge n}) = \mathbb{E}\big(Y_T I_{\{T \leq n\}} + Y_n I_{\{T > n\}}\big)$$
$$\geq x\mathbb{P}(T \leq n) - \mathbb{E}(Y_n^-),$$

whence $x\mathbb{P}(T \leq n) \leq \mathbb{E}(Y_0) + \mathbb{E}(Y_n^-)$.  ∎

Part (a) of the maximal inequality may be used to handle the maximum of a submartingale, and part (b) may be used as follows to handle its minimum. Suppose that $(Y, \mathcal{F})$ is a submartingale with finite means. Then $(-Y, \mathcal{F})$ is a supermartingale, and therefore

(4) $$\mathbb{P}\left(\min_{0 \leq k \leq n} Y_k \leq -x\right) \leq \frac{\mathbb{E}(Y_n^+) - \mathbb{E}(Y_0)}{x} \quad for \quad x > 0,$$

by (1b). Using (1a) also, we find that

$$\mathbb{P}\left(\max_{0 \leq k \leq n} |Y_k| \geq x\right) \leq \frac{2\mathbb{E}(Y_n^+) - \mathbb{E}(Y_0)}{x} \leq \frac{3}{x} \sup_k \mathbb{E}|Y_k|.$$

Sending $n$ to infinity (and hiding a minor 'continuity' argument), we deduce that

(5)
$$\mathbb{P}\left(\sup_k |Y_k| \geq x\right) \leq \frac{3}{x} \sup_k \mathbb{E}|Y_k|, \quad \text{for} \quad x > 0.$$

A slightly tighter conclusion is valid if $(Y, \mathscr{F})$ is a martingale rather than merely a submartingale. In this case, $(|Y_n|, \mathscr{F}_n)$ is a submartingale, whence (1a) yields

(6)
$$\mathbb{P}\left(\sup_k |Y_k| \geq x\right) \leq \frac{1}{x} \sup_k \mathbb{E}|Y_k|, \quad \text{for} \quad x > 0.$$

**(7) Example. Doob–Kolmogorov inequality (7.8.2).** Let $(Y, \mathscr{F})$ be a martingale such that $\mathbb{E}(Y_n^2) < \infty$ for all $n$. Then $(Y_n^2, \mathscr{F}_n)$ is a submartingale, whence

(8)
$$\mathbb{P}\left(\max_{0\leq k\leq n} |Y_k| \geq x\right) = \mathbb{P}\left(\max_{0\leq k\leq n} Y_k^2 \geq x^2\right) \leq \frac{\mathbb{E}(Y_n^2)}{x^2}$$

for $x > 0$, in agreement with (7.8.2). This is the major step in the proof of the convergence theorem (7.8.1) for martingales with bounded second moments.    ●

**(9) Example. Kolmogorov's inequality.** Let $X_1, X_2, \ldots$ be independent random variables with finite means and variances. Applying the Doob–Kolmogorov inequality (8) to the martingale $Y_n = S_n - \mathbb{E}(S_n)$ where $S_n = X_1 + X_2 + \cdots + X_n$, we obtain

(10)
$$\mathbb{P}\left(\max_{1\leq k\leq n} |S_k - \mathbb{E}(S_k)| \geq x\right) \leq \frac{1}{x^2} \text{var}(S_n) \quad \text{for} \quad x > 0.$$

This powerful inequality is the principal step in the usual proof of the strong law of large numbers (7.5.1). See Problem (7.11.29) for a simple proof not using martingales.    ●

The maximal inequality may be used to address the question of convergence in $r$th mean of martingales.

**(11) Theorem.** *Let $r > 1$, and let $(Y, \mathscr{F})$ be a martingale such that $\sup_n \mathbb{E}|Y_n^r| < \infty$. Then $Y_n \xrightarrow{r} Y_\infty$ where $Y_\infty$ is the (almost sure) limit of $Y_n$.*

This is not difficult to prove by way of Fatou's lemma and the theory of uniform integrability. Instead, we shall make use of the following inequality.

**(12) Lemma.** *Let $r > 1$, and let $(Y, \mathscr{F})$ be a non-negative submartingale such that $\mathbb{E}(Y_n^r) < \infty$ for all $n$. Then*

(13)
$$\mathbb{E}(Y_n^r) \leq \mathbb{E}((Y_n^*)^r) \leq \left(\frac{r}{r-1}\right)^r \mathbb{E}(Y_n^r).$$

**Proof.** Certainly $Y_n \leq Y_n^*$, and therefore the first inequality is trivial. Turning to the second, note first that
$$\mathbb{E}((Y_n^*)^r) \leq \mathbb{E}\left((Y_0 + Y_1 + \cdots + Y_n)^r\right) < \infty.$$

Now, integrate by parts and use the maximal inequality (2) to obtain

$$\mathbb{E}((Y_n^*)^r) = \int_0^\infty r x^{r-1} \mathbb{P}(Y_n^* \geq x)\, dx \leq \int_0^\infty r x^{r-2} \mathbb{E}(Y_n I_{\{Y_n^* \geq x\}})\, dx$$

$$= \mathbb{E}\left(Y_n \int_0^{Y_n^*} r x^{r-2}\, dx\right) = \frac{r}{r-1} \mathbb{E}\left(Y_n (Y_n^*)^{r-1}\right).$$

We have by Hölder's inequality that

$$\mathbb{E}\left(Y_n (Y_n^*)^{r-1}\right) \leq [\mathbb{E}(Y_n^r)]^{1/r} [\mathbb{E}((Y_n^*)^r)]^{(r-1)/r}.$$

Substituting this, and solving, we obtain

$$[\mathbb{E}((Y_n^*)^r)]^{1/r} \leq \frac{r}{r-1} [\mathbb{E}(Y_n^r)]^{1/r}. \qquad \blacksquare$$

**Proof of Theorem (11).** Using the moment condition, $Y_\infty = \lim_{n\to\infty} Y_n$ exists almost surely. Now $(|Y_n|, \mathcal{F}_n)$ is a non-negative submartingale, and hence $\mathbb{E}(\sup_k |Y_k|^r) < \infty$ by Lemma (12) and monotone convergence (5.6.12). Hence $\{Y_k^r : k \geq 0\}$ is uniformly integrable (Exercise (7.10.6)), implying by Exercise (7.10.2) that $Y_k \xrightarrow{r} Y_\infty$ as required. $\qquad \blacksquare$

---

### Exercise for Section 12.6

**1. Martingale laws of large numbers.** Let $M_n = \sum_{r=1}^n D_r$ be a zero-mean martingale with difference sequence $\{D_r : r \geq 1\}$, such that $\sigma_r^2 = \text{var}(D_r) < \infty$. Show the following.
(a) We have that $\mathbb{E}(D_r D_s) = 0$ for $r \neq s$.
(b) If $n^{-2} \sum_{r=1}^n \sigma_r^2 \to 0$ as $n \to \infty$, then $D$ satisfies the weak law of large numbers in that $n^{-1} M_n \xrightarrow{P} 0$.
(c) If $\sum_r \sigma_r^2/r^2 < \infty$, then $D$ satisfies the strong law of large numbers in that $n^{-1} M_n \xrightarrow{a.s.} 0$.

---

## 12.7 Backward martingales and continuous-time martingales

The ideas of martingale theory find expression in several other contexts, of which we consider two in this section. The first of these concerns backward martingales. We call a sequence $\mathcal{G} = \{\mathcal{G}_n : n \geq 0\}$ of $\sigma$-fields *decreasing* if $\mathcal{G}_n \supseteq \mathcal{G}_{n+1}$ for all $n$.

**(1) Definition.** Let $\mathcal{G}$ be a decreasing sequence of $\sigma$-fields and let $Y$ be a sequence of random variables which is adapted to $\mathcal{G}$. We call $(Y, \mathcal{G})$ a **backward** (or **reversed**) **martingale** if, for all $n \geq 0$,
(a) $\mathbb{E}|Y_n| < \infty$,
(b) $\mathbb{E}(Y_n \mid \mathcal{G}_{n+1}) = Y_{n+1}$.

Note that $\{(Y_n, \mathcal{G}_n) : n = 0, 1, 2, \ldots\}$ is a backward martingale if and only if the reversed sequence $\{(Y_n, \mathcal{G}_n) : n = \ldots, 2, 1, 0\}$ is a martingale, an observation which explains the use of the term.

**(2) Example. Strong law of large numbers.** Let $X_1, X_2, \ldots$ be independent identically distributed random variables with finite mean. Set $S_n = X_1 + X_2 + \cdots + X_n$ and let $\mathcal{G}_n = \sigma(S_n, S_{n+1}, \ldots)$. Then, using symmetry,

$$(3) \qquad \mathbb{E}(S_n \mid \mathcal{G}_{n+1}) = \mathbb{E}(S_n \mid S_{n+1}) = n\mathbb{E}(X_1 \mid S_{n+1}) = n\frac{S_{n+1}}{n+1}$$

since $S_{n+1} = \mathbb{E}(S_{n+1} \mid S_{n+1}) = (n+1)\mathbb{E}(X_1 \mid S_{n+1})$. Therefore $Y_n = S_n/n$ satisfies $\mathbb{E}(Y_n \mid \mathcal{G}_{n+1}) = Y_{n+1}$, whence $(Y, \mathcal{G})$ is a backward martingale. We shall see soon that backward martingales converge almost surely and in mean, and therefore there exists $Y_\infty$ such that $Y_n \to Y_\infty$ a.s. and in mean. By the zero–one law (7.3.15), $Y_\infty$ is almost surely constant, and hence $Y_\infty = \mathbb{E}(X_1)$ almost surely. We have proved the strong law of large numbers. ●

**(4) Backward-martingale convergence theorem.** *Let $(Y, \mathcal{G})$ be a backward martingale. Then $Y_n$ converges to a limit $Y_\infty$ almost surely and in mean.*

It is striking that no extra condition is necessary to ensure the convergence of backward martingales.

**Proof.** Note first that the sequence $\ldots, Y_n, Y_{n-1}, \ldots, Y_1, Y_0$ is a martingale with respect to the sequence $\ldots, \mathcal{G}_n, \mathcal{G}_{n-1}, \ldots, \mathcal{G}_1, \mathcal{G}_0$, and therefore $Y_n = \mathbb{E}(Y_0 \mid \mathcal{G}_n)$ for all $n$. However, $\mathbb{E}|Y_0| < \infty$, and therefore $\{Y_n\}$ is uniformly integrable by Example (7.10.13). It is therefore sufficient to prove that $Y_n$ converges almost surely. The usual way of doing this is via an upcrossings inequality. Applying (12.3.3) to the martingale $Y_n, Y_{n-1}, \ldots, Y_0$, we obtain that

$$\mathbb{E}U_n(a, b; Y) \le \frac{E((Y_0 - a)^+)}{b - a}$$

where $U_n(a, b; Y)$ is the number of upcrossings of $[a, b]$ by the sequence $Y_n, Y_{n-1}, \ldots, Y_0$. We let $n \to \infty$, and follow the proof of the martingale convergence theorem (12.3.1) to obtain the required result. ∎

Rather than developing the theory of backward martingales in detail, we confine ourselves to one observation and an application. Let $(Y, \mathcal{G})$ be a backward martingale, and let $T$ be a stopping time with respect to $\mathcal{G}$; that is, $\{T = n\} \in \mathcal{G}_n$ for all $n$. If $T$ is bounded, say $\mathbb{P}(T \le N) = 1$ for some fixed $N$, then the sequence $Z_N, Z_{N-1}, \ldots, Z_0$ defined by $Z_n = Y_{T \vee n}$ is a martingale with respect to the appropriate sequence of $\sigma$-fields (remember that $x \vee y = \max\{x, y\}$). Hence, by the optional sampling theorem (12.4.11a),

$$(5) \qquad \mathbb{E}(Y_T \mid \mathcal{G}_N) = Y_N.$$

**(6) Example. Ballot theorem (3.10.6).** Let $X_1, X_2, \ldots$ be independent identically distributed random variables taking values in $\{0, 1, 2, \ldots\}$, and let $S_n = X_1 + X_2 + \cdots + X_n$. We claim that

$$(7) \qquad \mathbb{P}\big(S_k \ge k \text{ for some } 1 \le k \le N \mid S_N = b\big) = \min\{1, b/N\},$$

whenever $b$ is such that $\mathbb{P}(S_N = b) > 0$. It is not immediately clear that this implies the ballot theorem, but look at it this way. In a ballot, each of $N$ voters has two votes; he or she allocates both votes either to candidate $A$ or to candidate $B$. Let us write $X_i$ for the number of votes allocated to $A$ by the $i$th voter, so that $X_i$ equals either $0$ or $2$; assume that the $X_i$ are independent. Now $S_k \geq k$ for some $1 \leq k \leq N$ if and only if $B$ is not always in the lead. Equation (7) implies

**(8)**    $\mathbb{P}\big(B \text{ always leads} \mid A \text{ receives a total of } 2a \text{ votes}\big)$

$$= 1 - \mathbb{P}\big(S_k \geq k \text{ for some } 1 \leq k \leq N \mid S_n = 2a\big)$$

$$= 1 - \frac{2a}{N} = \frac{p-q}{p+q}$$

if $0 \leq a < \frac{1}{2}N$, where $p = 2N - 2a$ is the number of votes received by $B$, and $q = 2a$ is the number received by $A$. This is the famous ballot theorem discussed after Corollary (3.10.6).

In order to prove equation (7), let $\mathcal{G}_n = \sigma(S_n, S_{n+1}, \dots)$, and recall that $(S_n/n, \mathcal{G}_n)$ is a backward martingale. Fix $N$, and let

$$T = \begin{cases} \max\{k : S_k \geq k \text{ and } 1 \leq k \leq N\} & \text{if this exists,} \\ 1 & \text{otherwise.} \end{cases}$$

This may not look like a stopping time, but it is. After all, for $1 < n \leq N$,

$$\{T = n\} = \{S_n \geq n, \ S_k < k \text{ for } n < k \leq N\},$$

an event defined in terms of $S_n, S_{n+1}, \dots$ and therefore lying in the $\sigma$-field $\mathcal{G}_n$ generated by these random variables. By a similar argument, $\{T = 1\} \in \mathcal{G}_1$.

We may assume that $S_N = b < N$, since (7) is obvious if $b \geq N$. Let $A = \{S_k \geq k$ for some $1 \leq k \leq N\}$. We have that $S_N < N$; therefore, if $A$ occurs, it must be the case that $S_T \geq T$ and $S_{T+1} < T + 1$. In this case $X_{T+1} = S_{T+1} - S_T < 1$, so that $X_{T+1} = 0$ and therefore $S_T/T = 1$. On the other hand, if $A$ does not occur then $T = 1$, and also $S_T = S_1 = 0$, implying that $S_T/T = 0$. It follows that $S_T/T = I_A$ if $S_N < N$, where $I_A$ is the indicator function of $A$. Taking expectations, we obtain

$$\mathbb{E}\left(\frac{1}{T} S_T \,\middle|\, S_N = b\right) = \mathbb{P}(A \mid S_N = b) \quad \text{if} \quad b < N.$$

Finally, we apply (5) to the backward martingale $(S_n/n, \mathcal{G}_n)$ to obtain

$$\mathbb{E}\left(\frac{1}{T} S_T \,\middle|\, S_N = b\right) = \mathbb{E}\left(\frac{1}{N} S_N \,\middle|\, S_N = b\right) = \frac{b}{N}.$$

The last two equations may be combined to give (7).                                    ●

In contrast to the theory of backward martingales, the theory of continuous-time martingales is hedged about with technical considerations. Let $(\Omega, \mathcal{F}, \mathbb{P})$ be a probability space. A *filtration* is a family $\mathcal{F} = \{\mathcal{F}_t : t \geq 0\}$ of sub-$\sigma$-fields of $\mathcal{F}$ satisfying $\mathcal{F}_s \subseteq \mathcal{F}_t$ whenever $s \leq t$. As before, we say that the (continuous-time) process $Y = \{Y(t) : t \geq 0\}$ is adapted to $\mathcal{F}$ if $Y(t)$ is $\mathcal{F}_t$-measurable for all $t$. If $Y$ is adapted to $\mathcal{F}$, we call $(Y, \mathcal{F})$ a *martingale* if $\mathbb{E}|Y(t)| < \infty$ for all $t$, and $\mathbb{E}(Y(t) \mid \mathcal{F}_s) = Y(s)$ whenever $s \leq t$. A random variable $T$ taking values in $[0, \infty]$ is called a *stopping time* (with respect to the filtration $\mathscr{F}$) if $\{T \leq t\} \in \mathscr{F}_t$ for all $t \geq 0$.

Possibly the most important type of stopping time is the first passage time $T(A) = \inf\{t : Y(t) \in A\}$ for a suitable subset $A$ of $\mathbb{R}$. Unfortunately $T(A)$ is not necessarily a stopping time. No problems arise if $A$ is closed and the sample paths $\Pi(\omega) = \{(t, Y(t; \omega)) : t \geq 0\}$ of $Y$ are continuous, but these conditions are over-restrictive. They may be relaxed at the price of making extra assumptions about the process $Y$ and the filtration $\mathscr{F}$. It is usual to assume in addition that:

(a) $(\Omega, \mathscr{F}, \mathbb{P})$ is complete,
(b) $\mathscr{F}_0$ contains all events $A$ of $\mathscr{F}$ satisfying $\mathbb{P}(A) = 0$,
(c) $\mathscr{F}$ is right-continuous in that $\mathscr{F}_t = \mathscr{F}_{t+}$ for all $t \geq 0$, where $\mathscr{F}_{t+} = \bigcap_{\epsilon > 0} \mathscr{F}_{t+\epsilon}$.

We shall refer to these conditions as the 'usual conditions'. Conditions (a) and (b) pose little difficulty, since an incomplete probability space may be completed, and the null events may be added to $\mathscr{F}_0$. Condition (c) is not of great importance if the process $Y$ has right-continuous sample paths, since then $Y(t) = \lim_{\epsilon \downarrow 0} Y(t + \epsilon)$ is $\mathscr{F}_{t+}$-measurable.

Here are some examples of continuous-time martingales.

**(9) Example. Poisson process.** Let $\{N(t) : t \geq 0\}$ be a Poisson process with intensity $\lambda$, and let $\mathscr{F}_t$ be the $\sigma$-field generated by $\{N(u) : 0 \leq u \leq t\}$. It is easily seen that

$$U(t) = N(t) - \lambda t,$$

$$V(t) = U(t)^2 - \lambda t,$$

$$W(t) = \exp[-\theta N(t) + \lambda t (1 - e^{-\theta})],$$

constitute martingales with respect to $\mathscr{F}$.

There is a converse statement. Suppose $N = \{N(t) : t \geq 0\}$ is an integer-valued non-decreasing process such that, for all $\theta$,

$$W(t) = \exp[-\theta N(t) + \lambda t (1 - e^{-\theta})]$$

is a martingale. Then, if $s < t$,

$$\mathbb{E}\big(\exp\{-\theta[N(t) - N(s)]\} \,\big|\, \mathscr{F}_s\big) = \mathbb{E}\left(\frac{W(t)}{W(s)} \exp[-\lambda(t - s)(1 - e^{-\theta})] \,\bigg|\, \mathscr{F}_s\right)$$

$$= \exp[-\lambda(t - s)(1 - e^{-\theta})]$$

by the martingale condition. Hence $N$ has independent increments, $N(t) - N(s)$ having the Poisson distribution with parameter $\lambda(t - s)$.  ●

**(10) Example. Wiener process.** Let $\{W(t) : t \geq 0\}$ be a standard Wiener process with continuous sample paths, and let $\mathscr{F}_t$ be the $\sigma$-field generated by $\{W(u) : 0 \leq u \leq t\}$. It is easily seen that $W(t)$, $W(t)^2 - t$, and $\exp[\theta W(t) - \frac{1}{2}\theta^2 t]$ constitute martingales with respect to $\mathscr{F}$. Conversely it may be shown that, if $W(t)$ and $W(t)^2 - t$ are martingales with continuous sample paths, and $W(0) = 0$, then $W$ is a standard Wiener process; this is sometimes called 'Lévy's characterization theorem'.  ●

Versions of the convergence and optional stopping theorems are valid in continuous time.

**(11) Convergence theorem.** *Let* $(Y, \mathcal{F})$ *be a martingale with right-continuous sample paths. If* $\mathbb{E}|Y(t)| \leq M$ *for some* $M$ *and all* $t$*, then* $Y_\infty = \lim_{t \to \infty} Y(t)$ *exists almost surely. If, in addition,* $(Y, \mathcal{F})$ *is uniformly integrable then* $Y(t) \overset{1}{\to} Y_\infty$.

**Sketch proof.** For each $m \geq 1$, the sequence $\{(Y(n2^{-m}), \mathcal{F}_{n2^{-m}}) : n \geq 0\}$ constitutes a discrete-time martingale. Under the conditions of the theorem, these martingales converge as $n \to \infty$. The right-continuity property of $Y$ may be used to fill in the gaps. ∎

**(12) Optional stopping theorem.** *Let* $(Y, \mathcal{F})$ *be a uniformly integrable martingale with right-continuous sample paths. Suppose that* $S$ *and* $T$ *are stopping times such that* $S \leq T$*. Then* $\mathbb{E}(Y(T) \mid \mathcal{F}_S) = Y(S)$.

The idea of the proof is to 'discretize' $Y$ as in the previous proof, use the optional stopping theorem for uniformly integrable discrete-time martingales, and then pass to the continuous limit.

---

## Exercises for Section 12.7

**1.**   Let $X$ be a continuous-time Markov chain with finite state space $S$ and generator $\mathbf{G}$. Let $\eta = \{\eta(i) : i \in S\}$ be a root of the equation $\mathbf{G}\eta' = \mathbf{0}$. Show that $\eta(X(t))$ constitutes a martingale with respect to $\mathcal{F}_t = \sigma(\{X(u) : u \leq t\})$.

**2.**   Let $N$ be a Poisson process with intensity $\lambda$ and $N(0) = 0$, and let $T_a = \min\{t : N(t) = a\}$, where $a$ is a positive integer. Assuming that $\mathbb{E}\{\exp(\psi T_a)\} < \infty$ for sufficiently small positive $\psi$, use the optional stopping theorem to show that $\mathrm{var}(T_a) = a\lambda^{-2}$. Show further that $T_a$ has characteristic function $\phi_a(s) = [\lambda/(\lambda - is)]^a$.

**3.**   Let $S_m = \sum_{r=1}^m X_r$, $m \leq n$, where the $X_r$ are independent and identically distributed with finite mean. Denote by $U_1, U_2, \ldots, U_n$ the order statistics of $n$ independent variables which are uniformly distributed on $(0, t)$, and set $U_{n+1} = t$. Show that $R_m = S_m/U_{m+1}$, $0 \leq m \leq n$, is a backward martingale with respect to a suitable sequence of $\sigma$-fields, and deduce that

$$\mathbb{P}(R_m \geq 1 \text{ for some } m \leq n \mid S_n = y) \leq \min\{y/t, 1\}.$$

**4.**   (a) Let $W$ be a standard Wiener process. Show that the following are martingales:

(i) $W(t)$,   (ii) $W(t)^2 - t$,   (iii) $W(t)^3 - 3tW(t)$,   (iv) $W(t)^4 - 6tW(t)^2 + 3t^2$.

(b) Let $T$ be the earliest time at which $W$ exits the interval $(-a, a)$. Show that $\mathbb{E}(T^n) < \infty$ for $n \geq 1$. Use the above martingales to show that $\mathbb{E}(T^2) = \frac{5}{3}a^4$. [You may use appropriate optional stopping theorems and the dominated convergence theorem.]

**5.**   Let $X$ and $Y$ be independent standard Wiener processes. Show that

$$M(t) = Y(t)X(t)^2 - \int_0^t Y(u)\, du$$

defines a martingale with respect to the natural filtration $\mathcal{F}_t = \sigma(\{X_v, Y_v : v \leq t\})$.

**6.**   **Brownian motion in a disc.** Let $Z(t) = z + X(t) + iY(t)$, where $X$ and $Y$ are independent standard Wiener processes and $|z| < 1$. Show that the first passage time $T$ to the unit circle has expected value $\mathbb{E}(T) = \frac{1}{2}(1 - |z|^2)$.

**7.**   **Kendall's taxicabs, Problem (11.8.16) revisited.** Let $X$ and $Y$ be independent Poisson processes with rate $\lambda$, and let $Q(t) = X(t) - Y(t)$. For positive integers $m, n$, find:

(a) the probability that $Q$ hits $n$ before $-m$,
(b) the expected time for $Q$ to hit either $-m$ or $n$.

**8.    Quadratic variation.** Let $N$ be a Poisson process with rate $\lambda$. Show the following.
(a) The predictable quadratic variation of $N$ over the interval $[0, t]$ equals $\lambda t$, which is to say that

$$\sum_{r=0}^{n-1} \mathbb{E}\left( \left[ N((r+1)t/n) - N(rt/n) \right]^2 \mid \mathcal{F}_{rt/n} \right) \to \lambda t \qquad \text{as } n \to \infty,$$

where $\mathcal{F}_u = \sigma(\{N(s) : s \le u\})$.
(b) The optional quadratic variation of $N(s) - \lambda s$ over $[0, t]$ equals $N(t)$. (See Exercise (12.2.4).)
(c) $M(t) = (N(t) - \lambda t)^2 - N(t)$ defines a martingale.

## 12.8  Some examples

**(1) Example.  Gambling systems.** In practice, gamblers do not invariably follow simple strategies, but they vary their manner of play according to a personal system. One way of expressing this is as follows. For a given game, write $Y_0, Y_1, \ldots$ for the sequence of capitals obtained by wagering one unit on each play; we allow the $Y_i$ to be negative. That is to say, let $Y_0$ be the initial capital, and let $Y_n$ be the capital after $n$ gambles each involving a unit stake. Take as filtration the sequence $\mathcal{F}$ given by $\mathcal{F}_n = \sigma(Y_0, Y_1, \ldots, Y_n)$. A general betting strategy would allow the gambler to vary her stake. If she bets $S_n$ on the $n$th play, her profit is $S_n(Y_n - Y_{n-1})$, since $Y_n - Y_{n-1}$ is the profit resulting from a stake of one unit. Hence the gambler's capital $Z_n$ after $n$ plays satisfies

$$(2) \qquad Z_n = Z_{n-1} + S_n(Y_n - Y_{n-1}) = Y_0 + \sum_{i=1}^{n} S_i(Y_i - Y_{i-1}),$$

where $Y_0$ is the gambler's initial capital. The $S_n$ must have the following special property. The gambler decides the value of $S_n$ in advance of the $n$th play, which is to say that $S_n$ depends only on $Y_0, Y_1, \ldots, Y_{n-1}$, and therefore $S_n$ is $\mathcal{F}_{n-1}$-measurable. That is, $(S, \mathcal{F})$ must be a predictable process.

The sequence $Z$ given by (2) is called the *transform* of $Y$ by $S$. If $Y$ is a martingale, we call $Z$ a *martingale transform*.

Suppose $(Y, \mathcal{F})$ is a martingale. The gambler may hope to find a predictable process $(S, \mathcal{F})$ (called a *system*) for which the martingale transform $Z$ (of $Y$ by $S$) is no longer a martingale. She hopes in vain, since all martingale transforms have the martingale property. Here is a version of that statement.

**(3) Theorem.** *Let $(S, \mathcal{F})$ be a predictable process, and let $Z$ be the transform of $Y$ by $S$. Then*:
(a) *if $(Y, \mathcal{F})$ is a martingale, then $(Z, \mathcal{F})$ is a martingale so long as $\mathbb{E}|Z_n| < \infty$ for all $n$,*
(b) *if $(Y, \mathcal{F})$ is a submartingale and in addition $S_n \ge 0$ for all $n$, then $(Z, \mathcal{F})$ is a submartingale so long as $\mathbb{E}(Z_n^+) < \infty$ for all $n$.*

**Proof.** From (2),

$$\mathbb{E}(Z_{n+1} \mid \mathcal{F}_n) - Z_n = \mathbb{E}\left[ S_{n+1}(Y_{n+1} - Y_n) \mid \mathcal{F}_n \right]$$
$$= S_{n+1}\left[ \mathbb{E}(Y_{n+1} \mid \mathcal{F}_n) - Y_n \right].$$

The last term is zero if $Y$ is a martingale, and is non-negative if $Y$ is a submartingale and $S_{n+1} \geq 0$.                                                                                                                          ■

A number of special cases are of value.

**(4)** *Optional skipping.* At each play, the gambler either wagers a unit stake or skips the round; $S$ equals either 0 or 1.

**(5)** *Optional stopping.* The gambler wagers a unit stake on each play until the (random) time $T$, when she gambles for the last time. That is,

$$S_n = \begin{cases} 1 & \text{if } n \leq T, \\ 0 & \text{if } n > T, \end{cases}$$

and $Z_n = Y_{T \wedge n}$. Now $\{T = n\} = \{S_n = 1, \ S_{n+1} = 0\} \in \mathcal{F}_n$, so that $T$ is a stopping time. It is a consequence of (3) that $(Y_{T \wedge n}, \mathcal{F}_n)$ is a martingale whenever $Y$ is a martingale, as established earlier.

**(6)** *Optional starting.* The gambler does not play until the $(T + 1)$th play, where $T$ is a stopping time. In this case $S_n = 0$ for $n \leq T$.                                                                      ●

**(7) Example. Likelihood ratios.** Let $X_1, X_2, \ldots$ be independent identically distributed random variables with common density function $f$. Suppose that it is known that $f(\cdot)$ is either $p(\cdot)$ or $q(\cdot)$, where $p$ and $q$ are given (different) densities; the statistical problem is to decide which of the two is the true density. A common approach is to calculate the *likelihood ratio*

$$Y_n = \frac{p(X_1) p(X_2) \cdots p(X_n)}{q(X_1) q(X_2) \cdots q(X_n)}$$

(assume for neatness for $q(x) > 0$ for all $x$), and to adopt the strategy:

**(8)**                        decide $p$ if $Y_n \geq a$,   decide $q$ if $Y_n < a$,

where $a$ is some predetermined positive level.

Let $\mathcal{F}_n = \sigma(X_1, X_2, \ldots, X_n)$. If $f = q$, then

$$\mathbb{E}(Y_{n+1} \mid \mathcal{F}_n) = Y_n \mathbb{E}\left(\frac{p(X_{n+1})}{q(X_{n+1})}\right) = Y_n \int_{-\infty}^{\infty} \frac{p(x)}{q(x)} q(x) \, dx = Y_n$$

since $p$ is a density function. Furthermore

$$\mathbb{E}|Y_n| = \int_{\mathbb{R}^n} \frac{p(x_1) p(x_2) \cdots p(x_n)}{q(x_1) q(x_2) \cdots q(x_n)} q(x_1) \cdots q(x_n) \, dx_1 \cdots dx_n = 1.$$

It follows that $(Y, \mathcal{F})$ is a martingale, under the assumption that $q$ is the common density function of the $X_i$. By an application of the convergence theorem, the limit $Y_\infty = \lim_{n \to \infty} Y_n$ exists almost surely under this assumption. We may calculate $Y_\infty$ explicitly as follows:

$$\log Y_n = \sum_{i=1}^{n} \log\left(\frac{p(X_i)}{q(X_i)}\right),$$

the sum of independent identically distributed random variables. The logarithm function is concave, so that

$$\mathbb{E}\left(\log\left(\frac{p(X_1)}{q(X_1)}\right)\right) < \log\left(\mathbb{E}\left(\frac{p(X_1)}{q(X_1)}\right)\right) = 0$$

by Jensen's inequality, Exercise (5.6.1). Applying the strong law of large numbers (7.5.1), we deduce that $n^{-1}\log Y_n$ converges almost surely to some point in $[-\infty, 0)$, implying that $Y_n \xrightarrow{\text{a.s.}} Y_\infty = 0$. (This is a case when the sequence $Y_n$ does not converge to $Y_\infty$ in mean, and $Y_n \neq \mathbb{E}(Y_\infty \mid \mathcal{F}_n)$.)

The fact that $Y_n \xrightarrow{\text{a.s.}} 0$ tells us that $Y_n < a$ for all large $n$, and hence the decision rule (8) gives the correct answer (that is, that $f = q$) for all large $n$. Indeed the probability that the outcome of the decision rule is ever in error satisfies $\mathbb{P}(Y_n \geq a \text{ for any } n \geq 1) \leq a^{-1}$, by the maximal inequality (12.6.6).                                                                    ●

**(9) Example. Epidemics.** A village contains $N + 1$ people, one of whom is suffering from a fatal and infectious illness. Let $S(t)$ be the number of susceptible people at time $t$ (that is, living people who have not yet been infected), let $I(t)$ be the number of infectives (that is, living people with the disease), and let $D(t) = N + 1 - S(t) - I(t)$ be the number of dead people. Assume that $(S(t), I(t), D(t))$ is a (trivariate) Markov chain in continuous time with transition rates

$$(s, i, d) \rightarrow \begin{cases} (s - 1, i + 1, d) & \text{at rate } \lambda si, \\ (s, i - 1, d + 1) & \text{at rate } \mu i; \end{cases}$$

that is to say, some susceptible becomes infective at rate $\lambda si$, and some infective dies at rate $\mu i$, where $s$ and $i$ are the numbers of susceptibles and infectives. This is the model of (6.12.4) with the introduction of death. The three variables always add up to $N + 1$, and therefore we may suppress reference to the dead, writing $(s, i)$ for a typical state of the process. Suppose we can find $\boldsymbol{\psi} = \{\psi(s, i) : 0 \leq s + i \leq N + 1\}$ such that $\mathbf{G}\boldsymbol{\psi}' = \mathbf{0}$, where $\mathbf{G}$ is the generator of the chain; think of $\boldsymbol{\psi}$ as a row vector. Then the transition semigroup $\mathbf{P}_t = e^{t\mathbf{G}}$ satisfies

$$\mathbf{P}_t \boldsymbol{\psi}' = \boldsymbol{\psi}' + \sum_{n=1}^{\infty} \frac{1}{n!} t^n \mathbf{G}^n \boldsymbol{\psi}' = \boldsymbol{\psi}',$$

whence it is easily seen (Exercise (12.7.1)) that $Y(t) = \psi(S(t), I(t))$ defines a continuous-time martingale with respect to the filtration $\mathcal{F}_t = \sigma(\{S(u), I(u) : 0 \leq u \leq t\})$.

Now $\mathbf{G}\boldsymbol{\psi}' = \mathbf{0}$ if and only if

**(10)**        $\lambda si \psi(s - 1, i + 1) - (\lambda si + \mu i)\psi(s, i) + \mu i \psi(s, i - 1) = 0$

for all relevant $i$ and $s$. If we look for a solution of the form $\psi(s, i) = \alpha(s)\beta(i)$, we obtain

**(11)**        $\lambda s\alpha(s - 1)\beta(i + 1) - (\lambda s + \mu)\alpha(s)\beta(i) + \mu\alpha(s)\beta(i - 1) = 0.$

Viewed as a difference equation in the $\beta(i)$, this suggests setting

**(12)**                              $\beta(i) = B^i$   for some $B$.

With this choice and a little calculation, one finds that

**(13)**                    $\alpha(s) = \prod_{k=s+1}^{N} \left(\frac{\lambda Bk - \mu(1 - B)}{\lambda B^2 k}\right)$

will do. With such choices for $\alpha$ and $\beta$, the process $\psi(S(t), I(t)) = \alpha(S(t))\beta(I(t))$ constitutes a martingale.

Two possibilities spring to mind. Either everyone dies ultimately (that is, $S(t) = 0$ before $I(t) = 0$) or the disease dies off before everyone has caught it (that is, $I(t) = 0$ before $S(t) = 0$). Let $T = \inf\{t : S(t)I(t) = 0\}$ be the time at which the process terminates. Clearly $T$ is a stopping time, and therefore

$$\mathbb{E}\big(\psi(S(T), I(T))\big) = \psi(S(0), I(0)) = \alpha(N)\beta(1) = B,$$

which is to say that

(14)
$$\mathbb{E}\left( B^{I(T)} \prod_{k=S(T)+1}^{N} \left( \frac{\lambda Bk - \mu(1 - B)}{\lambda B^2 k} \right) \right) = B$$

for all $B$. From this equation we wish to determine whether $S(T) = 0$ or $I(T) = 0$, corresponding to the two possibilities described above.

We have a free choice of $B$ in (14), and we choose the following values. For $1 \le r \le N$, define $B_r = \mu/(\lambda r + \mu)$, so that $\lambda r B_r - \mu(1 - B_r) = 0$. Substitute $B = B_r$ in (14) to obtain

(15)
$$\mathbb{E}\left( B_r^{S(T)-N} \prod_{k=S(T)+1}^{N} \left( \frac{k - r}{k} \right) \right) = B_r$$

(remember that $I(T) = 0$ if $S(T) \ne 0$). Put $r = N$ to get $\mathbb{P}(S(T) = N) = B_N$. More generally, we have from (15) that $p_j = \mathbb{P}(S(T) = j)$ satisfies

(16)
$$p_N + \frac{N - r}{N B_r} p_{N-1} + \frac{(N - r)(N - r - 1)}{N(N - 1)B_r^2} p_{N-2} + \cdots + \frac{(N - r)! \, r!}{N! \, B_r^{N-r}} p_r = B_r,$$

for $1 \le r \le N$. From these equations, $p_0 = \mathbb{P}(S(T) = 0)$ may in principle be calculated. ●

**(17) Example.** Our final two examples are relevant to mathematical analysis. Let $f : [0, 1] \to \mathbb{R}$ be a (measurable) function such that

(18)
$$\int_0^1 |f(x)| \, dx < \infty;$$

that is, $f$ is integrable. We shall show that there exists a sequence $\{f_n : n \ge 0\}$ of step functions such that $f_n(x) \to f(x)$ as $n \to \infty$, except possibly for an exceptional set of values of $x$ having Lebesgue measure 0.

Let $X$ be uniformly distributed on $[0, 1]$, and define $X_n$ by

(19)
$$X_n = k2^{-n} \quad \text{if} \quad k2^{-n} \le X < (k + 1)2^{-n}$$

where $k$ and $n$ are non-negative integers. It is easily seen that $X_n \uparrow X$ as $n \to \infty$, and furthermore $2^n(X_n - X_{n-1})$ equals the $n$th term in the binary expansion of $X$.

Define $Y = f(X)$ and $Y_n = \mathbb{E}(Y \mid \mathcal{F}_n)$ where $\mathcal{F}_n = \sigma(X_0, X_1, \dots, X_n)$. Now $\mathbb{E}|f(X)| < \infty$ by (18), and therefore $(Y, \mathcal{F})$ is a uniformly integrable martingale (see Example (12.3.9)). It follows that

$$(20) \qquad Y_n \to Y_\infty = \mathbb{E}(Y \mid \mathcal{F}_\infty) \quad \text{a.s. and in mean,}$$

where $\mathcal{F}_\infty = \sigma(X_0, X_1, X_2, \dots) = \sigma(X)$. Hence $Y_\infty = \mathbb{E}(f(X) \mid X) = f(X)$, and in addition

$$(21) \qquad Y_n = \mathbb{E}(Y \mid \mathcal{F}_n) = \mathbb{E}(Y \mid X_0, X_1, \dots, X_n) = \int_{X_n}^{X_n+2^{-n}} f(u) 2^n \, du = f_n(X)$$

where $f_n : [0, 1] \to \mathbb{R}$ is the step function defined by

$$f_n(x) = 2^n \int_{X_n}^{X_n+2^{-n}} f(u) \, du,$$

$x_n$ being the number of the form $k2^{-n}$ satisfying $x_n \le x < x_n + 2^{-n}$. We have from (20) that $f_n(X) \to f(X)$ a.s. and in mean, whence $f_n(x) \to f(x)$ for almost all $x$, and furthermore

$$\int_0^1 |f_n(x) - f(x)| \, dx \to 0 \quad \text{as} \quad n \to \infty. \qquad \bullet$$

**(22) Example.** This time let $f : [0, 1] \to \mathbb{R}$ be Lipschitz continuous, which is to say that there exists $C$ such that

$$(23) \qquad |f(x) - f(y)| \le C|x - y| \quad \text{for all } x, y \in [0, 1].$$

Lipschitz continuity is of course somewhere between continuity and differentiability: Lipschitz-continuous functions are necessarily continuous but need not be differentiable (in the usual sense). We shall see, however, that there must exist a function $g$ such that

$$f(x) - f(0) = \int_0^x g(u) \, du, \quad x \in [0, 1];$$

the function $g$ is called the *Radon–Nikodým derivative* of $f$ (with respect to Lebesgue measure).

As in the last example, let $X$ be uniformly distributed on $[0, 1]$, define $X_n$ by (19), and let

$$(24) \qquad Z_n = 2^n \big[ f(X_n + 2^{-n}) - f(X_n) \big].$$

It may be seen as follows that $(Z, \mathcal{F})$ is a martingale (with respect to the filtration $\mathcal{F}_n = \sigma(X_0, X_1, \dots, X_n)$). First, we check that $\mathbb{E}(Z_{n+1} \mid \mathcal{F}_n) = Z_n$. To this end note that, conditional on $X_0, X_1, \dots, X_n$, it is the case that $X_{n+1}$ is equally likely to take the value $X_n$ or the value $X_n + 2^{-n-1}$. Therefore

$$\mathbb{E}(Z_{n+1} \mid \mathcal{F}_n) = \tfrac{1}{2} 2^{n+1} \big[ f(X_n + 2^{-n-1}) - f(X_n) \big]$$
$$+ \tfrac{1}{2} 2^{n+1} \big[ f(X_n + 2^{-n}) - f(X_n + 2^{-n-1}) \big]$$
$$= 2^n \big[ f(X_n + 2^{-n}) - f(X_n) \big] = Z_n.$$

Secondly, by the Lipschitz continuity (23) of $f$, it is the case that $|Z_n| \le C$, whence $(Z, \mathcal{F})$ is a bounded martingale.

Therefore $Z_n$ converges almost surely and in mean to some limit $Z_\infty$, and furthermore $Z_n = \mathbb{E}(Z_\infty \mid \mathcal{F}_n)$ by Lemma (12.3.11). Now $Z_\infty$ is $\mathcal{F}_\infty$-measurable where $\mathcal{F}_\infty = \lim_{n \to \infty} \mathcal{F}_n = \sigma(X_0, X_1, X_2, \dots) = \sigma(X)$, which implies that $Z_\infty$ is a function of $X$, say $Z_\infty = g(X)$. As in equation (21), the relation

$$Z_n = \mathbb{E}\big(g(X) \mid X_0, X_1, \dots, X_n\big)$$

becomes

$$f(X_n + 2^{-n}) - f(X_n) = \int_{X_n}^{X_n + 2^{-n}} g(u)\, du.$$

This is an ('almost sure') identity for $X_n$, which has positive probability of taking any value of the form $k2^{-n}$ for $0 \le k < 2^n$. Hence

$$f\big((k+1)2^{-n}\big) - f(k2^{-n}) = \int_{k2^{-n}}^{(k+1)2^{-n}} g(u)\, du,$$

whence, by summing,

$$f(x) - f(0) = \int_0^x g(u)\, du$$

for all $x$ of the form $k2^{-n}$ for some $n \ge 1$ and $0 \le k < 2^n$. The corresponding result for general $x \in [0, 1]$ is obtained by taking a limit along a sequence of such 'dyadic rationals'. ●

---

## 12.9 Problems

1.    Let $Z_n$ be the size of the $n$th generation of a branching process with immigration in which the mean family size is $\mu$ ($\ne 1$) and the mean number of immigrants per generation is $m$. Show that

$$Y_n = \mu^{-n}\left\{ Z_n - m\frac{1 - \mu^n}{1 - \mu} \right\}$$

defines a martingale.

2.    In an age-dependent branching process, each individual gives birth to a random number of off-spring at random times. At time 0, there exists a single progenitor who has $N$ children at the subsequent times $B_1 \le B_2 \le \cdots \le B_N$; his family may be described by the vector $(N, B_1, B_2, \dots, B_N)$. Each subsequent member $x$ of the population has a family described similarly by a vector $(N(x), B_1(x), \dots, B_{N(x)}(x))$ having the same distribution as $(N, B_1, \dots, B_N)$ and independent of all other individuals' families. The number $N(x)$ is the number of his offspring, and $B_i(x)$ is the time between the births of the parent and the $i$th offspring. Let $\{B_{n,r} : r \ge 1\}$ be the times of births of individuals in the $n$th generation. Let $M_n(\theta) = \sum_r e^{-\theta B_{n,r}}$, and show that $Y_n = M_n(\theta)/\mathbb{E}(M_1(\theta))^n$ defines a martingale with respect to $\mathcal{F}_n = \sigma(\{B_{m,r} : m \le n, r \ge 1\})$, for any value of $\theta$ such that $\mathbb{E}M_1(\theta) < \infty$.

3.    Let $(Y, \mathcal{F})$ be a martingale with $\mathbb{E}Y_n = 0$ and $\mathbb{E}(Y_n^2) < \infty$ for all $n$. Show that

$$\mathbb{P}\left( \max_{1 \le k \le n} Y_k > x \right) \le \frac{\mathbb{E}(Y_n^2)}{\mathbb{E}(Y_n^2) + x^2}, \qquad x > 0.$$

**4.**    Let $(Y, \mathcal{F})$ be a non-negative submartingale with $Y_0 = 0$, and let $\{c_n\}$ be a non-increasing sequence of positive numbers. Show that

$$\mathbb{P}\left(\max_{1\leq k\leq n} c_k Y_k \geq x\right) \leq \frac{1}{x}\sum_{k=1}^{n} c_k \mathbb{E}(Y_k - Y_{k-1}), \qquad x > 0.$$

Such an inequality is sometimes named after subsets of Hájek, Rényi, and Chow. Deduce Kolmogorov's inequality for the sum of independent random variables. [Hint: Work with the martingale $Z_n = c_n Y_n - \sum_{k=1}^{n} c_k \mathbb{E}(X_k \mid \mathcal{F}_{k-1}) + \sum_{k=1}^{n}(c_{k-1} - c_k)Y_{k-1}$ where $X_k = Y_k - Y_{k-1}$.]

**5.**    Suppose that the sequence $\{X_n : n \geq 1\}$ of random variables satisfies $\mathbb{E}(X_n \mid X_1, X_2, \ldots, X_{n-1}) = 0$ for all $n$, and also $\sum_{k=1}^{\infty} \mathbb{E}(|X_k|^r)/k^r < \infty$ for some $r \in [1, 2]$. Let $S_n = \sum_{i=1}^{n} Z_i$ where $Z_i = X_i/i$, and show that

$$\mathbb{P}\left(\max_{1\leq k\leq n} |S_{m+k} - S_m| \geq x\right) \leq \frac{1}{x^r}\mathbb{E}(|S_{m+n} - S_m|^r), \qquad x > 0.$$

Deduce that $S_n$ converges a.s. as $n \to \infty$, and hence that $n^{-1}\sum_1^n X_k \xrightarrow{\text{a.s.}} 0$. [Hint: In the case $1 < r \leq 2$, prove and use the fact that $h(u) = |u|^r$ satisfies $h(v) - h(u) \leq (v-u)h'(u) + 2h((v-u)/2)$. Kronecker's lemma is useful for the last part.]

**6.**    Let $X_1, X_2, \ldots$ be independent random variables with

$$X_n = \begin{cases} 1 & \text{with probability } (2n)^{-1}, \\ 0 & \text{with probability } 1 - n^{-1}, \\ -1 & \text{with probability } (2n)^{-1}. \end{cases}$$

Let $Y_1 = X_1$ and for $n \geq 2$

$$Y_n = \begin{cases} X_n & \text{if } Y_{n-1} = 0, \\ nY_{n-1}|X_n| & \text{if } Y_{n-1} \neq 0. \end{cases}$$

Show that $Y_n$ is a martingale with respect to $\mathcal{F}_n = \sigma(Y_1, Y_2, \ldots, Y_n)$. Show that $Y_n$ does not converge almost surely. Does $Y_n$ converge in any way? Why does the martingale convergence theorem not apply?

**7.**    Let $X_1, X_2, \ldots$ be independent identically distributed random variables and suppose that $M(t) = \mathbb{E}(e^{tX_1})$ satisfies $M(t) = 1$ for some $t > 0$. Show that $\mathbb{P}(S_k \geq x \text{ for some } k) \leq e^{-tx}$ for $x > 0$ and such a value of $t$, where $S_k = X_1 + X_2 + \cdots + X_k$.

**8.**    Let $Z_n$ be the size of the $n$th generation of a branching process with family-size probability generating function $G(s)$, and assume $Z_0 = 1$. Let $\xi$ be the smallest positive root of $G(s) = s$. Use the martingale convergence theorem to show that, if $0 < \xi < 1$, then $\mathbb{P}(Z_n \to 0) = \xi$ and $\mathbb{P}(Z_n \to \infty) = 1 - \xi$.

**9.**    Let $(Y, \mathcal{F})$ be a non-negative martingale, and let $Y_n^* = \max\{Y_k : 0 \leq k \leq n\}$. Show that

$$\mathbb{E}(Y_n^*) \leq \frac{e}{e-1}\left\{1 + \mathbb{E}\left(Y_n(\log Y_n)^+\right)\right\}.$$

[Hint: $a \log^+ b \leq a \log^+ a + b/e$ if $a, b \geq 0$, where $\log^+ x = \max\{0, \log x\}$.]

**10.**    Let $X = \{X(t) : t \geq 0\}$ be a birth–death process with parameters $\lambda_i, \mu_i$, where $\lambda_i = 0$ if and only if $i = 0$. Define $h(0) = 0, h(1) = 1$, and

$$h(j) = 1 + \sum_{i=1}^{j-1} \frac{\mu_1\mu_2\cdots\mu_i}{\lambda_1\lambda_2\cdots\lambda_i}, \qquad j \geq 2.$$

Show that $h(X(t))$ constitutes a martingale with respect to the filtration $\mathcal{F}_t = \sigma(\{X(u) : 0 \leq u \leq t\})$, whenever $\mathbb{E}h(X(t)) < \infty$ for all $t$. (You may assume that the forward equations are satisfied.)

Fix $n$, and let $m < n$; let $\pi(m)$ be the probability that the process is absorbed at 0 before it reaches size $n$, having started at size $m$. Show that $\pi(m) = 1 - \{h(m)/h(n)\}$.

**11.** Let $(Y, \mathcal{F})$ be a submartingale such that $\mathbb{E}(Y_n^+) \le M$ for some $M$ and all $n$.

(a) Show that $M_n = \lim_{m \to \infty} \mathbb{E}(Y_{n+m}^+ \mid \mathcal{F}_n)$ exists (almost surely) and defines a martingale with respect to $\mathcal{F}$.

(b) Show that $Y_n$ may be expressed in the form $Y_n = X_n - Z_n$ where $(X, \mathcal{F})$ is a non-negative martingale, and $(Z, \mathcal{F})$ is a non-negative supermartingale. This representation of $Y$ is sometimes termed the 'Krickeberg decomposition'.

(c) Let $(Y, \mathcal{F})$ be a martingale such that $\mathbb{E}|Y_n| \le M$ for some $M$ and all $n$. Show that $Y$ may be expressed as the difference of two non-negative martingales.

**12.** Let $\pounds Y_n$ be the assets of an insurance company after $n$ years of trading. During each year it receives a total (fixed) income of $\pounds P$ in premiums. During the $n$th year it pays out a total of $\pounds C_n$ in claims. Thus $Y_{n+1} = Y_n + P - C_{n+1}$. Suppose that $C_1, C_2, \dots$ are independent $N(\mu, \sigma^2)$ variables and show that the probability of ultimate bankruptcy satisfies

$$\mathbb{P}\big(Y_n \le 0 \text{ for some } n\big) \le \exp\left\{-\frac{2(P - \mu)Y_0}{\sigma^2}\right\}.$$

**13. Pólya's urn.** A bag contains red and blue balls, with initially $r$ red and $b$ blue where $rb > 0$. A ball is drawn from the bag, its colour noted, and then it is returned to the bag together with a new ball of the same colour. Let $R_n$ be the number of red balls after $n$ such operations.

(a) Show that $Y_n = R_n/(n + r + b)$ is a martingale which converges almost surely and in mean.

(b) Let $T$ be the number of balls drawn until the first blue ball appears, and suppose that $r = b = 1$. Show that $\mathbb{E}\{(T + 2)^{-1}\} = \frac{1}{4}$.

(c) Suppose $r = b = 1$, and show that $\mathbb{P}(Y_n \ge \frac{3}{4} \text{ for some } n) \le \frac{2}{3}$.

**14.** Here is a modification of the last problem. Let $\{A_n : n \ge 1\}$ be a sequence of random variables, each being a non-negative integer. We are provided with the bag of Problem (12.9.13), and we add balls according to the following rules. At each stage a ball is drawn from the bag, and its colour noted; we assume that the distribution of this colour depends only on the current contents of the bag and not on any further information concerning the $A_n$. We return this ball together with $A_n$ new balls of the same colour. Write $R_n$ and $B_n$ for the numbers of red and blue balls in the urn after $n$ operations, and let $\mathcal{F}_n = \sigma(\{R_k, B_k : 0 \le k \le n\})$. Show that $Y_n = R_n/(R_n + B_n)$ defines a martingale. Suppose $R_0 = B_0 = 1$, let $T$ be the number of balls drawn until the first blue ball appears, and show that

$$\mathbb{E}\left(\frac{1 + A_T}{2 + \sum_{i=1}^{T} A_i}\right) = \frac{1}{2},$$

so long as $\sum_n \left(2 + \sum_{i=1}^{n} A_i\right)^{-1} = \infty$ a.s.

**15. Labouchere system.** Here is a gambling system for playing a fair game. Choose a sequence $x_1, x_2, \dots, x_n$ of positive numbers.

Wager the sum of the first and last numbers on an evens bet. If you win, delete those two numbers; if you lose, append their sum as an extra term $x_{n+1}$ ($= x_1 + x_n$) at the right-hand end of the sequence.

You play iteratively according to the above rule. If the sequence ever contains one term only, you wager that amount on an evens bet. If you win, you delete the term, and if you lose you append it to the sequence to obtain two terms.

Show that, with probability 1, the game terminates with a profit of $\sum_1^n x_i$, and that the time until termination has finite mean.

This looks like another clever strategy. Show that the mean size of your maximum deficit is infinite. (When Henry Labouchere was sent down from Trinity College, Cambridge, in 1852, his gambling debts exceeded £6000.)

**16.** Here is a martingale approach to the question of determining the mean number of tosses of a coin before the first appearance of the sequence HHH. A large casino contains infinitely many gamblers $G_1, G_2, \ldots$, each with an initial fortune of \$1. A croupier tosses a coin repeatedly. For each $n$, gambler $G_n$ bets as follows. Just before the $n$th toss he stakes his \$1 on the event that the $n$th toss shows heads. The game is assumed fair, so that he receives a total of \$$p^{-1}$ if he wins, where $p$ is the probability of heads. If he wins this gamble, then he *repeatedly* stakes his entire current fortune on heads, at the same odds as his first gamble. At the first subsequent tail he loses his fortune and leaves the casino, penniless. Let $S_n$ be the casino's profit (losses count negative) after the $n$th toss. Show that $S_n$ is a martingale. Let $N$ be the number of tosses before the first appearance of HHH; show that $N$ is a stopping time and hence find $\mathbb{E}(N)$.

Now adapt this scheme to calculate the mean time to the first appearance of the sequence HTH.

**17.** Let $\{(X_k, Y_k) : k \geq 1\}$ be a sequence of independent identically distributed random vectors such that each $X_k$ and $Y_k$ takes values in the set $\{-1, 0, 1, 2, \ldots\}$. Suppose that $\mathbb{E}(X_1) = \mathbb{E}(Y_1) = 0$ and $\mathbb{E}(X_1 Y_1) = c$, and furthermore $X_1$ and $Y_1$ have finite non-zero variances. Let $U_0$ and $V_0$ be positive integers, and define $(U_{n+1}, V_{n+1}) = (U_n, V_n) + (X_{n+1}, Y_{n+1})$ for each $n \geq 0$. Let $T = \min\{n : U_n V_n = 0\}$ be the first hitting time by the random walk $(U_n, V_n)$ of the axes of $\mathbb{R}^2$. Show that $\mathbb{E}(T) < \infty$ if and only if $c < 0$, and that $\mathbb{E}(T) = -\mathbb{E}(U_0 V_0)/c$ in this case. [Hint: You might show that $U_n V_n - cn$ is a martingale.]

**18.** The game 'Red Now' may be played by a single player with a well shuffled conventional pack of 52 playing cards. At times $n = 1, 2, \ldots, 52$ the player turns over a new card and observes its colour. Just once in the game he must say, just before exposing a card, "Red Now". He wins the game if the next exposed card is red. Let $R_n$ be the number of red cards remaining face down after the $n$th card has been turned over. Show that $X_n = R_n/(52 - n)$, $0 \leq n < 52$, defines a martingale. Show that there is no strategy for the player which results in a probability of winning different from $\frac{1}{2}$.

**19.** A businessman has a redundant piece of equipment which he advertises for sale, inviting "offers over £1000". He anticipates that, each week for the foreseeable future, he will be approached by one prospective purchaser, the offers made in week $0, 1, \ldots$ being £$1000 X_0$, £$1000 X_1, \ldots$, where $X_0, X_1, \ldots$ are independent random variables with a common density function $f$ and finite mean. Storage of the equipment costs £$1000c$ per week and the prevailing rate of interest is $\alpha$ ($> 0$) per week. Explain why a sensible strategy for the businessman is to sell in the week $T$, where $T$ is a stopping time chosen so as to maximize

$$\mu(T) = \mathbb{E}\left\{ (1 + \alpha)^{-T} X_T - \sum_{n=1}^{T} (1 + \alpha)^{-n} c \right\}.$$

Show that this problem is equivalent to maximizing $\mathbb{E}\{(1 + \alpha)^{-T} Z_T\}$ where $Z_n = X_n + c/\alpha$. Show that there exists a unique positive real number $\gamma$ with the property that

$$\alpha \gamma = \int_{\gamma}^{\infty} \mathbb{P}(Z_n > y) \, dy,$$

and that, for this value of $\gamma$, the sequence $V_n = (1 + \alpha)^{-n} \max\{Z_n, \gamma\}$ constitutes a supermartingale. Deduce that the optimal strategy for the businessman is to set a target price $\tau$ (which you should specify in terms of $\gamma$) and sell the first time he is offered at least this price.

In the case when $f(x) = 2x^{-3}$ for $x \geq 1$, and $c = \alpha = \frac{1}{90}$, find his target price and the expected number of weeks he will have to wait before selling.

**20.** Let $Z$ be a branching process satisfying $Z_0 = 1$, $\mathbb{E}(Z_1) < 1$, and $\mathbb{P}(Z_1 \geq 2) > 0$. Show that $\mathbb{E}(\sup_n Z_n) \leq \eta/(\eta - 1)$, where $\eta$ is the largest root of the equation $x = G(x)$ and $G$ is the probability generating function of $Z_1$.

**21. Matching.** In a cloakroom there are $K$ coats belonging to $K$ people who make an attempt to leave by picking a coat at random. Those who pick their own coat leave, the rest return the coats and try again at random. Let $N$ be the number of rounds of attempts until everyone has left. Show that $\mathbb{E}N = K$ and $\text{var}(N) \leq K$.

**22.** Let $W$ be a standard Wiener process, and define

$$M(t) = \int_0^t W(u)\, du - \tfrac{1}{3} W(t)^3.$$

Show that $M(t)$ is a martingale, and deduce that the expected area under the path of $W$ until it first reaches one of the levels $a\ (> 0)$ or $b\ (< 0)$ is $-\tfrac{1}{3} ab(a + b)$.

**23.** Let $W = (W_1, W_2, \ldots, W_d)$ be a $d$-dimensional Wiener process, the $W_i$ being independent one-dimensional Wiener processes with $W_i(0) = 0$ and variance parameter $\sigma^2 = d^{-1}$. Let $R(t)^2 = W_1(t)^2 + W_2(t)^2 + \cdots + W_d(t)^2$, and show that $R(t)^2 - t$ is a martingale. Deduce that the mean time to hit the sphere of $\mathbb{R}^d$ with radius $a$ is $a^2$.

**24.** Let $W$ be a standard one-dimensional Wiener process, and let $a, b > 0$. Let $T$ be the earliest time at which $W$ visits either of the two points $-a, b$. Show that $\mathbb{P}(W(T) = b) = a/(a + b)$ and $\mathbb{E}(T) = ab$. In the case $a = b$, find $\mathbb{E}(e^{-sT})$ for $s > 0$.

**25.** Let $(a_n)$ be a real sequence satisfying $a_n \in (0, 1)$, and let $\{U_n : n \geq 1\}$ be independent random variables with the uniform distribution on $(0, 1)$. Define

$$X_{n+1} = \begin{cases} (1 - a_n)X_n + a_n & \text{if } X_n > U_{n+1}, \\ (1 - a_n)X_n & \text{otherwise,} \end{cases}$$

where $X_0 = \rho \in (0, 1)$.
(a) Show that the sequence $X = \{X_n : n \geq 0\}$ is a martingale with respect to the filtration $\mathcal{F}_n = \sigma(X_0, X_1, \ldots, X_n)$, and that $X_n$ converges a.s. and in mean square to some $X_\infty$.
(b) Show that the infinite sum $\sum_{n=1}^\infty \mathbb{E}\{(X_{n+1} - X_n)^2 \mid \mathcal{F}_n\}$ converges a.s. and in mean to some random variable $A$ with $\mathbb{E}(A) = \mathbb{E}(X_\infty^2) - \rho^2$.
(c) Hence prove that $S = \sum_{n=0}^\infty a_n^2 X_n(1 - X_n)$ is a.s. finite.
(d) Deduce that, if $\sum_n a_n^2 = \infty$, then $X_\infty$ takes only the values 0 and 1. In this case, what is $\mathbb{P}(X_\infty = 1)$?

**26. Exponential inequality for Wiener process.** Let $W$ be a standard Wiener process, and show that

$$\mathbb{P}\left( \sup_{0 \leq t \leq T} W(t) \geq x \right) \leq \exp\{-\tfrac{1}{2} x^2/T\}, \qquad x > 0.$$

You may assume a version of Doob's maximal inequality (12.6.1) for continuous parameter submartingales.

**27. Insurance, Problem (8.10.7) revisited.** An insurance company receives premiums (net of costs) at rate $\rho$ per unit time. Claims $X_1, X_2, \ldots$ are independent random variables with the exponential distribution with parameter $\mu$, and they arrive at the times of a Poisson process of rate $\lambda$ (assume the usual independence, as well as $\lambda, \mu, \rho > 0$). Let $Y(t)$ be the assets of the company at time $t$ where $Y(0) = y > 0$. Show that

$$\mathbb{P}\big(Y(t) \leq 0 \text{ for some } t > 0\big) = \begin{cases} \left(1 - \dfrac{\theta}{\mu}\right) e^{-\theta y} & \text{if } \theta > 0, \\ 1 & \text{otherwise,} \end{cases}$$

where $\theta = \mu - (\lambda/\rho)$.

# 13

# Diffusion processes

*Summary.* An elementary description of the Wiener process (Brownian motion) is presented, and used to motivate an account of diffusion processes based on the instantaneous mean and variance. This leads to the forward and backward equations for diffusions. First-passage probabilities of the Wiener process are explored using the reflection principle. Interpretations of absorbing and reflecting barriers for diffusions are presented. There is a brief account of excursions, and of the Brownian bridge. The Itô calculus is summarized, and used to construct a class of diffusions which are martingales. The theory of financial mathematics based on the Wiener process is described, including option pricing and the Black–Scholes formula. Finally, there is a discussion of the links between diffusions, harmonic functions, and potential theory.

## 13.1 Introduction

Random processes come in many types. For example, they may run in discrete time or continuous time, and their state spaces may also be discrete or continuous. In the main, we have so far considered processes which are *discrete* either in time or space; our purpose in this chapter is to approach the theory of processes indexed by continuous time and taking values in the real line $\mathbb{R}$. Many important examples belong to this category: meteorological data, communication systems with noise, molecular motion, and so on. In other important cases, such random processes provide useful approximations to the physical process in question: processes in population genetics or population evolution, for example.

The archetypal diffusion process is the Wiener process $W$ of Example (9.6.13), a Gaussian process with stationary independent increments. Think about $W$ as a description of the motion of a particle moving randomly but continuously about $\mathbb{R}$. There are various ways of *defining* the Wiener process, and each such definition has two components. First of all, we require a *distributional* property, such as that the finite-dimensional distributions are Gaussian, and so on. The second component, not explored in Chapter 9, is that the sample paths of the process $\{W(t; \omega) : t \geq 0\}$, thought of as random functions on the underlying probability space $(\Omega, \mathcal{F}, \mathbb{P})$, are almost surely continuous. This assumption is important and natural, and of particular relevance when studying first passage times of the process.

Similar properties are required of a diffusion process, and we reserve the term 'diffusion' for a process $\{X(t) : t \geq 0\}$ having the strong Markov property and whose sample paths are almost surely continuous.

---

## 13.2  Brownian motion

Suppose we observe a container of water. The water may appear to be motionless, but this is an illusion. If we are able to approach the container so closely as to be able to distinguish individual molecules then we may perceive that each molecule enjoys a motion which is unceasing and without any apparent order. The disorder of this movement arises from the frequent occasions at which the molecule is repulsed by other molecules which are nearby at the time. A revolutionary microscope design enabled the Dutch scientist A. van Leeuwenhoek (1632–1723) to observe the apparently random motion of micro-organisms dubbed 'animalcules', but this motion was *biological* in cause. Credit for noticing that all sufficiently tiny particles enjoy a random movement of *physical* origin is usually given to the botanist R. Brown (1773–1858). Brown studied in 1827 the motion of tiny particles suspended in water, and he lent his name to the type of erratic movement thus observed. It was a major thrust of mathematics in the 20th century to model such phenomena, and this has led to the mathematical object termed the 'Wiener process', an informal motivation for which is presented in this section.

Brownian motion takes place in continuous time and continuous space. Our first attempt to model it might proceed by approximating to it by a discrete process such as a random walk. At any epoch of time the position of an observed particle is constrained to move about the points $\{(a\delta, b\delta, c\delta) : a, b, c = 0, \pm 1, \pm 2, \ldots\}$ of a three-dimensional 'cubic' lattice in which the distance between neighbouring points is $\delta$; the quantity $\delta$ is a fixed positive number which is very small. Suppose further that the particle performs a symmetric random walk on this lattice (see Problem (6.15.9) for the case $\delta = 1$) so that its position $\mathbf{S}_n$ after $n$ jumps satisfies

$$\mathbb{P}(\mathbf{S}_{n+1} = \mathbf{S}_n + \delta\boldsymbol{\epsilon}) = \tfrac{1}{6} \quad \text{if} \quad \boldsymbol{\epsilon} = (\pm 1, 0, 0), (0, \pm 1, 0), (0, 0, \pm 1).$$

Let us concentrate on the $x$ coordinate of the particle, and write $\mathbf{S}_n = (S_n^1, S_n^2, S_n^3)$. Then

$$S_n^1 - S_0^1 = \sum_{i=1}^{n} X_i$$

as in Section 3.9, where $\{X_i\}$ is an independent identically distributed sequence with

$$\mathbb{P}(X_i = k\delta) = \begin{cases} \tfrac{1}{6} & \text{if } k = -1, \\ \tfrac{1}{6} & \text{if } k = +1, \\ \tfrac{2}{3} & \text{if } k = 0. \end{cases}$$

We are interested in the displacement $S_n^1 - S_0^1$ when $n$ is large; the central limit theorem (5.10.4) tells us that the distribution of this displacement is approximately $N(0, \tfrac{1}{3}n\delta^2)$. Now suppose that the jumps of the random walk take place at time epochs $\tau, 2\tau, 3\tau, \ldots$ where $\tau > 0$; $\tau$ is the time between jumps and is very small, implying that a very large number of jumps occur in any 'reasonable' time interval. Observe the particle after some time $t$ ($> 0$)

has elapsed. By this time it has experienced $n = \lfloor t/\tau \rfloor$ jumps, and so its $x$ coordinate $S^1(t)$ is such that $S^1(t) - S^1(0)$ is approximately $N(0, \frac{1}{3}t\delta^2/\tau)$. At this stage in the analysis we let the inter-point distance $\delta$ and the inter-jump time $\tau$ approach zero; in so doing we hope that the discrete random walk may approach some limit whose properties have something in common with the observed features of Brownian motion. We let $\delta \downarrow 0$ and $\tau \downarrow 0$ in such a way that $\frac{1}{3}\delta^2/\tau$ remains constant, since the variance of the distribution of $S^1(t) - S^1(0)$ fails to settle down to a non-trivial limit otherwise. Set

(1) $$\tfrac{1}{3}\delta^2/\tau = \sigma^2$$

where $\sigma^2$ is a positive constant, and pass to the limit to obtain that the distribution of $S^1(t) - S^1(0)$ approaches $N(0, \sigma^2 t)$. We can apply the same argument to the $y$ coordinate and to the $z$ coordinate of the particle to deduce that the particle's position $\mathbf{S}(t) = (S^1(t), S^2(t), S^3(t))$ at time $t$ is such that the asymptotic distribution of the coordinates of the displacement $\mathbf{S}(t) - \mathbf{S}(0)$ is multivariate normal whenever $\delta, \tau \downarrow 0$, and (1) holds; furthermore, it is not too hard to see that $S^1(t)$, $S^2(t)$, and $S^3(t)$ are independent of each other.

We may guess from the asymptotic properties of this random walk that an adequate model for Brownian motion will involve a process $\mathbf{X} = \{\mathbf{X}(t) : t \geq 0\}$ taking values in $\mathbb{R}^3$ with a coordinate representation $\mathbf{X}(t) = (X^1(t), X^2(t), X^3(t))$ such that:

(a) $\mathbf{X}(0) = (0, 0, 0)$, say,

(b) $X^1$, $X^2$, and $X^3$ are independent and identically distributed processes,

(c) $X^1(s + t) - X^1(s)$ is $N(0, \sigma^2 t)$ for any $s, t \geq 0$,

(d) $X^1$ has *independent increments* in that $X^1(v) - X^1(u)$ and $X^1(t) - X^1(s)$ are independent whenever $u \leq v \leq s \leq t$.

We have not yet shown the existence of such a process $\mathbf{X}$; the foregoing argument only indicates certain plausible distributional properties without showing that they are attainable. However, properties (c) and (d) are not new to us and remind us of the Wiener process of Example (9.6.13); we deduce that such a process $\mathbf{X}$ indeed exists, and is given by $\mathbf{X}(t) = (W^1(t), W^2(t), W^3(t))$ where $W^1$, $W^2$, and $W^3$ are independent Wiener processes.

This conclusion is gratifying in that it demonstrates the existence of a random process which seems to enjoy at least some of the features of Brownian motion. A more detailed and technical analysis indicates some weak points of the Wiener model. This is beyond the scope of this text, and we able only to skim the surface of the main difficulty. For each $\omega$ in the sample space $\Omega$, $\{\mathbf{X}(t; \omega) : t \geq 0\}$ is a sample path of the process along which the particle may move. It can be shown that, in some sense to be discussed in the next section,

(a) the sample paths are continuous functions of $t$,

(b) almost all sample paths are nowhere differentiable functions of $t$.

Property (a) is physically necessary, but (b) is a property which *cannot* be shared by the physical phenomenon which we are modelling, since mechanical considerations, such as Newton's laws, imply that only particles with zero mass can move along routes which are nowhere differentiable. As a model for the local movement (over a short time interval) of particles, the Wiener process is poor; over longer periods of time the properties of the Wiener process are indeed very similar to experimental results.

A popular improved model for the local behaviour of Brownian paths is the so-called Ornstein–Uhlenbeck process. We close this section with a short account of this. Roughly, it is founded on the assumption that the velocity of the particle (rather than its position) undergoes a random walk; the ensuing motion is damped by the frictional resistance of the fluid. The result is a 'velocity process' with continuous sample paths; their integrals represent the sample paths of the particle itself. Think of the motion in one dimension as before, and write $V_n$ for the velocity of the particle after the $n$th jump. At the next jump the change $V_{n+1} - V_n$ in the velocity is assumed to have two contributions: the frictional resistance to motion, and some random fluctuation owing to collisions with other particles. We shall assume that the former damping effect is directly proportional to $V_n$, so that $V_{n+1} = V_n + X_{n+1}$; this is the so-called *Langevin equation*. We require that:

$$\mathbb{E}(X_{n+1} \mid V_n) = -\beta V_n \qquad \text{: frictional effect,}$$

$$\text{var}(X_{n+1} \mid V_n) = \sigma^2 \qquad \text{: collision effect,}$$

where $\beta$ and $\sigma^2$ are constants. The sequence $\{V_n\}$ is no longer a random walk on some regular grid of points, but it can be shown that the distributions converge as before, after suitable passage to the limit. Furthermore, there exists a process $V = \{V(t) : t \geq 0\}$ with the corresponding distributional properties, and whose sample paths turn out to be almost surely continuous. These sample paths do not represent possible routes of the particle, but rather describe the development of its velocity as time passes. The possible paths of the particle through the space which it inhabits are found by integrating the sample paths of $V$ with respect to time. The resulting paths are almost surely continuously differentiable functions of time.

---

## Exercise for Section 13.2

**1.    Total variation.** Show that the total variation of a standard Wiener process over any non-trivial interval does not exist.
[The total variation $V_{a,b}(f)$ of a function $f : \mathbb{R} \to \mathbb{R}$ over the bounded interval $[a, b]$ is defined as

$$V_{a,b}(f) = \sup \sum |f(s_{r+1}) - f(s_r)|,$$

where the supremum is taken over all increasing sequences $s_0 = a, s_1, s_2, \ldots, s_n = b$. You may find it useful that the quadratic variation of the Wiener process exists and is non-zero almost surely.]

---

## 13.3  Diffusion processes

We say that a particle is 'diffusing' about a space $\mathbb{R}^n$ whenever it experiences erratic and disordered motion through the space; for example, we may speak of radioactive particles diffusing through the atmosphere, or even of a rumour diffusing through a population. For the moment, we restrict our attention to one-dimensional diffusions, for which the position of the observed particle at any time is a point on the real line; similar arguments will hold for higher dimensions. Our first diffusion model is the Wiener process.

**(1) Definition.** A **Wiener process** $W = \{W(t) : t \geq 0\}$, starting from $W(0) = w$, say, is a real-valued Gaussian process such that:
  (a)  $W$ has independent increments (see Lemma (9.6.16)),
  (b)  $W(s + t) - W(s)$ is distributed as $N(0, \sigma^2 t)$ for all $s, t \geq 0$ where $\sigma^2$ is a positive constant,
  (c)  the sample paths of $W$ are continuous.

Clearly (1a) and (1b) specify the finite-dimensional distributions (fdds) of a Wiener process $W$, and the argument of Theorem (9.6.1) shows there exists a Gaussian process with these fdds. In agreement with Example (9.6.13), the autocovariance function of $W$ is given by

$$c(s, t) = \mathbb{E}\big([W(s) - W(0)][W(t) - W(0)]\big)$$
$$= \mathbb{E}\big([W(s) - W(0)]^2 + [W(s) - W(0)][W(t) - W(s)]\big)$$
$$= \sigma^2 s + 0 \qquad \text{if} \quad 0 \le s \le t,$$

which is to say that

(2) $$c(s, t) = \sigma^2 \min\{s, t\} \quad \text{for all } s, t \ge 0.$$

The process $W$ is called a *standard* Wiener process if $\sigma^2 = 1$ and $W(0) = 0$. If $W$ is non-standard, then $W_1(t) = (W(t) - W(0))/\sigma$ is standard. The process $W$ is said to have 'stationary' independent increments since the distribution of $W(s + t) - W(s)$ depends on $t$ alone. A simple application of Theorem (9.6.7) shows that $W$ is a Markov process.

The Wiener process $W$ can be used to model the apparently random displacement of Brownian motion in any chosen direction. For this reason, $W$ is sometimes called 'Brownian motion', a term which *we* reserve to describe the motivating physical phenomenon.

Does the Wiener process exist? That is to say, does there exist a probability space $(\Omega, \mathcal{F}, \mathbb{P})$ and a Gaussian process $W$ thereon, satisfying (1a, b, c)? The answer to this non-trivial question is of course in the affirmative, and we defer to Theorem (19) an explicit construction of such a process. The difficulty lies not in satisfying the distributional properties (1a, b) but in showing that this may achieved with *continuous* sample paths.

Roughly speaking, there are two types of statement to be made about diffusion processes in general, and the Wiener process in particular. The first deals with sample path properties, and the second with distributional properties.

Figure 13.1 is a diagram of a typical sample path. Certain distributional properties of continuity are immediate. For example, $W$ is 'continuous in mean square' in that

$$\mathbb{E}\big([W(s + t) - W(s)]^2\big) \to 0 \quad \text{as} \quad t \to 0;$$

this follows easily from equation (2).

Let us turn our attention to the distributions of a standard Wiener process $W$. Suppose we are given that $W(s) = x$, say, where $s \ge 0$ and $x \in \mathbb{R}$. Conditional on this, $W(t)$ is distributed as $N(x, t - s)$ for $t \ge s$, which is to say that the conditional distribution function

$$F(t, y \mid s, x) = \mathbb{P}\big(W(t) \le y \mid W(s) = x\big)$$

has density function, often called the *transition density*,

(3) $$f(t, y \mid s, x) = \frac{\partial}{\partial y} F(t, y \mid s, x)$$

which is given by

(4) $$f(t, y \mid s, x) = \frac{1}{\sqrt{2\pi(t - s)}} \exp\left(-\frac{(y - x)^2}{2(t - s)}\right), \qquad -\infty < y < \infty.$$

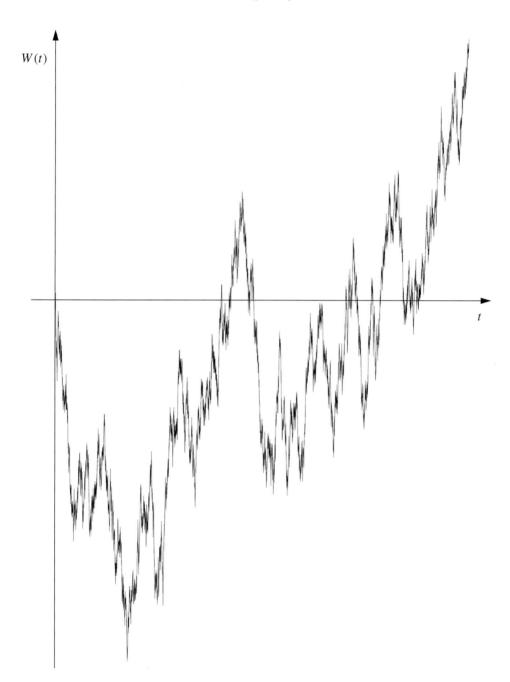

Figure 13.1. A typical realization of a Wiener process $W$. This is a scale drawing of a sample path of $W$ over the time interval $[0, 1]$. Note that the path is continuous but very spiky. This picture indicates the general features of the path only; the dense black portions indicate superimposed fluctuations which are too fine for this method of description. Any magnification of part of the path would reveal fluctuations of order comparable to those of the original path. This picture was drawn by a Monte Carlo method, using nearly 90,000 steps of a symmetric random walk.

This is a function of four variables, but just grit your teeth. It is easy to check that $f$ is the solution of the following differential equations.

**(5)** *Forward diffusion equation:* $\qquad \dfrac{\partial f}{\partial t} = \dfrac{1}{2} \dfrac{\partial^2 f}{\partial y^2}.$

**(6)** *Backward diffusion equation:* $\qquad \dfrac{\partial f}{\partial s} = -\dfrac{1}{2} \dfrac{\partial^2 f}{\partial x^2}.$

We ought to specify the boundary conditions for these equations, but we avoid this at the moment. Subject to certain conditions, (4) is the unique density function which solves (5) or (6). There is a good reason why (5) and (6) are called the *forward* and *backward* equations. Remember that $W$ is a Markov process, and use arguments similar to those of Sections 6.8 and 6.9. Equation (5) is obtained by conditioning $W(t + h)$ on the value of $W(t)$ and letting $h \downarrow 0$; (6) is obtained by conditioning $W(t)$ on the value of $W(s + h)$ and letting $h \downarrow 0$. You are treading in Einstein's footprints as you perform these calculations. The derivatives in (5) and (6) have coefficients which do not depend on $x, y, s, t$; this reflects the fact that the Wiener process is homogeneous in space and time, in that:

(a) the increment $W(t) - W(s)$ is independent of $W(s)$ for all $t \geq s$,

(b) the increments are stationary in time.

Next we turn our attention to diffusion processes which *lack* this homogeneity.

The Wiener process is a Markov process, and the Markov property provides a method for deriving the forward and backward equations. There are other Markov diffusion processes to which this method may be applied in order to obtain similar forward and backward equations; the coefficients in these equations will *not* generally be constant. The existence of such processes can be demonstrated rigorously, but here we explore their distributions only. Let $D = \{D(t) : t \geq 0\}$ denote a diffusion process. In addition to requiring that $D$ has (almost surely) continuous sample paths, we need to impose some conditions on the transitions of $D$ in order to derive its diffusion equations; these conditions take the form of specifying the mean and variance of increments $D(t + h) - D(t)$ of the process over small time intervals $(t, t + h)$. Suppose that there exist functions $a(t, x), b(t, x)$ such that:

$$\mathbb{P}\big(|D(t + h) - D(t)| > \epsilon \,\big|\, D(t) = x\big) = o(h) \quad \text{for all } \epsilon > 0,$$
$$\mathbb{E}\big(D(t + h) - D(t) \,\big|\, D(t) = x\big) = a(t, x)h + o(h),$$
$$\mathbb{E}\big([D(t + h) - D(t)]^2 \,\big|\, D(t) = x\big) = b(t, x)h + o(h).$$

The functions $a$ and $b$ are called the 'instantaneous mean' (or 'drift') and 'instantaneous variance' of $D$ respectively. Subject to certain other technical conditions (see Feller 1971, pp. 332–335), if $s \leq t$ then the conditional (or *transition*) density function of $D(t)$ given $D(s) = x$,

$$f(t, y \mid s, x) = \frac{\partial}{\partial y} \mathbb{P}\big(D(t) \leq y \,\big|\, D(s) = x\big),$$

satisfies the following partial differential equations.

**(7)** *Forward equation:*   $\dfrac{\partial f}{\partial t} = -\dfrac{\partial}{\partial y}[a(t,y)f] + \dfrac{1}{2}\dfrac{\partial^2}{\partial y^2}[b(t,y)f].$

**(8)** *Backward equation:*   $\dfrac{\partial f}{\partial s} = -a(s,x)\dfrac{\partial f}{\partial x} - \dfrac{1}{2}b(s,x)\dfrac{\partial^2 f}{\partial x^2}.$

It is a noteworthy fact that the density function $f$ is specified as soon as the instantaneous mean $a$ and variance $b$ are known; we need no further information about the distribution of a typical increment. This is very convenient for many applications, since $a$ and $b$ are often specified in a natural manner by the physical description of the process.

**(9) Example. The Wiener process.** If increments of any given length have zero means and constant variances then

$$a(t,x) = 0, \quad b(t,x) = \sigma^2,$$

for some $\sigma^2 > 0$. Equations (7) and (8) are of the form of (5) and (6) with the inclusion of a factor $\sigma^2$.                                                                                                            ●

**(10) Example. The Wiener process with drift.** Suppose a particle undergoes a type of one-dimensional Brownian motion, in which it experiences a drift at constant rate in some particular direction. That is to say,

$$a(t,x) = m, \quad b(t,x) = \sigma^2,$$

for some drift rate $m$ and constant $\sigma^2$. The forward diffusion equation becomes

$$\frac{\partial f}{\partial t} = -m\frac{\partial f}{\partial y} + \frac{1}{2}\sigma^2\frac{\partial^2 f}{\partial y^2}$$

and it follows that the corresponding diffusion process $D$ is such that $D(t) = \sigma W(t) + mt$ where $W$ is a standard Wiener process.                                                                                    ●

**(11) The Ornstein–Uhlenbeck process.** Recall the discussion of this process at the end of Section 13.2. It experiences a drift towards the origin of magnitude proportional to its dislacement. That is to say,

$$a(t,x) = -\beta x, \quad b(t,x) = \sigma^2,$$

and the forward equation is

$$\frac{\partial f}{\partial t} = \beta\frac{\partial}{\partial y}(yf) + \frac{1}{2}\sigma^2\frac{\partial^2 f}{\partial y^2}.$$

See Problem (13.8.4) for one solution of this equation.                                                                ●

**(12) Example. Feller's diffusion approximation to the branching process.** Diffusion models are sometimes useful as continuous approximations to discrete processes. In Section 13.2 we saw that the Wiener process approximates to the random walk under certain circumstances;

here is another example of such an approximation. Let $\{Z_n\}$ be the size of the $n$th generation of a branching process, with $Z_0 = 1$ and such that $\mathbb{E}(Z_1) = \mu$ and $\text{var}(Z_1) = \sigma^2$. A typical increment $Z_{n+1} - Z_n$ has mean and variance given by

$$\mathbb{E}(Z_{n+1} - Z_n \mid Z_n = x) = (\mu - 1)x,$$

$$\text{var}(Z_{n+1} - Z_n \mid Z_n = x) = \sigma^2 x;$$

these are directly proportional to the size of $Z_n$. Now, suppose that the time intervals between successive generations become shorter and shorter, but that the means and variances of the increments retain this proportionality; of course, we need to abandon the condition that the process be integer-valued. This suggests a diffusion model as an approximation to the branching process, with instantaneous mean and variance given by

$$a(t, x) = ax, \quad b(t, x) = bx,$$

and the forward equation of such a process is

(13)
$$\frac{\partial f}{\partial t} = -a \frac{\partial}{\partial y} (yf) + \frac{1}{2} b \frac{\partial^2}{\partial y^2} (yf).$$

Subject to appropriate boundary conditions, this equation has a unique solution; this may be found by taking Laplace transforms of (13) in order to find the moment generating function of the value of the diffusion process at time $t$.                                  ●

**(14) Example. A branching diffusion process.** The next example is a modification of the process of (6.12.15) which modelled the distribution in space of the members of a branching process. Read the first paragraph of (6.12.15) again before proceeding with this example. It is often the case that the members of a population move around the space which they inhabit during their lifetimes. With this in mind we introduce a modification into the process of (6.12.15). Suppose a typical individual is born at time $s$ and at position $x$. We suppose that this individual moves about $\mathbb{R}$ until its lifetime $T$ is finished, at which point it dies and divides, leaving its offspring at the position at which it dies. We suppose further that it moves according to a standard Wiener process $W$, so that it is at position $x + W(t)$ at time $s + t$ whenever $0 \le t \le T$. We assume that each individual moves independently of the positions of all the other individuals. We retain the notation of (6.12.15) whenever it is suitable, writing $N$ for the number of offspring of the initial individual, $W$ for the process describing its motion, and $T$ for its lifetime. This individual dies at the point $W(T)$.

We no longer seek complete information about the distribution of the individuals around the space, but restrict ourselves to a less demanding task. It is natural to wonder about the rate at which members of the population move away from the place of birth of the founding member. Let $M(t)$ denote the position of the individual who is furthest right from the origin at time $t$. That is,

$$M(t) = \sup\{x : Z_1(t, x) > 0\}$$

where $Z_1(t, x)$ is the number of living individuals at time $t$ who are positioned at points in the interval $[x, \infty)$. We shall study the distribution function of $M(t)$

$$F(t, x) = \mathbb{P}(M(t) \le x),$$

and we proceed roughly as before, noting that

**(15)**
$$F(t, x) = \int_0^\infty \mathbb{P}\big(M(t) \leq x \mid T = s\big) f_T(s)\, ds$$

where $f_T$ is the density function of $T$. However,

$$\mathbb{P}\big(M(t) \leq x \mid T = s\big) = \mathbb{P}\big(W(t) \leq x\big) \quad \text{if} \quad s > t,$$

whilst, if $s \leq t$, use of conditional probabilities gives

$$\mathbb{P}\big(M(t) \leq x \mid T = s\big)$$
$$= \sum_{n=0}^\infty \int_{-\infty}^\infty \mathbb{P}\big(M(t) \leq x \mid T = s,\ N = n,\ W(s) = w\big) \mathbb{P}(N = n) f_{W(s)}(w)\, dw$$

where $f_{W(s)}$ is the density function of $W(s)$. However, if $s \leq t$, then

$$\mathbb{P}\big(M(t) \leq x \mid T = s,\ N = n,\ W(s) = w\big) = \big[\mathbb{P}\big(M(t - s) \leq x - w\big)\big]^n,$$

and so (15) becomes

**(16)**
$$F(t, x) = \int_{s=0}^t \int_{w=-\infty}^\infty G_N[F(t - s, x - w)] f_{W(s)}(w) f_T(s)\, dw\, ds$$
$$+ \mathbb{P}\big(W(t) \leq x\big) \int_t^\infty f_T(s)\, ds.$$

We consider here only the Markovian case when $T$ is exponentially distributed, so that

$$f_T(s) = \mu e^{-\mu s} \quad \text{for} \quad s \geq 0.$$

Multiply throughout (16) by $e^{\mu t}$, substitute $t - s = u$ and $x - w = v$ within the integral, and differentiate with respect to $t$ to obtain

$$e^{\mu t}\left(\mu F + \frac{\partial F}{\partial t}\right) = \mu \int_{-\infty}^\infty G_N(F(t, v)) f_{W(0)}(x - v) e^{\mu t}\, dv$$
$$+ \mu \int_{u=0}^t \int_{v=-\infty}^\infty G_N(F(u, v)) \left(\frac{\partial}{\partial t} f_{W(t-u)}(x - v)\right) e^{\mu u}\, dv\, du$$
$$+ \frac{\partial}{\partial t} \mathbb{P}\big(W(t) \leq x\big).$$

Now differentiate the same equation twice with respect to $x$, remembering that $f_{W(s)}(w)$ satisfies the diffusion equations and that $\delta(v) = f_{W(0)}(x - v)$ needs to be interpreted as the Dirac $\delta$ function at the point $v = x$ to find that

**(17)**
$$\mu F + \frac{\partial F}{\partial t} = \mu G_N(F) + \frac{1}{2}\frac{\partial^2 F}{\partial x^2}.$$

Many eminent mathematicians have studied this equation; for example, Kolmogorov and Fisher were concerned with it in connection with the distribution of gene frequencies. It is difficult to extract precise information from (17). One approach is to look for solutions of the form $F(t, x) = \psi(x - ct)$ for some constant $c$ to obtain the following second-order ordinary differential equation for $\psi$:

**(18)** $$\psi'' + 2c\psi' + 2\mu H(\psi) = 0$$

where $H(\psi) = G_N(\psi) - \psi$. Solutions to (18) yield information about the asymptotic distribution of the so-called 'advancing wave' of the members of the process.                ●

Finally in this section, we show that Wiener processes exist. The difficulty is the requirement that sample paths be continuous. Certainly there exist Gaussian processes with independent normally distributed increments as required in (1a, b), but there is no reason in general why such a process should have continuous sample paths. We shall show next that one may construct such a Gaussian process with this extra property of continuity.

**(19) Theorem. Existence of the Wiener process.** *There exists a probability space that supports a random process $W = \{W(t) : t \geq 0\}$ satisfying the conditions of Definition (1).*

**Proof.** Let us restrict ourselves for the moment to the time interval $[0, 1]$, and suppose that $X$ is a Gaussian process on $[0, 1]$ with independent increments, such that $X(0) = 0$, and $X(s + t) - X(s)$ is $N(0, t)$ for $s, t \geq 0$. We shall concentrate on a certain countable subset $Q$ of $[0, 1]$, namely the set of 'dyadic rationals', being the set of points of the form $m2^{-n}$ for some $n \geq 1$ and $0 \leq m \leq 2^n$. For each $n \geq 1$, we define the process $X_n(t)$ by $X_n(t) = X(t)$ if $t = m2^{-n}$ for some integer $m$, and by linear interpolation otherwise; that is to say,

$$X_n(t) = X(m2^{-n}) + 2^n(t - m2^{-n})\left[X((m+1)2^{-n}) - X(m2^{-n})\right]$$

if $m2^{-n} < t < (m+1)2^{-n}$. Thus $X_n$ is a piecewise-linear and uniformly continuous function comprising $2^n$ line segments. Think of $X_{n+1}$ as being obtained from $X_n$ by repositioning the centres of these line segments by amounts which are independent and normally distributed. It is clear that

**(20)** $$X_n(t) \to X(t) \quad \text{for} \quad t \in Q,$$

since, if $t \in Q$, then $X_n(t) = X(t)$ for all large $n$. The first step is to show that the convergence in (20) is (almost surely) uniform on $Q$, since this will imply that the limit function $X$ is (almost surely) uniformly continuous on $Q$. Now

**(21)** $$X_n(t) = \sum_{j=1}^{n} Z_j(t)$$

where $Z_j(t) = X_j(t) - X_{j-1}(t)$ and $X_0(t) = 0$. This series representation for $X_n$ converges uniformly on $Q$ if

**(22)** $$\sum_{j=1}^{\infty} \sup_{t \in Q} |Z_j(t)| < \infty.$$

We note that $Z_j(t) = 0$ for values of $t$ having the form $m2^{-j}$ where $m$ is even. It may be seen by drawing a diagram that

$$\sup_{t \in Q} |Z_j(t)| = \max\{|Z_j(m2^{-j})| : m = 1, 3, \ldots, 2^j - 1\}$$

and therefore

**(23)**          $$\mathbb{P}\left(\sup_{t \in Q} |Z_j(t)| > x\right) \leq \sum_{m \text{ odd}} \mathbb{P}\big(|Z_j(m2^{-j})| > x\big).$$

Now

$$Z_j(2^{-j}) = X(2^{-j}) - \tfrac{1}{2}\big[X(0) + X(2^{-j+1})\big]$$
$$= \tfrac{1}{2}\big[X(2^{-j}) - X(0)\big] - \tfrac{1}{2}\big[X(2^{-j+1}) - X(2^{-j})\big],$$

and therefore $\mathbb{E}Z_j(2^{-j}) = 0$ and, using the independence of increments, $\text{var}(Z_j(2^{-j})) = 2^{-j-1}$; a similar calculation is valid for $Z_j(m2^{-j})$ for $m = 1, 3, \ldots, 2^j - 1$. It follows by the bound in Exercise (4.4.8) on the tail of the normal distribution that, for all such $m$,

$$\mathbb{P}\big(|Z_j(m2^{-j})| > x\big) \leq \frac{1}{x2^{j/2}}e^{-x^2 2^j}, \quad x > 0.$$

Setting $x = c\sqrt{j2^{-j}\log 2}$, we obtain from (23) that

$$\mathbb{P}\left(\sup_{t \in Q} |Z_j(t)| > x\right) \leq 2^{j-1}\frac{2^{-c^2 j}}{c\sqrt{j\log 2}}.$$

Choosing $c > 1$, the last term is summable in $j$, implying by the Borel–Cantelli lemma (7.3.10a) that

$$\sup_{t \in Q} |Z_j(t)| > c\sqrt{\frac{j\log 2}{2^j}}$$

for only finitely many values of $j$ (almost surely). Hence

$$\sum_j \sup_{t \in Q} |Z_j(t)| < \infty \quad \text{almost surely},$$

and the argument prior to (22) yields that $X$ is (almost surely) uniformly continuous on $Q$.

We have proved that $X$ has (almost surely) uniformly continuous sample paths on the set of dyadic rationals; a similar argument is valid for other countable dense subsets of $[0, 1]$. It is quite another thing for $X$ to be continuous on the entire interval $[0, 1]$, and actually this need not be the case. We can, however, extend $X$ by continuity from the dyadic rationals to the whole of $[0, 1]$: for $t \in [0, 1]$, define

$$Y(t) = \lim_{\substack{s \to t \\ s \in Q}} X(s),$$

the limit being taken as $s$ approaches $t$ through the dyadic rationals. Such a limit exists almost surely for all $t$ since $X$ is (almost surely) uniformly continuous on $Q$. It is not difficult to check that the extended process $Y$ is indeed a Gaussian process with covariance function $\mathrm{cov}(Y(s), Y(t)) = \min\{s, t\}$, and, most important, the sample paths of $Y$ are (almost surely) continuous.

Finally we remove the 'almost surely' from the last conclusion. Let $\Omega'$ be the subset of the sample space $\Omega$ containing all $\omega$ for which the corresponding path of $Y$ is continuous on $\mathbb{R}$. We now restrict ourselves to the smaller sample space $\Omega'$, with its induced $\sigma$-field and probability measure. Since $\mathbb{P}(\Omega') = 1$, this change is invisible in all calculations of probabilities. Conditions (1a) and (1b) remain valid in the restricted space.                                                    ∎

This completes the proof of the existence of a Wiener process on $[0, 1]$. A similar argument can be made to work on the time interval $[0, \infty)$, but it is easier either: (a) to patch together continuous Wiener processes on $[n, n+1]$ for $n = 0, 1, \ldots$, or (b) to use the result of Problem (9.7.18c).

---

## Exercises for Section 13.3

**1.**   Let $X = \{X(t) : t \geq 0\}$ be a simple birth–death process with parameters $\lambda_n = n\lambda$ and $\mu_n = n\mu$. Suggest a diffusion approximation to $X$.

**2.**   **Bartlett's equation.** Let $D$ be a diffusion with instantaneous mean and variance $a(t, x)$ and $b(t, x)$, and let $M(t, \theta) = \mathbb{E}(e^{\theta D(t)})$, the moment generating function of $D(t)$. Use the forward diffusion equation to derive *Bartlett's equation*:

$$\frac{\partial M}{\partial t} = \theta a \left( t, \frac{\partial}{\partial \theta} \right) M + \frac{1}{2} \theta^2 b \left( t, \frac{\partial}{\partial \theta} \right) M$$

where we interpret

$$g \left( t, \frac{\partial}{\partial \theta} \right) M = \sum_n \gamma_n(t) \frac{\partial^n M}{\partial \theta^n}$$

if $g(t, x) = \sum_{n=0}^{\infty} \gamma_n(t) x^n$.

**3.**   Write down Bartlett's equation in the case of the Wiener process $D$ having drift $m$ and instantaneous variance 1, and solve it subject to the boundary condition $D(0) = 0$.

**4.**   Write down Bartlett's equation in the case of an Ornstein–Uhlenbeck process $D$ having instantaneous mean $a(t, x) = -x$ and variance $b(t, x) = 1$, and solve it subject to the boundary condition $D(0) = 0$.

**5.**   **Bessel process.** Let $W_1(t)$, $W_2(t)$, $W_3(t)$ be independent Wiener processes. The positive $R = R(t)$ such that $R^2 = W_1^2 + W_2^2 + W_3^2$ is the three-dimensional *Bessel process*. Show that $R$ is a Markov process. Is this result true in a general number $n$ of dimensions?

**6.**   Show that the transition density for the Bessel process defined in Exercise (13.3.5) is

$$f(t, y \mid s, x) = \frac{\partial}{\partial y} \mathbb{P}\big(R(t) \leq y \mid R(s) = x\big)$$

$$= \frac{y/x}{\sqrt{2\pi(t-s)}} \left\{ \exp\left( -\frac{(y-x)^2}{2(t-s)} \right) - \exp\left( -\frac{(y+x)^2}{2(t-s)} \right) \right\}.$$

**7.**   If $W$ is a Wiener process and the function $g : \mathbb{R} \to \mathbb{R}$ is continuous and strictly monotone, show that $g(W)$ is a continuous Markov process.

**8.** Let $W$ be a Wiener process. Which of the following define martingales?

(a) $e^{\sigma W(t)}$,   (b) $cW(t/c^2)$,   (c) $tW(t) - \int_0^t W(s)\,ds$.

**9. Exponential martingale, geometric Brownian motion.** Let $W$ be a standard Wiener process and define $S(t) = e^{at+bW(t)}$. Show that:

(a) $S$ is a Markov process,

(b) $S$ is a martingale (with respect to the filtration generated by $W$) if and only if $a + \frac{1}{2}b^2 = 0$, and in this case $\mathbb{E}(S(t)) = 1$.

**10.** Find the transition density for the Markov process of Exercise (13.3.9a).

**11.** Verify that the transition density (13.3.4) of the standard Wiener process satisfies the Chapman–Kolmogorov equations in continuous time.

---

## 13.4  First passage times

We have often been interested in the time which elapses before a Markov chain visits a specified state for the first time, and we continue this chapter with an account of some of the corresponding problems for a diffusion process.

Consider first a standard Wiener process $W$. The process $W_1$ given by

**(1)** $$W_1(t) = W(t + T) - W(T), \quad t \geq 0,$$

is a standard Wiener process for any fixed value of $T$ and, conditional on $W(T)$, $W_1$ is independent of $\{W(s) : s < T\}$; the Poisson process enjoys a similar property, which in Section 6.8 we called the 'weak Markov property'. It is a very important and useful fact that this holds even when $T$ is a random variable, so long as $T$ is a stopping time for $W$. We encountered stopping times in the context of continuous-time martingales in Section 12.7.

**(2) Definition.** Let $\mathcal{F}_t$ be the smallest $\sigma$-field with respect to which $W(s)$ is measurable for each $s \leq t$. The random variable $T$ is called a **stopping time** for $W$ if $\{T \leq t\} \in \mathcal{F}_t$ for all $t$.

We say that $W$ has the 'strong Markov property' in that this independence holds for all stopping times $T$. Why not try to prove this? Here, we make use of the strong Markov property for certain particular stopping times $T$.

**(3) Definition.** The **first passage time** $T(x)$ to the point $x \in \mathbb{R}$ is given by

$$T(x) = \inf\{t : W(t) = x\}.$$

The continuity of sample paths is essential in order that this definition make sense: a Wiener process cannot jump over the value $x$, but must pass through it. The proof of the following lemma is omitted.

**(4) Lemma.** *The random variable $T(x)$ is a stopping time for $W$.*

**(5) Theorem.** *The random variable $T(x)$ has density function*

$$f_{T(x)}(t) = \frac{|x|}{\sqrt{2\pi t^3}} \exp\left(-\frac{x^2}{2t}\right), \quad t \geq 0.$$

Clearly $T(x)$ and $T(-x)$ are identically distributed. For the case when $x = 1$ we encountered this density function and its moment generating function in Problems (5.12.18) and (5.12.19); it is easy to deduce that $T(x)$ has the same distribution as $Z^{-2}$ where $Z$ is $N(0, x^{-2})$. In advance of giving the proof of Theorem (5), here is a result about the size of the maximum of a Wiener process. The process $M$ given by $M(t) = \sup\{W(s) : 0 \le s \le t\}$ is called the *maximum process* of $W$.

**(6) Theorem.** *The random variable $M(t)$ has the same distribution as $|W(t)|$. Thus $M(t)$ has density function*

$$f_{M(t)}(m) = \sqrt{\frac{2}{\pi t}}\, \exp\left(-\frac{m^2}{2t}\right), \qquad m \ge 0.$$

You should draw your own diagrams to illustrate the translations and reflections used in the proofs of this section. The following proof uses a method commonly called the 'reflection principle'.

**Proof of (6).** Suppose $m > 0$, and observe that

(7) $$T(m) \le t \quad \text{if and only if} \quad M(t) \ge m.$$

Then

$$\mathbb{P}\big(M(t) \ge m\big) = \mathbb{P}\big(M(t) \ge m,\ W(t) - m \ge 0\big) + \mathbb{P}\big(M(t) \ge m,\ W(t) - m < 0\big).$$

However, by (7),

$$\begin{aligned}
\mathbb{P}\big(M(t) \ge m,\ W(t) - m < 0\big) &= \mathbb{P}\big(W(t) - W(T(m)) < 0 \mid T(m) \le t\big)\mathbb{P}\big(T(m) \le t\big) \\
&= \mathbb{P}\big(W(t) - W(T(m)) \ge 0 \mid T(m) \le t\big)\mathbb{P}\big(T(m) \le t\big) \\
&= \mathbb{P}\big(M(t) \ge m,\ W(t) - m \ge 0\big)
\end{aligned}$$

since $W(t) - W(T(m))$ is symmetric whenever $t \ge T(m)$ by the strong Markov property; we have used sample path continuity here, and more specifically that $\mathbb{P}(W(T(m)) = m) = 1$. Thus

$$\mathbb{P}\big(M(t) \ge m\big) = 2\mathbb{P}\big(M(t) \ge m,\ W(t) \ge m\big) = 2\mathbb{P}\big(W(t) \ge m\big)$$

since $W(t) \le M(t)$. Hence $\mathbb{P}(M(t) \ge m) = \mathbb{P}(|W(t)| \ge m)$ and the theorem is proved on noting that $|W(t)|$ is the absolute value of an $N(0, t)$ variable. ∎

**Proof of (5).** This follows immediately from (7), since if $x > 0$ then

$$\begin{aligned}
\mathbb{P}\big(T(x) \le t\big) &= \mathbb{P}\big(M(t) \ge x\big) = \mathbb{P}\big(|W(t)| \ge x\big) \\
&= \sqrt{\frac{2}{\pi t}} \int_x^\infty \exp\left(-\frac{m^2}{2t}\right) dm \\
&= \int_0^t \frac{|x|}{\sqrt{2\pi y^3}} \exp\left(-\frac{x^2}{2y}\right) dy
\end{aligned}$$

by the substitution $y = x^2 t / m^2$. ∎

We are now in a position to derive some famous results about the times at which $W$ returns to its starting point, the origin. We say that '$W$ has a zero at time $t$' if $W(t) = 0$, and we write $\cos^{-1}$ for the inverse trigonometric function, sometimes written arc cos.

**(8) Theorem.** *Suppose $0 \le t_0 < t_1$. The probability that a standard Wiener process $W$ has a zero in the time interval $(t_0, t_1)$, is $(2/\pi) \cos^{-1} \sqrt{t_0/t_1}$.*

**Proof.** Let $E(u, v)$ denote the event

$$E(u, v) = \{W(t) = 0 \text{ for some } t \in (u, v)\}.$$

Condition on $W(t_0)$ to obtain

$$\mathbb{P}(E(t_0, t_1)) = \int_{-\infty}^{\infty} \mathbb{P}\big(E(t_0, t_1) \,\big|\, W(t_0) = w\big) f_0(w)\, dw$$

$$= 2 \int_{-\infty}^{0} \mathbb{P}\big(E(t_0, t_1) \,\big|\, W(t_0) = w\big) f_0(w)\, dw$$

by the symmetry of $W$, where $f_0$ is the density function of $W(t_0)$. However, if $a > 0$,

$$\mathbb{P}\big(E(t_0, t_1) \,\big|\, W(t_0) = -a\big) = \mathbb{P}\big(T(a) < t_1 - t_0 \,\big|\, W(0) = 0\big)$$

by the homogeneity of $W$ in time and space. Use (5) to obtain that

$$\mathbb{P}(E(t_0, t_1)) = 2 \int_{a=0}^{\infty} \int_{t=0}^{t_1-t_0} f_{T(a)}(t) f_0(-a)\, dt\, da$$

$$= \frac{1}{\pi \sqrt{t_0}} \int_{t=0}^{t_1-t_0} t^{-\frac{3}{2}} \int_{a=0}^{\infty} a \exp\left[ -\frac{1}{2} a^2 \left( \frac{t+t_0}{t t_0} \right) \right] da\, dt$$

$$= \frac{\sqrt{t_0}}{\pi} \int_0^{t_1-t_0} \frac{dt}{(t+t_0)\sqrt{t}}$$

$$= \frac{2}{\pi} \tan^{-1} \sqrt{\frac{t_1}{t_0} - 1} \qquad \text{by the substitution } t = t_0 s^2$$

$$= \frac{2}{\pi} \cos^{-1} \sqrt{t_0/t_1} \qquad \text{as required.} \qquad \blacksquare$$

The result of (8) indicates some remarkable properties of the sample paths of $W$. Set $t_0 = 0$ to obtain

$$\mathbb{P}\big(\text{there exists a zero in } (0, t) \,\big|\, W(0) = 0\big) = 1 \quad \text{for all } t > 0,$$

and it follows that

$$T(0) = \inf\{t > 0 : W(t) = 0\}$$

satisfies $T(0) = 0$ almost surely. A deeper analysis shows that, with probability 1, $W$ has infinitely many zeros in any non-empty time interval $[0, t]$; it is no wonder that $W$ has non-differentiable sample paths! The set $Z = \{t : W(t) = 0\}$ of zeros of $W$ is rather a large set; in fact it turns out that $Z$ has Hausdorff dimension $\frac{1}{2}$ (see Mandelbrot 1983 for an introduction to fractional dimensionality).

The proofs of Theorems (5), (6), and (8) have relied heavily upon certain symmetries of the Wiener process; these are similar to the symmetries of the random walk of Section 3.10. Other diffusions may not have these symmetries, and we may need other techniques for investigating their first passage times. We illustrate this point by a glance at the Wiener process with drift. Let $D = \{D(t) : t \geq 0\}$ be a diffusion process with instantaneous mean and variance given by

$$a(t, x) = m, \quad b(t, x) = 1,$$

where $m$ is a constant. It is easy to check that, if $D(0) = 0$, then $D(t)$ is distributed as $N(mt, t)$. It is not so easy to find the distributions of the sizes of the maxima of $D$, and we take this opportunity to display the usefulness of martingales and optional stopping.

**(9) Theorem.** *Let $U(t) = e^{-2m D(t)}$. Then $U = \{U(t) : t \geq 0\}$ is a martingale.*

Our only experience to date of continuous-time martingales is contained in Section 12.7.

**Proof.** The process $D$ is Markovian, and so $U$ is a Markov process also. To check that the continuous-martingale condition holds, it suffices to show that

**(10)**                  $$\mathbb{E}\big(U(t + s) \,\big|\, U(t)\big) = U(t) \quad \text{for all } s, t \geq 0.$$

However,

**(11)**
$$\mathbb{E}\big(U(t+s) \,\big|\, U(t) = e^{-2md}\big) = \mathbb{E}(e^{-2m D(t+s)} \mid D(t) = d)$$
$$= \mathbb{E}\big(\exp\{-2m[D(t+s) - D(t)] - 2md\} \,\big|\, D(t) = d\big)$$
$$= e^{-2md}\,\mathbb{E}\big(\exp\{-2m[D(t+s) - D(t)]\}\big)$$
$$= e^{-2md}\,\mathbb{E}(e^{-2m D(s)})$$

because $D$ is Markovian with stationary independent increments. Now, $\mathbb{E}(e^{-2m D(s)}) = M(-2m)$ where $M$ is the moment generating function of an $N(ms, s)$ variable; this function $M$ is given in Example (5.8.5) as $M(u) = e^{msu + \frac{1}{2}su^2}$. Thus $\mathbb{E}(e^{-2m D(s)}) = 1$ and so (10) follows from (11).  ∎

We can use this martingale to find the distribution of first passage times, just as we did in Example (12.5.6) for the random walk. Let $x, y > 0$ and define

$$T(x, -y) = \inf\{t : \text{either } D(t) = x \text{ or } D(t) = -y\}$$

to be the first passage time of $D$ to the set $\{x, -y\}$. It is easily shown that $T(x, -y)$ is a stopping time which is almost surely finite.

**(12) Theorem.** $\mathbb{E}\big(U[T(x, -y)]\big) = 1$ *for all $x, y > 0$.*

**Proof.** This is just an application of a version of the optional stopping theorem (12.7.12). The process $U$ is a martingale and $T(x, -y)$ is a stopping time. Therefore

$$\mathbb{E}\big(U[T(x, -y)]\big) = \mathbb{E}(U(0)) = 1.$$  ∎

**(13) Corollary.** *If $m < 0$ and $x > 0$, the probability that $D$ ever visits the point $x$ is*

$$\mathbb{P}\big(D(t) = x \text{ for some } t\big) = e^{2mx}.$$

**Proof.** By Theorem (12),

$$1 = e^{-2mx}\mathbb{P}\big(D[T(x, -y)] = x\big) + e^{2my}\big\{1 - \mathbb{P}\big(D[T(x, -y)] = x\big)\big\}.$$

Let $y \to \infty$ to obtain

$$\mathbb{P}\big(D[T(x, -y)] = x\big) \to e^{2mx}$$

so long as $m < 0$. Now complete the proof yourself. ∎

The condition of Corollary (13), that the drift be negative, is natural; it is clear that if $m > 0$ then $D$ almost surely visits all points on the positive part of the real axis. The result of (13) tells us about the size of the maximum of $D$ also, since if $x > 0$,

$$\Big\{\max_{t\geq0} D(t) \geq x\Big\} = \big\{D(t) = x \text{ for some } t\big\},$$

and the distribution of $M = \max\{D(t) : t \geq 0\}$ is easily deduced.

**(14) Corollary.** *If $m < 0$ then $M$ is exponentially distributed with parameter $-2m$.*

---

## Exercises for Section 13.4

**1.**   Let $W$ be a standard Wiener process and let $X(t) = \exp\{i\theta W(t) + \frac{1}{2}\theta^2 t\}$ where $i = \sqrt{-1}$. Show that $X$ is a martingale with respect to the filtration given by $\mathcal{F}_t = \sigma(\{W(u) : u \leq t\})$.

**2.**   Let $T$ be the (random) time at which a standard Wiener process $W$ hits the 'barrier' in space–time given by $y = at + b$ where $a \leq 0$ and $b > 0$; that is, $T = \inf\{t : W(t) = at + b\}$. Use the result of Exercise (13.4.1) to show that the density function of $T$ has Laplace transform given by

$$\mathbb{E}(e^{-\psi T}) = \exp\{|a|b - b\sqrt{a^2 + 2\psi}\}, \qquad \psi \geq 0.$$

You may assume that the conditions of the optional stopping theorem are satisfied.

**3.**   Let $W$ be a standard Wiener process, and let $T$ be the time of the last zero of $W$ prior to time $t$. Show that $\mathbb{P}(T \leq u) = (2/\pi)\sin^{-1}\sqrt{u/t}$ for $0 \leq u \leq t$.

**4.**   Let $X$ and $Y$ be independent standard Wiener processes, so that $(X, Y)$ is a planar Wiener process. Show that the value of $Y$ at the time of the first passage of $X$ to the level $x > 0$ has the Cauchy distribution with density $f(y) = x/[\pi(x^2 + y^2)]$.

**5. Inverse Gaussian density.**   Let $D(t) = W(t) - at$ where $W$ is a standard Wiener process and $a \leq 0$ is a *drift* parameter. Use the result of Exercise (13.4.2) to show that the density function of the first passage time by $D$ to the point $b > 0$ is

$$f(x) = \frac{b}{\sqrt{2\pi x^3}} \exp\left\{-\frac{(ax + b)^2}{2x}\right\}, \qquad x > 0.$$

**6.**   Show that the standard Wiener process $W(t)$ and its maximum process $M(t) = \sup\{W(s) : 0 \leq s \leq t\}$ satisfy

$$\mathbb{P}\big(M(t) \geq m, \ W(t) \leq w\big) = 1 - \Phi\big((2m - w)/\sqrt{t}\big), \qquad m \geq 0, \quad 0 \leq w \leq m,$$

where $\Phi$ is the $N(0, 1)$ distribution function.

**7.** Find the density function of $\sqrt{T(x)}$ where $T(x)$ is the first passage time of a standard Wiener process to the point $x$. Show that it is the same as that of $1/|Z|$ for a suitable normally distributed random variable $Z$.

**8. Continuation.** Show that $T(x)$ has the same distribution as $T(1)x^2$, and deduce that the process $\{T(x) : x \geq 0\}$ is self-similar. Hence or otherwise, show that $T(x)$ has a stable distribution with exponent $\frac{1}{2}$. [See Feller 1971, p. 174.]

**9. Feller's diffusion approximation to the branching process, Example (13.3.12) revisited.** Let $X = \{X(t) : t \geq 0\}$ be the diffusion model of Example (13.3.12) with $X(0) = x > 0$, and with instantaneous mean and variance $ax$ and $bx$ where $a < 0$ and $b > 0$. Let $Y = \int_0^T X(u)\,du$ be the integrated size of the accumulated population up to the extinction time $T = \inf\{t \geq 0 : X(t) = 0\}$. Show that the moment generating function $m(x, \psi) = \mathbb{E}_x(e^{-\psi Y})$ satisfies

$$\frac{1}{2}b\frac{d^2 m}{dx^2} + a\frac{dm}{dx} - \psi m = 0, \qquad \psi \geq 0.$$

Deduce that $Y$ has a 'first-passage' distribution of the type of Exercise (13.4.2), and specify which.

## 13.5 Barriers

Diffusing particles are rarely allowed to roam freely, but are often restricted to a given part of space; for example, Brown's pollen particles were suspended in fluid which was confined to a container. What may happen when a particle hits a barrier? As with random walks, two simple types of barrier are the *absorbing* and the *reflecting*, although there are various other types of some complexity.

We begin with the case of the Wiener process. Let $w > 0$, let $W$ be a standard Wiener process, and consider the shifted process $w + W(t)$ which starts at $w$. The Wiener process $W^a$ *absorbed* at 0 is defined to be the process given by

**(1)**
$$W^a(t) = \begin{cases} w + W(t) & \text{if } t < T, \\ 0 & \text{if } t \geq T, \end{cases}$$

where $T = \inf\{t : w + W(t) = 0\}$ is the hitting time of the position 0. The Wiener process $W^r$ *reflected* at 0 is defined as the process $W^r(t) = |w + W(t)|$.

Viewing the diffusion equations (13.3.7)–(13.3.8) as forward and backward equations, it is clear that $W^a$ and $W^r$ satisfy these equations so long as they are away from the barrier. That is to say, $W^a$ and $W^r$ are diffusion processes. In order to find their transition density functions, we might solve the diffusion equations subject to suitable boundary conditions. For the special case of the Wiener process, however, it is simpler to argue as follows.

**(2) Theorem.** *Let $f(t, y)$ denote the density function of the random variable $W(t)$, and let $W^a$ and $W^r$ be given as above.*
(a) *The density function of $W^a(t)$ is*
$$f^a(t, y) = f(t, y - w) - f(t, y + w), \quad y > 0.$$
(b) *The density function of $W^r(t)$ is*
$$f^r(t, y) = f(t, y - w) + f(t, y + w), \quad y > 0.$$

The function $f(t, y)$ is the $N(0, t)$ density function,

(3)
$$f(t, y) = \frac{1}{\sqrt{2\pi t}} \exp(-\tfrac{1}{2} y^2 / t).$$

**Proof.** Let $I$ be a subinterval of $(0, \infty)$, and let $I^r = \{x \in \mathbb{R} : -x \in I\}$ be the reflection of $I$ in the point 0. Then

$$\mathbb{P}\big(W^a(t) \in I\big) = \mathbb{P}\big(\{w + W(t) \in I\} \cap \{T > t\}\big)$$
$$= \mathbb{P}\big(w + W(t) \in I\big) - \mathbb{P}\big(\{w + W(t) \in I\} \cap \{T \le t\}\big)$$
$$= \mathbb{P}\big(w + W(t) \in I\big) - \mathbb{P}\big(w + W(t) \in I^r\big)$$

using the reflection principle and the strong Markov property. The result follows.
  The result of part (b) is immediate from the fact that $W^r(t) = |w + W(t)|$.  ∎

We turn now to the absorption and reflection of a *general* diffusion process. Let $D = \{D(t) : t \ge 0\}$ be a diffusion process; we write $a$ and $b$ for the instantaneous mean and variance functions of $D$, and shall suppose that $b(t, x) > 0$ for all $x$ ($\ge 0$) and $t$. We make a further assumption, that $D$ is *regular* in that

(4)              $\mathbb{P}\big(D(t) = y$ for some $t \mid D(0) = x\big) = 1$    for all $x, y \ge 0$.

Suppose that the process starts from $D(0) = d$ say, where $d > 0$. Placing an absorbing barrier at 0 amounts to killing $D$ when it first hits 0. The resulting process $D^a$ is given by

$$D^a(t) = \begin{cases} D(t) & \text{if } T > t, \\ 0 & \text{if } T \le t, \end{cases}$$

where $T = \inf\{t : D(t) = 0\}$; this formulation requires $D$ to have continuous sample paths.
  Viewing the diffusion equations (13.3.7)–(13.3.8) as forward and backward equations, it is clear that they are satisfied away from the barrier. The presence of the absorbing barrier affects the solution to the diffusion equations through the boundary conditions.
  Denote by $f^a(t, y)$ the density function of $D^a(t)$; we might write $f^a(t, y) = f^a(t, y \mid 0, d)$ to emphasize the value of $D^a(0)$. The boundary condition appropriate to an absorbing barrier at 0 is

(5)                                    $f^a(t, 0) = 0$    for all $t$.

It is not completely obvious that (5) is the correct condition, but the following rough argument may be made rigorous. The idea is that, if the particle is near to the absorbing barrier, then small local fluctuations, arising from the non-zero instantaneous variance, will carry it to the absorbing barrier extremely quickly. Therefore the chance of it being near to the barrier but unabsorbed is extremely small.
  A slightly more rigorous justification for (5) is as follows. Suppose that (5) does not hold, which is to say that there exist $\epsilon, \eta > 0$ and $0 < u < v$ such that

(6)                          $f^a(t, y) > \eta$    for    $0 < y \le \epsilon, \ u \le t \le v$.

There is probability at least $\eta\,dx$ that $0 < D^a(t) \leq dx$ whenever $u \leq t \leq v$ and $0 < dx \leq \epsilon$. Hence the probability of absorption in the time interval $(t, t+dt)$ is at least

(7)
$$\eta\,dx\mathbb{P}\big(D^a(t+dt) - D^a(t) < -dx \,\big|\, 0 < D^a(t) \leq dx\big).$$

The instantaneous variance satisfies $b(t,x) \geq \beta$ for $0 < x \leq \epsilon$, $u \leq t \leq v$, for some $\beta > 0$, implying that $D^a(t+dt) - D^a(t)$ has variance at least $\beta\,dt$, under the condition that $0 < D^a(t) \leq dx$. Therefore,

$$\mathbb{P}\big(D^a(t+dt) - D^a(t) < -\gamma\sqrt{dt} \,\big|\, 0 < D^a(t) \leq dx\big) \geq \delta$$

for some $\gamma, \delta > 0$. Substituting $dx = \gamma\sqrt{dt}$ in (7), we obtain $\mathbb{P}(t < T < t+dt) \geq (\eta\gamma\delta)\sqrt{dt}$, implying by integration that $\mathbb{P}(u < T < v) = \infty$, which is clearly impossible. Hence (5) holds.

**(8) Example. Wiener process with drift.** Suppose that $a(t,x) = m$ and $b(t,x) = 1$ for all $t$ and $x$. Put an absorbing barrier at $0$ and suppose $D(0) = d > 0$. We wish to find a solution $g(t,y)$ to the forward equation

(9)
$$\frac{\partial g}{\partial t} = -m\frac{\partial g}{\partial y} + \frac{1}{2}\frac{\partial^2 g}{\partial y^2}, \quad y > 0,$$

subject to the boundary conditions

(10) $$g(t,0) = 0, \qquad t \geq 0,$$
(11) $$g(0,y) = \delta_d(y), \qquad y \geq 0,$$

where $\delta_d$ is the Dirac $\delta$ function centred at $d$. We know from Example (13.3.10), and in any case it is easy to check from first principles, that the function

(12)
$$g(t, y \mid x) = \frac{1}{\sqrt{2\pi t}}\exp\left(-\frac{(y - x - mt)^2}{2t}\right)$$

satisfies (9), for all possible 'sources' $x$. Our target is to find a linear combination of such functions $g(\cdot, \cdot \mid x)$ which satisfies (10) and (11). It turns out that

(13) $$f^a(t,y) = g(t, y \mid d) - e^{-2md}g(t, y \mid -d), \quad y > 0,$$

is such a function; assuming the solution is unique (which it is), this is therefore the density function of $D^a(t)$. We may think of it as a mixture of the function $g(\cdot, \cdot \mid d)$ with source $d$ together with a corresponding function from the 'image source' $-d$, being the reflection of $d$ in the barrier at $0$.

It is a small step to deduce the density function of the time $T$ until the absorption of the particle. At time $t$, either the process has been absorbed, or its position has density function given by (13). Hence

$$\mathbb{P}(T \leq t) = 1 - \int_0^\infty f^a(t,y)\,dy = 1 - \Phi\left(\frac{mt+d}{\sqrt{t}}\right) + e^{-2md}\Phi\left(\frac{mt-d}{\sqrt{t}}\right)$$

by (12) and (13), where $\Phi$ is the $N(0, 1)$ distribution function. Differentiate with respect to $t$ to obtain

**(14)**
$$f_T(t) = \frac{d}{\sqrt{2\pi t^3}}\exp\left(-\frac{(d + mt)^2}{2t}\right), \quad t > 0.$$

It is easily seen that

$$\mathbb{P}(\text{absorption takes place}) = \mathbb{P}(T < \infty) = \begin{cases} 1 & \text{if } m \leq 0, \\ e^{-2md} & \text{if } m > 0. \end{cases} \qquad \bullet$$

Turning to the matter of a reflecting barrier, suppose once again that $D$ is a regular diffusion process with instantaneous mean $a$ and variance $b$, starting from $D(0) = d > 0$. A reflecting barrier at the origin has the effect of disallowing infinitesimal negative jumps at the origin and replacing them by positive jumps. A formal definition requires careful treatment of the sample paths, and this is omitted here. Think instead about a reflecting barrier as giving rise to an appropriate boundary condition for the diffusion equations. Let us denote the reflected process by $D^r$, and let $f^r(t, y)$ be its density function at time $t$. The reflected process lives on $[0, \infty)$, and therefore

$$\int_0^\infty f^r(t, y)\, dy = 1 \quad \text{for all } t.$$

Differentiating with respect to $t$ and using the forward diffusion equation, we obtain at the expense of mathematical rigour that

$$0 = \frac{\partial}{\partial t}\int_0^\infty f^r(t, y)\, dy$$

$$= \int_0^\infty \frac{\partial f^r}{\partial t}\, dy = \int_0^\infty \left(-\frac{\partial}{\partial y}(af^r) + \frac{1}{2}\frac{\partial^2}{\partial y^2}(bf^r)\right) dy$$

$$= \left[-af^r + \frac{1}{2}\frac{\partial}{\partial y}(bf^r)\right]_0^\infty = \left(af^r - \frac{1}{2}\frac{\partial}{\partial y}(bf^r)\right)\bigg|_{y=0}.$$

This indicates that the density function $f^r(t, y)$ of $D^r(t)$ is obtained by solving the forward diffusion equation

$$\frac{\partial g}{\partial t} = -\frac{\partial}{\partial y}(ag) + \frac{1}{2}\frac{\partial^2}{\partial y^2}(bg)$$

subject to the boundary condition

**(15)**
$$\left(ag - \frac{1}{2}\frac{\partial}{\partial y}(bg)\right)\bigg|_{y=0} = 0 \quad \text{for } t \geq 0,$$

as well as the initial condition

**(16)**
$$g(0, y) = \delta_d(y) \quad \text{for } y \geq 0.$$

**(17) Example. Wiener process with drift.** Once again suppose that $a(t, x) = m$ and $b(t, x) = 1$ for all $x, t$. This time we seek a linear combination of the functions $g$ given in (12) which satisfies equations (15) and (16). It turns out that the answer contains an image at $-d$ together with a continuous line of images over the range $(-\infty, -d)$. That is to say, the solution has the form

$$f^{\mathrm{r}}(t, y) = g(t, y \mid d) + Ag(t, y \mid -d) + \int_{-\infty}^{-d} B(x)g(t, y \mid x)\, dx$$

for certain $A$ and $B(x)$. Substituting this into equation (15), one obtains after some work that

**(18)**                        $A = e^{-2md}, \quad B(x) = -2me^{2mx}.$                        ●

---

## Exercises for Section 13.5

**1.** Let $D$ be a standard Wiener process with drift $m$ starting from $D(0) = d > 0$, and suppose that there is a reflecting barrier at the origin. Show that the density function $f^{\mathrm{r}}(t, y)$ of $D(t)$ satisfies $f^{\mathrm{r}}(t, y) \to 0$ as $t \to \infty$ if $m \geq 0$, whereas $f^{\mathrm{r}}(t, y) \to 2|m|e^{-2|m|y}$ for $y > 0$, as $t \to \infty$ if $m < 0$.

**2. Conditioned Wiener process.** Let $W$ be a standard Wiener process started at $W(0) = x \in (0, a)$, with absorbing barriers at $0$ and $a$. Let $C$ be the process $W$ conditioned on the event that $W$ is absorbed at $a$. Show that the instantaneous mean and variance of $C$, when $C(t) = c \in (0, a)$, are $1/c$ and $1$, respectively.

**3. Continuation.** Let $T$ be the first passage time by $C$ to the point $a$. Show that $\mathbb{E}_x(T) = \frac{1}{3}(a^2 - x^2)$ for $x \in (0, a)$.

---

## 13.6 Excursions and the Brownian bridge

This section is devoted to properties of the Wiener process conditioned on certain special events. We begin with a question concerning the set of zeros of the process. Let $W = \{W(t) : t \geq 0\}$ be a Wiener process with $W(0) = w$, say, and with variance-parameter $\sigma^2 = 1$. What is the probability that $W$ has no zeros in the time interval $(0, v]$ given that it has none in the smaller interval $(0, u]$? The question is not too interesting if $w \neq 0$, since in this case the probability in question is just the ratio

**(1)**                $\dfrac{\mathbb{P}(\text{no zeros in } (0, v] \mid W(0) = w)}{\mathbb{P}(\text{no zeros in } (0, u] \mid W(0) = w)}$

each term of which is easily calculated from the distribution of maxima (13.4.6). The difficulty arises when $w = 0$, since both numerator and denominator in (1) equal 0. In this case, it may be seen that the required probability is the limit of (1) as $w \to 0$. We have that this limit equals $\lim_{w \to 0}\{g_w(v)/g_w(u)\}$ where $g_w(x)$ is the probability that a Wiener process starting at $w$ fails to reach 0 by time $x$. Using symmetry and Theorem (13.4.6),

$$g_w(x) = \sqrt{\frac{2}{\pi x}} \int_0^{|w|} \exp(-\tfrac{1}{2}m^2/x)\, dm,$$

whence $g_w(v)/g_w(u) \to \sqrt{u/v}$ as $w \to 0$, which we write as

(2)        $\mathbb{P}\big(W \neq 0 \text{ on } (0,\, v] \,\big|\, W \neq 0 \text{ on } (0,\, u],\ W(0) = 0\big) = \sqrt{u/v}, \quad 0 < u \leq v.$

A similar argument results in

(3)        $\mathbb{P}\big(W > 0 \text{ on } (0,\, v] \,\big|\, W > 0 \text{ on } (0,\, u],\ W(0) = 0\big) = \sqrt{u/v}, \quad 0 < u \leq v,$

by the symmetry of the Wiener process.

An 'excursion' of $W$ is a trip taken by $W$ away from 0. That is to say, if $W(u) = W(v) = 0$ and $W(t) \neq 0$ for $u < t < v$, then the trajectory of $W$ during the time interval $[u,\, v]$ is called an *excursion* of the process; excursions are *positive* if $W > 0$ throughout $(u,\, v)$, and *negative* otherwise. For any time $t > 0$, let $t - Z(t)$ be the time of the last zero prior to $t$, which is to say that $Z(t) = \sup\{s : W(t - s) = 0\}$; we suppose that $W(0) = 0$. At time $t$, some excursion is in progress whose current duration is $Z(t)$.

(4) **Theorem.** *Let $Y(t) = \sqrt{Z(t)}\,\mathrm{sign}\{W(t)\}$, and $\mathcal{F}_t = \sigma(\{Y(u) : 0 \leq u \leq t\})$. Then $(Y, \mathcal{F})$ is a martingale, called the* excursions martingale.

**Proof.** Clearly $Z(t) \leq t$, so that $\mathbb{E}|Y(t)| \leq \sqrt{t}$. It suffices to prove that

(5)                                    $\mathbb{E}\big(Y(t) \,\big|\, \mathcal{F}_s\big) = Y(s) \quad \text{for} \quad s < t.$

Suppose $s < t$, and let $A$ be the event that $W(u) = 0$ for some $u \in [s,\, t]$. With a slight abuse of notation,

$$\mathbb{E}\big(Y(t) \,\big|\, \mathcal{F}_s\big) = \mathbb{E}\big(Y(t) \,\big|\, \mathcal{F}_s,\, A\big)\mathbb{P}(A \mid \mathcal{F}_s) + \mathbb{E}\big(Y(t) \,\big|\, \mathcal{F}_s,\, A^c\big)\mathbb{P}(A^c \mid \mathcal{F}_s).$$

Now,

(6)                                    $\mathbb{E}\big(Y(t) \,\big|\, \mathcal{F}_s,\, A\big) = 0$

since, on the event $A$, the random variable $Y(t)$ is symmetric. On the other hand,

(7)                            $\mathbb{E}\big(Y(t) \,\big|\, \mathcal{F}_s,\, A^c\big) = \sqrt{t - s + Z(s)}\ \mathrm{sign}\{W(s)\}$

since, given $\mathcal{F}_s$ and $A^c$, the current duration of the excursion at time $t$ is $(t - s) + Z(s)$, and $\mathrm{sign}\{W(t)\} = \mathrm{sign}\{W(s)\}$. Furthermore $\mathbb{P}(A^c \mid \mathcal{F}_s)$ equals the probability that $W$ has strictly the same sign on $(s - Z(s),\, t]$ given the corresponding event on $(s - Z(s),\, s]$, which gives

$$\mathbb{P}(A^c \mid \mathcal{F}_s) = \sqrt{\frac{Z(s)}{t - s + Z(s)}} \quad \text{by (3).}$$

Combining this with equations (6) and (7), we obtain $\mathbb{E}(Y(t) \mid \mathcal{F}_s) = Y(s)$ as required.    ∎

(8) **Corollary.** *The probability that the standard Wiener process $W$ has a positive excursion of total duration at least $a$ before it has a negative excursion of total duration at least $b$ is $\sqrt{b}/(\sqrt{a} + \sqrt{b})$.*

**Proof.** Let $T = \inf\{t : Y(t) \geq \sqrt{a}$ or $Y(t) \leq -\sqrt{b}\}$, the time which elapses before $W$ records a positive excursion of duration at least $a$ or a negative excursion of duration at least $b$. It may be shown that the optional stopping theorem for continuous-time martingales is applicable, and hence $\mathbb{E}(Y(T)) = \mathbb{E}(Y(0)) = 0$. However,

$$\mathbb{E}(Y(T)) = \pi\sqrt{a} - (1 - \pi)\sqrt{b}$$

where $\pi$ is the required probability.                                                             ∎

We turn next to the Brownian bridge. Think about a sample path of $W$ on the time interval $[0, 1]$ as the shape of a random string with its left end tied to the origin. What does it look like if you tie down its right end also? That is to say, what sort of process is $\{W(t) : 0 \leq t \leq 1\}$ conditioned on the event that $W(1) = 0$? This new process is called the 'tied-down Wiener process' or the 'Brownian bridge'. There are various ways of studying it, the most obvious of which is perhaps to calculate the fdds of $W$ conditional on the event $\{W(1) \in (-\eta, \eta)\}$, and then take the limit as $\eta \downarrow 0$. This is easily done, and leads to the next theorem.

**(9) Theorem.** *Let $B = \{B(t) : 0 \leq t \leq 1\}$ be a process with continuous sample paths and the same fdds as $\{W(t) : 0 \leq t \leq 1\}$ conditioned on $W(0) = W(1) = 0$. The process $B$ is a diffusion process with instantaneous mean $a$ and variance $b$ given by*

$$(10) \qquad\qquad a(t, x) = -\frac{x}{1 - t}, \quad b(t, x) = 1, \quad x \in \mathbb{R}, \ 0 \leq t \leq 1.$$

Note that the Brownian bridge has the same instantaneous variance as $W$, but its instantaneous mean increases in magnitude as $t \to 1$ and has the effect of guiding the process to its finishing point $B(1) = 0$.

**Proof.** We make use of an elementary calculation involving conditional density functions. Let $W$ be a standard Wiener process, and suppose that $0 \leq u \leq v$. It is left as an *exercise* to prove that, conditional on the event $\{W(v) = y\}$, the distribution of $W(u)$ is normal with mean $yu/v$ and variance $u(v - u)/v$. In particular,

$$(11) \qquad\qquad \mathbb{E}\big(W(u) \mid W(0) = 0, \ W(v) = y\big) = \frac{yu}{v},$$

$$(12) \qquad\qquad \mathbb{E}\big(W(u)^2 \mid W(0) = 0, \ W(v) = y\big) = \left(\frac{yu}{v}\right)^2 + \frac{u(v - u)}{v}.$$

Returning to the Brownian bridge $B$, after a little reflection one sees that it is Gaussian and Markov, since $W$ has these properties. Furthermore the instantaneous mean is given by

$$\mathbb{E}\big(B(t + h) - B(t) \mid B(t) = x\big) = -\frac{xh}{1 - t}$$

by (11) with $y = -x$, $u = h$, $v = 1 - t$; similarly the instantaneous variance is given by the following consequence of (12):

$$\mathbb{E}\big(|B(t + h) - B(t)|^2 \mid B(t) = x\big) = h + o(h).$$                                ∎

An elementary calculation based on equations (11) and (12) shows that

$$(13) \qquad\qquad \mathrm{cov}\big(B(s), B(t)\big) = \min\{s, t\} - st, \quad 0 \leq s, t \leq 1.$$

## Exercises for Section 13.6

**1.   Brownian meander.** Let $W$ be a standard Wiener process. Show that the conditional density function of $W(t)$, given that $W(u) > 0$ for $0 < u < t$, is $g(x) = (x/t)e^{-x^2/(2t)}$, $x > 0$.

**2.**   Show that the autocovariance function of the Brownian bridge is $c(s, t) = \min\{s, t\} - st$, $0 \le s, t \le 1$.

**3.**   Let $W$ be a standard Wiener process, and let $\widehat{W}(t) = W(t) - tW(1)$. Show that $\{\widehat{W}(t) : 0 \le t \le 1\}$ is a Brownian bridge.

**4.**   If $W$ is a Wiener process with $W(0) = 0$, show that $\widetilde{W}(t) = (1 - t)W(t/(1 - t))$ for $0 \le t < 1$, $\widetilde{W}(1) = 0$, defines a Brownian bridge.

   Deduce, with help from Corollary (13.4.14), that the probability of the Brownian bridge ever rising above the height $m > 0$ is $e^{-2m^2}$.

**5.**   Let $0 < s < t < 1$. Show that the probability that the Brownian bridge has no zeros in the interval $(s, t)$ is $(2/\pi) \cos^{-1} \sqrt{(t - s)/[t(1 - s)]}$.

## 13.7  Stochastic calculus

We have so far considered a diffusion process† $D = \{D_t : t \ge 0\}$ as a Markov process with continuous sample paths, having some given 'instantaneous mean' $\mu(t, x)$ and 'instantaneous variance' $\sigma^2(t, x)$. The most fundamental diffusion process is the standard Wiener process $W = \{W_t : t \ge 0\}$, with instantaneous mean 0 and variance 1. We have seen in Section 13.3 how to use this characterization of $W$ in order to construct more general diffusions. With this use of the word 'instantaneous', it may seem natural, after a quick look at Section 13.3, to relate increments of $D$ and $W$ in the infinitesimal form‡

**(1)**
$$dD_t = \mu(t, D_t)\, dt + \sigma(t, D_t)\, dW_t,$$

or equivalently its integrated form

**(2)**
$$D_t - D_0 = \int_0^t \mu(s, D_s)\, ds + \int_0^t \sigma(s, D_s)\, dW_s.$$

The last integral has the form $\int_0^t \psi(s)\, dW_s$ where $\psi$ is a random process. Whereas we saw in Section 9.4 how to construct such an integral for deterministic functions $\psi$, this more general case poses new problems, not least since a Wiener process is not differentiable. This section contains a general discussion of the stochastic integral, the steps necessary to establish it rigorously being deferred to Section 13.8.

   For an example of the infinitesimal form (1) as a modelling tool, suppose that $X_t$ is the price of some stock, bond, or commodity at time $t$. How may we represent the change $dX_t$ over a small time interval $(t, t + dt)$? It may be a matter of observation that changes in the price $X_t$ are proportional to the price, and otherwise appear to be as random in sign and magnitude

---

†For notational convenience, we shall write $X_t$ or $X(t)$ interchangeably in the next few sections.

‡Often called a 'stochastic differential equation', abbreviated to SDE.

as are the displacements of a molecule. It would be plausible to write $dX_t = bX_t \, dW_t$, or $X_t - X_0 = \int_0^t bX_s \, dW_s$, for some constant $b$. Such a process $X$ is called a *geometric Wiener process*, or *geometric Brownian motion*; see Example (13.9.9) and Section 13.10.

We have already constructed certain representations of diffusion processes in terms of $W$. For example we have from Problem (13.12.1) that $t W_{1/t}$ and $\alpha W_{t/\alpha^2}$ are Wiener processes. Similarly, the process $D_t = \mu t + \sigma W_t$ is a Wiener process with drift. In addition, Ornstein–Uhlenbeck processes arise in a multiplicity of ways, for example as the processes $U_i$ given by:

$$U_1(t) = e^{-\beta t} W(e^{2\beta t} - 1), \quad U_2(t) = e^{-\beta t} W(e^{2\beta t}), \quad U_3(t) = W(t) - \beta \int_0^t e^{-\beta(t-s)} W(s) \, ds.$$

(See Problem (13.12.3) and Exercises (13.7.4) and (13.7.5).) Expressions of this form enable us to deduce sample path properties of the process in question from those of the underlying Wiener process. For example, since $W$ has continuous sample paths, so do the $U_i$.

It is illuminating to start with such an expression and to derive a differential form such as equation (1). Let $X$ be a process which is a function of a standard Wiener process $W$, that is, $X_t = f(W_t)$ for some given $f$. Experience of the usual Newton–Leibniz chain rule would suggest that $dX_t = f'(W_t) \, dW_t$ but this turns out to be incorrect in this context. If $f$ is sufficiently smooth, a formal application of Taylor's theorem gives

$$X_{t+\delta t} - X_t = f'(W_t)(\delta W_t) + \tfrac{1}{2} f''(W_t)(\delta W_t)^2 + \cdots$$

where $\delta W_t = W_{t+\delta t} - W_t$. In the usual derivation of the chain rule, one uses the fact that the second term on the right side is $o(\delta t)$. However, $(\delta W_t)^2$ has mean $\delta t$, and something new is needed. It turns out that $\delta t$ is indeed an acceptable approximation for $(\delta W_t)^2$, and that the subsequent terms in the Taylor expansion are insignificant in the limit as $\delta t \to 0$. One is therefore led to the formula

**(3)**                                $dX_t = f'(W_t) \, dW_t + \tfrac{1}{2} f''(W_t) \, dt.$

Note the extra term over that suggested by the usual chain rule. Equation (3) may be written in its integrated form

$$X_t - X_0 = \int_0^t f'(W_s) \, dW_s + \int_0^t \tfrac{1}{2} f''(W_s) \, ds.$$

Sense can be made of this only when we have a proper definition of the stochastic integral $\int_0^t f'(W_s) \, dW_s$. Equation (3) is a special case of what is known as Itô's formula, to which we return in Section 13.9.

Let us next work with a concrete example in the other direction, asking for an non-rigorous interpretation of the stochastic integral $\int_0^t W_s \, dW_s$. By analogy with the usual integral, we take $t = n\delta$ where $\delta$ is small and positive, and we partition the interval $(0, t]$ into the intervals $(j\delta, (j+1)\delta], 0 \le j < n$. Following the usual prescription, we take some $\theta_j \in [j\delta, (j+1)\delta]$ and form the sum $I_n = \sum_{j=0}^{n-1} W_{\theta_j}(W_{(j+1)\delta} - W_{j\delta})$.

In the context of the usual Riemann integral, the values $W_{j\delta}$, $W_{\theta_j}$, and $W_{(j+1)\delta}$ would be sufficiently close to one another for $I_n$ to have a limit as $n \to \infty$ which is independent of the choice of the $\theta_j$. The Wiener process $W$ has sample paths with unbounded variation, and therein lies the difference.

Suppose that we take $\theta_j = j\delta$ for each $j$. It is easy to check that

$$2I_n = \sum_{j=0}^{n-1}(W_{(j+1)\delta}^2 - W_{j\delta}^2) - \sum_{j=0}^{n-1}(W_{(j+1)\delta} - W_{j\delta})^2 = W_t^2 - W_0^2 - Z_n$$

where $Z_n = \sum_{j=0}^{n-1}(W_{(j+1)\delta} - W_{j\delta})^2$. It is the case that

(4)                                       $\mathbb{E}\big((Z_n - t)^2\big) \to 0 \quad \text{as } n \to \infty,$

which is to say that $Z_n \to t$ in mean square (see Exercise (13.7.2)). It follows that $I_n \to \frac{1}{2}(W_t^2 - t)$ in mean square as $n \to \infty$, and we are led to the interpretation

(5)                                       $$\int_0^t W_s \, dW = \tfrac{1}{2}(W_t^2 - t).$$

This proposal is verified in Example (13.9.7).

The calculation above is an example of what is called an *Itô integral*. The choice of the $\theta_j$ was central to the argument which leads to (5), and other choices lead to different answers. In Exercise (13.7.3) is considered the case when $\theta_j = (j + 1)\delta$, and this leads to the value $\frac{1}{2}(W_t^2 + t)$ for the integral. When $\theta_j$ is the midpoint of the interval $[j\delta, (j + 1)\delta]$, the answer is the more familiar $W_t^2$, and the corresponding integral is termed the *Stratonovich integral*.

---

## Exercises for Section 13.7

**1.  Doob's $L_2$ inequality.** Let $W$ be a standard Wiener process, and show that

$$\mathbb{E}\Big(\max_{0\le s\le t} |W_s|^2\Big) \le 4\mathbb{E}(W_t^2).$$

**2.**  Let $W$ be a standard Wiener process. Fix $t > 0$, $n \ge 1$, and let $\delta = t/n$. Show that $Z_n = \sum_{j=0}^{n-1}(W_{(j+1)\delta} - W_{j\delta})^2$ satisfies $Z_n \to t$ in mean square as $n \to \infty$.

**3.**  Let $W$ be a standard Wiener process. Fix $t > 0$, $n \ge 1$, and let $\delta = t/n$. Let $V_j = W_{j\delta}$ and $\Delta_j = V_{j+1} - V_j$. Evaluate the limits of the following as $n \to \infty$:
(a) $I_1(n) = \sum_j V_j \Delta_j$,
(b) $I_2(n) = \sum_j V_{j+1}\Delta_j$,
(c) $I_3(n) = \sum_j \frac{1}{2}(V_{j+1} + V_j)\Delta_j$,
(d) $I_4(n) = \sum_j W_{(j+\frac{1}{2})\delta}\Delta_j$.

**4.**  Let $W$ be a standard Wiener process. Show that $U(t) = e^{-\beta t}W(e^{2\beta t})$ defines a stationary Ornstein–Uhlenbeck process.

**5.**  Let $W$ be a standard Wiener process. Show that $U_t = W_t - \beta \int_0^t e^{-\beta(t-s)}W_s \, ds$ defines an Ornstein–Uhlenbeck process.

## 13.8  The Itô integral

Our target in this section is to present a definition of the integral $\int_0^\infty \psi_s \, dW_s$, where $\psi$ is a random process satisfying conditions to be stated. Some of the details will be omitted from the account which follows.

Integrals of the form $\int_0^\infty \phi(s) \, dS_s$ were explored in Section 9.4 for deterministic functions $\phi$, subject to the following assumptions on the process $S$:

(a) $\mathbb{E}(|S_t|^2) < \infty$ for all $t$,

(b) $\mathbb{E}(|S_{t+h} - S_t|^2) \to 0$ as $h \downarrow 0$, for all $t$,

(c) $S$ has orthogonal increments.

It was required that $\phi$ satisfy $\int_0^\infty |\phi(s)|^2 \, dG(s) < \infty$ where $G(t) = \mathbb{E}(|S_t - S_0|^2)$.

It is a simple exercise to check that conditions (a)–(c) are satisfied by the standard Wiener process $W$, and that $G(t) = t$ in this case. We turn next to conditions to be satisfied by the integrand $\psi$.

Let $W = \{W_t : t \geq 0\}$ be a standard Wiener process on the complete probability space $(\Omega, \mathcal{F}, \mathbb{P})$. Let $\mathcal{F}_t$ be the smallest sub-$\sigma$-field of $\mathcal{F}$ with respect to which the variables $W_s$, $0 \leq s \leq t$, are measurable and which contains the null events $\mathcal{N} = \{A \in \mathcal{F} : \mathbb{P}(A) = 0\}$. We write $\mathcal{F} = \{\mathcal{F}_t : t \geq 0\}$ for the consequent filtration.

A random process $\psi$ is said to be *measurable* if, when viewed as a function $\psi_t(\omega)$ of both $t$ and the elementary event $\omega \in \Omega$, it is measurable with respect to the product $\sigma$-field $\mathcal{B} \otimes \mathcal{F}$; here, $\mathcal{B}$ denotes the Borel $\sigma$-field of subsets of $[0, \infty)$. The measurable process $\psi$ is said to be *adapted* to the filtration $\mathcal{F}$ if $\psi_t$ is $\mathcal{F}_t$-measurable for all $t$. It will emerge that adapted processes may be integrated against the Wiener process so long as they satisfy the integral condition

(1)
$$\mathbb{E}\left( \int_0^\infty |\psi_t|^2 \, dt \right) < \infty,$$

and we denote by $\mathcal{A}$ the set of all adapted processes satisfying (1). It may be shown that $\mathcal{A}$ is a Hilbert space (and is thus Cauchy complete) with the norm†

(2)
$$\|\psi\| = \sqrt{\mathbb{E}\left( \int_0^\infty |\psi_t|^2 \, dt \right)}.$$

We shall see that $\int_0^\infty \psi_s \, dW_s$ may be defined‡ for $\psi \in \mathcal{A}$.

We follow the scheme laid out in Section 9.4. The integral $\int_0^\infty \psi_s \, dW_s$ is first defined for a random step function $\psi_t = \sum_j C_j I_{(a_j, a_{j+1}]}(t)$ where the $a_j$ are constants and the $C_j$ are random variables with finite second moments which are $\mathcal{F}_{a_j}$-measurable. One then passes to limits of such step functions, finally checking that any process satisfying (1) may be expressed as such a limit. Here are some details.

---

†Actually $\| \cdot \|$ is not a norm, since $\|\psi\| = 0$ does not imply that $\psi = 0$. It is however a norm on the set of equivalence classes obtained from the equivalence relation given by $\psi \sim \phi$ if $\mathbb{P}(\psi = \phi) = 1$.

‡Integrals over bounded intervals are defined similarly, by multiplying the integrand by the indicator function of the interval in question. That $\psi$ be adapted is not really the 'correct' condition. In a more general theory of stochastic integration, the process $W$ is replaced by a so-called semimartingale, and the integrand $\psi$ by a locally bounded predictable process.

Let $0 = a_0 < a_1 < \cdots < a_n = t$, and let $C_0, C_1, \ldots, C_{n-1}$ be random variables with finite second moments and such that each $C_j$ is $\mathcal{F}_{a_j}$-measurable. Define the random variable $\phi_t$ by

$$\phi_t = \sum_{j=0}^{n-1} C_j I_{(a_j, a_{j+1}]}(t) = \begin{cases} 0 & \text{if } t \leq 0 \text{ or } t > a_n, \\ C_j & \text{if } a_j < t \leq a_{j+1}. \end{cases}$$

We call the function $\phi$ a 'predictable step function'. The stochastic integral $I(\phi)$ of $\phi$ with respect to $W$ is evidently to be given by

(3) $$I(\phi) = \sum_{j=0}^{n-1} C_j (W_{a_{j+1}} - W_{a_j}).$$

It is easily seen that $I(\alpha\phi^1 + \beta\phi^2) = \alpha I(\phi^1) + \beta I(\phi^2)$ for two predictable step functions $\phi^1, \phi^2$ and $\alpha, \beta \in \mathbb{R}$.

The following 'isometry' asserts the equality of the norm $\|\phi\|$ and the $L_2$ norm of the integral of $\phi$. As before, we write $\|U\|_2 = \sqrt{\mathbb{E}|U^2|}$ where $U$ is a random variable.

**(4) Lemma.** *If $\phi$ is a predictable step function, $\|I(\phi)\|_2 = \|\phi\|$.*

**Proof.** Evidently,

(5)

$$\mathbb{E}(|I(\phi)|^2) = \mathbb{E}\left(\sum_{j=0}^{n-1} C_j (W_{a_{j+1}} - W_{a_j}) \sum_{k=0}^{n-1} C_k (W_{a_{k+1}} - W_{a_k})\right)$$

$$= \mathbb{E}\left(\sum_{j=0}^{n-1} C_j^2 (W_{a_{j+1}} - W_{a_j})^2 + 2 \sum_{0 \leq j < k \leq n-1} C_j C_k (W_{a_{j+1}} - W_{a_j})(W_{a_{k+1}} - W_{a_k})\right).$$

Using the fact that $C_j$ is $\mathcal{F}_{a_j}$-measurable,

$$\mathbb{E}(C_j^2 (W_{a_{j+1}} - W_{a_j})^2) = \mathbb{E}[\mathbb{E}(C_j^2 (W_{a_{j+1}} - W_{a_j})^2 \mid \mathcal{F}_{a_j})] = \mathbb{E}(C_j^2)(a_{j+1} - a_j).$$

Similarly, by conditioning on $\mathcal{F}_{a_k}$, we find that the mean of the final term in (5) equals 0. Therefore,

$$\mathbb{E}(|I(\phi)|^2) = \sum_j \mathbb{E}(C_j^2)(a_{j+1} - a_j) = \mathbb{E}\left(\int_0^\infty |\phi(t)|^2 \, dt\right) = \|\phi\|^2. \qquad \blacksquare$$

Next we consider limits of sequences of predictable step functions. Let $\psi \in \mathcal{A}$. It may be shown that there exists a sequence $\phi = \{\phi^{(n)}\}$ of predictable step functions such that $\|\phi^{(n)} - \psi\| \to 0$ as $n \to \infty$. We prove this under the assumption that $\psi$ has continuous sample paths, although this continuity condition is not necessary.

**(6) Theorem.** *Let $\psi \in \mathcal{A}$ be a process with continuous sample paths. There exists a sequence $\phi = \{\phi^{(n)}\}$ of predictable step functions such that $\|\phi^{(n)} - \psi\| \to 0$ as $n \to \infty$.*

**Proof.** Define the predictable step function

$$
\phi_t^{(n)} = \begin{cases} n \displaystyle\int_{(j-1)/n}^{j/n} \psi_s \, ds & \text{for } \dfrac{j}{n} < t \le \dfrac{j+1}{n},\ 1 \le j < n^2, \\ 0 & \text{otherwise.} \end{cases}
$$

By a standard use of the Cauchy–Schwarz inequality,

$$
\int_{j/n}^{(j+1)/n} |\phi_t^{(n)}|^2 \, dt = n \left| \int_{(j-1)/n}^{j/n} \psi_s \, ds \right|^2 \le \int_{(j-1)/n}^{j/n} |\psi_s|^2 \, ds \quad \text{for } j \ge 1.
$$

Hence

(7)
$$
\int_T^\infty |\phi_s^{(n)}|^2 \, ds \le \int_{T-(2/n)}^\infty |\psi_s|^2 \, ds \quad \text{for } T \ge 0.
$$

Now,

(8)
$$
\int_0^\infty |\phi_s^{(n)} - \psi_s|^2 \, ds = \int_0^T |\phi_s^{(n)} - \psi_s|^2 \, ds + \int_T^\infty |\phi_s^{(n)} - \psi_s|^2 \, ds.
$$

Using the continuity of the sample paths of $\psi$, $|\phi_s^{(n)} - \psi_s| \to 0$ as $n \to \infty$, uniformly on the interval $[0, T]$, whence the penultimate term in (8) tends to 0 as $n \to \infty$. Since $|x + y|^2 \le 2(|x|^2 + |y|^2)$ for $x, y \in \mathbb{R}$, the last term in (8) is by (7) no greater than $4 \int_{T-(2/n)}^\infty |\psi_s|^2 \, ds$. We let $n \to \infty$ and then $T \to \infty$ in (8). Since $\psi \in \mathcal{A}$, it is the case that $\int_0^\infty |\psi_s|^2 \, ds < \infty$ almost surely, and therefore

$$
\int_0^\infty |\phi_s^{(n)} - \psi_s|^2 \, ds \to 0 \qquad \text{almost surely, as } n \to \infty.
$$

By the same argument used to bound the last term in (8),

$$
0 \le \int_0^\infty |\phi_s^{(n)} - \psi_s|^2 \, ds \le 4 \int_0^\infty |\psi_s|^2 \, ds,
$$

and it follows by the dominated convergence theorem that $\|\phi^{(n)} - \psi\| \to 0$ as $n \to \infty$. ∎

Let $\psi \in \mathcal{A}$ and let $\phi = \{\phi^{(n)}\}$ be a sequence of predictable step functions converging in $\mathcal{A}$ to $\psi$. Since $\phi^{(m)} - \phi^{(n)}$ is itself a predictable step function, we have that

$$
\begin{aligned}
\|I(\phi^{(m)}) - I(\phi^{(n)})\|_2 &= \|I(\phi^{(m)} - \phi^{(n)})\|_2 \\
&= \|\phi^{(m)} - \phi^{(n)}\| && \text{by Lemma (4)} \\
&\le \|\phi^{(m)} - \psi\| + \|\phi^{(n)} - \psi\| && \text{by the triangle inequality} \\
&\to 0 && \text{as } m, n \to \infty.
\end{aligned}
$$

Therefore the sequence $I(\phi^{(n)})$ is mean-square Cauchy convergent, and hence converges in mean square to some limit random variable denoted $I(\phi)$. It is not difficult to show as follows that $\mathbb{P}(I(\phi) = I(\rho)) = 1$ for any other sequence $\rho$ of predictable step functions converging in $\mathcal{A}$ to $\psi$. We have by the triangle inequality that

$$\|I(\phi) - I(\rho)\|_2 \le \|I(\phi) - I(\phi^{(n)})\|_2 + \|I(\phi^{(n)}) - I(\rho^{(n)})\|_2 + \|I(\rho^{(n)}) - I(\rho)\|_2.$$

The first and third terms on the right side tend to 0 as $n \to \infty$. By Lemma (4) and the linearity of the integral operator on predictable step functions, the second term satisfies

$$\|I(\phi^{(n)}) - I(\rho^{(n)})\|_2 = \|I(\phi^{(n)} - \rho^{(n)})\|_2 = \|\phi^{(n)} - \rho^{(n)}\|$$

$$\le \|\phi^{(n)} - \psi\| + \|\rho^{(n)} - \psi\|$$

which tends to zero as $n \to \infty$. Therefore $\|I(\phi) - I(\rho)\|_2 = 0$, implying as claimed that $\mathbb{P}(I(\phi) = I(\rho)) = 1$.

The (almost surely) unique such quantity $I(\phi)$ is denoted by $I(\psi)$, which we call the *Itô integral* of the process $\psi$. It is usual to denote $I(\psi)$ by $\int_0^\infty \psi_s \, dW_s$, and we adopt this notation forthwith. We define $\int_0^t \psi_s \, dW_s$ to be $\int_0^\infty \psi_s I_{(0,t]}(s) \, dW_s$.

With the (Itô) stochastic integral defined, we may now agree to write

**(9)**                                    $$dX_t = \mu(t, X_t) \, dt + \sigma(t, X_t) \, dW_t$$

as a shorthand form of

**(10)**                                    $$X_t = X_0 + \int_0^t \mu(s, X_s) \, ds + \int_0^t \sigma(s, X_s) \, dW_s.$$

A continuous process $X$ defined by (9), by which we mean satisfying (10), is called an *Itô process*, or a diffusion process, with infinitesimal mean and variance $\mu(t, x)$ and $\sigma(t, x)^2$. The proof that there exists such a process is beyond our scope. Thus we may define diffusions via stochastic integrals, and it may be shown conversely that all diffusions previously considered in this book may be written as appropriate stochastic integrals.

It is an important property of the above stochastic integrals that they define martingales†. Once again, we prove this under the assumption that $\psi$ has continuous sample paths.

**(11) Theorem.** *Let $\psi \in \mathcal{A}$ be a process with continuous sample paths. The process $J_t = \int_0^t \psi_u \, dW_u$ is a martingale with respect to the filtration $\mathcal{F}$.*

**Proof.** Let $0 < s < t$ and, for $n \ge 1$, let $a_0, a_1, \ldots, a_n$ be such that $0 = a_0 < a_1 < \cdots < a_m = s < a_{m+1} < \cdots < a_n = t$ for some $m$. We define the predictable step function

$$\psi_u^{(n)} = \sum_{j=0}^{n-1} \psi_{a_j} I_{(a_j, a_{j+1}]}(u),$$

with integral

$$J_v^{(n)} = \int_0^v \psi_u^{(n)} \, dW_u = \sum_{j=0}^{n-1} \psi_{a_j} (W_{a_{j+1} \wedge v} - W_{a_j \wedge v}), \quad v \ge 0,$$

---

†In the absence of condition (1), we may obtain what is called a 'local martingale', but this is beyond the scope of this book.

where $x \wedge y = \min\{x, y\}$. Now,

$$\mathbb{E}(J_t^{(n)} \mid \mathcal{F}_s) = \sum_{j=0}^{n-1} \mathbb{E}(\psi_{a_j}(W_{a_{j+1}} - W_{a_j}) \mid \mathcal{F}_s)$$

where

$$\mathbb{E}(\psi_{a_j}(W_{a_{j+1}} - W_{a_j}) \mid \mathcal{F}_s) = \psi_{a_j}(W_{a_{j+1}} - W_{a_j}) \qquad \text{if } j < m,$$

and

$$\mathbb{E}(\psi_{a_j}(W_{a_{j+1}} - W_{a_j}) \mid \mathcal{F}_s) = \mathbb{E}[\mathbb{E}(\psi_{a_j}(W_{a_{j+1}} - W_{a_j}) \mid \mathcal{F}_{a_j}) \mid \mathcal{F}_s] = 0 \qquad \text{if } j \geq m,$$

since $W_{a_{j+1}} - W_{a_j}$ is independent of $\mathcal{F}_{a_j}$ and has zero mean. Therefore,

**(12)** $$\mathbb{E}(J_t^{(n)} \mid \mathcal{F}_s) = J_s^{(n)}.$$

We now let $n \to \infty$ and assume that $\max_j |a_{j+1} - a_j| \to 0$. As shown in the proof of Theorem (6),

$$\| \psi^{(n)} I_{(0,s]} - \psi I_{(0,s]} \| \to 0 \quad \text{and} \quad \| \psi^{(n)} I_{(0,t]} - \psi I_{(0,t]} \| \to 0,$$

whence $J_s^{(n)} \to \int_0^s \psi_u \, dW_u$ and $J_t^{(n)} \to \int_0^t \psi_u \, dW_u$ in mean square. We let $n \to \infty$ in (12), and use the result of Exercise (13.8.5), to find that $\mathbb{E}(J_t \mid \mathcal{F}_s) = J_s$ almost surely. It follows as claimed that $J$ is a martingale. ∎

There is a remarkable converse to Theorem (11) of which we omit the proof.

**(13) Theorem.** *Let $M$ be a martingale with respect to the filtration $\mathcal{F}$. There exists an adapted random process $\psi$ such that*

$$M_t = M_0 + \int_0^t \psi_u \, dW_u, \qquad t \geq 0.$$

## Exercises for Section 13.8

In the absence of any contrary indication, $W$ denotes a standard Wiener process, and $\mathcal{F}_t$ is the smallest $\sigma$-field containing all null events with respect to which every member of $\{W_u : 0 \leq u \leq t\}$ is measurable.

**1.**   (a) Verify directly that $\int_0^t s \, dW_s = t W_t - \int_0^t W_s \, ds$.

   (b) Verify directly that $\int_0^t W_s^2 \, dW_s = \frac{1}{3} W_t^3 - \int_0^t W_s \, ds$.

   (c) Show that $\mathbb{E}\left( \left[ \int_0^t W_s \, dW_s \right]^2 \right) = \int_0^t \mathbb{E}(W_s^2) \, ds$.

**2.**   Let $X_t = \int_0^t W_s \, ds$. Show that $X$ is a Gaussian process, and find its autocovariance and autocorrelation function.

**3.** Let $(\Omega, \mathcal{F}, \mathbb{P})$ be a probability space, and suppose that $X_n \xrightarrow{\text{m.s.}} X$ as $n \to \infty$. If $\mathcal{G} \subseteq \mathcal{F}$, show that $\mathbb{E}(X_n \mid \mathcal{G}) \xrightarrow{\text{m.s.}} \mathbb{E}(X \mid \mathcal{G})$.

**4.** Let $\psi_1$ and $\psi_2$ be predictable step functions, and show that

$$\mathbb{E}\{I(\psi_1)I(\psi_2)\} = \mathbb{E}\left(\int_0^\infty \psi_1(t)\psi_2(t)\,dt\right),$$

whenever both sides exist.

**5.** Assuming that *Gaussian white noise* $G_t = dW_t/dt$ exists in sufficiently many senses to appear as an integrand, show by integrating the stochastic differential equation $dX_t = -\beta X_t\,dt + dW_t$ that

$$X_t = W_t - \beta \int_0^t e^{-\beta(t-s)} W_s\,ds,$$

if $X_0 = 0$.

**6.** Let $\psi$ be an adapted process with $\|\psi\| < \infty$. Show that $\|I(\psi)\|_2 = \|\psi\|$.

---

## 13.9 Itô's formula

The 'stochastic differential equation', or 'SDE',

**(1)**                                $dX = \mu(t, X)\,dt + \sigma(t, X)\,dW$

is a shorthand for the now well-defined integral equation

$$X_t = X_0 + \int_0^t \mu(s, X_s)\,ds + \int_0^t \sigma(s, X_s)\,dW_s,$$

solutions to which are called *Itô processes*, or *diffusions*. Under rather weak conditions on $\mu$, $\sigma$, and $X_0$, it may be shown that the SDE (1) has a unique solution which is a Markov process with continuous sample paths. The proof of this is beyond our scope.

We turn to a central question. If $X$ satisfies the SDE (1) and $Y_t = f(t, X_t)$ for some given $f : [0, \infty) \times \mathbb{R} \to \mathbb{R}$, what is the infinitesimal formula for the process $Y$?

**(2) Theorem. Itô's formula.** *If $dX = \mu(t, X)\,dt + \sigma(t, X)\,dW$ and $Y_t = f(t, X_t)$, where $f$ is twice continuously differentiable on $[0, \infty) \times \mathbb{R}$, then $Y$ is also an Itô process, given by*†

**(3)**   $dY = \left[f_x(t, X)\mu(t, X) + f_t(t, X) + \tfrac{1}{2}f_{xx}(t, X)\sigma^2(t, X)\right]dt + f_x(t, X)\sigma(t, X)\,dW.$

This formula, whose proof is beyond our scope, may be extended to cover multivariate diffusions. We do not prove Itô's formula at the level of generality of (2), instead specializing to the following special case when $X$ is the standard Wiener process. The differentiability assumption on the function $f$ may be weakened.

**(4) Theorem. Itô's simple formula.** *Let $f(s, w)$ be thrice continuously differentiable on $[0, \infty) \times \mathbb{R}$, and let $W$ be a standard Wiener process. The process $Y_t = f(t, W_t)$ is an Itô process with*

$$dY = \left[ f_t(t, W) + \tfrac{1}{2} f_{ww}(t, W) \right] dt + f_w(t, W) \, dW.$$

**Sketch proof.** Let $n \geq 1$, $\delta = t/n$, and write $\Delta_j = W_{(j+1)\delta} - W_{j\delta}$. The idea is to express $f(t, W_t)$ as the sum

$$
\text{(5)} \qquad f(t, W_t) - f(0, W_0) = \sum_{j=0}^{n-1} \left[ f\big((j+1)\delta, W_{(j+1)\delta}\big) - f(j\delta, W_{(j+1)\delta}) \right]
$$

$$
+ \sum_{j=0}^{n-1} \left[ f(j\delta, W_{(j+1)\delta}) - f(j\delta, W_{j\delta}) \right]
$$

and to use Taylor's theorem to study its behaviour as $n \to \infty$. We leave out the majority of details necessary to achieve this, presenting instead the briefest summary.

By (5) and Taylor's theorem, there exist random variables $\theta_j \in [j\delta, (j+1)\delta]$ and $\Omega_j \in [W_{j\delta}, W_{(j+1)\delta}]$ such that

$$
\text{(6)} \quad f(t, W_t) - f(0, W_0) = \sum_{j=0}^{n-1} f_t(\theta_j, W_{(j+1)\delta})\delta + \sum_{j=0}^{n-1} f_w(j\delta, W_{j\delta})\Delta_j
$$

$$
+ \tfrac{1}{2} \sum_{j=0}^{n-1} f_{ww}(j\delta, W_{j\delta})\Delta_j^2 + \tfrac{1}{6} \sum_{j=0}^{n-1} f_{www}(j\delta, \Omega_j)\Delta_j^3.
$$

We consider these terms one by one, as $n \to \infty$.

(i) It is a consequence of the continuity properties of $f$ and $W$ that

$$
\sum_{j=0}^{n-1} f_t(\theta_j, W_{(j+1)\delta})\delta \xrightarrow{\text{a.s.}} \int_0^t f_t(s, W_s) \, ds.
$$

(ii) Using the differentiability of $f$, one may see that $\sum_{j=0}^{n-1} f_w(j\delta, W_{j\delta})\Delta_j$ converges in mean square as $n \to \infty$ to the Itô integral $\int_0^t f_w(s, W_s) \, dW_s$.

(iii) We have that $\mathbb{E}(\Delta_j^2) = \delta$, and $\Delta_j^2$ and $\Delta_k^2$ are independent for $j \neq k$. It follows after some algebra that

$$
\sum_{j=0}^{n-1} f_{ww}(j\delta, W_{j\delta})\Delta_j^2 - \sum_{j=0}^{n-1} f_{ww}(j\delta, W_{j\delta})\delta \xrightarrow{\text{m.s.}} 0.
$$

This implies the convergence of the third sum in (6) to the integral $\tfrac{1}{2} \int_0^t f_{ww}(s, W_s) \, ds$.

(iv) It may be shown after some work that the fourth term in (6) converges in mean square to zero as $n \to \infty$, and the required result follows by combining (i)–(iv). ∎

**(7) Example.** (a) Let $dX = \mu(t, X) \, dt + \sigma(t, X) \, dW$ and let $Y = X^2$. By Theorem (2),

$$dY = (2\mu X + \sigma^2) \, dt + 2\sigma X \, dW = \sigma(t, X)^2 \, dt + 2X \, dX.$$

(b) Let $Y_t = W_t^2$. Applying part (a) with $\mu = 0$ and $\sigma = 1$ (or alternatively using Itô's simple formula (4)), we find that $dY = dt + 2W \, dW$. By integration,

$$\int_0^t W_s \, dW_s = \tfrac{1}{2}(Y_t - Y_0 - t) = \tfrac{1}{2}(W_t^2 - t)$$

in agreement with formula (13.7.5).                                                ●

**(8) Example. Product rule.** Suppose that

$$dX = \mu_1(t, X) \, dt + \sigma_1(t, X) \, dW, \quad dY = \mu_2(t, Y) \, dt + \sigma_2(t, Y) \, dW,$$

in the notation of Itô's formula (2). We have by Example (7) that

$$d(X^2) = \sigma_1(t, X)^2 \, dt + 2X \, dX,$$

$$d(Y^2) = \sigma_2(t, Y)^2 \, dt + 2Y \, dY,$$

$$d((X + Y)^2) = \big(\sigma_1(t, X) + \sigma_2(t, Y)\big)^2 \, dt + 2(X + Y)(dX + dY).$$

Using the representation $XY = \tfrac{1}{2}\{(X + Y)^2 - X^2 - Y^2\}$, we deduce the product rule

$$d(XY) = X \, dY + Y \, dX + \sigma_1(t, X)\sigma_2(t, Y) \, dt.$$

Note the extra term over the usual Newton–Leibniz rule for differentiating a product.    ●

**(9) Example. Geometric Brownian motion.** Let $Y_t = \exp(\mu t + \sigma W_t)$ for constants $\mu, \sigma$. Itô's simple formula (4) yields

$$dY = (\mu + \tfrac{1}{2}\sigma^2)Y \, dt + \sigma Y \, dW,$$

so that $Y$ is a diffusion with instantaneous mean $a(t, y) = (\mu + \tfrac{1}{2}\sigma^2)y$ and instantaneous variance $b(t, y) = \sigma^2 y^2$. As indicated in Example (12.7.10), the process $Y$ is a martingale if and only if $\mu + \tfrac{1}{2}\sigma^2 = 0$.                                        ●

---

## Exercises for Section 13.9

In the absence of any contrary indication, $W$ denotes a standard Wiener process, and $\mathcal{F}_t$ is the smallest $\sigma$-field containing all null events with respect to which every member of $\{W_u : 0 \le u \le t\}$ is measurable.

**1.** Let $X$ and $Y$ be independent standard Wiener processes. Show that, with $R_t^2 = X_t^2 + Y_t^2$,

$$Z_t = \int_0^t \frac{X_s}{R_s} \, dX_s + \int_0^t \frac{Y_s}{R_s} \, dY_s$$

is a Wiener process. [Hint: Use Theorem (13.8.11) and Example (12.7.10).] Hence show that the squared process $R^2$, called the *squared Bessel process*, satisfies

$$R_t^2 = 2 \int_0^t R_s \, dW_s + 2t.$$

Generalize this conclusion to $n$ dimensions.

**2.** Write down the SDE obtained via Itô's formula for the process $Y_t = W_t^4$, and deduce that $\mathbb{E}(W_t^4) = 3t^2$.

**3.** Show that $Y_t = t W_t$ is an Itô process, and write down the corresponding SDE.

**4. Wiener process on a circle.** Let $Y_t = e^{i W_t}$. Show that $Y = X_1 + i X_2$ is a process on the unit circle satisfying

$$dX_1 = -\tfrac{1}{2} X_1 \, dt - X_2 \, dW, \quad dX_2 = -\tfrac{1}{2} X_2 \, dt + X_1 \, dW.$$

**5.** Find the SDEs satisfied by the processes:
(a) $X_t = W_t / (1 + t)$,
(b) $X_t = \sin W_t$,
(c) [Wiener process on an ellipse] $X_t = a \cos W_t$, $Y_t = b \sin W_t$, where $ab \neq 0$.

**6.** Find a random function $X(t)$ satisfying

$$dX(t) = \left(\tfrac{1}{2} X + \sqrt{1 + X^2}\right) dt + \sqrt{1 + X^2} \, dW(t),$$

with $X(0) = 0$.

---

## 13.10 Option pricing

It was essentially the Wiener process which Bachelier proposed in 1900 as a model for the evolution of stock prices. Interest in the applications of diffusions and martingales to stock prices has grown astonishingly since the fundamental work of Black, Scholes, and Merton in the early 1970s. The theory of mathematical finance is now well developed, and is one of the most striking modern applications of probability theory. We present here one simple model and application, namely the Black–Scholes solution for the pricing of a European call option. Numerous extensions of this result are possible, and the reader is referred to one of the many books on mathematical finance for further details (see Appendix II).

The Black–Scholes model concerns an economy which comprises two assets, a 'bond' (or 'money market account') whose value grows at a continuously compounded constant interest rate $r$, and a 'stock' whose price per unit is a stochastic process $S = \{S_t : t \geq 0\}$ indexed by time $t$. It is assumed that any quantity, *positive or negative* real valued, of either asset may be purchased at any time†. Writing $M_t$ for the cost of one unit of the bond at time $t$, and normalizing so that $M_0 = 1$, we have that

**(1)** $$dM_t = r M_t \, dt \quad \text{or} \quad M_t = e^{rt}.$$

---

†No taxes or commissions are payable, and the possession of stock brings no dividends. The purchase of negative quantities of bond or stock is called 'short selling' and can lead to a 'short position'.

A basic assumption of the Black–Scholes model is that $S$ satisfies the stochastic differential equation

$$(2) \qquad dS_t = S_t(\mu\,dt + \sigma\,dW_t) \quad \text{with solution} \quad S_t = \exp\big((\mu - \tfrac{1}{2}\sigma^2)t + \sigma W_t\big),$$

where $W$ is a standard Wiener process and we have normalized by setting $S_0 = 1$. That is to say, $S$ is a geometric Brownian motion (13.9.9) with parameters $\mu, \sigma$; in this context, $\sigma$ is usually called the *volatility* of the price process.

The market permits individuals to buy so-called 'forward options' on the stock, such products being termed 'derivatives'. One of the most important derivatives, the 'European call option', permits the buyer to purchase one unit of the stock at some given future time and at some predetermined price. More precisely, the option gives the holder the right to buy one unit of stock at time $T$, called the 'exercise date', for the price $K$, called the 'strike price'; the holder is not *required* to exercise this right. The fundamental question is to determine the 'correct price' of this option at some time $t$ satisfying $t \leq T$. The following elucidation of market forces leads to an interpretation of the notion of 'correct price', and utilizes some beautiful mathematics†.

We have at time $T$ that:

(a) if $S_T > K$, a holder of the option can buy one unit of the stock for $K$ and sell immediately for $S_T$, making an immediate profit of $S_T - K$,

(b) if $S_T \leq K$, it would be preferable to buy $K/S_T$ ($\geq 1$) units of the stock on the open market than to exercise the option.

It follows that the value $\phi_T$ of the option at time $T$ is given by $\phi_T = \max\{S_T - K, 0\} = (S_T - K)^+$. The discounted value of $\phi_T$ at an earlier time $t$ is $e^{-r(T-t)}(S_T - K)^+$, since an investment at time $t$ of this sum in the bond will be valued at $\phi_T$ at the later time $T$. One might naively suppose that the value of the option at an earlier time is given by its expectation; for example, the value at time 0 might be $\phi_0 = \mathbb{E}(e^{-rT}(S_T - K)^+)$. The financial market does not operate in this way, and *this answer is wrong*. It turns out in general that, in a market where options are thus priced according to the mean of their discounted value, the buyer of the option can devise a strategy for making a certain profit. Such an opportunity to make a risk-free profit is called an *arbitrage opportunity* and it may be assumed in practice that no such opportunity exists. In order to define the notion of arbitrage more properly‡, we discuss next the concept of 'portfolio'.

Let $\mathcal{F}_t$ be the $\sigma$-field generated by the random variables $\{S_u : 0 \leq u \leq t\}$. A *portfolio* is a pair $\alpha = \{\alpha_t : t \geq 0\}$, $\beta = \{\beta_t : t \geq 0\}$ of stochastic processes which are adapted to the filtration $\mathcal{F} = \{\mathcal{F}_t : t \geq 0\}$. We interpret the pair $(\alpha, \beta)$ as a time-dependent portfolio comprising $\alpha_t$ units of stock and $\beta_t$ units of the bond at time $t$. The value at time $t$ of the portfolio $(\alpha, \beta)$ is given by the *value function*

$$(3) \qquad\qquad\qquad\qquad V_t(\alpha, \beta) = \alpha_t S_t + \beta_t M_t,$$

and the portfolio is called *self-financing* if

$$(4) \qquad\qquad\qquad\qquad dV_t(\alpha, \beta) = \alpha_t\,dS_t + \beta_t\,dM_t,$$

---

†It would in practice be a mistake to adhere over rigidly to strategies based on the mathematical facts presented in this section and elsewhere. Such results are well known across the market and their use can be disadvantageous, as some have found out to their cost.

‡Such a concept was mentioned for a discrete system in Exercise (6.6.3).

which is to say that changes in value may be attributed to changes in the market only and not to the injection or withdrawal of funds. Condition (4) is a consequence of the modelling assumption implicit in (2) that $S$ is an Itô integral. It is is explained slightly more fully via the following discretization of time. Suppose that $\epsilon > 0$, and that time is divided into the intervals $I_n = [n\epsilon, (n+1)\epsilon)$. We assume that prices remain constant within each interval $I_n$. We exit interval $I_{n-1}$ having some portfolio $(\alpha_{(n-1)\epsilon}, \beta_{(n-1)\epsilon})$. At the start of $I_n$ this portfolio has value $v_n = \alpha_{(n-1)\epsilon} S_{n\epsilon} + \beta_{(n-1)\epsilon} M_{n\epsilon}$. The self-financing of the portfolio implies that the value at the end of $I_n$ equals $v_n$, which is to say that

(5)
$$\alpha_{(n-1)\epsilon} S_{n\epsilon} + \beta_{(n-1)\epsilon} M_{n\epsilon} = \alpha_{n\epsilon} S_{n\epsilon} + \beta_{n\epsilon} M_{n\epsilon}.$$

Now,

(6)
$$v_{n+1} - v_n = \alpha_{n\epsilon} S_{(n+1)\epsilon} + \beta_{n\epsilon} M_{(n+1)\epsilon} - \alpha_{(n-1)\epsilon} S_{n\epsilon} - \beta_{(n-1)\epsilon} M_{n\epsilon}$$
$$= \alpha_{n\epsilon} (S_{(n+1)\epsilon} - S_{n\epsilon}) + \beta_{n\epsilon} (M_{(n+1)\epsilon} - M_{n\epsilon})$$

by (5). Condition (4) is motivated by passing to the limit $\epsilon \downarrow 0$.

We say that a self-financing portfolio $(\alpha, \beta)$ *replicates* the given European call option if its value $V_T(\alpha, \beta)$ at time $T$ satisfies $V_T(\alpha, \beta) = (S_T - K)^+$ almost surely.

We now utilize the assumption that the market contains no arbitrage opportunities. Let $t < T$, and suppose that two options are available at a given time $t$. Option I costs $c_1$ per unit and yields a (strictly positive) value $\phi$ at time $T$; Option II costs $c_2$ per unit and yields the same value $\phi$ at time $T$. We may assume without loss of generality that $c_1 \geq c_2$. Consider the following strategy: at time $t$, buy $-c_2$ units of Option I and $c_1$ units of Option II. The total cost is $(-c_2)c_1 + c_1 c_2 = 0$, and the value at time $T$ is $(-c_2)\phi + c_1\phi = (c_1 - c_2)\phi$. If $c_1 > c_2$, there exists a strategy which yields a risk-free profit, in contradiction of the assumption of no arbitrage. Therefore $c_1 = c_2$.

Assume now that there exists a self-financing portfolio $(\alpha, \beta)$ which replicates the European call option. At time $t$ ($< T$) we may either invest in the option, or we may buy into the portfolio. Since their returns at time $T$ are equal, they must by the argument above have equal cost at time $t$. That is to say, in the absence of arbitrage, the 'correct value' of the European call option at time $t$ is $V_t(\alpha, \beta)$. In order to price the option, it remains to show that such a portfolio exists, and to find its value function.

First we calculate the value function of such a portfolio, and later we shall return to the question of its existence. Assume that $(\alpha, \beta)$ is a self-financing portfolio which replicates the European call option. It would be convenient if its discounted value function $e^{-rt} V_t$ were a martingale, since it would follow that $e^{-rt} V_t = \mathbb{E}(e^{-rT} V_T \mid \mathcal{F}_t)$ where $V_T = (S_T - K)^+$. This is not generally the case, but the following clever argument may be exploited. Although $e^{-rt} V_t$ is not a martingale on the probability space $(\Omega, \mathcal{F}, \mathbb{P})$, it turns out that there exists an alternative probability measure $\mathbb{Q}$ on the measurable pair $(\Omega, \mathcal{F})$ such that $e^{-rt} V_t$ is indeed a martingale on the probability space $(\Omega, \mathcal{F}, \mathbb{Q})$. The usual proof of this statement makes use of a result known as the Cameron–Martin–Girsanov formula which is beyond the scope of this book. In the case of the Black–Scholes model, one may argue directly via the following 'change of measure' formula.

(7) **Theorem.** *Let $B = \{B_t : 0 \leq t \leq T\}$ be a Wiener process with drift $0$ and instantaneous variance $\sigma^2$ on the probability space $(\Omega, \mathcal{F}, \mathbb{P})$, and let $v \in \mathbb{R}$. Define the random variable*

$$\Lambda = \exp\left\{ \frac{v}{\sigma^2} B_T - \frac{v^2}{2\sigma^2} T \right\},$$

and the measure $\mathbb{Q}$ by $\mathbb{Q}(A) = \mathbb{E}(\Lambda I_A)$. *Then $\mathbb{Q}$ is a probability measure and, regarded as a process on the probability space $(\Omega, \mathcal{F}, \mathbb{Q})$, $B$ is a Wiener process with drift $\nu$ and instantaneous variance $\sigma^2$.*

**Proof.** That $\mathbb{Q}$ is a probability measure is a consequence of the fact that

$$\mathbb{Q}(\Omega) = \mathbb{E}(\Lambda) = e^{-\nu^2 T/(2\sigma^2)}\mathbb{E}(e^{(\nu/\sigma^2)B_T}) = e^{-\nu^2 T/(2\sigma^2)}e^{\frac{1}{2}(\nu/\sigma^2)^2\sigma^2 T} = 1.$$

The distribution of $B$ under $\mathbb{Q}$ is specified by its finite-dimensional distributions (we recall the discussion of Section 8.9). Let $0 = t_0 < t_1 < \cdots < t_n = T$ and $x_0, x_1, \ldots, x_n = x \in \mathbb{R}$. The notation used in the following is informal but convenient. The process $B$ has independent normal increments under the measure $\mathbb{P}$. Writing $\{B_{t_i} \in dx_i\}$ for the event that $x_i < B_{t_i} \leq x_i + dx_i$, we have that

$$\mathbb{Q}\big(B_{t_1} \in dx_1,\ B_{t_2} \in dx_2, \ldots,\ B_{t_n} \in dx_n\big)$$

$$= \mathbb{E}\left(\exp\left(\frac{\nu}{\sigma^2}B_T - \frac{\nu^2}{2\sigma^2}T\right)I_{\{B_{t_1}\in dx_1\}\cap\cdots\cap\{B_{t_n}\in dx_n\}}\right)$$

$$= \exp\left(\frac{\nu}{\sigma^2}x - \frac{\nu^2}{2\sigma^2}T\right)\prod_{i=1}^{n}\left\{\frac{1}{\sqrt{2\pi\sigma^2(t_i - t_{i-1})}}\exp\left(-\frac{(x_i - x_{i-1})^2}{2\sigma^2(t_i - t_{i-1})}\right)dx_i\right\}$$

$$= \prod_{i=1}^{n}\left\{\frac{1}{\sqrt{2\pi\sigma^2(t_i - t_{i-1})}}\exp\left(-\frac{(x_i - x_{i-1} - \nu(t_i - t_{i-1}))^2}{2\sigma^2(t_i - t_{i-1})}\right)dx_i\right\}.$$

It follows that, under $\mathbb{Q}$, the sequence $B_{t_1}, B_{t_2}, \ldots, B_{t_n}$ is distributed in the manner of a Wiener process with drift $\nu$ and instantaneous variance $\sigma^2$. Since this holds for all sequences $t_1, t_2, \ldots, t_n$ and since $B$ has continuous sample paths, the claim of the theorem follows. ■

With $W$ the usual standard Wiener process, and $\nu \in \mathbb{R}$, there exists by Theorem (7) a probability measure $\mathbb{Q}_\nu$ under which $\sigma W$ is a Wiener process with drift $\nu$ and instantaneous variance $\sigma^2$. Therefore, under $\mathbb{Q}_\nu$, the process $\widetilde{W}$ given by $\sigma \widetilde{W}_t = -\nu t + \sigma W_y$ is a standard Wiener process. By equation (2) and the final observation of Example (13.9.9), under $\mathbb{Q}_\nu$ the process

$$e^{-rt}S_t = \exp\big((\mu - \tfrac{1}{2}\sigma^2 - r)t + \sigma W_t\big) = \exp\big((\mu - \tfrac{1}{2}\sigma^2 - r + \nu)t + \sigma \widetilde{W}_t\big)$$

is a diffusion with instantaneous mean and variance $a(t, x) = (\mu - r + \nu)x$ and $b(t, x) = \sigma^2 x^2$. By Example (12.7.10), it is a martingale under $\mathbb{Q}_\nu$ if $\mu - r + \nu = 0$, and we set $\nu = r - \mu$ accordingly, and write $\mathbb{Q} = \mathbb{Q}_\nu$. The fact that there exists a measure $\mathbb{Q}$ under which $e^{-rt}S_t$ is a martingale is pivotal for the solution to this problem and its generalizations.

It is a consequence that, under $\mathbb{Q}$, $e^{-rt}V_t$ constitutes a martingale. This may be seen as follows. By the product rule of Example (13.9.8),

$$(8) \quad d(e^{-rt}V_t) = e^{-rt}\,dV_t - re^{-rt}V_t\,dt$$

$$= e^{-rt}\alpha_t\,dS_t - re^{-rt}\alpha_t S_t\,dt + e^{-rt}\beta_t(dM_t - rM_t) \quad \text{by (4) and (3)}$$

$$= \alpha_t e^{-rt}S_t\big((\mu - r)\,dt + \sigma\,dW_t\big) \quad\quad\quad\quad \text{by (1) and (2)}$$

$$= \alpha_t e^{-rt}S_t(-\nu\,dt + \sigma\,dW_t),$$

where $v = r - \mu$ as above. Under $\mathbb{Q}$, $\sigma W$ is a Wiener process with drift $v$ and instantaneous variance $\sigma^2$, whence $\sigma \widetilde{W} = -vt + \sigma W$ is a Wiener process with drift $0$ and instantaneous variance $\sigma^2$. By (8),

$$e^{-rt} V_t = V_0 + \int_0^t \alpha_u e^{-ru} S_u \sigma \, d\widetilde{W}_u$$

which, by Theorem (13.8.11), defines a martingale under $\mathbb{Q}$. Now $V_t$ equals the value of the European call option at time $t$ and, by the martingale property,

**(9)**                    $V_t = e^{rt}(e^{-rt} V_t) = e^{rt} \mathbb{E}_{\mathbb{Q}}(e^{-rT} V_T \mid \mathcal{F}_t)$

where $\mathbb{E}_{\mathbb{Q}}$ denotes expectation with respect to $\mathbb{Q}$. The right side of (9) may be computed via the result of Exercise (13.10.1), leading to the following useful form of the value† of the option.

**(10) Theorem. Black–Scholes formula.** *Let $t < T$. The value at time $t$ of the European call option is*

**(11)**                    $S_t \Phi(d_1(t, S_t)) - Ke^{-r(T-t)} \Phi(d_2(t, S_t))$

*where $\Phi$ is the $N(0, 1)$ distribution function and*

**(12)**    $d_1(t, x) = \dfrac{\log(x/K) + (r + \tfrac{1}{2}\sigma^2)(T - t)}{\sigma\sqrt{T - t}}$,    $d_2(t, x) = d_1(t, x) - \sigma\sqrt{T - t}$.

Note that the Black–Scholes formula depends on the price process through $r$ and $\sigma^2$ and not through the value of $\mu$. A similar formula may be derived for any adapted contingent claim having finite second moment.

The discussion prior to the theorem does not constitute a full proof, since it was based on the assumption that there exists a self-financing strategy which replicates the European call option. In order to prove the Black–Scholes formula, we shall show the existence of a self-financing replicating strategy with value function (11). This portfolio may be identified from (11), since (11) is the value function of the portfolio $(\alpha, \beta)$ given by

**(13)**                    $\alpha_t = \Phi(d_1(t, S_t))$,    $\beta_t = -Ke^{-rT} \Phi(d_2(t, S_t))$.

Let $\xi(t, x)$, $\psi(t, x)$ be smooth functions of the real variables $t$, $x$, and consider the portfolio denoted $(\xi, \psi)$ which at time $t$ holds $\xi(t, S_t)$ units in stock and $\psi(t, S_t)$ units in the bond. This portfolio has value function $W_t(\xi, \psi) = w(t, S_t)$ where

**(14)**                    $w(t, x) = \xi(t, x)x + \psi(t, x)e^{rt}$.

**(15) Theorem.** *Let $\xi$, $\psi$ be such that the function $w$ given by (14) is twice continuously differentiable. The portfolio $(\xi, \psi)$ is self-financing if and only if:*

**(16)**                    $x\xi_x + e^{rt}\psi_x = 0$,

**(17)**                    $\tfrac{1}{2}\sigma^2 x^2 \xi_x + x\xi_t + e^{rt}\psi_t = 0$,

*where $f_x$, $f_t$ denote derivatives with respect to $x$, $t$.*

---

†The value given in Theorem (10) is sometimes called the 'no arbitrage value' or the 'risk-neutral value' of the option.

**Proof.** We apply Itô's formula (13.9.2) to the function $w$ of equation (14) to find via (2) that

$$dw(t, S_t) = w_x(t, S_t) \, dS_t + \left\{ w_t(t, S_t) + \tfrac{1}{2}\sigma^2 S_t^2 w_{xx}(t, S_t) \right\} dt$$

whereas, by (4) and (1), $(\xi, \psi)$ is self-financing if and only if

**(18)**                    $$dw(t, S_t) = \xi(t, S_t) \, dS_t + r\psi(t, S_t)e^{rt} \, dt.$$

Equating coefficients of the infinitesimals, we deduce that $(\xi, \psi)$ is self-financing if and only if $\xi = w_x$ and $r\psi e^{rt} = w_t + \tfrac{1}{2}\sigma^2 x^2 w_{xx}$, which is to say that:

**(19)**                    $$\xi = \xi + \xi_x x + \psi_x e^{rt},$$

**(20)**                    $$r\psi e^{rt} = \xi_t x + \psi_t e^{rt} + r\psi e^{rt} + \tfrac{1}{2}\sigma^2 x^2(\xi_{xx} x + 2\xi_x + \psi_{xx} e^{rt}).$$

Differentiating (19) with respect to $x$ yields

**(21)**                    $$0 = \xi_{xx} x + \xi_x + \psi_{xx} e^{rt},$$

which may be inserted into (20) to give as required that

$$0 = \xi_t x + \psi_t e^{rt} + \tfrac{1}{2}\sigma^2 x^2 \xi_x. \qquad \blacksquare$$

Theorem (15) leads to the following characterization of value functions of self-financing portfolios.

**(22) Corollary. Black–Scholes equation.** *Suppose that $w(t, x)$ is twice continuously differentiable. Then $w(t, S_t)$ is the value function of a self-financing portfolio if and only if*

**(23)**                    $$\tfrac{1}{2}\sigma^2 x^2 w_{xx} + rxw_x + w_t - rw = 0.$$

The Black–Scholes equation provides a means for finding self-financing portfolios which replicate general contingent claims. One 'simply' solves equation (23) subject to the boundary condition imposed by the particular claim in question. In the case of the European call option, the appropriate boundary condition is $w(T, x) = (x - K)^+$. It is not always easy to find the solution, but there is a general method known as the 'Feynman–Kac formula', not discussed further here, which allows a representation of the solution in terms of a diffusion process. When the solution exists, the claim is said to be 'hedgeable', and the self-financing portfolio which replicates it is called the 'hedge'.

**Proof.** Assume that $w$ satisfies (23), and set

$$\xi = w_x, \quad \psi = e^{-rt}(w - xw_x).$$

It is easily checked that the portfolio $(\xi, \psi)$ has value function $w(t, S_t)$ and, via (23), that the pair $\xi, \psi$ satisfy equations (16) and (17).

Conversely, if $w(t, S_t)$ is the value of a self-financing portfolio then $w(t, x) = \xi(t, x)x + \psi(t, x)e^{rt}$ for some pair $\xi, \psi$ satisfying equations (16) and (17). We compute $w_x$ and compare with (16) to find that $\xi = w_x$. Setting $\psi = e^{-rt}(w - xw_x)$, we substitute into (17) to deduce that equation (23) holds. ■

**Proof of Theorem (10).** Finally we return to the proof of the Black–Scholes formula, showing first that the portfolio $(\alpha, \beta)$, given in equations (13), is self-financing. Set

$$\alpha(t, x) = \Phi(d_1(t, x)), \quad \beta(t, x) = -Ke^{-rT}\Phi(d_2(t, x)),$$

where $d_1$ and $d_2$ are given in (12). We note from (12) that

(24)                    $$d_2^2 = d_1^2 - 2\log(x/K) - 2r(T - t),$$

and it is straightforward to deduce by substitution that the pair $\alpha, \beta$ satisfy equations (16) and (17). Therefore, the portfolio $(\alpha, \beta)$ is self-financing. By construction, it has value function $V_t(\alpha, \beta)$ given in (11).

We may take the limit in (11) as $t \uparrow T$. Since

$$d_i(t, S_t) \to \begin{cases} -\infty & \text{if } S_T < K, \\ \infty & \text{if } S_T > K, \end{cases}$$

for $i = 1, 2$, we deduce that $V_T(\alpha, \beta) = (S_T - K)^+$ whenever $S_T \neq K$. Now $\mathbb{P}(S_T = K) = 0$, and therefore $V_T(\alpha, \beta) = (S_T - K)^+$ almost surely. It follows as required that the portfolio $(\alpha, \beta)$ replicates the European call option. ■

---

## Exercises for Section 13.10

In the absence of any contrary indication, $W$ denotes a standard Wiener process, and $\mathcal{F}_t$ is the smallest $\sigma$-field containing all null events with respect to which every member of $\{W_u : 0 \le u \le t\}$ is measurable. The process $S_t = \exp((\mu - \frac{1}{2}\sigma^2)t + \sigma W_t)$ is a geometric Brownian motion, and $r \ge 0$ is the interest rate.

**1.** (a) Let $Z$ have the $N(\gamma, \tau^2)$ distribution. Show that

$$\mathbb{E}\big((ae^Z - K)^+\big) = ae^{\gamma + \frac{1}{2}\tau^2}\Phi\left(\frac{\log(a/K) + \gamma}{\tau} + \tau\right) - K\Phi\left(\frac{\log(a/K) + \gamma}{\tau}\right)$$

where $\Phi$ is the $N(0, 1)$ distribution function.

(b) Let $\mathbb{Q}$ be a probability measure under which $\sigma W$ is a Wiener process with drift $r - \mu$ and instantaneous variance $\sigma^2$. Show for $0 \le t \le T$ that

$$\mathbb{E}_{\mathbb{Q}}\big((S_T - K)^+ \mid \mathcal{F}_t\big) = S_t e^{r(T-t)}\Phi(d_1(t, S_t)) - K\Phi(d_2(t, S_t))$$

where

$$d_1(t, x) = \frac{\log(x/K) + (r + \frac{1}{2}\sigma^2)(T - t)}{\sigma\sqrt{T - t}}, \quad d_2(t, x) = d_1(t, x) - \sigma\sqrt{T - t}.$$

**2.** Consider a portfolio which, at time $t$, holds $\xi(t, S)$ units of stock and $\psi(t, S)$ units of bond, and assume these quantities depend only on the values of $S_u$ for $0 \le u \le t$. Find the function $\psi$ such that the portfolio is self-financing in the three cases:

(a) $\xi(t, S) = 1$ for all $t, S$,

(b) $\xi(t, S) = S_t$,

(c) $\xi(t, S) = \int_0^t S_v \, dv$.

**3.** Suppose the stock price $S_t$ is itself a Wiener process and the interest rate $r$ equals 0, so that a unit of bond has unit value for all time. In the notation of Exercise (13.10.2), which of the following define self-financing portfolios?

(a) $\xi(t, S) = \psi(t, S) = 1$ for all $t$, $S$,

(b) $\xi(t, S) = 2S_t$, $\psi(t, S) = -S_t^2 - t$,

(c) $\xi(t, S) = -t$, $\psi(t, S) = \int_0^t S_s \, ds$,

(d) $\xi(t, S) = \int_0^t S_s \, ds$, $\psi(t, S) = - \int_0^t S_s^2 \, ds$.

**4.** An 'American call option' differs from a European call option in that it may be exercised by the buyer *at any time up to the expiry date*. Show that the value of the American call option is the same as that of the corresponding European call option, and that there is no advantage to the holder of such an option to exercise it strictly before its expiry date.

**5.** Show that the Black–Scholes value at time 0 of the European call option is an increasing function of the initial stock price, the exercise date, the interest rate, and the volatility, and is a decreasing function of the strike price.

## 13.11 Passage probabilities and potentials

In this final section, we study in a superficial way a remarkable connection between probability theory and classical analysis, namely the relationship between the sample paths of a Wiener process and the Newtonian theory of gravitation.

We begin by recalling some fundamental facts from the theory of scalar potentials. Let us assume that matter is distributed about regions of $\mathbb{R}^d$ where $d \geq 2$. According to the laws of Newtonian attraction, this matter gives rise to a function $\phi : \mathbb{R}^d \to \mathbb{R}$ which assigns a *potential $\phi(\mathbf{x})$* to each point $\mathbf{x} = (x_1, x_2, \ldots, x_d) \in \mathbb{R}^d$. In regions of space which are empty of matter, the potential function $\phi$ satisfies

**(1)** *Laplace's equation*:                                      $\nabla^2 \phi = 0$,

where the *Laplacian $\nabla^2 \phi$* is given by

**(2)**
$$\nabla^2 \phi = \sum_{i=1}^{d} \frac{\partial^2 \phi}{\partial x_i^2}.$$

It is an important application of Green's theorem that solutions to Laplace's equation are also solutions to a type of integral equation. We make this specific as follows. Let $\mathbf{x}$ lie in the interior of a region $R$ of space which is empty of matter, and consider a ball $B$ contained in $R$ with radius $a$ and centre $\mathbf{x}$. The potential $\phi(\mathbf{x})$ at the centre of $B$ is the average of the potential over the surface $\Sigma$ of $B$. That is to say, $\phi(\mathbf{x})$ may be expressed as the surface integral

**(3)**
$$\phi(\mathbf{x}) = \int_{\mathbf{y} \in \Sigma} \frac{\phi(\mathbf{y})}{|\Sigma|} \, dS,$$

where $|\Sigma|$ is the surface area of $B$ (for example, $|\Sigma| = 2\pi a$ in two dimensions and $4\pi a^2$ in three). Furthermore, $\phi$ satisfies (3) for all such balls if and only if $\phi$ is a solution to Laplace's equation (1) in the appropriate region.

We turn now to probabilities. Let $\mathbf{W}(t) = (W_1(t), W_2(t), \ldots, W_d(t))$ be a $d$-dimensional Wiener process describing the position of a particle which diffuses around $\mathbb{R}^d$, so that the $W_i$ are independent one-dimensional Wiener processes. We assume that $\mathbf{W}(0) = \mathbf{w}$ and that the $W_i$ have variance parameter $\sigma^2$. The vector $\mathbf{W}(t)$ contains $d$ random variables with joint density function

(4)
$$f_{\mathbf{W}(t)}(\mathbf{x}) = \frac{1}{(2\pi\sigma^2 t)^{d/2}} \exp\left(-\frac{1}{2\sigma^2 t}|\mathbf{x} - \mathbf{w}|^2\right), \quad \mathbf{x} \in \mathbb{R}^d.$$

Let $H$, $J$ be disjoint subsets of $\mathbb{R}^d$ which are 'nice' in some manner which we will not make specific. Suppose the particle starts from $\mathbf{W}(0) = \mathbf{w}$, and let us ask for the probability that it visits some point of $H$ before it visits any point of $J$. A particular case of this question might arise as follows. Suppose that $\mathbf{w}$ is a point in the interior of some closed bounded connected domain $D$ of $\mathbb{R}^d$, and suppose that the surface $\partial D$ which bounds $D$ is fairly smooth (if $D$ is a ball then $\partial D$ is the bounding spherical surface, for example). Sooner or later the particle will enter $\partial D$ for the first time. If $\partial D = H \cup J$ for some disjoint sets $H$ and $J$, then we may ask for the probability that the particle enters $\partial D$ at a point in $H$ rather than at a point in $J$ (as an example, take $D$ to be the ball of radius 1 and centre $\mathbf{w}$, and let $H$ be a hemisphere of $D$).

In the above example, the process was bound (almost surely) to enter $H \cup J$ at some time. This is not true for general regions $H$, $J$. For example, the hitting time of a point in $\mathbb{R}^2$ is almost surely infinite, and we shall see that the hitting time of a sphere in $\mathbb{R}^3$ is infinite with strictly positive probability if the process starts outside the sphere. In order to include all eventualities, we introduce the hitting time $T_A = \inf\{t : \mathbf{W}(t) \in A\}$ of the subset $A$ of $\mathbb{R}^d$, with the usual convention that the infimum of the empty set is $\infty$. We write $\mathbb{P}_\mathbf{w}$ for the probability measure which governs the Wiener process $\mathbf{W}$ when it starts from $\mathbf{W}(0) = \mathbf{w}$.

**(5) Theorem.** *Let $H$ and $J$ be disjoint 'nice' subsets† of $\mathbb{R}^d$ such that $H \cup J$ is closed, and let $p(\mathbf{w}) = \mathbb{P}_\mathbf{w}(T_H < T_J)$. The function $p$ satisfies Laplace's equation, $\nabla^2 p(\mathbf{w}) = 0$, at all points $\mathbf{w} \notin H \cup J$, with the boundary conditions*

$$p(\mathbf{w}) = \begin{cases} 1 & \text{if } \mathbf{w} \in H, \\ 0 & \text{if } \mathbf{w} \in J. \end{cases}$$

**Proof.** Let $\mathbf{w} \notin H \cup J$. Since $H \cup J$ is assumed closed, there exists a ball $B$ contained in $\mathbb{R}^d \setminus (H \cup J)$ with centre $\mathbf{w}$. Let $a$ be the radius of $B$ and $\Sigma$ its surface. Let $T = \inf\{t : \mathbf{W}(t) \in \Sigma\}$ be the first passage time of $\mathbf{W}$ to the set $\Sigma$. The random variable $T$ is a stopping time for $\mathbf{W}$, and it is not difficult to see as follows that $\mathbb{P}_\mathbf{w}(T < \infty) = 1$. Let $A_i = \{|\mathbf{W}(i) - \mathbf{W}(i-1)| \le 2a\}$ and note that $\mathbb{P}_\mathbf{w}(A_1) < 1$, whence

$$\mathbb{P}_\mathbf{w}(T > n) \le \mathbb{P}_\mathbf{w}(A_1 \cap A_2 \cap \cdots \cap A_n)$$
$$= \mathbb{P}_\mathbf{w}(A_1)^n \qquad \text{by independence}$$
$$\to 0 \qquad \qquad \text{as } n \to \infty.$$

---

†We do not explain the condition that $H$ and $J$ be 'nice', but note that sets with smooth surfaces that can support an electric potential and charge, such as balls or Platonic solids, are nice.

We now condition on the hitting point $W(T)$. By the strong Markov property, given $W(T)$, the path of the process after time $T$ is a Wiener process with the new starting point $\mathbf{W}(T)$. It follows that the (conditional) probability that $\mathbf{W}$ visits $H$ before it visits $J$ is $p(\mathbf{W}(T))$, and we are led to the following formula:

$$(6) \qquad p(\mathbf{w}) = \int_{\mathbf{y} \in \Sigma} \mathbb{P}_{\mathbf{w}}\big(T_H < T_J \mid \mathbf{W}(T) = \mathbf{y}\big) f_{\mathbf{w}}(\mathbf{y}) \, dS$$

where $f_{\mathbf{w}}$ is the conditional density function of $\mathbf{W}(T)$ given $\mathbf{W}(0) = \mathbf{w}$. Using the spherical symmetry of the density function in (4), we have that $W(T)$ is uniformly distributed on $\Sigma$, which is to say that

$$f_{\mathbf{w}}(\mathbf{y}) = \frac{1}{|\Sigma|} \quad \text{for all } \mathbf{y} \in \Sigma,$$

and equation (6) becomes

$$(7) \qquad p(\mathbf{w}) = \int_{\mathbf{y} \in \Sigma} \frac{p(\mathbf{y})}{|\Sigma|} \, dS.$$

This integral equation holds for any ball $B$ with centre $\mathbf{w}$ whose contents do not overlap $H \cup J$, and we recognize it as the characteristic property (3) of solutions to Laplace's equation (1). Thus $p$ satisfies Laplace's equation. The boundary conditions are derived easily.                            ■

Theorem (5) provides us with an elegant technique for finding the probabilities that $\mathbf{W}$ visits certain subsets of $\mathbb{R}^d$. The principles of the method are simple, although some of the ensuing calculations may be lengthy since the difficulty of finding explicit solutions to Laplace's equation depends on the boundary conditions (see Example (14) and Problem (13.12.12), for instance).

**(8) Example.** Take $d = 2$, and start a two-dimensional Wiener process $\mathbf{W}$ at a point $\mathbf{W}(0) = \mathbf{w} \in \mathbb{R}^2$. Let $H$ be a circle with radius $\epsilon$ $(> 0)$ and centre at the origin, such that $\mathbf{w}$ does not lie within the inside of $H$. What is the probability that $\mathbf{W}$ ever visits $H$?
**Solution.** We shall need two boundary conditions in order to find the appropriate solution to Laplace's equation. The first arises from the case when $\mathbf{w} \in H$. To find the second, we introduce a second circle $J$, with radius $R$ and centre at the origin, and suppose that $R$ is much larger than $\epsilon$. We shall solve Laplace's equation in polar coordinates,

$$(9) \qquad \frac{1}{r}\frac{\partial}{\partial r}\left( r\frac{\partial p}{\partial r} \right) + \frac{1}{r^2}\frac{\partial^2 p}{\partial \theta^2} = 0,$$

in the region $\epsilon \le r \le R$, and use the boundary conditions

$$(10) \qquad p(\mathbf{w}) = \begin{cases} 1 & \text{if } \mathbf{w} \in H, \\ 0 & \text{if } \mathbf{w} \in J. \end{cases}$$

Solutions to equation (9) having circular symmetry take the form

$$p(\mathbf{w}) = A \log r + B \quad \text{if} \quad \mathbf{w} = (r, \theta),$$

where $A$ and $B$ are arbitrary constants. We use the boundary conditions to obtain the solution

$$p_R(\mathbf{w}) = \frac{\log(r/R)}{\log(\epsilon/R)}, \qquad \epsilon \le r \le R,$$

and we deduce by Theorem (5) that $\mathbb{P}_\mathbf{w}(T_H < T_J) = p_R(\mathbf{w})$.

In the limit as $R \to \infty$, we have that $T_J \to \infty$ almost surely, whence

$$p_R(\mathbf{w}) \to \mathbb{P}_\mathbf{w}(T_H < \infty) = 1.$$

We conclude that $\mathbf{W}$ almost surely visits any $\epsilon$-neighbourhood of the origin regardless of its starting point. Such a process is called *recurrent* (or *persistent*) since its sample paths pass arbitrarily closely to every point in the plane with probability 1.                                      ●

**(11) Example.** We consider next the same question as Example (8) but in three dimensions. Let $H$ be the sphere with radius $\epsilon$ and centre at the origin of $\mathbb{R}^3$. We start a three-dimensional Wiener process $\mathbf{W}$ from some point $\mathbf{W}(0) = \mathbf{w}$ which does not lie within $H$. What is the probability that $\mathbf{W}$ visits $H$?

**Solution.** As before, let $J$ be a sphere with radius $R$ and centre at the origin, where $R$ is much larger than $\epsilon$. We seek a solution to Laplace's equation in spherical polar coordinates

**(12)**
$$\frac{\partial}{\partial r}\left(r^2 \frac{\partial p}{\partial r}\right) + \frac{1}{\sin\theta}\frac{\partial}{\partial\theta}\left(\sin\theta\frac{\partial p}{\partial\theta}\right) + \frac{1}{\sin^2\phi}\frac{\partial^2 p}{\partial\phi^2} = 0,$$

subject to the boundary conditions (10). Solutions to equation (12) with spherical symmetry have the form

**(13)**
$$p(\mathbf{w}) = \frac{A}{r} + B \quad \text{if} \quad \mathbf{w} = (r, \theta, \phi).$$

We use the boundary conditions (10) to obtain the solution

$$p_R(\mathbf{w}) = \frac{r^{-1} - R^{-1}}{\epsilon^{-1} - R^{-1}}.$$

Let $R \to \infty$ to obtain by Theorem (5) that

$$p_R(\mathbf{w}) \to \mathbb{P}(T_H < \infty) = \frac{\epsilon}{r}, \qquad r > \epsilon.$$

That is to say, $\mathbf{W}$ ultimately visits $H$ with probability $\epsilon/r$. It is perhaps striking that the answer is *directly* proportional to $\epsilon$.

We have shown that the three-dimensional Wiener process is *not* recurrent, since its sample paths do not pass through every $\epsilon$-neighbourhood with probability 1. This mimics the behaviour of symmetric random walks; recall from Problems (5.12.6) and (6.15.9) that the two-dimensional symmetric random walk is recurrent whilst the three-dimensional walk is transient.                                      ●

*Diffusion processes*

**(14) Example.** Let $\Sigma$ be the surface of the unit sphere in $\mathbb{R}^3$ with centre at the origin, and let

$$H = \left\{ (r, \theta, \phi) : r = 1, \ 0 \le \theta \le \tfrac{1}{2}\pi \right\}$$

be the upper hemisphere of $\Sigma$. Start a three-dimensional Wiener process $\mathbf{W}$ from a point $\mathbf{W}(0) = \mathbf{w}$ which lies in the *inside* of $\Sigma$. What is the probability that $\mathbf{W}$ visits $H$ before it visits $J = \Sigma \setminus H$, the lower hemisphere of $\Sigma$?

**Solution.** The function $p(\mathbf{w}) = \mathbb{P}_{\mathbf{w}}(T_H < T_J)$ satisfies Laplace's equation (12) subject to the boundary conditions (10). Solutions to (12) which are independent of $\phi$ are also solutions to the simpler equation

$$\frac{\partial}{\partial r}\left( r^2 \frac{\partial p}{\partial r} \right) + \frac{1}{\sin\theta}\frac{\partial}{\partial \theta}\left( \sin\theta \frac{\partial p}{\partial \theta} \right) = 0.$$

We abandon the calculation at this point, leaving it to the reader to complete. Some knowledge of Legendre polynomials and the method of separation of variables may prove useful. ●

We may think of a Wiener process as a continuous space–time version of a symmetric random walk, and it is not surprising that Wiener processes and random walks have many properties in common. In particular, potential theory is of central importance to the theory of random walks. We terminate this chapter with a brief but electrifying demonstration of this.

Let $G = (V, E)$ be a finite connected graph with vertex set $V$ and edge set $E$. For simplicity we assume that $G$ has neither loops nor multiple edges. A particle performs a random walk about the vertex set $V$. If it is at vertex $v$ at time $n$, then it moves at time $n + 1$ to one of the neighbours of $v$, each such neighbour being chosen with equal probability, and independently of the previous trajectory of the particle. We write $X_n$ for the position of the particle at time $n$, and $\mathbb{P}_w$ for the probability measure governing the $X_n$ when $X_0 = w$.

For $A \subseteq V$, we define the passage time $T_A = \inf\{n : X_n \in A\}$. Let $H$ and $J$ be disjoint non-empty sets of vertices. We see by conditioning on $X_1$ that the function

**(15)** $$p(w) = \mathbb{P}_w(T_H < T_J)$$

satisfies the difference equation

$$p(w) = \sum_{x \in V} \mathbb{P}_w(X_1 = x)p(x) \qquad \text{for } w \notin H \cup J.$$

This expresses $p(w)$ as the average of the $p$-values of the neighbours of $w$:

**(16)** $$p(w) = \frac{1}{d(w)} \sum_{x : x \sim w} p(x) \qquad \text{for } w \notin H \cup J,$$

where $d(w)$ is the degree of the vertex $w$ and we write $x \sim w$ to mean that $x$ and $w$ are neighbours. Equation (16) is the discrete analogue of the integral equation (7). The boundary conditions are given as before by (10).

Equations (16) have an interesting interpretation in terms of electrical network theory. We may think of $G$ as an electrical network in which each edge is a resistor with resistance 1 ohm. We connect a battery into the network in such a way that the points in $H$ are raised to the potential 1 volt and the points in $J$ are joined to earth. It is physically clear that this potential difference induces a potential $\phi(w)$ at each vertex $w$, together with a current along each wire. These potentials and currents satisfy a well-known collection of equations called Kirchhoff's laws and Ohm's law, and it is an easy consequence of these laws (*exercise*) that $\phi$ is the unique solution to equations (16) subject to (10). It follows that

$$(17) \qquad\qquad \phi(w) = p(w) \quad \text{for all} \quad w \in V.$$

This equality between first passage probabilities and electrical potentials is the discrete analogue of Theorem (5)†.

As a beautiful application of this relationship, we shall show that random walk on an infinite connected graph is recurrent if and only if the graph has *infinite* resistance when viewed as an electrical network.

Let $G = (V, E)$ be an infinite connected graph with countably many vertices and finite vertex degrees, and let 0 denote a chosen vertex of $G$. We may turn $G$ into an (infinite) electrical network by replacing each edge by a unit resistor. For $u, v \in V$, let $d(u, v)$ be the number of edges in the shortest path joining $u$ and $v$, and define $\Delta_n = \{v \in V : d(0, v) = n\}$. Let $R_n$ be the electrical resistance between 0 and the set $\Delta_n$. That is to say, $1/R_n$ is the current which flows in the circuit obtained by setting 0 to earth and applying a unit potential to the vertices in $\Delta_n$. It is a standard fact from potential theory that $R_n \leq R_{n+1}$, and we define the *resistance* of $G$ to be the limit $R(G) = \lim_{n\to\infty} R_n$.

**(18) Theorem.** *A random walk on the graph $G$ is recurrent if and only if $R(G) = \infty$.*

**Proof.** Since $G$ is connected with finite vertex degrees, a random walk on $G$ is an irreducible Markov chain on the countable state space $V$. It suffices therefore to show that the vertex 0 is a recurrent state of the chain. We write $\mathbb{P}_x$ for the law of the random walk started from $X_0 = x$.

Let $\phi_n$ be the potential function in the electrical network obtained from $G$ by earthing 0 and applying unit potential to all vertices in $\Delta_n$. Note that $0 \leq \phi_n(x) \leq 1$ for all vertices $x$ (this is an application of what is termed the *maximum principle*). We have from the above discussion that $\phi_n(x) = \mathbb{P}_x(T_{\Delta_n} < T_0)$, where $T_A$ denotes the first hitting time of the set $A$. Now $T_{\Delta_n}$ is at least the minimum distance from $x$ to $\Delta_n$, which is at least $n - d(0, x)$, and therefore $\mathbb{P}_x(T_{\Delta_n} \to \infty$ as $n \to \infty) = 1$ for all $x$. It follows that

$$(19) \qquad\qquad \phi_n(x) \to \mathbb{P}_x(T_0 = \infty) \quad \text{as } n \to \infty.$$

Applying Ohm's law to the edges incident with 0, we have that the total current flowing out of 0 equals

$$\sum_{x:x\sim 0} \phi_n(x) = \frac{1}{R_n}.$$

We let $n \to \infty$ and use equation (19) to find that

$$(20) \qquad\qquad \sum_{x:x\sim 0} \mathbb{P}_x(T_0 = \infty) = \frac{1}{R(G)}.$$

---

†A further account of the relationship between random walks and electrical networks may be found in Grimmett (2018, ch. 1).

where $1/\infty$ is interpreted as 0.

We have by conditioning on $X_1$ that

$$\mathbb{P}_0(X_n = 0 \text{ for some } n \geq 1) = \frac{1}{d(0)} \sum_{x:x \sim 0} \mathbb{P}_x(T_0 < \infty)$$

$$= 1 - \frac{1}{d(0)} \sum_{x:x \sim 0} \mathbb{P}_x(T_0 = \infty) = 1 - \frac{1}{d(0)R(G)}.$$

The claim follows.                                                                                   ∎

**(21) Theorem. Recurrence of two-dimensional random walk.** *Symmetric random walk on the two-dimensional square lattice $\mathbb{Z}^2$ is recurrent.*

**Proof.** It suffices by Theorem (18) to prove that $R(\mathbb{Z}^2) = \infty$. We construct a lower bound for $R_n$ in the following way. For each $r \leq n$, short out all the points in $\Delta_r$ (draw your own diagram), and use the parallel and series resistance laws to find that

$$R_n \geq \frac{1}{4} + \frac{1}{12} + \cdots + \frac{1}{8n - 4}.$$

This implies that $R_n \to \infty$ as $n \to \infty$, and the result is shown.                         ∎

**(22) Theorem. Transience of three-dimensional random walk.** *Symmetric random walk on the three-dimensional cubic lattice $\mathbb{Z}^3$ is transient.*

**Proof.** It is a non-trivial and interesting *exercise* to prove that $R(\mathbb{Z}^3) < \infty$. See the solution of Problem (6.15.9) for another method of proof.                                              ∎

---

## Exercises for Section 13.11

**1.** Let $G$ be the closed sphere with radius $\epsilon$ and centre at the origin of $\mathbb{R}^d$ where $d \geq 3$. Let $\mathbf{W}$ be a $d$-dimensional Wiener process starting from $\mathbf{W}(0) = \mathbf{w} \notin G$. Show that the probability that $\mathbf{W}$ visits $G$ is $(\epsilon/r)^{d-2}$, where $r = |\mathbf{w}|$.

**2.** Let $G$ be an infinite connected graph with finite vertex degrees. Let $\Delta_n$ be the set of vertices $x$ which are distance $n$ from 0 (that is, the shortest path from $x$ to 0 contains $n$ edges), and let $N_n$ be the total number of edges joining pairs $x$, $y$ of vertices with $x \in \Delta_n$, $y \in \Delta_{n+1}$. Show that a random walk on $G$ is recurrent if $\sum_i N_i^{-1} = \infty$.

**3.** Let $G$ be a connected graph with finite vertex degrees, and let $H$ be a connected subgraph of $G$. Show that a random walk on $H$ is recurrent if a random walk on $G$ is recurrent, but that the converse is not generally true.

---

## 13.12 Problems

**1.** Let $W$ be a standard Wiener process, that is, a process with independent increments and continuous sample paths such that $W(s+t) - W(s)$ is $N(0, t)$ for $t > 0$. Let $\alpha$ be a positive constant. Show that:
(a) $\alpha W(t/\alpha^2)$ is a standard Wiener process,
(b) $W(t + \alpha) - W(\alpha)$ is a standard Wiener process,
(c) the process $V$, given by $V(t) = tW(1/t)$ for $t > 0$, $V(0) = 0$, is a standard Wiener process.

**2.** Let $X = \{X(t) : t \geq 0\}$ be a Gaussian process with continuous sample paths, zero means, and autocovariance function $c(s, t) = u(s)v(t)$ for $s \leq t$ where $u$ and $v$ are continuous functions. Suppose that the ratio $r(t) = u(t)/v(t)$ is continuous and strictly increasing with inverse function $r^{-1}$. Show that $W(t) = X(r^{-1}(t))/v(r^{-1}(t))$ is a standard Wiener process on a suitable interval of time.
   If $c(s, t) = s(1 - t)$ for $s \leq t < 1$, express $X$ in terms of $W$.

**3.** Let $\beta > 0$, and show that $U(t) = e^{-\beta t} W(e^{2\beta t} - 1)$ is an Ornstein–Uhlenbeck process if $W$ is a standard Wiener process.

**4.** Let $V = \{V(t) : t \geq 0\}$ be an Ornstein–Uhlenbeck process with instantaneous mean $a(t, x) = -\beta x$ where $\beta > 0$, with instantaneous variance $b(t, x) = \sigma^2$, and with $U(0) = u$. Show that $V(t)$ is $N(ue^{-\beta t}, \sigma^2(1 - e^{-2\beta t})/(2\beta))$. Deduce that $V(t)$ is asymptotically $N(0, \frac{1}{2}\sigma^2/\beta)$ as $t \to \infty$, and show that $V$ is strongly stationary if $V(0)$ is $N(0, \frac{1}{2}\sigma^2/\beta)$.
   Show that such a process is the *only* stationary Gaussian Markov process with continuous autocovariance function, and find its spectral density function.

**5.** **Feller's diffusion approximation to the branching process.** Let $D = \{D(t) : t \geq 0\}$ be a diffusion process with instantaneous mean $a(t, x) = \alpha x$ and instantaneous variance $b(t, x) = \beta x$ where $\alpha$ and $\beta$ are positive constants. Let $D(0) = d$. Show that the moment generating function of $D(t)$ is

$$M(t, \theta) = \exp\left\{\frac{2\alpha d\theta e^{\alpha t}}{\beta\theta(1 - e^{\alpha t}) + 2\alpha}\right\}.$$

Find the mean and variance of $D(t)$, and show that $\mathbb{P}(D(t) = 0) \to e^{-2d\alpha/\beta}$ as $t \to \infty$.

**6.** Let $D$ be an Ornstein–Uhlenbeck process with $D(0) = 0$, and place reflecting barriers at $-c$ and $d$ where $c, d > 0$. Find the limiting distribution of $D$ as $t \to \infty$.

**7.** Let $X_0, X_1, \ldots$ be independent $N(0, 1)$ variables, and show that

$$W(t) = \frac{t}{\sqrt{\pi}}X_0 + \sqrt{\frac{2}{\pi}}\sum_{k=1}^{\infty}\frac{\sin(kt)}{k}X_k$$

defines a standard Wiener process on $[0, \pi]$.

**8.** Let $W$ be a standard Wiener process with $W(0) = 0$. Place absorbing barriers at $-b$ and $b$, where $b > 0$, and let $W^a$ be $W$ absorbed at these barriers. Show that $W^a(t)$ has density function

$$f^a(y, t) = \frac{1}{\sqrt{2\pi t}}\sum_{k=-\infty}^{\infty}(-1)^k \exp\left\{-\frac{(y - 2kb)^2}{2t}\right\}, \qquad -b < y < b,$$

which may also be expressed as

$$f^a(y, t) = \sum_{n=1}^{\infty}a_n e^{-\lambda_n t}\sin\left(\frac{n\pi(y + b)}{2b}\right), \qquad -b < y < b,$$

where $a_n = b^{-1}\sin(\frac{1}{2}n\pi)$ and $\lambda_n = n^2\pi^2/(8b^2)$.
   Hence calculate $\mathbb{P}(\sup_{0 \leq s \leq t}|W(s)| > b)$ for the unrestricted process $W$.

**9.** Let $D$ be a Wiener process with drift $m$, and suppose that $D(0) = 0$. Place absorbing barriers at the points $x = -a$ and $x = b$ where $a$ and $b$ are positive real numbers. Show that the probability $p_a$ that the process is absorbed at $-a$ is given by

$$p_a = \frac{e^{2mb} - 1}{e^{2m(a+b)} - 1}.$$

**10.** Let $W$ be a standard Wiener process and let $F(u, v)$ be the event that $W$ has no zero in the interval $(u, v)$.

(a) If $ab > 0$, show that $\mathbb{P}\big(F(0, t) \mid W(0) = a, W(t) = b\big) = 1 - e^{-2ab/t}$.

(b) If $W(0) = 0$ and $0 < t_0 \le t_1 \le t_2$, show that

$$\mathbb{P}\big(F(t_0, t_2) \mid F(t_0, t_1)\big) = \frac{\sin^{-1} \sqrt{t_0/t_2}}{\sin^{-1} \sqrt{t_0/t_1}}.$$

(c) Deduce that, if $W(0) = 0$ and $0 < t_1 \le t_2$, then $\mathbb{P}(F(0, t_2) \mid F(0, t_1)) = \sqrt{t_1/t_2}$.

**11.** Let $W$ be a standard Wiener process. Show that

$$\mathbb{P}\left(\sup_{0 \le s \le t} |W(s)| \ge w\right) \le 2\mathbb{P}(|W(t)| \ge w) \le \frac{2t}{w^2} \qquad \text{for } w > 0.$$

Set $t = 2^n$ and $w = 2^{2n/3}$ and use the Borel–Cantelli lemma to show that $t^{-1} W(t) \to 0$ a.s. as $t \to \infty$.

**12.** Let $\mathbf{W}$ be a two-dimensional Wiener process with $\mathbf{W}(0) = \mathbf{w}$, and let $F$ be the unit circle. What is the probability that $\mathbf{W}$ visits the upper semicircle $G$ of $F$ before it visits the lower semicircle $H$?

**13.** Let $W_1$ and $W_2$ be independent standard Wiener processes; the pair $\mathbf{W}(t) = (W_1(t), W_2(t))$ represents the position of a particle which is experiencing Brownian motion in the plane. Let $l$ be some straight line in $\mathbb{R}^2$, and let P be the point on $l$ which is closest to the origin O. Draw a diagram. Show that

(a) the particle visits $l$, with probability one,

(b) if the particle hits $l$ for the first time at the point R, then the distance PR (measured as positive or negative as appropriate) has the Cauchy density function $f(x) = d/\{\pi(d^2 + x^2)\}$, $-\infty < x < \infty$, where $d$ is the distance OP,

(c) the angle $\widehat{\text{POR}}$ is uniformly distributed on $[-\frac{1}{2}\pi, \frac{1}{2}\pi]$.

**14. Lévy's conformal invariance property.** Let $\phi(x + iy) = u(x, y) + iv(x, y)$ be an analytic function on the complex plane with real part $u(x, y)$ and imaginary part $v(x, y)$, and assume that

$$\left(\frac{\partial u}{\partial x}\right)^2 + \left(\frac{\partial u}{\partial y}\right)^2 = 1.$$

Let $(W_1, W_2)$ be the planar Wiener process of Problem (13.12.13) above. Show that the pair $u(W_1, W_2)$, $v(W_1, W_2)$ is also a planar Wiener process.

**15.** Let $M(t) = \max_{0 \le s \le t} W(s)$, where $W$ is a standard Wiener process. Show that $M(t) - W(t)$ has the same distribution as $M(t)$.

**16.** Let $W$ be a standard Wiener process, $u \in \mathbb{R}$, and let $Z = \{t : W(t) = u\}$. Show that $Z$ is a null set (i.e. has Lebesgue measure zero) with probability one.

**17.** Let $M(t) = \max_{0 \le s \le t} W(s)$, where $W$ is a standard Wiener process. Show that $M(t)$ is attained at exactly one point in $[0, t]$, with probability one.

**18. Sparre Andersen theorem.** Let $s_0 = 0$ and $s_m = \sum_{j=1}^{m} x_j$, where $(x_j : 1 \le j \le n)$ is a given sequence of real numbers. Of the $n!$ permutations of $(x_j : 1 \le j \le n)$, let $A_r$ be the number of permutations in which exactly $r$ values of $(s_m : 0 \le m \le n)$ are strictly positive, and let $B_r$ be the number of permutations in which the maximum of $(s_m : 0 \le m \le n)$ first occurs at the $r$th place. Show that $A_r = B_r$ for $0 \le r \le n$. [Hint: Use induction on $n$.]

**19. Arc sine laws.** For the standard Wiener process $W$, let $A$ be the amount of time $u$ during the time interval $[0, t]$ for which $W(u) > 0$; let $L$ be the time of the last visit to the origin before $t$; and let $R$ be the time when $W$ attains its maximum in $[0, t]$. Show that $A$, $L$, and $R$ have the same distribution function $F(x) = (2/\pi) \sin^{-1} \sqrt{x/t}$ for $0 \le x \le t$. [Hint: Use the results of Problems (13.12.15)–(13.12.18).]

**20.** Let $W$ be a standard Wiener process, and let $U_x$ be the amount of time spent below the level $x$ ($\ge 0$) during the time interval $(0, 1)$, that is, $U_x = \int_0^1 I_{\{W(t)<x\}} \, dt$. Show that $U_x$ has density function

$$f_{U_x}(u) = \frac{1}{\pi \sqrt{u(1-u)}} \exp\left(-\frac{x^2}{2u}\right), \qquad 0 < u < 1.$$

Show also that

$$V_x = \begin{cases} \sup\{t \le 1 : W_t = x\} & \text{if this set is non-empty,} \\ 1 & \text{otherwise,} \end{cases}$$

has the same distribution as $U_x$.

**21. Tanaka's example.** Let $\text{sign}(x) = 1$ if $x > 0$ and $\text{sign}(x) = -1$ otherwise.

(a) Show that $V_t = \int_0^t \text{sign}(W_s) \, dW_s$ defines a standard Wiener process, where $W$ is itself such a process.

(b) Deduce that $V$ is the solution of the SDE $dX = \text{sign}(X_t) \, dW$ with $W(0) = 0$.

(c) Use the result of Problem (13.12.16) with $u = 0$ to show that $-V$ is also a solution of this SDE.

(d) Use the results of Example (12.7.10) and Theorem (13.8.11) to deduce that any solution of the SDE has the fdds of the Wiener process. [You have shown that solutions of the SDE are not unique in a pathwise sense, but do have unique fdds.]

**22.** After the level of an industrial process has been set at its desired value, it wanders in a random fashion. To counteract this the process is periodically reset to this desired value, at times $0, T, 2T, \dots$. If $W_t$ is the deviation from the desired level, $t$ units of time after a reset, then $\{W_t : 0 \le t < T\}$ can be modelled by a standard Wiener process. The behaviour of the process after a reset is independent of its behaviour before the reset. While $W_t$ is outside the range $(-a, a)$ the output from the process is unsatisfactory and a cost is incurred at rate $C$ per unit time. The cost of each reset is $R$. Show that the period $T$ which minimises the long-run average cost per unit time is $T^*$, where

$$R = C \int_0^{T^*} \frac{a}{\sqrt{(2\pi t)}} \exp\left(-\frac{a^2}{2t}\right) \, dt.$$

**23.** An economy is governed by the Black–Scholes model in which the stock price behaves as a geometric Brownian motion with volatility $\sigma$, and there is a constant interest rate $r$. An investor likes to have a constant proportion $\gamma$ ($\in (0, 1)$) of the current value of her self-financing portfolio in stock and the remainder in the bond. Show that the value function of her portfolio has the form $V_t = f(t)S_t^\gamma$ where $f(t) = c \exp\{(1 - \gamma)(\frac{1}{2}\gamma\sigma^2 + r)t\}$ for some constant $c$ depending on her initial wealth.

**24.** Let $u(t, x)$ be twice continuously differentiable in $x$ and once in $t$, for $x \in \mathbb{R}$ and $t \in [0, T]$. Let $W$ be the standard Wiener process. Show that $u$ is a solution of the heat equation

$$\frac{\partial u}{\partial t} = \frac{1}{2} \frac{\partial^2 u}{\partial x^2}$$

if and only if the process $U_t = u(T - t, W_t)$, $0 \le t \le T$, has zero drift.

**25. Walk on spheres.**

(a) Let $W$ be the standard Wiener process in $d = 3$ dimensions starting at the origin 0, and let $T$ be its first passage time to the sphere with radius $r$ and centre at 0. Show that $W(T)$ is independent of $T$.

(b) Let $C$ be an open convex region containing 0, with smooth boundary $B$. For $x \in C$, let $M(x)$ be the largest sphere with centre $x$ that is inscribable in $B \cup C$, with radius $r(x)$. Define the process $\{R_n : n \geq 0\}$ thus. We set $R_0 = 0$; given $R_1, R_2, \ldots, R_n$ for $n \geq 0$, $R_{n+1}$ is uniformly distributed on the sphere $M_n := M(R_n)$. Now define the increasing sequence $\{T_n : n \geq 0\}$ by: $T_0 = 0$, $T_1$ is the first passage time of $W$ to $M(0)$; $T_2$ is the first subsequent time at which $W$ hits $M_1$, and so on.

  (i) Show that the sequences $(R_n)$ and $(W(T_n))$ have the same distributions.
  (ii) Deduce that $R_\infty = \lim_{n \to \infty} R_n$ has the same distribution as $W(T_B)$, where $T_B$ is the first passage time of $W$ to $B$.
  (iii) Let $d$ be the diameter of the smallest sphere that contains $B \cup C$. For $0 < a < d/2$, let $S(a) = \inf\{n : r(R_n) \leq a\}$ be the least $n$ at which $R_n$ is within distance $a$ of the boundary $B$. Show that $S(a)$ is a.s. finite, and moreover that $\mathbb{E}(S(a)) \leq d/a$.

# Appendix I

## Foundations and notation

*Summary.* Here is a digest of topics with which many readers will already be familiar, and which are necessary for a full understanding of the text.

### (A) Basic notation

The end of each example or subsection is indicated by the symbol ●; the end of each proof is indicated by ■.

The largest integer which is not larger than the real number $x$ is denoted by $\lfloor x \rfloor$, and the smallest integer not smaller than $x$ by $\lceil x \rceil$. We use the following symbols:

$$\mathbb{R} \equiv \text{the real numbers } (-\infty, \infty),$$
$$\mathbb{Z} \equiv \text{the integers } \{\ldots, -2, -1, 0, 1, 2, \ldots\},$$
$$\mathbb{C} \equiv \text{the complex plane } \{x + iy : x, y \in \mathbb{R}\}.$$

Here are two 'delta' functions.

*Kronecker $\delta$*: If $i$ and $j$ belong to some set $S$, define $\delta_{ij} = \begin{cases} 1 & \text{if } i = j, \\ 0 & \text{if } i \neq j. \end{cases}$

*Dirac $\delta$ function*: If $x \in \mathbb{R}$, the symbol $\delta_x$ represents a notional function with the properties
(a) $\delta_x(y) = 0$ if $y \neq x$,
(b) $\int_{-\infty}^{\infty} g(y)\delta_x(y)\, dy = g(x)$ for all integrable $g : \mathbb{R} \to \mathbb{R}$.

### (B) Sets and counting

In addition to the union and intersection symbols, $\cup$ and $\cap$, we employ the following notation:

*set difference*:      $A \setminus B = \{x \in A : x \notin B\}$,
*symmetric difference*:    $A \triangle B = (A \setminus B) \cup (B \setminus A) = \{x \in A \cup B : x \notin A \cap B\}$.

The *cardinality* $|A|$ of a set $A$ is the number of elements contained in $A$. The *complement* of $A$ is denoted by $A^c$.

The *binomial coefficient* $\binom{n}{r}$ is the number of distinct combinations of $r$ objects that can be drawn from a set containing $n$ distinguishable objects. The following texts treat this material in more detail: Halmos 1960, Ross 2013, and Rudin 1976.

## (C)  Vectors and matrices

The symbol $\mathbf{x}$ denotes the row vector $(x_1, x_2, \dots)$ of finite or countably infinite length. The transposes of vectors $\mathbf{x}$ and matrices $\mathbf{V}$ are denoted by $\mathbf{x}'$ and $\mathbf{V}'$ respectively. The determinant of a square matrix $\mathbf{V}$ is written as $|\mathbf{V}|$.

The following books contain information about matrices, their eigenvalues, and their canonical forms: Lipschutz 1974, Rudin 1976, and Cox and Miller 1965.

## (D)  Convergence

**(1) Limits inferior and superior.** We often use inferior and superior limits, and so we review their definitions. Given any sequence $\{x_n : n \geq 1\}$ of real numbers, define

$$g_m = \inf_{n \geq m} x_n, \qquad h_m = \sup_{n \geq m} x_n.$$

Then $g_m \leq g_{m+1}$ and $h_m \geq h_{m+1}$ for all $m$, whence the sequences $\{g_m\}$ and $\{h_m\}$ converge as $m \to \infty$. Their limits are denoted by '$\liminf_{n \to \infty} x_n$' and '$\limsup_{n \to \infty} x_n$' respectively. Clearly, $\liminf_{n \to \infty} x_n \leq \limsup_{n \to \infty} x_n$. The following result is very useful.

**(2) Theorem.** *The sequence $\{x_n\}$ converges if and only if* $\liminf_{n \to \infty} x_n = \limsup_{n \to \infty} x_n$.

**(3) Cauchy convergence.** The criterion of convergence (for all $\epsilon > 0$, there exists $N$ such that $|x_n - x| < \epsilon$ if $n \geq N$) depends on knowledge of the limit $x$. In many practical instances it is convenient to use a criterion which does not rely on such knowledge.

**(4) Definition.** The sequence $\{x_n\}$ is called **Cauchy convergent** if, for all $\epsilon > 0$, there exists $N$ such that $|x_m - x_n| < \epsilon$ whenever $m, n \geq N$.

**(5) Theorem.** *A real sequence converges if and only if it is Cauchy convergent.*

**(6) Continuity of functions.** We recall that the function $g : \mathbb{R} \to \mathbb{R}$ is *continuous* at the point $x$ if $g(x + h) \to g(x)$ as $h \to 0$. We often encounter functions which satisfy only part of this condition.

**(7) Definition.** The function $g : \mathbb{R} \to \mathbb{R}$ is called:
 (i)  **right-continuous** if $g(x + h) \to g(x)$ as $h \downarrow 0$ for all $x$,
 (ii) **left-continuous** if $g(x + h) \to g(x)$ as $h \uparrow 0$ for all $x$.

The function $g$ is continuous if and only if $g$ is both right- and left-continuous.

If $g$ is monotone then it has left and right limits, $\lim_{h \uparrow 0} g(x + h)$, $\lim_{h \downarrow 0} g(x + h)$, at all points $x$; these may differ from $g(x)$ if $g$ is not continuous at $x$. We write

$$g(x+) = \lim_{h \downarrow 0} g(x + h), \qquad g(x-) = \lim_{h \uparrow 0} g(x + h).$$

**(8) Infinite products.** We make use of the following result concerning products of real numbers.

**(9) Theorem.** *Let* $p_n = \prod_{i=1}^{n}(1 + x_i)$.
  (a) *If* $x_i > 0$ *for all* $i$, *then* $p_n \to \infty$ *as* $n \to \infty$ *if and only if* $\sum_i x_i = \infty$.
  (b) *If* $-1 < x_i \leq 0$ *for all* $i$, *then* $p_n \to 0$ *if and only if* $\sum_i |x_i| = \infty$.

**(10) Landau's notation.** Use of the O/o notation† is standard. If $f$ and $g$ are two functions of a real variable $x$, then we say that:

$$f(x) = o(g(x)) \text{ as } x \to \infty \quad \text{if} \quad \lim_{x \to \infty} f(x)/g(x) = 0,$$

$$f(x) = O(g(x)) \text{ as } x \to \infty \quad \text{if} \quad |f(x)/g(x)| < C \text{ for all large } x \text{ and some constant } C.$$

Similar definitions hold as $x \downarrow 0$, and for real sequences $\{f(n)\}$, $\{g(n)\}$ as $n \to \infty$.

**(11) Asymptotics.** We write

$$f(x) \sim g(x) \text{ as } x \to \infty \quad \text{if} \quad \lim_{x \to \infty} f(x)/g(x) = 1,$$

with a similar definition as $x \downarrow 0$, and for sequences $\{f(n)\}$, $\{g(n)\}$ as $n \to \infty$. When we write $f(x) \simeq g(x)$, we mean that $f(x)$ is approximately equal to $g(x)$, perhaps in some limiting sense.

   For more details about the topics in this section see Apostol 1974 or Rudin 1976.

## (E)  Complex analysis

We make use of elementary manipulation of complex numbers, the formula $e^{itx} = \cos(tx) + i\sin(tx)$, and the theory of complex integration. Readers are referred to Phillips 1957, Nevanlinna and Paatero 1969, and Rudin 1986 for further details.

## (F)  Transforms

An *integral transform* of the function $g : \mathbb{R} \to \mathbb{R}$ is a function $\tilde{g}$ of the form

$$\tilde{g}(\theta) = \int_{-\infty}^{\infty} K(\theta, x) g(x) \, dx,$$

for some 'kernel' $K$. Such transforms are very useful in the theory of differential equations. Perhaps most useful is the *Laplace transform*.

**(12) Definition.** The **Laplace transform** of $g$ is defined to be the function

$$\hat{g}(\theta) = \int_{-\infty}^{\infty} e^{-\theta x} g(x) \, dx \quad \text{where} \quad \theta \in \mathbb{C}$$

whenever this integral exists.

---

†Invented by Paul Bachmann in 1894.

As a special case of the Laplace transform of $g$, set $\theta = i\lambda$ for real $\lambda$ to obtain the *Fourier transform*

$$G(\lambda) = \widehat{g}(i\lambda) = \int_{-\infty}^{\infty} e^{-i\lambda x} g(x)\, dx.$$

Often, we are interested in functions $g$ which are defined on the half-line $[0, \infty)$, with Laplace transform

$$\widehat{g}(\theta) = \int_{0}^{\infty} e^{-\theta x} g(x)\, dx.$$

Such a transform is called 'one-sided'. We often think of $\widehat{g}$ as a function of a *real* variable $\theta$.

Subject to certain conditions (such as existence and continuity) Laplace transforms have the following important properties.

**(13) Inversion.** *The function $g$ may be retrieved from knowledge of $\widehat{g}$ by the 'inversion formula'.*

**(14) Convolution.** *If $k(x) = \displaystyle\int_{-\infty}^{\infty} g(x-y)h(y)\, dy$ then $\widehat{k}(\theta) = \widehat{g}(\theta)\widehat{h}(\theta)$.*

**(15) Differentiation.** *If $G : [0, \infty) \to \mathbb{R}$ and $g = dG/dx$ then $\theta\widehat{G}(\theta) = \widehat{g}(\theta) + G(0)$.*

It is sometimes convenient to use a variant of the Laplace transform.

**(16) Definition.** The **Laplace–Stieltjes transform** of $g$ is defined to be

$$g^*(\theta) = \int_{-\infty}^{\infty} e^{-\theta x}\, dg(x) \quad \text{where} \quad \theta \in \mathbb{C}$$

whenever this integral exists.

We do not wish to discuss the definition of this integral (it is called a 'Lebesgue–Stieltjes' integral and is related to the integrals of Section 5.6). You may think about it in the following way. If $g$ is differentiable then its Laplace–Stieltjes transform $g^*$ is defined to be the Laplace transform of its derivative $g'$, since in this case $dg(x) = g'(x)\, dx$. Laplace–Stieltjes transforms $g^*$ always receive an asterisk in order to distinguish them from Laplace transforms. They have properties similar to (13), (14), and (15). For example, (14) becomes the following.

**(17) Convolution.** *If $k(x) = \displaystyle\int_{-\infty}^{\infty} g(x-y)\, dh(y)$ then $k^*(\theta) = g^*(\theta)h^*(\theta)$.*

Fourier–Stieltjes transforms may be defined similarly.

More details are provided by Apostol 1974 and Hildebrand 1962.

## (G)  Difference equations

The sequence $\{u_r : r \geq 0\}$ is said to satisfy a *difference equation* if

**(18)**
$$\sum_{i=0}^{m} a_i u_{n+m-i} = f(n), \quad n \geq 0,$$

for some fixed sequence $a_0, a_1, \ldots, a_m$ and given function $f$. If $a_0 a_m \neq 0$, the difference equation is said to be of order $m$. The general solution of this difference equation is

$$u_n = \sum_{i=1}^{r} \sum_{j=0}^{m_i-1} c_{ij} n^j \theta_i^n + p_n$$

where $\theta_1, \theta_2, \ldots, \theta_r$ are the distinct roots of the polynomial equation

$$\sum_{i=0}^{m} a_i \theta^{m-i} = 0,$$

$m_i$ being the multiplicity of the root $\theta_i$, and $\{p_n : n \geq 0\}$ is any particular solution to (18). In general there are $m$ arbitrary constants, whose determination requires $m$ boundary conditions.

More details are provided by Hall 1983.

## (H)  Partial differential equations

Let $a = a(x, y, u)$, $b = b(x, y, u)$, and $c = (x, y, u)$ be 'nice' functions of $\mathbb{R}^3$, and suppose that $u(x, y)$ satisfies the partial differential equation

(19)
$$a \frac{\partial u}{\partial x} + b \frac{\partial u}{\partial y} = c.$$

The solution $u = u(x, y)$ may be thought of as a surface $\phi(x, y, u) = 0$ where $\phi(x, y, u) = u - u(x, y)$. The normal to $\phi$ at the point $(x, y, u(x, y))$ lies in the direction

$$\nabla \phi = \left( -\frac{\partial u}{\partial x}, -\frac{\partial u}{\partial y}, 1 \right).$$

Consider a curve $\{x(t), y(t), u(t) : t \in \mathbb{R}\}$ in $\mathbb{R}^3$ defined by $\dot{x} = a$, $\dot{y} = b$, $\dot{u} = c$. The direction cosines of this curve are proportional to the vector $(a, b, c)$, whose scalar product with $\nabla \phi$ satisfies

$$\nabla \phi.(a, b, c) = -a \frac{\partial u}{\partial x} - b \frac{\partial u}{\partial y} + c = 0,$$

so that the curve is perpendicular to the normal vector $\nabla \phi$. Hence any such curve lies in the surface $\phi(x, y, u) = 0$, giving that the family of such curves generates the solution to the differential equation (19).

For more details concerning partial differential equations, see Hildebrand 1962, Piaggio 1965, or O'Neil 1999.

# Appendix II

# Further reading

This list is neither comprehensive nor canonical. The bibliography lists books that are useful for mathematical background and further exploration.

**Probability theory.** Ross 2013, Hoel *et al.* 1971a, Grimmett and Welsh 2014, and Stirzaker (1999, 2003) are excellent elementary texts. There are many fine advanced texts, including Billingsley 1995, Breiman 1968, Chung 2001, Durrett 2019, Kallenberg 2002, and Shiryaev 1984. The probability section of Kingman and Taylor 1966 provides a concise introduction to the modern theory, as does Itô 1984. Moran 1968 is often useful at our level.

The two volumes of Feller's treatise (Feller 1968, 1971) are essential reading for incipient probabilists; the first deals largely in discrete probability, and the second is an idiosyncratic and remarkable encyclopaedia of the continuous theory. Blom *et al.* 1994 give a modern collection of problems of discrete probability in the spirit of Feller. The book of Stoyanov 1997 provides many cautionary examples.

The Stein–Chen method of proving distributional limits is discussed at length by Barbour *et al.* 1992.

**Markov chains.** We know of no account of Markov chains that is wholly satisfactory at this level, though various treatments have attractions. Billingsley 1995 proves the limit theorem by the coupling argument; Cox and Miller 1965 and Karlin and Taylor 1975 contain many examples; Ross 2013 is clear and to the point; Norris 1997 is an attractive and slightly more sophisticated account; Kemeny *et al.* 1976 deal extensively with links to potential theory. Chung 1960 and Freedman 1971b are much more advanced and include rigorous treatments of the continuous-time theory; these are relatively difficult books. Brémaud 1998 includes a pleasant selection of recent applications of the theory, and Kingman 1996 is a concise account of Poisson processes.

**Other random processes.** Our selection from the enormous list of books on these topics is necessarily ruthless. Karlin and Taylor (1975, 1981), Cox and Miller 1965, and Ross 1996 each look at several kinds of random processes and applications in an accessible way.

Time series, stationarity, and extensions thereof, are well covered by Brockwell and Davis 1991 and by Daley and Vere-Jones 1988.

Apart from by Cox 1962, renewal theory is seldom treated in isolation, and is often considered in conjunction with Markov chains and point processes; see Ross 2013, Feller 1971, and Karlin and Taylor (1975, 1981).

Queueing theory is treated in the above books also, in the form of examples involving Markov chains and renewal processes. Examples of excellent books dedicated to queues include Kelly 1979, Wolff 1989, and Asmussen 2003.

Martingale theory was expounded systematically by Doob 1953. The fine book of Williams 1991 provides an invaluable introduction to measure theory (for the probabilist) and to martingales in discrete time. Other fairly accessible books include those by Neveu 1975, Hall and Heyde 1980, and Kopp 1984.

Diffusion processes in general, and the Wiener process in particular, are often considered by authors in the context of continuous-parameter martingales and stochastic integration. Of the torrent of bulky volumes, we mention Revuz and Yor 1999, Øksendal 1998, and Rogers and Williams (2000a, 2000b). There are many books on financial mathematics, at several levels, and we mention only Baxter and Rennie 1996, Bingham and Kiesel 1998, Björk 1998, and Nielsen 1999.

For further reading around geometrical probability (otherwise known as integral, or stochastic, geometry), see Schneider and Weil 2008 and, at the more probabilistic end of the subject with applications, Mathai 1999.

# Appendix III

## History and varieties of probability

### History

Mathematical probability has its origins in games of chance, principally in games with dice and cards. Early calculations involving dice were included in a well-known and widely distributed poem entitled *De Vetula*, written in France around 1250 AD, (possibly by Richard de Fournival, a French cleric). Dice and cards continued as the main vessels of gambling in the fifteenth and sixteenth centuries, during which mathematics flowered as part of the Renaissance. A number of Italian mathematicians of this period (including Galileo) gave calculations of the number and proportion of winning outcomes in various fashionable games. One of them (G. Cardano) went so far as to write a book, *On games of chance*, sometime shortly after 1550. This was not published however until 1663, by which time probability theory had already had its official inauguration elsewhere.

It was around 1654 that B. Pascal and P. de Fermat generated a celebrated correspondence about their solutions of the problem of the points. These were soon widely known, and C. Huygens developed these ideas in a book published in 1657, in Latin. Translations into Dutch (1660) and English (1692) soon followed. The preface by John Arbuthnot to the English version (see Appendix IV) makes it clear that the intuitive notions underlying this work were similar to those commonly in force nowadays.

These first simple ideas were soon extended by Jacob (otherwise known as James) Bernoulli in *Ars conjectandi* (1713) and by A. de Moivre in *Doctrine of chances* (1718, 1738, 1756). These books included simple versions of the weak law of large numbers and the central limit theorem. Methods, results, and ideas were all greatly refined and generalized by P. Laplace in a series of books from 1774 to 1827. Many other eminent mathematicians of this period wrote on probability: Euler, Gauss, Lagrange, Legendre, Poisson, and so on.

However, as ever harder problems were tackled by ever more powerful mathematical techniques during the nineteenth century, the lack of a well-defined axiomatic structure was recognized as a serious handicap. In 1900, D. Hilbert included this as his sixth problem, though it may seem somewhat odd to modern eyes that he labelled probability as a physical science.

The resolution of the problem took just 33 years, starting with the work of E. Borel and H. Lebesgue. In his 1902 thesis, Lebesgue used his measures (non-negative, countably additive set functions on $\mathbb{R}$) to construct the Lebesgue integral, which, when abstracted, enabled the definition of the expected value of a random variable on a probability space $(\Omega, \mathcal{F}, \mathbb{P})$; see Section 5.6.

In 1909, Borel proposed his remarkable normal number theorem†; see Problem (9.7.14). A number $x$ is normal in base $b$ when the $b$ digits have equal limiting frequencies in its $b$-ary expansion, and absolutely normal if it is normal in every base $b > 1$. The normal number theorem asserts that almost every number in $[0, 1]$ is absolutely normal, that is, the Lebesgue measure of the set of such numbers is 1. Despite this, absolutely normal numbers are extremely difficult to display explicitly, a computable absolutely normal number being first demonstrated in 2002 by V. Becher and S. Figueira. In 1916, M. Sierpiński had shown an example of an absolutely normal number, but this is not known to be computable. Further non-computable absolutely normal numbers are G. Chaitin's constants $\Omega$, each being the halting probability of a type of universal Turing machine.

Later, M. Fréchet formalized measure theory in the context of general abstract spaces (metric and topological), discarding the earlier Euclidean models. This enabled P. Lévy (1925) to write a very influential book on probability, including accounts of characteristic functions, Lévy processes, and much more. To his later chagrin, his book just fell short of being the definitive foundation of the field.

Elsewhere (in Poland), H. Steinhaus, in three papers from 1923, used measure theory to axiomatize probability in the context of infinite sequences of independent events (or random variables). This enabled him to determine the convergence (or not) of random power series. An almost simultaneous axiomatic treatment of coin tossing based on measure theory was given by A. Łomnicki in 1923. Also, in 1930, O. Nikodým extended earlier work of J. Radon to obtain the Radon–Nikodým theorem on change of measure; see Example (12.8.22). This, in turn, enables a rigorous development of conditional expectation and probability in the context of probability measures on $\sigma$-fields.

Meanwhile, in Russia, S. Bernstein (1917, 1927), A. Khinchin and A. Kolmogorov (1925), A. Khinchin (1927), and E. Slutsky (1922), amongst others, had considered the logical and axiomatic foundations of probability, while proving on their way an enormous array of theorems, including, for example, central limit theorems, laws of large numbers, and laws of the iterated logarithm.

In what was simultaneously a conclusion to Hilbert's problem, and the beginning of an explosion in the growth of probability theory, in his 1933 book *Grundbegriffe der Wahrschein-lichkeitsrechnung*, written to aid roof repairs on his dacha, A. Kolmogorov provided the axioms which today underpin most mathematical probability. Indeed, for most people, to say "the theory of probability" is to denote Kolmogorov's theory, even though others exist. His remarkable book contains the definitive version of the strong law of large numbers, and the first general definition and exposition of conditional expectation.

## Varieties

It is necessary to have an interpretation of probability, for this is what suggests appropriate axioms and useful applications. The oldest interpretations of probability are as:

  (a)  an indication of relative frequency, and

  (b)  an expression of symmetry or fairness.

These views were natural given the origins of the subject. A well-made die is symmetrical

---

†Strictly speaking we should refer to the Borel–Faber–Hausdorff normal number theorem, Borel's original proof having been "unmendably faulty" (see Khoshnevisan 2006).

and is equally likely to show any face; an ill-made die is biased and in the long run shows its faces in different relative frequencies. (Recall Ambrose Bierce's definition of 'dice': dice are small polka-dotted cubes of ivory constructed like a lawyer to lie upon any side, commonly the wrong one.)

However, there are many chance events which are neither repeatable nor symmetrical, and from earliest times probabilists have been alert to the fact that applications might be sought in fields other than gambling. G. Leibniz considered the degree to which some statement had been proved, and many later authors concerned themselves with the theory of testimony. Indeed, Daston 1988 has argued that legal questions and concepts were among the primary catalysts for the development of probability; mathematicians simply used the obvious and natural symmetries of fair games to model the far more slippery concepts of equity and fair (judicial) expectation. Such ideas lead to more complicated interpretations of probability such as:

(c)  to what extent some hypothesis is logically implied by the evidence, and

(d)  the degree of belief of an individual that some given event will occur.

This last interpretation is commonly known as 'subjective probability', and the concept is extremely fissiparous. Since different schools of thought choose different criteria for judging possible reasons for belief, a wide variety of axiomatic systems have come into being.

However, by a happy chance, in many cases of importance, the axioms can be reasonably reduced to exactly the axioms (1.3.1) with which we have been concerned. And systems not so reduced have in general proved very intractable to extensive analysis.

Finally we note that (a)–(d) do not exhaust the possible intepretations of probability theory, and that there remain areas where interpretations are as yet unagreed, notably in quantum mechanics. The reader may pursue this in books on physics and philosophy; see Krüger *et al.* 1987 and Hájek 2012.

This is an opportune moment to mention a model of randomness in sequences that is essentially different from classical probability theories of the type discussed above. In mathematical probability, a 'random sequence' is one selected from the set of possible sequences according to some probability measure, the sequence itself is not otherwise examined. For example, in the probability space for $10^6$ rolls of a fair die, the outcome $\alpha$ of $10^6$ 1's is assigned the same probability as any arguably more typical sequence $\beta$ comprising roughly equal proportions of the numbers from 1 to 6. However, intuitively, $\alpha$ may appear to have no randomness, since it has none of the irregularity possessed by $\beta$. This leads to the concept of the 'descriptive complexity' of a sequence, as the length of the shortest computer program needed to generate it. A sequence may be considered to be random if it is shorter than any program that can produce it. When properly formulated, this yields a property termed *Kolmogorov complexity*, in acknowledgement of the extensive contributions to the theory made by Kolmogorov. For the rolled die above, it may be shown that 'most' of the possible $6^{10^6}$ possible sequences are 'nearly' Kolmogorov random. It is beyond our current scope to make this statement precise; see Downey and Hirschfeldt 2010.

# Appendix IV

## John Arbuthnot's Preface to
## *Of the laws of chance* (1692)

It is thought as necessary to write a Preface before a Book, as it is judg'd civil, when you invite a Friend to Dinner, to proffer him a Glass of Hock beforehand for a Whet: And this being maim'd enough for want of a Dedication, I am resolv'd it shall not want an Epistle to the Reader too. I shall not take upon me to determine, whether it is lawful to play at Dice or not, leaving that to be disputed betwixt the *Fanatick Parsons* and the *Sharpers*; I am sure it is lawful to deal with Dice as with other Epidemic Distempers; and I am confident that the writing a Book about it, will contribute as little towards its Encouragement, as Fluxing and Precipitates do to Whoring.

It will be to little purpose to tell my Reader, of how great Antiquity the playing at Dice is. I will only let him know that by the *AleæLudus*, the Antients comprehended all Games, which were subjected to the determination of mere Chance; this sort of Gaming was strictly forbid by the Emperor *Justinian, Cod. Lib. 3. Tit. 43.* under severe Penalties; and *Phocius Nomocan. Tit. 9. Cap. 27.* acquaints us, that the Use of this was altogether denied the Clergy of that time. *Seneca* says very well, *Aleator quantò in arte est melior tantò est nequior*; That by how much the one is more skilful in Games, by so much he is the more culpable; or we may say of this, as an ingenious Man says of Dancing, That to be extraordinary good at it, is to be excellent in a Fault†; therefore I hope no body will imagine I had so mean a Design in this, as to teach the Art of Playing at Dice.

A great part of this Discourse is a Translation from Mons. *Huygen's* Treatise, *De ratiociniis in ludo Aleæ*; one, who in his Improvements of Philosophy, has but one Superior‡, and I think few or no Equals. The whole I undertook for my own Divertisement, next to the Satisfaction of some Friends, who would now and then be wrangling about the Proportions of Hazards in some Cases that are here decided. All it requir'd was a few spare Hours, and but little Work for the Brain; my Design in publishing it, was to make it of general Use, and perhaps persuade a raw Squire, by it, to keep his Money in his Pocket; and if, upon this account, I should incur the Clamours of the Sharpers, I do not much regard it, since they are a sort of People the world is not bound to provide for.

You will find here a very plain and easy Method of the Calculation of the Hazards of Game, which a man may understand, without knowing the Quadratures of *Curves*, the Doctrine

---

†An apophthegm of Francis Bacon who attributes it to Diogenes.

‡Isaac Newton.

of *Series's*, or the Laws of *Concentripetation* of Bodies, or the Periods of the *Satellites* of *Jupiter*; yea, without so much as the Elements of *Euclid*. There is nothing required for the comprehending the whole, but common Sense and practical Arithmetick; saving a few Touches of *Algebra*, as in the first Three Propositions, where the Reader, without suspicion of Popery, may make use of a strong implicit Faith; tho' I must confess, it does not much recommend it self to me in these Purposes; for I had rather he would enquire, and I believe he will find the Speculation not unpleasant.

Every man's Success in any Affair is proportional to his Conduct and Fortune. Fortune (in the sense of most People) signifies an Event which depends on Chance, agreeing with my Wish; and Misfortune signifies such an one, whose immediate Causes I don't know, and consequently can neither foretel nor produce it (for it is no Heresy to believe, that Providence suffers ordinary matters to run in the Channel of second Causes). Now I suppose, that all a wise Man can do in such a Case is, to lay his Business on such Events, as have the most powerful second Causes, and this is true both in the great Events of the World, and in ordinary Games. It is impossible for a Die, with such determin'd force and direction, not to fall on such a determin'd side, only I don't know the force and direction which makes it fall on such a determin'd side, and therefore I call that Chance, which is nothing but want of Art; that only which is left to me, is to wager where there are the greatest number of Chances, and consequently the greatest probability to gain; and the whole Art of Gaming, where there is any thing of Hazard, will be reduc'd to this at last, *viz.* in dubious Cases to calculate on which side there are most Chances; and tho' this can't be done in the midst of Game precisely to an Unit, yet a Man who knows the Principles, may make such a conjecture, as will be a sufficient direction to him; and tho' it is possible, if there are any Chances against him at all, that he may lose, yet when he chuseth the safest side, he may part with his Money with more content (if there can be any at all) in such a Case.

I will not debate, whether one may engage another in a disadvantageous Wager. Games may be suppos'd to be a tryal of Wit as well as Fortune, and every Man, when he enters the Lists with another, unless out of Complaisance, takes it for granted, his Fortune and Judgment, are, at least, equal to those of his Play-Fellow; but this I am sure of, that false Dice, Tricks of *Leger-de-main, &c.* are inexcusable, for the question in Gaming is not, Who is the best Jugler?

The Reader may here observe the Force of Numbers, which can be successfully applied, even to those things, which one would imagine are subject to no rules. There are very few things which we know, which are not capable of being reduc'd to a Mathematical Reasoning; and when they cannot, it's a sign our Knowledge of them is very small and confus'd; and where a mathematical reasoning can be had, it's as great a folly to make use of any other, as to grope for a thing in the dark, when you have a Candle standing by you. I believe the Calculation of the Quantity of Probability might be improved to a very useful and pleasant Speculation, and applied to a great many Events which are accidental besides those of Games; only these Cases would be infinitely more confus'd, as depending on Chances which the most part of Men are ignorant of; and as I have hinted already, all the Politicks in the World are nothing else but a kind of Analysis of the Quantity of Probability in casual Events, and a good Politician signifies no more, but one who is dextrous at such Calculations; only the Principles which are made use of in the Solution of such Problems, can't be studied in a Closet, but acquir'd by the Observation of Mankind.

There is likewise a Calculation of the Quantity of Probability founded on Experience, to be made use of in Wagers about any thing; it is odds, if a Woman is *with Child*, but it shall be a *Boy*; and if you would know the just odds, you must consider the Proportion in the Bills that the Males bear to the Females†: The Yearly Bills of Mortality are observ'd to bear such Proportion to the live People as 1 to 30, or 26; therefore it is an even Wager, that one out of thirteen, dies within a Year (which may be a good reason, tho' not the true, of that foolish piece of superstition), because, at this rate, if 1 out of 26 dies, you are no loser. It is but 1 to 18 if you meet a *Parson* in the Street, that he proves to be a *Non-Juror‡*, because there is but 1 of 36 that are such. It is hardly 1 to 10, that a *Woman* of Twenty Years old has her *Maidenhead**, and almost the same Wager, that a *Town-Spark* of that Age has not been *clap'd*. I think a Man might venture some odds, that 100 of the *Gens d'arms* beats an equal Number of *Dutch Troopers*; and that an *English Regiment* stands its ground as long as another, making Experience our Guide in all these Cases and others of the like nature.

But there are no casual Events, which are so easily subjected to Numbers, as those of Games; and I believe, there the Speculation might be improved so far, as to bring in the Doctrine of the *Series's* and *Logarithms*. Since Gaming is become a Trade, I think it fit the Adventurers should be put upon the Square; and therefore in the Contrivance of Games there ought to be strict Calculation made use of, that they mayn't put one Party in more probability to gain them another; and likewise, if a Man has a considerable Venture; he ought to be allow'd to withdraw his Money when he pleases, paying according to the Circumstances he is then in: and it were easy in most Games to make Tables, by Inspection of which, a Man might know what he was either to pay or receive, in any Circumstances you can imagin, it being convenient to save a part of one's Money, rather than venture the loss of it all.

I shall add no more, but that a Mathematician will easily perceive, it is not put in such a Dress as to be taken notice of by him, there being abundance of Words spent to make the more ordinary sort of People understand it.

---

†Arbuthnot is here quoting the conclusions of John Graunt as set out in the book *Natural and Political Observations Made upon the Bills of Mortality* (1st edn 1662 Old Style or 1663 New Style; 5th edn 1676).

‡A 'Non-Juror' is one who refused to take an oath of allegiance to William and Mary in 1688.

*Karl Pearson has suggested that this may be a reference to a short-lived Company for the Assurance of Female Chastity.

# Appendix V

## Table of distributions

| | Mass/density function | Domain | Mean |
|---|---|---|---|
| **Bernoulli** | $f(1) = p, \; f(0) = q = 1 - p$ | $\{0, 1\}$ | $p$ |
| **Uniform (discrete)** | $n^{-1}$ | $\{1, 2, \ldots, n\}$ | $\frac{1}{2}(n+1)$ |
| **Binomial** bin$(n, p)$ | $\binom{n}{k} p^k (1-p)^{n-k}$ | $\{0, 1, \ldots, n\}$ | $np$ |
| **Geometric** | $p(1-p)^{k-1}$ | $k = 1, 2, \ldots$ | $p^{-1}$ |
| **Poisson** | $e^{-\lambda} \lambda^k / k!$ | $k = 0, 1, 2, \ldots$ | $\lambda$ |
| **Negative binomial** | $\binom{k-1}{n-1} p^n (1-p)^{k-n}$ | $k = n, n+1, \ldots$ | $np^{-1}$ |
| **Hypergeometric** | $\dfrac{\binom{b}{k}\binom{N-b}{n-k}}{\binom{N}{n}}, \; p = \dfrac{b}{N}, \; q = \dfrac{N-b}{N}$ | $(n - N + b)^+ \leq k \leq b \wedge n$ | $np$ |
| **Uniform (continuous)** | $(b-a)^{-1}$ | $[a, b]$ | $\frac{1}{2}(a+b)$ |
| **Exponential** | $\lambda e^{-\lambda x}$ | $[0, \infty)$ | $\lambda^{-1}$ |
| **Normal** $N(\mu, \sigma^2)$ | $\dfrac{1}{\sqrt{2\pi\sigma^2}} \exp\left\{ -\dfrac{(x-\mu)^2}{2\sigma^2} \right\}$ | $\mathbb{R}$ | $\mu$ |
| **Gamma** $\Gamma(\lambda, \tau)$ | $\dfrac{1}{\Gamma(\tau)} \lambda^\tau x^{\tau-1} e^{-\lambda x}$ | $[0, \infty)$ | $\tau\lambda^{-1}$ |
| **Cauchy** | $\dfrac{1}{\pi(1+x^2)}$ | $\mathbb{R}$ | $-$ |
| **Beta** $\beta(a, b)$ | $\dfrac{\Gamma(a+b)}{\Gamma(a)\Gamma(b)} x^{a-1}(1-x)^{b-1}$ | $[0, 1]$ | $\dfrac{a}{a+b}$ |
| **Doubly exponential** | $\exp(-x - e^{-x})$ | $\mathbb{R}$ | $\gamma$† |
| **Rayleigh** | $x e^{-\frac{1}{2}x^2}$ | $[0, \infty)$ | $\sqrt{\dfrac{\pi}{2}}$ |
| **Laplace** | $\frac{1}{2}\lambda e^{-\lambda|x|}$ | $\mathbb{R}$ | $0$ |

---

†The letter $\gamma$ denotes Euler's constant.

| Variance | Skewness | Characteristic function |
|---|---|---|
| $pq$ | $\dfrac{q-p}{\sqrt{pq}}$ | $q + pe^{it}$ |
| $\frac{1}{12}(n^2 - 1)$ | $0$ | $\dfrac{e^{it}(1 - e^{int})}{n(1 - e^{it})}$ |
| $np(1-p)$ | $\dfrac{1-2p}{\sqrt{np(1-p)}}$ | $(1 - p + pe^{it})^n$ |
| $(1-p)p^{-2}$ | $\dfrac{2-p}{\sqrt{1-p}}$ | $\dfrac{p}{e^{-it} - 1 + p}$ |
| $\lambda$ | $\lambda^{-\frac{1}{2}}$ | $\exp\{\lambda(e^{it} - 1)\}$ |
| $n(1-p)p^{-2}$ | $\dfrac{2-p}{\sqrt{n(1-p)}}$ | $\left(\dfrac{p}{e^{-it} - 1 + p}\right)^n$ |
| $\dfrac{npq(N-n)}{N-1}$ | $\dfrac{q-p}{\sqrt{npq}}\sqrt{\dfrac{N-1}{N-n}}\left(\dfrac{N-2n}{N-2}\right)$ | $\dfrac{\binom{N-b}{n}}{\binom{N}{n}}F(-n, -b; N-b-n+1; e^{it})$† |
| $\frac{1}{12}(b-a)^2$ | $0$ | $\dfrac{e^{ibt} - e^{iat}}{it(b-a)}$ |
| $\lambda^{-2}$ | $2$ | $\dfrac{\lambda}{\lambda - it}$ |
| $\sigma^2$ | $0$ | $e^{i\mu t - \frac{1}{2}\sigma^2 t^2}$ |
| $\tau\lambda^{-2}$ | $2\tau^{-\frac{1}{2}}$ | $\left(\dfrac{\lambda}{\lambda - it}\right)^{\tau}$ |
| — | — | $e^{-|t|}$ |
| $\dfrac{ab}{(a+b)^2(a+b+1)}$ | $\dfrac{2(b-a)}{a+b+2}\sqrt{\dfrac{a+b+1}{ab}}$ | $M(a, a+b, it)$† |
| $\frac{1}{6}\pi^2$ | $1.29857\ldots$ | $\Gamma(1 - it)$ |
| $2 - \dfrac{\pi}{2}$ | $\dfrac{2\sqrt{\pi}(\pi - 3)}{(4 - \pi)^{3/2}}$ | $1 + \sqrt{2\pi}it\left(1 - \Phi(-it)\right)e^{-\frac{1}{2}t^2}$† |
| $2\lambda^{-2}$ | $0$ | $\dfrac{\lambda^2}{\lambda^2 + t^2}$ |

---

† $F(a, b; c; z)$ is Gauss's hypergeometric function and $M(a, a+b, it)$ is a confluent hypergeometric function. The $N(0, 1)$ distribution function is denoted by $\Phi$.

# Appendix VI

# Chronology

A subset of the mathematicians, scientists and others mentioned in this book.

Aesop *c.*620–564 BC
Pythagoras *c.*570–495 BC, perhaps
Plato 428–348 BC
Diogenes 400–320 BC
Euclid 325–265 BC
Archimedes of Syracuse 288–212 BC
Seneca 4 BC–65
Luca Pacioli 1445–1514
Gerolamo Cardano 1501–1576
Gerardus Mercator 1512–1594
Francis Bacon 1561–1626
William Shakespeare 1564–1616
Galileo Galilei 1564–1642
Johannes Kepler 1571–1630
Pierre de Fermat 1601–1665
John Graunt 1620–1674
Blaise Pascal 1623–1662
Christiaan Huygens 1629–1695
Lorenzo Tonti 1630–1695
Antony van Leeuwenhoek 1632–1723
Samuel Pepys 1633–1703
Isaac Newton 1642–1727
Gottfried von Leibniz 1646–1716
William of Orange 1650–1702
Jacob [James] Bernoulli 1654–1705
Guillaume de L'Hôpital 1661–1704
John Arbuthnot 1667–1735
Abraham de Moivre 1667–1754
Pierre de Montmort 1678–1719
Brook Taylor 1685–1731
Nicholas Bernoulli 1687–1759
William Stukeley 1687–1765
James Stirling 1692–1770
Daniel Bernoulli 1700–1782
Thomas Bayes 1701–1761
Leonhard Euler 1707–1783
Georges Buffon 1707–1788
Adam Smith 1723–1790
Casanova de Seingalt 1725–1798
Edward Waring 1734–1798
Joseph-Louis Lagrange 1736–1813

Nicolas de Condorcet 1743–1794
Pierre-Simon de Laplace 1749–1827
Adrien-Marie Legendre 1752–1833
Heinrich Olbers 1758–1840
Thomas Malthus 1766–1834
Jean Fourier 1768–1830
Robert Brown 1773–1858
Carl Friedrich Gauss 1777–1855
Siméon Poisson 1781–1840
Friedrich Wilhelm Bessel 1784–1846
Georg Ohm 1789–1854
Augustin-Louis Cauchy 1789–1857
George Green 1793–1841
Irénée-Jules Bienaymé 1796–1878
Niels Abel 1802–1829
Carl Jacobi 1804–1851
Johann Dirichlet 1805–1859
Augustus De Morgan 1806–1871
William Makepeace Thackeray 1811–1863
Pierre Laurent 1813–1854
James Sylvester 1814–1897
George Boole 1815–1864
Karl Weierstrass 1815–1897
Pafnuti Chebyshov 1821–1894
Joseph Bertrand 1822–1900
Francis Galton 1822–1911
Leopold Kronecker 1823–1891
Gustav Kirchhoff 1824–1887
Georg Bernhard Riemann 1826–1866
Henry Smith 1826–1883
Morgan Crofton 1826–1915
Henry Watson 1827–1903
James Clerk Maxwell 1831–1879
Henry Labouchere 1831–1912
Lewis Carroll [Charles Dodgson] 1832–1898
Rudolf Lipschitz 1832–1903
John Venn 1834–1923
Simon Newcomb 1835–1909
Paul Bachmann 1837–1920
Josiah Willard Gibbs 1839–1903
Charles Peirce 1839–1914

William Whitworth 1840–1905
Ambrose Bierce 1842–1914
Rayleigh [John Strutt] 1842–1919
Friedrich Helmert 1843–1917
Hermann Schwarz 1843–1921
Georg Cantor 1845–1918
Gyula [Julius] Farkas 1847–1930
Vilfredo Pareto 1848–1923
Ferdinand Georg Frobenius 1849–1917
Charles Dow 1851–1902
Jules Henri Poincaré 1854–1912
Thomas Stieltjes 1856–1894
Edward Davis Jones 1856–1920
Andrei A. Markov 1856–1922
Alexander Liapunov 1857–1918
Karl Pearson 1857–1936
Ernesto Cesàro 1859–1906
Alfred Dreyfus 1859–1935
Otto Hölder 1859–1937
Alfred Whitehead 1861–1947
David Hilbert 1862–1943
Hermann Minkowski 1864–1909
Alfred Tauber 1866–1942
Felix Hausdorff 1868–1942
Johan Jensen 1869–1925
Louis Bachelier 1870–1946
Ernest Rutherford 1871–1937
George Udny Yule 1871–1951
Emile Borel 1871–1956
Paul Langevin 1872–1946
Bertrand Russell 1872–1970
Johan Steffensen 1873–1961
Andre Louis Cholesky 1875–1918
Henri Lebesgue 1875–1941
Francesco Cantelli 1875–1966
William Gosset [Student] 1876–1937
Tatiana Ehrenfest 1876–1964
Edmund Landau 1877–1938
Godfrey H. Hardy 1877–1947
Agner Erlang 1878–1929
Pierre Fatou 1878–1929
Maurice Fréchet 1878–1973
Guido Fubini 1879–1943
Albert Einstein 1879–1955
Paul Ehrenfest 1880–1933
Leonard Ornstein 1880–1941
Evgenii Slutsky 1880–1948
Norman Campbell 1880–1949
Sergei Bernstein 1880–1968
Oskar Perron 1880–1975
Harold Hurst 1880–1978

Antoni Marian Łomnicki 1881–1941
Leon Isserlis 1881–1966
Arthur Eddington 1882–1944
Hans Geiger 1882–1945
Harry Bateman 1882–1946
Waclaw Sierpiński 1882–1969
John Maynard Keynes 1883–1946
Eric Temple Bell 1883–1960
Henry T. H. Piaggio 1884–1967
John Edensor Littlewood 1885–1977
Paul Lévy 1886–1971
Johann Radon 1887–1956
Hugo Steinhaus 1887–1972
Otto Nikodým 1887–1974
George Pólya 1887–1985
Wilhelm Lenz 1888–1957
Sydney Chapman 1888–1970
Percy Daniell 1889–1946
Sewall Green Wright 1889–1988
Ronald Fisher 1890–1962
Emil Gumbel 1891–1966
Harold Jeffreys 1891–1989
Stefan Banach 1892–1945
Carlo Bonferroni 1892–1960
John B. S. Haldane 1892–1964
Paul Getty 1892–1976
Harald Cramér 1893–1985
Alexander Khinchin 1894–1959
Norbert Wiener 1894–1964
Heinz Hopf 1894–1971
Harold Hotelling 1895–1973
Rolf Herman Nevanlinna 1895–1980
Gábor Szegő 1895–1985
Carl Marius Christensen 1898–1973
Alfred Hitchcock 1899–1980
Joseph Berkson 1899–1982
Salomon Bochner 1899–1982
David van Dantzig 1900–1959
George Uhlenbeck 1900–1988
Antoni Zygmund 1900–1992
Ernst Ising 1900–1998
Abraham Wald 1902–1950
George Zipf 1902–1950
Paul Dirac 1902–1984
Leonard Tippett 1902–1985
John von Neumann 1903–1957
Andrei Nikolaevich Kolmogorov 1903–1987
Kaarlo Paatero 1903–1986
Sydney Goldstein 1903–1989
Bertha Swirles 1903–1999
Oliver Franks 1905–1992

Moritz Fenchel 1905–1988
Paul Hoel 1905–2000
William Feller 1906–1970
Bruno de Finetti 1906–1985
Eugene Lukacs 1906–1987
Andrew Berry 1906–1998
Raymond Paley 1907–1933
Conrad Palm 1907–1951
Michel Loève 1907–1979
Solomon Kullback 1907–1994
Gustav Elfving 1908–1984
Robert Cameron 1908–1989
Stanislaw Ulam 1909–1984
Paul Turán 1910–1976
Marshall Hall, Jr, 1910–1990
Maurice Bartlett 1910–2002
Wilfrid Stevens 1911–1958
Garrett Birkhoff 1911–1996
Shizuo Kakutani 1911–2004
William Martin 1911–2004
Boris Vladimirovich Gnedenko 1912–1995
Paul Erdős 1913–1996
Henry Lancaster 1913–2001
Anders Hald 1913–2007
Mark Kac 1914–1984
Wassily Hoeffding 1914–1991
Richard Leibler 1914–2003
George Dantzig 1914–2005
Wolfgang Doeblin 1915–1940
Nicholas Metropolis 1915–1999
Kiyosi Itô 1915–2008
Claude Shannon 1916–2001
Paul Halmos 1916–2006
Leonard Jimmie Savage 1917–1971
Patrick Moran 1917–1988
Norman Lloyd Johnson 1917–2004
Sten Malmquist 1917–2004
Kai Lai Chung 1917–2009
Richard Feynman 1918–1988
Frank Anscombe 1918–2001
Carl-Gustav Esseen 1918–2001
David Kendall 1918–2007
Harry Pollard 1919–1985
Erik Sparre Andersen 1919–2003
William Kruskal 1919–2005
Theodore Harris 1919–2006

David Blackwell 1919–2010
Edwin Hewitt 1920–1999
Joseph Doob 1920–2004
John Hammersley 1920–2004
Claude Ambrose Rogers 1920–2005
Lester Dubins 1920–2010
Charles Stein 1920–2016
Alfred Rényi 1921–1970
Gerd [Harry] Reuter 1921–1992
Walter Rudin 1921–2010
Monty Hall [Monte Halperin] 1921–2017
Edward Simpson 1922–2019
John Kelly, Jr 1923–1965
Dennis Lindley 1923–2013
Thomas Apostol 1923–2016
Pieter Kasteleyn 1924–1996
Lucien Le Cam 1924–2000
Samuel Karlin 1924–2007
Benoit Mandelbrot 1924–2010
James Jackson 1924–2011
George Marsaglia 1924–2011
Eugene Dynkin 1924–2014
Patrick Billingsley 1925–2011
James Laurie Snell 1925–2011
Jaroslav Hájek 1926–1974
John Kemeny 1926–1992
Frank Spitzer 1926–1992
Leo Breiman 1928–2005
Roland Dobrushin 1929–1995
Emanuel Parzen 1929–2016
Ruslan Leont'evich Stratonovich 1930–1997
Radha Laha 1930–1999
Wilfred Hastings 1930–2016
Hilton Miller 1931–2012
Jacques Neveu 1932–2016
Igor Vladimirovich Girsanov 1934–1967
Harry Kesten 1934–2019
Robert Ash 1935–2015
Fischer Black 1938–1995
David Freedman 1938–2008
Christopher Heyde 1939–2008
Catherine Doléans-Dade 1942–2004
Daniel Lunn 1942–2019
Marc Yor 1949–2014
Peter Hall 1951–2016

# Bibliography

Albertsen, K. (1995). *The extinction of families*. International Statistical Review 63, 234–239.

Aldous, D. and Fill, J. (2002/2014). *Reversible Markov chains and random walks on graphs*. https://www.stat.berkeley.edu/users/aldous/RWG/book.html.

Apostol, T. M. (1974). *Mathematical analysis* (2nd edn). Addison-Wesley, Reading, MA.

Applebaum, D. (2011). *Lévy processes and stochastic calculus* (2nd edn). Cambridge University Press.

Ash, R. B. and Doléans-Dade, C. A. (2000). *Probability and measure theory* (2nd edn). Academic Press, San Diego.

Asmussen, S. (2003). *Applied probability and queues* (2nd edn). Wiley, New York.

Athreya, K. B. and Ney, P. E. (1972). *Branching processes*. Springer, Berlin.

Barbour, A. D., Holst, L., and Janson, S. (1992). *Poisson approximation*. Oxford University Press.

Barndorff-Nielsen, O. E. and Shiryaev, A. (2015). *Change of time and change of measure* (2nd edn). World Scientific, Singapore.

Bass, R. F. (2011). *Stochastic processes*. Cambridge University Press.

Basu, S. and Dasgupta, A. (1997). *The mean, median, and mode of unimodal distributions: a characterization*. Theory of Probability and its Applications 41, 210–223..

Baxter, M. and Rennie, A. (1996). *Financial calculus*. Cambridge University Press.

Berger, A. and Hill, T. P. (2015). *An introduction to Benford's law*. Princeton University Press.

Bertoin, J. (1996). *Lévy processes*. Cambridge University Press.

Billingsley, P. (1995). *Probability and measure* (3rd edn). Wiley, New York.

Bingham, N. H. and Kiesel R. (1998). *Risk neutral valuation*. Springer, Berlin.

Björk, T. (1998). *Arbitrage theory in continuous time*. Oxford University Press.

Blom, G., Holst, L., and Sandell, D. (1994). *Problems and snapshots from the world of probability*. Springer, Berlin.

Breiman, L. (1968). *Probability*. Addison-Wesley, Reading, MA, reprinted by SIAM, 1992.

Brémaud, P. (1998). *Markov chains*. Springer, Berlin.

Brockwell, P. and Davis, R. (1991). *Time series: theory and methods* (2nd edn). Springer, Berlin.

Capinski, M. and Kopp, P. E. (2004). *Measure, integral and probability* (2nd edn). Springer, London.

Casanova de Seingalt (Giacomo Girolamo) (1922). *Memoirs* (trans. A. Machen), Vol. IV. Casanova Society, London.

Chatfield, C. (2003). *The analysis of time series* (6th edn). Chapman and Hall, London.

Chung, K. L. (1960). *Markov chains with stationary transition probabilities*. Springer, Berlin.

Chung, K. L. (2001). *A course in probability theory* (3rd edn). Academic Press, New York.

Chung, K. L. and Williams, R. J. (1990). *Introduction to stochastic integration*. Birkhäuser, Boston.

Clarke, L. E. (1975). *Random variables*. Longman, London.

Cox, D. R. (1962). *Renewal theory*. Longman, London.

Cox, D. R. and Miller, H. D. (1965). *The theory of stochastic processes*. Chapman and Hall, London.

Cox, D. R. and Smith, W. L. (1961). *Queues*. Chapman and Hall, London.

Daley, D. J. and Vere-Jones, D. (1988). *An introduction to the theory of point processes*. Springer, Berlin.

Daston, L. (1988). *Classical probability in the enlightenment*. Princeton University Press.

Doob, J. L. (1953). *Stochastic processes*. Wiley, New York.

Downey, R. G. and Hirschfeldt, D. R. (2010). *Algorithmic randomness and complexity*. Springer, New York.

Dubins, L. and Savage, L. (1965). *How to gamble if you must*. McGraw-Hill, New York, reprinted by Dover, 1976.

Dudley, R. M. (1989). *Real analysis and probability*. Wadsworth & Brooks/Cole, Pacific Grove, CA.

Durrett, R. T. (2019). *Probability: theory and examples* (5th edn). Cambridge University Press.

Embrechts, P. and Maejima, M. (2002). *Selfsimilar processes*. Princeton University Press.

Feller, W. (1957). *On boundaries and lateral conditions for the Kolmogorov differential equations*. Ann. Math. 65, 527–570.

Feller, W. (1968). *An introduction to probability theory and its applications,* Vol. 1 (3rd edn). Wiley, New York.

Feller, W. (1971). *An introduction to probability theory and its applications,* Vol. 2 (2nd edn). Wiley, New York.

Freedman, D. (1971a). *Brownian motion and diffusion*. Holden-Day, San Francisco, reprinted by Springer, 1983.

Freedman, D. (1971b). *Markov chains*. Holden-Day, San Francisco, reprinted by Springer, 1983.

Grimmett, G. R. (2018). *Probability on graphs* (2nd edn). Cambridge University Press.

Grimmett, G. R. and Stirzaker, D. R. (2020). *One thousand exercises in probability* (3rd edn). Oxford University Press.

Grimmett, G. R. and Welsh, D. J. A. (2014). *Probability, an introduction* (2nd edn). Oxford University Press.

Häggström, O. (2002). *Finite Markov chains with algorithmic applications*. Cambridge University Press.

Hájek, A. (2012). *Interpretations of probability, in The Stanford Encyclopedia of Philosophy, E. N. Zalta (ed.)*. https://plato.stanford.edu/entries/probability-interpret/.

Hald, A. (1990). *A history of probability and statistics and their applications before 1750*. Wiley, New York.

Hald, A. (1998). *A history of mathematical statistics from 1750 to 1930*. Wiley, New York.

Hall, M. (1983). *Combinatorial theory* (2nd edn). Wiley, New York.

Hall, P. and Heyde, C. C. (1980). *Martingale limit theory and its application*. Academic Press, New York.

Halmos, P. R. (1960). *Naive set theory*. Van Nostrand, Princeton, NJ.

Hildebrand, F. B. (1962). *Advanced theory of calculus*. Prentice Hall, Englewood Cliffs, NJ.

Hoel, P. G., Port, S. C., and Stone, C. J. (1971a). *Introduction to probability theory*. Houghton Mifflin, Boston.

Hoel, P. G., Port, S. C., and Stone, C. J. (1971b). *Introduction to stochastic processes*. Houghton Mifflin, Boston.

Itô, K. (1984). *Introduction to probability theory*. Cambridge University Press.

Janson, S. (2011). *Stable distributions*. https://arxiv.org/abs/1112.0220.

Kallenberg, O. (2002). *Foundations of modern probability* (2nd edn). Springer, Berlin.

Karatzas, I. and Shreve, S. E. (1991). *Brownian motion and stochastic calculus* (2nd edn). Springer, Berlin.

Karlin, S. and Taylor, H. M. (1975). *A first course in stochastic processes* (2nd edn). Academic Press, New York.

Karlin, S. and Taylor, H. M. (1981). *A second course in stochastic processes*. Academic Press, New York.

Kelly, F. P. (1979). *Reversibility and stochastic networks*. Wiley, New York.

Kemeny, J. G., Snell, J. L., and Knapp, A. W. (1976). *Denumerable Markov chains*. Springer, New York.

Khoshnevisan, D. (2006). *Normal numbers are normal*. Clay Mathematics Institute Annual Report, pp. 15, 27–31.

Kingman, J. F. C. (1996). *Poisson processes*. Oxford University Press.

Kingman, J. F. C. and Taylor, S. J. (1966). *Introduction to measure and probability*. Cambridge University Press.

Kopp, P. E. (1984). *Martingales and stochastic integrals*. Cambridge University Press.

Krüger, L. *et al.*, eds (1987). *The probabilistic revolution,* (2 vols). MIT Press, Cambridge, MA.

Laha, R. G. and Rohatgi, V. K. (1979). *Probability theory*. Wiley, New York.

Levin, D., Peres, Y., and Wilmer, E. (2017). *Markov chains and mixing times* (2nd edn). American Mathematical Society.

Lindvall, T. (1977). *A probabilistic proof of Blackwell's renewal theorem*. Annals of Probability 5, 482–485.

Lindvall, T. (2002). *Lectures on the coupling method*. Dover Books, Mineola, NY.

Liggett, T. M. (2010). *Continuous-time Markov processes*. American Mathematical Society, Providence RI.

Lipschutz, S. (1974). *Linear algebra,* Schaum Outline Series. McGraw-Hill, New York.

Loève, M. (1977). *Probability theory,* Vol. 1 (4th edn). Springer, Berlin.

Loève, M. (1978). *Probability theory,* Vol. 2 (4th edn). Springer, Berlin.

Lukacs, E. (1970). *Characteristic functions* (2nd edn). Griffin, London.

Mandelbrot, B. (1983). *The fractal geometry of nature*. Freeman, San Francisco.

Mathai, A. M. (1999). *An introduction to geometrical probability*. Gordon and Breach, Amsterdam.

Moran, P. A. P. (1968). *An introduction to probability theory*. Oxford University Press.

Musiela, M. and Rutkowski, M. (2008). *Martingale methods in financial modelling* (2nd edn). Springer, Berlin.

Nevanlinna, R. and Paatero, V. (1969). *Introduction to complex analysis*. Addison-Wesley, Reading, MA.

Neveu, J. (1975). *Discrete parameter martingales*. North-Holland, Amsterdam.

Nielsen, L. T. (1999). *Pricing and hedging of derivative securities*. Oxford University Press.

Norris, J. R. (1997). *Markov chains*. Cambridge University Press.

Øksendal, B. (1998). *Stochastic differential equations* (5th edn). Springer, Berlin.

O'Neil, P. V. (1999). *Beginning partial differential equations*. Wiley, New York.

Parzen, E. (1962). *Stochastic processes*. Holden-Day, San Francisco.

Phillips, E. G. (1957). *Functions of a complex variable*. Oliver and Boyd, Edinburgh.

Piaggio, H. T. H. (1965). *Differential equations*. Bell, London.

Pollard, D. (2002). *A user's guide to measure theoretic probability*. Cambridge University Press.

Prabhu, N. U. (1998). *Queues, insurance, dams, and data* (2nd edn). Springer, New York.

Revuz, D. and Yor, M. (1999). *Continuous martingales and Brownian motion* (3rd edn). Springer, Berlin.

Roberts, H. E. (1998). *Encyclopedia of comparative iconography*. Fitzroy Dearborn, Chicago & London.

Rogers, L. C. G. and Williams, D. (2000a). *Diffusions, Markov processes, and martingales*, Vol. 1 (2nd edn). Cambridge University Press.

Rogers, L. C. G. and Williams, D. (2000b). *Diffusions, Markov processes, and martingales*, Vol. 2 (2nd edn). Cambridge University Press.

Ross, S. (1996). *Stochastic processes* (2nd edn). Wiley, New York.

Ross, S. (2013). *A first course in probability* (9th edn). Prentice Hall, New York.

Rudin, W. (1976). *Principles of mathematical analysis* (3rd edn). McGraw-Hill, New York.

Rudin, W. (1986). *Real and complex analysis* (3rd edn). McGraw-Hill, New York.

Schneider, R. and Weil, W. (2008). *Stochastic and integral geometry*. Springer, Berlin.

Shiryaev, A. N. (1984). *Probability*. Springer, Berlin.

Shiryaev, A. N. (2012). *Problems in probability*. Springer, Berlin.

Stigler, S. M. (1980). *Stigler's law of eponymy*. Trans. N. Y. Acad. Sci. 39, 147–157. Reprinted in *Statistics on the table*, by Stigler, S. M. (1999), Harvard University Press.

Stigler, S. M. (1986). *The history of statistics*. Harvard University Press.

Stirzaker, D. R. (1999). *Probability and random variables*. Cambridge University Press.

Stirzaker, D. R. (2003). *Elementary probability* (2nd edn). Cambridge University Press.

Stoyanov, J. (1997). *Counterexamples in probability* (2nd edn). Wiley, New York.

Tankov, P. and Cont, R. (2003). *Financial modelling with jump processes*. Chapman and Hall/CRC, Boca Raton, FL.

Whittle, P. (2000). *Probability via expectation* (4th edn). Springer, New York.

Williams, D. (1991). *Probability with martingales*. Cambridge University Press.

Wolff, R. W. (1989). *Stochastic modelling and the theory of queues*. Prentice Hall, New York.

# Notation

| | | | |
|---|---|---|---|
| $\mathrm{bin}(n, p)$ | binomial distribution | $N(\mu, \sigma^2)$ | normal distribution |
| $c(n), c(t)$ | autocovariances | $N(t)$ | Poisson or renewal process |
| $\mathrm{cov}(X, Y)$ | covariance | $Q(t)$ | queue length |
| $d_{\mathrm{TV}}$ | total variation distance | $X, Y, Z, X(\omega)$ | random variables |
| $f, f_j, f_{ij}$ | probabilities | $\mathbf{X}, \mathbf{Y}, \mathbf{W}$ | random vectors |
| $f(x), f_X(\cdot)$ | mass or density functions | $\mathbf{V}(\mathbf{X})$ | covariance matrix |
| $f_{Y|X}(y \mid x)$ | conditional mass or density | $|\mathbf{V}|$ | determinant of $\mathbf{V}$ |
| $f_{X,Y}(x, y)$ | joint mass or density | $W(t), W_t, \mathbf{W}(t)$ | Wiener processes |
| $f'(t)$ | derivative of $f$ | $W, W_n$ | waiting times |
| $f * g$ | convolution | $\overline{X}$ | sample mean |
| $\widehat{g}$ | Laplace transform | $\mathcal{A}, \mathcal{B}, \mathcal{F}, \mathcal{G}, \mathcal{H}, \mathcal{I}$ | $\sigma$-fields |
| $g^*$ | Laplace–Stieltjes transform | $\mathcal{B}$ | Borel $\sigma$-field |
| $i$ | $\sqrt{-1}$ | $\delta_{ij}$ | Kronecker delta |
| $i, j, k, l, m, n, r, s$ | indices | $\delta(t)$ | Dirac delta |
| $m(\cdot), m^{\mathrm{d}}(\cdot)$ | mean renewal functions | $\eta$ | probability of extinction |
| $\max(\vee), \min(\wedge)$ | maximum, minimum | $\chi^2(\cdot)$ | chi-squared distribution |
| $p, p_i, p_{ij}, p(t), p_i(t)$ | probabilities | $\phi_X(t)$ | characteristic function |
| $x^+, x^-$ | $\max\{x, 0\}, -\min\{x, 0\}$ | $\mu$ | mean |
| $\lfloor x \rfloor$ | integer part of $x$ | $\mu_i$ | mean recurrence time |
| $\lceil x \rceil$ | least integer not less than $x$ | $\pi$ | stationary distribution |
| $\mathrm{var}(X)$ | variance | $\sigma$ | standard deviation |
| $\overline{z}$ | complex conjugate | $\rho(n)$ | autocorrelation |
| $|A|$ | cardinality of set $A$ | $\rho(X, Y)$ | correlation between $X$ and $Y$ |
| $A^c$ | complement of set $A$ | $\gamma$ | Euler's constant |
| $\mathbf{A}'$ | transpose or derivative of $\mathbf{A}$ | $\omega$ | elementary event |
| $B(a, b)$ | beta function | $\Gamma(t)$ | gamma function |
| $C(t), D(t), E(t)$ | current, total, excess life | $\Gamma(\lambda, t)$ | gamma distribution |
| $F(r, s)$ | $F$ distribution | $\Omega$ | sample space |
| $F(x), F_X(x)$ | distribution functions | $\Phi, \phi$ | $N(0, 1)$ distribution/density function |
| $F_{Y|X}(y \mid x)$ | conditional distribution | $\mathbb{C}$ | complex plane |
| $F_{X,Y}(x, y)$ | joint distribution | $\mathbb{E}$ | expectation |
| $G(s), G_X(s)$ | generating functions | $\mathbb{E}(\cdot \mid \mathcal{F})$ | conditional expectation |
| H, T | head, tail | $\mathbb{P}, \mathbb{Q}$ | probability measures |
| $I_A, I(A)$ | indicator of the event $A$ | $\mathbb{R}$ | real numbers |
| $J$ | Jacobian | $\mathbb{Z}$ | integers |
| $\log$ | natural logarithm | $\varnothing$ | empty set |
| $m_i$ | mean return time | $\| \cdot \|$ | norm |
| $M_X(t)$ | moment generating function | | |

# Index

Compiling an index is an occupation full as entertaining as that of darning socks, though by no means so advantageous to society.

Gilbert White, 1720–1793

Abbreviations used in this index: c.f. characteristic function; distn distribution; eqn equation; fn function; m.g.f. moment generating function; p.g.f. probability generating function; pr. process; r.v. random variable; r.w. random walk; s.r.w. simple random walk; thm theorem.